OXFORD MATHEMATICAL MONOGRAPHS

Von Neumann algebras
V.F.R. Jones

OXFORD MATHEMATICAL MONOGRAPHS

A. Belleni-Moranti: *Applied semigroups and evolution equations*
A. M. Arthurs: *Complementary variational principles* Second edition
M. Rosenblum and J. Rovnyak: *Hardy classes and operator theory*
J. W. P. Hirschfeld: *Finite projective spaces of three dimensions*
A. Pressley and G. Segal: *Loop groups*
D. E. Edmunds and W. D. Evans: *Spectral theory and differential operators*
Wang Jianhua: *The theory of games*
S. Omatu and J. H. Seinfeld: *Distributed parameter systems: theory and applications*
J. Hilgert, K. H. Hofmann, and J. D. Lawson: *Lie groups, convex cones, and semigroups*
S. Dineen: *The Schwarz lemma*
S. K. Donaldson and P. B. Kronheimer: *The geometry of four-manifolds*
D. W. Robinson: *Elliptic operators and Lie groups*
A. G. Werschulz: *The computational complexity of differential and integral equations*
L. Evens: *Cohomology of groups*
G. Effinger and D. R. Hayes: *Additive number theory of polynomials*
J. W. P. Hirschfeld and J. A. Thas: *General Galois geometries*
P. N. Hoffman and J. F. Humphreys: *Projective representations of the symmetric groups*
I. Györi and G. Ladas: *The oscillation theory of delay differential equations*
J. Heinonen, T. Kilpelainen, and O. Martio: *Non-linear potential theory*
B. Amberg, S. Franciosi, and F. de Giovanni: *Products of groups*
M. E. Gurtin: *Thermomechanics of evolving phase boundaries in the plane*
I. Ionescu and M. Sofonea: *Functional and numerical methods in viscoplasticity*
N. Woodhouse: *Geometric quantization* Second edition
U. Grenander: *General pattern theory*
J. Faraut and A. Koranyi: *Analysis on symmetric cones*
I. G. Macdonald: *Symmetric functions and Hall polynomials* Second edition
B. L. R. Shawyer and B.B. Watson: *Borel's methods of summability*
D. McDuff and D. Salamon: *Introduction to symplectic topology*
M. Holschneider: *Wavelets: an analysis tool*
Jacques Thévenaz: *G-algebras and modular representation theory*
Hans-Joachim Baues: *Homotopy type and homology*
P. D. D'Eath: *Black holes: gravitational interactions*
R. Lowen: *Approach spaces: the missing link in the topology–uniformity–metric triad*
Nguyen Dinh Cong: *Topological dynamics of random dynamical systems*
J. W. P. Hirschfeld: *Projective geometries over finite fields* Second edition
K. Matsuzaki and M. Taniguchi: *Hyperbolic manifolds and Kleinian groups*
David E. Evans and Yasuyuki Kawahigashi: *Quantum symmetries on operator algebras*

Quantum Symmetries on Operator Algebras

DAVID E. EVANS

School of Mathematics
University of Wales, Cardiff

and

YASUYUKI KAWAHIGASHI

Department of Mathematical Sciences
University of Tokyo

CLARENDON PRESS · OXFORD
1998

Oxford University Press, Great Clarendon Street, Oxford OX2 6DP
Oxford New York
Athens Auckland Bangkok Bogota Bombay
Buenos Aires Calcutta Cape Town Dar es Salaam
Delhi Florence Hong Kong Istanbul Karachi
Kuala Lumpur Madras Madrid Melbourne
Mexico City Nairobi Paris Singapore
Taipei Tokyo Toronto Warsaw
and associated companies in
Berlin Ibadan

Oxford is a trade mark of Oxford University Press

Published in the United States by
Oxford University Press Inc., New York

A catalogue record for this book is available from the British Library

Library of Congress Cataloging in Publication Data
(Data available)

ISBN 0 19 851175 2

Typeset by the authors
Printed in Great Britain by
Bookcraft Ltd., Midsomer Norton, Avon

to

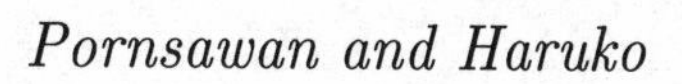

Pornsawan and Haruko

PREFACE

This book is aimed at graduate students and researchers in operator algebras and graduate students and researchers in the interface between operator algebras and statistical mechanics, algebraic, topological, and conformal field theory and low dimensional topological invariants. Indeed the book is based on several graduate courses given by us at our institutions. The first five chapters deal with the very basics of operator algebras – what we feel our graduate students need to know to do research in operator algebras and applications. This is supplemented by notes at the ends of those chapters giving more detailed surveys of contemporary work which is not surveyed in the rest of the book but helps to give the student a better picture of current research in the field. In fact the first three chapters (without the notes at the end of Chapters 2 and 3) have been the basis of undergraduate courses. With these prerequisites, Chapters 6–15 deal with the interactions mentioned above but from an operator algebraic point of view. Although we aim to introduce the reader to the research in operator algebras relevant to the directions of the last fifteen years in the above mentioned interface, our treatment is not exhaustive, especially with the scant attention paid to link invariants and quantum groups, because we already have ample literatures on these topics. Moreover the choice of topics included often reflects the eccentricities of the authors.

The theory of operator algebras was initiated by von Neumann and Murray as a tool for studying group representations and as a framework for quantum mechanics. The monographs of Bratteli and Robinson give an account of the theory of (non-commutative) operator algebras as a framework for quantum statistical mechanics. However, through the transfer matrix formalism, first introduced by Onsager and Kauffman in the two dimensional Ising model, non-commutative operator algebras have also become a framework for understanding classical statistical mechanics. Operator algebras also kept in touch with its roots in physics through the formalism of algebraic quantum field theory and especially the DHR theory of superselection sectors of Doplicher, Haag and Roberts, and K-theoretic methods. In particular the use of dimension groups of approximately finite dimensional C^*-algebras were used to provide dynamical systems with specified phase diagrams.

However, the study of operator algebras took a new turn with the introduction of subfactor theory. An outer group action on a factor can be recovered from the position of the fixed point algebra. An inclusion of a subfactor in a factor has a rich combinatorial structure which can be described by a quantization of a symmetry group or a paragroup in the language of Ocneanu. The theory of subfactors was initiated by Jones in 1981 who found that the minimal algebraic structure realizing this quantum symmetry was the Temperley–Lieb algebra or quantum $SU(2)$ at roots of unity. These algebras had already appeared in the work of Temperley and Lieb in the Potts model of statistical mechanics, and soon in 1984 provided a mechanism for providing new representations of braid groups and hence link invariants.

The introductory chapters allow us to discuss the basic structure of the fundamental building blocks of modern operator algebras, the matrix algebras and spaces of continuous functions on Hausdorff topological spaces (in fact matrix valued continuous functions on such spaces), the Cuntz algebras (Chapter 2), the irrational rotation algebra (notes to Chapter 3), and hyperfinite von Neumann algebras (Chapter 5). Chapter 6 discusses the Fermion algebra which is the basis for the treatment of the Ising model, our prototype IRF (interaction round a face) statistical mechanical model in Chapter 7, and our perspective of conformal field theory in Chapter 8. The bimodule approach of Ocneanu to paragroups is closely related to the IRF models of statistical mechanics (Chapter 7) and the sector approach of Longo, to algebraic quantum field theory of DHR and the conformal field theory ideas of Chapter 8. It is these interactions between subfactor theory in operator algebras, statistical mechanics, topological quantum field theory (TQFT; e.g. Chapter 12), rational conformal field theory (RCFT; e.g. Chapter 13), classification of modular invariants and common themes such as the orbifold construction which occupies us for the latter chapters.

This monograph is based on lectures given by the first author at RIMS, Kyoto University in 1990–91 and by the second author at University of Tokyo in the spring of 1993, as well as courses given by the first author at Warwick and Swansea in 1984–95. We thank D. Bisch, M. Choda, G. A. Elliott, T. Gannon, F. Goodman, U. Haagerup, K. C. Hannabuss, M. Izumi, V. F. R. Jones, T. Kohno, H. Kosaki, J. T. Lewis, R. Longo, A. Matsuo, A. Ocneanu, J. H. H. Perk, S. Popa, M. Takesaki, Y. Watatani, H. Wenzl, F. Xu, and S. Yamagami for helpful discussions and comments on the manuscript. We also thank E. J. Beggs, R. Bhat, C. Binnenhei, J. Bockenhauer, A. L. Carey, M. J. Gabriel, S. Goto, P. Johnson, T. Matsui, C. Pearce, P. Pinto, T. Sano, and N. Sato for reading preliminary drafts of this book and pointing out many typographical errors and ambiguous exposition. We especially thank M. Asaeda and K. Kawamuro for their careful reading of the manuscript and detailed comments. However any errors or misunderstandings that remain are due to the authors; updates on such corrections can be found under our home pages:

http://www.cf.ac.uk/uwcc/maths/evansde/

http://kyokan.ms.u-tokyo.ac.jp/~yasuyuki/

We are grateful to MEU Cymru for the Celtic braid diagrams at the beginning of each chapter and in the notes to Chapter 12 taken from their CD Rom Adnoddau Cymraeg, and to S. Winkler for drawing the toroidal orbifolds in the notes to Chapter 3.

Some of this book was written at the Fields Institute for Research in Mathematical Sciences during the programme on Operator Algebras and Applications in 1994–95. We thank the Fields Institute, the University of Tokyo and the University of Wales for their hospitality during our visits. We gratefully acknowledge the British Council, the Engineering and Physical Sciences Research Council, the Fields Institute for Research in the Mathematical Sciences, the Great Britain – Sasakawa Foundation, the Higher Education Funding Council for Wales, the Inamori Foundation, the Sumitomo Foundation, and the University of Wales for financial support during our work on this book.

Swansea D.E.E.
Tokyo Y.K.
18 September 1997

CONTENTS

1

OPERATOR THEORY BASICS

1.1 Introduction

This chapter introduces the most basic functional analysis of operators on Hilbert spaces. To avoid unnecessary complication, we try to minimize complexity in only proving what is absolutely essential to the general aims. Thus we restrict attention to separable Hilbert spaces and their operators – and not prove the more traditional aspects of an introductory functional analysis course such as a more general Banach space theory including the Hahn–Banach theorem, open mapping theorem, principle of uniform boundedness etc. In order to make much of the earlier chapters accessible to an undergraduate audience, we also avoid as much as possible measure theoretic considerations. Indeed the contents of the first three chapters are based on final year undergraduate courses.

1.2 Algebras of operators

We need to formulate a language for handling the operations of addition and multiplication of operators on a Hilbert space. We begin with some algebraic notions. An *algebra* is a set A endowed with three operations:

(1.2.1) scalar multiplication : λx, $\lambda \in \mathbb{C}$, $x \in A$

(1.2.2) vector addition : $x + y$, $x, y \in A$

(1.2.3) vector multiplication : xy, $x, y \in A$

with the following relations between these operations

(1.2.4) A is a vector space with respect to scalar multiplication and vector addition (relating (1.2.1) and (1.2.2)).

(1.2.5) A is a ring with respect to vector addition and vector multiplication (relating (1.2.2) and (1.2.3)).

(1.2.6) $\lambda(xy) = (\lambda x)y = x(\lambda y)$ for all $\lambda \in \mathbb{C}, x, y \in A$, (relating (1.2.1) and (1.2.3)).

An algebra A is said to be *commutative* if $xy = yx$ for all x, y in A. An algebra A is said to be *unital* if there exists a (necessarily unique) element e in A such that $ex = xe = x$ for all x in A. Then e is called the unit or identity of A and written 1 or 1_A. A map between unital algebras is said to be *unital* if it preserves the identities. A *homomorphism* π from an algebra A_1 to another algebra A_2 is a linear map which is also a ring homomorphism. An *isomorphism* is a bijective homomorphism. An *involution* on an algebra A is a map $x \to x^*$ on A satisfying

(1.2.7) $x \to x^*$ is conjugate linear.

(1.2.8) $(xy)^* = y^*x^*$, $x, y \in A$.

(1.2.9) $(x^*)^* = x$, $x \in A$.

The $*$-operation is also called the *adjoint operation*, x^* the *adjoint* of x, and an algebra with involution is called a $*$-*algebra*. An element x in a $*$-algebra is said to be *self-adjoint* or *hermitian* if $x = x^*$. The set of self adjoint elements of A is denoted by A_h. Each element x in A has a unique decomposition as $x = x_1 + ix_2$ with x_1 and x_2 in A_h; namely $x_1 = (x + x^*)/2$, $x_2 = (x - x^*)/2i$. Then A_h is a vector space over $\mathbb{R}$, but is not a subalgebra unless A is commutative. An element x in a $*$-algebra A is said to be *isometric* or an *isometry* if $x^*x = 1$, and *unitary* if both x and x^* are isometric. A $*$-homomorphism π between $*$-algebras is a homomorphism preserving adjoints : $\pi(x^*) = \pi(x)^*$. A $*$-*isomorphism* is a bijective $*$-homomorphism. A *normed (Banach) algebra* is an algebra A equipped with a map $\|\cdot\| : A \to \mathbb{R}^+$, such that as a vector space it is a normed (Banach) space and the norm is related to the multiplication by the property $\|xy\| \leq \|x\|\|y\|$, $x, y \in A$. A is a *normed (Banach)* $*$-*algebra* if moreover A has an involution which is isometric, i.e. $\|x^*\| = \|x\|$, $x \in A$. Similarly we define *automorphism, subalgebra,* $*$-*subalgebra, Banach subalgebra* etc.

Before we come to our prime examples, we need some more notation. If $T \in B(H_1, H_2)$, the bounded linear operators between Hilbert spaces H_1 and H_2, the adjoint T^* is given by the Riesz representation theorem as the unique operator in $B(H_2, H_1)$ such that $\langle Tf, g\rangle = \langle f, T^*g\rangle$, $f \in H_1$, $g \in H_2$. The map $T \to T^*$ is conjugate linear, anti–multiplicative, idempotent, isometric $\|T^*\| = \|T\|$ and $\|T^*T\| = \|T\|^2$, for $T \in B(H_1, H_2)$. Moreover T is isometric if and only if $\|Tx\| = \|x\|$ for all $x \in H_1$.

Example 1.1 If H is a Hilbert space, $B(H) = B(H, H)$ is a unital Banach $*$-algebra, with the additional property that

(1.2.10) $\|T^*T\| = \|T\|^2$, $T \in B(H)$.

Such an algebra is commutative only when H is zero or one dimensional.

Example 1.2 Let Ω be a compact Hausdorff space. Then $C(\Omega)$, the space of all complex valued continuous functions on Ω, is a commutative $*$-algebra under the pointwise operations:

(1.2.11) scalar multiplication $\qquad (\lambda f)(\omega) = \lambda f(\omega), \lambda \in \mathbb{C}, f \in C(\Omega)$.

(1.2.12) vector addition $(f+g)(\omega) = f(\omega) + g(\omega), f, g \in C(\Omega)$.

(1.2.13) vector multiplication $(fg)(\omega) = f(\omega)g(\omega), f, g \in C(\Omega)$.

(1.2.14) involution $(f^*)(\omega) = \overline{f(\omega)}, f \in C(\Omega)$.

$C(\Omega)$ is unital with identity 1, where $1(\omega) = 1$, $\omega \in \Omega$. We can define a norm $\|\cdot\|$ on $C(\Omega)$ by

$$\|f\| = \sup_{\omega \in \Omega} |f(\omega)|, \quad f \in C(\Omega), \tag{1.2.15}$$

which is finite since Ω compact. Then $C(\Omega)$ is a Banach $*$-algebra, where

$$\|f^* f\| = \|f\|^2, \quad f \in C(\Omega). \tag{1.2.16}$$

(Since $\|f^*f\| = \sup|(f^*f)(\omega)| = \sup|f(\omega)|^2 = \|f\|^2$.) Thus the norm is related to the $*$-multiplicative structure in a similar way to that in $B(H)$ (see 1.2.10).

To handle these two examples simultaneously, we make the following definition:

Definition 1.3 *A C^*-algebra or operator algebra is a Banach $*$-algebra A which is isometrically $*$-isomorphic to a (necessarily norm closed) $*$-subalgebra of $B(H)$ for some Hilbert space H.*

Thus a C^*-algebra on a Hilbert space H will be a subset of $B(H)$ which is closed under all the algebraic operations (scalar multiplication, vector addition, vector multiplication, and the involution) and under the norm topology. The following abstract characterization (i.e. without reference to any Hilbert space) is extremely important, but will not be proved here, since we will usually be able to concretely identify with *ad hoc* methods the algebras of interest to us as operator algebras:

Theorem 1.4 Gelfand–Naimark theorem. *If A is a Banach $*$-algebra where*

$$\|T^*T\| = \|T\|^2, \quad T \in A \tag{1.2.17}$$

then A is a C^*-algebra, i.e. there exists an isometric $*$-homomorphism π from A into $B(H)$ for some Hilbert space H.

In particular, if Ω is a compact Hausdorff space, then we should be able to construct a $*$-isometric embedding of $C(\Omega)$ into $B(H)$ for some Hilbert space H, since by Example 1.2, $C(\Omega)$ is a Banach $*$-algebra where $\|f^*f\| = \|f\|^2$, $f \in C(\Omega)$. Before we consider that, it is useful to first introduce the notion of direct sum of Hilbert spaces and C^*-algebras. Thus let $H_i, i = 1, 2, \ldots, N$ (where $1 \leq N \leq \infty$) be a sequence of Hilbert spaces. Then the space of sequences:

$$H = \left\{(\xi_i)_{i=1}^N \mid \xi_i \in H_i, \textstyle\sum_{i=1}^N \|\xi_i\|^2 < \infty\right\} \tag{1.2.18}$$

is a Hilbert space, for pointwise addition and scalar multiplication:

$$(\xi + \eta)_i = \xi_i + \eta_i, \quad (\lambda\xi)_i = \lambda\xi_i$$

$$\xi = (\xi_i), \quad \eta = (\eta_i) \in H, \quad \lambda \in \mathbb{C}$$

and inner product

$$\langle \xi, \eta \rangle = \sum_{i=1}^N \langle \xi_i, \eta_i \rangle . \tag{1.2.19}$$

We then write $\bigoplus_{i=1}^N H_i$ for H, called the Hilbert space direct sum (so that $\ell^2(N) = \bigoplus_{i=1}^N \mathbb{C}$), and $\bigoplus_{i=1}^N \xi_i$ for $(\xi_i)_{i=1}^N \in \bigoplus_{i=1}^N H_i$. The map $V_j : H_j \to \bigoplus_{i=1}^N H_i$ defined by

$$V_j : h \to (0, \ldots, 0, h, 0, \ldots) \quad j'\text{th position} \tag{1.2.20}$$

is an isometric embedding, so that we can identify H_j as a closed subspace of $\bigoplus_{i=1}^N H_i$. Then if $\{e_n^j\}_{n=1}^{m_j}$ is a complete orthonormal sequence for H_j, $(1 \leq j \leq N)$, then $\{V_j(e_n^j) \mid j, n\}$ is a complete orthonormal sequence for $\bigoplus_{i=1}^N H_i$, and so $\bigoplus_{i=1}^N H_i$ is separable. Next suppose A_i is a C^*-algebra, for each $i = 1, \ldots, N$ (where $1 \leq N \leq \infty$). Then what would be the appropriate C^*-direct sum $\bigoplus_{i=1}^N A_i$? Let us define the following space of sequences:

$$A = \left\{ (T_i)_{i=1}^N \mid T_i \in A_i, \sup_{i=1,\ldots,N} \|T_i\| < \infty \right\} . \tag{1.2.21}$$

Theorem 1.5 *A is a C^*-algebra under the pointwise operations (scalar multiplications, vector addition, multiplication, involution) for the norm*

$$\|(T_i)_{i=1}^N\| = \sup_{i=1,\ldots,N} \|T_i\| . \tag{1.2.22}$$

Moreover, if A_i can be represented (faithfully) on a Hilbert space H_i then A can be represented (faithfully) on the Hilbert space $\bigoplus_{i=1}^N H_i$.

Proof We leave it as an exercise that A is a Banach $*$-algebra for the above norm (1.2.22). Suppose without loss of generality that A_i is a C^*-algebra on a Hilbert space H_i. We will try to define a map

$$\pi : A \to B\left(\bigoplus_{i=1}^N H_i\right)$$

by $\pi(T)(\xi) = (T_i \xi_i)_{i=1}^N$, if $T = (T_i)_{i=1}^N \in A$, and $\xi = (\xi_i)_{i=1}^N \in \bigoplus_{i=1}^N H_i$. To show that this is well defined, note that for M finite:

$$\sum_{i=1}^M \|T_i \xi_i\|^2 \leq \sum_{i=1}^M \|T_i\|^2 \|\xi_i\|^2 \leq \|T\|^2 \sum_{i=1}^M \|\xi_i\|^2 \leq \|T\|^2 \|\xi\|^2 .$$

Hence $(T_i \xi_i)_{i=1}^N \in \bigoplus_{i=1}^N H_i$ and $\|(T_i \xi_i)\| \leq \|T\| \|\xi\|$. Thus $\pi(T)$ is a bounded linear map on $\bigoplus_{i=1}^N H_i$ and $\|\pi(T)\| \leq \|T\|$ for all $T \in A$. Now if $S \in B(H)$, then $\|S\| \geq \|S|_K\|$ for any subspace $K \subset H$. In particular, regarding H_j as a subspace of $\bigoplus_{i=1}^N H_i$, we see $\pi(T)|_{H_j} = T_j$, and so $\|\pi(T)\| \geq \|T_j\|$ for all j. Thus $\|\pi(T)\| \geq \sup \|T_j\|$, and so $\|\pi(T)\| = \|T\|$. □

We write $\bigoplus_{i=1}^N A_i$ for A, called the C^*-direct sum of the C^*-algebras A_i, and $\bigoplus_{i=1}^N T_i$ for $(T_i)_{i=1}^N \in \bigoplus_{i=1}^N A_i$. For example, if $A_i = \mathbb{C}$ ($= B(\mathbb{C})$) then we usually write ℓ^∞ for the C^*-direct sum $\bigoplus_{i=1}^\infty \mathbb{C} = \{(\xi_i) \mid \xi \in \mathbb{C}, \sup |\xi_i| < \infty\}$, which can be represented on the Hilbert space direct sum $\ell^2 = \bigoplus_{i=1}^\infty \mathbb{C}$. More generally, if $A_i = B(H_i)$ for each i, then we can embed $\bigoplus_{i=1}^N B(H_i)$ in $B\left(\bigoplus_{i=1}^N H_i\right)$. In particular if $H_1 = \mathbb{C}^n$, $H_2 = \mathbb{C}^m$ are finite dimensional Hilbert spaces, with canonical bases $e_1, \ldots, e_n$, and $f_1, \ldots, f_m$, then we can identify $\mathbb{C}^n \oplus \mathbb{C}^m$ with $\mathbb{C}^{n+m}$, such that $e_1, \ldots, e_n, f_1, \ldots, f_m$ in $\mathbb{C}^n \oplus \mathbb{C}^m$ (regarding $\mathbb{C}^n, \mathbb{C}^m \to \mathbb{C}^n \oplus \mathbb{C}^m$ as in (1.2.20)) are identified with the canonical basis in $\mathbb{C}^{n+m}$. Thus with these choices of bases we have identifications of $B(H_1)$, $B(H_2)$, $B(H_1 \oplus H_2)$ with the matrix algebras M_n, M_m, M_{m+n} respectively. Then the embedding

$$B(H_1) \bigoplus B(H_2) \to B\left(H_1 \bigoplus H_2\right)$$

of Theorem 1.5 is, with these identifications, the embedding of

$$M_n \bigoplus M_m \to M_{m+n}$$

given by

$$a \oplus b \to \begin{pmatrix} a & 0 \\ 0 & b \end{pmatrix}$$

where a, b are in M_n, M_m respectively. More generally, we can embed $M_{n_1} \bigoplus \cdots \bigoplus M_{n_p}$ isometrically in $M_{n_1 + \cdots + n_p}$ by

$$a_1 \oplus \cdots \oplus a_p \to \begin{pmatrix} a_1 & & & 0 \\ & \cdot & & \\ & & \cdot & \\ & & & \cdot \\ 0 & & & a_p \end{pmatrix}, a_i \in M_{n_i}\,. \tag{1.2.23}$$

With this construction of direct sums of operator algebras, we can now embed the algebra $C(\Omega)$ when Ω is a separable compact Hausdorff space as an algebra of operators on a separable Hilbert space:

Theorem 1.6 *If Ω is a separable, compact, Hausdorff space, then $C(\Omega)$ can be isometrically embedded as a C^*-subalgebra of $B(H)$ for a separable Hilbert space H.*

Proof Let $\{\omega_n\}_{n=1}^\infty$ be a countable dense subset of Ω. If $f \in C(\Omega)$, then f is bounded and so $(f(\omega_n))_{n=1}^\infty$ is a bounded sequence and so defines an element of $\bigoplus_{i=1}^\infty \mathbb{C}$ (the C^*-direct sum), which by Theorem 1.5, can be regarded as an operator, say $\pi(f)$, on $\ell^2 = \bigoplus_{i=1}^\infty \mathbb{C}$ (the Hilbert space direct sum). It is clear that $\pi : C(\Omega) \to B(\ell^2)$ is a $*$-homomorphism, which is isometric as

$$\|\pi(f)\| = \sup_{1\leq n<\infty} |f(\omega_n)| = \sup_{\omega\in\Omega} |f(\omega_n)| = \|f\|$$

where the second equality holds as f is continuous and $\{\omega_n\}_{n=1}^{\infty}$ is dense in Ω. □

Remark When Ω is not necessarily separable one could take a Borel measure μ on Ω, and form $L^2(\Omega,\mu)$, the space of complex valued measurable square integrable functions on Ω

$$\text{i.e.}\quad L^2(\Omega,\mu) = \left\{ f : \Omega \to \mathbb{C} \text{ measurable } \mid \int |f|^2 d\mu < \infty \right\}.$$

Then we can let $C(\Omega)$ act on $L^2(\Omega,\mu)$ by pointwise multiplication: $\pi(f)g = fg$. In the proof of Theorem 1.6, μ is counting measure on $\{x_n\}_{n=1}^{\infty}$ and zero on its complement $\Omega\backslash\{x_n\}_{n=1}^{\infty}$. We could also take μ to be counting measure on Ω, and this would lead to the space

$$\left\{ f : \Omega \to \mathbb{C} \mid \textstyle\sum_{\omega\in\Omega} |f(\omega)|^2 < \infty \right\} \tag{1.2.24}$$

Here we would have to be careful about the meaning of the summation, and unless Ω is countable this would lead to non-separable Hilbert spaces (see Exercises 1.3 – 1.6).

We have noted in Theorem 1.4, how to characterize C^*-algebras without reference to a Hilbert space. The following abstract characterization of the algebras $C(\Omega)$, where Ω is compact Hausdorff space, as precisely all commutative C^*-algebras will not be proved here, rather we will again make do with *ad hoc* methods for concrete identifications in the specific commutative C^*-algebras of interest to us.

Theorem 1.7 *If A is a commutative unital Banach $*$-algebra with the C^*-property of the norm, $\|f^*f\| = \|f\|^2$, $f \in A$, then there exists a compact Hausdorff space Ω such that A is isometrically $*$-isomorphic to $C(\Omega)$.*

We now consider the case of a singly generated commutative C^*-algebra. In fact, let us take a self adjoint element T in a unital C^*-algebra A. If $p \in \mathbb{C}[t]$, (the polynomials in an indeterminate t with complex coefficients), so that

$$p(t) = \textstyle\sum_{i=0}^{n} \lambda_i t^i \tag{1.2.25}$$

for some n, and $\lambda_0, \ldots, \lambda_n \in \mathbb{C}$, define

$$p(T) = \textstyle\sum_{i=0}^{n} \lambda_i T^i \tag{1.2.26}$$

where $T^0 = 1$, the unit of A. Then $p \to p(T)$ is a $*$-homomorphism from $\mathbb{C}[t]$ into A. Here we regard $\mathbb{C}[t]$ as a (commutative) unital algebra with operations

$$(\lambda p)(t) = \textstyle\sum_i \lambda\lambda_i t^i \tag{1.2.27}$$

$$(p+q)(t) = \sum_i (\lambda_i + \mu_i) t^i \tag{1.2.28}$$

$$(pq)(t) = \sum_k \left(\sum_{i+j=k} \lambda_i \mu_j \right) t^k \tag{1.2.29}$$

$$p^*(t) = \sum_i \overline{\lambda_i} t^i , \tag{1.2.30}$$

if $p(t) = \sum \lambda_i t^i$, $q(t) = \sum \mu_i t^i$, $\lambda \in \mathbb{C}$. Note that $p^*(T) = p(T)^*$ as T is self adjoint. Thus $B_0 = \{p(T) \mid p \in \mathbb{C}[t]\}$ is a commutative unital $*$-subalgebra of A. Hence (see Exercise 1.6), the closure B of B_0 in A is a commutative unital C^*-subalgebra of A. Then according to Theorem 1.7, there should exist a compact Hausdorff space Ω such that B is isometrically $*$-isomorphic to $C(\Omega)$. Rather than prove Theorem 1.7, we will in fact construct directly an isomorphism $B \cong C(\Omega)$, where Ω will not be just an abstract compact Hausdorff space, but will be a compact subset of the complex plane. Then under this identification the operator T in B will correspond to the continuous map

$$f(z) = z, \quad z \in \Omega \tag{1.2.31}$$

and the identity 1 of B to the unit 1 of $C(\Omega)$, namely

$$1(z) = 1, \quad z \in \Omega. \tag{1.2.32}$$

Note that (1.2.31) makes sense if $\Omega \subset \mathbb{C}$, and from now on, in a unital algebra we will write μ for $\mu 1$, if $\mu \in \mathbb{C}$. The problem we have to consider is how to identify Ω from the algebra B, or indeed how to recover Ω from $C(\Omega)$. We observe that if $\mu \in \mathbb{C}$, then there exists $g \in C(\Omega)$ such that

$$(\mu - f)g = 1 = g(\mu - f) \tag{1.2.33}$$

if and only if $\mu \notin \Omega$. For if $\mu \notin \Omega$, we take $g(z) = (\mu - z)^{-1}$, and if g exists, then

$$1 = (\mu - z)g(z) \tag{1.2.34}$$

for all z in Ω, so that $\mu \notin \Omega$. Thus

(1.2.35) $\quad \Omega = \{\mu \in \mathbb{C} \mid \mu - f \text{ does not have an inverse in } C(\Omega)\}$.

This allows us to recover Ω from $C(\Omega)$, and so we should be able to identify Ω as

(1.2.36) $\quad \{\mu \in \mathbb{C} \mid \mu - T \text{ does not have an inverse in } B\}$.

In the next section, we will consider the spectrum of an element T in a Banach algebra A, defined as in (1.2.36), and then later in Section 1.3 (Theorem 1.18), prove the identification of B with $C(\Omega)$ when T is a self adjoint element of a C^*-algebra.

Exercise 1.1 Let A be a normed algebra. Show that multiplication is jointly continuous, i.e. $(x, y) \to xy$ is continuous from $A \times A$ into A.

Exercise 1.2 Let $C_0(\mathbb{R})$ denote the $*$-algebra of continuous functions from $\mathbb{R}$ to $\mathbb{C}$ which vanish at infinity under the usual pointwise operations as in (1.2.11–14). (Here $f : \mathbb{R} \to \mathbb{C}$ is said to vanish at infinity if given $\varepsilon > 0$, there exists N such that for $|x| > N$, $|f(x)| < \varepsilon$). Show that $\|f\| = \sup_{x\in\mathbb{R}} |f(x)|$ defines a norm on $C_0(\mathbb{R})$, such that $C_0(\mathbb{R})$ becomes a Banach $*$-algebra with the C^*-property of the norm, but $C_0(\mathbb{R})$ is not unital.

Exercise 1.3 Let Ω be a set and g a map from Ω into $[0, \infty)$. Define

$$\textstyle\sum_{\omega\in\Omega} g(\omega) = \sup_F \sum_{\omega\in F} g(\omega) \tag{1.2.37}$$

where the supremum is taken over all finite subsets F of Ω. Show that if $\sum g(\omega) < \infty$, then the set $\{\omega \mid g(\omega) > 1/n\}$ is finite and so $\mathrm{supp}(g) = \{\omega \mid g(\omega) \neq 0\}$ is countable, and

$$\textstyle\sum_{\omega\in\Omega} g(\omega) = \sum_{\omega\in\mathrm{supp}(g)} g(\omega) \tag{1.2.38}$$

where the right hand side is the usual sum of an infinite series.

Exercise 1.4 Let Ω be a set and $\ell^2(\Omega)$ the space of all functions g from Ω into $\mathbb{C}$ such that $\sum_{\omega\in\Omega} |g(\omega)|^2 < \infty$. Show that ℓ^2 is a Hilbert space for the usual pointwise vector space operations and inner product

$$\langle f, g\rangle = \textstyle\sum_{\omega\in\Omega} f(\omega)\overline{g(\omega)}\,. \tag{1.2.39}$$

(Note that if $f, g \in \ell^2(\Omega)$, then $f, g, f + g$ and the summation of $\sum f(\omega)\overline{g(\omega)}$ are supported by some countable subset of Ω. Thus (1.2.39) converges from the existence of the inner product in $\ell^2(\mathbb{N})$. Similarly if (f_n) is sequence in $\ell^2(\Omega)$, then $\mathrm{supp}(f_n)$ is contained in the same countable subset of Ω and so completeness will follow from that of $\ell^2 = \ell^2(\mathbb{N})$.

Exercise 1.5 If Ω is a set and $\omega \in \Omega$, let e_ω be the element of $\ell^2(\Omega)$ given by $e_\omega(\omega') = \delta_{\omega\omega'}$. Show that $X = \{e_\omega \mid \omega \in \Omega\}$ is an orthonormal subset of $\ell^2(\Omega)$, where $\|e_\omega - e_{\omega'}\| = 2$ if $\omega \neq \omega'$. Deduce that if $\ell^2(\Omega)$ is separable then X is countable (and hence Ω is countable).

Exercise 1.6 If B_0 is a $*$-subalgebra of a Banach $*$-algebra A, show that the norm closure $\overline{B_0}$ is a Banach $*$-algebra.

Exercise 1.7 Let $\ell^\infty = \{(f_i)_{i=1}^\infty \mid f_i \in \mathbb{C}, \sup|f_i| < \infty\}$. Then $\pi : \ell^\infty \to B(\ell^2)$, given by $\pi(f)\xi = (f_i\xi_i)$, $f = (f_i) \in \ell^\infty$, $\xi = (\xi_i) \in \ell^2$, gives an isometric $*$-isomorphism of ℓ^∞ onto a C^*-subalgebra of $B(\ell^2)$. Suppose $S \in B(\ell^2)$ is such that $S\pi(f) = \pi(f)S$ for all $f \in \ell^\infty$. Show that there exists a unique $g \in \ell^\infty$ such that $S = \pi(g)$.

Exercise 1.8 Let M be a closed subspace of a Hilbert space H, so that (by the Riesz representation theorem for continuous linear functionals on M) we have a decomposition $H = M \oplus M^\perp$, i.e. every vector $h \in H$ has a unique decomposition $h = m + m^\perp$, where $m \in M$ and $m^\perp \in M^\perp$, the orthogonal subspace. Then $Ph = m$ defines a projection, i.e. an operator $P \in B(H)$ such that $P = P^* = P^2$. Show that $M \to P$ and $P \to PH$ defines a bijection between projections and closed subspaces of H.

1.3 Spectrum of an element

Let A be a unital algebra. We say that s in A is *invertible* (in A) if there exists t in A such that $st = ts = 1_A$. Then t is necessarily unique and written s^{-1}, and called the *inverse* of s. The set A^{-1} of invertible elements forms a group. If $s \in A$, the *resolvent* set of s, written $\rho_A(s)$, is defined to be the set

(1.3.1) $\quad \{\lambda \in \mathbb{C} \mid \lambda - s \text{ is invertible in } A\}.$

The *spectrum* of s, $\sigma_A(s)$ is defined as the complement of the resolvent, namely

(1.3.2) $\quad \sigma_A(s) = \{\lambda \in \mathbb{C} \mid \lambda - s \text{ is not invertible in } A\}.$

Later we shall see that if $s \in A \subset B$, where A and B are unital C^*-algebras with the *same unit* then $\rho_A(s) = \rho_B(s)$, $\sigma_A(s) = \sigma_B(s)$, so that for C^*-algebras we will eventually drop the subscripts and write $\sigma(s)$, $\rho(s)$. Before then we will also drop the subscripts when there is no confusion about which algebra we are thinking of s as lying in. We first aim to show that if x lies in a unital Banach algebra, then $\sigma(x)$ is compact, and so $\rho(x)$ is non-empty. If this is the case, we must have a method for proving the existence of inverses for $(\lambda - x)$ for some $\lambda \in \mathbb{C}$. To do that we will use power series expansions $\sum_{n=0}^{\infty} \lambda^{-n+1} x^n$, for large $|\lambda|$. To prove convergence of such series, it is convenient to consider first the following special case.

Lemma 1.8 *If x is in a unital Banach algebra A and $\|x\| < 1$, then $1 - x$ is invertible and*

(1.3.3) $\quad (1 - x)^{-1} = \sum_{i=0}^{\infty} x^i \quad (x^0 = 1).$

Proof We first show that the partial sums $s_m = \sum_{i=0}^{m} x^i$ are Cauchy. This is clear as if $m > n$ then

$$\|s_m - s_n\| = \left\|\sum_{i=n+1}^{m} x^i\right\| \leq \sum_{i=n+1}^{m} \|x^i\| \leq \sum_{i=n+1}^{m} \|x\|^i$$

and so as $\|x\| < 1$, $\{s_m\}$ is Cauchy. But A is complete and so s_m converges to an element y, say, in A. Then

(1.3.4) $\quad (1 - x)s_n = s_n(1 - x) = 1 - x^{n+1}.$

But multiplication is jointly continuous in A (Exercise 1.1) and $x^{n+1} = s_{n+1} - s_n \to 0$ as $n \to \infty$. Thus from (1.3.4) we see, letting $n \to \infty$, that

$$(1 - x)y = y(1 - x) = 1\,.$$

Thus $1 - x$ is invertible and $(1 - x)^{-1} = y$. □

ary 1.9 *If x is in a unital Banach algebra A, then*

(1.3.5) $\sigma(x) \subset \{\lambda \in \mathbb{C} \mid |\lambda| \leq \|x\|\}$

(1.3.6) $(\lambda - x)^{-1} = \sum_{m=0}^{\infty} \lambda^{-m-1} x^m$, *if* $|\lambda| > \|x\|$.

Proof Take $\lambda \in \mathbb{C}$, with $|\lambda| > \|x\|$. Then $\|\lambda^{-1}x\| < 1$, and so by Lemma 1.8, $1-\lambda^{-1}x$ is invertible and $(1-\lambda^{-1}x)^{-1} = \sum_{m=0}^{\infty} \lambda^{-m}x^m$. But $\lambda - x = \lambda(1-\lambda^{-1}x)$ and the invertible elements form a group. Thus $\lambda - x$ is invertible and

$$(\lambda - x)^{-1} = \lambda^{-1}(1 - \lambda^{-1}x)^{-1} = \sum_{m=0}^{\infty} \lambda^{-m-1}x^m .$$

□

Next we show that the resolvent set $\rho(x)$ is open (when x is an element of a Banach algebra), and that the map $\lambda \to R_\lambda = (\lambda - x)^{-1}$ from $\rho(x)$ into the invertible elements is analytic. A map f from an open subset U of $\mathbb{C}$ into a normed space X is *analytic* if the following exists in norm:

$$\lim_{\mu \to \lambda} (f(\mu) - f(\lambda))/(\mu - \lambda) \tag{1.3.7}$$

for each $\lambda \in U$. In which case the limit is denoted by $f'(\lambda)$ or $d/d\lambda\, f$, and called the derivative.

Theorem 1.10 *Let x be an element of a unital Banach algebra A. Then $\rho(x)$ is open and the map $\lambda \to R_\lambda$ is analytic on $\rho(x)$, and $d/d\lambda\, R_\lambda = -R_\lambda^2$. If $\lambda, \mu \in \rho(x)$, then $R_\lambda - R_\mu = (\mu - \lambda)R_\mu R_\lambda$, and in particular R_λ, R_μ commute.*

Proof If λ is in $\rho(x)$, then at least formally:

$$1/(\mu - x) = 1/[(\lambda - x) + (\mu - \lambda)] = 1/(\lambda - x)[1 - (\lambda - \mu)/(\lambda - x)]$$

$$= [1/(\lambda - x)]\textstyle\sum_{m=0}^{\infty} [(\lambda - \mu)/(\lambda - x)]^m .$$

Then for μ close to λ, $\|(\lambda - \mu)R_\lambda\| < 1$ (i.e. $|\lambda - \mu| < \|R_\lambda\|^{-1}$), we have $1 - (\lambda - \mu)R_\lambda$ is invertible by Lemma 1.8, and so since the invertible elements form a group, $\mu \in \rho(x)$, and

$$\mu - x = (\lambda - x) + (\mu - \lambda) = (\lambda - x)[1 - (\lambda - \mu)R_\lambda]$$

is invertible with

$$R_\mu = [1 - (\lambda - \mu)R_\lambda]^{-1}R_\lambda = \textstyle\sum_{m=0}^{\infty}(\lambda - \mu)^m R_\lambda^{m+1} \tag{1.3.8}$$

Moreover

$$\|R_\mu - R_\lambda\| = \|\textstyle\sum_{m=1}^{\infty}(\lambda - \mu)^m R_\lambda^{m+1}\| \leq \textstyle\sum_{m=1}^{\infty}|\lambda - \mu|^m \|R_\lambda\|^{m+1}$$

$$= |\lambda - \mu| \|R_\lambda\|^2 / [1 - |\lambda - \mu| \|R_\lambda\|] \, .$$

Hence $\lambda \to R_\lambda$ is clearly continuous. The resolvent identity for $\lambda, \mu \in \rho(x)$ is shown by:

$$\begin{aligned} R_\lambda - R_\mu &= [R_\mu(\mu - x)]R_\lambda - R_\mu[(\lambda - x)R_\lambda] \\ &= R_\mu[(\mu - x) - (\lambda - x)]R_\mu = (\mu - \lambda)R_\mu R_\lambda \, . \end{aligned}$$

Thus if $\lambda \neq \mu$, $R_\mu R_\lambda = (R_\lambda - R_\mu)/(\mu - \lambda)$ is symmetric in λ and μ, so $R_\lambda R_\mu = R_\mu R_\lambda$. Moreover, $(R_\mu - R_\lambda)/(\mu - \lambda) = -R_\mu R_\lambda \to -R_\lambda^2$ as $\mu \to \lambda$. Thus R is analytic and $d/d\lambda \, R_\lambda = -R_\lambda^2$. □

We define the *spectral radius* of an element x in unital algebra A to be the radius $r(x)$ of the smallest circle centred at the origin which completely encloses the spectrum. Thus

$$r(x) = \sup\{|\lambda| \mid \lambda \in \sigma(x)\} \, . \tag{1.3.9}$$

If A is a Banach algebra, then from Corollary 1.9, we have the estimate

$$r(x) \leq \|x\| \, . \tag{1.3.10}$$

We derive a precise formula for $r(x)$, but first a lemma.

Lemma 1.11 *If x is an element of a unital algebra A, then for any $p \in \mathbb{C}[t]$:*

$$\sigma(p(x)) = p(\sigma(x)) = \{p(\mu) \mid \mu \in \sigma(x)\} \tag{1.3.11}$$

Proof Take $\lambda \in \mathbb{C}$, and factorize the polynomial

$$\lambda - p(t) = b\textstyle\prod_{i=1}^n (a_i - t) \, .$$

where $a_i, b \in \mathbb{C}$. If $b = 0$, then the polynomial p is constant and the lemma is obvious. So suppose $b \neq 0$. Then

$$\lambda - p(x) = b\textstyle\prod_{i=1}^n (a_i - x) \, . \tag{1.3.12}$$

Now if $y_1, \ldots, y_n$ are commuting elements of A then the product $y_1 \cdots y_n$ is invertible if and only if all $y_1, \ldots, y_n$ are invertible. (If $y_1, \ldots, y_n$ are invertible, then so is $y_1 \cdots y_n$, since the invertible elements form a group. If $y_1 \cdots y_n$ is invertible then $y_1 \cdots y_n (y_1 \cdots y_n)^{-1} = 1$ and $(y_1 \cdots y_n)^{-1} y_2 \cdots y_n y_1 = 1$, and so y_1 has both right and left inverses, and hence is invertible). Thus $\lambda - p(x)$ is invertible if and only if $a_i - x$ is invertible for all i. Thus

$$\lambda \in \sigma(p(x)) \Leftrightarrow a_i \in \sigma(x) \quad \text{for some } i \Leftrightarrow \lambda \in p(\sigma(x)) \, ,$$

(for $\lambda = p(w)$ if and only if $\omega = a_i$ for some i). □

Theorem 1.12 *If x is an element of a unital Banach algebra A, then $\sigma(x)$ is compact, non-empty (if $A \neq \{0\}$), and*

$$r(x) = \lim_{n\to\infty} \|x^n\|^{1/n}. \tag{1.3.13}$$

Proof The spectrum $\sigma(x)$ is bounded by Corollary 1.9 and closed by Theorem 1.10, and hence is compact. Now by Lemma 1.11, $\sigma(x^n) = \sigma(x)^n$ and so if $\lambda \in \sigma(x)$, then $\lambda^n \in \sigma(x^n)$. Hence by (1.3.10) applied to x^n, $|\lambda^n| \leq r(x^n) \leq \|x^n\|$. Thus $|\lambda| \leq \|x^n\|^{1/n}$ for all $\lambda \in \sigma(x)$, and so $r(x) \leq \|x^n\|^{1/n}$, and hence

$$r(x) \leq \liminf \|x^n\|^{1/n}. \tag{1.3.14}$$

We know that R_λ is analytic in $|\lambda| > r(x)$, so that

$$R_\lambda = \textstyle\sum_{m=-1}^{\infty} b_m/\lambda^{m+1}, \quad |\lambda| > r(x) \tag{1.3.15}$$

for some coefficients $b_m \in A$, by A-valued analytic function theory. But by (1.3.6), we know that

$$R_\lambda = \textstyle\sum_{m=0}^{\infty} x^m/\lambda^{m+1}, \quad |\lambda| > \|x\|. \tag{1.3.16}$$

By comparing (1.3.15) and (1.3.16) and by uniqueness of power series expansions, we must have $b_{-1} = 0$ and $b_m = x^m$ for $m = 1, 2, \ldots$. Thus (1.3.16) also converges for $|\lambda| > r(x)$, and so x^m/λ^{m+1} is a bounded sequence for $|\lambda| > r(x)$. Thus if $|\lambda| > r(x)$, then $\|x^m/\lambda^{m+1}\| \leq \alpha$, for some α (which may depend on λ) and all m. Thus

$$\|x^m\|^{1/m} \leq \alpha^{1/m}|\lambda|^{(m+1)/m},$$

and so

$$\lim_{m\to\infty} \sup \|x^m\|^{1/m} \leq \lim_{m\to\infty} \alpha^{1/m}|\lambda|^{(m+1)/m} = |\lambda|.$$

Hence $\lim_{m\to\infty} \sup \|x^m\|^{1/m} \leq r(x)$, and consequently comparing with (1.3.14) we see that $\lim_{m\to\infty} \|x^m\|^{1/m}$ exists and equals $r(x)$. Now suppose that $\rho(x) = \mathbb{C}$, and $A \neq \{0\}$. Now by (1.3.6), we have for $|\lambda| > \|x\|$:

$$\|R_\lambda\| \leq \sum \|x^m\|/|\lambda^{m+1}| \leq \sum \|x\|^m/|\lambda|^{m+1}$$

$$= |\lambda|^{-1}(1 - \|x\|/|\lambda|)^{-1} \to 0 \text{ as } |\lambda| \to \infty.$$

Thus since R_λ is continuous, we see that R_λ is bounded on $\rho(x) = \mathbb{C}$, and is an everywhere defined bounded analytic function. Thus by the A-valued Liouville theorem, R is constant on $\mathbb{C}$. But $R_\lambda \to 0$ as $|\lambda| \to \infty$, so that $R_\lambda \equiv 0$ on $\mathbb{C}$. Hence $1 = R_\lambda(\lambda - x) = 0$, and so $A = \{0\}$, which is a contradiction. □

In a C^*-algebra, we can now use the C^*-property of the norm to show that for an element T, the norm of T is uniquely determined from the algebraic structure. An element T of a $*$-algebra is said to be *normal* if $T^*T = TT^*$ (e.g. all self adjoint and unitary elements are normal).

Corollary 1.13 *Let T be a normal element of unital C^*-algebra. Then $r(T) = \|T\|$.*

Proof $(T^*T)^m$ is self adjoint for all m, and so by the C^*-property of the norm, and normality:

$$\|(T^*T)^{2^n}\| = \|(T^*T)^{2^{n-1}}(T^*T)^{2^{n-1}}\| = \|(T^*T)^{2^{n-1}}\|^2$$

$$= \|T^*T\|^{2^n} \quad \text{inductively for all } n\,.$$

Hence

$$\|T^{2^n}\|^2 = \|(T^{2^n})^*T^{2^n}\| = \|(T^*T)^{2^n}\| = \|T^*T\|^{2^n} = \|T\|^{2^{n+1}}\,,$$

and

$$\|T\| = \|T^{2^n}\|^{1/2^n} = \lim_{m\to\infty} \|T^m\|^{1/m} = r(T)\,.$$

□

We have thus established already that the norm (at least for normal elements) in a C^*-algebra is fixed by the algebraic structure, since the spectral radius is a purely algebraic notion. This enables us to show, with another application of the C^*-property of the norm, that the norm of C^*-algebra is unique. If T is an arbitrary element of a unital C^*-algebra, then T^*T is self adjoint, and so $\|T^*T\| = r(T^*T)$. But $\|T\|^2 = \|T^*T\|$, and so $\|T\| = r(T^*T)^{1/2}$, which completely describes the norm in terms of the algebraic structure for any element. From this, it is a small matter to deduce that $*$-homomorphisms between C^*-algebras are automatically continuous.

Corollary 1.14 *Let α be a $*$-homomorphism between unital C^*-algebras A and B such that $\alpha(1_A) = 1_B$. Then α is automatically continuous, and is a contraction, $\|\alpha(x)\| \le \|x\|$, $x \in A$. Any $*$-isomorphism α between unital C^*-algebras is an isometric isomorphism, $\|\alpha(x)\| = \|x\|$, $x \in A$.*

Proof Take $y \in A$. Then if $\lambda - y$ is invertible, then so is $\lambda - \alpha(y)$, and so $\rho(y) \subset \rho(\alpha(y))$ or $\sigma(y) \supset \sigma(\alpha(y))$. Hence $r(y) \ge r(\alpha(y))$. Thus if $y = x^*x$ for $x \in A$, we see $r(y) \ge r(\alpha(y))$ implies $\|x\| = [r(x^*x)]^{1/2} \ge [r(\alpha(x^*x))]^{1/2} = [r(\alpha(x)^*\alpha(x))]^{1/2} = \|\alpha(x)\|$, and $\|x\| \ge \alpha(x)\|$. If α is an isomorphism, then $\alpha(1_A) = 1_B$ and we have $\sigma(y) = \sigma(\alpha(y))$, thus equality holds in the above, $\|x\| = \|\alpha(x)\|$, $x \in A$. Alternatively, if α and α^{-1} are contractions, then α is isometric.

□

Taking $A = B$, and α to be the identity map, we see that the norm on a C^*-algebra is unique. Next we aim to show that the spectrum of a self adjoint element x in a C^*-algebra A is real, i.e. $\sigma_A(x) \subset \mathbb{R}$. Now if x is a self adjoint operator on a Hilbert space H, it is rather easy and direct to show that $\sigma_{B(H)}(x) \subset \mathbb{R}$ (Lemma 1.22). We have yet to show that the spectrum of an element is independent of the C^*-algebra we think of the element as lying in. What we will do is to prove, by a Hilbert space free method, that if x is a self adjoint element of a unital C^*-algebra A, then $\sigma_A(x) \subset \mathbb{R}$. Then we will identify the C^*-algebra generated by x and 1 as the continuous functions on $\sigma_A(x)$ (Theorem 1.18), and then deduce from this that the spectrum is independent of the choice of algebra A (Corollary 1.19).

Theorem 1.15 *Let A be a unital C^*-algebra. Then:*

(1.3.17) *If x is invertible then $\sigma(x^{-1}) = \sigma(x)^{-1} = \{\lambda^{-1} \mid \lambda \in \sigma(x)\}$.*

(1.3.18) *If x is unitary then $\sigma(x) \subset \mathbb{T} = \{z \in \mathbb{C} \mid |z| = 1\}$.*

(1.3.19) *If x is self adjoint then $\sigma(x) \subset [-\|x\|, \|x\|] \subset \mathbb{R}$.*

Proof (1.3.17): Since x is invertible, then certainly $0 \in \rho(x)$. Thus, let $\lambda \neq 0$, $\lambda \in \mathbb{C}$. Then $\lambda - x = -\lambda x(\lambda^{-1} - x^{-1})$, and since the invertible elements form a group, we see $\lambda \in \rho(x)$ if and only if $\lambda^{-1} \in \rho(x^{-1})$.

(1.3.18): Note that $1^* = 1$ by uniqueness of the unit, and $\|1\|^2 = \|1^*1\| = \|1\|$, so that $\|1\| = 1$ if $A \neq \{0\}$. Then if x is unitary, $\|x\|^2 = \|x^*x\| = \|1\| = 1$, and so

$$\sigma(x) \subset \{\lambda \in \mathbb{C} \mid |\lambda| \leq 1\} \tag{1.3.20}$$

by (1.3.5). If x is unitary $x^{-1} = x^*$, and so

$$\sigma(x)^{-1} = \sigma(x^{-1}) = \sigma(x^*) \subset \{\lambda \mid |\lambda| \leq 1\} \tag{1.3.21}$$

by (1.3.17), and (1.3.20) applied to the unitary x^*. Combining (1.3.20) and (1.3.21) we see $\sigma(x) \subset \{\lambda \mid |\lambda| = 1\}$.

(1.3.19): Let x be self adjoint. The map taking $t \in \mathbb{R}$ to $(1 - it)/(1 + it)$ on the circle $\mathbb{T}$ leads us to consider $(1 - ix)(1 + ix)^{-1}$ which should be unitary if x is self adjoint. However, at this point we do not know that $1 + ix$ is invertible. Take $\alpha \in \mathbb{C}$, with Im $(\alpha) \neq 0$. We have to show $\alpha - x$ is invertible. But if β is real, non zero, then Im $(\beta\alpha) \neq 0$, and $\beta(\alpha - x) = \beta\alpha - \beta x$ and so $\alpha - x$ is invertible is equivalent to $\beta\alpha - \beta x$ being invertible. Thus by replacing x, α by $\beta x, \beta\alpha$ for β sufficiently small, we can assume $|\alpha|, \|x\| < 1$. Then $(1 + ix)$ is invertible by Lemma 1.8 and $u = (1 - ix)(1 + ix)^{-1}$ is unitary by Exercise 1.10. Now as Im $(\alpha) \neq 0$ (and $\alpha \neq i$), $(1 - i\alpha)/(1 + i\alpha) \notin \mathbb{T}$. Thus by (1.3.18) applied to the unitary u, we see that $(1 - i\alpha)/(1 + i\alpha) \in \rho(u)$ and so

$$-2i(\alpha - x)(1 + i\alpha)^{-1}(1 + ix)^{-1} = (1 - i\alpha)(1 + i\alpha)^{-1} - (1 - ix)(1 + ix)^{-1} \tag{1.3.22}$$

is invertible. Since the invertible elements form a group it is clear that $\alpha - x$ is invertible. We already know from (1.3.5) that $\sigma(x) \subset \{\lambda \in \mathbb{C} \mid |\lambda| \leq \|x\|\}$ and so (1.3.19) holds. □

If x is a self adjoint element of a unital C^*-algebra, we will identify the C^*-algebra generated by x and 1 with the continuous functions on $\sigma(x)$, in such a way that x,1 are identified with the functions $f(\lambda) = \lambda$, and $e(\lambda) = 1, \lambda \in \sigma(x)$, respectively. If this is so then the polynomial $\sum \lambda_i x^i$ must correspond to $\sum \lambda_i f^i$, (where $f^0 = e$). Hence if our claim is correct, then the polynomial functions on $\sigma(x)$ must be dense in all the continuous functions. We will do this in a constructive manner for an interval $[a, b]$, and then invoke Tietze's extension theorem:

Lemma 1.16 Tietze's extension theorem. *If X is a closed subset of a compact metric space M, and $g \in C(X)$, then there exists $h \in C(M)$ such that $h|_X = g$.*

Proof See (Dunford and Schwartz 1958), p.15–17. □

If Ω is a compact subset of the real line, and $p \in \mathbb{C}[t]$, we call $\lambda \to p(\lambda)$, $\lambda \in \Omega$ a *polynomial function* on Ω, and denote by $P(\Omega)$ the subalgebra of $C(\Omega)$ consisting of all polynomial functions. (In the notation of (1.2.26) and (1.2.31), $P(\Omega) = \{p(f) \mid p \in \mathbb{C}[t]\}$.)

Theorem 1.17 *If Ω is a compact subset of $\mathbb{R}$ then $P(\Omega)$ is dense in $C(\Omega)$, for the norm topology $\|g\| = \sup_{\lambda \in \Omega} |g(\lambda)|$ of uniform convergence.*

Proof We first prove density in the case when Ω is the unit interval. If $g \in C[0, 1]$, consider the Bernstein polynomial functions on $[0, 1]$:

$$B_n(g)(x) = \sum_{k=0}^{n} {}^nC_k x^k (1 - x)^{n-k} g\left(k/n\right) .$$

Then we leave it as an exercise on the binomial theorem for the reader to check:

(1.3.23) $B_n(e) = e$

(1.3.24) $B_n(f) = f$

(1.3.25) $B_n(f^2) = (1 - 1/n)f^2 + f/n$

where $e(\lambda) = 1$, $f(\lambda) = \lambda$ as usual. We will show that this is enough for us to be able to prove that $B_n(g) \to g$ for any $g \in C[0, 1]$. Now g is uniformly continuous, so given $\varepsilon > 0$, there exists $\delta > 0$ such that $|x - y| < \delta$, $x, y \in [0, 1]$ implies $|g(x) - g(y)| < \varepsilon$. For this δ, and a fixed x in $[0, 1]$ and a fixed integer n, define the following partition of the integers $\{0, 1, 2, \ldots, n\}$:

$$I = \{k \mid 0 \leq k \leq n, |k/n - x| < \delta\} \tag{1.3.26}$$

and its complement

$$J = \{k \mid 0 \leq k \leq n, 1 \leq |k/n - x|^2/\delta^2\} \tag{1.3.27}$$

Then by (1.3.23),

$$B_n(g)(x) - g(x) = \sum_{k=0}^{n} {}^nC_k x^k (1-x)^{n-k} (g(k/n) - g(x))$$

and so

$$\begin{aligned}
|B_n(g)(x) - g(x)| &\le \sum_{k=0}^{n} {}^nC_k x^k (1-x)^{n-k} |g(k/n) - g(x)| \\
&\le \sum_{k\in I} {}^nC_k x^k (1-x)^{n-k} \varepsilon \\
&\quad + \sum_{k\in J} {}^nC_k x^k (1-x)^{n-k} (k/n - x)^2 2\|g\|/\delta^2 \\
&\le \varepsilon \sum_{k=0}^{n} {}^nC_k x^k (1-x)^{n-k} \\
&\quad + 2\frac{\|g\|}{\delta^2} \sum_{k=0}^{n} {}^nC_k x^k (1-x)^{n-k} \left(x^2 - \frac{2kx}{n} + \frac{k^2}{n^2}\right) \\
&= \varepsilon + (2\|g\|/\delta^2) x(1-x)/n \quad \text{by (1.3.23) -- (1.3.25)} \\
&< \varepsilon + \|g\|/\delta^2 n < 2\varepsilon \quad \text{for } n > \|g\|/\varepsilon\delta^2 ,
\end{aligned}$$

i.e. $B_n(g) \to g$ uniformly on $[0,1]$.

Next consider the case of $\Omega = [a,b]$, a bounded interval. The map $\alpha : x \to xb + (1-x)a$ defines a homeomorphism of $[0,1]$ onto $[a,b]$. Then $\alpha^*(g) = g \circ \alpha$, $g \in C[a,b]$ defines a $*$-homomorphism with inverse $(\alpha^{-1})^*$, defined similarly. Since α is a polynomial function, it is clear that $\alpha^* P[a,b] \subset P[0,1]$, and similarly $(\alpha^{-1})^* P[0,1] \subset P[a,b]$ and so α^* maps $P[a,b]$ onto $P[0,1]$. Now α^* is isometric, as $\|\alpha^*(g)\| = \sup\{|g(\alpha(x))| \mid x \in [0,1]\} = \sup\{|g(y)| \mid y \in [a,b]\}$, as $\alpha[0,1] = [a,b]$. Thus, since $P[0,1]$ is dense in $C[0,1]$, we deduce that $P[a,b]$ is dense in $C[a,b]$. Finally, if Ω is an arbitrary compact subset of $\mathbb{R}$ choose a compact interval $[a,b]$ such that $\Omega \subset [a,b]$. If $g \in C(\Omega)$, choose by Tietze's extension theorem (Lemma 1.16) a continuous function h on $[a,b]$ extending g, and by what we have already established, a sequence of polynomial functions converging uniformly on $[a,b]$ to h. Restricting to Ω, we have a sequence of polynomial functions converging uniformly on Ω to g. This shows $\overline{P(\Omega)} = C(\Omega)$. □

We are now in a position to finally establish our topological spectral theorem.

Theorem 1.18 *Let T be a self adjoint element of a unital C^*-algebra A. Then there exists an unique isometric $*$-isomorphism π from $C(\sigma(T))$ onto $C^*(T)$, (where $C^*(T)$ is the smallest C^*-subalgebra containing T and 1_A, and is the closure of $\{P(T) \mid P \in \mathbb{C}[t]\}$) satisfying $\pi(f) = T$, $\pi(e) = 1_A$ where $f(\lambda) = \lambda$, $e(\lambda) = 1$, $\lambda \in \sigma(T)$.*

Remark Here $\sigma(T) = \sigma_A(T)$. However, since $C^*(T)$ is independent of which A we are considering T to sit inside, we can deduce (see Corollary 1.19) the independence of $\sigma_A(T)$ on A.

Proof of Theorem 1.18. For $p \in \mathbb{C}[t]$, regarded as a polynomial function on $\sigma(T) \subset \mathbb{R}$, define $\pi(p) = p(T) = \sum_{i=0}^n \lambda_i T^i$ if $p(t) = \sum_{i=0}^n \lambda_i t^i$, and $T^0 = 1_A$. Then as noted previously in Section 1.2, π is a homomorphism, which preserves involution, as $\sigma(T)$ is real by Theorem 1.15. We now show that π is isometric:

$$\begin{aligned}\|\pi(p)\|^2 &= \|\pi(p)^*\pi(p)\| = \|\pi(p^*p)\| \\ &= r(\pi(p^*p)) \text{by Corollary 1.13 applied to the self adjoint } \pi(p^*p) \\ &= \sup\{|\lambda| \mid \lambda \in \sigma(\pi(p^*p))\} \\ &= \sup\{|\lambda| \mid \lambda \in \sigma(p^*p)(T)\} \text{ as } \pi(p^*p) = (p^*p)(T) \\ &= \sup\{|\lambda| \mid \lambda \in (p^*p)(\sigma(T))\} \text{ as } \sigma(p^*p)(T) = (p^*p)(\sigma(T)) \text{ by Lemma 1.11} \\ &= \sup\{|p^*p(\mu)| \mid \mu \in \sigma(T)\} = \sup\{|p(\mu)|^2 \mid \mu \in \sigma(T)\} = \|p\|^2 .\end{aligned}$$

Thus $\|\pi(p)\| = \|p\|$, for all $p \in P(\sigma(T))$. Since π is isometric and $P(\sigma(T))$ is dense in $C(\sigma(T))$, we should be able to extend π by continuity to an isometric $*$-homomorphism from $C(\sigma(T))$ into A. It follows from Theorem 1.17 and (1.2.6) that π extends to an isometry from $C(\sigma(T))$ into A. Since π is a $*$-homomorphism on $P(\sigma(T))$, it now follows easily that π is a $*$-homomorphism of $C(\sigma(T))$ (onto $C^*(T)$). Uniqueness is clear, as if π_0 is another isometric $*$-homomorphism satisfying $\pi_0(f) = T$, $\pi_0(e) = 1$, then $\pi_0(p) = p(T) = \pi(p)$ for any $p \in P(\sigma(T))$ since π_0 is a homomorphism. Hence by density and continuity $\pi_0 = \pi$. □

Corollary 1.19 *Let A be a unital C^*-subalgebra of a unital C^*-algebra B, with $1_A = 1_B$, and T an element of A. Then $\sigma_A(T) = \sigma_B(T)$.*

Proof If $\lambda - T$ is invertible in A then $\lambda - T$ is invertible in B, and so $\rho_A(T) \subset \rho_B(T)$ or $\sigma_A(T) \supset \sigma_B(T)$. Next suppose T is self adjoint, so that we have isomorphisms $\pi_1 : C(\sigma_A(T)) \to C^*(T)$, and $\pi_2 : C(\sigma_B(T)) \to C^*(T)$ by Theorem 1.18. If p is a polynomial function on $\sigma_A(T)$ then clearly $\pi_1(p) = \pi_2(p|_{\sigma_B(T)})$ so that by continuity and density $\pi_1(g) = \pi_2(g|_{\sigma_B(T)})$ for all $g \in C(\sigma_A(T))$. Hence for all such g:

$$\|g\| = \|\pi_1(g)\| = \|\pi_2(g|_{\sigma_B(T)})\| = \|g|_{\sigma_B(T)}\| .$$

We claim that this forces $\sigma_A(T) = \sigma_B(T)$. If not take $\lambda \in \sigma_A(T)\backslash\sigma_B(T)$, and then by Tietze's extension theorem, we can find $g \in C(\sigma_A(T))$ such that $g|_{\sigma_B(T)} = 0$, and $g(\lambda) = 1$. Thus $\|g\| \geq 1$, and $\|g|_{\sigma_B(T)}\| = 0$, a contradiction. Hence $\sigma_A(T) = \sigma_B(T)$ for T self adjoint. Next let T be an arbitrary element of A, not necessarily self adjoint, and suppose $\lambda - T$ is invertible in B. Then by Exercise 1.9, $(\lambda - T)^*$ is invertible in B, and thus so is the self adjoint $(\lambda - T)^*(\lambda - T)$. Hence by the preceding self adjoint case, $(\lambda - T)^*(\lambda - T)$ is also invertible in A. Similarly $(\lambda - T)(\lambda - T)^*$ is invertible in A, and so $(\lambda - T)$ has both left and right inverses in A, and so is invertible in A. Thus $\rho_B(T) \subset \rho_A(T)$ as well. □

If T is a self adjoint element of a unital C^*-algebra, and f a continuous function on its spectrum we will also write $f(T)$ for the element $\pi(f)$ constructed in Theorem 1.18. This will extend our previous notation of Section 1.2 for $p(T)$ when p is a polynomial.

Example 1.20 Let us consider what the above means for a self adjoint element T in a matrix algebra M_n. Thus $T = [T_{ij}]$ where $T_{ij} = \overline{T_{ji}} \in \mathbb{C}$. The spectrum of T, being the zeros of the characteristic polynomial, is a finite subset $\{\lambda_1, \ldots, \lambda_r\}$ of real numbers where $1 \leq r \leq n$. (These are the distinct eigenvalues, $\lambda_i \neq \lambda_j, i \neq j$, so that they are not repeated according to any multiplicity). Then $C(\sigma(T))$ can be thought of as all maps from $\{\lambda_1, \ldots, \lambda_r\}$ into $\mathbb{C}$. Define $f_i(x) = 1$ if $x = \lambda_i$, and $f_i(x) = 0$ otherwise. Then $f_i \in C(\sigma(T))$, and $f_i = f_i^2 = \overline{f_i}$, for $i = 1, \ldots, r$. Then since $f \to f(T)$ is a $*$-homomorphism, $e_i = f_i(T)$ satisfies $e_i = e_i^2 = e_i^*$. Thus e_i is an orthogonal projection, and since $\sum f_i = 1, f_i f_j = \delta_{ij} f_i$, we see $\sum e_i = 1, e_i e_j = \delta_{ij} e_i$, and so the ranges of the e_i are mutually orthogonal, with $\mathbb{C}^n = \bigoplus_{i=1}^r e_i \mathbb{C}^n$. If $f \in C(\sigma(T))$, then $f = \sum_{i=1}^r f(\lambda_i) f_i$, and so $f(T) = \sum_{i=1}^r f(\lambda_i) e_i$. Let $\xi_1^i, \ldots \xi_{s(i)}^i$ be a complete orthonormal sequence for $e_i \mathbb{C}^n$, where s_i is the dimension of $e_i \mathbb{C}^n$. Then $\xi_1^1, \ldots, \xi_{s(1)}^1, \xi_1^2, \ldots, \xi_{s(2)}^2, \ldots, \xi_1^r, \ldots, \xi_{s(r)}^r$ will be a complete orthonormal sequence for $\mathbb{C}^n = \bigoplus_{i=1}^r e_i \mathbb{C}^n$. With respect to this ordered basis, e_i will have the diagonal matrix

$$O_{s(1)} \oplus \cdots \oplus O_{s(i-1)} \oplus 1_{s(i)} \oplus O_{s(i+1)} \oplus \cdots \oplus O_{s(r)} \in M_{s(1)} \oplus \cdots \oplus M_{s(r)} \subset M_n\,,$$

where $O_j, 1_j$ represent the zero and identity matrices respectively of M_j. Then $f(T) = \sum f(\lambda_i) e_i$ will have the diagonal matrix

$$f(\lambda_1) 1_{s(1)} \oplus \cdots \oplus f(\lambda_r) 1_{s(r)}\,.$$

For example, if $f(\lambda) = \lambda$ then $f(T) = T$ has the diagonal matrix $\lambda_1 1_{s(1)} \oplus \cdots \oplus \lambda_r 1_{s(r)}$. Thus Theorem 1.18 in this case shows us how to change our basis to diagonalize the self adjoint matrix T, i.e. there exists a unitary matrix u in M_n such that uTu^* is this diagonal matrix.

Example 1.21 Let Ω be a compact Hausdorff space and $g \in C(\Omega)$. Then it is clear (cf. (1.2.35)) that

$$\sigma(g) = \sigma_{C(\Omega)}(g) = \{g(\omega) \mid \omega \in \Omega\}\,. \tag{1.3.28}$$

Suppose g is self adjoint, i.e. g is real valued. Then $\pi : C(\sigma(g)) \to C(\Omega)$, given by $\pi(h) = h \circ g$, is an isometric $*$-homomorphism satisfying $\pi(f) = g$, if $f(\lambda) = \lambda$, and $\pi(e) = 1$, if $e(\lambda) = 1$. Thus by the uniqueness statement in Theorem 1.18, it is clear that

$$h(g) = h \circ g \tag{1.3.29}$$

if $h \in C(\sigma(g))$.

If T is a self adjoint operator, and f a continuous function on the spectrum, it follows from (1.3.28) and (1.3.29) that

$$\sigma(f(T)) = f(\sigma(T)) \tag{1.3.30}$$

thus generalising Lemma 1.11.

Exercise 1.9 Let x be an element of a unital $*$-algebra A. Show that x is invertible if and only if x^* is invertible. In which case $(x^{-1})^* = (x^*)^{-1}$.

Exercise 1.10 Let x be an element in a unital C^*- algebra, with $x = x^*$, $\|x\| < 1$. Show that $(1 - ix)(1 + ix)^{-1}$ is unitary.

Exercise 1.11 Verify (1.3.23–1.3.25).

Exercise 1.12 Regard $A = \mathbb{C}$ as a C^*-subalgebra of $B = \mathbb{C}^2$ by $\lambda \to (\lambda, 0)$. Let $e = 1_A \in A$. Show that $\sigma_A(e) \neq \sigma_B(e)$.

Exercise 1.13 Let $\{A_i\}, \{B_i\}$ be sequences of unital C^*-algebras, and $\{\pi_i\}$ a sequence of unital $*$-homomorphisms $\pi_i : A_i \to B_i$. Show that there is a $*$-homomorphism π from $\bigoplus_i A_i$ into $\bigoplus_i B_i$ such that $\pi\left(\bigoplus_i a_i\right) = \bigoplus_i \pi(a_i)$, if $\bigoplus_i a_i \in \bigoplus_i A_i$. We denote this π by $\bigoplus_i \pi_i$.

Exercise 1.14 If x is an element of a Banach algebra A, show that the series: $\sum_{n=0}^{\infty} x^n/n!$ converges to an element, say e^x, of A. If x and y are commuting elements of A (i.e. $xy = yx$) show that $e^x e^y = e^{x+y}$.

Exercise 1.15 Let T be a self adjoint element of a unital C^*-algebra A. Let exp denote the function $\exp : t \to e^t$ on $\sigma(T)$. Show that $\exp(T) = e^T$, where $\exp(T)$ is given by the spectral Theorem 1.18.

Exercise 1.16 Let T be a self adjoint element of a unital C^*-algebra A. Let f be a real valued continuous function on $\sigma(T)$ and g a continuous complex valued function of $\sigma(f(T))$. Show that $(g \circ f)(T) = g(f(T))$.

Exercise 1.17 Let T be a self adjoint element of a unital C^*-algebra A such that $\sigma(T) \subset (0, \infty)$. Show that there exists a self adjoint element R of A such that $e^R = T$.

Exercise 1.18 If A is an algebra, let $\overline{A} = \{(a, \lambda) \mid a \in A, \lambda \in \mathbb{C}\}$ with operations

$$\begin{aligned}
(a, \lambda) + (b, \mu) &= (a + b, \lambda + \mu) \\
\mu(a, \lambda) &= (\mu a, \mu\lambda) \\
(a, \lambda)(b, \mu) &= (ab + \lambda b + \mu a, \lambda\mu)
\end{aligned}$$

$a, b \in A, \lambda, \mu \in \mathbb{C}$. Show that $\overline{A}$ is a unital algebra and $i : a \to (a, 0)$ embeds A as an ideal in $\overline{A}$ such that $\overline{A}/A \cong \mathbb{C}$, i.e. we have a short exact sequence

$$0 \to A \to \overline{A} \to \mathbb{C} \to 0\,.$$

If A is a $*$-algebra, show that $\overline{A}$ is a $*$-algebra, if $(a, \lambda)^* = (a^*, \overline{\lambda})$.

Exercise 1.19 If A is a C^*-algebra (here we just need to know A is a Banach $*$-algebra, with the C^*-property of the norm). Show that there exists an unique C^*-norm on $\overline{A}$. (Hint: take $\|x\| = \sup\{\|xy\| \mid y \in A, \|y\| \leq 1\}, x \in \overline{A}$). Note that if A is a C^*-algebra on H, and $1_H \notin A$, we can identify $\overline{A}$ with $C^*(A, 1_H)$, the smallest C^*-subalgebra of $B(H)$ containing A and 1_H.

Exercise 1.20 Let A be a unital algebra.

(a) If $x \in A, \lambda \in \mathbb{C}, \lambda \neq 0$, show that $\sigma(\lambda x) = \lambda\sigma(x)$.

(b) If $x, y \in A$, show that $\sigma(xy) \cup \{0\} = \sigma(yx) \cup \{0\}$.

1.4 Positive operators, partial isometries and polar decomposition

Suppose $T \in B(H)$, for some Hilbert space H. We say that T is *positive* if $\langle Tx, x\rangle \geq 0$ for all $x \in H$. Since T is self adjoint if and only if $\langle Tx, x\rangle$ is real for all x, a positive operator is certainly self adjoint. If S, T are self adjoint, we write $S \geq T$ if $S - T$ is positive. Then $\geq$ is a partial ordering on the self adjoint operators. Now if $S \in B(H)$, then $T = S^*S \geq 0$ as $\langle Tx, x\rangle = \langle Sx, Sx\rangle \geq 0$, $x \in H$. Conversely we shall show in this section that if $T \in B(H)$ and $T \geq 0$, then there exists S in $B(H)$ such that $T = S^*S$. For this we will need the functional calculus developed in the previous section. Before doing that it is useful to refine the statement in Theorem 1.15, that the spectrum of a self adjoint element is real.

Lemma 1.22 *Let T be a self adjoint operator on a Hilbert space H, then*

$$\sigma(T) \subset \{\langle Tx, x\rangle \mid x \in H, \|x\| = 1\}^- . \tag{1.4.1}$$

Proof Take $\lambda \in \mathbb{C}$, with $\lambda \notin \{\langle Tx, x\rangle \mid x \in H, \|x\| = 1\}^- = M$. Then for some $\varepsilon > 0$,

$$\varepsilon < |\lambda - \langle Tx, x\rangle|, \quad \text{if } \|x\| = 1 .$$

i.e. if $\|x\| = 1$, then

$$\varepsilon < |\langle(\lambda - T)x, x\rangle| \leq \|(\lambda - T)x\| .$$

If $x \neq 0$, then by considering $x/\|x\|$, we see

$$\varepsilon\|x\| \leq \|(\lambda - T)x\| . \tag{1.4.2}$$

Thus $\mathrm{Ker}(\lambda - T) = 0$, and $\mathrm{Ran}(\lambda - T)$ is closed, as if $(\lambda - T)x_n \to y$ then

$$\|(\lambda - T)x_n - (\lambda - T)x_m\| = \|(\lambda - T)(x_n - x_m)\| \geq \varepsilon\|x_n - x_m\| .$$

Thus x_n is Cauchy and so converges to x, say, in H. Then since $\lambda - T$ is continuous, $y = (\lambda - T)x \in \mathrm{Ran}(\lambda - T)$. Now $\mathrm{Ran}(\lambda - T)^\perp = \mathrm{Ker}(\overline{\lambda} - T) = \{0\}$, using Exercise 1.21, which tells us that $\overline{\lambda} \notin M$ if $\lambda \notin M$, and what we have already established, that $\mathrm{Ker}(\mu - T) = \{0\}$ if $\mu \notin M$. Thus the closed subspace

$\operatorname{Ran}(\lambda - T) = H$, and $(\lambda - T)$ is invertible as a linear map from H into H. If $y \in H$, apply (1.4.2) to $x = (\lambda - T)^{-1}y$, then

$$\varepsilon\|(\lambda - T)^{-1}y\| \leq \|y\| \tag{1.4.3}$$

and so $(\lambda - T)^{-1}$ is bounded with $\|(\lambda - T)^{-1}\| \leq \varepsilon^{-1}$. □

In particular, this gives an elementary proof that the spectrum $\sigma_{B(H)}(T)$ is real (cf. Theorem 1.15; (1.3.19)). However, we preferred to give that abstract proof at that stage in order to be able to deduce from the spectral Theorem 1.18, in Corollary 1.19, that the spectrum is independent of the C^*-algebra to which we think of the element as belonging.

Corollary 1.23 *If T is a self adjoint operator on a Hilbert space H, then*

$$\|T\| = \sup\{|\langle Tx, x\rangle| \mid \|x\| = 1\}. \tag{1.4.4}$$

Proof By Corollary 1.13, Lemma 1.22, and the Schwarz inequality:

$$\|T\| = r(T) \leq \sup\{|\langle Tx, x\rangle| \mid \|x\| \leq 1\} \leq \|T\|.$$

□

Theorem 1.24 *Let T be an element of a C^*- algebra A which acts on a Hilbert space H with $1_H \in A$. Then the following conditions are equivalent:*

(1.4.5) $T = S^*S$ *for some* $S \in A$.

(1.4.6) $\langle Tx, x\rangle \geq 0$ *for all* $x \in H$.

(1.4.7) T *is self adjoint and* $\sigma(T) \subset [0, \infty)$.

Proof If (1.4.6) holds, then T is self adjoint as $R \in B(H)$ is self adjoint if and only if $\langle Rx, x\rangle$ is real for all $x \in H$. Then $\sigma(T) \subset [0, \infty)$ by Lemma 1.22, and so (1.4.7) holds.

If (1.4.7) holds, define $g \in C(\sigma(T))$ by $g(\lambda) = \sqrt{\lambda}, \lambda \in \sigma(T)$. Then $g = g^*$ and $g^2(\lambda) = \lambda$. Thus $S = g(T)$ is self adjoint and $S^2 = T$, so that (1.4.5) holds.

If (1.4.5) holds, then $\langle Tx, x\rangle = \langle Sx, Sx\rangle \geq 0$. □

If A is a unital C^*-algebra, we define the positive elements A_+ of A to be

$$A_+ = \{S^*S \mid S \in A\}. \tag{1.4.8}$$

From Theorem 1.24, it is clear that A_+ forms a *cone* in A_h, i.e:

$$\lambda A_+ \subset A_+, \quad \lambda \geq 0, \quad A_+ + A_+ \subset A_+. \tag{1.4.9}$$

For instance, if Ω is a compact Hausdorff space, then

$$C(\Omega)_+ = \{f \in C(\Omega) \mid f(\Omega) \subset [0, \infty)\}. \tag{1.4.10}$$

Note also that any $*$-homomorphism π between C^*-algebras preserves the positive cones: $\pi(A_+) \subset B_+$, by (1.4.8).

Also if T is self adjoint and $c \leq T \leq C$, where c, C are scalars, then it follows from Theorem 1.18 that

$$\|T\| \leq \max\{|c|, C\}. \tag{1.4.11}$$

Moreover, if T is self adjoint and m (respectively M) is the supremum (respectively infimum) of all real c (respectively C) such that $c \leq T$ (respectively $T \leq C$), then

$$\|T\| = \max\{|m|, M\}. \tag{1.4.12}$$

If T is self adjoint, we can write T as a difference of two positive operators as follows. Define g_+, g_- on $\sigma(T)$ by $g_+(\lambda) = \max(\lambda, 0)$, and $g_-(\lambda) = \max(-\lambda, 0)$, so that $\lambda = g_+(\lambda) - g_-(\lambda)$ and $T = g_+(T) - g_-(T)$. Since $g_\pm \geq 0, g_+g_- = 0$, it is clear that $T_\pm = g_\pm(T) = \pi(g_\pm) \geq 0$, and $T_+T_- = 0$.

It follows from the proof of Theorem 1.24 that if $T \in A$ is positive then there exists a positive S in A such that $T = S^2$. We now show uniqueness of such a positive square root S, and will denote this object by $\sqrt{T}$ or $T^{1/2}$. If S is any element of A, we will denote $(S^*S)^{1/2}$ by $|S|$.

Theorem 1.25 *Let T be a positive element of a C^*-algebra A which acts on a Hilbert space H, with $1_H \in A$. Then there exists an unique positive S in A such that $S^2 = T$. Moreover S is a norm limit of polynomials in T.*

Proof By the proof of Theorem 1.24, there exists a positive S in $C^*(T)$ such that $S^2 = T$. Let S_1 be any other positive element of A such that $S_1^2 = T$. Then $S_1 T = S_1 S_1^2 = S_1^2 S_1 = TS_1$. Since $S \in C^*(T)$, it follows that S and S_1 commute. Again by the proof of Theorem 1.24, choose positive R, R_1 in $C^*(S), C^*(S_1)$ respectively such that $R^2 = S, R_1^2 = S_1$. If $y = (S - S_1)x$, then

$$\begin{aligned}\|Ry\|^2 + \|R_1 y\|^2 &= \langle R^2 y, y\rangle + \langle R_1^2 y, y\rangle = \langle (S + S_1)y, y\rangle \\ &= \langle (S + S_1)(S - S_1)x, y\rangle = 0.\end{aligned}$$

Thus $R(S - S_1) = R_1(S - S_1) = 0$, and so $S(S - S_1) = S_1(S - S_1) = 0$. Then $\|(S - S_1)x\|^2 = \langle (S - S_1)x, (S - S_1)x\rangle = 0$ and so $S = S_1$. □

Having noted that any operator T can be uniquely expressed as $T_1 + iT_2$ where T_i are self adjoint, and that a self adjoint operator S can be expressed as a difference of two positive operators S_+, S_-, where $S_+S_- = 0$, it is natural to ask what is the operator analogue of the polar decomposition $e^{i\theta}r$ in $\mathbb{C}$. For this, we need to consider not only isometries, but partial isometries which are isometric on a closed subspace of a Hilbert space and zero on its orthogonal complement.

Definition 1.26 *A partial isometry on a Hilbert space H is a linear operator U from H into itself such that for some closed subspace M of H, $U|_M$ is an isometry and $U|_{M^\perp} = 0$. Then M is called the initial space of U, and $N = UH$ is called the final space of U. Note that N is closed since $\|Um\| = \|m\|$, for all $m \in M$, and M is complete.*

The following lemma implies that the initial and final spaces of a partial isometry are well defined:

Lemma 1.27 *If U is a partial isometry then U^*U (respectively UU^*) is the projection on the initial space M (respectively final space N), and U^* is a partial isometry with initial space N and final space M.*

Proof If $m_1, m_2 \in M$, $m_1^\perp, m_2^\perp \in M^\perp$, then $Um_i^\perp = 0$ and

$$\begin{aligned}\langle U^*U(m_1 + m_1^\perp), (m_2 + m_2^\perp)\rangle &= \langle Um_1, Um_2\rangle = \langle m_1, m_2\rangle \\ &= \langle P_M(m_1 + m_1^\perp), m_2 + m_2^\perp\rangle .\end{aligned}$$

Hence $U^*U = P_M$. But $U|_{M^\perp} = 0$ and $P_{M^\perp} = 1 - U^*U$, so $U(1 - U^*U) = 0$ or $U = UU^*U$. Thus $UU^*|_N = 1_N$. Moreover $UU^*|_{N^\perp} = 0$, as $\mathrm{Ker}(U^*) = N^\perp$ by Exercise 1.21, and consequently $P_N = UU^*$. Finally, since $\|U^*Ux\|^2 = \langle U^*Ux, U^*Ux\rangle = \langle UU^*Ux, Ux\rangle = \langle Ux, Ux\rangle = \|Ux\|^2$, for all $x \in H$, using $UU^*U = U$. $U^*|_N$ is an isometry and so U^* is a partial isometry since $U^*|_{N^\perp} = 0$. The final space of U^* is $U^*N = U^*UH = M$. □

Corollary 1.28 *If $U \in B(H)$, then the following conditions are equivalent:*

(a) *U is a partial isometry*
(b) *U^*U is a projection*
(c) *$UU^*U = U$*
(d) *U^* is a partial isometry*
(e) *UU^* is a projection*
(f) *$U^*UU^* = U^*$.*

Proof The following trail of implications is clear from Lemma 1.27.

$$(b) \Leftarrow (a) \Leftrightarrow (d) \Rightarrow (c) .$$

$(b) \Rightarrow (a)$: If $U^*U = P_M$, then $U^*U = 0$ on $M^\perp$ so $\langle Ux, Ux\rangle = \langle U^*Ux, x\rangle = 0$ for $x \in M^\perp$ and so $U = 0$ on $M^\perp$. Moreover for $x \in H$:

$$\langle U(U^*Ux), U(U^*Ux)\rangle = \langle U^*UU^*Ux, U^*Ux\rangle = \langle U^*Ux, U^*Ux\rangle$$

as U^*U is a projection, and so $U|_M$ is isometric.

$(c) \Rightarrow (b)$ If $UU^*U = U$ then $U^*(UU^*U) = U^*(U)$ and since $(U^*U)^* = U^*U$, U^*U is a projection.

$(a) \Rightarrow (c)$ follows from the proof of Lemma 1.27. The remainder is clear. □

Later, in Sections 3.2–3.4, we will develop the notion of dimension for closed subspaces of a Hilbert space H, whose associated projections lie in some fixed operator algebra A on H, by considering two subspaces as being equivalent when there is a partial isometry in A from one to the other. That is why we go into such pedantic detail here on properties of partial isometries, which is not really needed for the polar decomposition:

Theorem 1.29 Polar decomposition. *If T is a bounded linear operator on a Hilbert space H, then there exists a positive operator R and a partial isometry U on H such that $T = UR$.*

Proof Let $R = |T| = (T^*T)^{1/2}$. Then

$$\|Tx\|^2 = \langle Tx, Tx\rangle = \langle T^*Tx, x\rangle = \langle R^2x, x\rangle = \langle Rx, Rx\rangle = \|Rx\|^2,$$

so $\|Tx\| = \|Rx\|$, for $x \in H$. Then $URx = Tx$ is a well defined isometric operator on $\text{Ran}(R)$, which extends by continuity to $\overline{\text{Ran}(R)} = M$. If we define $U = 0$ on $M^\perp$, then this gives a partial isometry U on H such that $T = UR$. □

Note that the partial isometry U constructed in Theorem 1.29 has initial space $\overline{\text{Ran}(T^*)}$, since $\overline{\text{Ran}(R)} = \text{Ker}(R)^\perp = \overline{\text{Ran}(T^*)}$, (by Exercise 1.21, and $\|R(x)\| = \|Tx\|$, so that $\text{Ker}(R) = \text{Ker}(T)$) and final space $\overline{\text{Ran}(T)}$. In finite dimensions, the existence of a partial isometry from $\text{Ran}(T^*)$ onto $\text{Ran}(T)$ merely says that the row rank of the matrix T is the same as its column rank.

Exercise 1.21 If $S \in B(H)$, for some Hilbert space H, show that $\text{Ker}\, S = \text{Ran}(S^*)^\perp$.

Exercise 1.22 If M, N are closed subspaces of a Hilbert space H with associated projections P and Q, show that the following four conditions are equivalent:

$$\text{(i)}\quad M \subset N, \qquad \text{(ii)}\quad P \le Q, \qquad \text{(iii)}\quad P = QP, \qquad \text{(iv)}\quad P = PQ.$$

Exercise 1.23 Define the shift operator s on ℓ^2 by $(\xi_1, \xi_2, \xi_3, \ldots) \mapsto (0, \xi_1, \xi_2, \xi_3, \ldots)$. Show that s is isometric, and compute the operators s^* and ss^*.

Exercise 1.24 Let P be the orthogonal projection on a closed subspace M of a Hilbert space H, and T a bounded linear operator on H. Show that the following conditions are equivalent:

(i) T leaves M invariant (i.e. $TM \subset M$).
(ii) $PTP = TP$.
(iii) T^* leaves $M^\perp$ invariant.
(iv) $(1 - P)T^*(1 - P) = T^*(1 - P)$.

Deduce that the following are equivalent:

(v) T leaves M and $M^\perp$ invariant.
(vi) $TP = PT$.

Exercise 1.25 Let V be an isometry on a Hilbert space H, and $K = (VH)^\perp$. Show that $\{V^nK \mid n = 0, 1, 2, \ldots\}$ are closed, pairwise orthogonal subspaces. Prove that $L = (\bigoplus_{n=0}^\infty V^nK)^\perp$ is invariant under V, and $V|_L$ is unitary.

Exercise 1.26 If T is a positive operator on a Hilbert space, show that $\text{Ker}(T) = \text{Ker}(T^{1/2})$.

Exercise 1.27 Let T be an operator on a Hilbert space H. Show that there exists an unique positive operator R and a partial isometry U such that $\mathrm{Ker}(U) = \mathrm{Ker}(T)$ and $T = UR$.

Exercise 1.28 Let T be a self adjoint operator in a unital C^*-algebra A, such that $T^3 = T^2$. Show that T is a projection.

Exercise 1.29 If S is a positive element in a unital C^*-algebra A such that $S \geq 1$, show that S is invertible and $S^{-1} \leq 1$.

Exercise 1.30 If S, T are positive elements in a unital C^*-algebra A, such that $S \geq T$, and T is invertible, show that S is invertible and $S^{-1} \leq T^{-1}$.

Exercise 1.31 Show that if P and Q are projections on a Hilbert space H then $(PQ)^n$ and $(PQP)^n$ converge strongly as $n \to \infty$ to the projection $P \wedge Q$ on $PH \cap QH$.

1.5 Spectral theorem for unitaries

Let T be an element of a unital C^*-algebra A, and $C^*(\mathrm{T})$ the smallest unital C^*-subalgebra of A containing T, namely the closure of the polynomials in T and T^*. Clearly $C^*(\mathrm{T})$ is commutative if and only if T is normal, i.e. $TT^* = T^*T$. We have seen so far that if T is self adjoint $T = T^*$, then $\sigma(T)$ is real and $C^*(T)$ is canonically isomorphic to $C(\sigma(T))$. If T is normal, it is still true that $C^*(T)$ is canonically isomorphic to $C(\sigma(T))$. We shall not prove that fact here, but merely show how easy it is to modify the spectral Theorem 1.18 to show that $C^*(U)$ is canonically isomorphic to $C(\sigma(U))$ when U is unitary, $U^*U = UU^* = 1$. Thus consider a unitary U in a unital C^*-algebra A so that $\sigma(U) \subset \mathbb{T}$ by Theorem 1.12. If Ω is a closed subset of $\mathbb{T}$, then a Laurent polynomial $p \in \mathbb{C}[t, t^{-1}]$

(1.5.1) $p(t) = \sum_{|i| \leq N} a_i t^i$

defines a continuous function p on Ω, which we call a trigonometric polynomial on Ω. We let $P(\Omega)$ denote the totality of trigonometric polynomials on Ω, which is a $*$-subalgebra of $C(\Omega)$. Then

(1.5.2) $\pi : P(\sigma(U)) \to C^*(U)$

defined by $\pi(p) = p(U) = \sum_{|i| \leq N} a_i U^i$, for a trigonometric polynomial p on Ω as in (1.5.1), is a $*$-homomorphism from $P(\sigma(U))$ onto the $*$-algebra generated by U. If we are to follow the programme of Theorem 1.18, we need to establish

(1.5.3) $P(\Omega)$ is dense in $C(\Omega)$ for any closed $\Omega \subset \mathbb{T}$.

(1.5.4) $\pi : P(\sigma(U)) \to C^*(U)$ is isometric.

We prove the former and leave the latter, being a rather mild modification of the previous argument in Theorem 1.18. To show that $P(\Omega)$ is dense in $C(\Omega)$ it is enough to show this in the case when $\Omega = \mathbb{T}$ and then invoke Tietze's extension theorem. We will often think of a function on $\mathbb{T}$ as being a function on $\mathbb{R}$ of period 1, namely $x \to f(e^{2\pi i x})$, and vice versa. Then $\langle f, g \rangle = \int_0^1 f(x)\overline{g(x)}dx$ defines an inner product on $C(\mathbb{T})$, with corresponding norm $\|f\|_2 = [\int_0^1 |f(x)|^2 dx]^{1/2}$, such that $e_j(x) = e^{2\pi i j x}, x \in \mathbb{R}$, defines a sequence of orthonormal vectors in $C(\mathbb{T})$ with respect to this inner product. Since $\|f\|_2 \leq \|f\| = \sup\{|f(z)| \mid z \in \mathbb{T}\}$, then if our claim is correct that the trigonometric polynomials are dense in $C(\mathbb{T})$ for the uniform norm, they will also be dense in $C(\mathbb{T})$ for the $\|\cdot\|_2$ norm, and consequently in $\mathrm{L}^2(\mathbb{T})$, the completion of $C(\mathbb{T})$ with respect to $\|\cdot\|_2$. Then by Parseval's theorem, for any $f \in C(\mathbb{T})$,

$$\left\| f - \textstyle\sum_{|j|\leq n} \langle f, e_j \rangle e_j \right\|_2 \to 0 \text{ as } n \to \infty,$$

i.e. $\int_0^1 |f(x) - \sum_{|j|\leq n} \langle f, e_j \rangle e^{2\pi i j x}|^2 \, dx \to 0$, as $n \to \infty$.

In general this does not mean that

$$\left\| f - \textstyle\sum_{|j|\leq n} \langle f, e_j \rangle e_j \right\| \to 0, \text{ as } n \to \infty.$$

In fact there exists continuous f such that $\sum_{|j|\leq n} \langle f, e_j \rangle e^{2\pi i j x}$ does not converge everywhere by Du Bois–Reymond (see (Körner 1986), Theorem 18.1), although by Carleson (see (Körner 1986), Chapter 19) the Fourier series of a continuous function converges almost everywhere to the given function. However, what is true is that for a continuous function f the *Fourier series* $\sum \langle f, e_j \rangle e_j$ is *Cesaro summable* to f in the uniform norm $\|\cdot\|$, i.e. if $S_n = \sum_{|j|\leq n} \langle f, e_j \rangle e_j$, then the averages $(S_1 + \cdots + S_n)/n$ converge to f. In particular, $P(\mathbb{T})$ is dense in $C(\mathbb{T})$.

Theorem 1.30 Fejer's theorem. *Let $S_n = \sum_{|j|\leq n} \langle f, e_j \rangle e_j$ where $f \in C(\mathbb{T})$. Then $(S_1 + \cdots + S_n)/n \to f$ as $n \to \infty$ in the uniform norm.*

Proof We have

$$\begin{aligned} S_n(y) = \textstyle\sum_{|j|\leq n} \langle f, e_j \rangle e_j(y) &= \textstyle\sum_{|j|\leq n} \int_0^1 f(x) e^{-2\pi i j x} \, dx \; e^{2\pi i j y} \\ &= \int_0^1 f(x) D_n(y - x) \, dx \end{aligned} \tag{1.5.5}$$

where D_n is the *Dirichlet kernel* :

$$D_n(z) = \textstyle\sum_{|j|\leq n} e^{2\pi i j z} = e^{-2\pi i n z}(e^{2\pi(2n+1)z} - 1)/(e^{2\pi i z} - 1). \tag{1.5.6}$$

Then

$$\begin{aligned} 1/n \textstyle\sum_{i=1}^n S_i(y) &= 1/n \textstyle\sum_{i=1}^n \int_0^1 f(x) D_i(y - x) \, dx \\ &= \int_0^1 f(x) F_n(y - x) \, dx \end{aligned} \tag{1.5.7}$$

where F_n is the *Fejer kernel*:

$$F_n(z) = 1/n\sum_{i=1}^n D_i(z) = (1 - \cos 2n\pi z)/n(1 - \cos 2\pi z) \\ = (\sin n\pi z/\sin \pi z)^2/n\,. \tag{1.5.8}$$

Note that $F_n(z)$ is always positive (but this is not so for $D_n(z)$). Taking $f = e_0$, in (1.5.7), we see $1 = \int_0^1 F_n(y-x)\,dx$. Thus

$$\begin{aligned}\sum_{i=1}^n S_i(y)/n - f(y) &= \int_0^1 (f(x) - f(y))F_n(y-x)\,dx \\ &= \int_{-\frac{1}{2}}^{+\frac{1}{2}} (f(y+z) - f(y))F_n(z)\,dz\end{aligned} \tag{1.5.9}$$

where we have used periodicity of all functions considered. Since f is uniformly continuous, given $\varepsilon > 0$, there exists δ such that $|z| < \delta$ implies $|f(y+z) - f(y)| < \varepsilon$. When $1/2 \geq |z| \geq \delta, F_n(z) \leq n^{-1}(\sin \pi\delta)^{-2}$. Hence, from (1.5.9), we see

$$\begin{aligned}\left|\sum_{i=1}^n S_i(y)/n - f(y)\right| &\leq \int_{-\frac{1}{2}}^{+\frac{1}{2}} |f(y+z) - f(y)|F_n(z)\,dz \\ \leq \int_{|z|<\delta} |f(y+z) - f(y)|F_n(z)\,dz &+ \int_{\delta \leq |z| \leq \frac{1}{2}} |f(y+z) - f(y)|F_n(z)\,dz \\ \leq \varepsilon \int_{|z|<\delta} F_n(z)\,dz + \frac{4\|f\|}{n(\sin \pi\delta)^2} &\leq \varepsilon \int_{-\frac{1}{2}}^{+\frac{1}{2}} F_n(z)\,dz + \frac{4\|f\|}{n(\sin \pi\delta)^2} \\ = \varepsilon + 4\|f\|/n(\sin \pi\delta)^2 < 2\varepsilon, &\quad \text{for all } n \text{ large}.\end{aligned}$$

Thus $\sum_{i=1}^n S_i/n \to f$ uniformly. □

Following the programme for the spectral theorem proven in Section 1.3 for self adjoint operators, we can now deduce the analogous theorem for unitaries:

Theorem 1.31 *Let U be a unitary in a unital C^*-algebra A. Then there exists a unique isometric $*$-isomorphism π from $C(\sigma(U))$ onto $C^*(U)$ (where $C^*(U)$ is the smallest C^*-subalgebra containing U, and is the closure of the set of $p(U)$, $p \in \mathbb{C}[t, t^{-1}]$), satisfying $\pi(f) = U$, $\pi(e) = 1$ where $f(\lambda) = \lambda$, $e(\lambda) = 1$, $\lambda \in \sigma(U)$.*

Exercise 1.32 Let x be a self adjoint element of a unital C^*-algebra A. If $U = U(x)$ is the unitary $(1-ix)(1+ix)^{-1}$, show that $\sigma(U) \subset \mathbb{T}\backslash\{-1\}$. Conversely if U is a unitary in A, with $-1 \notin \sigma(U)$, show that $x = x(U) = i(1-U)(1+U)^{-1}$ is self adjoint, and that the maps $x \to U(x)$, and $U \to x(U)$ are mutually inverse.

Exercise 1.33 If x is a self adjoint element of a unital C^*-algebra, and $U = U(x)$ as in Exercise 1.32, show that $t \to (1-it)(1+it)^{-1}$ is a homeomorphism α of $\sigma(x)$ onto $\sigma(U)$, and so $\alpha^* : C(\sigma(U)) \to C(\sigma(x))$ given by $\alpha^* f = f \circ \alpha$, $f \in C(\sigma(U))$ is a $*$-isomorphism. Show that $C^*(x) = C^*(U)$, and if $\pi_1 : C(\sigma(x)) \to C^*(x)$, π_2 :

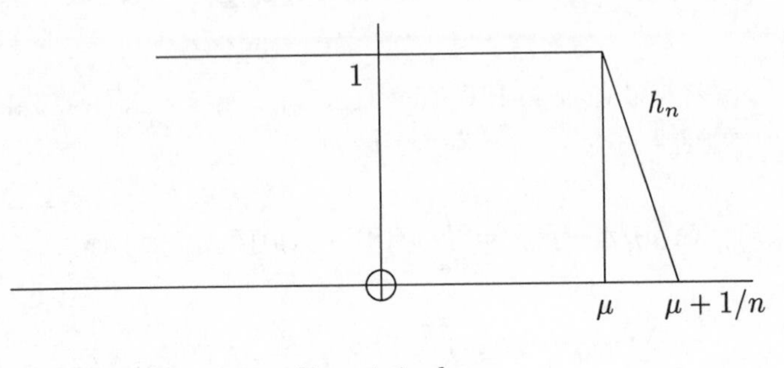

FIG. 1.1. h_n

$C(\sigma(U)) \to C^*(U)$ are the $*$-isomorphisms of Theorem 1.18 and 1.31 respectively, show that $\pi_1 = \pi_2 \circ \alpha^*$.

1.6 Spectral projections

We have seen that if T is a self adjoint matrix then the functional calculus, developed in Section 1.3, allows us to write $T = \sum_{i=1}^{r} \lambda_i e_i$, where $\{\lambda_i\}_{i=1}^{r}$ are the distinct eigenvalues of T, and $e_i = \mathrm{Ker}(T - \lambda_i)$ are the (orthogonal) eigenspaces. In fact $e_i = f_i(T)$, where f_i is the characteristic function of the singleton $\{\lambda_i\}$. We will now strive for an infinite dimensional analogue of $T = \sum \lambda_i e_i$, where the summation is replaced by a Riemann Stieltjes integral. The first obstacle is how to define $f(T)$, when f is the characteristic function of an interval, and so *not* continuous. In particular, if T is self adjoint, we wish to make sense of $e_\mu(T)$, where e_μ is the characteristic function $\chi_{(-\infty,\mu]}$. (More precisely e_μ is the restriction of $\chi_{(-\infty,\mu]}$ to the spectrum $\sigma(T) \subset \mathbb{R}$). Note that although e_μ is usually not continuous, it is upper semi continuous. Recall that a real valued function f on a metric space M is *upper semi continuous (u.s.c.)* if $f^{-1}(-\infty, a)$ is open for all real a. Note that an u.s.c. function on a compact metric space is bounded above. We also see that it is easy to approximate e_μ from above by a decreasing sequence of continuous functions, as in Fig. 1.1. For $x \in \sigma(T)$ define:

$$h_n(x) = \begin{cases} 1 & x \le \mu \\ 1 - (x-\mu)n & \mu \le x \le \mu + 1/n \\ 0 & \mu + 1/n \le x \end{cases}$$

Then $h_n \ge h_{n+1}$, and $\inf h_n(x) = e_\mu(x)$. Note that if a function f on metric space satisfies $f(x) = \inf h_n(x)$ (written $f = \inf h_n$), for a sequence h_n of continuous maps, $h_n \ge h_{n+1}$ then f is certainly u.s.c. Conversely:

Lemma 1.32 *Let f be a bounded u.s.c. real valued function on a compact subset $M \subset \mathbb{R}$, then there exists a sequence h_n of continuous functions on M such that $h_n \ge h_{n+1}$, and $\inf h_n = f$.*

Proof We can assume $f \le 0$. Let Z denote the functions on M which are restrictions from $\mathbb{R}$ of functions h, as in Fig. 1.2, where $a < b < c < d, e$ are

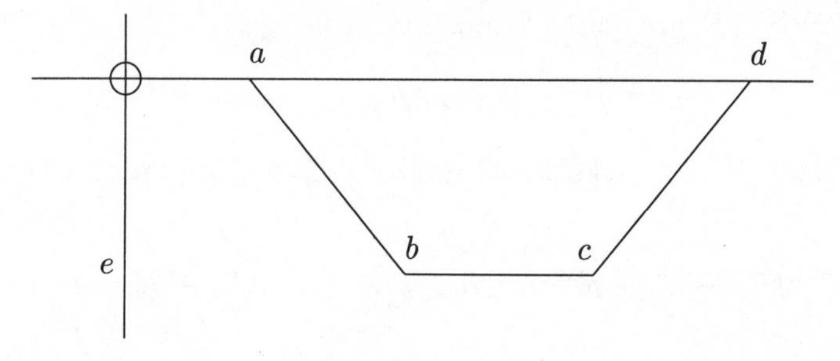

FIG. 1.2. h

rational $h = 0$ on $(-\infty, a]$ and $[d, \infty)$, $h(x) = e$, $x \in [b, c]$, and h is linear on $[a, b]$ and $[c, d]$. Then Z is countable, and if $F(x) = \inf\{g(x) \mid g \in Z, g \geq f\}$, then $0 \geq F(x) \geq f(x)$, as $0 \in Z$. We claim that $f(u) = F(u)$, for all $u \in M$. If $f(u) = 0$, then this is clear. Suppose $0 > f(u) = v$. Take $\varepsilon > 0$, such that $v + \varepsilon < 0$, and $v + \varepsilon$ is rational. Then $u \in f^{-1}(-\infty, v + \varepsilon) = U$ which is open, and so suppose $u \in (b, c) \subset [a, d] \subset U$, where $a < b < c < d$ are rational. Take $h \in Z$ such that h is as above where $e = v + \varepsilon$. Then $h \geq f$, and so $F(u) \leq h(u) = v + \varepsilon$, thus $F(u) \leq f(u) + \varepsilon$, for all $\varepsilon > 0$, and so $F(u) = f(u)$. Now $\{g \in Z \mid g \geq f\}$, being countable, can be enumerated $\{g_1, g_2, \ldots\}$. If we let $h_n = \min\{g_1, g_2, \ldots, g_n\}$, then h_n is continuous, $h_n \geq h_{n+1}$, and $\inf h_n = \inf g_n = f$. □

Thus if f is an u.s.c. function on $\sigma(T)$, bounded below by C, we can choose a sequence of continuous functions h_n on $\sigma(T)$, $h_n \geq h_{n+1} \geq C$, such that h_n converges to f. Then by the functional calculus for continuous functions, $h_n(T) \geq h_{n+1}(T) \geq C$. Thus to show that the sequence of self adjoint operators $h_n(T)$ converges, we need a lemma that decreasing sequences of self adjoint operators, bounded below, converge, in the following sense:

Definition 1.33 *A sequence T_n of bounded operators on a Hilbert space H converge strongly to bounded operator T on H if $\|T_n x - Tx\| \to 0$ as $n \to \infty$, for each $x \in H$.*

Recall that T_n converges to T in norm or uniformly if $\|T_n - T\| \to 0$. It is clear that norm convergence implies strong convergence. We will go into more detail on various forms of convergence for sequences of operators in Section 1.7.

Lemma 1.34 *Let B_n be a sequence of bounded self adjoint operators on a Hilbert space H such that $B_n \geq B_{n+1} \geq C$ for some self adjoint operator C. Then there exists an unique self adjoint B on H such that B_n converges strongly to B. Moreover $B = \inf B_n$, i.e. if $E \leq B_n$, then $E \leq B$, and if $D \leq B_n$ for all n, then $D \leq B$.*

Proof For each fixed $x \in H$, the sequence of real numbers $\langle B_n x, x\rangle$ is monotone decreasing and bounded below by $\langle Cx, x\rangle$. Hence $\lim_{n\to\infty}\langle B_n x, x\rangle$ exists. For

$n > m$, $Q(x,y) = \langle (B_m - B_n)x, y\rangle$, $x, y \in H$, defines a sesquilinear map such that $Q(x,x) \geq 0$. Hence by the Schwarz inequality,

$$|Q(x,y)|^2 \leq Q(x,x)Q(y,y), \quad x, y \in H. \tag{1.6.1}$$

Now $0 \leq B_m - B_n \leq C \leq \|C\|$, and so $\|B_m - B_n\| \leq \|C\|$ by (1.4.11). Then by (1.6.1),

$$\begin{aligned}|\langle (B_m - B_n)x, y\rangle|^2 &\leq \langle (B_m - B_n)x, x\rangle\langle (B_m - B_n)y, y\rangle \\ &\leq \langle (B_m - B_n)x, x\rangle \|(B_m - B_n)\|\|y\|^2 \\ &\leq \langle (B_m - B_n)x, x\rangle \|C\|\|y\|^2 .\end{aligned}$$

Hence $\|(B_m - B_n)x\|^2 \leq \langle (B_m - B_n)x, x\rangle\|C\| \to 0$ as $m, n \to \infty$. Thus $\lim_{n\to\infty} B_n(x)$ exists and is $B(x)$, say. Clearly B is linear. Now $-\|C\| \leq C \leq B_n \leq B_1 \leq \|B_1\|$, by (1.4.12), and so $\|B_n\| \leq \max\{\|B_1\|, \|C\|\}$, by (1.4.11). Thus $\|B(x)\| = \lim_{n\to\infty} \|B_n(x)\| \leq \max\{\|B_1\|, \|C\|\}\|x\|$, and so B is bounded. That B is self adjoint and is $\inf B_n$ is left as an exercise. □

Returning to our self adjoint operator T on a Hilbert space H, let f be a bounded u.s.c. function on $\sigma(T)$, say f is bounded below by C. Then if h_n is a decreasing sequence of continuous functions converging to f, we know by Lemma 1.34, that the decreasing sequence $h_n(T)$ of self adjoint operators converges strongly to some operator. Before we can define $f(T) = \lim_{n\to\infty} h_n(T)$, we need to check that this limit is independent of the choice of decreasing sequence converging to f. More generally, take f_1, f_2 u.s.c. functions on $\sigma(T)$, with families h_n^1, h_n^2 of continuous functions on $\sigma(T)$, such that $h_n^i \geq h_{n+1}^i$, and $f_i = \inf_n h_n^i$, for $i = 1, 2$, and $f_1 \geq f_2$. Then given $\varepsilon > 0$, and n, there exists N such that $h_N^2 < h_n^1 + \varepsilon$. This is because $X_m = \{x \in \sigma(T) \mid h_m^2(x) \geq h_n^1(x) + \varepsilon\}$ is closed, $X_{m+1} \subset X_m$, and $\cap X_m = \varnothing$. Thus by compactness $X_N = \varnothing$ for some N. Then by the functional calculus

$$h_N^2(T) \leq h_n^1(T) + \varepsilon .$$

But $\lim_{M\to\infty} h_M^2(T) \leq h_N^2(T)$ by Lemma 1.34, and so

$$\lim_{M\to\infty} h_M^2(T) \leq h_n^1(T) + \varepsilon .$$

Letting $\varepsilon \to 0$, we see again by Lemma 1.34 that

$$\lim_{M\to\infty} h_M^2(T) \leq \lim_{n\to\infty} h_n^1(T) .$$

In particular $f(T)$ is well defined for a bounded u.s.c function and if $f_2 \leq f_1$ are bounded u.s.c. functions then $f_2(T) \leq f_1(T)$. If $\Omega \subset \mathbb{R}$ is compact, let $U(\Omega)$

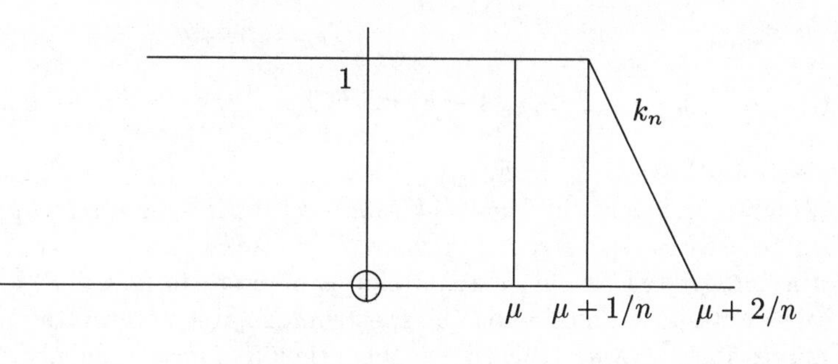

FIG. 1.3. k_n

denote the positive upper semi continuous functions on Ω. If $f, g \in U(\Omega)$, then $f + g, fg$ (defined by pointwise operations) are also in $U(\Omega)$. In fact if h_n, k_n are decreasing sequences of continuous functions converging to f, g respectively then $h_n + k_n, h_n k_n$ are decreasing sequences of continuous functions converging to $f+g, fg$ respectively, and so $f+g, fg$ are u.s.c. by the remark before Lemma 1.32. If $\Omega = \sigma(T)$, then

$$\begin{aligned}(f+g)(T) &= \lim_{n\to\infty}(h_n + k_n)(T) = \lim(h_n(T) + k_n(T)) \\ &= \lim h_n(T) + \lim k_n(T) = f(T) + g(T)\end{aligned}$$

and similarly $(fg)(T) = f(T)g(T)$, for $f, g \in U(\sigma(T))$. If $\Omega \subset \mathbb{R}$ is compact let $I(\Omega) = U(\Omega) - U(\Omega)$, which is an algebra (over $\mathbb{R}$). If $u, v, u', v' \in U(\sigma(T))$ and $u - v \geq u' - v'$, then $u + v' \geq u' + v$ and so $u(T) + v'(T) \geq u'(T) + v(T)$. Hence if $f \in I(\sigma(T))$, and $f = u - v$, where $u, v \in U(\sigma(T))$, then $f(T) = u(T) - v(T)$ is well defined, and if $f_1, f_2 \in I(\sigma(T))$ with $f_1 \geq f_2$ then $f_1(T) \geq f_2(T)$. Moreover $f \to f(T)$ is an algebra homomorphism from $I(\sigma(T))$ into the self adjoint operators on H.

Now for each $\mu \in \mathbb{R}$, the function e_μ which is the restriction of $\chi_{(-\infty,\mu]}$ to $\sigma(T)$ is in $U(\sigma(T))$ and $e_\mu^2 = e_\mu, e_\mu \leq e_\nu$ if $\mu \leq \nu$. Moreover if $\sigma(T) \subset [m, M]$ then $e_\mu = 0$ if $\mu < m$, $e_\mu = 1$ if $\mu \geq M$. Hence $E_\mu \equiv e_\mu(T)$ is an increasing family of projections such that $E_\mu = 0$ if $\mu < m$, $E_\mu = 1$ if $\mu \geq M$. Moreover E_μ is continuous from the right, in the sense that $\lim_{\varepsilon\to 0} \|E_{\mu+\varepsilon}x - E_\mu x\| = 0$ for all $\mu \in \mathbb{R}$, $x \in H$. For let k_n be the following continuous function on $\sigma(T)$:

$$k_n = \begin{cases} 1 & x \leq \mu + 1/n \\ 0 & x \geq \mu + 2/n \end{cases}$$

and linear on $[\mu + 1/n, \mu + 2/n]$, as in Fig. 1.3. Then $k_n \geq e_{\mu+1/n} \geq e_\mu$, and so $k_n(T) \geq E_{\mu+1/n} \geq E_\mu$. But $k_n \geq k_{n+1}$ and $k_n \to e_\mu$. Hence $E_\mu = \lim k_n(T)$. Also $\lim_{n\to\infty} E_{\mu+1/n}$ exists by Lemma 1.34. Therefore $\lim_{n\to\infty} E_{\mu+1/n} = E_\mu$. But for $0 < \varepsilon < 1/n$,

$$(E_{\mu+\varepsilon} - E_\mu)x = E_{\mu+\varepsilon}(1 - E_\mu)x = E_{\mu+\varepsilon}E_{\mu+1/n}(1 - E_\mu)x$$

and so

$$\|(E_{\mu+\varepsilon} - E_{\mu})x\| \le \|E_{\mu+1/n}(1 - E_{\mu})x\| = \|(E_{\mu+1/n} - E_{\mu})x\| \to 0 \text{ as } n \to \infty\,.$$

Here we have used $E_\mu E_\nu = E_{\min(\mu,\nu)}$.

We now claim that the family $\{E_\mu \mid \mu \in \mathbb{R}\}$, called the *spectral projections* of T, can be used to reconstruct T, in that $T = \int \lambda dE_\lambda$, where the right hand side is to be interpreted as a Riemann–Stieltjes integral. In fact $f(T) = \int f(\lambda)dE_\lambda$ for each continuous function on the spectrum, so first we need to make sense of the expression $\int f(\lambda)dE_\lambda$, when f is a continuous function defined on $[m, M]$ if $\sigma(T) \subset [m, M]$. Consider a sequence $\mu_0 < m < \mu_1 < \cdots < \mu_{n-1} < M \le \mu_n$, and $\lambda_k \in [\mu_{k-1}, \mu_k]$. If $\lim \sum_{k=1}^{n} f(\lambda_k)(E_{\mu_k} - E_{\mu_{k-1}})$ exists as $\max_{i=1}^{n} |\mu_i - \mu_{i-1}| \to 0$, we say that f is integrable with respect to $\{E_\lambda\}$ and write $\int f(\lambda)dE_\lambda$ for this limit. (Here, we only need an increasing family of projections.) We claim that any $f \in C(\sigma(T))$ is $\{E_\lambda\}$ integrable and $f(T) = \int f(\lambda)dE_\lambda$. Note that any $f \in C(\sigma(T))$ can be extended to a continuous function on an interval $[m, M]$ containing $\sigma(T)$. So we take any such extension to define $\int f(\lambda)dE_\lambda$, and so we are implicitly claiming that $\int f(\lambda)dE_\lambda$ (being equal to $f(T)$) is independent of choice of extension. Now f is uniformly continuous, so given $\varepsilon > 0$, there exists $\delta > 0$ such that if $|\lambda - \mu| < \delta$, then $|f(\lambda) - f(\mu)| < \varepsilon$. So if $\max |\mu_i - \mu_{i-1}| < \delta$, where $\mu_0 < m < \mu_1 < \cdots < \mu_{n-1} < M \le \mu_n$, and $\lambda_k \in [\mu_{k-1}, \mu_k]$ we have

$$-\varepsilon\chi_{(\mu_{k-1},\mu_k]} \le (f(\lambda) - f(\lambda_k))\chi_{(\mu_{k-1},\mu_k]} \le \varepsilon\chi_{(\mu_{k-1},\mu_k]}\,.$$

But $\chi_{(\alpha,\beta]} = \chi_{(-\infty,\beta]} - \chi_{(-\infty,\alpha]}$ if $\alpha < \beta$, and so by the functional calculus on $I(\sigma(T))$

$$-\varepsilon(E_{\mu_k} - E_{\mu_{k-1}}) \le (f(T) - f(\lambda_k))(E_{\mu_k} - E_{\mu_{k-1}}) \le \varepsilon(E_{\mu_k} - E_{\mu_{k-1}})\,.$$

But $\sum_{k=1}^{n}(E_{\mu_k} - E_{\mu_{k-1}}) = E_{\mu_n} - E_{\mu_0} = 1$ as $\mu_n \ge M, \mu_0 < m$. Therefore by addition:

$$-\varepsilon \le f(T) - \sum f(\lambda_k)(E_{\mu_k} - E_{\mu_{k-1}}) \le \varepsilon$$

and so $\|f(T) - \sum_{k=1}^{n} f(\lambda_k)(E_{\mu_k} - E_{\mu_{k-1}})\| < \varepsilon$, by 1.4.11. Summarizing, we have

Theorem 1.35 *Let T be a self adjoint operator on a Hilbert space H, with $\sigma(T) \subset [m, M]$. Then there exists a unique family $\{E_\mu \mid \mu \in \mathbb{R}\}$ of projections on H such that*

(a) $E_\mu \le E_\nu$ *or* $E_\mu = E_\nu E_\mu$ *if* $\mu \le \nu$.
(b) $E_\mu = 1$ *if* $\mu \ge M$, $E_\mu = 0$ *if* $\mu < m$.
(c) E_μ *is continuous from the right.*
(d) $T = \int \lambda dE_\lambda$.

Proof The only assertion which is now in need of proof is that of uniqueness. Take $\{F_\lambda\}$ to be any family of projections satisfying (*a*) – (*d*), and $\{E_\lambda\}$ the particular family $e_\mu(T)$ constructed previously. If $\mu \to F_\mu$ is increasing then

$(F_\nu - F_\mu)(F_{\nu'} - F_{\mu'}) = 0$ if $\mu \le \nu$, $\mu' \le \nu'$ and intervals $[\mu, \nu]$, $[\mu', \nu']$ are disjoint. Hence with the usual subdivisions,

$$\left[\sum_{k=1}^{n} \lambda_k (F_{\mu_k} - F_{\mu_{k-1}})\right]^r = \sum_{k=1}^{n} \lambda_k^r (F_{\mu_k} - F_{\mu_{k-1}})$$

and by continuity of multiplication, if

$$T = \lim \sum \lambda_k (F_{\mu_k} - F_{\mu_{k-1}}) \quad \text{then } T^r = \lim \sum \lambda_k^r (F_{\mu_k} - F_{\mu_{k-1}})$$

and so $p(T) = \int p(\lambda) dF_\lambda$ for any polynomial p. If f is continuous, choose a polynomial p such that $\|f - p\| < \varepsilon$, (uniformly on an interval containing $\sigma(T)$). Then $\|f(T) - p(T)\| < \varepsilon$, and if $\mu_0 < m < \mu_1 < \cdots < \mu_{n-1} < M < \mu_n$, and $\lambda_k \in [\mu_{k-1}, \mu_k]$, then

$$-\varepsilon \le \sum (f(\lambda_k) - p(\lambda_k))(F_{\mu_k} - F_{\mu_{k-1}}) \le \varepsilon$$

and so $\| \sum (f(\lambda_k) - p(\lambda_k))(F_{\mu_k} - F_{\mu_{k-1}}) \| \le \varepsilon$.
Hence

$$\begin{aligned} &\|f(T) - \sum f(\lambda_k)(F_{\mu_k} - F_{\mu_{k-1}})\| \\ &\le \|f(T) - p(T)\| + \| \sum (f(\lambda_k) - p(\lambda_k))(F_{\mu_k} - F_{\mu_{k-1}})\| \\ &\quad + \| \sum p(\lambda_k)(F_{\mu_k} - F_{\mu_{k-1}}) - p(T)\| \end{aligned}$$

and so f is $\{F_\lambda\}$ integrable and

$$\int f(\lambda) dF_\lambda = f(T).$$

Using right continuity of F_λ and h_n as in Fig. 1.1, we see

$$E_\mu(T) = \lim_{n \to \infty} h_n(T) = \lim_{n \to \infty} \int h_n(\lambda) dF_\lambda = F_\mu .$$

□

Example 1.36 At this point, one should re-examine Example 1.20 for a self adjoint matrix T, with finite spectrum. In the notation of that example,

$$e_\mu = \sum_{\lambda_i \le \mu} f_i$$

where $\{\lambda_1 < \lambda_2 < \cdots < \lambda_r\} = \sigma(T)$, and f_i is the characteristic function of the singleton $\{\lambda_i\}$. Hence $E_\mu = \sum_{\lambda_i \le \mu} e_i$, where $e_i = f_i(T)$ is the projection on $\mathrm{Ker}(\lambda_i - T)$. Then $E_{\lambda_i} - E_{\lambda_{i-1}} = e_i$, if $\lambda_0 < \lambda_1$ and

$$T = \sum_{i=1}^{r} \lambda_i e_i = \sum_{i=1}^{r} \lambda_i (E_{\lambda_i} - E_{\lambda_{i-1}}) .$$

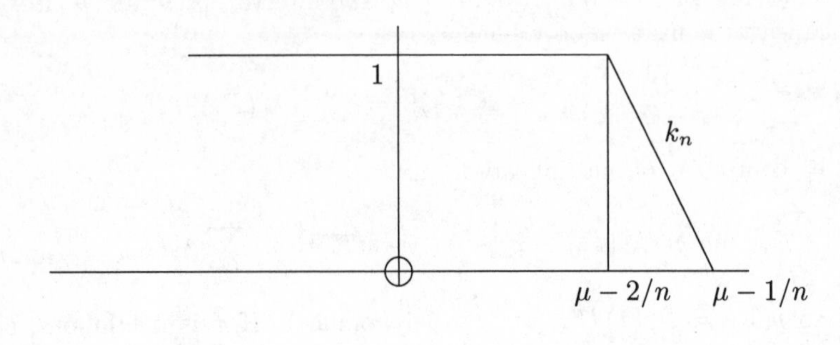

FIG. 1.4. k_n

In this example, we see that

$$\rho(T) \cap \mathbb{R} = \{\mu \mid E_\mu \text{ is constant in a neighbourhood of } \mu\}$$

and the projection on the eigenspace $\mathrm{Ker}(\lambda - T)$ is $E_\lambda - E_{\lambda-\varepsilon}$ for sufficiently small positive ε.

Now let T be a self adjoint operator on a Hilbert space H (not necessarily finite dimensional). Then, as in the proof that E is right continuous, we see that $\lim_{\varepsilon\searrow 0} E_{\mu-\varepsilon} \equiv E_{\mu-0}$ exists and defines a bounded self adjoint operator $E_{\mu-0}$. In fact, if f_μ is the restriction of $\chi_{(-\infty,\mu)}$ to $\sigma(T)$, then $f_\mu \in 1 - U(\sigma(T)) \subset I(\sigma(T))$, and we claim $f_\mu(T) = E_{\mu-0}$. Approximating f_μ from below by an increasing sequence k_n of continuous functions on $\sigma(T)$ as in Fig. 1.4:

$$k_n(x) = \begin{cases} 1 & x \le \mu - 2/n \\ 0 & x \ge \mu - 1/n \end{cases}$$

and k_n linear on $[\mu - 2/n, \mu - 1/n]$. Then

$$k_n \le e_{\mu-1/n} \le f_\mu, \quad \text{and} \quad k_n(T) \le E_{\mu-1/n} \le f_\mu(T)\,.$$

By definition $f_\mu(T) = \lim k_n(T)$, and so $f_\mu(T) = \lim_{n\to\infty} E_{\mu-1/n} = \lim_{\varepsilon\to 0} E_{\mu-\varepsilon}$. Since $f_\mu^2 = f_\mu$ and $e_\nu \le f_\mu$ if $\nu < \mu$, we see that $E_{\mu-0}$ is a projection and $E_\nu \le E_{\mu-0}$ if $\nu < \mu$.

Theorem 1.37 *Let $\{E_\mu\}$ be the spectral projections of a self adjoint operator T on a Hilbert space H. Then:*

(a) *If $\lambda \in \mathbb{R}$ then $\lambda \in \rho(T)$ if and only if E_μ is constant on some interval $[\lambda - \varepsilon, \lambda + \varepsilon]$ for some $\varepsilon > 0$.*
(b) *If $\lambda \in \mathbb{R}$, the projection on Ker$(\lambda - T)$ is $E_\lambda - E_{\lambda-0}$.*

Proof (*a*) Suppose $\lambda \in \rho(T) \cap \mathbb{R}$. Then $[\lambda-\varepsilon, \lambda+\varepsilon] \subset \rho(T)$ for some $\varepsilon > 0$. Then e_μ is constant for $\mu \in [\lambda-\varepsilon, \lambda+\varepsilon]$, being the restriction of $\chi_{(-\infty,\mu)}$ to $\sigma(T)$. Hence $E_\mu = e_\mu(T)$ is constant on the same interval. Conversely, suppose E_μ is constant on $[\lambda-\varepsilon,\lambda+\varepsilon]$. Take a continuous map g on $\mathbb{R}$ such that $g(\mu)=(\lambda-\mu)^{-1}$ for $\mu\notin[\lambda-\varepsilon,\lambda+\varepsilon]$. Then

$$(\lambda-T)g(T)=g(T)(\lambda-T)=\int g(\mu)(\lambda-\mu)dE_\mu=1\,,$$

as $g(\mu)(\lambda-\mu)=1$ outside $[\lambda-\varepsilon,\lambda+\varepsilon]$, and on that interval $[\lambda-\varepsilon,\lambda+\varepsilon]$, E_μ is constant. Thus $\lambda\in\rho(T)$ and $(\lambda-T)^{-1}=\int g(\lambda)dE_\mu$.

(*b*) $E_\lambda-E_{\lambda-0}=h_\lambda(T)$ where h_λ is the restriction of the characteristic function of the singleton $\{\lambda\}$ to $\sigma(T)$. Then $(\lambda-\mu)h_\lambda(\mu)=0$ for all μ and so $(\lambda-T)h_\lambda(T)=0$. Thus $(E_\lambda-E_{\lambda-0})(H)\subset\operatorname{Ker}(\lambda-T)$. Conversely, suppose $x\in\operatorname{Ker}(\lambda-T)$. Take $\nu<\lambda$, and define g continuous on $[m,M]\supset\sigma(T)$ such that $g(\mu)=(\lambda-\mu)^{-1}$ if $\mu\le\nu$. Then $\chi_{(-\infty,\nu]}(\mu)g(\mu)(\lambda-\mu)=\chi_{(-\infty,\nu]}(\mu)$. Hence $E_\nu g(T)(\lambda-T)=E_\nu$, and so $E_\nu x=0$ if $\nu<\lambda$, and $E_{\lambda-0}x=0$. Similarly $(1-E_\nu)x=0$ if $\nu>\lambda$ and so $E_\lambda x=x$. Thus $x\in(E_\lambda-E_{\lambda-0})H$. □

Exercise 1.34 Let χ be the characteristic function of a bounded interval (either open, closed or half open), and A_0 a self adjoint operator whose spectrum is disjoint from the endpoints of the interval. Show that the map $A\to\chi(A)$ is continuous on self adjoint operators at $A=A_0$.

1.7 Convergence and compact operators

We have already introduced, in the previous section, the notion of norm and strong convergence of operators. We now wish to investigate more closely different notions of convergence of not only operators on Hilbert space, but differing notions of convergence of sequences of vectors in a Hilbert space. This will lead us naturally to consider the *compact operator* on a Hilbert space. These are the norm closure of the finite rank operators, but can also be characterized as those operators which *improve* the convergence property of a sequence of vectors in the underlying Hilbert space. First let us consider different notions of convergence in a Hilbert space.

Definition 1.38 *If $\{f_n\}_{n=1}^\infty$ is a sequence in a Hilbert space H, we say that f_n converges to f in norm or strongly if $\|f_n-f\|\to0$ as $n\to\infty$, and weakly if $\langle f_n,g\rangle\to\langle f,g\rangle$, as $n\to\infty$, for each $g\in H$.*

Since $|\langle f_n-f,g\rangle|\le\|f_n-f\|\|g\|$, it is clear that norm convergence implies weak convergence. In finite dimensions, $H=\ell^2(N)$, with canonical complete orthonormal basis $\{e_i\}_{i=1}^N$, suppose $f_n=(f_n(i))$ converges weakly to $f=(f(i))$. Then $f_n(i)=\langle f_n,e_i\rangle\to\langle f,e_i\rangle=f(i)$ as $n\to\infty$. Thus $\|f_n-f\|^2=\sum_{i=1}^N|f_n(i)-f(i)|^2\to0$ as $n\to\infty$, and so weak convergence is equivalent to strong convergence in finite dimensions. In infinite dimensions this is not the case. In fact, let $\{e_i\}_{i=1}^\infty$ be an orthonormal sequence in an infinite dimensional Hilbert space. Then by Bessel,

$$\textstyle\sum_{n=1}^\infty|\langle g,e_n\rangle|^2\le\|g\|^2\,.$$

In particular, $\langle e_n, g\rangle \to 0$ as $n \to \infty$, and so $e_n \to 0$ weakly, but $\|e_n\| = 1$, and so e_n does *not* converge strongly to zero.

An operator T on H is said to be *completely continuous* or *compact* if it takes a (bounded) weakly converging sequence to a strongly converging one, i.e. if $f_n \to f$ weakly and $\sup\|f_n\| < \infty$ then $Tf_n \to Tf$ strongly. (Now any weakly convergent sequence is automatically bounded in norm, but we will not prove that here.) It is easily verified that the set of compact operators on H, denoted by $K(H)$, is a closed ideal of $B(H)$. (An *ideal* J in an algebra A is a subalgebra such that $AJ, JA \subset J$.) Now, if $\xi, \eta \in H$, let $T_{\xi,\eta}$ denote the linear map $f \to \langle f, \eta\rangle\xi = \phi_\eta(f)\xi$ where $\phi_\eta = \langle\cdot, \eta\rangle$. Then $(\xi, \eta) \to T_{\xi,\eta}$ is sesquilinear and bounded, in fact

$$\|T_{\xi,\eta}\| = \sup_{\|f\|=1} \|\phi_\eta(f)\xi\| = \|\xi\|\|\phi_\eta\| = \|\xi\|\|\eta\|\,.$$

Moreover for $\xi, \xi', \eta, \eta' \in H$, $S \in B(H)$, then

(1.7.1) $\quad T_{\xi,\eta}T_{\xi',\eta'} = \langle\xi', \eta\rangle T_{\xi,\eta'}$

(1.7.2) $\quad (T_{\xi,\eta})^* = T_{\eta,\xi}$

(1.7.3) $\quad ST_{\xi,\eta} = T_{S\xi,\eta}, T_{\xi,\eta}S = T_{\xi,S^*\eta}.$

Thus K_0= linear span $\{T_{\xi,\eta} \mid \xi, \eta \in H\}$ is a $*$-ideal in $B(H)$. The closure $K = \overline{K_0}$ in $B(H)$ is a closed $*$-ideal in $B(H)$. We claim that $K = K(H)$, so that the compact operators are precisely the norm closure of the finite rank operators. Note that if H is infinite dimensional then $K(H) \neq B(H)$, as we see from the example we used to distinguish weak and strong convergence of vectors, that a projection onto an infinite dimensional subspace of H is never compact. However, if $f_n \to f$ weakly, then $T_{\xi,\eta}(f_n) \to T_{\xi,\eta}(f)$ in norm, and so $K_0 \subset K(H)$. Since $K(H)$ is closed, $K \subset K(H)$.

Theorem 1.39 *Let T be a self adjoint operator on a Hilbert space H. Then T is compact if and only if $T = \sum_{i=1}^{\infty} \lambda_i e_i$* (norm convergence) *for some orthogonal sequence of finite dimensional projections e_i, and sequence λ_i of real numbers converging to zero.*

Proof Suppose T is compact. For $\varepsilon > 0$, take as in the proof of Theorem 1.37, a continuous map g such that $g(\mu) = 1/\mu$ if $\mu \leq -\varepsilon$. Then $g(\mu)\mu\chi_{(-\infty,-\varepsilon]} = \chi_{(-\infty,-\varepsilon]}$. Hence $E_{-\varepsilon} = g(T)TE_{-\varepsilon}$ is compact, since $K(H)$ is an ideal, and so $E_{-\varepsilon}$ is finite dimensional. Similarly, $1 - E_\varepsilon$ is finite dimensional. Then $x \to$ rank $E_x \in \mathbb{N}$ for $x \in (-\infty, -\varepsilon]$ is continuous from the right, monotone increasing, and so there must exist a finite number of points $-\infty = \gamma_0 < \gamma_1 < \gamma_2 < \cdots < \gamma_r < \gamma_{r+1} = -\varepsilon$ such that E_x is constant on $[\gamma_i, \gamma_{i+1})$, $i = 0, \ldots, r$, and E_x jumps at γ_i, $i = 1, 2, \ldots, r$, for $x \in (-\infty, -\varepsilon]$. (Similarly for $1 - E_x$ for $x \geq \varepsilon$.) Hence (cf. Example 1.36),

$$\int \lambda(\chi_{(-\infty,\varepsilon)}(\lambda) + \chi_{(\varepsilon,\infty)}(\lambda))dE_\lambda = \textstyle\sum_{i=1}^{n} \lambda_i P(\lambda_i)$$

where $P(\lambda) = E_\lambda - E_{\lambda-0}$, for some $\lambda_i, |\lambda_i| > \varepsilon$. Moreover (by considering Riemann approximations):

$$-\varepsilon \leq \int \lambda \chi_{[-\varepsilon,\varepsilon]} dE_\lambda \leq \varepsilon \,.$$

Hence $\|T - \sum_{i=1}^n \lambda_i P(\lambda_i)\| = \left\| \int \lambda \chi_{[-\varepsilon,\varepsilon]}(\lambda) dE_\lambda \right\| < \varepsilon$.

Conversely, if λ_i is a sequence of real numbers converging to zero, and e_i a sequence of finite dimensional orthogonal projections, put $\lambda_0 = 0$ and $e_0 =$ projection on $\left(\bigoplus_{\lambda_i \neq 0} e_i H\right)^\perp$. Then by Theorems 1.5, $\bigoplus_{i=0}^\infty \lambda_i e_i$ defines a bounded operator T on $\bigoplus_{i=1}^\infty e_i H$, identified with H. Then by the isometric property of this identification:

$$\left\|T - \textstyle\sum_{i=1}^r \lambda_i e_i\right\| = \max_{r+1 \leq i < \infty} |\lambda_i| \to 0 \text{ as } r \to \infty \,.$$

i.e. T is a norm limit of finite rank operators, hence compact. □

Corollary 1.40 *The compact operators on H is the norm closure of the finite rank operators.*

Proof Suppose T is compact. Then, since $K(H)$ is an ideal, so is $T^*T = S$. By Theorem 1.39,

$$T^*T = \textstyle\sum_{i=1}^\infty \lambda_i e_i$$

where $\lambda_i \geq 0, \sum e_i = 1, e_i$ are non zero finite rank projections. Now $\sigma(S) = \{\lambda_i \mid i = 1, 2, \ldots\}^-$ and $f \to \sum_{i=1}^\infty f(\lambda_i) e_i$ defines, by Theorem 1.5, an isometric $*$-homomorphism from $C(\sigma(S))$ into $B(H)$. Hence by uniqueness of the functional calculus (Theorem 1.18), we have

$$f(S) = \textstyle\sum_{i=1}^\infty f(\lambda_i) e_i \,.$$

In particular, $|T| = S^{1/2} = \sum \lambda_i^{1/2} e_i$. (Alternatively, one could get $S^{1/2} = \sum \lambda_i^{1/2} e_i$ using uniqueness of the square root, Theorem 1.25). Then another application of Theorem 1.39 shows that $|T|$ is compact, and it follows from the polar decomposition theorem that $T = U|T| = \lim \sum \lambda_i^{1/2} U e_i \in K$. □

Exercise 1.35 Let $f_n = (f_n(i))_{i=1}^\infty$ be a bounded sequence in ℓ^2. By a diagonalisation procedure, find a subsequence (n_j) such that $\lim_{j\to\infty} f_{n_j}(i)$ exists for each i. Show that $f(i) = \lim_{j\to\infty} f_{n_j}(i)$ defines an element $f = (f(i))$ in ℓ^2, and f_n converges to f weakly. Thus any bounded sequence in a Hilbert space has a weakly convergent subsequence.

1.8 The Calkin algebra and the Toeplitz extension

If J is an ideal (respectively $*$-ideal) in an algebra A, and $\pi : A \to A/J$, the quotient map $a \to a+J$, then we can endow A/J with unique algebraic operations such that π is a homomorphism (respectively $*$-homomorphism). In our case A, J will be C^*-algebras, and we will endow A/J with the quotient norm:

Lemma 1.41 *If Y is a closed subspace of a normed vector space X, then $\|x + Y\| = \inf\{\|x-y\| \mid y \in Y\}$ defines a norm on X/Y, which is complete if X is a Banach space.*

Proof If $\pi : X \to X/Y$ is the quotient map, and $\|\pi(x)\| = 0$, then there exists for each n, $y_n \in Y$ such that $\|x - y_n\| < 1/n$. Thus $y_n \to x$ and $x \in Y$ since Y is closed. Then $\pi(x) = 0$, and it is left to the reader to verify the remaining axioms for a norm. Let $\pi(x_n)$ be a Cauchy sequence in X/Y. By passing to a subsequence, we may assume $\|\pi(x_n) - \pi(x_{n+1})\| < 1/2^n$. Let $z_1 = x_1$, and inductively choose $z_n \in X$ such that $\pi(z_n) = \pi(x_n)$ and $\|z_n - z_{n+1}\| < 1/2^{n-1}$. Then $\sum(z_i - z_{i+1})$ is convergent, as X is complete, and so $\lim_{i\to\infty} z_i$ exists. Since $\|\pi(x)\| \le \|x\|$, π is continuous, and so $\lim_{i\to\infty}\pi(z_i) = \lim_{i\to\infty}\pi(x_i)$ exists. Thus X/Y is complete. □

Thus if J is a closed ideal in the normed algebra (respectively Banach algebra) A, then A/J is a normed algebra (respectively Banach algebra) as

$$\begin{aligned}\|\pi(x)\pi(y)\| = \|\pi(xy)\| &= \inf\{\|xy - j)\| \mid j \in J\} \\ &\le \inf\{\|(x-j_1)(y-j_2)\| \mid j_1, j_2 \in J\} \\ &\le \inf \|x - j_1\|\|y - j_2\| = \|\pi(x)\|\|\pi(y)\|\end{aligned}$$

using $(x-j_1)(y-j_2) - xy \in J$, if $j_1, j_2 \in J$. Similarly A/J is a normed $*$-algebra if J is a closed $*$-ideal in a normed $*$-algebra. Now if J is a closed $*$-ideal in a C^*-algebra A, is A/J a C^*-algebra? In view of Theorem 1.4, the crucial question is whether the norm of A/J has the C^*-property: $\|T^*T\| = \|T\|^2, T \in A/J$? This is indeed always the case. However, for simplicity we will only prove this in the case of ideal $K(H)$ in the C^*-algebra $B(H)$. Before giving a formal proof, let us look at the following simple example. Suppose e is a projection in a unital C^*-algebra A such that $ea = ae$ for all $a \in A$. Then $J = eA$ is a closed unital $*$-ideal in A, and A/J is $*$-isomorphic to $A(1-e)$, by identifying $\pi(a)$ with $a(1-e)$, $a \in A$. Now $A(1-e)$ is clearly a C^*-algebra, and if A/J is a C^*-algebra for the quotient norm, it would follow from uniqueness of the norm on a C^*-algebra (Corollary 1.14) that $\|\pi(a)\| = \|a(1-e)\|$. Thus if we could show $\|\pi(a)\| = \|a(1-e)\|$, then C^*-property of the norm on A/J would follow. In fact

$$\|\pi(a)^*\pi(a)\| = \|\pi(a^*a)\| = \|a^*a(1-e)\|,$$

and

$$\|\pi(a)\|^2 = \|a(1-e)\|^2 = \|(1-e)a^*a(1-e)\| = \|a^*a(1-e)\|$$

as $1-e$ is a projection commuting with a^*a. Now the ideal $K(H)$ does not have an identity if H is infinite dimensional, so the above argument could not apply directly. (If e is an identity for $K(H)$, then $T_{e\xi,\eta} = eT_{\xi,\eta} = T_{\xi,\eta}$, so that $e = 1_H$, which is impossible as 1_H is not completely continuous). However, $K(H)$ does have an approximate identity in the following sense. Let $\{e_i\}_{i=1}^{\infty}$ be a complete orthonormal sequence in the Hilbert space H, and let p_n denote the orthogonal projection on $\mathrm{lin}\{e_1, \ldots, e_n\}$. Thus $p_n = \sum_{i=1}^{n} T_{e_i,e_i} \in K(H)$, and $\lim_{n\to\infty} p_n\xi = \xi$ as $\{e_i\}$ is complete. Then

$$p_n T_{\xi,\eta} = T_{p_n\xi,\eta} \to T_{\xi,\eta} \quad \text{as } n \to \infty$$

for all $\xi, \eta \in H$, as $(\xi, \eta) \to T_{\xi,\eta}$ is continuous because $\|T_{\xi,\eta}\| = \|\xi\|\|\eta\|$. Thus by taking linear spans of $T_{\xi,\eta}$ and going to the closure, we see

$$p_n x - x,\ x p_n - x \to 0 \quad \text{as } n \to \infty$$

for all $x \in K(H)$. We say that $\{p_n\}$ is an *approximate identity* for $K(H)$. So as a substitute for the equality $\|\pi(a)\| = \|a(1-e)\|$, if e is an identity for the ideal, we have

Lemma 1.42 *For all $x \in B(H)$,*

$$\|\pi(x)\| = \lim_{n\to\infty} \|x(1-p_n)\|. \tag{1.8.1}$$

Proof If $x \in B(H)$, $xp_n \in K(H)$ as $K(H)$ is an ideal. Thus $\|\pi(x)\| \leq \|x - xp_n\|$ and so

$$\|\pi(x)\| \leq \lim\inf_{n\to\infty} \|x - xp_n\|.$$

Moreover, for any $j \in K(H)$:

$$\|x(1-p_n)\| \leq \|(x-j)(1-p_n)\| + \|j(1-p_n)\| \leq \|x-j\| + \|j(1-p_n)\|$$

and so $\lim\sup \|x(1-p_n)\| \leq \|x-j\| + \lim_{n\to\infty} \|j(1-p_n)\| = \|x-j\|$.
Thus

$$\lim\sup \|x(1-p_n)\| \leq \inf_j \|x-j\| \leq \lim\inf \|x(1-p_n)\|$$

and so $\lim_{n\to\infty} \|x(1-p_n)\|$ exists and is equal to $\|\pi(x)\|$. □

Corollary 1.43 *The norm on $B(H)/K(H)$ satisfies the C^*-property: $\|T^*T\| = \|T\|^2$, for $T \in B(H)/K(H)$.*

Proof Let $x \in B(H)$, then

$$\begin{aligned}\|\pi(x)\|^2 &= \lim_{n\to\infty} \|x(1-p_n)\|^2 = \lim_{n\to\infty} \|(1-p_n)x^*x(1-p_n)\| \\ &\leq \lim_{n\to\infty} \|x^*x(1-p_n)\|, \quad \text{as } \|1-p_n\| \leq 1 \\ &= \|\pi(x^*x)\| = \|\pi(x)^*\pi(x)\| \leq \|\pi(x)\|^2.\end{aligned}$$

Thus $\|\pi(x)\|^2 = \|\pi(x)^*\pi(x)\|$. □

The C^*-algebra $Q(H) = B(H)/K(H)$ is called the *Calkin algebra.* As an example of one of its C^*-subalgebras, consider the unilateral shift s on ℓ^2 (as in Exercise 1.23) defined by

$$s(\xi_1, \xi_2, \ldots) = (0, \xi_1, \xi_2, \ldots) \tag{1.8.2}$$

if $(\xi_i)_{i=1}^{\infty} \in \ell^2$. Thus $se_i = e_{i+1}$, $i = 1, 2, \ldots$ if $\{e_i\}_{i=0}^{\infty}$ is the canonical complete orthonormal basis. Then

$$s^*(\xi_1, \xi_2, \ldots) = (\xi_2, \xi_3, \ldots) \tag{1.8.3}$$

and

$$s^* e_i = \begin{cases} e_{i-1} & i \geq 1 \\ 0 & i = 0 \end{cases} .$$

Thus s is isometric, $s^*s = 1$, but not unitary as s^* has a kernel. In fact $ss^* = 1 - p_0$ where $p_0 = T_{e_0, e_0}$ is the projection on e_0. Thus $C^*(s)$ contains the rank one operator $p_0 = 1 - ss^*$. Moreover

$$T_{e_m, e_n} = s^m(1 - ss^*)s^{*n} \tag{1.8.4}$$

and since $(\xi, \eta) \to T_{\xi,\eta}$ is continuous and sesquilinear, we see that $K(\ell^2) \subset C^*(s)$. Then the quotient map $\pi : B(\ell^2) \to B(\ell^2)/K(\ell^2)$ restricts to $\pi : C^*(s) \to C^*(s)/K(\ell^2)$ and $U = \pi(s)$, is unitary as $p_0 \in K$ and $s^*s = 1 = ss^* + p_0$. Thus $C^*(s)/K(\ell^2) = C^*(U)$ which is a C^*-algebra by Corollary 1.43. Hence by Theorem 1.31, $C^*(s)/K(\ell^2) \cong C(\sigma(U))$. Note that here we do not need to employ the representation Theorem 1.4, since in the proof that $C^*(U)$ is isomorphic to $C(\sigma(U))$, all we used was the C^*-property of the norm. Also in any case we know $C(\sigma(U))$ can be represented on a Hilbert space by Theorem 1.6. We now show that $\sigma(U) = \mathbb{T}$.

If $t \in \mathbb{T}$, let F_t denote the unitary $1 \oplus t \oplus t^2 \oplus \cdots$ on ℓ^2 (as in the notation of Section 1.2), so that $F_t e_i = t^i e_i, i = 1, 2, \ldots$. Then $t \to F_t$ is a homomorphism from $\mathbb{T}$ into the group of unitaries on ℓ^2, so that $F_t^* = F_{\bar{t}}$. It is easy to verify that

$$F_t s F_t^* = ts\,. \tag{1.8.5}$$

Define a $*$-automorphism α_t of $B(\ell^2)$ by $\alpha_t(x) = F_t x F_t^*$, $x \in B(\ell^2)$, so that $t \to \alpha_t$ is a homomorphism from $\mathbb{T}$ into the group of $*$-automorphisms of $B(\ell^2)$. Then by (1.8.5), α_t leaves $C^*(s)$ invariant, and since $\alpha_t^{-1} = \alpha_{\bar{t}}$, it defines an automorphism of $C^*(s)$. Moreover $K(\ell^2)$ is an ideal, and so $\alpha_t(x) = F_t x F_t^*$ also defines a $*$-automorphism of $K(\ell^2)$. Hence if $\pi : C^*(s) \to C^*(s)/K(\ell^2)$ is the quotient map, there is for each $t \in \mathbb{T}$, a unique $*$-automorphism β_t of $C^*(s)/K(\ell^2)$ such that $\beta_t(\pi(x)) = \pi(\alpha_t(x))$, $x \in C^*(s)$. Since $\alpha_t(s) = ts$, we see $\beta_t(U) = tU$. But from the proof of Corollary 1.14, if β is an automorphism of a unital algebra, $\sigma(\beta(U)) = \sigma(U)$, and also $\sigma(tU) = t\sigma(U)$ by Exercise 1.20. Hence $\sigma(U) = t\sigma(U)$, for all $t \in \mathbb{T}$. But $C^*(s) \neq K(\ell^2)$, as 1_H is not completely continuous. Therefore, by Theorem 1.12, $\sigma(U) \neq \varnothing$. Hence since $\sigma(U)$ is closed under arbitrary rotations, $\sigma(U) = \mathbb{T}$.

So, summarizing, if s is the unilateral shift on ℓ^2, $C^*(s)$ contains the compact operators $K(\ell^2)$ and $C^*(s)/K(\ell^2) \cong C(\mathbb{T})$. This can be re-expressed as a short exact sequence:

$$0 \to K(\ell^2) \to C^*(s) \to C(\mathbb{T}) \to 0$$

or that $C^*(s)$ is an *extension* of the commutative algebra $C(\mathbb{T})$ by the compacts. This is called the *Toeplitz extension* of $C(\mathbb{T})$. The algebra $C^*(s)$ is in fact the unique algebra generated by a non-unitary isometry, Theorem 2.21.

Exercise 1.36 An *approximate identity* in a C^*-algebra A, is a set $\{e_\alpha : \alpha \in \Lambda\}$ of positive contractions in A, indexed by a net Λ such that $\lim_{\alpha\to\infty} \|xe_\alpha - x\| = 0$, for all $x \in A$, and $e_\alpha \leq e_\beta$ if $\alpha \leq \beta$. Let A be a C^*-algebra, and Λ the set of finite subsets of A, ordered by inclusion. If $\alpha \in \Lambda$, define $f_\alpha = \sum_{x\in\Lambda} xx^*$, and $e_\alpha = |\alpha| f_\alpha (1 + |\alpha| f_\alpha)^{-1} \in A$ where $|\alpha|$ denotes the cardinality of α, and $(1+|\alpha|f_\alpha)^{-1}$ is computed in A (see Exercises 1.18 and 1.19). If $x \in \alpha$, show that $(e_\alpha x - x)(e_\alpha x - x)^* \leq |\alpha|^{-1}$, and deduce using Exercise 1.30 that $\{e_\alpha : \alpha \in \Lambda\}$ is an approximate identity for A.

Exercise 1.37 Show that a separable C^*-algebra possesses a countable approximate identity.

Exercise 1.38 Let J be a norm closed $*$-ideal in a C^*-algebra A. Show that the quotient norm on A/J possesses the C^*-property.

Exercise 1.39 Let π be an injective $*$-homomorphism between the C^*-algebras A and B. Show that π is isometric. By Exercise 1.18, we can assume A, B, π are unital. By the C^*-property of the norm it is enough to show $\|\pi(x)\| = \|x\|$, if $x \in A_h$. We can thus take $A \cong C(\sigma(x))$, $B \cong C(\sigma(\pi(x)))$. Show that $\sigma(x) = \sigma(\pi(x))$.

Exercise 1.40 Let π be a $*$-homomorphism between C^*-algebras A and B. Deduce from Exercises 1.38 and 1.39 that $\pi(A)$ is a C^*-subalgebra of B.

Exercise 1.41 Show that the Calkin algebra $B(H)/K(H)$ is simple and does not have a representation on a separable Hilbert space.

1.9 Notes

Standard texts for the basics of the functional analysis of Hilbert spaces and their operators are (Bratteli and Robinson 1987), (Bratteli and Robinson 1981), (Dunford and Schwartz 1958), (Halmos 1967), (Kadison and Ringrose 1983), (Kadison and Ringrose 1986), (Reed and Simon 1972), (Riesz and Sz.-Nagy 1955). Exercise 1.34 is from (Kaplansky 1951a).

The spectral theorems for self-adjoint operators and unitary operators can be unified in a spectral theorem for normal operators N, $N^*N = NN^*$, so that $C^*(N, 1) \cong C(\sigma(N))$ under a $*$-isomorphism which takes $N, 1$ to the functions f, e respectively where $f(\lambda) = \lambda$, $e(\lambda) = 1$, $\lambda \in \sigma(N)$.

If $A = C(X)$ is a commutative unital C^*-algebra, then the spectrum X of A can be recovered from the characters of A, the homomorphisms $f \to f(x)$ from A into $\mathbb{C}$, or equivalently the maximal ideals of A via $\{f \in A \mid f(x) = 0\}$.

The proof given here that the image of the unilateral shift s in the Calkin algebra has spectrum the circle is non-standard. An operator T on a Hilbert space is *Fredholm* if kernels of T and T^* are both finite dimensional and T has closed image. By Atkinson's theorem (Murphy 1990), the Fredholm operators are precisely those that are invertible in the Calkin algebra. A direct computation then yields $\mathbb{T} = \{\lambda \in \mathbb{C} \mid \lambda - s \text{ is Fredholm}\}$.

2

C^*-ALGEBRA BASICS

2.1 Introduction

This chapter introduces the fundamental structure of C^*-algebras. In Section 2.2 we look at the most basic non-commutative building blocks, the matrix algebras, and in Section 2.3 their states described through density matrices. Analogously in infinite dimensions (going from $M_n = B(\mathbb{C}^n)$ to $B(H)$ where H is an infinite dimensional Hilbert space) we can identify in Section 2.4 the dual of the compact operators $K(H)^*$ with $T(H)$ the trace class operators, and $T(H)^*$ with $B(H)$ isometrically as Banach spaces. States on C^*-algebras give rise to representations by taking left multiplication on the Hilbert space formed as the completion of the algebra using the state to define a (pre-) inner product. This is the GNS construction of Section 2.5. There we also graphically describe all homomorphisms between finite dimensional C^*-algebras – the precursors of the Bratteli diagram, and path model to describe inductive limits of finite dimensional C^*-algebras in Section 2.9. Section 2.6 has a graphical description of finite dimensional C^*-algebras, Hilbert spaces and their tensor products, whilst the notion of inductive limits is formalized in Section 2.7 for Hilbert spaces and C^*-algebras. In Section 2.8 we generalize the construction of the Toeplitz extension, the unilateral shift, and the universal C^*-algebra of a unitary to the Cuntz algebras, and their Toeplitz extensions generated by isometries with orthogonal ranges. If matrix algebras M_n or matrix algebras over continuous functions $C(X, M_n) \cong M_n \otimes C(X)$ are the building blocks for finite C^*-algebras (where $x^*x = 1$ implies $xx^* = 1$), then the Cuntz algebras (or their tensor products with commutative algebras) are the building blocks for infinite algebras. They naturally contain inductive limits of finite dimensional C^*-algebras, indeed infinite tensor products of matrix algebras $F_n = \otimes_{\mathbb{N}} M_n$ and in some sense $\mathcal{O}_n$ is the C^*-algebra of the unilateral shift on these infinite tensor product algebras.

2.2 Matrix algebras

We have already looked in detail in Sections 1.3, 1.5 at commutative C^*-algebras, especially those generated by a single operator. Now we begin to look at another basic building block for operator algebras, namely the finite dimensional ones. In the first place, for each positive integer n, we have the matrix algebras $M_n = M(n)$ (Section 1.2), the $*$-algebra of all $n \times n$ matrices over $\mathbb{C}$, identified as the C^*-algebra of all (necessarily bounded) linear maps on a finite dimensional Hilbert space $\mathbb{C}^n$. This is a finite dimensional C^*-algebra, and non-commutative unless $n = 1$. More generally, if $\mathbf{n} = (n_1, \ldots, n_p)$ is a p-tuple of positive integers, $M(\mathbf{n}) = \bigoplus_{i=1}^p M(n_i)$ is a finite dimensional C^*-algebra. Recall from Theorem 1.5, that $\| \bigoplus_{i=1}^p T_i\| = \max_{i=1}^p \|T_i\|$, if $T_i \in M(n_i)$, and we can embed $M(\mathbf{n})$ as a C^*-subalgebra of $M(\sum_{i=1}^p n_i)$ by

$$\bigoplus_{i=1}^p T_i \to \begin{pmatrix} T_1 & & 0 \\ & \cdot & \\ & & \cdot \\ & & \cdot \\ 0 & & T_p \end{pmatrix} \tag{2.2.1}$$

Such a C^*-algebra is called a *multi-matrix algebra.* Note that $M(\mathbf{n})$ is non-commutative, unless $n_i = 1$ for all i. It is the case that every finite dimensional C^*-algebra is $*$-isomorphic, (and so necessarily isometrically $*$-isomorphic by Corollary 1.14) to a multi-matrix algebra. We shall not prove this, but refer the interested reader to (Takesaki 1979). In particular, every commutative finite dimensional C^*-algebra is isomorphic to $\mathbb{C}^p$, or $C(X)$, where X is a discrete set of p elements, for some finite p.

In an algebra A, we define *matrix units* indexed by a set Λ to be a non-zero set $\{e_{ij} \mid i, j \in \Lambda\}$ of elements of A such that

$$e_{ij}e_{kl} = e_{il}\delta_{jk} \,. \tag{2.2.2}$$

If moreover A is $*$-algebra, we say that they form $*$-*matrix units* if also:

$$e_{ij}^* = e_{ji} \tag{2.2.3}$$

Thus in a matrix algebra $M(n)$ we have the canonical $*$-matrix units

$$e_{ij} = [\delta_{ir}\delta_{js}]_{r,s=1}^n \,, \quad i, j = 1, \ldots, n, \tag{2.2.4}$$

consisting of 1 in the (i, j)th place and zeros elsewhere. Note that $e_{ij} = T_{e_i,e_j}$ in the notation of Section 1.7 or $e_{ij}e_l = \delta_{jl}e_i$, if e_i denotes the canonical basis of $\mathbb{C}^n$. It is convenient to be able to identify an algebra as a matrix algebra simply by writing down matrix units. Before doing that, note that a C^*-algebra A is said to be *simple* if every closed $*$-ideal in the algebra is either $\{0\}$ or the whole algebra.

Lemma 2.1 *A matrix algebra is simple.*

Proof Let $T = [T_{ij}]$ be a non zero element of an ideal J in $M(n), n \in \mathbb{N}$. If $T_{ij} \neq 0$, then from $T_{ij}e_{pq} = e_{pi}Te_{jq} \in J$, we see that J contains all matrix units, so that any non-zero ideal is equal to $M(n)$. □

Lemma 2.2 *If A is a $*$-algebra with $*$-matrix units $\{f_{ij} \mid i,j = 1,2,\ldots,n\}$ such that A=lin $\{f_{ij}\}$, then A is $*$-isomorphic to $M(n)$.*

Proof Define a homomorphism $\phi : M(n) \to A$ by $\phi(e_{ij}) = f_{ij}$. Since $A = \text{lin}\{f_{ij}\}$, ϕ is surjective and non-zero. Moreover by simplicity of $M(n)$, Lemma 2.1, the ideal Ker $\phi = \{0\}$, and so ϕ is a $*$-isomorphism. □

For later use, let us record:

Lemma 2.3 *Every ideal J in a multi-matrix algebra $M(\mathbf{n})$, $\mathbf{n} = (n_i) \in \mathbb{N}^p$ is of the form $\bigoplus_{i\in\Lambda} M(n_i)$, for some subset $\Lambda \subset \{1,\ldots,p\}$.*

Proof Regard $M(n_i)$ as a sub-algebra of $M(\mathbf{n})$ by embedding

$$t \to (0,\ldots,0,\underset{i\text{th place}}{t},0,\ldots), \quad t \in M(n_i)\,.$$

We see that $J \cap M(n_i)$ is an ideal of $M(n_i)$, so that $J \cap M(n_i) = \{0\}$ or $M(n_i)$ by Lemma 1.41. Let $\Lambda = \{i \mid J \cap M(n_i) = M(n_i)\}$. Now $J \supset \sum_{i=1}^{p} J \cap M(n_i)$, and if $j \in J$, then $j = \sum jp_i \in \sum J \cap M(n_i)$, if p_i denotes the unit of $M(n_i)$. Hence $J = \sum J \cap M(n_i) = \bigoplus_{i\in\Lambda} M(n_i)$. Conversely, it is clear that $\bigoplus_{i\in\Lambda} M(n_i)$ is an ideal for any subset $\Lambda \subset \{1,\ldots,p\}$. □

2.3 States

Take a matrix algebra $M(n)$. As a vector space $M(n)$ is isomorphic to $\mathbb{C}^{n^2}$. Thus taking the canonical matrix units as a canonical basis for $M(n)$, we have from $\ell^2(n^2)$ an inner product on $M(n)$ defined by

$$\langle A, B\rangle = \textstyle\sum_{i,j=1}^{n} A_{ij}\overline{B_{ij}} \tag{2.3.1}$$

if $A = [A_{ij}], B = [B_{ij}] \in M(n)$. Note that if tr denotes the trace or sum of the diagonal entries of a matrix, i.e. tr $C = \sum_{i=1}^{n} C_{ii}$, if $C = [C_{ij}] \in M(n)$, then

$$\langle A, B\rangle = \operatorname{tr} B^* A\,. \tag{2.3.2}$$

Thus if φ is a linear functional on $M(n)$, there exists by the Riesz–Frechet representation theorem a unique B in $M(n)$ such that for all A in $M(n)$:

$$\varphi(A) = \varphi_{B^*}(A) = \langle A, B^*\rangle = \operatorname{tr} BA\,. \tag{2.3.3}$$

The linear functionals which are positive on positive elements are of particular interest.

Definition 2.4 *A linear map T between C^*-algebras A and B is said to be positive if it preserves the positive cones, namely $T(A_+) \subset B_+$. A state on a C^*-algebra A is a positive linear functional.*

Let φ be a state on a unital C^*-algebra A. Then by the Cauchy–Schwarz inequality applied to the sesquilinear map $D : (a, b) \to \varphi(b^*a)$ we see that $|D(a,1)|^2 \leq D(a,a)D(1,1)$ or $|\varphi(a)|^2 \leq \varphi(a^*a)\varphi(1)$, for all $a \in A$. But $0 \leq a^*a \leq \|a^*a\|$ by Section 1.4, and so $\varphi(a^*a) \leq \varphi(1)\|a\|^2$. Hence $|\varphi(a)| \leq \varphi(1)\|a\|$, and so φ is continuous and $\|\varphi\| = \varphi(1)$. Thus any state on a unital C^*-algebra is automatically continuous. (The same is true for non-unital C^*-algebras and positive linear maps, but we will not bother to prove it here. See Exercise 2.5.) Note also that it follows from the polarization identity that $\varphi(b^*a) = D(a, b) = \overline{D(b,a)} = \varphi(\overline{a^*b})$ or $\varphi(b^*) = \overline{\varphi(b)}$ for all $b \in A$. Thus a state preserves adjoints. (This could also be deduced from the decompositions in Section 1.2 and Section 1.4 of $A = A_h + iA_h$, and $A_h = A_+ - A_+$, and also for positive linear maps between C^*-algebras.) A state on a C^*-algebra A is said to be normalized if $\|\varphi\| = 1$ (or $\varphi(1) = 1$ for a unital algebra).

States on commutative C^*-algebras can be concretely described as follows (Pedersen 1989). Suppose X is a compact Hausdorff space, and μ is a finite signed measure on X. Then

$$\varphi(f) = \int_X f(x)\, d\mu(x)\,, \quad f \in C(X) \tag{2.3.4}$$

defines a continuous linear functional on $C(X)$. Moreover, the *Riesz–Kakutani representation theorem* tells us that any continuous linear functional φ on $C(X)$ is of this form, for some such measure μ, and that φ is a state if and only if μ is a positive measure. Thus normalized states on a C^*-algebra should be regarded as analogous to probability measures on a (compact Hausdorff) topological space. Returning to matrix algebras, we have:

Proposition 2.5 *There is an isomorphism between elements B of $M(n)$ and linear functionals φ on $M(n)$ given by*

$$\varphi(A) = \operatorname{tr} AB\,, \quad A \in M(n)\,. \tag{2.3.5}$$

Then $\|\varphi\| = \operatorname{tr}|B|$, where $\|\varphi\|$ is computed using the C^*-norm on $M(n)$, and φ is a state if and only if B is positive.

Proof We have already identified $M(n)^*$ with $M(n)$ in (2.3.3). If $B = U|B|$ is the polar decomposition of B so that $U^*B = |B|$, then $\operatorname{tr}|B| = \operatorname{tr} U^*B = \varphi(U^*) \leq \|\varphi\|\|U^*\|$ so that $\operatorname{tr}|B| \leq \|\varphi\|$. Moreover, if $A \in M(n)$,

$$\begin{aligned}
&\varphi(A) = \operatorname{tr} AB = \operatorname{tr} AU|B|^{\frac{1}{2}}|B|^{\frac{1}{2}} = \langle AU|B|^{\frac{1}{2}}, |B|^{\frac{1}{2}}\rangle \\
&\leq [\langle AU|B|^{\frac{1}{2}}, AU|B|^{\frac{1}{2}}\rangle\langle |B|^{\frac{1}{2}}, |B|^{\frac{1}{2}}\rangle]^{\frac{1}{2}} = [\operatorname{tr}(|B|^{\frac{1}{2}}U^*A^*AU|B|^{\frac{1}{2}})\operatorname{tr}|B|]^{\frac{1}{2}} \\
&\leq \|U^*A^*AU\|^{\frac{1}{2}} \operatorname{tr}|B| \leq \|A\| \operatorname{tr}|B|\,,
\end{aligned}$$

so that $\operatorname{tr}|B| = \|\varphi\|$. Note that here we used positivity of the trace (which is the linear functional corresponding to $B = 1$). In fact if B is positive, say $B = D^*D$, for some $D \in M(n)$, then

$$\begin{aligned}\varphi(C^*C) &= \operatorname{tr} C^*CDD^* = \operatorname{tr} DC^*CD^* \\ &= \operatorname{tr} DC^*(DC^*)^* = \langle DC^*, DC^* \rangle \geq 0 .\end{aligned}$$

Conversely, suppose φ is a state. Now if $\xi = (\xi_i), \eta = (\eta_i) \in \mathbb{C}^n$, then $T_{\xi,\eta}$ has matrix $[\xi_i \overline{\eta}_j]$, and so tr $T_{\xi,\eta} = \langle \xi, \eta \rangle$. Hence $\varphi(T_{\xi,\eta}) = \operatorname{tr}(BT_{\xi,\eta}) = \operatorname{tr}(T_{B\xi,\eta}) = \langle B\xi, \eta \rangle$. But $T_{\xi,\xi} \geq 0$, and so $\langle B\xi, \xi \rangle = \varphi(T_{\xi,\xi}) \geq 0$, since φ is a state. Thus B is positive. □

A *convex set* X in a vector space V is a subset X of V such that the convex combination $\lambda x + (1-\lambda))y \in X$ if $x, y \in X$, $0 \leq \lambda \leq 1$. An *affine map (affine isomorphism)* between convex sets X_1 and X_2 is a map (respectively bijection) T from X_1 into X_2 which preserves convex combinations. Thus $T(\lambda x + (1-\lambda)y) = \lambda T(x) + (1-\lambda)T(y)$, if $x, y \in X_1$, $0 \leq \lambda \leq 1$. The set of normalized states on $M(n)$ is affinely isomorphic to the *density matrices* in $M(n)$, namely the positive matrices of trace one:

$$\{B \in M(n)_+ \mid \operatorname{tr} B = 1\} . \tag{2.3.6}$$

A positive matrix B has, by Example 1.21, a decomposition $B = \sum_{i=1}^r \lambda_i T_{e_i,e_i}$, where $\{e_i\}_{i=1}^r$ is an orthonormal sequence in $\mathbb{C}^n$, $1 \leq r \leq n$, $\lambda_i > 0$, and then tr $B = \sum \lambda_i$. Thus a normalized state

$$\varphi(A) = \textstyle\sum_{i=1}^r \lambda_i \langle Ae_i, e_i \rangle , \quad A \in M(n) \tag{2.3.7}$$

is a convex combination of vector states.

2.4 Hilbert–Schmidt and trace class operators

We briefly indicate how to take over the description of states on $M(n)$ in Section 2.3 to a description of states on the C^*-algebra K of compact operators. Not every operator x on an infinite dimensional Hilbert space has a well defined or finite trace $\sum_{i=1}^\infty x_{ii}$ if $x_{ij} = \langle xe_j, e_i \rangle$ is the matrix of x with respect to a complete orthonormal sequence $\{e_i\}$, (e.g. take $x = 1$). So in analogy with integration theory, regarding the trace as an integral or a measure, let us first define the Hilbert space of 'L^2-operators' for which tr $x^*x < \infty$.

If $x \in B(H)$, where H is a Hilbert space with complete orthonormal bases $\{e_i\}$ and $\{e'_i\}$, we have

$$\begin{aligned}\textstyle\sum_i \|xe_i\|^2 &= \textstyle\sum_i \sum_j |\langle xe_i, e'_j \rangle|^2 \text{ by Parseval} \\ &= \textstyle\sum_j \sum_i |\langle x^*e'_j, e_i \rangle|^2 = \sum_j \|x^*e'_j\|^2 .\end{aligned} \tag{2.4.1}$$

Hence

$$\|x\|_2 = \left[\textstyle\sum_i \|xe_i\|^2\right]^{\frac{1}{2}} \tag{2.4.2}$$

is independent of the choice of complete orthonormal basis $\{e_i\}$, and we define the Hilbert–Schmidt operators to be

$$HS(H) = \{x \in B(H) \mid \|x\|_2 < \infty\}\,. \tag{2.4.3}$$

If $x, y \in HS(H)$, i.e. $\{\|xe_i\|\}, \{\|ye_i\|\} \in \ell^2$ for a complete orthonormal basis $\{e_i\}$, then $\|(x+y)e_i\| \le \|xe_i\| + \|ye_i\|$ and $\ell^2(\mathbb{N})$ being closed under addition shows that $x + y \in HS(H)$ and $\|x+y\|_2 \le \|x\|_2 + \|y\|_2$.

Also if $x \in HS(H)$, $a \in B(H)$, we see from (2.4.1) and (2.4.2) that x^*, $|x|$, ax, $xa \in HS(H)$ and

$$\begin{gathered}\|x^*\|_2 = \|x\|_2 = \||x|\|_2 \\ \|ax\|_2 \le \|a\|\|x\|_2\,, \|xa\| \le \|x\|_2\|a\|\,.\end{gathered} \tag{2.4.4}$$

Thus $HS(H)$ is a $*$-ideal in $B(H)$, and $\|\cdot\|_2$ defines a norm on $HS(H)$ because $\|x\| \le \|x\|_2$, for if e is any unit vector in H, we can find a complete orthonormal basis $\{e_i\}$, such that $e_1 = e$, and so $\|xe\|^2 \le \|x\|_2^2$.

$HS(H)$ is indeed a Hilbert space, for if $a, b \in HS(H)$, then $|\langle ae_i, be_i\rangle| \le \|ae_i\|\|be_i\|$ and the existence of the inner product in $\ell^2(\mathbb{N})$ shows that $\sum\langle ae_i, be_i\rangle$ is absolutely convergent, and

$$\langle a, b\rangle = \textstyle\sum_i \langle ae_i, be_i\rangle \tag{2.4.5}$$

defines an inner product on $HS(H)$ such that $\|a\|_2 = \langle a, a\rangle^{\frac{1}{2}}$. (The inner product can be seen to be independent of the choice of complete orthonormal basis by polarizing (2.4.1) which also yields $\langle a, b\rangle = \langle b^*, a^*\rangle$.) Moreover $HS(H)$ is complete with respect to $\|\cdot\|_2$, for if x_n is a Cauchy sequence in $HS(H)$, then $\|x_n - x_m\| \le \|x_n - x_m\|_2$ shows that x_n is Cauchy in $B(H)$, and so converges in operator norm to x in $B(H)$, since $B(H)$ is complete. Now given $\varepsilon > 0$, there exists N such that $\|x_n - x_m\|_2 < \varepsilon$, for $n, m \ge N$. Hence for any M, $\sum_{i=1}^M \|(x_n - x_m)e_i\|^2 < \varepsilon^2$, and so letting $n \to \infty$, we see $\sum_{i=1}^M \|(x - x_m)e_i\|^2 \le \varepsilon^2$, and so $x_n - x \in HS(H)$, and $\|x - x_m\|_2 \le \varepsilon$. Since $HS(H)$ is a vector space, this means $x \in HS(H)$, and x_n converges to x in $HS(H)$.

Next, we define the 'integrable' or *trace class operators* $T(H)$ to be

$$\{x \in B(H) \mid \textstyle\sum_i \langle |x|e_i, e_i\rangle < \infty\} = \{x \in B(H) \mid |x|^{\frac{1}{2}} \in HS(H)\}\,. \tag{2.4.6}$$

If $x \in T(H)$, then the polar decomposition $x = u|x| = (u|x|^{1/2})(|x|^{1/2})$ shows that $x = ab$, for some $a, b \in HS(H)$, and by (2.4.1), $\sum\langle xe_i, e_i\rangle$ is absolutely convergent, to $\langle b, a^*\rangle$, for any complete orthonormal basis $\{e_i\}$. We define $\operatorname{tr} x = \sum\langle xe_i, e_i\rangle$. Next, if $x = ab$, for some $a, b \in HS(H)$, and $x = u|x|$ is the polar decomposition of x so that $u^*x = |x|$, then $|ab| = (u^*a)b \in \{xy \mid x, y \in HS(H)\}$, and so $\sum\langle |ab|e_i, e_i\rangle$ is absolutely convergent, i.e. $ab \in T(H)$. Thus

$$T(H) = \{xy \mid x, y \in HS(H)\}\,. \tag{2.4.7}$$

It remains to show that $T(H)$ is a Banach space for the norm $\|x\|_1 = \operatorname{tr}|x|$.

Note that in finite dimensions, $M(n)$ is a normed space for $\|x\|_1 = \mathrm{tr}|x|$, for by Proposition 2.5, $M(n)^*$ with the norm dual to the operator norm is isometrically isomorphic to $M(n)$ equipped with $\|\cdot\|_1$. In particular $\mathrm{tr}|x+y| \le \mathrm{tr}|x| + \mathrm{tr}|y|$, in finite dimensions certainly. We have to show that this situation persists in infinite dimensions, in that $T(H)$ is a vector space and $\|\cdot\|_1$ satisfies the triangle inequality. Thus let $x, y \in T(H)$, and $x+y = u|x+y|$, $x = v|x|$, $y = w|y|$ be the polar decompositions of $x+y, x, y$ respectively, so that $|x+y| = u^*(x+y)$. Then $|x+y| = u^*v|x| + u^*w|y| = (u^*v|x|^{1/2}|x|^{1/2} + u^*w|y|^{1/2}|y|^{1/2})$ is trace class by what we have shown already, and

$$\begin{aligned}\mathrm{tr}|x+y| &= \langle |x|^{\frac{1}{2}}, |x|^{\frac{1}{2}}v^*u\rangle + \langle |y|^{\frac{1}{2}}, |y|^{\frac{1}{2}}w^*u\rangle \\ &\le \||x|^{\frac{1}{2}}\|_2 \||x|^{\frac{1}{2}}v^*u\|_2 + \||y|^{\frac{1}{2}}\|_2 \||y|^{\frac{1}{2}}w^*u\|_2 \\ &\le \||x|^{\frac{1}{2}}\|_2^2 + \||y|^{\frac{1}{2}}\|_2^2 = \|x\|_1 + \|y\|_1 .\end{aligned}$$

Thus $T(H)$ is a normed vector space, and from (2.4.7) we see it is a $*$-ideal in $B(H)$. We defer a proof of completeness of $T(H)$ until Theorem 2.6.

Note that if $x \in T(H)$, with polar decomposition $x = u|x|$, then $|x|^{1/2} \in HS(H)$, and $x = u|x|^{1/2}|x|^{1/2} \in HS(H)$, since $HS(H)$ is an ideal in $B(H)$. Hence $T(H) \subset HS(H)$, and

$$\|x\|_2^2 = \|(u|x|^{\frac{1}{2}})|x|^{\frac{1}{2}}\|_2^2 \le \|u|x|^{\frac{1}{2}}\|^2 \||x|^{\frac{1}{2}}\|_2^2 \le \||x|^{\frac{1}{2}}\|^2 \|x\|_1 .$$

Hence $\|x\| \le \|x\|_2 \le \|x\|_1$, for $x \in T(H)$, as $\||x|^{\frac{1}{2}}\|^2 = \|x\|$, by the spectral theorem for x^*x.

Next, we show that every Hilbert–Schmidt operator on H is compact. For if $x \in HS(H)$, $\{e_i\}$ is any complete orthonormal basis in H, and $p_n = \sum_{i=1}^n T_{e_i,e_i}$ is the projection on $\mathrm{lin}_{i=1}^n\{e_i\}$, then

$$\|(x - xp_n)\|_2^2 = \textstyle\sum_i \|(x - xp_n)e_i\|^2 = \sum_{i>n} \|xe_i\|^2 \to 0 \quad \text{as } n \to \infty .$$

Hence the sequence of finite rank operators xp_n converges to x in $HS(H)$. This means that $HS(H)$ is the completion in the $\|\cdot\|_2$ norm of the finite rank operators, and since $\|x\| \le \|x\|_2$, it means that every Hilbert–Schmidt operator is a norm limit of finite rank operators and so is compact. Similarly, using completeness of $T(H)$, established below, and $\|x\|_1 \le \|x\|_2$, we see that the trace class operators are compact and are the completion of the finite rank operators with respect to the $\|\cdot\|_1$ norm. Recall from Corollary 1.43, that the compact operators $K(H)$ are the completion of the finite rank operators with respect to the operator norm. Thus if $\rho \in HS(H)$, then taking the polar decomposition $\rho = u|\rho|$, we have $|\rho|$ is Hilbert–Schmidt, hence compact and so by Theorem 1.39 has a decomposition

$$|\rho| = \textstyle\sum_i \lambda_i T_{e_i,e_i}$$

for some positive λ_i converging to zero and complete orthonormal basis $\{e_i\}$. Then $\sum_i \||\rho|e_i\|^2 = \lambda_i^2$ and so ρ being Hilbert–Schmidt (respectively trace class) is equivalent to $\sum_i \lambda_i^2 < \infty$ (respectively $\sum_i \lambda_i < \infty$). Now we can prove the infinite dimensional version of the duality in Proposition 2.5:

Theorem 2.6 *If H is a Hilbert space, $K(H)^*$ is isometrically isomorphic to $T(H)$ under the pairing $(x, \rho) = \operatorname{tr} \rho x, x \in K(H)$, $\rho \in T(H)$. Similarly $T(H)^* \cong B(H)$.*

Proof If $x \in B(H)$, $\rho \in T(H)$, then $\rho x \in T(H)$, and $x \to \operatorname{tr} \rho x$ is additive on $B(H)$. Moreover, if $\rho = u|\rho|$ is the polar decomposition of ρ, then

$$|\operatorname{tr} \rho x| = \langle xu|\rho|^{\frac{1}{2}}, |\rho|^{\frac{1}{2}}\rangle \leq \|xu|\rho|^{\frac{1}{2}}\|_2 \||\rho|^{\frac{1}{2}}\|_2 \leq \|x\|\|\rho\|_1 .$$

Thus for each $\rho \in T(H)$, there exists $\psi_\rho \in K(H)^*$ such that $\psi_\rho(x) = \operatorname{tr} \rho x, x \in K(H), \|\psi_\rho\| \leq \|\rho\|_1$. Conversely, if $\psi \in K(H)^*$, then for all $x \in HS(H) \subset K(H)$, $|\psi(x)| \leq \|x\|\|\psi\| \leq \|x\|_2\|\psi\|$. Hence $\psi \in HS(H)^*$, and so by the Riesz representation theorem for a continuous linear functional on a Hilbert space, there exists a unique $\rho \in HS(H)$ such that $\psi(x) = \operatorname{tr} \rho x$, $x \in HS(H)$. Let $\rho = u|\rho|$ be the polar decomposition of ρ so that $u^*\rho = |\rho| = \sum \lambda_i T_{e_i,e_i}$ for some complete orthonormal sequence $\{e_i\}$. Let $x = (\sum_{i=1}^n T_{e_i,e_i})u^* \in HS(H)$, then $x\rho = \sum_{i=1}^n \lambda_i T_{e_i,e_i}$, and so $\sum \lambda_i = \operatorname{tr}\rho x = \psi(x) \leq \|x\|\|\psi\| \leq \|\psi\|$. Hence $\rho \in T(H)$, and $\|\rho\|_1 \leq \|\psi\|$. Thus $\psi = \psi_\rho$ on $HS(H)$, and by continuity, we see $\psi = \psi_\rho$ on $K(H)$, the completion of $HS(H)$ with respect to the operator norm, and $\|\psi_\rho\| = \|\rho\|_1$. Similarly, we can show $T(H)^* = B(H)$. □

Corollary 2.7 *$T(H)$ is complete with respect to $\|\cdot\|_1$.*

It is now easy to see, as in Section 2.3, that there is affine isomorphism between normalized states on $K(H)$, and *density operators*, namely positive trace class operators ρ, of trace one, tr $\rho = 1$. Moreover, any positive trace class operator ρ has a decomposition $\rho = \sum \lambda_i T_{e_i,e_i}$, where $\lambda_i > 0$, $\sum \lambda_i = 1$, $\{e_i\}$ an orthonormal sequence, and so every normalized state ϕ is a convex combination of vector states (c.f. Exercise 2.2):

$$\phi(x) = \textstyle\sum_{i=1}^\infty \lambda_i \langle xe_i, e_i\rangle .$$

Exercise 2.1 Show that if $\rho \in T(H)$, $x \in B(H)$, then $\operatorname{tr} \rho x = \operatorname{tr} x\rho$. (Note, $\langle a, b\rangle = \langle b^*, a^*\rangle$, $a, b \in HS(H)$, shows $\operatorname{tr} ab = \operatorname{tr} ba$, if $a, b \in HS(H)$, and this has already been used in the text.)

Exercise 2.2 Suppose ρ is a positive trace class operator, with spectral decomposition $\rho = \sum_{i=1}^\infty \lambda_i T_{e_i,e_i}$ where $\{e_i\}$ is a complete orthonormal sequence, $\lambda_i > 0$. Show that $\rho_n = \sum_{i=1}^n \lambda_i T_{e_i,e_i}$ converges to ρ in $\|\cdot\|_1$ norm, and deduce that $\operatorname{tr} \rho x = \sum_{i=1}^\infty \lambda_i \langle xe_i, e_i\rangle$, $\quad x \in B(H)$.

Exercise 2.3 (a) If $x \in B(H)$, show that $\psi_x(\rho) = \operatorname{tr} \rho x$, $\rho \in T(H)$, defines an element ψ_x of $T(H)^*$ such that $\|\psi_x\| \leq \|x\|$.

(b) If $\xi, \eta \in H$, show that tr $T_{\xi,\eta} = \langle \xi, \eta\rangle$ and $\|T_{\xi,\eta}\|_1 = \|\xi\|\|\eta\|$.

(c) If $\psi \in T(H)^*$, show that there exists an unique $x \in B(H)$ such that $\psi(T_{\xi,\eta}) = \langle x\xi, \eta\rangle$, and $\|x\| = \|\psi\|$.

(d) Deduce that if $\psi \in T(H)^*$, there exists an unique $x \in B(H)$ such that $\psi(\rho) = \operatorname{tr} \rho x$, and $\|\psi\| = \|x\|$.

Exercise 2.4 (Powers–Størmer inequality). Show that if A and B are positive operators on a Hilbert space H, then $\|A - B\|_2^2 \leq \|A^2 - B^2\|_1$.

2.5 The Gelfand–Naimark–Segal construction

If we have a probability measure μ on a compact Hausdorff space X, or a normalized state $\varphi = \varphi_\mu$ on $C(X)$, we can form the Hilbert space $\mathrm{L}^2(\mu)$, as the completion of $C(X)$ with respect to the inner product $\langle f, g\rangle = \int f\bar{g}\, d\mu = \varphi(f\bar{g})$, $f, g \in C(X)$. There is then a canonical representation π of $C(X)$ on $\mathrm{L}^2(X, \mu)$ given by pointwise multiplication:

$$\pi(f)g = fg \tag{2.5.1}$$

for $f, g \in C(X)$, where g, fg are regarded as vectors in $\mathrm{L}^2(X, \mu)$. Moreover if $\Omega = 1$, the identity of $C(X)$ regarded as a vector in $\mathrm{L}^2(X, \mu)$, then

$$\varphi(f) = \int (f.1)1\, d\mu = \langle \pi(f)\Omega, \Omega\rangle\,, \tag{2.5.2}$$

$$\mathrm{L}^2(X, \mu) = \overline{\pi(C(X))\Omega}\,. \tag{2.5.3}$$

If φ is a normalized state on a (unital) C^*-algebra A, then there is an analogous construction of a triple $(H_\varphi, \pi_\varphi, \Omega_\varphi)$ of a Hilbert space H_φ obtained by completing A with respect to an inner product induced by φ, a distinguished unit vector Ω_φ, obtained by regarding the unit of A as an element of H_φ, and a representation of A as a C^*-algebra on H_φ such that

(2.5.4) $\varphi(x) = \langle \pi_\varphi(x)\Omega_\varphi, \Omega_\varphi\rangle$, $x \in A$

(2.5.5) $H_\varphi = \overline{\pi_\varphi(A)\Omega_\varphi}$.

Theorem 2.8 Gelfand–Naimark–Segal (GNS) decomposition. *If φ is a normalized state on a unital C^*-algebra A, then there exists a representation π_φ of A on a Hilbert space H_φ, with unit vector $\Omega_\varphi \in H_\varphi$, such that* (2.5.4) *and* (2.5.5) *hold.*

Proof We define a sesquilinear map on A by $(x, y) = \varphi(y^*x)$, $x, y \in A$. We first quotient A by the vectors of zero length (the functions μ-almost everywhere equal to zero in the commutative case). Let

$$N \equiv \{x \in A \mid (x, x) = 0\} = \{x \in A \mid (x, y) = 0, \forall y \in A\}$$

by the Cauchy–Schwarz inequality (1.4.11). Hence N is a subspace of A. Let $i : A \to A/N$ denote the quotient map. Then $\langle i(x), i(y)\rangle = (x, y)$, $x, y \in A$ defines a positive definite inner product on A/N and let H_φ denote its completion, $\Omega_\varphi =$

$i(1)$. If $a, x \in A$, then $0 \leq x^*a^*ax \leq \|a\|^2 x^*x$ by Exercise 1.29, and so $\|i(ax)\| \leq \|a\|\|i(x)\|$. Thus N is a left ideal and there is a well defined bounded linear operator $\pi_0(a)$ on A/N so that $\pi_0(a)i(x) = i(ax)$, $a, x \in A$. We let $\pi(a)$ denote its extension to a continuous linear map on H_φ. Since $\pi_0(ab) = \pi_0(a)\pi_0(b)$, $\pi_0(a+b) = \pi_0(a) + \pi_0(b)$, and

$$\begin{aligned} \langle \pi_0(a)i(x), i(y)\rangle &= \langle i(ax), i(y)\rangle = \varphi(y^*ax) \\ = \varphi((a^*y)^*x) &= \langle i(x), i(a^*y)\rangle = \langle i(x), \pi_0(a^*)i(y)\rangle, \quad a, x, y \in A \end{aligned}$$

it is clear that $\pi = \pi_\varphi$ is a $*$-representation. The remainder is clear as

$$\langle \pi_\varphi(x)\Omega_\varphi, \Omega_\varphi\rangle = \langle \pi(x)i(1), i(1)\rangle = \langle i(x), i(1)\rangle = \varphi(1^*x) = \varphi(x),$$

and $\pi_\varphi(A)\Omega_\varphi = A/N$. □

If φ, ψ are states on a C^*-algebra, we say that ψ is dominated by φ, written $\psi \leq \varphi$, if $\varphi - \psi$ is state. If S is a set of operators on a Hilbert space H, we let S' denote the *commutant* of S, namely $\{t \in B(H) \mid ts = st, s \in S\}$, or those operators commuting with S.

Lemma 2.9 *Let φ be a state on a (unital) C^*-algebra A. There exists an affine correspondence between positive contractions T in $\pi_\varphi(A)'$ and states on A dominated by φ given by*

(2.5.6) $\quad \varphi_T(a) = \langle \pi_\varphi(a)T\Omega_\varphi, \Omega_\varphi\rangle$, $a \in A$.

Proof Let T be a positive contraction in $\pi_\varphi(A)'$. Then by Theorem 1.25 and Exercise 2.6, there exists an operator S in $\pi_\varphi(A)'$ such that $T = S^*S$. Then

$$\begin{aligned} 0 &\leq \langle \pi_\varphi(a^*a)S\Omega_\varphi, S\Omega_\varphi\rangle = \langle T\pi_\varphi(a)\Omega_\varphi, \pi_\varphi(a)\Omega_\varphi\rangle \\ &\leq \langle \pi_\varphi(a)\Omega_\varphi, \pi_\varphi(a)\Omega_\varphi\rangle = \langle \pi_\varphi(a^*a)\Omega_\varphi, \Omega_\varphi\rangle \end{aligned}$$

shows that $0 \leq \varphi_T(a^*a) \leq \varphi(a^*a)$, and so φ_T is a state dominated by φ. Conversely suppose ω is a state dominated by φ. Then $0 \leq \omega(a^*a) \leq \varphi(a^*a)$ says that

$$0 \leq \|\pi_\omega(a)\Omega_\omega\|^2 < \|\pi_\varphi(a)\Omega_\varphi\|^2, \quad a \in A.$$

Hence there exists an unique contraction R from H_φ into H_ω such that $R\pi_\varphi(a)\Omega_\varphi = \pi_\omega(a)\Omega_\varphi$. Then $R\pi_\varphi(a)\pi_\varphi(b)\Omega_\varphi = R\pi_\varphi(ab)\Omega_\varphi = \pi_\omega(ab)\Omega_\omega = \pi_\omega(a)\pi_\omega(b)\Omega_\omega = \pi_\omega(a)R\pi_\varphi(b)\Omega_\varphi$, $a, b \in A$. Hence $R\pi_\varphi(a) = \pi_\omega(a)R$, $a \in A$, and so $T = R^*R$ is a positive contraction in $\pi_\varphi(A)'$. Then

$$\langle T\pi_\varphi(a)\Omega_\varphi, \Omega_\varphi\rangle = \langle R\pi_\varphi(a)\Omega_\varphi, R\Omega_\varphi\rangle = \langle \pi_\omega(a)\Omega_\omega, \Omega_\omega\rangle = \omega(a)$$

and so $\omega = \varphi_T$. □

Remark Note that from the proof of the above proposition, we see that the triple $(H_\varphi, \pi_\varphi, \Omega_\varphi)$ associated with a state φ is unique up to unitary equivalence, i.e. if π is a representation of A on a Hilbert space H with distinguished vector Ω such that $\varphi(a) = \langle \pi(a)\Omega, \Omega\rangle$, $a \in A$, $\overline{\pi(A)\Omega} = H$ then there exists an unique unitary U from H onto H_φ such that $U\Omega = \Omega_\varphi$, and $\pi(a) = U^*\pi_\varphi(a)U$, $a \in A$. We can thus refer to $(H_\varphi, \pi_\varphi, \Omega_\varphi)$ as the GNS triple or decomposition of a state φ. A representation π of a C^*-algebra A on a Hilbert space H is said to be *cyclic* if there exists a vector Ω in H such that $\overline{\pi(A)\Omega} = H$. In this case Ω is said be a *cyclic vector* for π. Two representations $\pi_i, i = 1, 2$ of a C^*-algebra A on Hilbert spaces H_i are said to be *unitarily equivalent* if there exists a unitary U from H_1 onto H_2 such that $\pi_1(a) = U^*\pi_2(a)U$, $a \in A$.

Let π be a representation of a C^*-algebra A on a Hilbert space H. Then π is said to be *non-degenerate* if $\xi \in H$, $\pi(a)\xi = 0$ for all $a \in A$ implies $\xi = 0$. Now in general if $K = \overline{\pi(A)H}$, then K is $\pi(A)$ invariant, and so from $\langle \pi(a)\xi, \pi(b)\xi'\rangle = \langle \xi, \pi(a^*b)\xi'\rangle$ we see that $\pi(A)K^\perp = \{0\}$, and π is non-degenerate on K. Thus $\pi = \pi_1 \oplus \pi_2$ on $H = K \oplus K^\perp$, where π_1 is non-degenerate and $\pi_2 = 0$.

Let us concretely identify the GNS triple of a state φ on $M(n)$. By (2.3.7), there exists an orthonormal sequence $\{e_i\}_{i=1}^r, 1 \le r \le n, \lambda_i > 0$ such that

$$\varphi(x) = \textstyle\sum_{i=1}^r \lambda_i \langle x e_i, e_i\rangle\,. \tag{2.5.7}$$

Let $H = \bigoplus_{i=1}^r \mathbb{C}^n$, $\Omega = \bigoplus_{i=1}^r \sqrt{\lambda_i} e_i$, $\pi(x) = \bigoplus_{i=1}^r x$, $x \in M(n)$. Then

$$\varphi(x) = \left\langle \textstyle\bigoplus_{i=1}^r x\sqrt{\lambda_i} e_i, \bigoplus_{i=1}^r \sqrt{\lambda_i} e_i \right\rangle = \langle \pi(x)\Omega, \Omega\rangle, \quad x \in M(n)\,.$$

Moreover $\pi(T_{\xi,e_i})\Omega = (0, \ldots, \sqrt{\lambda_i}\xi, 0, \ldots, 0)$ so that Ω is cyclic for π. Hence we can identify (π, H, Ω) as the GNS triple for φ. From this, it is a small matter to deduce that every representation of $M(n)$ (and similarly for K) is unitarily equivalent to a direct sum of identity representations.

Lemma 2.10 *Any non-degenerate representation of a C^*-algebra is a direct sum of cyclic representations.*

Proof Suppose $\pi = \pi_1 \oplus \pi_2$ on $H = H_1 \oplus H_2$, where π_1 is a direct sum of cyclic representations. Take Ω a non-zero vector in H_2, and let $K = \overline{\pi_2(A)\Omega}$. If $x \in A$, $\pi_2(x), \pi_2(x)^*$ leave K invariant and so by Exercise 2.2, $\pi_2(x)$ leaves $K^\perp \cap H_2$ invariant. Moreover π_2 is cyclic on K (if A is unital this is more than obvious, if A is not unital use an approximate identity, Exercise 1.36). In this way, using Zorn's lemma (or mere induction if dim $H < \infty$), we see that π is a direct sum $\oplus\pi_i$ of cyclic representations. Since H is separable, we see that we have at most a countable number of cyclic representations. □

Proposition 2.11 *Every non-degenerate representation π of $M(n)$ (or K) is unitarily equivalent to a direct sum of identity representations, $\pi = \bigoplus_{i=1}^N \mathrm{id}$, for some $1 \le N \le \infty$, where $\mathrm{id}\colon M_n \to M_n = B(\mathbb{C}^n)$ is the identity map.*

Proof By Lemma 2.10, we can assume that π possesses a cyclic vector Ω. Then the result follows from the uniqueness of the GNS decomposition of the state:

$\langle \pi(\cdot)\Omega, \Omega\rangle$, and the concrete identification of the GNS triple of any state on $M(n)$ described above (and similarly for K). □

Thus if $\pi : M(m) \to M(n)$ is a $*$-homomorphism, there is an integer k such that $km \leq n$, and a unitary u in $M(n)$ such that

$$\pi(x) = u(x \oplus \cdots \oplus x \oplus 0)u^* \tag{2.5.8}$$

in the notation of Section 1.2, where 0 denotes the zero of $M(n-km)$. This follows from Proposition 2.11 and the Remark on page 52. If $\pi : M(\mathbf{m}) \to M(n)$ is a $*$-homomorphism and $\mathbf{m} = (m_1, \ldots, m_p)$, let e_i denote the unit of $M(m_i)$. Then $\{\pi(e_i)\}_{i=1}^{p}$ are pairwise orthogonal projections in $M(n)$, and so by restricting attention to $\pi : M(m_i) \to \pi(e_i)M(n)\pi(e_i)$, we can find integers $(k_1, \ldots, k_p)$ such that $\sum_i k_i m_i \leq n$, and a unitary u in $M(n)$ such that

$$\pi(x_1 \oplus \cdots \oplus x_p) = u(x_1 \oplus \cdots \oplus x_1 \oplus x_2 \oplus \cdots \oplus x_2 \cdots \oplus x_p \oplus \cdots \oplus x_p \oplus 0)u^*$$

where on the right hand side k_1 x_1-summands, k_2 x_2-summands, x_p k_p-summands and where 0 denotes the zero of $M(n - \sum_{i=1}^{p} k_i m_i)$. From this, we easily deduce:

Proposition 2.12 *If $\pi : M(\mathbf{m}) \to M(\mathbf{n})$ is a $*$-homomorphism between multi-matrix C^*-algebras, where $\mathbf{m} = (m_1, \ldots, m_p)$, $\mathbf{n} = (n_1, \ldots, n_q)$, then there exist unique integers $[k_{ij}]_{i=1,j=1}^{q,p}$ and a unitary $u = (u_1, \ldots, u_q) \in M(\mathbf{n})$ such that $\sum_{i=1}^{p} k_{ji} m_i \leq n_j$, $j = 1, \ldots, q$ and*

$$\begin{aligned}
&\pi(x_1 \oplus \cdots \oplus x_p) \\
&= u_1(x_1 \overset{k_{11}}{\oplus} \cdots \oplus x_1 \oplus x_2 \overset{k_{12}}{\oplus} \cdots \oplus x_2 \oplus \cdots \oplus x_p \overset{k_{1p}}{\oplus} \cdots \oplus x_p \oplus 0)u_1^* \oplus \\
&\quad \vdots \\
&\oplus u_q(x_1 \overset{k_{q1}}{\oplus} \cdots \oplus x_1 \oplus x_2 \overset{k_{q2}}{\oplus} \cdots \oplus x_2 \oplus \cdots \oplus x_p \overset{k_{qp}}{\oplus} \cdots \oplus x_p \oplus 0)u_q^* .
\end{aligned} \tag{2.5.9}$$

Proof It only remains to show uniqueness. This follows because $k_{ji} = \mathrm{tr}(\pi(g_i)h_j)$ if h_j is the unit of $M(m_j)$, and g_i, a rank one-projection in $M(n_1)$. □

If $u = 1$ in (2.5.9), we say that π is a *canonical* or *standard homomorphism.* Thus the preceding proposition says that any $*$-homomorphism between multi-matrix algebras is unitarily equivalent to a canonical homomorphism. This can be used to describe all irreducible representations of $M(n)$ (or K). A representation π of a C^*-algebra A on a Hilbert space H is *irreducible* if it cannot be decomposed as a direct sum $\pi = \pi_1 \oplus \pi_2$ on $H = H_1 \oplus H_2$, where π_i are representations of A on proper subspaces of H. Equivalently (cf. the proof of Lemma 2.10), there does not exist a proper invariant closed subspace $K \subset H$ for π (i.e. such that $\pi(A)K \subset K$). It is clear that the identity representation of $M(n)$ (or K) is irreducible, since $T_{\xi,\eta}f = \langle f, \eta\rangle\xi$ shows that every non-zero vector is cyclic, and so there are certainly no proper invariant subspaces.

Corollary 2.13 *Every irreducible representation of $M(n)$ (or $K(H)$) is unitarily equivalent to the identity representation.*

Proof Suppose π is an irreducible representation. Then π is unitarily equivalent to $\bigoplus_{i=1}^{N} id$, for some N, by Proposition 2.12. If $N \neq 1$, then π is certainly reducible. □

In particular, every automorphism α of $M(n)$ ($K(H)$) is implemented by a unitary u in $M(n)$ (respectively $B(H)$), i.e. $\alpha(x) = uxu^*$, $x \in M(n)$ (respectively $K(H)$), since we can regard α as an irreducible representation. For completeness, note the following:

Proposition 2.14 *Let π be a representation of a C^*-algebra A on a Hilbert space H. Then the following conditions are equivalent:*

(2.5.10) *π is irreducible.*

(2.5.11) *There does not exist a proper closed subspace K of H such that $\pi(A)K \subset K$.*

(2.5.12) $\pi(A)' = \mathbb{C}1$.

(2.5.13) *Every non-zero vector in H is cyclic for π.*

Proof We have already observed in the remarks before Corollary 2.13 that (2.5.10)⇔(2.5.11)⇔(2.5.13). Moreover if $\pi = \pi_1 \oplus \pi_2$ on $H = H_1 \oplus H_2$, then E_1 the projection on H_1 lies in $\pi(A)'$ and so (2.5.12)⇒(2.5.10). Suppose (2.5.10) holds. If E is a projection in $\pi(A)'$, then $EH \oplus (1-E)H$ reduces π, and so either $E = 0$ or 1. If $T \in \pi(A)'$, we can assume T is self adjoint. Then by Section 1.6, $T = \int \lambda dE_\lambda$, where $E_\lambda \in \pi(A)'$ since E_λ is a strong limit of polynomials in T. Thus $E_\lambda = 0$ or 1 for each λ, and since E is increasing and right continuous there is a μ such that $E_\lambda = 0, \lambda < \mu, E_\lambda = 1, \lambda \geq \mu$. Then $T = \int \lambda dE_\lambda = \mu 1$. Hence (2.5.12) holds. □

We can describe geometrically those states whose GNS representations are irreducible. If φ is an element of a convex set C, we say that φ is *extreme* if whenever $\varphi = \lambda\varphi_1 + (1-\lambda)\varphi_2$ where $\varphi_1, \varphi_2 \in C, 0 \leq \lambda \leq 1$, then $\varphi = \varphi_1 = \varphi_2$. The extreme points of the convex set of normalized states are called *pure states.* Moreover, a state φ is *extremal* if whenever ψ is another state as dominated by φ, $\psi \leq \varphi$, then there exists $\lambda \geq 0$ such that $\psi = \lambda\varphi$.

Proposition 2.15 *Let φ be a normalized state on a C^*-algebra A, with GNS triple (π, H, Ω). Then the following conditions are equivalent:*

(2.5.14) *φ is an extremal state.*

(2.5.15) *φ is a pure state.*

(2.5.16) *π is irreducible.*

Proof That (2.5.14)⇔(2.5.16) follows from Lemma 2.9. Suppose (2.5.15) holds, and ψ is a non-zero state dominated by φ. (For simplicity, let us take A unital.)

If $\varphi - \psi$ is a state, then $\|\varphi - \psi\| = \varphi(1) - \psi(1)$, and so we can assume $1 = \varphi(1) \neq \psi(1)$. Then

$$\varphi = \psi(1)\psi/\psi(1) + (1 - \psi(1))((\varphi - \psi)/\varphi(1) - \psi(1))$$

expresses φ as a convex combination of states, and so $\psi/\psi(1) = \varphi$ since φ is pure. Conversely if (2.5.14) holds, and $\varphi = \lambda\varphi_1 + (1 - \lambda)\varphi_2$, where φ_i are normalized states, then $\varphi \geq \lambda\varphi_1$ and so φ_1 is a multiple of φ. Since $\varphi_1(1) = 1 = \varphi(1)$, we must have $\varphi = \varphi_1$. □

Exercise 2.5 Let φ be a positive linear functional on a C^*-algebra A.

(a) Suppose there exists $x_n \in A_+$, $\|x_n\| \leq 1$, and $\varphi(x_n) \geq n^2$. Show $x = \sum x_n/n^2 \in A_+$ and $\varphi(x) \geq \sum_{n=1}^{m} \varphi(x_n)/n^2 \geq m$. Deduce that $\sup\{\varphi(x) \mid x \in A_+, \|x\| \leq 1\} < \infty$ and φ is bounded.

(b) If $\{e_\lambda\}$ is an approximate identity for A, show that $M = \lim \varphi(e_\lambda)$ exists and $0 \leq M \leq \|\varphi\|$. Deduce from the Schwarz inequality $|\varphi(xe_\lambda)|^2 \leq \varphi(x^*x)\varphi(e_\lambda^2)$, that $\|\varphi\| = M$, and that the normalized states on A form a convex set.

(c) If N denotes the closed left ideal $\{x \in A \mid \varphi(x^*x) = 0\}$, $i : A \to A/N$ the quotient map, H the completion of A/N with respect to the inner product $\langle i(x), i(y)\rangle = \varphi(y^*x)$, $x, y \in A$, and π the representation of A on H such that $\pi(a)i(x) = i(ax)$. Show that $\Omega = \lim i(e_\lambda)$ exists, $i(x) = \pi(x)\Omega$, $H = \overline{\pi(A)\Omega}$, and $\varphi(x) = \langle \pi(x)\Omega, \Omega\rangle$, $x \in A$.

Exercise 2.6 If S is a family of operators on a Hilbert space H, closed under involution (i.e. $x \in S \Rightarrow x^* \in S$), show that S' is a C^*-algebra, closed under weak limits.

Exercise 2.7 A non-zero projection p in a C^*-algebra A is minimal if whenever q is a non-zero projection in A with $q \leq p$, then $q = p$. Thus the minimal projections in $B(H)$ are the rank one projections $\xi \otimes \overline{\xi}$, as ξ runs over the unit vectors of H. If α is an automorphism of $B(H)$, show that α acts on minimal projections as $\alpha(\xi \otimes \overline{\xi}) = u\xi \otimes \overline{u\xi}$ for a unitary linear operator on H and that $\alpha = \mathrm{Ad}(u)$, i.e. is spatial.

2.6 Graphical representation of operators and tensor products

The vector space $M_{m,n}$ of $m \times n$ complex matrices consists of arrays $W = [W_{\xi,\eta}]$ where $\xi \in \{1, 2, \ldots, m\}$, $\eta \in \{1, 2, \ldots, n\}$. Alternatively, we could take $\xi \in I, \eta \in J$, where I and J are any indexing sets of cardinality $|I| = m, |J| = n$. We write $M_{I,J}$ for the corresponding vector space, and $M_I = M_{I,I} (= M_{|I|})$. Graphically, we draw Fig. 2.1 where ξ, η label edges, the two notations being useful in different contexts, and refer to such a w as a *face operator*. Strictly speaking, we should ensure that our notation is not obscured or becomes confused by rotations of the figures in the plane. So we should orient the edges which carry the labelling of the indexing sets as in Fig. 2.1. In this section and elsewhere, where no ambiguity will arise as to which is the incoming or outgoing index in a matrix, we will omit the orientation of the edges in our figures.

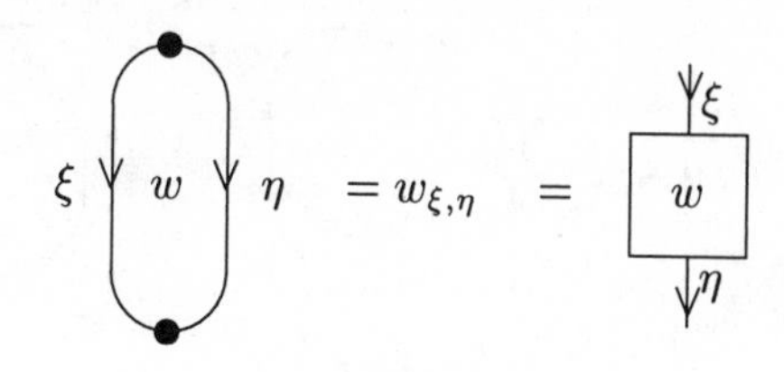

FIG. 2.1. Face operator w

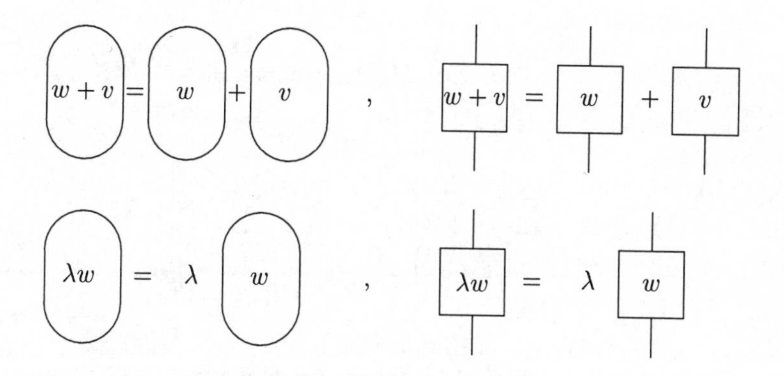

FIG. 2.2. Vector space operations on face operators

Vector space operations are then given by Fig 2.2 with involution $w^* \in M_{J,I}$ if $w \in M_{I,J}$ (see Fig. 2.3). Matrix multiplication $(wv)_{\xi\eta} = \sum_\phi w_{\xi\phi} v_{\phi\eta}$ becomes Fig. 2.4, and we write the right hand sides as Fig 2.5 respectively or Fig. 2.6 with internal edges summed, i.e. Fig. 2.7. Then $M_{I,*} = \mathbb{C}^I$ is identified with $\ell^2(I)$ with inner product $((M_{*,I} = (M_{I,*})^*) \times M_{*,I} \to \mathbb{C})$ as in Fig. 2.8. If $u \in \ell^2(I), v \in \ell^2(J)$, then $uv^* \in M_{I,J} = \mathrm{End}(\ell^2(I), \ell^2(J))$, is the rank one operator $T_{u,v}$ and we have Fig. 2.9 where the internal edge is unique. Then $M_{I,J} = \mathrm{End}(\ell^2(I), \ell^2(J))$ acts on $\ell^2(J)$ by Fig. 2.10. The trace of an operator $w \in M_I$ is as in Fig. 2.11, so that $\mathrm{tr}\, ab = \mathrm{tr}\, ba$ is expressed graphically as in Fig. 2.12, e.g. $\mathrm{tr}\, uv^* = v^*u \in \mathbb{C}$, if $u, v \in \ell^2(I)$.

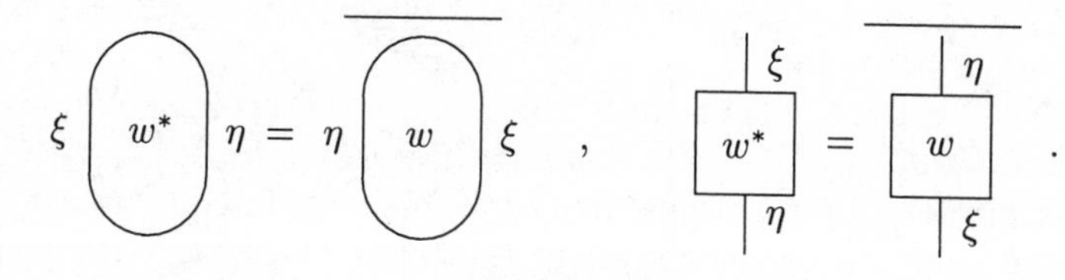

FIG. 2.3. Involution on face operators

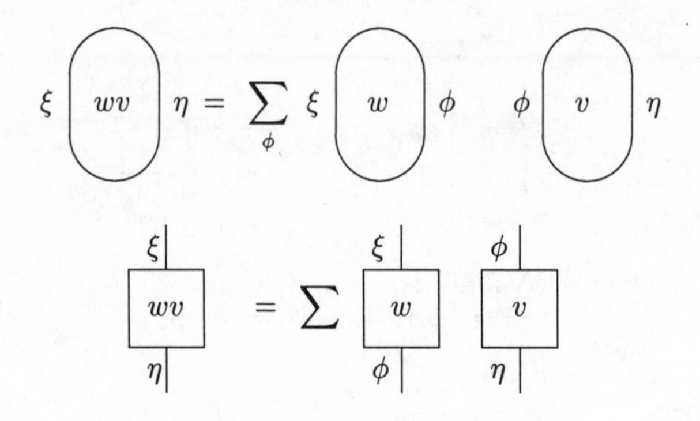

FIG. 2.4. Matrix product wv

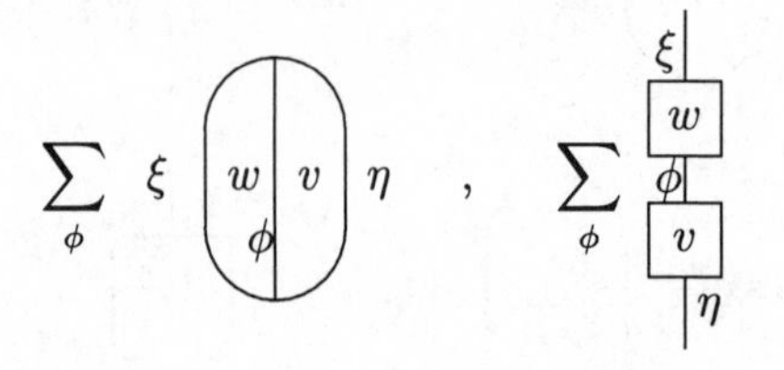

FIG. 2.5. Matrix product wv

We have seen how to add or take direct sums $\oplus$ of inner product spaces, algebras etc., in Section 1.2. Now we wish to multiply or take tensor products $\otimes$ of inner product spaces, algebras etc. For the time being it is sufficient to take the tensor product $V_1 \otimes V_2$ of *finite dimensional* vector spaces V_1, V_2, which may be inner product spaces, matrix algebras etc. Thus let I, J, I', J' be finite sets with cardinality m, n, m', n' respectively. If $A \in M_{I,J}, B \in M_{I',J'}$ we let $A \otimes B$ denote the matrix of $M_{I \times I', J \times J'}$ given by

$$(A \otimes B)_{(i,r),(j,s)} = A_{ij} B_{rs} \quad i \in I, r \in I', j \in J, s \in J' ,$$

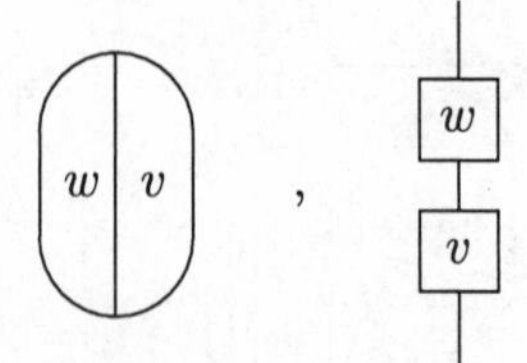

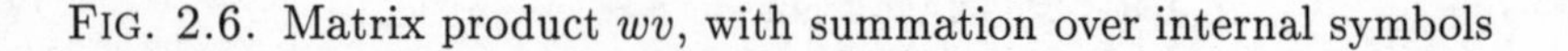

FIG. 2.6. Matrix product wv, with summation over internal symbols

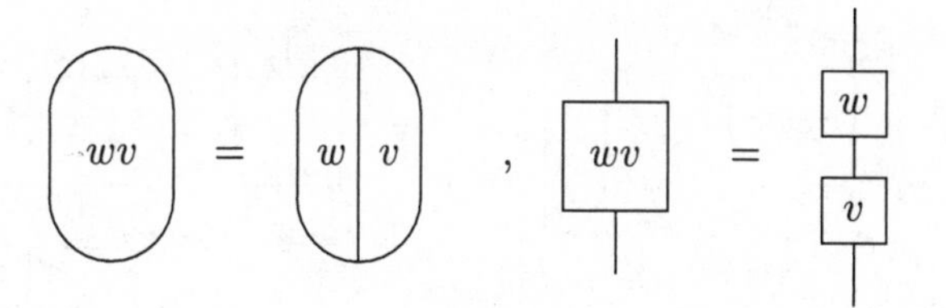

FIG. 2.7. Abbreviated diagram of matrix product

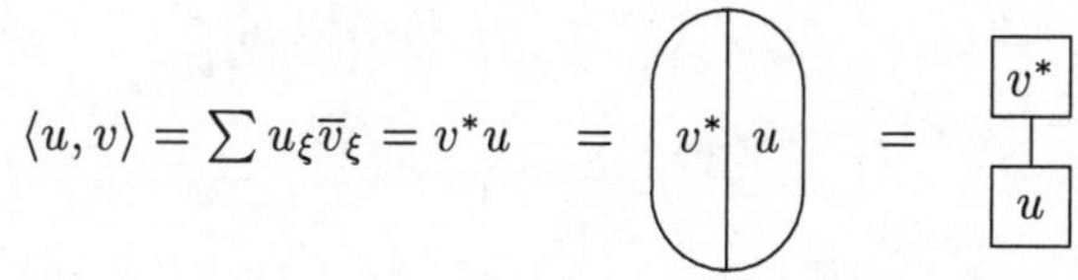

$$\langle u, v\rangle = \sum u_\xi \overline{v}_\xi = v^* u \quad = \quad v^* \mid u \quad = \quad \begin{matrix} v^* \\ u \end{matrix}$$

FIG. 2.8. Scalar product of face vectors

i.e. graphically as Fig. 2.13 where we use the notation of Fig. 2.14 for convenience. We regard $M_{(mm',nn')}$ as being matrices indexed by (i,r) and (j,s) where $1 \leq i \leq m$, $1 \leq r \leq m'$, $1 \leq j \leq n$, $1 \leq s \leq n'$. If $A \in M_{(m,n)}$, $B \in M_{(m',n')}$, we let $A \otimes B$ denote the matrix of $M_{(mm',nn')}$ given by

$$A \otimes B = \begin{pmatrix} AB_{11} & AB_{12} \dots AB_{1n'} \\ AB_{21} & AB_{22} \quad\quad \vdots \\ \vdots & \\ AB_{m'1} & \dots \quad AB_{m',n'} \end{pmatrix} \tag{2.6.1}$$

i.e. $(A \otimes B)_{(i,r),(j,s)} = A_{ij} B_{rs}$, where i, j, r, s are as above. The tensor product $\otimes$ satisfies

(2.6.2) $A \to A \otimes B, B \to A \otimes B$ are linear, i.e.

$$(\lambda A_1 + \mu A_2) \otimes B = \lambda(A_1 \otimes B) + \mu(A_2 \otimes B) \quad \text{etc.},$$

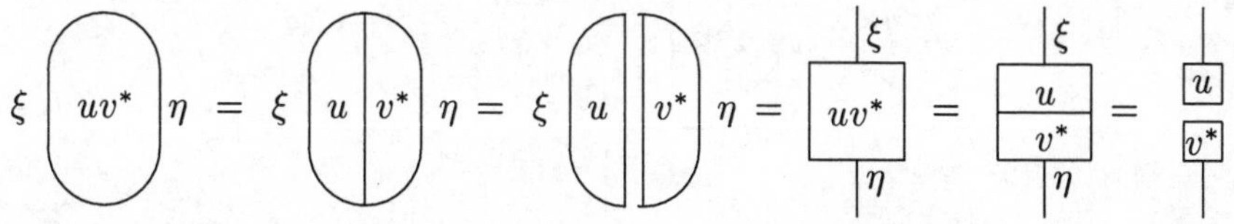

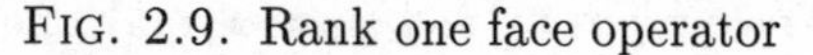

FIG. 2.9. Rank one face operator

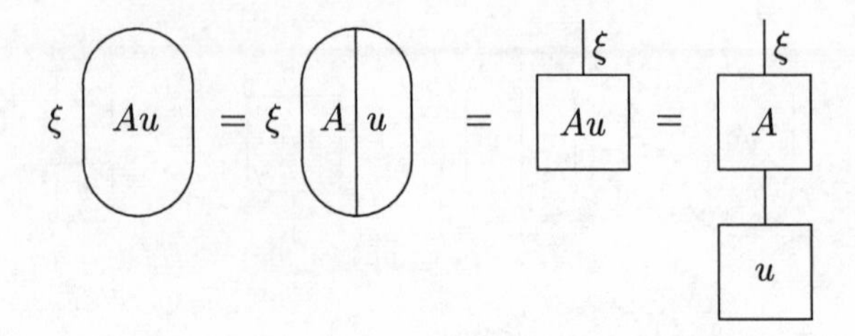

FIG. 2.10. Face operator acting on vector

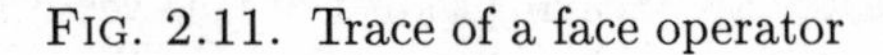

FIG. 2.11. Trace of a face operator

(2.6.3) $(A \otimes B)^* = A^* \otimes B^*$,

(2.6.4) $(A \otimes B)(C \otimes D) = AC \otimes BD$.

This can be seen graphically as in Fig. 2.15, where the reader is left to insert the indices. It will be seen that

$$M_{I,J} \otimes M_{I',J'} \equiv \operatorname{lin}\{A \otimes B \mid A \in M_{I,J}, B \in M_{I',J'}\} \tag{2.6.5}$$

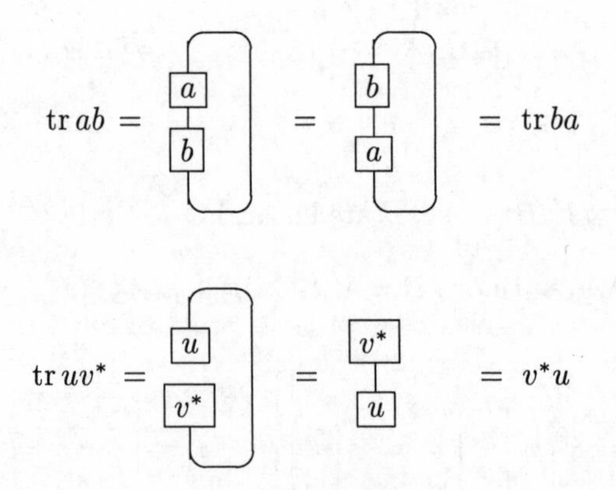

FIG. 2.12. Trace property and trace of rank one face operator

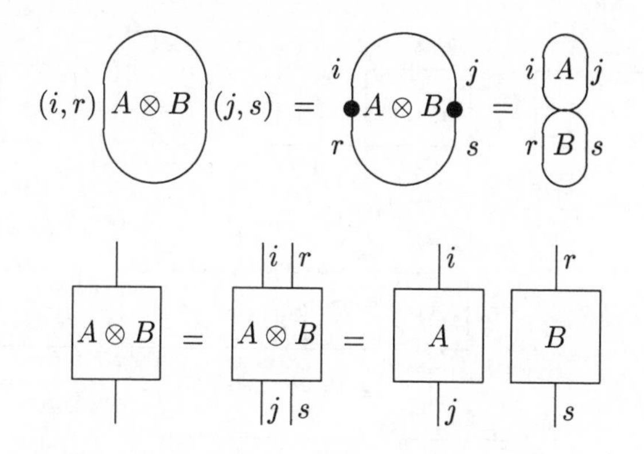

FIG. 2.13. Tensor product of face operators

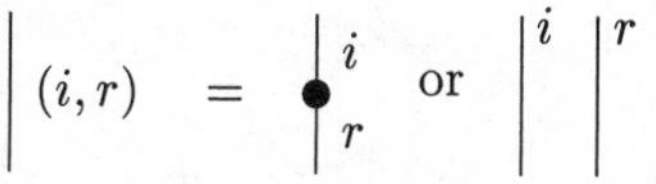

$(i, r) \quad = \quad i \;\; r \quad$ or $\quad i \;\; r$

FIG. 2.14. Labels for tensor products

exhausts all of $M_{(I\times I', J\times J')}$. For example, we have $\ell^2(I)\otimes\ell^2(I') = \ell^2(I\times I')$, where if $\{e_i\}, \{e_j\}$, are canonical bases for $\ell^2(I), \ell^2(I')$ respectively then $e_i\otimes e_j$ are identified with the canonical basis $e_{(i,j)}$ for $\ell^2(I\times I')$, and

$$\langle e\otimes f, e'\otimes f'\rangle = \langle e, e'\rangle\langle f, f'\rangle\,, \quad e, e'\in\ell(I),\ f, f'\in\ell^2(I') \tag{2.6.6}$$

using the canonical inner products in $\ell^2(I)$, $\ell^2(I')$, $\ell^2(I\times I')$. Thus we identify $M_{I,J}\otimes M_{I',J'}$ with $M_{I\times I',J\times J'}$ where if $\{e_{ij}\}$ and $\{e_{rs}\}$ are canonical matrix units for $M_{I,J}, M_{I',J'}$ respectively, then $\{e_{ij}\otimes e_{rs}\}$ are canonical matrix units $e_{(i,r)(j,s)}$ of $M_{I\times I',J\times J'}$ with respect to the bases $\{e_i\otimes e_r\}$ and $\{e_j\otimes e_s\}$. In fact

$$T_{e\otimes f, e'\otimes f'} = T_{e,e'}\otimes T_{f,f'}\,. \tag{2.6.7}$$

The labelling set(s) could be the edges of a graph. For example Fig. 2.16 could be the edges between the same pair of vertices p and q. If also I' are the edges of Fig. 2.17 then $I\times I'$ can be identified with the edges of Fig. 2.18. Similarly suppose J and J' are the edges of Fig. 2.19 respectively, so that $J\times J'$ is identified with Fig. 2.20(a). It is not necessary that $q = q'$. Then an element of $M_{I\times I',J\times J'}$ is written as Fig. 2.20(b) where $\alpha,\beta,\gamma,\delta$ are edges of I, I', J, J' respectively. Thus associated with each step in Fig. 2.21 with $I = I'$ of cardinality $n_1, J = J'$ of cardinality n_2, etc. we have matrix algebras $M_{n_1} = M_I$, $M_{n_1n_2} = M_{I\times J}$, $M_{n_1n_2n_3} = M_{I\times J\times K}$, ... which we can put inside each other as follows, by $T\to T\otimes 1$ or graphically as in Fig 2.22.

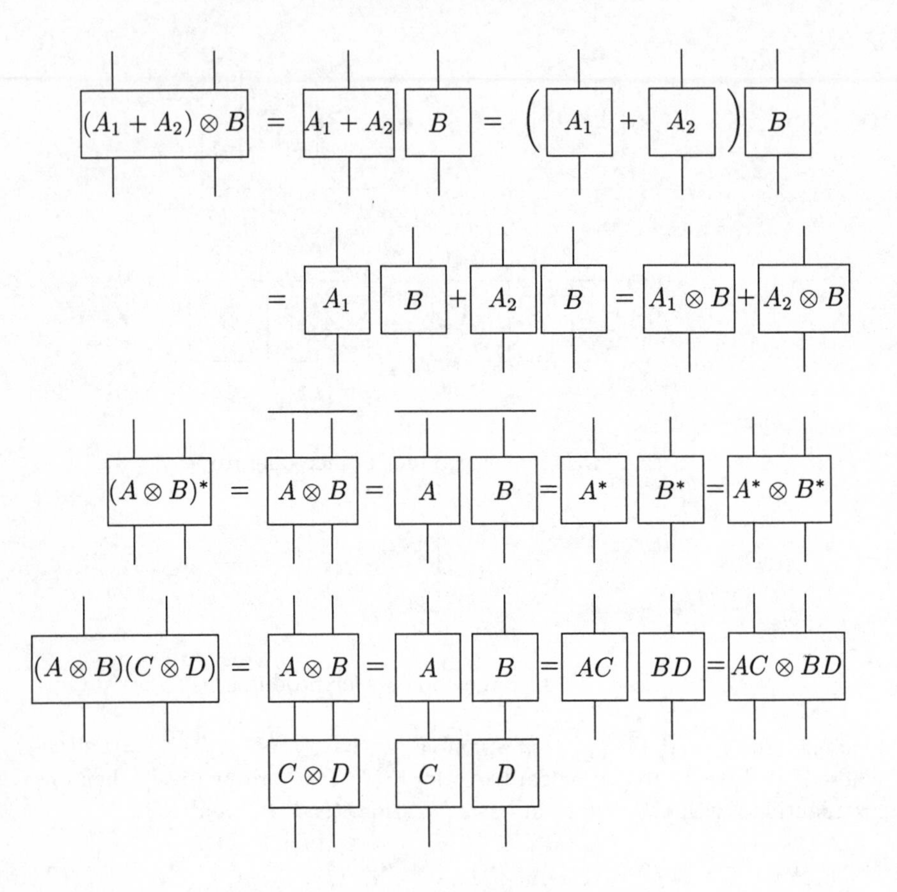

FIG. 2.15. Tensor product identities

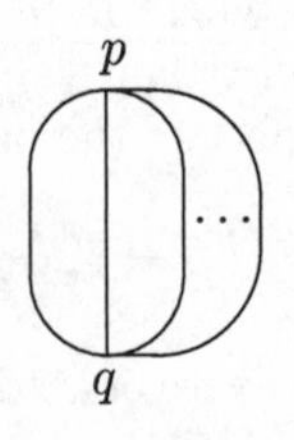

FIG. 2.16. Labels I for M_I

We would also take Pascal's triangle as in Fig. 2.40, and let I_α denote the paths from $*$ to a vertex α. Because of the branching rule, the cardinalities of indexing sets I_α are binomial coefficients. We will look at these more complicated path algebras in Section 2.9, after considering inductive limits in Section 2.7.

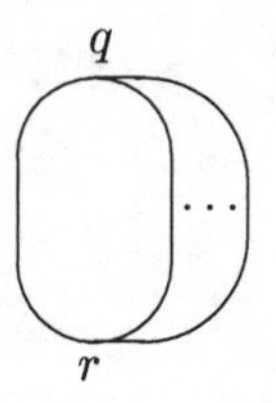

FIG. 2.17. I'

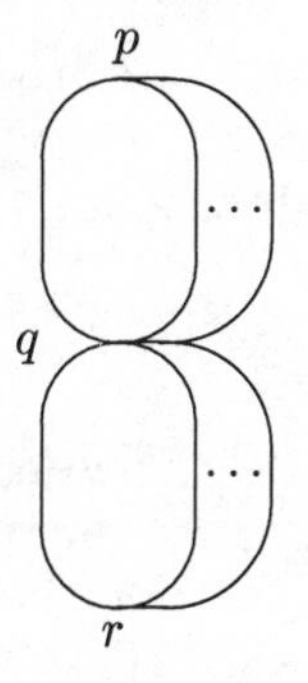

FIG. 2.18. $I \times I'$

Exercise 2.8 If H is a Hilbert space, we let $\overline{H}$, the conjugate space, be the Hilbert space consisting of vectors $\{\xi \mid \xi \in H\}$, where $\overline{\xi} + \overline{\eta} = \overline{\xi + \eta}$, $\lambda \cdot \overline{\xi} = \overline{\overline{\lambda}\xi}$, $\langle \overline{\xi}, \overline{\eta} \rangle = \langle \eta, \xi \rangle$, $\xi, \eta \in H$. Show that $e \otimes \overline{f} \to T_{e,f}$ identifies $\mathbb{C}^m \otimes \overline{\mathbb{C}^m}$ with M_m.

Exercise 2.9 We show how to define tensor products of infinite dimensional vector spaces. To avoid confusion, we denote algebraic procedure by $\odot$ and reserve $\otimes$ for completions. If Λ is a set, the free vector space generated by Λ is the (weak) direct sum $\bigoplus_{\lambda \in \Lambda}$, where we identify a point $\mu \in \Lambda$ with $(v_\lambda)_{\lambda \in \Lambda}$, $v_\lambda = \delta_{\mu\lambda}$. If V_1 and V_2 are two vector spaces, let U denote the free vector space generated by the Cartesian product $V_1 \times V_2$, and W the subspace generated by

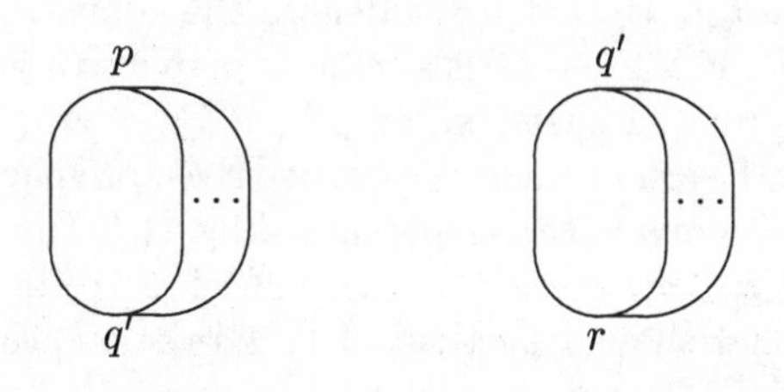

FIG. 2.19. J and J'

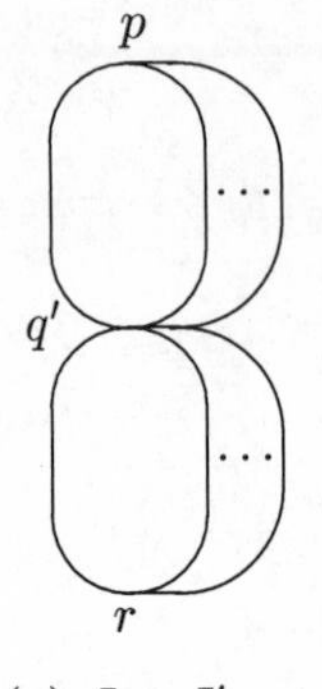

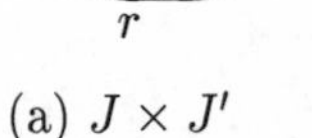

(a) $J \times J'$

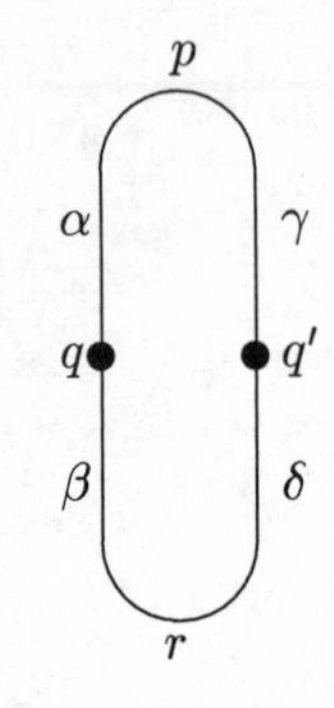

(b) Face operator in $M_{I\times I',J\times J'}$

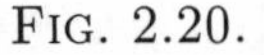

FIG. 2.20.

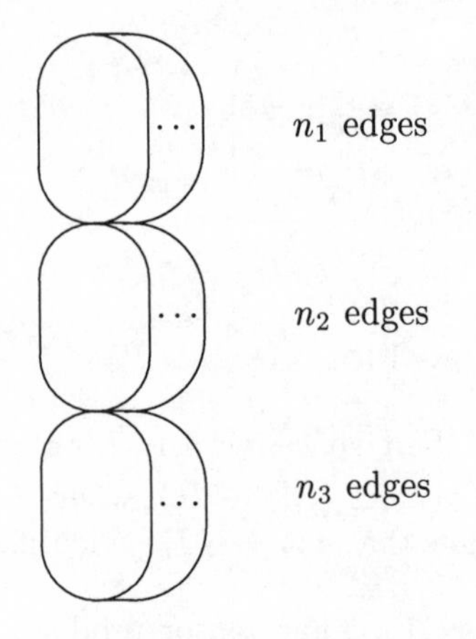

FIG. 2.21. $M_{I\times J\times K}$

$$(\alpha_1 a_1 + \alpha_2 a_2, b_1) - \alpha_1(a_1, b_1) - \alpha_1(a_2, b_1)$$
$$(a_1, \beta_1 b_1 + \beta_2 b_2) - \beta_1(a_1, b_1) - \beta_2(a_1, b_2)$$

$\alpha_i, \beta_j \in \mathbb{C}$, $a_i \in V_1$, $b_i \in V_2$. Let $a \otimes b$ denote the equivalence class of (a, b) in $U/W \equiv V_1 \odot V_2$. If $\psi : V_1 \times V_2 \to W$ is a bilinear map into a vector space W, show that there exists an unique linear map $\psi : V_1 \times V_2 \to W$ such that $\phi(a \otimes b) = \psi(a, b)$. If $T_i : V_i \to W_i$ are linear maps between vector spaces for $i = 1, 2$, show that there exists an unique linear map denoted by $T_1 \otimes T_2 : V_1 \odot V_2 \to W_1 \odot W_2$ such that $(T_1 \otimes T_2)(a \otimes b) = T_1 a \otimes T_2 b$, $a \in V_1$, $b \in V_2$. If V_1, V_2 etc. are finite dimensional, show that these definitions of $V_1 \odot V_2$, $a \otimes b$, $T_1 \otimes T_2$ are consistent with those in Section 2.6.

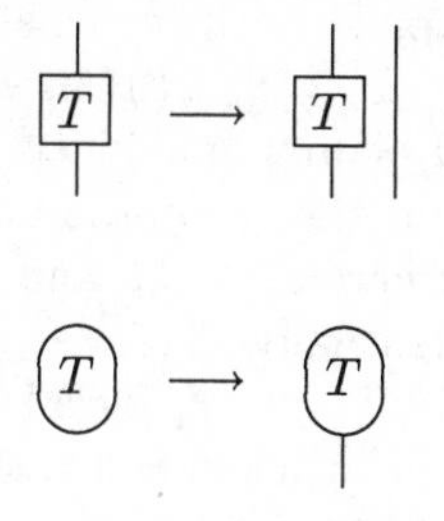

FIG. 2.22. $T \to T \otimes 1$

Exercise 2.10 Let V_1 and V_2 be inner product spaces. Show that there exists an unique sesquilinear map $\langle\, ,\, \rangle$ on $(V_1 \odot V_2) \times (V_1 \odot V_2)$ such that

$$\langle a \otimes b, a' \otimes b' \rangle = \langle a, a' \rangle \langle b, b' \rangle \,,$$

$a, a' \in V_1$, $b, b' \in V_2$. If $\{e_i\}_{i=1}^{N}$ is an orthonormal set in V_2 show that

$$\langle \textstyle\sum_{i=1}^{N} a_i \otimes e_i, \sum_{i=1}^{N} a_j \otimes e_j \rangle = \sum_{i=1}^{N} \|a_i\|^2 \,.$$

Deduce that $\langle\, ,\, \rangle$ defines an inner product on $V_1 \odot V_2$. If H_1 and H_2 are Hilbert spaces, let $H_1 \otimes H_2$ denote the Hilbert space completion of $H_1 \odot H_2$. If H is a Hilbert space, show that $\xi \otimes \overline{\eta} \to T_{\xi,\eta}$ defines a unitary from $H \otimes \overline{H}$ onto $HS(H)$.

Exercise 2.11 If H is a Hilbert space, embed $H \odot \overline{H}$ in $B(H)$ by $f \otimes \overline{g} \to T_{f,g}$, i.e. $(f \otimes \overline{g})\, h = \langle h, g \rangle f$, $f, g, h \in H$. Show that if $z \in H \odot \overline{H}$, then

$$\|z\|_1 = \inf \left\{ \textstyle\sum_i \|f_i\|\, \|g_i\| \mid z = \sum_i f_i \otimes \overline{g}_i \right\} \tag{2.6.8}$$

where $\|\cdot\|_1$ denotes the trace class norm. This means that $T(H)$, the trace class operators is the *projective tensor product* $H \otimes^{\vee} \overline{H}$, the completion of $H \odot \overline{H}$ with respect to the projective norm (2.6.8).

2.7 Inductive limits

We need the concept of inductive limits of sets, particularly of Hilbert spaces and of C^*-algebras. Thus let $\{Z_n \mid n = 1, 2, \ldots\}$ be a sequence of sets, with connecting maps $\theta_n : Z_n \to Z_{n+1}$, and define by composition, maps $\theta_{nm} : Z_n \to Z_m$, for $m \geq n$, such that $\theta_{ml}\theta_{nm} = \theta_{nl}$ if $l \geq m \geq n$, and $\theta_{mm} = id_{Z_m}, \theta_{m,m+1} = \theta_m$.

The inductive limit Z_∞ can be defined as the space of equivalence classes of the relation $\sim$ on the disjoint union

$$\textstyle\bigsqcup_{n=1}^{\infty} Z_n = \{(z, n) \mid z \in Z_n, n = 1, 2, \ldots\}$$

where $\sim$ is the smallest equivalence relation generated by $(z, n) \sim (\theta_n(z), n+1)$. More precisely, the relation $\sim$ is given by $(x, n) \sim (y, m)$ if and only if x in Z_n

and y in Z_m are eventually mapped to the same element, i.e. if and only if there exists $N > m, n$ such that $\theta_{nN}(x) = \theta_{mN}(y)$ in Z_N. We write $Z_\infty = \lim_{\rightarrow} Z_n$, or $\lim_{\rightarrow}(Z_n, \theta_n)$, if there could be any ambiguity about which connecting maps to take. Alternatively, one could take the product space $\Pi_{n=1}^{\infty} Z_n$, or the space of all sequences $z = (z_n)_{n=1}^{\infty}$, where $z_n \in Z_n$, and then consider the subset $\mathcal{Z}$ of all sequences (z_n) such that eventually $z_{n+1} = \theta_n(z_n)$. Then define a relation $\approx$ on Z by $(z_n) \approx (z'_n)$ if and only if $z_n = z'_n$ for all large enough n. Then Z_∞ can also be identified with $\mathcal{Z}/\approx$, the space of equivalence classes of $\approx$.

For example if A is a set with subsets $A_n \subset A_{n+1} \subset A$ such that $A = \bigcup_{n=1}^{\infty} A_n$, we can identify A with $\lim_{\rightarrow}(A_n, \theta_n)$, where $\theta_n : A_n \to A_{n+1}$ is the inclusion map.

If [ ] denotes the equivalence class of the enclosed element, define $\theta_{n\infty} : Z_n \to Z_\infty$ by

$$\theta_{n\infty}(z) = [(z, n)], \quad \text{if } z \in Z_n . \tag{2.7.1}$$

Thus if $W_n = \theta_{n\infty}(Z_n)$, then $W_n \subset W_{n+1}$, and $Z_\infty = \bigcup_{n=1}^{\infty} W_n$. If each θ_n is injective then so are θ_{mn} if $\infty \geq n \geq m$, where $\theta_{\infty\infty} = \mathrm{id}_{Z_\infty}$. If we have a commuting diagram of maps

$$\begin{array}{ccccccccc} Z_1 & \xrightarrow{\theta_1} & Z_2 & \xrightarrow{\theta_2} & Z_3 & \xrightarrow{\theta_3} & Z_4 & \to & \cdots \\ \downarrow \phi_1 & & \downarrow \phi_2 & & \downarrow \phi_3 & & \downarrow \phi_4 & & \\ Z'_1 & \xrightarrow[\theta'_1]{} & Z'_2 & \xrightarrow[\theta'_2]{} & Z'_3 & \xrightarrow[\theta'_3]{} & Z'_4 & \to & \cdots \end{array} \tag{2.7.2}$$

we can define a unique map $\phi_\infty : Z_\infty \to Z'_\infty$ such that

$$\phi_\infty \theta_{n\infty} = \theta'_{n\infty} \phi_n . \tag{2.7.3}$$

If each ϕ_n is surjective (respectively injective) then ϕ_∞ is also surjective (respectively injective). If each Z_n is a group (respectively vector space, inner product space, algebra, $*$-algebra etc.), and $\{\theta_n\}$ are group homomorphisms (respectively linear transformations, linear isometries, algebra homomorphisms, $*$-algebra homomorphisms etc.) then Z_∞ can be endowed with a unique group structure (respectively vector space, inner product, algebraic, $*$-algebraic structure etc.) such that $\theta_{n\infty} : Z_n \to Z_\infty$ all become group homomorphisms (respectively linear transformations, linear isometries, algebra homomorphisms, $*$-algebra homomorphisms etc.). For example, if the Z_n are inner product spaces, define

$$\theta_{n\infty}(x) + \theta_{m\infty}(y) = \theta_{N\infty}(\theta_{nN}(x) + \theta_{mN}(y)) , \tag{2.7.4}$$

$$\lambda \cdot \theta_{n\infty}(x) = \theta_{n\infty}(\lambda x) , \tag{2.7.5}$$

$$\langle \theta_{n\infty}(x), \theta_{m\infty}(y) \rangle = \langle \theta_{nN}(x), \theta_{mN}(y) \rangle_{Z_n} , \tag{2.7.6}$$

if $x \in Z_n$, $y \in Z_m$, $N > m, n$, $\lambda \in \mathbb{C}$. By abuse of notation, we will write Z^0_∞ for the algebraic direct limit $\lim_\to(Z_n, \theta_n)$, Z_∞, for its Hilbert space completion $\overline{\lim_\to}(Z_n, \theta_n)$, and in the situation of (2.7.2), where ϕ_i are contractions between Hilbert spaces, we let ϕ_∞ again denote the extension of the contraction ϕ_i between Z^0_∞ and Z'^0_∞ to that between Z_∞ and Z'_∞.

If $\{Z_n\}$ are C^*-algebras, and θ_n are $*$-homomorphisms, we can define an unique norm $\|\cdot\|$ on the $*$-algebraic inductive limit

$$\lim_{\to}(Z_n, \theta_n) = \bigcup_{n=1}^{\infty} \theta_{n\infty}(Z_n), \tag{2.7.7}$$

such that

(2.7.8) $\|x^*x\| = \|x\|^2$, $x \in \bigcup_{n=1}^\infty \theta_{n\infty}(Z_n)$

(2.7.9) $\theta_{n\infty}(Z_n)$ is complete with respect to this norm, i.e. $\theta_{n\infty}(Z_n)$ is a C^*-algebra.

It is enough for our purposes to prove this when each Z_n is a finite dimensional C^*-algebra, i.e. a multi-matrix C^*-algebra. To see this note that $J = \mathrm{Ker}\,\theta_{n\infty}$ is a $*$-ideal in Z_n, and so by Lemma 2.3 is also a multi-matrix algebra. If $Z_n = M(\mathbf{n})$, $\mathbf{n} = (n_1, \ldots n_p)$, $J = \bigoplus_{i\in\Lambda} M(n_i)$, then $Z_n/J \cong \bigoplus_{i\notin\Lambda} M(n_i)$ is another multi-matrix algebra, and so since $\theta_{n\infty}(Z_n) \cong Z_n/\mathrm{Ker}\,\theta_{n\infty}$ as $*$-algebras, $\theta_{n\infty}(Z_n)$ can be equipped with a norm $\|\cdot\|_n$ under which $\|x^*x\|_n = \|x\|_n^2$, $x \in \theta_{n\infty}(Z_n)$. By the uniqueness of the C^*-norm on a finite dimensional C^*-algebra, the restriction of $\|\cdot\|_{n+1}$ on $\theta_{n+1,\infty}(Z_{n+1})$ to $\theta_{n\infty}(Z_n)$ is $\|\cdot\|_n$. We can thus consistently define a norm $\|\cdot\|$ on $\bigcup_{n=1}^\infty \theta_{n\infty}(Z_n)$ such that its restriction to $\theta_{n\infty}(Z_n)$ is $\|\cdot\|_n$, and (2.7.8), (2.7.9) hold. (See also Exercise 2.12.) Again, we let Z^0_∞ denote the $*$-algebraic inductive limit $\lim(Z_n, \theta_n)$, $Z_\infty = \overline{\lim_\to}(Z_n, \theta_n)$ for the completion of $Z_\infty = \bigcup_{n=1}^\infty \theta_{n\infty}(Z_n)$ with respect to this norm, and in the C^*-situation of (2.7.2) where ϕ_i are $*$-homomorphisms (hence contractions by Corollary 1.14) between C^*-algebras, we again let ϕ_∞ denote the extension of ϕ_∞ between Z^0_∞ and Z'^0_∞ to a $*$-homomorphism between Z_∞ and Z'_∞.

An inductive limit of finite dimensional C^*-algebras is called an *approximately finite dimensional C^*-algebra* or AF *algebra*. For example, consider a sequence $A_n = M(\mathbf{m_n})$ of multi-matrix algebras, where $M(\mathbf{m_n})$ is embedded in $M(\mathbf{m_{n+1}})$ via a standard homomorphism with multiplicity matrix Λ_n, represented graphically by Fig. 2.23. Here we draw $\lambda^{(n)}_{ji}$ lines between the ith node in the upper row and the jth node in the lower row, with i and j representing simple subalgebras or minimal central projections in $M(\mathbf{m_n})$ and $M(\mathbf{m_{n+1}})$ respectively. It is convenient to adjoin an additional stage, the unital embedding

$$\mathbb{C} \to M(\mathbf{m_1}) \tag{2.7.10}$$

represented by the row matrix $(1, \ldots, 1) = \lambda^{(1)}$. The diagram of an AF algebra obtained by concatenating the multiplicity graphs is called a *Bratteli diagram* of

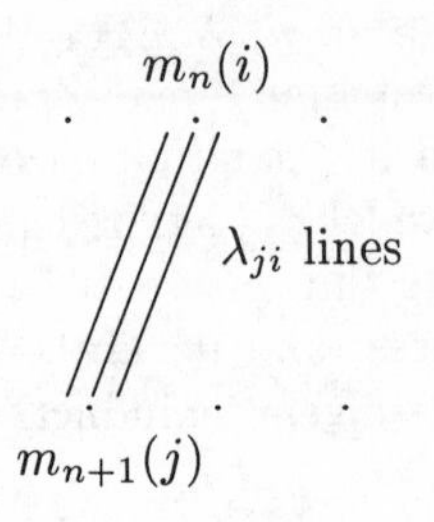

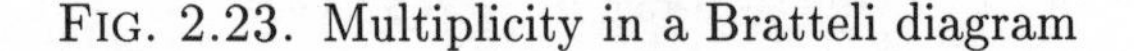
FIG. 2.23. Multiplicity in a Bratteli diagram

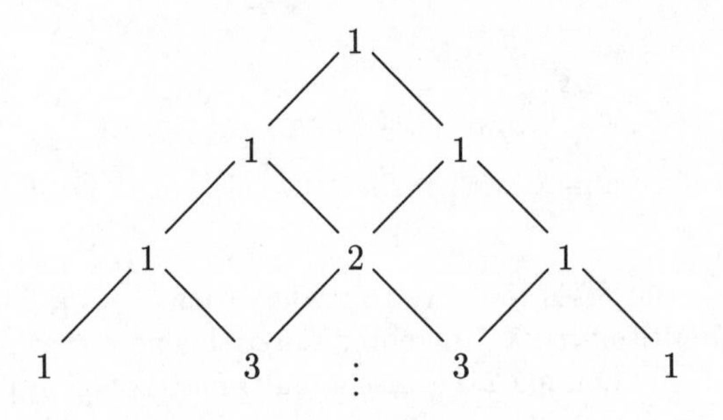

FIG. 2.24. Pascal's triangle

the AF algebra. An algebra may have many different Bratteli diagrams associated with it. As an example consider the AF algebra given in Fig. 2.24, i.e. the inductive limit F of $\mathbb{C} \to \mathbb{C} \oplus \mathbb{C} \to \mathbb{C} \oplus M_2 \oplus \mathbb{C} \to \ldots$. Note that if the mappings $M(\mathbf{m_n}) \to M(\mathbf{m_{n+1}})$ are all unital (equivalently $\lambda^{(n)}\mathbf{m_n} = \mathbf{m_{n+1}}$) then we do not need to record the size of the matrix algebra $M(m_n(i))$ at the ith node of the nth row, as they are uniquely determined from the diagram by

$$\mathbf{m_{n+1}} = \lambda^{(n)}\mathbf{m_n} = \lambda^{(n)}\lambda^{(n-1)}\cdots\lambda^{(1)}\,1\,. \tag{2.7.11}$$

We show that we can associate with any AF algebra a Bratteli diagram, and that two AF algebras which give rise to the same Bratteli diagram are isomorphic. In the first place an AF algebra A is obtained as an inductive limit from (A_n, θ_n) where each A_n is isomorphic to a multi-matrix algebra $M(\mathbf{m_n})$, via an isomorphism ψ_n. By Proposition 2.11, the homomorphism $\theta_n : A_n \to A_{n+1}$ is inner conjugate to standard homomorphism from $M(\mathbf{m_n})$ into $M(\mathbf{m_{n+1}})$ represented by a multiplicity graph $\lambda^{(n)}$, i.e. there exists u_{n+1} in $M(\mathbf{m_{n+1}})$ such that Fig. 2.25 commutes.

Inductively, we construct a sequence of unitaries w_n in $M(\mathbf{m_n})$ such that Fig. 2.26 commutes. This is clear at the initial case $n = 1$, if we take $w_1 = 1, w_2 = u_2$. The construction of w_{n+1} can be made first by observing that if w_n is a unitary in $M(\mathbf{m_n})$ then we can find a unitary w_n' in $M(\mathbf{m_{n+1}})$ such that

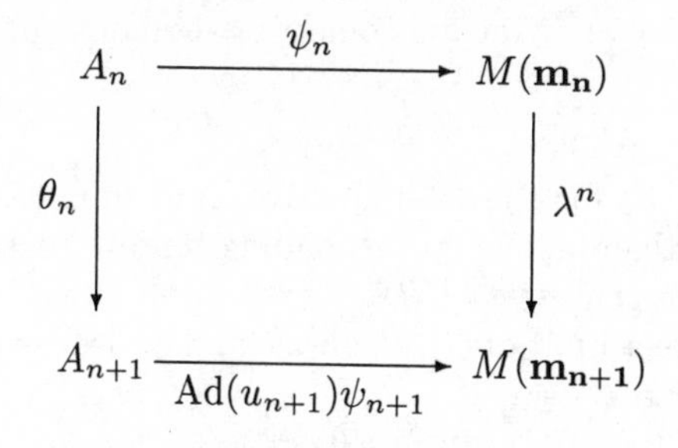

FIG. 2.25.

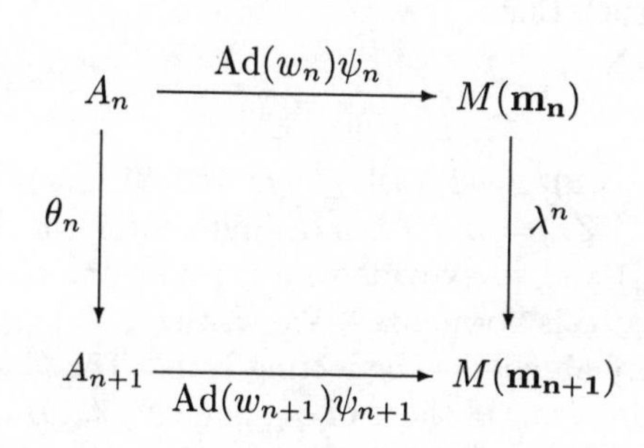

FIG. 2.26.

$\lambda^n \mathrm{Ad}(w_n) = \mathrm{Ad}(w'_n)\lambda^n$ (for λ^n unital, take $w'_n = \lambda^n(w_n)$). Then from Fig. 2.25 we see that

$$\text{(2.7.12)} \quad \lambda^n \mathrm{Ad}(w_n)\psi_n = \mathrm{Ad}(w'_n)\lambda^n\psi_n = \mathrm{Ad}(w'_n)\mathrm{Ad}(u_{n+1})\psi_{n+1}\theta_n$$

and so we take $w_{n+1} = w'_n u_{n+1}$.

As an example of C^*-inductive limits, we will construct infinite tensor products of finite dimensional C^*-algebras. For each $i = 1, 2, \ldots$, let A_i be a multimatrix C^*-algebra, and define $Z_n = A_1 \otimes \cdots \otimes A_n$. Since $\|T \otimes 1_{A_{n+1}}\| = \|T\|$

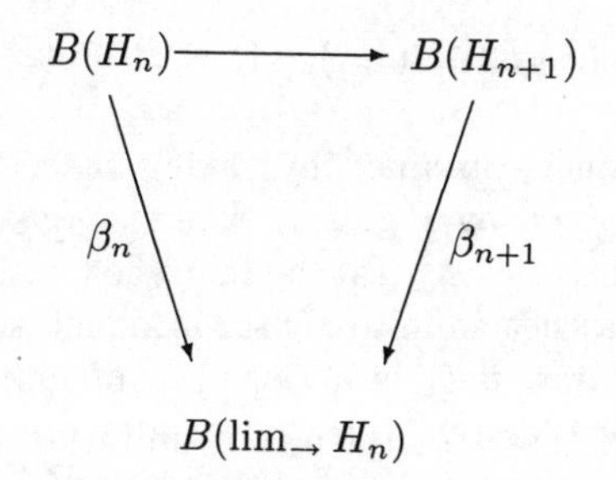

FIG. 2.27.

(see Theorem 1.5), we can define an isometric $*$-homomorphism $\theta_n : Z_n \to Z_{n+1}$ by

$$\theta_n(T) = T \otimes 1_{A_{n+1}}, \quad T \in Z_n\,. \tag{2.7.13}$$

We then write $\bigotimes_{i=1}^{\infty} A_i$ for the C^*-inductive limit $\lim_{\to}(A_1 \otimes \cdots \otimes A_n, \theta_n)$. If each A_i is a matrix algebra, the corresponding infinite tensor product $\bigotimes_{i=1}^{\infty} A_i$ is called *uniformly hyperfinite* or *UHF*.

If (Z_n, θ_n) is a direct limit of C^*-algebras, and ϕ_n is a sequence of normalized states on Z_n such that

$$\phi_{n+1} \circ \theta_n = \phi_n \tag{2.7.14}$$

we can define by (2.7.2) an unique linear functional ϕ on the $*$-algebraic direct limit $\bigcup_{n=1}^{\infty} \theta_{n\infty}(Z_n)$ such that

$$\phi \circ \theta_{n\infty} = \phi_n\,. \tag{2.7.15}$$

If $\theta_{n\infty}(y) = z$, then $|\phi(z)| = |\phi_n(y)| \leq \|y\|$ and so $|\phi(z)| \leq \|z\| = \inf\{\|y\| \mid \theta_{n\infty}(y) = z\}$ since $\theta_{n\infty}(Z_n) = Z_n/J$ has the quotient norm. Thus ϕ is a contraction on $\bigcup_{n=1}^{\infty} \theta_{n\infty}(Z_n)$ and so extends to a state on the completion Z_∞. From (2.7.15), it is clear that ϕ is normalized. We write $\phi_\infty = \lim(\phi_n, \theta_n)$ or $\lim_{\to} \phi_n$, if there is no ambiguity about the connecting maps. The GNS decomposition of ϕ_∞ can be expressed in terms of those of ϕ_n, namely (π_n, H_n, Ω_n) as follows. We define an isometry $V_n : H_n \to H_{n+1}$ by $V_n \pi_n(a)\Omega_n = \pi_{n+1}(\theta_n(a))\Omega_{n+1}$, $a \in Z_n$, and homomorphisms $\alpha_n : B(H_n) \to B(H_{n+1})$ by $\alpha_n(x) = V_n x V_n^*$, $x \in B(H_n)$. We then have a commutative diagram:

$$\begin{array}{ccccc} Z_1 & \xrightarrow{\theta_1} & Z_2 & \xrightarrow{\theta_2} & Z_3 \to \dots \\ \downarrow \pi_1 & & \downarrow \pi_2 & & \downarrow \pi_3 \\ B(H_1) & \xrightarrow[\alpha_1]{} & B(H_2) & \xrightarrow[\alpha_2]{} & B(H_3) \to \dots \end{array} \tag{2.7.16}$$

and so by going to the C^*-inductive limits we have a homomorphism

$$\pi_\infty : \lim_{\to}(Z_n, \theta_n) \to \lim_{\to}(B(H_n), \alpha_n) \to B(\lim_{\to}(H_n, V_n))\,, \tag{2.7.17}$$

where the final inclusion is obtained by taking the C^*-inductive limit of the commutative diagram given by Fig. 2.27 where $\beta_n(x) = V_{n\infty} x V_{n\infty}^*$. If $\Omega_\infty = V_{i\infty}\Omega_i$, then the GNS triple of ϕ_∞ can be identified with $(\pi_\infty H_\infty, \Omega_\infty)$.

Using this, we construct infinite tensor product states on infinite tensor product C^*-algebras. First, if ξ_i is a sequence of unit vectors in a sequence of Hilbert spaces H_i, we let $\bigotimes_{i=1}^{\infty}(H_i, \xi_i)$ denote the Hilbert space inductive limit of $\{\bigotimes_{i=1}^{n} H_i \mid \theta_n : \xi \to \xi \otimes \xi_n\}$, and write $\bigotimes_{i=1}^{\infty} \xi_i = \xi_1 \otimes \xi_2 \otimes \cdots$ for $\theta_{n\infty}(\xi_1 \otimes \cdots \otimes \xi_n)$. If ψ_i is a sequence of normalized states on a sequence of finite dimensional C^*-algebras A_i, we let $\bigotimes_{i=1}^{\infty} \psi_i$ denote $\lim_{\to}(\psi_1 \otimes \cdots \otimes \psi_n)$

on $\lim A_1 \otimes \cdots \otimes A_n = \bigotimes_{i=1} A_i$, and call such a state a product state. Note that $\psi_1 \otimes \cdots \otimes \psi_n$ is a state and its GNS triple can be identified with $(\pi_1 \otimes \cdots \otimes \pi_n, H_1 \otimes \cdots \otimes H_n, \xi_1 \otimes \cdots \otimes \xi_n)$ if (π_i, H_i, ξ_i) is the GNS triple of ψ_i, and that of $\bigotimes_{i=1}^{\infty} \psi_i$ can be identified with $(\bigotimes_{i=1}^{\infty} \pi_i, \bigotimes_{i=1}^{\infty}(H_i, \xi_i), \bigotimes_{i=1}^{\infty} \xi_i)$ if $\bigotimes_{i=1}^{\infty} \pi_i = \lim_{\rightarrow} \bigotimes_{i=1}^{n} \pi_i$ is constructed as in (2.7.17).

Exercise 2.12 Let $\theta_n : Z_n \to Z_{n+1}$ be a sequence of $*$-homomorphisms between C^*-algebras. Show that Ker $\theta_{n\infty} = \bigcap_{m \geq n}$Ker θ_{nm}, and so is a closed two sided $*$-ideal in Z_n. Show that Ker $\theta_{n\infty} = \theta_n^{-1}$Ker $(\theta_{n+1,\infty})$, and deduce that there is an injective $*$-homomorphism between the C^*-algebras $Z_n/$Ker $\theta_{n\infty} \to Z_{n+1}/$Ker $\theta_{n+1,\infty}$, induced by θ_n. Assuming we know that an injective $*$-homomorphism between C^*-algebras is isometric (see Exercise 1.39), show that we can equip the $*$-algebraic direct limit $\lim(Z_n, \theta_n) = \bigcup \theta_{n\infty}(Z_n)$ with an unique norm $\|\cdot\|$ such that

(a) $\|x^*x\| = \|x\|^2$, $x \in \bigcup \theta_{n\infty}(Z_n)$

(b) $\theta_{n\infty}(Z_n)$ is complete with respect to this norm,

i.e. $\theta_{n\infty}(Z_n)$ is a C^*-algebra.

2.8 The Cuntz algebras

In Section 1.8 we constructed the Toeplitz extension, namely the short exact sequence

$$0 \to K \to C^*(s) \to C(\mathbb{T}) \to 0 \tag{2.8.1}$$

where s is a non-unitary isometry, indeed the unilateral shift on ℓ^2, $C^*(s)$ contains the compacts K on ℓ^2, and $C(\mathbb{T})$ the complex valued continuous functions on the circle are generated by a unitary u with full spectrum $\sigma(u) = \mathbb{T}$. Here we look at generalizations of this, where we replace the single non-unitary isometry s, and relations $s^*s = 1, ss^* \neq 1$ by n-isometries $t_1, \ldots, t_n$ which satisfy

$$t_i^* t_j = \delta_{ij} 1 \tag{2.8.2}$$

$$\textstyle\sum_{i=1}^{n} t_i t_i^* \lneq 1 \tag{2.8.3}$$

and replace the unitary u, with relations $u^*u = 1 = uu^*$ with n-isometries $s_1, \ldots, s_n$ satisfying

$$s_i^* s_j = \delta_{ij} 1 \tag{2.8.4}$$

$$\textstyle\sum_{i=1}^{n} s_i s_i^* = 1\,. \tag{2.8.5}$$

If $\mathcal{T}_n$ (respectively $\mathcal{O}_n$) denotes the C^*-algebra $C^*(t_1, \ldots, t_n)$ generated by $t_1, \ldots, t_n$ (respectively $C^*(s_1, \ldots, s_n)$ generated by $s_1, \ldots, s_n$) then we will show that $\mathcal{T}_n$ has a natural representation where t_i are 'shift operators', on a separable

Hilbert space, such that $\mathcal{T}_n$ contains the compact operators so that $\mathcal{T}_n/K \cong \mathcal{O}_n$. Thus we have a short exact sequence generalizing (2.8.1):

$$0 \to K \to \mathcal{T}_n \to \mathcal{O}_n \to 0\,. \tag{2.8.6}$$

We will also see that $\mathcal{O}_n$ carries a natural action of the circle group $\mathbb{T}$, such that its fixed point algebra is isomorphic to a C^*-infinite tensor product $\bigotimes_1^\infty M_n$, and use this to deduce simplicity of $\mathcal{O}_n$, if $n \geq 2$.

First, let $H = \mathbb{C}^n$ be an n-dimensional Hilbert space, $1 \leq n \leq \infty$, and let

$$F(H) = \bigoplus_{m=0}^\infty (\otimes^m H) \tag{2.8.7}$$

where $\otimes^m H = H \otimes \cdots \otimes H$, m copies, and $\otimes^0 H$ is a one-dimensional Hilbert space spanned by a unit vector Ω, called the vacuum. $F(H)$ is sometimes called *full Fock space.* Note that if $n = 1$, then $F(\mathbb{C}) = \ell^2$, on which we have defined the unilateral shift $s(\xi_1, \xi_2, \ldots) = (0, \xi_1, \ldots)$. More generally, we can define n shift operators on $F(\mathbb{C}^n)$ as follows. If $\xi \in \mathbb{C}^n$, then

$$\langle \xi \otimes f, \xi \otimes f \rangle = \langle \xi, \xi \rangle \langle f, f \rangle, \quad f \in \otimes^m \mathbb{C}^n\,. \tag{2.8.8}$$

Hence there is an operator $t^m(\xi) : \otimes^m \mathbb{C}^n \to \otimes^{m+1}\mathbb{C}^n$ such that

$$t^m(\xi) f = \xi \otimes f\,, \quad t^0(\xi)\Omega = \xi \quad \text{and} \quad \|t^m(\xi)\| = \|\xi\|\,.$$

Then define $t(\xi) = \bigoplus_{m=0}^\infty t^m(\xi) \in B(F(H^n))$ as in Section 1.8 when $n = 1$. If u is a unitary on $\mathbb{C}^n$, let F_u denote the unitary $\bigoplus_{m=0}^\infty (\otimes^m u)$ on $F(H^n)$. It is easy to check that

(2.8.9) $t(\xi) f = \xi \otimes f$, $f \in \otimes^m H$, $m \geq 1$

(2.8.10) $t(\xi)\Omega = \xi$

(2.8.11) $F_u t(\xi) F_u^* = t(u\xi)$

(2.8.12) $t(\xi)^* t(\eta) = \langle \eta, \xi \rangle 1$

(2.8.13) $\sum_{i=1}^n t(e_i) t(e_i)^* + T_{\Omega,\Omega} = 1$

for any complete orthonormal set $\{e_i\}$ in $\mathbb{C}^n$. Thus writing $t_i = t(e_i)$ we get n isometries satisfying (2.8.2) and (2.8.3), where if $n = 1$ we recover the unilateral shift on ℓ^2. Note that it follows from (2.8.13) that $\mathcal{T}_n = C^*(t_1, \ldots, t_n)$ contains $K(F(\mathbb{C}^n))$, the compact operators on $F(\mathbb{C}^n)$. For if $\sum = \{1, 2, \ldots, n\}$, $\mu = (\mu_1, \ldots, \mu_m) \in \sum^m$, let $t_\mu = t_{\mu_1} \cdots t_{\mu_m}$, $e_\mu = e_{\mu_1} \otimes \cdots \otimes e_{\mu_m} \in \otimes^m H$, if $e_1, \ldots, e_n$ is the canonical basis for $\mathbb{C}^n$. Then (2.8.13) says that $\mathcal{T}_n$ contains the rank one operator $T_{\Omega,\Omega}$, the projection on Ω. In fact

$$t_\mu \left[1 - \textstyle\sum_{i=1}^n t_i t_i^*\right] t_{\mu'}^* = T_{e_\mu, e_{\mu'}}\,. \tag{2.8.14}$$

Thus $\mathcal{T}_n$ contains every rank one operator on $F(\mathbb{C}^n)$, and hence contains the compact operators. Then the quotient map $\pi : B(F(\mathbb{C}^n)) \to B(F(\mathbb{C}^n))/K(F(\mathbb{C}^n))$

restricts to $\pi : \mathcal{T}_n \to \mathcal{T}_n/K(F(\mathbb{C}^n))$, and it follows from (2.8.12) and (2.8.13) that $s_i = \pi(t_i)$ are isometries satisfying (2.8.4) and (2.8.5). Thus $\mathcal{T}_n/K(F(\mathbb{C}^n)) = C^*(s_1, \ldots, s_n) \equiv \mathcal{O}_n$, which is a C^*-algebra by Corollary 1.43. Let $U(n)$ denote the group of unitaries on $\mathbb{C}^n$. If $u \in U(n)$, define a $*$-automorphism α_u of $B(F(\mathbb{C}^n))$ by $\alpha_u(x) = F_u x F_u^*$, $x \in B(F(\mathbb{C}^n))$, so that $u \to \alpha_u$ is a homomorphism from $U(n)$ into the group of $*$-automorphisms of $B(F(\mathbb{C}^n))$. Then by (2.8.11), α_u leaves $T_n = C^*(t_1, \ldots, t_n)$ invariant, and since $\alpha_u^{-1} = \alpha_{\overline{u}}$, it defines an automorphism of T_n. Moreover, $K(F(\mathbb{C}^n))$ is an ideal and so $\alpha_u(x) = F_u x F_u^*$ also defines a $*$-automorphism of $K(F(\mathbb{C}^n))$. Hence if $\pi : \mathcal{T}_n \to \mathcal{T}_n/K(F(\mathbb{C}^n))$ is the quotient map, there is for each $u \in U(n)$, a unique $*$-automorphism β_u of $\mathcal{T}_n/K(F(\mathbb{C}^n))$ such that $\beta_u(\pi(x)) = \pi\alpha_u(x), x \in T_n$. Since $\alpha_u t(\xi) = t(u\xi)$, we see $\beta_u s(\xi) = s(u\xi)$, if $s(\xi) = \sum_{i=1}^n \xi_i s_i = \pi(t(\xi))$, $\xi = (\xi_i)$.

Now because of relation (2.8.4), any monomial in $\{s_i, s_j^*\}$ is of the form $s_\mu s_\nu^*$, $\mu, \nu \in \bigcup_m \sum^m$, where $s_\mu = s_{\mu_1} \cdots s_{\mu_m}$ if $\mu = (\mu_1, \ldots, \mu_m) \in \sum^m$, $s_\varnothing = 1$. Hence $\mathcal{O}_n$ is generated as a Banach space by

$$\{s_\mu s_\nu^* \mid \mu, \nu \in \textstyle\bigcup_m \sum^m\} \tag{2.8.15}$$

If $t \in \mathbb{T} \subset U(n)$, $\beta_t(s_\mu s_\nu^*) = t^{|\mu|-|\nu|} s_\mu s_\mu^*$ where $|\mu| = m$, if $\mu \in \sum^m$. Hence we see that β is *strongly continuous* in the sense that $t \to \beta_t(x)$ is norm continuous on $\mathbb{T}$ for each $x \in \mathcal{O}_n$. (A similar computation shows that α is strongly continuous on T_n). Let F_n denote the fixed point algebra

$$\{x \in \mathcal{O}_n \mid \beta_t(x) = x, \text{ for all } t \in \mathbb{T}\} \tag{2.8.16}$$

which is a C^*-subalgebra of $\mathcal{O}_n$. Since β is strongly continuous we can form a Riemann integral $P(x) = \int_{\mathbb{T}} \beta_t(x)\, dt$, $x \in \mathcal{O}_n$, and see that P defines a positive norm one bounded linear operator on $\mathcal{O}_n$, whose range is F_n. In particular $P(s_\mu s_\nu^*) = s_\mu s_\nu^* \delta_{|\mu|,|\nu|}$, and so we see from (2.8.15) that F_n is generated as a Banach space by

$$\{s_\mu s_\nu^* \mid \mu, \nu \in \textstyle\sum^m, \quad m = 0, 1, 2, \ldots\}\,. \tag{2.8.17}$$

Now again from (2.8.4) we see that $\{s_\mu s_\nu^* \mid \mu, \nu \in \sum^m\}$ form a set of $n^m \times n^m$ matrix units in $\mathcal{O}_n$, and so by Lemma 2.2 linearly generate an algebra F_n^m which is isomorphic to $M_{\Sigma^m} = M_{n^m} = M_n \otimes \cdots \otimes M_n$ (m factors), where $s_\mu s_\nu^*$ corresponds to $e_{\mu_1,\nu_1} \otimes \cdots \otimes e_{\mu_m,\nu_m}$ if $\mu = (\mu_1, \ldots, \mu_m)$, $\nu = (\nu_1, \ldots, \nu_m)$ and $\{e_{ij} \mid i, j = 1, \ldots, n\}$ are canonical matrix units in M_n. (See (2.5.9) and the remark immediately preceding that.) From

$$s_\mu s_\nu^* = \textstyle\sum_i s_\mu s_i s_i^* s_\nu^* \tag{2.8.18}$$

we see that the embedding $F_n^m \subset F_n^{m+1}$ corresponds to $\otimes^m M_n \to \otimes^{m+1} M_n$ given by $x \to x \otimes 1_n$, where 1_n denotes the unit of M_n. Hence the fixed point algebra $F_n \cong \bigotimes_{\mathbb{N}} M_n$, a UHF algebra. We now show that both F_n and $\mathcal{O}_n$ are simple C^*-algebras.

Lemma 2.16 *F_n is simple.*

Proof Let J be a proper closed $*$-ideal in F_n. Then $J \cap F_n^m$ is an ideal in F_n^m and so by Lemma 2.1, $J \cap F_n^m = \{0\}$ or F_n^m. If $J \cap F_n^m = F_n^m$ for one m, say M, then $J \cap F_n^{m'} = F_n^{m'}$ for all $m' \geq M$, and so $J = F_n$. Suppose $J \cap F_n^m = \{0\}$ for all m. Now by Exercise 1.38, F_n/J is a C^*-algebra, and let $\pi : F_n \to F_n/J$ denote the quotient map. Now $J \cap F_n^m = \{0\}$ means that π is injective on F_n^m, and hence isometric on F_n^m by uniqueness of the C^*-norm on a finite dimensional C^*-algebra. Hence π is isometric on $F_n = \overline{\bigcup_m F_n^m}$, and consequently π is injective or $J = \{0\}$. □

Now on any matrix algebra $M(p)$ we have the normalized trace $\tau[A_{ij}] = (1/p)\sum A_{ii}$. Hence by 2.7 there is an unique normalized state τ on F_n such that $\tau|F_n^m$ is the normalized trace on $\otimes^m M_n$, and which has the *trace* property $\tau(ab) = \tau(ba)$, $a, b \in F_n$. In fact $\tau = \bigotimes_1^\infty (1/n\,\mathrm{tr})$, the product state formed by taking the normalized trace $1/n$ tr$(\cdot)$ on M_n. A state φ on a C^*-algebra A is said to be *faithful* if $\varphi(x^*x) > 0$ for any non-zero x in A.

Corollary 2.17 *The trace τ on F_n is faithful.*

Proof Let (π, H, Ω) denote the GNS triple for τ on F_n. Suppose $\tau(x^*x) = 0$. Then (cf. the proof of Theorem 2.8) $\tau(yx) = 0$ for all $y \in F_n$. Consequently using the trace property

$$\langle \pi(x)\pi(z)\Omega, \pi(z')\Omega\rangle = \tau(z'^*xz) = \tau(zz'^*x) = 0 \quad \text{for all } z, z' \in F_n\,.$$

Since π is cyclic, this means $x \in \mathrm{Ker}\,\pi = \{0\}$, as F_n is simple by Lemma 2.16. Hence τ is faithful. □

Corollary 2.18 *The state $\varphi = \tau \circ P$ is faithful on $\mathcal{O}_n$.*

Proof Suppose $\tau P(x^*x) = 0$ for some $x \in \mathcal{O}_n$. Now $P = \int \beta_t(\cdot)\,dt$ is positive, and so $P(x^*x) = 0$ since τ is faithful. If we cheat somewhat, and assume that $\mathcal{O}_n$ has a faithful representation on a Hilbert space H, then $0 = \langle P(x^*x)\xi, \xi\rangle = \int \langle \alpha_t(x^*x)\xi, \xi\rangle\,dt = \int \|\alpha_t(x)\xi\|^2\,dt$ for all $\xi \in H$ and consequently $\|\alpha_t(x)\xi\| = 0$ and so $x = 0$. □

Corollary 2.19 *$\mathcal{O}_n$ is simple if $n \geq 2$.*

Proof Let $\sigma(x) = \sum_{i=1}^n s_i^* x s_i/n$, $x \in \mathcal{O}_n$. It is easy to check that $\lim_{m\to\infty} \sigma^m(x) = \tau \circ P(x)$, if $n \geq 2$, first for $x = s_\mu s_\nu^*$, and then for arbitrary $x \in \mathcal{O}_n$. Let J be a non-zero closed $*$-ideal in $\mathcal{O}_n$, and x a non-zero element of J. Then since σ leaves ideals invariant,

$$\tau \circ P(x^*x) = \lim_{m\to\infty} \sigma^m(x^*x) \in J\,.$$

Hence since $\tau \circ P$ is faithful, J contains a non-zero scalar, and so $J = \mathcal{O}_n$ since J is an ideal. □

Not only is $\mathcal{O}_n$ simple but it is the universal C^*-algebra generated by n isometries s_i with range projectors summing to the unit.

Theorem 2.20 *If $\{r_i\}$ is a sequence of n isometries with $\sum r_i r_i^* = 1$ and $n \geq 2$ then there exists a unique $*$-isomorphism from $C^*(r_i)$ onto $\mathcal{O}_n$ taking r_i to s_i $i = 1, \ldots, n$.*

Proof Consider the C^*-algebra $A = C(\mathbb{T}, C^*(r_i))$ and put $\hat{r}_i = U \otimes r_i$ if U is the canonical generator of $C(\mathbb{T})$, there exists a strongly continuous automorphism group $\{\alpha_t : t \in \mathbb{T}\}$ of A such that $\alpha_t(\hat{r}_i) = t\hat{r}_i$ induced by rotation on the spectrum of U. Let $\hat{s}_i = \hat{r}_i \oplus s_i$ so that $C^*(\hat{s}_i)$ also has a dynamics $\{\beta_t : t \in \mathbb{T}\}$ such that $\beta_t(\hat{s}_i) = t\hat{s}_i$. Then by the argument of Corollary 2.19, $C^*(\hat{s}_i)$ is simple. The natural maps $C^*(\hat{s}_i) \to C^*(\hat{r}_i) \to C^*(r_i)$ and $C^*(\hat{s}_i) \to C^*(s_i)$ take $\hat{s}_i$ to r_i and s_i respectively. Here we have taken the restriction of projection maps from $C^*(\hat{r}_i) \oplus C^*(s_i)$ to the coordinates, and map $C^*(\hat{r}_i) \to C^*(r_i)$ by any point evaluation $C^*(s_i) \leftarrow C^*(\hat{s}_i) \to C^*(\hat{r}_i) \to C^*(r_i)$. So by simplicity of $C^*(\hat{s}_i)$, all the maps are isomorphisms, and we have a $*$-isomorphism of $C^*(s_i)$ onto $C^*(r_i)$ taking s_i to r_i. □

Using this we can deduce the universal property of the Toeplitz algebras – although of course they are not simple.

Theorem 2.21 *Let $\{t_i\}_{i=1}^n$, $\{r_i\}_{i=1}^n$ be finite sequences of isometries with $\sum_{i=1}^n t_i t_i^* \lneq 1$, and $\sum_{i=1}^n r_i r_i^* \lneq 1$ and $n \geq 1$. Then there exists a unique $*$-isomorphism of $C^*(t_1, \ldots, t_n)$ onto $C^*(r_1, \ldots, r_n)$ taking t_i onto r_i, $i = 1, \ldots, n$.*

Proof If t_i are represented on a Hilbert space H, we can by considering $t_i \otimes 1$ on $H \otimes \ell^2(\mathbb{Z})$ assume that the range of $1 - \sum_{i=1}^n t_i t_i^*$ is infinite dimensional. (This dilation is necessary in the Toeplitz extension, where $1 - \sum_{i=1}^n t_i t_i^*$ is the rank one projection in the vacuum). Then adjoin an isometry t_{n+1} so that $\sum_{i=1}^{n+1} t_i t_i^* = 1$ and similarly an isometry r_{n+1} so that $\sum_{i=1}^{n+1} r_i r_i^* = 1$. Since $n + 1 \geq 2$, we can apply Theorem to obtain an isomorphism of $C^*(t_1, \ldots, t_{n+1})$ onto $C^*(r_1, \ldots, r_{n+1})$ taking t_i to r_i $i = 1, \ldots, n + 1$. We then restrict. □

Exercise 2.13 Show that if $x \in \mathcal{O}_n$ is non-zero and positive there exists $y \in \mathcal{O}_n$ such that $y^* x y = 1$.

2.9 Path algebra model of an AF algebra

Here we describe the path model of an AF algebra $A = \lim_{\to} A_n$ constructed from multi-matrix algebras A_n under canonical or standard unital embeddings. In the first place, let us consider the unital embedding

$$\mathbb{C} \to M_n \tag{2.9.1}$$

which according to our convention of 2.7 is described by the graph or Bratteli diagram given by Fig. 2.28 where the vertices p, q represent the units of $\mathbb{C}, M_n$ respectively, and the n edges joining p to q represents the multiplicity of the

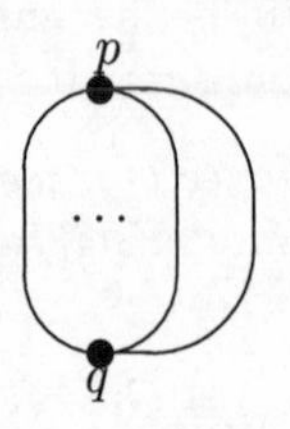

FIG. 2.28. Bratelli diagram of $\mathbb{C} \subset M_X$

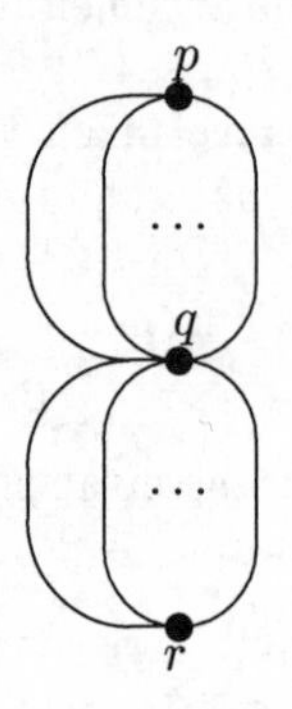

FIG. 2.29. Bratelli diagram of $\mathbb{C} \subset M_X \subset M_Y$

embedding of $\mathbb{C}$ in M_n. Now M_n= End($\ell^2(X)$) where X is any set of cardinality n, generated by matrix units $\{f_{\mu\nu} \mid \mu, \nu \in X\}$. It is convenient to use the n-edges of the graph in Fig. 2.28 for the indexing set X, i.e. Path (p, q) the set of edges from p to q and write A_{pq}= End ($\ell^2(\text{Path}(p, q))$). The embedding (2.9.1) or Fig. 2.28 is then

$$1 \to \sum_{\mu} f_{\mu\mu} \tag{2.9.2}$$

where the summation is over $\mu \in \text{Path}(p, q)$. Next consider the unital standard embeddings:

$$\mathbb{C} \to M_n \to M_{nk} \tag{2.9.3}$$

described by the Bratteli diagram given by Fig. 2.29 where now the vertices p, q, r denote the units of $\mathbb{C}, M_n, M_{nk}$ respectively and there are n edges, Path (p, q) from p to q, and k edges Path (q, r) from q to r representing the multiplicities of the corresponding embeddings. Then Path (p, r), the set of (minimal length!) paths from p to r in the Bratteli diagram can be identified with the Cartesian product Path $(p, q)\times$ Path (q, r), and consequently we identify

$$\ell^2(\text{Path}\,(p, r)) = \ell^2(\text{Path}\,(p, q)) \otimes \ell^2(\text{Path}\,(q, r)) \tag{2.9.4}$$

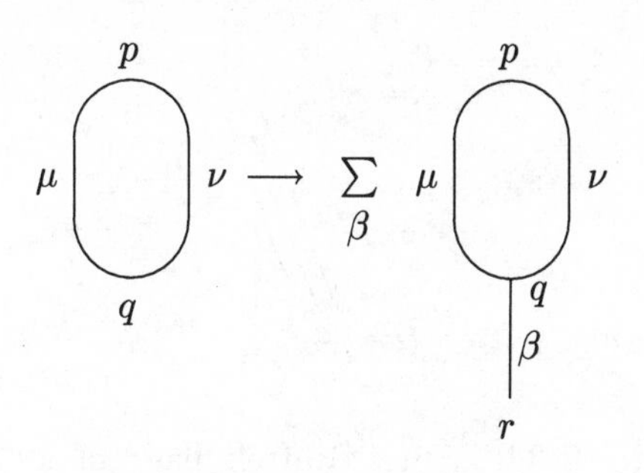

FIG. 2.30. $f_{\mu\nu} \to \sum_\beta f_{\mu\beta,\nu\beta}$

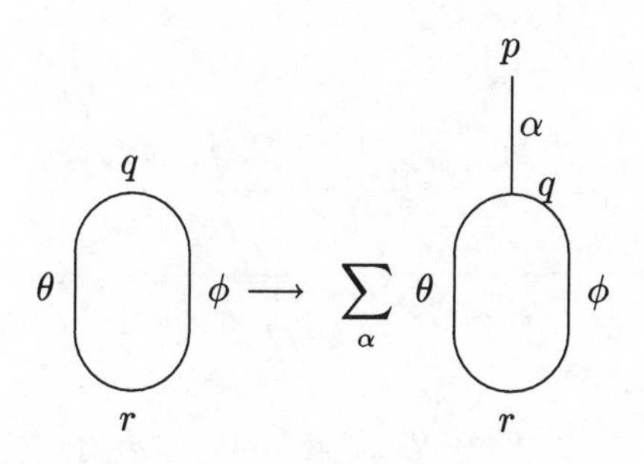

FIG. 2.31. $f_{\theta\phi} \to \sum_{\alpha \in \mathrm{Path}(p,q)} f_{\alpha\theta,\alpha\phi}$

and

$$A_{pr} = A_{pq} \otimes A_{qr} \tag{2.9.5}$$

if $A_{vw} = \mathrm{End}\ \ell^2(\mathrm{Path}\,(v,w))$, using

$$f_{\mu\theta,\nu\phi} = f_{\mu\nu} \otimes f_{\theta\phi} \tag{2.9.6}$$

for $\mu,\nu \in$ Path (p,q), $\theta,\phi \in$ Path (q,r). Under the identification (2.9.5) and (2.9.6) the embedding of $M_n = A_{pq} = \mathrm{End}\ (\ell^2(\mathrm{Path}\ (p,q)))$ in $M_{nk} = A_{pr} = \mathrm{End}\ (\ell^2(\mathrm{Path}\ (p,q)))$ is

$$\begin{aligned} f_{\mu\nu} &\to f_{\mu\nu} \otimes 1 = \textstyle\sum_{\beta \in \mathrm{Path}(q,r)} f_{\mu\nu} \otimes f_{\beta\beta} \\ \text{i.e.}\quad f_{\mu\nu} &\to \textstyle\sum_\beta f_{\mu\beta,\nu\beta}\,. \end{aligned} \tag{2.9.7}$$

Thus the matrix unit corresponding to the pair of edges μ, ν is taken to the sum obtained by extending the end point of these edges in all possible ways (see Fig. 2.30). What is the meaning of the algebra $M_k = A_{qr} = \mathrm{End}\ [\ell^2((\mathrm{Path}\ (q,r))]$ in terms of the embeddings of $M_n \to M_{nk}$ or of the path picture? Clearly it is the relative commutant of M_n in M_{nk}:

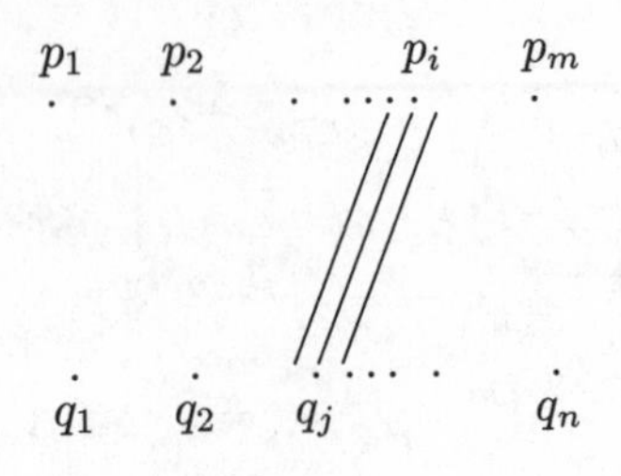

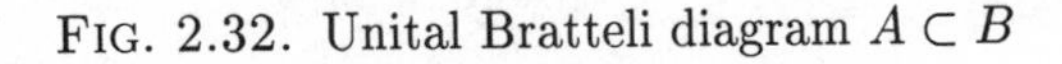

FIG. 2.32. Unital Bratteli diagram $A \subset B$

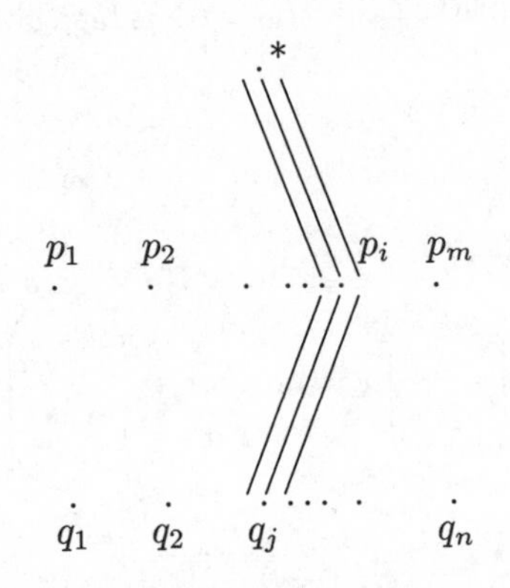

FIG. 2.33. Unital Bratteli diagram $\mathbb{C} \subset A \subset B$

$$A'_{pq} \cap A_{pr} = A_{qr} \tag{2.9.8}$$

and A_{qr} is embedded in A_{pr} (cf. (2.9.7)) by

$$f_{\theta\phi} \to \sum_{\alpha \in \mathrm{Path}(p,q)} f_{\alpha\theta,\alpha\phi} \,. \tag{2.9.9}$$

The matrix unit corresponding to the pair of edges θ, ϕ in Path (q, r) is taken to the sum obtained by extending the initial vertex of the edges in all possible ways (see Fig 2.31).

Consider now the general case of a unital standard embedding of multi matrix algebra $A \to B$ represented by the Bratteli diagram given by Fig. 2.32, where $(p_1, \ldots, p_m)$ and $(q_1, \ldots, q_n)$ are the minimal central projections of A, B respectively, and the Λ_{ji} lines are drawn from vertex p_i to vertex q_j to represent the multiplicity of the embedding of A_{pi} in B_{qj}. If $A_{pi} = \mathrm{End}\ \ell^2(X_i)$ we complete the Bratteli diagram to the embeddings $\mathbb{C} \to A \to B$ represented by Fig. 2.33 where $*$ denotes the unit of $\mathbb{C}$, and X_i can be taken to be Path $(*, p_i)$ by what we have discussed in Fig. 2.28. Since we have a unital embedding we can identify

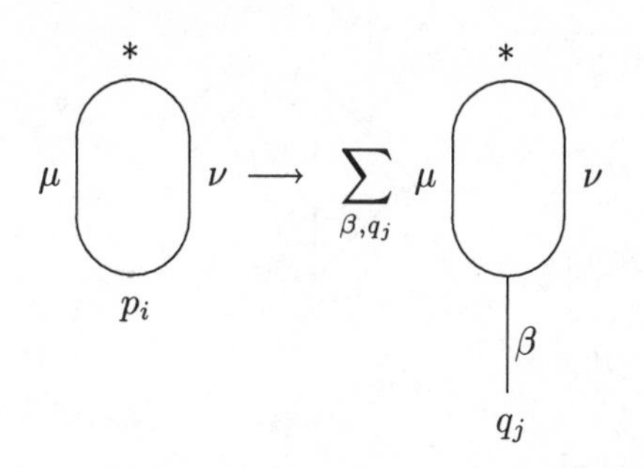

FIG. 2.34. $f_{\mu\nu} \to \sum_{\beta \in \mathrm{Path}(p_i, q_j)} f_{\mu\beta,\nu\beta}$

$$B_{q_j} = \mathrm{End}\, \ell^2(Y_j) \tag{2.9.10}$$

where $Y_j = \sqcup_i$ Path $(*, p_i)\times$ Path (p_i, q_j) can be identified with Path $(*, q_j)$, the set of (minimal) paths in the Bratteli diagram from $*$ to q_j, and where Path (p_i, q_j) denotes the edges from p_i to q_j. In this notation, matrix units of A_i are embedded in B_j by

$$f_{\mu\nu} \to \sum_{\beta \in \mathrm{Path}(p_i, q_j)} f_{\mu\beta,\nu\beta}, \quad \mu, \nu \in \mathrm{Path}(*, p_i)\,. \tag{2.9.11}$$

The embedding of $f_{\mu\nu}$ in B is obtained by a further summation in (2.9.11) over j (see Fig. 2.34) to the summation obtained by extending the terminal vertex of the edges in all possible ways.

We can now define the path model of the AF algebra associated with a Bratteli diagram, with standard embeddings. Recall from Section 2.7 that the AF algebra associated with a Bratteli diagram is unique and any AF algebra is isomorphic to such an algebra with standard embeddings. Suppose in the first case that the Bratteli diagram describes unital embeddings. We have a sequence $\Omega[m]$ of finite sets (each to represent the minimal central projections of a finite dimensional C^*-algebra $A_n \cong M[\mathbf{m_n}]$) and a multiplicity graph or matrix $\Lambda^n = [\lambda_{ij}^{(n)}]$ whose rows are labelled by $j \in \Omega[n]$ and columns by $i \in \Omega[n+1]$, where $\lambda_{ij}^{(n)}$ is the multiplicity of the embedding of the simple algebra at vertex i into that at vertex j. In the unital case we adjoin $A_0 = \mathbb{C}, \Omega[0] = \{*\}, \Lambda^{(1)} = (1, \ldots, 1)$ ($|\Omega(1)|$ copies of 1), then $m_n = \lambda^{(n)}\lambda^{(n-1)} \cdots \lambda^{(1)}.1$. The Bratteli diagram consists of the graph with vertices $\sqcup_{n \geq 0}\Omega[n]$, and $\lambda_{ji}^{(n)}$ edges oriented between vertex i in $\Omega[n]$ and j in $\Omega[n+1]$, as indicated in Fig. 2.35.

The vertices $\Omega[n]$ denote the nth level of the Bratteli diagram. For $m < n$, $i \in \Omega[m]$, $j \in \Omega[n]$, let Path (i, j) denote the space of paths in the Bratteli diagram from i to j. (Such paths have length $n - m$, since we have oriented the Bratteli diagram.) For $\mu \in$ Path (i, j), i (respectively j) is called the initial (respectively terminal) vertex of μ. Then let $\Omega[m, n] = \sqcup_{i \in \Omega[m], j \in \Omega[n]}$ Path (i, j), the space of paths from level m to level n as in Fig. 2.36.

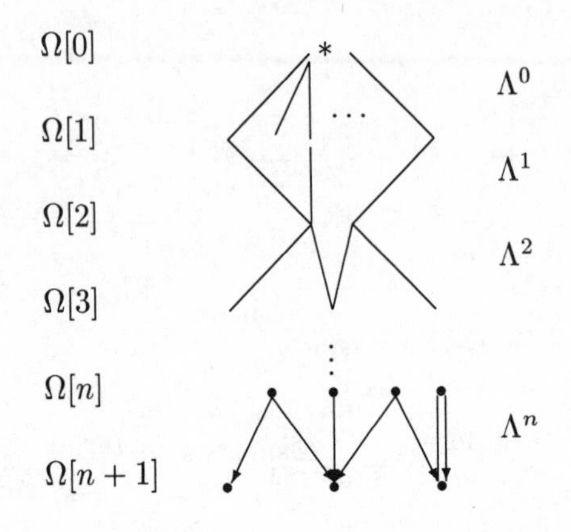

FIG. 2.35. Unital Bratteli diagram

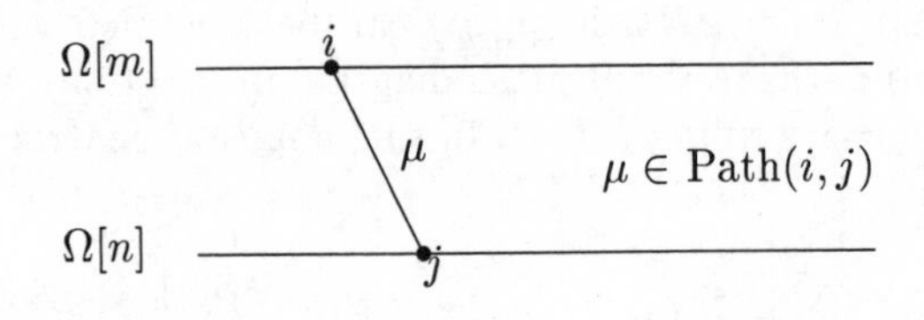

FIG. 2.36. $\Omega[m, n]$

We let Path $(i, i) = \{i\}$, Path $(i, i') = \varnothing$, if $i, i' \in \Omega[m]$, $i \neq i'$, $\Omega[m, m] = \Omega[m]$. For $i \in \Omega[m], j \in \Omega[n]$, let

$$A_{ij} = \mathrm{End}\, \ell^2(\mathrm{Path}\,(i, j)) \tag{2.9.12}$$

generated by matrix units $f_{\mu,\nu}$, $\mu, \nu \in$ Path (i, j) and

$$A[m, n] = \oplus A_{ij} = \oplus \mathrm{End}\, \ell^2(\mathrm{Path}\,(i, j)) \tag{2.9.13}$$

where the summation is over all i at level m, j at level n. Thus $A[m, n]$ is generated by matrix units $f_{\mu,\nu}$ where $\mu, \nu \in \Omega[m, n]$ with same initial vertex and same terminal vertex. If $[m, n] \subset [m', n']$ we embed $A[m, n]$ in $A[m', n']$ by

$$f_{\mu,\nu} \to \textstyle\sum_{\alpha,\beta} f_{\alpha\mu\beta,\alpha\nu\beta} \tag{2.9.14}$$

for $\mu, \nu \in$ Path (i, j), where the summation is over all $\alpha \in$ Path (i', i), $\beta \in$ Path (j, j') and i' on level m', j' on level n', as in Fig. 2.37.

Remark When considering a matrix algebra, or more generally $M_{I,J}$ or a path algebra $A[m, n]$, one can emphasize $(f_{\mu,\nu})$, the matrix units, or the matrices or *face operators* $T = [T_{\mu\nu}]$ $(\mu, \nu \in \Omega[m, n]$, in either case). In the face operator picture, we have addition involution and multiplication as in Figs 2.2–2.7 with

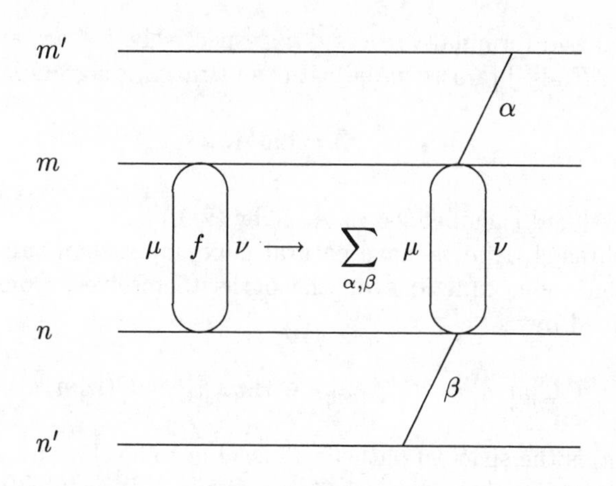

FIG. 2.37. $A[m,n] \subset A[m',n']$

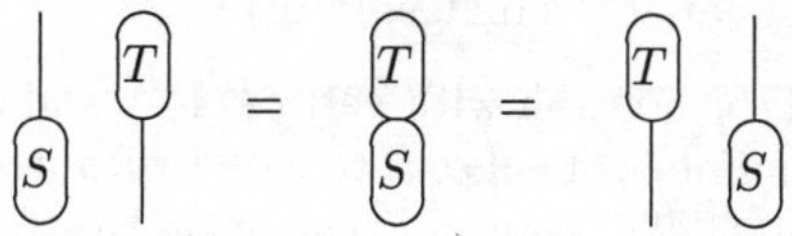

FIG. 2.38. $A[0,m]' \cap A[0,n] = A[m,n]$

labels ξ, η in $\Omega[m,n]$. For example with multiplication the right hand side of Fig. 2.4 is interpreted to be zero if there is no ϕ, i.e. the initial vertices of ξ and η must coincide, otherwise the (ξ,η) entry of wv is zero. Then (extending the graphical formulae Fig. 2.22 for $T \to 1 \otimes T \otimes 1$ of Section 2.6), we embed an arbitrary element T (a face operator) of the path algebra $A[m,n]$ in $A[m',n']$ with the same picture of Fig. 2.37.

Then by what we show in (2.9.8)

$$A[0,m]' \cap A[0,n] = A[m,n]\,. \tag{2.9.15}$$

In terms of face operators, this is Fig. 2.38.

In particular the centre of $A_n = A[0,n]$ is identified with $A[n,n] = C(\Omega[m]) = \mathbb{C}^{\Omega[m]}$, and the minimal central projections of $A[0,n]$ can be identified with $\Omega(n) : f_{i,i} \leftrightarrow i$. Then

$$A[m,n] = C(\Omega[m])' \cap C(\Omega[n])' \cap \operatorname{End} \ell^2\,(\Omega[m,n])$$

where $\Omega[m], \Omega[n]$ act on $\ell^2\,(\Omega[m,n])$ by

$$i\gamma = \delta_{ii'}\gamma\,, \quad j\gamma = \delta_{jj'}\gamma$$

if γ has initial and terminal vertices i', j' respectively, i, i' are at level m, j, j' at level n. The AF algebra associated with the Bratteli diagram is then

$$A = \lim_{\rightarrow} A_n$$

where $A_n = A[0,n]$ is embedded in A_{n+1} by (2.9.14).

The algebras A, $A[m,n]$ have natural maximal abelian subalgebras, the algebras of continuous functions on the paths themselves. Consider $C[m,n] \subset A[m,n]$, defined by

$$C[m,n] = C^*\{f_{\mu\mu} \mid \mu \in \Omega[m,n]\} = C(\Omega[m,n]) \tag{2.9.16}$$

where $\Omega[m,n]$ is the space of paths from level m to level n and $0 \leq m \leq n \leq \infty$. If $m' \leq m$, $n \leq n'$, then the embedding $C(\Omega[m,n]) \to C(\Omega[m',n'])$ (induced from $A[m,n] \to A[m',n']$) is the transpose of the canonical chopping off maps from $\Omega[m',n']$ to $\Omega[m,n]$. If $\Omega = \Omega[0,\infty)$ is the space of semi-infinite paths, and $\Omega_m = \Omega[0,m]$, then $\Omega = \lim_{\leftarrow} \Omega_m$ a *direct limit* over finite indices, and $C(\Omega) = \lim_{\rightarrow} C[m,n]$ is an AF subalgebra of A.

Example 2.22 Let Γ be a graph with vertices $\Gamma^{(0)}$ and edges $\Gamma^{(1)}$. Assume Γ is locally finite, i.e. the number of edges associated with a vertex is finite. Let I be a finite set of vertices of $\Gamma^{(0)}$. We can construct a Bratteli diagram $(\Gamma, I)^\wedge$ (and hence an AF algebra $A(\Gamma, I)$) as the space of semi-infinite paths in Γ beginning at I. Thus

$$(\Gamma, I)^\wedge = \{(w_n) \mid w_n \in \Gamma^{(1)}, t(w_n) = i(w_{n+1}), n = 0, 1, 2, \ldots \\ i(w_0) = *, t(w_0) \in I\}$$

where i and t denote the initial and terminal vertices of an edge. Here as usual we have adjoined some vertex $*$ at the beginning of the diagram – which can sometimes be identified with a vertex of Γ. If there are no multiple edges the Bratteli diagram is

$$(\Gamma, I)^\wedge = \{(v_n) \mid v_n \in \Gamma^{(0)}, (v_n, v_{n+1}) \in \Gamma^{(1)}, n = 0, 1, 2, \ldots \\ v_0 = *, v_1 \in I\} \tag{2.9.17}$$

as usual we have adjoined some vertex $*$ at the beginning of the diagram – which can sometimes be identified with a vertex of Γ.

Example 2.23 Consider the graph given by Fig. 2.39 with distinguished vertex 0. The Bratteli diagram $(A_{\infty,\infty}, 0)^\wedge$ is Pascal's triangle of Fig. 2.40. Consider the following projections g_i and partial isometries v_i in the path algebra $A(A_{\infty,\infty})$. Define g_i to be the projection in $A[i-1,i]$ as projection on the edges from NE to SW as in Fig. 2.41, and v_i to be the partial isometry in $A[i-1,i+1]$ with

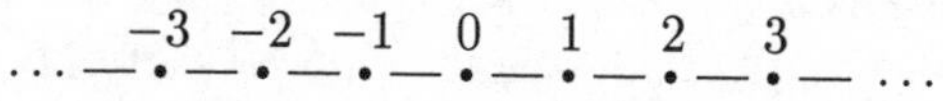

$$\cdots \underset{-3}{\bullet} \; \underset{-2}{\bullet} \; \underset{-1}{\bullet} \; \underset{0}{\bullet} \; \underset{1}{\bullet} \; \underset{2}{\bullet} \; \underset{3}{\bullet} \cdots$$

FIG. 2.39. $A_{\infty,\infty}$

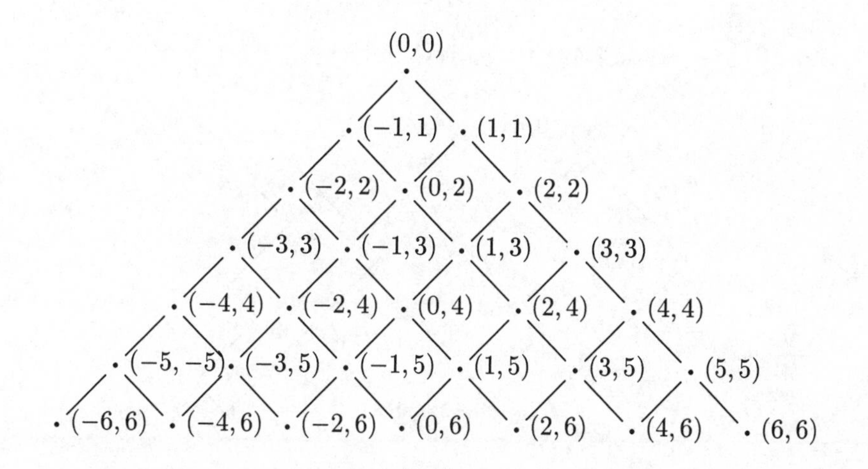

FIG. 2.40. Pascal's triangle $(A_{\infty,\infty}, 0)^\wedge$

initial support $g_{i+1}(1-g_i)$ and final support $g_i(1-g_{i+1})$ which flips paths in the diamonds as in Fig. 2.42 and is zero on other paths from level $i-1$ to $i+1$. Then it is clear that $\{g_i\}$ and $\{v_i\}$ generate $A(A_{\infty,\infty})$ and that the following relations are satisfied:

$$\begin{array}{ll} g_i^2 = g_i^* = g_i & [g_i, g_j] = 0 \\ [v_i, g_j] = 0 & \text{if} \quad j = i, i+1 \\ [v_i, v_j^*] = 0 & \text{if} \quad |i-j| \geq 2 \end{array}$$
$$v_i^* v_i = g_{i+1}(1-g_i)\,,\; v_i v_i^* = g_i(1-g_{i+1})$$

The algebra $A(A_{\infty,\infty})$ is actually the universal C^*-algebra for these generators and relations. Thus let $\{G_i, V_i\}$ be families of operators in a C^*-algebra satisfying the same relations. We can define a $*$-homomorphism $\pi : A(A_{\infty,\infty}) \to C^*(G_i, V_i)$ as follows. For a standard path γ as in Fig. 2.43, we

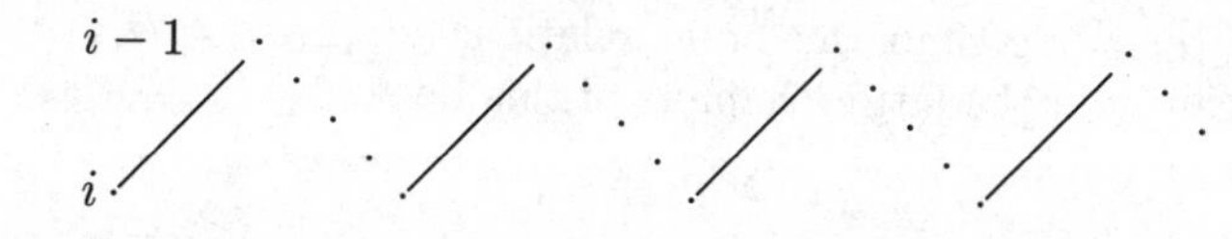

FIG. 2.41. Support of projection g_i

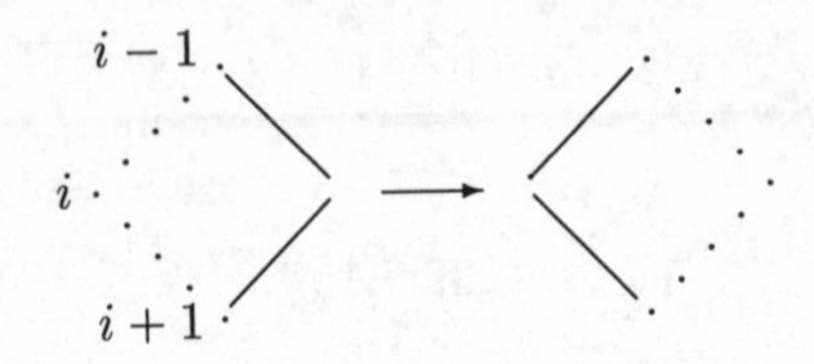

FIG. 2.42. $v_i : g_{i+1}(1-g_i) \to g_i(1-g_{i+1})$

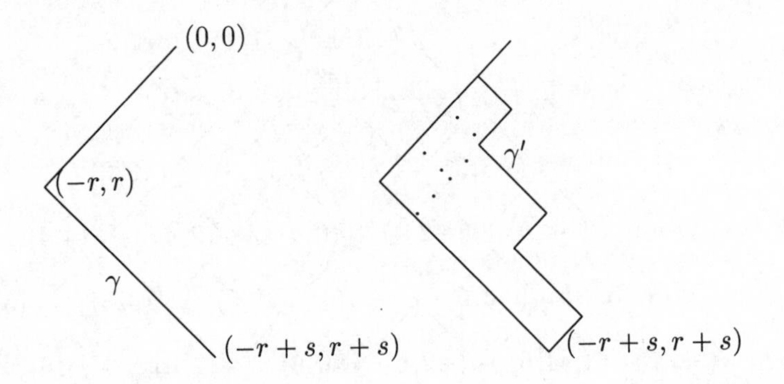

FIG. 2.43. Paths γ and γ' in $A^{\wedge}_{\infty,\infty}$

map $e_{\gamma,\gamma} \to G_1 \cdots G_r(1-G_{r+1}) \cdots (1-G_{r+s})$. Any other path γ' with the same initial and terminal vertex as γ is obtained by flipping, in a natural order, a certain set of diamonds S with $\prod_{i \in S} v_i$, the product taken with respect to the natural order. Then map

$$e_{\gamma',\gamma} \to \prod_{i \in S} V_i \prod_{i=1}^{r} G_i \prod_{j=1}^{s} (1 - G_{r+j})$$

with the product over $\prod_{i \in S} V_i$ in the same order. Note the infinitesimal braid/Temperley–Lieb relations

$$v_i^2 = 0\,; \quad v_i v_{i\pm1} v_i = 0 \tag{2.9.18}$$

which signify the Yang–Baxter equations:

$$R_i(s)R_{i+1}(s+t)R_i(t) = R_{i+1}(t)R_i(s+t)R_{i+1}(t), \text{ if } R_i(s) = 1 + sv_i$$

The 'exponentiated' version of relations 2.9.18, the Temperley–Lieb relations $e_i^2 = e_i$, $e_i e_{i\pm1} e_i = \tau e_i$ and the braid relations $\sigma_i \sigma_{i+1} \sigma_i = \sigma_{i+1} \sigma_i \sigma_{i+1}$ will feature significantly in the latter chapters of this book.

2.10 Notes

If A, B are C^*-algebras represented faithfully on Hilbert spaces H and K, we can represent the algebraic tensor product $A \odot B$ on $H \otimes K$. The completion in $B(H \otimes K)$ is a C^*-algebra independent of the choice of representations of A and B. It is called the *spatial* or *minimal* tensor product, as any other choice of norm on $A \odot B$ for which the completion is a C^*-algebra is automatically greater, and there exists a maximal one such. The algebra A is said to be nuclear if the minimal norm is equal to the maximal norm on $A \odot B$ for any other C^*-algebra B. The notion was introduced in (Takesaki 1964) who showed that the free group C^*-algebra $C^*(\mathbb{F}_n)$ is not nuclear if $n \geq 2$. AF algebras are clearly nuclear as are commutative C^*-algebras. Indeed $C_0(X) \otimes A \cong C_0(X, A)$, the A-valued continuous functions on a locally compact Hausdorff space vanishing at infinity with the supremum norm $\|f\| = \sup\{\|f(x)\| \mid x \in X\}$. The class of nuclear C^*-algebras is closed under direct limits, crossed products by amenable group, and stable isomorphism. If J is an ideal in a C^*-algebra A, then A is nuclear if and only if J and A/J are both nuclear. (See Notes on Chapter 3 for detailed references on nuclearity.) The tensor product $A \otimes B$ is simple if and only if A and B are simple and the tensor product is the minimal one.

Our treatment of inductive limits is influenced by that of (Effros 1981). (Bratteli 1972) used the picture of multiplicity graphs to describe AF algebras and showed that two AF algebras which give rise to the same Bratteli diagram are isomorphic. The Cuntz algebras were shown in (Cuntz 1977) to be simple and universal. The construction of the Cuntz algebra from full Fock space is from (Evans 1980b). The proof here of Corollary 2.19 follows that of (Evans 1980b), (and the step in the proof of Theorem 2.20 on page 75 is from (de Schreye and van Daele 1981)). The path model of an AF algebra is due to (Evans 1984a), (Evans 1985b). The algebraic description of the $A(A_{\infty,\infty})$, the gauge-invariant Fermion algebra, is from (Connes and Evans 1989) (see also (Evans and Gould 1994a)).

The C^*-algebra $\mathcal{O}_n$ is the C^*-algebra of the full shift (Evans 1980b), (Cuntz and Krieger 1980); namely $K \otimes \mathcal{O}_n \cong C^* \left((\prod' \mathbb{Z}/n) \rtimes (\oplus \mathbb{Z}/n) \rtimes \mathbb{Z} \right)$ where $\prod' \mathbb{Z}/n$ is the restricted product of all sequences $(x_i)_{i=-\infty}^{\infty}$ such that $x_i = 0$ for $i << 0$, with inductive limit topology of the product topologies on $\prod_{i=j}^{\infty} \mathbb{Z}/n$, the weak product $\bigoplus_{i=-\infty}^{\infty} \mathbb{Z}/n$ acts by finitely many changes of co-ordinates and $\mathbb{Z}$ as the shift. The crossed product C^*-algebra $A \rtimes G$ where the discrete (amenable) group G acts by automorphisms on a C^*-algebra A is the universal algebra generated by A and a unitary group implementing the action. (If the group is not discrete, we have to take the completion of $C_c(G, A)$ the A-valued functions on G with a convolution twisted by the action and if the group is not amenable, we need to take care to possibly distinguish between the regular or reduced crossed product $A \rtimes G$ formed on $L^2(G) \otimes H$, if $A \subset B(H)$ as in the first Remark on page 225 and the universal construction). If G acts by homeomorphisms on a locally compact Hausdorff space X, and the action is *free* ($g \cdot x = x, x \in G, g \in G$ implies $g =$ identity) we can form the crossed product as a groupoid C^*-algebra generated by arrows or transitions $x \longrightarrow g \cdot x$. That is we take functions on arrows and use the (partial) compositions of arrows to convolve functions.

$$(fg)(a) = \textstyle\sum_{a=bc} f(b)g(c)\,.$$

(For non-discrete G, we would replace the sum by an integral). If Ω is discrete, the delta functions on arrows are partial isometries. So for the cyclic group actions by translation on itself $C(\mathbb{Z}/n) \rtimes (\mathbb{Z}/n) \cong M_{\mathbb{Z}/n} = M_n$, generated by matrix units $e_{i,i+j}$ associated with arrows $i \longrightarrow i+j$. In this way $C(\prod_1^\infty (\mathbb{Z}/n)) \rtimes \oplus(\mathbb{Z}/n)$ is the UHF algebra $M_{\mathbb{Z}_n^\infty}$. The crossed product $C(\mathbb{T}) \rtimes (\mathbb{Z}/n)$, where $\mathbb{Z}/n$ acts on $\mathbb{T}$ by rotation through $e^{2\pi i/n}$, picks up a copy of $C(\mathbb{Z}/n) \rtimes (\mathbb{Z}/n) \cong M_n$ for each point of the orbit space $\mathbb{T}/(\mathbb{Z}/n) \cong \mathbb{T}$, each orbit being identified with $\mathbb{Z}/n$. Gluing these matrix algebras along the orbit space, we get $C(\mathbb{T}) \rtimes (\mathbb{Z}/n) \cong C(\mathbb{T}, M_n) \cong C(\mathbb{T}) \otimes M_n$.

Groupoids can be formed from equivalence relations on a set Ω by drawing an arrow from $x \longrightarrow y$ if x is equivalent to y. The algebra $A(A_{\infty,\infty})$ of Section 2.9 is the C^*-algebra for the following equivalence relation on the space of paths in the Bratteli diagram $\hat{A}_{\infty,\infty}$. A path is an infinite word in two symbols a, b representing edges / and \ respectively (see Fig. 3.11). Then $A(A_{\infty,\infty})$ is the C^*-algebra for the equivalence relation which identifies two paths where one word is obtained from the other by permuting finitely many letters. The partial isometry g_i arises from the arrow which reverses the ab in the $i - i+1$ letters of a word to ba.

The Cuntz algebras and their Toeplitz constructions for full shifts can be generalized to other shifts or subshifts (Bunce and Deddens 1975), (Cuntz and Krieger 1980), (Evans 1982a), (Kumjian *et al.* preprint 1996) as follows. Let G be a directed graph which is locally finite in the sense that each vertex has a finite number of incoming edges, but may possibly be infinite. Let $P_n(G)$ denote the paths of length n in G and define $F_G = \oplus_{n=0}^\infty \ell^2(P_n(G))$. For each edge f in G, define a partial isometry on F_G by a shift action:

$$t_f e_p = \begin{cases} e_{fp} & \text{if } r(f) = s(p) \\ 0 & \text{otherwise}. \end{cases}$$

where r, s refer to the range and source of a path, and $\{e_h \mid h \in P_n(G)\}$ is a canonical orthonormal basis for $\ell^2(P_n(G))$. Then $\{t_f \mid f \in G^{(1)}\}$ satisfy the relations

$$t_f^* t_f = \sum_{g:s(g)=r(f)} t_g t_g^* + \Omega \otimes \overline{\Omega} \tag{2.10.1}$$

where $\ell^2(P_0(G))$ is defined to be a one dimensional space spanned by a vacuum vector Ω. Then $\mathcal{T}_G \equiv C^*(t_f \mid f \in G^{(1)})$ contains the compact operators $K = K(F_G)$, and we have the exact sequence

$$0 \to K \to \mathcal{T}_G \to \mathcal{O}_G \to 0 \tag{2.10.2}$$

where $\mathcal{O}_G$ is the *Cuntz–Krieger* algebra generated by partial isometries $\{s_f \mid f \in G^{(1)}\}$ satisfying the exact relations:

$$s_f^* s_f = \sum_{g:s(g)=r(f)} s_g s_g^*. \tag{2.10.3}$$

Example 2.24 If

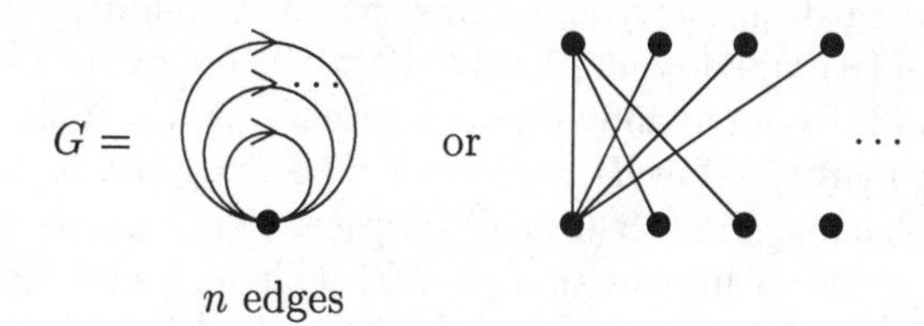

the full bipartite graph on $2n$ vertices, then $\mathcal{T}_G \cong \mathcal{T}_n$, $\mathcal{O}_G \cong \mathcal{O}_n$ the Cuntz–Toeplitz algebra and the Cuntz algebra respectively.

Example 2.25 If

$$G = \bullet \longrightarrow \bullet \longrightarrow \bullet \longrightarrow \bullet \longrightarrow \bullet \longrightarrow \cdots$$

then $\mathcal{T}_G \cong \mathcal{T}$, $\mathcal{O}_G \cong K$ where $\mathcal{T}$ is the standard Toeplitz algebra generated by a non-unitary isometry and K the compact operators on a separable infinite dimensional Hilbert space.

Example 2.26 If

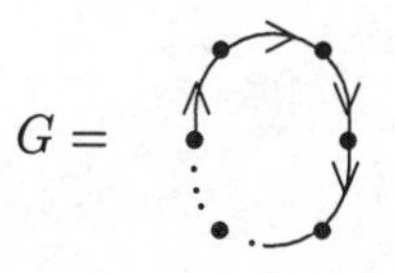

then $\mathcal{T}_G \cong M_I(\mathcal{T})$, $\mathcal{O}_G \cong M_I(C(\mathbb{T}))$, where $\mathcal{T}$ is the standard Toeplitz algebra generated by a non-unitary isometry, and I is the vertex set of G.

Although it is not always so easy to give a crossed product description of these algebras, in terms of groups on spaces, there is a clear description of these algebras via groupoids. We construct this groupoid on the Cantor space of semi-infinite paths $x = (x_i)$ in the graph, and draw an arrow $x \xrightarrow{n} y$ if there is some $n \in \mathbb{Z}$ such that $x_i = y_{i+n}$ for $i >> 0$. We then compose arrows $x \xrightarrow{n} y$, $y \xrightarrow{m} z$ to give $x \xrightarrow{n+m} z$, and need to keep track of the shift, by integer n, as the shift may not be free.

Now a crossed product $C_0(X) \rtimes G$ is simple if and only if the action is *free* ($g \cdot x = x$ for some $x \in X$, $g \in G$, implies $g =$ identity) and *minimal* (no closed invariant subsets). Minimality for these groupoid C^*-algebras $\mathcal{O}_G$ can be described as follows (Cuntz 1981b), (Kumjian *et al.* preprint 1996), at least in the presence of the following condition (k):

(k) No vertex has only one loop passing through it, i.e. there are either none or more than one such loop.

The C^*-algebra $\mathcal{O}_G$ is simple if and only if G is cofinal, i.e. for any given infinite path on the graph and a vertex of the graph, there is a path from that vertex onto the given path. In (k) it is not assumed that there exists a loop. Indeed the following dichotomy in the presence of cofinality is a conjecture of Kumjian based on (Kumjian *et al.* preprint 1996): There exists a loop if and only if the C^*-algebra $\mathcal{O}_G$ is purely infinite, otherwise $\mathcal{O}_G$ is AF. A simple C^*-algebra is said to be *purely infinite* if every hereditary C^*-subalgebra contains an infinite projection. (A C^*-subalgebra B of a C^*-algebra A is said to be *hereditary* if $0 \leq a \leq b$, $a \in A$, $b \in B$ implies that $a \in B$). (Zhang 1990) has shown that a purely infinite C^*-algebra A is either stable (i.e. $A \cong A \otimes K$) or unital. A unital simple C^*-algebra A is *purely infinite* if and only if for each positive $x \in A$, there exists $y \in A$ such that $y^*xy = 1$ (see e.g. Exercise 2.13).

Suppose Λ_n is the graph of Example 2.26 where n is the number of vertices. Winding the graph Λ_{nm} around Λ_n m times induces a homomorphism from $\mathcal{T}_{\Lambda_n}$ into $\mathcal{T}_{\Lambda_{nm}}$ and $\mathcal{O}_{\Lambda_n}$ into $\mathcal{O}_{\Lambda_{nm}}$. Taking the inductive limit of $\mathcal{O}_{\Lambda_{p_i}}$ for a sequence of integers $p_i | p_{i+1}$ we have the *Bunce–Deddens algebra* isomorphic to $C(\mathbb{T}) \rtimes G$ where G is the group of rotations through $\{e^{2\pi i/p_j} \mid j\}$. If $p_j \uparrow \infty$, then this group is dense, the action minimal (and free) and so the crossed product is simple. Alternatively, to see simplicity, the state $\varphi = \int \alpha_t \, dt$ is a faithful trace if $\{\alpha_t \mid t \in \mathbb{T}\}$ is the group of rotations on $C(\mathbb{T})$ which extends to $C(\mathbb{T}) \rtimes G$ (as the G and $\mathbb{T}$ actions commute, G being a subgroup of the abelian $\mathbb{T}$). For the dense set of $t \in G$, α_t is inner, hence by the argument of Corollary 2.19, $C(\mathbb{T}) \rtimes G$ is simple.

3

K-THEORY

3.1 Introduction

The bijection between projections and subspaces of a Hilbert space H allows us to recognize a dimension function on projections. This dimension function is in terms of its rank, i.e. the dimension of the corresponding Hilbert subspace K, a number n in $\{0, 1, 2, \ldots, N = \dim H\}$ (when the Hilbert space is separable). This is really a comparison of the subspace K in terms of one dimensional or minimal subspaces $K = \mathbb{C}e_1 \oplus \cdots \oplus \mathbb{C}e_n$. Algebraically this gives a decomposition of the projection f associated with K as $f = f_1 + \cdots + f_n$, where f_i is the rank one projection $e_i \otimes \bar{e}_i = T_{e_i,e_i}$. Then $v_i = T_{e_i,e_1} = e_i \otimes \bar{e}_1$ is a partial isometry taking f_1 to f_i: $v_i v_i^* = f_i$, $v_i^* v_i = f_1$. This is a comparison theory amongst projections in $B(H)$. For projections in a C^*-algebra, we would identify projections e and f for which there is a partial isometry v from e to f: $vv^* = f$, $v^*v = e$. In general, we would not have minimal projections, but could still think of the equivalence classes (of Murray–von Neumann) as a set of dimensions generalizing $\{1, 2, \ldots, N\}$.

Just as one goes from the (partial) semigroup $\{0, 1, 2, \ldots, N\}$ to $\mathbb{Z}$, we can form a group $K_0(A)$ if we first turn the equivalence classes of projections into a semigroup. For inductive limits of finite dimensional C^*-algebras, the AF algebras, these K_0 groups, or dimension groups are a complete invariant – just as they are for the simple infinite Cuntz algebras $\mathcal{O}_n$. This raises the question of how far one can go in attempting to classify C^*-algebras via K-theoretic invariants.

3.2 Equivalence of projections

We are now going to introduce the idea of dimension of a subspace of a Hilbert space, relative to a C^*-algebra, through an equivalence relation on the projections in a C^*-algebra.

Let A be a $*$-algebra, and $P(A)$ denote the projections in A, i.e. $P(A) = \{e \in A \mid e = e^2 = e^*\}$. We define an equivalence relation $\sim$ on $P(A)$, by $e \sim f$ if there exists $v \in A$ such that $e = v^*v$, $f = vv^*$ and say that e and

f are *Murray–von Neumann equivalent.* (Note that if A is a C^*-algebra, v is necessarily a partial isometry by Corollary 1.28.) It is clear that $\sim$ is symmetric ($e = e^*e = ee^*$), reflexive, and transitive, for if $e \sim f$ ($e = v^*v, f = vv^*$), $f \sim g$ ($f = w^*w, g = ww^*$) then if $u = wv$, $u^*u = v^*w^*wv = v^*fv = v^*vv^*v = e^2 = e$. Similarly $uu^* = g$, and so $e \sim g$. We let $D(A) = P(A)/\sim$ be the space of equivalence classes, where D stands for *Dimension*.

For example, if $A = M_n$, and e, f are two projections in A, then the following are equivalent:

(3.2.1) $e \sim f$

(3.2.2) $\operatorname{rank} e = \operatorname{rank} f \in \{0, 1, 2, \ldots, n\}$.

Hence $D(M_n) = \{0, 1, 2, \ldots, n\}$. Beginning with $D(A)$, we will construct a group $K_0(A)$ depending only on the algebra A. For example, for M_n, we will go from $\{0, 1, 2, \ldots, n\}$ to the integers $\mathbb{Z}$. However, first let us examine the connection between $\sim$ and a certain unitary equivalence. If e, f are projections in a unital $*$-algebra A, we say that e is *unitarily equivalent* to f, written $e \approx f$, if there exists a unitary u in A such that $e = ufu^*$. It is clear that $\approx$ is an equivalence relation, stronger than $\sim$. For if $e \approx f$, where $e, f \in P(A)$, u a unitary in A, then if $v = fu^*$, $e = v^*v$, $f = vv^*$, and so $e \sim f$. Moreover if $e \approx f$, then $e = ufu^*$ shows that $1 - e = u(1 - f)u^*$, and so $1 - e \approx 1 - f$. Thus

$$\begin{array}{ccc} e \approx f & \Leftrightarrow & 1 - e \approx 1 - f \\ \Downarrow & & \Downarrow \\ e \sim f & & 1 - e \sim 1 - f \end{array} \tag{3.2.3}$$

Now we claim that in general $e \sim f$ does not necessarily imply $1 - e \sim 1 - f$, and so in general $e \sim f$ does not imply $e \approx f$. For take $A = B(\ell^2)$, and s to be the unilateral shift on ℓ^2, $s(\xi_1, \xi_2, \ldots) = (0, \xi_1, \xi_2, \ldots)$. Then $s^*s = 1$, $ss^* \neq 1$. Thus $1 \sim ss^*$, but if $1 - 1 \sim 1 - ss^*$, then $0 \sim p$, where p is the rank one projection on $\{(\xi, 0, \ldots) : \xi \in \mathbb{C}\}$. But if $0 = v^*v$, $p = vv^*$, then $v = 0$ which is impossible. Thus in general $\sim$ and $\approx$ are distinct equivalence relations. However, in finite dimensions, if A is a multi-matrix algebra, then it is clear from (3.2.1) and (3.2.2) that $\sim$ and $\approx$ are the same. In general, if e, f are projections in a unital C^*-algebra A then

$$e \sim f \text{ and } 1 - e \sim 1 - f \Leftrightarrow e \approx f\,. \tag{3.2.4}$$

For suppose $e = v^*v$, $f = vv^*$, $1 - e = w^*w$, $1 - f = ww^*$, then $u = v + w$ is a unitary taking e to f.

Now in order to try and turn $D(A)$ into a semigroup, and subsequently a group, we would like to make sense of $[e] + [f]$, if $e, f \in P(A)$, and $[\ ]$ denotes the equivalence class in $P(A)/\sim$. We could try to define

$$[e] + [f] = [e + f]\,.$$

However, in general $e + f$ is not a projection (e.g. if $e = f \neq 0$), and (for a C^*-algebra) is only a projection when the range of e is orthogonal to the range of f, i.e. when $ef = 0$. (Alternatively, it would be enough to find for $e, f \in P(A)$, $e', f' \in P(A)$, such that $e' \sim e$, $f' \sim f$ and e' is orthogonal to f'). In general, we will use a matrix trick to go to 2×2 matrices, where $\begin{pmatrix} e & 0 \\ 0 & 0 \end{pmatrix}$ and $\begin{pmatrix} 0 & 0 \\ 0 & f \end{pmatrix}$ are orthogonal projections, and try to define

$$[e] + [f] = \left[\begin{pmatrix} e & 0 \\ 0 & f \end{pmatrix} \right] .$$

To make sense of this, we must introduce projections in matrix algebras over A. Consequently, it is useful to know that a matrix algebra over a C^*-algebra is a C^*-algebra.

If A is an algebra, and n a positive integer, then $M_n(A) = \{[a_{ij}]_{i,j=1}^n \mid a_{ij} \in A\}$ is an algebra under the usual matrix operations. If A is a $*$-algebra, then $M_n(A)$ is also a $*$-algebra for the involution $[a_{ij}]^* = [a_{ji}^*]$. If A is a C^*-algebra (on a Hilbert space H) we would like to put a norm on $M_n(A)$ so that it becomes a C^*-algebra. We claim that $M_n(B(H))$ acts on $H \oplus \cdots \oplus H$ (n-copies), and in fact we can identify $M_n(B(H))$ with $B(H \oplus \cdots \oplus H)$. If $A = [a_{ij}] \in M_n(B(H))$, $\xi = (\xi_i) \in H \oplus \cdots \oplus H$, define $A \cdot \xi = \left(\sum_{j=1}^n a_{ij}\xi_j\right)_{i=1}^n \in H \oplus \cdots \oplus H$. Then

$$\left\|\sum_{j=1}^n a_{ij}\xi_j\right\|^2 \leq \left(\sum_{j=1}^n \|a_{ij}\|\|\xi_j\|\right)^2 \leq \sum_{j=1}^n \|a_{ij}\|^2 \sum_{j=1}^n \|\xi_j\|^2 .$$

Hence

$$\|A\xi\|^2 = \sum_{i=1}^n \left\|\sum_{j=1}^n a_{ij}\xi_j\right\|^2 \leq \sum_{i,j=1}^n \|a_{ij}\|^2 \sum_{j=1}^n \|\xi_j\|^2 .$$

Thus $\xi \to A \cdot \xi$ defines a bounded linear map on $H \oplus \cdots \oplus H$ and so we have a linear map φ from $M_n(B(H))$ into $B(H \oplus \cdots \oplus H)$. Conversely, take $T \in B(H \oplus \cdots \oplus H)$, and let p_i denote the projection of $H \oplus \cdots \oplus H$ onto the ith coordinate so that $H_i \cong p_i(H \oplus \cdots \oplus H) \cong H$. Define $T_{ij} = p_i T | H_j$. Then $T \to [T_{ij}]$ defines a linear map from $B(H \oplus \cdots \oplus H)$ onto $M_n(B(H))$ which is inverse to the map φ described above. Thus $M_n(B(H)) \cong B(H \oplus \cdots \oplus H)$, as $*$-algebras. This allows us to define a norm on $M_n(B(H))$ such that it becomes a C^*-algebra. If A is a C^*-subalgebra of $B(H)$, then $M_n(A)$ is a C^*-subalgebra of $M_n(B(H))$. For if $A_m = [A_m(i,j)] \in M_n(A)$, and $A_m \to A = [A(i,j)]$ in $M_n(B(H))$, then $A_m(i,j) = p_i A_m | H_j \to p_i A | H_j = A(i,j)$. Thus A complete means that $A(i,j) \in A$, or $A \in M_n(A)$. Hence $M_n(A)$ is a C^*-algebra.

If A is a $*$-algebra, we let $P_n(A)$ (respectively $D_n(A)$) denote $P(M_n(A))$ (respectively $D(M_n(A))$), the projections (respectively the dimensions) in $M_n(A)$. If $\varphi : A \to B$ is a $*$-homomorphism between $*$-algebras, then $\varphi(P(A)) \subset P(B)$, and if $e \sim f$ in A, then $\varphi(e) \sim \varphi(f)$ in B, for if $e = v^*v$, $f = vv^*$, then $\varphi(e) = \varphi(v)^*\varphi(v)$, $\varphi(f) = \varphi(v)\varphi(v)^*$. Thus φ induces a map $[e] \to [\varphi(e)]$ from $D(A)$ into $D(B)$. In particular, if A is a $*$-algebra, we have a $*$-homomorphism

$\psi_n : M_n(A) \to M_{n+1}(A)$ given by $a \to \begin{pmatrix} a & 0 \\ 0 & 0 \end{pmatrix}$, $a \in M_n(A)$. Thus we have an induced map $\psi_n : D(M_n(A)) \to D(M_{n+1}(A))$, such that $\psi_n[e] = [e \oplus 0_1]$, where if $x \in M_n(A)$, $y \in M_p(A)$, we write $x \oplus y = \begin{pmatrix} x & 0 \\ 0 & y \end{pmatrix} \in M_{n+p}(A)$, and $0_p, 1_p$ denote the zero and unit of $M_p(A)$. We let $D_\infty(A) = \lim_{\to} (D_n(A), \psi_n)$, and claim that $D_\infty(A)$ is an abelian semigroup with identity. If A is a unital C^*-algebra, $K_0(A)$ is the enveloping or Grothendieck group of the semigroup $D_\infty(A)$. (If A is non unital, this is not quite the whole story, see Exercises 3.7, 3.8). Now if A is an AF algebra, the construction of $K_0(A)$ is less cumbersome, because the embeddings $\psi_n : D_n(A) \to D_{n+1}(A)$ are injective, and the semigroup $D_\infty(A)$ has the cancellation property (i.e. if $\alpha + \beta = \gamma + \beta$ in $D_\infty(A)$, then $\alpha = \gamma$). So in the next section, before proceeding any further, we will check these claims. See also Exercises 4.5 and 4.6.

Exercise 3.1 If A is a $*$-algebra, let $M_\infty(A)$ denote the $*$-algebra of infinite matrices $[a_{ij}]_{i,j=1}^\infty$, where $a_{ij} \in A$, with only finitely many non-zero entries, endowed with the usual matrix operations. If $\psi_n : M_n(A) \to M_{n+1}(A)$ is the map $x \to x \oplus 0_1$, show that $\lim_{\to}(M_n(A), \psi_n)$ is $*$-isomorphic to $M_\infty(A)$ and $D_\infty(A)$ can be identified with $D(M_\infty(A))$.

Exercise 3.2 Let A be a unital C^*-algebra.

(a) Take $e, f \in P(A)$ such that $e = v^*v$, $f = vv^*$. Show that $u = \begin{pmatrix} v^* & 1-e \\ 1-f & v \end{pmatrix}$ is a unitary in $M_2(A)$ such that $e \oplus 0_1 = u(f \oplus 0_1)u^*$.

(b) If $\psi_n : M_n(A) \to M_{n+1}(A)$ is as in Exercise 3.1, show that ψ_n preserves the relation $\approx$, and that we can identify $\lim_{\to}(P_n(A)/\approx)$ with $\lim_{\to}(P_n(A)/\sim)$.

3.3 K_0 for AF algebras

We first prove some technical lemmas, useful for understanding K_0 of an AF algebra.

Lemma 3.1 *If e, f are projections in a unital C^*-algebra A with $\|e - f\| < 1$ then $e \approx f$.*

Proof Let $A \subset B(H)$, $T = fe$ and $T = u|T|$ be the polar decomposition of T in $B(H)$ in Theorem 1.29. Now $\|e - efe\| = \|e(e - f)e\| < 1$, and so efe is invertible in eAe by Lemma 1.8. Consequently $|T| = (efe)^{1/2}$ is invertible in eAe. From $e = e(efe)(efe)^{-1}$, we see $\operatorname{Ran} e = \operatorname{Ran} T^*$, and similarly $\operatorname{Ran} f = \operatorname{Ran} T$. Also from the remarks after Theorem 1.29, the initial space of u is $\operatorname{Ran} T^*$ and the final space of u is $\operatorname{Ran} T$. Thus $u^*u = e$, $uu^* = f$. But $T = u|T|$, and so $u = ue = u|T|\,|T|^{-1} = T|T|^{-1} \in A$. Thus $e \sim f$ and from $\|(1-e) - (1-f)\| < 1$ we see $1 - e \sim 1 - f$ and so $e \approx f$. □

Let A be a unital C^*-algebra which is the norm closure of an increasing sequence of C^*-algebras $A_n \subset A_{n+1}$, where A_n is (isomorphic to) a multi-matrix C^*-algebra, and $1_{A_n} = 1_A$. Let $B = \bigcup_{n=1}^\infty A_n$, a dense $*$-subalgebra of A.

Lemma 3.2 *If $e \in P(A)$ and $\varepsilon > 0$, there exists $f \in P(B)$ such that $\|e-f\| < \varepsilon$.*

Proof By density take a self adjoint $x \in A_n$ for sufficiently large n such that $\|x - e\| < \varepsilon^2 \leq 1$. Then $\|x\| \leq \|x - e\| + \|e\| < 2$ and $x^2 - x = x(x - e) + (x - e)e + e - x$ so that $\|x^2 - x\| < 4\varepsilon^2$. Thus if $\lambda \in \sigma(x)$, $|\lambda(\lambda - 1)| < 4\varepsilon^2$, and so either $|\lambda| < 2\varepsilon$ or $|1 - \lambda| < 2\varepsilon$. Thus $\sigma(x) \subset [-2\varepsilon, 2\varepsilon] \cup [1 - 2\varepsilon, 1 + 2\varepsilon]$. Let f be the function on $\sigma(x)$ such that

$$f(\lambda) = \begin{cases} 1 & \text{if } \lambda \in [1 - 2\varepsilon, 1 + 2\varepsilon] \\ 0 & \text{otherwise} \end{cases} .$$

Then $|f(t) - t| \leq 2\varepsilon$, for $t \in \sigma(x)$. Thus $f = f(x)$ is a projection in A_n such that $\|f - e\| < \|f(x) - x\| + \|x - e\| \leq 2\varepsilon + \varepsilon^2$. □

Lemma 3.3 *If u is a unitary in the unital AF algebra A, then there exists a unitary v in B such that $\|u - v\| < \varepsilon$.*

Proof Choose $x \in A_n$ such that $\|u-x\| = \varepsilon < 1$, and $\|x\| < 1$. Then $\|1-u^*x\| = \|u^*(u - x)\| = \|u - x\| < 1$, and so u^*x, and hence x is invertible by Lemma 1.8. Thus $|x|$ is invertible and so the partial isometry $v = x|x|^{-1} \in A_n$ is invertible, and hence is a unitary. Thus

$$\begin{aligned} &\|u - v\| \leq \|u - x\| + \|x - v\| \leq \|u - x\| + \|v(|x| - 1)\| \\ &\leq \|u - x\| + \||x| - 1\| \leq \|u - x\| + \|x^*x - 1\| \\ &\leq \|u - x\| + \|(x^* - u^*)u\| + \|u^*(x - u)\| \leq 3\|u - x\| = 3\varepsilon . \end{aligned}$$

Here we have used that $|\lambda^{1/2} - 1| \leq |\lambda - 1|$ for $0 \leq \lambda \leq 1$, and so $\||x| - 1\| \leq \|x^*x - 1\|$. □

Proposition 3.4 *Let $A = \overline{\cup A_n}$ be a unital AF algebra where $A_n \subset A_{n+1}$ is an increasing sequence of multi-matrix algebras, $1_{A_n} = 1_A$ and $B = \cup A_n$. Then:*

(a) *If e is a projection in A, then there exists a projection f in B such that $e \approx f$.*
(b) *If $e, f \in P(B)$ such that $e \approx f$ in A, then $e \approx f$ in B.*
(c) *If $e, e' \in P_m(A)$, $f \in P_n(A)$ such that $e \oplus f \sim e' \oplus f$, then $e \approx e'$.*
(d) *The maps $\psi_n : D_n(A) \to D_{n+1}(A)$, and consequently $\psi_{n,\infty} : D_n(A) \to D_\infty(A)$ are injective.*

Proof (a) Choose by Lemma 3.2 a projection f in B such that $\|e - f\| < 1$. Then $e \approx f$ by Lemma 3.1.
(b) Suppose u is a unitary in A such that $e = ufu^*$. By Lemma 3.3, take a unitary v in B such that $\|u - v\| < 1/2$. Then $\|e - vfv^*\| < 1$. Hence by Lemma 3.1, there exists a unitary w in B such that $e = wvfv^*w^*$. Hence $e \approx f$ in B.
(c) By replacing f by $f \oplus 0_p$ for sufficiently large p we can assume that $e \oplus f \approx e' \oplus f$, using the 2×2 matrix trick of Exercise 3.2. Then by (a), choose projections e_0, e_0' in $P_m(B)$, $f_0 \in P_n(B)$ such that $e \approx e_0$, $e' \approx e_0'$, $f \approx f_0$. Then

$e_0 \oplus f_0 \approx e_0' \oplus f_0$ in A, and so $e_0 \oplus f_0 \approx e_0' \oplus f_0$ in B, by (b). Then by finite dimensional considerations, $e_0 \approx e_0'$ in B, and so $e \approx e'$.
(d) This follows from (c). □

Remark Consequently if $e, e' \in P_m(B)$ (respectively $P_m(A)$) the following conditions are equivalent:

(3.3.1) $e \oplus f \sim e' \oplus f$ in $M_{m+n}(A)$ for some $f \in P_n(A)$

(3.3.2) $e \approx e'$ in $M_m(B)$ (respectively $M_m(A)$).

So now if $A = \overline{\cup A_n}$ is a unital AF algebra, and $D_\infty(A) = \lim_\rightarrow D_n(A)$, we write $[e]$ for the equivalence class of $e \in P_\infty(A) = \lim P_m(A) \subset M_\infty(A)$ (see Exercise 3.1). Note that by Proposition 3.4 there is no possibility of confusion here. We now claim that $D_\infty(A)$ is an abelian semigroup with identity, and possesses the cancellation law. First of all we have maps $P_n(A) \times P_m(A) \to P_{n+m}(A)$ given by $(e, f) \to e \oplus f$. If $e \sim e'$, $f \sim f'$ then $e \oplus f \sim e' \oplus f'$. So we have maps

$$\begin{gathered} D_n(A) \times D_m(A) \to D_{n+m}(A) \to D_\infty(A) \\ ([e], [f]) \to [e \oplus f]\,. \end{gathered}$$

We claim that the maps $D_n(A) \times D_m(A) \to D_\infty(A)$ are compatible under the maps $D_n(A) \to D_{n+1}(A)$, $D_m(A) \to D_{m+1}(A)$. For if $e \in P_n(A)$, $f \in P_m(A)$ then $e \oplus 0_1 \oplus f \oplus 0_1 \approx e \oplus f \oplus 0_2$ in $M_{m+n+2}(A)$ via a permutation matrix; i.e. a change of basis. Consequently

$$[e \oplus 0_1 \oplus f \oplus 0_1] = [e \oplus f \oplus 0_2] \text{ in } D_{n+m+2}(A)$$

or

$$[e \oplus 0_1 \oplus f \oplus 0_1] = [e \oplus f] \text{ if } D_\infty(A)\,.$$

Thus we have a commutative diagram

$$\begin{array}{ccc} D_n(A) \times D_m(A) & & \\ \downarrow & \searrow & \\ & & D_\infty(A) \\ & \nearrow & \\ D_{n+1}(A) \times D_{m+1}(A) & & \end{array}$$

which gives a well defined product: $D_\infty(A) \times D_\infty(A) \to D_\infty(A)$. It is easy to check that this makes $D_\infty(A)$ an abelian semigroup with identity $[0_1]$. To see that $D_\infty(A)$ is abelian, note that $e \oplus f \approx f \oplus e$ via a permutation matrix and so $[e \oplus f] = [f \oplus e]$. Moreover, $[e \oplus 0_m] = [e]$ and so $[0_1] = [0_m]$ is an identity. We leave the reader to check associativity of the product. Finally it is clear from the Remark on page 94 that $D_\infty(A)$ has the cancellation property.

Let $K_0(A)$ denote the enveloping or Grothendieck group, constructed as follows, in the same way as one goes from $\mathbb{N}$ to $\mathbb{Z}$. Take any abelian semigroup S with identity 0, which has the cancellation property. Let $K_0(S) = S \times S/R$ where R is the equivalence relation on S given by $(\alpha, \beta)R(\alpha', \beta')$ if $\alpha + \beta' = \alpha' + \beta$. Let $\pi : S \times S \to K_0(S)$ be the quotient map, and define an addition on $K_0(S)$ by

$$\pi(\alpha, \beta) + \pi(\alpha', \beta') = \pi(\alpha + \alpha', \beta + \beta') . \tag{3.3.3}$$

This addition is well defined and with this operation $K_0(S)$ is an abelian group with identity $\pi(0,0)$, and inverse operation given by $-\pi(\alpha, \beta) = \pi(\beta, \alpha)$. Moreover $\gamma(s) = \pi(s, 0)$ embeds S as a subsemigroup $K_0(S)_+$ of $K_0(S)$. We usually identify S with $\gamma(S)$. If A is a unital AF algebra, we write $K_0(A) = K_0(D_\infty(A))$, and $K_0(A)_+ = K_0(D_\infty(A))_+ = D_\infty(A)$. (For the modifications when A is not necessarily AF or unital see Exercises 3.3–3.8). If S_1 and S_2 are abelian semigroups with identities (and the cancellation property), and $\varphi : S_1 \to S_2$ is a homomorphism, then there is an induced homomorphism denoted by $K_0(\varphi)$ such that $K_0(\varphi) \circ \gamma = \gamma \circ \varphi$. Thus if $\theta : A \to B$ is a $*$-homomorphism between (unital AF-C)*-algebras A and B, we have induced $*$-homomorphisms $\theta_n : M_n(A) \to M_n(B)$ such that $\theta_n[a_{ij}] = [\theta(a_{ij})]$, $[a_{ij}] \in M_n(A)$. Thus there are induced maps (Section 3.2) $\theta_n : D_n(A) \to D_n(B)$, $\theta_n[e] = [\theta_n(e)]$ and a commutative diagram:

$$\begin{array}{ccccccc}
D_1(A) & \to & D_2(A) & \to & D_3(A) & \to & \dots \\
\downarrow \theta_1 & & \downarrow \theta_2 & & \downarrow \theta_3 & & \\
D_1(B) & \to & D_2(B) & \to & D_3(B) & \to & \dots
\end{array} \tag{3.3.4}$$

Hence there is an induced map $\theta_\infty : D_\infty(A) \to D_\infty(B)$, $\theta_\infty[e] = [\theta_n(e)]$, $e \in P_m(A)$, which is a homomorphism as $\theta_m(e) \oplus \theta_n(f) = \theta_{m+n}(e \oplus f)$, if $e \in P_m(A)$, $f \in P_n(A)$. Hence there is an induced homomorphism from $K_0(A)$ into $K_0(B)$, denoted by $K_0(\theta) = K_0(\theta_\infty)$ such that

$$K_0(\theta)([e] - [f]) = [\theta_m(e)] - [\theta_n(f)] \tag{3.3.5}$$

if $e \in P_m(A)$, $f \in P_n(A)$. K_0 is functorial in that if A, B, C are (unital AF-C)*-algebras and $\varphi : A \to B$, $\psi : B \to C$ are $*$-homomorphisms then $K_0(\psi\varphi) = K_0(\psi)K_0(\varphi)$, and $K_0(\mathrm{id}) = \mathrm{id}$. Thus if θ is an isomorphism between (unital AF-C)*-algebras, then $K_0(\theta)$ is a group isomorphism. Thus K_0 could be used to distinguish between $*$-algebras (see e.g. Theorem 3.8 below).

Example 3.5 (a) If $A = M_m$, then $M_n(M_m) = M_{mn}$. As already observed in Section 3.2, we can identify $D(M_p)$ with $I_p = \{0, 1, 2, \dots, p\}$, by $[e] \to \operatorname{rank} e$, $e \in P(M_p)$. So we can identify $D_n(A) = D(M_{mn})$ with $I_{mn} = J_n$ via a map θ_n. Figure 3.1 is a commutative diagram as rank $(e \oplus 0_1) = \operatorname{rank} e$, and so we have an

$$\begin{array}{ccccccc} D_1 & \to & D_2 & \to & D_3 & \to & \dots \\ \downarrow \theta_1 & & \downarrow \theta_2 & & \downarrow \theta_3 & & \\ J_1 & \to & J_2 & \to & J_3 & \to & \dots \end{array}$$

FIG. 3.1. Computation of $K_0(M_m)$

induced injective map $\theta_\infty : D_\infty = \cup D_n \to \mathbb{N} = \cup J_n$, which is a homomorphism as rank $(e \oplus f) = \operatorname{rank} e + \operatorname{rank} f$. Hence we can identify $D_\infty(A)$ with $\mathbb{N}$ as semigroups and consequently

$$(K_0(M_m), K_0(M_m)_+, [1]) = (\mathbb{Z}, \mathbb{N}, m) \tag{3.3.6}$$

where [1] denotes the equivalence class of the identity.

(b) If A, B are (unital AF-C)*-algebras, then there are identifications $P(M_m(A)) \times P(M_m(B)) \to P(M_m(A \oplus B))$ which preserve $\sim$, and so $D_m(A) \times D_m(B)$ can be identified with $D_m(A \oplus B)$. These identifications are compatible in going from m to $m+1$, and so we can identify $D_\infty(A) \times D_\infty(B)$ with $D_\infty(A \oplus B)$ as semigroups, and consequently we can identify $(K_0(A \oplus B), K_0(A \oplus B)_+)$ with $(K_0(A) \oplus K_0(B), K_0(A)_+ \oplus K_0(B)_+)$.

(c) In particular, if $M(\mathbf{n}) = M(n_1) \oplus \cdots \oplus M(n_p)$ is a multi-matrix algebra

$$(K_0(M(\mathbf{n}), K_0(M(\mathbf{n})_+, [1]) = (\mathbb{Z}^p, \mathbb{Z}^p_+, \mathbf{n}) \tag{3.3.7}$$

where $\mathbb{Z}^p_+ = \mathbb{N}^p$.

Let $\theta : M(\mathbf{m}) \to M(\mathbf{n})$ be a canonical homomorphism given by a matrix $[k_{ij}]^{q,p}_{i=1,j=1}$, $\mathbf{m} = (m_1, \dots m_p)$, $\mathbf{n} = (n_1, \dots, n_q)$ as in Section 2.5. We claim that $K_0(\theta) : \mathbb{Z}^p \to \mathbb{Z}^q$ is given by multiplication by the $q \times p$ matrix $[k_{ij}]$. Consider first $q = p = 1$, so that $\theta(x) = (x \oplus \cdots \oplus x \oplus 0)$. Then rank $\theta(e) = k \operatorname{rank} e$, if $e \in P(M(m))$, i.e. $K_0(\theta)x = kx$, $x \in \mathbb{Z}$. The general case is similar with obvious adjustments.

Suppose $\varphi_i : A_i \to A_{i+1}$ is a sequence of unital $*$-homomorphisms between multi-matrix algebras A_i, and $B_i = \varphi_{i,\infty}(A_i) \subset A_\infty$, the C^*-direct limit of A_i. Since $\varphi_i(1_{A_i}) = 1_{A_{i+1}}$, we see that $1_{B_i} = \varphi_{i,\infty}(1_{A_i}) = \varphi_{i+1,\infty}(1_{A_{i+1}}) = 1_{B_{i+1}}$, and so $1_{B_i} = 1_A$.

Theorem 3.6 *In the above situation*

$$K_0(A_\infty) = \lim_{\to}(K_0(A_i), K_0(\varphi_i)) \tag{3.3.8}$$

$$K_0(A_\infty)_+ = \lim_{\to}(K_0(A_i)_+, K_0(\varphi_i)) \,. \tag{3.3.9}$$

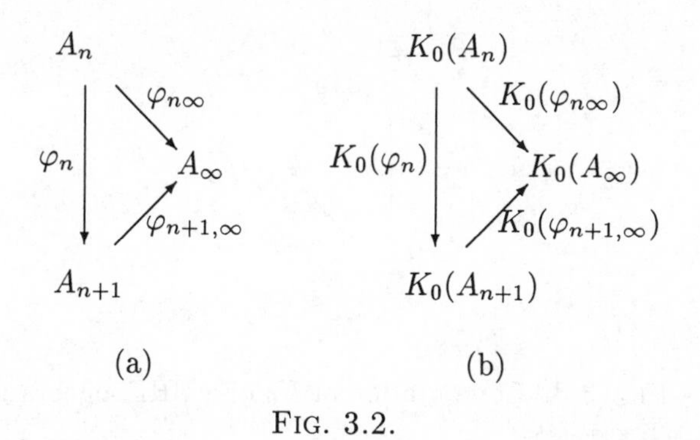

FIG. 3.2.

Proof From Fig. 3.2(a) we get Fig. 3.2(b). So if $\psi_n = K_0(\varphi_{n\infty})$, this gives an induced map $\psi_\infty : \lim_\rightarrow (K_0(A_n), K_0(\varphi_n)) \rightarrow K_0(A_\infty)$. To see surjectivity of ψ_∞, take $e \in P_\infty(A_\infty)$. Then by Proposition 3.4(a), there exists $f \in P_\infty(A_n)$ for some n, such that $e \approx f$. Then $f = \varphi_{n\infty}(g)$ for some projection g in A_n, and so $[e] = K_o(\varphi_{n\infty})[g]$. Thus ψ_∞ is surjective. To see injectivity, let us assume for simplicity that $K_0(\varphi_n)$ are all injective. (Exercise for the general case!) Suppose $\psi_\infty([e] - [f] = 0$, for $e, f \in P_\infty(A_n)$ some n. Then $\psi_\infty([e] - [f]) = 0$ means that $[\varphi_{n\infty}(e)] = [\varphi_{n\infty}(f)]$ in $K_0(A_\infty)$. Then by Proposition 3.4(b), $[\varphi_n(e)] = [\varphi_n(f)]$ in $K_0(B_n)$. Since $K_0(\varphi_n)$ is injective $[e] - [f] = 0$ in $K_0(A_n)$, and so ψ_∞ is injective. This establishes (3.3.8) and (3.3.9). □

Example 3.7 Take $q_1, q_2, \ldots$ to be a sequence of positive integers, and $p_i = q_1 \cdots q_i$, so that $q_{i+1} = p_{i+1}/p_i$, where $p_1, p_2, \ldots$ is a sequence of positive integers so that $p_i | p_{i+1}$. We embed $M_{p_i} \rightarrow M_{p_{i+1}}$ by

$$x \rightarrow \begin{vmatrix} x & & & & \\ & \cdot & & & \\ & & \cdot & & \\ & & & \cdot & \\ & & & & x \end{vmatrix}, q_{i+1} \text{ copies down the diagonal.} \tag{3.3.10}$$

or identifying $M_{p_{i+1}}$ with $M_{p_i} \otimes M_{q_{i+1}}$ by $\psi_i : x \rightarrow x \otimes 1_{q_{i+1}}$. Let $M[p_i]_{i=1}^\infty = \lim_\rightarrow M(p_i) = \bigotimes_{i=1}^\infty M_{q_i}$. Then by Theorem 3.6, we have

$$K_0(M[p_i]) = \varinjlim(K_0(M_{p_i}), K_0(\psi_i)) = \varinjlim(\mathbb{Z}, q_{i+1})\,. \tag{3.3.11}$$

Thus we need to compute the direct limit of

$$\mathbb{Z} \underset{q_1}{\rightarrow} \mathbb{Z} \underset{q_2}{\rightarrow} \mathbb{Z} \rightarrow \ldots \tag{3.3.12}$$

Let $\mathbb{Z}[1/p_i]$ denote the group of rationals $\{a/p_i \mid a \in \mathbb{Z}, i = 1, 2, \ldots\}$. Then we have the commutative diagram of Fig. 3.3 where $\theta_n(x) = x/p_n$. Taking the direct

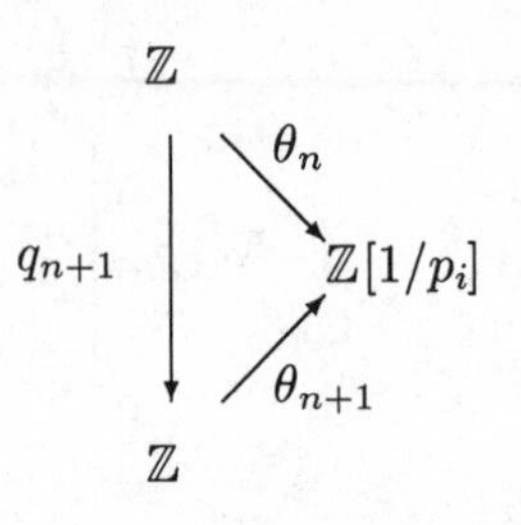

FIG. 3.3. Computation of K_0 of a UHF algebra

limit of the injective homomorphisms, we get an isomorphism $\theta_\infty : K_0(M[p_i]) \to \mathbb{Z}[1/p_i]$ such that $K_0(M[p_i])_+$ is identified with $\mathbb{Z}[1/p_i]_+ = \{a/p_i \mid a \in \mathbb{N}, i = 1, 2, \ldots\}$. In $K_0(M(p_i))$, $[1]$ corresponds to p_i, hence the class of the identity in $K_0(M[p_i])$ corresponds to $\theta_i(p_i) = 1$. With this computation, it is a small matter to classify *UHF algebras* up to isomorphism:

Theorem 3.8 Glimm *Let $p_1, p_2, \ldots, p'_1, p'_2, \ldots$ be sequences of positive integers such that $p_i | p_{i+1}$, $p'_i | p'_{i+1}$. Then $M[p_i] \cong M[p'_i]$ if and only if for all i there exists j such that $p_i | p'_j$ and $p'_i | p_j$.*

Proof Suppose $\alpha : M[p_i] \to M[p'_i]$ is a $*$-isomorphism. Then α is unital, and so $K_0(\alpha) : (K_0(M[p_i]), [1]) \to (K_0(M[p'_i]), [1])$ is a group isomorphism. Thus $\alpha_* \equiv K_0(\alpha) : \mathbb{Z}[1/p_i] \to \mathbb{Z}[1/p'_i]$ takes 1 to 1. Now

$$\begin{aligned} &\alpha_*(1/p_i) + \cdots + \alpha_*(1/p_i) \quad p_i - \text{terms} \\ &= \alpha_*(1/p_i + \cdots + 1/p_i) = \alpha_*(1) = 1\,. \end{aligned}$$

Hence $\alpha_*(1/p_i) = 1/p_i$. Suppose $a \in \mathbb{N}$, then

$$\begin{aligned} \alpha_*(a/p_i) &= \alpha_*(1/p_i + \cdots + 1/p_i) \quad a - \text{terms} \\ &= \alpha_*(1/p_i) + \cdots + \alpha_*(1/p_i) = a/p_i\,. \end{aligned}$$

Hence $\alpha_*(x) = x$, for all $x \in \mathbb{Z}[1/p_i]$, and so $\mathbb{Z}[1/p_i] = \mathbb{Z}[1/p'_i]$ as sets. If we take $1/p_i \in \mathbb{Z}[1/p_i]$, then $1/p_i = a/p'_{j_1}$ for some $a \in \mathbb{N}$, j_1, and so $p_i | {p'}_{j_1}$. Similarly $p'_i | p_{j_2}$. Take $j \geq j_1, j_2$ then $p_i | p'_j$ and $p'_i | p_j$.

Conversely suppose for each i there exists j such that $p_i | p'_j$, $p'_i | p_j$. Let $n(1) = 1$, then $p_{n(1)} | p'_{m(1)}$ say, and $p'_{m(1)} | p_{n(2)}, \ldots$. In this way we construct sequences of integers $n(1), n(2), \ldots, m(1), m(2), \ldots$ such that $n(i), m(i) \to \infty$ as $i \to \infty$, and $p_{n(i)} | p'_{m(i)}$, $p'_{m(i)} | p_{n(i+1)}$. We can thus form a commutative diagram of Fig. 3.4 consisting of canonical homomorphisms. Taking direct limits, we have $\theta_\infty : M[p_i] \to M[p'_i]$, and $\psi_\infty : M[p'_i] \to M[p_i]$, which are easily seen to be inverse to each other. Hence $M[p_i]$ and $M[p'_i]$ are isomorphic C^*-algebras. □

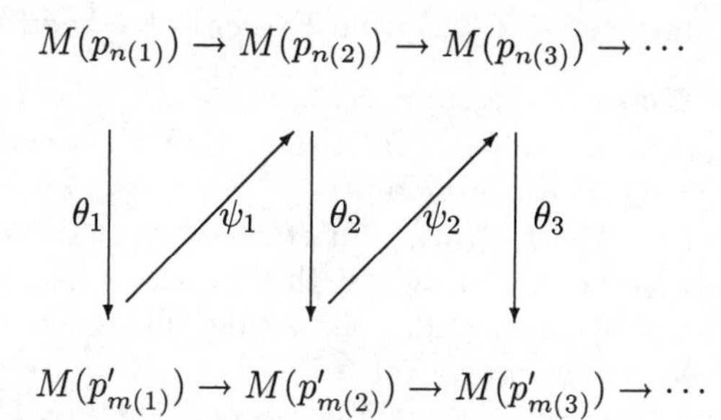

FIG. 3.4. Intertwining of UHF algebras

In particular, we have thus constructed an uncountable number of non-isomorphic simple unital separable C^*-algebras. For simplicity of UHF algebras, see Lemma 2.16.

Exercise 3.3 Let A be a $*$-algebra, and $\psi_n : D_n(A) \to D_{n+1}(A)$ the maps as in Sections 2.3 and 2.4. Show that if $[e \oplus 0_\infty]$ denotes $\psi_{n\infty}[e] \in D_\infty(A)$, $e \in P_n(A)$, then

$$[e \oplus 0_\infty] + [f \oplus 0_\infty] = [e \oplus f \oplus 0_\infty]$$

makes $D_\infty(A)$ into an abelian semigroup with identity.

Exercise 3.4 If S is an abelian semigroup with identity, show that $(\alpha, \beta)R(\alpha', \beta')$ if $\alpha + \beta' + \gamma = \alpha' + \beta + \gamma$ for some γ defines an equivalence relation on $S \times S$ such that $K_0(S) = S \times S/R$ becomes an abelian group with addition as in (3.3.3), and $\gamma : S \to S \times S/R$, given by $\gamma(s) = \pi(s, 0)$, $s \in S$ where π is the quotient map is a homomorphism. If A is a unital $*$-algebra, let $K_0(A) = K_0(D_\infty(A))$, $K_0(A)_+ = \gamma[D_\infty(A)]$.

Exercise 3.5 If $\theta : A \to B$ is a $*$-homomorphism between unital $*$-algebras, show how to define an unique homomorphism $K_0(\theta) : K_0(A) \to K_0(B)$ such that $K_0(\theta)\gamma[\varphi_{n\infty}(e)] = \gamma[\varphi_{n\infty}(\theta_n(e))]$, $e \in P_n(A)$, if $\theta_n : M_n(a) \to M_n(B)$, is $\theta_n[a_{ij}] = [\theta(a_{ij})]$.

Exercise 3.6 Let $A = \overline{\cup A_n}$ be a unital C^*-algebra, with $A_n \subset A_{n+1}$ unital C^*-subalgebras such that $1_{A_n} = 1_A$. Show that $K_0(A) \cong \lim_{\to}(K_0(A_n), K_0(i_n))$, if $i_n : A_n \to A_{n+1}$ are the inclusion maps.

Exercise 3.7 If A is a $*$-algebra, and $\pi : \tilde{A} \to \mathbb{C}$ is the quotient map $\pi(a, \lambda) = \lambda$, we define $K_0(A) =$ Kernel of $K_0(\pi) : K_0(\tilde{A}) \to K_0(\mathbb{C})$, using Exercises 3.4, 3.5 to define K_0 for unital algebras, and homomorphisms between unital algebras. If φ is a $*$-homomorphism between $*$-algebras A and B, extend φ to a unital homomorphism $\tilde{\varphi}$ between $\tilde{A}$ and $\tilde{B}$ (see Exercises 1.18 and 1.19). Show that $K_0(\tilde{\varphi})$ between $K_0(\tilde{A})$ and $K_0(\tilde{B})$ restricts to a homomorphism $K_0(\varphi)$ between $K_0(A)$ and $K_0(B)$, such that if ψ is a $*$-homomorphism between B and another $*$-algebra C then $K_0(\psi\varphi) = K_0(\psi)K_0(\varphi)$.

Exercise 3.8 If A is a unital $*$-algebra, show that $\tilde{A} \cong A \oplus \mathbb{C}$ as $*$-algebras and deduce that the definitions of $K_0(A)$ from Exercises 3.4 and 3.7 are consistent.

Exercise 3.9 Let A be a C^*-algebra
(a) If e is an *idempotent* (i.e. $e = e^2$) in A show that there exists a projection f in A such that $Ae \cong Af$ as left A-modules.
(b) If e, f are idempotents in $M_n(A)$ for some $n \geq 1$, show that $M_n(A)e \cong M_n(A)f$ as left A-modules if and only if there exists a, b in $M_n(A)$ such that $e = ab$, $f = ba$. If e, f are projections show that this is equivalent to e and f being Murray–von Neumann equivalent.
(c) If R is a ring and E a (right) R-module, we say that E is *finitely generated* if there exist finitely many elements $e_1, \ldots, e_n$ in E such that $E = \sum_{i=1}^{N} Re_i$, and E is *projective* if any surjective R-module homomorphism $M \to E$ has a right inverse $E \to M$.
If A is a unital C^*-algebra, show that E is a finitely generated projective A-module if and only if there is a projection e in $M_n(A)$ for some $n \geq 1$ such that $eM_n(A)e \cong E$ (as A-modules). Thus $[e] \to [M_n(A)e]$ is an isomorphism from $P_\infty(A)/\sim$ into isomorphism classes of finitely generated projective A-modules equipped with the natural addition (of isomorphism classes) of modules.

3.4 Classification of AF algebras

In the previous section we showed that K_0 as an ordered group, together with the class of the unit [1] is a complete invariant for UHF algebras. Here we extend this to have a complete K-theoretic invariant for AF algebras. The invariant will be $K_0(A)$ as an ordered group, together with the distinguished subset of its positive cone, $\Gamma(A) = \{[e] \mid e \in P(A)\}$ called the *scale* of A. For unital algebras this invariant is essentially the same as $(K_0(A), [1_A])$.

Suppose first that A and B are unital AF algebras and that $\varphi : K_0(A) \to K_0(B)$ is an ordered group isomorphism, which identifies $[1_A]$ with $[1_B]$ and

$$A_1 \to A_2 \to \cdots \to A \tag{3.4.1}$$

$$B_1 \to B_2 \to \cdots \to B \tag{3.4.2}$$

are unital inclusions, with A_i, B_i isomorphic to the multi-matrix algebras $M(\mathbf{p_i})$, $M(\mathbf{q_i})$ respectively. Let Λ_i, Θ_i denote the multiplicity graphs of the embeddings $A_i \to A_{i+1}$, $B_i \to B_{i+1}$ respectively, so that at the dimension group level we identify

$$K_0(A_i) \to K_0(A_{i+1}) \text{ with } \mathbb{Z}^{|\mathbf{p_i}|} \to^{\Lambda_i} \mathbb{Z}^{|\mathbf{p_{i+1}}|}$$

and

$$K_0(B_i) \to K_0(B_{i+1}) \text{ with } \mathbb{Z}^{|\mathbf{q_i}|} \to^{\Theta_i} \mathbb{Z}^{|\mathbf{q_{i+1}}|}$$

respectively. Thus at the dimension group level (3.4.1) and (3.4.2) become

$$\mathbb{Z}^{|\mathbf{p_1}|} \overset{\Lambda_1}{\to} \mathbb{Z}^{|\mathbf{p_2}|} \overset{\Lambda_2}{\to} \mathbb{Z}^{|\mathbf{p_3}|} \overset{\Lambda_3}{\to} \cdots \to K_0(A) \tag{3.4.3}$$

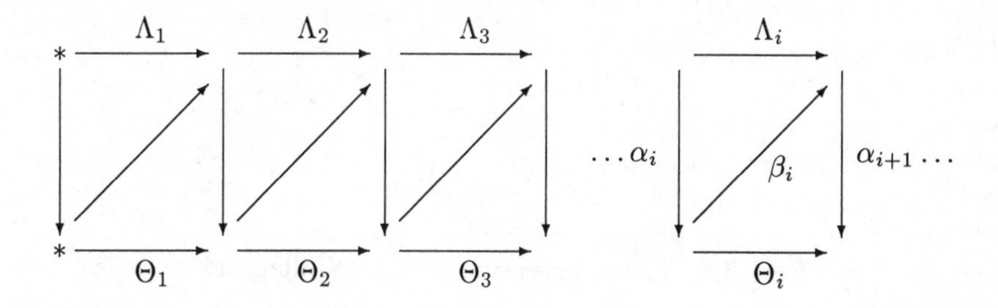

FIG. 3.5. Intertwining Bratteli diagrams

$$\mathbb{Z}^{|\mathbf{q_1}|} \stackrel{\Theta_1}{\to} \mathbb{Z}^{|\mathbf{q_2}|} \stackrel{\Theta_2}{\to} \mathbb{Z}^{|\mathbf{q_3}|} \stackrel{\Theta_3}{\to} \cdots \to K_0(B) \tag{3.4.4}$$

and φ takes $K_0(A) = \lim_{\to} \mathbb{Z}^{|\mathbf{p_i}|}$ into $K_0(B) = \lim_{\to} \mathbb{Z}^{|\mathbf{q_i}|}$. Since $\mathbb{Z}^{|\mathbf{p_i}|}$ is finitely generated we can assume by relabelling the initial algebra in (3.4.2) that φ takes $K_0(A_1) = \mathbb{Z}^{|\mathbf{p_1}|}$ into $K_0(B_1) = \mathbb{Z}^{|\mathbf{q_2}|}$. Similarly by telescoping in (3.4.1) we can assume that φ^{-1} takes $K_0(B_1) = \mathbb{Z}^{|\mathbf{q_2}|}$ into $K_0(A_2) = \mathbb{Z}^{|\mathbf{p_2}|}$. We thus construct a commutative diagram Fig. 3.5 where each triangle is a commuting square of non-negative matrices, and where α_i is (the restriction of) φ from $\mathbb{Z}^{|\mathbf{p_i}|} = K_0(M(\mathbf{p_i}))$ into $\mathbb{Z}^{|\mathbf{q_i}|} = K_0(M(\mathbf{q_i}))$, and β_i is (the restriction of) φ^{-1} from $\mathbb{Z}^{|\mathbf{q_i}|} = K_0(M(\mathbf{q_i}))$ into $\mathbb{Z}^{|\mathbf{p_{i+1}}|} = K_0(M(\mathbf{p_{i+1}}))$. Extend the diagram one rectangle to the left in Fig. 3.5 if necessary so that $A_1 = \mathbb{C} = B_1$ and $\alpha_1 = 1$. Since $\Lambda_i = \beta_i\alpha_i$, and $\Theta_i = \alpha_{i+1}\beta_i$ we can identify the path algebra of $\alpha_1\beta_1\alpha_2\beta_2\cdots\alpha_i\beta_i = \Lambda_1\Lambda_2\cdots\Lambda_i$ with $M(\mathbf{p_i})$ and $\beta_1\alpha_2\beta_2\alpha_3\cdots\beta_i\alpha_{i+1} = \Theta_1\Theta_2\cdots\Theta_i$ with $M(\mathbf{q_i})$. In this way the path models of both (3.4.1) for A and (3.4.2) for B are identified with the path model for $\alpha_1\beta_1\alpha_2\beta_2\cdots$. Hence A is isomorphic to B.

To complete the proof in the general (not necessarily unital) case, we only need the extra understanding of what the path model is for an AF algebra as in (3.4.1), where the embeddings are not necessarily unital. The unital extension $M(\tilde{\mathbf{p}})$ of $M(\mathbf{p})$ can be identified with $M(\mathbf{p}\oplus 1) = M(\mathbf{p})\oplus\mathbb{C}$. If $\varphi : M(\mathbf{p}) \to M(\mathbf{q})$ is a $*$-homomorphism with multiplicity matrix Λ then the unital extension $\tilde{\varphi}$ (see Exercise 1.18) from $M(\mathbf{p}\oplus 1)$ to $M(\mathbf{q}\oplus 1)$ $\tilde{\varphi} : a \oplus \alpha \to (\varphi(a) + \alpha(1-\varphi(1))\oplus\alpha$ has multiplicity matrix $\tilde{\Lambda} = \begin{pmatrix} \Lambda & \mathbf{p}-\Lambda\mathbf{q} \\ 0 & 1 \end{pmatrix}$. An extra vertex 1 is introduced at each level, and $(p - \Lambda q)_i$ additional edges are added from 1 in the upper level to $p(i)$ so that the total number of edges leading into $p(i)$ is now $p(i)$, and one edge joins the two new vertices. Thus no edge leads into 1 on the lower level. The path model for

$$\mathbb{C} = M(\mathbf{p_1}) \stackrel{\Lambda_1}{\to} M(\mathbf{p_2}) \stackrel{\Lambda_2}{\to} M(\mathbf{p_3}) \stackrel{\Lambda_3}{\to} \cdots \tag{3.4.5}$$

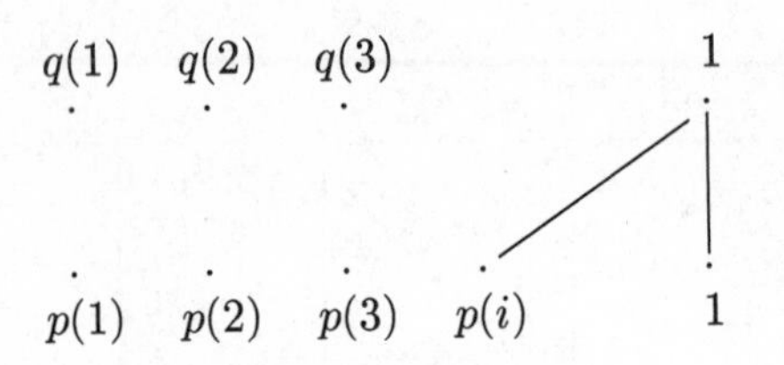

FIG. 3.6. Unital extension of Bratteli diagram

is then obtained as follows. First from the path model for

$$\mathbb{C} = M(\mathbf{p_1} \oplus 1) \xrightarrow{\tilde{\Lambda}_1} M(\mathbf{p_2} \oplus 1) \xrightarrow{\tilde{\Lambda}_2} M(\mathbf{p_3} \oplus 1) \xrightarrow{\tilde{\Lambda}_3} \cdots \tag{3.4.6}$$

and for $m \leq n$, define

$$A[m,n] = \bigoplus_{i,j} A_{ij}$$

where the summation is over all vertices i at level m for the Bratteli diagram of (3.4.5) (corresponding to minimal central projection of $M(\mathbf{p_m} \oplus 1)$) and vertices j at level n corresponding to minimal central projections of $M(\mathbf{p_n})$ and A_{ij} are as defined in (2.9.12). Then $A[0,m] \cong M(\mathbf{p_m})$, generated by matrix units $f_{u,v}$, where $u, v \in \text{Path}(*, j)$, j a minimal central projection of $M(\mathbf{p_n})$, and the paths are taken from the Bratteli diagram of (3.4.6) (not (3.4.5)) so that the paths may stray outside the vertices and edges of the Bratteli diagram of (3.4.5) but always end at a vertex of the original Bratteli diagram. Moreover

$$A[0,m]' \cap A[0,n] = A[m,n] \text{ if } 0 \leq m \leq n\,. \tag{3.4.7}$$

Theorem 3.9 *Let A, B be AF algebras and $\Phi : K_0(A) \to K_0(B)$ an ordered group homomorphism taking the scale $\Gamma(A)$ of A into the scale $\Gamma(B)$ of B. Then there exists a homomorphism $\varphi : A \to B$ such that $K_0(\varphi) = \Phi$. If Φ is a scale preserving ordered group isomorphism then φ can be chosen to be an isomorphism.*

Proof If $\varphi : K_0(M(\mathbf{p})) \to K_0(M(\mathbf{q}))$ is a map represented by a matrix Λ, then scale preserving is equivalent to $p \geq \Lambda q$. We construct a commuting diagram as in Fig. 3.5, and then notice that the functorial property $(\varphi\psi)^{\sim} = \tilde{\varphi}\tilde{\psi}$ means that we have commuting diagrams as in Fig. 3.7 and then complete the argument as in the unital case. □

Exercise 3.10 Let $A = \lim_{\to}(A_n, \phi_n)$ be an inductive limit of C^*-algebras. If J is an ideal in A, let $J_n = \{x \in A_n \mid \phi_{n\infty}(x) \in J\}$. Show that J_n is an ideal in A_n and $J = \overline{\cup_n \phi_{n\infty}(J_n)} \cong \lim_{\to}(J_n, \phi_n|_{J_n})$. In particular an inductive limit of simple C^*-algebras is simple (cf. Lemma 2.16).

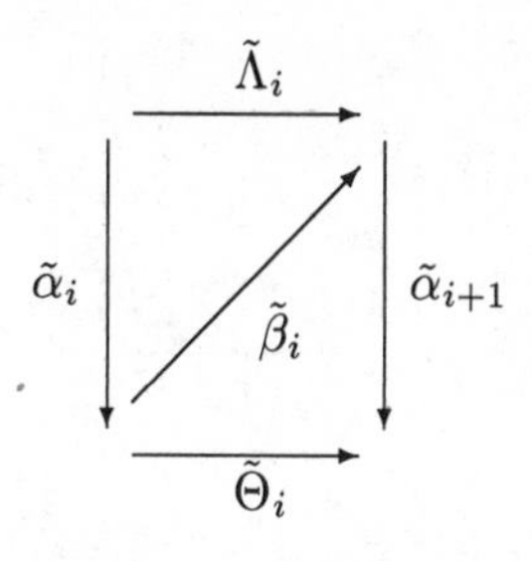

FIG. 3.7. Intertwining Bratteli diagrams for unital extensions

Exercise 3.11 Let A be an AF algebra with Bratteli diagram D. Discard, as we may, all vertices of D and edges incident to them from the second level onwards, which have no descendants. Let Λ be a subgraph of D such that
(a) if a vertex v lies in Λ, and v has a descendant in D at the next level, then the descendants and all edges from v to the descendants lie in D.
(b) if a vertex v lies in D, such that every ancestor of v lies in Λ then v and the edges from the ancestors to v lie in Λ.
Show that Λ is the Bratteli diagram of an ideal J of A, and that every ideal of A arises in this way. Show that A/J is AF with Bratteli diagram $D\backslash\Lambda$ obtained by removing all the vertices of Λ and all edges containing at least one vertex of Λ. Deduce that A is simple if and only if for each vertex v there is some lower level where each vertex is a descendant of v.

Exercise 3.12 Let A be a separable C^*-algebra. Show that A is AF if and only if given $\varepsilon > 0$, $f_1, \ldots, f_r$ in A, there exists a finite dimensional C^*-subalgebra B of A and $g_1, \ldots, g_r$ in B such that $\|f_i - g_i\| < \varepsilon$, $i = 1, \ldots, r$. If moreover, B_0 is a given finite dimensional C^*-subalgebra of A, B can be chosen to contain B_0.

3.5 Intertwiners and connections

In the previous section we came across filtered homomorphisms of path algebras via identifications of paths. These are the simplest examples of intertwiners or connections between path algebras. Before we look at a description of the most general filtered homomorphism between path algebras, let us look at an illuminating example – the fixed point algebras of UHF algebras under (product) compact group actions.

Example 3.10 Let $\pi_j : G \to M_{n(j)}$, be a unitary representation of a compact group G, and $\alpha_g = \bigotimes_{j=1}^{\infty} \operatorname{Ad} \pi_j$, the product type action of G on the UHF algebra

$$A = \bigotimes_{j=1}^{\infty} M_{n(j)} = \varinjlim A_m \tag{3.5.1}$$

where $A_m = \bigotimes_{j=1}^{m} M_{n(j)}$. The Bratteli diagram of this tower of algebras is given by singletons $\Omega[m]$ and graphs $\mu^{(m)}$ with $n(m)$ edges connecting $\Omega[m]$ and

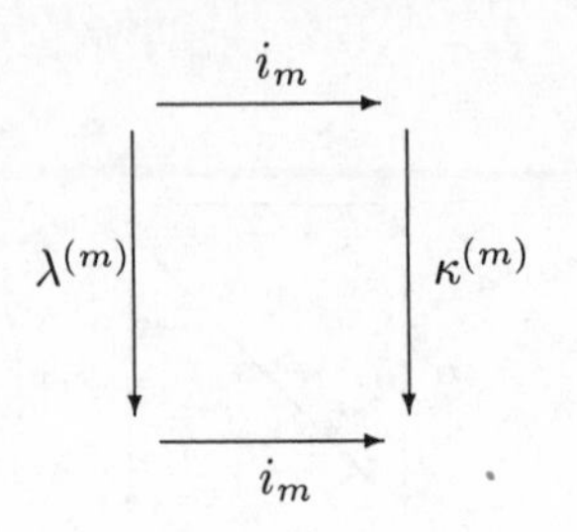

FIG. 3.8. Bratteli diagram for $\left(\otimes_{j=1}^{\infty} M_{n(j)}\right)^G \subset \otimes_{j=1}^{\infty} M_{n(j)}$

$\Omega[m+1]$. The fixed point algebra A^G is AF, being the C^*-inductive limit of the fixed point algebras A_m^G. Let $\{\chi_\alpha\}$ denote the irreducible characters of G, where $\chi_0 = \chi^{(0)}$ is the trivial character, and $\chi^{(j)}$ the character of π_j. For each $m = 0, 1, 2 \ldots$, let

$$\chi^{(0)}\chi^{(1)} \cdots \chi^{(m)} = \textstyle\sum_\alpha a_{m\alpha}\chi_\alpha \tag{3.5.2}$$

be the decomposition of the character $\chi^{(0)} \cdots \chi^{(m)}$ corresponding to the representation $\pi^{(m)} = \bigoplus_{j=0}^{m} \pi_{n(j)}$ into irreducible characters χ_α where $a_{m\alpha}$ are positive integers. By Schur's lemma, the fixed point algebra A_m^G has the following decomposition into simple components:

$$A_m^G \cong \oplus_\alpha M_{a_{m\alpha}} . \tag{3.5.3}$$

The multiplicity $\kappa_{\alpha\beta}^{(m)}$ of the embedding of the simple components $M_{a_{m\alpha}}$ into $M_{a_{m+1,\beta}}$ is determined by the decomposition of $\chi^{(m+1)}\chi_\alpha$ into irreducible characters

$$\chi^{(m+1)}\chi_\alpha = \textstyle\sum_\beta \kappa_{\alpha\beta}^{(m)} \chi_\beta . \tag{3.5.4}$$

Thus $A^G \cong A(\kappa)$, but to understand the embedding $\varphi : A(\kappa) \to A(\mu)$ we need the Clebsch–Gordan coefficients which expresses the equality (3.5.4) at the representation level. Let $\Omega'[m]$ denote the vertices at level m of the Bratteli diagram of $A(\kappa)$, i.e., the irreducible representations of G which figure in (3.5.3). Then define i_m to be the graph with $\dim \pi_\alpha$ edges connecting α in $\Omega'[m]$ to the singleton $\Omega[m]$. Then the *connection* or intertwiner describing the embedding of $A(\kappa)$ in $A(\mu)$ is the Clebsch–Gordan coefficient U in $(i_{m+1}\kappa^{(m)}, \lambda^{(m)} i_m)$:

$$\pi^{(m+1)} \otimes \pi_\alpha \cong U\left(\textstyle\bigoplus_\beta \pi_\beta \otimes 1_{\kappa_{\alpha\beta}^{(m)}}\right) U^* \tag{3.5.5}$$

i.e. we identify the path algebras along $\llcorner\!\!\rightarrow$ and $\urcorner\!\!\downarrow$ in Fig. 3.8 via the unitary U, which is not necessarily a diagonal identification as in Fig. 3.5. All filtered embeddings of path algebras can be described by such unitary intertwiners.

First, we consider intertwiners between path algebras of sequences of graphs which have the same initial and same terminal vertices, in a similar fashion to how we constructed the path algebras themselves in the first place. If $\lambda_1, \lambda_2, \ldots$ and $\mu_1, \mu_2, \ldots$ are two sequences of graphs such that

$$t(\lambda_n) = i(\lambda_{n+1}), \quad t(\mu_n) = i(\mu_{n+1}) \tag{3.5.6}$$
$$i(\lambda_n) = i(\mu_n), \quad t(\lambda_n) = t(\mu_n) \tag{3.5.7}$$

we can define (λ_r, μ_r) in

$$\bigoplus_{i,j} \mathrm{End}\left(l^2\left(\mathrm{Path}_{\lambda_r}(i,j)\right), l^2\left(\mathrm{Path}_{\mu_r}(i,j)\right)\right)$$

to be the intertwiners between $(\lambda_r, \lambda_r) \equiv A(\lambda)[r, r+1]$ and $A(\mu)[r, r+1]$ generated by matrix units (μ, ν), with $\mu \in \mathrm{Path}_{\lambda_r}(i,j)$, $\nu \in \mathrm{Path}_{\mu_r}(i,j)$ with same initial and same terminal vertices i and j respectively. In particular, concatenating graphs, we have intertwiners

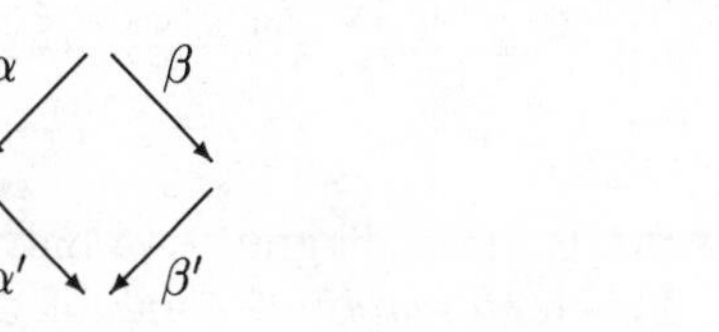

(3.5.8)

where $\alpha, \alpha', \beta, \beta'$ are edges of $\lambda_r, \lambda_{r+1}, \mu_r, \mu_{r+1}$ respectively (where of course we do not really need (3.5.7) for $n = r$).

Consider a filtered $*$-homomorphism φ between path algebras $A(\lambda)$ and $A(\mu)$ filtered in the sense that $A(\lambda)_n$ is taken to $A(\mu)_n$ with this restriction denoted by φ_n. Now the $*$-homomorphism φ_n is up to an inner automorphism of $A(\mu)_n$ a path endomorphism of the type (2.9.11). More precisely, there is a graph i_n between the nth levels of $A(\lambda)$ and $A(\mu)$, and a unitary intertwiner $u_n \in \left(i_n \lambda^{(n)} \cdots \lambda^{(0)}, \mu^{(n)} \cdots \mu^{(0)}\right)$ such that

$$\varphi_n = \mathrm{Ad}(u_n) i_n . \tag{3.5.9}$$

Then since $\mu^{(n)} \varphi_n = \varphi_{n+1} \lambda^{(n)}$, $w_{n+1} = i_n(u_n^*) u_{n+1}$ is an intertwiner in

$$\left(i_{n+1} \lambda^{(n+1)} \cdots \lambda^{(0)}, \mu^{(n+1)} i_n \lambda^{(n)} \cdots \lambda^{(0)}\right)$$

which commutes with

$$\left(\lambda^{(n)} \cdots \lambda^{(0)}, \lambda^{(n)} \cdots \lambda^{(0)}\right) = A(\lambda)_n .$$

Hence by the extension of (3.4.7) to intertwiners, $w_{n+1} \in \left(i_{n+1} \lambda^{(n+1)}, \mu^{(n+1)} i_n\right)$ and u_n of Fig. 3.9 implements φ_n as in (3.5.9).

Consider the special case of Example 3.10 where $\pi_j \equiv \pi$. Then if χ is the character of π, (3.5.2) and (3.5.4) reduce to

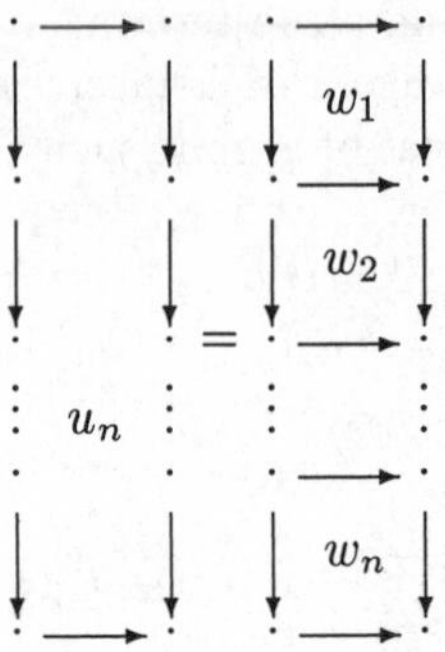

FIG. 3.9. Intertwiner u_n

$$\chi^m = \sum_\alpha a_{m\alpha}\chi_\alpha\,, \tag{3.5.10}$$

$$\chi\chi_\alpha = \sum_\beta k_{\alpha\beta}\chi_\beta\,. \tag{3.5.11}$$

To determine the Bratteli diagram, we only need to know (3.5.10) for $m = 1$ and (3.5.11). The *representation graph* of G corresponding to π is the graph $\Gamma_\pi = \Gamma_\pi(G)$ with vertices $\{\chi_\alpha\}$ and $k_{\alpha\beta}$ edges between χ_α and χ_β. Then Γ_π is connected if and only if π is faithful. It is clear that $A(\kappa) = A(\Gamma_\pi(G))$. Thus $A^G \cong A(\Gamma_\pi)$, where the graph Γ_π has distinguished vertex labelled χ_0.

Let H be a closed subgroup of G. The embedding of A^G_m in A^H_m is determined as follows. Let $\{\sigma_\mu\}$ denote the irreducible characters of H, σ_0 the trivial representation and $\sigma = \chi|H$. Then

$$A^H_m \cong \bigoplus\nolimits_\mu M_{b_{m\mu}} \tag{3.5.12}$$

if

$$\sigma^m = \sum\nolimits_\mu b_{m\mu}\sigma_\mu\,. \tag{3.5.13}$$

The multiplicity $q_{\alpha\mu}$ of the embedding of $M_{a_{m\alpha}}$ in $M_{b_{m\mu}}$ is given by the decomposition of $\chi_\alpha|H$ into irreducible characters of H:

$$\chi_\alpha|H = \sum\nolimits_\mu q_{\alpha\mu}\sigma_\mu \tag{3.5.14}$$

with $q_{0\mu} = \delta_{o\mu}$.

Let $G = SU(2)$, $\pi : G \to M_2$ the standard representation and $H = \mathbb{T}$, so that $(\pi|\mathbb{T})(z) = \begin{pmatrix} z & 0 \\ 0 & \bar{z} \end{pmatrix}$, $z \in \mathbb{T}$. Let $\{\chi_i\}_{i=0}^\infty$ be the irreducible characters of $SU(2)$, where χ_0 is the trivial character and $\chi = \chi_1$ the character of π. Then by the Clebsch–Gordan formula we have

$$\chi_1\chi_i = \chi_{i-1} + \chi_{i+1}\,, \quad i = 0, 1, 2, \ldots \tag{3.5.15}$$

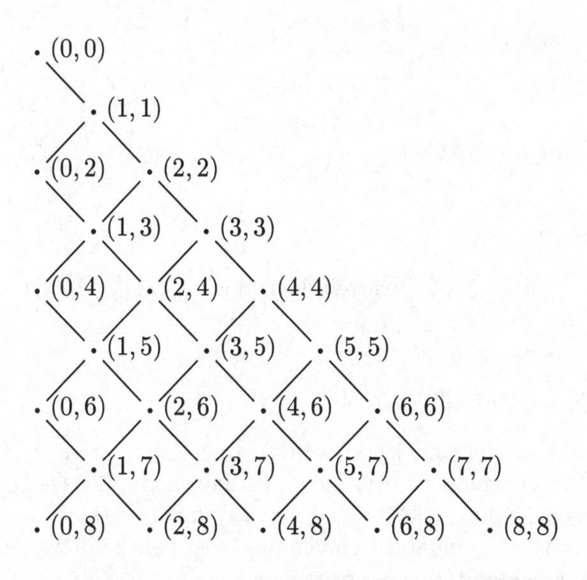

FIG. 3.10. $\hat{A}_\infty$, a Bratteli diagram for $(\otimes_{\mathbb{N}} M_2)^{SU(2)}$

if $\chi_{-1} = 0$. Thus the representation graph Γ_χ is A_∞, with distinguished vertex $* = 0$, and so $(\bigotimes_{\mathbb{N}} M_2)^{SU(2)} \cong A(A_\infty)$, with Bratteli diagram as in Fig. 3.10.

Let $\{\sigma_i\}_{i=-\infty}^{\infty}$ be the irreducible characters of $\mathbb{T}$, where $\sigma_i(z) = z^i$. Then if $\sigma = \chi_1|\mathbb{T}$ we have

$$\sigma = \sigma_1 + \sigma_{-1}$$

and

$$\sigma\sigma_i = \sigma_{i-1} + \sigma_{i+1}\,, \quad i \in \mathbb{Z}\,. \tag{3.5.16}$$

Thus Γ_σ is identified with $A_{\infty,\infty}$ as in Fig. 2.39 with distinguished vertex $* = 0$, and so $(\bigotimes_{\mathbb{N}} M_2)^{\mathbb{T}} \cong A(A_{\infty,\infty})$ with Bratteli diagram as in Fig 2.40.

The continuous functions on the space of paths of a Bratteli diagram embed in the AF algebra of the diagram. Thus we have $C(\prod \mathbb{Z}/2) \to \otimes M_2$, and $C(\Omega) \to (\otimes M_2)^{\mathbb{T}}$ where Ω is the path space of Pascal's triangle. Note that $C(\prod \mathbb{Z}/2) = \otimes \mathbb{C}^2 \subset \otimes M_2$ is gauge invariant and indeed $C(\prod \mathbb{Z}/2) = C(\Omega)$. This means the path spaces of the two Cantor sets $\prod \mathbb{Z}/2$ and Ω should be identified. This we do in Fig. 3.11. First label edges / , \ in $\hat{A}_{\infty,\infty}$ by a and b, and those () in $\prod \mathbb{Z}/2$ by a and b. Then a path in either Cantor set is an infinite word as in Fig. 3.11 (iii), which is a Bratteli diagram for $C(\Omega) \cong C(\prod \mathbb{Z}/2)$. This identifies with the Cantor construction $\vdash\!\overset{a}{\text{—}}\!+\!\text{—}\!+\!\overset{b}{\text{—}}\!\dashv$ of ignoring the middle third and selecting the left a or right b interval at each stage.

3.6 Traces on AF algebras

A trace on a $*$-algebra A is a linear map τ on A such that

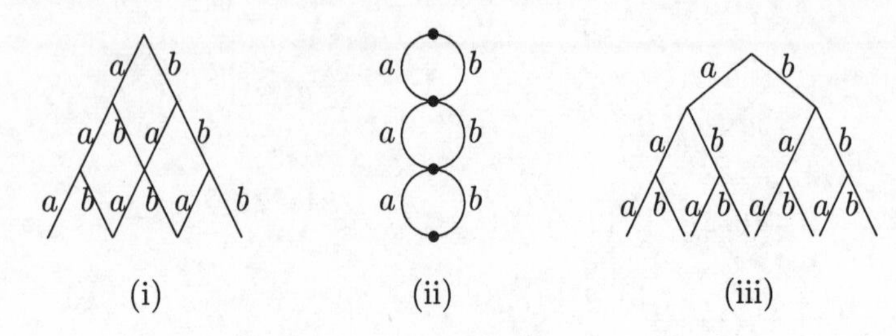

FIG. 3.11. Bratteli diagrams for $C\left(\prod_{\mathbb{N}}(\mathbb{Z}/2)\right)$

(3.6.1) τ is positive, i.e. $\tau(a^*a) \geq 0$.

(3.6.2) $\tau(ab) = \tau(ba)$, all $a, b \in A$.

If e, f are equivalent projections in A, then from $e = v^*v$, $f = vv^*$ we see $\tau(e) = \tau(f)$ for any trace τ on A, and so we have a map $\overline{\tau} : D(A) \to [0, \infty)$. Now if τ is a trace on A, then $\tau_n[a_{ij}] = \sum_{i=1}^n \tau(a_{ii})$ defines a trace τ_n on $M_n(A)$, such that $\tau_{n+1}|M_n(A) = \tau_n$ under the embedding $a \to a \oplus 0_1$ and, $M_n(A) \to M_{n+1}(A)$. Thus there is a commutative diagram

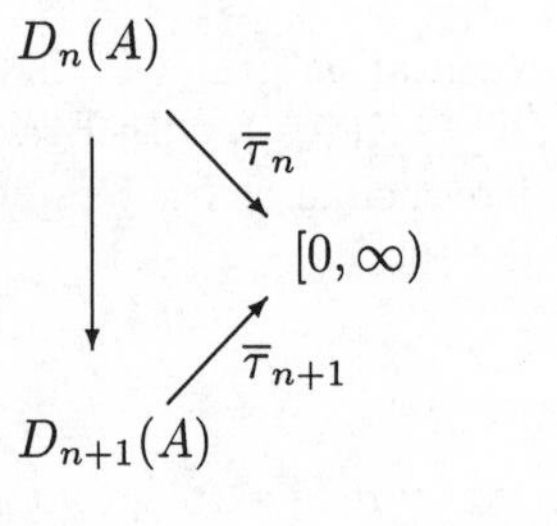

(3.6.3)

and so we have an induced map $\tau_* : D_\infty(A) \to [0, \infty)$. Since $\tau_{n+m}(e \oplus f) = \tau_n(e) + \tau_m(f)$, $e \in P_n(A)$, $f \in P_m(A)$ we see that τ_* is a homomorphism from $D_\infty(A) \to [0, \infty)$, and so extends to a homomorphism also denoted by τ_* from $K_0(A)$ into $\mathbb{R}(= K_0([0, \infty))$, which is positive in that $\tau_*(K_0(A)_+) \subset [0, \infty)$. A *state* on $K_0(A)$ is a homomorphism ψ from $K_0(A)$ into $\mathbb{R}$, which is positive, $\psi(K_0(A_+)) \subset [0, \infty)$. It is said to be normalized if A is unital and $\psi[1] = 1$. We will show that on a unital AF C^*-algebra every normalized state on $K_0(A)$ arises from an unique normalized trace on A. As an application of this, we will deduce a solution to the Hausdorff moment problem.

Proposition 3.11 *If A is a unital AF C^*-algebra, and ψ is a state on $K_0(A)$, there exists an unique trace τ on A such that $\psi = \tau_*$.*

Proof Case (a) Suppose $A = M(\mathbf{n})$, $\mathbf{n} = (n_1, \ldots, n_p)$. A homomorphism $\psi : K_0(A) = \mathbb{Z}^p \to \mathbb{R}$ is given by a sequence $(\lambda_i)_{i=1}^p \in \mathbb{R}^p$, such that $\psi(n_i) = \sum_{i=1}^p \lambda_i n_i$. If ψ is positive on $K_0(A)_+ = \mathbb{N}^p$, then $\lambda_i \geq 0$. Define $\tau\left(\bigoplus_{i=1}^p x_i\right) =$

$\sum_{i=1}^{p} \lambda_i \mathrm{tr}(x_i)$, where $x_i \in M(n_i)$ and tr is the un-normalized trace on $M(m)$, i.e. $\mathrm{tr}[a_{ij}] = \sum_{i=1}^{m} a_{ii}$, $[a_{ij}] \in M(m)$. Then τ is a trace on $M(\mathbf{n})$, and if e_i is a rank one projection in $M(n_i)$, then $\psi[e_i] = \lambda_i = \tau_*[e_i]$. From this $\tau_* = \psi$, and uniqueness of τ follows.
Case (b) Let $\varphi_n : A_n \to A_{n+1}$ be unital $*$-homomorphisms between unital multi-matrix algebras, and $A = \lim_{\to} A_n$, the C^*-direct limit. So we have a commutative diagram 3.6.4:

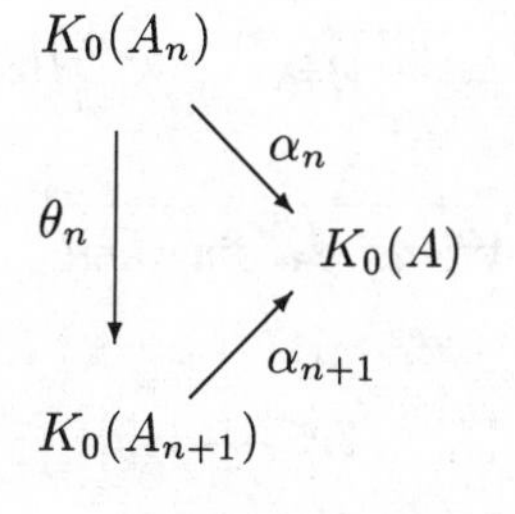

(3.6.4)

where $\theta_n = K_0(\varphi_n)$, $\alpha_n = K_0(\varphi_{n\infty})$. Then $\psi_n \equiv \psi \circ \alpha_n$ is a state on $K_0(A_n)$, and so by case (a) there exists an unique trace τ_n on A_n such that $(\tau_n)_* = \psi \circ \alpha_n$. Then $(\tau_n)_* = \psi \circ \alpha_n = \psi \alpha_{n+1} \theta_n = (\tau_{n+1})_* (\varphi_n)_*$. This says that $(\tau_n)_* = (\tau_{n+1} \varphi_n)_*$ and so by uniqueness in case (a) $\tau_n = \tau_{n+1} \circ \varphi_n$. Since ψ is normalized, $\tau_n(1) = 1$, and so taking direct limits (see Section 2.6) $\tau = \lim_{\to} \tau_n$ is a trace on A. From $\tau_* \alpha_n[e] = \tau(\varphi_{n\infty}(e)) = \tau_n(e) = (\tau_n)_*[e] = \psi \circ \alpha_n[e]$, we see that $\tau_* = \psi$. Uniqueness follows from uniqueness in case (a). □

As an example consider the AF algebra $A(A_{\infty,\infty}) \cong (\bigotimes_{\mathbb{N}} M_2)^{\mathbb{T}}$, the *current algebra*, associated with Pascal's triangle of Fig 2.24, i.e. the inductive limit F of $\mathbb{C} \to \mathbb{C} \oplus \mathbb{C} \to \mathbb{C} \oplus M_2 \oplus \mathbb{C} \to \cdots$.

Theorem 3.12 *$K_0(F) \cong \mathbb{Z}[t]$ and $K_0(F)_+$ corresponds to*

$$\{0\} \cup \{P \in \mathbb{Z}[t] \mid P(t) > 0, \text{ for all } t \in (0,1)\}.$$

Proof Here $\mathbb{Z}[t]$ denotes the ring of all finite polynomials in an indeterminate t with integer coefficients and usual operations (1.2.27–1.2.29). We first note the following:

(a) If ${}^nC_m = 0$ if $m < 0$, then

$$t^p = \sum_{k=0}^{n} {}^{n-p}C_{k-p} t^k (1-t)^{n-k},$$
$$(1-t)^p = \sum_{k=0}^{n} {}^{n-p}C_k t^k (1-t)^{n-k}.$$

(b) If $e_k^n(t) = t^k(1-t)^{n-k}$, $n = 0, 1, 2, \ldots$, $k = 0, \ldots, n$, then for a fixed n, $\{e_k^n\}_{k=0}^{n}$ are linearly independent.

(c) $\mathrm{lin}\{e_k^n \mid n, k\} = \mathbb{Z}[t]$.

(d) $e_k^n = e_k^{n+1} + e_{k+1}^{n+1}$.

If F_n is the multi-matrix algebra associated with the nth row of Fig. 2.24, so that $F_1 = \mathbb{C}$, $F_2 = \mathbb{C} \oplus \mathbb{C}$, etc., and $\alpha_n : F_n \to F_{n+1}$ is the canonical homomorphism associated with (3.4.5), let $\psi_n = K_0(\alpha_n) : \mathbb{Z}^n \to \mathbb{Z}^{n+1}$. Define $\theta_n : \mathbb{Z}^{n+1} \to \mathbb{Z}[t]$ by

$$\theta_n(m_i)_{i=0}^n = \textstyle\sum_{k=0}^n m_k e_k^n \,. \tag{3.6.5}$$

In terms of the path algebra description of Figs 2.40–2.43, we are mapping

$$[e_{\gamma,\gamma}] = \left[\textstyle\prod_{i=1}^r g_i \prod_{j=1}^s (1 - g_{r+j})\right] \to t^r(1-t)^s$$

and in terms of the word description of paths $a \to t$, $b \to 1-t$ (cf. Exercise 3.14). Then θ_n is injective by (b). We have a diagram:

$$\begin{array}{ccc} \mathbb{Z}^n & \xrightarrow{\theta_{n-1}} & \mathbb{Z}[t] \\ {\scriptstyle\psi_n}\downarrow & & \downarrow \\ \mathbb{Z}^{n+1} & \xrightarrow{\theta_n} & \mathbb{Z}[t] \end{array} \tag{3.6.6}$$

This diagram is commutative because

$$\begin{array}{ccc} \underset{i\text{th place}}{(0,\ldots,m_i,\ldots,0)} & \longrightarrow & m_i e_i^{n-1} \\ \downarrow & & \downarrow \\ \underset{i-(i+1)\text{ places}}{(0,\ldots,m_i,m_i,0,\ldots,0)} & \longrightarrow & m_i e_i^n + m_i e_{i+1}^n \end{array} \tag{3.6.7}$$

where we use (d). We thus have an injective map $\theta_\infty : \lim_\to(\mathbb{Z}^n, \psi_n) \to \mathbb{Z}[t]$, which is surjective by (c). Hence by Theorem 3.6 $K_0(F) = \mathbb{Z}[t]$. Also by Theorem 3.6, to identify $K_0(F)_+$, we need to compute $\lim_\to(\mathbb{N}^n, \psi_n) \subset K_0(F)$. Thus, using (3.6.5), we need to know, when given $P \in \mathbb{Z}[t]$, can we write

$$P = \textstyle\sum_{k=0}^n b_k e_k^n \,, \text{ where } b_k \in \mathbb{N}, \quad 0 \le k \le n \,, \tag{3.6.8}$$

for some n? It is clear that any P of the form (3.6.8) is either 0 or satisfies $P(t) > 0$ for $t \in (0,1)$. Conversely, suppose $P \in \mathbb{Z}[t]$ is such that $P(t) > 0$ for $t \in (0,1)$. We say $P \in \mathbb{Z}[t]$ is strictly positive, written $P > 0$, if $P \neq 0$, and P is of the form (3.6.8). We now make the following claims:

(e) If $P, Q > 0$ then $PQ > 0$. If

$$P = \textstyle\sum_{k=0}^{m} \lambda_k t^k (1-t)^{m-k}\,, \quad Q = \sum_{k=0}^{n} \mu_k t^k (1-t)^{n-k}$$

then $PQ = \sum_{j=0}^{m+n} \left(\sum_{k+l=j} \lambda_k \mu_l\right) t^j (1-t)^{m+n-j}$. If $\lambda_k, \mu_l \geq 0$, then $\sum \lambda_k \mu_l \geq 0$, and so $PQ > 0$.

(f) Let $p \in \mathbb{Z}[t]$, such that $P(t) > 0$ for $t \in [0,1]$. Suppose $P = \sum_{p=0}^{m} a_p t^p$, $a_p \in \mathbb{Z}$, and take

$$P_n = \textstyle\sum_{p=0}^{m} a_p t(t-1/n)\cdots(t-(p-1)/n)/(1-1/n)\cdots(1-(p-1)/n)\,.$$

Then $P_n \to P$ uniformly on $[0,1]$ so that $P_n(t) > 0$ on $[0,1]$ for large n. Now for $n \geq m$:

$$P = \textstyle\sum_{p=0}^{m} a_p t^p = \sum_{p=0}^{m} a_p \sum_{k=0}^{n} {}^{n-p}C_{k-p} t^k (1-t)^{n-k} \text{ (by (a))} = \sum_{k=0}^{n} b_k e_k^n$$

where $b_k = \sum_{p=0}^{m} a_p \, {}^{n-p}C_{k-p} \in \mathbb{Z}$, $0 \leq k \leq n$. Now $b_k = {}^nC_k P_n(k/n) > 0$ for n large. Hence $P > 0$.

(g) Take $P \in \mathbb{Z}[t]$ such that $P(t) > 0$ for $t \in (0,1)$. Then $P = t^m(1-t)^n Q$ where $m, n \geq 0$, and $Q(t) > 0$ for $t \in [0,1]$. Hence $Q \geq 0$ by (f), and so $P > 0$ by (e). □

Now by Proposition 3.11, to study normalized traces on F, it is enough to look at normalized states on $K_0(F) = \mathbb{Z}[t]$. Now $\mathbb{Z}[t] = R$ is a ring with a distinguished subset R_+ (Theorem 3.12) satisfying

$$R_+ + R_+ \subset R_+\,, \quad R_+ R_+ \subset R_+\,, \quad R = R_+ - R_+\,. \tag{3.6.9}$$

The latter property is correct for the dimension group of any AF algebra. Alternatively, if $P \in R$, then by compactness, choose $N \in \mathbb{N}$ such that $P(t) < N$ for $t \in [0,1]$. Then $P = N - (N - P)$ where $N, N - P \in R_+$. For convenience let us write $P \leq Q$, $(P, Q \in R)$ if $Q - P \in R_+$, and $\psi \leq \varphi$, ψ, φ states on R, if $\varphi - \psi$ is a state.

We claim that there is a bijection between states φ on $K_0(F) = R$ and measures μ on $[0,1]$ such that

$$\varphi(P) = \int_0^1 P(\theta)\, d\mu(\theta)\,. \tag{3.6.10}$$

It is clear that any measure μ gives rise to a state φ on R, using the characterization of R_+. We give an indication of a proof of the converse. First note that if φ is a state such that $\varphi(1) = 0$ then $\varphi = 0$. For if $P \in R_+$, then by the preceding, there exists $N \in \mathbb{N}$ such that $0 \leq P \leq N$. Hence $0 \leq \varphi(P) \leq \varphi(N) = N\varphi(1) = 0$. Hence $\varphi = 0$, on $R = R_+ - R_+$. Now as in Section 2.5, we say that a state φ

on R is *extremal* if whenever ψ is another state dominated by φ, $\psi \leq \varphi$, then there exists $\lambda \geq 0$ such that $\psi = \lambda\varphi$. Then as in the proof of Proposition 2.15, a normalized state is pure amongst the normalized states, if and only if it is extremal.

Lemma 3.13 *φ is an extremal normalized state on R if and only if φ is a non-zero positive ring homomorphism from R into $\mathbb{R}$.*

Proof Let φ be an extremal normalized state on R. Take non-zero $P \in R_+$ such that $0 \leq P < N$ some $N \in \mathbb{N}$. Take $g \in R_+$, then $N - P \geq 0$, $g \geq 0$ shows $(N-P)g \geq 0$ or $0 \leq Pg \leq Ng$. Hence $0 \leq \varphi(Pg) \leq N\varphi(g)$. Let $\psi(g) = \varphi(Pg)/N$, then ψ is a state dominated by φ, and since φ is extremal we deduce $\psi = \lambda\varphi$ for some $\lambda \geq 0$, i.e. $\varphi(Pg) = (N\lambda)\varphi(g)$, for all $g \in R = R_+ - R_+$. Taking $g = 1$, we see $\varphi(P) = N\lambda$, and $\varphi(Pg) = \varphi(P)\varphi(g)$ for all $P, g \in R = R_+ - R_+$, i.e. φ is a positive ring homomorphism. The proof of the converse is omitted (see Proposition I.2 of (Handelman 1985), (Vershik and Kerov 1987), (Wassermann 1981) — its basic idea goes back to Gelfand theory). See also Lemma 9.36. □

The next question to arise is: what are the ring homomorphisms on $R = \mathbb{Z}[t]$ which are positive on R_+? Take any ring homomorphism $\varphi : \mathbb{Z}[t] \to \mathbb{R}$. Let $\varphi(t) = \theta \in \mathbb{R}$. Then if $P = \sum_{i=0}^n \lambda_i t^i$, $\varphi(P) = \varphi\left(\sum \lambda_i t^i\right) = \sum \lambda_i \varphi(t)^i = \sum \lambda_i \theta^i = P(\theta)$. If φ is positive, then since $t, 1-t \geq 0$, we see $\theta = \varphi(t)$, $1 - \theta = \varphi(1-t) \geq 0$, i.e. $\theta \in [0,1]$. Conversely, if $\theta \in [0,1]$, then $\varphi_\theta(P) = P(\theta)$ is clearly a positive ring homomorphism, using the identification R_+ of Theorem 3.12. So we have a bijection between extreme normalized traces τ_θ on F, extremal normalized states φ_θ on R and the unit interval $[0,1]$.

One can show, again the proof is omitted (see (Handelman 1985), (Vershik and Kerov 1987), (Wassermann 1981)), that if φ is any normalized state on R, then it is an average of pure ones, i.e. there exists an unique measure μ on $[0,1]$ such that

$$\varphi = \int_0^1 \varphi_\theta \, d\mu(\theta)$$

or

$$\varphi(P) = \int_0^1 \varphi_\theta(P) \, d\mu(\theta) = \int_0^1 P(\theta) \, d\mu(\theta) \,. \tag{3.6.11}$$

Now if τ is a trace on $F = \overline{\cup F_n}$, it is enough to know that the sequence $\tau_n = \tau | F_{n+1}$. Now $F_{n+1} = \bigoplus_{s=0}^n N_n(s)$ where $N_n(s) = M({}^nC_s)$. By the proof of Proposition 3.11, any trace τ_n on F_{n+1} is a positive linear combination of canonical traces on $N_n(s)$. Thus to specify τ it is enough to know the value $\alpha_\tau(n,s)$ of τ on a rank one projection in $N_n(s)$, $0 \leq s \leq n$. To ensure compatibility of τ_n under $F_{n+1} \subset F_{n+2}$ one needs

$$\alpha_\tau(n,s) = \alpha_\tau(n+1,s) + \alpha_\tau(n+1,s+1) \tag{3.6.12}$$

which arises from the embedding:

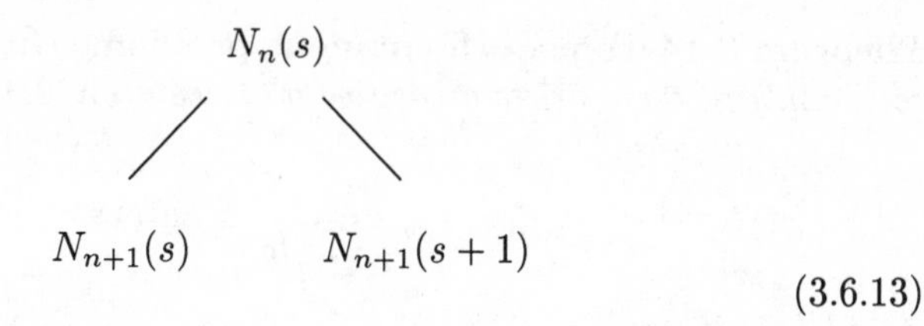

(3.6.13)

Conversely any family $\{\alpha(n,s) \mid 0 \leq s \leq n, n = 0,1,2,\ldots\}$ of positive real numbers satisfying $\alpha(0,0) = 1$ and the compatibility condition (3.6.12) gives rise to a normalized trace on F (since one can construct a compatible sequence of normalized traces on F_n). Now to specify $\{\alpha(n,s)] \mid 0 \leq s \leq n, n = 0,1,2,\ldots\}$ one only need be given $\alpha(n,n)$, since all other $\alpha(n,s)$ can be computed from (3.6.12). Thus to specify a trace is to associate a positive real number to each node in (3.4.5) such that if we take any small triangle in that diagram

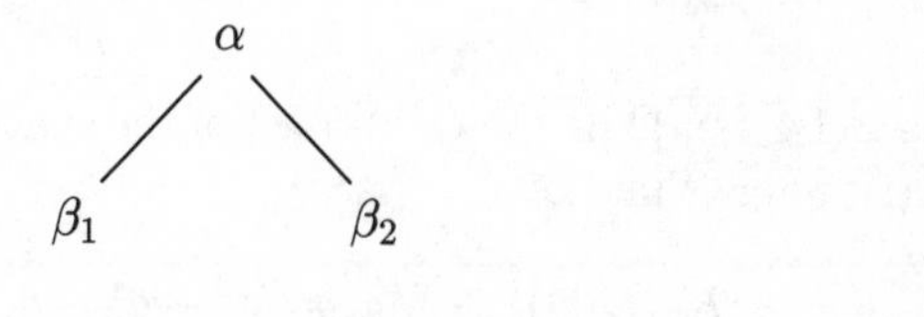

(3.6.14)

then $\alpha = \beta_1 + \beta_2$, where we have inserted positive reals α, β_1, β_2 representing traces of rank-one projections. Moreover to do this it is enough to have a sequence $\{a_i\}$ of positive real numbers along the right-hand extreme diagonal such that when we use rule (3.6.14) to complete the diagram,

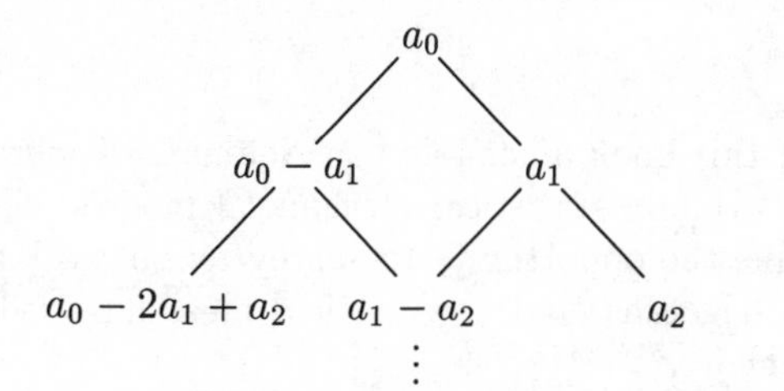

(3.6.15)

then we only get positive real numbers at each node. The numbers obtained in this way are

$$\sum_{i=0}^{k} a_{s+i}\, {}^{k}C_i(-1)^i \geq 0\,, \quad 0 \leq k \leq s\,, \quad s = 0,1,2,\ldots. \tag{3.6.16}$$

So we must have a correspondence between normalized traces on F, normalized states φ on $\mathbb{Z}[t]$, sequences $\{a_i\}$ of real numbers satisfying (3.6.16), ($a_0 = 1$), and probability measures μ on $[0,1]$. Now under these correspondences, a_n = trace of a rank one projection in $N_n(n) = \varphi(t^n) = \int_0^1 t^n \, d\mu(t)$, where the penultimate equality follows because under the identification of Theorem 3.12, a rank one projection, in $N_n(n)$ corresponds to t^n. From all this we deduce:

Theorem 3.14 Hausdorff moment problem. *Given a sequence $\{a_i\}_{i=0}^{\infty}$ of real numbers, there exists a positive measure μ on $[0,1]$ such that*

$$a_i = \int_0^1 t^i \, d\mu(t)$$

if and only if

$$\sum_{i=0}^{k} a_{s+i} \, {}^kC_i (-1)^i \geq 0 \,, \quad 0 \leq k \leq s \,, \quad s = 0, 1, 2, \ldots .$$

Exercise 3.13 Let X be a compact Hausdorff space. Show that $C(X)$ is AF if and only if X is *totally disconnected* (i.e. X has a basis of closed and open sets). In which case show that $K_0(C(X)) \cong C(X, \mathbb{Z})$ the continuous integer valued functions on X, with K_0^+ identified with $f \in C(X, \mathbb{Z})$ such that $f(x) \geq 0$, for all $x \in X$.

Exercise 3.14 Let Ω be the Cantor set of the space of paths in Pascal's triangle, Fig. 3.11(i). Show that

$$K_0(C(\Omega)) \cong \mathbb{Z}[a, b]/\langle a^2 + ab - a, b^2 + ba - b \rangle$$

where a, b are two non-commuting indeterminates and $\langle \; \rangle$ denotes the ideal generated by the two polynomials $a^2 + ab - a$ and $b^2 + ba - b$. What is the ordering on $\mathbb{Z}[a, b]/\langle a^2 + ab - a, b^2 + ba - b \rangle$? Show that the natural inclusion $C(\Omega) \to A(A_{-\infty,\infty})$ at the K_0 level takes $\mathbb{Z}[a, b]/\langle a^2 + ab - a, b^2 + ba - b \rangle$ into $\mathbb{Z}[t]$ by $a \to t$, $b \to 1 - t$.

3.7 Notes

In the latter part of this book amenable von Neumann algebras, their orbifold automorphisms and orbifold statistical mechanical models will play key roles. In these notes we take the opportunity to survey recent work on the classification of amenable C^*-algebras, and how orbifold ideas have led to striking new phenomena in this theory.

K-theory for commutative C^*-algebras was introduced via vector bundles. The isomorphism classes of vector bundles over a compact topological space X form a semigroup under addition, from which one can form the enveloping group $K^0(X)$. This can be identified with $K_0(C(X))$, since the space of continuous sections $\Gamma(E)$ of a bundle E is a finitely generated projective module over $C(X)$, and every such module arises in this way. Moreover (Exercise 3.9) if A is a unital C^*-algebra, every finitely generated projective module over A is isomorphic to $M_n(A)e$ for a projection e in a matrix algebra over A. We refer to (Blackadar 1986), (Douglas 1980), (Fillmore 1996), (Rosenberg 1994), for historical comments on the role of K-theory in topology.

In operator algebras, the type I, II, III classification of von Neumann algebras is essentially K-theoretic. If M is a von Neumann algebra (i.e. a C^*-algebra on a Hilbert space H closed under taking strong limits) and is a factor (i.e. trivial centre) then the Murray–von Neumann equivalence classes of projections (in the algebra itself without considering matrices over the algebra) is order isomorphic to one of the following possibilities only, labelled types I, II, III respectively:

$$\begin{array}{lll} \{0,1,2,\ldots,n\} & n \leq \infty & \mathrm{I}_n \\ [0,1] & & \mathrm{II}_1 \\ [0,\infty] & & \mathrm{II}_\infty \\ \{0,\infty\} & & \mathrm{III} \end{array} \tag{3.7.1}$$

On taking Grothendieck completions, we get the K-groups $\mathbb{Z}, \mathbb{R}$ or $\{0\}$ for types I, II, III factors respectively.

(Glimm 1960) introduced the class of UHF algebras and classified them by the generalized integers. (Dixmier 1967) introduced the matroid C^*-algebras, inductive limits of (not necessary unital) embeddings of matrix algebras and classified them. Since a rank one dimension group is of the form $\mathbb{Z}[1/p_i]$, a matroid C^*-algebra is stably isomorphic to a UHF algebra $\bigotimes_{\mathbb{N}} M_{p_i}$. (Bratteli 1972) introduced the class of AF algebras and classified them by their Bratteli diagrams. Two diagrams give rise to isomorphic algebras if and only if we can go from one diagram to the other by the process of telescoping (or the reverse process) in a finite number of stages.

Although K-theory had already been formulated for Banach algebras previously (see e.g. (Taylor 1975)), it was not until (Elliott 1976) that a K-theoretic classification of AF algebras was given in terms of K_0 (usually referred to as a dimension group in the AF context), i.e. K_0 as a scaled ordered group is a complete invariant. This invariant (and the theory of AF algebras) played an important role in the later development of operator algebras. The characterizsation of which ordered abelian groups could arise as the dimension group of an AF algebra was completed by (Effros *et al.* 1981). Indeed an abelian group can arise as the dimension group of an AF algebra (unique by Elliott's classification up to stable $*$-isomorphism) if and only if it is countable, with subsemigroup G_+, such that G is *partially ordered* (i.e. $G_+ \cap -G_+ = \{0\}$), *unperforated* (i.e. if $g \in G$, $ng \in G_+$ for some $n \in \mathbb{N}$, then $g \in G_+$) and satisfies the *Riesz interpolation property* (i.e. if $x_i, y_j \in G$, $x_i \leq y_j$ for $i,j = 1,2$, then there exists $z \in G$ such that $x_i \leq z \leq y_j$ for all i,j). Amongst its many striking applications is the realization of any closed subset of the real numbers as the temperature state space of a C^*-dynamical system (Bratteli *et al.* 1980).

Our treatment of K-theory or dimension groups of AF-algebras is based on that of (Effros 1981). Standard expositions for the K-theory of C^*-algebras are (Taylor 1975), (Blackadar 1986) and (Wegge-Olsen 1993).

The description of filtered homomorphisms between path algebras is due to (Ocneanu 1988); the exposition here is taken from (Evans and Kawahigashi 1994a). This notion of connection will be a key feature in constructing subfactors in Chapters 10 onwards (when we embed one AF algebra in another and take the weak closures) and in understanding the paragroup invariant of a subfactor

$$\begin{array}{ccccc} K_0(J) & \to & K_0(A) & \to & K_0(A/J) \\ \uparrow & & & & \downarrow \\ K_1(A/J) & \leftarrow & K_1(A) & \leftarrow & K_1(J) \end{array}$$

FIG. 3.12. Six-term exact sequence of K-theory

$N \subset M$. This involves extending the inclusion in a canonical way $N \subset M \subset M_1 \subset M_2, \ldots$ and computing (in the finite index case) an inclusion of $M' \cap M_i$ in $N' \cap M_i$ of finite dimensional algebras, or a filtered inclusion of $\bigcup_i (M' \cap M_i)$ in $\bigcup_i (N' \cap M_i)$.

The characterization of the dimension group of the GICAR algebra is due to (Renault 1980), although the characterization of polynomials in $\mathbb{Z}(t)$ which are strictly positive in the interval $(0, 1)$ is classical (see e.g. (Karlin and Studden 1966) p.126). Replacing $[0, 1]$ by closed subsets $F \subset [0, 1]$ and ordering $\mathbb{Z}[t]$ by $\{0\} \cup \{P \in \mathbb{Z}[t] \mid P(t) > 0$, for all $t \in (0, 1)$ in a neighbourhood of $F\}$ also yields dimension groups – such groups were used by (Bratteli *et al.* 1980) in their work on equilibrium states. We will find concrete constructions of some AF algebras with such dimension groups with $F = [0, \lambda]$, and some $\lambda < 1$ in Section 9.7. The proof given here of the solution to the Hausdorff moment problem is due to (Evans 1982a).

We have defined K_0 in terms of projections (or idempotents in a Banach algebra). Similarly we can define K_1 in terms of unitaries (or invertible elements in a Banach algebra). More precisely, if A is unital we embed the unitary group $U_n(A)$ of $M_n(A)$ in U_{n+1} by $u \to u \oplus 1$, and write $U_\infty(A) = \lim_\to U_n(A)$, and $U_\infty(A)_0$ for the connected component of the unit, then $K_1(A) = U_\infty(A)/U_\infty(A)_0 = \pi_0(U_\infty(A))$. Looking at the exact sequence of Exercise 1.18 and since $K_1(\mathbb{C}) = 0$, we would now be led to put $K_1(A) \cong K_1(\tilde{A})$ for non-unital algebras. One can define higher K-groups by $K_i(A) = \pi_{i-1}(U_\infty(A))$, $i \geq 1$. However (in our complex situation) there is a periodicity of Bott: $K_i(A) \cong K_{i+2}(A)$ which essentially leaves us with K_0 and K_1. In fact $K_i(A) \cong K_{1-i}(SA)$, $i = 0, 1$, if $SA = C_0(\mathbb{R}) \otimes A$ is the *suspension* of A. We also have the continuity $K_1(\lim_\to A_n) \cong \lim_\to K_1(A_n)$ so that $K_1(A) \cong K_1(A \otimes K) \cong \pi_0(U_1(A \otimes K))$. As the unitary group of a finite dimensional C^*-algebra is clearly connected, $K_1(A) = 0$ for any AF algebra A. Surprisingly, as we shall consider briefly below, this is often the only obstruction towards an operator algebra being AF.

As usual in algebraic topology, computation of K-groups is facilitated by exact sequences. In the first place one has the six-term exact sequence of Fig. 3.12 relating the K-theory of an ideal J in a C^*-algebra A, with that of A and that of the quotient A/J. Here the horizontal maps are those induced by the natural maps $J \to A \to A/J$. The vertical boundary map $\partial = K_1(A/J) \to K_0(J)$ is the *index* map, generalizing the notion of Fredholm index (kernel – cokernel) of an invertible element of the Calkin algebra in the extension $0 \to K(H) \to B(H) \to$

$B(H)/K(H) \to 0$. The other boundary map $\partial : K_0(A/J) \to K_1(J)$ is closely related to Bott periodicity. If e is a projection in A/J, lift to an element x in A. Then $t \to e^{2\pi i t x} + J$ is a loop in A/J, hence an element of $K_1(C(\mathbb{T}) \otimes A/J) = K_1(S(A/J))$ which by the index map can be taken to $K_0(SJ) \cong K_1(J)$. Note that for the very special exact sequence $0 \to A \to \tilde{A} \to \mathbb{C} \to 0$, the six term exact sequence breaks up to $K_1(A) \cong K_1(\tilde{A})$ and the short exact sequence $0 \to K_0(A) \to K_0(\tilde{A}) \to K_0(\mathbb{C}) \to 0$ of Exercise 3.7.

(Cuntz 1981a) used the Toeplitz extension $\mathcal{T}_n$ of a Cuntz algebra $\mathcal{O}_n$ of 2.8.6 to compute the K-theory of $\mathcal{O}_n$:

$$K_0(\mathcal{O}_n) \cong \mathbb{Z}/(n-1)\,, \quad K_1(\mathcal{O}_n) = \{0\}\,. \tag{3.7.2}$$

Here K_0 is generated by the class of the unit (or of $s_i s_i^*$; note that $\sum_{i=1}^n s_i s_i^* = 1$ implies $n[1] = [1]$). This means the algebras $\mathcal{O}_n$ are all non-isomorphic – this had already been established by (Paschke and Salinas 1979), (Pimsner and Popa 1978) using Ext groups.

The Ext theory is a homology theory dual to that of the (cohomology) K-theory. This homology theory had been described by (Atiyah 1970), for commutative C^*-algebras in terms of pairs of representations of the algebra intertwined by a Fredholm operator modulo the compacts. A concrete presentation of this theory appeared with the work of BDF (Brown *et al.* 1973), (Brown *et al.* 1977) on extensions of C^*-algebras. This led the way to the KK-theory of Kasparov ($KK(A, B)$ is a bifunctor on separable C^*-algebras, which is contravariant in the first variable, covariant in the second so that $KK(A, \mathbb{C}) \cong K^0(A)$, $KK(\mathbb{C}, B) \cong K_0(B)$ with a bilinear pairing $KK(A, B) \times KK(B, C) \to KK(A, C)$ extending the duality between homology and cohomology), and to the E-theory of (Connes and Higson 1990) defined as $E(A, B) = [[SA \otimes K, SB \otimes K]]$, where $[[-, -]]$ denotes homotopy classes of asymptotic homomorphisms and a bilinear pairing extending composition of homomorphisms, which is better suited for non-amenable C^*-algebras.

Baum and Connes have introduced geometrically defined K-groups, which specialize to discrete groups, Lie groups and foliations. Using elliptic operators, they construct a homomorphism from their geometric K-groups to the algebraically defined K-groups of the C^*-algebras of the corresponding system, and conjecture that this is an isomorphism. Progress on this conjecture would shed light on a number of problems. It would illuminate the construction of the discrete series of semi-simple Lie groups. For discrete groups, the conjecture would imply the Novikov conjecture on homotopy invariance of higher signatures (Cappell 1976) and conjectures of Gromov, Lawson and Rosenberg (Rosenberg 1983) concerning topological obstructions to the existence of a Riemannian metric of positive scalar curvature. See (Baum *et al.* 1994) for a recent review on the conjecture. Connes has provided a framework for studying a foliation F on a manifold V through non-commutative algebras, which are to be regarded as a substitute for the impractical continuous functions on the singular space of leaves. The algebras, which are constructed from the holonomy groupoid, provide, for example, a means of extending the Atiyah–Singer index theorem to differential operators elliptic along the leaves. If the foliation possesses a holonomy invariant measure, one can construct a finite von Neumann algebra and a real valued index theorem (Connes 1979), (Connes 1982a). When such a measure does not exist, one can only construct canonically a traceless C^*-algebra (a topological rather than a measure theoretic object) $C^*(V, F)$ of the foliation, but (Connes and Skandalis 1984) could still derive an index theorem for longitudinal elliptic operators, taking values in K_0 of $C^*(V, F)$.

If we fix C^*-algebras A, B consider the problem of classifying all extensions

$$0 \to A \to E \to B \to 0 \tag{3.7.3}$$

up to some notion of equivalence. In the original BDF set up, $A = K$, $B = C(X)$:

$$0 \to K \to E \to C(X) \to 0\,.$$

Extensions can be added as follows. Given extensions:

$$0 \to A \to E_i \xrightarrow{\pi_i} B \to 0$$

we define E to be the C^*-algebra of matrices $t = \begin{pmatrix} e_1 & x \\ y & e_2 \end{pmatrix}$, where $x, y \in A$, $e_i \in E_i$ such that $\pi_1(e_1) = \pi_2(e_2)$. Then define the extension

$$0 \to M_2(A) \to E \xrightarrow{\pi} B \to 0$$

by $\pi(t) = \pi_i(e_i)$. If $A \cong M_2(A)$, e.g. if $A = K$, or just *stable* $A \cong A \otimes K$, we then have an extension of A by B, and $\mathrm{Ext}(A, B)$ will be a semigroup but not necessarily a group if A, B are not nuclear. The invertible elements of $\mathrm{Ext}(A, B)$ are precisely those for which the quotient map in (3.7.3) has a completely positive lifting, and the neutral element arises from the (class of) of extensions with a lifting which is a $*$-homomorphism (i.e. the *split extensions*). As an example let us go back and examine again the extensions of K by $C(\mathbb{T})$ starting with the Toeplitz extension of Section 1.8:

$$0 \to K \to C^*(s) \to C(\mathbb{T}) \to 0\,.$$

If we embed $\ell^2(\mathbb{N})$ in $\ell^2(\mathbb{Z}) \cong L^2(\mathbb{T})$ in the natural way, define the *Toeplitz operator* $T_f = pM_f|_{\ell^2(\mathbb{N})}$ where p is the projection on $\ell^2(\mathbb{N})$ and M_f is the multiplication operator on $L^2(\mathbb{T})$, for $f \in C(\mathbb{T})$. If $f(z) = z^n$, then $T_f = s^n$ or s^{*-n} if $n \geq 0$ or $n \leq 0$ respectively so that $C^*(s) = C^*(T_f \mid f \in C(\mathbb{T}))$ by Theorem 1.30. Then $f \to T_f$ is a completely positive (see Section 4.2) lifting of the quotient map $C^*(s) \to C(\mathbb{T})$.

We will use in the study of phase transition in the two dimensional Ising model, the following fact for Toeplitz operators. The Toeplitz operator T_f is invertible if and only if f is invertible and has zero winding number. We indicate why this should be so. From the six-term exact sequence for ideals we have $K_1(C^*(s)) \to K_1(C(\mathbb{T})) \to K_0(K) = \mathbb{Z}$. Take $f \in C(\mathbb{T})$. If T_f is invertible, then its image f in $C(\mathbb{T})$ must be invertible and its K_1 class $[f]$ must map to zero in $K_0(K)$, i.e. f must have winding number zero. Conversely if f is invertible then T_f must be Fredholm. An operator in $B(H)$ is invertible in the Calkin algebra if and only if it is Fredholm. By (Coburn 1967), if $f \in C(\mathbb{T})$ is invertible then $\dim \mathrm{Ker}(\mathrm{T_f}) \cdot \dim \mathrm{Ker}(\mathrm{T_f^*}) = 0$. But if f has winding number zero, then T_f must have Fredholm index zero by homotopy invariance and so T_f and T_f^* have no kernels and T_f is invertible.

The lifting s of the generator of $C(\mathbb{T})$ is Fredholm of index -1. In a similar way, we can identify $\mathrm{Ext}\, C(\mathbb{T}) \cong \mathbb{Z}$ through the Fredholm index of a lifting. To see this consider first more general extensions

$$0 \to K \to E \to B \to 0 .$$

Since K is an ideal in E, an arbitrary pointwise lifting of $e \to b$ allows one to multiply elements of $K(H)$ ($bK \subset K, Kb \subset K$), i.e. give us elements of $B(H)$ (well defined only modulo K due to the arbitrariness of the lifting). Similarly being given an extension (3.7.3) is equivalent to being given a homomorphism from B into $M(A)/A$ where $M(A)$ is the universal *multiplier algebra* of A. The multiplier algebra $M(A)$ can be concretely identified (for any faithful representation of A on a Hilbert space) as the unital C^*-algebra of operators T such that $TA, AT \subset A$. For example $M(K(H)) = B(H)$ and $M(C_0(\Omega)) = C_b(\Omega)$, all continuous bounded functions on a local compact space Ω, so that $M(C_0(\Omega)) = C(\beta\Omega)$, where $\beta\Omega$ is the Stone–Cech compactification. In particular any extension of K by $C(\mathbb{T})$ gives a map from $C(\mathbb{T})$ into the Calkin algebra, where the Fredholm index can be computed. Without delving into the multiplier algebra but going to the six-term exact sequence, we have $K_1(C(\mathbb{T})) \xrightarrow{\partial} K_0(K) \to 0$ $(= K_1(K))$ and the boundary map ∂ is the Fredholm index.

Actually the Toeplitz extension is closely involved in Bott periodicity. If we take the standard Toeplitz extension and consider the ideal $C_0(\mathbb{R})$ in $C(\mathbb{T})$ of functions vanishing at one point and its preimage $J_0 = C^*(T_f \mid f \in C_0(\mathbb{R}))$ in J, we have the exact sequence

$$0 \to K \to J_0 \to C_0(\mathbb{R}) \to 0 . \tag{3.7.4}$$

The algebra J_0 is contractible in that $K_*(J_0) = 0$, so that $K_i(C_0(\mathbb{R})) \cong K_{i-1}(K)$. In fact by tensoring this exact sequence (3.7.4) with any C^*-algebra A not only could (Cuntz 1984) show Bott periodicity for K_* but that any functor F from C^*-algebras to abelian groups which is half exact (i.e. if J is an ideal in A, then $F(J) \to F(A) \to F(A/J)$ is exact) and homotopy invariant satisfies a periodicity: $F(S^2(A \otimes K)) \cong F(A \otimes K)$. If moreover $F(\mathbb{C}) = \mathbb{Z}, F(S\mathbb{C}) = 0$, and F is stable and continuous, then $F = K_0$, at least on a bootstrap class (Cuntz 1984). The E theory of (Connes and Higson 1990) is stable ($E(A \otimes K, B) =$

$E(A, B \otimes K) = E(A, B)$) and half exact in each variable (whilst KK is only split exact, i.e. in general for non-nuclear algebras, we need $A \to A/J$ to be split before KK is (split) exact in each variable). Thus Bott periodicity holds, indeed $E(S^2A, B) = E(SA, SB) = E(A, S^2B)$. Moreover (Connes and Higson 1990) have shown that any bivariant functor F from C^*-algebras to abelian groups which is homotopic, half exact, and stable factors through E. In particular, there is a natural map $KK(A, B) \to E(A, B)$ which is an isomorphism for nuclear algebras.

The Kasparov groups can be understood from the UCT (Universal Coefficient Theorem) of (Rosenberg and Schochet 1987):

$$0 \to \mathrm{Ext}^1_{\mathbb{Z}}(K_*(A), K_*(B)) \xrightarrow{\delta} KK_*(A, B) \xrightarrow{\gamma} \mathrm{Hom}(K_*(A), K_*(B)) \to 0$$

where δ is of degree 1, γ is of degree 0 as mappings of graded groups, and Hom means graded homomorphisms. This is true at least for the bootstrap class $\mathcal{N}$ of amenable C^*-algebras, which contains the type I C^*-algebras, and is closed under taking direct limits, ideals, quotients, extensions, crossed products by $\mathbb{R}$ or $\mathbb{Z}$. The K-theory of tensor products can be computed from the Künneth short exact sequence of (Rosenberg and Schochet 1987):

$$0 \to K_*(A) \otimes K_*(B) \to K_*(A \otimes B) \to \mathrm{Tor}_1^{\mathbb{Z}}(K_*(A), K_*(B)) \to 0$$

where $\mathrm{Tor}(G, H) = \mathrm{Ker}(H \otimes L \to G \otimes M)$ if $L \subset M$ are free abelian groups such that $M/L \cong H$. This is true at least if A is in the bootstrap class $\mathcal{N}$ and B is *σ-unital* (i.e. has a countable approximate unit), and either $K_*(A)$ or $K_*(B)$ is finitely generated.

Following the seminal work of Cuntz on the computation of the K-theory of $\mathcal{O}_n$, (Pimsner and Voiculescu 1980b) produced an exact sequence for computing the K-theory of a crossed product $A \rtimes_\alpha \mathbb{Z}$ of a C^*-algebra A by an automorphism α:

$$\begin{array}{ccccc} K_0(A) & \xrightarrow{1_* - \alpha_*} & K_0(A) & \to & K_0(A \rtimes \mathbb{Z}) \\ \uparrow & & & & \downarrow \\ K_1(A \rtimes \mathbb{Z}) & \leftarrow & K_1(A) & \xleftarrow{1_* - \alpha_*} & K_1(A) \end{array} \tag{3.7.5}$$

Again this exact sequence can be proven via Toeplitz extensions. If A is represented on a Hilbert space H, let A act on $H \oplus H \oplus \cdots$ by $\bigoplus_{i=0}^{\infty} \alpha^i$ and consider the algebra $J(A)$ generated by A and the unilateral shift. Then we have an exact sequence

$$0 \to A \otimes K \to J(A) \to A \rtimes \mathbb{Z} \to 0\,. \tag{3.7.6}$$

The Pimsner–Voiculescu exact sequence amounts to showing that $K_*(A) \cong K_*(J(A))$ and identifying the maps $K_*(A) = K_*(A \otimes K) \to K_*(J(A)) \cong K_*(A)$

with id $- \alpha_*$ and $K_*(A) \cong K_*(J(A)) \to K_*(A \rtimes \mathbb{Z})$ with the natural maps arising from the inclusion $A \to A \rtimes \mathbb{Z}$. The exact sequence can also be deduced from Connes' Thom isomorphism (Connes 1981) $K_i(A \rtimes \mathbb{R}) \cong K_{1-i}(A)$ (which includes Bott periodicity for the case of trivial actions) by considering the mapping torus $M_\alpha = \{f \in C(\mathbb{R}, A) \mid f(x+1) = \alpha(f(x))\}$, the exact sequence $0 \to SA \to M_\alpha \to A \to 0$, the Morita equivalence between $A \rtimes_\alpha \mathbb{Z}$ and $M_\alpha \rtimes \mathbb{R}$ so that $K_i(A \rtimes_\alpha \mathbb{Z}) \cong K_i(M_\alpha \rtimes \mathbb{R}) \cong K_{1-i}(M_\alpha)$. Connes' Thom isomorphism can be deduced from an analogue of (3.7.6) where the K-theory of the middle algebra is trivial.

The Pimsner–Voiculescu exact sequence can be used to compute the K-theory of the Cuntz–Krieger algebras $\mathcal{O}_A$. For finite graphs with incidence matrix A (Cuntz and Krieger 1980), we can represent the stable algebra $\mathcal{O}_A \otimes K$ as a crossed product $\overline{F}_A \rtimes \mathbb{Z}$ of the stable AF algebra $\overline{F}_A$ with constant embedding A to obtain (cf. (3.7.2) for the full shift):

$$K_1(\mathcal{O}_A) \cong \mathbb{Z}^I/(1-A^t)\mathbb{Z}^I, \quad K_1(\mathcal{O}_A) \cong \mathrm{Ker}(1-A^t) \tag{3.7.7}$$

if I is the vertex set of the graph. In the case of infinite graphs it is more convenient to use the dual version (Paschke 1981) of the Pimsner–Voiculescu exact sequence with $\mathbb{T}$ actions (Pask and Raeburn 1996) to obtain the same result. Since $(\overline{F_A}, \sigma_A)$ is an invariant of conjugacy of the subshift, (X_A, σ_A), this gives an interpretation of the Bowen–Franks invariant $\mathbb{Z}^I/(1-A^t)\mathbb{Z}^I$ of conjugacy. Note that the Williams problem in classical dynamical systems is equivalent to showing that $(K_0(\overline{F}_A), \sigma_{A*})$ is a complete invariant of topological conjugacy of (X_A, σ_A). The dynamical system $(\mathcal{O}_A \otimes K, \sigma_t)$ with the dual gauge action is an invariant of flow equivalence of the shifts (the conjugacy of the one parameter flows built on the shifts) – best seen by groupoid considerations as for the Morita equivalence of the irrational rotation and Kronecker foliations described below. The sign of the determinant of $1 - A$ is not an invariant of $\mathcal{O}_A$ – as follows by the classification by Rørdam of (inductive limits of direct sums of matrix algebras over) Cuntz–Krieger algebras which we will delve into below, (Rørdam 1995a).

In a tour de force, (Connes 1976a) classified amenable (= injective = hyperfinite) factors. A von Neumann algebra M on a (separable) Hilbert space H is injective if there exists a projection of norm one from $B(H)$, the bounded linear operators on H, onto M. Amongst other things Connes showed that a factor is amenable (in the sense of the vanishing of certain cohomology groups) if and only if it is hyperfinite (can be expressed as an inductive limit of finite dimensional algebras for the weak topology) if and only if it is injective. Already from the work of Murray and von Neumann, there is an unique hyperfinite factor of type I_n ($B(\ell^2(n))$) and type II_1 ($R = \otimes_{\mathbb{N}} M_2$ with the completion taken in the weak operator topology in the trace representation). But now Connes could show uniqueness of the hyperfinite II_∞ ($R_1 = R \otimes B(\ell^2(\infty))$) and III_λ, $0 \leq \lambda < 1$, with the III_1 case being completed by (Haagerup 1987)). The III_λ factors are given by the crossed product of an automorphism on R_1 scaling the trace by

λ. The III_0 factors are classified by their flow of weights. (Jones 1983) in the early 1980s began a study of subfactors, primarily of the hyperfinite II_1 factor. The classification of amenable subfactors is now a major field with two aspects: analytic and combinatorial. As we shall see in the remainder of this book, the latter aspect has applications and connections with quantum groups, link and manifold invariants, statistical mechanics, conformal field theory etc.

On the other hand there has been remarkable progress since 1988 in the classification of amenable C^*-algebras. A C^*-algebra is amenable (in the sense of the vanishing of certain cohomology groups) if and only if its double dual or enveloping von Neumann algebra is amenable or equivalently if it is nuclear (has an unique tensor product with any other C^*-algebra or there is a unique way of putting a C^*-norm on the algebraic tensor product of the algebra with any other C^*-algebra so that the completion is again a C^*-algebra), (Choi and Effros 1976a), (Choi and Effros 1976b), (Choi and Effros 1977), (Choi and Effros 1978) (Effros and Lance 1977), (Connes 1976a), (Connes 1978), (Connes 1975a), (Haagerup 1983). The study of (separable) amenable C^*-algebras may not appear at first to be as clean as that of amenable von Neumann algebras. However around 1988 some significant but at the time apparently unrelated breakthroughs occurred. Whilst there exists an unique subfactor of index two of the hyperfinite II_1 factor R (and hence an unique outer action of the group $\mathbb{Z}/2$ on R) it was not until (Blackadar 1990), (see also (Kumjian 1988)) that a symmetry was found on a Pauli algebra (the infinite tensor product $\bigotimes_{\mathbb{N}} M_2$ completed in the norm topology) whose fixed point algebra was not AF. This involved writing the Pauli algebra in a non-standard way as an inductive limit of direct sums of matrix algebras over the circle. Independently (Putnam 1990a) approximated the crossed products of minimal actions on Cantor sets by direct sums of matrix algebras over the circle. The class of amenable C^*-algebras is much larger than the class of AF algebras (whose K_1 always vanishes), but there is much evidence to suggest that amenable separable (simple) C^*-algebras may be represented as inductive limits of elementary building blocks. Blackadar's construction was followed up by a construction of (Evans and Kishimoto 1991) to obtain a C^*-algebra A which is not AF (K_1 is a non-zero torsion group), whose tensor product with a UHF algebra is UHF. Tensoring with the UHF algebra destroys the K_1 obstruction for the original algebra to be AF, and the question arises as to the possible classification of amenable separable C^*-algebras by K-theoretic invariants. Also independently, (Dadarlat and Nemethi 1990) had classified inductive limits of direct sums of matrices over continuous functions on CW-complexes, up to shape equivalence (in this case it means the approximating diagrams as in Fig. 3.5 are valid up to homotopy) by K_0 and K_1.

(Elliott 1995) has proposed a subdivision of amenable simple (separable) C^*-algebras into three cases where the K-theoretic invariants K_0, K_1, T^+ (= topological convex cone of traces) with the pairing $K_0^+ \times T^+ \to \mathbb{R}$ should provide complete (stable) isomorphism invariants. The group $K_0(A)$ contains two natural subgroups $K_0^+ \cap (-K_0^+) \subset K_0^+ - K_0^+$. In the simple case they must be either 0 or K_0. Combining with the cone T^+ of positive traces on the algebra (defined on the *Pedersen ideal*, the smallest dense two sided ideal), which has a pairing with K_0, the subdivision is:

Case 1 $K_0^+ = 0, T^+ \neq 0$.

Case 2 K_0 ordered group $\neq 0, T^+ \neq 0$.

Case 3 $K_0^+ = K_0, T^+ = 0$.

Not only is this subdivision disjoint, it is exhaustive (Elliott 1995). Case 2 and Case 3 include the finite AF algebras and infinite Cuntz algebras respectively, for which we have already noted that K_0 is a complete invariant. Note that K_0 of an AF algebra is not only ordered $K_0^+ \cap -K_0^+ = 0$, but it is weakly unperforated. (Villadsen, in press) has however an example of a simple amenable separable C^*-algebra (an inductive limit of matrix algebras over tori) which has perforation, there is an element g with $ng > 0$, for some positive integer n but g itself is not positive. However, classification results by K-theoretic invariants now go far beyond the AF and Cuntz algebras and we will briefly review recent developments. In the context of Case 3, note that in a purely infinite simple C^*-algebra, the map $[p] \to K_0$ is an isomorphism from equivalence classes of projections in the algebra itself (Cuntz 1981a). In particular $K_0 = K_0^+$. Case 1 includes the stable projectionless C^*-algebras of (Blackadar 1980) with the first classification result for unital projectionless algebras due to (Jiang and Su (preprint 1996)) and Elliott (unpublished). Recall that for non-unital algebras, we adjoin an identity to A, and define $K_0(A)$ from an exact sequence:

$$0 \to K_0(A) \to K_0(A + \mathbb{C}1) \to K_0(\mathbb{C}1) \to 0\,.$$

Here it is possible (see e.g. the stable projectionless C^*-algebras of (Blackadar 1980)) that $K_0(A)_+$ (= [projections in matrices over A]) = 0, so that the equality $K_0(A) = K_0(A)_+ - K_0(A_+)$ may fail for non-unital algebras. In Case 1 the invariant consists of the countable abelian groups K_0 and K_1, together with the cone T^+ in duality with K_0. (Elliott 1995) has exhausted all possibilities for the invariant (i.e. K_* can be arbitrary countable abelian groups) and an arbitrary simplical cone with zero pairing with K_0. There are examples where the pairing can be non-zero, and the question arises as to whether the pairing can be arbitrary. The range of the invariant in Case 2 is still not totally clear. However, the range of the invariant for inductive limits of circle algebras (direct sums of matrix algebras over continuous functions on the circle) is that of K_0, K_1 arbitrary torsion free countable groups, with K_0 weakly unperforated, has the Riesz property and extreme rays of T^+ yield extreme rays in the cone of positive functionals on K_0.

The programme to classify finite amenable separable C^*-algebras has so far involved at least two parts. One is to write the C^*-algebra as an inductive limit of simpler building blocks, e.g. crossed products of continuous functions on a Cantor set by a minimal action (Putnam 1990a), the irrational rotation algebras (Elliott and Evans 1993) and the crossed products of Fermion algebras by Bogoliubov automorphisms (Bratteli *et al.* 1993), (Bratteli *et al.* 1995) are $A\mathbb{T}$, that is inductive limits of circle algebras (direct sums of matrix algebras over continuous functions on the circle). The second is the classification of such inductive limit C^*-algebras by K-theoretic invariants.

This recent work on inductive limit of homogeneous and sub-homogeneous algebras is primarily oriented towards a better understanding of non-AF amenable C^*-algebras, but has succeeded in solving problems concerning AF algebras which had been unresolved since formulated by (Effros 1981) around 1978. For each integer $p > 1$ there exists a C^*-algebra A_p which is not an inductive limit of finite dimensional C^*-algebras, such that $A_p \otimes (\bigotimes_{\mathbb{N}} M_p)$ is isomorphic to the UHF algebra $\bigotimes_{\mathbb{N}} M_p$ (Evans and Kishimoto 1991), there exist compact actions on UHF algebras whose fixed point algebras are not AF (Evans and Kishimoto 1991) (the latter was as already noted above shown by (Blackadar 1990), see also (Kumjian 1988), for $\mathbb{Z}/2$ which initiated the explosion of work on inductive limits of type I algebras). In particular any such action of a compact group on an AF algebra with non-AF fixed point algebra cannot be locally representable in the sense of (Handelman and Rossman 1985). Evans pointed out to Elliott in 1989 that the main obstruction for A_p being AF seemed to be its $K_1 = \mathbb{Z}/p$ and an apparent similarity between the proof of the AF property of $A_p \otimes (\bigotimes_{\mathbb{N}} M_p)$ and the p-torsion of $K_1(A_p)$. Tensoring with the algebra $\bigotimes_{\mathbb{N}} M_p$ destroyed this obstruction. Elliott began to investigate whether this K_1 obstruction was the only obstruction for such an inductive limit algebra being AF. Since then there has been remarkable progress on this problem and indeed it is already known that $K_*(A) = K_0 \oplus K_1 (\cong K_0(C(\mathbb{T}) \otimes A)$ and so equipped with an order structure) together with the space of trace states classifies large classes of algebras. This began with the classification of inductive limits of (finite direct sums of) matrix algebras over

- the circle (assuming that the limit algebra has real rank zero – self adjoint elements are approximable by ones with finite spectrum) using K_* (Elliott 1993a);
- the interval (assuming that the limit algebra is simple) but also including the simplex of tracial states in the invariant (which reduces to a singleton in the case of real rank zero) (Elliott 1993b).

The classification was extended to real rank zero C^*-algebras which are inductive limits of direct sums of matrix algebras over continuous functions on graphs (possibly non-Hausdorff) in the thesis of (Su 1992). Increasing the dimension increases the difficulties in extending these results. However a combination of the work of (Gong, preprint a) and (Dadarlat 1995) shows that simple C^*-algebras of real rank zero that can be represented as inductive limits $A = \lim_{\to} A_n$, where each A_n is a finite direct sum of subhomogeneous algebras of the form $eM_m(C(X))e$, where X is finite CW complex and e a projection in $M_m(C(X))$, are classified by the ordered scaled K_*, as long as there is slow dimension growth (e.g sup dim $C(X) < \infty$). Indeed all such algebras can be described by inductive limits where $\dim X \leq 3$. However this invariant is not a complete invariant when the algebra is not simple, even when $\dim X \leq 2$ (Gong, preprint 1994).

In the non-simple case, mod-p K-theory is the extra ingredient needed to produce complete invariants. Put $K_*(A;\mathbb{Z}/p) = K_*(A \otimes A_p)$ for any nuclear C^*-algebra A_p such that $K_0(A_p) = \mathbb{Z}/p$, $K_1(A_p) = 0$. Although the Cuntz algebra $\mathcal{O}_{p+1}$ has the correct K-theory it can as an infinite algebra play havoc with ordering, so when working with finite algebras it is more natural to take I_p the ideal in the dimension drop algebra vanishing at one end point:

$$I_p = \underline{0 \qquad M_p \qquad \mathbb{C}} \tag{3.7.8}$$

where $K_1(I_p) = 0$, $K_0(I_p) = \mathbb{Z}/p$. The unital $\tilde{I}_p$ (see (3.7.9)) were used in (Evans and Kishimoto 1991) to produce the torsion in K_1 of the non-AF algebra A_p such that $A_p \otimes (\bigotimes_{\mathbb{N}} M_p) \cong \bigotimes_{\mathbb{N}} M_p$. Then we have (Dadarlat and Loring 1996a): $[I_p, A \otimes K] \cong [[I_p, A \otimes K]] \cong [[SI_p, SA \otimes K]] \cong KK(I_p, A) \cong KK^1(\mathbb{C}, I_p \otimes A) \cong K_0(SI_p \otimes A) = K_0(A;\mathbb{Z}/p)$. Here $[-,-]$ denotes the Grothendieck group of homotopy classes of homomorphisms, and the first isomorphism in this chain involves the semi-projectivity of I_p (Loring 1996) – a stability result for I_p in terms of generators and relations. In this way $K_*(A;\mathbb{Z} \oplus \mathbb{Z}/p) = [\tilde{I}_p, A \otimes C(\mathbb{T}) \otimes K]$ has a distinguished semigroup, the classes of morphisms, and determines an ordering on $\underline{K}(A) = \oplus_{p=0}^{\infty} K_*(A;\mathbb{Z}/p)$, (Dadarlat and Loring 1996a). With input from the work of (Eilers 1996), (Dadarlat and Loring 1996b) equip $\underline{K}(A)$ with the additional Bockstein operations $K_*(A) \rightleftharpoons K_*(A;\mathbb{Z}/p)$, $K_*(A;\mathbb{Z}/pq) \rightleftharpoons K_*(A;\mathbb{Z}/p)$ and produce a multi-coefficient UCT (valid in the same context as the UCT):

$$0 \to \operatorname{Pext}(K_*(A), K_*(B)) \to KK(A,B) \to \operatorname{Hom}(\underline{K}(A), \underline{K}(B)) \to 0$$

which permits a classification of real rank zero C^*-algebras which are inductive limits of subhomogeneous algebras with slow dimension growth (Dadarlat and Gong in press). The *pure extensions* $\operatorname{Pext}(K_*(A), K_*(B)) \subset \operatorname{Ext}^1_{\mathbb{Z}}(K_*(A), K_*(B))$ are the locally trivial extensions. (An extension $0 \to H \to K \to G \to 0$ is *pure* if every element of G of finite order can be lifted to an element of K of the same order).

One of the ingredients which has led to new phenomena in the theory of operator algebras are the non-commutative orbifolds – non-commutative C^*-algebras associated with certain singular orbifolds. (Blackadar 1990) obtained an exotic symmetry on the Pauli algebra $F_2 = \bigotimes_{\mathbb{N}} M_2$, by expressing this algebra in a non-standard and surprising way as an inductive limit of algebras $C(\mathbb{T}) \otimes M_{4^n}$ using winding operations (both negative and positive) on the circle $\mathbb{T}$. Kumjian considered the Bunce–Deddens algebra, namely the crossed product $C(\mathbb{T}) \rtimes G_2$ of the circle $\mathbb{T}$ by dyadic rotations G_2. This algebra $C(\mathbb{T}) \rtimes G_2$ is not AF, but by throwing in the flip $z \to \bar{z}$ on $\mathbb{T}$, one obtains an AF algebra $(C(\mathbb{T}) \rtimes G_2) \rtimes (\mathbb{Z}/2)$ (Kumjian 1988). Taking a $(\mathbb{Z}/2)^\wedge$ dual action one obtains an AF algebra and a symmetry whose fixed point algebra, essentially a Bunce–Deddens algebra, is not AF but is of real rank zero in the sense of (Brown and Pedersen 1991) which we now discuss. An AF algebra certainly contains many projections, indeed is generated linearly by them as a Banach space since a finite dimensional

C^*-algebra has this property. In fact a commutative C^*-algebra $C(X)$ (the continuous functions on a compact Hausdorff space) is AF if and only if X is totally disconnected. Thus the AF property is one non-commutative generalization of a space being totally disconnected or zero dimensional, but there are others. One such is the notion of real rank zero (Brown and Pedersen 1991). Real rank zero is a weaker notion than being approximately finite dimensional (but for commutative C^*-algebras, they coincide). A unital C^*-algebra is of real rank zero if the invertible self adjoint elements are dense in all self adjoint elements; equivalently one of the following conditions hold:

- self adjoint elements of finite spectrum are dense in all self adjoint elements;
- hereditary subalgebras have approximate identities consisting of projections.

The Bunce–Deddens algebra has non-trivial K_1, (and so cannot be AF for this reason), but for real rank zero algebras built up (as the Bunce–Deddens algebra and its crossed product with $\mathbb{Z}/2$ are) from certain subhomogeneous algebras on the interval or the circle, the K_1-obstruction is the only obstruction to the algebra being AF. Through these examples, one is naturally led to consider the orbifold $\mathbb{T}/(\mathbb{Z}/2)$, represented or described via the non-commutative C^*-algebra $C(\mathbb{T}) \rtimes (\mathbb{Z}/2)$ which can be identified with the algebra of M_2-valued functions on a compact interval which are diagonal (or lie in subalgebras isomorphic to $\mathbb{C}^2$) at the endpoints.

$$\mathbb{C}^2 \overset{M_2}{\text{———————}} \mathbb{C}^2 \tag{3.7.9}$$

The given action of $\mathbb{Z}/2$ on $\mathbb{T}$ has (generically) two point orbits (away from the singular points ± 1 which are both fixed under the action):

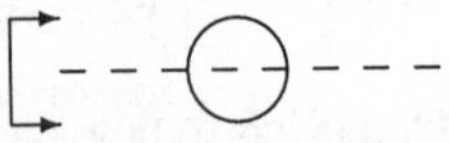

So locally, $\mathbb{Z}/2$ either acts on a two-point space interchanging the points, or fixes a singleton. Then in the former case $C\{:\} \rtimes (\mathbb{Z}/2)$ is given by two diagonal entries and the transition between the two which generate M_2, whilst in the latter case $C\{\cdot\} \rtimes (\mathbb{Z}/2)$, it is only the transition which is significant, and generates the universal algebra of a self adjoint unitary, or $C^*((\mathbb{Z}/2)^\wedge)$. Gluing these together along the orbit space (the interval) we get $C(\mathbb{T}) \rtimes (\mathbb{Z}/2)$ is identified with M_2 valued functions on the interval $[-1, 1]$ which are diagonal at the end points. The spectrum of this algebra can be represented diagrammatically by:

$$: \text{———————} : \tag{3.7.10}$$

The original algebra $C(\mathbb{T})$ embeds in the crossed product:

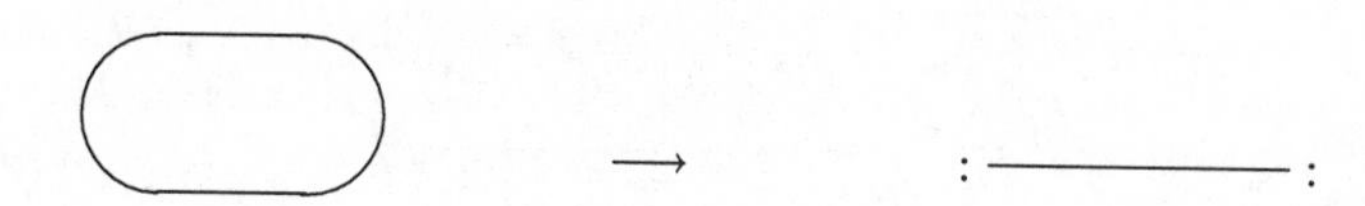

in a natural way. The crossed product $C(\mathbb{T}) \otimes M_{2^n}$ by the flip $z \to \bar{z}$ on $\mathbb{T}$ can also be represented by (3.7.10); it is the algebra of $M_{2^{n+1}}$ valued functions on the unit interval, whose values at the endpoints lie in subalgebras isomorphic to $M_{2^n} \oplus M_{2^n}$. Kumjian's algebra $C(\mathbb{T}) \rtimes G_2 \rtimes (\mathbb{Z}/2)$ can be viewed as an inductive limit of algebras $(C(\mathbb{T}) \otimes M_{2^n}) \rtimes (\mathbb{Z}/2)$. Thus began a study of inductive limits of subhomogeneous algebras on intervals, circles, etc. using folding and winding operations on the interval and circle respectively to construct embeddings.

The construction of the non-AF algebra A, such that

$$A_p \otimes (\bigotimes_{\mathbb{N}} M_p) \cong \bigotimes_{\mathbb{N}} M_p \,, \tag{3.7.11}$$

also uses inductive limits of matrix valued functions on the interval, with constraints or singularities at the endpoints. We sketch the construction when $p = 2$. This time the basic building block consists of algebras

$$A_n = M_{2^{n-1}} \overset{M_{2^n}}{\rule{4cm}{0.4pt}} M_{2^{n-1}}$$

consisting of M_{2^n} valued functions on the unit interval $[0,1]$, which at the end points lie in unital subalgebras isomorphic to $M_{2^{n-1}}$. Then A_n is embedded in A_{n+1} by a folding operation:

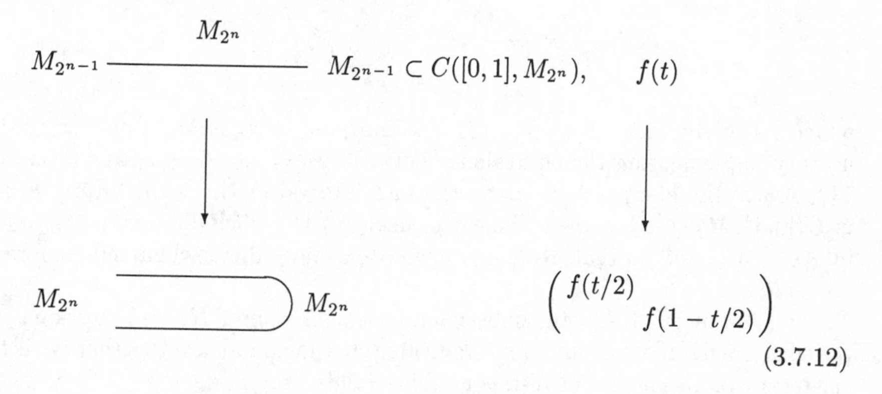

(3.7.12)

We remark that the position of the subalgebras $M_{2^{n-1}}$ in M_{2^n} is not important in defining A_n, i.e. changing their position will give an isomorphic algebra. To see this consider the following algebras:

$$A_n^{(i)} = \frac{M_{2^n}}{B^{(i)} \qquad\qquad C^{(i)}} \tag{3.7.13}$$

where $B^{(i)} \cong C^{(i)} \cong M_{2^{n-1}}$ are unital subalgebras, $i = 0, 1$. The subalgebras at the endpoints will be conjugate by unitaries u_0, u_1 in M_{2^n}, $B^{(1)} = u_0 B^0 u_0^*$, $C^{(1)} = u_1 C^{(0)} u_1^*$. Choose a continuous path of unitaries $\{u_s \mid s \in [0,1]\}$ in M_{2^n} connecting u_0 to u_1. Then $\operatorname{Ad} u_s$ will define an isomorphism of $A_n^{(0)}$ onto $A_n^{(1)}$. In particular, the subalgebras at the end points in (3.7.13) could be taken to be in the same position, but if we insisted on that choice, then we could not keep the simple form of the folding embedding in (3.7.12).

To see that $A \otimes F_2$ is AF, where $A = \lim_{\rightarrow} A_n$ and F_2 is the Pauli algebra $\bigotimes_{\mathbb{N}} M_2$, we need to approximate $f_1, \ldots, f_r$ in A_n by a finite dimensional subalgebra in $A_{n+m} \otimes M_2$ for m large (Exercise 3.12). By uniform continuity, given $\varepsilon > 0$, $\|f_i(s) - f_i(t)\| < \varepsilon$, for all $i = 1, \ldots, r$, and $|s - t|$ sufficiently small. Thus after sufficiently many foldings the embedded functions $\tilde{f}_i$ in A_{n+m} satisfy $\|\tilde{f}_i(s) - \tilde{f}_i(t)\| < \varepsilon$ for all $i = 1, 2, \ldots, r$ and all s, t. Letting $N = n + m$, consider $F_i(t) = \tilde{f}_i(0)$ in $C([0,1], M_{2^N})$. Then $F_i(0) = \tilde{f}_i(0)$ certainly lies in the correct subalgebra (isomorphic to $M_{2^{N-1}}$) at the left hand endpoint. But $F_i(1) = \tilde{f}_i(0)$ will not necessarily lie in the correct subalgebra (isomorphic to $M_{2^{N-1}}$) at the right hand endpoint, say $M_{2^{N-1}} \otimes 1 \subset M_{2^N}$. But $\tilde{f}_i(0) \in M_{2^{N-1}} \otimes M_2 \otimes 1_2$ a subalgebra of M_{2^N} certainly unitarily equivalent to $M_{2^{N-1}} \otimes 1_2 \otimes M_2$. Then $F_i \otimes 1_2$ will lie in:

$$\frac{M_{2^{N-1}} \otimes M_2 \otimes M_2}{M_{2^{N-1}} \otimes M_2 \otimes 1_2 \qquad\qquad M_{2^{N-1}} \otimes M_2 \otimes 1_2}$$

which by the remark before (3.7.13) is isomorphic to $A_n \otimes M_2$. Connecting this unitary implementing the equivalence between $M_{2^{N-1}} \otimes M_2 \otimes 1_2$ and $M_{2^{N-1}} \otimes 1_2 \otimes M_2$ to the identity back along the unit interval $[0,1]$, we can adjust $F_i \otimes 1$ in $C([0,1], M_{2^N}) \otimes M_2$ to an element g_i in $A_N \otimes M_2$ which approximates $\tilde{f}_i \otimes 1_2$ in $A_N \otimes M_2$, and so that $g_1, \ldots, g_r$ generate a finite dimensional subalgebra of $A_N \otimes M_2$.

The K-theory of A_n (and subsequently $K_*(A) = \lim_{\rightarrow} K_*(A_n)$) can be computed from the ideal of functions vanishing at the endpoints, together with the six-term exact sequence of K-theory. This yields:

$$K_0(A_n) \cong \mathbb{Z}\,, \quad K_1(A_n) \cong \mathbb{Z}/2 \quad \text{and} \quad K_0(A) \cong \mathbb{Z}[1/2]\,, \quad K_1(A) \cong \mathbb{Z}/2\,.$$

The generator of $K_1(A_n) \cong \mathbb{Z}/2$ is given by the unitary:

$$t \to \begin{pmatrix} e^{2\pi it} & & & & \\ & 1 & & & \\ & & \cdot & & \\ & & & \cdot & \\ & & & & 1 \end{pmatrix} \otimes 1_2 \in M_{2^{n-1}} \otimes M_2 \,.$$

The 2-torsion in $K_1(A_n)$ can be seen as follows. If U is any unitary in A_n, then since the interval $[0,1]$ and the unitary group of M_{2^n} are connected, let W be a continuous map from $[0,1]^2$ into $M_{2^{n-1}} \otimes M_2$ such that $W(0,t) = U(t)$, $W(1,t) = 1$. Then $V_s(t) = W(s,t) \otimes 1_2$ defines a path $\{V_s \mid s \in [0,1]\}$ of unitaries in the algebra:

$$\begin{array}{ccc} & M_{2^{n-1}} \otimes M_2 \otimes M_2 & \\ \hline M_{2^{n-1}} \otimes M_2 \otimes 1_2 & & M_{2^{n-1}} \otimes M_2 \otimes 1_2 \end{array}$$

which using the remark (3.7.13) on page 128 is isomorphic to $A_n \otimes M_2$ (cf. the proof that $A \otimes F_2$ is AF) with $V_0 = U \otimes 1_2$, $V_1 = 1_{A_n} \otimes 1_2$. Hence $2[U] = 0$ in $K_1(A_n)$. From the expression for the generator we see that under the isomorphism $K_1(A_n) \cong \mathbb{Z}/2$, the inclusion $A_n \to A_{n+1}$ induces the identity map on $\mathbb{Z}/2$, and so $K_1(A) \cong \mathbb{Z}/2$. Thus there is at least a K_1-obstruction to A being an AF algebra. Tensoring with $F_2 \cong \bigotimes_{\mathbb{N}} M_2$ destroys this K_1 as $K_1(A \otimes F_2) \cong \lim_{\to} K_1(A \otimes M_{2^n}) \cong \lim_{\to} K_1(A) \cong 0$ as $K_1(A \otimes M_{2^n}) \cong K_1(A)$ with $K_1(A \otimes M_{2^n}) \to K_1(A \otimes M_{2^{n+1}})$ corresponding to multiplication by 2. We have noted that $A \otimes F_2$ is AF. To show that $A \otimes F_2$ is isomorphic to F_2, it is enough by Elliott's classification of AF algebras by K_0 (Theorem 3.9), to show that $K_0(A \otimes F_2) \cong K_0(F_2)$. This can be established from $K_0(A) \cong \mathbb{Z}[1/2]$.

A unital C^*-algebra A has real rank $\leq n$ if given $\varepsilon > 0$, $(x_1, x_2, \ldots, x_{n+1})$ in A_h^{n+1}, there exists $(y_1, y_2, \ldots, y_{n+1})$ in A_h^{n+1} such that $\|x_i - y_i\| < \varepsilon$ and $\sum_{i=1}^{n+1} y_i^2$ is invertible. Thus, as we have already said, A is of real rank zero if the invertible self adjoint elements are dense in all self adjoint elements. If A is commutative, say $C(X)$, then A being of real rank zero coincides with the property of the algebra being AF or X being totally disconnected. In fact $RR(C(X)) = \dim X$, the covering dimension of X. Brown and Pedersen based their definition on the notion of topological stable rank of (Rieffel 1983a): a unital C^*-algebra has topological stable rank $\leq n$ if given $\varepsilon > 0$, $(a_1, a_2, \ldots, a_n) \in A^n$, there exists $(b_1, \ldots, b_n) \in A^n$ with $\|a_i - b_i\| < \varepsilon$ and $\sum_{i=1}^{n} b_i^* b_i$ is invertible. (Herman and Vaserstein 1984) have shown that (for C^*-algebras) topological stable rank is the same as Bass stable rank and (Villadsen (preprint 1997)) has shown that for each $n \in \mathbb{N}$, there is a simple C^*-algebra of topological stable rank n. Moreover, A is of stable rank 1 if and only if the invertible elements are dense, $\operatorname{tsr} C(X) = [\dim(X)/2] + 1$ where [ ] denotes the integer part and $RR(A) \leq 2(\operatorname{tsr} A) - 1$. It is natural to ask when a C^*-algebra such as

$$\lim_{\rightarrow} \oplus (C(X_i) \otimes M_{m_i}) \tag{3.7.14}$$

is of real rank zero, or even what the real rank of $C(X) \otimes M_m$ is? (Beggs and Evans 1991) have shown that if X is a compact Hausdorff space then $\dim(X) \leq n(2m-1)$ if and only if $RR(C(X) \otimes M_m) \leq n$. This should be contrasted with the situation for topological stable rank (Rieffel 1983a), where

$$\operatorname{tsr}(B \otimes M_m) = \{(\operatorname{tsr}(B) - 1)/m\} + 1 \tag{3.7.15}$$

and $\{x\}$ denotes the least integer greater than x. In particular, for m large and $RR(C(X))$ finite

$$RR(C(X) \otimes M_m) = \begin{cases} 0 & \text{if } \dim(X) = 0 \\ 1 & \text{otherwise} \end{cases} .$$

That $RR(C(X) \otimes M_m) = 0$ if and only if $\dim(X) = 0$ was shown by (Brown and Pedersen 1991).

(Dadarlat *et al.* 1992) have shown that, when the dimension of X_i in (3.7.14) is uniformly bounded, and the inductive limit C^*-algebra is simple then it is automatically of topological stable rank one. (Note that if the algebra is simple, then necessarily $m_i \to \infty$, and so the topological stable rank is at most two by (3.7.15)). It is not however necessarily of real rank zero, even when each X_i is the circle. (Blackadar *et al.* 1992a) have considered inductive limits $\lim_{\rightarrow} \oplus(C(\mathbb{T}) \otimes M_{m_i})$ using winding embeddings, and have shown that the inductive limit is of real rank zero if and only if the number of standard ± 1 times around embeddings is asymptotically small compared with the total number of embeddings as $i \to \infty$. Thus we can produce simple C^*-algebras of the form $\lim_{\rightarrow} \oplus(C(\mathbb{T}) \otimes M_{m_i})$ with topological stable rank one but non-zero real rank – the invertible elements are dense, but the invertible self adjoint elements are not dense in the self adjoint elements.

Other exotic symmetries have been analysed on non-commutative tori. These systems give rise to non-commutative toroidal orbifolds. (Albeverio and Høegh-Krohn 1980) classified (faithful) ergodic actions of a compact abelian group G on unital C^*-algebras using skew bicharacters on the dual group. If $G = \mathbb{T}^2$, the bicharacters are $((m,n),(m',n')) \to q^{mn'-m'n}$ for some $q = e^{2\pi i\theta} \in \mathbb{T}$ with corresponding rotation C^*-algebra A_q which for our purposes we will accept to be defined as the universal C^*-algebra generated by two unitaries U and V satisfying the commutation relation $VU = qUV$ (or the multiplicative commutator $[V^m U^n, V^{m'} U^{n'}]$ is the bicharacter $q^{mn'-m'n}$). Now $VUV^{-1} = qU$; $\operatorname{Ad} V$ leaves $C^*(U) \cong C(\mathbb{T})$ invariant and acts by rotation through θ on the circle, and A_q can be identified with the crossed product $C(\mathbb{T}) \times_\theta \mathbb{Z}$. The ergodic $\mathbb{T}^2$ action α on A_q is then $(t_1, t_2) : U \to t_1 U$, $V \to t_2 V$, $(t_1, t_2) \in \mathbb{T}^2$. The non-commutative 2-torus or rotation algebra is a classic example of a non-AF amenable algebra. When θ is irrational, A_q is simple and is the unique C^*-algebra generated by two unitaries satisfying the above commutation relation (contrast with $C(\mathbb{T}^2)$

when $\theta = 0$). When θ is irrational, simplicity and universality can be shown as for the Cuntz algebras in Corollary 2.19 (cf. also the proof in the notes to Chapter 2 of simplicity of the Bunce–Deddens algebra). We can consider the trace $\tau = \int\int \alpha_{s,t}\, ds\, dt$, with the ergodic α action which is approximately inner as $\alpha_{q,1} = \mathrm{Ad}\, V$, $\alpha_{1,q^{-1}} = \mathrm{Ad}\, U$ and $\{q^n\}$ is dense in $\mathbb{T}$.

The stable algebra $A_q \otimes K$ is isomorphic to the C^*-algebra of the Kronecker foliation on the two torus $C^*(\mathbb{T}^2, F_\theta)$, the crossed product $C(\mathbb{T}^2) \rtimes \mathbb{R}$ where $\mathbb{R}$ acts on $\mathbb{T}^2$ via a flow along lines of slope θ which decomposes the manifold $\mathbb{T}^2$ into a (singular) family or foliation by leaves. Since the dynamical systems $(\mathbb{T}, \mathbb{Z})$, $(\mathbb{T}^2, \mathbb{R})$ of the irrational rotation and Kronecker flow are minimal the orbit spaces $\mathbb{T}/\mathbb{Z}$ and $\mathbb{T}^2/\mathbb{R}$ are singular – the quotient topology is the trivial topology. The stable isomorphism of $C(\mathbb{T}) \rtimes_\theta \mathbb{Z}$ and $C(\mathbb{T}^2) \rtimes \mathbb{R}$ can be understood by first considering the following simpler situation. Consider the algebras M_I and M_J. Their representation theory is seen to be equivalent via the use of the $M_I - M_J$ bimodule $V = M_{I,J}$ of $I \times J$ matrices. This has two inner products $\langle x, y\rangle_{M_I} = xy^*$, $\langle x, y\rangle_{M_J} = y^*x$, which are M_I and M_J valued respectively which are compatible with the bimodule structure. The inner products and their compatibility is what is needed for $M_{I\sqcup J} = \begin{pmatrix} M_I & V \\ V^* & M_J \end{pmatrix}$ to form a simple C^*-algebra such that cutting down to the corners we get M_I and M_J. In groupoid language, M_I and M_J are generated by arrows from sets I and J respectively. The groupoids are the equivalence relation on $I \times I$, $J \times J$ respectively of all arrows $a \longrightarrow b$ and $i \longrightarrow j$ $a, b \in I$, $i, j \in J$. The bimodule V is the space $M_{I,J}$ using all arrows which start from I and end in J:

$$I \times J \qquad a \longrightarrow i \qquad a \in I,\, i \in J\,.$$

Composing these arrows on $I \times J$ with arrows from $I \times I$ and $J \times J$ on the left and right respectively, gives the bimodule structure, and composing two arrows from $I \times J$ using the reverse of one gives the two inner products:

$$\begin{array}{llll} a \xrightarrow{I\times I} b & b \xrightarrow{I\times J} i & i \xrightarrow{J\times J} j & \text{bimodule structure} \\ a \xrightarrow{I\times J} i & i \xleftarrow{J\times I} b & & \text{inner products} \\ j \xleftarrow{J\times I} c & c \xrightarrow{I\times J} k\,. & & \end{array}$$

Here $a, b, c \in I$, $i, j, k \in J$. We regard the dynamical system $(\mathbb{T}, \mathbb{Z})$ as arrows in $\mathbb{T}$:

$\mathbb{T}$ — x ⟶ $x + m\theta$

beginning in an arbitrary point $x \in \mathbb{T}$ and ending at a point $(x+m\theta)$ on the orbit. The dynamical system $(\mathbb{T}^2, \mathbb{R})$ is similarly described by arrows in $\mathbb{T}^2$ beginning at an arbitrary point z in $\mathbb{T}^2$ and ending at time t along the orbit. So we now take the imprimitivity bimodule given by arrows beginning at x in $\mathbb{T}$ ($= \mathbb{T} \times \{1\} \subset \mathbb{T}^2$) and ending at z along the flow line after time t as in Fig. 3.13. There is a natural

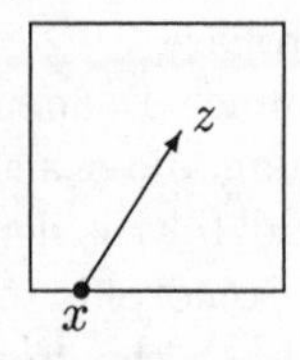

FIG. 3.13. $C(\mathbb{T}) \rtimes \mathbb{Z} - C(\mathbb{T}^2) \rtimes \mathbb{R}$ bimodule

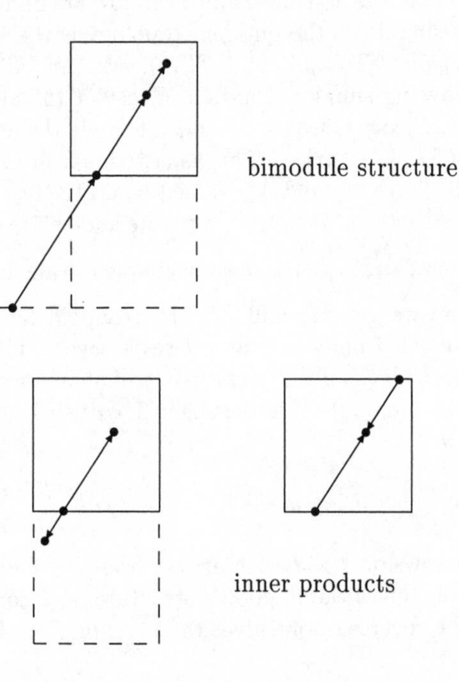

FIG. 3.14. Structure of $C(\mathbb{T}) \rtimes \mathbb{Z} - C(\mathbb{T}^2) \rtimes \mathbb{R}$ bimodule

$C(\mathbb{T}) \rtimes \mathbb{Z} - C(\mathbb{T}^2) \rtimes \mathbb{R}$ bimodule structure and $C(\mathbb{T}) \rtimes \mathbb{Z}$ and $C(\mathbb{T}^2) \rtimes \mathbb{R}$ valued inner products from composing arrows as in Fig. 3.14. The existence of such an $A-B$ bimodule V is the meaning of Morita equivalence between $A = C(\mathbb{T}) \rtimes \mathbb{Z}$ and $B = C(\mathbb{T}^2) \rtimes \mathbb{R}$. Tensoring with V produces an equivalence between modules ($X_A \to X_A \otimes_A V$, ${}_BY \to V_B \otimes_B Y$ – such constructions will be discussed in detail in Section 9.2 onwards) and hence identifies the K-theory $K_*(A) \cong K_*(B)$. More than that, the algebras A and B are stably isomorphic $A \otimes K \cong B \otimes K$ ($\cong B$). This is because A and B are both cut downs of $C = \begin{pmatrix} A & V \\ V^* & B \end{pmatrix}$, and by (Brown 1977), this procedure of cutting-down induces a stable isomorphism.

The K-theory of A_q can be computed from the Pimsner–Voiculescu six term

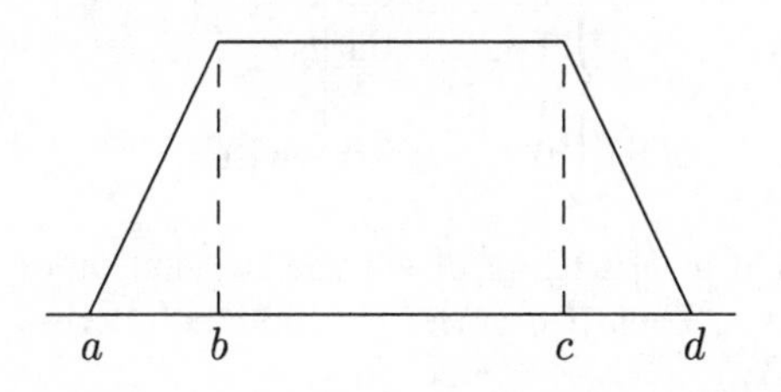

FIG. 3.15. Construction of a Rieffel projection

exact sequence of K-theory (3.7.5) as $K_0(A_q) \cong \mathbb{Z}^2$, $K_1(A_q) \cong \mathbb{Z}^2$. If $q = 1$, $K_0(C(\mathbb{T}^2))$ is generated by the trivial bundle and the Bott element, whilst $K_1(C(\mathbb{T}^2))$ is generated by the coordinate unitary maps $(z_1, z_2) \to z_i$, $i = 1, 2$, i.e. U and V. When q is rational (or more precisely when θ is a rational multiple of 2π, say r/s with r, s relatively prime), A_q can be identified with a homogeneous algebra over $\mathbb{T}^2$, with values in M_s. This bundle is not trivial unless $q = 1$. Indeed A_q is not isomorphic to $A_{q'}$ unless $q = q'$ or $q = \overline{q}'$, the rational case being due to (Høegh-Krohn and Skjelbred 1981) (see also (Rieffel 1983b)) and the irrational case to (Rieffel 1981), (Pimsner and Voiculescu 1980a). All rotation algebras have $K_0 \cong \mathbb{Z}^2$, and $K_1 \cong \mathbb{Z}^2$, but the irrational ones (the simple case) are indeed distinguished by K_0 as these are not simply groups but ordered groups (Rieffel 1981), (Pimsner and Voiculescu 1980a). The ordering on $K_0(A_q) \cong \mathbb{Z} + \theta\mathbb{Z}$ is the inherited ordering from $\mathbb{R}$, so that we can recover the angle θ (modulo some obvious identifications with $\mathbb{Z} \pm \theta$) from the ordering. Alternatively the ordering is given by the half plane in $\mathbb{Z}^2$ at slope θ with the x-axis. The unique trace τ on A_q actually implements an isomorphism $\tau_* : K_0(A_q) \to \mathbb{Z} + \theta\mathbb{Z} \subset \mathbb{R}$.

To construct the non-trivial *Rieffel projections* which have irrational trace, we will represent A_q on $L^2[0,1]$ as follows. Take U to be multiplication by $e^{2\pi ix}$ on $L^2[0,1]$. The characteristic function $\chi_{[b,c]}$ of an interval $[b, c]$ in $[0, 1]$ will be a projection, but not an element of $C^*(\mathrm{U})$. To modify this to get a continuous function, first take a real valued function F, $0 \leq F \leq 1$ initially defined on an interval $[a, b]$ as indicated in Fig. 3.15. Then

$$\begin{pmatrix} F & \sqrt{F(1-F)} \\ \sqrt{F(1-F)} & 1-F \end{pmatrix}$$

is a projection which we can take to act on $L^2[a, b] \oplus L^2[c, d]$ where $0 < b - a = d - c$. The Rieffel projection is then

$$e = \chi_{[b,c]} + \begin{pmatrix} F & \sqrt{F(1-F)} \\ \sqrt{F(1-F)} & 1-F \end{pmatrix}$$

on $L^2[0, 1]$. Looking at the diagonal parts of this operator e, which is evidently a projection by construction, we let

$$f = \begin{cases} F & \text{on } [a,b] \\ 1 & \text{on } [b,c] \\ 1-F & \text{on } [c,d] \\ 0 & \text{otherwise} \end{cases}$$

where we have identified $[a,b]$ with $[c,d]$ via the translation operator V_1 by distance $\beta = c - a \equiv d - b$. Then if g is the *continuous* function $\sqrt{F(1-F)}\chi_{[a,b]}$ we have

$$e_\beta = V_1^{-1} g + f + g V_1 \in C^*(U, V_1) \equiv A_\beta\,,$$

as $V_1 U V_1^{-1} = e^{2\pi i\beta} U$. Thus if $r' - s'\theta \in (\mathbb{Z} + \theta\mathbb{Z}) \cap [0,1]$, and V is translation by θ, $V_1 = V^{s'}$ is translation by $s'\theta$ and

$$e_\beta = V^{-s'} g + f + g V^{s'} \in A_\theta$$

is a projection with trace $\beta = r' - s'\theta$, and K_0 class $(r', -s')$.

Next suppose we have $\begin{pmatrix} r' & r \\ s' & s \end{pmatrix} \in SL(2, \mathbb{Z})$ with $s, s' > 0$ and $r/s < \theta < r'/s'$. Then $s\beta + s'\beta' = 1$, where $\beta = r' - s'\theta$, $\beta' = s\theta - r > 0$. We can then form the Rieffel projection e_β in $A_\theta = C^*(U,V)$ and rotate on the spectrum of U by $\alpha_{\exp 2\pi i/s, 1}$ to get s projections $\{e_\beta^i\}$. Since $s\beta < 1$, we can choose the base of the original Rieffel projection e_β to have length $< 1/s$ so that the s projections are orthogonal. We would like U and V respectively to be approximately multiplication and cyclic permutation matrices with respect to this tower of projections. For s large, U will be approximately a multiplication operator, because $e_\beta \le E^U$, where E^U is the spectral projection of U for the interval $[a,d]$ of Fig. 3.15, and U is approximately constant on E^U for s large.

For V to approximately rotate the projections, we need to estimate the difference $V e_\beta V^{-1} - \alpha_{\exp(2\pi i r/s),1}(e_\beta)$. Note that as $r's - rs' = 1$, r, s are relatively prime, so rotating by the powers of $\exp(2\pi i r/s)$ gives the same set of powers of $\exp(2\pi i/s)$, the sth roots of unity, traced out in a different order. Now $\operatorname{Ad} V = \alpha_{q,1}$, and to control

$$\begin{aligned} &\left\|\alpha_{q,1}(e_\beta) - \alpha_{\exp(2\pi i r/s),1}(e_\beta)\right\| \\ &\le \left\|\alpha_{q,1}(f) - \alpha_{\exp(2\pi i r/s),1}(f)\right\| + 2\left\|\alpha_{q,1}(g) - \alpha_{\exp(2\pi i r/s),1}(g)\right\| \end{aligned}$$

it is enough to estimate $\|\alpha_{q,1}(f) - \alpha_{\exp(2\pi i r/s),1}(f)\|$ and hence all we need to control is $|\text{maximum slope of } f| \cdot |\theta - r/s|$. Now in the construction of the Rieffel projection in Fig. 3.15 there was some freedom in assigning the lengths ℓ, x of the intervals $[c,d]$ and $[a,b]$ respectively. The total area $\ell + x = \beta$ constrains $x \le \beta$. In order to fit in q orthogonal copies the total length of $[a,d]$ must be less than $1/s$, i.e. $\ell + 2x \le 1/s$. Since $\ell + x = \beta$, this area constraint gives a further constraint on $x \le 1/s - \beta$. Thus we are forced that the slope $x^{-1} \ge \max\{\beta^{-1}, (1/s - \beta)^{-1}\}$. In order to maximize our chances of controlling the estimate we need, we take the slope to be the least that it has to be, namely slope (in $[a,b]$)$\equiv \max\{\beta^{-1}, (1/s -$

$\beta)^{-1}\} = \max\{1/\beta, s/s'\beta'\}$. If we let $\gamma \equiv (\theta - r/s)/(r'/s' - \theta) = r'\beta'/r\beta$ then $|\text{slope}||\theta - r/s| \leq (1/r', \gamma/r')$. Thus we are in business if we can control γ. This control is possible because if θ is irrational, there are infinitely many matrices $\begin{pmatrix} r' & r \\ s' & s \end{pmatrix}$ in $SL(2, \mathbb{Z})$ such that $1/4 < \gamma < 4$. For example if θ is in the golden mean, and we take r/s, r'/s' successive approximants in the continued fraction expansion, then

$$(r/s = i) < [(r + r')/(s + s') = j] < \theta < [(r + 2r')/(s + 2s') = k] < (r'/s' = \ell)$$

yields $(j - i)/(\ell - j) < \gamma < (k - i)/(\ell - k)$. Using $r's - rs' = 1$, this estimate is just $s'/s < \gamma < 2s'/s$. Since for the golden mean the ratio s'/s of successive denominators in the continued fraction expansion r_n/s_n converges to $(1+\sqrt{5})/2$, $(s_n = s_{n-1} + s_{n-2})$, we find that γ is kept away uniformly from 0 and ∞. By reversing the roles of U, V (more precisely taking the Fourier transform $U \to V$, $V \to U^{-1}$) we find also s' orthogonal Rieffel projections $\{e^j_{\beta'}\}$ such that V is now approximately diagonal with scalar entries and U is approximately a permutation matrix. These two towers will not be orthogonal to each other, but at least in K_0, $\sum_{i=1}^{s} e^i_\beta + \sum_{j=1}^{s'} e^j_{\beta'} = 1$, (i.e. $s\beta + s'\beta' = 1$). Now by (Rieffel 1983b), $K_0(A_q)$ has cancellation, which means we can find a unitary w in A_q such that $e + we'w = 1$, if $e = \sum e^i_\beta$, $e' = \sum e^j_{\beta'}$. We can do even better than this, as $e_\beta \in C^*(U, V^{s'})$, so that (as $\sum e^i_\beta$ is even more periodic) $e \in C^*(U^s, V^{s'})$. Similarly $e' \in C^*(U^s, V^{s'})$, so we can even find a unitary w in the rotation algebra $C^*(U^s, V^{s'}) \cong A_{ss'\theta} = A_{\gamma/1+\gamma}$ (as $\gamma/(1 + \gamma) = ss'\theta$ mod $\mathbb{Z}$). Since $s\theta, s'\theta$ are approximately integral U, V almost commute with $U^s, V^{s'}$. If we choose f smooth, so that e_β (and similarly $e_{\beta'}$) is smooth with respect to the $\mathbb{T}^2$ action, then we can also choose the conjugating unitary w to be a smooth function of $U^s, V^{s'}$. This smooth function can be controlled by a continuous field argument on the family of C^*-algebras $\{A_{\gamma/1+\gamma} \mid \gamma\}$, and so U, V almost commute with w. This means with respect to the modified orthogonal towers, U, V have approximately the form as before in each tower, one diagonal, one permutation matrix and reversed roles in the second tower. Note that the 'permutation' matrices are actually of the form

$$\begin{pmatrix} 0 & & * \\ 1 \cdot & & \\ \cdot & \cdot & \\ & & \cdot \\ 0 & & 1\,0 \end{pmatrix}$$

with $*$ a unitary with the same K_1 class as V in the first tower and U in the second tower. Since these K_1 classes of V and U are non-trivial, what we actually have is an approximation by

$$M_s(C(\mathbb{T})) \oplus M_{s'}(C(\mathbb{T})) \,. \tag{3.7.16}$$

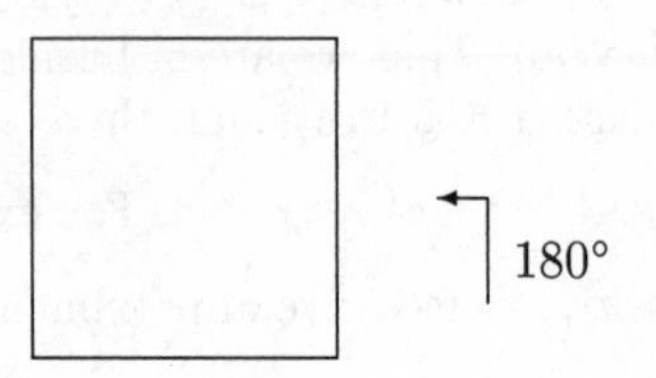

FIG. 3.16. $\mathbb{Z}/2$ action on $\mathbb{T}^2$

Higher dimensional simple non-commutative tori (generated by unitaries U_i which commute up to some scalars $U_l U_m = e^{2\pi i\theta_{l,m}} U_m U_l$) are also known generically to be $A\mathbb{T}$ algebras (Boca, preprint 1995), and all simple 3-tori are $A\mathbb{T}$ (Elliott and Lin 1996), (Lin 1996).

The structure of A_q as a circle algebra (3.7.16) also has some consequences for A_q previously shown by direct methods which can now be deduced from the general theory of circle algebras. A_q has real rank zero, i.e. invertible self adjoint elements are dense in the self adjoint elements – first shown by (Choi and Elliott 1990), (Haagerup, preprint 1991), and (Blackadar *et al.* 1992b) and now follows from (Blackadar *et al.* 1992a). A_q has topological stable rank one, i.e. invertible elements are dense – first shown by (Riedel 1985), (Putnam 1990b) and now follows from (Dadarlat *et al.* 1992). Moreover there exists an embedding of A_q into an AF algebra which is the identity on K_0 – first shown by (Pimsner and Voiculescu 1980a), but now follows from the classification result of (Elliott 1993a) for circle algebras (map $K_0(A_q) \to \mathbb{Z} + \theta\mathbb{Z}$, $K_1(A_q) \to 0$, and lift; alternatively map the circle algebras to interval algebras, where again K_1 has been destroyed).

To understand the structure of A_q as a circle algebra and $A_q \rtimes \mathbb{Z}_2 = C(\mathbb{T}) \rtimes \mathbb{Z} \rtimes (\mathbb{Z}/2)$, $C(\text{Cantor}) \rtimes \mathbb{Z} \rtimes (\mathbb{Z}/2)$ as AF algebras, the notion of non-commutative orbifold is again fruitful. The flip $(z_1, z_2) \to (\overline{z}_1, \overline{z}_2)$ on the classical or commutative 2-torus $\mathbb{T}^2$ gives rise to the orbifold $\mathbb{T}^2/(\mathbb{Z}/2)$ which is topologically a sphere. The flip corresponds geometrically on $\mathbb{T}^2 = \mathbb{R}^2/\mathbb{Z}^2$ to rotation through 180° in Fig. 3.16. To identify the quotient space $\mathbb{T}^2/(\mathbb{Z}/2)$ we only need the triangle in Fig. 3.17, with the identifications of the edges as shown. Folding, one obtains a tetrahedron, the four vertices corresponding to the fixed points of the flip on $\mathbb{T}^2$. Hence in the crossed product $C(\mathbb{T}^2) \rtimes (\mathbb{Z}/2)$ we pick up a copy of M_2 for each point of the sphere except that at the four singular points we are restricted to $\mathbb{C}^2$, i.e. M_2 valued functions on the sphere, restricted to $\mathbb{C}^2$ at four distinguished points.

A non-commutative toroidal orbifold arises as the fixed point algebra of the non-commutative torus $A_q = C^*(U, V)$ under the symmetry $U, V \to U^{-1}, V^{-1}$ respectively (which preserves the relation $VU = qUV$). Flipping generators in A_q means conjugating $(z_i) \to (\overline{z}_i)$ for the classical torus $\mathbb{T}^2$. In general $A_q^{\mathbb{Z}/2}$, $A_q \rtimes (\mathbb{Z}/2)$ can be referred to as non-commutative toroidal orbifolds. Note that the flip on $\mathbb{T}^2$ induces a transformation on the space of leaves of the Kronecker

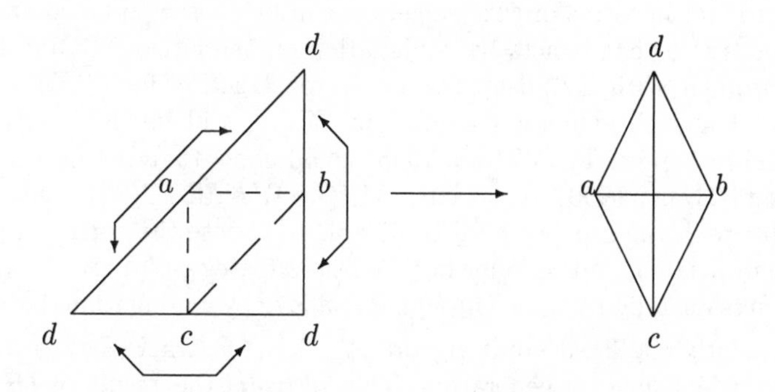

FIG. 3.17. Toroidal orbifold, $\mathbb{T}^2/(\mathbb{Z}/2) = S^2$

FIG. 3.18. Non-commutative toroidal orbifold

foliation F_θ (i.e. leaves are taken to leaves under the symmetry), and hence a singular foliation of the 2-sphere S^2. The fixed point algebra $B_q = A_q^{\mathbb{Z}/2}$ bears the same relation to this singular foliation (see the diagrams in Fig. 3.18), that the non-commutative torus A_q does to the Kronecker foliation F_θ; $C^*(\mathbb{T}^2, F_\theta)^{\mathbb{Z}/2}$ is Morita equivalent to B_q. If q is not a root of unity, it is natural to ask, in view of the situation with exotic symmetries on AF algebras obtained by flipping one circle, whether B_q is AF or not.

The rotation algebra A_θ is itself a crossed product $C(\mathbb{T}) \rtimes_\theta \mathbb{Z}$ where $\mathbb{Z}$ acts on the first circle $\mathbb{T}$ (the spectrum of U) by rotation through the angle θ ($VUV^{-1} = \exp(2\pi i\theta)U$). The first indication that such a crossed product could be a circle algebra was found by (Putnam 1990a), who showed that $C(X) \rtimes \mathbb{Z}$ is a circle algebra when $\mathbb{Z}$ is a minimal action on a Cantor set X. (Bratteli *et al.* 1995) showed that if there is also present a symmetry on X with a fixed point in such a way that $\mathbb{Z} \rtimes (\mathbb{Z}/2)$ acts on X (so that $\mathbb{Z}/2$ flips $\mathbb{Z}$, $x \to -x$), then the crossed product $C(X) \rtimes \mathbb{Z} \rtimes (\mathbb{Z}/2)$ is AF. Indeed the decomposition of the first crossed product in terms of circle algebras can be chosen to be compatible with the symmetry so that it acts by conjugation on the circles. Taking the further crossed product with $\mathbb{Z}/2$ destroys the K_1 obstruction for $C(X) \rtimes \mathbb{Z} \rtimes (\mathbb{Z}/2)$ being AF. For such algebras classified by K_*, K_1 will be the only obstruction to an algebra being AF. (Walters 1995) could show that the decomposition of (Elliott and Evans 1993) $A_\theta = \lim_\to M_s(C(\mathbb{T})) \oplus M_{s'}(C(\mathbb{T}))$ could be chosen compatible with the flip (we need to be able to choose the correcting unitary w to be flip invariant), and so following the same strategy, $C(\mathbb{T}) \rtimes_\theta \mathbb{Z} \rtimes (\mathbb{Z}/2)$ is AF where $\mathbb{Z}$ acts on $\mathbb{T}$ by rotation through θ and $\mathbb{Z}/2$ by conjugation, i.e. $A_\theta \rtimes (\mathbb{Z}/2)$ (or equivalently the fixed point algebra $A_\theta^{\mathbb{Z}/2}$) is AF, where $\mathbb{Z}/2$ is the previous symmetry which inverts generators. This recovers the result of (Bratteli and Kishimoto 1992). The dual action of $(\mathbb{Z}/2)^\wedge$ on the AF algebra $A_\theta \rtimes (\mathbb{Z}/2)$ has a fixed point algebra which is essentially isomorphic to A_θ by Pontryagin–Takai duality and hence is not AF.

If $q^2 \neq 1$, then the algebra B_q is generated by $A = U + U^{-1}$ and $B = V + V^{-1}$, (Bratteli *et al.* 1991b). It is reasonable to try and relate the AF (totally disconnected) structure of $A_\theta^{\mathbb{Z}/2} = C^*(U+U^{-1}, V+V^{-1})$ with the structure of the Cantor spectra of the Hamiltonians $H = U + U^{-1} + \lambda(V + V^{-1})$, $\lambda \neq 0$, when θ is irrational. Represent A_θ (for θ irrational) on $L^2(\mathbb{T})$, generated by multiplication and translation operators $Uh(z) = zh(z)$, $Vh(z) = h(\exp(2\pi i\theta)z)$. (The norm closure of the algebra generated by U and V generates A_θ, and so depends heavily on θ whilst the weak closure is always $B(L^2(\mathbb{T}))$. In this representation the operator $U + U^{-1} + \lambda(V + V^{-1})$ is the almost Mathieu operator:

$$f \to f(n+1) + f(n-1) + 2\lambda\cos(2\pi n\theta) f(n)$$

which (generically at least) has a Cantor spectrum (Bellisard and Simon 1982), (Simon 1982), (Choi *et al.* 1990) for θ irrational, $\lambda \neq 0$. Since the fixed point algebra under the symmetry which inverts U and V is generated by $U + U^{-1}$, $V + V^{-1}$, it is the natural C^*-algebra to Hamiltonians $U + U^{-1} + \lambda(V + V^{-1})$.

The non-commutative torus A_q is the universal C^*-algebra generated by two unitaries U and V satisfying the relation $VU = qUV$. In the irrational case ($q^n \neq 1$, for all integral $n \neq 0$), A_q is simple and is indeed the unique C^*algebra generated by unitaries U and V satisfying this commutation relation. The non-commutative orbifolds B_q are analogously described as follows (Bratteli *et al.* 1991b). There is an action of $SL(2, \mathbb{Z})$ on A_q, where the matrix $\begin{pmatrix} -1 & 0 \\ 0 & -1 \end{pmatrix}$ corresponds to the flip $U \to U^{-1}$, $V \to V^{-1}$. Since $\begin{pmatrix} -1 & 0 \\ 0 & -1 \end{pmatrix}$ generates the centre of $SL(2, \mathbb{Z})$, there is an action of $PSL(2, \mathbb{Z})$, on the fixed point algebra $B_q = A_q^{\mathbb{Z}/2}$. The generators $S = \begin{pmatrix} 0 & -1 \\ 1 & 0 \end{pmatrix}$, $T = \begin{pmatrix} 1 & 1 \\ 0 & 1 \end{pmatrix}$ of $PSL(2, \mathbb{Z})$ act on B_q as follows. S merely interchanges A and B, but T acts less trivially by fixing A, but taking B to $q^{1/2}(1 - q^2)^{-1}[A, B]_q$ where $[X, Y]_q$ denotes the twisted or q-commutator

$XY - qYX$. Similarly T^{-1} fixes A and takes B to $q^{-1/2}(1-q^2)^{-1}[A,B]_{q^{-1}}$. Hence $TT^{-1}(B) = B$ gives that

$$B = -(q-q^{-1})^2[A,[A,B]_{q^{-1}}]_q$$

or

$$BA^2 + A^2B = (q+q^{-1})ABA - (q-q^{-1})^2B\,. \tag{3.7.17}$$

Similarly

$$A = -(q-q^{-1})^2[B,[B,A]_{q^{-1}}]_q$$

or

$$AB^2 + A^2B = (q+q^{-1})BAB - (q-q^{-1})^2A\,. \tag{3.7.18}$$

A further relation is

$$\begin{aligned} BABA = {} & (q^2+1-q^{-2})ABAB - (q+q^{-1})A^2B^2 \\ & + (q+q^{-1}-(q^3+q^3))(A^2+B^2-2) \end{aligned} \tag{3.7.19}$$

In the irrational case ($q^n \neq 1$, for all non-zero integral n), B_q is simple and is the unique C^*-algebra generated by two self adjoint operators A and B (identified with $U+U^{-1}, V+V^{-1}$ respectively) satisfying the three independent relations (3.7.17), (3.7.18) and (3.7.19). Note that if $q^4 \neq 1$, then the third relation shows that the unit of B_q is in the polynomial algebra generated by A and B. This actually only occurs when $q^4 \neq 1$. A simpler relation between the unit and a polynomial expression in A and B can be deduced as follows. If U, V satisfy $UV = qVU$ then U^2, V satisfy $VU^2 = q^2U^2V$. Replacing A, B, q in (3.7.18) with $U^2 + U^{-2} = A^2 - 2$, B and q^2 we obtain:

$$\begin{aligned} 2(q^2-q^{-2})^2 1 = {} & A^2B^2 + B^2A^2 - (q^2+q^{-2})BA^2B \\ & + 2(q-q^{-1})^2B^2 + (q^2-q^{-2})^2A^2\,. \end{aligned}$$

Again this can be solved for the unit unless $q^4 = 1$.

In the rational case (more precisely if $q^n = 1$, for some integral n, but $q^2 \neq 1$), B_q is the enveloping C^*-algebra of $P(U,V)^{\mathbb{Z}/2}$, where $P_q = P(U,V)$ denotes the $*$-subalgebra of A_q generated by U and V (i.e. the polynomial algebra in U and V). The enveloping C^*-algebra of $P(U,V)^{\mathbb{Z}/2}$ does not exist when $q^2 = 1$.

To understand the K-theory of the non-commutative orbifold B_q, it is illuminating to first consider the rational case (Bratteli *et al.* 1992). In this case, with $q = \exp(2\pi ir/s)$, with r, s relatively prime integers, the algebra B_q can be identified with a subalgebra of the C^*-algebra $C(S^2, M_s)$ of continuous functions from the 2-sphere S^2 into M_s, which at four distinct points $\omega_0, \omega_1, \omega_2, \omega_3$ in S^2 commute with an associated projection E_i in M_s of dimension:

$\dim(E_i) = (s-1)/2$, when s is odd.

$\dim(E_0) = (s-2)/2$, $\dim(E_i) = s/2$, for $i = 1,2,3$ when s is even.

Thus when $q = e^{2\pi ir/s}$, the isomorphism class of B_q depends on s but not on r, although the non-commutative tori A_q and $A_{q'}$ are isomorphic if and only if $q' = q$ or $q' = q^{-1}$.

Using the six-term exact sequence for $C^*(SL(2,\mathbb{Z}))$ (Natsume 1985), (Kumjian 1990), computes $K_0(B_q) = \mathbb{Z}^6$ if $q^2 \neq 1$, ($\mathbb{Z}^5$ when $q = -1$, $\mathbb{Z}^2$ when $q = 1$), $K_1(B_q) = 0$ with the latter consistent with B_q being AF, as K_1 always vanishes for an AF algebra. These K-groups can easily be computed (Bratteli *et al.* 1992) when q is a root of unity using the following simple description of B_q in that case. The algebra $A_{\exp(2\pi ir/s)}$ is a homogeneous algebra over the 2-torus $\mathbb{T}^2$ with fibre M_s (Høegh-Krohn and Skjelbred 1981), (Rieffel 1981), but the obstruction to isomorphism for the non-commutative tori is eliminated for the non-commutative orbifolds by the behaviour of the four singular points. Consequently, the bundle for $B_{\exp(2\pi ir/s)}$ is trivial. The K-theory of $B_{\exp(2\pi ir/s)}$ can then be computed using the ideal of functions which vanish at all the ω_i, together with the six-term short exact sequence of K-theory, Fig. 3.12. More informally to see $K_0(B_q) = \mathbb{Z}^6$ if $q^2 \neq 1$, a projection in B_q will give an element in $\mathbb{Z}^8$, say $\{(d_1^i, d_2^i) \mid i = 0,1,2,3\}$ with three constraints $d_1^i + d_2^i =$ constant, which gives us $\mathbb{Z}^5$; the additional element of $K_0(B_q)$ is provided by the Bott element on the 2-sphere. (If $p \in C(S^2, M_s)$ is a projection, then $(d_1^i, d_2^i) = (\dim(p(\omega_i)E_i), \dim(p(\omega_i)(1-E_i)))$ is in $\mathbb{Z}^8$, where the dimension is computed in M_s, with $d_1^i + d_2^i =$ constant due to the homotopy invariance of $\dim f(\omega)$ for $\omega \in S^2$).

To understand the traces of B_q, it is convenient to bring the crossed product $C_q = A_q \times (\mathbb{Z}/2)$ into the game. When $q = e^{2\pi ir/s}$ and r and s are relatively prime positive integers, C_q is isomorphic to a subalgebra of the C^*-algebra $C(S^2, M_{2s})$ of continuous functions from the 2-sphere S^2 into the algebra M_{2s} determined as follows. There are four distinct points ω_0, ω_1, ω_2, ω_3 in S^2 and an orthogonal projection E in M_{2s} of dimension s, such that the subalgebra consists of those functions $f \in C(S^2, M_{2s})$ such that $f(\omega_i)$ commutes with E for $i = 0,1,2,3$. For any q, the algebraic crossed product $P_q \times (\mathbb{Z}/2)$ is the $*$-algebra generated by three unitaries U, V and W satisfying the relations $VU = qUV$, $WU = U^{-1}W$, $WV = V^{-1}W$, and $W^2 = 1$, its completion is the crossed product $C_q = A_q \times (\mathbb{Z}/2)$. There is a canonical continuous tracial state τ on $A_q \times (\mathbb{Z}/2)$ given by $\tau(q^{nm/2}U^nV^m) = 1$ if $n = m = 0$, and 0 otherwise, $\tau(q^{nm/2}U^nV^mW) = 0$. There are four other trace functionals τ_{p_1,p_2} where $p_1, p_2 \in \{\text{even, odd}\}$ defined on $P_q \times (\mathbb{Z}/2)$ by: $\tau_{p_1p_2}(q^{nm/2}U^nV^m) = 0$, $\tau_{p_1p_2}(q^{nm/2}U^nV^mW) = 1$, if parity$(n) = p_1$ and parity$(m) = p_2$, and 0 otherwise. When q is not a root of unity, the space of trace functionals on $P_q \times (\mathbb{Z}/2)$ is five dimensional spanned by τ and the $\tau_{p_1p_2}$, with the only continuous trace functionals on C_q (respectively A_q) being scalar multiples of τ (respectively $\tau|A_q$). It is however remarkable, when q is a root of unity, that the other four trace functionals are continuous and extend to C_q. The algebraically defined trace functionals $\tau_{p_1p_2}$ are related to the following geometrically defined trace functionals $\tilde{\tau}_{p_1p_2}$ which are clearly *continuous*. There is a convenient labelling of the four singular points ω_i by $\{\text{even, odd}\}^2$, i.e. $\omega_{p_1p_2}$. Define four continuous trace functionals on C_q by: $\tilde{\tau}_{p_1p_2}(f) = \mathrm{Tr}_{2s}((2E-1)f(\omega_{p_1p_2}))/s$, where Tr_{2s} is the unnormalized trace on M_{2s}, and E is the rank s projection in M_{2s} as above. Then, adopting a proper sign convention, the algebraically defined trace functionals are related to the geometrically defined trace functionals by

$$\tau_{p_1p_2} = (-1)^{p_1p_2}\tilde{\tau}_{p_1p_2} \text{ if } s \text{ is even and}$$

$$\tau_{nm} = (-1)^{rnm}\left\{\textstyle\sum_{p_1p_2\in\mathbb{Z}/2}(-1)^{np_1+mp_2}\tilde{\tau}_{p_1p_2}\right\} \text{ when } s \text{ is odd}.$$

In particular the trace functionals $\tau_{p_1p_2}$ are continuous when q is not a root of unity. The five trace functionals τ and $\tau_{p_1p_2}$ separate the following five projections in C_q written down by (Nest, preprint 1989):

$$1, (1+W)/2, (1+UW)/2, (1+VW)/2, (1+q^{1/2}UVW)/2$$

These projections, together with the Rieffel projection, show that $K_0(C_q)$ contains $\mathbb{Z}^6$, even when q is not a root of unity.

The four spurious trace functionals on $P_q \times (\mathbb{Z}/2)$, and in particular the fact that they are unbounded only in the rational case, have been exploited to show the following. The $\mathbb{T}^2$ action on the non-commutative torus A_q gives rise to a family of seminorms $\|\delta_1^m\delta_2^n(\cdot)\|$ from the corresponding derivations δ_1, δ_2 on the smooth subalgebra A_q^∞ and $B_q^\infty = A_q^\infty \cap B_q$. Then an analysis of the spurious traces shows that the linear span of projections in B_q^∞ is *not* dense in B_q^∞ for the natural topology for the seminorms. However, for irrational q, both A_q and B_q have real rank zero – the linear span of projections is dense for the norm topology.

(Nest, preprint 1989) has computed the cyclic cohomology of these non-commutative toroidal orbifolds. In the course of this he has shown that all derivations on the smooth elements of B_q (smooth with respect to the $\mathbb{T}^2$ torus action on the rotation algebra A_q) are approximately inner if q is not a root of unity (cf. (Powers and Sakai 1975) for the conjectured structure of derivations on AF algebras).

The classification of amenable infinite C^*-algebras has already reached a more advanced stage, or at least a cleaner statement of the results. This has culminated in Kirchberg classifying by K_*, the simple, purely infinite, separable amenable unital C^*-algebras which satisfy the UCT. As in the finite theory, inductive limits of elementary infinite building blocks were first classified. The first natural building blocks are the Cuntz algebras $\mathcal{O}_n$. These have torsion in $K_0(\mathcal{O}_n) = \mathbb{Z}/(n-1)$ with trivial K_1 (cf. matrix algebras as the basic building blocks for AF algebras). The Cuntz algebra $\mathcal{O}_n$ can be described as the crossed product $\bigotimes_{\mathbb{N}} M_n$, by the unilateral shift or the stable algebra as the crossed product $(K \otimes (\bigotimes_{\mathbb{N}} M_n)) \rtimes \mathbb{Z}$ by an automorphism which scales the trace by $1/n$. A non-commutative Rohlin property was established in (Bratteli *et al.* 1993) for the quasi-free shift on the Fermion algebra. This led to the result that the Cuntz algebra $\mathcal{O}_2$ is isomorphic to its tensor product with the Fermion algebra. The Rohlin property for the shift on $\bigotimes_{\mathbb{N}} M_2$ yields the same on $\bigotimes_{\mathbb{N}} M_{2n} = (\bigotimes_{\mathbb{N}} M_2) \otimes (\bigotimes_{\mathbb{N}} M_n)$. (Bratteli *et al.*

1993) obtained a weak Rohlin property on $\bigotimes_{\mathbb{N}} M_m$ for arbitrary m through the embedding of the GICAR in the UHF algebra which respected the shifts (Connes and Evans 1989), see Remark on page 343. (Kishimoto 1996) then showed that this was enough to obtain the usual Rohlin property. The decomposition of the crossed products of the Fermion algebra by the shift (Bratteli *et al.* 1993) or by Bogoliubov automorphisms (Bratteli *et al.* 1993) is also dependent on this Rohlin property – generically all such crossed products are isomorphic. (Rørdam 1993) gave a K-theoretical classification of inductive limits of arbitrary sequences of finite direct sums of matrix algebras of the Cuntz algebras $\mathcal{O}_m$ where m is even. Such algebras are of real rank zero and K_1 zero. (Rørdam 1995a) then increased the class of classifiable algebras with the building blocks composed of crossed products of certain AF algebras by shifts. In particular (and this is very different from factors), for every $\lambda \in (0,1)$ there exists a stable simple AF algebra with unique trace and an automorphism scaling the trace by λ, such that the crossed product is isomorphic to $\mathcal{O}_2 \otimes K$. (Bratteli *et al.* (to appear a)) extended Rørdam's classification to a class of simple algebras of real rank zero which are inductive limits of finite direct sums of matrix algebras over $\mathcal{O}_n \otimes C(\mathbb{T})$ for which K_1 may be non-zero. This step is analogous to the generalization in (Elliott 1993a) from matrix algebras over $\mathbb{C}$ to matrix algebras over $C(\mathbb{T})$ i.e. from AF algebras to real rank zero inductive limits of finite direct sums of matrix algebras over $C(\mathbb{T})$. Amongst other things, this gives:

- If A_α, A_β are irrational rotation C^*-algebras, then $\mathcal{O}_n \otimes A_\alpha \cong \mathcal{O}_n \otimes A_\beta$ if $n \geq 2$.

This follows because A_α, A_β are inductive limits of direct sums of matrix algebras over $C(\mathbb{T})$ by (Elliott and Evans 1993), and both tensor products have the same K-theory by the Künneth formula. The ordering on $K_0(\mathcal{O}_n \otimes A_\alpha)$ is now trivial, everything is positive by (Cuntz 1981a), and so the angle α cannot be recovered from $K_0(\mathcal{O}_n \otimes A_\alpha)$ as was done from the ordering on $K_0(A_\alpha)$.

- If n, m are positive integers, and $k-1$ is the greatest common divisor of $n-1$ and $m-1$, then

$$\mathcal{O}_n \otimes \mathcal{O}_m \cong \mathcal{O}_k \otimes \mathcal{O}_k \cong \mathcal{O}_k \otimes B_k$$

  where B_k is the Bunce–Deddens algebra $C(\mathbb{T}) \rtimes G_k$, and G_k is the subgroup of $\mathbb{T}$ generated by $\{2\pi/k^r \mid r \in \mathbb{N}\}$ and acts on $\mathbb{T}$ by translation.

- If $n-1$ and $m-1$ are relatively prime, then $\mathcal{O}_n \otimes \mathcal{O}_m \cong \mathcal{O}_2$ and $\mathcal{O}_2 \otimes A \cong \mathcal{O}_2$, for any simple unital C^*-algebra A which is an inductive limit of the class of (Bratteli *et al.* (to appear a)).

Of course, these results all now follow from Kirchberg's classification, which is built on his study of exact C^*-algebras, those algebras A for which $B \to A \otimes_{\min} B$ preserves short exact sequences. Any subalgebra of a nuclear C^*-algebra is exact, and (Kirchberg 1995) has shown that any exact, separable C^*-algebra can be embedded in $\mathcal{O}_2$. (Phillips, preprint 1994) independently obtained the same classification beginning with the following two results announced by Kirchberg at the Geneva ICM satellite in the summer of 1994:

(a) $A \otimes \mathcal{O}_2 \cong \mathcal{O}_2$ if A is simple, separable, unital and amenable;
(b) $A \otimes \mathcal{O}_\infty \cong A$ if A is also purely infinite.

The algebra $\mathcal{O}_\infty$ plays a significant role in the theory of Case 3, as we can see from (b), with $K_0(\mathcal{O}_\infty) \cong \mathbb{Z}, K_1(\mathcal{O}_\infty) \cong 0$. If $\mathcal{O}_\infty$ is a unit for a simple, amenable, unital separable A in the sense that $A \otimes \mathcal{O}_\infty \cong A$, then A is automatically purely infinite (Kirchberg 1995). The question then arises as to whether there is an infinite dimensional, simple, unital, amenable C^*-algebra B with unique trace and

$$K_0(B) \cong K_0(\mathbb{C}) \cong \mathbb{Z}$$

as scaled ordered groups and

$$K_1(B) \cong K_1(\mathbb{C}) \cong 0\,.$$

In fact (Jiang and Su (preprint 1996)) have constructed such a C^*-algebra $\mathcal{C}$ which is projectionless (no projections other than 0 or 1) and is an inductive limit of dimension drop algebras of the form

$$M_p \overset{M_n}{\text{———————}} M_q$$

where $p|n$, $q|n$ but p, q are relatively prime so that the building blocks themselves are essentially projectionless. With the classification of simple inductive limits of such dimension drop algebras, the algebra $\mathcal{C}$ is the unique C^*-algebra in this class with prescribed K-theory, unique trace and moreover $A \otimes \mathcal{C} \cong A$ for all infinite dimensional, separable, simple unital AF algebras. This is a strengthening of (3.7.11), where now since $\mathcal{C}$ is projectionless, it is real rank one. However, there are examples of infinite dimensional, unital, stably finite, amenable, simple C^*-algebras A for which $A \ncong A \otimes \mathcal{C}$ (Gong, Jiang and Su).

The classification of (Elliott 1976) for AF algebras by the ordered (scaled) group $K_0(A)$ was extended to classifying limit inner actions of a compact group G on AF algebras (Fack and Maréchal 1979), (Fack and Marèchal 1981), (Kishimoto 1977), (Handelman and Rossmann 1984), (Handelman and Rossman 1985) (i.e. where $A = \lim_\rightarrow A_n$, and the action of G is given by inner unitary actions on finite dimensional C^*-algebras A_n), and to limit actions of $\mathbb{Z}/2$ (Elliott and Su (to appear)), (Su, (to appear)) (i.e. where $A = \lim_\rightarrow A_n$ and $\mathbb{Z}/2$ leaves finite dimensional C^*-algebras A_n globally invariant). (Elliott and Su (to appear)), (Su, (to appear)) showed that limit actions of $\mathbb{Z}/2$ on AF algebras are classified by the inclusion of the dimension groups $K_0(A)$ in $K_0(A \rtimes_\alpha \mathbb{Z}/2)$, together with the class of the unit of A, and the class of the special element, that of unit of the fixed point algebra embedded as a corner of the crossed product, as well as the actions and dual actions on the dimension groups. (Evans and Su, in press)

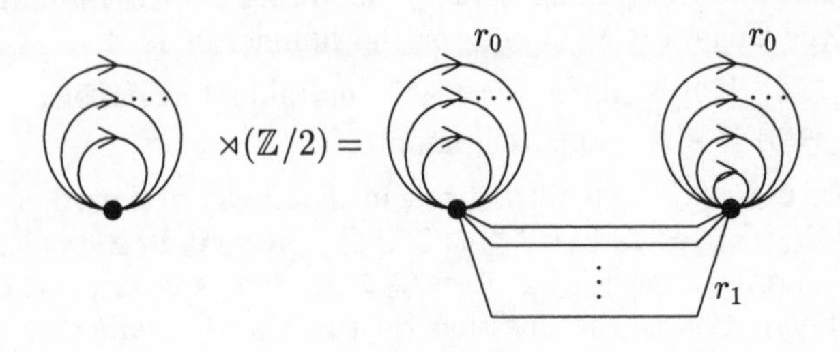

FIG. 3.19. $\mathcal{O}_n \rtimes \mathbb{Z}_2$ orbifold

classified some inductive limits of symmetries on such unital C^*-algebras that can be expressed as inductive limits of finite direct sums of matrix algebras over even Cuntz algebras. For simple algebras the invariant is the inclusion of $K_0(A)$ in $K_0(A \rtimes_\alpha \mathbb{Z}/2)$, decorated with the actions and dual actions on the K-groups. In particular, they could classify actions $s_i \to \pm s_i$, (see (Cuntz and Evans 1981)) on a Cuntz algebra $\mathcal{O}_n = C^*(s_1, \ldots, s_n)$, for n even.

If $V = 1_{r_0} \oplus (-1_{r_1})$ is a self adjoint unitary in M_n of signature $r = r_0 - r_1 = \operatorname{tr} V$, the fixed point algebra $M_n^{\mathrm{Ad}(V)} \cong M_{r_0} \oplus M_{r_1}$ where $r_0 = (n-r)/2$, $r_1 = (n+r)/2$. The fixed point algebra $\otimes^m M_n$ under $\otimes^m \mathrm{Ad}(V)$ has similarly two components with sizes $(n^m - r^m)/2$ and $(n^m + r^m)/2$ as $\otimes^m M_n \cong M_{n^m}$ and the signature of V^m is $\operatorname{tr} V^m = (\operatorname{tr} V)^m = r^m$. This suggests when $m \to \infty$ that what is significant is n^∞ to classify the algebra and $|r|^\infty$ to classify the action – which is precisely the classification of $\otimes_{\mathbb{N}} \mathrm{Ad}\, V$ on $\otimes_{\mathbb{N}} M_n$ by (Fack and Maréchal 1979). The fixed point algebra is an AF algebra with constant embeddings $\begin{pmatrix} r_0 & r_1 \\ r_1 & r_0 \end{pmatrix} = A$. The automorphism $\tau(V)$ on $\mathcal{O}_n$ restricts to $\otimes_{\mathbb{N}} \mathrm{Ad}\, V$ on the canonical UHF algebra, and commutes with the shift. Thus the fixed point algebra $\mathcal{O}_n$ is generated by the AF algebra with constant embeddings A and the shift, i.e. $\mathcal{O}_n^{\mathbb{Z}/2} \cong \mathcal{O}_A$. (If $r_1 = 0$, $\tau = \mathrm{id}$, we have to be more careful and $\mathcal{O}_n^{\mathbb{Z}/2} \cong \mathcal{O}_n \neq \mathcal{O}_{n \oplus n}$ but it is always the case that $\mathcal{O}_n \rtimes (\mathbb{Z}/2) \cong \mathcal{O}_A$.) Thus just by considering $K_0(\mathcal{O}_n \rtimes (\mathbb{Z}/2))$ we have that the absolute value of $\det(1 - A) = (n-1)(r-1)$ is an invariant.

Finally note that these symmetries are essentially orbifold actions (if $r \geq 0$) To see this, write $V = 1_r \oplus \sigma_{n-r}$ where σ_{n-r} is a Pauli matrix of signature zero. Then V leaves fixed r of the edges of the graph Fig. 3.19 of $\mathcal{O}_n$, and interchanges the remaining $(n-r)$. The vertex v then bifurcates into two new ones v_1 and v_2. Edges which are fixed are replaced with two edges which take v_i to v_i, and a pair of interchanged edges are replaced with a pair from v_i to v_j, $i \neq j$, and another pair from v_i to v_i, to obtain Fig. 3.19. This construction generalizes. If $\pi : G \to M_n$ is a unitary representation of a compact group G, then we know by Example 3.10 that the fixed point algebra $\left(\bigotimes_{\mathbb{N}} M_n\right)^G$ under the product action $\bigotimes_{\mathbb{N}} \mathrm{Ad}\, \pi$ has an AF fixed point algebra with Bratteli diagram Γ_π, vertices indexed by $\hat{G}$, and transitions given by the fusion rules of tensoring with π, but initial vertex the trivial representation. Thus $\mathcal{O}_n \rtimes_\pi G$ is isomorphic to a cut down (by the projection arising from the initial vertex) of a Cuntz–Krieger algebra $\mathcal{O}_A$, where A is the possibly infinite graph, vertices indexed by $\hat{G}$ and transitions given by π. Then (3.7.7) may be used to compute the K-theory of $\mathcal{O}_n \rtimes G$, e.g. $K_0(\mathcal{O}_2 \rtimes SU(2)) \cong \mathbb{Z}(t)/(1-t)\mathbb{Z}(t) \cong \mathbb{Z}$, $K_1(\mathcal{O}_2 \rtimes SU(2)) \cong \mathrm{Ker}(1-t) = 0$, when $\pi : SU(2) \to M_2$ is the fundamental representation, so that by Kirchberg, we would have $\mathcal{O}_2 \rtimes SU(2) \cong \mathcal{O}_\infty$.

This work on the classification of amenable subfactors and that on the combinatorial aspects of amenable subfactors have benefited from mutual interaction – e.g. the orbifold construction (Evans 1985a), (Evans and Kawahigashi 1994a). (Evans and Kawahigashi 1993) have constructed new subfactors by introducing an orbifold method into the theory of subfactors which also enables one to identify some new solvable models in statistical mechanics predicted by (Di Francesco and Zuber 1990a), see Chapter 13. Rohlin properties of automorphisms on AF and Cuntz algebras in C^*-algebras theory (Bratteli *et al.* 1993), (Bratteli *et al.* 1993), (Kishimoto 1995), (Kishimoto 1996), (Elliott *et al.* (to appear a)), (Evans and Kishimoto 1997) is built on Connes' work on Rohlin properties of automorphisms of factors. It is also striking to note the following result of (Bratteli *et al.* 1989): if α is an action of a compact abelian group on a separable prime C^*-algebra A, such that also the fixed point subalgebra A^α is prime, then the following conditions are equivalent:

- for each $g \in G$, either $\alpha_g = 1$ or α_g is properly outer;
- there exists a faithful irreducible representation of A which is also irreducible on A^α;
- there exists a faithful irreducible representation of A which is covariant.

The second condition is noteworthy in two respects. It involves only the fixed point subalgebra $A^\alpha \subset A$, not the action α itself. This is not evident in the case of the other two conditions. Second, while a representation verifying the third condition is required to be covariant, a representation verifying the second condition must in fact be as far as possible from being covariant. Thus the study of properties of actions on C^*-algebras through inclusions of algebras already has some firm foundation. The projectionless algebra $\mathcal{C}$ of (Jiang and Su (preprint 1996)) has been proposed by those authors as a C^*-analogue of the hyperfinite II_∞ factor – techniques involved in proving $\mathcal{C} \otimes \mathcal{C} \cong \mathcal{C}$ go back to the von Neumann algebra theory.

Similarly analogues the work of Dye and Krieger on orbit equivalence of measure theoretic dynamical systems in a von Neumann algebra setting are being found in topological dynamical systems in a C^*-setting. Dye showed that two ergodic finite measure preserving transformations are orbit equivalent, and Krieger showed that the von Neumann algebra crossed product $L^\infty(X,\mu) \rtimes_T \mathbb{Z}$ is a complete invariant of an ergodic non-singular transformation T on a measure space (X,μ) up to orbit equivalence. In the measure theoretic (respectively topological) setting, *orbit equivalence* means that there is a bimeasurable (respectively bicontinuous) bijection F of one space to the other taking orbits to orbits – or at least for almost all x: $F(T_1^{\mathbb{Z}}(x)) = T_2^{\mathbb{Z}}(x)$ in the measure theoretic case. Building on the work of (Putnam 1990a), with his Berg technique for constructing Rohlin towers for homeomorphisms of Cantor sets, (Giordano *et al.* 1995) showed that the ordered $(K_0(C_0(X) \rtimes \mathbb{Z})/\mathrm{Inf}, [1])$ is a complete invariant of orbit equivalence, for a homeomorphism T of a Cantor set X. Here Inf denotes the *infinitesimal elements* $a \in K_0$ such that $-pu \leq qa \leq pu$ for all $p, q \in \mathbb{N}$, some non-zero $u \in K_0^+$. The quotient group has the induced ordering, whilst

$$K_0(C(X) \rtimes \mathbb{Z}) \cong C(X, \mathbb{Z})/(1 - T_*)C(X, \mathbb{Z}) \tag{3.7.20}$$

by (3.7.5). The ordering on (3.7.20) is the natural one from $C(X, \mathbb{Z})$, but (Putnam 1990a) is needed to prove this (Giordano *et al.* 1995). Note that $K_0(C(X) \rtimes \mathbb{Z})$ is a complete invariant for $C(X) \rtimes \mathbb{Z}$ which is an $A\mathbb{T}$ algebra, by (Putnam 1990a) and (Elliott 1993a), ($K_1(C(X) \rtimes \mathbb{Z}) \cong \mathbb{Z}$ always), whilst the C^*-algebra $C(X) \rtimes \mathbb{Z}$ is a complete invariant of a stronger form of orbit equivalence.

4

POSITIVITY AND SEMIGROUPS

4.1 Introduction

A *one-parameter semigroup* on a Banach space X is a family $\{T_t \mid t \geq 0\}$ of bounded linear operators on X satisfying $T_t T_s = T_{t+s}$ for $t, s \in \mathbb{R}_+$, with $T_0 = \mathrm{id}$. We will be interested in strongly continuous semigroups where $\lim_{t \to s} \|(T_t - T_s)x\| = 0$, $x \in X$, and their *infinitesimal generators*

$$L = \frac{d}{dt}\bigg|_{t=0} T_t \,. \tag{4.1.1}$$

In particular, we will be concerned with the structure of L, and the problem of integrating (4.1.1), or when we can form a semigroup (maybe with certain positivity preserving properties) from an operator L. Typically X will be a Hilbert space, or a C^*-algebra A, or the trace class operators (or more generally the predual of a von Neumann algebra – see Section 5.2).

More precisely if $\{T_t\}$ is a strongly continuous one-parameter semigroup on a Banach space X

$$L(x) = \frac{d}{dt}\bigg|_{t=0} T_t(x) = \lim_{t \downarrow 0} \left(\frac{T_t - 1}{t} \right)(x) \tag{4.1.2}$$

where the domain of L, namely $D(L)$, consisting of those x in X for which the limit in (4.1.2) exists, is T_t-invariant and

$$\frac{d}{dt} T_t = L T_t = T_t L \,. \tag{4.1.3}$$

We will write $T_t = e^{tL}$. If L is bounded (which is equivalent to $\{T_t\}$ being norm continuous, $\lim_{t \downarrow 0} \|T_t - 1\| = 0$), then $T_t = \sum_{n=0}^{\infty} t^n L^n / n!$.

Moreover X may be a partially ordered with respect to a positive cone X_+:

$$X_+ + X_+ \subset X_+ \,, \quad \mathbb{R}_+ \cdot X_+ \subset X_+ \,, \quad X_+ \cap (-X_+) = \{0\} \tag{4.1.4}$$

and we will consider semigroups which preserve the positive cone $T_t(X_+) \subset X_+$ and refer to these semigroups as positive semigroups. Actually for C^*-algebras, we will need a stronger concept of positivity, namely complete positivity, which we will introduce in Section 4.2. One parameter groups arise naturally in the description of time development in physical systems. In classical mechanics one has a manifold M, with a flow $S_t : M \to M$, a group of say diffeomorphisms, and

$$(T_t f)(\omega) = f(S_t\omega), \quad T_t : C_0(M) \to C_0(M).$$

The joint continuity of S_t is equivalent to strong continuity of T. For example if phase space $M = \mathbb{R}^n \times \mathbb{R}^n$, J a complex structure on M, H the Hamiltonian, then $S_t(x) = x_t$ is the solution of $(d/dt)x_t = J\nabla H(x_t)$, $x_0 = x$. One-parameter semigroups can arise as time development of open systems as follows. Restricting to a subsystem, represented by a C^*-subalgebra B, via a *conditional expectation* (a unital projection of norm one) $N : A \to B$, one obtains that if α_t represents reversible dynamics on A then $T_t = N \circ \alpha_t|_B$ describes the irreversible dynamics on B:

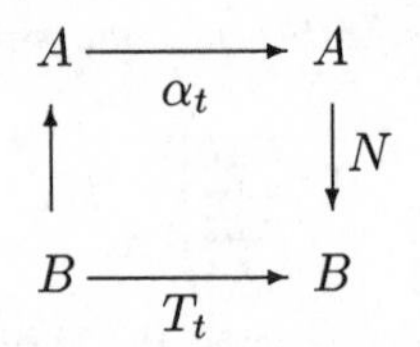

Then T_t is necessarily completely positive as α_t certainly is, and N is by a theorem of (Tomiyama 1957), (e.g. $N = 1 \otimes \varphi$, φ a state of B_1 if $A = B \otimes B_1$). For example if $\{\alpha_t\}$ is a group of $*$-automorphisms of a C^*-algebra A with infinitesimal generator δ then $\alpha_t(xy) = \alpha_t(x)\alpha_t(y)$, $\alpha_t(x^*) = \alpha_t(x)^*$ show that if $x, y \in D(\delta)$ then $xy, x^* \in D(\delta)$ and

$$\delta(xy) = \delta(x)y + x\delta(y), \quad \delta(x^*) = \delta(x)^*.$$

Such a δ will be called a *derivation.* It is analogous to a self adjoint operator on a Hilbert space, the infinitesimal generator of a one-parameter group of unitaries which preserves all the structure of a Hilbert space. This would represent reversible dynamics in quantum systems. For example, if $A = B(\mathcal{H})$, and $U_t = e^{iHt}$ is a strongly continuous one-parameter group of unitaries on $\mathcal{H}$, satisfying Schrödinger's equation $(d/dt)U_t x = iHU_t x$, where H is now the quantized Hamiltonian, then $\alpha_t(x) = U_t x U_t^*$, $x \in B(\mathcal{H})$. Thus

$$\delta(x) = iHx - ixH = i[H, x] = i\ \mathrm{ad}(H)(x).$$

In quantum statistical mechanics, the commutative algebra $C_0(M)$ is replaced by a non-commutative C^*-algebra and reversible Markovian dynamics is described by a group $\{T_t\}$ of $*$-automorphisms.

4.2 Positivity in matrix C^*-algebras

This section concerns positive linear maps T between C^*-algebra A and B. If either A or B is commutative, then such maps are automatically completely positive – Proposition 4.2. Certain Schwarz-type inequalities for positive maps are deduced.

If A is a $*$-algebra, and n is a positive integer, $M_n(A)$ denotes the $*$-algebra of all $n \times n$ matrices over A under the natural operations. Let T be a linear map between $*$-algebras A and B; and let T_n denote the product mapping $T \otimes 1_n$ from $M_n(A)$ into $M_n(B)$ where 1_n denotes the identity mapping on $M_n(\mathbb{C})$. Then T_n acts elementwise on each matrix over $A : T_n : [a_{ij}] \mapsto [T(a_{ij})]$. Suppose B is a C^*-algebra. Then T_n is *positive* if and only if $T_n(a^*a) \geq 0$ for each a in $M_n(A)$ in which case we say T is *n-positive.* We say T is *completely positive* if T is n-positive for all $n \geq 1$. It is useful to study the order structure of matrix C^*-algebras more closely:

Lemma 4.1 *Let A be a C^*-algebra, and $a = [a_{ij}]$ be an element of $M_n(A)$. The following conditions are equivalent:*

(i) $a \geq 0$;

(ii) *a is a finite sum of matrices, each of the form $[b_i^* b_j]$ where $b_1, \ldots, b_n \in A$;*

(iii) *$\sum_{i,j} a_i^* a_{ij} a_j \geq 0$, for all sequences $a_1, \ldots, a_n$ in A.*

Thus a linear map T from A into another C^*-algebra B is completely positive if and only if $\sum_{i,j} b_i^* T(a_i^* a_j) b_j \geq 0$, for all sequences $a_1, \ldots, a_n$ in A, $b_1, \ldots, b_n$ in B.

If A is commutative, then the above three conditions are also equivalent to:

(iv) *$\sum_{i,j} a_{ij} \overline{\lambda}_i \lambda_j \geq 0$, for all sequences $\lambda_1, \ldots, \lambda_n$ in $\mathbb{C}$.*

Proof (i)⇒(ii) If $a \in [a_{ij}] \in M_n(A)$, $a^*a = \sum_{p=1}^n [a_{pi}^* a_{pj}]$

(ii)⇒(iii) is trivial;

(iii)⇒(i): If we represent A on a Hilbert space H, we can decompose H into cyclic orthogonal subspaces. Thus we can assume A has a cyclic vector $f \in H$. Then

$$\sum_{i,j} \langle a_{ij} a_j f, a_i f \rangle = \left\langle \left(\sum_{i,j} a_i^* a_{ij} a_j \right) f, f \right\rangle \geq 0 ,$$

for all $a_1, \ldots, a_n$ in A. Thus, since f is cyclic, $\sum_{i,j} \langle a_{ij} f_j, f_i \rangle \geq 0$ for all $f_1, \ldots, f_n$ in H. That is, $[a_{ij}]$ is positive.

(iv)⇒(iii): Represent A as $C_0(X)$, the continuous functions vanishing at infinity on a locally compact Hausdorff space X. Then

$$\begin{aligned}
&\sum_{i,j} a_{ij} \overline{z}_i z_j \geq 0 , \text{ for all } z_i, \ldots, z_n \in \mathbb{C} , \\
\Rightarrow &\sum_{i,j} a_{ij}(x) \overline{z}_i z_j \geq 0 , \text{ for all } z_i, \ldots, z_n \in \mathbb{C}, x \in X , \\
\Rightarrow &[a_{ij}(x)] \geq 0 \text{ (in } M_n(\mathbb{C})) , \text{ for all } x \in X , \\
\Rightarrow &\sum_{i,j} a_{ij}(x) \overline{a_i(x)} a_j(x) \geq 0 , \text{ for all } a_1, \ldots, a_n \in A , x \in X ,
\end{aligned}$$

$$\Rightarrow \sum_{i,j} a_{ij} a_i^* a_j \geq 0\,, \text{ for all } a_1, \ldots, a_n \in A\,.$$

(i,ii)⇒(iv) is trivial. □

Proposition 4.2 *Let A, B be C^*-algebras, with either A or B commutative. Then any positive linear map from A into B is completely positive.*

Proof Suppose first that B is commutative and $[a_{ij}] \in M_n(A)$ is positive. Then

$$\sum a_{ij} \overline{z}_i z_j \geq 0\,, \text{ for all } z_1, \ldots, z_n \in \mathbb{C}.$$

Then if T is any positive map from A into B, and

$$T\left(\sum a_{ij} \overline{z}_i z_j\right) \geq 0\,, \text{ for all } z_1, \ldots, z_n \in \mathbb{C},$$

hence

$$\sum T(a_{ij}) \overline{z}_i z_j \geq 0\,, \text{ for all } z_1, \ldots, z_n \in \mathbb{C}.$$

The conclusion follows from Lemma 4.1(iv).
Suppose next that A is commutative. By going to the second dual, we can assume that A is a von Neumann algebra and that the given positive linear map T from A into B is ultraweakly continuous. We represent A as $L^\infty(\Omega, \mu)$ for some localizable measure space (Ω, μ) with predual $L^1(\Omega, \mu)$, and we take B to act on a Hilbert space H. Then for all $f, g \in H$, the map $a \mapsto \langle T(a)f, g\rangle$ is ultraweakly continuous on $L^\infty(\Omega, \mu)$. Hence there exists $h(f, g)$ in $L^1(\Omega, \mu)$ such that $\langle T(a)f, g\rangle = \langle a, h(f, g)\rangle$. Moreover $f, g \mapsto h(f, g)$ is sesquilinear, and $h(f, f) \geq 0$ since T is positive. Let $f_1, \ldots, f_n$ be elements of H; then for all $z_1, \ldots, z_n$ in $\mathbb{C}$:

$$\sum z_i \overline{z}_j h(f_i, f_j) = h\left(\sum z_i f_i, \sum z_i f_j\right) \geq 0\,,$$

$$\Rightarrow \sum z_i \overline{z}_j h(f_i, f_j)(\omega) \geq 0 \text{ for almost all } \omega \text{ in } \Omega\,, \tag{4.2.1}$$

$$\Rightarrow [h(f_i, f_j)(\omega)] \geq 0 \quad \text{a.e.}$$

Then, for all $a_1, \ldots, a_n$ in $L^\infty(\Omega, \mu)$, $\sum \langle T(a_i^* a_j) f_j, f_i\rangle = \sum \langle a_i^* a_j, h(f_j, f_i)\rangle = \int_\Omega \sum \overline{a_i(\omega)} a_j(\omega) h(f_j, f_i)(\omega)\, d\mu(\omega) \geq 0$, by (4.2.1). □

It is worth while, especially in connection with generators of positive semigroups in Section 4.7, to look closely at Schwarz-type inequalities for positive maps. Suppose A is a C^*-algebra on a Hilbert space H, and $C \in A^+$. Then for all y in A, $\varepsilon > 0$, $y^*(C + \varepsilon)^{-1} y \in A$ and increases monotonically as $\varepsilon \downarrow 0$. Hence $\lim_{\varepsilon \downarrow 0} y^*(C + \varepsilon)^{-1} y$ exists strongly, if $y^*(C + \varepsilon)^{-1} y$ is bounded above for all $\varepsilon > 0$. We write this limit as $y^* C^{-1} y$. If V is a vector subspace of A and $y^* C^{-1} y$ exists for all y in V, then $\lim_{\varepsilon \downarrow 0} y^*(C + \varepsilon)^{-1} x$ exists for all x, y in V and we write this as $y^* C^{-1} x$.

Lemma 4.3 *Let A be a $*$-algebra, H a Hilbert space and $w : A \to B(H)$ an $(n+1)$-positive $*$-linear map, where $n \geq 2$. Then if $y, x_i \in A$, $i = 1, 2, \ldots, n$, we have*

$$[w(x_i^* x_j)] \geq \left[w(x_i^* y) w(y^* y)^{-1} w(y^* x_j)\right] \text{ in } M_n\left(B(H)\right).$$

In particular

$$\|w(y^* y)\| \, [w(x_i^* x_j)] \geq [w(x_i^* y) w(y^* x_j)] .$$

Proof Put $x_0 = y$. Then if $\xi_0, \xi_1, \ldots, \xi_n \in H$,

$$\textstyle\sum_{i,j=0}^n \langle w(x_i^* x_j)\xi_j, \xi_i\rangle \geq 0 \text{ since } w \text{ is } (n+1)\text{-positive.}$$

We now fix $\xi_1, \ldots, \xi_n$ and put $\xi_0 = -\left[w(y^* y) + \varepsilon\right]^{-1} \sum_{i=1}^n w(y^* x_i)\xi_i$. Hence

$$\begin{aligned}
&\textstyle\sum_{i,j=1}^n \langle w(x_i^* x_j)\xi_j, \xi_i\rangle \\
&\geq \textstyle\sum_{i,j=1}^n \big\langle w(x_i^* y) \left\{2[w(y^* y) + \varepsilon]^{-1}\right. \\
&\qquad \left. -[w(y^* y) + \varepsilon]^{-2}\right\} w(y^* y) w(y^* x_j)\xi_j, \xi_i\big\rangle \\
&\geq \textstyle\sum_{i,j=1}^n \left\langle w(x_i^* y)[w(y^* y) + \varepsilon]^{-1}[w(y^* x_j)]\xi_j, \xi_i\right\rangle .
\end{aligned}$$

□

In particular, if T is an $(n+1)$-positive linear map between C^*-algebras then

$$\|T\| \, [T(x_i^* x_j)] \geq [T(x_i)^* T(x_j)] \text{ for all } x_1, \ldots, x_n . \tag{4.2.2}$$

A linear map T satisfying this inequality is said to be $(n + 1/2)$-positive. In particular, we say T is sesqui-positive if it is $1\frac{1}{2}$ positive. Thus $(n+1)$-positive $\Rightarrow$ $(n + 1/2)$-positive $\Rightarrow$ n-positive.

Corollary 4.4 *Let T be a positive linear map from a C^*-algebra A into another C^*-algebra B. If a is a normal element of A, then*

$$\|T\| T(a^* a) \geq T(a)^* T(a) . \tag{4.2.3}$$

More generally:

$$\|T\| T(a^* a + a a^*) \geq T(a)^* T(a) + T(a) T(a)^* \tag{4.2.4}$$

for all a in A.

Proof If C is the commutative C^*-algebra generated by a normal element a, then the restriction of T to C is completely positive, by Proposition 4.2. Hence we can apply the Schwarz inequality of Lemma 4.3. If a is an arbitrary element of A, we can apply (4.2.3) to the self-adjoint elements $a + a^*$ and $i(a - a^*)$. Then (4.2.4) follows by addition. □

Remark We know from Section 2.2 that if T is a state on a unital C^*-algebra A, then $\|T\| = T(1)$. If T is now a completely positive map between unital C^*-algebras A then it is still the case that $\|T\| = \|T(1)\|$, but the proof is a bit more tricky. From (4.2.2), we see that $\|T\|T(1) \geq T(u)^*T(u)$ for all unitaries u. That this is enough to deduce $\|T\| = \|T(1)\|$ is a consequence of the unit ball of A being the closed convex hull of its unitaries.

Corollary 4.5 *Let T be a positive contraction between C^*-algebras A and B, and a, a self-adjoint element of A, such that $T(a^2) = T(a)^2$. Then*

$$T(ab + ba) = T(a)T(b) + T(b)T(a) \tag{4.2.5}$$

and

$$T(aba) = T(a)T(b)T(a) \tag{4.2.6}$$

for all b in A.

Proof Fix φ, a state on B, and consider the sesquilinear form D on A.

$$D : (x, y) \mapsto \varphi[T(xy^* + y^*x) - T(x)T(y)^* - T(y)^*T(x)] .$$

By Corollary 4.4, we have $D(x, x) \geq 0$ for all x in A. However $D(a, a) = 0$, by assumption, and so $D(a, x) = 0$ by the Cauchy–Schwarz inequality applied to D; hence (4.2.5) holds. Then (4.2.6) follows from the Jordan identity $2aba = [[a, b]_+, a]_+ - [a^2, b]_+$, if $[a, b]_+ = ab + ba$. □

Exercise 4.1 Show that a linear map T between C^*-algebras is $(n + 1/2)$-positive if and only if T_n is sesqui-positive.

Exercise 4.2 Show that if T is a positive linear map between C^*-algebras, then T is bounded and $\|T\| = \sup \|T(u)\|$, where u is a positive approximate identity for A.

4.3 Positive-definite kernels

Throughout this section X denotes a set and H a Hilbert space; a map $K : X \times X \to B(H)$ is called a *kernel* and the set of such kernels is a vector space denoted by $K(X; H)$. A kernel K in $K(X; H)$ is said to be *positive-definite* if for each positive n and each choice of elements $x_1, \ldots, x_n$ in X, the matrix $[K(x_i, x_j)]$ is positive. Clearly a positive-definite kernel is hermitian symmetric $K(x, y)^* = K(y, x)$.

Let H' be a Hilbert space, let V be a map from X into $B(H, H')$, and put $K(x, y) = V(x)^*V(y)$. Then K is positive-definite as in (4.2.1)(a) (ii). The principal result of this section is that a kernel K is positive-definite if and only if it can be expressed in this form. Let K be a kernel in $K(X; H)$. Let H_V be a Hilbert space and $V : X \to B(H, H_V)$ a map such that $K(x, y) = V(x)^*V(y)$ for all x, y in X. Then V is said to be a *Kolmogorov decomposition* of K; if $H_V = \vee\{V(x)h \mid x \in X, h \in H\}$ then V is said to be *minimal*. Two Kolmogorov decompositions V and V' are said to be *equivalent* if there is a unitary mapping $U : H_V \to H_V$, such that $V'(x) = UV(x)$ for all x in X. A minimal Kolmogorov decomposition is universal in the following sense:

Lemma 4.6 *Let K be in $K(X;H)$ and let V be a minimal Kolmogorov decomposition of K. Then to each Kolmogorov decomposition V' of K there corresponds a unique isometry $W : H_V \to H_{V'}$ such that $V'(x) = WV(x)$ for all x in X. Moreover, if V' is minimal then W is unitary.*

Proof Since V is minimal the set of elements of the form $\sum_j V(x_j)h_j$ is dense in H_V. The map $W(\sum V(x_j)h_j) = \sum V'(x_j)h_j$ is well-defined and isometric since

$$\langle V(y)k, V(x)h\rangle = \langle K(x,y)k, h\rangle = \langle V'(y)k, V'(x)h\rangle ,$$

and hence it extends by continuity to an isometry $W : H_V \to H_{V'}$. The rest is routine. □

We have yet to show the existence of a Kolmogorov decomposition for an arbitrary positive-definite kernel; we remedy this by constructing a decomposition canonically associated with the kernel. We employ a Hilbert space of H-valued functions spanned by those of the form $x \mapsto K(x,y)h$, using the positivity of K to get an inner product. For this purpose it is convenient to reformulate the definition of a positive-definite kernel. But first we need another definition. Let $F_0 = F_0(X;H)$ denote the vector space of H-valued functions on X having finite support; let $F = F(X;H)$ denote the vector space of all H-valued functions on X. We identify F with a subspace of the algebraic anti-dual F_0' of F_0 by defining the pairing $p, f \mapsto (p,f)$ of F and F_0 by

$$(p,f) = \sum_{x\in X} \langle p(x), f(x)\rangle \, . \tag{4.3.1}$$

Given K in $K(X;H)$ we define the associated convolution operator $K : F_0(X;H) \to F(X;H)$ by

$$(Kf)(x) = \sum_{x\in X} K(x,y)f(y) \, . \tag{4.3.2}$$

Thus the kernel K in $K(X;H)$ is *positive-definite* if and only if the associated convolution operator $K : F_0(X;H) \to F(X;H)$ is positive:

$$(Kf,f) \geq 0 \, , \text{ for all } f \text{ in } F_0(X;H) \, .$$

Next we need a vector-space result:

Lemma 4.7 *Let V be a complex vector space, and let V' be its algebraic anti-dual, with the pairing $V' \times V \to \mathbb{C}$ written $v', v \mapsto (v', v)$. Let $A : V \to V'$ be a linear mapping such that $(Av, v) \geq 0$ for all v in V. Then there is a well-defined inner product on the image space AV given by $\langle Av_1, Av_2\rangle = (Av_1, v_2)$.*

Proof The sesquilinear form $v_1, v_2 \mapsto a(v_1, v_2) = (Av_1, v_2)$ is positive, so that the Schwarz inequality holds:

$$|a(v_1, v_2)|^2 \leq a(v_1, v_1)a(v_2, v_2).$$

It follows that the set $V_A = \{v \in V \mid a(v, v) = 0\}$ coincides with the subspace $\ker A$, and so the natural projection $\pi : V \to V/\ker A$ carries the form $a(\cdot, \cdot)$ into an inner-product $\langle \cdot, \cdot \rangle_A$ on $V/\ker A$ given by $\langle \pi(v_1), \pi(v_2) \rangle_A = a(v_1, v_2)$. The vector-space isomorphism $A' : V/\ker A \to AV$ given by $A' \circ \pi = A$ carries the inner product $\langle \cdot, \cdot \rangle_A$ into an inner product $\langle \cdot, \cdot \rangle$ on AV, given by

$$\langle Av_1, Av_2 \rangle = \langle A'\pi(v_1), A'\pi(v_2) \rangle = \langle \pi(v_1), \pi(v_2) \rangle_A = a(v_1, v_2) = (Av_1, v_2).$$

□

Proposition 4.8 *For each positive-definite K in $K(X; H)$ there exists a unique Hilbert space $R(K)$ of H-valued functions on X such that*

(a) $R(K) = \vee\{K(\cdot, x)h \mid x \in X, h \in H\}$,
(b) $\langle f(x), h \rangle = \langle f, K(\cdot, x)h \rangle$ *for all f in $R(K)$, x in X and h in H.*

Proof Since the kernel K is positive-definite the associated convolution operator K of $F_0 = F_0(X; H)$ into F' defined in (4.3.2) satisfies the hypotheses of Lemma 4.7. Let $\overline{KF_0}$ be the completion of KF with respect to the norm obtained from the inner product $\langle Kf_1, Kf_2 \rangle = (Kf_1, f_2)$, and identify KF_0 with a dense subset of $\overline{KF_0}$. For each x in X and h in H define the function h_x in F_0 by putting $h_x(y) = h$ if $y = x$ and $h_x(y) = 0$ otherwise; then $(Kh_x)(y) = K(y, x)h$. Define K_x on H by $K_x h = Kh_x$ for all x in X and h in H; then $\|K_x h\| \leq \|K(x, x)\|^2 \|h\|$, so that K_x is a bounded linear map from H into $\overline{KF_0}$. A straightforward calculation shows that on KF_0 we have $K_x^* f = f(x)$. The mapping of $\overline{KF_0}$ into the space of all H-valued functions on X which sends f into the function $x \mapsto K_x^* f$ is linear, injective and compatible with the identification of KF_0 with a dense subset of $\overline{KF_0}$. Thus we can regard $\overline{KF_0}$ as a Hilbert space $R(K)$ of H-valued functions on X. We have proved that $R(K)$ satisfies (a) and (b); the uniqueness assertion clearly holds. □

$R(K)$ is called the *reproducing-kernel Hilbert space* determined by K.

Corollary 4.9 *A kernel has a Kolmogorov decomposition if and only if it is positive-definite.*

Proof If K is a positive-definite kernel, take $V(x) = K_x : H \to R(K)$ as in the proof of Proposition 4.8; then $K(x, y) = V(x)^* V(y)$. Thus $(K, R(K))$ is a Kolmogorov decomposition of K; from Proposition 4.8 it is minimal. □

The set $K^+(X; H)$ of positive-definite kernels in $K(X; H)$ forms a cone; we define the induced partial ordering: put $K \geq K'$ if and only if $K - K'$ is in $K^+(X; H)$. The next result says that R is functorial:

Proposition 4.10 *Let K and K' be positive-definite kernels; then $K \geq K'$ if and only if there is a (necessarily unique) contraction $C : R(K) \to R(K')$ such that $K'_x = CK_x$ for all x in X.*

Proof Let K, K' be in $K^+(X; H)$. Then $K \geq K'$ if and only if $(Kv, v) \geq (K'v, v)$ for all v in $F_0(X; H)$; this holds if and only if $\langle Kv, Kv\rangle \geq \langle K'v, K'v\rangle$ for all v in $F_0(X; H)$. This is the case if and only if there is a contraction $C : R(K) \to R(K')$ such that $Kv = CK'v$ for all v in $F_0(X; H)$. The result follows by considering the generating set $\{h_x \mid h \in H, x \in X\}$ in $F_0(X; H)$, since $K_x h = Kh_x = CK'h_x = CK'_x h$ for all x in X and h in H. Putting this result together with Corollary 4.9 we have: □

Corollary 4.11 *Let K and K' be positive-definite kernels with Kolmogorov decompositions V and V' respectively. Then $K \geq K'$ if and only if there is a positive contraction T in $B(H_V)$ such that*

$$K'(x, y) = V(x)^* T V(y)$$

for all x, y in X.

Proposition 4.12 *Let K be in $K^+(X; H)$; then for each $\varepsilon > 0$ and each z in X we have*

$$K(\cdot, \cdot) \geq K(\cdot, z)(\varepsilon + K(z, z))^{-1} K(z, \cdot) .$$

In particular, the Schwarz inequality holds:

$$K(\cdot, \cdot)\|K(z, z)\| \geq K(\cdot, z)K(z, \cdot) .$$

Proof Let V be a minimal Kolmogorov decomposition for K; then we have

$$K(x, z)(\varepsilon + K(z, z))^{-1} K(z, y) = V(x)^* V(z)(\varepsilon + V(z)^* V(z))^{-1} V(z)^* V(y)$$

for all x, y, z in X. Thus by Proposition 4.10 it is enough to show that the operator

$$W = (\varepsilon + V(z)^* V(z))^{-1/2} V(z)^*$$

is a contraction. But

$$\begin{aligned} WW^* &= (\varepsilon + V(z)^* V(z))^{-1/2} V(z)^* V(z)(\varepsilon + V(z)^* V(z))^{-1/2} \\ &= (\varepsilon + V(z)^* V(z))^{-1} V(z)^* V(z) \leq 1 \end{aligned}$$

by the spectral theorem. □

4.4 Positive-definite functions

The principal results in this section are two well-known representation theorems: the Naimark–Sz.-Nagy characterization of positive-definite functions on groups (Corollary 4.14) and the Stinespring decomposition for completely positive maps on Banach $*$-algebras (Corollary 4.15). We exploit the existence and uniqueness of minimal Kolmogorov decompositions for certain functions on semigroups with involution.

Let S be a semigroup, and let $J : S \to S$ be a map of S into itself such that (i) $J^2 = \mathrm{id}_S$, (ii) $J(ab) = J(b)J(a)$ for all a, b in S; then J is said to be an *involution.* An element a of a semigroup with involution (S, J) is said to be an *isometry* if $J(s)J(a)at = J(s)t$ for all s, t in S. The set S_J of isometries in (S, J) is a subsemigroup. For example, when S is a group let $J(a) = a^{-1}$ for all a in S; then $S_J = S$. When S is a $*$-algebra with unit, let $J(a) = a^*$ then $S_J = \{a \in S \mid a^*a = 1\}$ so that the elements of S_J are isometries in the usual sense, and $S_J \cap J(S_J)$ are the unitaries.

Let H be a Hilbert space and let (S, J) be a semigroup with involution; then a function $T : S \to B(H)$ is said to be *positive-definite* if the kernel $a, b \mapsto T(J(a)b)$ is positive-definite. A *Kolmogorov decomposition* for a positive-definite function is a Kolmogorov decomposition for its associated kernel. For example, let G be a group, and $\pi : S \to B(H_\pi)$ a unitary representation of G. Let $W : H \to H_\pi$ be an isometry; then the function $T(g) = W^*\pi(g)W$ is positive-definite and has a Kolmogorov decomposition V where $V(g) = U(g)W$. We shall see that every positive-definite function on a group can be put in this form.

Lemma 4.13 *Let (S, J) be a semigroup with involution, let $T : S \to B(H)$ be a positive-definite function on S, and let V be a minimal Kolmogorov decomposition for T. Then there exists a unique homomorphism ϕ of S_J into the semigroup of isometries on H_V, such that $\phi(b)V(a) = V(ba)$ for all b in S_J and all a in S. It follows that $T(J(a)bc) = V(a)^*\phi(b)V(c)$ for all b in S_J and all a, c in S, and that the restriction of ϕ to $S_J \cap J(S_J)$ is a $*$-map: $\phi(b)^* = \phi(Jb)$.*

Proof For all a, c in S we have $V(ba)^*V(bc) = T(J(ba)bc) = T(J(a)c) = V(a)^*V(c)$ whenever b is in S_J. Hence, by Proposition 4.8, the minimality of V entails the existence of a unique isometry $\phi(b) : H_V \to H_V$, such that

$$\phi(b)V(c) = V(bc) \tag{4.4.1}$$

for all c in S. It follows from (4.4.1) that $\phi(b)\phi(b') = \phi(bb')$ for all b, b' in S_J. Now suppose that b is in $S_J \cap J(S_J)$; then for all a, c in S we have

$$\begin{aligned} V(a)^*\phi(b)^*V(c) &= [\phi(b)V(a)]^*\, V(c) = V(ba)^*V(c) \\ &= T(J(ba)c) = T(J(a)J(b)c) = V(a)^*\phi(Jb)V(c)\,, \end{aligned}$$

so that $\phi(b)^* = \phi(Jb)$ by uniqueness. □

Corollary 4.14 *Let G be a group, and let $T : G \to B(H)$ be a positive-definite function on G. Then there exists a Hilbert space H_π, a unitary representation $\pi : G \to B(H_\pi)$ and a map V in $B(H, H_\pi)$ such that*

$$T(g) = V^*\pi(g)V \tag{4.4.2}$$

for all g in G. If the decomposition (4.4.4) is minimal then it is unique up to unitary equivalence.

Let A be a $*$-algebra with involution $J(a) = a^*$. A completely positive map $T : A \to B(H)$ is linear and positive-definite; if V is a minimal Kolmogorov decomposition for T then $V : A \to B(H, H_V)$ is linear. For example, let $W : H \to H_W$ be an isometry, and let A be a $*$-subalgebra of $B(H_W)$. Then $T : A \to B(H)$ given by $T(a) = W^*aW$ is completely positive. Moreover, a $*$-representation $\pi : A \to B(H)$ of a $*$-algebra A is completely positive. All completely positive maps from C^*-algebras into $B(H)$ are compositions of these two fundamental examples:

Corollary 4.15 *Let A be a Banach $*$-algebra with bounded approximate identity, and let T be a completely positive map from A into $B(H)$. Then there exists, uniquely up to unitary equivalence, a Hilbert space H_V, a $*$-representation π of A on H_V, and a map V in $B(H, H_V)$, such that*

$$T(a) = V^*\pi(a)V \tag{4.4.3}$$

for all a in A, and

$$H_V = \vee\{\pi(a)Vh \mid a \in A, h \in H\}\,. \tag{4.4.4}$$

Proof Let V be a minimal Kolmogorov decomposition for T, and let A' denote the unital Banach $*$-algebra obtained from A by adjoining an identity. Then A is an ideal in A' and

$$V(xa)^*V(bc) = T((ba)^*(bc)) = T(a^*c) = V(a^*)V(c)$$

for all a, c in A and b in $U(A')$, the unitaries of A'. As in Lemma 4.13 let $\phi : U(A') \to B(H_V)$ be the $*$-homomorphism such that $\phi(b)V(c) = V(bc)$ for all b in $U(A')$, c in A. Then for a in A' we have $a = \sum_{i=1}^n z_i u_i$, where $z_1, \ldots, z_n$ are complex numbers and $u_1, \ldots, u_n$ are in $U(A')$; put $\pi'(a) = \sum_{i=1}^n z_i\phi(u_i)$. Then $\pi'(a)V(c) = V(ac)$ for all c in A, so that π' is a well-defined $*$-homomorphism from A' into $B(H_V)$. Let π denote the restriction of π' to A. It follows (cf. Exercise 4.2) that T is bounded and hence so is $V(\cdot)$, since $\|V(x)\|^2 = \|T(x^*x)\|$ for all x in A. We identify $B(H, H_V)$ with the dual of the space of trace class operators from H_V into H. Let $\{u_\gamma\}$ be a bounded approximate identity for A, then the net $\{V(u_\gamma)\}$ is bounded in $B(H, H_V)$ and so has a weak* limit V say. We see that $\pi(a)V = \lim \pi(a)V(u_\gamma) = \lim V(au_\gamma) = V(a)$ for all a in A. The result follows. □

The Stinespring representation theorem can be used to obtain a description of completely positive normal maps:

Theorem 4.16 *Let A be a von Neumann algebra on a Hilbert space H, and K another Hilbert space. If ψ is a completely positive ultraweakly continuous map* *from A into $B(K)$, then there exist $\{A_i, i \in X\}$ in $B(K,H)$ such that, for all x in A,*

$$\psi(x) = \textstyle\sum_i A_i^* x A_i\,. \tag{4.4.5}$$

Proof By the Stinespring decomposition, we can assume that ψ is a normal representation with cyclic vector f. Then since $\langle \psi(\cdot) f, f\rangle$ is a normal state on A, there exist vectors $\{f_i \mid i \in \mathbb{N}\}$ in H such that $\sum \|f_i\|^2 < \infty$, and $\langle \psi(x)f, f\rangle = \sum \langle x f_i, f_i\rangle$ for all x in A. Since $\|x f_i\| \le \|\psi(x) f\|$ for all x in A, there exist contractions A_i from K into H such that $A_i \psi(x) f = x f_i$. Then, for all x, z in A, we have

$$\begin{aligned}\langle \psi(x)\psi(z)f, \psi(z)f\rangle &= \langle \psi(z^* x z) f, f\rangle = \textstyle\sum_i \langle z^* x z f_i, f_i\rangle \\ = \textstyle\sum_i \langle x z f_i, z f_i\rangle &= \textstyle\sum_i \langle x A_i \psi(z) f, A_i \psi(z) f\rangle = \textstyle\sum_i \langle A_i^* x A_i \psi(z) f, \psi(z) f\rangle\,.\end{aligned}$$

Since f is a cyclic vector for ψ, we have $\psi(x) = \sum_i A_i^* x A_i$ for all x in A; the series converges in the ultraweak topology. □

It is apt at this point to discuss the intimate relationship between positive-definite functions on groups and completely positive maps on algebras, and in particular the relationship between the Naimark–Sz.-Nagy representation and the Stinespring decomposition. In the first place, consider a unital Banach $*$-algebra A, and let G denote a subgroup of its group of unitaries such that $\mathrm{Lin}(G) = A$. As we have already essentially noted, a completely positive map on A restricts to a positive-definite function on G. Conversely, if T is a linear map on A such that its restriction to G is positive-definite, then T is completely positive. For if a_i, $i = 1, \ldots, n$ are elements of A, then there exists complex numbers z_{ip} and elements g_p of G, $p = 1, \ldots, m$, such that $a_i = \sum_p z_{ip} g_p$, since $A = \mathrm{Lin}(G)$. From the linearity of T we have

$$T(a_i^* a_j) = \textstyle\sum_{p,q} \overline{z_{ip}} T(g_p^{-1} g_q) z_{jq}; \tag{4.4.6}$$

regarding the right-hand side as a matrix-element of the product of three matrices, we see that $[T(a_i^* a_j)]$ is a positive matrix since $[T(g_p^{-1} g_q)]$ is. Moreover, T is a homomorphism if and only if its restriction to G is a unitary representation. Thus the restriction map takes the Stinespring decomposition into the Naimark–Sz.-Nagy representation.

This connection can be taken further. Suppose G is a locally compact group, and T is a strongly continuous positive-definite function on G (acting on a Hilbert space H, say). Then it is easy to verify that

$$T'(f) = \int_G f(g) T(g)\, dg\,, \tag{4.4.7}$$

where dg is a left-invariant Haar measure on G, defines a completely positive map T' of the Banach $*$-algebra $L^1(G)$ into $B(H)$. Moreover it can be shown, using the existence of an approximate identity for $L^1(G)$, that each completely positive map on $L^1(G)$ arises in this way; $T'$is a homomorphism of $L^1(G)$ if and only if T is a unitary representation of G. Thus the (minimal) Naimark–Sz.-Nagy representation of T on G (Corollary 4.14),

$$T(g) = V^*U(g)V\,, \tag{4.4.8}$$

gives the (minimal) Stinespring decomposition on $L^1(G)$ (Corollary 4.15),

$$T'(f) = V^*U'(f)V\,, \tag{4.4.9}$$

and vice versa.

If S is an abelian semigroup, let $\gamma : S \to K(S)$ be the canonical homomorphism of S into the Grothendieck group of S as in Section 3.3.

Exercise 4.3 Let $T : S \to B(H)$ be a homomorphism of an abelian semigroup S into the semigroup of isometries on a Hilbert space H. Show that there is a positive-definite function T' on $K(S)$ such that $T'_{\gamma(t)-\gamma(s)} = T_s^*T_t$, for all (s,t) in $S \times S$.

Exercise 4.4 Let S be an abelian semigroup and let $T : S \to B(H)$ be a homomorphism such that $T_0 = 1$. Then T is said to have a *unitary dilation in the strong sense* if there exists a Hilbert space H_V, an isometry $V : H \to H_V$, and a unitary representation $U : K(S) \to B(H_V)$ of the Grothendieck group of S such that $VT_s = U_{\gamma(s)}V$. Show that T has a unitary dilation in the strong sense if and only if T_s is an isometry for all s in S.

Exercise 4.5 (a) Show that the following are equivalent for an abelian semigroup S:

(i) $\gamma(S) \cap [-\gamma(S)] = \{0\}$.

(ii) If s, t, u, v are in S and $s + u = v$ and $u = t + v$, then $s + w = w$ for some w in S.

(b) Let S be an abelian semigroup for which (a) holds, and let $T : K(S) \to B(H)$ satisfy

(i) $T_0 = 1$, (ii) $T_k^* = T_{-k}$ for all k in $K(S)$,

(iii) $T_kT_{k'} = T_{k+k'}$ whenever k, k' and $k + k'$ are not in $[-\gamma(S)]$.

Show that T is positive-definite if and only if T_k is a contraction for each k in $K(S)$; in which case T has a unitary dilation.

Exercise 4.6 (a) Show that the following are equivalent for an abelian semigroup S:

(i) $\gamma(S) \cup [-\gamma(S)] = K(S)$.

(ii) Whenever s, t are in S there exist u, v, w in S such that either $t + u = v$, $s + u = w + v$, or $t + u = v + w$, $s + u = v$.

In this case S is said to be *totally ordered.*

(b) Let S be a totally ordered abelian semigroup, and let $T : S \to B(H)$ be a homomorphism satisfying

(i) $T_0 = 1$, (ii) $\|T_s\| \leq 1$,

and the cancellation law:

(iii) if $h + s = h + t$ then $T_s = T_t$.

Show that there is a unique positive-definite function T' on $K(S)$ such that $T'_{\gamma(s)} = T_s$ and $T'_{-\gamma(s)} = T_s^*$ for all s in S; hence T has a unitary dilation.

Exercise 4.7 Let H be a Hilbert space; for each positive integer n, let H_n denote the n-fold tensor product $\bigotimes^n H$. Show that there is a unitary action of S_n, the group of all permutations on n symbols, on H_n given by

$$\pi(f_1 \otimes \cdots \otimes f_n) = f_{\pi^{-1}(1)} \otimes \cdots \otimes f_{\pi^{-1}(n)}\,,$$

for π in S_n, $f_1, \ldots, f_n$ in H

Exercise 4.8 Show that

$$P_n = (n!)^{-1} \sum_{\pi \in S_n} \pi\,,$$

$$Q_n = (n!)^{-1} \sum_{\pi \in S_n} \varepsilon(\pi)\pi\,,$$

(where $\varepsilon(\pi)$ denotes the signature of the permutation π), are projections on H_n. The range of P_n is the space H_n^s of symmetric tensors of degree n, and Q_n is the projection from H_n onto the space H_n^a of antisymmetric tensors of degree n over H, with $H_0^s = H_0^a = H_0$ spanned by the unit vacuum vector Ω. Symmetric (or Boson) Fock space $F^s(H)$ is then defined by

$$F^s(H) = \bigoplus_{n=0}^{\infty} H_n^s\,.$$

Anti-symmetric (or Fermion) Fock space $F^a(H)$ is defined by

$$F^a(H) = \bigoplus_{n=0}^{\infty} H_n^a\,.$$

Let T be a contraction between Hilbert spaces H and K; show that T_n intertwines the actions of S_n on H_n and $K_n : T_n\pi = \pi T_n$ for all π in S_n, and so $T_n = \bigotimes^n T$ maps H_n^s into K_n^s, and H_n^a into K_n^a. Thus $F(T)$ induces a contraction $F^s(T) : F^s(H) \to F^s(K)$ (respectively $F^a(T) : F^a(H) \to F^a(K)$). Then F^s and F^a inherit properties $F(TS) = F(T)F(S)$, $F(T^*) = F(T)^*$ of the functor F of Section 2.8.

Exercise 4.9 Let h be a vector in the Hilbert space H, and let h_n denote the n-fold tensor product $h \otimes \cdots \otimes h$ which lies in H_n^s, with $h_0 = \Omega$. Then $\langle h_n, k_n \rangle = \langle h, k \rangle^n$ for all h, k in H. Show that $h \mapsto h_n$ is a minimal Kolmogorov decomposition of the positive-definite kernel $h, k \mapsto \langle h, k \rangle^n$ on $H \times H$. Define $\mathrm{Exp} : H \to F^s(H)$ by

$$\mathrm{Exp}(h) = \bigoplus_{n=0}^{\infty} (n!)^{-1/2} h_n \,.$$

Show that $\langle \mathrm{Exp}(h), \mathrm{Exp}(k) \rangle = \exp\langle h, k \rangle$, and:

$$\frac{d^n}{dt^n} \mathrm{Exp}(th)|_{t=0} = (n!)^{1/2} h_n \,.$$

so that $\mathrm{Exp} : H \to F^s(H)$ is a minimal Kolmogorov decomposition for the positive-definite kernel $h, k \mapsto \exp\langle h, k \rangle$ on $H \times H$. Show that $\{\mathrm{Exp}(h) \mid h \in H\}$ is a linearly independent total set of vectors for $F^s(H)$.

Exercise 4.10 Show that there is a natural identification of $F^s(H \oplus K)$ with $F^s(H) \otimes F^s(K)$ under which $\mathrm{Exp}(h \oplus k) = \mathrm{Exp}(h) \otimes \mathrm{Exp}(k)$, and $F^s(S \oplus T) = F^s(S) \otimes F^s(T)$.

Exercise 4.11 Consider the linearly independent total set of normalized vectors

$$C(h) = \mathrm{Exp}(2^{-1/2} h) \exp\left(-\|h\|^2/4\right); \quad h \in H \,.$$

Show that $C(\cdot)$ is a minimal Kolmogorov decomposition for the positive-definite kernel

$$h, k \mapsto \exp\left(\|h - k\|^2/4\right) \exp(i\mathrm{Im}\langle h, k \rangle/2) \,. \tag{4.4.10}$$

Thus $F^s(H)$ can be identified with the reproducing kernel Hilbert space for the kernel (4.4.10). Show that the map

$$\omega : (h, k) \mapsto \exp(i\mathrm{Im}\langle h, k \rangle/2) \tag{4.4.11}$$

defines a multiplier.

A *multiplier* b on a group G is a map from $G \times G$ into the unit circle such that

a) $b(g, 0) = b(0, g) = 1$,

b) $b(g, g')b(g + g', g'') = b(g, g' + g'')b(g', g'')$ for all g, g', g'' in G,

[i.e. a circle-valued normalized 2-cocycle].

A *b-representation* of a group G with multiplier b is a map U from G into the unitary operators on some Hilbert space such that

c) $U(0) = 1$,

d) $U(g)U(g') = U(g + g')b(g, g')$, for all g, g' in G.

A *projective representation* is a b-representation for some multiplier b. Properties (a) and (b) of a multiplier are consistency conditions for the existence of b-representations; e.g. (b) reflects the associative law.

Since $\{C(h) \mid h \in H\}$ is a linearly independent total set of normalized vectors, there is a well-defined unitary $W(h)$, for each h in H, such that

$$W(h)C(k) = C(h+k)\omega(k,h)$$

for all k in H. Moreover, $W(h)$ obeys the canonical commutation relations:

$$W(h)W(k) = W(h+k)\omega(h,k)\,.$$

A *representation of the* CCR (*canonical commutation relations*) is a projective representation of a Hilbert space H with multiplier ω given by (4.4.11). The C^*-algebra generated by a representation W of the CCR (the norm-closed linear span of the unitaries $\{W(h) \mid h \in H\}$) is denoted by $W(H)$. The representation of the CCR defined above is called the *Fock representation.*

Let H, K be Hilbert spaces; a representation W of the CCR over H on $B(K)$ is said to be K-*cyclic* if there exists a V in $B(K, H_W)$ such that $H_W = \vee\{W(h)Vk \mid h \in H, k \in K\}$. Let (W, V) be a K-cyclic representation of the CCR over H, and define a map $M : H \to B(K)$ by $h \mapsto M(h) = V^*W(h)V$. Then M is called the *generating function* of (W, V).

Exercise 4.12 Let H, K be Hilbert spaces, and M a map from H into $B(K)$. Show that there exists a K-cyclic representation (W, V) having M as its generating function if and only if the kernel $(h,k) \mapsto M(k-h)\omega(k,h)$ is positive-definite on $H \times H$. In this case (W, V) is uniquely determined up to unitary equivalence. [If $M(k-h)\omega(k,h) = V(h)^*V(k)$, then $V(h+h'')^*V(h'+h'')\omega(h',h'')\overline{\omega(h,h'')} = V(h)^*V(h')$; thus, by the uniqueness of the minimal Kolmogorov decomposition, there exists a well-defined unitary $W(h'')$ such that $W(h'')V(h) = V(h+h'')\omega(h,h'')$]. Thus the Fock representation of the CCR is determined by the generating functional $h \mapsto \langle W(h)\Omega, \Omega\rangle = \exp(-\|h\|^2/4)$.

Exercise 4.13 The *Schrödinger representation* of the CCR over $\mathbb{C}$, is defined on $L^2(\mathbb{R})$ as follows: $(W(x,y)g)(s) = e^{ix(2s+y)/2}g(s+y)$ for g in $L^2(\mathbb{R})$. Verify that this defines a representation of the CCR over $\mathbb{C}$. Using the cyclic vector $\Omega(s) = \pi^{-1/4}e^{-s^2/2}$; show that the Schrödinger representation has the same generating functional (4.4.10) as the Fock representation; so that the representations are unitarily equivalent. Show that the Schrödinger representation on $L^2(\mathbb{R})$ is irreducible.

Exercise 4.14 For each $\lambda \geq 1$, show using Exercise 4.12 that there exists a cyclic representation W_λ of the CCR over H, acting on a Hilbert space $F_\lambda(H)$, with cyclic vector Ω_λ, and generating functional μ_λ given by $\mu_\lambda(h) = \exp(-\lambda\|h\|^2/4)$. Since $\mu_\lambda(h \oplus k) = \mu_\lambda(h)\mu_\lambda(k)$, identify $F_\lambda(H \oplus K)$ with $F_\lambda(H) \otimes F_\lambda(K)$, and $W_\lambda(h \oplus k)$ with $W_\lambda(h) \otimes W_\lambda(k)$, and hence $W_\lambda(H \oplus K)$ with the spatial C^*-tensor product $W_\lambda(H) \otimes W_\lambda(K)$.

Exercise 4.15 Let J be a conjugation on H (that is, an antilinear map satisfying $J^2 = 1$ and $\langle Jh, Jh' \rangle = \langle h', h \rangle$ for all h, h' in H). Given $\lambda \geq 1$, choose $\alpha, \beta \geq 0$ such that $\alpha^2 + \beta^2 = \lambda$, $\alpha^2 - \beta^2 = 1$, and put $W_\lambda(H) = W(\alpha h) \otimes W(\beta Jh)$. Then W_λ, defined on $F_\lambda(H) = F(H) \otimes F(H)$, is a representation of the CCR with $\langle W_\lambda(H)\Omega_\lambda, \Omega_\lambda \rangle = \exp(-\lambda \|h\|^2/4)$ if $\Omega_\lambda = \Omega \otimes \Omega$.

Exercise 4.16 Let H, K be Hilbert spaces, A a linear map from H into K, and f a map from H into $\mathbb{C}$. Show that there exists a completely positive map $T : W(H) \to W(K)$ such that $T[W(h)] = W(Ah)f(h)$ for all h in H, if and only if the kernel $h, k \mapsto f(k-h)\omega(k,h)/\omega(Ak, Ah))$ is positive-definite on $H \times H$.

Exercise 4.17 Let $\lambda \geq 1$ be fixed, and $T : H \to K$ a contraction between Hilbert spaces. Show that

(a) there is a completely positive map $W_\lambda(T) : W_\lambda(H) \to W_\lambda(K)$ of C^*-algebras such that $W_\lambda(T)[W_\lambda(h)] = W_\lambda(Th)e^{-\lambda\{\|h\|^2 - \|Th\|^2\}/4}$ for all h in H.

(b) W_λ is functorial: $W_\lambda(ST) = W_\lambda(S)W_\lambda(T)$, $W_\lambda(T) = 1$ and $W_\lambda(S \oplus T) = W_\lambda(S) \otimes W_\lambda(T)$,

(c) $W_\lambda(0)$ is the state determined by μ_λ, and is invariant under $W_\lambda(T) : \mu_\lambda \circ W_\lambda(T) = \mu_\lambda$.

Exercise 4.18 Recall that to each contraction $H \to K$ there corresponds a contraction $F(T) : F(H) \to F(K)$ such that

$$\begin{aligned} F(T)W(h)\Omega &= F(T)\mathrm{Exp}(2^{-1/2}h)e^{-\|h\|^2/4} \\ &= \mathrm{Exp}(2^{-1/2}Th)e^{-\|h\|^2/4} = C(Th)e^{-\{\|h\|^2 - \|Th\|^2\}/4} \\ &= W(Th)\Omega e^{-\{\|h\|^2 - \|Th\|^2\}/4} = W(T)[W(h)]\Omega\,. \end{aligned}$$

Analogously for $\lambda \geq 1$: Let H, K be Hilbert spaces; for each contraction $T : H \to K$, show that $\|W_\lambda(T)[x]\Omega_\lambda\|^2 \leq \|x\Omega_\lambda\|^2$. Deduce that there is a contraction $F_\lambda(T) : F_\lambda(H) \to F_\lambda(K)$ such that

(a) $F_\lambda(T)W_\lambda(h)\Omega = W_\lambda(Th)\Omega_\lambda e^{-\lambda\{\|h\|^2 - \|Th\|^2\}/4}$ for all h in H.

(b) F_λ is functorial: $F_\lambda(ST) = F_\lambda(S)F_\lambda(T)$, $F_\lambda(1) = 1$, and $F_\lambda(T)^* = F_\lambda(T^*)$, $F_\lambda(S \oplus T) = F_\lambda(S) \otimes F_\lambda(T)$, and $F_\lambda(0)$ is the projection on the vacuum.

(c) The map $x \mapsto W_\lambda(T)[x] - F_\lambda(T)(x)F_\lambda(T)^*$ from $W_\lambda(H)$ into $B(F_\lambda(K))$ is completely positive.

(d) $W_\lambda(T)$ is a homomorphism if and only if T is an isometry and $W_\lambda(T) = F_\lambda(T)(\cdot)F_\lambda(T)^*$ if and only if T is a co-isometry.

(e) Let $T : H \to K$ be a contraction; then there exists a Hilbert space L and isometries $V_1 : H \to L$ and $V_2 : K \to L$ such that $T = V_2^* V_1$. Show the following Stinespring decomposition for $W_\lambda(T)$: $W_\lambda(T) = F_\lambda(V_2)^*(W_\lambda(V_1)[\cdot])F_\lambda(V_2)$.

(f) $\mu_\lambda(W_\lambda(T)[x]y) = \mu_\lambda(xW_\lambda(T^*)[y])$, for all x in $W_\lambda(H)$ and y in $W_\lambda(K)$.

Exercise 4.19 Consider for fixed $n \in \mathbb{N}$, the permutation group S_n, and denote by $\pi_i \in S_n$, $i = 1, \ldots, n-1$, the transposition between i and $i+1$. Suppose $T_i \in$

$B(H)$ for $i = 1, \ldots, n-1$ where H is a Hilbert space, are self adjoint contractions satisfying $T_i T_{i+1} T_i = T_{i+1} T_i T_{i+1}$, $i = 1, 2, \ldots, n-2$, $T_i T_j = T_j T_i$ if $|i - j| \geq 2$. If $\sigma = \pi_{i(1)} \cdots \pi_{i(k)} \in S_n$ is a reduced word, define $\phi(\sigma) = T_{i(1)} \cdots T_{i(k)}$. Show that ϕ is completely positive. (S_n is the group with generators π_i, $i = 1, \ldots, n-1$ and relations $\pi_i \pi_{i+1} \pi_i = \pi_{i+1} \pi_i \pi_{i+1}$, $\pi_i^2 = 1$, $\pi_i \pi_j = \pi_j \pi_i$, $|i - j| \geq 2$.)

Exercise 4.20 Let x be a contraction in a Cuntz algebra $\mathcal{O}_n$ with generators $(s_1, \ldots, s_n)$. Show that there exists a completely positive unital linear map ϕ on $\mathcal{O}_n$ such that $\phi(s_i) = x s_i$. If x is a *co-isometry* ($xx^* = 1$), show that ϕ is unique and given by

$$\phi\left(s_{i_1} \cdots s_{i_r} s_{j_t}^* \cdots s_{j_1}^*\right) = (x s_{i_1}) \cdots (x s_{i_r})(x s_{j_t})^* \cdots (x s_{j_1})^* .$$

Let $s(f) = \sum f_i s_i$, where f is a vector in $\ell^2(n)$. Then in particular there is, for each unit vector ξ in $\ell^2(n)$, a unique completely positive map, φ_ξ on $\mathcal{O}_n$ such that $\varphi_\xi(s(f)) = \langle f, \xi \rangle$. Indeed φ_ξ is the *Cuntz state*:

$$\varphi_\xi(s(f_1) \cdots s(f_r) s(g_t)^* \cdots s(g_1)^*) = \prod_{i=1}^{r} \langle f_i, \xi \rangle \prod_{j=1}^{t} \langle \xi, g_j \rangle .$$

More generally, if $v_i \in B(H)$ with $\sum_{i=1}^n v_i v_i^* = 1$, show that there exists a completely positive unital linear map from $\mathcal{O}_n$ into $B(H)$ taking $s_{i_1} \cdots s_{i_r} s_{j_t}^* \cdots s_{j_1}^*$ into $v_{i_1} \cdots v_{i_r} v_{j_t}^* \cdots v_{j_1}^*$.

4.5 One-parameter semigroups on Banach spaces

Let $\{T_t\}$ be a strongly continuous one-parameter semigroup on a Banach space X with infinitesimal generator L. One can deduce from the submultiplicativity $\|T_{t+s}\| \leq \|T_t\| \|T_s\|$ that $\omega_0 = \lim_{t \to \infty} t^{-1} \log \|T_t\|$ exists. Then $\omega_0 \in [-\infty, \infty)$ is called the *spectral type* of the semigroup and if $\omega > \omega_0$ then $\|T_t\| \leq M e^{\omega t}$ for some finite M. In fact

$$\omega_0 = \inf\{\omega \mid \|T_t\| \leq M e^{\omega t} \text{ for some } M, \text{ all } t > 0\} . \tag{4.5.1}$$

If f is a complex continuous function on $[0, \infty)$, with $\int_0^\infty |f(t)| e^{\omega_0 t}\, dt < \infty$, then we can define

$$T_f(x) = \int_0^\infty f(t) T_t(x)\, dt \tag{4.5.2}$$

as a Riemann integral. From

$$T_s T_f x = \int_0^\infty f(t - s) T_t(x)\, dt \tag{4.5.3}$$

we can deduce that if f is smooth then $T_f(x) \in D(L)$ and

$$L T_f(x) = -T_{f'}(x) - x . \tag{4.5.4}$$

Consequently, by considering a sequence of such an f approaching a delta function at zero, we see that the domain of L is dense in X. That L is closed follows from

$$T_t(x) - x = \int_0^\infty T_s L(x)\,ds\,, \quad x \in D(L)\,.$$

Moreover,the C^n-vectors:

$$D(L^n) = \{x \mid t \to T_t(x) \text{ is } C^n\} \tag{4.5.5}$$

are dense in X. However, the *analytic vectors*

$$\Big\{x \mid t \to T_t(x) \text{ is analytic}\Big\} = \Big\{x \in C^\infty \mid \sum_{n=0}^{\infty} t^n \|L^n x\|/n! < \infty \text{ for } t > 0\Big\} \tag{4.5.6}$$

are dense for strongly continuous groups, but this can fail for semigroups (Nelson 1960), e.g. translation to the right on $L^2(0,\infty)$. Taking $f(t) = e^{\lambda t}$, $\operatorname{Re}\lambda > \omega$ in (4.5.4) we see that

$$(\lambda - L)\int_0^\infty e^{-\lambda t} T_t(x)\,dt = x\,. \tag{4.5.7}$$

Thus $\lambda \in \rho(L)$, the resolvent of L,

$$(\lambda - L)^{-1} = \int_0^\infty e^{-\lambda t} T_t\,dt \tag{4.5.8}$$

and the spectrum of L, $\sigma(L)$, is contained in a half plane $\{\lambda \mid \operatorname{Re}\lambda \le \omega_0\}$. If $\rho_T = \sup\{\operatorname{Re} z \mid z \in \sigma(L)\}$, then

$$\rho_T \le \omega_0\,. \tag{4.5.9}$$

However, equality can fail even for Hilbert spaces, C^*-algebras, and some positive semigroups in ordered Banach spaces (Zabzyk 1975), (Gelfand and Naimark 1943), (Greiner *et al.* 1981), but equality does hold for positive semigroups on C^*-algebras or L^2-spaces (Gelfand and Naimark 1943), (Naimark 1972). The inequality (4.5.9) can be strengthened (Hille and Philips 1957) to

$$\sigma(T_t) \supset \exp t\sigma(L)\,. \tag{4.5.10}$$

If ρ denotes the spectral radius, then

$$\rho(T_t) \ge \exp t\rho_T \tag{4.5.11}$$

and $\rho(T_t) = \exp t\omega_0$ follows from $\omega_0 = \lim_{t\to\infty} t^{-1}\log\|T_t\|$, $\rho(T_t) = \lim_{n\to\infty}\|T_t^n\|^{1/n}$ and the semigroup law. The inclusion in (4.5.10) can be proper, indeed the spectrum of L can be empty (Hille and Philips 1957), (Zabzyk 1975).

In particular, the spectrum of L being bounded does not imply that L is bounded, (but this is not so for the generator of a group of isometries as we shall see in Section 4.6), for this one need only take translation to the right on $L^2(0,1)$. Moreover $\sigma(L) = \varnothing$ can occur for contraction semigroups of positive maps, but $\sigma(L) \neq \varnothing$ for positive bounded groups (see also Section 4.6).

We have noted that if $\{T_t\}$ is strongly continuous one-parameter semigroup with infinitesimal generator L, and $\|T_t\| \leq Me^{\omega t}$, for $\omega > \omega_0$, then $\operatorname{Re}\lambda > \omega_0$ implies that

$$(\lambda - L)^{-1} = \int_0^\infty e^{-\lambda t} T_t \, dt \, .$$

Similarly,

$$(\lambda - L)^{-n} = \frac{1}{n!} \int_0^\infty e^{-\lambda t} t^{n-1} T_t \, dt \tag{4.5.12}$$

and so $\|(\lambda - L)^{-n}\| \leq M(\operatorname{Re}\lambda - \omega)^{-n}$.

The formula $(1 - tL/n)^{-n} \to e^{tL}$ allows us to reconstruct of one-parameter semigroups from their infinitesimal generators:

Theorem 4.17 Hille–Yosida–Phillips . *Let L be a closed densely defined linear operator on a Banach space X. Then L is the infinitesimal generator of a strongly continuous one-parameter semigroup $\{T_t\}$ if, and only if, the following two conditions hold for some $\omega_0 \geq 0$:*

(4.5.13) $\lambda \in \rho(L)$ *for all large* $\lambda > \omega_0$.

(4.5.14) $\|(\lambda - L)^{-n}\| \leq M(\lambda - \omega_0)^{-n}$ *for such* λ.

Then $T_t = \lim_{n\to\infty}(1 - tL/n)^{-n} = \lim_{\lambda\to\infty} \exp -t\lambda[1 - \lambda(\lambda - L)^{-1}]$.

This can be improved for contraction semigroups, where $\|(\lambda - L)^{-n}\| \leq \lambda^{-n}$:

Theorem 4.18 Hille–Yosida, Lumer–Phillips. *A closed densely defined linear operator L on a Banach space X generates a strongly continuous contraction semigroup if, and only if, one of the following two equivalent conditions* (4.5.15) *or* (4.5.16) *holds:*

(4.5.15a) $\lambda \in \rho(L)$, *for all* λ *large and positive.*

(4.5.15b) $\|(\lambda - L)^{-1}\| \leq \lambda^{-1}$ *for such* λ.

(4.5.16a) $(\lambda - L)X = X$ *for one* $\lambda > 0$.

(4.5.16b) $\lambda\|x\| \leq \|(\lambda - L)x\|$, λ *large and positive, all* x *in* $D(L)$.

For a densely defined linear operator L, condition (4.5.16b) *is also equivalent to either of the following:*

(4.5.16c) *for all* $x \in D(L)$, *there exists* $\phi \in X^*$, $\phi \neq 0$ *such that* $\phi(x) = \|x\|$ *and* $\operatorname{Re}(\phi L(x)) \leq 0$

(4.5.16d) *for all* $x \in D(L)$, $\phi \in X^*$ *with* $\phi(x) = \|x\|$, $\|\phi\| = 1$, *then* $\operatorname{Re}(\phi L(x)) \leq 0$.

In this case L is said to be dissipative. A dissipative operator is closable and $\overline{L}$ is dissipative.

In practice, one often appeals to

Theorem 4.19 *A closed densely defined operator L on a Banach space generates a one-parameter group of isometries if, and only if, both the following conditions hold:*

(4.5.17) $\pm L$ *are dissipative*

(4.5.18) L *possesses a dense set of analytic vectors.*

We will be interested in semigroups of positive maps. Note that if e^{tL} is strongly continuous on an ordered Banach space X where X_+ is a closed cone then

$$e^{tL} \geq 0\,, t \geq 0 \Leftrightarrow (\lambda - L)^{-1} \geq 0 \quad \lambda >> 0 \tag{4.5.19}$$

as

$$(\lambda - L)^{-1} = \int_0^\infty e^{-\lambda t} T_t \, dt\,, \quad e^{tL} = \lim_{n \to \infty} (1 - tL/n)^{-n}\,.$$

Moreover, if $\{e^{tL_i}\}$, $i = 1, 2$ are strongly continuous one-parameter semigroups, then under suitable conditions (Kato 1966) we have the Trotter product formula:

$$e^{t(L_1 + L_2)} = \lim_{n \to \infty} \left(e^{tL_1/n} e^{tL_2/n} \right)^n\,. \tag{4.5.20}$$

Thus

$$e^{tL_i} \geq 0\,, t \geq 0\,, i = 1, 2 \Rightarrow e^{t(L_1 + L_2)} \geq 0\,, t \geq 0\,. \tag{4.5.21}$$

For example, consider the generator δ of a strongly continuous one-parameter group of $*$-automorphisms of a C^*-algebra A. Then $\|e^{\delta t}\| = 1$, and so by the Hille–Yosida theorem $\pm\sqrt{\lambda} \in \rho(\delta)$ and $\|(\sqrt{\lambda} \pm \delta)^{-1}\| \leq \lambda^{-1/2}$ for $\lambda > 0$. But $(\lambda - \delta^2) = (\sqrt{\lambda} - \delta)(\sqrt{\lambda} + \delta)$, and so $\lambda \in \rho(\delta^2)$, and $\|(\lambda - \delta^2)^{-1}\| \leq \|(\sqrt{\lambda} - \delta)^{-1}\| \|(\sqrt{\lambda} + \delta)^{-1}\| \leq \lambda^{-1}$ for $\lambda > 0$. This means by the Hille–Yoshida theorem that δ^2 generates a contraction semigroup, which is positive as $(\lambda - \delta^2)^{-1} = (\sqrt{\lambda} - \delta)^{-1}(\sqrt{\lambda} + \delta)^{-1} \geq 0$. (It is clear that δ^2 is closed.) Note that this could also be established from the formula

$$e^{t\delta^2} = \int_{-\infty}^{\infty} \frac{e^{-s^2/4t}}{4\pi t} e^{s\delta} \, ds \geq 0\,.$$

Thus the Trotter product formula indicates, that (formally at least), if $\delta_0, \delta_1, \delta_2, \ldots$ are a sequence of derivations then $L = \delta_0 + \delta_1^2 + \delta_2^2 + \cdots$ 'generates' a (completely) positive semigroup.

4.6 Spectral subspaces

Suppose that T_t is a strongly continuous group of isometries on a Banach space X. Then

$$f \to T_f = \int_{-\infty}^{\infty} f(t) T_t \, dt \tag{4.6.1}$$

is multiplicative for convolution on $\mathcal{S} = \mathcal{S}(\mathbb{R})$, Schwartz space, or $f \to T_{\hat{f}} \equiv \beta(f)$ is multiplicative for pointwise product on $\mathcal{S}$, where ˆ denotes Fourier transform. If X is a Hilbert space, then β lifts to a map from $B(\mathbb{R})$, the bounded Borel functions on $\mathbb{R}$ into X, and $P(A) = \beta(\chi_A)$ $(A \subset \mathbb{R})$ is a projection valued measure such that

$$\beta(f) = \int_{-\infty}^{\infty} f(s) \, dP(s) \,. \tag{4.6.2}$$

So comparing (4.6.1) with (4.6.2), then

$$T_t = \int_{-\infty}^{\infty} e^{ist} \, dP(s) \tag{4.6.3}$$

or $T_t = e^{iHt}$, where $H = \int s \, dP(s)$ is self adjoint, which is *Stone's theorem.* For an arbitrary Banach space, such a projection valued measure does not exist, but analogues of

$$\text{spectral subspaces} \quad X^T(A) = P(A)X \tag{4.6.4}$$

and

$$\text{spectrum} \quad \sigma(H) \tag{4.6.5}$$

do exist. If $\{T_t\}$ is a strongly continuous group of isometries on a Banach space X (and if we formally think of $T_f = \int \hat{f}(s) \, dP(s)$), then we define (Arveson 1974), (Borchers 1974) the *spectral subspace* of a closed set $A \subset \mathbb{R}$ as

$$X^T(A) = \cap \{ \operatorname{Ker} T_f \mid f \in \mathcal{S}, \operatorname{supp} \hat{f} \cap A = \varnothing \} \tag{4.6.6}$$

and the *spectrum* of $\{T_t\}$ as

$$\sigma(T) = \cap \{ \text{closed } A \mid X^T(A) = X \} \,. \tag{4.6.7}$$

The above would make sense for a locally compact abelian group G, but when $G = \mathbb{R}$, the spectrum of T, $\sigma(T)$, is the same as the spectrum of the infinitesimal generator L (Evans 1976b). Thus as $\sigma(T) \neq \varnothing$ (Arveson 1974), (Borchers 1974) we deduce that $\sigma(L) \neq \varnothing$. Moreover since L is bounded if, and only if, $\sigma(T)$ is bounded (Olesen 1974), we deduce that the generator of a strongly continuous one-parameter group of isometries is bounded, if and only if its generator has bounded spectrum.

As an application of the power of the theory of spectral subspaces, consider $\{\alpha_t\}$ a *weakly continuous* group of $*$-automorphisms of a von Neumann algebra M on a Hilbert space $\mathcal{H}$ (i.e. $t \to \langle \alpha_t(x)\zeta, \eta \rangle$ is continuous for all $x \in M$

$\zeta, \eta \in \mathcal{H}$). One can define spectral subspaces as for strongly continuous groups. Suppose

$$\bigcap_{t \in \mathbb{R}} M[t, \infty)\mathcal{H} = \{0\}. \tag{4.6.8}$$

This is the case for example if either

(4.6.9) $\alpha_t = e^{\delta t}$ is norm continuous, in which case as $\sigma(\delta) = \text{support}\, M(\cdot)$, we have $M[t, \infty) = \{0\}$ if $t > \|\delta\|$.

(4.6.10) $\alpha_t = V_t(\cdot)V_t^*$ where V_t is some unitary group on $\mathcal{H}$ with spectrum (of Vor its generator) in a half line.

Then $P[t, \infty) = \bigcap_{s<t} M[s, \infty)\mathcal{H}$ is a projection valued measure such that $U_s = \int e^{ist}\, dP(t) \in M$, and $\alpha_s = U_s(\cdot)U_s^*$ (Arveson 1974). Thus for (4.6.9) we see that if δ is a bounded derivation, then $H = \int s\, dP(s) \in M$ is bounded. Consequently $\delta = i\text{ad}\, H$, and so every bounded derivation on a von Neumann algebra is inner (Kadison 1966), (Sakai 1966). Borchers' theorem (Borchers 1966) is recovered from (4.6.10) in that if a weakly continuous group of $*$-automorphisms on a von Neumann algebra can be implemented by a unitary group with positive generator, then the generator can be chosen affiliated to the algebra.

4.7 Positive and completely positive one-parameter semigroups on C^*-algebras

Recall from Section 4.2 that a bounded linear map T on a C^*-algebra A is completely positive if, and only if

$$[T(x_i^* y^* y x_j)] \geq [T(x_i^* y^* y) T(y^* y)^{-1} T(y^* y x_j)] \tag{4.7.1}$$

for all $y, x_1, \ldots, x_n \in A$, $n = 1, 2, \ldots$. If $T_t = e^{tL}$ is a completely positive norm continuous semigroup, then by differentiation of (4.7.1):

$$[L(x_i^* y^* y x_j) + x_i^* L(y^* y) x_j] \geq [L(x_i^* y^* y) x_j + x_i^* L(y^* y x_j)]. \tag{4.7.2}$$

In particular if the maps T_t are unital ($T_t(1) = 1$ for all t if and only if $L(1) = 0$) then

$$[L(a_i^* a_j)] \geq [L(a_i)^* a_j + a_i^* L(a_j)], \quad a_1, \ldots, a_n \in A \tag{4.7.3}$$

(cf. the derivation law for generators of homomorphisms).
It is worth examining such inequalities more carefully:

Lemma 4.20 *Let $L : B \to A$ be a bounded symmetric linear map between C^*-algebras A and B, with $B \subset A \subset B(H)$. Then the following conditions are equivalent:*

(a) *For all b in B, the kernels*

$$(s, t) \to L(s^* b^* b t) + s^* L(b^* b) t - L(s^* b^* b) t - s^* L(b^* b t)$$

are positive definite on $B \times B$.

(b) *The following holds for all n in $\mathbb{N}$:*

$$\sum_{i,j}\langle L(b_i^* b_j)\xi_j, \xi_i\rangle \geq 0$$

for all $b_1, \ldots, b_n \in A$, $\xi_1, \ldots, \xi_n \in H$ which satisfy $\sum_{i=1}^n b_i \xi_i = 0$.

(c) *The following holds for all n in $\mathbb{N}$:*

$$\sum_{i,j} a_i^* L(b_i^* b_j) a_j \geq 0$$

for all $b_1, \ldots, b_n \in B$, $a_1, \ldots, a_n \in A$ which satisfy $\sum_{i=1}^n b_i a_i = 0$.

(d) *The following holds for all n in $\mathbb{N}$:*

$$f[L(b_i^* b_j)] \geq 0$$

for all $f \in M_n(A)_+^*$, $b_1, \ldots, b_n \in A$ which satisfy $f[b_i^* b_j] = 0$.

Proof (d) $\Rightarrow$ (b) is trivial.

(a) $\Rightarrow$ (c) Let $a_1, \ldots, a_n$ in A and $b_1, \ldots, b_n$ in B satisfy $\sum_{i=1}^n a_i b_i = 0$. Then for all a in A we have

$$\sum_{i,j=1}^n b_i^* \{L(a_i^* a^* a a_j) + a_i^* L(a^* a) a_j - L(a_i^* a^*) a_j - a_i^* L(a^* a a_j)\} b_j \geq 0\,;$$

thus

$$\sum_{i,j=1}^n b_i^* L(a_i^* a^* a a_j) b_j \geq 0\,.$$

Taking a to be an approximate identity for A, we have

$$\sum_{i,j=1}^n b_i^* L(a_i^* a_j) b_j \geq 0\,.$$

(c) $\Rightarrow$ (a) Suppose $c_1, \ldots, c_n, e_1, \ldots, e_n$ in A, and $f_1, \ldots, f_n$ in B are arbitrary. Define

$$a_i = \begin{cases} c_i\,, & 1 \leq i \leq n\,, \\ c_{i-n} e_{i-n}\,, & n < i \leq 2n\,. \end{cases}$$

and

$$b_i = \begin{cases} -e_i f_i\,, & 1 \leq i \leq n\,, \\ f_{i-n}\,, & n \leq i \leq 2n\,. \end{cases}$$

Then $\sum_{i=1}^{2n} a_i b_i = 0$, so that $\sum_{i,j=1}^{2n} b_i^* L(a_i^* a_j) b_j \geq 0$; substituting for a_i and b_i, we have

$$\begin{aligned} &\sum_{i,j=1}^n f_i^* L(e_i^* c_i^* c_j e_j) f_j + \sum_{i,j=1}^n f_i^* e_i^* L(c_i^* c_j) e_j f_j \\ &\geq \sum_{i,j=1}^n f_i^* L(e_i^* c_i^* c_j) e_j f_j + \sum_{i,j=1}^n f_i^* e_i^* L(c_i^* c_j e_j) f_j\,. \end{aligned}$$

Thus the kernels

$$(s_1, s_2), (t_1, t_2) \mapsto L(s_1^* s_2^* t_2 t_1) + s_1^* L(s_2^* t_2) t_1 - L(s_1^* s_2^* t_2) t_1 - s_1^* L(s_2^* t_2 t_1)$$

are positive-definite on $B^2 \times B^2$ and so (a) certainly holds.

(b) $\Rightarrow$ (c) Let $b_1, \ldots, b_n \in B$, $a_1, \ldots, a_n \in A$ satisfy $\sum_{i=1}^n b_i a_i = 0$. Then $\sum_{i=1}^n b_i(a_i\eta) = 0$ for all $\eta \in H$. Hence, by (b)

$$\left\langle \sum_{i,j} a_i^* L(b_i^* b_j) a_j \eta, \eta \right\rangle = \sum_{i,j} \langle L(b_i^* b_j)(a_i\eta)(a_i\eta)\rangle \geq 0\,,$$

and so (c) holds.

(a) $\Rightarrow$ (b) Take $b_1, \ldots, b_n \in B$, $f \in M_n(A)^*_+$ with $f[(b_i^* b_j)] = 0$. Let

$$b_0 = \begin{pmatrix} b_1 \ldots b_n \\ 0 \end{pmatrix} \in M_n$$

so that $f(b_0^* b_0) = 0$ and so $f(x b_0) = 0 = f(b_0^* x)$ for all $x \in M_n(A)$ by the Schwarz inequality. In particular

$$f([x_i b_j]) = 0 = f([b_i^* x_j]) \text{ for all } x_1, \ldots, x_n \in A\,.$$

But by (a) for any $b \in B$:

$$[L(b_i^* b^* b b_j) + b_i^* L(b^* b) b_j)] \geq [L(b_i^* b^* b) b_j] + [b_i^* L(b^* b b_j)]\,.$$

Then applying the state $f([L(b_i^* b^* b b_j)] \geq 0$ and, hence, $f([L(b_i^* b_j)]) \geq 0$ by taking an approximate identity for B, so that (d) holds. □

A map satisfying one of the equivalent conditions of Lemma 4.20 is said to be *conditionally completely positive.* Conditions (a) and (c) can be regarded as the infinitesimal versions of (4.2.2) and (4.2.1) (iii) respectively. In the following, we naturally regard the double dual of a C^*-algebra as a unital C^*-algebra (indeed a von Neumann algebra – see Section 5.2).

Lemma 4.21 *Let B be a C^*-subalgebra of a C^*-algebra A and $L : B \to A$ a symmetric bounded linear map. Then the following conditions are equivalent:*

(a) *L is conditionally completely positive and $1L^{**}(1)1 \leq 0$, where 1 denotes the identity of B^{**}.*
(b) *The kernel $(s,t) \to L(s^*t) - L(s)^*t - s^*L(t)$ is positive definite on $B \times B$.*

Proof Suppose (a) holds. Then taking b in Lemma 4.20 (a) to be an approximate identity for B, converging to 1 in B^{**} we see that (b) holds. Conversely suppose (b) holds. Then clearly as in the proof of Lemma 4.20 ($a \Rightarrow c$) L is conditionally completely positive. Moreover for all self adjoint b in B, $L(b^2) \geq L(b)b + bL(b)$, and so $L^{**}(1) \geq L^{**}(1)1 + 1L^{**}(1)$ and finally $0 \geq 1L^{**}(1)1$. □

Theorem 4.22 *Let L be a bounded symmetric linear map on a C^*-algebra $\mathcal{A} \subset B(H)$. Then the following conditions are equivalent:*

(a) *e^{tL} is positive for all positive t.*

(b) $(\lambda - L)^{-1}$ *is positive for all large positive* λ.

(c) *For any commutative* C^**-subalgebra* B *of* A, *the map* $L|_B : B \to A$ *is conditionally completely positive.*

(d) $L(x^2y^2) + xL(y^2)x \geq L(xy^2)x + xL(xy^2)$ *for all commuting pairs of self adjoint elements* x, y *in* A.

(e) $f(L(b^*b)) \geq 0$ *for all* $f \in A_+^*$, $b \in A$ *which satisfy* $f(b^*b) = 0$.

(f) $\langle L(b^*b)\xi, \xi \rangle \geq 0$ *for all* $b \in A$, $\xi \in H$ *which satisfy* $b\xi = 0$.

(g) $a^*L(b^*b)a \geq 0$ *for all* $a, b \in A$ *which satisfy* $ba = 0$.

If the algebra is unital, then these conditions are also equivalent to:

(h) $L(x^2) + xL(1)x \geq L(x)x + xL(x)$ *for all self-adjoint* x *in* A.

(i) $L(1) + u^*L(1)u \geq L(u^*)u + u^*L(u)$ *for all unitaries* u *in* A.

Proof (d)⇒(h) and (e)⇒(f) are trivial.

(g)⇒(b) Let λ be greater than $\|L\|$. In order to show that $(\lambda - L)^{-1} \geq 0$, it is enough to show that $x \geq 0$ whenever x is self-adjoint and $(\lambda - L)x \geq 0$. Let $x = x^+ - x^-$ with x^+ and x^- positive and $x^+x^- = 0$. Then, by (g), we have $x^-L(x^+)x^- \geq 0$, so that

$$\begin{aligned} 0 \leq x^-[(1 - \lambda^{-1}L)(x)]x^- &= x^-xx^- - \lambda^{-1}x^-L(x)x^- \\ &= -(x^-)^3 - \lambda^{-1}x^-L(x^+)x^- + \lambda^{-1}x^-L(x^-)x^- . \end{aligned}$$

Thus $0 \leq (x^-)^3 \leq \lambda^{-1}x^-L(x^-)x^-$, and so $\|x^-\|^3 \leq \lambda^{-1}\|L\|\|x^{-1}\|^3$, since $\|a\| \leq \|b\|$ whenever $0 \leq a \leq b$. Hence $x^- = 0$, since $\lambda^{-1}\|L\| < 1$.

(b)⇒(a) We have $e^{tL} = \lim_{n\to\infty}(1 - tL/n)^{-n}$.

(c)⇒(d) by Lemma 4.20.

(a)⇒(c) has been observed when deducing (4.7.2), using Proposition 4.2.

(d),(h)⇒(e) and (f)⇒(g) as in the proof of Lemma 4.20.

(a)⇔(i) Let $K = -L(1)/2$, and put $L''(x) = Kx + xK$. Then $e^{tL''}(x) = e^{tK}xe^{tK}$, so that $\{e^{tL''} \mid t \in \mathbb{R}\}$ is a group of positive maps. Applying the Lie–Trotter formula (Kato 1966) to $L' = L + L''$, we have $e^{tL'} \geq 0$ for all $t \geq 0$ if and only if $e^{tL} \geq 0$ for all $t \geq 0$. By this reduction we assume $L(1) = 0$.

(a)⇒(i) Since $e^{tL} \geq 0$ and $e^{tL}(1) = 1$ for all $t \geq 0$, we have $\|e^{tL}\| = 1$ for all $t \geq 0$. Hence $e^{tL}(u^*)e^{tL}(u) \leq 1$ for all $t \geq 0$; differentiating this inequality at $t = 0$, we have $L(u^*)u + u^*L(u) \leq 0$ for all unitaries u in A.

(i)⇒(a) Since we have assumed that $e^{tL}(1) = 1$ for all $t \geq 0$, it is enough (by Section 4.3) to prove that e^{tL} is a contraction for all $t \geq 0$. By Section 4.1, this is the case if $\lim_{t\downarrow 0}(\|1 + tL\| - 1)/t \leq 0$. Moreover

$$\|1 + tL\| = \sup\{\|u + tL(u)\| \mid u \text{ unitary}\}$$

(see Section 4.2). But if u is unitary and $t \geq 0$, we have

$$\begin{aligned}\|u + tL(u)\|^2 &= \|1 + t[L(u^*)u + u^*L(u)] + t^2L(u)^*L(u)\| \\ &\leq \|1 + t^2L(u)^*L(u)\| \leq 1 + t^2\|L\|^2 .\end{aligned}$$

Thus $\|1 + tL\| \leq [1 + t^2\|L\|^2]^{1/2}$, and so

$$\lim_{t\downarrow 0}(\|1 + tL\| - 1)/t \leq \lim_{t\downarrow 0}[1 + t^2\|L\|^2]^{1/2} - 1)/t = 0;$$

hence e^{tL} is a contraction for each $t \geq 0$. □

We recall from Section 4.5 that generators of strongly continuous contraction semigroups are dissipative, and that dissipative maps are closable. Thus everywhere defined dissipative maps are bounded. It is thus natural to ask whether maps satisfying say condition (d) of Theorem 4.22 are automatically dissipative. This includes the case of L being a derivation (Sakai 1966).

Theorem 4.23 *Let L be a self adjoint linear map on a unital C^*-algebra A, with the following property:*

$$\text{if } y \text{ is in } A_+\,, f \text{ is in } A_+^*\,, \text{and } f(y) = 0\,, \text{then } f(L(y)) \geq 0\,. \tag{4.7.4}$$

Then L is bounded, and so e^{tL} is positive for all $t \geq 0$.

Proof The map $x \mapsto L(x) - [L(1)x + xL(1)]/2$ satisfies condition (4.7.4) whenever L does, so we may assume that $L(1) = 0$. We will show that, in this case, L is dissipative on A_h (in the sense of Section 4.5):

$$\lambda\|x\| \leq \|\lambda x - Lx\| \text{ for all } x \text{ in } A_h \text{ and } \lambda > 0\,. \tag{4.7.5}$$

In order to prove this for some self adjoint x, we may assume that there exists a positive f in A^* such that $f(x) = \|x\|$ and $\|f\| = 1$. Then $f(\|x\| - x) = 0$, and so $f(L(\|x\| - x)) \geq 0$; that is, we have $f(L(x)) \leq 0$. Let λ be strictly positive, then $\lambda f(x) \leq f(\lambda x - Lx) \leq \|f\|\|\lambda x - Lx\|$. Hence $\lambda\|x\| \leq \|f\|\|\lambda x - Lx\|$ for all self adjoint x in A. It follows that L is closed on A_h, and so L is bounded: Let $\{f_n \in A_h\}$ be a sequence satisfying $f_n \to 0$, $Lf_n \to g$; then for all h in A_h, and $\lambda > 0$, we have

$$\lambda\|\lambda f_n + h\| \leq \|(\lambda - L)(\lambda f_n + h)\|\,.$$

Letting $n \to \infty$, we have $\lambda\|h\| \leq \|\lambda(h - g) - L(h)\|$; as $\lambda \to \infty$ we have $\|h\| \leq \|h - g\|$ for all h in A_h. Hence $g = 0$. It then follows that e^{tL} is positive for all $t \geq 0$. Alternatively, this follows from (4.7.5) which shows that $(1 - \lambda^{-1}L)^{-1}$ is a contraction for all $\lambda > \|L\|$, and hence is positive since it preserves the identity (see Section 4.3). □

The results listed in Theorem 4.22 relate mainly to the Jordan structure of a C^*-algebra, but they will be used to prove a result about its C^*-structure in Theorem 4.26.

Theorem 4.24 *Let L be a densely defined symmetric linear map on an a unital C^*-algebra A. Then the following conditions are equivalent:*

(a) *L generates a strongly continuous positive semigroup.*
(b) *$(\lambda - L)D = A$ for all large λ.*
$fL(a) \geq 0$, for all $f \in A_+^*$, $a \in D(L)_+$ satisfying $f(a) = 0$.
(c) *$(\lambda - L)D = A$ for all large λ.*
λ large, $a \in D$, $(\lambda - L)a \geq 0 \Rightarrow a \geq 0$.
(d) *$(\lambda - L)^{-1}$ exists and is positive for all large λ.*

If the C^*-algebra A is not unital, then (d) does not necessarily imply that L is a generator (Batty and Davies 1983). We can improve Theorem 4.24 for positive contraction semigroups.

Theorem 4.25 *Let L be a bounded symmetric linear map on a C^*-algebra A. Then the following conditions are equivalent:*

(a) *e^{tL} is a positive contraction for all positive t.*
(b) *$\lambda(\lambda - L)^{-1}$ is a positive contraction for all large positive λ.*
(c) *$L^{**}(1_{A^{**}}) \leq 0$, and for any commutative C^*-subalgebra B of A, $L|_B$ is conditionally completely positive.*
(d) *For any commutative C^*-subalgebra B of A, the kernel $(s,t) \to L(s^*t) - L(s)^*t - s^*L(t)$ is positive-definite on $B \times B$.*
(e) *$L(x^2) \geq L(x)x + xL(x)$ for all self adjoint x in A.*

Theorem 4.26 *Let L be a bounded symmetric linear map on a C^*-algebra A. Then the following conditions are equivalent:*

(a) *$e^{tL}(x^*x) \geq e^{tL}(x)^*e^{tL}(x)$, $t \geq 0$, for all x in A.*
(b) *$L(x^*x) \geq L(x)^*x + x^*L(x)$ for all x in A.*

Proof (a)$\Rightarrow$(b) This follows by differentiating.

(b)$\Rightarrow$(a) Suppose (b) holds; adjoin an identity 1 to A, and extend L to the enlarged algebra by putting $L(1) = 0$. Then, by Theorem 4.25, e^{tL} is positive on the enlarged algebra for all $t \geq 0$. Fix x in A and define

$$f(t) = e^{tL}(x^*x) - e^{tL}(x)^*e^{tL}(x), \quad t \geq 0.$$

Then $f'(t) = Le^{tL}(x^*x) - [Le^{tL}(x)^*]e^{tL}(x) - e^{tL}(x)^*[Le^{tL}(x)]$, so that

$$f(t) - e^{tL}f(0) = \int_0^t \frac{d}{ds}[e^{(t-s)L}f(s)]\,ds$$
$$= -\int_0^t e^{(t-s)L}Lf(s)\,ds + \int_0^t e^{(t-s)L}\frac{d}{ds}f(s)\,ds$$

$$= \int_0^t e^{(t-s)L}\{L[e^{sL}(x)^*e^{sL}(x)] - [Le^{sL}(x)^*]e^{sL}(x) - e^{sL}(x)^*[Le^{sL}(x)]\}ds.$$

But, by hypothesis,

$$L[e^{sL}(x)^*e^{sL}(x)] \geq [Le^{sL}(x)^*]e^{sL}(x) + e^{sL}(x)^*[Le^{sL}(x)]$$

for all x in A and $s \geq 0$. Moreover, $e^{(t-s)L}$ is positive for $0 \leq s \leq t$; hence $f(t) \geq e^{Lt}f(0) = 0$ for all $t \geq 0$. □

Any bounded derivation of a C^*-algebra is implemented in the weak closure, see Section 4.6. Analogously for bounded generators of dynamical semigroups:

Theorem 4.27 *Let $L : A \to A$ be a bounded symmetric linear map on a C^*-algebra A, with A faithfully represented on a Hilbert space. Then the following conditions are equivalent:*

(a) *e^{tL} is completely positive for all $t \geq 0$.*
(b) *L is conditionally completely positive.*
(c) *There exists a norm continuous semigroup G_t in A'' such that $x \to e^{tL}(x) - G_t(x)G_t^*$ is completely positive from A into A''.*
(d) *There exists $K \in A''$ such that $\psi \equiv L - K(\cdot) - (\cdot)K^*$ is completely positive.*

Exercise 4.21 Let

$$\sigma_0 = \begin{pmatrix} 1 & 0 \\ 0 & 1 \end{pmatrix}, \quad \sigma_1 = \begin{pmatrix} 1 & 0 \\ 0 & -1 \end{pmatrix}, \quad \sigma_2 = \begin{pmatrix} 0 & 1 \\ 1 & 0 \end{pmatrix}, \quad \sigma_3 = \begin{pmatrix} 0 & i \\ -i & 0 \end{pmatrix}$$

denote the Pauli matrices in M_2, and L a linear map on M_2 such that $L\sigma_i = \lambda_i\sigma_i$, with $\lambda_0 = 0$ (i.e. $L(1) = 0$). Show that:

(a) $\{e^{tL} \mid t \geq 0\}$ is a positive semigroup if and only if $\lambda_i \geq 0$, $i = 1, 2, 3$.
(b) $\{e^{tL} \mid t \geq 0\}$ is a semigroup of sesqui-positive maps if and only if $\lambda_i \geq 0$ and $4\lambda_i\lambda_j - t_k^2 \geq 0$ for $i \neq j \neq k \neq i$, where $t_i = -\lambda_i + \sum_{j\neq i}\lambda_j$, $i = 1, 2, 3$.
(c) $\{e^{tL} \mid t \geq 0\}$ is a semigroup of 2-positive maps if and only if $t_i \geq 0$, $i = 1, 2, 3$.

4.8 Notes

Sections 4.2–4.5 are taken from (Evans and Lewis 1977a).

(Størmer 1963) showed that a positive map from an arbitrary C^*-algebra into a commutative C^*-algebra is completely positive; he used a slightly different method from the one given here. That any positive map from a commutative C^*-algebra into an arbitrary C^*-algebra is completely positive was shown by (Naimark 1943a), (Naimark 1943b), and by (Stinespring 1955). The Schwarz inequality (4.2.3) in Corollary 4.4 was first obtained for self adjoint elements by (Kadison 1952), who used an entirely different method. Corollary 4.4 and its proof were first recorded by (Størmer 1963), along with the Schwarz inequalities of Proposition 4.12 for completely positive maps. For other Schwarz-type inequalities, with various positivity assumptions, see (Araki 1960), (Choi 1974), (Evans 1976a), (Lieb and Ruskai 1974). Corollary 4.5 is due to Broise (unpublished), and is recorded by (Størmer 1980). The proof given here is due to (Evans and Høegh-Krohn 1978) and uses an observation of (Evans 1977).

The main result of Section 4.3 is Proposition 4.8. For scalar-valued kernels on $\mathbb{Z} \times \mathbb{Z}$, it was proved by (Kolmogorov 1941); he showed that a kernel is the correlation kernel of a stochastic process if and only if it is positive-definite (Parthasarathy and Schmidt 1972). For operator-valued kernels, versions of Corollary 4.9, with various restrictive assumptions on X, can be found in the literature: (Payen 1964), (Kunze 1967), (Ponomarenko 1956), (Allen *et al.* 1975). The idea of using the image–space rather than the quotient-space (Naimark 1943a) goes back to (Aronszajn 1950); it has been exploited by (Halmos 1967) and (Schrader and Uhlenbrock 1975) for Hilbert space dilations, and by (Kunze 1967) and in group representation theory.

The dilation theorem for positive-definite functions on groups (Corollary 4.14) is due to (Naimark 1943b); it was extended to $*$-semigroups by (Sz.-Nagy 1955). The canonical decomposition of a completely positive scalar-valued map (that is, of a state) on a C^*-algebra is known as the GNS construction (Gelfand and Naimark 1943), (Segal 1947). It was extended by (Stinespring 1955) to operator-valued completely positive maps on unital C^*-algebras. The relationship between the Stinespring decomposition for algebras and the Naimark dilation for groups has been described several times in the literature (see (Suciu 1967)). If G is a locally compact group, there is a canonical bijection between completely positive maps on $L^1(G)$ and those on $C^*(G)$, the enveloping C^*-algebra of $L^1(G)$. If G is abelian, $C^*(G)$ can be identified via the Fourier transform with $C_0(\hat{G})$, the continuous functions vanishing at infinity on $\hat{G}$, the dual of G.

The theory of dilations of continuous semigroups began with (Cooper 1947) who discovered the dilation in Exercise 4.4; it is interesting to note that his motivation came from quantum mechanics (Cooper 1950a), (Cooper 1950b). The result of Exercise 4.5 (b), on the dilation of semigroups of contractions, is due to (Sz.-Nagy 1953); it is a powerful tool in Hilbert space theory (Sz.-Nagy and Foias 1970).

Exercise 4.6 (b) comes from (Sz.-Nagy 1955), who discovered the connection between positive-definite functions on $\mathbb{Z}$ and $*$-semi-groups of contractions indexed by $\mathbb{N}$. This method was generalized by (Mlak 1966) and (Suciu 1973). The construction of a unitary dilation of a contraction semigroup contracting strongly to zero is due to (Lax and Phillips 1967); this method can be modified to give an alternative proof of Exercise 4.5 (b) ((Sz.-Nagy and Foias 1970), section 1.10.2)

(Kraus 1971) obtained the canonical decomposition of a normal completely positive map on the von Neumann algebra of all bounded operators on a Hilbert space. (Choi 1975) showed that if, in Theorem 4.16, H and K are finite-dimensional the decomposition can be chosen so that the cardinality of the set X is at most $\dim(H)\dim(K)$.

The fundamental paper on Fock space is by (Cook 1951). The character-zation of a generating functional of the CCR is due to (Araki 1960) and to Segal 1961) independently – see Exercise 4.12. The extremal universally invari-ant states (whose generating functionals are of the form μ_λ in Exercise 4.14) were introduced by (Segal 1962).

Quasi-free dynamical semigroups associated with representations of the CCR were investigated in the thesis of (Thomas 1971); see also (Lewis and Thomas 1975a). In the algebraic context they were studied by (Davies 1972a), (Davies 1972b), (Davies 1976a) and also by (Demoen *et al.* 1977), (Emch 1972), (Emch *et al.* 1978), (Evans and Lewis 1976b), and (Lindblad 1976c). Necessity in Exercise 4.16 was proved by (Evans and Lewis 1976b), whilst sufficiency was shown by Demoen *et al.* 1977). In fact, (Demoen *et al.* 1977) introduce the multiplier $h, k) \to \omega(h,k)/\omega(Ah, Ak)$, and use it to construct a CCR algebra $W_A(H)$ over H; they exploit the fact that the function f of Exercise 4.16 gives rise to a completely positive map if and only if it is a generating functional of a state of he algebra $W_A(H)$.

Dilations of quasi-free dynamical semigroups induced by contraction semi-groups can be found in the FKM model (Ford *et al.* 1965), (Thomas 1971), Lewis and Thomas 1975a), (Lewis and Thomas 1975b). They have been stud-ed in detail by (Davies 1972a), (Emch 1976), (Emch *et al.* 1978), (Evans and Lewis 1976b). We will study completely positive quasi–free maps in some detail n Section 6.8.

Exercise 4.19 is from (Bozejko and Speicher 1994) and Exercise 4.20 is from Bratteli *et al.* 1986b) and (Popescu 1989).

Standard texts on one-parameter semigroups are (Hille and Philips 1957), Kato 1966), (Davies 1976b). Theorems 4.17 and 4.18 are due to (Feller 1953), Miyadera 1952), (Phillips 1952), (Hille 1948), (Hille 1952), (Yosida 1948), Lumer and Philips 1961), (Batty 1978), whilst Theorem 4.19 is taken from Bratteli and Robinson 1987).

The main content of Theorem 4.22 is showing that (g) implies (a), and is due to (Evans and Lewis 1977a) and (Evans and Hanche-Olsen 1979). The remainder of the equivalences of (a) – (g) follow from Lemma 4.20 and its proof, together with remarks in (Evans 1984b). In the unital case, that (i) implies (a) is due to (Tsui 1977), who also observed that (h) implies (i) is implicit in the work of Lindblad 1976a).

Theorems 4.23, 4.26 are due to (Evans and Hanche-Olsen 1979), (Evans 1976a). Theorem 4.23 is an improvement on the work of (Kishimoto 1976); we use (Sullivan 1975) for a proof of a result of (Lumer and Philips 1961): a densely defined dissipative linear map is closeable. (Kishimoto 1976) has shown that if L is a symmetric linear map on a dense domain D in a C^*-algebra A, satisfying

$$L(x^2) \geq L(x)x + xL(x), \quad x \in D \tag{4.8.1}$$

and the domain D has the property that if $x \in D \cap A_+$, then $x^{1/2} \in D$, then L is automatically bounded. Theorem 4.26 was first proved for identity-preserving semigroups on unital C^*-algebras by (Lindblad 1976a) with a different method. Theorem 4.24 is due to (Arendt *et al.* 1982) and (Batty *et al.* 1983).

The concept of conditionally completely positive maps was introduced by the first author – see (Evans and Lewis 1977a); Lemma 4.20 is built on the work of the first author – see (Evans and Lewis 1977a) and (Lindblad 1976b).

The canonical decomposition $L = \psi + K(\cdot) + (\cdot)K^*$, in Theorem 4.27(a) of norm-continuous semigroups of completely positive normal maps on a von Neumann algebra was first obtained independently by (Gorini *et al.* 1976) for finite dimensional matrix algebras, and by (Lindblad 1976a) for hyperfinite von Neumann algebras. The final result Theorem 4.27 for an arbitrary C^*-algebra is due to (Christensen and Evans 1979). For the analogous result for semigroups of positive-definite functions on groups, see (Parthasarathy and Schmidt 1972). For earlier work on the generators of dynamical semigroups, and dissipativity, see (Kossakowski 1972a), (Kossakowski 1972b), (Kossakowski 1973), and (Ingarden and Kossakowski 1975).

By (Lindblad 1976b), (Evans and Lewis 1977a), the following are equivalent for a von Neumann algebra M:
(a) whenever M is faithfully represented as a von Neumann algebra in a Hilbert space H, we have $H^1(M, B(H)) = 0$
(b) whenever M is faithfully represented as a von Neumann algebra on a Hilbert space H, and $L = M \to B(H)$ is a conditionally completely positive ultraweakly continuous $*$–linear map, there exists K in $B(H)$ such that $L - K^*(\cdot) - (\cdot)K$ is completely positive.

If M is a von Neumann algebra on a Hilbert space H, it is known that $H^1(M, B(H)) = 0$ if: (i) M is type I or hyperfinite (Johnson 1972), (Ringrose 1972): (ii) M has a cyclic vector on H (e.g. if M is properly infinite) (Christensen 1982b). It is conjectured that $H^1(M, B(H)) = 0$ for all von Neumann algebras.

A representation W of the CCR is said to be *non-singular* if the map $t \to W(th)$ is weakly continuous on $\mathbb{R}$ for each h in H (or equivalently if W is strongly continuous on all finite dimensional subspaces). In this case, by Stone's theorem there is for each $h \in H$, a self adjoint operator $R(h)$ called a *field operator*, such that $W(th) = \exp itR(h)$. The universally invariant representations W_λ are non-singular. It is sometimes instructive to regard the field operators $R(h)$ as the random variables of a non-commutative probability theory, the generating functional of a non-singular representation $\mu(h) = \langle \exp iR(h)\Omega, \Omega\rangle$ is analogous to the characteristic function of a probability distribution.

The field operators satisfy, at least formally, the commutation relation

$$R(h)R(k) - R(k)R(h) = -i\mathrm{Im}\langle h, k\rangle 1$$

as a consequence of W satisfying the canonical commutation relations. Define the annihilation operator $c(h)$ by

$$c(h) = 2^{-1/2}(R(h) + iR(ih))$$

and the creation operator $c^*(h) = c(h)^*$ by $c^*(h) = 2^{-1/2}(R(h) - iR(ih))$, so that again formally at least

$$W(h) = \exp(i2^{-1/2}c^*(h))\exp(i2^{-1/2}c(h))\exp -\|h\|^2/4 . \tag{4.8.2}$$

In the Fock representation, c and c^* may be described in terms of the left creation operators on full Fock space

$$c^*(h) = P_s(N!)^{1/2}t(h)P_s \tag{4.8.3}$$

where P_s is the projection on symmetric Fock space, and N is the number operator, $Nf = nf$, $N!f = n!f$ and $f \in H_n$ and $t(h)$ is the Toeplitz operator on full Fock space of Section 2.8.

What is crucial here in the CCR theory is the imaginary part of the inner product. More generally all we need is a *symplectic form* σ on a real vector space, i.e. a bilinear form σ on H which is skew symmetric $\sigma(x,y) = -\sigma(y,x)$, for each x, y in H. For example, if K is a complex inner product space, then $\sigma(x,y) = \mathrm{Im}\langle x,y\rangle$ is *non-degenerate symplectic form* in that $\sigma(x,y) = 0$ for all y only when $x = 0$. Then if (H,σ) is a non-degenerate symplectic space, there is an unique C^*-algebra $W(H,\sigma)$ generated by unitaries $\{W(f) \mid f \in H\}$ satisfying

$$W(f)W(g) = \exp(i\sigma(f,g))\,W(f+g)\,, \quad f,g \in H\,.$$

Moreover $W(H,\sigma)$ is simple. The uniqueness and simplicity follows using the same arguments of Chapters 2 and 3 for that of the Cuntz algebra, the Bunce–Deddens algebra and the irrational rotation algebra. For the theory of CCR algebras for degenerate symplectic spaces, see (Manuceau *et al.* 1973).

We can analogously define annihilation and creation operators on Fermion Fock space, but first some notation. Let $f_1, \ldots, f_n$ lie in the Hilbert space H, and define $f_1 \wedge \cdots \wedge f_n$ by

$$f_1 \wedge \cdots \wedge f_n = (n!)^{1/2} Q_n(f_1 \otimes \cdots \otimes f_n)\,.$$

Then we have

$$\begin{aligned} &\langle f_1 \wedge \cdots \wedge f_n, g_1 \wedge \cdots \wedge g_n\rangle = (n!)\langle Q_n(f_1 \otimes \cdots \otimes f_n), g_1 \otimes \cdots \otimes g_n\rangle \\ &= \textstyle\sum_\pi \varepsilon(\pi)\langle f_{\pi^{-1}(1)} \otimes \cdots \otimes f_{\pi^{-1}(n)}, g_1 \otimes \cdots \otimes g_n\rangle \\ &= \textstyle\sum_\pi \varepsilon(\pi)\langle f_{\pi^{-1}(1)}, g_1\rangle \cdots \langle f_{\pi^{-1}(1)}, g_n\rangle = \det(\langle f_i, g_j\rangle)\,. \end{aligned}$$

Thus the map $(f_i)_{i=1}^n \mapsto f_1 \wedge \cdots \wedge f_n$ of H^n into H_n^a is a minimal Kolmogorov decomposition for the positive-definite kernel $((f_i),(g_i)) \mapsto \det(\langle f_i, g_j\rangle)$ on $H^n \times H^n$. To avoid confusion, we may add subscripts s or a to our annihilation, creation operators when considering Bosons or Fermions, but usually we can dispense with these. So define

$$c^*(h) = P_a(N!)^{1/2}t(h)P_a\,, \quad c(h) = P_a(N!)^{1/2}t(h)^*P_a\,. \tag{4.8.4}$$

In this case $c(h)$ and $c^*(h) = c(h)^*$ are bounded operators; indeed they satisfy the canonical anti-commutation relations

$$c(f)c^*(g) + c^*(g)c(f) = \langle g, f\rangle 1 .$$

To see this note that $h_1 \wedge \cdots \wedge h_n$ is in the domain of $c^*(f)$, for $h_1, \ldots, h_n \in H$ and $c^*(f)h_1 \wedge \cdots \wedge h_n = f \wedge h_1 \wedge \cdots \wedge h_n$. Now let f be a unit vector in H, and put $M = \{f\}^\perp$. Then $c^*(f)$ maps $\wedge^n M$ isometrically onto $f \wedge (\wedge^n H)$ and annihilates $f \wedge (\wedge^{n-1} H)$, the orthogonal complement of $\wedge^n M$ in $\wedge^n H$. Thus $c^*(f)$ maps $F^a(M)$ isometrically onto $F^a(H) \ominus F^a(M)$ and annihilates $F^a(H) \ominus F^a(M)$. That is $c(f)c^*(f) + c(f)c^*(f) = 1$, or more generally $c(f)c^*(f) + c(f)c^*(f) = \langle f, f\rangle 1$ for all f in H. So by polarization

$$c(f)c^*(g) + c^*(g)c(f) = \langle g, f\rangle 1 , \quad f, g \in H .$$

We also have

$$c(f)c(g) + c(g)c(f) = 0 ,$$

since $f \wedge g + g \wedge f = 0$ for all f, g in H. This particular representation of the canonical anti-commutation relations is called the *Fock representation.*

The CCR algebras, being non-separable, are natural homes for counterexamples. For example, if P, Q are the position and momentum operators on $L^2(\mathbb{R})$ obtained from the Schrödinger representation of $W(\mathbb{C})$ in Exercise 4.13, then the Hamiltonian $H = P^2/2 + V(Q)$ do not induce $*$-automorphisms of $W(\mathbb{C})$ when $V \in W(\mathbb{C})$, or $V \in L^\infty \cap L^1$ except when the potential is trivial (Fannes and Verbeure 1974). The structure of all KMS states of a quasi-free Fermion dynamics is simply described in Chapter 6. The structure of KMS states of a quasi-free Boson dynamics can be quite subtle (Petz 1990).

The Boson and Fermion set-ups are most elegantly related through the stochastic calculus of (Hudson and Parthasarathy 1986). For this one needs the one particle space to be infinite dimensional and we take it to be $H = L^2(R_+)$ (otherwise F_a is finite dimensional but F_s is not). If h is a self adjoint operator on H, let $dF(h)$ denote the infinitesimal generator of the unitary group $F(e^{ith})$, e.g. $N = dF(1)$ (on Boson or Fock space with subscripts a, s if there is a risk of confusion). The *annihilation* and *creation processes* and the *gauge processes* are the processes (families of operators) given by

$$c^\#(t) = c^\#(\chi_{[0,t]}), \quad \Lambda(t) = dF(\chi_{[0,t]})$$

– we have such processes for Bosons and Fermions, but again we omit for the time being the subscripts a or s. Let $J(t) = F_s(-\chi_{[0,t]} + \chi_{[t,\infty]})$, $t \geq 0$ be the self adjoint unitary on Boson Fock space. Then the stochastic integrals $Jdc_s^\#$ behave like Fermions – $dc_a^\#$. By uniqueness of the Fock representation for Fermions $F_s(H)$ can be identified with Fermion Fock space, in such a way that $dc_a^\# = Jdc_s^\#$, $dc_s^\# = Jdc_a^\#$, and the gauge processes are identified.

The study of dynamical semigroups has been taken up by Powers and subsequently Arveson in the context of analysing the structure of one-parameter semigroups of endomorphisms of $B(H)$. The first examples are those induced by CAR or CCR algebras in the Fock representations of one-parameter semigroups of isometries on $L^2(0,\infty)\otimes\ell^2(n)$ obtained by translation in the interval $(0,\infty)$. These only depend on n, up to a natural notion of cocycle conjugacy. Such semigroups are spatial in that there is an intertwining semigroup of isometries U_t on H such that $U_t x = \alpha_t(x)U_t$ for $x \in B(H)$, $t \geq 0$. The intertwining semigroup of isometries is not unique, indeed the degree of freedom can be used to define an invariant, an index, which reduces to n in the case of CAR or CCR flows. There are semigroups on $B(H)$ which are not spatial, i.e. do not possess intertwining semigroups of isometries (Powers 1987), and spatial semigroups which are not cocycle conjugate to CAR or CCR flows (Powers, preprint 1994). However the latter flows satisfy a stronger notion of being completely spatial, and such semigroups are classified up to cocycle conjugacy by their index (Arveson 1989a). This work leads to a study of continuous tensor products of Hilbert spaces, e.g. (Arveson 1989a), (Arveson 1989c).

The three families of Hilbert spaces, and operators $F_a(H), c_a(h), h \in H)$, $(F(H), t(h), h \in H)$, $(F_s(H), c_s(h), h \in H)$, and corresponding relations can be extrapolated to give a one-parameter family indexed by $q \in [-1,+1]$ so that $q = -1, 0, +1$ correspond to the above three cases – Fermion, Cuntz–Toeplitz and Boson respectively (Bozejko and Speicher 1991). We will consider operators $c(f)$, $f \in H$ satisfying the q-commutation relations

$$c(f)c^*(g) - qc^*(g)c(f) = \langle g, f\rangle 1 \tag{4.8.5}$$

on a twisted Fock space. Start with Fock space $F(H) = \bigoplus_{n=0}^{\infty}\bigotimes^n H$, and let $\mathcal{F}$ denote the subspace linearly spanned by product vectors. Define linear maps, $c^*(f)$ and $c(f)$ on $\mathcal{F}$ by

$$c^*(f)\Omega = f \qquad c(f)\Omega = 0$$
$$c^*(f)h_1\otimes\cdots\otimes h_n = f\otimes h_1\otimes\cdots\otimes h_n$$
$$c(f)h_1\otimes\cdots\otimes h_n = \textstyle\sum_{i=1}^{n} q^{i-1}\langle h_i, f\rangle h_1\otimes\cdots\otimes h_i^{\vee}\otimes\cdots\otimes h_n$$

where $h_i^{\vee}$ means that h_i is omitted.

Then $c(f), c^*(g)$ satisfy relation (4.8.5) on $\mathcal{F}$. These operators are adjoint i.e. $\langle c^*(f)\xi, \eta\rangle = \langle \xi, c(f)\eta\rangle$, $\xi,\eta \in \mathcal{F})$ with respect to the symmetric bilinear form $\langle\,,\,\rangle_q$ defined on $\mathcal{F}$ via

$$\langle f\otimes g, h\rangle_q = \langle g, c(f)h\rangle_q\,, \quad g,h\in\mathcal{F}\,, \quad f\in H$$

which is related to the usual positive definite inner product on $F(H)$ by

$$\langle \xi, \eta\rangle_q = \langle \xi, P_q\eta\rangle\,, \quad \xi,\eta\in\mathcal{F}$$

where

$$P_q = \bigoplus_{n=0}^{\infty} P_q^n$$

and P_q^n is the operator on H_n given by

$$P_q^n = \sum_{\pi \in S_n} q^{i(\pi)} \pi$$

and $i(\pi)$ denotes the number of inversions of $\pi \in S_n$, i.e.

$$i(\pi) = \#\{(i,j) \in \{1,\ldots,n\}^2 \mid i < j\,, \pi(i) > \pi(j)\}\,.$$

Graphically, if a permutation $\pi \in S_n$ is drawn as n-straight lines between n vertices on each of two parallel rows

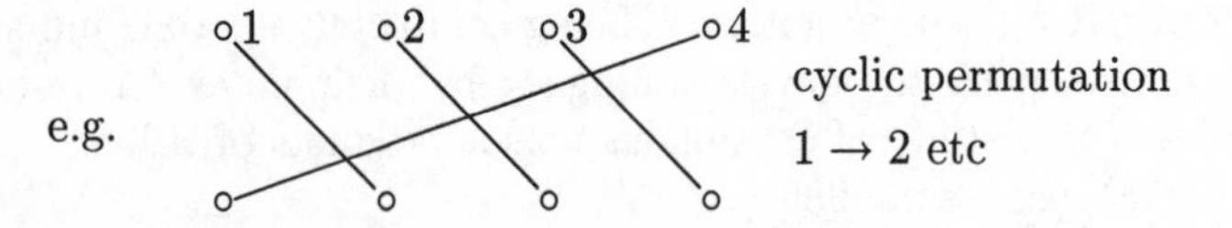

then $i(\pi)$ counts the number of crossings. For $q \in [-1,1]$, P_q is positive since $\pi \to q^{i(\pi)}$ is positive definite. Equivalently (cf. Theorem 4.27) $\pi \to -i(\pi)$ is conditionally negative definite, i.e.

$$\sum \varphi_\sigma \overline{\varphi_\pi} i(\pi^{-1}\sigma) \leq 0 \tag{4.8.6}$$

for each $\varphi : S_n \to \mathbb{C}$ such that $\sum \varphi_\pi = 0$. Let $K = \ell^2(\Delta)$, where $\Delta = \{(i,j) \mid i,j = 1,2,\ldots,n\}$ and V be the representation of S_n on K induced by $(i,j) \to (\pi(i),\pi(j))$. Then

$$i(\pi) = \langle (1 - V_\pi)\Omega, \Omega \rangle$$

where $\Omega = \chi_{\Delta^+}$, $\Delta^+ = \{(i,j) \mid i < j\}$, so that (4.8.6) clearly holds.

Thus $\langle\, ,\,\rangle_q$ is a positive scalar product, which is degenerate only when $q = \pm 1$. Let $F_q(H)$ denote the completion of $\mathcal{F}$ with respect to this scalar product. Note that

$$\langle f_1 \otimes \cdots \otimes f_n, g_1 \otimes \cdots \otimes g_n \rangle = \det_q \langle f_i, g_j \rangle$$

where $\det_q[a_{ij}] = \sum_\pi q^{i(\pi)} a_{1\pi(1)} \ldots a_{n\pi(n)}$ with $\det_{-1} = \det$, $\det_{+1} =$ the permanent, $\det_0[a_{ij}] = \prod_i a_{ii}$. Then F_q is the orthogonal sum of the reproducing kernel Hilbert spaces for the positive definite functions

$$(f_i),(g_i) \to \det_q[\langle f_i, g_j \rangle] \text{ on } H^n \times H^n\,, \quad n = 0,1,2,\ldots\,.$$

When $q = 0$, $P_0 = 1$ thus $F_0(H) = F(H)$, and $c^*(f)$ identifies with $t(f)$. If P_s, P_a are the projections of $F(H)$, on symmetric and anti-symmetric Fock space, then $P_{+1} = (N!)^{1/2} P_s$, $P_{-1} = (N!)^{1/2} P_a$, so that $F_{\pm 1}(H)$ can be identified with

$F_s(H)$ and $F_a(H)$ respectively, and c^* with the usual creation operators. The degeneracy in $\langle\ ,\ \rangle_q$ at $q = \pm 1$ forces extra relations in those cases, namely

$$\begin{array}{ll} q = 1 & c(f)c(g) = c(g)c(f) \\ q = -1 & c(f)c(g) = -c(g)c(f) \end{array} .$$

The operators $c(f)$ are bounded on F_q if and only if $q < 1$, in which case

$$\|c(f)\| = \begin{cases} \dfrac{1}{\sqrt{1-q}}\|f\| & 0 \leq q < 1 \\ \|f\| & q \leq 0 \end{cases} . \tag{4.8.7}$$

Let $\mathcal{E}^q$ denote the universal C^*-algebra generated by operators $c(\xi)$, anti-linear in $\xi \in H$ (for a fixed Hilbert space H) satisfying relations (4.8.5). We have a representation of $\mathcal{E}^q$ on F_q, called the *Fock representation*, uniquely determined up to unitary equivalence as the representation for $\mathcal{E}^q$ for which there is a cyclic vector Ω annihilated by $\{c(\xi) \mid \xi \in H\}$. If $q = 0$, this gives the Cuntz–Toeplitz algebra of Section 2.8. It is known (Jørgensen *et al.* 1994) that if $d = \dim H$ is finite, and greater than 1, and $|q| < \sqrt{2} - 1$ then the Fock representation is faithful, and $\mathcal{E}^q \cong \mathcal{E}^0$. If e_i is an orthonormal basis for H, let $c_i = c(e_i)$, $t_i = t(e_i)$. Then a positive element $r \in \mathcal{E}^0$ is constructed so that

$$r^2 = p + q \sum_{i,j=1}^{d} (t_i r t_j)(t_j r t_i)^*$$

where $p = \sum t_i t_i^*$ (so that $1 - p$ is the projection on the vacuum in the Fock representation). Then $c_i \to t_i^* r$ gives an isomorphism between $\mathcal{E}^q$ and $\mathcal{E}^0$. (Dykema and Nica 1993) construct a unitary U from twisted Fock space to the usual full Fock space such that $U\mathcal{R}^q U^* \supset \mathcal{R}^0$ for all $q \in (-1, 1)$ where $\mathcal{R}^q$ denotes $\mathcal{E}^q$ in its Fock representation. Thus $\mathcal{R}^q$ contains the compact operators $K(F_q)$. Equality $U\mathcal{R}^q U^* = \mathcal{R}^0$ is shown for all

$$q^2 < \sum_{k=-\infty}^{\infty} (-1)^k |q|^{k^2} = \prod_{k=1}^{\infty} \frac{1 - |q|^k}{1 + |q|^k},$$

which extends the range $|q| < \sqrt{2} - 1$.

5

VON NEUMANN ALGEBRA BASICS

5.1 Introduction

The aim of this chapter is to present the basic theory of von Neumann algebras with minimal preparation. The general theory of von Neumann algebras often requires several hundred pages in standard textbooks, and this has been an obstacle for non-operator algebraists to grasp the theory for applications in other fields such as low-dimensional topology or quantum group theory. We present only what is needed in the remainder of the book, and often give *ad hoc* definitions and make extra assumptions rather than work in full generality. (When we give a non-standard corner-cutting definition, we also explain what the standard definition is.)

Section 5.2 deals with basic definitions and properties of von Neumann algebras. We study the most fundamental class of von Neumann algebras in this book, the II_1 factors, in Section 5.3. We define a notion of a crossed product for a von Neumann algebra and an action of a group on it in Section 5.4. We deal only with finite groups here, which from a subfactor theory viewpoint are the classical objects before passing to the 'quantization'. We also define the tensor product of von Neumann algebras. Section 5.5 contains the theory of coupling constant of II_1 factors and their continuous dimension. This work of Murray–von Neumann was done in the very early days of the theory. These ideas are essential for subfactor theory in the second half of this book. We present a very brief review of the Tomita–Takesaki theory for more general von Neumann algebras in Section 5.6. This is not logically necessary for most of the remainder of this book, but we will occasionally use this theory in some later sections because it sometimes gives a conceptually better framework.

5.2 Definitions and basic properties

Let H be a Hilbert space and $B(H)$ the space of all bounded linear operators on H. (For most applications, it is enough to work on separable Hilbert spaces.) If a subalgebra of $B(H)$ is closed under the adjoint operation $*$ and norm topology, it is a C^*-algebra, as we already know. We consider here a $*$-subalgebra of $B(H)$

which is closed in a topology weaker than the norm topology. So we make the following definition (cf. strong and weak convergence of vectors in a Hilbert space in Definition 1.38).

Definition 5.1 *We define the following four topologies on $B(H)$.*

1. *A net x_i in $B(H)$ converges to $x \in B(H)$ in the strong operator topology if $\|x_i\xi - x\xi\| \to 0$ for all $\xi \in H$.*
2. *A net x_i in $B(H)$ converges to $x \in B(H)$ in the weak operator topology if $\|\langle x_i\xi, \eta\rangle - \langle x\xi, \eta\rangle\| \to 0$ for all $\xi, \eta \in H$.*
3. *A net x_i in $B(H)$ converges to $x \in B(H)$ in the σ-strong operator topology if $\sum_{j=1}^{\infty} \|x_i\xi_j - x\xi_j\|^2 \to 0$ for all $\xi_j \in H$ such that $\sum_{j=1}^{\infty} \|\xi_j\|^2 < \infty$.*
4. *A net x_i in $B(H)$ converges to $x \in B(H)$ in the σ-weak operator topology if $\sum_{j=1}^{\infty} \langle (x_i - x)\xi_j, \eta_j\rangle \to 0$ for all $\xi_j, \eta_j \in H$ such that $\sum_{j=1}^{\infty} \|\xi_j\|^2 < \infty$ and $\sum_{j=1}^{\infty} \|\eta_j\|^2 < \infty$.*

Actually we have two more commonly used topologies. They are the strong* operator topology and the σ-strong* operator topology, but they are not necessary in this book. Still, having four topologies may look quite complicated, but they are not so complicated as they seem at first. The prefix 'σ-' does not cause much difference in this book by Proposition 5.2 and the Kaplansky density theorem (Theorem 5.14). The use of nets rather than sequences is also not essential in this book again by the Kaplansky density theorem and the fact that we usually work on separable Hilbert spaces. The subtle differences among the various topologies do cause much difficulty in the more analytic aspects of von Neumann algebra theory, but they are essentially irrelevant in most parts of this book. One has to realize, however, that the norm topology is different from the above topologies in an essential way.

We often drop the adjective 'operator'. So if we say that x_i converges to x strongly or that x is in the weak closure of A, we actually mean that x_i converges to x in the strong operator topology and that x is in the closure of A in the weak operator topology.

Proposition 5.2 *The σ-strong topology and the strong topology coincide on the unit ball of $B(H)$ with respect to the norm. Similarly, the σ-weak topology and the weak topology coincide on the unit ball of $B(H)$.*

Proof Suppose we have $x_i \to 0$ in the strong topology with $\|x_i\| \le 1$. Then for $\{\xi_j\}$ in H with $\sum_{j=1}^{\infty} \|\xi_j\|^2 < \infty$, we get

$$\lim_i \sum_{j=1}^{\infty} \|x_i\xi_j\|^2 = \sum_{j=1}^{\infty} \lim_i \|x_i\xi_j\|^2 = 0.$$

The other statement is proved similarly. □

Recall the following basic definition, which has already played a role in the characterizing states on a C^*-algebra majorized by a fixed state (Lemma 2.9).

Definition 5.3 *For a subset S of $B(H)$, we define*

$$S' = \{x \in B(H) \mid xy = yx \text{ for all } y \in S\}.$$

The set S' is called the commutant *of S.*

The following theorem (the double commutant theorem of von Neumann) is *the* most fundamental theorem in von Neumann algebra theory.

Theorem 5.4 *For a $*$-subalgebra M of $B(H)$ containing the identity operator 1, the following conditions are equivalent:*

1. *$M'' = M$;*
2. *M is σ-strongly closed;*
3. *M is σ-weakly closed;*
4. *M is strongly closed;*
5. *M is weakly closed.*

Proof It is clear that Condition 1 implies Condition 5. It is then enough to prove that Condition 2 implies Condition 1, because the other implications are clear.

Suppose that we have $x \in M''$. We will then prove that x is contained in the σ-strong closure of M. A neighbourhood of x in the σ-strong topology is of the following form for some $\varepsilon > 0$ and $\xi_j \in H$ with $\sum_{j=1}^\infty \|\xi_j\|^2 < \infty$:

$$\{y \in B(H) \mid \sum_{j=1}^\infty \|(x-y)\xi_j\|^2 < \varepsilon^2\}.$$

Suppose first that ξ_1 is the only non-zero vector in $\{\xi_j\}$ for simplicity. We then need to show that there exists $y \in M$ such that $\|(x-y)\xi_1\| < \varepsilon$. We set

$$K = \overline{M\xi_1} = \overline{\{a\xi_1 \mid a \in M\}} \subset H.$$

Let p be the projection onto this subspace K. For any $a \in M$, we have $pap = ap$, so we get $pa^*p = a^*p$. Because a is an arbitrary element in a $*$-subalgebra M, we have $pap = pa$ for any $a \in M$. (Here we used the condition that M is closed under the $*$-operation in an essential way.) This shows $p \in M'$ and hence $xp = px$. This gives $x\xi_1 = xp\xi_1 = px\xi_1$, and hence $x\xi_1 \in K$. This implies that there exists $y \in M$ such that $\|(x-y)\xi_1\| < \varepsilon$.

We next consider the general case. For the given $\{\xi_j\}$, we set $\tilde{\xi} = (\xi_1, \xi_2, \ldots)$ in $\tilde{H} = \bigoplus_{i=1}^\infty H$ which is a countable direct sum of copies of H. For any element $a \in B(H)$, we have an operator $\tilde{a} = \bigoplus_{i=1}^\infty a$ on $\tilde{H}$ defined by

$$\tilde{a}(\eta_1, \eta_2, \ldots) = (a\eta_1, a\eta_2, \ldots).$$

In this way, $\tilde{M} = \{\tilde{a} \mid a \in M\}$ is a $*$-subalgebra of $B(\tilde{H})$. We then easily get $\tilde{M}'' = \widetilde{M''}$, and hence $\tilde{x} \in \tilde{M}''$. Applying the above argument to this case, we get $\tilde{y} \in \tilde{M}$ with $\|(\tilde{x} - \tilde{y})\tilde{\xi}\| < \varepsilon$, which means $y \in M$ and

$$\sum_{j=1}^{\infty} \|(x-y)\xi_j\|^2 < \varepsilon^2.$$

□

Definition 5.5 *If a $*$-subalgebra M of $B(H)$ containing 1 satisfies one (and hence all) of the conditions in Theorem 5.4, then we say that M is a* von Neumann algebra.

Theorem 5.4 also means the following. If S is a subset $B(H)$ closed under the $*$-operation, then S'' is the smallest von Neumann algebra in $B(H)$ containing S. We call S'' the *von Neumann algebra generated by S*.

Note that any topology in Definition 5.1 is weaker than the norm topology and hence that any von Neumann algebra is also automatically a C^*-algebra. It is, however, rather useless to regard a von Neumann algebra as a C^*-algebra except for some elementary aspects of the theory. This is because von Neumann algebras are rather unusual as C^*-algebras as we will see later. Also note that the C^*-algebras we are interested in are separable, and a von Neumann algebra separable for the norm topology is finite dimensional.

Example 5.6 Let (X, μ) be a measure space. Any L^∞ function acts on $H = L^2(X, \mu)$ by multiplication. In this way, we can regard $L^\infty(X, \mu)$ as a subalgebra of $B(H)$. This is a von Neumann algebra. So this is also a commutative C^*-algebra and hence an algebra of continuous functions of some compact space, but this topological space is highly disconnected. (Such a topological space is called hyperstonean. See III.1 of (Takesaki 1979).)

This example is very fundamental, and we regard a general von Neumann algebra as a 'non-commutative analogue' of $L^\infty(X, \mu)$.

We need the following easy observation next.

Proposition 5.7 *Let A be a C^*-algebra. For any $x \in A$, there exist four unitaries $u_i \in A$, $i = 1, 2, 3, 4$ and scalars $c_i \in \mathbb{C}$ such that $x = \sum_{i=1}^4 c_i u_i$.*

Proof Because any element x is a linear combination of two self-adjoint elements, we may assume that x is self-adjoint and then we prove that x is a linear combination of two unitaries. We may assume that $\|x\| \le 1$ and then write

$$x = \frac{1}{2}(x + i\sqrt{1-x^2}) + \frac{1}{2}(x - i\sqrt{1-x^2}).$$

□

Based on this proposition, we get the following theorem.

Theorem 5.8 *Let M be a von Neumann algebra and x a self-adjoint element in M. Then any spectral projection of x belongs to M. Furthermore, for any element $x \in M$ with polar decomposition $x = uh$, we have $u, h \in M$.*

Proof Let

$$x = \int_a^b \lambda \, dE_\lambda$$

be the spectral decomposition of x (for some a, b). Then for any unitary $u \in M'$, we get

$$\int_a^b \lambda \, d(uE_\lambda u^*) = uxu^* = x = \int_a^b \lambda \, dE_\lambda.$$

By uniqueness of the spectral decomposition, this implies that $uE_\lambda u^* = E_\lambda$ for any λ and any unitary $u \in M'$. Proposition 5.7 then implies $E_\lambda \in M'' = M$.

The other statement similarly follows from uniqueness of the polar decomposition. □

We next characterize a good subclass of bounded linear functionals on a von Neumann algebra M as follows.

Theorem 5.9 *Let M be a von Neumann algebra on a Hilbert space H. For $\phi \in M^*$, the following conditions are equivalent:*

1. *The functional ϕ is σ-weakly continuous.*
2. *The functional ϕ is σ-strongly continuous.*
3. *There exist vectors $\xi_j, \eta_j \in H$, $j = 1, 2, 3, \ldots$, such that*

$$\sum_{j=1}^{\infty} \|\xi_j\|^2 < \infty, \quad \sum_{j=1}^{\infty} \|\eta_j\|^2 < \infty, \quad \text{and} \quad \phi(x) = \sum_{j=1}^{\infty} (x\xi_j, \eta_j).$$

Proof It is trivial that Condition 1 implies Condition 2 and that Condition 3 implies Condition 1. It is thus enough to prove that Condition 2 implies Condition 3.

So we assume that ϕ is σ-strongly continuous and we will prove that ϕ is of the form of Condition 3. For any $\varepsilon > 0$, there exists $\delta > 0$ and vectors $\xi_j \in H$ with $\sum_{j=1}^{\infty} \|\xi_j\|^2 < \infty$ such that if $\sum_{j=1}^{\infty} \|x\xi_j\|^2 < \delta$, then we get $|\phi(x)|^2 < \varepsilon$. This implies that

$$|\phi(x)|^2 \leq \frac{\varepsilon}{\delta} \sum_{j=1}^{\infty} \|x\xi_j\|^2.$$

Let $\tilde{H}$ be a countable direct sum of copies of H, and let

$$K = \{(x\xi_1, x\xi_2, \ldots) \mid x \in M\} \subset \tilde{H}.$$

Let $\tilde{\phi}$ be the continuous linear functional on K defined by

$$\tilde{\phi}(x\xi_1, x\xi_2, \ldots) = \phi(x),$$

and extend it to a continuous linear functional on $\tilde{H}$, again denoted by $\tilde{\phi}$.

Then we have a vector $\eta = (\eta_1, \eta_2, \ldots) \in \tilde{H}$ such that

$$\tilde{\phi}(\zeta_1, \zeta_2, \ldots) = \sum_{j=1}^{\infty} (\zeta_j, \eta_j),$$

which gives the desired conclusion. □

Furthermore, if ϕ is a state, we can take $\xi_j = \eta_j$ in condition 3 with a little bit more work.

Definition 5.10 *A functional ϕ on M satisfying one (hence all) of the conditions in Theorem 5.9 is said to be* normal. *The set of all normal linear functionals on M is denoted by M_* and called the* predual *of M.*

The reason for the notation M_* and the name 'predual' comes from the following theorem.

Theorem 5.11 *Let M be a von Neumann algebra. Then the predual M_* is a Banach space, and the dual of the predual of M, $(M_*)^*$, is naturally isomorphic to M. Furthermore, the σ-weak topology on M coincides with the weak* topology on $(M_*)^*$ as a Banach space.*

Proof First note that $(B(H)_*)^* = B(H)$ by Theorems 5.9 and 2.6. Set

$$\begin{aligned} M^\perp &= \{\phi \in B(H)_* \mid \phi|_M = 0\}, \\ M^{\perp\perp} &= \{x \in B(H) \mid \phi(x) = 0, \text{ for all } \phi \in M^\perp\}. \end{aligned}$$

We first prove that $M = M^{\perp\perp}$. It is trivial that $M \subset M^{\perp\perp}$. Suppose $x \notin M$. Because M is σ-weakly closed, there exist $\phi \in B(H)_*$ and $\varepsilon > 0$ such that if $|\phi(x) - \phi(y)| < \varepsilon$, then $y \notin M$. This shows that if $y \in M$, then $|\phi(x) - \phi(y)| \geq \varepsilon$. Because M is a linear space, we get $\phi(y) = 0$ for all $y \in M$ and $|\phi(x)| \geq \varepsilon$. This means that we have $\phi \in M^\perp$ and $x \notin M^{\perp\perp}$. Thus we have proved $M = M^{\perp\perp}$.

We next prove that $B(H)_*/M^\perp$ is isomorphic to M_*. For $\phi \in B(H)_*$, we have a natural restriction $\phi|_M \in M_*$, and this gives a well-defined map from $B(H)_*/M^\perp$ to M_*. This is surjective by Theorem 5.9. We next show that this restriction preserves the norm.

It is trivial that $\|\phi|_M\| \leq \|\phi\|$. Conversely, suppose $\phi \in B(H)_*$ has $\|\phi\| = 1$ in the quotient Banach space $B(H)_*/M^\perp$. This means that the distance between ϕ and $M^\perp$ is 1 in $B(H)_*$. The Hahn–Banach theorem (see (Pedersen 1989)) gives an $x \in (B(H)_*)^* = B(H)$ such that $\|x\| = 1$, $\phi(x) = 1$ and $x|_{M^\perp} = 0$. This shows $x \in M^{\perp\perp} = M$ and $\|\phi|_M\| \geq 1$. Thus the restriction map gives an isometric isomorphism and we have $(M_*)^* = (B(H)_*/M^\perp)^* = M^{\perp\perp} = M$.

Finally, if we have $x_i \to x$ σ-weakly in M, we get $\phi(x_i) \to \phi(x)$ for all $\phi \in B(H)_*$, and then we have $\phi(x_i) \to \phi(x)$ for all $\phi \in M_* = B(H)_*/M^\perp$. This means that $x_i \to x$ in the weak* topology. We have converse implications for all these, so we get the desired conclusion. □

For a von Neumann algebra $L^\infty(X,\mu)$, we can show that $L^\infty(X,\mu)_*$ is given by $L^1(X,\mu)$. (Recall that $L^1(X,\mu)^* = L^\infty(X,\mu)$. It can be shown that M_* is the only Banach space with $(M_*)^* \cong M$ for a von Neumann algebra M.) (This is essentially because a state ϕ on M is normal if and only if for any increasing net x_i of self-adjoint elements of M, bounded above we have $\phi(\vee x_i) = \vee\phi(x_i)$ – a notion which only depends on the order structure.) Note that $L^\infty(X,\mu)^*$ is much bigger than $L^1(X,\mu)$ (unless $L^\infty(X,\mu)$ is finite dimensional).

The above theorem shows that the σ-weak topology on a von Neumann algebra is quite natural although it may not look so at first glance. This theorem also gives the following corollary.

Corollary 5.12 *Let M be a von Neumann algebra. Then the unit ball of M is σ-weakly compact.*

Proof This follows from Theorem 5.11 and the Banach–Alaoglu theorem which states that the unit ball in any dual Banach space is compact in the weak* topology. (See Theorem 1.6.5 of (Kadison and Ringrose 1983), for example, for the Banach–Alaoglu theorem.) □

We also need another characterization of normal linear functionals on a von Neumann algebra.

Theorem 5.13 *Let M be a von Neumann algebra. For $\phi \in M^*$, the following conditions are equivalent:*

1. *The functional ϕ is normal.*
2. *The functional ϕ is weakly continuous on the unit ball, with respect to the norm, of M.*
3. *The functional ϕ is strongly continuous on the unit ball, with respect to the norm, of M.*

Proof We obtain the conclusion from Proposition 5.2, Theorem 5.9, the Hahn–Banach theorem, and the Krein–Smulian theorem, which states that a convex set in a dual space of a Banach space is weak* closed if and only if its intersection with every bounded ball is weak* closed. (See page 55 of (Pedersen 1979), for example, for the Krein–Smulian theorem.) □

The following is the famous density theorem of Kaplansky.

Theorem 5.14 *Let A be a $*$-subalgebra of $B(H)$, and x an element in the weak closure of A. Then there exists a net x_i in A such that $\|x_i\| \le \|x\|$ and $x_i \to x$ σ-strongly.*

This is one of the fundamental theorems in the theory of operator algebras, but we need this theorem in this book only in an indirect way, so we only sketch an outline of a proof as follows. (See Theorem 5.3.5 of (Kadison and Ringrose 1983) or Theorem II.4.8 of (Takesaki 1979), for example, for a complete proof.)

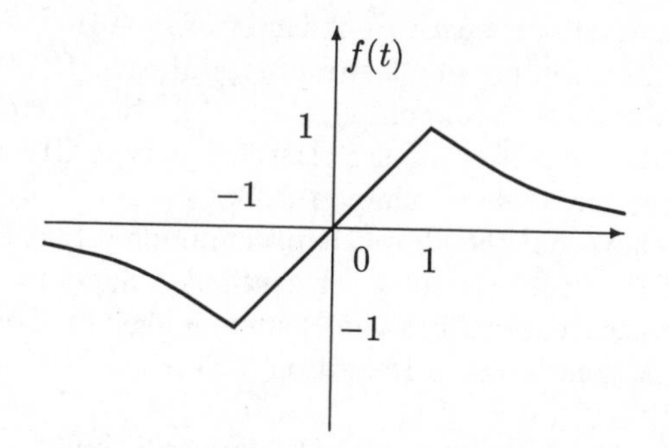

FIG. 5.1. The function f for the Kaplansky density theorem

First we reduce the general case to the case $x = x^*$. We may assume $\|x\| = 1$. Choose a net (x_i) converging to x weakly. We may assume that each x_i is self-adjoint. Then applying a function f as in Fig. 5.1, we can prove $f(x_i) \to f(x) = x$ σ-weakly, and this is enough for our later purpose.

Finally we remark that the second dual space A^{**} of any C^*-algebra A can naturally be regarded as a von Neumann algebra, (e.g. $K(H)^{**} = B(H)$, for the compact operator $K(H)$). We omit its proof here. See section III.2 of (Takesaki 1979).

5.3 II$_1$ factors

We next study a subclass of von Neumann algebras called II$_1$ factors. This is the class of von Neumann algebras we will work on in most of the remainder of the book. First we give a definition.

Definition 5.15 *An infinite dimensional von Neumann algebra M is called a* II$_1$ factor *if it has a unique σ-weakly continuous linear functional* tr *on M satisfying the following properties:*

1. $\mathrm{tr}(1) = 1$.
2. $\mathrm{tr}(x^*x) \geq 0$ *for all* $x \in M$.
3. *If* $\mathrm{tr}(x^*x) = 0$, *then we have* $x = 0$.
4. $\mathrm{tr}(xy) = \mathrm{tr}(yx)$ *for* $x, y \in M$.

Such a functional tr *is called a* trace *on M.*

Actually, this is not a standard definition in the theory of operator algebras. The standard definition for a von Neumann algebra M being a *factor* is that the centre $M \cap M'$ is $\mathbb{C}$. (This is also equivalent to the condition that any σ-weakly continuous representation is faithful. It is also equivalent to simplicity in the sense that the algebra M does not have a non-trivial σ-weakly closed two-sided ideal.) The prefix 'II$_1$-' is used because it is one of the types classified by Murray–von Neumann. They had types I$_n$ ($n = 1, 2, 3, \ldots, \infty$), II$_1$, II$_\infty$, and III. (A simple description of this classification is as follows. A factor of type I$_n$ is a matrix algebra $M_n(\mathbb{C})$. A factor of type I$_\infty$ is the algebra of all bounded linear operators $B(H)$ on a Hilbert space H. A factor of type II_1 is as above. A factor of type II_∞ is a tensor product of a factor of type II_1 and $B(H)$ – see Section 5.4 for a tensor product of von Neumann algebras. All other factors are called type III.) The type III case has been further classified as type III_λ with $0 \le \lambda \le 1$ by (Connes 1973). (See the notes to Chapter 3.)

We can easily show that the above Definition implies that the centre $M \cap M'$ is $\mathbb{C}$ as follows. If the centre is not $\mathbb{C}$, it contains a non-trivial projection p by Theorem 5.8, as its centre itself is a von Neumann algebra. Then we get another tr' satisfying all the conditions in Definition 5.15 by

$$\mathrm{tr}'(x) = \mathrm{tr}(xp + 2x(1-p))/\mathrm{tr}(2-p).$$

The name 'trace' comes from the fact that it behaves like a trace in ordinary linear algebra. We have already used the symbol Tr for traces on $B(H)$, but the lower case here means that we have $\mathrm{tr}(1) = 1$. (Note that we get $\mathrm{Tr}(1) = +\infty$ unless the Hilbert space is finite dimensional.) Strictly speaking, the trace in the above definition is called a finite faithful normalized trace, but we will just call it a trace in this book, as is often done in the literature on II_1 factors. Some of the conditions in Definition 5.15 are actually redundant, but we do not care about that here.

First, we note that the algebra $B(H)$ for an infinite dimensional Hilbert space H is not a II_1 factor. Suppose that $B(H)$ had a trace in the above sense. Let $H = H_1 \oplus H_2$ be a decomposition of H into a direct sum of two subspaces H_1, H_2 which are isomorphic to H. Let u, v be isometries from H onto H_1, H_2, respectively. Then we have

$$1 = \mathrm{tr}(1) = \mathrm{tr}(uu^* + vv^*) = \mathrm{tr}(u^*u + v^*v) = \mathrm{tr}(1) + \mathrm{tr}(1) = 2,$$

which is a contradiction. However, $B(H)$ does have exotic (unbounded) traces other than the normal one Tr – the Dixmier trace as in (Connes 1994).

The following is a fundamental example of a II_1 factor.

Example 5.16 We have an increasing sequence

$$M_2(\mathbb{C}) \subset M_2(\mathbb{C}) \otimes M_2(\mathbb{C}) \subset M_2(\mathbb{C}) \otimes M_2(\mathbb{C}) \otimes M_2(\mathbb{C}) \subset \cdots$$

given by embeddings $x \mapsto x \otimes 1$. Let A be the union of this increasing sequence. On the nth tensor power of $M_2(\mathbb{C})$, we define $\mathrm{tr}(x) = \mathrm{Tr}(x)/2^n$, where Tr is the ordinary trace in linear algebra. Then it is easy to see that this tr is compatible with the above embeddings and thus gives a well-defined functional on A. We use the GNS-representation with respect to tr as in Theorem 2.8. (Here A is not a C^*-algebra, because it is not norm closed, but this does not matter for the GNS-construction.)

That is, we put an inner product on A by $\langle x, y\rangle = \mathrm{tr}(y^*x)$ and get a Hilbert space H as the completion of A with respect to this inner product. For $a \in A$, we have $\pi(a) \in B(H)$ as the extension of the left multiplication by a on A.

Dropping π for simplicity, we regard A as a $*$-subalgebra of $B(H)$. Let M be the weak closure of A. Writing $\hat{1}$ for the image of $1 \in A$ in H, we can extend tr to M by $\mathrm{tr}(x) = \langle x\hat{1}, \hat{1}\rangle$. It is easy to see that this tr satisfies all the conditions in Definition 5.15 and thus M is a II_1 factor. (Uniqueness of the trace on M follows from uniqueness on A, which then follows from uniqueness on $M_{2^n}(\mathbb{C})$.)

The above example is given by a weak closure of a union of an increasing sequence of finite dimensional operator algebras. We have the following definition for such von Neumann algebras.

Definition 5.17 *A von Neumann algebra M is said to be* hyperfinite *if M is a weak closure of a union of an increasing sequence of finite dimensional von Neumann algebras.*

Note that finite dimensional von Neumann algebras and finite dimensional C^*-algebras give the same class – they are direct sums of matrix algebras. The name AFD, an abbreviation of 'approximately finite dimensional', is also often used instead of 'hyperfinite'. These two names mean the exactly same condition in Definition 5.17.

The II_1 factor in Example 5.16 is thus called a *hyperfinite II_1 factor*. Actually, we have the following uniqueness Theorem due to Murray–von Neumann, so the algebra in Example 5.16 is often called *the* hyperfinite II_1 factor.

Theorem 5.18 *All hyperfinite II_1 factors are isomorphic.*

This is one of the most fundamental theorem in von Neumann algebra theory, and plays an essential role in the second half of this book, but a proof of this theorem is rather technical, so we just sketch an outline of a proof here. (See Theorem 12.2.1 of (Kadison and Ringrose 1986) for a complete proof, for example.)

We may and do choose a weakly dense sequence $\{x_n\}$ in M. (The hyperfiniteness assumption shows existence of such a sequence.) We inductively choose an increasing sequence of finite dimensional algebras $\{A_n\}$ in M so that each A_n is isomorphic to $M_{2^k}(\mathbb{C})$ for some k and the union $\bigcup_{n=1}^{\infty} A_n$ is weakly dense in M as follows.

Suppose we have chosen the sequence up to A_n. By the hyperfiniteness, we can find a finite dimensional B which 'almost' contains A and $x_1, x_2, \dots, x_{n+1}$. (cf. Exercise 3.12 for AF C^*-algebras). Because M is a II_1 factor, it is possible to find a sufficiently large k and a subalgebra C of M such that C is isomorphic to $M_{2^k}(\mathbb{C})$ and C 'almost' contains B. Then we can 'rotate' C by 'a small amount' to get A_{n+1} which actually contains A_n. The sequence $\{A_n\}$ obtained in this way gives the desired uniqueness.

In particular, if we take a weak closure of any UHF algebra in its GNS representation with respect to trace, we obtain the same algebra. Taking weak closures means much information is lost – but the main point of subfactor theory is that the position of one hyperfinite II_1 factor embedded as a subalgebra of another has a rich combinatorial structure.

Some II_1 factors turn out to be hyperfinite though they do not seem to be so at first glance. For example, take the irrational rotation C^*-algebra A_θ. This has a unique trace, and the weak closure in the GNS-representation with respect to the trace gives a II_1 factor. This is actually hyperfinite and follows from the realization of A_θ as an inductive limit of direct sums of matrix algebras over continuous functions on the circle – see (Elliott and Evans 1993) and notes to Chapter 3.

Hyperfinite von Neumann algebras are of fundamental importance, and have been studied extensively. There are continuously many isomorphism classes of non-hyperfinite II_1 factors, but our readers do not have to know about them to read the remainder of the book.

Hyperfiniteness is closely related to amenability of groups. Let M be a von Neumann algebra generated by the left regular representation of a discrete group G on $\ell^2(G)$. If G is 'sufficiently non-commutative', we can show that M is a II_1 factor. (See Exercise 5.1.) This M is a hyperfinite II_1 factor if and only if G is amenable.

By our Definition 5.5, a von Neumann algebra M has a Hilbert space on which it acts, from the beginning. However, we often want to change the Hilbert space of a given von Neumann algebra. The most important case of such changes is given by the GNS-representation with respect to the trace (Theorem 2.8) as follows.

Let M be a II_1 factor with trace tr. Then M acts on M itself by left multiplication. We put an inner product on M by $\langle x, y\rangle = \mathrm{tr}(y^*x)$, which is positive definite by conditions 2, 3 in Definition 5.15. We denote the completion of M with respect to this inner product by $L^2(M)$. We denote the image of $x \in M$ in $L^2(M)$ by $\hat{x}$.

The notation $L^2(M)$ comes from the following analogy. The algebra $L^\infty[0,1]$ is a von Neumann algebra on the Hilbert space $L^2[0,1]$. The map $f \in L^\infty[0,1] \mapsto \int_0^1 f(x)\, dx$ satisfies all the four conditions in Definition 5.15. If we apply the above procedure of the GNS-construction to this map, which is a state on $L^\infty[0,1]$, we get $L^2[0,1]$ from $L^\infty[0,1]$. We regard the above procedure for II_1 factors as a 'non-commutative analogue' of this procedure for $L^\infty[0,1]$.

We have a natural representation π of M on $L^2(M)$ defined by $\pi(x)\hat{y} = \widehat{xy}$. (Strictly speaking, this defines $\pi(x)$ on a dense subspace, and it is easy to see that we can extend $\pi(x)$ to a bounded linear operator on the entire $L^2(M)$.)

Theorem 5.19 *The above GNS representation π of M in the trace has the following properties:*

1. *The map π is injective.*
2. *The image $\pi(M)$ is a von Neumann algebra.*
3. *The maps π and π^{-1} are both σ-weakly continuous.*

Proof 1. It is easy to see that if $\pi(x) = 0$, then $\mathrm{tr}(x^*x) = 0$, and hence $x = 0$ by Condition 3 in Definition 5.15.

2. It is easy to see that $\pi(M)$ is a $*$-subalgebra of $B(L^2(M))$. Let a be an element of the σ-weak closure of $\pi(M)$ with $\|a\| = 1$. There exists a net $(\pi(x_i))$ σ-weakly converging to a such that $\|\pi(x_i)\| \le 1$ and $x_i \in M$ by the Kaplansky density theorem (Theorem 5.14). It is easy to see $\|x_i\| = \|\pi(x_i)\|$, thus we may assume that (x_i) converges σ-weakly to some $x \in M$ by Corollary 5.12. Then it is easy to see that $(\pi(x_i))$ converges to $\pi(x)$ σ-weakly, and this implies that $a = \pi(x) \in \pi(M)$. This shows that $\pi(M)$ is closed in the σ-weak topology, hence it is a von Neumann algebra.

3. The restriction of π to the unit ball of M is a σ-weakly continuous injective map onto the unit ball of $\pi(M)$. By Corollary 5.12, the converse of this map is also σ-weakly continuous. Then for any $\phi \in M_*$, the map $\phi\cdot\pi^{-1}$ is σ-weakly continuous on the unit ball of $\pi(M)$. By Theorem 5.13, this gives σ-weak continuity of π^{-1}. □

The above theorem essentially says that we can 'identify' M acting on the original Hilbert space with $\pi(M)$ acting on $L^2(M)$. This representation π is called the *standard representation.*

Next we have the following Radon–Nikodym type theorem.

Theorem 5.20 *Let M be a II_1 factor with trace* tr. *Suppose $\phi \in M_*$ and $0 \le \phi(x^*x) \le \mathrm{tr}(x^*x)$ for all $x \in M$. Then there exists $a \in M$ with $0 \le a \le 1$ and $\phi(x) = \mathrm{tr}(ax)$.*

Proof Let $I = \{a \in M \mid 0 \le a \le 1\}$ and define a map Φ sending $a \in I$ to $\phi_a \in M_*$ defined by $\phi_a(x) = \mathrm{tr}(ax)$. We claim that this is a continuous map from I with σ-weak topology to M_* with weak topology as a Banach space.

If we have $a_i \to a$ in I in the σ-weak topology, we get $a_i x \to ax$ in the σ-weak topology for all $x \in M$. This shows that we have $\mathrm{tr}(a_i x) \to \mathrm{tr}(ax)$ for all $x \in$M, and this implies $\phi_{a_i}(x) \to \phi_a(x)$ for all $x \in M = (M_*)^*$. Consequently, $\phi_{a_i} \to \phi_a$ in the weak topology of M_*. The set I is σ-weakly compact by Corollary 5.12, thus the image $\Phi(I)$ is weakly closed.

Suppose ϕ satisfying the condition of the theorem does not belong to $\Phi(I)$. Because $\Phi(I)$ is convex, the (real) Hahn–Banach theorem (see (Pedersen 1989)) gives an $x \in (M_*)^* = M$ with $x = x^*$ satisfying $\phi_a(x) < \phi(x)$ for all $a \in I$. Let g be the characteristic function of the half interval $[0, \infty)$. Set $p = g(x)$. We have $x = px - (1-p)(-x)$ and thus we get

$$\mathrm{tr}(px) = \phi_p(x) < \phi(x) \le \phi(px) = \phi(pxp) \le \mathrm{tr}(pxp) = \mathrm{tr}(px),$$

which is a contradiction. □

The reason that the above theorem is said to be of Radon–Nikodym type is as follows. As we have seen earlier, a II_1 factor is regarded as a non-commutative analogue of a probability measure space. Then the above theorem can be regarded as an analogue of the classical Radon–Nikodym theorem. (A normal state plays the role of a measure.) Our assumption $0 \le \phi(x^*x) \le \mathrm{tr}(x^*x)$ is stronger than just absolute continuity, so we get a bounded a in the conclusion, which is also stronger than in the classical Radon–Nikodym theorem.

Next we study comparison of projections in II_1 factors. First we make the following definition.

Definition 5.21 *Let M be a von Neumann algebra. For projections $e, f \in M$, we say that e and f are equivalent, written as $e \sim f$, if there exists a partial isometry $u \in M$ with $u^*u = e$, $uu^* = f$.*

For projections $e, f \in M$, we write $e \prec f$, if there exists a projection $e_0 \in M$ with $e_0 \sim e$ and $e_0 \leq f$.

Note that this definition of equivalence of projections is the same as that which is used in K-theory in Chapter 3. Historically, the above definition is due to Murray–von Neumann, and is much older than K-theory.

We first have the following lemma.

Lemma 5.22 *Let e, f be two non-zero projections in a II_1 factor M on a Hilbert space H. Then there exist two non-zero projections $e_0, f_0 \in M$ such that $e_0 \leq e$, $f_0 \leq f$, and $e_0 \sim f_0$.*

Proof Let p be the projection onto the closure of MfH. For any $x \in M$ and $x' \in M'$, we have $pxp = xp$ and $px'p = x'p$. These imply that $p \in M \cap M' = \mathbb{C}$ and hence that $p = 0, 1$. Because $f \neq 0$, we have $p = 1$. Because $e \neq 0$, we get $eMf \neq 0$. (If $eMf = 0$, then we would have $eMfH = 0$, while $e \neq 0$ and $p = 1$, which is a contradiction.) Thus there exists an $x \in M$ with $exf \neq 0$. Let $uh = exf$ be its polar decomposition. Setting $u^*u = f_0$ and $uu^* = e_0$, we get the conclusion. □

Then we can prove the following fundamental theorem for II_1 factors.

Theorem 5.23 *Let M be a II_1 factor. We have the following.*

1. *If $e \prec f$ and $f \prec e$, then $e \sim f$.*
2. *For two projections $e, f \in M$, one of the following holds.*
   - $e \prec f$, $e \not\sim f$.
   - $e \sim f$.
   - $f \prec e$, $e \not\sim f$.
3. *For two projections $e, f \in M$, the three cases in 2 are equivalent to the following three cases, respectively.*
   - $\mathrm{tr}(e) < \mathrm{tr}(f)$.
   - $\mathrm{tr}(e) = \mathrm{tr}(f)$.
   - $\mathrm{tr}(e) > \mathrm{tr}(f)$.

Proof 1. This is proved in the exactly same way as in the standard proof of the Cantor–Bernstein theorem for comparison of cardinality in set theory.

2. Let $\{e_i\}$ and $\{f_i\}$ be families of mutually orthogonal projections with $e_i \leq e$, $f_i \leq f$, $e_i \sim f_i$. Zorn's lemma implies that we have a maximal pair $\{e_i\}$ $\{f_i\}$ of such families. Lemma 5.22 and maximality imply that at least one of $e - \sum_i e_i$ and $f - \sum_i f_i$ is zero. (Note that $\sum_i e_i$, $\sum_i f_i$ are in M because M is strongly closed.) Thus we have one of the listed conditions.

3. The conditions 2, 3, 4 of the trace in Definition 5.15 show that each of the three conditions in 2 implies the corresponding condition in 3. This shows the desired equivalence. □

The next theorem shows the continuous nature of II$_1$ factors.

Theorem 5.24 *Let M be a II$_1$ factor with* tr. *Then the set*

$$T = \{\mathrm{tr}(p) \mid p \text{ is a projection in } M\}$$

is equal to $[0, 1]$.

Proof If the set T is finite, it is easy to see that it must be of the form $T = \{0, 1/n, 2/n, \ldots, 1\}$ by Theorem 5.23 and then M contains a subalgebra A isomorphic to $M_n(\mathbb{C})$. Since M is isomorphic to $M_n(\mathbb{C}) \otimes (M_n(\mathbb{C})' \cap M)$, we know that $(M_n(\mathbb{C})' \cap M)$ is isomorphic to $\mathbb{C}$, which contradicts infinite dimensionality of M.

This implies that T is dense in $[0, 1]$, since if r is in T, then positive multiples of r less than 1 are also in T. For $r \in [0, 1]$, choose a decreasing sequence $\{r_n\}_n \subset T$ converging to r. Theorem 5.23 implies that we have a decreasing sequence $\{p_n\}_n \subset M$ with $\mathrm{tr}(p_n) = r_n$. We have $p = \lim_n p_n \in M$ and σ-weak continuity of the trace implies that $r \in T$. □

At the end of this section, we study conditional expectations on II$_1$ factors.

Theorem 5.25 *Let M be a II$_1$ factor with* tr, *and N a von Neumann subalgebra. There exists a unique map E_N from M onto N with the following properties:*

1. $E_N(x^*x) \geq 0$ *for all* $x \in M$.
2. $E_N(x) = x$ *for all* $x \in N$.
3. $\mathrm{tr}(xy) = \mathrm{tr}(E_N(x)y)$ *for all* $x \in M$, $y \in N$.
4. $E_N(axb) = aE_N(x)b$ *for all* $x \in M$, $a, b \in N$.
5. $E_N(x^*) = E_N(x)^*$ *for all* $x \in M$.
6. $\|E_N(x)\| \leq \|x\|$ *for all* $x \in M$.
7. *The map E_N is σ-weakly continuous.*

Proof Choose $x \in M$ with $0 \leq x \leq 1$. We define $\phi_x \in N_*$ by $\phi_x(y) = \mathrm{tr}(xy)$. Then we have

$$\phi_x(y^*y) = \mathrm{tr}(xy^*y) = \mathrm{tr}(yxy^*) \leq \mathrm{tr}(y^*y).$$

By Theorem 5.20, there exists $x_0 \in N$ such that $\mathrm{tr}(xy) = \phi_x(y) = \mathrm{tr}(x_0 y)$ and $0 \leq x_0 \leq 1$. We set $E(x) = x_0$. By extending this definition linearly, we get a well-defined map E. We now have $\mathrm{tr}(xy) = \mathrm{tr}(E(x)y)$ for all $x \in M$ and $y \in N$. So Properties 1, 2, 3 are valid, and it is clear that Property 3 gives uniqueness of E.

4. If $x \in M$ and $a, b, y \in N$, we get

$$\mathrm{tr}(E(axb)y) = \mathrm{tr}(axby) = \mathrm{tr}(xbya) = \mathrm{tr}(E(x)bya) = \mathrm{tr}(aE(x)by),$$

which gives the desired property.

Table 5.1 *Analogy between probability spaces and II_1 factors*

probability space	II_1 factor
$L^\infty(X,\mu)$	M
$L^2(X,\mu)$	$L^2(M)$
$L^1(X,\mu)$	M_*
measurable subset	projection
measure	tr, normal state
σ-subfields	von Neumann subalgebra
conditional expectation	conditional expectation

5. Positivity of the trace gives $\overline{\mathrm{tr}(x^*)} = \mathrm{tr}(x)$ for all $x \in M$. Then we have

$$\begin{aligned}\mathrm{tr}(E(x^*)y) &= \mathrm{tr}(x^*y) = \overline{\mathrm{tr}(y^*x)} = \overline{\mathrm{tr}(xy^*)} \\ &= \overline{\mathrm{tr}(E(x)y^*)} = \mathrm{tr}(yE(x)^*) = \mathrm{tr}(E(x)^*y),\end{aligned}$$

for all $x \in M$ and $y \in N$, which completes a proof.

6. By 1, we have $E((E(x)-x)^*(E(x)-x)) \geq 0$. Expanding this and using 3, 4, we get $E(x^*x) \geq E(x)^*E(x)$.

7. Suppose that $x_i \to 0$ σ-weakly in M. We will show that $E(x_i) \to 0$ σ-weakly in N. It is enough to work on $L^2(N)$. That is, we have to prove $\sum_{j=1}^\infty \langle E(x_i)\xi_j, \eta_j\rangle \to 0$ for $\xi_j, \eta_j \in L^2(N)$ as in Definition 5.1. This is same as asking $\sum_{j=1}^\infty \langle x_i\xi_j, \eta_j\rangle \to 0$, and ξ_j, η_j are in $L^2(M)$. So we are done. □

Definition 5.26 *The map E_N in Theorem 5.25 is called the* conditional expectation from M onto N.

The term 'conditional expectation' again comes from an analogy with measure theory (or probability theory). A von Neumann subalgebra is a non-commutative analogue of a σ-subfield in measure theory. A measurable function on a σ-field is not measurable with respect to a σ-subfield in general, so we have to modify the function to get a measurable function with respect to a σ-subfield. This process is the classical conditional expectation.

We list in Table 5.1 the analogy between probability spaces and II_1 factors (or more generally, von Neumann algebras).

Exercise 5.1 Let G be a countable group and π its left regular representation on $\ell^2(G)$. That is, we have

$$(\pi_g\xi)(h) = \xi(g^{-1}h), \quad \xi \in \ell^2(G), g, h \in G.$$

Let $L(G) = \pi(G)''$. (This is called the *group von Neumann algebra.*)

We further assume that any conjugacy class in G contains infinite elements unless it has only the identity of G. (Such a G is called an *ICC group*, where 'ICC' stands for infinite conjugacy class.) Then prove that $L(G)$ is a II_1 factor. Also find discrete groups satisfying this property.

5.4 Crossed products and tensor products

We study basic facts on crossed products by actions of finite groups on von Neumann algebras. From an algebraic viewpoint, this is an analogue of a semi-direct product of groups. In this section, we deal with only finite groups for simplicity. (Also note that there are brief discussions on C^*-algebraic crossed products in the notes to Chapters 2 and 3.)

Let M be a von Neumann algebra. We denote its $*$-automorphism group by $\mathrm{Aut}(M)$. Let G be a finite group and α be a group homomorphism from G into $\mathrm{Aut}(M)$. Such an α is called an *action* of G on M. We write α_g for the image of $g \in G$. An automorphism α of M is called *inner* if it is given by $\alpha(x) = uxu^*$ for some unitary $u \in M$. An automorphism which is not inner is called *outer*. An action α of G on M is called *outer* if α_g is outer for all $g \in G \setminus \{1\}$.

Let M be a von Neumann algebra on a Hilbert space H and α an action of a finite group G. Let $\{H_g\}$ be copies of H labelled by G, and set $\tilde{H} = \bigoplus_{g\in G} H_g$. We define operators $\pi_\alpha(x)$ and u_g on $\tilde{H}$ for $x \in M$ and $g \in G$ as follows, where $(\xi(h))_h \in \tilde{H}$:

$$(\pi_\alpha(x)\xi)(h) = \alpha_h^{-1}(x)\xi(h),$$
$$(u_g\xi)(h) = \xi(g^{-1}h).$$

Then it is easy to see that $u_g\pi_\alpha(x)u_g^* = \pi_\alpha(\alpha_g(x))$. We denote the von Neumann algebra generated by $\pi_\alpha(M)$ and $\{u_g\}_{g\in G}$ by $M \rtimes_\alpha G$ and call it a *crossed product of M by an action α*. It can be shown that the isomorphism class of the crossed product does not depend on which Hilbert space M is represented on. When no confusion arises, we often just write $M \rtimes G$ without specifying α. Using π_a, we often identify M and its image. So we usually regard M as a subalgebra of $M \rtimes_\alpha G$. Then it is easy to see that an element in $M \rtimes_\alpha G$ can be uniquely written as $\sum_{g\in G} x_g u_g$ with $x_g \in M$. It is possible to define crossed products for actions of locally compact groups which are not finite, but then we will not have such a clear description of elements in the crossed product, which causes considerable technical difficulty. See Chapter 13 of (Kadison and Ringrose 1986) for more on crossed products.

The basic idea in making the crossed product is adding universal unitaries u_g to the von Neumann algebra M so that the action α_g is given by $\mathrm{Ad}(u_g)$ on M. The unitaries u_g are called the *implementing unitaries*.

We first present a lemma on outerness of an automorphism.

Lemma 5.27 *Let α be an outer automorphism of a II_1 factor M. If we have $a \in M$ such that $a\alpha(x) = xa$ for all $x \in M$, then $a = 0$.*

Proof By taking adjoints, we also get $\alpha(x)a^* = a^*x$ for all $x \in M$. Thus we obtain $aa^* \in M \cap M' = \mathbb{C}$. If $a \neq 0$, we may assume that $aa^* = 1$ by multiplying a by a scalar, if necessary. Then a is a partial isometry and $\mathrm{tr}(a^*a) = \mathrm{tr}(aa^*) = 1$, so a is a unitary. We thus have $\alpha(x) = a^*xa$ for a unitary $a \in M$, which is a contradiction. □

Theorem 5.28 *Let α be an action of a finite group G on a II_1 factor M. Then the action α is outer if and only if $M' \cap (M \rtimes_\alpha G) = \mathbb{C}$.*

Proof Suppose $\alpha_g = \mathrm{Ad}(u)$ for some $g \in G$, $g \neq 1$, and a unitary $u \in M$. Then it is easy to see that $u^* u_g \in M' \cap (M \rtimes_\alpha G)$.

Conversely, suppose that α is outer. If $\sum_{g\in G} x_g u_g \in M' \cap (M \rtimes_\alpha G)$, then we have $x_g \alpha_g(x) = x x_g$ for all $g \in G$. For $g = 1$, this implies $x_1 \in \mathbb{C}$. For $g \neq 1$, this implies $x_g = 0$ by Lemma 5.27. □

Any finite group G has an outer action on the hyperfinite II_1 factor as we will see in Section 6.7.

We next define a tensor product of von Neumann algebras. Let M, N be von Neumann algebras on Hilbert spaces H, K respectively. For $x \in M$ and $y \in N$, the operator $x \otimes y$ makes a sense naturally as a bounded linear operator on $H \otimes K$. We denote by $M \otimes N$ the von Neumann algebra generated by these operators. It is called the *tensor product* of M and N. It is known that the isomorphism class of $M \otimes N$ depends only on isomorphism classes of M and N, i.e. it does not depend on the choice of representations of M and N on Hilbert spaces.

We next give the celebrated theorem of Tomita on commutants of tensor products. We omit its proof here, because we will not use this theorem in this book. A proof can be found in Theorem 11.2.16 in (Kadison and Ringrose 1986) or Theorem 5.9 in (Takesaki 1979).

Theorem 5.29 *Let M, N be von Neumann algebras, then*

$$(M \otimes N)' = M' \otimes N'.$$

Exercise 5.2 Let M be a II_1 factor and α be an outer action of a finite group G on M. Prove that $M \rtimes_\alpha G$ is also a II_1 factor.

Exercise 5.3 Let M be a II_1 factor and α be an outer action of a finite group G on M. Compute the conditional expectation E from $M \rtimes_\alpha G$ onto M.

Exercise 5.4 Let M, N be II_1 factors. Show that $M \otimes N$ is also a II_1 factor and that the conditional expectation $E : M \otimes N \to M \otimes \mathbb{C}$ is given by $E(x \otimes y) = \mathrm{tr}(y)x$, for $x \in M$, $y \in N$.

5.5 Coupling constants

We study the coupling constant theory of Murray–von Neumann from today's viewpoint. This is an essential tool for the study of subfactors in the second half of this book. Let M be a II_1 factor on a Hilbert space H. Our aim is to define the 'dimension' (or 'rank') $\dim_M H$ of H over M normalized so that we have $\dim_M L^2(M) = 1$.

First we study the standard representation.

Proposition 5.30 *Let M be a II_1 factor with the standard representation on $L^2(M)$. Define an operator J on $L^2(M)$ by $J(\hat{x}) = \widehat{x^*}$ for $x \in M$. (It is clear that this defines an anti-linear bounded operator on $L^2(M)$.) Then $M' = JMJ$*

Proof First note that

$$Jx J\hat{y} = (Jx)\widehat{y^*} = J\widehat{xy^*} = \widehat{yx^*}$$

for $x, y \in M$. Thus JxJ is identified with the right multiplication by x^*. (We also write $\hat{y}x^*$ for $\widehat{yx^*}$.)

It is trivial that we have $JMJ \subset M'$, so take $a \in M'$. Without loss of generality, we may assume $0 \leq a \leq 1$. Set $\phi(x) = \langle xa\hat{1}, \hat{1}\rangle$ for $x \in M$. This gives an element in M_*. We have

$$\phi(x^*x) = \langle x^*xa\hat{1}, \hat{1}\rangle = \langle ax\hat{1}, x\hat{1}\rangle = \langle a\hat{x}, \hat{x}\rangle \geq 0,$$

and

$$0 \leq \phi(x^*x) \leq \|a\|\|\hat{x}\|^2 \leq \mathrm{tr}(x^*x).$$

By Theorem 5.20, we get $y \in M$ such that $\phi(x) = \mathrm{tr}(xy)$. This implies that

$$\langle a\hat{1}, \widehat{x^*}\rangle = \langle xa\hat{1}, \hat{1}\rangle = \phi(x) = \mathrm{tr}(xy) = \langle \hat{y}, \widehat{x^*}\rangle,$$

for all $x \in M$, and thus we have $a\hat{1} = \hat{y}$. Then for all $x \in M$, we get

$$a\hat{x} = ax\hat{1} = xa\hat{1} = x\hat{y} = \hat{x}y,$$

which means that a is given by right multiplication by y. Thus we are done. □

Let M be a II_1 factor on a Hilbert space H. We make the following definition.

Definition 5.31 *Suppose the action of M on H is isomorphic to the action of M on $(\bigoplus_{j=1}^n L^2(M))p$ by left multiplication, where $p = (p_{jk})$ is some projection in $M_n(\mathbb{C}) \otimes M$ with $p_{jk} \in M$. We then define $\dim_M H = \sum_{j=1}^n \mathrm{tr}(p_{jj})$ and call it the* coupling constant *of M in H.*

If the action of M on H is not of this form, we set the coupling constant to be ∞.

It is easy to see that this definition is well-defined and $\dim_M$ is additive. Our definition here is not standard at all. For standard definitions and more detailed study on the coupling constant, see Chapter V of (Takesaki 1979) or section 3.2 of (Goodman *et al.* 1989), where the condition in the above definition is proved from a weaker definition.

Our next aim is to prove the fundamental theorem here, Theorem 5.33. Before that, we need a lemma.

Lemma 5.32 *Let M be a II_1 factor on a Hilbert space H.*

1. *For a positive integer n, the algebra $M_n(\mathbb{C}) \otimes M$ on $\oplus_{j=1}^n H$ is also a II_1 factor.*
2. *If p is a non-zero projection in M, then pMp is also a II_1 factor on pH.*

3. *If p is a non-zero projection in M' and $\dim_M H < \infty$, then Mp is also a II_1 factor on pH.*

Proof 1. First note that the algebra $M_n(\mathbb{C}) \otimes M$ naturally acts on $\oplus_{j=1}^n H$. For $(x_{ij})_{i,j=1,\dots,n} \in M_n(\mathbb{C}) \otimes M$, the trace $\mathrm{tr}(x_{ij}) = \sum_{j=1}^n \mathrm{tr}(x_{jj})/n$ naturally gives the unique trace on the algebra.

2. One can show that pMp is a von Neumann algebra on pH. It is trivial that $\mathrm{tr}_p(pxp) = \mathrm{tr}(pxp)/\mathrm{tr}(p)$ gives a trace on pMp, so we have to prove that this is the unique trace on pMp.

Set

$$T = \{\mathrm{tr}(p) \mid pMp \text{ is a II}_1 \text{ factor}\}.$$

This is well defined by Theorem 5.23. By a similar consideration to the above proof of 1, we know that all the rational numbers between 0 and 1 are in T. (Note that we have Theorem 5.24.) Then σ-weak continuity of trace gives $T = [0, 1]$.

3. It is easy to see that M' is given by right multiplication by $q(M_n(\mathbb{C}) \otimes M)q$ for some projection $q \in M_n(\mathbb{C}) \otimes M$. (See the proof of Proposition 5.30.) Then the map $x \in M \mapsto xp \in Mp$ is an isomorphism. □

The assumption $\dim_M H < \infty$ in 3 is actually redundant. This is because the map $x \in M \mapsto xp \in Mp$ is actually an isomorphism in general, if M is a factor. Note that if M is hyperfinite in Lemma 5.32, then the resulting factors are also hyperfinite.

Theorem 5.33 *Suppose* $\dim_M H \in (0, \infty)$. *Then the following hold:*

1. *The commutant M' of M in $B(H)$ is a II_1 factor.*
2. $\dim_{M'} H = (\dim_M H)^{-1}$.
3. *For a non-zero projection $e' \in M'$,* $\dim_M(e'H) = \mathrm{tr}_{M'}(e') \dim_M H$.
4. *For a non-zero projection $e \in M$, we get*

$$\dim_{eMe}(eH) = \mathrm{tr}_M(e)^{-1} \dim_M H.$$

The symbol $\mathrm{tr}_{M'}$ *means the unique normalized trace on the II_1 factor M'.*

Proof 1. This follows from Lemma 5.32.

2. This follows from the above description of M'.

3. As in the proof of Part 3 of of Lemma 5.32, we can identify M' with the right multiplication algebra by $q(M_n(\mathbb{C}) \otimes M)q$ for some projection $q \in M_n(\mathbb{C}) \otimes M$. Then we can identify e' with right multiplication by some projection $e \in M_n(\mathbb{C}) \otimes M$ with $e \leq q$. Then we can identify $e'H = (\bigoplus_{j=1}^n L^2(M))e$. We now have

$$\dim_M e'H = \mathrm{tr}_{q(M_n(\mathbb{C})\otimes M)q}(e) \dim_M H = \mathrm{tr}_{M'}(e') \dim_M H.$$

4. First we claim that $(eMe)'$ on eH is equal to $eM'e$. It is trivial that $eM'e \subset (eMe)'$. Take any $x \in (eMe)'$. We assume that $\mathrm{tr}(e) \geq 1/2$ for simplicity and choose a projection $f \in M$ with $f \leq e$ and $\mathrm{tr}(f) = 1/2$. We choose a partial

isometry $u \in M$ so that $uu^* = f$ and $u^*u = 1 - f$ and set $x' = fx + u^*xu$. A direct computation shows $x' \in M'$. We then have $f \in eMe$, $x, ex' \in (eMe)'$. We also have $f(ex') = fx' = fx$, which implies $x = ex' \in eM'e$, because the map $y \in (eMe)' \mapsto fy$ is an isomorphism as in the proof of Lemma 5.32 3. Thus we now have $(eMe)' = eM'e$.

By the claim, 2, and 3 of this theorem, we now have

$$\begin{aligned}\dim_{eMe} eH &= (\dim_{(eMe)'} eH)^{-1} = (\dim_{eM'e} eH)^{-1}\\ &= (\mathrm{tr}_M(e)\dim_{M'} H)^{-1} = \mathrm{tr}_M(e)^{-1}\dim_M H.\end{aligned}$$

□

5.6 Tomita–Takesaki theory for hyperfinite von Neumann algebras

In this section, we present the fundamentals of the Tomita–Takesaki theory for hyperfinite von Neumann algebras. The Tomita–Takesaki theory works without the hyperfiniteness assumption, but this assumption makes the proofs considerably easier and clearer.

Let M be a hyperfinite von Neumann algebra with a cyclic and separating vector $\xi \in H$. (A vector $\xi \in H$ is said to be *cyclic* if $M\xi$ is dense in H. It is said to be *separating* if $x\xi = 0$ for $x \in M$ implies x_0.) In general, we may not find such a vector, but it is known that for all the von Neumann algebras acting on a separable Hilbert space, we can find such a vector after changing the Hilbert space, if necessary. (For example, if the von Neumann algebra has a faithful state, we can use a GNS-representation.) So we assume existence of such a vector and fix one vector ξ. If M is a II_1 factor, we can find such a vector if and only if the representation of M on H is standard.

Our aim is to get the *modular conjugation* operator J and a *modular automorphism group* σ_t with appropriate properties.

We define two anti-linear operators S_0 and F_0 on H by

$$\begin{aligned}S_0(x\xi) &= x^*\xi, \quad x \in M,\\ F_0(y\xi) &= y^*\xi, \quad y \in M',\end{aligned}$$

where the domains of S_0 and F_0 are $M\xi$ and $M'\xi$, respectively. For $x \in M$ and $y \in M'$, we have

$$\begin{aligned}\langle S_0x\xi, y\xi\rangle &= \langle x^*\xi, y\xi\rangle = \langle \xi, xy\xi\rangle = \langle \xi, yx\xi\rangle\\ &= \langle y^*\xi, x\xi\rangle = \langle F_0y\xi, x\xi\rangle,\end{aligned}$$

which means that both S_0 and F_0 are closable. We set $F = S_0^*$, $S = F_0^*$, and $\Delta = FS$. We also let $S = J\Delta^{1/2}$ be the polar decomposition of S. We call Δ a *modular operator* of M and J a *modular conjugation* of M. (Both operators depend on ξ.) Note that Δ is non-singular and positive and that J is an anti-linear unitary operator with $J^2 = 1$. (In general, Δ is not bounded.)

First we study a finite dimensional M. Let tr be a faithful trace, and ϕ be any faithful state on M. Then we have a positive invertible operator $h \in M$ with $\phi(x) = \mathrm{tr}(xh) = \mathrm{tr}(h^{1/2}xh^{1/2})$. Let π be the GNS-representation (Theorem 2.8) of M with respect to tr on H_{tr}. Then the vector $h^{1/2}$ gives a cyclic separating vector for $\pi(M)$ on H_{tr}. It is easy to see that the modular operator Δ with respect to $h^{1/2} \in H_{\mathrm{tr}}$ is given by $\Delta(\eta) = h\eta h^{-1}$ for $\eta \in H_{\mathrm{tr}}$. This implies $\Delta^{it}\pi(x)\Delta^{-it} = \pi(h^{it}xh^{-it})$ for $x \in M$ and $t \in \mathbb{R}$.

We next work on hyperfinite M. Let M_k be an increasing sequence of finite dimensional von Neumann algebras such that $A = \bigcup_{k=1}^{\infty} M_k$ is weakly dense in M. Our ξ, S, F, Δ are as above.

First, let H' be the domain of S with inner product

$$\langle \eta, \zeta \rangle' = \langle \eta, \zeta \rangle + \langle S\zeta, S\eta \rangle.$$

Because S is a closed operator, this makes H' a Hilbert space. It is easy to see $A\xi$ is a dense linear subspace of H'.

For $\eta, \zeta \in H'$, we have

$$|\langle \eta, \zeta \rangle| \le \|\eta\| \, \|\zeta\| \le \|\eta\|' \, \|\zeta\|',$$

where $\|\eta\|' = \langle \eta, \eta \rangle'^{1/2}$. Thus we have a linear operator T on H' with $\langle T\eta, \zeta \rangle' = \langle \eta, \zeta \rangle$.

Let $p_k \in M_k'$ be the projection of H onto $M_k\xi$. We have a von Neumann algebra $M_{k p_k}$ on $p_k H$. Then we see that π_k defined by $\pi_k(x) = xp_k$ for $x \in M_k$ is a $*$-isomorphism of M_k on H onto $M_{k p_k}$ on $p_k H$. (Note that ξ is a cyclic and separating vector for $M_{k p_k}$ on $p_k H$.) Thus for $x \in M_k$, we can define an antilinear operator S_k by $S_k(x\xi) = x^*\xi$ for $x \in M_k$. By the above consideration for finite dimensional cases, we have $\Delta_k^{it} M_{k p_k} \Delta_k^{-it} = M_{k p_k}$ for $t \in \mathbb{R}$ and $\Delta_k = S_k^* S_k$.

It is easy to see that $S_k = S|_{M_k\xi}$. We set H_k' to be the linear space $M_k\xi$ with inner product $\langle \eta, \zeta \rangle' = \langle \eta, \zeta \rangle + \langle S_k\zeta, S_k\eta \rangle$, for $\eta, \zeta \in M_k\xi$. Then H_k' is a Hilbert subspace of H'. As above, we have a linear operator T_k on H_k' with $\langle T_k\eta, \zeta \rangle' = \langle \eta, \zeta \rangle$ for $\eta, \zeta \in H_k'$. We have the following lemma.

Lemma 5.34 *Let $q_k \in B(H')$ be the projection of H' onto H_k'. Then the operators $\tilde{T}_k = T_k q_k + (1 - q_k)$ satisfy $\|T\eta - \tilde{T}_k\eta\|' \to 0$ for all $\eta \in H'$.*

Proof Because $\bigcup_{k=1}^{\infty} H_k' = A\xi$ is dense in H' as we have seen above, the projections q_k converge to 1 on H' strongly. For $\eta, \zeta \in H_k'$, we have $\langle T\eta, \zeta \rangle = \langle T_k\eta, \zeta \rangle$. This gives $\|T\eta - \tilde{T}_k\eta\|' \to 0$ for all $\eta \in H'$. □

Next we extend the modular operator Δ_k to the entire space H by $\tilde{\Delta}_k = \Delta_k p_k + (1 - p_k)$. For each k, the operator Δ_k is positive and invertible.

We cite without proof a general theorem in functional analysis. (See Theorem VIII.20 in (Reed and Simon 1972), for example.)

Theorem 5.35 *Let $\{D_n\}$ be a sequence of (possibly unbounded) positive non-singular operators, D a positive non-singular operator. Assume that $(D_n + 1)^{-1} \to (D + 1)^{-1}$ strongly.*

Then for any bounded continuous complex valued function f on $(0,\infty)$, we have $f(D_n) \to f(D)$ strongly.

Lemma 5.36 *For each $t \in \mathbb{R}$, we have $\|\tilde{\Delta}_k^{it}\eta - \Delta^{it}\eta\| \to 0$ for all $\eta \in H$.*

Proof It is enough to prove that $(\tilde{\Delta}_k + 1)^{-1} \to (\Delta + 1)^{-1}$ strongly by Theorem 5.35. First note that the range of $(\Delta + 1)^{-1}$ is equal to the domain of Δ, which is contained in the domain of S. For any η, ζ in the domain of S, we have

$$\begin{aligned}\langle (\Delta+1)^{-1}\eta, \zeta\rangle' &= \langle (\Delta+1)^{-1}\eta, \zeta\rangle + \langle S\zeta, S(\Delta+1)^{-1}\eta\rangle \\ &= \langle (\Delta+1)^{-1}\eta, \zeta\rangle + \langle \Delta(\Delta+1)^{-1}\eta, \zeta\rangle \\ &= \langle (\Delta+1)(\Delta+1)^{-1}\eta, \zeta\rangle \\ &= \langle \eta, \zeta\rangle = \langle T\eta, \zeta\rangle',\end{aligned}$$

which implies $T = (\Delta + 1)^{-1}$ on the domain of S. By the same argument, we have $T_k = (\Delta_k + 1)^{-1}$ and hence $\tilde{T}_k|_{M_k\xi} = (\tilde{\Delta}_k + 1)^{-1}|_{M_k\xi}$. By Lemma 5.34, we have

$$\begin{aligned}\|(\tilde{\Delta}_k+1)^{-1}\eta - (\Delta+1)^{-1}\eta\| &\le \|(\tilde{\Delta}_k+1)^{-1}\eta - (\Delta+1)^{-1}\eta\|' \\ &\le \|\tilde{T}_k\eta - T\eta\|' \to 0,\end{aligned}$$

where $\eta \in A\xi$ and k is sufficiently large. Because $A\xi$ is dense in H, we have the conclusion. □

By the isomorphism π_k defined above, we can define an automorphism σ_t^k of M_k for $t \in \mathbb{R}$ by $\pi_k(\sigma_t^k(x)) = \Delta_k^{it}\pi_k(x)\Delta_k^{-it}$. We then have the following lemma.

Lemma 5.37 *For $x \in A$, the sequence $\{\sigma_t^k(x)\}_k$ defined for sufficiently large k converges to $\Delta^{it}x\Delta^{-it}$ strongly.*

Proof For $x \in A$ and $\eta \in A\xi$, we have

$$\sigma_t^k(x)\eta = \pi_k(\sigma_t^k(x))\eta = \Delta_k^{it}x\Delta_k^{-it}\eta = \tilde{\Delta}_k^{it}x\tilde{\Delta}_k^{-it}\eta,$$

for sufficiently large k. This and Lemma 5.36 imply

$$\|\sigma_t^k(x)\eta - \Delta^{it}x\Delta^{-it}\eta\| \to 0$$

for each $x \in A$, $\eta \in A\xi$, and $t \in \mathbb{R}$. This gives the conclusion. □

Lemma 5.37 shows that $\Delta^{it}A\Delta^{-it}$ is contained in M for any $t \in \mathbb{R}$ and thus $\Delta^{it}M\Delta^{-it}$ is contained in M. By symmetry, we have $\Delta^{it}M\Delta^{-it} = M$ and thus $\sigma_t(x) = \Delta^{it}x\Delta^{-it}$ gives a one-parameter automorphism group of M. This σ_t is called a *modular automorphism group* of M (with respect to ξ). For the construction of the modular automorphism in the general setting, see section 2.5 of (Bratteli and Robinson 1987), Chapter 9 of (Kadison and Ringrose 1986), or Chapter 1 of (Takesaki 1983).)

We state the following characterization of the modular automorphism group without proof.

Theorem 5.38 *For the modular automorphism group σ_t as above, we have the following two properties:*

1. *We have $\phi \cdot \sigma_t = \phi$, where ϕ is the vector state given by the cyclic separating vector ξ.*
2. *For any $x, y \in M$, there exists a bounded continuous function $F(z)$ on $\bar{D}$, where*
$$D = \{z \in \mathbb{C} \mid 0 \leq \operatorname{Im} z \leq 1\},$$
such that $F(z)$ is holomorphic on D and
$$F(t) = \phi(\sigma_t(x)y), \quad F(t+i) = \phi(y\sigma_t(x)), \qquad t \in \mathbb{R}.$$

Conversely, any one-parameter automorphism group σ_t satisfying the above two properties coincides with the modular automorphism group.

The two conditions in the above theorem is called the *KMS condition*, where 'KMS' stands for Kubo–Martin–Schwinger. The KMS condition can be equivalently formulated as follows. For any $x \in M$, and y entire for σ, we have

$$\varphi(yx) = \varphi(x\sigma_{-i}(y)).$$

In the case when φ is a trace, the modular group $\sigma_t = \mathrm{id}$, and the KMS condition in general is a modification of the trace condition $\operatorname{tr}(yx) = \operatorname{tr}(xy)$. See Theorems 9.2.13 and 9.2.16 of (Kadison and Ringrose 1986) or Theorem 4.1 of (Takesaki 1983) for a proof of this theorem, for example.

In the above setting, we can also prove $JMJ = M'$. We then say that the representation of M on H is *standard.*

More precisely, this is the KMS condition at inverse temperature -1. More generally, a state φ is a KMS state at inverse temperature β or KMS_β state for a one-parameter group α_t (strongly continuous for a C^*-algebra, and weakly continuous for a von Neumann algebra) if φ is a KMS state at inverse temperature -1 for $\alpha_{-\beta t}$, or equivalently if

$$\varphi(yx) = \varphi(x\alpha_{\beta i}(y))$$

for all x and all entire y. For example, the *Gibbs state* $\operatorname{tr}(e^{-\beta H}\cdot)$ is the (unique) KMS_β state for the group $e^{iHt}(\cdot)e^{-iHt}$ on a matrix algebra as

$$\operatorname{tr}(e^{-\beta H}yx) = \operatorname{tr}\left(e^{-\beta H}xe^{-\beta H}ye^{\beta H}\right)$$

for all x, y. The role of KMS_β states as equilibrium states is discussed in the introduction to Chapter 7. The case $\beta = \infty$ corresponds to the notion of ground state and will be significant in Chapters 6 and 7 (see Exercise 5.5).

At the end of this section, we briefly sketch structure of positive cones associated to the standard representation. First, we need a definition.

Definition 5.39 *Let P be a convex cone in a Hilbert space H. The dual cone P^0 of P is defined by*

$$P^0 = \{\eta \in H \mid \langle \eta, \zeta \rangle \geq 0, \zeta \in P\}.$$

If $P = P^0$, the cone P is said to be self-dual.

We again work on a von Neumann algebra M with a cyclic separating vector $\xi \in H$. We have Δ, J as above. Then we have the following theorem, whose proof is again omitted here. See, for example, section 2.5.4 of (Bratteli and Robinson 1987) for a proof.

Theorem 5.40 *Setting*

$$P^{\sharp} = \overline{\{x^* x \xi \mid x \in M\}}, \quad P^{\flat} = J P^{\sharp}, \quad P^{\natural} = \overline{\Delta^{1/4} P^{\sharp}},$$

we get the following:

1. *The convex cones $P^{\sharp}$ and $P^{\flat}$ are mutually dual.*
2. *The cone $P^{\natural}$ is self-dual.*
3. *For $x \in M$, we have $xJxJP^{\natural} \subset P^{\natural}$.*
4. *To each positive normal functional ω on M, we have a unique $\eta \in P^{\natural}$ such that ω is the vector state ω_η associated to η. We also have*

$$\|\eta - \zeta\|^2 \leq \|\omega_\eta - \omega_\zeta\| \leq \|\eta - \zeta\| \, \|\eta + \zeta\|, \quad \eta, \zeta \in P^{\natural}.$$

Here the cone $P^{\natural}$ is called the *natural cone.*

Spectral analysis of the modular automorphism group gives much information on the structure of M. In particular, the classification of type III factors into type III_λ in (Connes 1973) is based on such an analysis.

Exercise 5.5 Let α_t be a strongly continuous one parameter group on a C^*-algebra A, with infinitesimal generator δ, and φ a state on A. Show that the following conditions are equivalent (if they are satisfied φ is said to be a *ground state* on A):

(a) $-i\varphi(a^*\delta(a)) \geq 0$ for all a in the domain of δ.

(b) if a, b in A are α-entire then $z \to \varphi(a\alpha_z(b))$ is uniformly bounded on $\{z \in \mathbb{C} \mid \operatorname{Im} z \geq 0\}$.

(c) if a, b are in A, $t \in \mathbb{R} \to \varphi(a\alpha_t(b))$ has an extension to a continuous function on $\operatorname{Im} z \geq 0$, which is analytic and bounded in $\operatorname{Im} z > 0$.

(d) If $\hat{f}$ is a C^∞ function with compact support in $(-\infty, 0)$, then $\varphi(\alpha_f(a)^*\alpha_f(a)) = 0$ for all a in A.

(e) φ is α invariant and if $e^{ih_\varphi}\pi_\varphi(a)\Omega_\varphi = \pi_\varphi(\alpha_t(a))\Omega_\varphi$ is the corresponding one parameter unitary representation on H_φ, then $h_\varphi \geq 0$.

If φ is a ground state, then $e^{ith_\varphi} \in \pi_\varphi(A)''$ by Borchers' theorem (see Section 4.6). We denote the set of ground states by KMS_∞, or regard such states as satisfying the KMS condition at inverse temperature $\beta = \infty$. Similarly we can define ceiling states for inverse temperature $\beta = -\infty$. Note that if ω_α is $\mathrm{KMS}_{\beta_\alpha}$ state where $-\infty \leq \beta_\alpha \leq \infty$ and $\omega_\alpha \to \omega$ in the weak $*$-topology (i.e. $\omega_\alpha(a) \to \omega(a)$ for a in A) as $\beta_\alpha \to \beta$, then ω is a KMS_β state.

5.7 Notes

The results in Sections 5.2–5.5 are classical, and many of them are due to Murray–von Neumann. These can be found in any standard text book on operator algebras such as (Bratteli and Robinson 1987), (Bratteli and Robinson 1981), (Dixmier 1969), (Kadison and Ringrose 1983), (Kadison and Ringrose 1986), (Pedersen 1979), (Sakai 1971), (Strătilă and Zsido 1979), (Takesaki 1979).

A von Neumann algebra has been also characterized abstractly as a C^*-algebra M with $M = X^*$ for some Banach space X by S. Sakai. (See (Sakai 1971).)

Hyperfiniteness has been characterized by a more intrinsic property 'injectivity', or 'amenability' as is often called today, by (Connes 1976a). This is one of the deepest theorems in the theory of operator algebras. The classification of hyperfinite factors has been completed by (Connes 1976a), (Connes 1985b) and (Haagerup 1987). The notes to Chapter 3 contain a discussion on amenable C^*-algebras.

An outer action of a finite group on the hyperfinite factor is known to be unique for each finite group by (Connes 1977) and (Jones 1980b). See Section 15.4 for a brief review on classification of group actions.

Theorem 5.29 was first proved by M. Tomita. The Tomita–Takesaki theory was first published in (Takesaki 1970). Here we followed (Longo 1978) in Section 5.6 with an extra assumption of hyperfiniteness. Theorem 5.38 was obtained in (Takesaki 1970), while the uniqueness of the modular automorphisms and the invariance in this theorem had been independently obtained in (Winnink 1969). The name KMS condition came from a formulation of this condition in (Haag *et al.* 1967). The inequality in Theorem 5.40 is due to (Araki 1963), which is a generalization of the Powers–Størmer inequality, Exercise 2.4, in (Powers and Størmer 1970). See (Bratteli and Robinson 1987), (Kadison and Ringrose 1986), (Pedersen 1979), (Strătilă 1981) and (Takesaki 1983) for more on the Tomita–Takesaki theory. Exercise 5.5 is from (Bratteli and Robinson 1981).

6

THE FERMION ALGEBRA

6.1 Introduction

This chapter introduces the Fermion algebra as the main technical tool for studying in Chapter 7 the two dimensional Ising model as the prototype lattice model in statistical mechanics. It will also feature in Chapter 8 as a means of constructing representations of chiral algebras, particularly the Virasoro algebra, in conformal field theories which can appear in critical statistical mechanical models.

Ignoring most of the structure, the Fermion algebra is just an infinite tensor product of 2×2 matrix algebras. Such infinite tensor products of matrix algebras, UHF algebras, are studied in Sections 6.2 and 6.3, as well as their representations on infinite tensor products of Hilbert spaces. Subtleties about completions of such infinite tensor products of Hilbert spaces lead to inequivalent representations of UHF algebras, and criteria to decide when infinite tensor automorphisms are inner or weakly inner in the trace representation. In the Fermion algebra setting, these will translate as trace class and Hilbert–Schmidt conditions. The Fermion algebra is the algebra associated with the canonical anti-commutation relations, CAR. There are three natural ways of making this notion precise. The one we mainly adopt is the self-dual formalism of (Araki 1970a), (Araki 1987), where we treat annihilation and creation operators on an equal footing. Here $B(f \oplus g) = a^*(f) + a(Jg)$ if a is a representation of the CAR in its most primitive form, the complex formalism $\{a(f), a^*(g)\} = \langle g, f\rangle$, $\{a(f), a(g)\} = 0$, for a conjugation J on a Hilbert space and where $\{x, y\} = xy + yx$ denotes the *anti-commutator* or Jordan product. The self dual CAR algebra A^{SDC} is then generated by the operator valued map B satisfying $\{B(f), B(g)^*\} = \langle f, g\rangle$, $B(f \oplus g)^* = B(Jg \oplus Jf)$. The Fermion algebra A^{CAR} generated by the operator valued conjugate linear map a will also feature, but we will not make much reference to the third (real) formalism, the Clifford algebra.

Although the Fermion algebra A^F is isomorphic as a C^*-algebra to the Pauli algebra A^P, the UHF algebra which is the infinite tensor product of 2×2 matrix algebras, we have to be careful about this isomorphism, in particular for the gradings – Section 6.5. In Section 6.6 we introduce the quasi-free states on the Fermion algebra. They can be regarded as analogues of Gaussian distributions in classical probability theory in that all the n-point functions can be computed once one knows the two-point functions. In particular such a state φ_S is completely determined by an operator S on the underlying one-particle Hilbert space. Similarly a distinguished class of $*$-endomorphisms, the Bogoliubov $*$-endomorphisms, of the Fermion algebra are those induced by isometries U on the one-particle space commuting with the conjugation so that $B(f) \to B(Uf)$, extends to a $*$-endomorphism denoted by $\tau(U)$. As we have already hinted, such $*$-automorphisms are inner or weakly inner when a trace class or Hilbert–Schmidt condition is satisfied (either $U - 1$ is in such a class and $\det U = 1$ or $U + 1$ is in such a class and $\det(-U) = -1$). The implementing unitary however must be either even or odd and the two conditions reflect this dichotomy. This is discussed in Sections 6.7–6.9, as is the question of constructing explicit implementers for Bogoliubov endomorphisms, (or indeed completely positive maps) and their infinitesimal versions, quasi-free $*$-derivations extending $B(f) \to B(Hf)$ for skew adjoint operators H which commute with the conjugation. We also discuss equivalence of quasi-free states on the Fermion algebra where Hilbert–Schmidt criteria arise.

Although we have to be careful to distinguish the Fermion and Pauli algebras, we can identify their even subalgebras, and consider them as subalgebras of a mutual enveloping algebra. Even states on the Fermion algebra such as quasi-free states are naturally transformed to states on the Pauli algebra. This is how our states in the C^*-setting of the Ising model will appear in Chapter 7. We will need criteria for understanding equivalence of restrictions of quasi-free states to the Fermion algebra, and criteria for the purity of the even state on a Pauli algebra arising from a (pure) quasi-free state on the Fermion algebra. This purity in the Pauli algebra is not automatic and will be the key point in observing the phase transition in the Ising model in our C^*-setting. Section 6.10 contains an analysis of the Pauli states and ground states for dynamics in the Fermion, Pauli and even settings, arising from quasi-free dynamics.

6.2 Infinite tensor products of Hilbert spaces

To understand better the various representations encountered in different product representations and states of UHF algebras, we first need to look closely at the construction of an infinite tensor product of Hilbert spaces.

We can form the infinite tensor product of a sequence of vector spaces $V_i, i = 1, 2, \ldots$ as in Exercise 6.2 to be the quotient of the free vector space generated by the Cartesian product $\prod_{i=1}^{\infty} V_i$ by the subspace generated by $\lambda x + \mu y - z$, where $x = \{x_i\}$, $y = \{y_i\}$, $z = \{z_i\}$, $\lambda, \mu \in \mathbb{C}$, and $x_i = y_i = z_i$, except for one index j, where $\lambda x_j + \mu y_j = z_j$. We let $\bigotimes_{i=1}^{\infty} x_i$ denote the equivalence class of $\{x_i\}$ in the quotient, written $\bigodot_{i=1}^{\infty} X_i$. If H_i is a sequence of Hilbert spaces, there will be problems in putting a natural inner product on the infinite tensor product through:

$$\left\langle \bigotimes_{i=1}^{\infty} x_i, \bigotimes_{i=1}^{\infty} y_i \right\rangle = \prod_{i=1}^{\infty} \langle x_i, y_i \rangle \tag{6.2.1}$$

due to the possible non-convergence of the right hand side. At least if $\Omega_i \in H_i$ is a sequence of unit vectors then $\prod_{i=1}^{\infty} \langle \Omega_i, \Omega_i \rangle$ converges. Then, for such a given sequence $\Omega = \{\Omega_i\}$, which infinite tensors $\bigotimes_{i=1}^{\infty} x_i$ permit convergence of $\prod_{i=1}^{\infty} \langle x_i, \Omega_i \rangle$? Certainly, we can start with the subspace of $\bigodot_{i=1}^{\infty} H_i$, spanned by tensors $\bigotimes_{i=1}^{\infty} x_i$, such that $x_i = \Omega_i$ except for finitely many i. Denote this subspace by $\bigodot_{i=1}^{\infty}(H_i, \Omega_i)$, and its completion with respect to the inner product (6.2.1), is denoted by $\bigotimes_{i=1}^{\infty}(H_i, \Omega_i)$. Note that $\bigotimes_{i=1}^{\infty}(H_i, \Omega_i) = \lim_{\rightarrow}(H_1 \otimes \cdots \otimes H_n, \xi \rightarrow \xi \otimes \Omega_{n+1})$ and if $\bigotimes_{i=1}^{\infty} x_i \in \bigodot_{i=1}^{\infty} H_i$ with $x_i = \Omega_i$ for $i > N$, then $\bigotimes_{i=1}^{\infty} x_i = x_1 \otimes x_2 \otimes \cdots$ is the image of $x_1 \otimes \cdots \otimes x_N$ in $\lim_{\rightarrow} H_1 \otimes \cdots \otimes H_n$.

Lemma 6.1 *Let $\{z_i\}$ be a sequence of non-zero vectors in $\{H_i\}$. Then the sequence $x_n = z_1 \otimes z_2 \otimes \cdots \otimes z_n \otimes \Omega_{n+1} \otimes \Omega_{n+2} \otimes \cdots$ converges in $\bigotimes_{i=1}^{\infty}(H_i, \Omega_i)$ to a non-zero vector, denoted by $\bigotimes_{i=1}^{\infty} z_i = z_1 \otimes z_2 \otimes \cdots$ if:*
(6.2.2) $\sum_{i=1}^{\infty} | \|z_i\|^2 - 1| < \infty$, *and*
(6.2.3) $\sum_{i=1}^{\infty} |\langle z_i, \Omega_i \rangle - 1| < \infty$.

Proof If $m > n$, from $x_n - x_m = z_1 \otimes \cdots \otimes z_n \otimes (\Omega_{n+1} \otimes \cdots \otimes \Omega_m - z_{n+1} \otimes \cdots \otimes z_m) \otimes \Omega_{m+1} \otimes \Omega_{m+2} \otimes \cdots$ we have

$$\begin{aligned}
\|x_n - x_m\|^2 &= \prod_{i=1}^{n} \|z_i\|^2 \|\Omega_{n+1} \otimes \cdots \otimes \Omega_m - z_{n+1} \otimes \cdots \otimes z_m\|^2 \\
&= \prod_{i=1}^{n} \|z_i\|^2 \left(1 + \prod_{n+1}^{m} \|z_i\|^2 - 2 \prod_{n+1}^{m} \mathrm{Re}\langle z_i, \Omega_i \rangle\right) \\
&\rightarrow 0 \quad \text{as} \quad n, m \rightarrow \infty
\end{aligned}$$

since by assumption, all $z_i, \langle z_i, \Omega_i \rangle \neq 0$ for large i, so that

$$\prod_{n+1}^{m} \|z_i\|^2, \prod_{n+1}^{m} \mathrm{Re}\langle z_i, \Omega \rangle \rightarrow 1 \quad \text{as } n, m \rightarrow \infty .$$

□

We say that the sequence $\{z_i\}$ or the tensor $\bigotimes_{i=1}^{\infty} z_i$ is c_0 if $\sum_{i=1}^{\infty} | \|z_i\|^2 - 1| < \infty$. Define a relation $\approx$ on c_0 sequences $\{z_i\}$ and $\{z_i'\}$ of vectors in $\{H_i\}$ by (6.2.3), namely $\{z_i\} \approx \{z_i'\}$ if and only if $\sum_{i=1}^{\infty} |\langle z_i, z_i' \rangle - 1| < \infty$.

Lemma 6.2 *$\approx$ defines an equivalence relation on c_0 sequences in $\{H_i\}$.*

Proof The only property which is not immediate is transitivity. Suppose that f, g, h are c_0 sequences such that $f \approx g$, $g \approx h$. We will sketch a proof in the case $\|g_i\| = 1$, for all i, as this will be sufficient for our purposes. Now

$$\langle f_i, h_i \rangle - 1 = \langle f_i - \langle f_i, g_i \rangle g_i, h_i \rangle + \langle f_i, g_i \rangle \langle g_i, h_i \rangle - 1$$

$$= \langle f_i - \langle f_i, g_i \rangle g_i, h_i - \langle h_i, g_i \rangle g_i \rangle$$
$$+ \langle f_i, g_i \rangle \langle g_i, h_i \rangle - 1, \quad \text{using} \quad \|g_i\| = 1 .$$

Thus

$$\begin{aligned} |\langle f_i, h_i \rangle - 1| &\le \|f_i - \langle f_i, g_i \rangle g_i\| \|h_i - \langle h_i, g_i \rangle g_i\| + |\langle f_i, g_i \rangle \langle g_i, h_i \rangle - 1| \\ &= \left[\|f_i\|^2 - |\langle f_i, g_i \rangle|^2\right]^{\frac{1}{2}} \left[\|h_i\|^2 - |\langle h_i, g_i \rangle|^2\right]^{\frac{1}{2}} + |\langle f_i, g_i \rangle \langle g_i, h_i \rangle - 1| . \end{aligned}$$

Convergence of $\sum |\langle f_i, h_i \rangle - 1|$ now follows from that of $\sum | \, \|f_i\|^2 - 1|$, $\sum | \, \|h_i\|^2 - 1|$, $\sum |\langle f_i, g_i \rangle - 1|$ and $\sum |\langle g_i, h_i \rangle - 1|$. □

It is clear that the following subset of the algebraic tensor product $\bigodot_{i=1}^{\infty} H_i$:

$$\left\{ \bigotimes_{i=1}^{\infty} z_i \mid \sum_{i=1}^{\infty} | \, \|z_i\|^2 - 1| < \infty, \{z_i\} \approx \{\Omega_i\} \right\}$$

can be identified with a total set of vectors for $\bigotimes_{i=1}^{\infty} (H_i, \Omega_i)$. From this, and the fact that $\approx$ is an equivalence relation, we see that if $\{\Omega_i\}$ and $\{\Omega_i'\}$ are two sequences of unit vectors, with $\{\Omega_i\} \approx \{\Omega_i'\}$, we can identify $\bigotimes_{i=1}^{\infty} (H_i, \Omega_i)$ with $\bigotimes_{i=1}^{\infty} (H_i, \Omega_i')$. If $\{H_i\}$, and $\{K_i\}$ are two sequences of Hilbert spaces, then $\bigodot B(H_i, K_i)$ maps $\bigodot_{i=1}^{\infty} H_i$ into $\bigodot_{i=1}^{\infty} K_i$ by the universal property of tensor products (see Exercise 6.2 and Section 2.6):

$$\bigotimes_{i=1}^{\infty} T_i \bigotimes_{i=1}^{\infty} x_i = \bigotimes_{i=1}^{\infty} T_i x_i \tag{6.2.4}$$

$x_i \in H_i$, $T_i \in B(H_i, K_i)$. Thus if $\{\Omega_i\}$ is a sequence of unit vectors in $\{H_i\}$, and T_i isometries from H_i to K_i, then (6.2.4) defines an unique isometry also denoted by $\bigotimes_{i=1}^{\infty} T_i$ from $\bigotimes_{i=1}^{\infty} (H_i, \Omega_i)$ to $\bigotimes_{i=1}^{\infty} (K_i, T_i \Omega_i)$. If $\{x_i\}$ is a sequence in $\{H_i\}$ satisfying $\sum_{i=1}^{\infty} | \, \|x_i\|^2 - 1| < \infty$, and $\{x_i\} \approx \{\Omega_i\}$, then $\sum_{i=1}^{\infty} | \, \|T_i x_i\|^2 - 1| < \infty$ and $\{T_i x_i\} \approx \{T \Omega_i\}$.

Note also that if $\{\Omega_i\}$ and $\{\Omega_i'\}$ are two sequences of unit vectors in $\{H_i\}$ with

$$\sum_{i=1}^{\infty} ||\langle \Omega_i, \Omega_i' \rangle| - 1| < \infty , \tag{6.2.5}$$

and letting $|\langle \Omega_i, \Omega_i' \rangle| = \alpha_i \langle \Omega_i, \Omega_i' \rangle$, where $\alpha_i \in \mathbb{T}$, we see that $\{\alpha_i \Omega_i\} \approx \{\Omega_i'\}$. Hence we see that there are canonical identifications

$$\bigotimes_{i=1}^{\infty} (H_i, \Omega_i') \to \bigotimes_{i=1}^{\infty} (H_i, \alpha_i \Omega_i) \to \bigotimes_{i=1}^{\infty} (H_i, \Omega_i) \tag{6.2.6}$$

using $\{\Omega_i'\} \approx \{\alpha_i \Omega_i\}$ and the unitary $\bigotimes_i \overline{\alpha_i}$ respectively.

Two representations π_1 and π_2 of a C^*-algebra A are said to be *algebraically equivalent* if Ker π_1 = Ker π_2, which is the case if and only if there exists an isomorphism $\alpha = \pi_1(A) \to \pi_2(A)$ such that $\alpha\pi_1 = \pi_2$. They are said to be *quasi-equivalent* if there exists an isomorphism $\alpha : \pi_1(A)'' \to \pi_2(A)''$ such that $\alpha\pi_1 = \pi_2$, and *unitarily equivalent* if there exists a unitary $u : H_{\pi_1} \to H_{\pi_2}$ such that $\mathrm{Ad}(u)\pi_1 = \pi_2$. Thus unitary equivalence implies quasi-equivalence which implies algebraic equivalence, and quasi-equivalent irreducible representations are unitarily equivalent since every isomorphism $\alpha : B(H_{\pi_1}) \to B(H_{\pi_2})$ is spatial by Exercise 2.7.

Exercise 6.1 Provide the details of the proof of Lemma 6.2 when $\|g_i\| \not\equiv 1$.

Exercise 6.2 If V_i is a sequence of vector spaces and $T : \prod_{i=1}^{\infty} V_i \to V$ is a multilinear mapping into a vector space V, then there exists an unique linear map $\hat{T} : \bigotimes_{i=1}^{\infty} V_i \to V$ such that $\hat{T}\left(\bigotimes_{i=1}^{\infty} x_i\right) = T(x_i)$.

6.3 Product states and automorphisms of UHF algebras

The aim of this section is to discuss implementability of product automorphisms and equivalence of product states on a UHF algebra. Let A_i be a sequence of finite dimensional matrix algebras and $A = \bigotimes_{i=1}^{\infty} A_i$.

Theorem 6.3 *Let $\{u_i\}$ be a sequence of unitaries in $\{A_i\}$ with $1 \in \sigma(u_i)$. Then $\bigotimes_{i=1}^{\infty} \mathrm{Ad}(u_i)$ is inner if and only if $\lim_{n\to\infty} u_1 \otimes \cdots \otimes u_n \otimes 1_{n+1} \otimes 1_{n+2} \otimes \cdots = u$ exists in $A = \bigotimes_{i=1}^{\infty} A_i$ (where $1_n = 1_{A_n}$). In this case $\bigotimes_{i=1}^{\infty} \mathrm{Ad}(u_i) = \mathrm{Ad}(u)$.*

Proof Suppose $\bigotimes_{i=1}^{\infty} \mathrm{Ad}(u_i) = \mathrm{Ad}(u)$, for some $u \in A$. (Note that by rotating, we could always move to the situation where $1 \in \sigma(u_i)$). Let $v_n = u_1 \otimes u_2 \otimes \cdots \otimes u_n \otimes 1_{n+1} \otimes \cdots$, so that $v_n^{-1}u \in (A_1 \otimes \cdots \otimes A_n \otimes 1)^c = 1 \otimes A_{n+1} \otimes A_{n+2} \otimes \cdots$ (see (2.9.15)). Choose unitaries w_n in $A_1 \otimes \cdots \otimes A_n = B_n$ such that $w_n \to u$ as in Lemma 3.3. If φ_1, φ_2 are states on B_n and B_n^c respectively, then

$$\begin{aligned}|\varphi_2(v_n^{-1}u) - \varphi_1(v_n^{-1}w_n)| &= |\varphi_1 \otimes \varphi_2(v_n^{-1}u - v_n^{-1}w_n)| \\ &\le \|v_n^{-1}u - v_n^{-1}w_n\| \le \|u - w_n\| .\end{aligned}$$

Hence taking $\lambda_n = \varphi_1(v_n^{-1}w_n)$, for a sequence of states φ_1 on B_n, we see

$$\|v_n^{-1}u - \lambda_n\| = \sup_{\varphi_2} \|\varphi_2(v_n^{-1}u) - \lambda_n\| \le \|u - w_n\| ,$$

and

$$\|u - \lambda_n v_n\| \to 0 . \tag{6.3.1}$$

But $1 \in \sigma(u_i)$, and so if $m < n$,

$$\begin{aligned}|\lambda_m - \lambda_n| &\le \|\lambda_m u_{m+1} \otimes \cdots \otimes u_n - \lambda_n 1_{m+1} \otimes \cdots \otimes 1_n\| \\ &= \|\lambda_m v_m - \lambda_n v_n\| \to 0 \quad \text{as} \quad m, n \to \infty .\end{aligned}$$

Thus $\{\lambda_n\}$ is convergent to some λ, $|\lambda| = 1$ (see (6.3.1)). Hence $\{v_n\}$ is also convergent and $\lim_{\rightarrow} v_n = \lambda^{-1}u$. □

Corollary 6.4 *Let λ_i be a sequence of complex numbers on the circle. Then $\bigotimes_{i=1}^{\infty} \mathrm{Ad}\begin{pmatrix} 1 & 0 \\ 0 & \lambda_i \end{pmatrix}$ is inner on the Pauli algebra $\bigotimes_{i=1}^{\infty} M_2$ if and only if $\prod_{i=1}^{\infty} \lambda_i$ is convergent. In particular $\bigotimes_{i=1}^{\infty} \mathrm{Ad}\begin{pmatrix} 1 & 0 \\ 0 & \lambda \end{pmatrix}$ is outer if $\lambda \in \mathbb{T}\backslash\{1\}$.*

Proof Let $u_i = \begin{pmatrix} 1 & 0 \\ 0 & \lambda_i \end{pmatrix}$, and $x_n = u_1 \otimes u_2 \otimes \cdots \otimes u_n$. Then for $m > n$,

$$x_n - x_m = u_1 \otimes u_2 \otimes \cdots \otimes u_n \otimes (1 \otimes 1 \otimes \cdots \otimes 1 - u_{n+1} \otimes \cdots \otimes u_m)$$

so that the result follows from

$$\begin{aligned} \|x_n - x_m\| &= \|1 \otimes 1 \otimes \cdots \otimes 1 - u_{n+1} \otimes u_{n+2} \otimes \cdots \otimes u_m\| \\ &= \max\{|1 - \lambda_{i_1}\lambda_{i_2} \cdots \lambda_{i_r}| \mid n \leq i_1 < \cdots < i_r \leq m\}. \end{aligned}$$

□

Remark Recall from Section 6.2 that we defined quasi-equivalence and unitary equivalence of two representations π_1 and π_2 of a C^*-algebra A. The following conditions are equivalent:

(a) π_1 and π_2 are quasi-equivalent, i.e. there exists an isomorphism τ from $\pi_1(A)''$ onto $\pi_2(A)''$ taking $\pi_1(a)$ to $\pi_2(a)$, for $a \in A$.

(b) π_1 and π_2 have the same normal states, or normal folia. The *normal folium* $\mathcal{F}(\pi)$ of a representation π is the set of π-*normal states* of A of the form $a \to \rho(\pi(a))$ for a normal state $\rho \in \pi(A)''_*$.

(c) π_1 is unitarily equivalent to a subrepresentation of a multiple of π_2 and vice versa.

(d) π_1 and π_2 are unitarily equivalent up to multiplicity.

That (a) implies (b) is clear as if ρ is in $\pi_1(A)''_*$, then $\rho \cdot \tau$ is in $\pi_2(A)''_*$. If (b) holds, then writing $\pi_1 = \oplus_\alpha \pi_\alpha$ as a sum of cyclic representations with cyclic vector ξ_α, then $\langle \pi_1(\cdot)\xi_\alpha, \xi_\alpha \rangle \in \mathcal{F}(\pi_1) = \mathcal{F}(\pi_2)$ so that $\langle \pi_1(\cdot)\xi_\alpha, \xi_\alpha \rangle = \sum_n \langle \pi_2(\cdot)\xi_n^\alpha, \xi_n^\alpha \rangle$ where $\sum_n \|\xi_n^\alpha\|^2 = 1$. We then argue as in Proposition 2.11 to get that π_1 is unitarily equivalent to a subrepresentation of a multiple of π_2 (and vice versa). A Cantor–Bernstein argument shows that (c) implies (d), and (d) implies (a) is clear as a representation is clearly quasi-equivalent to any multiple.

The antithesis of quasi-equivalence is *disjointness*. The following conditions are equivalent for two representations π_1 and π_2 of a C^*-algebra:

(e) π_1 and π_2 are *disjoint*, written $\pi_1 \not\sim \pi_2$, in the sense that they do not posses unitarily equivalent subrepresentations.

(f) π_1 and π_2 have no quasi-equivalent subrepresentations.

Moreover two representations are quasi-equivalent if and only if each has no subrepresentation disjoint from the other representation. Factor representations (i.e. representations which generate factors) are either quasi-equivalent or disjoint. Irreducible representations are either quasi-equivalent or disjoint, since irreducible representations have no non-trivial subrepresentations. We define disjointness, quasi-equivalence, equivalence of states in terms of their cyclic representations.

Theorem 6.5 *Let $\{K_n\}$ and $\{L_n\}$ be sequences of density operators in $\{A_n\}$, and $\rho = \bigotimes_{i=1}^{\infty} \rho_{K_i}$, $\rho' = \bigotimes_{i=1}^{\infty} \rho_{L_i}$ be the corresponding product states on A. Consider the following conditions:*
(6.3.2) *The states ρ and ρ' are quasi-equivalent on A.*
(6.3.3) *The states ρ and ρ' are unitarily equivalent on A.*
(6.3.4) $\sum_{i=1}^{\infty}(1 - \operatorname{tr} K_i^{\frac{1}{2}} L_i^{\frac{1}{2}}) < \infty$.
All three conditions are equivalent if the following holds:
(6.3.5) *Either all ρ_{K_i}, ρ_{L_i} are pure, or all are faithful.*

Proof We give the proof in the case $K_i, L_i > 0$, for all i. The GNS triplet of ρ_{K_i} identifies with $(H_i = A_i K_i^{1/2} = A_i, \pi_i, K_i^{1/2})$, where $\langle a, b\rangle = \operatorname{tr} ab^*$, $a, b \in H_i$, and $\pi_i(a)b = ab$ for $a \in A_i, b \in H_i$. Thus the GNS triplet of $\bigotimes_{i=1}^{\infty} \rho_{K_i}$ then identifies with $(\bigotimes_{i=1}^{\infty}(H_i, K_i^{1/2}), \bigotimes_{i=1}^{\infty} \pi_i, \bigotimes_{i=1}^{\infty} K_i^{1/2})$, and so condition (6.3.4) holds if and only if $\{K_i^{1/2}\} \approx \{L_i^{1/2}\}$. Hence by Section 6.2, if (6.3.4) holds then ρ and ρ' are unitarily equivalent. Next, suppose ρ and ρ' are quasi-equivalent. Then $\bigotimes_{i=1}^{\infty} \rho_{K_i}$ is a KMS state for $\bigotimes_{i=1}^{\infty} A_i$ with respect to $\{\bigotimes_{i=1}^{\infty} \operatorname{Ad}(K_i^{it}) \mid t \in \mathbb{R}\}$ and so is faithful on the weak closure.

We can thus regard the quasi-equivalent states ρ and ρ' as faithful states on the same von Neumann algebra M, with vector representatives $\xi_\rho, \xi_{\rho'}$ in the natural positive cone (Section 5.6). Then

$$\|\xi_\rho - \xi_{\rho'}\| \le \|\rho - \rho'\| \le \|\xi_\rho - \xi_{\rho'}\| \|\xi_\rho + \xi_{\rho'}\| = (1 - \langle \xi_\rho, \xi_{\rho'}\rangle)^{\frac{1}{2}} \cdot 2 < 2$$

as $\langle \xi_\rho, \xi_{\rho'}\rangle > 0$, since ρ and ρ' are faithful. Restricting to $M_1 = \bigotimes_{i=1}^{N} A_i$, we see that $\bigotimes_{i=1}^{N} K_i^{1/2}, \bigotimes_{i=1}^{N} L_i^{1/2}$ can be identified with the natural cone representatives of $\bigotimes_{i=1}^{N} \rho_{K_i}, \bigotimes_{i=1}^{N} \rho_{L_i}$. Thus

$$\begin{aligned}\left\| \bigotimes_{i=1}^{N} K_i^{\frac{1}{2}} - \bigotimes_{i=1}^{N} L_i^{\frac{1}{2}} \right\| &\le \|\rho|_{M_1} - \rho'|_{M_1}\| \quad \text{by Proposition 2.5} \\ &\le \|\rho - \rho'\| < 2 \quad \text{by the above.}\end{aligned}$$

Hence

$$\prod_{i=1}^{N} \operatorname{tr}(K_i^{\frac{1}{2}} L_i^{\frac{1}{2}}) = \left\langle \bigotimes_{i=1}^{N} K_i^{\frac{1}{2}}, \bigotimes_{i=1}^{N} L_i^{\frac{1}{2}} \right\rangle \ge (1 - \|\rho - \rho'\|/2) > 0$$

and so $\sum_{i=1}^{\infty}(1 - \operatorname{tr} K_i^{1/2} L_i^{1/2}) < \infty$. Since unitary equivalence implies quasi-equivalence, the remainder is clear. □

Corollary 6.6 *If* $\{K_i\}$ *and* $\{u_i\}$ *are sequences of density operators and unitaries in* $\{A_i\}$ *respectively, such that* (6.3.5) *holds, then* $\bigotimes_{i=1}^{\infty} \mathrm{Ad}(u_i)$ *is unitarily implementable in* $\bigotimes_{i=1}^{\infty} \rho_{K_i}$ *if and only if*

$$\sum_{i=1}^{\infty}(1 - \mathrm{tr} K_i^{\frac{1}{2}} u_i K_i^{\frac{1}{2}} u_i^*) < \infty .$$

Proof This follows because $\bigotimes_{i=1}^{\infty} \mathrm{Ad}(u_i)$ is implementable in $\bigotimes_{i=1}^{\infty} \rho_{K_i}$ if and only if $\bigotimes_{i=1}^{\infty} \rho_{K_i}$ is unitarily equivalent to $\bigotimes_{i=1}^{\infty} \rho_{u_i K_i u_i^*}$. □

Theorem 6.7 *If* u_i *is a unitary in* A_i *commuting with a density operator* K_i *in* A_i*, then* $\bigotimes_{i=1}^{\infty} \mathrm{Ad}(u_i)$ *is weakly inner in* $\bigotimes_{i=1}^{\infty} \rho_{K_i}$ *if and only if*

$$\sum_{i=1}^{\infty}(1 - |\mathrm{tr}(u_i K_i)|) < \infty . \tag{6.3.6}$$

Proof If (6.3.6) holds, then by rotating u_i we can assume that $\mathrm{tr}\, u_i K_i \geq 0$, and

$$\sum_{i=1}^{\infty}(1 - \mathrm{tr}\, u_i K_i) < \infty . \tag{6.3.7}$$

Then by Lemma 6.1,

$$u_1 K_1^{\frac{1}{2}} \otimes u_2 K_2^{\frac{1}{2}} \otimes \cdots \otimes u_n K_n^{\frac{1}{2}} \otimes K_{n+1}^{\frac{1}{2}} \otimes K_{n+2}^{\frac{1}{2}} \cdots$$

converges in $\bigotimes_{i=1}^{\infty}(H_i, K_i^{1/2})$ to $\bigotimes_{i=1}^{\infty} u_i K_i^{1/2}$. Now the strong topology on $M = A''$, coincides with the metric topology given by the L^2-norm $\|x\|_2 = \langle x\Omega, x\Omega\rangle$, $\Omega = \bigotimes_{i=1}^{\infty} K_i^{1/2}$, since $\|xy'\Omega\| = \|y'x\Omega\| \leq \|y'\|\|x\Omega\|$ for $y' \in M'$ and Ω is cyclic for M' if $K_i > 0$. Thus (6.3.7) is equivalent to the existence of the strong limit u of $u_1 \otimes u_2 \otimes \cdots u_n \otimes 1 \otimes \cdots$ in M. Similarly, the strong limit of $u_1^* \otimes \cdots u_n^* \otimes 1 \otimes \cdots$ exists, hence u must be unitary, and $\sigma \equiv \bigotimes_{i=1}^{\infty} \mathrm{Ad}\, u_i = \mathrm{Ad}\, u$.

Conversely suppose that there exists a unitary u in M such that $\sigma = \mathrm{Ad}\, u$. We identify $(\bigotimes_{i=1}^{\infty}(A_i, \rho_{K_i}), \bigotimes_{i=1}^{\infty} \mathrm{Ad}\, u_i)$ with

$$\left(\left(\bigotimes_{i=1}^{n} A_i\right) \otimes \left(\bigotimes_{i=n+1}^{\infty}(A_i, \rho_{K_i})\right), \left(\bigotimes_{i=1}^{n} \mathrm{Ad}\, u_i\right) \otimes \left(\bigotimes_{i=n+1}^{\infty} \mathrm{Ad}\,(u_i)\right)\right)$$

and (M, σ, Ω) with $(M_n \otimes M^n, \sigma_n \otimes \sigma^n, \Omega_n \otimes \Omega^n)$ in the obvious notation. And so $1 \otimes \sigma^n(x)w = w(1 \otimes x)$, for some w in M, all x in M^n. Take a slice map $\varphi \otimes 1 : M_n \otimes M^n \to 1 \otimes M^n$ where φ is a state of M^n. Then $\sigma^n(x)v = vx$ for some $v \neq 0$ in M^n, and all x in M^n. Then $v^*\sigma^n(x) = xv^*$, $x \in M^n$, and so $v^*vx = v^*\sigma^n(x)v = xv^*v, x \in M^n$. Thus $v^*v = \alpha 1, \alpha \in \mathbb{R}$, as M^n is a factor. Similarly $vv^* = \beta 1, \beta \in \mathbb{R}$, but $\alpha = \beta$ as $\|v^*v\| = \|vv^*\|$. Hence $v^n = \alpha^{-1/2}v$ is a unitary in M^n such that $\sigma^n = \mathrm{Ad}(v^n)$. Since M is a factor we can by rotating v^n if necessary assume that

$$u = u_1 \otimes \cdots \otimes u_n \otimes v^n .$$

Since u is in $(\bigotimes_{i=1}^{\infty} A_i)''$,

$$\left\langle u \bigotimes_{i=1}^{\infty} K_i^{\frac{1}{2}}, a_1 K_1^{\frac{1}{2}} \otimes \cdots \otimes a_N K_N^{\frac{1}{2}} \otimes K_{N+1}^{\frac{1}{2}} \otimes \cdots \right\rangle > 0$$

for some $a_1, \ldots, a_N$. Hence for $n \geq N$:

$$\prod_{i=1}^{N} \operatorname{tr}(a_i^* u_i K_i) \prod_{i=N+1}^{n} \operatorname{tr}(u_i K_i) \langle v^n \Omega^n, \Omega^n \rangle > 0$$

and so $\prod_{i=N+1}^{n} \operatorname{tr} u_i K_i$ is convergent and $\sum_{i=1}^{\infty} 1 - |\operatorname{tr}(u_i K_i)| < \infty$. □

6.4 Basics of the Fermion algebra

If H is a complex Hilbert space, consider the C^*-algebra A generated by the range of a conjugate linear map a satisfying the *canonical anti-commutation relations (CAR)* :

$$a(f)a(g) + a(g)a(f) = 0 \tag{6.4.1}$$

$$a(f)a^*(g) + a^*(g)a(f) = \langle g, f \rangle 1 \tag{6.4.2}$$

for $f, g \in H$, where $a^*(g) = a(g)^*$. Since $a(f)a(f)^* \leq \|f\|^2$ we have $\|a(f)\| \leq \|f\|$ and a is automatically bounded. One particular representation (Chapter 4, Notes) is as follows on the anti-symmetric or Fermion Fock space $F(H) = \bigoplus_{n=1}^{\infty} \wedge^n H$:

$$a^*(f) g_1 \wedge \cdots \wedge g_n = f \wedge g_1 \wedge \cdots \wedge g_n , \quad f, g_i \in H .$$

We claim that A is uniquely determined by the relations (6.4.1) and (6.4.2) and is isomorphic to $\bigotimes^{\dim H} M_2$. Note that Fermion Fock space has dimension $\sum_n C_n^{\dim H} = 2^{\dim H}$, so that if H is finite dimensional, the Fock representation would automatically be irreducible. This is also the case when H is infinite dimensional, see (6.6.4)

First suppose $H = \mathbb{C}$, then A is generated by a single operator $c = c(f)$, $\|f\| = 1$, with $c^*c + cc^* = 1$, $c^2 = c^{*2} = 0$. Define $e_{ij} \in A$ by

$$\begin{pmatrix} e_{11} & e_{12} \\ e_{21} & e_{22} \end{pmatrix} = \begin{pmatrix} cc^* & c \\ c^* & c^*c \end{pmatrix} .$$

Then $\{e_{ij}\}$ are $*$-matrix units for A, the $*$-algebra generated by c, and $A = \operatorname{lin}\{e_{ij}\}$. Hence by Lemma 2.2, $A \cong M_2$. In general, if $\{f_i\}$ are unit vectors in H, then each $c_i = c(f_i)$ generates a 2×2 matrix algebra, which however do not pairwise commute even when f_i are orthogonal (as $c_i c_j = -c_j c_i, i \neq j$). To obtain commuting families of 2×2 matrix algebras, we proceed as follows.

If u is a unitary on H, then $F(u) = \bigoplus_{r=0}^{\infty} \wedge^r u$ is a unitary on Fock space such that

$$F(u)c(f)F(u)^* = c(uf), \quad f \in H. \tag{6.4.3}$$

Taking $u_t = e^{it}, t \in \mathbb{R}$,

$$F(u_t) = 1 \oplus e^{it} \oplus e^{i2t} \oplus e^{i3t} \oplus \cdots = \exp iNt, \tag{6.4.4}$$

if $N = 0 \oplus 1 \oplus 2 \oplus \cdots$ the *number operator*, $N|_{\wedge^r} = r$.

Thus $e^{iNt}c^*(f)e^{-iNt} = e^{it}c^*(f)$, and so differentiating $Nc^*(f) - c^*(f)N = c^*(f)$ with suitable interpretation of the left hand side as N is unbounded if dim H is infinite. Putting $t = \pi$, $V = F(-1) = \exp iN\pi = \bigoplus_r (-1)^r$. Then V is a self adjoint unitary anti-commuting with $c^*(f)$. We claim that $N = \sum_i c_i^* c_i$, where $c_i = c(f_i)$ if f_i is a complete orthonormal sequence for H. Let $N_0 = \sum c_i^* c_i$. Then

$$N_0 c^*(f) - c^*(f) N_0 = c^*(f)$$

as

$$\begin{aligned}\sum_i c^*(f_i)c(f_i)c^*(f) &= -\sum_i c^*(f_i)c^*(f)c(f_i) + \sum_i c^*(f_i)\langle f, f_i\rangle \\ &= c^*(f)N_0 + c^*(f)\end{aligned}$$

at least on finite linear combinations of anti-symmetric tensors. Thus

$$\begin{aligned}N_0 g_1 \wedge \cdots \wedge g_r &= N_0 c^*(g_1)c^*(g_2)\cdots c^*(g_r)\Omega \\ &= c^*(g_1)N_0 c^*(g_2)\cdots c^*(g_r)\Omega + c^*(g_1)c^*(g_2)\cdots c^*(g_r)\Omega \\ &= r g_1 \wedge \cdots \wedge g_r \quad \text{inductively}.\end{aligned}$$

Consequently $N = N_0$. Taking H finite dimensional,

$$\begin{aligned}e^{iN\pi} = \exp\left(i\pi \textstyle\sum_j c_j^* c_j\right) &= \textstyle\prod_j \exp(i\pi c_j^* c_j) \quad (\text{as } \{c_j^* c_j\} \text{ mutually commute}) \\ &= \textstyle\prod_j (1 - 2c_j^* c_j)\end{aligned}$$

anti-commutes with each c_j (or all $c(f), f \in H$). We are now in a position to show

Proposition 6.8 *Let H be a Hilbert space, and let c, c' be representations of the CAR over H generating $A(H), A'(H)$ C^*-algebras respectively. Then there exists a (necessarily unique) $*$-isomorphism $\beta : A(H) \to A'(H)$ such that $\beta(c(h)) = c'(h), h \in H$. Moreover, $A(H)$ is $*$-isomorphic to $\bigotimes^{\dim H} M_2$.*

Proof Let $\{f_i\}$ be a complete orthonormal sequence for H, $c_j = c(f_j)$, $u_j = 1 - 2c_j^* c_j$. Then u_j is a self adjoint unitary commuting with $c_k, k \neq j$ and anti commuting with c_j. Thus

$$\begin{pmatrix} c_1c_1^* & c_1 \\ c_1^* & c_1^*c_1 \end{pmatrix}, \quad \begin{pmatrix} c_2c_2^* & u_1c_2 \\ c_2^*u_1 & c_2^*c_2 \end{pmatrix}$$

are two commuting families of matrix units. Proceeding in this way $v_i = u_1 \cdots u_i$ anti-commutes with c_j, $j \leq i$, and commutes with c_j, $j > i$. Then

$$(e_{pq}^r) = \begin{pmatrix} c_rc_r^* & v_{r-1}c_r \\ c_r^*v_{r-1} & c_r^*c_r \end{pmatrix} \quad r = 1, 2, \ldots \tag{6.4.5}$$

are commuting families of 2×2 matrices. Thus

$$\begin{aligned} e_{11}^r &= c_rc_r^* & e_{12}^r &= \prod_{i=1}^{r-1}(1 - 2c_i^*c_i)c_r \\ e_{21}^r &= \prod_{i=1}^{r-1}(1 - 2c_i^*c_i)c_r^* & e_{22}^r &= c_r^*c_r \end{aligned} \tag{6.4.6}$$

$$c_r = \prod_{i=1}^{r-1}(1 - 2e_{11}^r)e_{12}^r\,. \tag{6.4.7}$$

This means that the $*$-algebra generated by $\{c_j \mid j \leq m\}$ coincides with the $*$-algebra generated by $\{e_{pq}^r \mid r \leq m, p, q = 1, 2\}$ which is isomorphic to $\otimes^m M_2$. By the boundedness of a, this is enough to show the Proposition. □

We let $A(H) = A^F(H) = A^{\text{CAR}}(H)$ denote *the* C^*-algebra generated by the range of a conjugate linear map c satisfying (6.4.1) and (6.4.2), which is unique in the sense of the above proposition. Note that what we have actually done is the following. Suppose $H = K \oplus L$ where K is finite dimensional. Then if c^1, c^2 are representations of the CAR over K, L respectively, $N = \sum_j (c_j^1)^* c_j^1$ the number operator for K, $V = e^{i\pi N} \in A(K)$, then

$$f \oplus g \to c^1(f) \otimes 1 + V \otimes c^2(g), \quad f \in K, g \in L \tag{6.4.8}$$

is a representation of the CAR on $K \oplus L$, so that $A(K \oplus L) \cong A(K) \otimes A(L)$. Proceeding inductively in this way taking K to be one-dimensional, we obtain $A(\mathbb{C} \oplus \mathbb{C}^n) \cong A(\mathbb{C}) \otimes A(\mathbb{C}^n) \cong \otimes^{n+1} M_2$ and the formulae (6.4.5).

Let J be any complex conjugation on H (i.e. a conjugate linear unitary involution $J^2 = 1, Ji = -iJ$, $\langle Jf, Jg \rangle = \langle g, f \rangle$ $f, g \in H$). Then $f \to c^*(Jf)$ is a representation of the canonical anticommutation relations over H, if c is. More generally, if U is a linear operator on H, V conjugate linear on H with

$$V^*U + U^*V = 0\,, \quad U^*U + V^*V = 1\,, \tag{6.4.9}$$

then $f \to c(Uf) + c^*(Vf)$ is a representation of the canonical anticommutation relations on H. To handle annihilation and creation operators on an equal footing it is convenient to introduce the self dual formalism. Let $K = H \oplus H$, regarded as a complex Hilbert space. Then for J a complex conjugation on H,

$$\Gamma(f \oplus g) = Jg \oplus Jf \quad f, g \in H \tag{6.4.10}$$

is a complex conjugation on K. If c is a representation of the CARs over H, then

$$B(f \oplus g) = c^*(f) + c(Jg), \quad f \oplus g \in K \tag{6.4.11}$$

defines a complex linear map B on K satisfying

$$\begin{aligned} B(f)B(g)^* + B(g)^*B(f) &= \langle f, g \rangle \\ B(f)^* &= B(\Gamma f) \quad f, g \in K. \end{aligned} \tag{6.4.12}$$

The *self dual CAR algebra* $A^F(K, \Gamma) = A^{\mathrm{SDC}}(K, \Gamma)$ is the unital C^*-algebra generated by the range of a complex linear map B on a Hilbert space K, with complex conjugation Γ satisfying the *self dual canonical anti-commutation relations* (6.4.12).

A *basis projection* P on K is a projection P on K satisfying $\Gamma P \Gamma = 1 - P$. [For example, if $K = H \oplus H$, $\Gamma = \begin{pmatrix} 0 & J \\ J & 0 \end{pmatrix}$, then $P = \begin{pmatrix} 1 & 0 \\ 0 & 0 \end{pmatrix}$ is a basis projection.] Then we can identify $B(f)$, $B(\Gamma f)$, $f \in PK$ with creation and annihilation operators $c^*(f), c(f)$ respectively in the CAR algebra $A(PK)$ over PK, so that

$$A^{\mathrm{SDC}}(K, \Gamma) \cong A^{\mathrm{CAR}}(PK) \cong \otimes^{\dim K/2} M_2 . \tag{6.4.13}$$

Remark Note that a basis projection exists if and only if K is even dimensional (i.e. either finite dimensional and even, or infinite dimensional). First take an orthonormal basis for K, fixed by Γ. [If $\{e_1, \ldots, e_k\}$ is an orthonormal set in K, fixed by Γ, and $f \in \{e_1, e_2, \ldots, e_k\}^\perp$ is non-zero, then either $f + \Gamma f$ or $i(f - \Gamma f)$ is non-zero and orthogonal to $e_1, \ldots, e_k$. So in finite dimensions the claim is clear.] Then if $\{e_1, e_2, \ldots, e_{2n}\}, n \leq \infty$ is a Γ-fixed, $\Gamma e_j = e_j$ orthonormal basis for K, let P be the projection on $\overline{\mathrm{lin}}\{e_j + ie_j \mid j\}$, so that $1 - P$ is the projection on $\overline{\mathrm{lin}}\{\Gamma(e_j + ie_j) = e_j - ie_j \mid j\}$.

Take an orthonormal basis $\{e_i\}$ for K fixed by Γ. Under the identification of $A^{\mathrm{SDC}}(\mathbb{C}e_1 + \mathbb{C}e_2)$ with $A^{\mathrm{CAR}}(P(\mathbb{C}e_1 + \mathbb{C}e_2))$ with M_2 using (6.4.5) or (6.4.6), then

$$B(e_1) = \frac{1}{\sqrt{2}} \begin{pmatrix} 0 & 1 \\ 1 & 0 \end{pmatrix}, \quad B(e_2) = \frac{1}{\sqrt{2}} \begin{pmatrix} 0 & i \\ -i & 0 \end{pmatrix}$$

so that in particular $A(\mathbb{C}e_1) \cong \mathbb{C}^2$. Then, from the analysis of (6.4.5), we see that if $K_n = \mathrm{lin}\{e_1, \ldots, e_n\}$,

$$A(K_{2n}) \cong A(K_{2n-2}) \otimes A(\mathbb{C}e_{2n-1} + \mathbb{C}e_{2n}) .$$

Now $A(\mathbb{C}e_{2n-1} + \mathbb{C}e_{2n}) \cong M_2$, and the subalgebra generated by a Γ invariant vector is a commutative algebra isomorphic to $\mathbb{C}^2$. In this way, we see that

$$A(K_{2n}) \cong \otimes^n M_2 \cong M_{2^n} \tag{6.4.14}$$

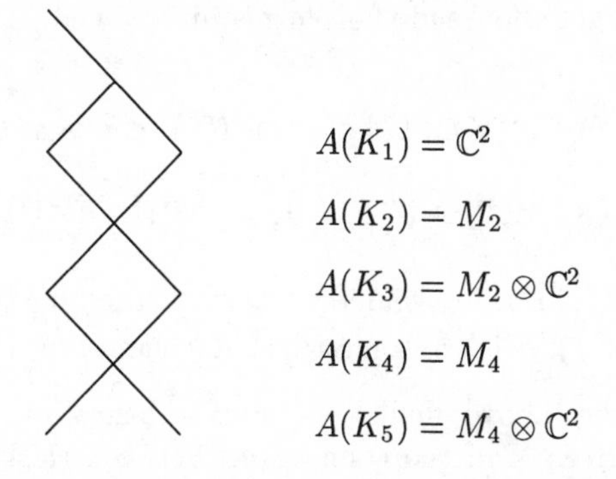

FIG. 6.1. Bratteli diagram for $A(K_n) \subset A(K_{n+1})$

$$A(K_{2n+1}) \cong (\otimes^n M_2) \oplus (\otimes^n M_2) \cong M_{2^n} \oplus M_{2^n} \tag{6.4.15}$$

and the Bratteli diagram of the inclusions $A(K_1) \subset A(K_2) \subset \cdots$ is given by Fig. 6.1. Thus if K is odd dimensional, we should take $A(K)$ to be the universal algebra generated by the self dual anti-commutation relations. If K is even dimensional $A(K)$, the C^*-algebra generated by the self dual CARs is simple, and is uniquely determined by those relations up to isomorphism.

Exercise 6.3 If Γ is a conjugation on a Hilbert space K and $h \in K$, show that

$$\|B(h)\| = 2^{-\frac{1}{2}} \left[\|h\|^2 + \left(\|h\|^4 - |\langle h, \Gamma h \rangle|^2 \right)^{\frac{1}{2}} \right]^{\frac{1}{2}}.$$

Exercise 6.4 Let A be the $*$-algebra generated by a map B on an odd dimensional Hilbert space with orthonormal basis of Γ-invariant vectors $e_1, \ldots, e_{2n+1}$ where Γ is a conjugation. Show that $z = B(e_1) \cdots B(e_{2n+1})$ generates the centre of A and $z^* = (-1)^n z$, $z^* z = 2^{-2n+1}$.

Exercise 6.5 Let L be a real inner product space. The *Clifford algebra* Cliff(L) is the (universal) C^*-algebra generated by the range of a real linear map γ satisfying $\gamma(h)^* = \gamma(h)$, $\gamma(h)^2 = \|h\|^2$, $h \in L$. Let H be a complex Hilbert space, and L be H viewed as a real inner product space (using Re $\langle\, , \rangle$) with multiplication by i on L as an orthogonal (i.e. isometric) transformation of square -1, also denoted by i, i.e. i is a *complex structure* on L. Let $K = H \oplus H$, $P = 1 \oplus 0$, and Γ a conjugation on K such that $\Gamma P \Gamma = 1 - P$. Identify H with PK, in the natural way. Show that Cliff$(L) \cong A^{\mathrm{CAR}}(H) \cong A^{\mathrm{SDC}}(K, \Gamma)$ under the identifications:

$$\begin{aligned}
a(f) &= (\gamma(f) + i\gamma(if))/\sqrt{2} = B(\Gamma f) \quad f \in H = PK \\
\gamma(f) &= (a(f) + a^*(f))/\sqrt{2} = (B(f) + B(\Gamma f))/\sqrt{2}, \quad f \in H = PK \\
B(f) &= a^*(Pf) + a(P\Gamma f) \\
&= (\gamma(P(f + \Gamma f)) + i\gamma(iP(-f + \Gamma f))/\sqrt{2} \quad f \in K.
\end{aligned}$$

Exercise 6.6 Bogoliubov endomorphisms are defined in the CAR, self dual, Clifford formalism by

(a) $c(f) \to c(Uf) + c^*(Vf)$, U linear on H, V conjugate linear on H,

$$V^*U + U^*V = 0\,, \qquad U^*U + V^*V = 1\,.$$

(b) $\gamma(f) \to \gamma(Tf)$, T is an isometry transformation on L.
(c) $B(f) \to B(Sf)$, S is an isometry on K commuting with Γ.

Show that the above define $*$-endomorphisms on $A^{\mathrm{CAR}}(H)$, Cliff(L), $A^{\mathrm{SDC}}(K)$ respectively and the connections between (U, V), T and S defining the same endomorphisms are

$$\begin{aligned} U &= \text{ complex linear part of } T = (T - iTi)/2 = PSP \\ V &= \text{ conjugate linear part of } T = (T + iTi)/2 = PS\Gamma P \\ T &= U + V = P(S + \Gamma)P \\ S &= U + V\Gamma + \Gamma V + \Gamma U \Gamma = (1 - J)(T + T\Gamma + \Gamma T + \Gamma T \Gamma)/2 \end{aligned}$$

We will usually write $\tau(S)$ for the corresponding *Bogoliubov endomorphism*, or $\alpha(U)$ in the case $V = 0$, (so that U is an isometry).

6.5 The Pauli and Fermion algebras as graded algebras

A *graded algebra* is an algebra A with a symmetry θ (a period two automorphism). The eigenspaces of θ split the algebra into even and odd parts $A_\pm = \{x \in A \mid \theta(x) = \pm x\}$, so that $A = A_+ + A_-$. As C^*-algebras, the Pauli algebra $A^P = \bigotimes_{\mathbb{Z}} M_2$ and the Fermion algebra $A^F = A^{\mathrm{CAR}}(\ell^2(\mathbb{Z}))$ are isomorphic. However, we are not merely interested in these objects as C^*-algebras, but also in their quasi-local structure, and gradings. In fact, these two algebras are better considered as distinct C^*-subalgebras of a larger C^*-algebra $\hat{A}$. Before we examine this, we first introduce a little more notation.

If $1 \le L \le \infty$ let I_L denote the intervals of integers $[-L, L]$, and $A_L^P = \bigotimes_{I_L} M_2$, $A_L^F = A^{\mathrm{CAR}}(\ell^2(I_L))$ so that $A_L^P \subset A_{L+1}^P$, $A_L^F \subset A_{L+1}^F$, $A^P = \overline{\bigcup A_L^P}$, $A^F = \overline{\bigcup A_L^F}$.

The matrix algebra M_2 is generated by the *Pauli matrices* (cf. (7.10.2)):

$$\sigma_x = \begin{pmatrix} 1 & 0 \\ 0 & -1 \end{pmatrix}, \qquad \sigma_y = \begin{pmatrix} 0 & i \\ -i & 0 \end{pmatrix}, \qquad \sigma_z = \begin{pmatrix} 0 & 1 \\ 1 & 0 \end{pmatrix},$$

so that σ_α are self adjoint unitaries $\sigma_\alpha^2 = 1$, and $\sigma_x \sigma_y = i\sigma_z$ (or $\sigma_\alpha \sigma_\beta = i\sigma_\gamma$ if (α, β, γ) is any cyclic permutation of (x, y, z)). Then

$$\begin{pmatrix} e_{11} & e_{12} \\ e_{21} & e_{22} \end{pmatrix} = \frac{1}{2} \begin{pmatrix} 1 + \sigma_x & \sigma_z - i\sigma_y \\ \sigma_z + i\sigma_y & 1 - \sigma_x \end{pmatrix}$$

are matrix units and so (cf. Lemma 2.2) A_L^P is the unique C^*-algebra generated by self adjoint unitaries $\sigma_x^m, \sigma_y^m, \sigma_z^m$, $m \in I$, which satisfy the mixed commutation relations $\sigma_x^m \sigma_y^m = i\sigma_z^m$, $\sigma_\alpha^n \sigma_\beta^m = \sigma_\beta^m \sigma_\alpha^n$ $n \neq m$.

The *-algebras A_L^P and A_L^F are all graded, the gradings being given by period two $*$-automorphisms, which with abuse of notation we all denote by θ:

Grading in the Pauli algebra A_L^P : $\theta(\sigma_z^i) = \sigma_z^i$, $\theta(\sigma_x^i) = -\sigma_x^i$, $\theta(\sigma_y^i) = -\sigma_y^i$, $i \in I_L$ (i.e. a rotation through π of all spins around the z axis). If $L < \infty$, θ is inner, in fact $\theta = \mathrm{Ad}(u)$, where $u = \prod_{i \in I_L} \sigma_z^i$, but if $L = \infty$, $\bigotimes_{\mathbb{Z}} \mathrm{Ad}(\sigma_z^i)$ is outer by Corollary 6.4. Then A_{L+}^P is generated by σ_z^i, $-L \leq i \leq L$ and $\sigma_x^i \sigma_x^{i+1}$, $-L \leq i < L$ so that $A_+^P = \overline{\bigcup A_{L+}^P}$.

Grading in the Fermion algebra A_L^F: Here θ is the quasi-free automorphism given by the operator -1 on the one-particle space $\ell^2(I_L)$. Thus $\theta(c(f)) = -c(f)$, $f \in \ell^2(I_L)$ or $\theta(c_i) = -c_i$, $i \in I_L$.

The Jordan–Wigner transformation is an isomorphism $\alpha_L : A_L^P \to A_L^F$ $L < \infty$ constructed in Section 6.4, which identifies these two algebras:

$$\sigma_z^j = 2c_j^* c_j - 1, \quad \sigma_x^j = TS_j(c_j + c_j^*), \quad \sigma_y^j = TS_j i(c_j - c_j^*) \tag{6.5.1}$$

where $T = \prod_{k=-L}^{0} \sigma_z^k$ and $TS_j = \prod_{k=-L}^{j-1} \sigma_z^k$. Then $\alpha_L \theta = \theta \alpha_L$, so that we can identify A_{L+}^P with A_{L+}^F using α_L. Since the tail T depends on L, it is clear that the identifications of A_L^P with A_L^F, and A_{L+1}^P with A_{L+1}^F are *not compatible* with the embeddings $A_L^F \subset A_{L+1}^F$ and $A_L^P \subset A_{L+1}^P$, i.e. Fig. 6.2(a) is *not* commutative. However, since $T^2 = 1$, the tail T disappears in a product $\sigma_x^j \sigma_x^{j+1}$. In fact, under the identification α_L,

$$\sigma_x^j \sigma_x^{j+1} = (c_j - c_j^*)(c_{j+1} + c_{j+1}^*) \tag{6.5.2}$$

Thus since A_{L+}^P is generated by σ_z^i and $\sigma_x^j \sigma^{j+1}$ it is clear that the restriction of α_L to the even subalgebra is *not* L-dependent, i.e. the diagram in Fig. 6.2(a) is commutative when restricted to even algebras. Then $\beta = \lim_{L \to \infty} \alpha_L|_{A_{L+}^P}$ gives an isomorphism of A_+^P with A_+^F, which preserves some of the quasi-local structure, in that it at least identifies A_{L+}^P with A_{L+}^F.

However, for a one sided lattice, the situation is rather different. In that case, the corresponding Jordan–Wigner isomorphism $\gamma_L : \bigotimes_1^L M_2 \to A^{\mathrm{CAR}}(\ell^2[1, l])$ is canonical, in the sense that Fig. 6.2(b) is commutative. This is because in this situation, of a one-sided lattice, the corresponding tail $TS_j = \prod_{k=1}^{j-1} \sigma_z^k$ is not L-dependent. Thus $\gamma = \lim_{L \to \infty} \gamma_L$ gives an isomorphism of the one-sided Pauli algebra $\bigotimes_1^\infty M_2$ with $A^{\mathrm{CAR}}(\ell^2[1, \infty])$, which identifies $\bigotimes_1^L M_2$ with $A^{\mathrm{CAR}}(\ell^2[1, L])$.

In the two-sided case, we need to make sense of $\prod_{k=-\infty}^{j-1} \sigma_z^k$ or $\prod_{k=-\infty}^{0} \sigma_z^k$. This product does not converge in the algebra A^P. However, to study the self adjoint unitary $u_- = \prod_{k=-L}^{0} \sigma_z^k$ if $L < \infty$, it is almost the same to consider the

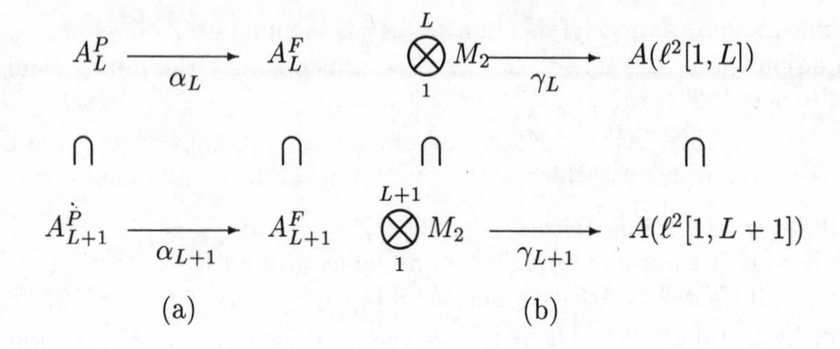

FIG. 6.2. Jordan–Wigner transformation

period two inner automorphism $\theta_- = \mathrm{Ad}(u_-)$ which it defines on A_L^P. Then θ_- is rather similar to θ except that it reverses the spins σ_x and σ_y on only one side:

$$\theta_-(\sigma_z^j) = \sigma_z^j \,,\, \theta_-(\sigma_\alpha^j) = \begin{cases} \sigma_\alpha^j & j \geq 1 \\ -\sigma_\alpha^j & j < 1 \end{cases} ,j \in I_L, \alpha = x, y \,. \tag{6.5.3}$$

Now we can define θ_- on $A_\infty^P = A^P$ using the formulae (6.5.3) for all $j \in \mathbb{Z}$. The only difference is that now θ_- is outer on A^P. In fact $\theta_- = \bigotimes_{-\infty}^{\infty} \mathrm{Ad}(u_j)$ where $u_j = \sigma^z =$ if $j < 1$, and $u_j = 1$ if $j \geq 1$. Similarly, we define the outer automorphism θ_- on A^F by $\theta_-(c_j) = c_j$ if $j \geq 1$, $-c_j$ if $j < 1$, the quasi-free automorphism induced by the unitary $u_- = (-1) \oplus 1$ on $\ell^2(\mathbb{Z}) = \ell^2(-\infty, 0] \oplus \ell^2[1, \infty)$.

We construct the crossed product $\hat{A} = A^F \rtimes_{\theta_-} (\mathbb{Z}/2)$ which is generated by A^F and a self adjoint unitary $T \in \hat{A}$ satisfying

$$T^2 = 1\,, \quad T^* = T\,, \quad \theta_-(a) = TaT\,, \quad a \in A^F\,. \tag{6.5.4}$$

Since θ_- is outer, $\hat{A}$ is uniquely determined by these relations (6.5.4) (see the Remark on page 225). Then using (6.5.1) the Pauli spin algebra A^P can be identified with the C^*-subalgebra of $\hat{A}$ generated by

$$\sigma_z^j = 2c_j^* c_j - 1\,, \quad \sigma_x^j = TS_j(c_j + c_j^*)\,, \quad \sigma_y^j = TS_j i(c_j - c_j^*) \tag{6.5.5}$$

where

$$S_j = \begin{cases} \prod_{k=1}^{j-1} \sigma_z^k & \text{if } j > 1 \\ 1 & \text{if } j = 1 \\ \prod_{k=j}^{0} \sigma_z^k & \text{if } j < 1 \end{cases} . \tag{6.5.6}$$

Then we have Fig. 6.3. In this way the even and odd spaces of A^P and A^F are related by

$$\begin{array}{ccc} \hat{A} & = & A^F \rtimes \mathbb{Z}/2 \\ \cup & & \cup \\ A^P & & A^F \\ \cup & & \cup \\ A^P_+ & = & A^F_+ \end{array}$$

FIG. 6.3. $A^P \rtimes \mathbb{Z}/2 = A^F \rtimes \mathbb{Z}/2$

$$A^P_+ = A_+ , \quad A^P_- = TA_- \tag{6.5.7}$$

so that

$$A^F = A_+ + A_- , \quad A^P = A_+ + TA_- \tag{6.5.8}$$

where we write $A_\pm = A^F_\pm$. The automorphisms θ and θ_- on A^P and A^F extend consistently to automorphisms θ and θ_- on $\hat{A}$ by $\theta(T) = \theta_-(T) = T$. Note also that using (6.5.1) it is easy to see that we can identify $A^F \rtimes_{\theta_-} (\mathbb{Z}/2)$ and $A^P \rtimes_{\theta_-} (\mathbb{Z}/2)$.

An even state on A^P or A^F is one which is invariant under θ. There is a bijective affine correspondence between even states φ on A^P and even states $\hat{\varphi}$ on A^F. We will construct even states φ_β on A^P using the fact that the Ising Hamiltonian will be even, and it will turn out that the corresponding even states $\hat{\varphi}_\beta$ on A^F are quasi-free. We will use the embeddings of Fig. 6.3 to study properties of φ_β in terms of the corresponding quasi-free states on A^F in Chapter 7.

Remark If T is any self adjoint unitary implementing θ_- then $\hat{A} = C^*(A, T) \cong A$. Now $T\sigma^z_j \notin A_j \equiv C^*(\sigma^j_\alpha) \cong M_2$ as θ_- is outer and so is a non-trivial central self adjoint unitary in $C^*(A_j, T)$, with spectral projections $E_j, 1 - E_j$, for $j < 0$. Then $C^*(A_j, T) = A_j E_j \oplus A_j(1 - E_j) \cong M_2 \oplus M_2$. In this way $C^*(A_{-j} \otimes \cdots \otimes A_{-1}, T) \cong M_{2^j} \oplus M_{2^j}$ with embeddings given by the Bratteli diagram of Fig. 6.4. (This is the diagram of the inclusions $A(K_1) \subset A(K_3) \subset A(K_5) \subset \cdots$ of Fig. 6.1). Note that if $A \subset B(H)$, then we can embed $A \subset B(H \oplus H)$ by $a \to a \oplus \theta_-(a)$, and then $T = \begin{pmatrix} 0 & 1 \\ 1 & 0 \end{pmatrix}$ is a self adjoint unitary implementing θ_-, (see (7.11.12), (7.11.13)).

Remark It follows easily from the Jordan–Wigner description that if Γ is a conjugation on a Hilbert space K and E is a Γ-invariant projection, not finite dimensional and odd, then

$$A(K)_+ \cap A((1 - E)K)' = A(K)_+ \cap A((1 - E)K)'_+ = A(EK)_+$$

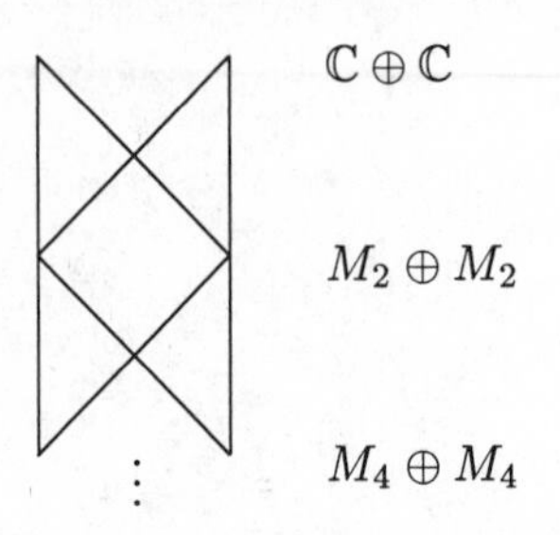

FIG. 6.4. Bratteli diagram for $A(K) \rtimes (\mathbb{Z}/2)$

or $A(K)_+ \cong A(EK)_+ \otimes A(1-E)K)_+$ under the natural inclusions $A(EK)$, $A((1-E)K) \subset A(K)$ (cf. (2.9.15)).

6.6 Quasi-free states

We now introduce the class of quasi-free states on CAR algebras. They can be regarded as analogues of Gaussian distributions in classical probability in that all the n-point functions can be computed once one knows the two-point functions, (see (6.6.3) below).

First, if H is a complex Hilbert space, and φ is any state on the CAR algebra $A(H)$, one can see from the Riesz representation theorem and the anti-commutation relations (6.4.1) and (6.4.2) that there is an unique operator R on H satisfying

$$0 \leq R \leq 1 \tag{6.6.1}$$
$$\varphi(a^*(f)a(g)) = \langle Rf, g \rangle, \quad f, g \in H. \tag{6.6.2}$$

The (gauge-invariant) *quasi-free states* are those which are completely determined from the operator R in the following sense. If R is a positive contraction, then there is an unique state, denoted by ω_R, and called a (gauge-invariant) quasi-free state on $A^{\mathrm{CAR}}(H)$ satisfying

$$\omega_R\,(a^*(f_n)\cdots a^*(f_1)a(g_1)\cdots a(g_m)) = \delta_{mn} \det\left[\langle Rf_i, g_j \rangle\right] \tag{6.6.3}$$

Moreover, ω_R is a pure state (i.e. π_{ω_R} is irreducible) if and only if R is a projection. Purity of ω_0 follows from

$$\omega_0\,(a^*(f_n)\cdots a^*(f_1)a(g_1)\cdots a(g_m)) = 0 \text{ if } n+m > 0\,, \tag{6.6.4}$$

as any state dominated by a multiple of ω_0 also annihilates these monomials, which together with the unit 1 generate the Fermion algebra as a Banach space. In this case, one can identify the GNS Hilbert space of ω_0 with the space of anti-symmetric tensors $\bigoplus_{n=0}^{\infty} \wedge^n H$ if $\wedge^0 H$ is a one dimensional Hilbert space spanned by a unit vector Ω, the cyclic vector Ω_{ω_0} of the GNS triple with Ω, and the representation $\pi_{\omega_0} = \pi_0$ is given by

$$\pi_0(a^*(f))g_1 \wedge \cdots \wedge g_m = f \wedge g_1 \wedge \cdots \wedge g_n \tag{6.6.5}$$

Similarly, in the self dual picture, one has a notion of quasi-free state. Take K to be a complex Hilbert space, with complex conjugation Γ. Then if φ is any normalized state on $A^{\mathrm{SDC}}(K, \Gamma)$ there is an unique operator S on K satisfying:

$$1 \geq S^* = S \geq 0 \tag{6.6.6}$$

$$S + \Gamma S \Gamma = 1 \tag{6.6.7}$$

$$\varphi(B(f)^* B(g)) = \langle Sg, f \rangle , \quad f, g \in K . \tag{6.6.8}$$

To define a quasi-free state on the self dual algebra, it is useful to use the Pfaffian notation. If $I = \{i_1 < \cdots < i_r\}$ is a finite ordered set with cardinality $|I| = r$ we let D_I denote the set of all subsets of I with the induced ordering. If $J, K \in D_I$, $J = \{j_1, \ldots, j_s\}$, $K = \{k_1, \ldots, k_l\}$ with $I = J \sqcup K$ (i.e. $I = J \cup K$, $J \cap K = \varnothing$), let $\varepsilon(J, K)$ denote the signature of the permutation

$$\begin{pmatrix} i_1 \ldots i_r \\ j_1 \ldots j_s k_1 \ldots k_l \end{pmatrix}$$

and $\overline{\varepsilon}(J, K)$ denote the signature of the permutation

$$\begin{pmatrix} i_1 \ldots i_{2n} \\ j_1 k_1 j_2 k_2 \ldots j_n k_n \end{pmatrix}$$

if $|J| = |K| = |I|/2$. In this case

$$\overline{\varepsilon}(J, K) = (-1)^{n(n-1)/2} \varepsilon(J, K)$$

If $a_{ij} \in \mathbb{C}$ for $i, j \in I$, with $|I| = 2n$ even, let the *Pfaffian* $\mathrm{Pf}[a_{ij}]$ be

$$\mathrm{Pf}[a_{ij}] = \sum \overline{\varepsilon}(J, K) a_{j_1 k_1} a_{j_2 k_2} \cdots a_{j_n k_n}$$

where the summation is over all disjoint J, K in D_I with $J \cup K = I$, $J = \{j_1, \ldots, j_n\}$, $K = \{k_1, \ldots, k_n\}$ and $j_m < k_m$, $m = 1, \ldots, n$. Moreover we let $\mathrm{Pf}[a_{ij}] = 1$ if $I = \varnothing$, and $\mathrm{Pf}[a_{ij}] = 0$ if $|I|$ odd.

Let $Q(K, \Gamma)$ denote the bounded linear operators on K satisfying (6.6.6) and (6.6.7). If $S \in Q(K, \Gamma)$, there is an unique state on $A^{\mathrm{SDC}}(K)$, denoted by φ_S and called a *quasi-free state*, satisfying

$$\varphi_S(B(f_1) \cdots B(f_m)) = \mathrm{Pf}[\varphi_S(B(f_i)B(f_j))] \tag{6.6.9}$$

and

$$\varphi_S(B(f)^*B(g)) = \langle Sg, f\rangle\,. \tag{6.6.10}$$

Moreover φ_S is pure if and only if S is a projection (in fact a basis projection by (6.6.6) and (6.6.7)).

If H is a complex Hilbert space, with conjugation γ, and $K = H \oplus H$ with conjugation $\Gamma = \begin{pmatrix} 0 & \gamma \\ \gamma & 0 \end{pmatrix}$, and R a positive contraction on H, then $S = (1-R) \oplus \gamma R \gamma$ is in $Q(K,\Gamma)$. Under the identification of $A^{\mathrm{CAR}}(H)$ with $A^{\mathrm{SDC}}(K,\Gamma)$, ω_R corresponds to φ_S.

The existence of φ_S is seen as follows. First if S is a positive contraction on K, let P_S denote the operator on $K \oplus K$ given by the matrix

$$P_S = \begin{pmatrix} S & S^{\frac{1}{2}}(1-S)^{\frac{1}{2}} \\ S^{\frac{1}{2}}(1-S)^{\frac{1}{2}} & 1-S \end{pmatrix} \tag{6.6.11}$$

If S is in $Q(K,\Gamma)$, then P_S is a basis projection on $(\hat{K},\hat{\Gamma})$ where $\hat{K} = K \oplus K$, $\hat{\Gamma} = \Gamma \oplus (-\Gamma)$. Then we know how to define the Fock state φ_{P_S} on $A^{\mathrm{SDC}}(\hat{K},\hat{\Gamma})$ by (6.6.5) on $A^{\mathrm{CAR}}(P_S\hat{K})$. Then the restriction of φ_{P_S} to $A^{\mathrm{SDC}}(K,\Gamma)$ is the quasi-free state φ_S. Here we give an alternative constructive proof of this existence which simultaneously provides us with a $\mathbb{N}$-grading of the GNS Hilbert space into n-particle spaces for $n = 0, 1, 2, \ldots$.

First, in the complex formalism, if H is a Hilbert space the monomials

$$\{a(g_n)\cdots a(g_1)a^*(f_1)\cdots a^*(f_m) \mid f_i, g_j \in H\} \tag{6.6.12}$$

$$\{a^*(f_m)\cdots a^*(f_1)a(g_1)\cdots a(g_n) \mid f_i, g_i \in H\} \tag{6.6.13}$$

as n, m vary separately span a dense subspace of the Fermion algebra. The Fock state ω_0 and *anti-Fock state* ω_1 have particularly simple expressions on the appropriate monomials:

$$\omega_0(a^*(f_m)\cdots a^*(f_1)a(g_1)\cdots a(g_n)) = \delta_{nm}\delta_{n0} \tag{6.6.14}$$

$$\omega_1(a(g_n)\cdots a(g_1)a^*(f_1)\cdots a^*(f_m)) = \delta_{nm}\delta_{n0} \tag{6.6.15}$$

although ω_0, ω_1 on the inappropriate monomial is not so simple (cf. (6.6.3)) and

$$\omega_0(a(g_n)\cdots a(g_1)a^*(f_1)\cdots a^*(f_m)) = \det[\langle f_i, g_j\rangle]\delta_{nm}\,. \tag{6.6.16}$$

To express a monomial with all creation operators on the right as a linear combination of those with all creation operators on the left, take $g_1, \ldots, g_n$ and $f_1, \ldots, f_m$, elements of the Hilbert space H, and then use the CAR to move each creation operator one at a time to the left to obtain

$$\begin{aligned} &a(g_n)\cdots a(g_1)a^*(f_1)\cdots a^*(f_m) \\ &= \sum \varepsilon a^*(f(\sigma'))a(g(\rho'))\ \det\left[\langle f_i, g_j\rangle \mid i \in \sigma, j \in \rho\right] \end{aligned} \tag{6.6.17}$$

where the summation extends over all subsets σ of $\{1,2,\ldots,m\}$, ρ of $\{1,2,\ldots,n\}$ in increasing order, and if σ', respectively ρ' in increasing order is the complement of σ, respectively ρ in $\{1,2,\ldots,m\}$, respectively $\{1,2,\ldots,n\}$, then ε is the product of $(-1)^{(m-p)(n-p)}$ where $|\sigma|=|\rho|=p$, and the signatures of permutations $(1,2,\ldots,m)\to(\sigma,\sigma')$, and $(1,2,\ldots,n)\to(\rho,\rho')$. Note that (6.6.16) follows from this expression (6.6.17).

The Wick ordered product, normally ordered with respect to the quasi-free state ω_R, can then be defined by

$$\begin{aligned} &: a^*(f_1)\ldots a^*(f_m)a(g_n)\ldots a(g_1) :_R \\ &= \sum \varepsilon a^*(f(\sigma'))a(g(\rho')) \det\left[\omega_R(a^*(f_i)a(g_j)) \mid i\in\sigma, j\in\rho\right] \end{aligned} \qquad (6.6.18)$$

where the summation is as described for (6.6.17). Note that (6.6.16) follows from the expressions (6.6.17) and (6.6.14).

We will see in the Remark on page 232 how to normal order $: a^{\#}(f_1)\cdots a^{\#}(f_n) :$ where $a^{\#}$ denotes either a or a^*, but in the ordering $a^{\#}(f_1)\cdots a^{\#}(f_n)$ it is not necessary to have all the creation operators on the left. We will however be guided by (6.6.17) to get a normal ordering in the self dual formalism.

Thus let Γ be a conjugation on a Hilbert space K. Let S be any operator on K. Define an indefinite inner product on K by

$$\langle f,g\rangle_S = \langle Sf,g\rangle\,, \quad f,g\in K\,. \qquad (6.6.19)$$

so that $\langle f,g\rangle_S=\langle\Gamma g,\Gamma f\rangle_{\Gamma S^*\Gamma}$. If $I=\{i_1<\cdots<i_r\}$ is a finite ordered set and $\{f_i \mid i\in I\}\subset K$, we let

$$B(f_I)=B(f_{i_1})\cdots B(f_{i_r})\,, \quad B(f_\varnothing)=1 \qquad (6.6.20)$$

and

$$\varphi_S(B(f_I))=\mathrm{Pf}[\langle f_j,\Gamma f_i\rangle_S \mid i,j\in I]\,. \qquad (6.6.21)$$

Then define the Wick ordered product by

$$: B(f_I) :=: B(f_I) :_S = \textstyle\sum_{J\sqcup K=I}(-1)^{|K|/2}\varepsilon(J,K)B(f_J)\varphi_S(B(f_K)) \qquad (6.6.22)$$

where the summation is over all disjoint J,K in D_I with $J\cup K=I$.

Lemma 6.9 *If S,T are in $Q(K,\Gamma)$:*

(a) $B(f_I)=\sum_{J\sqcup K=I}\varepsilon(J,K) : B(f_J) :_S \varphi_S(B(f_K))$ (6.6.23)

(b) $B(f) : B(f_I) :_S =: B(f)B(f_I) :_S$

$$+\sum_{s=1}^{r}(-1)^{s+1}\varphi_S(B(f)B(f_{i_s})) : B(f_{i_1})\cdots\widehat{B(f_{i_s})}\cdots B(f_{i_r}) :_S \qquad (6.6.24)$$

where $\hat{}$ *over an element means that element is omitted.*

(c) $: B(f_{i_1}) \dots B(f_{i_r}) :_S$ *is an anti – symmetric function of* $(i_1, \dots, i_r)$.

(d) $: B(f_I) :_T = \sum_{J \sqcup K = I} \varepsilon(J, K) : B(f_J) :_S \mathrm{Pf}[\langle f_i, \Gamma f_j \rangle_{T-S} \mid i, j \in K]$.

(6.6.25)

Proof First (b) is shown by examining the definition of $: B(f)B(f_I) :$. Then (a) is shown by induction on $|I|$. If (a) holds for $|I| = n$, multiply this formula by $B(f)$ on the left, and then use (b) to express $B(f) : B(f_J) :$ in terms of Wick products. Next (c) follows inductively from (a), and (d) follows from the definition of $: :_T$, (a) for $: :_S$ and Pfaffian expansions. □

Lemma 6.10 *If* $n \geq 1$, *then* $((f_i)_{i=1}^n, (g_i)_{i=1}^n) \to \det[\langle f_i, g_j \rangle_S]$ *is positive-definite on* $K^n \times K^n$.

Proof If $n = 1$, then $(f, g) \to \langle f, g \rangle_S = \langle S^{1/2} f, S^{1/2} g \rangle$ is clearly positive definite on $K \times K$. It only remains to show that if $A_{ij} \in M_n(\mathbb{C})$ for $i, j = 1, \dots, m$, and $[A_{ij}]$ is positive in $M_m(M_n(\mathbb{C}))$ then $[\det(A_{ij})]$ is positive in $M_n(\mathbb{C})$ (for then consider $(f_r^i)_{r=1}^n \in K^n$, $i = 1, \dots, m$ and $A_{ij} = [\langle f_r^i, f_s^j \rangle]_{r,s=1}^n$). Let $[A_{ij}] = B^* B$, where $B = [B_{ij}] \in M_m(M_n(\mathbb{C}))$. Then, taking n tensor factors:

$$[A_{ij} \otimes \cdots \otimes A_{ij}] = \sum_{r_1, \dots, r_n = 1}^{m} [(B_{r_1 i} \otimes \cdots B_{r_n i})^* (B_{r_1 j} \otimes \cdots \otimes B_{r_n j})] \geq 0 .$$

But $\det A_{ij} = \wedge^n A_{ij}$, and so cutting down to $\wedge^n \mathbb{C}^n$, $[\det(A_{ij})] \geq 0$. □

Let (C_n, F_S^n) denote the minimal Kolmogorov decomposition of the positive definite kernel

$$((f_i), (g_j)) \to \det[\langle f_i, g_j \rangle_S] \quad \text{on } K^n \times K^n ,$$

so that

$$\langle C_n(f_1, \dots, f_n), C_n(g_1, \dots, g_n) \rangle = \det[\langle f_i, g_j \rangle_S] . \qquad (6.6.26)$$

Then $C_n(f_1, \dots, f_n)$ is an anti-symmetric function of $f_1, \dots, f_n$. Let F_S^n denote the linear span of $\{C_n(f_1, \dots, f_n) \mid f_i \in K\}$. Define $F_S = \bigoplus_{n=0}^{\infty} F_S^n$ where F_S^0 is a one-dimensional Hilbert space spanned by a unit vector $\Omega = \Omega_S$. If $f \in K$ define

$$\hat{C}_n(f, f_1, \dots, f_n) = C_{n+1}(f, f_1, \dots, f_n)$$

$$+ \sum_{i=1}^{n} (-1)^{i+1} \langle f_i, \Gamma f \rangle_S C_{n-1}(f_1, \dots, \hat{f}_i, \dots, f_n).$$

Then easy computations show that

$$\langle \hat{C}_{n+1}(f, f_1, \ldots, f_n), C_m(g_1, \ldots, g_m)\rangle$$
$$= \langle C_n(f_1, \ldots, f_n), \hat{C}_{m+1}(\Gamma f, g_1, \ldots, g_m)\rangle$$

for all $f_1, \ldots, f_n, g_1, \ldots, g_m, n, m$. Hence there is a well defined operator $\pi_0(f) : F_S \to F_S$ such that

$$\pi_0(f)C_n(f_1, \ldots, f_n) = \hat{C}_n(f, f_1, \ldots, f_n) \tag{6.6.27}$$

and $\pi_0(f)^* \supset \pi_0(\Gamma f)$. Moreover

$$\pi_0(g)\pi_0(f)C_n(f_1, \ldots, f_n) = \pi_0(g)C_{n+1}(f, f_1, \ldots, f_n)$$
$$+ \sum_{i=1}^{n} (-1)^{i+1}\langle f_i, \Gamma f\rangle_S \pi_0(g)C_{n-1}(f_1, \ldots, \hat{f}_i, \ldots, f_n)$$
$$= C_{n+2}(g, f, f_1, \ldots, f_n) + \langle f, \Gamma g\rangle_S C_n(f_1, \ldots, f_n)$$
$$+ \sum_{i=1}^{n} (-1)^{i}\langle f_i, \Gamma g\rangle_S C_n(f, f_1, \ldots, \hat{f}_i, \ldots, f_n)$$
$$+ \sum_{i=1}^{n} (-1)^{i+1}\langle f_i, \Gamma f\rangle_S C_n(g, f_1, \ldots, \hat{f}_i, \ldots, f_n)$$
$$+ \sum_{i=1}^{n} (-1)^{i+1}\langle f_i, \Gamma f\rangle_S \sum_{j<i} (-1)^{j+1}\langle f_j, \Gamma g\rangle_S C_{n-2}(f_1, \ldots, \hat{f}_j, \ldots, \hat{f}_i, \ldots, f_n)$$
$$+ \sum_{i=1}^{n} (-1)^{i+1}\langle f_i, \Gamma f\rangle_S \sum_{j>i} (-1)^{j}\langle f_j, \Gamma g\rangle C_{n-2}(f_1, \ldots, \hat{f}_i, \ldots, \hat{f}_j, \ldots, f_n)$$
$$= C_{n+2}(g, f, f_1, \ldots, f_n) + \langle f, \Gamma g\rangle_S C_n(f_1, \ldots, f_n)\,.$$

But

$$\langle f, \Gamma g\rangle_S + \langle g, \Gamma f\rangle_S = \langle f, \Gamma g\rangle_S + \langle f, \Gamma g\rangle_{1-S} = \langle f, \Gamma g\rangle$$

and C_{n+2} is anti-symmetric, hence

$$\pi_0(g)\pi_0(f) + \pi_0(f)\pi_0(g) = \langle f, \Gamma g\rangle 1$$

on F_S. Taking $\Gamma g = f$ and using $\pi_0(f)^* \supset \pi_0(\Gamma f)$ we see that $\pi_0(f)$ is bounded as

$$\|\pi_0(f)\xi\|^2 + \|\pi_0(\Gamma f)\xi\|^2 = \|f\|^2\|\xi\|^2$$

for $\xi \in F_S$.

Let $\pi(f)$ be the closure of $\pi_0(f)$. Then by uniqueness of the self dual CAR algebra, there exists an unique representation $\pi = \pi_S$ of $A^{\mathrm{SDC}}(K,\Gamma)$ on F_S such that $\pi(B(f)) = \pi(f)$. Moreover we claim:

$$\pi(: B(f_1)\cdots B(f_n) :)\Omega = C_n(f_1,\dots,f_n)\,, \quad f_i \in K\,. \tag{6.6.28}$$

Assume, inductively that this is so for $n-1$. Then

$$\pi(: B(f_1)\cdots B(f_n) :)\Omega = \pi(B(f_1))\pi(: B(f_2)\cdots B(f_n) :)\Omega$$

$$-\sum_{i=2}^{n}(-1)^i\langle f_i,\Gamma f_1\rangle_S\pi(: B(f_2)\cdots B(\hat{f}_i)\cdots B(f_n) :) \quad \text{by (6.6.24)}$$

$$= \pi(B(f_1))C_{n-1}(f_2,\dots,f_n) - \sum_{i=2}^{n}(-1)^i\langle f_i,\Gamma f_1\rangle_S C_{n-2}(f_2,\dots,\hat{f}_i,\dots,f_n)$$

$$= C_n(f_1,\dots,f_n) \text{ by definition of } \pi(f_1) = \pi(B(f_1))\,.$$

Thus (π_S, F_S, Ω_S) is a cyclic representation of the self dual CAR algebra A. Define the state φ_S on A by $\varphi_S(x) = \langle\pi_S(x)\Omega_S,\Omega_S\rangle$ for $x \in A$ which extends (6.6.21). Our construction now allows us to deduce:

Theorem 6.11 *Let S be a bounded operator on K satisfying*

(a) $1 \geq S^* = S \geq 0\,,$ (b) $S + \Gamma S\Gamma = 1.$

Then there is an unique state φ_S on $A^{\mathrm{SDC}}(K,\Gamma)$ such that

(c) $\varphi_S(B(f_1)\cdots B(f_n)) = \mathrm{Pf}[\varphi_S(B(f_i)B(f_j))]$

(d) $\varphi_S(B(f)^*B(g)) = \langle Sg, f\rangle$

(e) $\varphi_S(: B(f_1)\cdots B(f_m) :^* : B(g_1)\cdots B(g_n) :) = \det[\langle g_i, f_j\rangle_S]\delta_{nm}$

for $f, g, f_i, g_j \in K$.
There is a grading $F_S = \bigoplus_{n=0}^{\infty} F_S^n$ of the GNS Hilbert space of φ_S such that the GNS cyclic vector Ω_S spans F_S^0 and if π_S is the GNS cyclic representation, then $(F_S^n, (f_1,\dots,f_n) \to \pi_S(: B(f_1)\dots B(f_n)) : \Omega_S)$ is the minimal Kolmogorov decomposition of the positive-definite kernel $((f_i),(g_j)) \to \det[\langle f_i, g_j\rangle_S]$.

Remark If S is a basis projection, then using the identification of $A^{\mathrm{SDC}}(K,\Gamma)$ with $A^{\mathrm{CAR}}(SK)$ one has (see (6.4.11))

$$B(f) = a^*(Sf) + a(S\Gamma f)\,.$$

Then one can easily show inductively that in the GNS or Fock representation of $A^{\mathrm{CAR}}(SK)$ with vacuum vector Ω, one has

$$B(f_I)\Omega = \textstyle\sum_{J\sqcup K=I}\varepsilon(J,K)\varphi_S(B(f_K))B((1-S)f_J)\Omega$$

$$= \sum_{J \sqcup K = I} \varepsilon(J,K) \mathrm{Pf}\left[\langle Sf_i, f_j \rangle \mid i,j \in K\right] a^*(Sf_J)\Omega$$

so that (cf. (6.6.23))

$$: B(f_I) : \Omega = : a^*(Sf_I) : \Omega \,. \tag{6.6.29}$$

This means that the n particle spaces F_S^n coincide for the complex and self lual formalism, and that (6.6.29) naturally leads to the definition of (6.6.22) (or 6.6.23)) for the Wick ordered product $: B(f_I) :_S$.

Remark When working in the complex formalism, the use of the conjugate Iilbert space is often more natural than taking a complex conjugation. As an example we give a concrete construction of the quasi free state ω_R in the CAR ormalism where R is a positive contraction on a Hilbert space H using the conjugate Hilbert space $\overline{H}$. For an operator R on H with $0 \leq R \leq 1$, take positive operators α, β on H such that $\alpha^2 + \beta^2 = 1$ and $\beta^2 = R$ (cf. (6.6.11)). Let $V = F(-1)$ be the self adjoint unitary on $F(H)$ such that V fixes the vacuum Ω and $Va(f) = -a(f)V$ for all f in H. Then

$$\pi'_R : a(f) \to \pi_0\left(a(\alpha f)\right) \otimes V + 1 \otimes \pi_0\left(a(\overline{\beta f})\right)^* ; \quad f \in H$$

lefines a representation π'_R of the CAR algebra $A(H)$ on $F(H) \otimes F(\overline{H})$. We lefine the quasi-free state ω_R on $A(H)$ by

$$\omega_R(x) = \langle \pi'_R(x)\Omega \otimes \Omega, \Omega \otimes \Omega \rangle; \quad x \in A(H) \tag{6.6.30}$$

o that

$$\omega_R\left[a^*(f_n) \cdots a^*(f_1) a(g_1) \cdots a(g_m)\right] = \det\left[\langle Rf_i, g_j \rangle\right] \delta_{nm} \,.$$

What we have actually done in (6.6.30) is define ω_R through φ_{P_S}, where P_S is as in (6.6.11), if $S = (1-R) \oplus R$ on $H \oplus H$, and Γ the conjugation $\begin{pmatrix} 0 & J \\ J & 0 \end{pmatrix}$ if J is a conjugation on H commuting with R.

We denote the GNS decomposition of ω_R by $(\pi_R, F_R(H), \Omega_R)$, so that $F_R(H)$ can be identified with $[\pi'_R A(H)\Omega \otimes \Omega]^-$, Ω_R with $\Omega \otimes \Omega$ and π_R with a subrepresentation of π'_R. We can decompose H as $H_1 \oplus H_2$ where $H_1 = \mathrm{Ker}\left[|1-2R| - 1\right]$ and $H_2 = H \ominus H_1$. Then $RH_i \subset H_i$ for $i = 1,2$ and if $R_i = R|_{H_i}$ we have $R_1^2 = R_1$. The quasi-free state ω_R is said to have *no Fock part* if $H_1 = 0$, or equivalently if $0 < R < 1$, and to be *completely Fock* if $H_2 = 0$, or equivalently if $R^2 = R$. (There are analogous notions in the self dual formalism). If R has no Fock part, the representation π'_R is already cyclic with cyclic vector Ω_R. Moreover in this situation, the state ω_R is KMS at inverse temperature $\beta \neq \infty$, for the quasi-free group of automorphisms $A(e^{iht})$ where the self-adjoint operator h on H is given by $\coth(\beta h/2) = 1 - 2R$. In the completely Fock case when R is a projection itself, we would have $\alpha = 1 - R$, $\beta = R$. We will usually then express ω_R in terms of ω_0 as follows. If R is a projection on H, take a conjugation J on

H, commuting with R, and define the Bogoliubov representation π_R of $A(H)$ on $F(H)$ by

$$\pi_R : a(f) \to \pi_0(a((1-R)f)) + \pi_0(a^*(JRf)) . \qquad (6.6.31)$$

Then $(F(H), \pi_R, \Omega)$ can be identified with the GNS decomposition of the pure quasi-free state $\omega_R = \langle \pi_R(\cdot)\Omega, \Omega\rangle$ on $A(H)$. We usually identify $A(H)$ with its Fock representation, so that π_0 is the identity representation. This can be understood in terms of $A(H \oplus H, \Gamma)$, if Γ is the conjugation $\begin{pmatrix} 0 & J \\ J & 0 \end{pmatrix}$, as follows. The map $a(f) \to a((1-R)f) + a^*(JRf)$ is the Bogoliubov automorphism $\tau(V)$ in the self dual formalism (Exercise 6.6) where $V = \begin{pmatrix} 1-R & R \\ R & 1-R \end{pmatrix}$ is a self adjoint unitary commuting with Γ. Then $VEV = S$ if $E = 1 \oplus 0$, $S = (1-R) \oplus R$ denotes the basis projections. Thus $\varphi_E \circ \tau(V) = \varphi_S$.

If we identify $A(H)$ with $\otimes^{\dim H} M_2$ as in (6.4.5), then ω_0 is identified with the product state $\otimes^{\dim H} \omega_{\left(\begin{smallmatrix} 1 \\ 0 \end{smallmatrix}\right)}$.

The argument for purity of φ_0 following (6.6.4) shows that if φ is a state on $A^{SDC}(K, \Gamma)$, such that $\varphi = \varphi_P$ on two point functions $\{B(f)B(g) \mid f, g \in K\}$ for a basis projection P then $\varphi = \varphi_P$.

Exercise 6.7 Let Γ be a conjugation on a Hilbert space K, P a basis projection and Q a projection commuting with Γ. Let $R = \pi_P(A(QK))''$. Show that the following three conditions are equivalent:

(a) Ω_P is cyclic for R; $(1-Q) \wedge (1-P) = 0$; $(1-Q) \wedge P = 0$

as are:

(b) Ω_P is separating for R; $Q \wedge (1-P) = 0$; $Q \wedge P = 0$.

Exercise 6.8 Let α be an automorphism of the Fermion algebra A^{SDC} such that $\varphi_S \circ \alpha$ is a quasi-free state whenever φ_S is. Show that if A^{SDC} is infinite dimensional then α is a Bogoliubov automorphism $\tau(U)$. (Hint: When is a convex combination of quasi-free states quasi-free ?)

Exercise 6.9 If Γ is a conjugation on a Hilbert space H, and $S \in Q(H, \Gamma)$ show that φ_S is not factorial unless

(i) $S^{1/2}(1-S)^{1/2}$ is Hilbert–Schmidt.

(ii) $\mathrm{Ker}(2S-1)$ is odd dimensional.

Exercise 6.10 If E is the projection on a finite dimensional subspace with orthonormal basis $f_1, \ldots, f_n$ show that

$$\omega_E = \langle (\,\cdot\,) f_1 \wedge \cdots \wedge f_n, f_1 \wedge \cdots \wedge f_n \rangle .$$

This shows the existence of a quasi-free Fock state ω_E for any projection E, not necessarily finite dimensional, and hence ω_S for any positive contraction S, by restriction.

6.7 Implementability of quasi-free automorphisms and derivations and equivalence of quasi-free states

In this section we consider the problem of implementing quasi-free automorphisms and quasi-free derivations by unitaries or self adjoint operators respectively in the Fermion algebra itself – or in some quasi-free representation. Since the Bogoliubov automorphisms act transitively on Fock quasi-free states, this is related to the question of equivalence of quasi-free states. We start with the definitions of quasi-free derivations and quasi-free homomorphisms – which are not necessarily $*$-maps, as we will have need for these in Section 7.14 when we discuss dynamics in the one dimensional quantum model associated with the two dimensional classical Ising model. Let K be a complex Hilbert space with distinguished conjugation Γ, and $A_0 = A_0^F$ the dense $*$-subalgebra of A^F generated by $B(f)$, $f \in K$.

Lemma 6.12 *(i) Let U be a bounded linear operator on K such that $\Gamma U^* \Gamma U = 1$. Then $\tau(U)B(f) = B(Uf)$ defines a unique homomorphism $\tau(U)$ of $A_0 = A_0^F$. It is a $*$-homomorphism if and only if U is isometric and $U\Gamma = \Gamma U$, in which case it extends to a *-homomorphism of $A = A^F$.*

(ii) Let H be a bounded linear operator on K such that $\Gamma H^* \Gamma = -H$. Then $\delta(H)B(f) = B(Hf)$ defines a unique derivation on A_0. It is a $*$-derivation if and only if H is skew adjoint, $H^* = -H$.

Proof (i) follows from the universal property enjoyed by A_0. (ii) follows from (i) by defining on $A_0 : \delta(H) = (d/dt)_{t=0}\tau(e^{Ht})$. □

If $H \in B(K)$, let $e(H) = (H - \Gamma H^* \Gamma)/2$. Then e is a projection from $B(K)$ onto $D = \{H \in B(K) \mid \Gamma H^* \Gamma = -H\}$. Moreover, since $\text{tr}\ \Gamma H^* \Gamma = \text{tr}\ H$, for any trace class H, e takes $T(K)$ onto $E = \{H \in T(K) \mid H = -\Gamma H^* \Gamma,\ \text{tr}\ H = 0\}$.

Proposition 6.13 *The map $f \otimes \overline{g} \to B(f)B(g)^*$ extends to a self adjoint bounded linear map b from $T(K)$ into A such that*

(a) $b(H)^* = b(H^*)$.

(b) $\tau(U)b(H) = b(UH\Gamma U^* \Gamma)$ *for all bounded U on K such that* $\Gamma U^* \Gamma U = 1$.

(c) $\delta(H_1)b(H) = b([H_1, H])$ *for all bounded H_1 on K such that* $\Gamma H_1^* \Gamma = -H_1$.

(d) $b(H) = b(e(H)) + \text{tr}\ H$ *for any trace class H on K.*

Moreover, if H is an operator on K such that $\Gamma H^* \Gamma = -H$, then $\delta(H)$ extends to a bounded (inner) derivation on A if and only if H is trace class. In this case

(e) $2\delta(H)(a) = [b(H), a] \quad a \in A$.

(f) $\varphi_S[b(H)] = -\text{tr}\ SH$.

(g) $\|H\|_1/2 = \|b(H)\|$ *if H is skew adjoint.*

Proof The existence of the symmetric map $H \to b(H)$ on the finite rank operators on K satisfying $b(f \otimes \overline{g}) = B(f)B(g)^*$ is clear. Now if $H = \sum_i f_i \otimes \overline{g}_i$ is a finite rank operator, we see that

$$\|b(H)\| \leq \textstyle\sum_i \|B(f_i)\| \, \|B(g_i)\| \leq \sum_i \|f_i\| \, \|g_i\| \, .$$

Thinking of $T(K)$ as the projective tensor product $K \otimes^\vee \overline{K}$, so that

$$\|H\|_1 = \inf \left\{ \textstyle\sum_i \|f_i\| \, \|g_i\| \mid H = \sum_i f_i \otimes \overline{g}_i \right\}$$

for $H \in T(K)$ (Exercise 2.11), we have $\|b(H)\| \leq \|H\|_1$ for all finite rank operators on K. Hence by linearity and continuity we can uniquely extend the map $H \to b(H)$ to a linear contraction from $T(K)$ into A. Then

$$\begin{aligned} \tau(U)(B(f)B(g)^*) &= \tau(U)\,(B(f)B(\Gamma g)) = B(Uf)B(U\Gamma g) \\ &= B(Uf)B(\Gamma U \Gamma g)^* = b(Uf \otimes \overline{\Gamma U \Gamma g}) = b(Uf \otimes \overline{g}(\Gamma U \Gamma)^*) \end{aligned}$$

so that (b) holds by linearity and continuity. Then (c) follows by differentiating

$$\tau(e^{H_1 t}) b(H) = b(e^{H_1 t} H e^{-H_1 t}) \, .$$

If $H = f \otimes \overline{g}$, then $2\alpha(H) = f \otimes \overline{g} - \Gamma g \otimes \overline{\Gamma f}$, so that

$$\begin{aligned} 2b(H) - 2b(\alpha(H)) &= 2B(f)B(g)^* - [B(f)B(g)^* - B(\Gamma g)B(\Gamma f)^*] \\ &= B(f)B(g)^* + B(g)^*B(f) = \langle f, g \rangle = \operatorname{tr} H \end{aligned}$$

and (d) holds by linearity and continuity. Also,

$$\begin{aligned} & B(f)B(g)^*B(h) - B(h)B(f)B(g)^* \\ &= -B(f)B(h)B(g)^* + \langle h, g \rangle B(f) - B(h)B(f)B(g)^* \\ &= -B(f)B(\Gamma h)^*B(g)^* + \langle h, g \rangle B(f) - B(h)B(f)B(g)^* \\ &= B(\Gamma h)^*B(f)B(g)^* - \langle f, \Gamma h \rangle B(g)^* + \langle h, g \rangle B(f) - B(h)B(f)B(g)^* \\ &= B(\langle h, g \rangle f - \langle f, \Gamma h \rangle \Gamma g) = B(2e(f \otimes \overline{g})h) \, . \end{aligned}$$

Thus if H is any trace class operator on K, we see

$$[b(H), B(h)] = B(2e(H)h), \quad h \in K$$

by linearity and continuity. Hence in this case $\delta(e(H))$ is bounded and implemented by $b(H)/2$. Also

$$\varphi_S(B(f)B(g)^*) = \langle S\Gamma g, \Gamma f \rangle = \langle (1 - S)f, g \rangle = \operatorname{tr}[(1 - S)\,(f \otimes \overline{g})] \, ,$$

so that (f) holds by linearity and continuity.

Now let H be a self adjoint operator on K with $\Gamma H\Gamma = -H$, and $\delta(H)$ bounded. Let $H = \int \lambda\, dE(\lambda)$ be the spectral decomposition of H. By $H = -\Gamma H\Gamma$, and uniqueness of the spectral decomposition we see that $\Gamma E(\lambda)\Gamma = E(-\lambda)$. Then $E_+ \equiv E(0,\infty)$ is in general not a basis projection but $\Gamma E_+\Gamma \perp E_+$, with $\Gamma E_+\Gamma = E_- \equiv E(-\infty, 0)$, and $\Gamma E_0\Gamma = E_0$ if $E_0 = E\{0\}$. Let $f_1, \ldots, f_n$ be arbitrary orthogonal unit vectors in E_+K, and take $f_{n+1}, \ldots, f_m$ orthogonal unit vectors in E_+K, such that $Hf_i \in \text{lin span}\,\{f_1, \ldots, f_m\}$ if $1 \le i \le n$. Then

$$\{B(f_i), B(f_j)\} = \langle f_i, \Gamma f_j\rangle = 0 \tag{6.7.1}$$

as $\Gamma E_+\Gamma \perp E_+$. In particular $B(f_i)^2 = 0$. Now

$$\begin{aligned}
&\delta(H)[B(f_1)\cdots B(f_n)]\\
&= B(Hf_1)B(f_2)\cdots B(f_n) + B(f_1)B(Hf_2)B(f_3)\cdots B(f_n)\\
&\quad + \cdots + B(f_1)\cdots B(f_{n-1})B(Hf_n)\\
&= B\left(\textstyle\sum_{i=1}^m \langle Hf_1, f_i\rangle f_i\right) B(f_2)\cdots B(f_n)\\
&\quad + B(f_1)B\left(\textstyle\sum_{i=1}^m \langle Hf_2, f_i\rangle f_i\right) B(f_3)\cdots B(f_n)\\
&\quad + \cdots + B(f_1)\cdots B(f_{n-1})B\left(\textstyle\sum_{i=1}^m \langle Hf_n, f_i\rangle f_i\right)
\end{aligned}$$

using $Hf_j = \sum_{i=1}^m \langle Hf_j, f_i\rangle f_i$. Hence by (6.7.1), we see

$$\delta(H)[B(f_1)\cdots B(f_n)]B(f_{n+1})\cdots B(f_m) = \sum_{i=1}^n \langle Hf_i, f_i\rangle B(f_1)\cdots B(f_m).$$

But $HE_+ \ge 0$, and $\|\prod_i B(f_i)\| = 1$, so that

$$0 \le \textstyle\sum_i \langle Hf_i, f_i\rangle \le \left\|\delta(H)|_{A(E_+K)}\right\| \le \|\delta(H)\|\,.$$

Hence H is trace class and $\|H\|_1 = 2\|HE_+\|_1 \le 2\|\delta(H)\|$.

Now let $\{f_j\}$ be a complete orthonormal basis of eigenvectors for H on $E_+ = \sum_{\lambda>0} E_\lambda = \Gamma E_+\Gamma$, with eigenvalues $\{\lambda_j > 0\}$, so that $\{\Gamma f_j\}$ is a complete orthonormal basis of eigenvectors for H on $E_- = \sum_{\lambda<0} E_\lambda = \Gamma E_+\Gamma$, with eigenvalues $\{-\lambda_j\}$. Thus

$$\begin{aligned}
b(H) &= \textstyle\sum_j \lambda_j B(f_j)B(f_j)^* - \sum_j \lambda_j B(\Gamma f_j)B(\Gamma f_j)^*\\
&= \textstyle\sum_j \lambda_j [B(f_j)B(f_j)^* - B(f_j)^*B(f_j)]\,.
\end{aligned}$$

But

$$\begin{aligned}
-B(f)B(f)^* - B(f)^*B(f) &\le B(f)B(f)^* - B(f)^*B(f)\\
&\le B(f)B(f)^* + B(f)^*B(f)
\end{aligned}$$

so that

$$-\|f\|^2 \le B(f)B(f)^* - B(f)^*B(f) \le \|f\|^2$$

and

$$\|B(f)B(f)^* - B(f)^*B(f)\| \leq \|f\|^2 .$$

Hence $\|b(H)\| \leq \sum \lambda_i = \|H\|_1/2$. Now let $S = E_+ + E_0/2$, so that S is a positive contraction with $\Gamma S \Gamma = 1 - S$. Then $\varphi_S[b(H)] = -\text{tr } SH = -\text{tr}(HE_+)$. Hence $\|H\|_1/2 = \|HE_+\|_1 \leq \|b(H)\|$, as $\|\varphi_S\| = 1$, since φ_S is a state. □

We will call a projection F_0 such that F_0 is orthogonal to $\Gamma F_0 \Gamma$ a *partial basis projection.*

We will need the notion of determinant of certain operators on an infinite dimensional Hilbert space H. Just as we restricted the definition of the trace to the integrable or trace class elements $T(H)$, we will restrict the domain of the determinant. Operators with a well defined determinant will be those $X = 1 + A$ which are trace class perturbations of the identity, $A \in T(H)$. First note that if H is finite dimensional, then $\wedge^k A$ on $\wedge^k H$ has eigenvalues $\{\lambda_{i_1}\lambda_{i_2}\cdots\lambda_{i_k} \mid i_1 < \cdots < i_k\}$ if A is normal and has eigenvalues $\{\lambda_i\}$. Thus

$$\text{tr } \wedge^k A = \sum_{i_1 < \ldots < i_k} \lambda_{i_1} \cdots \lambda_{i_k} .$$

Hence $\det(1 + A) = \prod_i (1 + \lambda_i) = \sum_k \text{tr } \wedge^k A$. If A is trace class (but not necessarily normal), then $\| \wedge^k A\| \leq \|A\|_1^k/k!$ and so we can use

$$\det(1 + A) = \sum_{k=0}^{\infty} \text{tr } \wedge^k A \tag{6.7.2}$$

to define a *determinant.* Then $A \to \det(1 + A)$ is continuous for the trace class norm (indeed $|\det(1 + A) - \det(1 + B)| \leq \|A - B\|_1 \exp(\|A\|_1 + \|B\|_1 + 1))$. Since this determinant coincides with the usual one in finite dimensions, det is multiplicative on $1 + T(H)$. Moreover $X \in 1 + T(H)$ is invertible if and only if $\det X \neq 0$. Furthermore, an element $X \in 1 + T(H)$ is invertible if and only if there exists A in $T(H)$ such that $X = e^A$. (If X is unitary, A can be chosen skew, cf. Lemma 6.14 below.) Then by continuity we have $\det e^A = \exp \text{tr } A$ which could be used as an alternative definition of the determinant.

Now if Γ is a conjugation on K, and A is a trace class operator, then it is clear that

$$\text{tr } \Gamma A \Gamma = \overline{\text{tr } A} = \text{tr } A^* .$$

It is then just as clear from (6.7.2) that

$$\det(1 + \Gamma B \Gamma) = \overline{\det(1 + B)} = \det(1 + B^*) .$$

Hence if U is a unitary on K commuting with Γ such that $1 - U$ is trace class then

$$\det(U)^2 = \det(\Gamma U \Gamma)\det U = \det U^* \det U = \det U^* U = 1 ,$$

and so $\det U = \pm 1$.

In fact suppose $U-1=\sum(\lambda-1)E_\lambda$ is the spectral decomposition of a unitary U on K commuting with Γ such that $U-1$ is compact. Then $\Gamma E_\lambda \Gamma = E_{\overline{\lambda}}$ as $\Gamma U \Gamma = U$, $\sigma(U)=\overline{\sigma(U)}$ and

$$\sigma(U)\backslash\{\pm 1\} = \{e^{\pm i\theta_n} \mid 0<\theta_n<\pi\} .$$

Then if $e_n = E^U\{e^{i\theta_n}\}$,

$$U = \sum_n \left(e^{i\theta_n} e_n + e^{-i\theta_n}\Gamma e_n \Gamma\right) + E_1 - E_{-1} \tag{6.7.3}$$

and

$$\|1-U\|_1 = \sum_n 2|1-e^{-i\theta_n}| + \dim E_{-1} = \sum_n 4\sin(\theta_n/2) + \dim E_{-1} .$$

Remark Note that $\operatorname{tr}|1-U|^p<\infty$, $1\le p<\infty$ is equivalent to $\{\theta_n\}$ being in ℓ^p and $E_{-1}=\operatorname{Ker}(U+1)$ being finite dimensional. If $p=1$, then

$$\det U = (-1)^{\det E_{-1}}$$

and this can be used to define det if $p>1$, e.g. $p=2$.

When $\det U=1$, it is most convenient to take logarithms. More precisely:

Lemma 6.14 *Let U be a unitary on K such that $\Gamma U\Gamma = U$, $1-U$ is trace class and $\det U = 1$. Then there exists a skew adjoint trace class operator H on K commuting with Γ such that $U=e^H$.*

Proof Adopting the previous notation, since $\dim E_{-1}$ is even, there exists a projection F_0 such that $F_0+\Gamma F_0\Gamma = E_{-1}$. Let $H=\sum i\theta_n(e_n-\Gamma e_n\Gamma)+i\pi(f_0-\Gamma f_0\Gamma)$. Then H is skew adjoint, is trace class as $\{\theta_n\}$ is ℓ^1, and commutes with Γ. Moreover $e^H=U$ by (6.7.3). □

An inner automorphism $\operatorname{Ad}(U)$ of a graded algebra is said to be *even*, respectively *odd*, if U can be chosen to be even, respectively odd. Note that an inner automorphism of a simple C^*-algebra is automatically either even or odd (cf. Section 7.7).

Theorem 6.15 *Let U be a unitary on a complex Hilbert space K, commuting with Γ. Then $\tau(U)$ on $A=A^{\mathrm{F}}(K,\Gamma)$ is inner if and only if one of the following conditions hold:*

(6.7.4) $1-U$ *is trace class and* $\det(U)=+1$.

(6.7.5) $1+U$ *is trace class and* $\det(-U)=-1$.

In which case (6.7.4) *(respectively* (6.7.5)*) holds if and only if $\tau(U)$ is even (respectively odd).*

Proof We first show sufficiency of conditions (6.7.4) and (6.7.5). Suppose $1-U$ is trace class and $\det(U)=1$. Then by Lemma 6.14 there exists a trace class skew adjoint operator H on K commuting with Γ such that $U=e^H$. Then by

Proposition 6.13, $\delta(H)$ is a bounded inner skew derivation on K, implemented by $b(H)/2$. Hence $\tau(U) = \tau(e^H) = e^{\delta(H)} = \operatorname{Ad} e^{b(H)/2}$ is inner and even. On the other hand, suppose e is a Γ fixed unit vector in K. Then $V = \sqrt{2}B(e)$ is a self adjoint odd unitary such that

$$\begin{aligned} VB(h)V &= B(e)^*B(h)B(e)/2 = -B(h)B(e)^*B(e)/2 + \langle h, e\rangle B(e)/2 \\ &= B(-h + 2\langle h, e\rangle e) = B(I_e h) \end{aligned}$$

and $I = I_e$ denotes the inversion or reflection $-1 + 2e \otimes \bar{e}$. Thus $1 + I$ is rank one, $\det(-I) = -1$, and $\tau(I)$ is implementable by the odd V. Hence if $1 + U$ is trace class and $\det(-U) = -1$, then $1 - IU$ is trace class and $\det IU = 1$. Hence by the preceding $\tau(IU)$ is inner and even, and $\tau(U)$ is inner and odd. We have thus already shown sufficiency of conditions (6.7.4) and (6.7.5).

Next suppose that $\tau(U)$ is inner. If K is of finite odd dimension, and $\{e_i\}$ a Γ invariant orthonormal basis then $z = \prod_i B(e_i)$ is central, and $\tau(U)(z) = (\det U)z$. Since central elements are fixed by inner automorphisms, $\det U = 1$ in this case.

Suppose then that K is not of finite odd dimension, and $\tau(U) = \operatorname{Ad} Q$ where $Q \in A^F$ is unitary and is either even or odd by the comment before Theorem 6.15. We have already noted that the reflection I_e is implemented by the odd $B(h)$, so that multiplying U by I_e takes case (6.7.4) to case (6.7.5). We can thus assume Q is even. We first show that $1 - U$ is compact and then eventually reduce to the conditions of (6.7.4) that $1 - U$ is trace class and $\det U = 1$. Let E_k be a sequence of Γ-invariant finite dimensional projections increasing to 1, and take unitary $Q_k \in A_+^F(E_k K, \Gamma)$ converging to Q in $A_+^F(K, \Gamma)$ in norm by Lemma 3.3 so that $Q_k^* \to Q^*$. Let h_n be a sequence of vectors in K weakly convergent to zero. Then

$$\begin{aligned} &\|B(U-1)h_n\| = \|\operatorname{Ad}(Q)B(h_n) - B(h_n)\| \\ &= \|\operatorname{Ad}(Q)B(E_k h_n) - B(E_k h_n)\| + \|(\operatorname{Ad} Q - \operatorname{Ad} Q_k)B((1-E_k)h_n)\| \\ &\leq 2\|B(E_k h_n)\| + \|\operatorname{Ad} Q - \operatorname{Ad} Q_k\| \, \|h_n\| \to 0 \text{ as } n \to \infty \end{aligned}$$

where we have used $\operatorname{Ad} Q_k | A(1 - E_k K) = \mathrm{id}$, $E_k h_n \to 0$ as $n \to \infty$, since E_k is finite dimensional and $\operatorname{Ad} Q - \operatorname{Ad} Q_k \to 0$. Since $\|B(h)\| \geq \|h\|/2$, this shows that $U - 1$ takes weakly convergent sequences to strongly convergent ones and hence is compact.

Let $E = E^U\{1\}$ be the spectral projection of U corresponding to the eigenvalue 1. If E is not finite dimensional and odd, then as $\operatorname{Ad} Q = \mathrm{id}$ on $A(EK)$, and Q is even, we have $Q \in A((1-E)K)$ by the Remark on page 225. We can in any case assume E is finite rank. Let e_l be a complete orthogonal set of eigenvectors for U with eigenvalues $\exp i\alpha_l$, $|\alpha_l| \leq \pi$. Take E_n the projection on the space spanned by $\{e_l \mid |\alpha_l| > 1/n \text{ or } \alpha_l = 0\}$. Then $E_n \nearrow 1$, and take Q_n unitary in $A_+^F(E_n K, \Gamma)$, $Q_n \to Q$. Then

$$\|\operatorname{Ad}(Q)(a) - a\| = \|(\operatorname{Ad} Q - \operatorname{Ad} Q_n)(a)\| \leq \|\operatorname{Ad} Q - \operatorname{Ad} Q_n\| \, \|a\|$$

for all $a \in A^F((1-E_n)K)$, where $[Q_n, a] = 0$. In particular for $a = B(e_{l_1})\cdots B(e_{l_m})$, $l_j > n$, $\|a\| = 1$, $\mathrm{Ad}(Q)(a) = \exp\left(i\sum_j \alpha_{l_j}\right) a$ so that

$$\left|\exp\left(i\textstyle\sum_{j=1}^m \alpha_{l_j}\right) - 1\right| \to 0$$

for $l_j > n$, and all m, as $n \to \infty$. This means $\sum_l |\alpha_l| < \infty$ and $U - 1$ is of trace class. Taking the self adjoint trace class operator

$$H = \sum_{0<\beta<\pi} \beta\left(E(e^{i\beta}) - E(e^{-i\beta})\right)$$

(cf. (6.7.3)) so that $\Gamma H \Gamma = -H$, $\tau(e^{iH})$ is implemented by $Q(e^{iH})$, with $\det(e^{iH}) = 1$. Replacing U by Ue^{-iH}, we reduce to the case where $U = 1 - 2E$, E is a projection of finite rank n, with Γ-invariant basis $e_1, \ldots, e_n$. If n is even $\tau(1-2E)$ is implemented by $B(e_1)\cdots B(e_n)$, and $\det(1-2E) = 1$. If n is odd, we replace Q by $QB(e_2)\cdots B(e_n)$ to reduce to the case $n = 1$. Again using the Remark on page 225, $Q \in A_+(K) \cap A((1-E)K)' = A_+(EK)$ implements a non-trivial automorphism on $A(EK) = \mathbb{C} \oplus \mathbb{C}$ which is impossible. Thus n is even and $\det U = 1$. □

We will next consider the related question of implementing quasi-free automorphisms and derivations in quasi-free states by operators not in the Fermion algebra itself but in the von Neumann algebra generated in such representations.

For example (Blattner 1958), cf. Theorem 6.7, has shown that $\tau(U)$ is weakly inner in the trace representation (i.e. $\tau(U) = \mathrm{Ad}\, Q$, $Q \in A(H)'' \subset B(H_{\mathrm{tr}})$ in the trace representation) if and only if either

$$\begin{aligned} &1 - U \text{ is Hilbert} - \text{Schmidt}\,, \quad \det U = 1 \\ &1 + U \text{ is Hilbert} - \text{Schmidt}\,, \quad \det\,(-U) = -1 \end{aligned}$$

in which case Q is even, odd respectively. Here of course we use the Remark on page 239 for the case $p = 2$ to make sense of the determinant.

If the quasi-free state in question is pure, then of course weak implementability in this sense just means being implemented by an operator in the GNS Hilbert Fock space F, as the representation is now irreducible $\pi(A^F)'' = B(F)$. This irreducibility means that every bounded operator is a limit of polynomials in annihilation and creation operators, although in a specific example it may not be immediately apparent what is a canonical way of doing so. Thus in the Fock representation on $F_a(H)$:

$$F(1+X) = \sum_I \frac{1}{|I|!} c^*(Xf_I)c(f_I) \tag{6.7.6}$$

where the summation is over all finite subsets I if $\{f_j \mid j \in J\}$ is a complete orthonormal basis for H, and $c(f_I) = c(f_{i_1})\cdots c(f_{i_n})$, if $I = \{i_1, \ldots, i_n\}$, e.g.

$c^*(Xf_I)c(f_I)c^*(f_i)\Omega = c^*(Xf_i)\Omega, c^*(f_i)\Omega, 0$ if $I = \{i\}, \varnothing$ or otherwise respectively. Here in the complex formalism, Fermion Fock space $F_a(H)$ we have the annihilation operators $c(f)$, $f \in H$ satisfying the canonical anti-commutation relations (6.4.1) and (6.4.2).

Note that if H is finite dimensional then $\det(1+X) = \wedge^{|J|}(1+X)$ so that using (6.7.6) (cf. Exercise 6.11) we have

$$\det(1+X) = \langle F(1+X)c^*(f_J)\Omega, c^*(f_J)\Omega\rangle = \textstyle\sum_p \det(pXp) \tag{6.7.7}$$

where the summation is over all projections $p = p_I$ on the span of $\{f_j \mid j \in I\}$ as I runs over the subsets of J. Indeed

$$\det(1+\lambda X) = \sum \lambda^{|I|}\det(pXp) \tag{6.7.8}$$

is an entire function of λ, for X trace class (Ruijsenaars 1977), (Ringrose 1971) and is easily verified if X is normal.

It will not immediately be clear when expressions such as the right hand side of (6.7.6) converge to bounded operators – note that the left hand side is bounded only when $1+X$ is contractive. Let us look at the operators of second quantization more carefully. Second quantization takes operators on H to those on $F_a(H)$, more precisely if T (respectively h) is a contraction on H (self adjoint operator on H) then $F(T)$ (respectively $dF(h)$) is a contraction (respectively self adjoint operator) so that

$$F(\exp ith) = \exp itdF(h)\,.$$

Moreover $F(T)c(f) = c(Tf)F(T)$, so that if U is unitary $F(U)$ implements the Bogoliubov automorphism $\alpha(U) : c(f) \to c(Uf)$ and $dF(h)$ implements the quasi-free derivation $d\alpha(h)$ (the infinitesimal generator of $\alpha(e^{ith})$) so that $[dF(h), c(f)] = c(hf) = d\alpha(h)(c(f))$. Thus $\alpha(U) = \operatorname{Ad} F(U)$, $d\alpha(h) = \operatorname{ad} i\, dF(h)$. If A is trace class, then we have that $dF(A)$ is bounded and indeed is in the Fermion algebra $dF(A) = \sum_n c^*(Af_n)c(f_n) \equiv c^*Ac$ by Proposition 6.13 for any complete orthonormal sequence. An exponentiated version of dF is what is appearing in (6.7.6). We can write this as

$$: \exp dF(X) : = \sum_n \sum_i \; : [c^*(Xf_i)c(f_i)]^n : /n!$$

where the double dots : : are indicating a normal ordering of moving creation operators to the left of annihilation operators. We will come back to look at this more carefully.

We can of course similarly consider $c^*Ac^* \equiv \sum c^*(Af_i)c^*(Jf_i)$, $cAc \equiv \sum c(JAf_i)c(f_i)$ if J is a conjugation on H. These will turn out to be defined as unbounded self adjoint operators for Hilbert–Schmidt A (not necessarily trace class), as follows from the estimate

$$\|cAcP_l\| \leq l\|A\|_2\,, \qquad \|c^*Ac^*P_l\| \leq (l+2)\|A\|_2 \tag{6.7.9}$$

f P_l is the projection on the l-particle space. We write $c_i = c(f_i)$, $cAc = \sum_{i,j}\overline{A}_{ij}c_ic_j$ where $\|A\|^2 = \sum_{i,j}|\overline{A}_{ij}|^2$, and represent the annihilation oper-tors as a shift on anti-symmetric functions in the finite particle spaces by $_if(j_1,\ldots,j_l) = (l+1)^{1/2}f(i,j_1,\ldots,j_l)$ if $f \in \wedge^{l+1}$, so that $c_if \in \wedge^l$. Then or anti-symmetric f, g in l and $(l-2)$ particle spaces respectively, we have

$$\begin{aligned}\langle cAcf, g\rangle &= \textstyle\sum_{i,j}\overline{A}_{ij}\,\langle c_ic_jf, g\rangle = [l(l-1)]^{\frac{1}{2}}\sum_{i,j}\overline{A}_{ij}\sum_\alpha f_{ij\alpha}\overline{g}_\alpha \\ &\leq l\textstyle\sum_{i,j}|\overline{A}_{ij}|\left(\sum_\alpha|f_{ij\alpha}|^2\right)^{\frac{1}{2}}\left(\sum_\alpha|g_\alpha|^2\right)^{\frac{1}{2}} \\ &\leq l\left(\textstyle\sum_{i,j}|\overline{A}_{ij}|^2\right)^{\frac{1}{2}}\left(\sum_{i,j}\sum_\alpha|f_{ij\alpha}|^2\right)^{\frac{1}{2}}\left(\sum_\alpha|g_\alpha|^2\right)^{\frac{1}{2}} \\ &= l\|A\|_2\,\|f\|\,\|g\|\,.\end{aligned}$$

Thus cAc is bounded on $\wedge^l$ and $\|cAcP_l\| \leq l\|A\|_2$. Consequently $\|c^*Ac^*P_l\| = \|cA^*cP_{l+2}\|$ implies $\|c^*Ac^*P_l\| \leq (l+2)\|A\|_2$.

This will be the main idea in implementing quasi-free automorphisms and lerivations in Fock states from Hilbert–Schmidt rather than trace class esti-nates. Such Fock states arise naturally as follows. For physical reasons, we are nterested in finding self adjoint operators with positive spectrum (which may ıot necessarily be the case with $dF(h)$) which implement the quasi-free dynam-cs $\alpha(e^{iht})$. If we take P_+ and $P \equiv P_-$ to be the spectral projections of $[0,\infty)$ nd $(-\infty, 0]$ respectively, we can work with the Fock state ω_{P_-} (instead of ω_0 as ve have done in this discussion so far). The corresponding representation of the anonical anti-commutation relations is given by (6.6.31), namely

$$c' : f \to c(P_+f) + c^*(JP_-f)$$

vhere J is a conjugation commuting with P, $JP = PJ$. The quasi-free dynamics now implemented by $dF(|h|) = dF(h_{++}) - dF(h_{--})$ if J commutes with h. Iere we decompose an operator a on H, as a matrix $a_{\sigma\sigma'} = P_\sigma aP_{\sigma'}$. If h is trace lass but not necessarily commuting with P, then $K \equiv \tau(V)dF(h) = c'^*hc'$ nplements the dynamics $t \to c'(e^{ith}f) = \tau(V)c(e^{ith}f)$ if V is the Bogoli-bov unitary $\begin{pmatrix} P_+ & P_- \\ P_- & P_+\end{pmatrix}$. If we distinguish the creation operator c^* on P_+ and P_- in creating states of particles and anti-particles as a^* and b^* respectively o that $c'(g) = a(P_+g) + b^*(JP_-g)$, then with our previous convention for $^*Ac, c^*Ac^*, cAc$ we have

$$K = \left(a^*\; b\right)h\left(a\; b^*\right)^t = a^*h_{++}a + bh_{--}b^* + a^*h_{+-}b^* + bh_{-+}a\,.$$

we subtract $\langle K\Omega, \Omega\rangle = \operatorname{tr} h_{--}$ from K to get an implementer which annihilates ne vacuum state ω_Ω, we find the implementer

$$dF_{P_-}(h) = dF(h_{++}) - dF(h^t_{--}) + a^*h_{+-}b^* + bh_{-+}a$$

where x^t is the transpose Jx^*J as $dF(h^t_{--}) = -bh_{--}b^* + \operatorname{tr} h_{--}$ from $b^*(f)b(g) = -b(g)b(f)^* + \langle f, g\rangle$. (Note that we now see $dF_{P_-}(h) = dF(|h|)$ in the special case when P_+, P_- are the spectral projections.) The right hand side now makes sense for any bounded h with h_{+-}, h_{-+} Hilbert–Schmidt. If h is self adjoint, $dF_{P_-}(h)$ is essentially self adjoint (being symmetric and having the finite particle vectors as a dense set of analytic vectors, it has a self adjoint closure by Nelson's theorem (Nelson 1960)). If $o_P(H)$ denotes the Lie algebra of bounded operators h with off-diagonal h_{+-}, h_{-+} Hilbert–Schmidt then we have a map $h \to dF_{P_-}(h)$ from $o_P(H)$ such that if h is self adjoint

$$\operatorname{Ad}\left(\exp it\, dF_{P_-}(h)\right)(c'(f)) = c'(e^{ith} f)\,.$$

However dF_{P_-} does not preserve commutators, as even though dF did preserve commutators, we adjusted dF by $\operatorname{tr} h_{--}$. So

$$dF_{P_-}([h,k]) = [dF_{P_-}(h), dF_{P_-}(k)] + \operatorname{tr}[h,k]_{--}$$

where the correction is the *Schwinger term*

$$S(h,k) = \operatorname{tr}[h,k]_{--} = \operatorname{tr}(h_{-+}k_{+-} - k_{-+}h_{+-}) \tag{6.7.10}$$

which is well defined when the off-diagonal terms of h and k are Hilbert–Schmidt. So by considering first trace class operators and then using an approximating argument we can deduce for all $h, k \in o_P(H)$:

$$[dF_{P_-}(h), dF_{P_-}(k)] = dF_{P_-}[h,k] + \operatorname{tr}(h_{-+}k_{+-} - k_{-+}h_{+-})\,. \tag{6.7.11}$$

This cocycle is not however the most fundamental. There is another construction yielding one-half of the expression $\operatorname{tr}(h_{-+}k_{+-} - k_{-+}h_{+-})$ as the correction by taking a real formulation – we replace $F_a(H)$ by $F_a(H_+)$, instead of distinguishing a and b we essentially identify them, the *neutral* situation where we identify the particle and the anti-particle with real *Majorana Fermions*.

Suppose there is a conjugation Γ on H such that $\Gamma P = (1-P)\Gamma$. Note that previously our conjugation commuted with P. Then with $P_+ = 1 - P$, $P_- = P$

$$B(f) = c^*(P_+f) + c(\Gamma P_- f) \quad f \in H\,. \tag{6.7.12}$$

satisfies the self dual form of the canonical anti-commutation relations

$$\{B(f), B(g)^*\} = \langle f, g\rangle\,, \quad B(f)^* = B(\Gamma f)\,.$$

Again as in Proposition 6.13, if h is self adjoint, trace class and commutes with Γ, $\sum B(hf_n)B(f_n)^*/2$ implements the quasi-free derivation $d\tau(h)$. To annihilate the vacuum state we now subtract $\operatorname{tr} h_{--}/2$. Let $o_P(H,\Gamma)$ denote the elements h of $o_P(H)$, which satisfy $\Gamma h^*\Gamma = -h$. Then for $h \in o_P(H,\Gamma)$,

$$dQ_{P_-}(h) = \left[dF(h_{++}) - dF(h^T_{--}) + c^* h_{+-} c^* + c h_{-+} c\right]/2 \qquad (6.7.13)$$

yields an operator which for h self adjoint is self adjoint (essentially self-adjoint on finite particle vectors) satisfying

$$[dQ_{P_-}(h), dQ_{P_-}(k)] = dQ_{P_-}[h,k] + \text{tr}(h_{-+}k_{+-} - k_{-+}h_{+-})/2\,. \qquad (6.7.14)$$

To understand this formalism better, we will consider the related problem of implementing Bogoliubov $*$-automorphisms in the complex CAR formalism. After all $\text{Ad}(\exp itdQ_{P_-}(h))$ will implement $\tau(e^{ith})$. We work with the CAR algebra $A(H)$ built over a separable Hilbert space H. We will show that a Bogoliubov $*$-automorphism $\alpha(U)$ is implementable in a Fock representation π_P if and only if U_{+-}, U_{-+} are Hilbert–Schmidt. We will give a constructive discussion of this, beginning with the case where $U_{\varepsilon\varepsilon}$ have no kernels, i.e. $\ker(U_{\varepsilon\varepsilon}) = 0$.

We first write the unitary U as

$$\begin{pmatrix} 1 & U_{+-}U_{--}^{-1} \\ 0 & 1 \end{pmatrix} \begin{pmatrix} U_{++} - U_{+-}U_{--}^{-1}U_{-+} & 0 \\ 0 & U_{--} \end{pmatrix} \begin{pmatrix} 1 & 0 \\ U_{--}^{-1}U_{-+} & 1 \end{pmatrix} = U\,.$$

The outer factors are implemented by

$$\exp dF_P \begin{pmatrix} 0 & -U_{+-}U_{--}^{-1} \\ 0 & 0 \end{pmatrix}, \quad \exp dF_P \begin{pmatrix} 0 & 0 \\ -U_{--}^{-1}U_{-+} & 0 \end{pmatrix}$$

respectively. The middle factor commutes with P, but in general not one of the three components is unitary.

For this reason suppose S is an operator on a Hilbert space H with a left inverse S^{-1} (i.e. $S^{-1}S = 1$). Then let $\alpha(S)$ denote the induced endomorphism of the local algebra A_0, the $*$-algebra generated by $\{c(f) \mid f \in H\}$, such that

$$\alpha(c(f)) = c(Sf)\,, \qquad \alpha(c^*(f)) = c^*(S^{-1*}f)\,,$$

$f \in H$ (so that α is symmetric if and only if S is isometric, cf. Lemma 6.12).

We have already seen that the quasi-free derivation $d\alpha \begin{pmatrix} A_{++} & 0 \\ 0 & A_{--} \end{pmatrix}$ is implemented in the state ω_P by $dF_P(A_{++} \oplus A_{--}) = dF_0(A_{++} \oplus (-A^t_{--}))$, where $A^t = JA^*J$ and J is the conjugation commuting with P. Similarly, the quasi-free automorphism $\alpha(A_{++} \oplus A_{--})$ is implemented in ω_P by

$$\begin{aligned} F_P(A_{++} \oplus A_{--}) &= F_0(A_{++} \oplus (A^t_{--})^{-1})\,. \\ &= :\exp dF_0\left((A_{++} - 1) \oplus ((A^t_{--})^{-1} - 1)\right): \quad \text{by (6.7.6)} \\ &= :\exp dF_P\left((A_{++} - 1) \oplus (1 - A_{--}^{-1})\right): \end{aligned}$$

as $-((A^t_{--})^{-1} - 1)^t = 1 - A_{--}^{-1}$. Thus the middle factor is implemented by

$$F_P\left(-U_{++} + U_{+-}U_{--}^{-1}U_{-+} \oplus U_{--}\right)$$

$$= :\exp dF_P\left(U_{++} - 1 - U_{+-}U_{--}^{-1}U_{-+} \oplus (1 - U_{--}^{-1})\right):$$

Combining, the Bogoliubov automorphism $\alpha(U)$ is implemented by

$$\exp dF_P\begin{pmatrix} 0 & \Lambda_{+-} \\ 0 & 0 \end{pmatrix} : \exp dF_P\begin{pmatrix} \Lambda_{++} & 0 \\ 0 & \Lambda_{--} \end{pmatrix} : \exp dF_P\begin{pmatrix} 0 & 0 \\ \Lambda_{-+} & 0 \end{pmatrix}$$
$$= :\exp dF_P(\Lambda): \qquad (6.7.15$$

where $\Lambda =$

$$\begin{pmatrix} U_{++} - 1 - U_{+-}U_{--}^{-1}U_{-+} & U_{+-}U_{--}^{-1} \\ U_{--}^{-1}U_{-+} & 1 - U_{--}^{-1} \end{pmatrix} = (1 + (U-1)(1-P))^{-1}(U-1)\,.$$

By the unitarity of U, we have the relations

$$\begin{aligned} \Lambda_{--} &= 1_{--} - U_{--}^{-1} && = 1_{--} - U_{--}^* + U_{-+}^* U_{++}^{*\,-1} U_{+-}^* \\ \Lambda_{-+} &= U_{--}^{-1} U_{-+} && = -U_{-+}^* U_{++}^{*\,-1} \\ \Lambda_{+-} &= U_{+-} U_{--}^{-1} && = -U_{++}^{*\,-1} U_{+-}^* \\ \Lambda_{++} &= U_{++} - 1_{++} - U_{+-} U_{--}^{-1} U_{-+} && = -1_{++} + U_{++}^{*\,-1} \end{aligned} \qquad (6.7.16$$

$$\begin{aligned} U_{++} - \Lambda_{+-}U_{-+} &= 1_{++} + \Lambda_{++} \\ \Lambda_{-+} - U_{-+} + \Lambda_{--}U_{-+} &= 0 \\ U_{--} + \Lambda_{-+}^* U_{+-} &= 1_{--} - \Lambda_{--}^* \\ U_{+-} + \Lambda_{+-}^* + \Lambda_{++}^* U_{+-} &= 0\,. \end{aligned}$$

So far we have been rather cavalier with these computations, let us be a bi more precise regarding what we have so far. We know from Proposition 6.13 tha if S is a projection on a Hilbert space H, then the map

$$f \otimes \overline{g} \to \pi_S[c^*(f)c(g)]$$

extends to a bounded linear map dF_S from $T(H)$ into $\pi_S(A(H))$ such that fo any complete orthonormal sequence h_i in H

$$dF_S(L) = \textstyle\sum_{i=1}^{\infty} \pi_S\left[c^*(Lh_i)c(h_i)\right]\,.$$

We know we can also make sense of $\exp dF_S(L)\varphi$ for certain Hilbert–Schmid operators L on H and certain vectors φ in $F(H)$ – see the estimates (6.7.9). T understand how this may depend continuously on L, it is useful to have furthe estimates such as (6.7.18) below.

First suppose that S_1 and S_2 are finite-dimensional projections on H wit $S_i \leq S$. Let $L_i \in B(H)$ with $L_i = (1-S)L_iS_i$, $i = 1, 2$. Let h_j be a complet orthonormal sequence set for $\mathrm{lin}\{S_iH \mid i = 1,2\}$. Then if J is a conjugation commuting with S

$$dF_S(L_i) = \textstyle\sum_j c^*(L_ih_j)c^*(Jh_j)$$

and since $\langle L_ih_j, Jh_k\rangle = 0$, for all i,j,k, we have, using (6.7.7) or (6.7.8),

$$\begin{aligned}
&\langle \exp dF_S(L_1)\Omega, \exp dF_S(L_2)\Omega\rangle \\
&= \sum_{n=0}^{\infty} \frac{1}{(n!)^2} \sum_{|I|=|K|=n} \langle c^*(L_1h_I)c^*(Jh_I)\Omega, c^*(L_2h_K)c^*(Jh_K)\Omega\rangle \\
&= \det(1 + L_2^*L_1)\,. \qquad (6.7.17)
\end{aligned}$$

Thus

$$\begin{aligned}
\|\exp dF_S(L_1)\Omega - \exp dF_S(L_2)\Omega\|^2 &= \det(1 + L_1^*L_1) + \det(1 + L_2^*L_2) \\
&\quad - \det(1 + L_1^*L_2) - \det(1 + L_2^*L_1)\,.
\end{aligned}$$

Using the continuity property of the determinant, we see that the map $L \to \exp dF_S(L)\Omega$ defined on the finite-rank operators from SH into $(1-S)H$ extends to a continuous map from $HS(SH, (1-S)H)$ into $F(H)$.

More generally consider two projections S_1, S_2 on H with $S_1 \leq S_2$. We can assume that our conjugation J commutes with both S_1 and S_2. Then with similar arguments to the above we have

$$\langle \exp dF_{S_1}(L_1)\Omega, \exp dF_{S_2}(L_2)\Omega\rangle = \det(1 + L_2^*L_1) \qquad (6.7.18)$$

for all Hilbert–Schmidt operators L_i on H with $L_i = (1 - S_i)L_iS_i$. Then if S is a projection on H, take a sequence of projections $S_n \nearrow S$, and a conjugation J to commute with all S_n. Then if L_n are Hilbert–Schmidt operators on H with $L_n = (1 - S_n)L_nS_n$ and $L_n \to L$ in Hilbert–Schmidt norm, we have

$$\lim_{n\to\infty} \exp dF_{S_n}(L_n)\Omega = \exp dF_S(L)\Omega\,. \qquad (6.7.19)$$

Finally, for any projection S on H, and any finite-rank operator L from SH into $(1 - S)H$, and any finite sequence $f_1, \ldots, f_n$ in H, we have

$$\exp\,[dF_S(L)]\, c^*(f_1)\cdots c^*(f_n)\Omega = c^*(Lf_1)\cdots c^*(Lf_n)\exp\,[dF_S(L)]\,\Omega\,. \quad (6.7.20)$$

In this way we can define $\exp[dF_S(L)]\varphi$ for all φ in

$$c^*(H)\Omega = \{c^*(f_1)\cdots c^*(f_n)\Omega \mid f_i \in H, \quad n = 0,1,2,\ldots\}\,,$$

such that (6.7.20) holds.

Let Λ be a bounded operator on H with matrix $[\Lambda_{\sigma\sigma'} = S_\sigma\Lambda S_{\sigma'}]$, if $S_+ = 1 - S$, $S_- = S$, such that Λ_{-+} is Hilbert–Schmidt. We will define an operator $:\exp dF_S(\Lambda):$ with domain $c^*(H)\Omega$. Here the dots will indicate normal ordering with respect to the Fock state and will give a precise meaning for $\sum : dF_S(\Lambda)^n : /n!$. Let $\{f_i\}$ and $\{g_i\}$ be complete orthonormal bases for S_+H and S_-H, respectively. If $\varphi \in c^*(H)\Omega$, we define

$$: \exp dF_S(\Lambda) : \varphi = \sum_{L=0}^{\infty} \exp\left(dF_S(\Lambda_{+-})\right) U_L \varphi \tag{6.7.21}$$

where

$$U_L = \sum_{j+k+l=L} O_{jkl}/j!k!l!$$

and where

$$O_{jkl} = \sum_{|\alpha|=j, |\beta|=k, |\gamma|=\ell} a^*(\Lambda_{++} f_\alpha) b^*(-J\Lambda_{--} J g_\beta) b(g_\beta) b(J\Lambda_{-+} f_\gamma) a(f_\gamma) a(f_\alpha)\,.$$

Here the summation is over all subsets $\alpha = (p_1, \ldots, p_j)$ etc. (not necessarily in increasing order) so that for example $a^*(\Lambda_{++} f_\alpha) = a^*(\Lambda_{++} f(p_1)) \cdots a^*(\Lambda_{++} f(p_j))$, and a, b denote c on H_+, H_- respectively as usual. Then O_{jkl} maps $c^*(H)\Omega$ into itself, and $O_{jkl}\varphi = 0$ for large j, k, l for $\varphi \in c^*(H)\Omega$. In (6.7.20) $\exp dF_S(\Lambda_{+-})$ is defined on $c^*(H)\Omega$. Hence (6.7.21) is well defined.

This gives a precise meaning for the expression which, up to a normalization which we will sort out in a moment in (6.7.22) and (6.7.23), is the implementing unitary in the case in which $U_{\sigma\sigma}$ have no kernels. The general case is discussed in Exercise 6.12. The Bogoliubov automorphism $\alpha(U)$ on $A = A^{\mathrm{F}}(H)$ maps

$$\begin{aligned} a(f) \to a'(f) &= a(U_{++} f) + b^*(JU_{-+} f) \\ b(Jg) \to b'(Jg) &= b(JU_{--} g) + a^*(U_{+-} g) \end{aligned}$$

and by uniqueness of the Fock representation is implementable in π_P if and only if there is a corresponding vacuum vector, i.e. if there is a non-zero vector Ω' in F such that

$$a'(f)\Omega' = b'(Jg)\Omega' = 0, \quad f \in H_+, g \in H_-\,.$$

In this case, normalizing $\|\Omega'\| = 1$, the unitary $\mathcal{U}$ implementing $\alpha(U)$ in ω_P is defined by $\mathcal{U}\Omega = \Omega'$, and more generally:

$$\mathcal{U} x \Omega = \alpha(U)(x)\Omega'\,, \tag{6.7.22}$$

for x local. Note that $: \exp dF_P(\Lambda) : \Omega$ is not normalized (generally) so that using (6.7.17), we should put

$$\Omega' = \det(1_{++} + \Lambda_{+-}^* \Lambda_{+-})^{-\frac{1}{2}} : \exp dF_P(\Lambda) : \Omega\,. \tag{6.7.23}$$

Conversely, suppose $\alpha(U)$ is implementable in the state ω_P, and suppose $U_{\sigma\sigma}^*$ may have kernels. Take $\{f_j^\sigma\}_{j=1}^\infty$, a complete orthonormal basis for $H_\sigma = P_\sigma H$,

$\sigma = \pm$ in such a way that $\{f_j^\sigma\}_{j=1}^{L_\sigma}$ and $\{f_j^\sigma\}_{j=L_\sigma+1}^{\infty}$ are orthonormal bases for $\ker U_{\sigma\sigma}^*$ and $(\ker U_{\sigma\sigma}^*)^\perp = \overline{\operatorname{Ran} U_{\sigma\sigma}}$ respectively. We will show $L_\sigma < \infty$. If ρ, τ are finite ordered subsets of positive integers, we let as usual

$$a^*(f_\rho) = a^*(f_{\rho_1}) \cdots a^*(f_{\rho_m}), \quad b^*(Jf_\tau) = b^*(Jf_{\tau_1}) \cdots b^*(Jf_{\tau_n})$$

so that

$$\phi_{\rho,\tau} = a^*(f_\rho^+) b^*(Jf_\tau^-)\Omega$$

will be a complete orthonormal basis for the anti-symmetric Fock space $F_a(H_+ \oplus H_-)$. With respect to this basis, we can write

$$\Omega' = \textstyle\sum_{\rho,\tau} \alpha_{\rho,\tau} \phi_{\rho,\tau} .$$

From the behaviour of the new vacuum vector Ω' in the presence of the Bogoliubov automorphism we have

$$a^*(f_i^+)\Omega' = b^*(Jf_j^-)\Omega' = 0 , \quad i = 1, \ldots, L_+, \, j = 1, \ldots, L_- .$$

This means that

$$\alpha_{\rho,\tau} = 0 \tag{6.7.24}$$

f there exists just one i such that $i \leq L_+$ and $i \notin \rho$. Since ρ is always finite his forces $L_+ < \infty$ (and similarly $L_- < \infty$). Define $P^{m',n'}$ to be the projection $P^{m',n'} \phi_{\rho,\tau} = \delta_{m',m}\delta_{n,n'}\phi_{\rho,\tau}$ if $m = |\rho|$, $n = |\tau|$. Thus by (6.7.24),

$$P^{m,n} = 0 \text{ if } m < L_+ \text{ or } n < L_- . \tag{6.7.25}$$

Now $a'(f)\Omega = 0$ so that from $P^{L_+ + m, L_- + n} a'(f) = 0$, $m, n \geq 0$, we have

$$a(U_{++}f)P^{L_+ + m+1, L_- + n}\Omega' = -b^*(JU_{-+}f)P^{L_+ + m, L_- + n-1}\Omega' . \tag{6.7.26}$$

Then again from (6.7.24) if $P^{L_+ + m, L_- + n-1}\Omega' = 0$, then

$$P^{L_+ + m+1, L_- + n}\Omega' = 0.$$

So from (6.7.25) we have $P^{L_+ + m+k, L_- + k}\Omega' = 0$, $m > 0, k \geq 0$ and similarly $P^{L_+ + k, L_- + m+k} = 0$. Now

$$P^{L_+, L_-}\Omega' = \beta \prod_{i=1}^{L_+} a^*(f_i^+) \prod_{j=1}^{L_-} b^*(Jf_j^-)\Omega$$

o that if $\beta = 0$, then we would deduce $P^{L_+ + m, L_- + m}\Omega' = 0$ for all $m \geq 0$ and we would have shown $\Omega' = 0$. Thus $\beta \neq 0$ and we can write

$$P^{L_+ + 1, L_- + 1}\Omega' = \sum_{k=L_+ + 1}^{\infty} \sum_{l=L_- + 1}^{\infty} H_{kl} a^*(f_k^+) b^*(Jf_l^-) P^{L_+, L_-}\Omega' \tag{6.7.27}$$

and define a Hilbert–Schmidt operator $H : \overline{\mathrm{Ran}(U_{--})} \to \overline{\mathrm{Ran}(U_{++})}$ by $H = \sum H_{kl} f_k^+ \otimes \overline{f_l^-}$. Then from (6.7.26) with $m = 0, n = 1$ we have

$$a(U_{++}f)P^{L_+ +1, L_- +1}\Omega' = b^*(JU_{-+}f)P^{L_+, L_-}\Omega' .$$

Comparing this with (6.7.27) we see $H^*U_{++} = -U_{-+}$ on $\left(\ker U_{++}^*\right)^\perp$, and similarly $HU_{--} = U_{+-}$ on $(\ker U_{--})^\perp$. Hence U_{-+} and U_{+-} are Hilbert–Schmidt.

Finally, note that $\left\{U^* f_j^+\right\}_{j=1}^{L+}$ and $\left\{U^* f_j^-\right\}_{j=1}^{L-}$ are orthonormal bases for $\ker U_{--}$ and $\ker U_{++}$ respectively.

Thus what we have done is obtain the necessary and sufficient conditions in the complex CAR formalism for a Bogoliubov unitary U to be implementable in a Fock state ω_P, namely $[P, U]$ should be Hilbert–Schmidt. In the self dual formalism, this generalizes to

Theorem 6.16 *Let U be a unitary on a Hilbert space H commuting with a conjugation Γ, and P and P' basis projections in $Q(H, \Gamma)$. Then*

(a) $\tau(U)$ is implementable in φ_P if and only if $[U, P]$ is Hilbert–Schmidt.

(b) φ_P and $\varphi_{P'}$ are equivalent if and only if $P - P'$ is Hilbert–Schmidt.

Remark If P_1, P_2 are basis projections with $\|P_1 - P_2\| < 1$, then $X = P_1P_2 + (1 - P_1)(1 - P_2)$ has polar decomposition $U_{12}R$, where $R = |X|^{1/2} = (1 - (P_1 - P_2)^2)^{1/2}$, commutes with Γ, P_1, P_2. Then U_{12} is unitary taking P_2 to $U_{12}P_2U_{12}^* = P_1$.

Proof of Theorem 6.16 Now $\tau(U)$ is implementable in φ_P if and only if $\varphi_P \circ \tau(U) = \varphi_{U^*PU}$ is equivalent to φ_P. Thus (b) implies (a) as $U^*PU - P = U^*(PU - UP)$.

(b) We will give the details only for showing that if $P - P'$ is Hilbert–Schmidt then φ_P and $\varphi_{P'}$ are unitarily equivalent. Suppose $P - P'$ is Hilbert–Schmidt. Then $P' \wedge (1 - P)$ is the eigenspace of $P' - P$ for eigenvalue 1 and so is of finite dimension. By rotation we can eliminate this and reduce to the case $P' \wedge (1 - P) = P \wedge (1 - P') = 0$. More precisely, if $e_1, \ldots, e_n$ is a complete orthonormal basis for $P' \wedge (1 - P)$, then $\prod_{j=1}^n (B(e_j) + B(e_j)^*)$ is a unitary implementing a unitary $W = 1+$ a finite rank operator, where $(-1)^n U$ interchanges $P' \wedge (1 - P)$ and $P \wedge (1 - P')$, and fixes $P \wedge P'$, $(1 - P) \wedge (1 - P')$. Replacing P' by $P'' = WP'W^*$ we have $P'' \wedge (1 - P) = 0 = P \wedge (1 - P'')$. Then using the preceding Remark with $P_1 = P$, $P_2 = P''$, $R^{-1} - 1 = ((1 - (P - P'')^2)^{1/2} - 1$ is trace class and $U_{12} - 1 \doteq U_{12}R(R^{-1} - 1) + R - 1$ is Hilbert–Schmidt. Then $U_{12} + U_{12}^* = 2h^{-1}(1 - (P - P'')^2) > 0$, and so $\|U_{12} - 1\| < \sqrt{2}$. We can then write $U_{12} = e^H$ where H is skew adjoint Hilbert–Schmidt, $\Gamma H \Gamma = H$, $\|H\| < \pi/2$. Then as we shall see in Section 6.9 $\tau(U_{12})$ is implementable by $\exp dQ_P(H)$. □

Exercise 6.11 Let Γ be a conjugation on a Hilbert space K, and $S_1, S_2 \in Q(K, \Gamma)$. Show that φ_{S_1} and φ_{S_2} are quasi-equivalent states if and only if $S_1^{1/2} - S_2^{1/2}$ is Hilbert–Schmidt.

Exercise 6.12 Let U be a unitary on a Hilbert space H, P a projection on H such that $[U,P]$ is Hilbert–Schmidt. Let $\{f_i^\sigma \mid i=1,\ldots,L_\sigma\}$ be a complete orthonormal basis for $\ker(U_{\sigma\sigma}=P_\sigma UP_\sigma)$, $\sigma=+,-$ where $P_+=1-P$, $P_-=P$. If ρ, (respectively τ) is a subset of $\{1,\ldots,L_-\}$ (respectively $\{1,\ldots,L_+\}$ with complement $\overline{\rho}$ (respectively $\overline{\tau}$), identified as subsets of $\{1,\ldots,L_-+L_+\}$ ordered in the natural way so that $\rho\sqcup\overline{\rho}=\{1,\ldots,L_+\}$, let $\mathrm{sgn}(\rho,\tau)$ be the signature of the permutation of $\{1,\ldots,L_++L_-\}$ into the order $(\overline{\rho},\overline{\tau},\rho,\tau)$. Consider the operator

$$\mathcal{U}=\textstyle\sum_{\rho,\tau}\mathrm{sgn}(\rho,\tau)a^*(Uf_\rho^-)b^*(JUf_\tau^+)\tilde{\mathcal{U}}b(Jf_{\overline{\rho}}^-)a(f_{\overline{\tau}}^+)$$

where $\tilde{\mathcal{U}}$ is now defined as in (6.7.22) and (6.7.23) for U replaced by $(-1)^{L_++L_-}U$. Show that on the vacuum vector Ω, $\mathcal{U}$ reduces to

$$\begin{aligned}\mathcal{U}\Omega=&\det(1_{--}+\Lambda_{+-}^*\Lambda_{+-})^{-\frac{1}{2}}a^*(Uf^-_{\{1,\ldots,L_-\}})b^*(JUf^+_{\{1,\ldots,L_+\}})\\ &\exp dF_P(\Lambda_{+-})\Omega\,.\end{aligned}$$

and that $\mathcal{U}$ is unitary and implements the Bogoliubov automorphism $\alpha(U)$ in the quasi-free state ω_P.

6.8 Completely positive quasi-free maps

So far we have discussed quasi-free endomorphisms and quasi-free states, and the implementability of quasi-free automorphisms. These notions can be extended to the realm of quasi-free completely positive maps. These we will define via the process of normal ordering, which we have already utilized to construct quasi-free states. After that we will discuss an analogue, for completely positive maps, of unitary implementation of automorphisms. More generally let T be a completely positive contraction on a C^*-algebra A, which leaves a state φ on A invariant. Let (π,H,ξ) be the GNS decomposition of φ. Then by the Schwarz inequality for a completely positive map there is a well-defined contraction $F=F(T)$ say on H such that

$$F\left(\pi(x)\xi\right)=\pi\left(T(x)\right)\xi\,,\quad\text{for all } x \text{ in } A. \tag{6.8.1}$$

We are interested in showing that

$$x\to\pi\left(T(x)\right)-F\pi(x)F^* \text{ is completely positive} \tag{6.8.2}$$

from A into $B(H)$. This is an analogue of unitary implementation of automorphisms, and we will particularly be concerned with this concept for quasi-free completely positive maps. An operator F such that (6.8.2) holds is called a *dissipator* for T in the representation π.

Lemma 6.17 *Let T and T' be completely positive contractions on a C^*-algebra A, having a common invariant separating state φ such that $\varphi[T(x)y]$ $=\varphi[xT'(y)]$, for all x,y in A. Then if $F=F(T)$, the map $x\to\pi(T(x))-F\pi(x)F^*$ is completely positive from A into $B(H)$.*

Proof If φ is separating, then ξ is cyclic for $\pi(A)'$ and the set of vector states $\{\omega_{a\xi} \mid a \in \pi(A)', \|a\xi\| = 1\}$ is dense in all vector states on $\pi(A)$. If $a \in \pi(A)'$, then $\omega_{a\xi} \leq \|a\|^2\omega_\xi$ on $\pi(A)$, so that the normal folium $\mathcal{F}(\varphi) = \{\psi \in A_+^* \mid \psi \leq \alpha\varphi$, some $\alpha\}^-$. Hence T, T' induce maps on the predual of $\pi(A)''$. It follows that there exist completely positive normal maps $\hat{T}$ and $\hat{T}'$ on the von Neumann algebra $B = \pi(A)''$ such that $\hat{T}(\pi(x)) = \pi(T(x))$ and $\hat{T}'(\pi(x)) = \pi(T'(x))$ for all x in A. Then $\langle \hat{T}(a)b\xi, \xi\rangle = \langle a\hat{T}'(b)\xi, \xi\rangle$ for all a, b in B. The vector state $b \to \langle b\xi, \xi\rangle; b \in B$, is a faithful normal state on B with modular automorphism group $\{\sigma_t \mid t \in \mathbb{R}\}$ say. Then $\hat{T}$ and $\hat{T}'$ commute with the modular automorphism group σ_t. In fact let S be the closure of the map $b\xi \to b^*\xi; b \in B$. Then since $\hat{T}$ and $\hat{T}'$ are $*$-maps, we see that $F(\hat{T})$ and $F(\hat{T}')$ commute strongly with S. But $F(\hat{T}') = F(\hat{T})^*$ and so $F(\hat{T})$ commutes strongly with both S and S^* and hence also with the modular operator $\Delta = (S^*S)^{1/2}$. Thus $\hat{T}$ commutes with the modular automorphism group $\sigma_t = \Delta^{it}(\cdot)\Delta^{-it}$ since ξ is a separating vector for B. Similarly $\hat{T}'\sigma_t = \sigma_t\hat{T}'$ for all t in $\mathbb{R}$. Let B_e denote the σ-weakly dense $*$-subalgebra of entire elements of B. Then $\sigma_z(a)^* = \sigma_{\bar{z}}(a^*)$ and $\langle\sigma_z(a)\xi, \xi\rangle = \langle a\xi, \xi\rangle$ for all a in B_e, z in $\mathbb{C}$. Moreover $\hat{T}$ and $\hat{T}'$ leave B_e invariant and $\hat{T}\sigma_z(a) = \sigma_z\hat{T}(a), \hat{T}'\sigma_z(a) = \sigma_z\hat{T}'(a)$ for all a in B_e, z in $\mathbb{C}$. We will use the KMS condition in the form: $\langle ab\xi, \xi\rangle = \langle b\sigma_{-i}(a)\xi, \xi\rangle$, for all $a \in B_e$, $b \in B$. Then for all x, y and B_e we have

$$
\begin{aligned}
&\langle x^*x\hat{T}'(y)\xi, \hat{T}'(y)\xi\rangle = \langle \hat{T}'(y^*)x^*x\hat{T}'(y)\xi, \xi\rangle = \langle x\hat{T}'(y)\sigma_{-i}[\hat{T}'(y^*)x^*]\xi, \xi\rangle \\
&= \langle \sigma_{-i/2}\left\{\sigma_{i/2}(x)\sigma_{i/2}[\hat{T}'(y)]\sigma_{-i/2}[\hat{T}'(y^*)]\sigma_{-i/2}(x^*)\right\}\xi, \xi\rangle \\
&= \langle \sigma_{i/2}(x)\hat{T}'(\sigma_{i/2}y)\hat{T}'[\sigma_{i/2}(y)^*]\sigma_{i/2}(x)^*\xi, \xi\rangle \\
&\leq \langle \sigma_{i/2}(x)\hat{T}'[\sigma_{i/2}(y)\sigma_{i/2}(y)^*]\sigma_{i/2}(x)^*\xi, \xi\rangle \text{ by the Kadison -- Schwarz inequality} \\
&= \langle \hat{T}'\left\{\sigma_{-i/2}[\sigma_i(y)y^*]\right\}\sigma_{-i/2}(x^*)\sigma_{-i/2}(x)\xi, \xi\rangle \\
&= \langle T'[\sigma_i(y)y^*]x^*x\xi, \xi\rangle = \langle \sigma_i(y)y^*\hat{T}(x^*x)\xi, \xi\rangle = \langle y^*\hat{T}(x^*x)y\xi, \xi\rangle
\end{aligned}
$$

□

Let $T : H \to K$ be a contraction between Hilbert spaces H and K intertwining with conjugations Γ_H and Γ_K. (Usually we will drop the subscripts on conjugations and refer to them all by Γ). We wish to construct in a quasi-free fashion completely positive maps from $A(H)$ into $A(K)$ extending T in the sense that $B(f)$ must be mapped onto $B(Tf)$. Moreover we wish this construction to be functorial and to reduce to the study of quasi-free states for the case $T = 0$, and to quasi-free $*$-homomorphisms for isometric T. Thus if T is an isometry we define $A(T) = \tau(T)$ to be the unique $*$-homomorphism from $A(H)$ into $A(K)$ such that $A(T)(B(f)) = B(Tf)$, for all f in H. Projections at the Hilbert space level will give rise to conditional expectations at the C^*-level. Thus let H and K be Hilbert spaces, and R in $Q(H, \Gamma)$. Consider the $*$-homomorphism π of $A(K \oplus H)$ into $A(K) \otimes B(F_R(H))$ given by

$$\pi(B(k \oplus h)) = B(k) \otimes V_R + 1 \otimes \pi_R(B(h)), \quad k \oplus h \in K \oplus H$$

where V_R is the parity unitary on $F_R(H)$ such that $V_R\Omega_R = \Omega_R$, $V_R B(f) V_R = -B(f)$, $f \in H$. We can then define a projection N from $A(K \oplus H)$ onto $A(K)$ by

$$x \to 1 \otimes \tilde{\varphi}_R(\pi(x)); \quad x \in A(K \oplus H) \tag{6.8.3}$$

where $\tilde{\varphi}_R$ is the vector state $\Omega_R \otimes \overline{\Omega}_R$ on $B(F_R(H))$. Then for any operator S in $Q(K, \Gamma)$ we have

$$\varphi_{S \oplus R} = \varphi_S \circ N\,. \tag{6.8.4}$$

Now let $T : H \to K$ be an arbitrary contraction between Hilbert spaces H and K. Then if U denotes the unitary on $K \oplus H$, (with conjugation $\Gamma_K \oplus \Gamma_H$):

$$U = \begin{bmatrix} -(1 - TT^*)^{\frac{1}{2}} & T \\ T^* & (1 - T^*T)^{\frac{1}{2}} \end{bmatrix} \quad \text{on } K \oplus H$$

and W_1' and W_2 are the isometries $h \to (0, h)$, $h \in H$; $k \to (k, 0)$, $k \in K$ respectively, one has the unitary dilation $T = W_2^* U W_1'$ for T (cf. Chapter 4). In particular $T = W_2^* W_1$ is a decomposition of T into a co-isometry and an isometry where

$$W_1 h = U W_1' h = Th \oplus (1 - T^*T)^{\frac{1}{2}} h; \quad h \in H\,.$$

If T intertwines with the conjugations, then so do W_1', U and W_2. If R is in $Q(H, \Gamma)$, we define the completely positive map $A_R(T)$ by composition:

$$A_R(T) = N \circ A(W_1)$$

where N is the conditional expectation in (6.8.3) of $A(K \oplus H)$ onto $A(K)$.

In the first place note that $A_R(T)$ maps the quasi-free states of $A(K)$ into those of $A(H)$. Let S be an operator in $Q(K, \Gamma)$. Then

$$\varphi_S A_R(T) = \varphi_S N A(W_1) = \varphi_{S \oplus R} A(W_1) = \varphi_Q$$

where $Q = W_1^*(S \oplus R) W_1 = T^*ST + (1 - T^*T)^{1/2} R (1 - T^*T)^{1/2}$.

Thus if $TR = ST$ we have $\varphi_S A_R(T) = \varphi_R$. In fact even more is true:

$$\varphi_S[A_R(T)(x)y] = \varphi_R[x A_S(T^*)(y)] \tag{6.8.5}$$

for all x in $A(H)$, y in $A(K)$, since

$$\begin{aligned} \varphi_S[A_R(T)(x)y] &= \varphi_S[NA(W_1)(x)y] = \varphi_S N[A(W_1)(x)A(W_2)(y)] \\ &= \varphi_{S \oplus R}[A(U)A(W_1')(x)A(W_2)(y)] = \varphi_{S \oplus R}\left[A(W_1')(x)A(U^{-1})A(W_2)(y)\right] \end{aligned}$$

$= \varphi_R[xA_S(T^*)(y)]$.

Now let T be an arbitrary contraction between Hilbert spaces H and K intertwining with the conjugations and take R in $Q(H,\Gamma)$. Then if $D = (1 - T^*T)^{1/2}$ and $f_1, \ldots, f_n$, are elements of H, we have from the definitions:

$$A_R(T)[B(f_1)\cdots B(f_n)] = \textstyle\sum_\sigma \varepsilon B(Tf_\sigma)\varphi_R(B(Df_{\sigma'}))\,. \tag{6.8.6}$$

The summation extends over all subsets $\sigma = \{i_1, \ldots, i_p\}$ of $\{1, \ldots, n\}$ in increasing order; and if $\sigma' = \{j_1, \ldots, j_{n-p}\}$ is the complement of σ in $\{1, \ldots, n\}$, in increasing order, the ε is the signature of the permutation $(1, \ldots, n) \to (\sigma, \sigma')$. Assume now that there exists an S in $Q(K,\Gamma)$, and $TR = ST$. Then simplifications take place. If T is a contraction between Hilbert spaces H and K which intertwines with operators R and S in $Q(H,\Gamma)$, $Q(K,\Gamma)$, and with the conjugations, we have

$$A_R(T)\,[:B(h_1)\cdots B(h_n):_R] =: B(Th_1)\cdots B(Th_n):_S \tag{6.8.7}$$

for all $h_1, \ldots, h_n$ in H. (However, recall from the definition that $A_R(T)$ in no way depends on S). This follows from the definition of normal ordering (6.6.22), the definition of the $A_R(T)$ and Pfaffian identities.

From this we can conclude that if T' is another contraction from K into another Hilbert space L, and that if there exists an operator Q in $Q(L,\Gamma)$, $T'S = QT'$, $T'\Gamma = \Gamma T'$, then

$$A_S(T')A_R(T) = A_R(T'T)\,. \tag{6.8.8}$$

We now study the contraction $F_{R,S}(T)$ from $F_R(H)$ into $F_S(K)$ given by

$$F_{R,S}(T)\,(\pi_R(x)\Omega_R) = \pi_S\,(A_R(T)(x))\,\Omega_S; \quad x \in A(H) \tag{6.8.9}$$

If $H = K$ and $R = S$, we write $F_R(T)$ for $F_{R,R}(T)$. If R and S have no Fock part then Ω_R and Ω_S are separating, and we are in the situation of Lemma 6.17. In general let

$$R_n = 1/n + (1 - 2/n)R \quad \text{and} \quad S_n = 1/n + (1 - 2/n)S\,.$$

Then R_n in $Q(H,\Gamma)$ and S_n in $Q(K,\Gamma)$ have no Fock parts, $TR_n = S_nT$, and $R_n \to R$, $S_n \to S$ as $n \to \infty$. Thus we conclude from Lemma 6.17 that

$$x \to \pi_S\,(A_R(T)(x)) - F\,(\pi_R(x))\,F^* \tag{6.8.10}$$

is completely positive from A into $B(F_S(K))$, where $F = F_{R,S}(T)$. In what follows we will often dispense with the symbols π_R and π_S when taking representations.

If T is a co-isometry we have $A_R(T)(1) - F_{R,S}(T)1F_{R,S}(T)^* = 0$ and so from the preceding paragraphs, we have $A_R(T) = F(\cdot)F^*$. Conversely suppose $A_R(T) = F(\cdot)F^*$ holds for some contraction T. Then $1 = F_{R,S}(T)F_{R,S}(T^*) = F_S(TT^*)$, and by considering the action on $\pi_S(B(h))$ it follows that T is a co-isometry. We summarize these results as follows.

Theorem 6.18 *Let T be a contraction between Hilbert spaces H and K, with R, S operators in $Q(H,\Gamma)$ and $Q(K,\Gamma)$ respectively such that $TR = ST$, and $T\Gamma = \Gamma T$. Then there exists an unique completely positive unital map $A_R(T)$ from $A(H)$ into $A(K)$ such that*

$$A_R(T) : B(f_1) \cdots B(f_n) :_R = : B(Tf_1) \cdots B(Tf_n) :_S$$

for all $f_1, \ldots, f_n$ in H. Moreover:

(i) $\varphi_S[A_R(T)(x)y] = \varphi_R[xA_S(T^*)y]$, $x \in A(H)$, $y \in A(K)$.

(ii) The map $x \to \pi_S(A_R(T)(x)) - F(\pi_R(x))F^*$ is completely positive from $A(H)$ into $B(F_S(K))$, and where F is the contraction from $F_R(H)$ into $F_S(K)$ given by $\pi_R(x)\Omega_R \to \pi_S(A_R(T)(x))\Omega_S$. Also $\pi_S(A_R(T)(x)) = F(\pi_R(x))F^*$, for all x in $A(H)$ if and only if T is a co-isometry, and $A_R(T)$ is a homomorphism if and only if T is an isometry.

Restricting to the complex formalism, suppose T is a contraction between Hilbert spaces H and K, which intertwines with projections R, S on H and K, respectively, we let $A_R(T)$ denote the completely positive unital map from $A(H)$ into $A(K)$ such that

$$\begin{aligned} &A_R(T)\,[: a^*(f_1) \cdots a^*(f_n)a(g_1) \cdots a(g_m) :_R] \\ &= : a^*(Tf_1) \cdots a^*(Tf_n)a(Tg_1) \cdots a(Tg_m) :_S \end{aligned} \tag{6.8.11}$$

where double dots indicate normal ordering with respect to the states ω_R and ω_S (6.6.18). Thus $A_R(0) = \omega_R$, and if T is an isometry, $A_R(T)$ is a homomorphism independent of R, denoted by $A(T) = \alpha(T)$. Furthermore there is a well-defined contraction $F = F_{R,S}(T)$ from $F(H)$ into $F(K)$ such that

$$F\pi_R(x)\Omega = \pi_S(A_R(T)(x))\Omega$$

for all x in $A(H)$. If $H = K, R = S$, we write $F_R(T) = F_{R,R}(T)$, and if $R = S = 0$, $F(T) = F_0(T)$ is the usual Fock space contraction as defined in Exercise 4.8.

Next we will consider A_R at the infinitesimal level – at least in the complex formalism. Recall that differentiating one-parameter semigroups of isometries led to quasi-free derivations. Let $\{T_t \mid t \geq 0\}$ be a strongly continuous contraction semigroup on a Hilbert space H, which commutes with an operator R on H, $0 \leq R \leq 1$. If B denotes the infinitesimal generator of the semigroup T_t, let $dA_R(B)$ denote the infinitesimal generator of the strongly continuous contraction semigroup $A_R(e^{Bt})$ at the C^*-level. For which B is the induced semigroup $A_R(e^{Bt})$ norm continuous? If $A_R(e^{Bt})$ is norm continuous, then B is certainly bounded since $A_R(e^{Bt})(a(f)) = a(e^{Bt}f)$. Now a bounded operator B on H gives rise to a contraction semigroup $\{e^{Bt} \mid t \geq 0\}$ if and only if B is dissipative; i.e. $\mathrm{Re}(B) \leq 0$. It was shown in Proposition 6.13 that if B is skew-adjoint or equivalently if e^{Bt} is a unitary group, then $A(e^{Bt})$ is a norm continuous group of $*$-automorphisms if and only if B is of trace class.

Theorem 6.19 *Let $\{e^{Bt} \mid t \geq 0\}$ be a strongly continuous contraction semigroup on a Hilbert space H, which commutes with an operator R on H, $0 \leq R \leq 1$. Then $\{A_R(e^{Bt}) \mid t \geq 0\}$ is a norm continuous semigroup of completely positive maps on $A^{CAR}(H)$ if and only if B is of trace class.*

Proof From (6.8.11) we see that the $*$-algebra $A_0(H)$ generated by $\{a(h) \mid h \in H\}$ is contained in the domain of $dA_R(B)$ and on the local algebra A_0, $h_i \in H$:

$$dA_R(B)[: a^{\#}(h_1) \cdots a^{\#}(h_n) :_R]$$
$$= \sum_{i=1}^{n} : \prod_{j=1}^{i-1} a^{\#}(h_j)\, a^{\#}(Bh_i) \prod_{j=i+1}^{n} a^{\#}(h_j) :_R . \tag{6.8.12}$$

It follows from (6.8.12) that $dA_R(B)$ is additive on dissipative B which commute with R (at least on the local algebra A_0).

Let B be of trace class. Then

$$B = \textstyle\sum_n i\mu_n h_n \otimes \overline{h}_n - \sum_n \lambda_n g_n \otimes \overline{g}_n$$

where $\{h_n\}$ and $\{g_n\}$ are two orthonormal sets in H, $\mu_n \in \mathbb{R}$, $\lambda_n > 0$ (since B is dissipative), and $\sum |\mu_n|$, $\sum \lambda_n < \infty$. Since $\mathrm{Re}(B)$ commutes with R, one can choose the sequence $\{g_n\}$ such that there exists a sequence of real numbers r_n, $0 \leq r_n \leq 1$, with $Rg_n = r_n g_n$. Then if θ is the grading on $A(H)$, we have

$$\begin{aligned}
dA_R(B)(x) &= i\textstyle\sum_n \mu_n [a^*(h_n)a(h_n), x] \\
&+ \textstyle\sum_n \lambda_n \big[2\,[(1-r_n)a^*(g_n) + r_n a(g_n)]\,\theta(x)\,[(1-r_n)a(g_n) + r_n a^*(g_n)] \\
&- \{((1-r_n)a^*(g_n) + r_n a(g_n))\,((1-r_n)a(g_n) + r_n a^*(g_n))\,, x\}\big] \\
&+ \textstyle\sum_n 2\lambda_n (r_n - r_n^2)\,[(a^*(g_n) - a(g_n))\,\theta(x)\,(a(g_n) - a^*(g_n)) - x]
\end{aligned}$$

for all x in $A(H)$. The first term, being given by the Hamiltonian part of the system, is familiar (Proposition 6.13). In the Fock case, $R = 0$, this expression can be verified directly. The general case can be verified as follows. It is enough to consider the case B is real. Let J be a conjugation on H such that $Jg_n = g_n$. If p is the projection on $\mathrm{Ker}\, B$, we see that $A_{(1-p)R}(e^{Bt}) = A_R(e^{Bt})$, if $t \geq 0$. Thus we can assume that $JR = RJ$ as well as $Je^{Bt} = e^{Bt}J$. Let E be the purification projection P_R (6.6.11). Then $(J \oplus J)E = E(J \oplus J)$. Let π_E be the Bogoliubov automorphism of $A(H \oplus H)$ given by $\pi_E a(f) = a((1-E)f) + a^*(E(J \oplus J)f)$. Then $A_E(e^{tB} \oplus e^{tB}) = \pi_E^{-1} A_0(e^{tB} \oplus e^{tB})\pi_E$. Hence we can obtain the form of $dA_E(B \oplus B) = \pi_E^{-1}\, dA_0(B \oplus B)\pi_E$ using the orthonormal set $\{g_n \oplus 0, 0 \oplus g_m\}$

which diagonalizes $B\oplus B$. But $dA_R(B)$ is the restriction of $dA_E(B\oplus B)$ to $A(H)$, and so some calculations yield the stated decomposition.

Conversely suppose $dA_R(B)$ is bounded. Let H_0 be a finite dimensional subspace in H with complete orthonormal set $e_1,\ldots,e_n,e_{n+1},\ldots,e_m$ such that $Be_i \in H_0$ for $i = 1,\ldots,n$. Then we see as in Proposition 6.13 that $[dA_R(B)(y)]x = \sum_{i=1}^n \langle Be_i, e_i\rangle yx$, where $y = a(e_1)\cdots a(e_n)$, and $x = a(e_{n+1})\cdots a(e_m)$, and:

$$\textstyle\sum_{i=1}^n \langle (B+B^*)e_i, e_i\rangle [yx + (yx)^*] = [dA_R(B)(y)]x + x^* dA_R(B)(y^*)\,.$$

But $\|yx + (yx)^*\| \geq 1$, so that

$$\left|\textstyle\sum_i \langle (B+B^*)e_i, e_i\rangle\right| \leq 2\|dA_R(B)\| < \infty\,.$$

Hence since $B+B^* \leq 0$ we see that $B+B^*$ is of trace class, with $-\mathrm{tr}(B+B^*) \leq 2\|dA_R(B)\|$. It follows from the first part of the theorem that $dA_R(B+B^*)$ is bounded, (as well as $dA_R(B)$ by hypothesis). Thus from (6.8.12) we see that $dA_R(B-B^*)$ is bounded and so $B-B^*$ is of trace class by Proposition 6.13. Hence B is of trace class. □

Next we consider the completely positive analogue of implementability, namely producing a *dissipator* in the sense of (6.8.2). We will do this for $A_R(T)$ in the complex formalism, by dilating T to a unitary U, use the implementing criteria and constructive procedure of (6.7.15) or (6.7.23) for U, and then project back to produce a dissipator for $A_R(T)$.

Let P and R be projections on a Hilbert space H, and T a contraction on H such that $TR = RT$. We will be interested in constructing first a non-zero contraction $\mathcal{F}$ on $F(H)$ such that both maps

$$y \to \pi_P\left(A_R(T)(y)\right) - \mathcal{F}\pi_P(y)\mathcal{F}^* \quad \text{and} \quad y \to \pi_P\left(A_R(T^*)(y)\right) - \mathcal{F}^*\pi_P(y)\mathcal{F}$$

are completely positive from $A(H)$ into $B(F(H))$.

Let K denote the Hilbert space $H\oplus H$, and W the embedding $h \to h\oplus 0$ of H in K. If U denotes the unitary

$$\begin{pmatrix} T & -(1-TT^*)^{\frac{1}{2}} \\ (1-T^*T)^{\frac{1}{2}} & T^* \end{pmatrix} \tag{6.8.13}$$

on K, so that $T = W^*UW$. Then $[U, P\oplus R]$ being Hilbert–Schmidt is equivalent to $TP-PT$ being Hilbert–Schmidt, and the following four operators being trace class:

$$\begin{gathered} P(1-T^*T)(1-R)P, \quad P(1-TT^*)(1-R)P, \\ (1-P)(1-T^*T)R(1-P), \quad (1-P)(1-TT^*)R(1-P)\,. \end{gathered}$$

We let $\mathcal{C}(P,R)$ denote the contractions T on H such that $TR = RT$ and which satisfy these trace class and Hilbert–Schmidt conditions. Then $\mathcal{C}(P,R)$ is a $*$-

semigroup. Note that a unitary T lies in $\mathcal{C}(P,0)$ if and only if $TP - PT$ is Hilbert–Schmidt, which is the condition of Theorem 6.16(a) that the quasi-free automorphism $A(T)$ is unitarily implementable in the quasi-free state ω_P. We denote such unitaries on H by $\mathcal{O}(P)$. Moreover, the operator $T = 0$ lies in $\mathcal{C}(P,R)$ if and only if $P-R$ is Hilbert–Schmidt, which is the condition of Theorem 6.16(b) that the quasi-free states ω_R and ω_P are quasi-equivalent.

Let P, R be projections on H, and T a contraction in $\mathcal{C}(P,R)$. Then $U \in \mathcal{O}(P \oplus R)$, so there exists by Theorem 6.16 a unitary $\mathcal{U}$ on $F(K)$ such that

$$\pi_{P\oplus R}\left(A(U)(x)\right) = \mathcal{U}\left(\pi_{P\oplus R}(x)\right)\mathcal{U}^* \tag{6.8.14}$$

for all x in $A(K)$. Moreover since $WP = (P \oplus R)W$, there exists by Theorem 6.18 an isometry $F = F_{P,P\oplus R}(W)$ from $F(H)$ into $F(K)$ such that

$$y \to \pi_{P\oplus R}\left(A(W)(y)\right) - F\left(\pi_P(y)\right)F^* \tag{6.8.15}$$

is completely positive from $A(H)$ into $B(F(K))$, and

$$\pi_P\left(A_{P\oplus R}(W^*)(x)\right) = F^*\left(\pi_{P\oplus R}(x)\right)F \tag{6.8.16}$$

for all x in $A(K)$. In fact F is defined by

$$F\pi_P(y)\Omega = \pi_{P\oplus R}\left(A(W)(y)\right)\Omega$$

and all y in $A(H)$. Let J_1 and J_2 be conjugations on H, commuting with P and R, respectively. For convenience, we will always take $J_1 \oplus J_2$ to be our conjugation which commutes with $P \oplus R$. Then $WJ_1 = (J_1 \oplus J_2)W$. Hence $A(W)\pi_P = \pi_{P\oplus R}A(W)$ (as well as $\pi_P A_{P\oplus R}(W^*) = A_0(W^*)\pi_{P\oplus R}$) and so $F = F(W)$ the usual Fock space isometry $\bigoplus_{n=0}^{\infty} \wedge^n W$ as defined in the Notes to Chapter 4.

Then

$$\begin{aligned}\pi_P A_R(T)\pi_P^{-1} &= \pi_P A_{P\oplus R}(W^*)A(U)A(W)\pi_P^{-1} \\ &= A_0(W^*)\pi_{P\oplus R}A(U)\pi_{P\oplus R}^{-1}A(W)\end{aligned} \tag{6.8.17}$$

which extends to a normal completely positive map on $A(H)'' = B(F(H))$, using $A_0(W^*)\pi_{P\oplus R} = F^*\pi_P(\cdot)F$, $\pi_{P\oplus R}A(U) = \mathcal{U}\pi_{P\oplus R}(\cdot)\mathcal{U}^*$, and that $A(W)$ $A(H) \to A(H \oplus H)$ is essentially spatial $x \to x \otimes 1$.

We let $\mathcal{F}$ denote the contraction $F^*\mathcal{U}F$ on $F(H)$, which is uniquely determined by T, P, and R, up to a phase. Thus for any finite sequence $y_1, \ldots, y_n$ in $A(H)$:

$$\begin{aligned}[\pi_P A_R(T)(y_i^* y_j)] &= [\pi_P(A_{P\oplus R}(W^*)A(U)A(W)(y_i^* y_j)] \quad \text{by (6.8.8)} \\ &= [F^*\pi_{P\oplus R}(A(U)A(W)(y_i^* y_j))F] \quad \text{by (6.8.16)} \\ &= [F^*\mathcal{U}\pi_{P\oplus R}(A(W)(y_i^* y_j))\mathcal{U}^* F] \quad \text{by (6.8.14)} \\ &\geq [F^*\mathcal{U}F\pi_P(y_i^* y_j)F^*\mathcal{U}^* F] \quad \text{by (6.8.15)}.\end{aligned}$$

Hence $y \to \pi_P A_R(T)(y) - \mathcal{F}\pi_P(y)\mathcal{F}^*$ is completely positive from $A(H)$ into $B(F(H))$. Moreover $\mathcal{F} \neq 0$ since we shall see $\langle \mathcal{F}\Omega, \Omega\rangle \neq 0$ from (6.8.14).

We investigate the contraction $\mathcal{F}$ a little closer. If S is an operator on K (respectively H), we let $[S_{\sigma\sigma'}]$ denote the matrix of S with respect to the decomposition $Q_- = Q = P \oplus R$, $Q_+ = 1 - Q$ (respectively $P_- = P$, $P_+ = 1 - P$). We make the following assumption on (T, P, R) in order to cut down the notation:

$$U_{++},\, U_{--} \text{ are invertible.} \tag{6.8.18}$$

We let $\Lambda = \Lambda(U)$ be the bounded operator on K with matrix $\Lambda_{\sigma\sigma'}$ given as in (6.7.16), so that $\Lambda_{+-}(U)$, $\Lambda_{-+}(U)$ are Hilbert–Schmidt. We define the operator $\Lambda = \Lambda(T)$ on H by

$$\Lambda(T) = W^*\Lambda(U)W \tag{6.8.19}$$

so that $\Lambda_{+-}(T)$, $\Lambda_{-+}(T)$ are also Hilbert–Schmidt. Note that if $R = 0$, then $T_{--} = U_{--}$, and it is easily verified that (6.8.19) reduces to

$$\begin{aligned} &\Lambda_{--} = 1_{--} - T_{--}^{-1}, \qquad &&\Lambda_{-+} = T_{--}^{-1}T_{-+}, \\ &\Lambda_{+-} = T_{+-}T_{--}^{-1}, \qquad &&\Lambda_{++} = T_{++} - 1_{++} - T_{+-}T_{--}^{-1}T_{-+}. \end{aligned} \tag{6.8.20}$$

We now return to our system (T, P, R), where $T \in \mathcal{C}(P, R)$ as usual and (6.8.18) holds. Then by (6.7.23) we have, up to a phase θ, that

$$\mathcal{U}\varphi = \theta \det\left[1 + \Lambda_{+-}^*(U)\Lambda_{+-}(U)\right]^{-\frac{1}{2}} : \exp dF_{P\oplus R}(\Lambda(U)) : \varphi$$

for all φ in $a^*(K)\Omega$.

However $\mathcal{F} = F(W)^*\mathcal{U}F(W)$ by definition, and $A_0(W^*) = F(W)^*(\cdot)\, F(W)$ by Theorem 6.18. We can thus deduce:

Theorem 6.20 *Let P and R be projections on a Hilbert space H, and T a contraction in $\mathcal{C}(P, R)$ such that* (6.8.18) *holds. Then $x \to \pi_P A_R(T)\pi_P^{-1}(x)$ extends to a completely positive normal map on $A(H)'' = B(F(H))$.*

Moreover there exists a contraction $\mathcal{F}$ on $F(H)$ with $\langle \mathcal{F}\Omega, \Omega\rangle \neq 0$, given on $a^*(H)\Omega$ by

$$\mathcal{F} = \det\left[1 + \Lambda_{+-}^*(U)\Lambda_{+-}(U)\right]^{-\frac{1}{2}} : \exp dF_P(\Lambda(T)) : , \tag{6.8.21}$$

where $\Lambda(U)$, $\Lambda(T)$ are given by (6.7.16) and (6.8.19), such that the maps

$$x \to \pi_P\left(A_R(T)(x)\right) - \mathcal{F}\pi_P(x)\mathcal{F}^* \quad \text{and} \quad x \to \pi_P\left(A_R(T^*)(x)\right) - \mathcal{F}^*\pi_P(x)\mathcal{F}$$

are completely positive from $A(H)$ into $B(F(H))$. If $R = 0$, then (6.8.21) *reduces to*

$$\mathcal{F} = \det\left[1 + T_{--}^{*-1}P(1 - T^*PT)PT_{--}^{-1}\right]^{-\frac{1}{2}} : \exp dF_P(\Lambda(T)) : , \tag{6.8.22}$$

where $\Lambda(T)$ is given by (6.8.20).

In defining $\mathcal{F}$ previously, there was a phase ambiguity. We now remove this by defining $\mathcal{F} = \mathcal{F}(T)$ such that (6.8.21) holds. If we wish to emphasize the dependence on P we write $\mathcal{F}(T, P)$. We will produce dissipators for some quasi-free semigroups with the aid of:

Lemma 6.21 *Let P, R be projections on H with $P - R$ Hilbert–Schmidt, and let α be the automorphism of $B(F(H))$ such that $\alpha\pi_R = \pi_P$. Then for any contraction T on H which commutes with R, the contraction $\mathcal{F}(T)$ is up to a phase θ given by $\theta\mathcal{F}(T) = \alpha(F_R(T))$. If P and R are both finite dimensional, and* (6.8.18) *holds, then $\theta = \det U_{--}/|\det U_{--}|$.*

Proof Since $P \oplus R - R \oplus R$ is Hilbert–Schmidt, there exists by Theorem 6.16(b) an automorphism β of $B(F(K))$ such that $\beta\pi_{R\oplus R} = \pi_{P\oplus R}$. Let α (respectively β) be implemented by a unitary V_1 (respectively V_2). Then for $x \in A(K)$,

$$\begin{aligned} F(W)^*V_2^*\pi_{R\oplus R}(x)V_2F(W) &= F(W)^*\pi_{P\oplus R}(x)F(W) \\ = \pi_P A_{P\oplus R}(W^*)(x) &= \pi_P A_{R\oplus R}(W^*)(x) \\ = V_1^*\pi_R A_{R\oplus R}(W^*)(x)V_1 &= V_1^*F(W^*)\pi_{R\oplus R}(x)F(W)V_1\,. \end{aligned}$$

Hence there exists a unitary u such that

$$u\pi_{R\oplus R}(x)V_2F(W) = \pi_{R\oplus R}(x)F(W)V_1$$

for all x in $A(K)$. In particular $u \in \pi_{R\oplus R}(A(K))' = \mathbb{C}$ and absorbing this into V_1: $V_2F(W) = F(W)V_1$. Clearly we have, up to a phase $\mathcal{U} = \beta\,[F_{R\oplus R}(U)]$. Hence up to a phase

$$\begin{aligned} \mathcal{F}(T) = F(W)^*\mathcal{U}F(W) &= F(W)^*V_2^*F_{R\oplus R}(U)V_2F(W) \\ &= V_1^*F(W)^*F_{R\oplus R}(U)F(W)V_1 = V_1^*F_R(T)V_1\,. \end{aligned} \tag{6.8.23}$$

Suppose P and R are both finite dimensional and that (6.8.18) holds, and let θ be the phase in (6.8.23). Then

$$\theta^{-1}\langle\alpha F_R(T)\Omega,\Omega\rangle = \langle\mathcal{F}(T)\Omega,\Omega\rangle = \det\left[1 + \Lambda_{+-}^*(U)\Lambda_{+-}(U)\right]^{-\frac{1}{2}}\,.$$

Let $\tilde{\pi}_R, \tilde{\pi}_P$ denote the unique extensions of π_R, π_P, respectively, to automorphisms of $B(F(H))$ so that $\alpha = \tilde{\pi}_P\tilde{\pi}_R^{-1}$. Suppose $\pi, \pi_{R\oplus R}$ are implemented by unitaries $u_R, u_{R\oplus R}$ respectively. Then for all $x \in A(K)$:

$$\begin{aligned} \omega_{P\oplus R}(x) = \omega_P \circ A_{R\oplus R}(W^*)(x) &= \omega_P(u_R^*[\pi_R A_{R\oplus R}(W^*)\pi(x)]u_R) \\ &= \omega_P[u_R^*F(W)^*\pi_{R\oplus R}(x)F(W)u_R]\,. \end{aligned}$$

Hence

$$\langle\alpha(F_R(T))\Omega,\Omega\rangle = \omega_P\left[\tilde{\pi}_R^{-1}(F_R(T))\right] = \omega_P\left[u_R^*F(W)^*F_{R\oplus R}(U)F(W)u_R\right]$$

$$= \omega_P \left[u_R^* F(W)^* \tilde{\pi}_{R\oplus R}(F(U)) F(W) u_R \right] = \omega_{P\oplus R}(F(U))$$
$$= \det(U_{--}) \qquad \text{by Exercise 6.10}.$$

Thus

$$\theta = \det\left[1 + U_{--}^{*-1} U_{+-}^* U_{+-} U_{--}^{-1}\right]^{\frac{1}{2}} \det(U_{--})$$
$$= \left\{\det\left[1 + U_{--}^{*-1} U_{+-}^* U_{+-} U_{--}^{-1}\right] \det(U_{--}^*) \det(U_{--})\right\}^{\frac{1}{2}} \det(U_{--})/\det|U_{--}|$$
$$= \det(U_{--})/\det|U_{--}|$$

ısing $U_{--}^* U_{--} + U_{+-}^* U_{+-} = 1_{--}$. □

We can now combine this to obtain dissipators for symmetric dynamical emigroups:

Theorem 6.22 *Let L be a negative bounded self-adjoint operator on a Hilbert pace H, and P a projection on H, such that PLP is trace class. Then $\{\pi_P A_0(e^{tL})\pi_P^{-1} \mid t \geq 0\}$ extends to a weakly continuous semigroup of completely ositive normal maps on $A(H)'' = B(F(H))$. Moreover there exists a strongly ontinuous self-adjoint contraction semigroup G_t on $F(H)$ such that*

$$x \to \pi_P A_0(e^{tL})(x) - G_t \pi_P(x) G_t$$

s completely positive from $A(H)$ into $B(F(H))$ for all positive t.

'roof If PLP is of trace class, we have PL is Hilbert–Schmidt and it is verified rom the power series expansion $e^{tL} = \sum_n (tL)^n/n!$ that $e^{tL} \in C(P,0)$ for all ≥ 0. Let P_n be a sequence of finite-rank projections on H such that $P_n \nearrow P$. We an assume that our conjugation J commutes with all P_n. Let $G_t^n = \pi_{P_n}(F(e^{tL}))$ or $t \geq 0$. Then $\{G_t^n \mid t \geq 0\}$ is a strongly continuous self-adjoint contraction emigroup on $F(H)$. Let t_0 be such that $0 \leq t \leq t_0 \Rightarrow \|e^{tL} - 1\| < 1/16$. Then $0 \leq t \leq t_0$, it is clear that the systems $(e^{tL}, P, 0)$, $(e^{tL}, P_n, 0)$ $n = 1, 2, \ldots$, all atisfy (6.8.18). Moreover, by Lemma 6.21, we have for $0 \leq t \leq t_0$:

$$\theta \mathcal{F}(e^{tL}, P_n) = G_t^n, \tag{6.8.24}$$

here the phase $\theta = \det(P_n e^{tL} P_n)/\det|P_n e^{tL} P_n| \equiv 1$ since $e^{tL} \geq 0$. We show for $\in H$:

$$\mathcal{F}(e^{tL}, P)h = \lim_{n\to\infty} \mathcal{F}(e^{tL}, P_n)h \tag{6.8.25}$$

niformly in $0 \leq t \leq t_0$.

It is easily seen from (6.8.20) that $\Lambda(e^{tL}, P_n) \to \Lambda(e^{tL}, P)$ uniformly strongly s $n \to \infty$, for $0 \leq t \leq t_0$. Thus from (6.7.21) and the continuity property 5.7.19) it is enough to show that as $n \to \infty$:

$$\Lambda_{+-}(e^{tL}, P_n) \to \Lambda_{+-}(e^{tL}, P) \tag{6.8.26}$$

and

$$(P_n e^{tL} P_n)^{-1}(1 - e^{tL} P_n e^{tL})(P_n e^{tL} P_n)^{-1}$$
$$\to (P e^{tL} P)^{-1}(1 - e^{tL} P e^{tL})(P e^{tL} P)^{-1} \qquad (6.8.27)$$

uniformly in Hilbert–Schmidt and trace norm, respectively. Now

$$\Lambda_{+-}(e^{tL}, P_n) = -(1 - P_n) e^{tL} P_n (P_n e^{tL} P_n)^{-1}$$
$$= -(1 - P) e^{tL} P (P_n e^{tL} P_n)^{-1} - (1 - P_n) P (e^{tL} - 1) P (P_n e^{tL} P_n)^{-1} .$$

Thus approximating $(1 - P)e^{tL} \in HS(H)$ and $P(e^{tL} - 1)P \in T(H) \subset HS(H)$ uniformly by finite-rank operators and using $(P_n e^{tL} P_n)^{-1} \to (P e^{tL} P)^{-1}$, it is established that (6.8.26) holds uniformly for $0 \le t \le t_0$. Similarly we have

$$(P_n e^{tL} P_n)^{-1}(1 - e^{tL} P_n e^{tL})(P_n e^{tL} P_n)^{-1}$$
$$= (P_n e^{tL} P_n)^{-1} P (1 - e^{2tL}) P (P_n e^{tL} P_n)^{-1}$$
$$+ (P_n e^{tL} P_n)^{-1} P e^{tL} (1 - P) e^{tL} P (P_n e^{tL} P_n)^{-1}$$
$$+ (P_n e^{tL} P_n)^{-1} P (e^{tL} - 1) P (1 - P_n) P (e^{tL} - 1) P (P_n e^{tL} P_n)^{-1}$$

and note that $P(1 - e^{2tL})P$, $Pe^{tL}(1 - P)e^{tL}P \in T(H)$, and $P(e^{tL} - 1)P \in T(H) \subset HS(H)$, in order to establish (6.8.27).

Now define $\{G_t \mid t \ge 0\}$ by $G_t = [\mathcal{F}(e^{tL/n}, P)]^n$ if $0 \le t/n < t_0/2$. Then by (6.8.24) and (6.8.25), G_t is a strongly continuous, self-adjoint contraction semigroup on $F(H)$ satisfying the claim of the theorem. We know by Theorem 6.19 that $\{\pi_P A_0(e^{tL}) \pi_P^{-1} \mid t \ge 0\}$ extends to a semigroup of completely positive normal maps on $B(F(H))$. It only remains to show the weak continuity of this extension. Using (6.8.17) it is enough to check that the unitary family $\mathcal{U}_t \in B(F(K))$, $0 \le t \le t_0$, given on $a^*(K)\Omega$ by

$$\mathcal{U}_t = \det\left[1 + \Lambda_{+-}(U_t)^* \Lambda_{+-}(U_t)\right]^{1/2} : \exp dF_{P \oplus 0}(\Lambda(U_t)) :$$

is weakly continuous in t, where U_t is the family of unitaries on K dilating e^{tL} pointwise. The proof is similar to that of (6.8.25). □

Exercise 6.13 Let T be a contraction between Hilbert spaces H and K, with R, S operators on H and K respectively such that $0 \le R \le 1$, $0 \le S \le 1$ and $TR = ST$. Show that $A_R(T)$ is pure (amongst the completely positive unital maps from $A(H)$ into $B(H_{\omega_S})$) if and only if $TT^* = 1$ and $R^2 = R$, i.e. $A_R(T)$ is a conditional expectation with respect to a Fock state.

6.9 Spatiality of quasi-free one-parameter automorphism groups

The purpose of this section is to implement quasi-free derivations in the self dual formalism by self adjoint operators.

In the complex formalism, we have implemented the quasi-free derivation $d\tau(h)$ in the Fock state ω_{P_-} by $dQ_{P_-}(h)$, at least when h is bounded self adjoint and $[P, h]$ is Hilbert–Schmidt if $P_- = P$, $P_+ = 1 - P$. We are now going to implement quasi-free derivations in the self dual formalism, when H is self adjoint, but the off diagonal terms H_{+-}, H_{-+} are Hilbert–Schmidt, beginning with the case of H trace class, then when H is Hilbert–Schmidt, next when H is bounded, and finally the general case.

We are going to show that for any skew adjoint H with Hilbert–Schmidt off diagonals that there is a unique skew adjoint operator $dQ_P(H)$ on H_P such that the unitary operator $e^{tdQ_P(H)}$ implements $\tau(e^{tH})$ in the Fock state φ_P, Ω_P is in its domain and $\langle dQ_P(H)\Omega_P, \Omega_P\rangle = 0$. Moreover any vector in the linear space D spanned by $F_n(PK)$ for all $n \in \mathbb{N}$ (states with a finite particle number) is an analytic vector of the self adjoint operator $dQ_P(H)$ and we have

$$dQ_P(H_1) + dQ_P(H_2) = dQ_P(H_1 + H_2) \tag{6.9.1}$$

$$dQ_P(\lambda H) = \lambda dQ_P(H) \quad \lambda \in \mathbb{R}, \tag{6.9.2}$$

$$[dQ_P(H_1), dQ_P(H_2)] = dQ_P([H_1, H_2]) + s_P(H_1, H_2)\,1\,, \tag{6.9.3}$$

$$[dQ_P(H), \pi_P(B(h))] = \pi_P(B(Hh)) \tag{6.9.4}$$

on D where s_P is the Schwinger term

$$s_P(H_1, H_2) = \mathrm{tr}\,(PH_2(1-P)H_1P - PH_1(1-P)H_2P)\,/2\,. \tag{6.9.5}$$

Some notation is useful to keep track of the topologies and the estimates. The set of unitary operators U on K such that $[\Gamma, U] = 0$ and $PU(1-P)$ is Hilbert–Schmidt (equivalently $(1-P)UP = \Gamma PU(1-P)\Gamma$ is Hilbert–Schmidt) is denoted by $\mathcal{O}_P(K,\Gamma)$. The P-norm topology (respectively P-strong topology) is given by the norm $\|U - U'\| + \|P(U-U')(1-P)\|_2$ (respectively semi-norms $\|(U - U')\xi\| + \|P(U-U')(1-P)\|_2$, $\xi \in K$). Let $o_P(K,\Gamma)$ denote the real Banach space of bounded skew adjoint operators H on K, such that $\Gamma H \Gamma = H$, and $PH(1-P)$ is trace class equipped with the P-norm. We already have an expression for implementing $dQ(H)$ when H is trace class, namely $b(H)/2$. Relaxing the condition on H means we have to take the normal ordered implementer, i.e. we put:

$$dQ_P(H) = \pi_P\left(b_P(H)\right)/2 = \pi_P(b(H)/2 - \varphi_P\left(b(H)\right)/2) \tag{6.9.6}$$

Before we tackle the main aim of this section of implementers for quasi-free derivations in the self dual formalism, we first put what is available from the theory of implementing quasi-free automorphisms in a setting which will allow us to use the implementers Q_t of the one-parameter unitary group $\tau(e^{tH})$ with carefully chosen phase so that $Q_t = e^{tdQ_P(H)}$ is a strongly continuous one-parameter group with the required infinitesimal generator. To keep track of our procedure, we will need closed expressions for the one-point functions $\varphi_P(Q_t)$.

So to begin the discussion we consider unitaries implementing equivalences between quasi-free states $\varphi_P, \varphi_{P'}$ on the self dual Fermion algebra (which exists when and only when $P - P'$ is Hilbert–Schmidt) and a computation of their one-point functions. If $P - P'$ is trace class, we can be precise about an expression for a unitary Q implementing the equivalence between the states φ_P and $\varphi_{P'}$ and that of the one-point function $\varphi_P(Q)$.

Let $\sin\alpha = |P - P'|$ where $0 \leq \alpha \leq \pi/2$, so that α is trace class. Since $(P - P')^2$ commutes with P and P', α commutes with P and P' (as well as Γ). Let

$$E_0 = P \wedge P' + (1 - P) \wedge (1 - P')$$
$$E_{\pi/2} = P \wedge (1 - P') + (1 - P) \wedge P'$$

the spectral projections of α for the eigenvalues 0 and $\pi/2$ respectively. If we let $P_0 = P(1 - E_0 - E_{\pi/2})$, $P_0' = P'(1 - E_0 - E_{\pi/2})$, then

$$P_0 \wedge P_0' = P_0 \wedge (1 - P_0') = (1 - P_0) \wedge P_0' = (1 - P_0) \wedge (1 - P_0') = 0\,.$$

Moreover $(1-P_0)P_0'P_0$ has a polar decomposition uh where u is a partial isometry from P_0 to $1 - P_0$. With respect to the matrix units $(e_{ij}) = \begin{pmatrix} P_0 & u \\ u^* & 1 - P_0 \end{pmatrix}$, P_0 and P_0' become

$$\begin{pmatrix} 1 & 0 \\ 0 & 0 \end{pmatrix}, \quad \begin{pmatrix} c^2 & cs \\ cs & s^2 \end{pmatrix}, \tag{6.9.7}$$

respectively, where $c^2 + s^2 = 1$. Writing $c = \cos\beta$, then $\alpha = \beta \oplus \beta$. Then

$$R \equiv e^{iH} = \begin{pmatrix} c & s \\ -s & c \end{pmatrix}, \tag{6.9.8}$$

if $H \equiv i\alpha \begin{pmatrix} 0 & 1 \\ -1 & 0 \end{pmatrix}$ takes P_0 to P_0', i.e. $P_0' = RP_0R^*$. (We could define $H = H(P/P')$ without reference to matrix units and (6.9.7) as follows. Now $[P, P']^2 = -\sin^2\alpha\cos^2\alpha$, and we take the partial isometry $(\sin\alpha\cos\alpha)^{-1}[P, P']$ in the polar decomposition of $[P, P']$ and let

$$H(P/P') = i\alpha(\sin\alpha\cos\alpha)^{-1}[P, P']\,.)$$

Let $e_1, \ldots, e_n$ be an orthonormal basis for $P \wedge (1 - P')$ so that $n < \infty$ as we have assumed $P - P'$ is trace class. Then $\Gamma e_1, \ldots, \Gamma e_n$ is an orthonormal basis for $\Gamma(P \wedge (1-P'))\Gamma = (1-P) \wedge P'$. Let U be the unitary on H defined by $U = 1$ on $1 - E_{\pi/2}$, and $Ue_j = \Gamma e_j$, $U\Gamma e_j = e_j$. Define

$$\hat{R}(P'/P) = e^{iH(P'/P)}U$$
$$Q = \exp\left[\tfrac{i}{2}b(H(P'/P))\right] \textstyle\prod_{j=1}^n \left[B(e_j) - B(\Gamma e_j)\right]\,.$$

Then $\hat{R}$ is an unitary with the same parity as n and

$$\hat{R}P\hat{R}^* = P', \quad QB(f)Q^* = B((-1)^n\hat{R}f), \quad \varphi_P(QaQ^*) = \varphi_{P'}(a)\,.$$

(In particular, when $[P,P']=0$, $H(P'/P)=0$, $Q=\prod_{j=1}^{n}[B(e_j)-B(\Gamma e_j)]$ we recover Exercise 6.10.)

We claim that the value of Q is given in the quasi-free state φ_P using the positive quartic root:

$$\varphi_P(Q)=(\det\cos\alpha)^{\frac{1}{4}}\,. \tag{6.9.9}$$

If $n\neq 0$, then the right hand side is zero, but so also is the left hand side using the quasi-free rule (6.6.9) and $\langle Pe_k,g\rangle=\langle P\Gamma e_k,g\rangle=\langle Pf,g\rangle=0$ for $g=e_j,\Gamma e_j,k\neq j$, and $f\in(1-E_0)K$. We can thus assume $E_{\pi/2}=0$. Now $Q\in A((1-E_0)K,\Gamma)$ so that we can replace φ_P on $A(K)$ on the left hand side by $\varphi_{P(1-E_0)}$ on $A((1-E_0)K,\Gamma)$ or reduce to the case $E_0=0$. Then we can find a basis projection E, and a unitary u commuting with P and P' such that $uEu^*=1-E$. Then $\operatorname{tr}EH(P'/P)=\operatorname{tr}H(P'/P)/2=0$. We identify $b(H(P'/P))/2$ in $A^F(K,\Gamma)$ with $c^*EH(P'/P)c=dF(EH(P'/P))$ in $A^F(EK)$, and φ_P on $A^F(K,\Gamma)$ with ω_0 on $A^F(EK)$. Then

$$\varphi_P(e^{ib(H(P'/P)/2})=\omega_0\left(\exp[ic^*EH(P'/P)c]\right)\,.$$

If $H=\sum\lambda_j f_j\otimes\overline{f_j}$ is finite rank with f_j orthonormal and λ_j real, then

$$\begin{aligned}\exp ic^*Hc&=\textstyle\prod_j\exp i\lambda_j c^*(f_j)c(f_j)\\&=\textstyle\prod_j(1+\mu_j c^*(f_j)c(f_j))\,,\quad \mu_j=e^{i\lambda_j}-1\,.\end{aligned}$$

Hence (cf. (6.7.7) and (6.7.8)), with $\mu_K=\prod_{j\in K}\mu_j$,

$$\begin{aligned}\omega_0(e^{ic^*Hc})&=\textstyle\sum_n\sum_{|K|=n}\mu_K\\&=\det[E+(e^{iH}-1)(1-P)E]\\&=\exp[\operatorname{tr}\,\log(1+(e^{iH}-1)(1-P)E)]\\&=\exp\operatorname{tr}_E\log[1+(1-P)(e^{iH}-1)(1-P)]\\&=\exp\operatorname{tr}_{E(1-P)}\log(1-P)e^{iH}(1-P)\\&=\exp\operatorname{tr}_{E(1-P)}\log\cos\alpha\quad\text{(see (6.9.8))}\\&=\exp\frac{1}{4}\operatorname{tr}\,\log\cos\alpha=\det(\cos\alpha)^{\frac{1}{4}}\,.\end{aligned} \tag{6.9.10}$$

and so (6.9.9) follows by continuity.

Now suppose U is a unitary commuting with Γ such that $[U,P]$ is trace class. Let $P'=UPU^*$. By the above, $P'=\hat{R}P\hat{R}$, where $\hat{R}$ is implemented by $\mathcal{U}$. Then since $\left[\hat{R}^*U,P\right]=0$, $\tau\left(\hat{R}^*U\right)$ leaves φ_P invariant, and so implemented by a unitary $\mathcal{V}$ such that $\mathcal{V}\Omega_P=\Omega_P$. Hence $\tau(U)=\tau(\hat{R})\tau(\hat{R}^*U)$ is implemented by $\mathcal{W}=\mathcal{U}\mathcal{V}$ and

$$\varphi_P(\mathcal{W})=\varphi_P(\mathcal{U})=\det(\cos\alpha)^{1/4}\,. \tag{6.9.11}$$

But $\sin^2\alpha=[P,UPU^*]$, so that

$$\det \cos^2 \alpha = \det[PU^*PUP + (1-P)U^*(1-P)U(1-P)]$$

and up to a phase, U is implemented by $\mathcal{U}$ such that

$$\varphi_P(\mathcal{U}) = \det(PU^*P)^{1/2}\,. \tag{6.9.12}$$

In fact

$$\begin{aligned}(\det \cos\alpha)^{1/4} &= (\det{}_P(\cos\alpha\, P))^{1/2} = \left(\det{}_P(P\hat{R}^*P)\right)^{1/2} \\ &= \det{}_P(PU^*PU\hat{R}^*P)^{1/2} = (\det{}_P PU^*P)^{1/2}\det{}_P(U\hat{R}^*P)^{1/2}\,,\end{aligned}$$

where $\det_P(U\hat{R}^*P)^{1/2}$ is a phase factor in $\mathbb{T}$.

So we now begin to apply this theory of implementing a single Bogoliubo automorphism to a one-parameter unitary group e^{tH}. In particular, suppose H is skew adjoint commuting with Γ, such that $[H,P]$ is trace class. Then e^{tH} is a unitary commuting with Γ such that $[e^{tH},P]$ is trace class. Hence by the abov $\tau(e^H)$ is implemented by a unitary $\mathcal{W}$ such that

$$\begin{aligned}\varphi_P(\mathcal{W}) &= \det{}_P\left(Pe^{-H}P\right)^{\frac{1}{2}} \\ &= \exp\left\{-\int_0^1 \operatorname{tr}\left[\left(Pe^{-tH}\right)^{-1}\frac{d}{dt}\left(Pe^{-tH}P\right)\right]dt\right\} \\ &= \exp\left\{-\int_0^1 \operatorname{tr}\left[\left(Pe^{-tH}P\right)^{-1}\left(PHe^{-tH}\right)\right]dt/2\right\}.\end{aligned} \tag{6.9.13}$$

Here the integration is along any path in $\mathbb{C}$ avoiding the values of t such tha $Pe^{-tH}P$ has non-zero kernel on PK. The function $t \to \det_P(Pe^{tH}Pe^{-tH}P)$ i holomorphic on $\mathbb{C}$, with isolated zeros, so that ker $Pe^{-tH}P$ is zero except a isolated points without finite accumulation points. This forces the existence c the domain of integration in (6.9.13).

However, we know if H is trace class then $e^{b(H)/2}$ implements $\tau(e^H)$, an there was a phase fudging in the unitary $\mathcal{W}$ of (6.9.11). In fact we have that $\mathcal{W}$ constructed above is $\pm e^{b(H)/2}$ or that

$$\varphi_P(e^{b(H)/2}) = \det{}_P(Pe^{-H}P)^{\frac{1}{2}}\,. \tag{6.9.14}$$

To see this we can compute the right hand side of

$$\frac{d}{dt}\varphi_P(e^{tb(H)/2}) = \varphi_P(e^{tb(h)/2}b(H))/2$$

as $\alpha(t)\varphi_P\left(e^{tb(H)/2}\right)$, where $\alpha(t) = -\operatorname{tr}(Pe^{-H}P)^{-1}(PHe^{-H}P)$, and integrate.

If H is skew adjoint trace class, P a basis projection, then with $dQ_P(H)$ a in (6.9.6) we deduce from (6.9.14) and (6.9.13) that

$$\varphi_P\left(e^{dQ_P(H)}\right) = \exp\left\{-\int_0^1 \operatorname{tr}\left[(Pe^{-tH}P)^{-1}PH(1-P)e^{-tH}P\right]\,dt/2\right\} \tag{6.9.15}$$

where the integration is along any path avoiding the isolated values where $Pe^{-tH}P$ has non-zero kernel, and the right hand side is defined to be 0 if $Pe^{-H}P$ has a non-zero kernel.

So we now have an expression when H is trace class for the one point function $\varphi_P(e^{tdQ_P(H)})$ of the implementer – the skew adjoint operator $dQ_P(H)$ which we already had constructed in Section 6.7 for trace class H. Next we look at the Hilbert–Schmidt case.

So let H be a skew adjoint Hilbert–Schmidt operator satisfying $\Gamma H = H\Gamma$. If P is a basis projection, then $e^{tH} \in \mathcal{O}_P(K,\Gamma)$ for all t. Let $1-E_n$ be the spectral projection of iH for $[-1/n, 1/n]$, so that $E_n = \Gamma E_n \Gamma$ is of finite dimension, and $H_n \equiv HE_n$, is of finite rank. Then for fixed t, the points $-1 \notin \sigma(e^{-t(H_n-H_m)})$, and $0 \notin \sigma(Pe^{-t(H_n-H_m)}P)$ for large n, m and so by (6.9.15) we have the formula:

$$\left\|\left(e^{tb_P(H_n)/2} - e^{tb_P(H_m)/2}\right)\Omega_P\right\|^2 \tag{6.9.16}$$

$$\begin{aligned} &= 2\left(1 - \operatorname{Re}\langle e^{tb_P(H_n-H_m)/2}\Omega_P, \Omega_P\rangle\right) \\ &= 2\left[1 - \operatorname{Re}\exp\left\{(-1/2)\int_0^t \operatorname{tr}\left[(Pe^{-s(H_n-H_m)}P)^{-1}P(H_n-H_m)\right.\right.\right. \\ &\qquad\qquad \left.\left.\left.(1-P)e^{-s(H_n-H_m)}P\right]\,ds\right\}\right]. \end{aligned} \tag{6.9.17}$$

If $Pe^{-tH}P$ is invertible, taking a path $\{Pe^{-sH}P\}$ of invertible elements from P to $Pe^{-tH}P$, we can for large n assume $Pe^{-sH_n}P$ are all invertible. Then (6.9.15) holds for H as (uniformly):

$$\begin{aligned} (Pe^{-sH_n}P)^{-1} &\to (Pe^{-sH}P)^{-1} \quad \text{in norm}, \\ PH_n(1-P) &\to PH(1-P) \quad \text{in Hilbert – Schmidt norm}, \\ (1-P)e^{-sH_n}P &\to (1-P)e^{-sH}P \quad \text{in Hilbert – Schmidt norm}. \end{aligned}$$

In particular both $PH(1-P)$ and $(1-P)e^{-sH}P$ are Hilbert–Schmidt so the integrand in (6.9.15) is trace class. Hence (6.9.17) $\to 0$ as $n, m \to \infty$ by continuity of the determinant, and

$$e^{tb_P(H_n)/2}\pi_P(a)\Omega_P = \pi_P\left[\tau(e^{tH_n})(a)\right]e^{tb_P(H_n)/2}\Omega_P \tag{6.9.18}$$

is Cauchy for a in the Fermion algebra, and so $e^{tb_P(H_N)/2}$ converges $*$-strongly to a one-parameter group of unitaries whose infinitesimal generator we denote by $dQ_P(H)$. By continuity it is seen from (6.9.18) that $\exp tdQ_P(H)$ implements $\tau(e^{tH})$ in π_P for H Hilbert–Schmidt, since we already know it does so for finite rank H (Proposition 6.13) and the following holds:

$$e^{tdQ_P(H)}\pi_P(a)\Omega_P = \pi_P\left[\tau(e^{tH})(a)\right]e^{tdQ_P(H)}\Omega_P\,. \tag{6.9.19}$$

It then follows from

$$e^{tdQ_P(H)}B(h) = B(e^{tH})e^{tQ_P(H)}$$

that any vector $h_1\wedge\cdots\wedge h_n = B(h_1)\cdots B(h_n)\Omega_P \in F_n(PK)$ is analytic provided H is in $o_P(K,\Gamma)$.

Now $f_H(t) = f(t)$, the function of the right hand side of (6.9.15), is holomorphic in t, as $(1-P)e^{sH}P$ is an entire family of Hilbert–Schmidt operators, and $(Pe^{sH}P)^{-1}$ is holomorphic for norm topology where defined. Hence if $U(s) = e^{sdQ_P(H)}$, $U(s)\Omega$ is differentiable at $s=0$, as can be seen by considering $\|((U(s)-1)/s-(U(t)-1)/t)\,\Omega\|$ and

$$\begin{aligned}\langle (U(s)-1)\Omega,(U(t)-1)\Omega\rangle &= \langle U(s+t)+1-U(s)-U(t)\Omega,\Omega\rangle\\ &= f(s+t)+1-f(s)-f(t)\,.\end{aligned}$$

Then by differentiating $\langle U(s)U(t_1)\Omega, U(t_2)\Omega\rangle = f(s+t_1-t_2)$ with respect to t_1 and t_2 we have

$$\langle U(s)dQ_P(H)\Omega, dQ_P(H)\Omega\rangle = -f''(s)\,. \tag{6.9.20}$$

Thus by iteration, Ω_P is the domain of $dQ_P(H)^n$ and

$$\langle e^{sQ_P(H)}dQ_P(H)^n\Omega, dQ_P(H)^n\Omega\rangle = (-1)^n f^{(2n)}(s)\,.$$

In particular

$$\textstyle\sum_n |z|^n \|dQ_P(H)^n\Omega\| < \infty \quad \text{for some } z\neq 0$$

follows from analyticity of f, and so Ω_P is an analytic vector for $dQ_P(H)$ and $e^{tdQ_P(H)}\Omega_P$ is analytic in a neighbourhood of $t=0$.

To show additivity of $dQ_P(\cdot)$ we identify $dQ_P(H)\Omega_P$ as a bilinear expression in $\pi_P(B(h)B(k))\Omega_P$ (cf.(6.7.13)) and then differentiate (6.9.19). We have for $h_1,\ldots h_n\in PK$

$$\begin{aligned}&\langle e^{tQ_P(H)}\Omega_P, \pi_P(B(h_1))\cdots\pi_P(B(h_n))\Omega_P\rangle\\ &= \langle \Omega_P, \pi_P(B(e^{-tH}h_1)\cdots B(e^{-tH}h_n))\Omega_P\rangle\,.\end{aligned} \tag{6.9.21}$$

We express $B(e^{-tH}h_1)\cdots B(e^{-tH}h_n)\equiv B(e^{-tH}h_I)$ in terms of normal ordered products using (6.6.23):

$$\sum \varepsilon : B(e^{-tH}h_J) :_{1-P} \varphi_{1-P}(B(e^{-tH}h_K))\,. \tag{6.9.22}$$

Examining the possibilities shows that the only possibility for the derivative of (6.9.21) not to vanish at $t=0$ is when $n=2$ (as $: B(f_J) :_{1-P} \Omega_P = 0$ if at least one $f_j\in PK$ in a sequence (f_i) in K, and similarly $\varphi_{1-P}(B(f_K))=0$ for such a sequence, so that for a term in (6.9.22), the derivative of their product vanishes at $t = 0$ if $|J| + |K| \geq 2$). Computing explicitly for $n = 2$:

$$\begin{aligned}\langle dQ_P(H)\Omega_P, B(h_1)B(h_2)\Omega_P\rangle &= \frac{d}{dt}\varphi_{1-P}\left(B(e^{-tH}h_2)^* B(e^{-tH}h_1)^*\right) \\ &= -\langle \Gamma(1-P)HPh_1, h_2\rangle .\end{aligned}$$

Thus by Parseval:

$$\langle dQ_P(H_1)\Omega_P, dQ_P(H_2)\Omega_P\rangle = -\mathrm{tr}(PH_2(1-P)H_1P)/2 .$$

Note that in fact, in accordance with (6.7.13) and (6.9.6):

$$dQ_P(H)\Omega_P = \textstyle\sum_i B((1-P)HPh_i)B(h_i)\Omega_P/2$$

when h_i is a complete orthonormal sequence for PK.

Next we show the cocycle relation (6.9.3). We have from (6.9.20) at $t = 0$:

$$\|dQ_P(H)\Omega_P\|^2 = -f''(0) = \mathrm{tr}\, PH(1-P)HP/2 ,$$

so that by linearity already established $H \to dQ_P(H)\Omega_P$ is continuous for the seminorm $\|(1-P)HP\|_2$. Since $\tau(e^{Ht})$ is continuous in H, it follows from (6.9.19) that $dQ_P(H)\varphi$ is P-norm continuous in $H \in o_P(K,\Gamma)$ for all φ in D_0.

We then deduce the commutation relation (6.9.3), at least as quadratic forms where we interpret the matrix value of the commutator given by $\langle [dQ_P(H_1), dQ_P(H_2)]\varphi_1, \varphi_2\rangle$ as

$$\langle dQ_P(H_1)\varphi_1, dQ_P(H_2)\varphi_2\rangle - \langle dQ_P(H_2)\varphi_1, dQ_P(H_1)\varphi_2\rangle$$

for $\varphi_1, \varphi_2 \in D_0$. Since we already have the form of the Schwinger term for finite rank H in (6.7.14), we could deduce it for arbitrary $H_1, H_2 \in o_P(K,\Gamma)$, by approximation using continuity of $dQ_P(H)\Omega_P$ in H for the semi-norm $\|PH(1-P)\|_2$.

After considering the case of H, Hilbert–Schmidt, let us consider next the case H is bounded and in $o_P(K,\Gamma)$.

Then define

$$H_n = E_nHE_n + \Gamma E_nHE_n\Gamma + PH(1-P) + (1-P)HP$$

where E_n is a sequence of finite dimensional projections increasing to P. Then H_n are Hilbert–Schmidt and $H_n \to H$ in P-strong topology. If we take the limit of (6.9.15) for H_n, we have on the one hand

$$\lim \left\langle e^{t dQ_P(H_n)}\Omega_P, \Omega_P\right\rangle = f_H(t) .$$

On the other hand, as $[e^{tH}, P]$ is Hilbert–Schmidt, $\tau(e^{tH})$ is implementable by a unitary Q_t in π_P, and by (6.9.14),

$$|\langle Q_t\Omega_P, \Omega_P\rangle| = \det\left[Pe^{tH}Pe^{-tH}P\right]^{\frac{1}{4}} = |f_H(t)|\,.$$

Thus by a suitable phase choice, we can arrange $\langle Q_t\Omega_P, \Omega_P\rangle = f_H(t)$, or

$$\langle Q_t\Omega_P, \Omega_P\rangle = \lim\langle e^{tdQ_P(H_n)}\Omega_P, \Omega_P\rangle\,.$$

Then by a previous argument as in (6.9.21) and (6.9.22), we deduce

$$\lim_{n\to\infty}\left\langle e^{tdQ_P(H_n)}\Omega_P, \psi\right\rangle = \langle Q_t\Omega_P, \psi\rangle$$

for $\psi \in D_0$, and t small. This means $e^{tdQ_P(H_n)}\Omega_P \to Q_t\Omega_P$ strongly, and hence $e^{tdQ_P(H_n)} \to Q_t$ strongly using the covariance property (6.9.19). This means that Q_t is a strongly continuous one parameter group of unitaries implementing $\tau(e^{tH})$, satisfies (6.9.19), with generator $dQ_P(H)$ say.

These methods can be extended to cover the case of the infinitesimal generator H of a strongly continuous one-parameter group $U(t)$ in $\mathcal{O}_P(K,\Gamma)$.

Take $U_{12}(t) \in \mathcal{O}_P(K,\Gamma)$ continuous in the Hilbert–Schmidt norm, satisfying

$$U_{12}(t)PU_{12}(t)^* = U(t)PU(t)^*$$

for small t, and implemented by $Q(U_{12}(t))$. Then $U_{12}(t)^*U(t)$ commutes with P and so implemented by Q_t^1 strongly continuous in t. Then $Q_t \equiv Q(U_{12}(t))Q_t^1$ implements $U(t)$, for small t. We can extend Q to $\mathbb{R}$ by multiplication $Q_t = (Q_{t/n})^n$, to a strongly continuous family. By irreducibility, Q is a multiplier representation – and by triviality of continuous 2-cocycles on $\mathbb{R}$, we can adjust Q to be a strongly continuous one-parameter group of unitaries, with infinitesimal generator $dQ_P(H)$ say. Similar approximation arguments to what already has been used show that (by approximating $PHP + (1-P)H(1-P)$ on spectral projections of finite intervals) that we can adjust $dQ_P(H)$ by a phase if necessary so that Ω_P is in the domain of $dQ_P(H)$, $\langle dQ_P(H)\Omega_P, \Omega_P\rangle = 0$, $e^{tdQ_P(H)}$ implements $\tau(e^{tH})$, and (6.9.15) holds. Then (6.9.1)–(6.9.4) follow as in the case H is Hilbert–Schmidt.

Remark For non-Fock states φ_S, the quantization $dQ_S(H) \equiv dQ_{P_S}(H \oplus 0)$ has a Schwinger term

$$\begin{aligned} s_S(H_1, H_2) &\equiv s_{P_S}(H_1 \oplus 0, H_2 \oplus 0) \\ &= \operatorname{tr}\{SH_2(1-S)H_1S + DH_2(1-S)H_1D \\ &\quad -SH_1(1-S)H_2S - DH_1(1-S)H_2D\}/2 \end{aligned}$$

where $D = S^{1/2}(1-S)^{1/2}$. This will appear in the non-Fock Ramond state in Section 8.3.

xercise 6.14 Let K be a Hilbert space equipped with a conjugation Γ, and a basis projection. Show that $o_P(K,\Gamma)$ is closed under commutators. If $\in o_P(K,\Gamma)$, show that $\{e^{tH} \mid t \in \mathbb{R}\}$ is a norm-continuous one-parameter sub-roup of $\mathcal{O}_P(K,\Gamma)$, and that every norm-continuous one-parameter subgroup of $_P(K,\Gamma)$ arises in this way. Show that any such norm-continuous one-parameter roup is continuous for the P-norm topology.

xercise 6.15 Suppose H is a skew adjoint operator on a Hilbert space K ommuting with a conjugation Γ, such that PHP is skew adjoint, and the closure $(1-P)HP$ is in the Hilbert–Schmidt class. Show that e^{tH} is a continuous one-arameter subgroup of $\mathcal{O}_P(K,\Gamma)$ relative to P-strong topology.

.10 Fock states on the even algebra and the Pauli algebra

iven a quasi-free Fock state φ_E on the Fermion algebra, we can form an even ate φ_E^P on the Pauli algebra, whose restriction to the even subalgebra is quasi-ee Fock $\varphi_E^P|_{A_+} = \varphi_E|_{A_+}$. Although a quasi-free Fock state on the Fermion gebra is pure, the corresponding even state on the Pauli algebra is not nec-ssarily so, the obstruction to purity is a certain mod 2 index. Suppose π is a ock representation of A^F on a Fock space, regarded as a space of antisymmetric nsors. Then F splits as $F_+ \oplus F_-$ where F_+ and F_- are the subspaces of even nd odd tensors respectively. Then $\pi|_{A_+}$ leaves $F_\pm$ invariant, and the mod 2 dex enters because these two representations are not equivalent and is needed determine when the restriction of equivalent Fock states on the Fermi algebra, main equivalent when restricted to the even subalgebra.

We first consider the following general situation. Let A be a unital C^*-algebra ith two commuting symmetries α and β. Let $\hat{A}$ be the crossed product of A the β-action of $\mathbb{Z}/2$, which is generated by A and T in $\hat{A}$ satisfying $T^2 = 1$, $^* = T$, $Ta = \beta(a)T$, $a \in A$. Let A be graded by α, $B = A_+ + TA_-$ a *-subalgebra of $\hat{A}$, and extend α, β to $*$-automorphisms $\hat{\alpha}, \hat{\beta}$ on $\hat{A}$ fixing T. ippose there is an odd unitary U in A. For an even state φ of A, there exists unique even state φ^B of B which is an extension of the restriction of φ to A_+. 'e determine criteria for deducing when φ^B is pure in terms of φ on A and its striction to A_+. We will then apply these criteria to the case $A = A^F$, $\alpha = \theta$, $= \theta_-$, $B = A^P$, $U = 2^{-1/2}(c_i + c_i^*)$ for a fixed i.

emma 6.23 *If φ is α-invariant and pure, then $\varphi|A_+$ is pure.*

roof Taking $\varphi_+ \to \varphi = \varphi_+ \circ (1+\alpha)/2$, and $\varphi \to \varphi_+ = \varphi|_{A^+}$ gives an affine jection between states φ_+ of A^+ and α-invariant states φ of A. Hence φ_+ pure if and only if φ is an extremal α-invariant state. In particular, if φ is -invariant and pure, then φ_+ is pure. □

ote that $\mathrm{Ad}\, U$ leaves A_+ invariant as U is odd.

emma 6.24 *If an α-invariant state φ of A is pure, then φ_+ and $\varphi_- \equiv \varphi_+ \circ$ $\mathrm{d}\, U$ on A_+ are disjoint.*

Proof Let $(H_\varphi, \pi_\varphi, \xi_\varphi)$ be the GNS triple of φ. Due to α-invariance of φ,

$$H_\varphi = H_+ \oplus H_-\,, \quad H_\pm = \overline{\pi_\varphi(A_\pm)\xi_\varphi}\,. \tag{6.10.1}$$

Then $H_{\varphi\pm}$, $\pi_{\varphi\pm}$ and $\xi_{\varphi\pm}$ can be identified with $H_\pm$, $\pi_\pm = (\pi_\varphi|A_+)|H_\pm$ an $\xi_+ = \xi_\varphi$, $\xi_- = \pi_\varphi(U)\xi_\varphi$. By Lemma 6.23, φ_+ and hence $\varphi_- = \varphi_+ \circ \operatorname{Ad} U$ are pur and so $\pi_\pm$ are irreducible. We now derive a contradiction assuming that they ar equivalent. Since $\pi_\pm$ are irreducible, there exists a unitary $W_0 \in \pi_+(A_+)''$, $\operatorname{Ad} W$ on $\pi_+(A_+)$ implementing $\operatorname{Ad} U$ on A_+. Since $\pi_+ \sim \pi_-$, there exists a unitar $W \in \pi_\varphi(A_+)''$ on $\pi_\varphi(A_+)$ implementing $\operatorname{Ad} U$ on A_+. Since $(\operatorname{Ad} W)W = W$, W has to commute with $\pi_\varphi(U)$. For $a \in A_-$, $aU^* \in A_+$, and

$$\begin{aligned}(\text{Ad } W)\pi_\varphi(a) &= (\text{Ad } W)(\pi_\varphi(aU^*)\pi_\varphi(U)) = \{(\text{Ad } W)\pi_\varphi(aU^*)\}\pi_\varphi(U) \\ &= \pi_\varphi(\{(\text{Ad } U)(aU^*)\}U) = \pi_\varphi((\text{Ad } U)a)\,.\end{aligned} \tag{6.10.2}$$

Therefore $\operatorname{Ad} W$ coincides with $\operatorname{Ad}\pi_\varphi(U)$ on $\pi_\varphi(A)$ and hence on $\pi_\varphi(A)''$. Sinc π_φ is irreducible, $W = c\pi_\varphi(U)$ for some complex number c of modulus 1. O the other hand, by α-invariance of φ, α can be extended to an isomorphism of $\pi_\varphi(A)''$. By $W \in \pi_\varphi(A_+)''$, we have $\overline{\alpha}(W) = W$, whilst $\alpha(U) = -U$. Thi contradicts with $W = c\pi_\varphi(U)$, $W^*W = 1$. □

Theorem 6.25 *Assume that φ is an α-invariant pure state of A. Then φ^B not pure if and only if*

(1) *φ and $\varphi \circ \beta$ are equivalent and*

(2) *$\varphi_+ = \varphi|A_+$ and $\varphi_+ \circ \beta$ are not equivalent.*

If φ^B is not pure, it is a mixture of two inequivalent pure states.

The proof is divided into several lemmas.

Lemma 6.26 *Let A be a C^*-subalgebra of a C^*-algebra B and $\varepsilon : B \to A$ conditional expectation. If φ_1 and φ_2 are quasi-equivalent states of A, then $\varphi_1 \circ$ and $\varphi_2 \circ \varepsilon$ are quasi-equivalent on B.*

Proof If φ is a state, the normal folium (see the Remark on page 214) $\mathcal{F}(\varphi) = \{\psi \circ \pi_\varphi \mid \psi \in \pi_\varphi(A)''_*\}$ is the norm-closed convex hull of the states $\varphi(a^* \cdot a)$ a a runs over the domain A of φ. Then $\varphi_1 \sim \varphi_2$ if and only if $\mathcal{F}(\varphi_1) = \mathcal{F}(\varphi_2$ if and only if $\varphi_1 \in \mathcal{F}(\varphi_2)$ and $\varphi_2 \in \mathcal{F}(\varphi_1)$. Then if φ_2 is a norm limit convex combinations of $\varphi_1(a^* \cdot a)$, $(a \in A)$, then $\varphi_2 \circ \varepsilon$ is a norm limit of conve combinations of $\varphi_1(a^*\varepsilon(\cdot)a) = \varphi_1 \circ \varepsilon(a^* \cdot a)$. Hence $\varphi_2 \in \mathcal{F}(\varphi_1)$ implies tha $\varphi_2 \circ \varepsilon \in \mathcal{F}(\varphi_1 \circ \varepsilon)$.

Lemma 6.27 *Let φ be an α-invariant pure state of A. The states $\varphi_+ = \varphi|A$ and $\varphi_+ \circ \beta \circ \operatorname{Ad} U$ of A_+ are equivalent if and only if*

(1) *φ is equivalent to $\varphi \circ \beta$ and*

(2) *φ_+ is not equivalent to $\varphi_+ \circ \beta$.*

Proof If $\varphi_+ \sim \varphi_+ \circ \beta \circ \operatorname{Ad} U$, then the states $\varphi = \varphi_+ \circ p$ and $\varphi \circ \beta \circ \operatorname{Ad} U = \varphi_+ \circ \beta \circ \operatorname{Ad} U \circ p$ of A are equivalent by Lemma 6.26 where $p = (1+\alpha)/2$. Since $\operatorname{Ad} U$ is inner on A, we obtain $\varphi \sim \varphi \circ \beta$. (Note that $[\alpha, \operatorname{Ad} U] = 0$ due to $\alpha(U) = -U$.)

Now assume that $\varphi \sim \varphi \circ \beta$. Since $\pi_\varphi | A_+ = \pi_+ \oplus \pi_-$ and $\pi_\pm$ are disjoint by Lemmas 6.23 and 6.24, we have two alternatives:

(A) $\pi_+ \sim \pi_+ \circ \beta$ disjoint from $\pi_- \sim \pi_- \circ \beta$,
(B) $\pi_+ \sim \pi_- \circ \beta$ disjoint from $\pi_- \sim \pi_+ \circ \beta$.

The desired conclusion $\pi_+ \sim \pi_- \circ \beta$ $(\sim \pi_+ \circ \beta \circ \operatorname{Ad} U)$ holds in and only in case (B), which is characterized by $\varphi \sim \varphi \circ \beta$ and π_+ disjoint from $\pi_+ \circ \beta$. □

Lemma 6.28 *If φ_+ and $\varphi_+ \circ \beta \circ \operatorname{Ad} U$ are disjoint, then φ^B is pure.*

Proof Let H_φ^β, π_φ^β and ξ_φ^β be the GNS triplet for the state $\varphi \circ \beta$. Let

$$\hat{H}_\varphi = H_\varphi \oplus H_\varphi^\beta, \tag{6.10.3}$$

$$\begin{aligned} &\hat{\pi}_\varphi(a + Tb)\left[\pi_\varphi(a_1)\xi_\varphi \oplus \pi_\varphi^\beta(a_2)\xi_\varphi^\beta\right] \\ &= \pi_\varphi(aa_1 + \beta(ba_2))\xi_\varphi \oplus \pi_\varphi^\beta(aa_2 + \beta(ba_1))\xi_\varphi^\beta . \end{aligned} \tag{6.10.4}$$

Then the restriction of $\hat{\pi}_\varphi(B)$ to

$$H_\varphi^B = H_+ \oplus H_-^\beta, \quad H_\pm^\beta \equiv \overline{\pi_\varphi^\beta(A_\pm)\xi_\varphi^\beta} \tag{6.10.5}$$

(denoted by π_φ^B) yields the cyclic representation of B associated with φ^B.

Since $\pi_\varphi^B | A_+ \sim \pi_+ \oplus \pi_- \circ \beta$ and the assumption implies the disjointness of π_+ and $\pi_- \circ \beta$ (due to $\pi_- \circ \beta = \pi_+ \circ \operatorname{Ad} U \circ \beta = \pi_+ \circ \beta \circ \operatorname{Ad} \beta(U) \sim \pi_+ \circ \beta \circ \operatorname{Ad} U$ because of $U^* \beta(U) \in A_+$), any $x \in \pi_\varphi^B(A_+)'$ is of the form

$$x = \lambda 1_+ \oplus \mu 1_- . \tag{6.10.6}$$

If $x \in \pi_\varphi^B(A)'$, x must commute with $\pi_\varphi^B(TU)$ which connects H_+ with H_-^β and hence $\lambda = \mu$. This proves that π_φ^B is irreducible. □

Lemma 6.29 *If φ_+ is equivalent to $\varphi_+ \circ \beta \circ \operatorname{Ad} U$, then φ^B is a non-trivial mixture of two inequivalent pure states.*

Proof Since π_+ is irreducible and $\varphi_+ \sim \varphi_+ \circ \beta \circ \operatorname{Ad} U$, there exists a unitary $W_0 \in \pi_+(A_+)''$, $\operatorname{Ad} W_0$ on $\pi_+(A_+)$ implementing $\beta \circ \operatorname{Ad} U$ on A_+. By $\pi_+ \sim \pi_+ \circ \beta \circ \operatorname{Ad} U$, there exists a unitary $W \in \pi_\varphi^B(A_+)''$, such that $\operatorname{Ad} W$ on $\pi_\varphi^B(A_+)$ implements $\beta \circ \operatorname{Ad} U$ on A_+. Since $(\operatorname{Ad} W)W = W$, W has to commute with $\pi_\varphi^B(TU)^*$. Let

$$V = W \pi_\varphi^B(TU)^* . \tag{6.10.7}$$

For any $a_- \in TA_-$, $a_- = (a_- TU)(TU)^*$ and hence

$$(\operatorname{Ad} W)\pi_\varphi^B(a_-) = \left[(\operatorname{Ad} W)\pi_\varphi^B(a_- TU)\right] \pi_\varphi^B(TU)^*$$

$$= \pi_\varphi^B \left(\left[(\hat{\beta} \circ \mathrm{Ad}\, U)(a_- TU) \right] (TU)^* \right) = \pi_\varphi^B \left((\hat{\beta} \circ \mathrm{Ad}\, U)(a_-) \right), \quad (6.10.8)$$

because $(\hat{\beta} \mathrm{Ad}\, U)(TU) = \hat{\beta}(UT) = \beta(U)T = TU$. Therefore V is in the centre $\pi_\varphi^B(B)' \cap \pi_\varphi^B(B)''$. It is non-trivial because it connects H_+ and H_+^β. Thus φ^B is not pure.

Since the restriction of π_φ^B to A_+ is a sum of two equivalent irreducible representations, $\pi_\varphi^B(A_+)'$ is isomorphic to the algebra of 2×2 matrices and $\pi_\varphi^B(B)'$ has to be its non-trivial $*$-subalgebra, which must be a two-dimensional commutative algebra, coinciding with the centre of $\pi_\varphi^B(B)''$. Then φ^B is a non-trivial mixture of two inequivalent pure states. □

If π is the Fock representation of $A^{\mathrm{CAR}}(H)$ on anti-symmetric tensors $F = F_a(H)$. Then F splits as $F_+ \oplus F_-$ where F_+ and F_- are the subspaces of even and odd tensors respectively, and $\pi|_{A_+}$ leaves $F_\pm$ invariant and gives two representations $\pi_\pm$ of A_+ on $F_\pm$, which by Lemmas 6.23 and 6.24 are irreducible and disjoint. (Note that if $H = \mathbb{C}^n$ is finite dimensional, then $A_+ \cong M_{2^{n-1}} \oplus M_{2^{n-1}}$ with π_+, π_- being the two disjoint representations on the factors $M_{2^{n-1}}$. If H is infinite dimensional then under the identification $A_+ \cong \lim_\rightarrow (M_{2^{n-1}} \oplus M_{2^{n-1}}) \cong \bigotimes_{\mathbb{N}} M_2$, $\varphi_\pm$ correspond to $\bigotimes_{\mathbb{N}} \omega_{\left(\begin{smallmatrix} 1 \\ 0 \end{smallmatrix}\right)}$, $\bigotimes_{\mathbb{N}} \omega_{\left(\begin{smallmatrix} 0 \\ 1 \end{smallmatrix}\right)}$ which are disjoint by Lemma 6.28.) Now if E is finite rank projection on H, we can identify by Exercise 6.10 the GNS triple for the quasi-free state ω_E with (F, π, Ω_E) if

$$\Omega_E = f_1 \wedge \cdots \wedge f_n = c^*(f_1) \cdots c^*(f_n)\Omega$$

if $(f_1, \ldots, f_n)$ is a complete orthonormal basis for the range of E. Hence $(\pi_{\omega_E})_\pm$ are identified with $\pi_\pm$ or $\pi_\mp$ according as to n is even or odd, i.e. $\omega_E|_{A_+}$ is equivalent to $\omega_0|_{A_+}$ if and only if $\dim E$ is even. Note that if E is a projection then $\omega_0 \sim \omega_E$ if and only if E is of finite rank, Theorem 6.16(b). If E is a projection, then we see from Lemma 6.26 that $\omega_0|_{A_+} \sim \omega_E|_{A_+}$ implies that $\omega_0 \sim \omega_E$. Thus $\omega_0|_{A_+} \sim \omega_E|_{A_+}$ if and only if E is of finite rank and even dimension. This is generalized in Theorem 6.30 to a criterion for equivalence of arbitrary quasi-free states in the self dual formalism. To deal with the parity we introduce the following $\mathbb{Z}/2$–index between two basis projectors E_1, E_2 for which $E_1 - E_2$ is in the Hilbert–Schmidt class:

$$\sigma(E_1, E_2) = (-1)^{\dim E_1 \wedge (1 - E_2)}. \quad (6.10.9)$$

By $\Gamma[E_1 \wedge (1 - E_2)]\Gamma = (1 - E_1) \wedge E_2$, σ is symmetric in E_1 and E_2.

Theorem 6.30 *The restriction of the Fock states φ_{E_1} and φ_{E_2} on A^F to the even part A_+^F are equivalent if and only if*

(i) *$E_1 - E_2$ is in the Hilbert–Schmidt class and*

(ii) *$\sigma(E_1, E_2) = 1$.*

Proof Let φ_{1+} and φ_{2+} denote the restrictions of φ_{E_1} and φ_{E_2} to A_+^F. If φ_{1+} and φ_{2+} are equivalent, then φ_{E_1} and φ_{E_2} must be equivalent by Lemma 6.26. In view of Theorem 6.16, this implies that the condition (i) is a necessary condition.

We now assume that (i) holds. We follow the proof of Theorem 6.16. There exists a Bogoliubov transformation U which does not change $E_j(1 - E_{\pi/2})$ ($j = 1, 2$) and changes $E_2 E_{\pi/2}$ to $E_1 E_{\pi/2}$, where

$$E_{\pi/2} = E_1 \wedge (1 - E_2) + (1 - E_1) \wedge E_2 \,, \tag{6.10.10}$$

namely

$$E_2' \equiv U^* E_2 U = E_2(1 - E_{\pi/2}) + E_1 E_{\pi/2} \,. \tag{6.10.11}$$

The Bogoliubov automorphism of A^F induced by this U is inner and the implementing unitary operator $\hat{Q}(U)$ belongs to A_σ^F, $\sigma = +$ or $-$ according to whether the $\mathbb{Z}/2$-index (6.10.9) is even or odd. Since $\|E_2' - E_1\| < 1$ we are in the situation of the Remark on page 250 so that we can find H_{12} a skew adjoint Hilbert–Schmidt operator commuting with Γ such that $\exp H_{12}$ takes E_2' to E_1. Hence the Bogoliubov automorphism $\tau(\exp H_{12})$ is implemented by the *even* $\exp dQ_P(H_{12})$. Therefore the cyclic representation π_{2+} associated with φ_{2+} (which is the restriction of the cyclic representation π_2 of A^F associated with φ_{E_2} to the subalgebra A_+^F and to the subspace $(A_+^F \xi_2)^-$ where ξ_2 is the cyclic vector for φ_{E_2}) is equivalent to the representation π_{1+} or π_{1-} through the unitary operator $\exp(dQ_P(H_{12}))\, \pi_1(\hat{Q}(U))$ according to whether the $\mathbb{Z}/2$-index is $+1$ or -1, where $\pi_{1\pm}$ are the restrictions of $\pi_1 | A_+^F$ to the closure of $\pi_1(A_\pm^F)\xi_1$, ξ_1 being the cyclic vector for φ_{E_1}. In view of Lemma 6.24, π_{2+} is equivalent to or disjoint from π_{1+} according to whether the $\mathbb{Z}/2$-index is $+1$ or -1. □

Remark We give concrete realizations of the state φ_E^P on the Pauli algebra A^P arising from the quasi-free state φ_E on A^F, or at least in the cases of interest to us where φ_E and $\varphi_E \circ \theta_-$ are equivalent on A^F, i.e. $E - \theta_- E \theta_-$ is Hilbert–Schmidt. First we identify the GNS triple for $\hat{\varphi}_E$ the θ_- invariant extension of φ_E to $\hat{A}$. The cyclic representation $(\hat{\pi}, \hat{A})$ on $\hat{H}$ (cyclic vector $\hat{\Omega}$) restricts to a direct sum of four irreducible representations (π_{ij}, H_{ij}) of A_+:

$$H_{11} = \left[\hat{\pi}\left(A_+^F\right)\hat{\Omega}\right]^-, \quad H_{12} = \left[\hat{\pi}\left(A_-^F\right)\hat{\Omega}\right]^-,$$
$$H_{21} = \left[\hat{\pi}\left(A_+^F T\right)\hat{\Omega}\right]^-, \quad H_{22} = \left[\hat{\pi}\left(A_-^F T\right)\hat{\Omega}\right]^-.$$

Then $B(\hat{H})$ is graded by $H_{11} \oplus H_{21}$ being even and $H_{12} \oplus H_{22}$ being odd, with $\hat{\pi}(T)$ being even as usual. The Fermion algebra A^F acts on $H_i \equiv H_{i1} \oplus H_{i2}$ by two equivalent representations associated to φ_E and $\varphi_{\theta_- E \theta_-}$ for $i = 1, 2$ respectively, intertwined by unitary $w_{12} : H_2 \to H_1$. Since both $\hat{\pi}(T) w_{12}^*$ and $w_{12}\hat{\pi}(T)$ implement θ_- in the irreducible representation of $\hat{\pi}|_{A^F}$ on H_1, we can arrange $\hat{\pi}(T) w_{12}^* = w_{12}\hat{\pi}(T)$ on H_1.

The self adjoint unitary $w = \begin{pmatrix} 0 & w_{12} \\ w_{12}^* & 0 \end{pmatrix} \in \hat{\pi}(\hat{A})'$ on $H_1 \oplus H_2$ gives the decomposition:

$$\hat{H} = H^+ \oplus H^- \qquad H^\pm = (1 \pm w)\hat{H} \tag{6.10.12}$$

$$\hat{\Omega} = 2^{-\frac{1}{2}}(\Omega^+ + \Omega^-) \qquad \Omega^\pm = 2^{-\frac{1}{2}}(1 \pm w)\hat{\Omega} \tag{6.10.13}$$

$$\hat{\pi} = \hat{\pi}^+ \oplus \hat{\pi}^- \tag{6.10.14}$$

$$\hat{\varphi}_E = (\hat{\varphi}_{E_+} + \hat{\varphi}_{E_-})/2, \quad \varphi_{E\pm} = \omega_{\Omega^\pm} \circ \hat{\pi}. \tag{6.10.15}$$

The GNS triple of $\hat{\varphi}_{E\pm}$ on $\hat{A}$ can be identified with $(\hat{\pi}^\pm, H_1, \hat{\Omega})$ if we define

$$\hat{\pi}^\pm(a_1 + a_2 T) = [\hat{\pi}(a_1) \pm \hat{\pi}(a_2 T)w]\,|_{H_1}, \quad a_i \in A^F$$

where $\hat{\pi}_\pm(A^F) = \hat{\pi}(A^F)|H_1 \cong \pi_E(A^F)$ are already irreducible.

Now $\hat{\pi}(A^F)|_{H_1}$ is irreducible, so take $\hat{\pi}(a_n) \to \hat{\pi}(T)w_{12}^* = w_{12}\hat{\pi}(T)$ on H_1. Since $\hat{\pi}(a_n)|_{H_1} = w_{12}\hat{\pi}(a_n)|_{H_2}w_{12}^*$, we have

$$\hat{\pi}(a_n) \to \hat{\pi}(T)w \text{ on } H_1 \oplus H_2\,. \tag{6.10.16}$$

Since $w = \pm 1$ on $H^\pm$ we get $\hat{\pi}(Ta_n) \to \pm 1$ on $H^\pm$, and so $\hat{\pi}^\pm$ are disjoint irreducible representations of $\hat{A}$. However, $\hat{\pi}^\pm$ are not necessarily disjoint on $A^P (\subset \hat{A})$. We have to distinguish two cases.

case (a) $\sigma(E, \theta_- E \theta_-) = 1$.

Here w_{12} interchanges π_{11} with π_{21} and π_{12} with π_{22} and so is even. Thus

$$\hat{\varphi}_{E\pm}(a_- T) = \pm \langle \hat{\pi}(a_-)\hat{\pi}(T)w\Omega, \Omega \rangle = 0$$

for odd a_- in A_-^F as both $\hat{\pi}(T)$ and w are even. Hence $\hat{\varphi}_{E\pm}|_{A^P} = \varphi_E^P$. The GNS triple of φ_E^P can be identified with $(H, \pi^\pm, \Omega)$, if (H, π, Ω) is the GNS space of φ_E, and

$$\pi^\pm(a_+ + a_- T) = \pi(a_+) \pm \pi(a_-)v \tag{6.10.17}$$

and v is an (even) unitary implementing θ_- in φ_E, $a_\pm \in A_\pm^F$.

case (b) $\sigma(E, \theta_- E \theta_-) = -1$.

Here w_{12} interchanges π_{11} with π_{22} and π_{12} with π_{21}, and so is odd. In this case, if we look at the odd part of (6.10.16) we have that $\hat{\pi}(Ta_{n-}) \to \pm 1$ on $H^\pm$ where a_{n-} is the odd part of a_n, and so not only are $\hat{\pi}^\pm$ disjoint irreducible representations of $\hat{A}$, they are disjoint irreducible representations of A^P. Thus

$$\varphi_E^P = (w_\Omega \circ \pi^+ + w_\Omega \circ \pi^-)/2$$

is the decomposition of φ_E^P into disjoint pure states with GNS triples $\left(\pi^\pm, H_+, \hat{\Omega}\right)$, where if (H, π, Ω) is the GNS triple of φ_E with even subspace H_+, and (an odd) v implements θ_- in φ_E, then for $a_\pm \in A_\pm^F$

$$\pi^\pm(a_+ + a_- T) = \pi(a_+) \pm \pi(a_-)v\,. \tag{6.10.18}$$

Theorem 6.31 *The index $\sigma(E_1, E_2)$ is continuous with respect to the norm topology in E_1 and E_2.*

Proof $E_1 \wedge (1-E_2)$ is the eigenprojection of $E_1 - E_2$ belonging to the eigenvalue 1. Let $E_1^0 - E_2^0$ be in the Hilbert–Schmidt class and 4ε (> 0) be the distance of 1 from $\sigma(E_1^0 - E_2^0)\backslash\{1\}$. Then there exists a neighbourhood $\mathcal{N}$ of (E_1^0, E_2^0) such that for any $(E_1, E_2) \in \mathcal{N}$, $\sigma(E_1 - E_2) \subset [-1, 1-3\varepsilon] \cup [1-\varepsilon, 1]$. Then

$$E = (2\pi i)^{-1} \int_{|z-1|=2\varepsilon} (z - (E_1 - E_2))^{-1}\, dz \tag{6.10.19}$$

must be the spectral projection of $E_1 - E_2$ for the interval $[1-\varepsilon, 1]$. Since it is of finite dimension for (E_1^0, E_2^0) and is continuous in the norm topology by (6.10.19), $\dim E$ is finite and independent of $(E_1, E_2) \in \mathcal{N}$.

Let $(E_1, E_2) \in \mathcal{N}$ be a pair of basis projections, such that $E_1 - E_2$ is Hilbert–Schmidt. We claim that the multiplicity of an eigenspace F_λ of $E_1 - E_2$ belonging to an eigenvalue λ not equal to 0 or ± 1 must be even. Due to the constancy of the dimension of E it will then follow that the multiplicity of the eigenvalue 1 of $E_1 - E_2$ is even or odd according to whether the multiplicity of the eigenvalue 1 of $E_1^0 - E_2^0$ is even or odd. Namely we have constancy of the $\mathbb{Z}/2$-index $\sigma(E_1, E_2) = \sigma(E_1^0, E_2^0)$ in $\mathcal{N}$.

If $[E_1, E_2] = U|[E_1, E_2]|$ is the polar decomposition then $|[E_1, E_2]|^2 = -[E_1, E_2]^2 = (E_1 - E_2)^2(1 - (E_1 - E_2)^2)$ with kernel the eigenprojectors $F_0, F_{\pm 1}$ of $E_1 - E_2$ for eigenvalues $0, \pm 1$. Since $[E_1, E_2]$ is skew adjoint, so is U, and U commutes with $|[E_1, E_2]|$ and $[E_1, E_2]$ and $U^2 = -1 + \text{Proj Ker}[E_1, E_2] = -1 + F_0 + F_1 + F_{-1}$. In fact $(E_1 - E_2)^2$ commutes with E_1, E_2 and Γ, and so $(E_1 - E_2)^2$ commutes with U. From $E_1[E_1, E_2] = E_1 E_2 (1 - E_1) = [E_1, E_2](1 - E_1)$, we have $E_1 U |[E_1, E_2]| = U|[E_1, E_2]|(1 - E_1) = U(1 - E_1)|[E_1, E_2]|$ so that $E_1 U = U(1 - E_1)$ as U vanishes on $\text{Ran}|[E_1, E_2]|^\perp = \text{Ker}|[E_1, E_2]|$ and similarly $E_2 U = U(1 - E_2)$. Thus $(E_1 - E_2)U = -(E_1 - E_2)U$ and $(\Gamma U, E_1 - E_2) = 0$, and $\Gamma U (E_1 - E_2) U^* \Gamma = (E_1 - E_2) U U^*$. Since $(\Gamma U)^2 = U^2 = -1 + F_0 + F_1 + F_{-1}$, we have that if $\lambda \neq 0, \pm 1$, that $J = \Gamma U|_{F_\lambda K}$ is an anti-unitary operator on $F_\lambda K$, ($[\Gamma U, F_\lambda] = 0$) $J^2 = -1$. The existence of this complex structure on $F_\lambda K$ makes $\dim F_\lambda$ even. In fact if $\{e_i\}_{i=1}^n$ is any basis for F_λ, then $J_{ij} = \langle J e_i, e_j \rangle$ is a non-singular, anti-symmetric matrix, with $\det J = \det J^T = \det(-J) = (-1)^n \det J$ so that n must be even, if T is the transpose. □

Remark One may give an interpretation of the index map (6.10.9) in terms of K_1 of a certain Banach algebra. To see how this goes we let $H_\mathbb{R}$ denote a real separable Hilbert space, H a complex separable Hilbert space, J a fixed complex structure on $H_\mathbb{R}$ or H (i.e. $J^2 = -1$, $J^* = -J$) and Q the ideal of Hilbert–Schmidt operators on $H_\mathbb{R}$ or H (or indeed any symmetrical normed ideal in the algebra of bounded operators on $H_\mathbb{R}$ or H in the sense of (Simon 1979)).

Write $B_Q(H_\mathbb{R})$ (resp $B_Q(H)$) for the subalgebra of the bounded operators on $H_\mathbb{R}$ (resp H) consisting of those A which almost commute with J in the sense that $AJ - JA \in Q$. Then $B_Q(H_\mathbb{R})$ and $B_Q(H)$ are Banach algebras in the norm

$\|A\| = \|A\|_\infty + \|AJ - JA\|_Q$ if $\|\cdot\|_\infty$ denotes the operator norm in $B(H_\mathbb{R})$ or $B(H)$. Denote by $G_Q(H_\mathbb{R})$ and $G_Q(H)$ the respective groups of invertible elements of these algebras. Let $\mathcal{O}_Q$ (Resp $\mathcal{U}_Q$) denote the orthogonal (resp unitary) subgroup of $G_Q(H_\mathbb{R})$ (resp $G_Q(H)$). Then $\mathcal{O}_Q$ and $\mathcal{U}_Q$ are deformation retracts of the respective groups of invertible elements and hence have the same homotopy type as those groups. To determine this homotopy type one argues as follows (the case of $\mathcal{O}_Q$ only is sketched, $\mathcal{U}_Q$ is similar).

$\mathcal{O}_Q$ acts transitively by conjugation on the space X consisting of all complex structures differing from J by an element of Q. Moreover this action is jointly continuous when X is equipped with the metric topology: $\|J_1 - J_2\|_Q$. As the isotropy subgroup of $J \in X$ is clearly the group $\mathcal{U}$ of unitary operators on $H_\mathbb{R}$ (equipped with complex structure J) we have $\mathcal{O}_Q/\mathcal{U}$ homeomorphic to X. Furthermore the quotient map $\mathcal{O}_Q \to X$ is a locally trivial principal fibration (using the fact that $\mathcal{O}_Q$ is a Banach Lie group) and a standard argument exploiting contractibility of the fibre $\mathcal{U}$ shows that $\mathcal{O}_Q$ and X have the same homotopy type. On the other hand the homotopy type of X is known from the proof of Bott periodicity (see for example (Milnor 1963)) to be that of $\mathcal{O}(\infty)/\mathcal{U}(\infty) = \lim_\to \mathcal{O}(n)/\mathcal{U}(n)$ (where the right hand side of this equality means the inductive limit of these homogeneous spaces as $n \to \infty$ and $\mathcal{O}(\infty)$ and $\mathcal{U}(\infty)$ denote the stable orthogonal and unitary groups respectively). In particular $\pi_0(\mathcal{O}_Q) \cong \mathbb{Z}/2$.

For $\mathcal{U}_Q$ one obtains $\pi_0(\mathcal{O}_Q) \cong \mathbb{Z}$ (provided J satisfies (6.10.20) below) and $\mathcal{U}_Q$ has the homotopy type of the space of Fredholm operators on H. The connected components of $\mathcal{O}_Q$ and $\mathcal{U}_Q$ are separated by the following homomorphisms. Define $i_\mathcal{O} : \mathcal{O}_Q \to \mathbb{Z}/2$ by

$$i_\mathcal{O}(R) = \dim_\mathbb{C} \ker(JR + RJ) (\mathrm{mod}\, 2) , \quad R \in \mathcal{O}_Q$$

and $i_\mathcal{U} : \mathcal{U}_Q \to \mathbb{Z}$ by

$$\begin{aligned} i_\mathcal{U}(U) &= \dim\ \ker(P_+ U P_+) - \dim\ \ker(P_+ U^* P_+) \\ &= \text{Fredholm index of } P_+ U P_+ , \quad U \in \mathcal{U}_Q \end{aligned}$$

where $J = i(P_+ - P_-)$ is the spectral decomposition of J, and for surjectivity of $i_\mathcal{U}$ we need

$$P_\pm \text{ both have infinite dimensional range.} \tag{6.10.20}$$

Both $i_\mathcal{O}$ and $i_\mathcal{U}$ are well defined as $JR + RJ$ and $P_+ U P_+$ are Fredholm operators for all $R \in \mathcal{O}_Q$ and $U \in \mathcal{U}_Q$ respectively.

These index maps have another interpretation, in terms of the groups $K_1(B_Q(H_\mathbb{R}))$ and $K_1(B_Q(H))$. To determine $K_1(B_Q(H_\mathbb{R}))$ it is sufficient to find π_0 of the unitary group of $B_Q(H_\mathbb{R}) \otimes M_n$ for each n, when M_n denotes the $n \times n$ matrices (over $\mathbb{R}$). This latter problem is easily solved as

$$B_Q(H_\mathbb{R}) \otimes M_n \cong B_Q(H_\mathbb{R} \otimes \mathbb{R}^n)$$

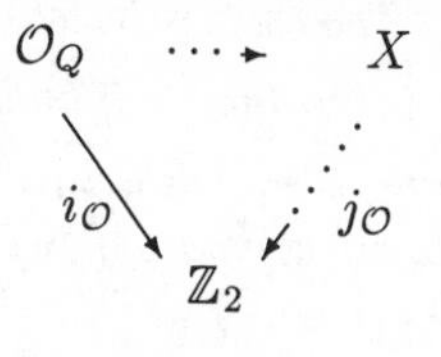

FIG. 6.5.

where $H_{\mathbb{R}} \otimes \mathbb{R}^n$ is equipped with the complex structure $J \otimes 1_n$ ($1_n = n \times n$ identity matrix). So π_0 of the group of invertibles of $B_Q(H_{\mathbb{R}}) \otimes M_n$ is $\mathbb{Z}/2$, independently of n. Thus $K_1(B_Q(H_{\mathbb{R}})) \cong \mathbb{Z}/2$. Similarly $K_1(B_Q(H)) \cong \mathbb{Z}$ (cf.(Carey 1985) and (Carey 1984)).

To connect up with (6.10.9) we observe that $i_{\mathcal{O}}$ factors through X via the commutative diagram Fig. 6.5 where the horizontal arrow is the quotient map. To give a simple expression for $j_{\mathcal{O}}$ we introduce

$$\tilde{H} = H_{\mathbb{R}} \oplus H_{\mathbb{R}}\,; \Gamma = \begin{pmatrix} 0 & 1 \\ 1 & 0 \end{pmatrix} \;:\; \tilde{H} \to \tilde{H}$$

and regard $\tilde{H}$ as a complex Hilbert space with complex structure $J \oplus -J$. Then one may define a map $R \to \begin{pmatrix} T_1 & T_2 \\ T_2 & T_1 \end{pmatrix}$, where $T_1 = (R - JRJ)/2$, $T_2 = (R + JRJ)/2$ which is an isomorphism of $\mathcal{O}_Q$ with the group U_Γ of unitary operators U on $\tilde{H}$ commuting with Γ and such that $P_+UP_- + P_-UP_+ \in Q$ where $P_- = 1 - P_+$,$P_+ = \begin{pmatrix} 1 & 0 \\ 0 & 0 \end{pmatrix}$ and $\Gamma P_- \Gamma = P_+$. Thus under this map, J corresponds to P_+ and $X \cong \mathcal{O}_Q/U$ is now homeomorphic to the space of all basis projections P on $\tilde{H}$ satisfying $P - P_+ \in Q$ (i.e. the corresponding homogeneous space of U_Γ). It is now easy to check that

$$j_{\mathcal{O}}(P) = \sigma(P_+, P).$$

In Chapter 7 we will discuss ground states (zero temperature states or $\beta = \infty$ in the KMS theory) for dynamics on the Pauli algebra arising from quasi-free dynamics on the Fermion algebra. Here we will make a preliminary investigation of the ground states of quasi-free dynamics on the Fermion algebra and the even algebra. This will be needed for our treatment of Ising model ground states (and the XY-model (Exercise 7.1)) in moving back and forth between ($A^P \supset A_+$) and $A^F \supset A_+$).

First, we show some technical lemmas which stand in a more general situation.

Lemma 6.32 *Let A be a C^*-algebra, with grading θ, so that there exists an odd unitary W in A, and ω, an even state on A with GNS triple (H, π, Ω). Let $\pi_\pm$ be the restriction of the representation π of A_+ to subspaces $H_\pm \equiv \overline{\pi(A_\pm)\Omega}$.*

Assume that π_+ is irreducible. Then π_- is irreducible too and the following hold

(1) The representation π of A is irreducible if and only if π_+ and π_- are disjoint

(2) If π_+ and π_- are not disjoint, then π is a direct sum of two disjoint irreducibl representations of A and ω is an average of two disjoint pure states $\omega_\pm$ whic are interchanged by $\theta : \omega_\pm \circ \theta = \omega_\mp$.

Proof The representation π_- is irreducible as $\pi_- = \operatorname{Ad}\pi(W^*)\pi_+\operatorname{Ad}W$.
(1) Suppose $\pi_+ \not\sim \pi_-$. If $C \in \pi(A)'$, then $C = C_+ \oplus C_-$ on $H_+ \oplus H_-$, a $\pi|_{A_+} = \pi_+ \oplus \pi_-$, with the off diagonal terms of C vanishing by disjointness and $C_\pm$ scalar by irreducibility of $\pi_\pm$. Then from $\pi(a_-)C_+ = C_-\pi(a_-)$, we hav $C_+ = C_-$ (unless $H_- = 0$, in which case $C = C_+$) so that C is a scalar, and is irreducible.
(2) If $\pi_+ \sim \pi_-$, then we have a unitary u from H_+ to H_- such that $u\pi_+(a) = \pi_-(a)u$, for all $a \in A$. We will replace this u by U in $\pi(A_-)^-$:

Lemma 6.33 *There exists a unitary U in $\pi(A_-)^{-w}$, the weak closure, such th $U\xi = u\xi$, for all ξ in H_+.*

Proof We can assume $\pi(a_-) \neq 0$ for some $a_- \in A_-$; otherwise $H_- = 0$, an then $H_+ = 0$ also by equivalence. The map $x \to x \oplus uxu^*$ gives an isomorphisn from $B(H_+) = \pi_+(A_+)''$ to $\pi(A_+)''$. Take ξ_i be a complete orthonormal set i H_+, and e_{ij} the corresponding matrix units in $\pi(A_+)''$, i.e. $e_{ij} = (\xi_i \otimes \overline{\xi}_j) \oplus (u\xi_i \otimes \overline{u\xi_j})$. Fix i, k and a_- such that $\langle \pi(a_-)\xi_i, u\xi_k \rangle \neq 0$, so that $u_i = e_{ik}\pi(a_-)e_{ii} \neq 0$ Hence since e_{ii} is a minimal projection, $u_i^* u_i$ is a non-zero scalar multiple of e_i so by rescaling a_-, we can ensure $u_i^* u_i = e_{ii}$, and $u_i u_i^* = e_{ii}$, and $u_i \xi_j = \delta_{ij} u\xi$ Then $u_j \equiv e_{ji} u_i e_{ij} \in \pi(A_-)^{-w}$ has the mutually orthogonal e_{jj} as initial an final projections (since this is so for $j = i$) with $u_j \xi_j = u\xi_j$. We can then defin $U = \sum u_j \in \pi(A_-)^{-w}$ so that U and u coincide on H_+, and U^* and u^* coincid on H_-. Then $u^*u - 1 = 0$ on H_+, $uu^* - 1 = 0$ on H_-. But $u^*u, uu^* \in \pi(A_+)''$ and so by the isomorphisms $\pi(A_+)'' \cong \pi_\pm(A)''$ already mentioned for $+$, we hav U unitary.

We return to the proof of Lemma 6.32. Since u intertwines π_+ with π_-, w have $U\pi(a_+)\xi_+ = U\pi_+(a_+)\xi_+ = \pi_-(a_+)U\xi_+ = \pi(a_+)U\xi_+$, for all $a_+ \in A$ $\xi_+ \in H_+$. Hence $U^*\pi(a_+)U - \pi(a_+) \in \pi(A_+)'' \cong \pi_+(A_+)''$ vanishes on H hence is zero. Thus $U \in \pi(A_+)'$. If $a_- \in A_-$, then $\pi(a_-)U^* \in \pi(A_+)''$, an so $[U, \pi(a_-)U^*] = 0$, i.e. $U \in \pi(A_-)'$, and indeed $U \in \mathcal{Z}(\pi(A)'')$. Now as U $\pi(A_-)^{-w}$, $U^2 \in \pi(A_+)''$, and also commutes with $\pi(A_+)$ and so is a scalar $\pi(A_+)'' \cong \pi_+(A_+)'' \cong B(H_+)$ is a factor. By rotation, we can arrange $U^2 =$ Let $V(\theta)$ be the unitary ± 1 on $H_\pm$ respectively, implementing θ in π, so th extending the grading from $\pi(A)$ to $B(H_\pi)$ via $\operatorname{Ad}V(\theta)$, U is odd. Let (c the Remark on page 275) $\Omega_\pm = 2^{-1/2}(1 \pm U)\Omega$, $H^\pm = \overline{\pi(A)\Omega_\pm}$. Then $H^\pm$ a orthogonal subspaces as $(1+U^*)(1-U) = 0$, and $U \in \mathcal{Z}(\pi(A)'')$, and $\Omega_\pm$ are un vectors as $(\Omega \in H_+) \perp (U\Omega \in H_-)$. Now $\pi(A_+)''(\cong B(H_+))$ has multiplici two, with $H_\pm$ non trivial invariant spaces. Hence A_+ acts irreducibly on H Then $\omega_\pm = \omega_{\Omega_\pm} \circ \pi$ are pure states of A and A_+, as π is irreducible for A, ar

A_+ acts irreducibly on $H^{\pm}$. Now $V(\theta)$ interchanges $\Omega_\pm$ so that θ interchanges $\omega_\pm$, i.e. $\omega_\pm \circ \theta = \omega_\pm$. Moreover, $\omega = \omega_\Omega \circ \pi = \omega_{U\Omega} \circ \pi$ as $U \in \mathcal{Z}(\pi(A)'')$, so that $\omega = (\omega_+ + \omega_-)/2$. If $\overline{\omega} = \omega_\Omega$ then

$$\omega_\pm(a_+ + a_-) = \omega(a_+) \pm \overline{\omega}(\pi(a_-)U)\,, \quad a_\pm \in A_\pm$$

as U is odd in the centre of $\pi(A)''$. We can identify (as in the Remark on page 275) the GNS triple for $\omega_\pm$ with $(H_+, \rho_\pm, \Omega)$ where

$$\rho_\pm(a_+ + a_-) = \pi_+(a_+) \pm \pi_-(a_-)U\,, \quad a_\pm \in A_\pm\,. \tag{6.10.21}$$

Thus $\rho_\pm(A)'' = \pi_+(A_+)'' = B(H_+)$ and so again $\rho_\pm$ are irreducible, as we already know since $\omega_\pm$ are pure states of A. □

Lemma 6.34 *Let $\varphi = (\psi + \psi\cdot\theta)/2$ be a state on A, with ψ a pure state on A not equivalent to $\psi\cdot\theta$. If (π, A, Ω) is the GNS triple of φ and $\pi_\pm$ the representation π restricted to the subspaces $H_\pm = \overline{\pi(A_\pm)\Omega}$, then π_+ and π_- are irreducible.*

Proof We first show that π_+ is irreducible (and then use the argument of Lemma 6.32, to deduce that π_- is irreducible in the presence of an odd unitary in A – although this is not necessary). Let $\varphi_+ = \varphi|_{A_+}$, and μ a state of A_+ majorized by φ_+, i.e. $\mu \le \lambda\varphi_+$ for some $\lambda > 0$. Put $\overline{\mu} = \mu \circ (1+\theta)/2$ on A, then if $a = a_+ + a_-$, $a_\pm \in A_\pm$,

$$\begin{aligned}\overline{\mu}(a^*a) &= \mu(a_+a_+^*) + \mu(a_-^*a_-)\\ &\le \lambda\varphi_+(a_+a_+^*) + \lambda\varphi_+(a_-^*a_-) = \lambda\varphi(a^*a)\,.\end{aligned}$$

Since ψ and $\psi\cdot\theta$ are inequivalent pure states and $\overline{\mu} \le \lambda(\psi + \psi\cdot\theta)/2$ then from Lemma 2.9, we must have $\overline{\mu} = \gamma\psi + (1-\gamma)\psi\cdot\theta$ for some $\gamma \in [0,1]$. But as $\overline{\mu}$ is even, γ is forced to be $1/2$ and so $\overline{\mu} = \varphi$ and so $\mu = \varphi_+$. Consequently φ_+ is a pure state. □

An invariant state φ on A gives rise to a strongly continuous one-parameter group of unitaries $U_\varphi(t)$ in the GNS representation $(H_\varphi, \pi_\varphi, \Omega_\varphi)$ implementing α_t:

$$U_\varphi(t)\pi_\varphi(a)U_\varphi(t)^* = \pi_\varphi(\alpha_t(a))\,, \quad a \in A\,, \quad U_\varphi(t)\Omega_\varphi = \Omega_\varphi\,.$$

Then φ is a *ground state* if $U_\varphi(t) = \exp ith_\varphi$ has a positive generator $h_\varphi \ge 0$ or equivalently

$$-i\varphi(a^*\delta(a)) \ge 0$$

for all a in the domain of the generator δ of α_t (Exercise 5.5). Note that it follows from this characterization that an extremal ground state is a pure state (i.e. an extremal state) and hence is irreducible (cf. an extremal KMS state is factorial).

Let (A, α_t) be a C^*-dynamical system, with θ a grading commuting with α, gives rise to another C^*-dynamical system (A_+, α_t). By restriction a ground state φ of (A, α_t) gives rise to a ground state for (A_+, α_t), *contained* in the former in the sense that the associated GNS representation of the restricted state is a sub-representation of the restriction of the GNS representation π_φ to A_+. We next describe all possibilities for containment of representations when restricting in this fashion, and apply this in the situation of the Fermion algebra in Theorem 6.38.

Theorem 6.35 *(1) The correspondence by containment between irreducible ground state representations of* (A, α_t) *and those of* (A_+, α_t) *is either* (i) *one-to-two,* (ii) *one-to-one, or* (iii) *two-to-one.*

(2) For a given pure ground state φ *of* (A, α_t)*, case* (iii) *occurs if and only if* φ *and* $\varphi \circ \theta$ *are disjoint.*

(3) If φ *and* $\varphi \circ \theta$ *are disjoint pure states, then there exists a* θ*-invariant pure ground state* φ_0 *giving rise to the same representation. Case* (i) *occurs if and only if the infimum of the spectrum of the restriction of* h_φ *to* $\pi_{\varphi_0}(A_-)\Omega_{\varphi_0}$ *is an eigenvalue.*

(4) The restriction of a pure ground state φ *of* (A, α_t) *to* A_+ *is not pure if and only if* $\varphi \neq \varphi \circ \theta$ *and* φ *and* $\varphi \circ \theta$ *are equivalent. This can happen only in case* (i).

(5) If ψ *is a pure ground state of* (A_+, α_t)*, and the associated representation of* A_+ *is contained in a ground state representation of* (A, α_t)*, then case* (iii) *occurs if and only if the even extension of* ψ *to* A *is not pure.*

Proof (a) If φ is an even pure ground state of A, then $\varphi_+ = \varphi|_{A_+}$ is a pure ground state with both π_+, π_- irreducible by Lemmas 6.23 and 6.24. If ξ is an eigenvector of $h_\varphi|_{H_-}$ belonging to the infimum of the spectrum, then $\langle \pi_-(\cdot)\xi, \xi \rangle$ is a ground state and both $\pi_\pm$ are ground state representations. Conversely suppose $\langle \pi_-(\cdot)\xi, \xi \rangle$, $\xi \in H_-$ is a ground state representation, so that ξ belongs to the eigenspace of the eigenvalue at the infimum of the spectrum of a self adjoint operator implementing α in π_-. Since π_- is irreducible, the generator of any unitary group implementing α in π_- is unique up to translation. Then ξ must belong to the eigenspace at the infimum of the spectrum of $h_\varphi|_{H_-}$. As $\pi_\pm$ are disjoint by Lemma 6.24, π_φ contains either two or one ground state representations depending on whether the infimum of $\sigma(h_\varphi|_{H_-})$ belongs to the point spectrum. If (π_1, H_1) an irreducible ground state representation of (A, α_t) contains either π_+ or π_-, then there exists $\xi \in H_1$, $\xi \in H_+$ or H_- such that $\psi_1 = \langle \pi_1(\cdot)\xi_1, \xi_1 \rangle$ and $\psi(\cdot) = \langle \pi(\cdot)\xi, \xi \rangle$ coincide on A_+. But ψ is even and so $\psi = (\psi_1 + \psi_1 \circ \theta)/2$. Since ψ is pure, $\psi_1 = \psi$, and so π_1 is unique up to unitary equivalence, being equivalent to π. Thus the correspondence is either one-to-two or two-to-one.

(b) Suppose φ is a pure ground state inequivalent to $\varphi \circ \theta$. The state $\overline{\varphi} \equiv (\varphi + \varphi \circ \theta)/2$ on A gives rise by Lemma 6.34 to two irreducible representations π_+, π_- on A_+ which are equivalent by Lemma 6.32. Decomposing $\overline{\varphi}$ by Lemma 6.32, its GNS representation is a direct sum of mutually disjoint irreducible representations $\pi^\pm$, and $\overline{\varphi} = (\omega_+ + \omega_-)/2$ as a mixture of inequivalent pure states interchanged by θ. Since φ is pure, $\varphi = \omega_+$ or ω_-, say ω_+. As the dynamics is

graded, if $\omega_+ (= \varphi)$ is a ground state then so is ω_-. Then the disjoint ground state representations $\pi^\pm$ both contain π_+. If an irreducible ground state representation contains π_+ and π_-, then by the argument of (a), it is a subrepresentation of $\pi_{\overline{\varphi}}$ and so coincides with π_φ or $\pi_{\varphi\circ\theta}$ (π_+ or π_- in the notation of Lemma 6.32). This case gives a two-to-one correspondence.
(c) Suppose φ is a pure ground state $\varphi \neq \varphi \circ \theta$, but φ equivalent to $\varphi \circ \theta$. Let $V(\theta)$ implement θ in π_φ, where $(\pi_\varphi, H_\varphi, \Omega_\varphi)$ is the GNS triple for φ. Since π_φ is irreducible and $\theta^2 = 1$, we can arrange $V(\theta)^2 = 1$. Again since α is graded, $\alpha\theta\alpha^{-1} = \theta$, and π is irreducible, we have

$$U_\varphi(t)V(\theta)U_\varphi(t)^{-1} = c(t)V(\theta)$$

for some scalar $c(t)$. Squaring, we get $c(t)^2 = 1$, and so we have by continuity $c(t) \equiv 1$ and $V(\theta)$ and $U_\varphi(t)$ commute. We can define

$$\Omega_0 = \|\Omega_\varphi + V(\theta)\Omega_\varphi\|^{-1} (\Omega_\varphi + V(\theta)\Omega_\varphi)$$

as $\varphi \neq \varphi \circ \theta$ so that $\Omega_\varphi \neq -V(\theta)\Omega_\varphi$. Then Ω_φ is both $V(\theta)$ and $U_\varphi(t)$ invariant, and as $h_\varphi \geq 0$, $\varphi_0 = \langle \pi_\varphi(\cdot)\Omega_0, \Omega_0\rangle$ is an even ground state representation, with $\pi_{\varphi_0} = \pi_\varphi$. We now get into the situation of (a) with (H, π, Ω) the GNS triple of φ, with associated grading $H = H_+ \oplus H_-$ coming from θ and $\pi|_{A_+} = \pi_+ \oplus \pi_-$. Then φ is vector state for a vector $\Omega = \Omega^+ \oplus \Omega^-$, where $\Omega^\pm \neq 0$ (otherwise $\varphi = \varphi \circ \theta$). Again since $V(\theta)$ and $U_\varphi(t)$ commute, $\Omega^\pm$ give pure ground states on A. Restricting to A_+, we write $\psi_\pm = \langle \pi_\pm(\cdot)\Omega^\pm, \Omega^\pm\rangle$ on A_+. Then $\varphi_+ = \|\Omega^+\|^2 \psi_+ + \|\Omega^-\|^2 \psi_-$ is a mixture of two inequivalent pure states, and so is not pure. □

Lemma 6.36 *The graded extension $\overline{\omega}$ of a pure ground state ω of A_+ to A is a ground state of A if $\overline{\omega}$ is not pure.*

Proof Since ω is α_t invariant, and α_t is graded, $\overline{\omega}$ is α_t invariant. By Lemma 6.32 $\overline{\omega}$ is a mixture of two inequivalent pure states $\omega_\pm$. Thus $\overline{\omega} = \overline{\omega} \circ \alpha_t$ is also a mixture of the inequivalent pure states $\omega_\pm \circ \alpha_t$, so that $\omega_\pm \circ \alpha_t \in \{\omega_\pm\}$ for each t, and $\omega_\pm \circ \alpha_t = \omega_\pm$ by continuity. Let $U_t^\pm = U_{\omega_\pm}(t)$ implementing α_t in $\omega_\pm$, which by (6.10.21) can be identified with U_t on $H_\pm$, where U_t implements α_t in the ground state $\omega|_{A_+}$. Thus $\omega_\pm$ are pure ground states of A, and $\overline{\omega}$ is itself a ground state. □

We now restrict attention to the quasi-free formalism, so that L is a self adjoint operator on Hilbert space K anti-commuting with a conjugation Γ, and consider the quasi-free dynamics $\tau(e^{iLt})$ on $A^F = A^{SDC}(K, \Gamma)$ and its restriction to the even algebra A_+^F.

Proposition 6.37 *Let $\alpha_t = \tau(e^{iLt})$ be a quasi-free dynamics on $A = A^{SDC}(K, \Gamma)$, E_0 the eigenprojection of L for the eigenvalue 0 and $E_\pm$ the spectral projections for $(0, \infty)$ and $(-\infty, 0)$ respectively. Then all ground states of (A, α_t) are given by*

$$\varphi(a_1 a_0) = \varphi_{E_+}(a_1)\varphi_0(a_0) \tag{6.10.22}$$

where $a_0 \in A^F_{E_0}$, $a_1 \in A^F_{1-E_0}$, φ_0 *is an arbitrary state of* $A^F_{E_0}$, *and* φ_{E_+} *the Fock state for* $A^F_{1-E_0}$ *given by* E_+. *In particular, there exists a unique ground state if and only if* 0 *is not an eigenvalue of* L.

Proof Case (i) $E_0 = 0$.
Now φ is a ground state if $\varphi(\alpha_f(a)^*\alpha_f(a)) = 0$ for all $a \in A$, $f \in C^\infty$, $\operatorname{supp}\hat{f} \subset (-\infty, 0)$. In particular if $a = B(h)$, $\alpha_f(B(h)) = B(\hat{f}(L)h)$ then $\varphi(B(h)^*B(h)) = 0$ for all $h \in 1 - E_+$. Thus as in (6.6.4), φ must be the Fock state φ_{E_+}. It is clear that φ_{E_+} is a ground state by inspecting the spectrum of $h_\varphi = \bigoplus_{m=0}^\infty \bigotimes^m L$ restricted to antisymmetric tensors on E_+ where $L|_{E_+} \geq 0$.

Case (ii) $E_0 \neq 0$.
Let $K_0 = E_0 K$, $K_1 = (1 - E_0)K$ which are Γ invariant, $A^F_0 = A^F(E_0 K)$, $A^F_1 = A^F((1 - E_0)K)$. As before $\varphi(B(h)^*B(h)) = 0$ wherever $h \in E_- K$ so that $\varphi(B(h)^*a) = \varphi(aB(h)) = 0$, $a \in A^F$, $h \in E_- K$ by the Schwarz inequality. If $h \in K_1$, write $h = h_+ + h_-$, $h_\pm \in E_\pm K$, so that $B(h) = B(\Gamma h_+)^* + B(h_-)$ with $\Gamma h_+, h_- \in E_- K_1$. Thus in (6.10.22) it is enough to consider $a_1 = B(h_1)^* \cdots B(h_m)^* B(k_1) \cdots B(k_n)$ with $h_i, h_j \in E_- K_1$. Then equality in (6.10.22) follows as both sides vanish when $n + m \neq 0$. Conversely to see that (6.10.22) defines a ground state let (K_i, π_i, Ω), $i = 0, 1$ be the GNS triples for φ_0 on A^F_0 and φ_{E_+} on A^F_1 respectively. If V implements θ in K_1 as in (6.4.8) fixing Ω_1, then $(K_1 \otimes K_0, B(h_0 + h_1) \to \pi_1(B(h_1)) \otimes 1 + V \otimes B(h_0), \Omega_1 \otimes \Omega_0)$ will be the GNS triple of a state φ on A^F satisfying (6.10.22) which is a ground state as α_t is implemented by $e^{ith} \otimes 1$ where $h \geq 0$ as in case (i). □

Restricting to the even subalgebra:

Theorem 6.38 (1) *There exists a unique ground state for* (A^F_+, α_t) *if and only if one of the following (mutually exclusive) conditions (A) and (B) is satisfied:*

(A) $E_0 = 0$ *and the infimum of the positive part of the spectrum of* L *is not an eigenvalue of* L.

(B) dim $E_0 = 1$.

In this case, the unique ground state is the restriction of any ground state of (A^F, α_t). *More explicitly for* $a_\pm$ *in* $A^F_{1-E_0,\pm}$ *and* $E_0 h = h$

$$\varphi(a_+ + a_- B(h)) = \varphi_{E_+}(a_+) . \tag{6.10.23}$$

(2) *If* dim $E_0 > 1$, *the set of all ground states of* (A^F_+, α) *coincides with the set of restrictions of all ground states of* (A^F, α) *to* A^F_+.

(3) *If* $E_0 = 0$ *and the infimum* λ *of the positive part of the spectrum of* L *is an eigenvalue with eigenprojection* E^λ, *then an extremal ground state* φ *of* (A^F_+, α) *is either the restriction of* φ_{E_+} *to* A^F_+ *or the Fock state* φ_h *defined by*

$$\varphi_h(a) = \varphi_{E_+}(B(h)^* a B(h))/\langle h, h\rangle = \varphi_{E_+ - P(h) + \Gamma P(h)\Gamma}(a) \qquad (6.10.24)$$

where h is any non-zero vector satisfying $E^\lambda h = h$ and $P(h)$ is the orthogonal projection onto the one-dimensional space spanned by h. The representations of all φ_h are equivalent and disjoint from that obtained from the restriction of φ_{E_+} to A_+^F.

Proof Suppose ψ is a pure ground state of A_+^F. We show that $\overline{\psi}$, the even extension to A^F is a ground state. If Δ is a closed subset of $\mathbb{R}$, the spectral subspace (Section 4.6) $A^F(\Delta) = \left\{x \in A^F \mid \alpha_f(x) = 0, \operatorname{supp}\hat{f} \cap \Delta = \varnothing\right\}$, $A(\Delta_1)A(\Delta_2) \subset A(\overline{\Delta_1 + \Delta_2})$. If h_1, $h_2 \in E^L(-\infty, \varepsilon]$, $E^L(-\infty, 0]$ respectively, then $B(h_1), B(h_2) \in A(-\infty, \varepsilon]$, $A(-\infty, 0]$ respectively, and $B(h_2)B(h_1) \in A_+(-\infty, \varepsilon]$.

A ground state ψ of A_+^F satisfies $\psi(a^* a) = 0$, $a \in A_+^F(-\infty, \varepsilon]$ and $\varepsilon > 0$. Thus $\pi_{\overline{\psi}}(B(h_2)B(h_1))\,\Omega = 0$, if $\left(K, \pi_{\overline{\psi}}, \Omega\right)$ is the GNS triple for $\overline{\psi}$. Suppose $\pi_{\overline{\psi}}(B(h))\Omega = 0$, for all $h \in E^L(-\infty, 0)$, in which case we argue as in Proposition 6.37 to deduce (6.10.22) for $\overline{\psi}$ and show $\overline{\psi}$ is a ground state of A^F. Alternatively, we suppose $\xi \equiv \pi_{\overline{\psi}}(B(h))\Omega \neq 0$, and $\langle B(h_2)\xi, \xi\rangle = 0$ for all $h_2 \in E^L(-\infty, 0]$. Let $E' \in \pi_{\overline{\psi}}(A^F)'$ be the projection on $\overline{\pi_{\overline{\psi}}(A^F)\xi}$. Then $\pi(B(h_1)E'\Omega = E'\pi(Bh_1)\Omega = E'\xi = \xi$ so that $E'\Omega \neq 0$ and cyclic for A^F in $\overline{\pi_{\overline{\psi}}(A^F)\xi}$. Then $\varphi = \langle \pi_{\overline{\psi}}(\cdot)E'\Omega, E'\Omega\rangle/\|E'\Omega\|^2$ has a ground state representation (as $\langle \pi_{\overline{\psi}}(\cdot)\xi, \xi\rangle/\|\xi\|^2$ is a ground state of A^F).

We deduce from Lemma 6.32 that $\overline{\psi}$ either is pure or $\overline{\psi} = (\omega_+ + \omega_-)/2$ with $\omega_\pm$ disjoint pure states with representations $\pi^\pm$. If $\overline{\psi}$ is pure, then $\varphi = \overline{\psi}$ as $E' \in \pi_{\overline{\psi}}(A^F)')$ and $\pi_{\overline{\psi}}$ is a ground state representation. In the alternative situation $\overline{\psi} = (\omega_+ + \omega_-)/2$, then $E' \in \pi_{\overline{\psi}}(A^F)' = (\pi_{\omega_+} \oplus \pi_{\omega_-})(A^F)'$ can only be $1 \oplus 0$, $0 \oplus 1$ or $1 \oplus 1$, i.e. $\varphi = \omega_\pm$ or $\overline{\psi}$. If $\varphi = \overline{\psi}$ we are as in the situation already treated. If $\varphi = \omega_+$, then π^+ is a ground state representation and so is $\pi^- \cong \pi^+ \circ \theta$ as α_t commutes with θ. Hence $\overline{\psi} = (\omega_+ + \omega_-)/2$ has a ground state representation as we asserted.

We next discuss three cases according to the dimension of E_0 being zero, one or greater than one.

Case (a) $E_0 = 0$.

Here φ_{E_+} is according to Proposition 6.37 the unique ground state of (A^F, α_t), and contains disjoint representations $\pi_\pm$ of A_+^F on even and odd parts of Fock space. The representation π_+ is the GNS representation of the ground state $\varphi_{E_+}|A_+$. To see whether or not π_- is a ground state representation we consider the spectrum of h_φ for $\varphi = \varphi_{E_+}$ on H_-, the anti-symmetric odd tensors on E_+H. Since $h_\varphi|H_-$ is the restriction of $\bigoplus_{n=0}^\infty \bigotimes^{2n+1} L$ to H_-, the infimum λ of the spectrum of h_φ occurs in the one particle space and is the infimum of the positive spectrum of L, and λ will be in the point spectrum of h_φ if and only if λ is in the point spectrum of L. In this case $\lambda > 0$ as we have already assumed $E_0 = 0$. Thus if λ is not in the point spectrum of L we have an unique ground state representation on A_+ by Theorem 6.35, and then as $E^{h_\varphi}(0) = \mathbb{C}\Omega$ the vacuum vector, there is only one ground state, namely the restriction of φ_{E_+} to A_+. Next consider the case where $\lambda > 0$ is in the point spectrum of L. Then a pure ground state with representation π_- is a vector state for $\xi \in E^{h_\varphi}(\lambda)H_-$ with $\xi = \pi(B(h_\lambda))\Omega$ for $h_\lambda \in E^L(\lambda)$. Then $\pi(B(h))\xi = 0$ whenever $E'h = 0$, for the basis projection $E' = E_+ - P(h_\lambda) + \Gamma P(h_\lambda)\Gamma$. Hence the vector state ω_ξ is the Fock state $\varphi_{E'}$.

Case (b) dim $E_0 = 1$.

Since $E_0 \equiv \mathbb{C}h_0$ is Γ-invariant and $\Gamma^2 = 1$ we can choose h_0 so that $\Gamma h_0 = h_0$, as in the Remark on page 220. If $\|h_0\|^2 = 2$, then $B(h_0)$ is a self adjoint unitary, $A_0^F = A^F(E_0) \cong \mathbb{C} \oplus \mathbb{C}$ with two pure states $\varphi_\pm$ interchanged by θ. Thus by Proposition 6.37 (A^F, α_t) has two pure ground states for $\varphi_0 = \varphi_\pm$, interchanged by θ, and disjoint as $B(h_0) \to \pm 1$ in their respective representations. Then A^F has a unique even ground state which contains by Lemma 6.32 a unique ground state representation of A_+^F, and both $\varphi_\pm$ restrict to the unique ground state of A_+^F. Note that $H_{\varphi_\pm}$ can be identified with the even part of the Fock space for φ on A^F. Uniqueness follows from $E^{h_\varphi}(0) = \mathbb{C}\Omega$.

Case (c) dim $E_0 > 1$.

In this case $A_{0+} \neq \mathbb{C}$ and states on A_0 which are distinct on A_{0+} give rise to distinct ground states of A_+. □

Exercise 6.16 Let Γ be a conjugation on a Hilbert space H which commutes with an action of a compact group. If $S \in Q(K, \Gamma)$ is G-invariant, let φ_S^G denote the restriction of φ_S to the fixed point algebra $A(K, \Gamma)^G$.

(a) If E_1 and E_2 are G-invariant basis projections, show that $\varphi_{E_1}^G$ and $\varphi_{E_2}^G$ are equivalent if and only if $E_1 - E_2$ is Hilbert–Schmidt and $\det_{E_1 \wedge (1-E_2)}(g) = 1$ for all g in G. ($E_1 \wedge (1 - E_2)$ has finite dimensional range if $E_1 - E_2$ is Hilbert–Schmidt, which is G-invariant.)

(b) If $S_1, S_2 \in Q(K, \Gamma)$ are G-invariant with no Fock parts ($0 < S_i < 1$), show that $\varphi_{S_1}^G$ and $\varphi_{S_2}^G$ are quasi-equivalent if and only if $S_1^{1/2} - S_2^{1/2}$ is Hilbert–Schmidt.

6.11 Quasi-free endomorphisms

In discussing fusion rules via endomorphisms of the (even) Fermi algebra we will have need to understand the decomposition of $\pi_P \circ \tau(U)$ on A^F or A_+^F into its irreducible components when π_P is an irreducible Fock representation and $\tau(U)$ is a quasi-free endomorphism. We will be working in the self dual formalism, so that K is a Hilbert space equipped with a conjugation Γ, and P is a basis projection.

If U is an isometry on K commuting with Γ, and of finite co-rank, dim Ker $U^* \equiv m_U < \infty$, let $n_U = \dim \operatorname{Ker}(U^*|_{PK}) \leq m_U/2$.

Proposition 6.39

$$\pi_P \circ \tau(U) \cong \bigoplus^{2^{n_U}} \pi_{U^*PU} .$$

Proof If $\{h_j : j \in J\}$ is an orthonormal basis for $\operatorname{Ker} U^* \cap PK$, then $\Omega_I = \underline{\pi_P(B(h_I))\Omega_P}$, for $I \subset J$ ordered give orthogonal invariant subspaces $H_I = \overline{\pi_P\tau(U)(A)\Omega_I}$, $H_P = \oplus H_I$, with $\omega_{\Omega_I} = \langle \pi_P \circ \tau(U)(\cdot)\Omega_I, \Omega_I\rangle = \varphi_P \circ \tau(U)$, for all I, so that $\pi_P \circ \tau(U)$ on H_I are all equivalent. □

Here π_{U^*PU} may not be irreducible but is the GNS representation of $\varphi_{U^*PU} = \varphi_P \circ \tau(U)$.

Theorem 6.40

$$\pi_P \circ \tau(U) \cong \begin{cases} 2^{m_U/2}\pi_P & m_U \text{ even} \\ 2^{(m_U-1)/2}(\pi^+ \oplus \pi^-) & m_U \text{ odd} \end{cases}$$

where $\pi^\pm$ are disjoint irreducible representations.

Proof case (a) $m_U = 2$.
Take, using the Remark on page 220, $e_\pm$ an orthonormal basis of $\operatorname{Ker} U^*$ such that $e_+ = \Gamma e_-$. Put $\lambda_\pm = \langle Pe_\pm, e_\pm\rangle$ so that $\lambda_+ + \lambda_- = 1$ and $E_\pm = e_\pm \otimes \overline{e}_\pm$, $UU^* = 1 - E_+ - E_-$. If $\lambda_- = 0$, then $e_+ \in PK$, $e_- \in (1-P)K$, U^*PU is a basis projection and $\pi_P \circ \tau(U) \cong \pi_{U^*PU} \oplus \pi_{U^*PU}$ is a sum of two irreducible Fock representations. Thus we can assume $\lambda_\pm \neq 0$. In this case we claim $n_U = 0$ (recall $n_U \leq (m_U)/2$, so that $n_U = 0$ or 1 if $m_U = 2$), and so $\pi_P \circ \tau(U)$ is cyclic and can be identified with π_{U^*PU} which will not be irreducible but a mixture of two Fock states. Suppose $v \in \operatorname{Ker}(U^*|_{PK})$. Then $v = \mu_+e_+ + \mu_-e_-$, $\mu_\pm \in \mathbb{C}$, where $\Gamma v \in (1-P)K$. Then $E_+PE_- = E_+(1 - \Gamma P\Gamma)E_- = \Gamma(E_-PE_+)\Gamma$ using $E_+E_- = 0$, $\Gamma E_\pm \Gamma = E_\mp$. Hence $\langle Pe_-, e_+\rangle = \langle E_+PE_-e_-, e_+\rangle = -\langle \Gamma(E_-PE_+)\Gamma e_-, e_+\rangle = -\langle e_-, PE_+e_+\rangle$ as $\Gamma e_\pm = e_\mp$, and so $\langle Pe_\mp, e_\pm\rangle = 0$. Then $0 = \langle P\Gamma v, \Gamma v\rangle = |\mu_+|^2\lambda_- + |\mu_-|^2\lambda_+$ so that $\mu_\pm = 0$ as $\lambda_\pm > 0$, and $v = 0$.

Let

$$P_\pm = U^*PU + \lambda_\pm^{-1}U^*P(E_\mp - E_\pm)PU \tag{6.11.1}$$

which are basis projections. Then

$$\varphi_P \circ \tau(U) = \lambda_+\varphi_+ + \lambda_-\varphi_-$$

where

$$\varphi_\pm = \varphi_{P_\pm} = \lambda_\pm^{-1}\left\langle \pi_P \circ \tau(U)(\cdot)\pi_P(B(e_\mp)B(e_\pm))\Omega_P, \Omega_P\right\rangle . \tag{6.11.2}$$

Since $P_+ - P_- = (\lambda_+\lambda_-)^{-1}U^*P(E_- - E_+)PU$ is finite rank, φ_{P_+} and φ_{P_-} are equivalent Fock states.

Thus if $\lambda_+\lambda_- = 0$ then $\varphi_P \circ \tau(U)$ is pure but $(\pi_P \circ \tau(U), \Omega_P)$ is not cyclic. If $\lambda_+\lambda_- \neq 0$, then $\varphi_P \circ \tau(U)$ is a mixture of two equivalent pure states, and

$\pi_P \circ \tau(U)$ is cyclic. In both cases $\pi_P \circ \tau(U)$ is a mixture of two equivalent irreducible Fock representations.

case (b) $m_U = 1$.
Take g_0 a Γ-invariant unit vector in $\operatorname{Ker} U^*$, and put $G_0 = g_0 \otimes \overline{g}_0$, $S = U^*PU$. Now $\Gamma G_0 P G_0 \Gamma = G_0(1-P)G_0 = G_0 - G_0PG_0$ as $\Gamma G_0 \Gamma = G_0$. Hence $G_0PG_0 = G_0/2$, $\langle P g_0, g_0\rangle = 1/2$. Hence $E = 4(S - S^2) = 4U^*PG_0PU$ is a rank one projection $e_0 \otimes \overline{e}_0$, where $e_0 = 2U^*Pg_0$, and $\Gamma E \Gamma = E$ as $U^*G_0 = 0$. Then $ES = E/2 = SE$ and $F = S - E/2$ is a projection with $1 - F - \Gamma F \Gamma = E$. Thus $S = F + E/2 = F + (1 - F - \Gamma F \Gamma)/2 = (1 + F - \Gamma F \Gamma)/2$. A projection F such that $F \perp \Gamma F \Gamma$ and $1 - F - \Gamma F \Gamma$ is rank one is called a partial basis projection with Γ-codimension 1. Since such a projection is a basis projection on $(F + \Gamma F \Gamma)K$, we can form the Fock representation $(\mathcal{F}_F, \pi_F, \Omega_F)$ of $A((\Gamma + \Gamma F \Gamma)K, \Gamma)$. We can define irreducible representations $\pi_{(F,\pm e_0)} = \pi_{(F,\pm)}$ of $A(K,\Gamma)$ on $\mathcal{F}_F$ by

$$\pi_{(F,\pm e_0)}(B(h)) = \frac{1}{\sqrt{2}}\langle h, \pm e_0\rangle V(\theta) + \pi_F\left(B(Fh + \Gamma F \Gamma h)\right) \tag{6.11.3}$$

where $V(\theta)$ is the parity operator. Then defining $\varphi_{(F,\pm e_0)} = \langle \pi_{(F,\pm e_0)}(\cdot)\Omega_F, \Omega_F\rangle$, the quasi-free state $\varphi_F = \left(\varphi_{(F,+)} + \varphi_{(F,-)}\right)/2$ is a decomposition into disjoint pure states. Now from $G_0PG_0 = G_0/2$ we have $\operatorname{Ker}(U^* \cap PK) = 0$ i.e. $n_U = 0$, and so by the previous remarks, $\varphi_P \circ \tau(U)$ $(= \varphi_S)$ has a GNS triple $(H_P, \pi_P \circ \tau(U), \Omega_P)$. Thus $\pi_P \circ \tau(U)$ is a mixture of two disjoint irreducible representations.

Now consider the general cases. First choose a Γ-invariant orthonormal basis $\{e_n\}_{n=1}^\infty$ of K, $\Gamma e_n = e_n$ by the Remark on page 220.

If m_U is even, choose $\{f_n\}_{n=1}^{m_U}$ a Γ-invariant orthonormal basis of $\operatorname{Ker} U^*$, and write $f_{m_U+n} = Ue_n$, $n = 1, 2, \ldots$ so that $U = \sum_{n=1}^\infty f_{m_U+n} \otimes \overline{e}_n$. Define

$$U_0 = \textstyle\sum_{n=1}^\infty f_n \otimes \overline{e}_n\,, \qquad U_1 = \sum_{n=1}^\infty e_{n+2} \otimes \overline{e}_n$$

so that $m_{U_i} = i$, $U = U_0 U_2^{m_U/2}$, and $P_0 = U_0^* P U_0$ is a basis projection as U_0 is unitary. Then $\pi_P \circ \tau(U) = \pi_{P_0} \circ \tau(U_2)^{m_U/2}$ and we can inductively use step (a).

If m_U is odd, let $\{f_i\}_{i=0}^{m_U-1}$ be a Γ-invariant orthonormal basis for $\operatorname{Ker} U^*$, and $f_{m_U-1+n} = Ue_n$, $n = 1, 2, \ldots$, so that $U = \sum_{n=1}^\infty f_{m_U-1+n} \otimes \overline{e}_n$. Define isometries

$$U_1 = \textstyle\sum_{n=1}^\infty f_n \otimes \overline{e}_n\,, \qquad U_2 = \sum_{n=0}^\infty f_{n+2} \otimes \overline{f}_n$$

so that $m_{U_i} = i$, and $U = U_2^{(m_U-1)/2} U_1$. We first use the general even case to decompose $\pi_P \circ \tau(U_2^{(n_U-1)/2}) = \bigoplus_{j=1}^{2^{(m_U-1)/2}} \pi_{P_j}$ where π_{P_j} are equivalent Fock representations and then decompose $\pi_{P_j} \circ \tau(U_1)$ as a sum of disjoint non-Fock irreducible representations using case (b). □

Note that $\pi_{(F,e_0)}|_{A^+}$ is irreducible. For if $x \in \pi_{(F,e_0)}(A^+)'$ then $x \in \pi_F(A(F + \Gamma F \Gamma)^+)' = C^*(V(\theta))$ by Section 6.10, as a Fock representation on the even algebra is a sum of two disjoint irreducible representations on even and odd tensors. But $[V(\theta), \pi_{(F,e_0)}(B(e_0)B(f))] = \sqrt{2}\pi_F(B(f))$ for all $f \in (F + \Gamma F \Gamma)K$, so that $V(\theta) \notin \pi_F(A(F+\Gamma F\Gamma)^+)'$, so that $x \in \mathbb{C}$. The representations $\pi_{(F,\pm e_0)}$ become equivalent on $A(K)^+$ as on that subalgebra they are interchanged by $V(\theta)$. Thus restricting Theorem 6.40 to the even subalgebra we get

Corollary 6.41

$$\pi_P \circ \tau(U)|_{A^+} \cong \begin{cases} 2^{m_U/2}(\pi_+ \oplus \pi_-) & m_U \text{ even} \\ 2^{(m_U+1)/2}\pi & m_U \text{ odd} \end{cases}$$

where $\pi_\pm$ are the disjoint irreducible representations on even and odd tensors and π is an irreducible representation which is not quasi-free.

6.12 Notes

Von Neumann formalized the basics of infinite tensor products of Hilbert spaces (von Neumann 1961). The infinite tensor product construction is fundamental in the theory of C^*-algebras and W^*-algebras, providing simple C^*-algebras, (the UHF algebras) and type I, II, III factors by completion in product states. The types of product states on UHF algebra $\bigotimes_{j=1}^\infty M_{p_j}$ can be described as follows: (Araki 1963), (Bures 1963), (Moore 1967). Suppose ρ_{K_j} is a state on M_{p_j} whose density matrix is diagonal with eigenvalues $(k_j(1),\ldots,k_j(p_j))$ in decreasing order. Then let $\rho = \bigotimes_{j=1}^\infty \rho_{K_j}$ on $\bigotimes_{j=1}^\infty M_{p_j}$. Then ρ is of type I if and only if $\sum_j k_j(1) < \infty$, of type II_1 if and only if $\sum_{i,j} \left|1-(k_j(i)p_j)^{1/2}\right|^2/p_j < \infty$, and $n_j < \infty$, for all j and $n_j > 1$ for infinitely many j. Suppose $k_j(1) \geq a > 0$ for some a, all j (which is automatic if p_j is uniformly bounded) then ρ is of type III if and only if $\sum_{j,i} k_j(i)|k_j(1)/k_j(i)-1|_c^2 = \infty$ for some (and hence all) positive c, where $|x|_i = \min(|x|, c)$. The proofs of Theorems 6.7 and 6.5 are from (Connes 1973) and (Araki *et al.* 1984).

The structure of the states $\omega = \bigotimes_{j=1}^\infty \rho_{K_j} \circ p$ on the Cuntz algebra $\mathcal{O}_n$ if $p : \mathcal{O}_n \to \bigotimes_1^\infty M_n$ is the canonical projection is more complex than the product states $\rho = \bigotimes_{j=1}^\infty \rho_{K_j}$ (Araki *et al.* 1984), e.g. unlike the product states, they are not always factorial. Indeed ω is not factorial if and only if $\sum_{j=1}^\infty (1 - \mathrm{tr}\, K_j^{1/2} K_{j+s}^{1/2}) < \infty$ for some $s > 0$ and ρ is of type I.

We refer to (Bratteli and Robinson 1981), (Kadison and Ringrose 1986) for full details concerning quasi-equivalence and disjointness in the Remark on page 214.

The isomorphism between the Pauli and Fermi algebras of Jordan–Wigner was formulated in the bilateral case by Araki, who was the first to regard the Pauli and Fermi algebras as a sub-C^*-algebra of the universal algebra, the crossed product by the symmetry θ_-, and to only identify the even subalgebras. It was initially used by (Araki and Barouch 1983) in the XY-model, and quickly taken up by (Araki and Evans 1983) in the Ising model. The constructive formulation of quasi-free states here is the self dual setting of that in (Evans and Lewis 1984) for the Clifford formalism with normal ordering taken from (Evans 1979b). The result of Exercise 6.8 is due to (Hugenholtz and Kadison 1975) in the complex formalism and (Wolfe 1975) in the self dual formalism. (Shale and Stinespring 1965) characterized when quasi-free Bogoliubov automorphisms are inner (Blattner 1958), characterized when they are weakly inner in the trace representation. (Powers and Størmer 1970) showed when two quasi-free states are equivalent in the complex formalism, whilst (Araki 1970a) extended it to the self dual case.

For expositions on determinants, see (Ringrose 1971), (Simon 1977b), (Simon 1979).

The dictionary between Bogoliubov endomorphisms of Exercise 6.6 is taken from (Bratteli 1982).

Our theory of implementability of quasi-free derivations and automorphisms is based on (Berezin 1966), (Fredenhagen 1977), (Araki 1970a), (Carey and Ruijsenaars 1987), (Powers and Størmer 1970), (Ruijsenaars 1977), (Sato *et al.* 1977–1980), (Araki 1987), see also (Bongaarts 1970), (Chadam 1968), (Hochstenbach 1975), (Klaus and Scharf 1977), (Labontè 1974), (Lundberg 1976), (Kristensen *et al.* 1967), (Schrader and Uhlenbrock 1975). Explicit formula for implementers go back to (Friedrichs 1953). The Schwinger term $s_P(h,k)$ can be written as

$$s_P(h,k) = \mathrm{tr}(h_{-+}k_{+-} - k_{-+}h_{+-})/2 = \mathrm{tr}(F[F,h][F,k])/8$$

if $F = 2P - 1$, and satisfies the *cyclic* condition $s(h,k) = -s(k,h)$ and the *Hochschild* condition $s(h_1h_2,h_3) - s(h_1,h_2h_3) + s(h_3h_1,h_2) = 0$. This is an example of a cyclic 1-cocycle (Connes 1994). In the course of attempting to introduce a Chern character of an element of K-homology, Connes was led to initiate a purely algebraic theory of cyclic cohomology.

For analogous ideas in the CCR quasi-free theory of implementability of derivations and automorphisms, and equivalence of states, see for example (Ruijsenaars 1977) and (van Daele and Verbeure 1971).

The general theory of quasi-free completely positive maps is taken from (Evans 1979b), (Evans 1980a) – see also (Hugenholtz and Kadison 1975) and Chapter 4 for Fock completely positive maps $A_0(\cdot)$ in the complex formalism). (Davies 1978) showed that a trace class operator generated a norm continuous completely positive semigroup $A_0(e^{tB})$; the converse and the general setting of Theorem 6.19 is taken from (Evans 1979b).

(Araki and Evans 1983) initiated the theory of Fock states on the even and Pauli algebra, from which Theorems 6.25, 6.30, 6.31 are taken. In particular they introduced the mod 2 index needed to understand even Fock quasi-free states.

The Remark on page 277 is taken from (Carey and Evans 1988). Theorems 6.35 and 6.38 are taken from (Araki 1986) and (Araki and Matsui 1985) on the study of ground states on the even and Pauli algebras. Lemma 6.26 is due to (Strătilă and Voiculescu 1978).

Theorem 6.30, and in particular the $\mathbb{Z}/2$-index of (6.10.9) has also been successfully used in (Araki and Matsui 1985) to help find the number of pure ground states in the one-dimensional XY-model – see Exercise 7.1. The generalization of (Araki and Evans 1983) from $\mathbb{Z}/2$ to an arbitrary compact group is due to (Matsui 1987a), Exercise 6.16. We also refer the reader to (Baker 1978, 1979), (Bake

1980), (Baker and Powers 1983a), (Baker and Powers 1983b), (Matsui 1987a), (Matsui 1987b) for other work on the equivalence of restrictions of states to fixed point algebras (e.g. the gauge invariant CAR algebra).

Section 6.11 is based on (Binnenhei 1995), (Böckenhauer 1996a), where the question of decomposing quasi-free endomorphisms in the Fock representation of the Fermi algebra and its even part is considered.

7

THE ISING MODEL

7.1 Introduction

Statistical mechanics aims to study the global properties of classical or quantum systems, which may have a large number of building blocks and interactions, through a small number of global concepts such as volume, temperature, equilibrium states, critical exponents.

We may be particularly interested in critical phenomena and phase transition such as a liquid/gas transition or a change in magnetic properties. Consider for example a metal bar suspended in a magnetic field. Apart from diamagnetic effects (where currents are induced by Lenz's law to oppose the field) there may be a (greater) opposite paramagnetic effect (as in aluminium) where there is a permanent magnetic moment present in the material which the field tends to line up. At lower temperatures, there is a greater tendency towards order, whilst at higher temperatures this has to compete with the disordering power of thermal motion. In a ferromagnet, such as iron, this paramagnetic effect is exaggerated by the particles also lining up by their own magnetic field and interactions. An external magnetic field H may induce a magnetization $m(H)$ in the iron bar, which as the field is switched off, $H \searrow 0$, leaves a residual magnetization $m(H) \searrow m^*$. At high temperatures, where disorder is dominant m^* may be zero (paramagnetism), whilst at low temperatures where order is dominant m^* may be non zero (ferromagnetism) (see Fig. 7.1). In Fig. 7.1 $m(H)$ and m^* may depend on the temperature T, whilst m^* may be non-zero only below some critical temperature T_c (see Fig. 7.2).

Before studying lattice models exhibiting such behaviour, let us recall the essence of Newtonian mechanics. The position of a single particle is described by a point in $\mathbb{R}^3$, the positions of N particles by a point $q = (q_i)$ in *configuration space* $\mathbb{R}^n$, $n = 3N$. By Newton, if $m_{3i-2} = m_{3i-1} = m_{3i}$ is the mass of the ith particle, then $m_j \ddot{q}_j$ are the components of the forces on the particles, which in a conservative system are $-\partial V/\partial q_j = \dot{p}_j$ for some potential $V = V(q)$, where p_j is the jth momentum $m_j \dot{q}_j$. If $x_j = q_j$, $x_{j+n} = p_j$, $j = 1, 2, \ldots, n$, $x = (x_i) \in \mathbb{R}^{2n}$ then the Hamiltonian $H = H(x) = \sum_{j=1}^{n} x_{n+j}^2 m_j/2 + V(x_1, \ldots, x_n)$ satisfies the

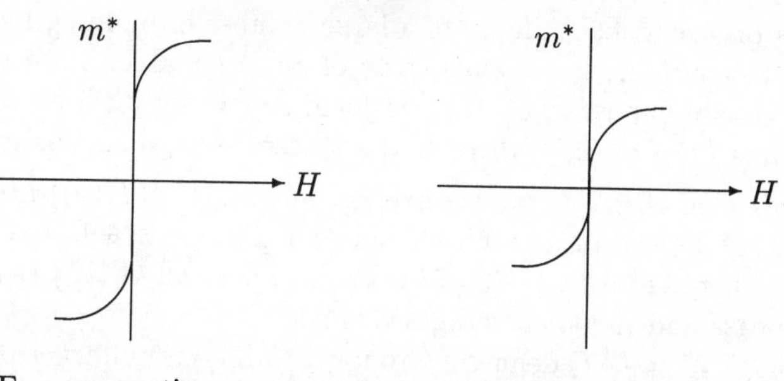

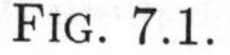

FIG. 7.1.

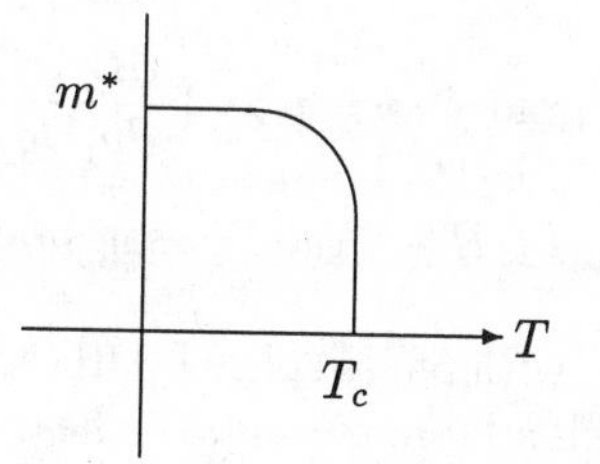

FIG. 7.2. Magnetization m^*

first order equations

$$\dot{x} = J \text{ grad } H(x) \tag{7.1.1}$$

where $J = \begin{pmatrix} 0 & 1 \\ -1 & 0 \end{pmatrix}$ is a complex structure on $\mathbb{R}^{2n}$, i.e. J is orthogonal, $J^* J = 1$, and $J^2 = -1$. In Hamiltonian mechanics we have *phase space* $\Gamma = \mathbb{R}^{2n}$ and a differential equation (7.1.1) is *Hamiltonian* if $H : \Gamma \to \mathbb{R}$ and J is a complex structure on Γ. Assume that H is sufficiently smooth so that $t \to T_t$ is a flow on Γ (i.e. $T_0 = \text{id}, T_{t+s} = T_t T_s, t, s \in \mathbb{R}$), where $T_t x = x(t)$ is the (unique) solution of (7.1.1) subject to $x(0) = x$. By Liouville's theorem the flow T_t preserves Lebesgue measure on Γ. The energy function H is constant on orbits, $HT_t = H$, and let μ_E be the measure induced, by Lebesgue measure on Γ, on $\Omega_E = H^{-1}(E)$, the surface of constant energy $E \in \mathbb{R}$; μ_E is preserved by the flow.

The above was sketched for genuine coordinates (q_j), of particles; coordinates in configuration or phase space could also be generalized ones, e.g. to describe the location of a thin rod in Euclidean space needs only five coordinates: the location of one fixed point of the rod in $\mathbb{R}^3$ and two angles to describe the orientation of the rod.

An *observable* is a real valued function on phase space, such as the energy H. With a gas or a magnet made up of a large number of building blocks, we need to be able to consider the average value of an observable. Since the flow takes place on constant energy levels Ω_E, we could simply for fixed energy E take the space average $\langle f\rangle_E = \int_{\Omega_E} f\, d\mu_E$. If the system is ergodic this is the same as Boltzmann's idea of taking a time average $\lim_{t\to\infty}\int_0^t f(T_s(x))\,ds/t$. In general, the latter may depend on the initial condition $T_0(x) = x \in \Gamma$, but if the system is ergodic (where an orbit $\{T_t(x) \mid t \in \mathbb{R}\}$ is almost all of Ω_E) then this should not be the case and the two averages coincide.

Note that $U_t f = f\circ T_t$ defines a strongly continuous unitary group on $\mathrm{L}^2(\Omega_E)$ (the completion of $C_c(\Omega_E)$ with respect to $\langle f, g\rangle = \int f\bar{g}\, d\mu_E$). Then $\int_0^t U_s\, ds/t$ converges strongly as $t\to\infty$, to the projection P on $\{f \mid U_s f = f,$ for all $s\}$, which in the ergodic case are the constant functions so that $Pf = \langle f, 1\rangle/\langle 1, 1\rangle$, the space average. The generator of U_t is

$$\begin{aligned}\left.\frac{d}{dt}\right|_{t=0} U_t f &= \operatorname{grad} f\cdot\dot{x} = i\sum_j\left(\frac{\partial H}{\partial p_j}\frac{\partial f}{\partial q_j} - \frac{\partial H}{\partial q_j}\frac{\partial f}{\partial p_j}\right)\\ &= i\{f, H\}, \quad \text{the Poisson bracket.}\end{aligned}\tag{7.1.2}$$

We think of our system, with phase space Γ_S, Hamiltonian H_S interacting and in thermal equilibrium with a large system, the heat bath with phase space Γ_B, Hamiltonian H_B, containing a large number N of particles at some temperature T – so large that its temperature is unaffected by any interchange of heat with the small system. As a heat bath, we can take an ideal gas, satisfying Boyle's law $PV/N =$ constant if $P =$ pressure, $V =$ volume, at constant temperature. Thus we can use PV/N to define a temperature scale $PV/N = kT$, where k is Boltzmann's constant. Then for $\gamma = (q,p) \in \mathbb{R}^{6N} = \Gamma_B$,

$$H_B(\gamma) = \sum \frac{1}{2m}p_i^2 + \infty\cdot\chi_{\Lambda^c}(q) \tag{7.1.3}$$

where the gas is restricted to be in a box Λ, and Λ^c denotes the complement.

Then for $H(\gamma) = H_S(\gamma_S) + H_B(\gamma_B), \gamma = (\gamma_S, \gamma_B) \in \Gamma_S\times\Gamma_B = \Gamma$, a central limit theorem for identically distributed random variables shows that if F is an observable for the system S then

$$\lim_{N\to\infty}\langle F\rangle_{H=3N/2\beta} = \frac{\int_{\Gamma_S} F(\gamma_S)e^{-\beta H_S(\gamma_S)}d\gamma_S}{\int_{\Gamma_S} e^{-\beta H_S(\gamma_S)}d\gamma_S}\,. \tag{7.1.4}$$

Thus from the *microcanonical ensemble*, μ_E and *microcanonical average* $\mu_E(f)$, we are led to the *canonical ensemble* or *Gibbs ensemble* the measure $Z^{-1}\exp(-\beta H_S(\gamma_S))$ on phase space, where

$$Z = \int_{\Gamma_S}\exp(-\beta H_S(\gamma_S))\, d\gamma_S \tag{7.1.5}$$

s the *partition function*, and *canonical average* or *Gibbs average* (7.1.4).

Suppose $H(\alpha)$ are a family of Hamiltonians indexed by some external parameters α such as the external magnetic field or geometrical parameters of some
onstraining vessel. If $-\partial H/\partial q$ are the components of force, then $-\partial H/\partial \alpha_j$ will
imilarly be the (generalized) force opposing the change in α, with average value
$-\mathbb{E}_\beta(\partial H/\partial\alpha) = \beta^{-1}\partial/\partial\alpha \log Z$, e.g. $-\partial H/\partial V$ is the pressure, the force tending
o oppose the change in volume V. For a perfect gas, (7.1.3), we can compute
$Z = V^N(2\pi m/\beta)^{3N/2}$ so that the pressure $P = \beta^{-1}\partial/\partial V \log Z = N/\beta V$, and
omparing with Charles' law $P = kTN/V$ we identify β in the Gibbs ensemble
f the ideal gas with $1/kT$, essentially the inverse temperature. We will interpret
β as $1/kT$ or the inverse temperature for any system.

The *internal energy* (energy contained in the system) is $U = \mathbb{E}(H)$, whilst
he work done on the system against the generalized forces resisting change is
he *work differential*

$$W = \mathbb{E}(dH) = -\textstyle\sum_j \mathbb{E}(\partial H/\partial\alpha_j)d\alpha_j\,. \tag{7.1.6}$$

The heat absorbed by the system, the *heat differential* is $q = dU - W$. Then
$\beta q = dI$ a perfect differential where

$$I = -\mathbb{E}(\log\rho) = \beta\mathbb{E}(H) + \log Z \quad \text{if } \rho = e^{-\beta H}/Z\,. \tag{7.1.7}$$

Thus $\int_{(\beta',\alpha')}^{(\beta,\alpha)} q/T$ is independent of the path from (β',α') to (β,α) in the (β,α)
lane and there exists a function S such that $S(T,\alpha) - S(T',\alpha') = \int q/T$. This
atter point had been observed empirically and S is called the *entropy*. Thus the
ntropy $S = kI = -k\mathbb{E}(\log\rho) = k\beta U + k\log Z$, and $U - TS = -kT\log Z$ is
alled the (*Helmholtz*) *free energy* f, so that $e^{-\beta f} = Z = \int e^{-\beta H}$.

Phase transitions will be related to the existence of singularities in some
erivative of the free energy f, and equilibrium states determined by minimiza-
ion of the free energy $f = U - TS$ = internal energy − temperature·entropy. In a
hysical system, nature minimizes the right-hand side (for a fixed temperature).
o at zero (or low) temperature the internal energy is minimized. At infinite (or
igh) temperature the entropy is dominant and needs to be maximized. Thus
here are two forces which may be competing, e.g. in a ferromagnet, one trying to
roduce order and align the building blocks and one trying to produce disorder
nd randomness. Which aspect wins depends on the model and of course the
emperature – with some critical temperature describing a phase transition.

We will model such critical phenomena in lattices L (a graph) usually the
quare lattices $\mathbb{Z}^2$, but for expositional purposes it is also useful to look at the
ne dimensional lattice $\mathbb{Z}$. At each site or vertex of L we have a variable – a spin
r magnetization. In the Ising model we simply have either a positive or negative
rientation or value represented by $+1$ or -1. Thus a *state* of the Ising model
s a distribution of pluses and minuses over the lattice L, so any configuration
s represented by a point in configuration space, the compact Hausdorff space
$P = \{\pm 1\}^L$. The Q-state Potts model is a generalization of the Ising model

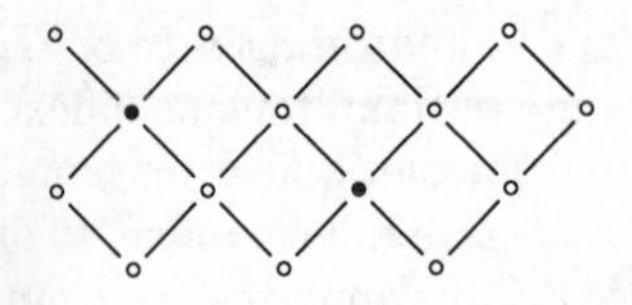

FIG. 7.3. An Ising model configuration

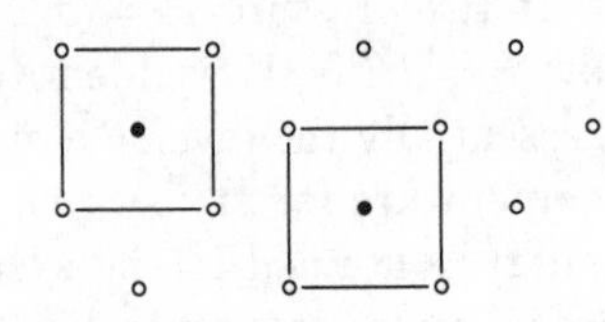

FIG. 7.4. A Hard square configuration

where the spin at any lattice site can take on Q-values or directions where Q i an integer. The state space of the Ising model could also be used to describe lattice gas where $+$ means an occupied site $\bullet$ and $-$ unoccupied $\circ$ as in Fig. 7.3

To represent particles with non-zero volume, a similar state space can b used where we draw a square around each occupied site as in Fig. 7.4. To ensur distinct particles do not occupy the same portion of space, i.e. the squares i Fig. 7.4 should not overlap, we should only consider these states in the Isin state space such that the nearest neighbour in L of every occupied site with $\bullet$ i unoccupied $\circ$. Thus the state space of the hard square model are the distribution σ of the vertices of the graph of Fig. 7.5 over the vertices of the lattice L, suc that if α and β are nearest neighbours in L, then the corresponding values σ and σ_β are joined in the graph. Such a state space P could be defined for an graph G – but if G contains some multiple edges (two vertices joined by mor than one edge) then we should consider distributions of the edges of G on L. Fo the Dynkin diagram A_3 with vertices labelled as $\{\bullet, \pm\}$ and square lattice L on obtains two copies of the Ising model as in Fig. 7.6 by placing the frozen spin on the even or odd sublattices of L.

This graph may be generalized to those in Fig. 7.7 for the Q-state Pott model, or the Dynkin diagrams A_n of Figure 10.8, for the models of (Andrew *et al.* 1984). Thinking of this as an oriented graph with basic transitions

$$\circ \rightleftarrows \bullet$$

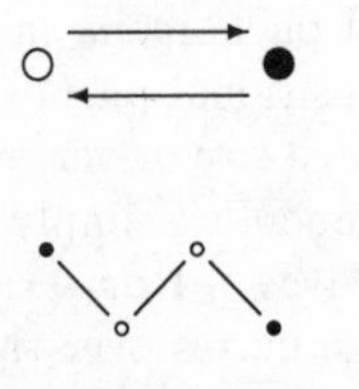

FIG. 7.5.

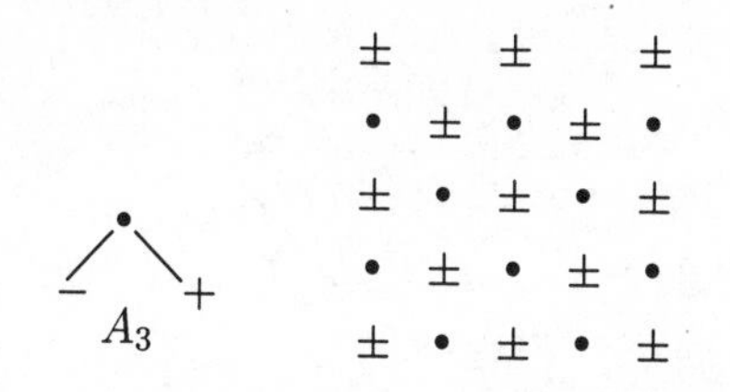

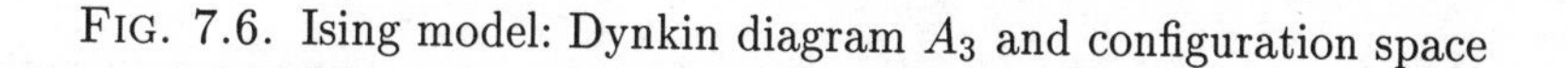

FIG. 7.6. Ising model: Dynkin diagram A_3 and configuration space

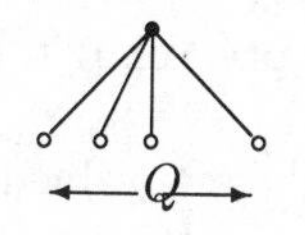

FIG. 7.7. Q-state Potts model

we can generalize to the Weyl alcove $\mathcal{A}^{(n)}$ of the level k-integrable representations of the Kač–Moody algebra $A_{N-1}^{(1)} = su(N)^\wedge$. For $SU(3)$ instead of $SU(2)$ (see Fig. 13.18) we replace

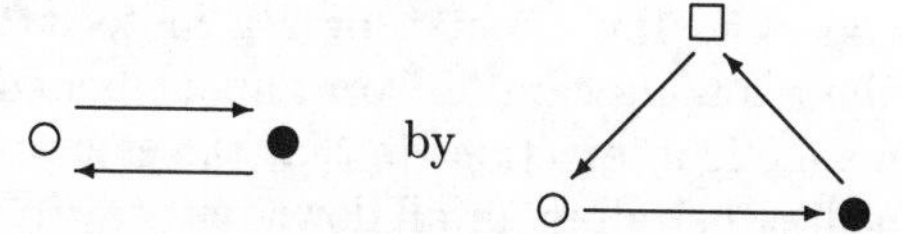

i.e. transitions generated by 3 vectors (instead of 2) which add to zero. Let

$$P_{++}^{(n)} = \{\lambda = \textstyle\sum_{i=1}^{N-1} \lambda_i \Lambda_i \mid \lambda_i \geq 1 \,, \sum_{i=1}^{N-1} \lambda_i \leq n-1\} \tag{7.1.8}$$

where the Λ_is are the $N-1$ weights of the fundamental representation and $n = k + N$. The vertices of the graph are the elements of $P_{++}^{(n)}$ and its oriented edges are given by N vectors e_i defined by

$$\begin{aligned} e_1 &= \Lambda_1 \,, \\ e_i &= \Lambda_i - \Lambda_{i-1} \,, \quad i = 1, 2, \ldots, N-1 \,, \\ e_N &= \Lambda_{N-1} \,. \end{aligned} \tag{7.1.9}$$

These models were introduced and studied by (Date *et al.* 1987). We can also label our states by $\left\{(\gamma_i)_{i=1}^{N-1} \mid \gamma_i \geq 0, \sum_{i=1}^{N-1} \gamma_i \leq l\right\}$ where $l = n - N$ is the *level* $(\lambda_i = \gamma_i + 1)$, or under the transformation $(\gamma_i)_{i=1}^{N-1} \to (\gamma_1 + \gamma_2 + \cdots + \gamma_{N-1}, \gamma_2 + \cdots + \gamma_{N-1}, \ldots, \gamma_{N-2} + \gamma_{N-1}, \gamma_{N-1}) = (\alpha_i)_{i=1}^{N-1}$ with partitions or Young tableaux

$$\{(\alpha_i)_{i=1}^{N} \mid l \geq \alpha_1 \geq \cdots \geq \alpha_{N-1} \geq \alpha_N = 0\} \tag{7.1.10}$$

where there is a constraint in the number of rows given by the rank N (in fact $N-1$) and in the number of columns given by the level l. The procedure

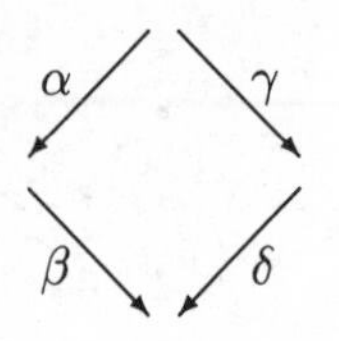

FIG. 7.8. Local configuration

of moving along an edge e_i can then be interpreted as adding or subtracting boxes to a Young tableau as in Fig. 11.75. The unoccupied state corresponds to $(1, 1, \ldots, 1)$, $(0, 0, \ldots, 0)$ or the empty Young tableaux in these three descriptions, which we often denote by $*$ or 0.

The classical observables are the real valued functions on configuration space, such as the energy or Hamiltonian H. To completely specify the model, we have to describe the Hamiltonian. The Ising Hamiltonian with nearest neighbour interactions is

$$H(\sigma) = - \sum_{(\alpha,\beta)\in \mathrm{L}^{(1)}} J\sigma(\alpha)\sigma(\beta) \tag{7.1.11}$$

where σ is a configuration in P, and the sum is over nearest neighbours α, β in L. (We will first take the Hamiltonian over a finite lattice and then take the thermodynamic limit, taking care to worry about boundary conditions.) If $J > 0$, one has ferromagnetic interactions in that the energy is minimized when all the spins are aligned, either all up or all down, either $\sigma(\alpha) \equiv 1$ or $\sigma(\alpha) \equiv -1$ respectively.

Similarly for more general graphs Γ, the energy associated with any face of the lattice is a function of the shape of the boundary and the distribution of the edges $\Gamma^{(1)}$ along it. The energy of a configuration σ is then the sum of the energy of the faces. For the square lattice L of Fig. 7.3 each face has a boundary of the form of Fig. 7.8. The energy associated with a configuration where $\alpha\beta$, $\gamma\delta$ are paths of length 2 in Γ with the same initial and same terminal vertices is denoted $H(\alpha, \beta, \gamma, \delta)$. Strictly speaking it could depend also on the position of the face in Fig. 7.8 in the lattice L. The energy of the configuration is then $\sum_{\text{faces}} H(\alpha, \beta, \gamma, \delta)$. Writing

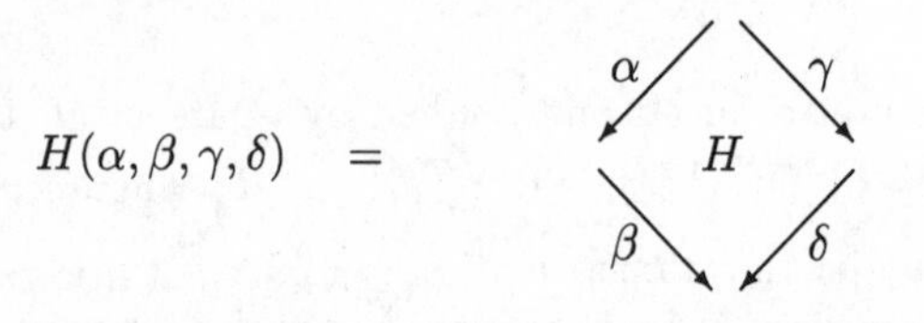

(7.1.12)

we see that a (translation invariant) energy function on the square lattice for the graph Γ is precisely the same as certain elements of the path AF algebra built from Γ, namely those living between levels n and $n + 2$ in $A(\Gamma)[n, n + 2]$. The Boltzmann weight $e^{-\beta H}$, thought of as an element of the path algebra, is called a *face operator*.

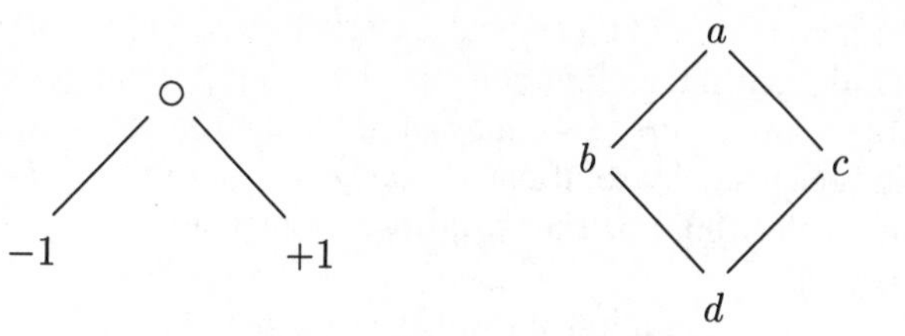

FIG. 7.9. Ising state space A_3 and energy $H(a,b,c,d) = -J(ad+bc)$

In the case of the Ising model, then the edges in $\alpha, \beta, \gamma, \delta$ (Fig. 7.8) can be described in terms of vertices of A_3, as in Fig. 7.9, with Hamiltonian $H(a,b,c,d) = -J(ad+bc)$, $a,b,c,d \in \{-1,0,1\}$ the 3 vertices of A_3.

To begin our discussion on equilibrium states for lattice models, both classical and quantum, let us first consider finite classical system, with an energy function H on a finite configuration space P of states. Then in analogy with classical mechanics discussed above, we are led to consider the Gibbs state, the measure μ on P which assigns a (Boltzmann) weight $Z^{-1}\exp(-\beta H(\sigma))$ to a state σ, where Z, the partition function, is a normalizing factor $Z = \sum_\sigma \exp(-\beta H(\sigma))$, and again β is to be interpreted as the inverse temperature. This is the unique measure μ on P which satisfies the variational principle, i.e. minimizes the free energy $= U - \beta^{-1}S$ where $U = E(H)$ the internal energy $\sum H(\sigma)\mu(\sigma)/\sum \mu(\sigma)$, and S is the entropy $= -\sum \mu(\sigma)\log \mu(\sigma)$.

We will take as equilibrium states for infinite classical lattice models, those that satisfy a variational principle; or equivalently at least for spin systems those that in a certain sense (the DLR equations of (Dobrushin 1968), (Lanford and Ruelle 1969)) give Gibbs states on finite subsystems. For us, with an emphasis in this book on a more non-commutative framework, the corresponding notions in quantum statistical mechanics are easier to explain. Suppose we have Hamiltonians $H_\Lambda = \sum_{X\subset\Lambda}\Phi(X)$ for finite subsets $\Lambda \subset \mathbb{Z}^d$, and some potential function Φ, where $\Phi(X) \in \bigotimes_X M_n \subset \bigotimes_{\mathbb{Z}^d} M_n$ is self adjoint. Then under a suitable smoothness hypothesis on Φ (e.g. if Φ is translation invariant, containing only one and two body interactions of finite range as in the Ising model) then $\sigma_t = \lim_{\Lambda\to\infty} e^{itH_\Lambda}(\cdot)e^{-itH_\Lambda}$ exists on the UHF algebra $\bigotimes_{\mathbb{Z}^d} M_n$ and defines a dynamics. Equilibrium states for the dynamics σ_t at inverse temperature β then correspond to KMS states (Section 5.6), solutions of a variational principle (minimizing the (free energy − T entropy) as in (7.1.13)) or the Gibbs condition (the non-commutative analogue of the DLR equations). A (faithful) state φ satisfies the *Gibbs condition* if for all finite subsets Λ of $\mathbb{Z}^d$, if we perturb the state φ by $\beta H_{\partial\Lambda} = \beta\sum_{X\cap\Lambda\neq\varnothing, X\cap\Lambda^c=\varnothing}\Phi(X)$ (remove the interaction across the boundary $\partial\Lambda$) then the perturbed state is a product state $\varphi_\Lambda \otimes \psi_{\Lambda^c}$, where ψ_{Λ^c} is a state on $\bigotimes_{\Lambda^c} M_n$, and φ_Λ is the *Gibbs state* $\mathrm{tr}(e^{-\beta H_\Lambda}\cdot)/\mathrm{tr}(e^{-\beta H_\Lambda})$ on $\bigotimes_\Lambda M_d$, the non-commutative or quantum analogue of the classical Gibbs state (7.1.4). These conditions are equivalent if the potential Φ is sufficiently smooth.

Consider a lattice model built from a graph Γ, a two dimensional lattice L

and Boltzmann weights. The classical observables are $C(P)$, the commutative C^*-algebra of all continuous functions on the compact configuration space, and at each inverse temperature β we may be interested in the simplex K_β of equilibrium states, given say by solutions to the DLR equations of (Dobrushin 1968), (Lanford and Ruelle 1969) or the variational principle:

$$\text{minimize (energy} - T \text{ entropy)} \tag{7.1.13}$$

where T is the temperature. In the algebraic approach, one uses the transfer matrix formalism to transform the setting to that of a one-dimensional quantum model, represented by a non-commutative C^*-algebra, an AF algebra (possibly $A(\Gamma)$ or its two sided version or the C^*-algebra generated by the local transfer matrices, essentially the operators in (7.1.12)). For each inverse temperature β one looks for a map $F \to F_\beta$ from (local) classical observables in $C(P)$ to the quantum algebra A such that one can recover the classical expectation values or correlation functions from a knowledge of the quantum ones alone:

$$\mu(F) = \varphi_\mu(F_\beta) \tag{7.1.14}$$

For the nearest neighbour two dimensional Ising model, on $L = \mathbb{Z}^2$, the quantum algebra A is the Pauli algebra, whilst the local transfer matrices generate the even part A_+. Spatial translation by $\mathbb{Z}^2$ in the classical model $\bigotimes_{\mathbb{Z}^2} \mathbb{C}^2$ corresponds to spatial translation in $A^P = \bigotimes_{\mathbb{Z}} M_2$ where $\mathbb{Z}^2$ has been compressed to $\{0\} \times \mathbb{Z}$, together with an evolution $\{T^n(\cdot)T^{-n} \mid n \in \mathbb{Z}\}$ in the orthogonal transfer direction, arising from the transfer matrix T.

Let us fix some boundary conditions, and then for each inverse temperature β let φ_β denote the corresponding *state* on A. [In general, the map φ_μ in (7.1.14) is not a state; positivity of φ_μ is related to reflection positivity of μ (Section 7.14)]. Then if β_c denotes the inverse critical temperature of Onsager, there exist (Theorem 7.6) automorphisms $\{\nu_\beta \mid \beta \neq \beta_c\}$ of A which do not depend on the boundary conditions, and real analytic in $\beta \neq \beta_c$ such that

$$\varphi_\beta = \begin{cases} \varphi_\infty \circ \nu_\beta & \beta > \beta_c \\ \varphi_0 \circ \nu_\beta & \beta < \beta_c \end{cases} . \tag{7.1.15}$$

The structure at zero and infinite temperatures is accessible. In fact $\varphi_0 = \otimes_{\mathbb{Z}} \omega_\Omega$ where $\Omega = \begin{pmatrix} 1 \\ 1 \end{pmatrix} / \sqrt{2}$ is the disordered state, and for $+$ or $-$ boundary conditions $\varphi_\infty^+ = \otimes_{\mathbb{Z}} \omega_{\left(\begin{smallmatrix} 1 \\ 0 \end{smallmatrix}\right)}, \varphi_\infty^- = \otimes_{\mathbb{Z}} \omega_{\left(\begin{smallmatrix} 0 \\ 1 \end{smallmatrix}\right)}$ respectively, and $\varphi_\infty = \frac{1}{2}(\varphi_\infty^+ + \varphi_\infty^-)$ for free or periodic boundary conditions. Thus with free or periodic boundary conditions, we conclude from (7.1.15) that φ_β is pure for $0 \leq \beta < \beta_c$ (in fact also for $\beta = \beta_c$ by different methods) and is a mixture of two inequivalent pure states $\varphi_\beta^\pm$ for $\beta > \beta_c$. Moreover for $\beta < \beta_c < \beta'$, there exists an endomorphism $\alpha : A \to \pi(A_+)''$ where π denotes the GNS representation of φ_β, such that $\varphi_{\beta'}^+ = \overline{\varphi}_\beta \circ \alpha$ and $\overline{\varphi}_\beta$ denotes the natural extension of φ_β to $\pi(A)''$. Since the family $\{\nu_\beta \mid \beta \neq \beta_c\}$

can be chosen (real) analytic in β, we can conclude that $\langle F\rangle^{\pm}$ is real analytic in $\beta \neq \beta_c$, where $\langle\rangle^{\pm}$ denote the classical states corresponding to + and − boundary conditions respectively. This will follow from $\langle F\rangle^{\pm} = \phi^{+}_{\infty} \circ \nu_\beta(F_\beta)$. Moreover, $\lim_{M\to\infty} T_M^{-it}(\cdot)T_M^{it} = \alpha_t$, where T is the transfer matrix, defines a dynamics on A_+ which can be lifted to a strongly continuous one-parameter group on A which has an unique ground state for $\beta < \beta_c$ and two extremal ground states $\phi^{\pm}_{\beta}$ for $\beta > \beta_c$.

A detailed analysis of the quantum model provides the following for the classical model. Let

$$\psi = \begin{cases} \langle\rangle, & \beta < \beta_c \quad m = 2(K_1^* - K_2) \\ \langle\rangle^{\pm}, & \beta > \beta_c \quad m = 4(K_2 - K_1^*) \end{cases} \tag{7.1.16}$$

where $K_i = \beta J_i$ are the interaction parameters in (7.1.11), $\sinh 2K \sinh K^* = 1$, and $\langle\ \rangle$ denotes the equilibrium state for free or periodic boundary conditions, $\langle\ \rangle^{\pm}$ for fixed $\pm$ boundary conditions. Then

$$\lim_{l\to\infty} e^{|l|m} \left[\psi(F_1\tau_{(l,0)}(F_2)) - \psi(F_1)\psi(F_2)\right] = 0$$

for all F_1, F_2 local. For any $\varepsilon > 0$, there exist $F_{\varepsilon,1}$, $F_{\varepsilon,2}$ such that

$$\lim_{l\to\infty} e^{|l|(m+\varepsilon)} \left[\psi(F_1\tau_{(l,0)}F_2) - \psi(F_1)\psi(F_2)\right] = \infty\,. \tag{7.1.17}$$

We consider in Section 7.2 phase transition in terms of differentiability of the free energy for classical lattice models. We do this first in quite general settings, using the powerful GKS correlation inequalities and then more specifically in Sections 7.3 and 7.5 in the Ising model where we can actually make some computations and obtain some closed expressions, although in two dimensions not that of the free energy in a non-zero magnetic field.

In Sections 7.4–7.5 we discuss the transfer matrices of the two dimensional nearest neighbour Ising models, in Section 7.6 their use to derive various thermodynamic quantities, and in Sections 7.7–7.9, particularly (7.1.15), the C^*-algebra approach to phase transition in the two dimensional Ising model. At the heart of all this is the high temperature – low temperature duality of Kramers and Wannier, which we go into in some detail in Section 7.11. The Temperley–Lieb algebra is introduced in Section 7.10, as being generated by face operators in certain statistical mechanical models including the Q-state Potts model and the IBF model. It will play a significant role in the subfactor theory of later chapters. Sections 7.12 and 7.13 discuss the analyticity of expectation values $\langle\ \rangle_\beta$, $\langle\ \rangle^{\pm}_{\beta}$, through that of $\{\nu_\beta \mid \beta \neq \beta_c\}$, and a computation of the magnetization respectively, Section 7.14 contains an analysis of the ground states of the dynamics on the Pauli algebra arising from the transfer matrix, $\lim_{M\to\infty} T_M^{-it}(\cdot)T_M^{it}$, and (7.1.16) and (7.1.17). Finally Section 7.15 discusses the Bost–Connes model of spontaneously broken symmetry related to the distribution of prime numbers.

7.2 Equilibrium states

Consider $e_1 = 1 = 1 \oplus 1$, $e_+ = \sigma_x = 1 \oplus (-1)$, a basis for a copy of $\mathbb{C}^2 = C\{\pm 1\}$ in M_2, so that $e_1(\sigma) = 1$, $e_+(\sigma) = \sigma$, $\sigma \in \{\pm 1\}$. Then if Λ is a finite set, we identify $C\{\pm 1\}^\Lambda \cong \otimes_\Lambda C\{\pm 1\} = \otimes_\Lambda \mathbb{C}^2$ (where if X, Y are finite sets $C(X \times Y) \cong C(X) \otimes C(Y)$, under the identification $(f \otimes g)(x,y) = f(x)g(y)$ $f \in C(X)$, $g \in C(Y)$, $(x,y) \in X \times Y$). Since e_1, e_+ are basis vectors for $\mathbb{C}^2$ a basis for $\otimes_\Lambda \mathbb{C}^2$ is $e_A = (\otimes_A e_+) \otimes (\otimes_{\Lambda \setminus A} e_1)$, $A \subset \Lambda$. Then if $\sigma \in \{\pm 1\}^\Lambda$ $e_A(\sigma) = \prod_{x \in A} \sigma(x) \equiv \sigma_A$. By the basis property of e_As, any Hamiltonian or any function in $C\{\pm 1\}^\Lambda$ can be decomposed uniquely as

$$H(\sigma) = -\textstyle\sum_{A \subset \Lambda} J_A \sigma_A\,, \quad \sigma \in P(\Lambda) = \{\pm 1\}^\Lambda\,.$$

The term $-J_A\sigma_A$ represents the interaction between the points of $A \subset \Lambda$; so that for the Ising model $J_A = 0$ if $|A| > 2$. For $J_A \geq 0$, the Hamiltonian favours alignment. If $\sigma_i = \sigma_j$ is more probable than $\sigma_i \neq \sigma_j$ then $\langle \sigma_i \sigma_j \rangle = \mathbb{P}(\sigma_i = \sigma_j) - \mathbb{P}(\sigma_i \neq \sigma_j) \geq 0$. At zero temperature, with perfect alignment $\langle \sigma_i \sigma_j \rangle = 1$ the term $J_{(k,l)}$ appears as $\beta J_{(k,l)}$ in the Hamiltonian, so increasing J should have the same effect as lowering the temperature, so if $\langle \sigma_i \sigma_j \rangle \nearrow 1$ as $T \searrow 0$, then $\langle \sigma_i \sigma_j \rangle$ increases as $J_{(k,l)} \nearrow$, or $\partial/\partial J_{(k,l)} \langle \sigma_i \sigma_j \rangle \geq 0$.

Theorem 7.1 GKS inequalities. *Suppose $H(\sigma) = -\sum J_A \sigma_A$, with ferromagnetic interactions $J \geq 0$. Then for all $B, C \subset \Lambda$*

(7.2.1) $\langle \sigma_B \rangle \geq 0$

(7.2.2) $\beta^{-1} (\partial/\partial J_C) \langle \sigma_B \rangle = \langle \sigma_B \sigma_C \rangle - \langle \sigma_B \rangle \langle \sigma_C \rangle \geq 0$.

Proof Configuration space $P(\Lambda) = (\mathbb{Z}/2)^\Lambda$ is an abelian group. If we identify $P(\Lambda)$ with subsets of Λ, by $\sigma \to \sigma^{-1}(-1)$, multiplication corresponds to symmetric difference $BC = (B \cup C) \setminus (B \cap C)$; $\varnothing = \mathrm{id}$, $B^{-1} = B$ or $B^2 = \varnothing$. With $B \subset \Lambda$ and $\sigma \in P(\Lambda)$ identified with $A \subset \Lambda$, define $A(B) \equiv \sigma_B = (-1)^{|A \cap B|} = B(A)$. The family $\{A(\cdot) \mid A \in P\}$ identifies $P(\Lambda) = (\mathbb{Z}/2)^\Lambda$ with its group dual $P(\Lambda)^\wedge$ the characters of $P(\Lambda)$. Schur's orthogonality relations for X_1, X_2 homomorphisms from a finite abelian group $G \to \mathbb{T}$:

$$\langle X_1, X_2 \rangle = \textstyle\sum_g X_1(g) \overline{X_2(g)} = |G| \delta_{X_1, X_2}$$

in this context says

$$\langle A, B \rangle = \textstyle\sum_{E \subset \Lambda} A(E) B(E) = 2^{|\Lambda|} \delta_{A,B}\,.$$

Let $B \subset \Lambda$ be non empty then

$$\begin{aligned} Z\langle \sigma_B \rangle &= \sum_{E \subset \Lambda} B(E) \exp\left(\beta \sum_A J_A A(E) \right) \\ &= \sum_{E \subset \Lambda} \sum_{n=0}^{\infty} B(E) \frac{\beta^n}{n!} \sum_{A_1, \ldots, A_n} J_{A_1} \cdots J_{A_n} A_1(E) \cdots A_n(E) \end{aligned}$$

$$= \sum_{n=1}^{\infty} \frac{\beta^n}{n!} \langle B, B \rangle \sum_{A_1 \cdots A_n = B} J_{A_1} \cdots J_{A_n}$$

by Schur's orthogonality relations. This gives $\langle \sigma_B \rangle \geq 0$, if $J_A \geq 0$, $\beta \geq 0$. We readily show $\beta^{-1}\partial/\partial J_C \langle \sigma_B \rangle = \langle \sigma_B \sigma_C \rangle - \langle \sigma_B \rangle \langle \sigma_C \rangle$, by straightforward differentiation. Then

$$\begin{aligned} & Z_\Lambda^2 \{\langle \sigma_B \sigma_C \rangle - \langle \sigma_B \rangle \langle \sigma_C \rangle\} \\ &= \textstyle\sum_{F,G} [(BC)(G) - B(F)C(G)] \exp\left(\beta \sum J_A(A(F) + A(G))\right) \\ &= \textstyle\sum_E (1 - B(E)) \sum_G D(G) \exp(\beta \sum_A J'_A A(G)) \text{ putting } E = FG, D = BC \end{aligned}$$

≥ 0 by (a) for $\langle \sigma_D \rangle' \geq 0$ where $J'_A = J_A(1 + A(E)) \geq 0$. □

In a finite lattice Λ, with zero magnetic field and local Hamiltonian

$$H_\Lambda(\sigma) = \textstyle\sum_{x,y\, n.n} - J\sigma(x)\sigma(y)$$

with a summation over nearest neighbours $(n.n)$ we start considering the thermodynamic limit (as $|\Lambda| \to \infty$) of

$$E_\Lambda(f) = \langle f \rangle_\Lambda = \frac{\sum_\sigma f(\sigma) \exp(-\beta H(\sigma))}{\sum_\sigma \exp(-\beta H(\sigma))} = \textstyle\sum_{\sigma \in P(\Lambda)} f(\sigma) \mathbb{P}_\beta(\sigma)$$

vhere $\mathbb{P}_\beta(\sigma)$ is the Gibbs probability

$$\mathbb{P}_\beta(\sigma) = \exp\left(-\beta H(\sigma)\right)/Z\,, \quad Z = \textstyle\sum_{\sigma \in P(\Lambda)} \exp\left(-\beta H(\sigma)\right)\,.$$

The Gibbs ensemble gives a state, a positive linear functional on $C(P(\Lambda))$. We could consider $f \in C(P(\Lambda))$, depending only on configurations inside some fixed ubset $\Lambda_0 \subset \Lambda$ (i.e. use the embedding $C(P(\Lambda_0)) \subset C(P(\Lambda))$ by $f \to f \otimes$ $\in C(P(\Lambda_0)) \otimes C(P(\Lambda \backslash \Lambda_0)) \cong C(P(\Lambda))$, or extend f to $\tilde{f}$ in $P(\Lambda)$ by setting $\tilde{f}(a, b) = f(a)$, if $a \in P(\Lambda_0)$, $b \in P(\Lambda \backslash \Lambda_0)$.) We then compute $\langle f \rangle_\Lambda$, letting $\Lambda| \to \infty$ to obtain a state on $C(P(\mathbb{Z}^2))$.

More generally we could impose boundary conditions, e.g. fix all spins to e positive (or all negative) outside Λ. If there exists a phase transition, at low nough temperatures, then the effect of these two operations could be different. If ve compute $\langle f \rangle^{\pm}$ with $\pm$ boundary conditions respectively, then $\lim_{\Lambda \to \infty} \langle \sigma(x) \rangle_\Lambda^+$ xists, and for low enough temperatures is strictly positive by Peierls' argument. nstead of embedding $C(P(\Lambda_0))$ in $C(P(\Lambda))$ in the above way, we could also have ushed $f \to f \otimes e_A$ for A some subset of $\Lambda \backslash \Lambda_0$ i.e. prescribing the configuration o be $+$ in $A \subset \Lambda \backslash \Lambda_0$. Alternatively one can keep the same embedding on observbles but change the Hamiltonian to take account of boundary effects in $\Lambda \backslash \Lambda_0$. ince we are primarily interested in the Ising model and its nearest neighbour nteractions, it is enough for us to consider the boundary $\partial \Lambda_0$ of Λ_0, and modify the Hamiltonian H to $H^b_\Lambda = H_\Lambda + b_{\partial\Lambda}$ where $b_{\partial\Lambda}$ is some interaction across the boundary. Then with

$$E^b_\Lambda(f) = \langle f\rangle^b_\Lambda = \frac{\sum_\sigma f(\sigma)\exp(-\beta H^b(\sigma))}{\sum_\sigma \exp(-\beta H^b(\sigma))}$$

we would take $\lim_{\Lambda\to\infty} E^b_\Lambda(f)$ where $f \in C(P(\Lambda_0)) \subset C(P(\Lambda))$, $\Lambda_0 \subset \Lambda$, to define a state, $E^b_\beta = E^b$ a positive linear function on $C(P(\mathbb{Z}^2))$.

In particular, we have the magnetization $\langle\sigma(x)\rangle^+_\Lambda$, $x \in \Lambda$ with + boundary conditions $\left(b^+_{\partial\Lambda} = -J\sum_{x\in\partial\Lambda}\sigma(x)\right)$. Then Peierls' celebrated estimate (Peierls 1936) says that for fixed $a \in (0,1)$, there exists β_0 such that

$$\langle\sigma(x)\rangle^+_\Lambda \geq a$$

for all $\beta > \beta_0$, independent of x in Λ. From this one gets a non-zero spontaneous magnetization, or at least

$$\lim_{|\Lambda|\to\infty} \inf \frac{1}{|\Lambda|}\sum_x \langle\sigma(x)\rangle^+ > 0\,.$$

In fact $\Lambda \to \langle\sigma(x)\rangle^+_\Lambda$ is monotone decreasing, as a consequence of the GKS inequalities when we switch on a positive magnetization in the complement of Λ. Define

$$H^\lambda_\Lambda = H^+_\Lambda - \lambda\sum_{x\in\Lambda\setminus\Lambda'}\sigma(x)\,.$$

Then $\langle\sigma_B\rangle^+_{\Lambda'}(\lambda)$ is increasing in λ by the GKS inequalities, and at $\lambda = 0$, $\lambda = \infty$ we obtain $\langle\sigma_B\rangle^+_\Lambda$, $\langle\sigma_B\rangle^+_{\Lambda'}$ respectively. (Moreover, there exists a critical temperature T_c, such that $\langle\sigma\rangle^+ > 0$ for $T < T_c$, as in Fig. 7.2.)

Clustering is also a consequence of the finite range property of the interaction and these correlation inequalities. If A, B are finite subsets of $\mathbb{Z}^2$ then

$$\lim_{x\to\infty} \langle\sigma_{A+x}\sigma_B\rangle^+ = \langle\sigma_A\rangle^+\langle\sigma_B\rangle^+\,. \tag{7.2.3}$$

If $A \subset \Lambda_1$, $B \subset \Lambda_2$, then by GKS inequalities

$$\langle\sigma_{A+x}\sigma_B\rangle^+ \leq \langle\sigma_{A+x}\sigma_B\rangle^+_{(\Lambda_1+x)\cup\Lambda_2} = \langle\sigma_{A+x}\rangle^+_{\Lambda_1}\langle\sigma_B\rangle^+_{\Lambda_2}$$

where the last equality holds if $|x|$ is sufficiently large that no element of B has a nearest neighbour in $A + x$. Thus $\lim\sup\langle\sigma_{A+x}\sigma_B\rangle^+ \leq \langle\sigma_A\rangle^+\langle\sigma_B\rangle^+$. On the other hand $\lim\inf\langle\sigma_{A+x}\sigma_B\rangle^+ \geq \langle\sigma_A\rangle^+\langle\sigma_B\rangle^+$ follows from the GKS inequality (7.2.2).

The existence of a residual or spontaneous magnetization is related to (non) differentiability of the free energy. The free energy exists under quite general hypotheses. With + boundary conditions and an external magnetic field H:

$$H^+_\Lambda = \sum_{x,y\in\Lambda,n.n} -J\sigma(x)\sigma(y) - J\sum_{x\in\partial\Lambda}\sigma(x) - H\sum_{x\in\Lambda}\sigma(x)$$

the one point function is $\langle\sigma_A\rangle^+_{\Lambda,H}$ is monotone increasing for $H \geq 0$ by another application of the GKS inequalities. So in particular and using Peierls' estimate for $\beta > \beta_0$:

$$\lim_{H\searrow 0}\inf \lim_{|\Lambda|\to\infty}\inf \sum_{x\in\Lambda} \frac{1}{|\Lambda|}\left\langle \sigma(x)^+_{\Lambda,H}\right\rangle \geq a > 0\,.$$

These limits actually exist and give the *residual magnetization*, which is non-zero for low enough temperatures. We can relate this to differentiability of the free energy. Let us again return to the situation of an arbitrary but specified boundary condition labelled by b, and the corresponding partition function and free energy per lattice site in finite volume:

$$Z^b = \textstyle\sum_{\sigma\in P(\Lambda)} \exp -\beta H^b(\sigma)\,, \quad f^b = -\dfrac{1}{\beta|\Lambda|}\log Z^b\,.$$

We compute

$$\begin{aligned}\frac{\partial f^b}{\partial H} &= -\frac{1}{\beta|\Lambda|}\frac{(Z^b)'}{Z^b} = \textstyle\sum_\sigma\sum_x \sigma(x)\exp\left(-\beta H^b(\sigma)\right)/\beta|\Lambda|Z^b \\ &= -\textstyle\sum_{x\in\Lambda}\langle\sigma(x)\rangle^b_\Lambda = -m^b_\Lambda(H) \qquad (7.2.4)\end{aligned}$$

where $m^b_\Lambda(H)$ denotes the average magnetization. Differentiating again:

$$\frac{\partial^2 f^b}{\partial H^2} = -\frac{\beta}{|\Lambda|}\left\langle\left(\textstyle\sum_x \sigma(x) - \left\langle\sum_y \sigma(y)\right\rangle^b\right)^2\right\rangle^b \leq 0$$

so that $H \to f^b(H)$ is concave. We will show that $f(H) = \lim_{\Lambda\to\infty} f^b_\Lambda(H)$ exists and is independent of the boundary condition. Then f will automatically be concave. A concave function is continuous and possesses both left and right derivatives D^+f, D^-f and $D^-f \geq D^+f$, with strict inequality at only a countable number of points. In the thermodynamic limit $\langle\sigma(x)\rangle^\pm = \langle\sigma(y)\rangle^\pm$ for all x, y and (cf. (7.2.4))

$$-D^\pm f(0) = \langle\sigma(x)\rangle^\pm_{H=0}\,.$$

In fact $[-D^-f(0), -D^+f(0)] = \{\langle\sigma(x)\rangle^b \mid b \text{ boundary conditions}\}$. We will not prove this, it depends on having a suitable interaction or restrictions on J_A which includes the Ising Hamiltonian. We will however show $D^+f(0) \neq D^-f(0)$ in the Ising model at low temperatures.

We could define a phase transition to exist at (β, H_0) if $D^+f(H_0) \neq D^-f(H_0)$, i.e. if the free energy is not differentiable at H_0. In this case, we say that f has a first order phase transition in H (at H_0). It turns out (under suitable conditions on the interaction) that f is differentiable away from zero, i.e. $H \neq 0$, and $\lim_{H\searrow 0}\partial f/\partial H = D^+f(0)$. There is no phase transition at $H = 0$, if and only if the free energy is differentiable at 0, or $\langle\sigma(x)\rangle^b$ is independent of the boundary condition or $\lim_{H\searrow 0}\partial f/\partial H = 0$.

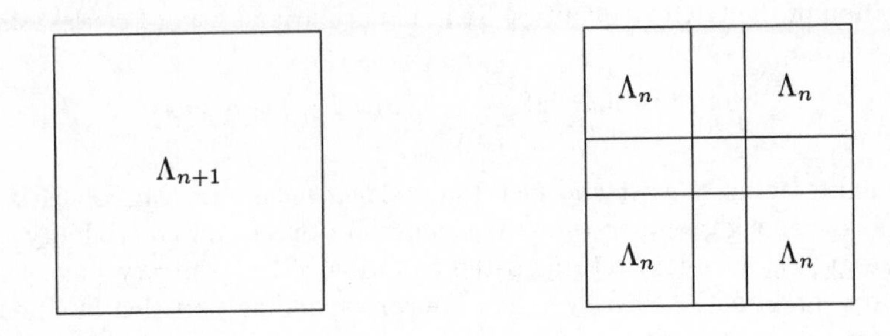

FIG. 7.10. Division of sublattice Λ_{n+1}

To show that the free energy is not differentiable at $H = 0$ for the two dimensional Ising model, let Λ_n be a nested sequence of finite lattices increasing to cover $\mathbb{Z}^2$, and put

$$f_{2n} = f^+_{\Lambda_n}(H)\,, \quad f_{2n+1} = f^-_{\Lambda_n}(H)\,.$$

Then $f_n \to f$ the free energy, the f_n are differentiable and $f'_{2n}(0) = -m^+_{\Lambda_n}(0)$, $f'_{2n+1}(0) = -m^-_{\Lambda_n}(0)$. Now $\langle\sigma(x)\rangle^+_{\Lambda,H=0} = -\langle\sigma(x)\rangle^-_{\Lambda,H=0}$ so that if $\langle\sigma(x)\rangle^+_{\Lambda,H=0} > a > 0$ for $\beta > \beta_0$ by Peierls' estimate then

$$f'_{2n}(0) < -a < a < f'_{2n+1}(0)\,.$$

But if f_n is a sequence of concave functions converging pointwise to f then

$$f'_+(x) \le \liminf\,(f_n)'_+(x) \le \limsup\,(f_n)'_-(x) \le f'_-(x)\,.$$

Hence as $a \neq 0$, $f'_+(0) \neq f'_-(0)$ and the free energy f is not differentiable at $H = 0$.

We have still to show that $\lim_{\Lambda\to\infty} f^b_\Lambda(H)$ exists and is independent of boundary condition b, as long as $\|b_{\partial\Lambda}\|/|\Lambda| \to 0$ as $|\Lambda| \to \infty$ when

$$H^b = -\sum_{x,y\in\Lambda} J\sigma(x)\sigma(y) - \sum_{x\in\Lambda} H\sigma(x) + b_{\partial\Lambda}\,.$$

We consider first the case $b_{\partial\Lambda} = 0$, and square boxes Λ_n in $\mathbb{Z}^2$ of size $2n$ (Fig. 7.10). Then Λ_{n+1} consists of four sublattices each of size Λ_n, and a cross X_n. Then

$$H(\Lambda_{n+1}) = H^{(1)}(\Lambda_n) + H^{(2)}(\Lambda_n) + H^{(3)}(\Lambda_n) + H^{(4)}(\Lambda_n) + H(X_n)$$

where $H^{(i)}(\Lambda_n)$ is the Hamiltonian of region i of size Λ_n, and $H(X_n)$ that of the cross X_n. Ignoring X_n altogether then $Z(\Lambda_{n+1}) = Z(\Lambda_n)^4$, so that $f(\Lambda_{n+1})$

$f(\Lambda_n)$. When we do not ignore $H(X_n)$ we use Lemma 7.2 below to compare $f(\Lambda_{n+1})$ with $f(\Lambda_n)$. So with $H_1 = H(\Lambda_{n+1})$, $H_2 = \sum_i H^i(\Lambda_n)$, we have with $H(X_n) = -J\sum_{X_n} \sigma(x)\sigma(y)$:

$$|f(\Lambda_{n+1}) - f(\Lambda_n)| \leq \|H(X_n)\| / |\Lambda_{n+1}| \leq |J| 2^n 4/(2^{n+1})^2 = |J|/2^n .$$

Thus $\sum_n f(\Lambda_{n+1}) - f(\Lambda_n)$ is convergent, and so is $\lim f(\Lambda_n)$. By considering boxes of size $2^n L$ and using similar arguments one can deduce that $\lim_{\Lambda\to\infty} f(\Lambda)$ exists (for boxes Λ of size L, as $L \to \infty$), say f.

Introducing a boundary term, $H^b_\Lambda = H_\Lambda + b_{\partial\Lambda}$, we have $|f_\Lambda - f^b_\Lambda| \leq \|H - H^b\|/|\Lambda| = \|b_{\partial\Lambda}\|/|\Lambda|$. Thus as long as $\lim_{\Lambda\to\infty} \|b_{\partial\Lambda}\|/|\Lambda| = 0$, $\lim_{\Lambda\to\infty} f^b_\Lambda$ exists and coincides with $\lim_{\Lambda\to\infty} f_\Lambda$. For example, with + or − boundary conditions

$$b^{\pm}_{\partial\Lambda} = \mp\textstyle\sum_{x\in\partial\Lambda} J\sigma(x)$$

$\|b^{\pm}_{\partial\Lambda}\| \leq |J||\partial\Lambda| = |J|4L$, if Λ is of size L. Then $\|b^{\pm}_{\partial\Lambda}\| / |\Lambda| = |J|4L/L^2 = 4|J|/L \to 0$ as $L \to \infty$. Similarly for periodic boundary conditions.

Lemma 7.2 *If $H_1, H_2 \in C(P(\Lambda))$ are self adjoint, and if the free energy per site $f_i = -\log(Z_i)/\beta|\Lambda|$, where Z_i is the partition function $\sum_\sigma e^{-\beta H_i(\sigma)}$ then $|f_1 - f_2| \leq \|H_1 - H_2\|/|\Lambda|$.*

Proof From

$$-\|H_1 - H_2\| + H_2(\sigma) \leq H_1(\sigma) \leq H_2(\sigma) + \|H_1 - H_2\|$$

we have

$$e^{-\beta H_2(\sigma)} e^{\beta\|H_1 - H_2\|} \leq e^{-\beta H_1(\sigma)} \leq e^{-\beta H_2(\sigma)} e^{\beta\|H_1 - H_2\|} .$$

Hence summing over states:

$$Z_2 e^{\beta\|H_1 - H_2\|} \leq Z_1 \leq Z_2 e^{\beta\|H_1 - H_2\|} .$$

Taking logarithms we have $|f_1 - f_2| \leq \|H_1 - H_2\| / |\Lambda|$. □

Thus the free energy per lattice site exists in the thermodynamic limit and is independent of boundary conditions. Without computing $f(H)$, when $H \neq 0$, we have seen using Peierls' estimate that $f^+(0) \neq f^-(0)$ or the free energy is not differentiable at zero. (Onsager 1941) computed $f(0)$ in the two dimensional Ising model by the transfer matrix formalism, but in higher dimension no explicit computation of $f(H)$ exists even if $H = 0$. However in higher dimensions, one can still prove the existence of a phase transition in a nearest neighbour Ising model by putting all the interactions to be zero except in parallel two dimensional planes, using Peierls' estimate there and then turning on the interactions in the new dimensions and exploiting the GKS inequalities.

7.3 One dimensional Ising model

We illustrate the reduction of dimension ($C(P) \to A$) in the case of the one dimensional nearest neighbour Ising model, reducing from $C\{\pm\}^{\mathbb{Z}} = \otimes_{\mathbb{Z}}\mathbb{C}^2$ to a single matrix algebra M_2 on one site.

Starting with the configurations on a finite interval $P_L = \{\sigma = (\sigma(i))_{i=-L}^{L}$ $\sigma(i) \in \{\pm 1\}\}$, the Ising Hamiltonian is

$$H(\sigma) = -\textstyle\sum_{i=-L}^{L} J\sigma(i)\sigma(i+1)\,, \quad J > 0 \tag{7.3.1}$$

with say periodic boundary conditions $\sigma(L+1) = \sigma(-L)$. The first thing to compute is the partition function

$$Z = \sum_{\sigma} \exp\left(-\beta H(\sigma)\right) \tag{7.3.2}$$

where the sum is over all configurations in P_L. Since

$$\exp\left(-\beta H(\sigma)\right) = \textstyle\prod_{i=-L}^{L} \exp\left(\beta J\sigma(i)\sigma(i+1)\right) \tag{7.3.3}$$

we can express the partition function in terms of a 2×2 matrix, by writing

$$T_{\sigma,\sigma'} = \exp\left(\beta J\sigma\sigma'\right) \tag{7.3.4}$$

where $\sigma, \sigma' \in \{\pm 1\}$ are configurations on one site. Then

$$Z = \textstyle\sum_{\sigma} T(\sigma(-L), \sigma(-L+1)) \cdots T(\sigma(L-1), \sigma(L)) T(\sigma(L), \sigma(-L))\,. \tag{7.3.5}$$

Because of our periodic boundary conditions, this can be interpreted as the trace of a certain power of T, namely $Z = \mathrm{tr}(T^{2L+1})$, where the trace is computed in the 2×2 matrices. Next, let F be a local observable, say on P_ℓ where $\ell < L$ (ℓ is thought of as being constant, whilst L will eventually go to infinity in the thermodynamic limit). We now claim that the expectation value

$$\langle F\rangle_\beta^L = \textstyle\sum_{\sigma} F(\sigma) \exp\left(-\beta H(\sigma)\right) / \sum_{\sigma} \exp\left(-\beta H(\sigma)\right) \tag{7.3.6}$$

can be expressed as

$$\langle F\rangle_\beta^L = \mathrm{tr}\left(T^{2L+1} F_\beta\right) / \mathrm{tr} T^{2L+1} \tag{7.3.7}$$

where F_β is an element of M_2, depending on β and linearly on F but not on L. To see this, consider $F = \prod_{i=-\ell}^{\ell} F_i$ where F_i is a function on the ith site alone. Then as we did for the partition function, we write

$$\begin{aligned}
&\textstyle\sum F(\sigma) \exp\left(-\beta H(\sigma)\right) \\
&= \textstyle\sum_{\sigma} T(\sigma(-L), \sigma(-L+1) \\
&\quad \cdots F_{-l}(\sigma(-l)) T(\sigma(l), \sigma(-l+1)) F_{-l+1}(\sigma(-l+1))
\end{aligned}$$

$$\cdots F_l(\sigma(l))T(\sigma(l)\sigma(l+1))\cdots T(\sigma(L-1),\sigma(L))T(\sigma(L),\sigma(-L))$$
$$= \text{Trace}\, T^L F_\beta T^{L+1} \tag{7.3.8}$$

if $F_\beta = T^{-l}F_{-l}TF_{-l+1}\cdots TF_lT^l$, and $F_i = F_i(\sigma_x)$, $\sigma_x = \begin{pmatrix} 1 & 0 \\ 0 & -1 \end{pmatrix}$. Now letting $L \to \infty$ in (7.3.7) we see that $\langle F\rangle_\beta = \langle F_\beta\Omega, \Omega\rangle$, where $\Omega = \begin{pmatrix} 1 \\ 1 \end{pmatrix}/\sqrt{2}$ is the unit vector corresponding to the largest eigenvalue of T. Actually, this only holds for non-zero temperatures. If $\beta = \infty$, one should be more careful. For finite β, the transfer matrix T has a non-degenerate largest eigenvalue. (This can be computed explicitly in this situation or notice that T has strictly positive entries, and so one can invoke the Perron–Frobenius theorem. The transfer matrix T is $e^K + e^{-K}\sigma_z$, $K = \beta J$. (We adopt the convention that $\sigma_z = \begin{pmatrix} 0 & 1 \\ 1 & 0 \end{pmatrix}$, for the Pauli matrices). Multiplying T by a scalar does not alter the value of (7.3.7) so we could replace it by $1 + e^{-2K}\sigma_z$, which becomes the identity matrix at $\beta = \infty$ (for ferromagnetic interactions $J > 0$) and so the largest eigenspace is doubly degenerate and $\langle F\rangle_\infty = \lim_{L\to\infty}\langle F\rangle_\infty^L = \text{tr}(F_\infty)$. Thus we have

$$\langle F\rangle_\beta = \varphi_\beta(F_\beta) \tag{7.3.9}$$

where

$$\varphi_\beta = \begin{cases} \omega_\Omega & \beta \neq \infty \\ \left(\omega_{\left(\begin{smallmatrix}1\\0\end{smallmatrix}\right)} + \omega_{\left(\begin{smallmatrix}0\\1\end{smallmatrix}\right)}\right)/2 & \beta = \infty \end{cases} \tag{7.3.10}$$

Thus φ_β is the pure state corresponding to an equal weight distribution for any non-zero temperature, and a mixture of the two pure states corresponding to spin up and spin down at zero temperature.

The situation at zero temperature can be handled more elegantly as follows. We write

$$T = e^K + e^{-K}\sigma_z = (2\sinh 2K)^{\frac{1}{2}} e^{K^*\sigma_z} \tag{7.3.11}$$

if $K = \beta J$ and K^* is given by $\sinh 2K \sinh 2K^* = 1$, so that $K^{**} = K$. We can then replace T by $e^{K^*\sigma_z}$ in (7.3.7) and then regard K^* as the parameter instead of β or K. Then $K^* = 0$ corresponds to zero temperature.

To compute the magnetization, we should insert a magnetic field:

$$H(\sigma) = -J\textstyle\sum_{i=-L}^{L}\sigma(i)\sigma(i+1) - H\sum_{i=-L}^{L}\sigma(i)\,.$$

Again the partition function for periodic boundary conditions $\sigma(L+1) = \sigma(-L)$ is $Z = \text{tr}\, T^{2L+1}$, where

$$T_{\sigma,\sigma'} = \exp(K\sigma\sigma' + B(\sigma+\sigma')/2) \tag{7.3.12}$$
$$K = \beta J\,, \quad B = \beta H\,, \quad (\sigma,\sigma') \in \{\pm 1\}\,.$$

Then T has eigenvalues

$$\lambda_{\pm} = e^K \cosh B \pm (e^{2K} \sinh^2 B + e^{-K})^{\frac{1}{2}}$$

with $\lambda_+ > \lambda_-$, and

$$\begin{aligned} -\beta f &= \lim_{N\to\infty} \frac{1}{2L+1} \log Z = \lim_{N\to\infty} \frac{1}{2L+1} \log\left(\lambda_+^{2L+1} + \lambda_-^{2L+1}\right) \\ &= \log \lambda_+ = \log\left(e^K \cosh B + (e^{2K} \sinh^2 B + e^{-K})^{\frac{1}{2}}\right). \end{aligned}$$

This leads to a magnetization per spin

$$M(T,H) = -\frac{\partial}{\partial H} f(T,H) = \sinh B(\sinh^2 B + e^{-4K})^{\frac{1}{2}}$$

$\to 0$ as $H \to 0$, so there is no residual magnetization or phase transition in the one dimensional nearest neighbour Ising model.

In the presence of a magnetic field the eigenvectors of T are

$$z_+ = \begin{pmatrix} \cos\varphi \\ \sin\varphi \end{pmatrix} \qquad z_- = \begin{pmatrix} -\sin\varphi \\ \cos\varphi \end{pmatrix}$$

if $\cos\varphi = (1+\mu)/[2(1+\mu^2)]^{1/2}$, $\sin\varphi = (1-\mu)/[2(1+\mu^2)]^{1/2}$, $\mu = [\lambda_+ - e^{K+B}]e^K$. Then the expectation value $\langle f\rangle_\beta = \varphi_\beta(F_\beta)$ where $\varphi_\beta = \omega_{z_+}$, for $\beta \neq \infty$. In particular if $A = \{i_1 \leq i_2 \leq \cdots \leq i_p\}$,

$$\langle \sigma_A \rangle = \varphi_\beta\left((\sigma_A)_\beta\right) = \langle (\sigma_A)_\beta\, z_+, z_+\rangle = \langle \sigma_x T^{i_2-i_1} \sigma_x \cdots T^{i_p - i_{p-1}} \sigma_x z_+, z_+\rangle .$$

Thus

$$\begin{aligned} \langle \sigma_i \rangle &= \sin 2\varphi, \quad (= 0 \text{ if } H = 0) \\ \langle \sigma_i \sigma_j \rangle &= (\sin 2\varphi)^2 + r^{j-i} \cos^2 2\varphi, \end{aligned}$$

if $r = \lambda_-/\lambda_+ < 1$. Thus

$$|\langle \sigma_i \sigma_j\rangle - \langle \sigma_i\rangle\langle\sigma_j\rangle| = r^{j-i} \cos^2 2\varphi \to 0 \text{ as } j - i \to \infty .$$

Typically, for large x one has in a lattice model the behaviour:

$$\langle \sigma_0 \sigma_x \rangle - \langle \sigma_0 \rangle^2 \sim |x|^{-\eta'} \exp -|x|/\xi ,$$

where the scale factor ξ is called the *correlation length* ξ. The correlation length ξ diverges at criticality (where assuming $\langle\sigma_0\rangle = 0$ at T_c) there is a power law

$$\langle \sigma_0 \sigma_x \rangle \sim |x|^\eta$$

and η is called a *critical exponent.* Other critical exponents arise in the asymptotic behaviour of other thermodynamic functions.

Thus the one dimensional Ising model has a correlation length of $(\log \lambda_+/\lambda_-)^{-1}$.

7.4 The transfer matrix formalism for the two dimensional Ising model

In this section we begin our C^*-treatment of the two dimensional Ising model by discussing the general framework and showing why quasi-free states are relevant with the technical details provided in later sections. We already have a guide from the one dimensional discussion of the previous section.

We will reduce the configurations on a finite box

$$\Lambda_{LM} = \{(i,j) \mid |i| \leq M, |j| \leq L\} \tag{7.4.1}$$

to a single row $\{(i,0) \mid |i| \leq M\}$ picking up a two by two matrix algebra for each site on this row. We thus reduce from $C(P_{LM})$, where P_{LM} are the configurations $\{\pm\}^{\Lambda_{LM}}$, to $\bigotimes_{-M}^{M} M_2 = A_M^P$, a finite dimensional Pauli algebra. More precisely we have the local Hamiltonian, with periodic boundary conditions:

$$H^{LM}(\sigma) = -\sum_{i=-M}^{M}\sum_{j=-L}^{L} \left(J_1\sigma(i,j)\sigma(i+1,j) + J_2\sigma(i,j)\sigma(i,j+1)\right) \tag{7.4.2}$$

We break up a configuration $\sigma \in P_{LM}$ as a column

$$\sigma = \begin{pmatrix} \sigma^L \\ \vdots \\ \sigma^{-L} \end{pmatrix} \tag{7.4.3}$$

if $\sigma^j = (\sigma(-M,j),\ldots,\sigma(M,j)) \in \{\pm 1\}^{2M+1}$ denotes the configuration in σ along the jth row, for $|j| \leq L$. We then have the decomposition

$$H^{LM}(\sigma) = \sum_{j=-L}^{L} S(\sigma^j) + \sum_{j=-L}^{L} I(\sigma^j, \sigma^{j+1}) \tag{7.4.4}$$

in terms of the internal energies of the rows, and interaction energies between neighbouring rows if

$$S(\overline{\sigma}) = -J_2\sum_{j=-M}^{M}\overline{\sigma}(j)\overline{\sigma}(j+1) \tag{7.4.5}$$

$$I(\overline{\sigma},\overline{\sigma}') = -J_1\sum_{j=-M}^{M}\overline{\sigma}(j)\overline{\sigma}'(j) \tag{7.4.6}$$

and $\overline{\sigma},\overline{\sigma}' \in \{\pm 1\}^{2M+1}$ are row configurations. The transfer matrix $T = T_M$ is the symmetric $2^{2M+1} \times 2^{2M+1}$ array

$$T(\overline{\sigma},\overline{\sigma}') = \exp\left(-\beta\left\{[S(\overline{\sigma}) + S(\overline{\sigma}')]/2 + I(\overline{\sigma},\overline{\sigma}')\right\}\right) \tag{7.4.7}$$

if $\overline{\sigma},\overline{\sigma}' \in \{\pm 1\}^{2M+1}$, identified with an element of the Pauli algebra $A_M^P = C^*(\sigma_\alpha^i \mid |i| \leq M)$, so that

$$T = (2\sinh 2K_1)^{M+\frac{1}{2}} V^{\frac{1}{2}} W V^{\frac{1}{2}} \tag{7.4.8}$$

if

$$V = \exp K_2 \textstyle\sum_{j=-M}^{M} \sigma_x^j \sigma_x^{j+1}\,, \quad W = \exp K_1^* \textstyle\sum_{j=-M}^{M} \sigma_z^j \tag{7.4.9}$$

and

$$K_j = \beta J_j\,, \quad \sinh 2K_1 \sinh 2K_1^* = 1\,. \tag{7.4.10}$$

The contribution V (a diagonal matrix) comes from interactions along horizontal rows, and $(2\sinh 2K_1)^{M+1/2}W$ comes from the transfer direction, along vertical columns (cf. 7.3.11).

Then as in (7.3.5), the partition function $Z = \sum \exp(-\beta H(\sigma))$ can be identified with $\operatorname{tr} T^{2L+1}$, and if F is a local observable on Λ_{lm}, where $l < L$, $m < M$ we can identify $\sum F(\sigma)\exp(-\beta H(\sigma))$ with $\operatorname{tr}(T^{2L+1}F_\beta^M)$, where F_β^M is an operator in $\bigotimes_{-M}^{M} M_2$ independent of L, and the trace is computed in $\bigotimes_{-M}^{M} M_2$. Thus

$$\langle F\rangle_\beta^{LM} = \operatorname{tr} T^{2L+1} F_\beta^M / \operatorname{tr} T^{2L+1}\,. \tag{7.4.11}$$

For example, if $F = \prod_{i=-l}^{l} F_i$, where $F_i = F_i(\sigma^i)$ is a function of the ith row alone, then $F_\beta = T^{-l}\hat{F}_{-l}T\hat{F}_{-l-1}\cdots T\hat{F}_l T^{-l}$ where $\hat{F}_i = \hat{F}_i(\sigma_x^{-m},\ldots,\sigma_x^m)$ is a diagonal matrix. Now T_M has strictly positive entries, as long as K_1 is finite or K_1^* is non-zero. So by the Perron–Frobenius theorem (cf. Theorem 10.3), it has a unique unit eigenvector Ω^M associated with the largest eigenvalue, with $\Omega^M(\overline{\sigma}) > 0$, $\overline{\sigma} \in \{\pm 1\}^{2M+1}$. Thus letting $L \to \infty$, we pick out this eigenspace associated with the largest eigenvalue:

$$\langle F\rangle_\beta^M = \langle F_\beta^M \Omega^M, \Omega^M\rangle\,. \tag{7.4.12}$$

Now as $M \to \infty$, F_β^M is eventually constant, and as we shall see below the states $\varphi_\beta^M = \langle\,\cdot\,\Omega_M, \Omega_M\rangle$ on $\bigotimes_{-M}^{M} M_2$ converge to a state φ_β on the infinite Pauli algebra $A^P = \bigotimes_{-\infty}^{\infty} M_2$.

The structure of the state φ_β at infinite and zero temperatures is clear. The factor $(2\sinh 2K_1)^{M+1/2}$ from (7.4.8) becomes irrelevant in (7.4.11). We can thus regard K_1^* and K_2 as independent parameters. Then the extreme temperatures $\beta = 0$ and $\beta = \infty$ correspond to $K_2 = 0$, and $K_1^* = 0$ respectively. Having removed the scalar from (7.4.8), the transfer matrix T is then essentially W when $K_2 = 0$, and V when $K_1^* = 0$. The largest eigenspace of W is spanned by $\bigotimes_{-M}^{M}\begin{pmatrix}2^{-\frac12}\\ 2^{-\frac12}\end{pmatrix}$ whilst that of V is degenerate, spanned by $\bigotimes_{-M}^{M}\begin{pmatrix}1\\0\end{pmatrix}$ and $\bigotimes_{-M}^{M}\begin{pmatrix}0\\1\end{pmatrix}$. Thus

$$\varphi_0 = \bigotimes_{-\infty}^{\infty} \omega_\Omega\,, \tag{7.4.13}$$

and is pure whilst

$$\varphi_\infty = \left(\otimes_{-\infty}^{\infty}\omega_{\left(\begin{smallmatrix}1\\0\end{smallmatrix}\right)} + \otimes_{-\infty}^{\infty}\omega_{\left(\begin{smallmatrix}0\\1\end{smallmatrix}\right)}\right)\Big/2 \tag{7.4.14}$$

is a mixture of two inequivalent pure states, φ_∞^+, φ_∞^- respectively.

The question is how to pull out the information at zero and infinite temperature to the regions $(0, T_c)$ and (T_c, ∞) respectively, to see how the structure persists? For this it is better not to work with the Pauli algebra A^P alone, but introduce an auxiliary algebra, the Fermi algebra A^F over $\ell^2(\mathbb{Z})$ generated by the annihilation and creation operators $\{c_j, c_j^* \mid j \in \mathbb{Z}\}$ satisfying the canonical anti-commutation relations.

We have symmetries, both denoted by θ_-, of the Pauli and Fermi algebras as in Section 6.5, so that if $\hat{A}$ denotes the crossed product $A^P \rtimes_{\theta_-} (\mathbb{Z}/2)$ (equivalently $A^F \rtimes_{\theta_-} (\mathbb{Z}/2)$) we have

$$\begin{array}{ccc} \hat{A} & = & \hat{A} \\ \cup & & \cup \\ A^P & & A^F \\ \cup & & \cup \\ A_+^P & = & A_+^F \end{array} \qquad (7.4.15)$$

The state φ_β on A^P, being manufactured from the even transfer matrix is even, $\varphi_\beta = \varphi_\beta \circ \theta$, and so we can transform it quite naturally to an even state $\varphi_\beta^F = \varphi_\beta^F \circ \theta$ on A^F such that $\varphi_\beta^F \left| A_+^F \right. = \varphi_\beta \big| A_+^P$. The state φ_β^F is quasi-free, being constructed from the 'quadratic' Hamiltonian $\log T$ (see (7.4.22)). It is in fact a Fock state and so is always pure for any temperature. Thus the manifestation of the phase transition in the Fermi algebra picture is not immediately clear. To give a precise description of the quasi-free state φ_β^F it is convenient to adopt the self-dual formalism of Section 6.4, so that A^F is generated by the range of a linear map B on $\ell^2 \oplus \ell^2$ given by

$$B(h) = \textstyle\sum_{-\infty}^{\infty} (c_j^* f_j + c_j g_j) \qquad h = f \oplus g\,,\ f = (f_j)\,,\ g = (g_j)\,. \qquad (7.4.16)$$

Here B satisfies

$$\{B(h_1)^*, B(h_2)\} = \langle h_2, h_1 \rangle\,, \quad B(h)^* = B(\Gamma h) \qquad (7.4.17)$$

where

$$\Gamma(f \oplus g^*) = (g \oplus f^*)\,. \qquad (7.4.18)$$

and $(g^*)_i = \overline{g}_i$.

To see that φ_β^F is quasi-free, note that from (7.4.9) and Section 6.5 we have

$$V = \exp K_2 b^M(H_2)\,, \quad W = \exp K_1^* b^M(H_1)\,, \qquad (7.4.19)$$

where

$$H_2 = \frac{1}{2}\begin{pmatrix} -(U^* + U) & U^* - U \\ U - U^* & U^* + U \end{pmatrix}, \quad H_1 = \begin{pmatrix} 1 & 0 \\ 0 & -1 \end{pmatrix}, \qquad (7.4.20)$$

$$(Uf)_n = f_{n+1}\,, \quad (U^*f)_n = f_{n-1}\,, \quad f = (f_n) \in \ell^2(\mathbb{Z})\,. \tag{7.4.21}$$

We use the notation

$$B^M(h) = B(P_M h)\,, \quad h \in K_M\,,$$

where P_M is the orthogonal projection on

$$K_M = \left\{ f \oplus g \in \ell^2 \oplus \ell^2 \mid f_j = g_j = 0 \text{ if } |j| > M \right\}\,,$$

and

$$b^M(H) = \textstyle\sum_{m,n} B^M(e_m)\langle He_n, e_m\rangle B^M(e_n)^*$$

for any complete orthonormal basis $\{e_n\}$ for K_M. Then $H_j^* = H_j = -\Gamma H_j \Gamma$. Let H^M, H be the unique self adjoint operators satisfying

$$\begin{aligned} e^{2H^M} &= e^{K_2 H_2^M} e^{2K_1^* H_1^M} e^{K_2 H_2^M} \\ e^{2H} &= e^{K_2 H_2} e^{2K_1^* H_1} e^{K_2 H_2} \end{aligned}$$

where $H_j^M = P_M H_j | K_M$. Then H^M, H are self adjoint operators anti-commuting with Γ.

Now $e^{b(H^M)}, e^{K_2 b(H_2^M)/2}, e^{K_1^* b(H_1^M)}$ implement the quasi-free automorphisms $\tau(e^{2HM})$, $\tau(e^{K_2 H_2^M})$, $\tau(e^{2K_1^* H_1^M})$ respectively. Hence up to a scalar, the operators $e^{b(H^M)}$ and $e^{K_2 b(H_2^M)/2} e^{K_1^* b(H_1^M)} e^{K_2 b(H_2^M)/2}$ coincide. Since both are positive operators, the scalar is certainly positive, (and is indeed 1 by an application of the Baker–Hausdorff formula). Thus the transfer matrix T_M is given, up to a scalar, by

$$V^{\frac{1}{2}} W V^{\frac{1}{2}} = e^{b^M(H^M)}\,. \tag{7.4.22}$$

Let E_-^M, E_0^M, E_+^M be the spectral projection of H^M for $(-\infty, 0), \{0\}, (0, \infty)$ respectively. From $H^M = -\Gamma H^M \Gamma$ we see that $\Gamma E_\pm^M \Gamma = E_\mp^M$ and $\Gamma E_0^M \Gamma = E_0$. Let $\{e_\nu\}$ be a complete orthonormal basis of eigenvectors for H^M on $E_+^M K_M$ so that

$$H^M e_\nu = \lambda_\nu e_\nu\,, \quad \lambda_\nu > 0\,.$$

Then $b_\nu = B(e_\nu)$ satisfy the canonical anti-commutation relations:

$$b_\nu^* b_\mu + b_\mu b_\nu^* = \delta_{\mu\nu}\,, \quad b_\mu b_\nu + b_\nu b_\mu = 0\,,$$

and

$$b(H^M) = \textstyle\sum_\nu \lambda_\nu (1 - 2b_\nu^* b_\nu)\,.$$

The eigenspace of $b(H^M)$ corresponding to the largest eigenvalue consists of vectors Ω satisfying $b_\nu \Omega = 0$. Since the transfer matrix was constructed as a matrix with strictly positive entries, we have already observed that this is one dimensional. Now consider $\{a_\nu\}_{\nu\in\Lambda}$, $\{c_\alpha\}_{\alpha\in\Theta}$ as two representations of the canonical anti-commutation relations on two Hilbert spaces F_1 and F_2 with vacuum vectors Ω_1, Ω_2 respectively. If γ is an operator anti-commuting with $\{a_\nu\}$, then $\{a_\nu \otimes 1, \gamma \otimes c_\alpha\}$ with $\Omega_1 \otimes F_2$ as vacuum vectors for $\{a_\nu \otimes 1\}$. This is one dimensional only when $\Theta = \varnothing$. Thus we are forced to conclude that $E_0^M = 0$. Consequently $E_\pm^M$ are basis projections, and the vector state $\langle \cdot \Omega^M, \Omega^M\rangle$ of $A_{M+}^P = A_{M+}^F$ satisfies

$$\langle B^M(h)^* B^M(h)\Omega^M, \Omega^M\rangle = 0\,, \quad h \in E_+^M K_M\,.$$

Thus the extended even state on A_M^F must be the quasi-free state $\varphi_{E_+^M}$.

7.5 Diagonalization of the two dimensional Ising transfer matrix

We will explicitly diagonalize in this section the transfer matrix for periodic boundary conditions.

In this case the transfer matrix (7.4.8), with scalar factor removed, becomes $T = V^{1/2} W V^{1/2}$ where

$$V = \exp K_2 \left(\textstyle\sum_{j=1}^{M-1}(c_j - c_j^*)(c_{j+1} + c_{j+1}^*) - \Gamma(c_M - c_M^*)(c_1 + c_1^*)\right) \tag{7.5.1}$$

$$W = \exp K_1^* \left(\textstyle\sum_{j=1}^{M}(2c_j^* c_j - 1)\right) \tag{7.5.2}$$

and $\Gamma = (-1)^N$, when N is the number operator; the term $-\Gamma(c_M - c_M^*)(c_1 + c_1^*) = \sigma_x^M \sigma_x^1$ arising from periodic boundary conditions. (In dealing with finite systems, it is as well to treat $[1, M]$ instead of the interval $[-M, M]$ to ease the notation). Splitting Γ into its even and odd parts, let

$$V_\pm = \exp K_2 \left(\textstyle\sum_{j=1}^{M-1}(c_j - c_j^*)(c_{j+1} + c_{j+1}^*) \pm (c_M - c_M^*)(c_1^* + c_1)\right) \tag{7.5.3}$$

and $T^\pm = V_\pm^{1/2} W V_\pm^{1/2}$. Then Γ commutes with T, and $T = T^+|_{F_-} \oplus T^-|_{F_+}$, the restrictions of T^+ and T^- to the $-$ and $+$ eigenspaces of Γ respectively. Representing the Fermion algebra on Fock space F, these are the even and odd vectors in Fock space, F_+ and F_- respectively. So we only need to diagonalize $T^\pm$ separately and pick out the relevant eigenvectors in $F_\pm$. We easily write

$$\textstyle\sum_{i=1}^{M-1}(c_i - c_i^*)(c_{i+1} + c_{i+1}^*) \pm (c_M - c_M^*)(c_1 + c_1^*) = b(K^\pm) \tag{7.5.4}$$

if

$$K^\pm = \begin{pmatrix} -U^\pm & -U^\pm \\ U^\pm & U^\pm \end{pmatrix} \tag{7.5.5}$$

and $U^\pm$ are the operators on $\ell^2(M)$ given by

$$\left(U^{\pm}f\right)_j = f_{j+1} \quad j = 1, \ldots, M-1; \quad \left(U^{\pm}f\right)_M = \pm f_1 \,. \tag{7.5.6}$$

But $(c_i - c_i^*)(c_{i+1} + c_{i+1}^*)$ is self adjoint, and so $b(K^{\pm}) = b(H^{\pm})$ if

$$H^{\pm} = \frac{1}{2}\left(K^{\pm} + (K^{\pm})^*\right) = \frac{1}{2}\begin{pmatrix} -U^{\pm} - U^{\pm *} & -U^{\pm} + U^{\pm *} \\ U^{\pm} - U^{\pm *} & U^{\pm} + U^{\pm *} \end{pmatrix} . \tag{7.5.7}$$

To diagonalize $T^{\pm}$, we first diagonalize $U^{\pm}$. If M is even define

$$\begin{aligned} S^{+} &= \{\ell\pi/M, \ell = 0, \pm 2, \pm 4, \ldots, \pm(M-2), M\} \\ S^{-} &= \{\ell\pi/M, \ell = \pm 1, \pm 3, \ldots, \pm(M-1)\} \,. \end{aligned} \tag{7.5.8}$$

Then $f_q^{\pm} = M^{-1/2}\sum_m e^{iqm} e_m$, $q \in S^{\pm}$ are eigenvectors for $U^{\pm}$, if e_i are the canonical basis vectors for ℓ^2, with $U^{\pm} f_q^{\pm} = e^{iq} f_q^{\pm}$, $q \in S^{\pm}$. With respect to these bases $f_q^{\pm}$, $H^{\pm} = \bigoplus_q H^{\pm}(q)$, where

$$H^{\pm}(q) = \begin{pmatrix} -\cos q & -i \sin q \\ i \sin q & \cos q \end{pmatrix}, \quad q \in S^{\pm} \,. \tag{7.5.9}$$

Writing $\theta_q = B\begin{pmatrix} 0 \\ f_q^{\pm} \end{pmatrix}$, so that $B\begin{pmatrix} f_q^{\pm} \\ 0 \end{pmatrix} = \theta_{-q}^*$, we have

$$\begin{aligned} b(H^{\pm}) &= \textstyle\sum_q \cos q \,(\theta_{-q}^*\theta_{-q} - \theta_q\theta_q^*) + i\sum_q \sin q \,(\theta_q\theta_{-q} + \theta_{-q}^*\theta_q^*) \\ &= \textstyle\sum_q \cos q \,(\theta_{-q}^*\theta_{-q} + \theta_q^*\theta_q - 1) + \sum_q i \sin q \,(\theta_q\theta_{-q} + \theta_{-q}^*\theta_q^*) \end{aligned} \tag{7.5.10}$$

as $\sum_q \cos\omega_q^{\pm} = \mathrm{tr}(U^{\pm} + U^{\pm *})/2 = 0$. Hence

$$\begin{aligned} b(H^{+}) = &\sum_{q \geq 0, q \neq 0, \pi} 2\left[\cos q \,(\theta_{-q}^*\theta_{-q} + \theta_q^*\theta_q) + i \sin q \,(\theta_{-q}^*\theta_q^* + \theta_q\theta_{-q})\right] \\ &+ 2\sum_{q = 0, \pi} e^{i\omega_q^{+}}\left[\theta_q^*\theta_q - 1/2\right] \end{aligned} \tag{7.5.11}$$

$$b(H^{-}) = \sum_{q > 0} 2\left[\cos q \,(\theta_{-q}^*\theta_{-q} + \theta_q^*\theta_q) + i \sin q \,(\theta_q\theta_{-q} + \theta_{-q}^*\theta_q^*)\right] .$$

We can make the coefficients real, by inserting a factor in the definition $f_q^{\pm} = M^{-1/2}e^{-\pi i/4}\sum_m e^{iqm} e_m$. Consequently $V_{\pm} = \exp K_2(b(H^{\pm})) = \prod_{q \geq 0} V^{\pm}(q)$ if

$$\begin{aligned} V^{\pm}(q) = \exp 2K_2 &\left[\cos q \,(\theta_{-q}^*\theta_{-q} + \theta_q^*\theta_q)\right. \\ &\left. + \sin q \,(\theta_{-q}^*\theta_q^* + \theta_q\theta_{-q})\,\right], \quad q \neq 0, \pi \end{aligned} \tag{7.5.12}$$

and

$$V^{+}(q) = \exp 2K_2 \left[e^{iq}\left(\theta_q^*\theta_q - 1/2\right)\right], \quad q = 0, \pi . \tag{7.5.13}$$

Moreover, $W = \exp K_1^* \left(\sum_j 2c_j^* c_{j-1}\right) = \prod_{q\geq 0} W^{\pm}(q)$ if

$$W^{\pm}(q) = \exp 2K_1^*(\theta_q^*\theta_q + \theta_{-q}^*\theta_{-q} - 1), \quad q \neq 0, \pi \tag{7.5.14}$$

and

$$W^{+}(q) = \exp 2K_1^*\left(\theta_q^*\theta_q - 1/2\right), \quad q = 0, \pi . \tag{7.5.15}$$

Then $T^{\pm} = V_{\pm}^{1/2} W V_{\pm}^{1/2} = \prod_{q\geq 0} T^{\pm}(q)$ where it only remains to diagonalize $T^{\pm}(q) = V_{\pm}^{1/2}(q) W_{\pm}(q) V_{\pm}^{1/2}(q)$.

Here $T^{\pm}(q) \in C^*(\theta_q, \theta_{-q}) \cong M_4$ for $q \neq 0, \pi$, and $T^{+}(0) \in C^*(\theta_0) \cong M_2$, $T^{+}(\pi) \in C^*(\theta_\pi) \cong M_2$ so that $T^{-} \in \bigotimes_q M_4$ where $q \in S^{-}$, $q > 0$, $q \neq \pi$, $T^{+} \in M_2 \otimes \bigotimes_q M_4 \otimes M_2$, $q \in S^{+}$, $q > 0$. The grading operator Γ also splits as a product $\otimes_q \Gamma_q$, $\Gamma_q = (-1)^{N_q}$, where N_q is the local number operator $N_q = \theta_q^*\theta_q + \theta_{-q}^*\theta_{-q}$, $q \neq 0, \pi$, $N_0 = \theta_0^*\theta_0$, $N_\pi = \theta_\pi^*\theta_\pi$. In particular,

$$\begin{aligned} T^{+}(0) &= \exp\left(-2(K_1^* - K_2)\left(\theta_0^*\theta_0 - 1/2\right)\right), \\ T^{+}(\pi) &= \exp\left(-2(K_1^* + K_2)\left(\theta_\pi^*\theta_\pi - 1/2\right)\right) \end{aligned} \tag{7.5.16}$$

which have spectra $\exp(-2(K_1^* - K_2)(\pm 1/2))$ and $\exp(-2(K_1^* + K_2)(\pm 1/2))$ respectively, where the $+$ and $-$ signs correspond to even and odd eigenvectors.

To diagonalize the 4×4 matrix $T^{\pm}(q)$, we represent θ_q, θ_{-q} on $F(\mathbb{C}^2) = C\Omega \oplus \mathbb{C}^2 \oplus (\mathbb{C}^2 \wedge \mathbb{C}^2) = \mathbb{C}\Phi_0 \oplus (\mathbb{C}\Phi_q \oplus \mathbb{C}\Phi_{-q}) \oplus \mathbb{C}\Phi_{-qq}$ where $\Phi_0 = \Omega$ is the vacuum, $\Phi_{\pm q} = \theta_{\pm q}^*\Phi_0$ and $\Phi_{-qq} = \theta_{-q}^*\theta_q^*\Phi_0$. Then Φ_q, Φ_{-q} are odd eigenvectors of $T^{\pm}(q)$, being eigenvectors of $\theta_q\theta_{-q}$, $\theta_{-q}^*\theta_q^*$, with eigenvalue zero and eigenvectors of N_q with eigenvalue 1. Thus $\Phi_{\pm q}(q \neq 0, \pi)$ are odd eigenvectors of $T^{\pm}(q)$ with eigenvalue $\exp(2K_2 \cos q)$. To diagonalize $T^{\pm}(q)$, on $\mathbb{C}\Phi_0 \oplus \mathbb{C}\Phi_{-qq}$, we write (with respect to the basis Φ_{-qq}, Φ_0):

$$N_q = \begin{pmatrix} 2 & 0 \\ 0 & 0 \end{pmatrix} = 1 + \sigma_x, \quad \theta_{-q}^*\theta_q^* + \theta_q\theta_{-q} = \begin{pmatrix} 0 & 1 \\ 1 & 0 \end{pmatrix} = \sigma_z \tag{7.5.17}$$

so that $W^{\pm}(q) = \exp(-2K_1^*\sigma_x)$, and

$$V_{\pm}^{1/2}(q) = \exp K_2\left[\cos q\,(1 + \sigma_x) + \sin q\,\sigma_z\right] = \exp(K_2 \cos q) \exp K_2\sigma \tag{7.5.18}$$

if $\sigma = \cos q\,\sigma_x + \sin q\,\sigma_z$, with $\sigma^2 = 1$. By considering the determinant, the eigenvalues of $T^{\pm}(q)$ on this two dimensional subspace are $\exp(2K_2 \pm \varepsilon_q)$ (as $\det \exp(\alpha\sigma) = 1$); where ε_q is the positive root of

$$\cosh \varepsilon_q = \operatorname{tr}\left[\exp(-2K_1^*\sigma_x)\exp(-2K_2\sigma)\right]/2$$

$$= \operatorname{tr}\left[(\cosh 2K_1^* - \sinh 2K_1^* \sigma_x)(\cosh 2K_2 - \sinh 2K_2 \sigma)\right]/2$$

$$\text{i.e. } \cosh \varepsilon_q = \cosh 2K_1^* \cosh 2K_2 - \sinh 2K_1^* \sinh 2K_2 \cos q \tag{7.5.19}$$

using $\operatorname{tr} \sigma_x = \operatorname{tr} \sigma_z = \operatorname{tr} \sigma_x \sigma_z = 0$.

Elementary manipulations show that the eigenvectors belonging to the eigenvalues $\exp(2K_2 \pm \varepsilon_q)$ are respectively

$$\Psi_0 = \cos \varphi_q \Phi_0 + \sin \varphi_q \Phi_{-qq}\,, \quad \Psi_{-qq} = -\sin \varphi_q \Phi_0 + \cos \varphi_q \Phi_{-qq} \tag{7.5.20}$$

where φ_q is defined modulo $\pi/2$ by

$$\tan 2\varphi_q = 2C_q/(B_q - A_q) \tag{7.5.21}$$

and $\operatorname{sgn} 2\varphi_q = \operatorname{sgn} q$, where $\exp(2K_2 \cos q) \begin{pmatrix} A_q & C_q \\ C_q & B_q \end{pmatrix}$ is the matrix of $T^{\pm}(q)$ on this two dimensional space, i.e.

$$\begin{aligned} A_q &= \exp(-2K_1^*)(\cosh K_2 + \sinh K_2 \cos q)^2 + \exp(2K_1^*)(\sinh K_2 \sin q)^2 \\ B_q &= \exp(-2K_1^*)(\sinh K_2 \sin q)^2 + \exp(2K_1^*)(\cosh K_2 - \sinh K_2 \cos q)^2 \\ C_q &= (2 \sinh K_2 \sin q)(\cosh 2K_1^* \cosh K_2 - \sinh 2K_1^* \sinh K_2 \cos q)\,. \end{aligned}$$

With the Bogoliubov transformation

$$\xi_q = \cos \varphi_q \theta_q + \sin \varphi_q \theta_{-q}^*\,, \quad \xi_{-q} = \cos \varphi_q \theta_q - \sin \varphi_q \theta_{-q}^* \tag{7.5.22}$$

we have

$$T(q) = \exp(2K_2 \cos q) \exp\left[-\varepsilon_q(\xi_q^* \xi_q + \xi_{-q}^* \xi_{-q} - 1)\right]$$

for $q \neq 0, \pi$, and

$$\begin{aligned} T(0)T(\pi) &= \exp[2K_2(\cos 0 + \cos \pi)] \\ &\quad \times \exp\left[-\varepsilon_0\left(\xi_0^* \xi_0 - 1/2\right) - \varepsilon_\pi\left(\xi_\pi^* \xi_\pi - 1/2\right)\right] \end{aligned} \tag{7.5.23}$$

so that

$$T^{\pm} = (2 \sinh 2K_1)^{\frac{M}{2}} \exp\left[-\textstyle\sum_{\text{all } q} \varepsilon_q \left(\xi_q^* \xi_q - 1/2\right)\right] \tag{7.5.24}$$

using again $\sum \cos q = 0$. Thus the eigenvalues of the transfer matrix T are

$$(2 \sinh 2K_1)^{\frac{M}{2}} \exp\left(-\sum \varepsilon_q \left(\alpha_q - 1/2\right)\right) \tag{7.5.25}$$

where $\alpha_q \in \{0, 1\}$, the summation is over all even (respectively odd) q so that $\sum \alpha_q$ is odd (respectively even).

7.6 Computation of thermodynamic quantities in the two dimensional Ising model

We will use the explicit diagonalization of the two dimensional Ising model obtained in the last section to compute in this section various thermodynamic quantities, particularly correlation length, free energy and specific heat.

We can compute the correlation length in a similar way to what we did in Section 7.3 for the one dimensional case. To facilitate the discussion here we assume isotropic interactions $K_1 = K_2 = K$ at this point. Again what is important are the largest eigenvalues. Note that ε_0 is given by the solution to $\cosh \varepsilon_0 = \cosh 2(K - K^*)$ so that $\varepsilon_0 = \pm 2(K - K^*)$. The correct sign is $\varepsilon_0 = 2(K^* - K)$, obtained by the consideration after (7.5.16). The sign of ε_0 is crucial here as the case $K = K^*$ determines the critical temperature T_c (or inverse critical temperature β_c) and the temperature ranges $T > T_c$, $T < T_c$ correspond to $K^* > K$, $K^* < K$ respectively. For large M, $\varepsilon_{2j\pi/M} \sim \varepsilon_{(2j+1)\pi/M}$ except for $j = 0$ due to the sign of ε_0 changing at criticality so that $\varepsilon_{\pi/M} \sim \varepsilon_0$, $T > T_c$, $\varepsilon_{\pi/M} \sim -\varepsilon_0$ for $T < T_c$. From (7.5.25) the largest eigenvalues are $\lambda^+_{\max} = \exp\left(\sum_{q \text{ even}} |\varepsilon_q|/2\right)$ for T^+ and $\lambda^-_{\max} = \exp\left(\sum_{q \text{ odd}} \varepsilon_q/2\right)$ for T^- (up to scalars $(2\sinh 2K_1)^{M/2}$), with the former only available for the transfer matrix T for the temperature $T < T_c$. All other eigenvalues are obtained by multiplying by a factor of at most $\exp(-|\varepsilon_0|)$. Thus for $T > T_c$,.the maximum eigenvalue is strictly non-degenerate and this gives a mass m, the inverse of the correlation length as

$$m = 2(K^* - K)\,, \quad T > T_c\,.$$

For $T < T_c$, however the maximum eigenvalue of the transfer matrix is asymptotically degenerate, and

$$\frac{\lambda^+_{\max}}{\lambda^-_{\max}} \sim 1 - O(e^{-dM})\,, \quad T < T_c\,, \text{ as } M \to \infty\,.$$

To get the correlation length, we have to look at the next largest eigenvalue λ_2 which is asymptotically non-degenerate with the largest two eigenvalues

$$\frac{\lambda_2}{\lambda^-_{\max}} \sim e^{4(K^*-K)}$$

giving a mass

$$m = 4(K - K^*)\,, \qquad T < T_c\,.$$

It is these principles of asymptotic degeneracy or non degeneracy which yields the phase transition. We will study this behaviour in an operator algebraic formulation of the Ising model in Section 7.14 in terms of states on the Pauli algebra, which will be ground states for dynamics generated by transfer matrices. The mass m will appear in the GNS representation of these states.

In particular, the largest eigenvalue is

$$\lambda = (2\sinh 2K_1)^{\frac{M}{2}} \exp\left(\sum_{q\ \mathrm{even}} \varepsilon_q/2\right).$$

Taking logarithms

$$\frac{1}{M}\log\lambda_1(M) = \frac{1}{2}\log(2\sinh 2K_1) + \frac{1}{2M}\sum_{q\ \mathrm{even}} \varepsilon_q$$

and as $M \to \infty$, the right hand side converges to

$$\frac{1}{2}\log(2\sinh 2K_1) + \frac{1}{2\pi}\int \cosh^{-1}(\cosh 2K_1^* \cos 2K_2 - \sinh 2K_1^* \sinh 2K_2 \cos\theta)\, d\theta\,.$$

Substituting

$$\cosh^{-1}|z| = \frac{1}{\pi}\int_0^\pi \log 2(z - \cos\psi)\, d\psi$$

the free energy per site, as $M \to \infty$ becomes

$$\begin{aligned} f &= kT\left[\frac{1}{2}\log(2\sinh 2K_1) + \frac{1}{2\pi}\int_0^\pi \varepsilon_q\, dq\right] \\ &= \log 2 + \frac{1}{2\pi^2}\int_0^\pi\int_0^\pi \log[\cosh 2K_1 \cosh 2K_2 - \sinh 2K_1 \cos p \\ &\qquad - \sinh 2K_2 \cos q]\, dp\, dq\,. \end{aligned}$$

We will examine this in the isotropic case $K = K_1 = K_2$, so that

$$\begin{aligned} -\beta f &= \frac{1}{2}\log(2\sinh 2K) + \frac{1}{2}\log 2 \\ &\quad + \frac{1}{2\pi^2}\int_0^\pi\int_0^\pi \log 2(\cosh 2K \coth 2K - \cos\theta - \cos\psi)\, d\theta\, d\psi \\ &= \log 2 \\ &\quad + \frac{1}{2\pi^2}\int_0^\pi\int_0^\pi \log\left[\cosh^2 2K - \sinh 2K(\cos\theta_1 + \cos\theta_2)\right]\, d\theta_1\, d\theta_2\,. \end{aligned}$$

With $\beta = 1/kT$, $K = \beta J$ the internal energy

$$\begin{aligned} U &= -kT^2\partial/\partial T(f/kT) = J\partial/\partial K(f/kT) \\ &= -J\coth 2K\Bigg[1 + (\sinh^2 2K - 1) \\ &\qquad \frac{1}{\pi^2}\int_0^\pi\int_0^\pi \frac{d\theta_1\, d\theta_2}{\cosh^2 2K - \sinh 2K(\cos\theta_1 + \cos\theta_2)}\Bigg]. \end{aligned}$$

The only problem in the integrand is when θ_i are small:

$$\begin{aligned}
&\cosh^2 2K - \sinh 2K(\cos\theta_1 + \cos\theta_2) \\
&= (1 - \sinh 2K)^2 + (2 - \cos\theta_1 - \cos\theta_2)\sinh 2K \\
&\approx (1 - \sinh 2K)^2 + \frac{1}{2}\left(\theta_1^2 + \theta_2^2\right)\sinh 2K
\end{aligned}$$

and here the contribution to the integral is

$$\begin{aligned}
&\frac{1}{\pi^2}\int\!\!\int_{\theta_i \text{ small}} \frac{d\theta_1\, d\theta_2}{(1 - \sinh 2K)^2 + (\theta_1^2 + \theta_2^2)(\sinh 2K)/2} \\
&= \frac{2}{\pi}\int \frac{r\, dr}{\delta + r^2(\sinh 2K)/2}, \quad \delta = (1 - \sinh 2K)^2
\end{aligned}$$

giving a divergent contribution $-2\log|\delta|/\pi \sinh 2K$ but compensated by the factor $(\sinh^2 2K - 1)$ in the above expansion for U. Thus U is continuous even when $\sinh 2K - 1 = 0$, $(\delta = 0)$. If K_c is the unique solution to $\sinh 2K_c = 1$, $K = \beta J$, $K_c = \beta_c J$, and β_c is the inverse critical temperature, then U is continuous even at K_c and the internal energy

$$U \approx -J\coth 2K_c\,[1 + A(K - K_c)\log|K - K_c|]$$

where A is a constant. Thus the specific heat $C = \partial U/\partial T \approx \log|K - K_c|$ has a logarithmic divergence. More explicitly,

$$\begin{aligned}
C &= -J\coth 2K\left[1 + (2\tanh^2 2K - 1)\frac{2}{\pi}K(k)\right] \\
&= \frac{2k}{\pi}(K\coth 2K)^2\Big\{2K(k) - 2E(k) \\
&\qquad -2(1 - \tanh^2 2K)\left[\pi/2 + (2\tanh^2 2K - 1)K(k)\right]\Big\}
\end{aligned}$$

where

$$K(k) = \int_0^{\frac{\pi}{2}} (1 - k^2\sin^2\theta)^{-\frac{1}{2}}\, d\theta \sim \log[4(1 - k^2)^{-\frac{1}{2}}] \quad \text{as } k \to 1$$

$$E(k) = \int_0^{\frac{\pi}{2}} (1 - k^2\sin^2\theta)^{\frac{1}{2}}\, d\theta$$

are complete elliptic integrals of the first and second kind respectively, and $k = 2\sinh 2K/\cosh^2 2K$ (Fig. 7.11).

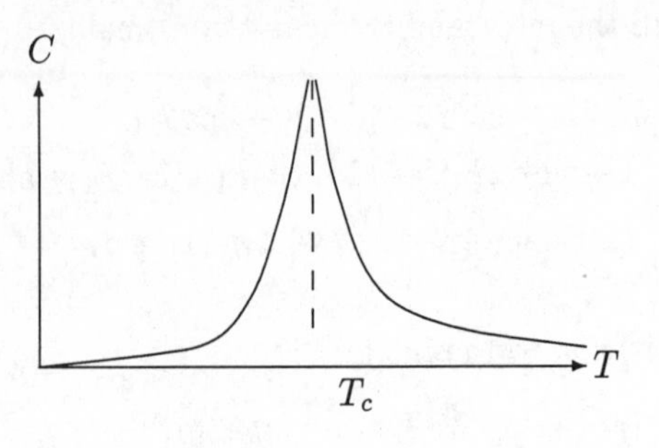

FIG. 7.11. Specific heat C

7.7 C^*-algebraic approach to phase transition in the two dimensional Ising model

With the explicit diagonalization of the transfer matrix in the previous section, we can begin to study the states φ_β on A^P of Pauli spins on a one-dimensional lattice which gives rise to the thermodynamic limit of the Gibbs ensemble in the two dimensional Ising model, with either periodic or free boundary conditions. We will show that the representation of the Pauli algebra associated with the state is factorial above and at the known critical temperature of Onsager, while it has a two dimensional centre below the critical temperature.

In the limit of $M \to \infty$, U^M and $(U^*)^M$ (restrictions of U and U^* to the subspace $H_M = \{f \in \ell^2 \mid f_j = 0, |j| > M\}$, which are no longer unitary) tend strongly to U and U^*. Hence H_j^M tends to H_j, $(j = 1, 2)$ and H^M to H defined by

$$e^{2H} = e^{K_2 H_2} e^{2K_1^* H_1} e^{K_2 H_2}, \quad H^* = H. \tag{7.7.1}$$

Due to the property $\Gamma H_j \Gamma = -H_j$, H also satisfies $\Gamma H \Gamma = -H$.

We need the following result:

Lemma 7.3 *H does not have a point spectrum at* 0.

Before going into its proof, we describe its consequence. Since H does not have the point spectrum at 0, Lemma 7.5 implies

$$\lim_{M\to\infty} E_\pm^M = E_\pm, \tag{7.7.2}$$

where E_- and E_+ are spectral projections of H for $(-\infty, 0)$ and $(0, \infty)$. This means that the limit of φ_M^+ as $M \to \infty$ is the restriction of the Fock state φ_{E_-} on A^F to $A_+^F = A_+^P$.

The automorphism θ on A^{PM} can be implemented by a unitary operator $U(\theta) = \prod_{j=-M}^{M} \sigma_z^j$. By θ-invariance of H^M, Ω^M is $U(\theta)$ invariant and hence $\langle a\Omega^M, \Omega^M \rangle = 0$ if $\theta(a) = -a$, $a \in A^{PM}$. Therefore φ_β is also θ-invariant and is determined by its restriction to A_+^P, i.e. $\varphi_\beta(a) = \varphi_\beta(a_+)$, $a_+ = (a+\theta(a))/2 \in A_+^P$ for any $a \in A^P$. Therefore we obtain the following:

Lemma 7.4 *φ_β is the unique θ-invariant extension, to A^P, of a state φ_β^+ on $A_+^P = A_+^F$, where φ_β^+ is the restriction, to A_+^F, of the Fock state φ_{E_-} on A^F and E_- is the spectral projection of H (given by (7.7.1)) for $(-\infty, 0)$.*

Proof of Lemma 7.3 The unitary operator U has the spectral decomposition $U = \int_0^{2\pi} e^{i\theta}\, dE_U(\theta)$ with a simple Lebesgue spectral measure on $[0, 2\pi]$. Thus $H = \int H(\theta)\, dE_U(\theta)$ with θ-dependent 2×2 matrix $H(\theta)$, given by

$$2H(\theta) = -\gamma(\theta)V(\theta)\,, \tag{7.7.3}$$

$$V(\theta) = \begin{pmatrix} \cos\vartheta(\theta) & -i\sin\vartheta(\theta) \\ i\sin\vartheta(\theta) & -\cos\vartheta(\theta) \end{pmatrix}, \tag{7.7.4}$$

where $\gamma(\theta) \geq 0$ is determined by

$$\cosh 2K_1^* \cosh 2K_2 - \sinh 2K_1^* \sinh 2K_2 \cos\theta = \cosh\gamma(\theta)\,, \tag{7.7.5}$$

and $\delta(\theta) \equiv \vartheta(\theta) - \theta$ is determined by

$$\cos\delta(\theta) = (\sinh\gamma(\theta))^{-1}(\cosh 2K_1^* \sinh 2K_2 - \sinh 2K_1^* \cosh 2K_2 \cos\theta)\,, \tag{7.7.6}$$

$$\sin\delta(\theta) = (\sinh\gamma(\theta))^{-1} \sinh 2K_1^* \sin\theta\,. \tag{7.7.7}$$

The right-hand side of (7.7.5) is not 1 except for discrete values of θ satisfying $\cos\theta = 1$ (and only if $K_1^* = K_2$) due to

$$\begin{aligned} |\sinh 2K_1^* \sinh 2K_2 \cos\theta| + 1 &\leq \sinh 2K_1^* \sinh 2K_2 + 1 \\ &\leq (\sinh^2 2K_1^* + 1)^{\frac{1}{2}} (\sinh^2 2K_2 + 1)^{\frac{1}{2}} = \cosh 2K_1^* \cosh 2K_2\,. \end{aligned} \tag{7.7.8}$$

Thus $\delta(\theta)$ is well defined (modulo 2π) by (7.7.6) and (7.7.7) for all θ if $K_1^* \neq K_2$ and for $\theta \neq 0$ (modulo 2π) if $K_1^* = K_2$. (If $K_1^* = K_2$, $\gamma(\theta) = 0$ for $\theta = 0$ and any value of $\vartheta(0)$ leads to the same $H(0)$.) Since the matrix part of (7.7.3) and (7.7.4) is self-adjoint unitary, $H(\theta)$ does not have an eigenvalue 0. □

Lemma 7.5 *Let $\lim A_n = A$, E_n and E be the spectral projections of A_n and A for an (infinite or finite) interval (b, a) either including or not including the eigenprojections of a and/or b. If a and b are not eigenvalues of A, then*

$$\lim E_n = E\,. \tag{7.7.9}$$

Proof It is enough to treat the case of $b = -\infty$, a finite because the case of $a = \infty$, b finite will follow by considering $-A_n \to -A$ and the case of a and b finite will follow by taking the product of projections for the two infinite cases. For any given vector ψ and $\varepsilon > 0$, there exists ϕ belonging to the A-spectral subspace for the complement of the interval $(a - \delta, a + \delta)$ for some $\delta > 0$ and satisfying $\|\psi - \phi\| < \varepsilon$. Since E_n is uniformly bounded, it is enough to prove $\lim E_n\phi = E\phi$ for such ϕ. However, on the A-spectral subspace for the complement of the interval $(a-\delta, a+\delta)$, we can apply Exercise 1.34 to the characteristic function of the interval (b,a) (the endpoints a and/or b included or not included according to the definition of E_n) and obtain $\lim E_n\phi = E\phi$. □

We can write $E_\beta = E_- = (1-V_\beta)/2$, where V is given by (7.7.4). The states φ_0 and φ_∞ correspond to infinite and zero temperatures ($\beta = 0$, $\beta = \infty$ respectively) as follows. The region $\beta > \beta_c$ corresponds to $K_1^* < K_2$, and $\beta < \beta_c$ to $K_1^* > K_2$. We will regard K_1^* and K_2 as independent parameters. Then $K_2 = 0$, $K_1^* > 0$ corresponds to $\beta = 0$, and $K_1^* = 0$, $K_2 > 0$ to $\beta = \infty$. To be more precise, in these cases, V, δ, ϑ, are given by the following:
Case (A). $K_2 = 0$, $K_1^* > 0$, $(\beta = 0)$. Here $\gamma(\theta) = 2K_1^*$, $\delta(\theta) = \pi - \theta$, $\vartheta(\theta) = \pi$,

$$V_0(\theta) = \begin{pmatrix} -1 & 0 \\ 0 & 1 \end{pmatrix}, \quad E_0 = (1-V_0)/2\,.$$

Then the even state φ_0 on A^P corresponding to the quasi-free state $\varphi_0^F = \varphi_{E_0}$ on A^F is the product state,

$$\varphi_0 = \bigotimes_{-\infty}^{\infty} \omega_\Omega\,,$$

where ω_Ω is the vector state on M_2 given by $\Omega = 2^{-\frac{1}{2}}\begin{pmatrix}1\\1\end{pmatrix}$. Note that $\sigma_z = \begin{pmatrix} 0 & 1 \\ 1 & 0 \end{pmatrix}$ has eigenvectors Ω and $2^{-\frac{1}{2}}\begin{pmatrix}1\\-1\end{pmatrix}$ with eigenvalues 1 and -1 respectively. Thus the eigenspace of $W = \exp\left\{K_1^* \sum_{j=-M}^{M} \sigma_z^j\right\}$ corresponding to the largest eigenvalue is non-degenerate, and spanned by $\bigotimes_{-M}^{M} \Omega$. The transfer matrix T_M in the case when $K_2 = 0$ is a scalar multiple of W and so the same applies to T_M.

Case (B). $K_1^* = 0$, $K_2 > 0$, $(\beta = \infty)$. Here $\gamma(\theta) = 2K_2$, $\delta(\theta) = 0$, $\vartheta(\theta) = \theta$,

$$V_\infty(\theta) = \begin{pmatrix} \cos\theta & -i\sin\theta \\ i\sin\theta & -\cos\theta \end{pmatrix}, \quad E_\infty = (1-V_\infty)/2\,.$$

The even state φ_∞ on A^P corresponding to the quasi-free state $\varphi_\infty^F = \varphi_{E_\infty}$ on A^F is the state

$$\varphi_\infty = \frac{1}{2}\left(\bigotimes_{-\infty}^{\infty} \omega_{\left(\begin{smallmatrix}1\\0\end{smallmatrix}\right)} + \bigotimes_{-\infty}^{\infty} \omega_{\left(\begin{smallmatrix}0\\1\end{smallmatrix}\right)}\right)$$

Note that $\sigma_x = \begin{pmatrix} 1 & 0 \\ 0 & -1 \end{pmatrix}$ has eigenvectors $\begin{pmatrix}1\\0\end{pmatrix}$ and $\begin{pmatrix}0\\1\end{pmatrix}$ with eigenvalue 1 and -1 respectively. Thus the eigenspace of $V = \exp\left\{K_2 \sum_{-M}^{M-1} \sigma_x^j \sigma_x^{j+1}\right\}$ corresponding to the largest eigenvalue is doubly degenerate and spanned by $\bigotimes_{-M}^{M}\begin{pmatrix}1\\0\end{pmatrix}$ and $\bigotimes_{-M}^{M}\begin{pmatrix}0\\1\end{pmatrix}$ (corresponding to all spins up and all spins down respectively). The transfer matrix T_M in the case when $K_1^* = 0$ is a scalar multiple of V, and so the same applies to this T_M.

Consider the states φ_β on the C^*-algebra of Pauli spins on a one-dimensional lattice (infinitely extended in both directions) which give rise to the thermodynamic limit of the Gibbs ensemble in the two-dimensional Ising model at inverse temperature β for free or periodic boundary conditions.

Theorem 7.6 *(a) There exists a family $\{\nu_\beta \mid \beta \neq \beta_c\}$ of automorphisms of the Pauli algebra A^P such that*

$$\varphi_\beta = \begin{cases} \varphi_\infty \circ \nu_\beta & \beta > \beta_c \\ \varphi_0 \circ \nu_\beta & \beta < \beta_c \end{cases} .$$

(b) The states φ_β are pure for $0 \leq \beta < \beta_c$ and a non-trivial mixture of two inequivalent pure states for $\beta > \beta_c$.

We already understand the picture at zero and infinite temperatures. Thus the second statement of the above theorem will follow from the first away from the critical temperature. At criticality we are obliged to use Theorem 6.25, which gives the criterion for when an even state φ^P of A^P is pure in terms of the associated even state φ^F of A^F. In fact this argument works also away from criticality. When applying Theorem 6.25 to the Ising model, we have to decide when the Fock states φ_{E_β} and $\varphi_{\theta_- E_\beta \theta_-}$ are equivalent on A^F and their restrictions on A_+. Now φ_{E_β} and $\varphi_{\theta_- E_\beta \theta_-}$ are equivalent if and only if $E_\beta - \theta_- E_\beta \theta_-$ is Hilbert–Schmidt, Theorem 6.16 (b).

Lemma 7.7 *$E_\beta - \theta_- E_\beta \theta_-$ is Hilbert–Schmidt if and only if $\beta \neq \beta_c$.*

Proof Denoting $\overline{E}_\beta = 1 - E_\beta$

$$\|E_\beta - \theta_- E_\beta \theta_-\|_{HS}^2 = 4 \lim_{\varepsilon \to 0} \int_{-\pi}^{\pi} \int_{-\pi}^{\pi} \frac{\operatorname{tr} E_\beta(\theta_1)\overline{E}_\beta(\theta_2)}{|1 - e^{i(\theta_1 - \theta_2) - \varepsilon}|^2} \frac{d\theta_1}{2\pi} \frac{d\theta_2}{2\pi} \tag{7.7.10}$$

where $\operatorname{tr}(E_\beta(\theta_1)\overline{E}_\beta(\theta_2)) = \{2 - \operatorname{tr} V(\theta_1)V(\theta_2)\}/4 = \{1 - \cos(\vartheta(\theta_1) - \vartheta(\theta_2))\}/2$. We consider this in detail for criticality only (as we shall see below in Lemma 7.15, that $E_\beta - \theta_- E_\beta \theta_-$ is even trace class for $\beta \neq \beta_c$); i.e. $K_1^* = K_2 = K$.

For $\cos\theta \neq 1$ (i.e. $\theta \neq 0 \bmod 2\pi$), $\cos\vartheta(\theta)$ and $\sin\vartheta(\theta)$ are real analytic in θ, and hence the integrand of (7.7.10) is integrable except possibly near the point $\theta_1 = \theta_2 = 0$. As $\theta \to 0$, $\gamma(\theta) \to +0$ with

$$\lim_{\theta \to 0} (\gamma(\theta)/\theta)^2 = \sinh^2 2K . \tag{7.7.11}$$

Therefore $\delta(\theta) \to \pm\pi/2 \pmod{2\pi}$ as $\theta \to \pm 0$ by (7.7.6) and (7.7.7). Thus the contribution from $\theta_1\theta_2 > 0$ in (7.7.10) in a neighbourhood of $\theta_1 = \theta_2 = 0$ is finite while the contribution from $\theta_1\theta_2 < 0$ is $+\infty$ due to

$$\int_0^\alpha d\theta_1 \int_{-\alpha}^0 d\theta_2 \left\{(1 - \cos(\theta_1 - \theta_2))^2 + \sin^2(\theta_1 - \theta_2)\right\}^{-1} = \infty .$$

□

We deduce using condition (a) of Theorem 7.6 that φ_{β_c} is pure, but we have to do more to decide what the situation is away from β_c. To use (b), we need the criterion of Theorem 6.30 to decide when the restriction of two Fock states to the even algebra are equivalent. To deal with this, we introduced in Section 6.8. the following (symmetric) $\mathbb{Z}/2$ index between two basis projections E_1 and E_2 for which $E_1 - E_2$ is Hilbert–Schmidt:

$$\sigma(E_1, E_2) = (-1)^{\dim E_1 \wedge (1-E_2)} . \tag{7.7.12}$$

Theorem 7.6(b) then follows from:

$$\sigma(E_\beta, \theta_- E_\beta \theta_-) = \begin{cases} 1 & \beta < \beta_c \\ -1 & \beta > \beta_c \end{cases} . \tag{7.7.13}$$

An evaluation of this index can be explicitly achieved when $K_2 = 0$, $K_1^* > 0$, $(\beta = 0)$ and $K_1^* = 0$, $K_2 > 0$, $(\beta = \infty)$. Alternatively, since we can compute the states φ_0 and φ_∞ explicitly, and we know the former is pure and the latter not, then σ at these temperatures must be what we claim it is by Theorem 6.30 in reverse. The index σ is continuous in the norm topology (Theorem 6.31), and E_β is norm continuous in K_1^* and K_2 (regarded as independent parameters) in the regions $K_1^* > K_2 \geq 0$ (or $\beta < \beta_c$) and $K_2 > K_1^* \geq 0$ (or $\beta > \beta_c$). Thus an explicit computation at zero and infinite temperatures is enough. However, in Section 7.8 we give a precise description of $E_\beta \wedge (1 - \theta_- E_\beta \theta_-)$. We postpone a discussion on continuity of σ to later (see Remark on page 334), but note continuity of E_β as follows.

Let the functions δ, ϑ, V and the projections E_β for another parameter $K_1'^*$ and K_2' be denoted by δ', ϑ', V' and $E_{\beta'}$. Then

$$\|E_\beta - E_{\beta'}\| = 2^{-1} \sup_\theta \|V(\theta) - V'(\theta)\| . \tag{7.7.14}$$

For $K_1^* \neq K_2$, $\gamma(\theta)$ is real analytic in K_1^*, K_2 and θ, and so are $\cos\delta(\theta)$ and $\sin\delta(\theta)$ due to $\gamma(\theta) \neq 0$. Hence $V(\theta)$ is uniformly continuous in K_1^*, K_2 and θ over a compact set. In particular, (7.7.14) tends to 0 as $(K_1'^*, K_2')$ tends to (K_1^*, K_2), and hence E is continuous in the parameter K_1^* and K_2 relative to the norm topology except for $K_1^* = K_2$.

To further exploit the explicit and simple information available at zero and infinite temperature, we begin with the duality between high and low temperatures used to locate the critical temperature (Kramers and Wannier 1941). Mathematically, this duality is effected by the following automorphism κ on A_+:

$$\kappa\left(\sigma_z^j\right) = \sigma_x^j \sigma_x^{j+1} , \quad \kappa\left(\sigma_x^j \sigma_x^{j+1}\right) = \sigma_z^{j+1} . \tag{7.7.15}$$

(See the remark of (Onsager 1941), page 123 on Kramers–Wannier duality). Roughly speaking κ interchanges the role of V and W in the transfer matrix T

Recall that σ_z^j and $\sigma_x^j\sigma_x^{j+1}$ generate A_+. Then κ^2 is the restriction of the shift on $\bigotimes_{-\infty}^{\infty} M_2$ to A_+, but we will see in Corollary 7.11 that κ does not extend to an automorphism of A^P. However, κ does extend to an automorphism of A^F.

Let U be the shift on ℓ^2:

$$(Uf)_k = f_{k+1}\,, \quad f = (f_k) \in \ell^2\,, \tag{7.7.16}$$

identified with multiplication by $e^{-i\theta}$ on $\mathrm{L}^2(\mathbb{T})$. Let

$$W = \frac{i}{2}\begin{pmatrix} 1-U^* & 1+U^* \\ -1-U^* & U^*-1 \end{pmatrix}. \tag{7.7.17}$$

Note that $W^2 = \begin{pmatrix} U^* & 0 \\ 0 & U^* \end{pmatrix}$, so that $\tau(W)^2 = \tau(W^2)$ is the Bogoliubov automorphism on the CAR algebra induced by the shift, or $\tau^2(c_j) = c_{j+1}$.

Lemma 7.8 *The restriction of the Bogoliubov automorphism $\tau(W)$ from A^F to A_+^F is κ.*

Proof If $\tau = \tau(W)$, we have $\tau(c_j^*) = i(c_j^* - c_{j+1}^* - c_j - c_{j+1})/2$, $\tau(c_j) = i(c_j^* + c_{j+1}^* - c_j + c_{j+1})/2$. Then $\tau(c_j - c_j^*) = i(c_{j+1}^* + c_{j+1})$, $\tau(c_j + c_j^*) = i(c_j^* - c_j)$ so that $\tau\left((c_j - c_j^*)(c_{j+1} + c_{j+1}^*)\right) = 2c_{j+1}^*c_{j+1} - 1$. Since $\tau^2(c_j) = c_{j+1}$, we see $\tau(2c_j^*c_j - 1) = (c_j - c_j^*)(c_{j+1} + c_{j+1}^*)$. □

We now extend the Kramers–Wannier automorphism κ to A^F by putting $\kappa = \tau(W)$. Also note that

$$W^*\begin{pmatrix} 1 & 0 \\ 0 & -1 \end{pmatrix} W = \frac{1}{2}\begin{pmatrix} -(U+U^*) & U^*-U \\ U-U^* & U+U^* \end{pmatrix}. \tag{7.7.18}$$

This means that κ takes the infinite temperature state φ_0^F to the zero temperature state φ_∞^F:

$$\varphi_0^F \circ \kappa = \varphi_\infty^F \tag{7.7.19}$$

as one would expect from (7.7.15) and (7.4.8), (7.4.9).

We now define

$$U_\beta = e^{-i\vartheta} \tag{7.7.20}$$

where ϑ is as defined in (7.7.6) and (7.7.7), and

$$W_\beta = \frac{i}{2}\begin{pmatrix} 1-U_\beta^* & 1+U_\beta^* \\ -(1+U_\beta^*) & U_\beta^*-1 \end{pmatrix}. \tag{7.7.21}$$

Then

$$W_\beta^*\begin{pmatrix} 1 & 0 \\ 0 & -1 \end{pmatrix} W_\beta = \frac{1}{2}\begin{pmatrix} -(U_\beta+U_\beta^*) & U_\beta^*-U_\beta \\ U_\beta-U_\beta^* & U_\beta+U_\beta^* \end{pmatrix} = -V_\beta\,. \tag{7.7.22}$$

This means that $E_\beta = W_\beta^* E_0 W_\beta$, and if $\gamma_\beta = \tau(W_\beta)$, the Bogoliubov automorphism induced by W_β, then

$$\varphi_0^F \circ \gamma_\beta = \varphi_\beta^F , \tag{7.7.23}$$

and

$$\varphi_\infty^F \circ \delta_\beta = \varphi_\beta^F , \tag{7.7.24}$$

if $\delta_\beta = \kappa^{-1}\gamma_\beta = \tau(W^* W_\beta)$. We will show that $\{\gamma_\beta|_{A_+} \mid 0 \leq \beta < \beta_c\}$ and $\{\delta_\beta|_{A_+} \mid \beta < \beta_c\}$ extend to automorphisms $\{\nu_\beta \mid \beta \neq \beta_c\}$ of A^P such that

$$\left.\begin{array}{ll} \varphi_0 \circ \nu_\beta = \varphi_\beta , & 0 \leq \beta < \beta_c \\ \varphi_\infty \circ \nu_\beta = \varphi_\beta , & \beta > \beta_c \end{array}\right\} . \tag{7.7.25}$$

We consider the problem of deciding when an automorphism of the even algebra A_+ extends to an automorphism of the Pauli algebra A^P.

Let C be a unital C^*-algebra graded by a symmetry θ. We say that an automorphism ν of C is graded if $\nu(C_\pm) \subset C_\pm$. An inner automorphism of C is said to be even (respectively odd) if it is implemented by an even (respectively odd) unitary. Note that if C is simple, then a graded inner automorphism on C is always either even or odd. For then if $\nu = \mathrm{Ad}(u)$, $u \in C$, we have $\nu = \theta\nu\theta$ since ν is graded, and so $\mathrm{Ad}\,\theta(u) = \mathrm{Ad}(u)$ on C. Since C is simple, this implies $\theta(u) = \lambda u$ for some $\lambda \in \mathbb{T}$. Letting $u = a + b$, where a, b are even and odd respectively, we see that $a - b = \lambda(a + b)$, or $a(1 - \lambda) = b(1 + \lambda) = 0$. Hence either $b = 0$, $\lambda = 1$ and u is even, or $a = 0$, $\lambda = -1$ and u is odd. We need something stronger than this:

Lemma 7.9 *Let u be a self adjoint unitary in a graded C^*-algebra such that C_+ is simple and $uC_+u = C_+$. Then u is either even or odd.*

Proof Let $u = a + b$, where a, b are even and odd respectively. We have to show that either a or b is zero. Now a, b are self adjoint and $(a + b)x(a + b) \in C_+$, for all $x \in C_+$. This means

$$axb + bxa = 0 , \quad \text{for all } x \in C_+ . \tag{7.7.26}$$

In particular $ab + ba = 0$, and since u is unitary we have $a^2 + b^2 = 1$. From (7.7.26) with $x = a$ we get $a^2b + ba^2 = 0$. Then using $a^2 = 1 - b^2$ we have $(1 - b^2)b + b(1 - b^2) = 0$, or $b = b^3$. Then $(ab)^*ab = ba^2b = b(1 - b^2)b = 0$, hence $ab = 0 = ba$. But then using (7.7.26), $b(axb + bxa) = 0$, for all $x \in C_+$ implies that $b^2xa = 0$ for all $x \in C_+$, or $(1 - a^2)xa = 0$ for all $x \in C_+$. But C_+ is simple and so either $a^2 = 1$ or $a = 0$, i.e. by $a^2 + b^2 = 1$ either $b = 0$ or $a = 0$. □

We now consider the following general situation (cf. Section 6.10). Let A be a unital C^*-algebra, with α, β two commuting automorphisms such that $\alpha^2 = \beta^2 = 1$. We grade A by α, and suppose there is a self adjoint odd unitary fixed by β. Let $\hat{A}$ be the crossed product of A by the β-action of $\mathbb{Z}/2$ which is generated by A and a $T \in \hat{A}$ satisfying $T^2 = 1$, $T^* = T$, $Ta = \beta(a)T$, $a \in \hat{A}$. Let $B = A_+ + TA_-$

which is a C^*-subalgebra of $\hat{A}$. Extend α, β to $\hat{A}$ by $\hat{\alpha}(a + Tb) = \alpha(a) + T\alpha(b)$, $\hat{\beta}(a + Tb) = \beta(a) + T\beta(b)$, $a, b \in A$. We grade $\hat{A}, B$ by $\hat{\alpha}$ and $\hat{\alpha}|_B$ respectively, so that $B_+ = A_+$, $B_- = TA_-$. If ν is a graded automorphism of A, we give a criterion when $\nu|A_+$ extends to an automorphism of B. We will then apply these criteria to the case $A = A^F$, $\alpha = \theta$, $\beta = \theta_-$, $B = A^P$, $U = c_i + c_i^*$ for any $i \geq 1$, and ν a quasi-free automorphism of A^F.

Proposition 7.10 *Let ν be a graded automorphism of A, where A_+ is simple. If $\nu|A_+$, extends to an automorphism $\tilde{\nu}$ of B, then $\tilde{\nu}$ must be graded.*

Proof Let $\sigma = TU \in TA_-$, so that σ is a self adjoint unitary in B, and $B = A_+ + \sigma A_+$. If $\nu|A_+$ extend to an automorphism $\tilde{\nu}$ of B, then ν, $\mathrm{Ad}(\sigma)$ leave A_+ invariant and $\nu\,\mathrm{Ad}(\sigma)\nu^{-1} = \mathrm{Ad}(\tilde{\nu}(\sigma))$ on A_+. Hence by Lemma 7.9, $\tilde{\nu}(\sigma)$ is either odd or even. If $\tilde{\nu}(\sigma)$ is odd, then $\tilde{\nu}$ is graded. If $\tilde{\nu}(\sigma)$ is even, then $\tilde{\nu}(B) \subset B_+$ which as $\tilde{\nu}$ is an automorphism means $B_+ = B$, and again $\tilde{\nu}$ is graded. □

Corollary 7.11 *The Kramers–Wannier automorphism $\kappa : A_+ \to A_+$ does not extend to an automorphism of A^P.*

Proof Suppose κ extends to a graded automorphism $\tilde{\kappa}$ of A^P. Then $\varphi_0 \circ \kappa = \varphi_\infty$ on A_+ means that $\varphi_0 \circ \tilde{\kappa} = \varphi_\infty$ on A^P, since φ_0 and φ_∞ are even states. But this is impossible as φ_0 is pure and φ_∞ is not. □

Note that since κ extends to an automorphism of A^F, it follows from Corollary 7.11 that the Jordan–Wigner transformation which identifies A_+^P with A_+^F in Section 6.5 cannot be extended to an isomorphism between A^P and A^F (although A^P and A^F are isomorphic C^*-algebras).

If ν is a graded automorphism of A, which extends to an automorphism $\hat{\nu}$ of $\hat{A}$, then $\beta\nu\beta\nu^{-1}(x) = T\hat{\nu}(T)x\hat{\nu}(T)T$, for all $x \in A$. In particular, if $\hat{\nu}$ is graded, then $T\hat{\nu}(T)$ is in $\hat{A}_+$. Note that by the argument of Proposition 7.10 if $\hat{A}_+$ is simple, then $\hat{\nu}$ must be graded. In the converse direction we have

Lemma 7.12 *Let ν be a graded automorphism of A, where A_+ is simple, and $\beta\nu\beta\nu^{-1}$ is an inner even automorphism of A Then ν extends to a graded automorphism of $\hat{A}$, leaving B invariant, and given by*

$$\hat{\nu}(a + Tb) = \nu(a) + Tv\nu(b)\,, \quad a, b \in A \tag{7.7.27}$$

where v is a unitary in A_+ such that $v\beta(v) = 1$, $\beta\nu\beta\nu^{-1} = \mathrm{Ad}(v)$ on A.

Proof Suppose $\beta\nu\beta\nu^{-1} = \mathrm{Ad}(v)$, for some v unitary in A_+. If $\gamma = \beta\nu\beta\nu^{-1}$, we have $\gamma\beta\gamma\beta = 1$. Therefore for $x \in A$: $x = \gamma\beta\gamma\beta(x) = v\beta(v)x\beta(v)^*v^*$. But A_+ is simple and so we must have $v\beta(v) \in \mathbb{T}$. By rotating v we may assume $v\beta(v) = 1$. Define $\hat{\nu} : \hat{A} \to \hat{A}$ by (7.7.27). We use $\nu\beta(x) = Tv\nu(x)v^*T$, $x \in A$, and $TvT = v^*$ to check that $\hat{\nu}$ is an automorphism. For $a, b, a_1, b_1, a_2, b_2 \in A$,

$$\hat{\nu}[(a + Tb)^*] = \hat{\nu}[a^* + T\beta b^*] = \nu(a^*) + Tv\nu\beta(b^*)$$

$$= \nu(a)^* + \nu(b^*)v^*T = [\hat{\nu}(a+Tb)]\,, \tag{7.7.28}$$

$$\begin{aligned}
\hat{\nu}\,[(a_1+Tb_1)(a_2+Tb_2)] &= \hat{\nu}\,[(a_1a_2+\beta(b_1)b_2+Tb_1a_2+T\beta(a_1)b_2)] \\
&= \nu(a_1a_2)+\nu(\beta(b_1)b_2)+Tv\nu(b_1a_2)+Tv\nu(\beta(a_1)b_2) \\
&= \nu(a_1)\nu(a_2)+Tv\nu(b_1)Tv\nu(b_2)+Tv\nu(b_1)\nu(a_2)+\nu(a_1)Tv\nu(b_2) \\
&= [\nu(a_1)+Tv\nu(b_1)][\nu(a_2)+Tv\nu(b_2)] = \hat{\nu}(a_1+Tb_1)\hat{\nu}(a_2+Tb_2)\,.
\end{aligned} \tag{7.7.29}$$

Thus $\hat{\nu}$ is an automorphism, and if v is in A_+, it is clear that $\hat{\nu}$ is graded and leaves B invariant. □

We now apply the criterion of Lemma 7.12 for extending automorphisms from the even algebra A_+^P to the Pauli algebra A^P to deduce:

Theorem 7.13 *The Bogoliubov automorphisms* $\{\tau(W^*W_\beta)|_{A_+} \mid \beta > \beta_c\}$ *and* $\{\tau(W_\beta)|_{A_+} \mid 0 \le \beta < \beta_c\}$ *extend to graded automorphisms* $\{\nu_\beta \mid \beta \ne \beta_c\}$ *of the Pauli algebra* A^P *such that*

$$\varphi_0 \circ \nu_\beta = \varphi_\beta\,, \quad 0 \le \beta < \beta_c\,, \tag{7.7.30}$$

$$\varphi_\infty \circ \nu_\beta = \varphi_\beta\,, \quad \beta > \beta_c\,. \tag{7.7.31}$$

The Bogoliubov automorphisms

$$\{\tau(W^*W_\beta)|_{A_+} \mid 0 \le \beta < \beta_c\} \quad \text{and} \quad \{\tau(W_\beta)|_{A_+} \mid \beta > \beta_c\}$$

do not extend to automorphisms of the Pauli algebra.

First we need some lemmas, and notation. Recall that the quasi-free automorphism θ_- is induced by a unitary, which by abuse of notation we also denote by θ_- on $\ell^2(\mathbb{Z}) \oplus \ell^2(\mathbb{Z})$ where $\theta_- = u_- \oplus u_-$, and $u_- = (-1) \oplus 1$ on $\ell^2(\mathbb{Z}) = \ell^2(-\infty, 0] \oplus \ell^2[1, \infty)$. Now put $p_\pm = (1 \pm \theta_-)/2)$, $q_\pm = (1 \pm u_-)/2$.

Lemma 7.14 *The following operators are trace class:*

$$\delta - u_-\delta u_-\,, \quad \text{for } \beta > \beta_c\,, \tag{7.7.32}$$

$$\vartheta - u_-\vartheta u_-\,, \quad \text{for } 0 \le \beta < \beta_c\,. \tag{7.7.33}$$

Proof If $z_i = \tanh K_i = e^{-2K_i^*}$, and $z_i^* = \tanh K_i^* = e^{-2K_i}$, then (7.7.5), (7.7.6) and (7.7.7) can be solved to get

$$e^{2i\delta(\omega)} = \frac{(1 - z_2 z_1^* e^{i\omega})(1 - z_1^* e^{-i\omega}/z_2)}{(1 - z_2 z_1^* e^{-i\omega})(1 - z_1^* e^{i\omega}/z_2)}\,. \tag{7.7.34}$$

Taking logarithms

$$i\delta(\omega) = \frac{1}{2}\log\frac{(1 - z_2 z_1^* e^{i\omega})(1 - z_1^* e^{-i\omega}/z_2)}{(1 - z_2 z_1^* e^{-i\omega})(1 - z_1^* e^{i\omega}/z_2)}$$

$$= \frac{1}{2}\Big[\log(1 - z_2 z_1^* e^{i\omega}) + \log(1 - z_1^* e^{-i\omega}/z_2)$$

$$- \log(1 - z_2 z_1^* e^{-i\omega}) - \log(1 - z_1^* e^{i\omega}/z_2)\Big]$$

$$= \sum_{n=1}^{\infty} \frac{1}{2n}([-(z_2 z_1^*)^n + (z_1^*/z_2)^n]e^{in\omega} + [-(z_1^*/z_2)^n + (z_2 z_1^*)^n]e^{-in\omega}).$$

Then for $\beta > \beta_c$ (i.e. $z_1^* < z_2 < 1$), the Fourier coefficients $\{k_n\}$ of $i\delta$ are given by

$$k_n = \frac{1}{2n}[(z_1^*/z_2)^n - (z_2 z_1^*)^n] = -k_{-n}, \quad n > 0, \tag{7.7.35}$$

and $\delta - u_- \delta u_- = 2(q_+ \delta q_- + q_- \delta q_+)$. Since δ is real, it is enough to show that $q_+ \delta q_-$ is trace class if $\beta > \beta_c$. Let $\{e^{-ik\omega} \mid k = 0, 1, 2, \ldots\}$ and $\{e^{-ik\omega} \mid k = 1, \ldots\}$ be complete orthonormal bases for $q_- \ell^2$ and $q_+ \ell^2$ respectively. Then the matrix of $iq_+ \delta q_-$ with respect to these bases is $\{k_{r+s+1} \mid r, s = 0, 1, 2, \ldots\}$. Thus with this identification of $q_- \ell^2$ and $q_+ \ell^2$ with $l^2(\mathbb{N}) = l^2_+$, and if $\chi_\lambda = \{\lambda^i\}_{i=0}^{\infty} \in l^2_+$, for $0 \leq \lambda < 1$, we have

$$iq_+ \delta q_- = \{k_{r+s+1} \mid r, s\} = \frac{1}{2}\int_{z_2 z_1^*}^{z_1^*/z_2} \chi_\lambda \otimes \overline{\chi}_\lambda \, d\lambda \tag{7.7.36}$$

which is trace class for $0 \leq z_1^* < z_2$, or $\beta > \beta_c$. We have from (7.7.34) that

$$e^{2i\vartheta(\omega)} = \frac{(1 - z_2 z_1^* e^{-i\omega})(1 - z_2 e^{i\omega}/z_1^*)}{(1 - z_2 z_1^* e^{i\omega})(1 - z_2 e^{-i\omega}/z_1^*)}, \tag{7.7.37}$$

so that $i\vartheta$ has Fourier coefficients $\{h_n\}$ given by

$$h_n = -\frac{1}{2n}[(z_2/z_1^*)^n + (z_2 z_1^*)^n] = -h_{-n}, \quad n > 0. \tag{7.7.38}$$

Thus

$$iq_+ \vartheta q_- = -\frac{1}{2}\left[\int_0^{z_2/z_1^*} \chi_\lambda \otimes \overline{\chi}_\lambda \, d\lambda + \int_0^{z_2 z_1^*} \chi_\lambda \otimes \overline{\chi}_\lambda \, d\lambda\right], \tag{7.7.39}$$

which is trace class for $0 \leq z_2 < z_1^*$ or $0 \leq \beta < \beta_c$. □

Lemma 7.15 *The following operators are trace class for $\beta \neq \beta_c$:*

$$W^* W_\beta - \theta_- W^* W_\beta \theta_-, \tag{7.7.40}$$
$$W_\beta - \theta_- W_\beta \theta_-. \tag{7.7.41}$$

Proof We have

$$W^*W_\beta = \frac{1}{2}\begin{pmatrix} 1+UU_\beta^* & 1-UU_\beta^* \\ 1-UU_\beta^* & 1+UU_\beta^* \end{pmatrix}. \tag{7.7.42}$$

Thus $W^*W_\beta - \theta_- W^*W_\beta\theta_-$ being trace class is equivalent to $UU_\beta^* - u_- UU_\beta^* u_-$ being trace class. But $UU_\beta^* = e^{i\delta}$ as $\vartheta(\theta) - \theta = \delta(\theta)$, and

$$e^{i\delta} - u_- e^{i\delta} u_- = i\int_0^1 u_- e^{i(1-s)\delta} u_- [\delta - u_-\delta u_-] e^{is\delta}\, ds. \tag{7.7.43}$$

Thus by (7.7.32) in Lemma 7.14 we see that

$$W^*W_\beta - \theta_- W^*W_\beta\theta_-, \quad \beta_c < \beta \le \infty \tag{7.7.44}$$

are trace class. Since $U_\beta^* = e^{i\vartheta}$, we see in a similar manner using (7.7.33) of Lemma 7.14 that

$$W_\beta - \theta_- W_\beta\theta_-, \quad 0 \le \beta \le \beta_c \tag{7.7.45}$$

are trace class. Now $W_\infty = W$, and a direct computation shows that

$$W^* - \theta_- W^*\theta_- \tag{7.7.46}$$

is trace class. The lemma now follows from the identity

$$W^*W_\beta - \theta_- W^*W_\beta\theta_- = W^*(W_\beta - \theta_- W_\beta\theta_-) + (W^* - \theta_- W^*\theta_-)\theta_- W_\beta\theta_-, \tag{7.7.47}$$

and the operators in (7.7.44), (7.7.45), and (7.7.46) being trace class. □

Remark It follows from Lemma 7.7 that the operators in (7.7.41) (and hence in (7.7.40) using (7.7.47) are Hilbert–Schmidt, but this is not enough for our approach here. Note that it also follows from Lemma 7.7 that $W^*W_{\beta_c} - \theta_- W^*W_{\beta_c}\theta_-$ and $W_{\beta_c} - \theta_- W_{\beta_c}\theta_-$ are not Hilbert–Schmidt.

We recall the following from Section 6.7. A Bogoliubov automorphism $\tau(v)$ on A^F is inner if and only if one of the following conditions hold:

$$1 - v \text{ is trace class and } \det v = 1, \tag{7.7.48}$$

$$1 + v \text{ is trace class and } \det(-v) = -1. \tag{7.7.49}$$

Moreover if (7.7.48) holds then $\tau(v)$ is even, and if (7.7.49) holds then $\tau(v)$ is odd. If a unitary v commutes with Γ, and $1 - v$ is trace class, then $\det(v) = \pm 1$. Moreover the map $\omega \to \det(1-\omega)$ is continuous on the trace class operators. We apply these considerations to the unitaries $\theta_- W^*W_\beta\theta_- W_\beta^* W$ and $\theta_- W_\beta\theta_- W$ for $\beta \ne \beta_c$.

Lemma 7.16

$$\det(\theta_- W^*W_\beta\theta_- W_\beta^* W) = \begin{cases} 1 & \beta_c < \beta \le \infty \\ -1 & 0 \le \beta < \beta_c \end{cases} \tag{7.7.50}$$

$$\det(\theta_- W_\beta\theta_- W_\beta^*) = \begin{cases} 1 & 0 \le \beta < \beta_c \\ -1 & \beta_c < \beta \le \infty \end{cases} \tag{7.7.51}$$

roof By Lemma 7.15, we see that

$$1 - \theta_- W^* W_\beta \theta_- W_\beta^* W = (W^* W_\beta - \theta_- W^* W_\beta \theta_-) W_\beta^* W \tag{7.7.52}$$

nd

$$1 - \theta_- W_\beta \theta_- W_\beta^* = (W_\beta - \theta_- W_\beta \theta_-) W_\beta^* \tag{7.7.53}$$

re trace class if $\beta \neq \beta_c$.

We now treat K_1^* and K_2 (or z_1 and z_2^*) as independent parameters. From 7.7.14) we see that W_β is norm continuous in the region $z_1^* \neq z_2$. Then we ave from (7.7.42), (7.7.43), (7.7.36), (7.7.39), (7.7.52), (7.7.53), and (7.7.47) hat $1 - \theta_- W^* W_\beta \theta_- W_\beta^* W$ and $1 - \theta_- W_\beta \theta_- W_\beta^*$ are continuous in z_1^* and z_2 in he trace class norm when $z_1^* \neq z_2$. Hence using continuity of the determinant, t is enough to compute the determinants in the cases $z_1^* = 0$, $z_2 > 0$, $(\beta = \infty)$, nd $z_2 = 0$, $z_1^* > 0$, $(\beta = 0)$, which is an easy exercise. □

roof of Theorem 7.6 (a). We now apply Lemma 7.12 to the automorphisms

$$\nu_\beta = \begin{cases} \tau(W^* W_\beta), & \beta > \beta_c \\ \tau(W_\beta), & 0 \leq \beta < \beta_c \end{cases}$$

sing Lemmas 7.12 and 7.15 and (7.7.48) to see that $\nu_\beta|_{A_+}$ extend to graded utomorphisms of A^P also denoted by ν_β. Then (7.7.25) follows from (7.7.23) nd (7.7.24). Finally, it is now clear using Corollary 7.11 that the automorphisms

$$\{\tau(W^* W_\beta|_{A+} \mid 0 \leq \beta < \beta_c\} \quad \text{and} \quad \{\tau(W_\beta)|_{A+} \mid \beta > \beta_c\}$$

o not extend to A^P. □

.8 Determinants and the index

n Section 7.7 two approaches were developed for understanding the structure of he states φ_β for the two-sided Pauli algebra A^P. The former gave necessary and ufficient criteria for deciding when a quasi-free Fock state φ_E on A^F gave an ven pure state φ_E^P on A^P such that $\varphi_{E|A_+}^P = \varphi_{E|A_+^F}$. This involved a discussion f when the restriction of two quasi-free Fock state φ_E and φ_F to the even algebra vere equivalent (in particular if $E = E_\beta$, $F = \theta_- E_\beta \theta_-$), and Hilbert–Schmidt stimates for $E - F$, and the $\mathbb{Z}/2$ index of $\sigma(E, F) = \dim E \wedge (1 - F) \bmod 2$. he latter gave criteria for when a Bogoliubov automorphism $\tau(U)$ could be ransformed to a graded automorphism on the Pauli algebra A^P. This involved race class estimates, for $1 - \theta_- U \theta_- U^*$, and a computation of determinants, et$(\theta_- U \theta_- U^*)$. Here we show how the computations of the index σ and deteriinant are related, and also how they are related to deciding when a restricted tate φ_S is a factor state, if $S = p_+ E p_+$, in terms of Hilbert–Schmidt estimates, nd the index: $\dim \operatorname{Ker}(2S - 1) \bmod 2$.

emma 7.17 *Let E be a basis projection, and U a unitary commuting with Γ uch that $1 - U$ is trace class. Then $\sigma(E, UEU^*) = \det U$.*

Proof It is enough to construct a canonical unitary U_0 commuting with Γ such that $1 - U_0$ is trace class $U_0 E U_0^* = UEU^*$, and $\sigma(E, U_0 E U_0^*) = \det U_0$. For then $U^* U_0$ commutes with the basis projection E, and so $\det U^* U_0 = 1$ or $\det U = \det U_0$. Then $\sigma(E, UEU^*) = \sigma(E, U_0 E U_0^*) = \det U_0 = \det U$. Consider the complex structures $J = i(1 - 2E)$, $J_1 = UJU^* = i(1 - 2E_1)$ if $E_1 = UEU^*$. Define $t = \left[(2 - JJ_1 - J_1 J)^{1/2} - K(2 + JJ_1 + J_1 J)^{1/2}\right]/2$ where K is the partial isometry in the polar decomposition of $J_1 J - JJ_1$, and $\ker K = \ker(J_1 J - JJ_1)$. Some algebra shows that $tJt^* = J_1$ on the orthogonal complement of $\ker K$. Now on $\ker K$, where J and J_1 are simultaneously diagonalizable, we have $V_{\sigma\mu} = \{x \mid Jx = \sigma i x, J_1 x = \mu i x\}$, $\sigma, \mu = \pm$. On $V_{++} \oplus V_{--}$ $J = J_1$, and $t = 1$ so that $tJt^* = J_1$ here again. It only remains to handle $V_{+-} \oplus V_{-+}$. But $V_{+-} = \mathrm{Range}((1 - E) \wedge E_1)$ and $V_{-+} = \mathrm{Range}(E \wedge (1 - E_1))$ which have the same dimension, and so let $S : V_{+-} \to V_{-+}$ be unitary and then $S_1 = \begin{pmatrix} 0 & S^* \\ S & 0 \end{pmatrix}$ on $V_{+-} \oplus V_{-+}$ takes $J = i \oplus (-i)$ to $J_1 = (-i) \oplus i$. Then $U_0 = t + S_1$ is a unitary, commuting with Γ, satisfying $U_0 J U_0^* = J_1$ or $U_0 E U_0^* = UEU^*$. Moreover, $1 - U_0$ is trace class, and since $\det(t) = 1$, we have $\det U_0 = (-1)^{\dim V_{+-}} = \sigma(E_0, U_0 E U_0^*)$. □

Remark From this one sees the connection between the index and determinant computations of Section 7.7. Note first that if E_1 and E_2 are basis projections with $E_1 - E_2$ Hilbert–Schmidt, then

$$\sigma(E_1, E_2) = \sigma(WE_1W^*, WE_2W^*) \tag{7.8.1}$$

for any unitary W commuting with Γ. We have $E_\beta = W_\beta^* E_0 W_\beta$ (7.7.22) and so

$$\begin{aligned}
\sigma(E_\beta, \theta_- E_\beta \theta_-) &= \sigma(W_\beta^* E_0 W_\beta, \theta_- W_\beta^* E_0 W_\beta \theta_-) \\
&= \sigma(E_0, W_\beta \theta_- W_\beta^* E_0 W_\beta \theta_- W_\beta^*) \text{ by (7.8.1)} \\
&= \sigma(E_0, \theta_- W_\beta \theta_- W_\beta^* E_0 W_\beta \theta_- W_\beta^* \theta_-) \text{ as } E_0 = \theta_- E_0 \theta_- \\
&= \det \theta_- W_\beta \theta_- W_\beta^* \quad \text{for } \beta \neq \beta_c
\end{aligned}$$

by Lemma 7.17, and $1 - \theta_- W_\beta \theta_- W_\beta^*$ being trace class for $\beta \neq \beta_c$ by Lemma 7.12. Note also that, with $W = W_\infty$,

$$\begin{aligned}
\det(\theta_- W_\beta \theta_- W_\beta^*) &= \det(W^* \theta_- W_\beta \theta_- W_\beta^* W) \\
&= \det(W^* \theta_- W \theta_- \theta_- W^* W_\beta \theta_- W_\beta^* W) \\
&= \det(W^* \theta_- W \theta_-) \det(\theta_- W^* W_\beta \theta_- W_\beta^* W) \\
&= -\det(\theta_- W^* \theta_- W_\beta^* W)
\end{aligned}$$

as $\det(W^* \theta_- W \theta_-) = -1$. This relates Lemma 7.16 with (7.7.13).

Remark Note also that joint continuity of the relative index $\sigma(E_1, E_2)$ is closely related to Lemma 7.17, at least for $E_1 - E_2$ Hilbert–Schmidt and for the trace class norm topology. (In the case of the Ising model, this is really a

one needs by the estimates of Lemma 7.15.) First observe, that if E_i are basis projections, with $E_i - E_j$ Hilbert–Schmidt, then

$$\sigma(E_1, E_3) = \sigma(E_1, E_2)\sigma(E_2, E_3)\,. \tag{7.8.2}$$

To check this, it is convenient to have a little more notation. If E is a basis projection, let π_E denote the GNS representation of the quasi-free state φ_E and $\pi_E^{\pm}$ the *disjoint* representations of A_+ obtained by restricting π_E to even and odd tensors respectively Lemma 6.24. Then by Theorem 6.30, $\sigma(E_i, E_j) = 1$ if and only if $\pi_{E_1}^+ \sim \pi_{E_j}^+$. Moreover $\sigma(E_i, E_j) = -1$, if and only if $\pi_{E_i}^+ \sim \pi_{E_j}^-$ if and only if $\pi_{E_i}^- \sim \pi_{E_j}^+$. Then (7.8.2) follows from these observations.

Now take E_0, F_0 basis projections with $E_0 - F_0$ Hilbert–Schmidt. We show that $(E, F) \to \sigma(E, F)$ is continuous with respect to the trace class norm at E_0, F_0. From (7.8.2),

$$\sigma(E, F) = \sigma(E, E_0)\sigma(E_0, F_0)\sigma(F_0, F) \tag{7.8.3}$$

and it is enough to show that $\sigma(F_0, F) = 1$ if F is close to F_0 in the trace class norm. Now if F, F_0 are basis projections with $\|F_0 - F\| < 1$, and $\|F_0 - F\|_1 < \infty$, we claim that there exists a unitary U commuting with Γ such that $F = UF_0U^*$ and

$$\|U - 1\| < 2\|F - F_0\|\,, \quad \|U - 1\|_1 \leq \|F - F_0\|_1\,. \tag{7.8.4}$$

Let v be the partial isometry in the polar decomposition of EF_0F, then $\Gamma v \Gamma$ is the partial isometry in the polar decomposition of $\Gamma F F_0 F \Gamma = (1-F)(1-F_0)(1-F)$. Then $U = v + \Gamma v \Gamma$ is a unitary commuting with Γ and has all the other properties required. Hence $\sigma(F_0, F) = \sigma(F_0, UF_0U^*) = \det U$ by Lemma 7.17. But for F close enough to F_0 in the trace class norm, $\det(U) = 1$ by (7.8.4) and the continuity of the determinant.

Next we show how our knowledge of the value of the index $\sigma(E_\beta\theta_- E_\beta\theta_-) = (-1)^{\dim(1-E_\beta)\wedge(\theta_- E_\beta \theta_-)}$ can be improved by calculating the subspace $(1 - E_\beta) \wedge \theta_- E_\beta \theta_-$ directly. As we have already seen in Section 7.7, this information is not needed to show the phase transition, but will be employed in Section 7.12 to compute the spontaneous magnetization.

Proposition 7.18 $(1 - E_\beta) \wedge \theta_- E_\beta \theta_- = 0$ *if* $\beta < \beta_c$*, and spanned by*

$$2^{-\frac{1}{2}}\left((1 + U_\beta)(q_- e^{-i\delta} q_-)^{-1} e_0, (1 - U_\beta)(q_- e^{-i\delta} q_-)^{-1} e_0\right) \quad \text{for } \beta > \beta_c\,. \tag{7.8.5}$$

Proof The unitary $v = 2^{-\frac{1}{2}}\begin{pmatrix} 1 & 1 \\ 1 & -1 \end{pmatrix}$ commutes with θ_- and $vE_\beta v = \begin{pmatrix} 1 & -U_\beta^* \\ -U_\beta & 1 \end{pmatrix}$. We will compute $(1 - vE_\beta v) \wedge (\theta_- vE_\beta v\theta_-)$. A vector $h = (f, g)$ is in the range of this projection if and only if

$$vE_\beta vh = 0\,, \quad v(1 - E_\beta)v\theta_- h = 0$$

or equivalently, $f = U_\beta^* g$, $u_- f = -U_\beta^* u_- g$, or $(U_\beta^* + u_- U_\beta^* u_-)g = 0$, and $(U_\beta + u_- U_\beta u_-)f = 0$. These equations are equivalent to

$$q_+ U_\beta^* q_+ g = q_- U_\beta^* q_- g = 0 \tag{7.8.6}$$

$$q_+ U_\beta q_+ f = q_- U_\beta q_- f = 0\,. \tag{7.8.7}$$

Now above $\beta_c, U_\beta = Ue^{-i\delta}$ where $e^{i\delta}$ has winding number zero, and U is the bilateral shift, or $e^{i\theta}$. Below β_c, U_β has winding number zero. Hence

$$q_\pm e^{-i\delta} q_\pm \quad \text{for} \quad \beta > \beta_c$$

and

$$q_\pm U_\beta q_\pm \quad \text{for} \quad \beta < \beta_c$$

are invertible (see Chapter 3, Notes). Now $q_+ U q_+$ is injective, and $q_- U q_-$ has kernel spanned by e_0. Hence

$$q_- U_\beta q_- = q_- U q_- e^{-i\delta} q_- \quad \beta > \beta_c$$

has kernel spanned by $(q_- e^{-i\delta} q_-)^{-1} e_0$, $q_- U_\beta q_-$ is injective for $\beta < \beta_c$ and $q_+ U_\beta q_+$ is injective for $\beta < \beta_c$.

Hence $(1 - vE_\beta v) \wedge (\theta_- v E_\beta v \theta_-) = 0$ if $\beta < \beta_c$, and is spanned by

$$2^{-\frac{1}{2}} \left((q_- e^{-i\delta} q_-)^{-1} e_0, U_\beta (q_- e^{-i\delta} q_-)^{-1} e_0 \right) \quad \text{for } \beta > \beta_c\,. \tag{7.8.8}$$

Consequently, $(1 - E_\beta) \wedge \theta_- E_\beta \theta_- = 0$ if $\beta < \beta_c$, and spanned by (7.8.5) for $\beta > \beta_c$. □

In computing the magnetization in Section 7.13, we will have need for a more explicit expression for $\left[q_- e^{-i\delta} q_-\right]^{-1}$.

7.9 The half lattice

Next we relate properties of the state φ_β of the full Pauli algebra $A^P = \bigotimes_{-\infty}^{\infty} M_2$ with that of $\overline{\varphi}_\beta$ on the half Pauli algebra $A[1, \infty) = \bigotimes_1^\infty M_2$. More generally, let E be a basis projection on H, and $S = p_+ E p_+$ (where $p_\pm = (1 \pm \theta_-)/2$ as usual). Now if S is a self adjoint contraction, with $S + \Gamma S \Gamma = 1$, then the quasi-free state φ_S is not factorial unless [Exercise 6.9]:

$$S^{\frac{1}{2}}(1 - S)^{\frac{1}{2}} \text{ is Hilbert} - \text{Schmidt} \tag{7.9.1}$$

$$\ker(2S - 1) \text{ is odd dimensional}\,. \tag{7.9.2}$$

Moreover if E is a basis projection, φ_E on A^P is not pure if and only if [Theorem 6.25]:

$$E - \theta_- E \theta_- \text{ is Hilbert} - \text{Schmidt} \tag{7.9.3}$$

$$\sigma(E, \theta_- E \theta_-) = -1\,. \tag{7.9.4}$$

Proposition 7.19 *If E is a basis projection of $H = l^2 \oplus l^2$, then φ_E is a pure state on A^P if and only if φ_S is a factorial state on $A^F[1,\infty)$ where $S = p_+ E p_+$.*

Proof It is easy to see that (7.9.1) and (7.9.3) are equivalent. If $x \in H$, let $x_\pm = p_\pm x$, and $E_{\sigma\mu} = p_\sigma E p_\mu$ if $\sigma, \mu = \pm$. Then x is in the range of $E \wedge (1 - \theta_- E \theta_-)$ if and only if

$$E_{++}x_+ + E_{+-}x_- = x_+ , \quad E_{-+}x_+ + E_{--}x_- = x_-$$

and $E_{++}x_+ - E_{+-}x_- = 0$, $-E_{-+}x_+ + E_{--}x_- = 0$, i.e. if and only if $2E_{++}x_+ = x_+$, $x_- = 2E_{-+}x_+$. But $S = E_{++}$ and so

$$\dim E \wedge (1 - \theta_- E \theta_-) = \dim \ker(2S - 1) \tag{7.9.5}$$

and so the proposition follows. □

Remark If $\mathcal{S}$ denotes set of self adjoint contractions on S on $l^2[1,\infty) \oplus l^2[1,\infty)$ with $S + \Gamma S \Gamma = 1$, $S^{1/2}(1-S)^{1/2}$ Hilbert–Schmidt, then the map $j : S \to \dim \ker(2S-1) \bmod 2$ from $\mathcal{S}$ to $\mathbb{Z}/2$ is continuous in the norm topology, because $j(S) = \sigma(P_S, \theta_- P_S \theta_-)$ by (7.9.5), where P_S is the basis projection in (6.6.11), and because of the continuity of σ and $S \to P_S$.

Next we show how the automorphism method of Section 7.7 could be adopted to the one-sided or half lattice. We have $\theta_- = u_- \oplus u_-$, $p_\pm = (1 \pm \theta_-)/2$, $q_\pm = (1 \pm u_-)/2$ as usual. Now $E_\beta = W_\beta E_0 W_\beta^*$, and so

$$p_+ E_\beta p_+ = p_+ W_\beta E_0 W_\beta^* p_+ = (p_+ W_\beta p_+)(p_+ E_0 p_+)(p_+ W_\beta^* p_+) + \text{trace class} \tag{7.9.6}$$

by Lemma 7.12. But

$$p_+ W_\beta p_+ = \frac{i}{2} \begin{pmatrix} q_+ - q_+ U_\beta^* q_+ & q_+ + q_+ U_\beta^* q_+ \\ -q_+ - q_+ U_\beta^* q_+ & -q_+ + q_+ U_\beta^* q_+ \end{pmatrix} \tag{7.9.7}$$

and is invertible for $0 \le \beta < \beta_c$, as in that range U_β has winding number zero and so

$$-\frac{i}{2} \begin{pmatrix} q_+ - (q_+ U_\beta^* q_+)^{-1} & -q_+ - (q_+ U_\beta^* q_+)^{-1} \\ q_+ + (q_+ U_\beta^* q_+)^{-1} & -q_+ + (q_+ U_\beta^* q_+)^{-1} \end{pmatrix} \tag{7.9.8}$$

is seen to be an inverse for $p_+ W_\beta p_+$ if $0 \le \beta < \beta_c$. Let $X = p_+ W_\beta p_+$ with polar decomposition $X = V|X|$. Then $X - p_+$ trace class implies $X^*X - p_+$ is trace class, and so $(|X| - p_+) = (|X|^2 - p_+)(|X| + p_+)^{-1}$ is trace class. Hence the quasi-free states for $|X| p_+ E_0 p_+ |X|$ and $p_+ E_0 p_+$ are equivalent, and by (7.9.6), the quasi-free states for $p_+ E_\beta p_+$ and $X p_+ E_0 p_+ X^*$ are equivalent. Hence $\varphi_{S_\beta} = \varphi_{p_+ E_\beta p_+}$ is equivalent to $\varphi_{|X| E_0 |X|} \circ \tau(V)$ which is equivalent to $\varphi_{p_+ E_0 p_+} \circ \tau(V) = \varphi_S \circ \tau(V)$. Thus φ_{S_β} is equivalent to $\varphi_{S_0} \circ \tau(V)$ for a unitary V, when $0 < \beta < \beta_c$. Similarly φ_{S_β} is equivalent to $\varphi_{S_\infty} \circ \tau(V)$ for a unitary V when $\beta > \beta_c$.

We conclude this section with some remarks on the Kramers–Wannier endomorphism for one-sided systems. As we have already noted in Section 6.5, the half Pauli algebra $A^P[1,\infty) = \bigotimes_1^\infty M_2$ and the half Fermion algebra $A^F[1,\infty) = C^*(c_i \mid i \in \mathbb{N})$ can be naturally identified since no infinite tail appears in the corresponding Jordan–Wigner correspondence. Let us abbreviate either algebra by B. The Kramers–Wannier endomorphism $\kappa : B_+ \to B_+$

$$\kappa(\sigma_z^j) = \sigma_x^j \sigma_x^{j+1}\,, \quad \kappa\left(\sigma_x^j \sigma_x^{j+1}\right) = \sigma_z^{j+1} \quad j \geq 1 \tag{7.9.9}$$

thus extends to an endomorphism κ of B such that

$$\kappa(c_j) = \frac{i}{2}(c_j^* + c_{j+1}^* - c_j + c_{j+1})\,, \quad j \geq 1\,. \tag{7.9.10}$$

In the Pauli picture, κ becomes

$$\kappa(\sigma_x^j) = \sigma_x^j \sigma_x^{j+1}\,, \quad \kappa(\sigma_x^j) = i\textstyle\prod_{k=1}^j \sigma_z^k \sigma_x^1\,. \tag{7.9.11}$$

Thus κ is graded (it is in fact a Bogoliubov endomorphism) on B, but $\kappa^2 \neq \sigma$, if σ is the unilateral shift on $B = \bigotimes_1^\infty M_2$. In fact

$$\kappa^2(\sigma_x^j) = \sigma_z^1 \sigma_x^{j+1}\,, \quad \kappa^2(\sigma_y^j) = \sigma_z^1 \sigma_y^{j+1}\,, \quad \kappa^2(\sigma_z^j) = \sigma_z^{j+1}$$

but of course $\kappa^2|B_+ = \sigma|B_+$. The question then arises of whether there does exist an endomorphism μ of B such that $\mu^2 = \sigma$, and $\mu|B_+ = \kappa|B_+$. Note that there certainly does not exist a *graded* endomorphism μ of B such that $\mu^2 = \sigma$, and $\mu|B_+ = \kappa|B_+$ for the following reason. Suppose such a μ exists. Then from $\overline{\varphi}_0 \circ \kappa = \overline{\varphi}_\infty$, $\overline{\varphi}_\infty \circ \kappa = \overline{\varphi}_0$ on B_+ we would have $\overline{\varphi}_0 \circ \mu = \overline{\varphi}_\infty$, $\overline{\varphi}_\infty \circ \mu = \overline{\varphi}_0$ on B. Now $\overline{\varphi}_\infty = (\overline{\varphi}_\infty^+ + \overline{\varphi}_\infty^-)/2$, and so $\overline{\varphi}_0 = \overline{\varphi}_\infty \circ \mu = (\overline{\varphi}_\infty^+ \circ \mu + \overline{\varphi}_\infty^- \circ \mu)/2$. Since $\overline{\varphi}_0$ is pure, we deduce $\overline{\varphi}_\infty^+ \circ \mu = \overline{\varphi}_\infty \circ \mu$, and so $\overline{\varphi}_\infty^+ = \overline{\varphi}_\infty^+ \circ \sigma = \overline{\varphi}_\infty^- \circ \mu^2 = \overline{\varphi}_\infty^- \circ \mu^2 = \overline{\varphi}_\infty^- \circ \sigma = \overline{\varphi}_\infty^-$ as $\overline{\varphi}_\infty^\pm$ are shift invariant. This contradicts $\overline{\varphi}_\infty^+ \neq \overline{\varphi}_\infty^-$.

Here is another viewpoint of the unilateral order–disorder transformation. Let $\mathbb{Z}/2$ act on $\mathbb{C}^2$ by transposition α. Then the crossed product $\mathbb{C}^2 \rtimes (\mathbb{Z}/2) \cong M_2$, and using Takesaki duality

$$\left(\left((\mathbb{C}^2 \rtimes_\alpha (\mathbb{Z}/2)\right) \rtimes_{\hat{\alpha}} (\mathbb{Z}/2)^\wedge\right) \rtimes_{\hat{\hat{\alpha}}} (\mathbb{Z}/2)^{\wedge\wedge}\Big) \cong M_2 \otimes M_2$$

where $((\mathbb{Z}/2)^\wedge, \hat{\alpha})$, $\left((\mathbb{Z}/2)^{\wedge\wedge}, \hat{\hat{\alpha}}\right)$ are the dual, double dual actions respectively. In this way

$$\bigotimes_1^\infty M_2 \cong \cdots (((\mathbb{C}^2 \rtimes (\mathbb{Z}/2)) \rtimes (\mathbb{Z}/2)) \rtimes (\mathbb{Z}/2)) \cdots$$

Now $\mathbb{C}^2 \rtimes (\mathbb{Z}/2)$ is generated by unitaries u_1, u_2 satisfying $u_i^2 = 1$, $u_1 u_2 = -u_2 u_1$ and $\bigotimes_1^\infty M_2$ is generated by self adjoint unitaries $u_1, u_2, \ldots$ satisfying

$$u_i u_j = u_j u_i\,, \quad |i - j| > 1\,, \quad u_i u_{i+1} = -u_{i+1} u_i \tag{7.9.12}$$

for $i, j \geq 1$. We concretely identify B with $C^*(u_j \mid j \in \mathbb{N})$ by

$$\{\sigma_x^j\}_{j=1}^{\infty} = \{u_1, u_1u_3, u_1u_3u_5, u_1u_3u_5u_7, \ldots\} . \tag{7.9.13}$$

$$\{\sigma_z^j\}_{j=1}^{\infty} = \{u_2, u_4, u_6, u_8, \ldots\} . \tag{7.9.14}$$

By the universal property of a crossed product, there exists a unique endomorphism $\nu : B \to B$ which sends u_i to u_{i+1}. Then $\nu|B_+ = \kappa|B_+$ since $\{\sigma_x^j\sigma_x^{j+1}\}_{j=1}^{\infty} = \{u_3, u_5, u_7, \ldots\}$. Note that

$$\nu\left(\sigma_x^j\right) = \textstyle\prod_{i=1}^{j} \sigma_z^i \tag{7.9.15}$$

so that ν is not graded, in fact $\nu(B) \subset B_+$. Moreover $\nu^2(\sigma_x^j) = \sigma_x^1\sigma_x^{j+1}$ so that $\nu^2 \neq \sigma$.

Remark The ν constructed above is a high temperature–low temperature duality between free boundary conditions and periodic boundary conditions, in that

$$\overline{\varphi}_{\infty}^{+} = \overline{\varphi}_0 \circ \nu \quad \text{on } B . \tag{7.9.16}$$

To get a bilateral version of this duality, we need to make sense of the bilateral version of (7.9.15), namely

$$\nu(\sigma_x^j) = \textstyle\prod_{-\infty}^{j} \sigma_z^i . \tag{7.9.17}$$

Now φ_0 is θ_- invariant, and so θ_- is implemented by a unitary $U(\theta_-)$ in the GNS space of φ_0. We can thus define ν from A^P into the even part of $\pi_0(A^P)''$ where π_0 denotes the GNS representation of φ_0 by

$$\nu(\sigma_x^j) = U(\theta_-)\textstyle\prod_1^j \sigma_z^i \tag{7.9.18}$$

$$\nu(\sigma_z^j) = \sigma_x^j\sigma_x^{j+1} . \tag{7.9.19}$$

Then $\nu|A_+^P$ is the Kramers–Wannier endomorphism and

$$\varphi_{\infty}^{+} = \varphi_0 \circ \nu \quad \text{on } A^P . \tag{7.9.20}$$

We shall return to this aspect of duality in Section 7.11.

7.10 The Temperley–Lieb algebra

The Ising model, the Pauli algebra, Pauli matrices and relations (7.9.12)–(7.9.14) can be generalized to a Q-state standard Potts model, replacing the graph of Fig. 7.6 by that of Fig. 7.7.

The transfer matrix formalism for the Q-state Potts Hamiltonian

$$H(\sigma) = -\textstyle\sum_{i,j \ n \cdot n} J\delta(\sigma_i, \sigma_j) \tag{7.10.1}$$

on the two dimensional lattice $\mathbb{Z}^2$ leads to the following algebraic set up. Define $Q \times Q$ unitary matrices U, V by

$$U = \begin{bmatrix} 1 & 0 & 0 & \cdots & 0 \\ 0 & q & 0 & \cdots & 0 \\ 0 & 0 & q^2 & \cdots & 0 \\ \cdot & \cdot & \cdot & & \cdot \\ \cdot & \cdot & \cdot & & \cdot \\ \cdot & \cdot & \cdot & & \cdot \\ 0 & 0 & 0 & \cdots & q^{Q-1} \end{bmatrix} \qquad V = \begin{bmatrix} 0 & 1 & 0 & \cdots & 0 & 0 \\ 0 & 0 & 1 & \cdots & 0 & 0 \\ 0 & 0 & 0 & \cdots & 0 & 0 \\ \cdot & \cdot & \cdot & & \cdot & \cdot \\ \cdot & \cdot & \cdot & & \cdot & \cdot \\ \cdot & \cdot & \cdot & & \cdot & \cdot \\ 0 & 0 & 0 & \cdots & 0 & 1 \\ 1 & 0 & 0 & \cdots & 0 & 0 \end{bmatrix} \tag{7.10.2}$$

where $q = e^{2\pi i/Q}$, so that $VU = qUV$, and U, V generate M_Q. Define a sequence of unitaries W_i in $M_Q \otimes M_Q \otimes M_Q \otimes \cdots$ by

$$W_{2i+1} = 1 \otimes \cdots 1 \otimes U^{-1} \otimes U \otimes 1 \otimes \cdots \tag{7.10.3}$$

where the factor U appears as the $(i+1)$th factor and

$$W_{2i} = 1 \otimes \cdots 1 \otimes V \otimes 1 \otimes \cdots \tag{7.10.4}$$

with V appearing as the ith factor so that

$$W_j^Q = 1\,, \quad W_i W_{i+1} = q W_{i+1} W_i\,, \quad W_i W_j = W_j W_i\,, \quad |i-j| > 1\,. \tag{7.10.5}$$

The symmetric group S_Q acts on $\mathbb{C}^Q$ by permuting basis vectors and so induces a product action on $F_Q = \otimes_{\mathbb{N}} M_Q$. In particular, there is a $\mathbb{Z}/Q$ action on F_Q by $\bigotimes_i \mathrm{Ad}(V) = \prod_i \mathrm{Ad}(W_{2i})$. The Temperley–Lieb operators in F_Q are the spectral projections of W_i corresponding to eigenvalue 1, namely

$$e_i = \left(1 + W_i + W_i^2 + \cdots + W_i^{Q-1}\right)/Q\,. \tag{7.10.6}$$

In particular

$$e_{2i-1} = \cdots 1 \otimes \underset{i-(i+1)\text{ factors}}{g} \otimes 1 \otimes \cdots \tag{7.10.7}$$

$$e_{2i} = \cdots 1 \otimes \underset{i\text{th factor}}{f} \otimes 1 \otimes \cdots \tag{7.10.8}$$

where g, f are projections in $M_Q \otimes M_Q$ and M_Q respectively given by

$$g = \textstyle\sum_{i=1}^{Q} E_{ii} \otimes E_{ii}\,, \qquad f = \sum_{i,j=1}^{Q} E_{ij}/Q \tag{7.10.9}$$

if $\{E_{ij} \mid i, j = 1, 2, \ldots, Q\}$ are matrix units for M_Q. In the Ising case:

$$e_{2i} = (1+\sigma_z^i)/2\,, \quad e_{2i+1} = (1+\sigma_x^i\sigma_x^{i+1})/2\,. \tag{7.10.10}$$

The transfer matrix for the two dimensional Q-state Potts model is then (cf. (7.4.9)) described as

$$V = \exp K_2 \sum e_{2i+1}\,, \qquad W = \exp K_1^* \sum e_{2i} \tag{7.10.11}$$

with Kramers–Wannier duality being the shift $e_i \to e_{i+1}$. Here the family $\{e_i\}$ satisfy the relations that non-nearest neighbours commute:

$$e_i e_j = e_j e_i \qquad |i-j| > 1 \tag{7.10.12}$$

whilst nearest neighbours satisfy

$$e_i e_{i\pm 1} e_i = \tau e_i \tag{7.10.13}$$

where $\tau^{-1} = Q$.

The transfer matrix formalism in the ABF for the Dynkin diagrams, A_ℓ, is similarly described with the aid of projections $\{e_i\}$ satisfying the same relations (7.10.12) and (7.10.13) with $\tau^{-1} = 4\cos^2\pi/(\ell+1)$, see Section 11.8.

We will see in Section 9.3, that an infinite family $\{e_i\}$ of projections satisfying (7.10.11) exist if and only if

$$\tau^{-1} \in [4,\infty) \cup \{4\cos^2\pi/(\ell+1) \mid \ell = 3,4,\cdots\}\,.$$

Note that $2 = 4\cos^2(\pi/4)$, $3 = 4\cos^2(\pi/6)$.

In the Q-state Potts model the family $\{e_i\}$ is S_Q invariant, and the $\{W_i\}$ generate the $\mathbb{Z}/Q$ fixed point algebra of F_Q:

$$C^*(e_i) \subset F_Q^{S_Q} \subset F_Q^{\mathbb{Z}/Q} = C^*(W_i)\,. \tag{7.10.14}$$

Moreover

$$\text{for } Q=2: \quad C^*(e_i) = F_2^{S_2} = F_2^{\mathbb{Z}/2} \tag{7.10.15}$$

$$\text{for } Q=3: \quad C^*(e_i) = F_3^{S_3} \underset{\neq}{\subset} F_3^{\mathbb{Z}/3} \tag{7.10.16}$$

and both inclusions in (7.10.14) are strict for $Q \geq 4$. This formulation is related to that of generalized Clifford algebras as follows. The formulae

$$\Gamma_i = W_1 W_2 \cdots W_i\,, \qquad W_i = \Gamma_{i-1}^{-1}\Gamma_i \tag{7.10.17}$$

is the generalized Jordan–Wigner transformation required between (7.10.5) and

$$\Gamma_i^n = 1\,, \quad \Gamma_i\Gamma_j = q\Gamma_j\Gamma_i\,, \quad i<j\,. \tag{7.10.18}$$

(For the bilateral infinite sequences, one treats the infinite tail as in Section 6.5.)

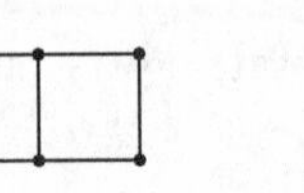
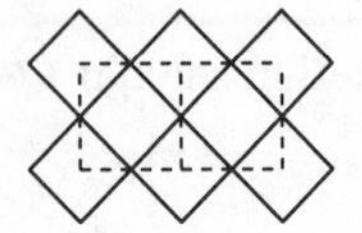

(a) lattice L (b) medial lattice L' (c) braided dual

FIG. 7.12.

(Temperley and Lieb 1971) showed that at criticality, the local Potts partition function on a planar graph L can be identified with the partition function of an ice type or six-vertex model on a dual or medial graph L'. This identification corresponds algebraically to a representation of the Temperley–Lieb relations in a Pauli algebra F_2:

$$e_n \to (1-t)e^n_{++}e^{n+1}_{--} + te^n_{--}e^{n+1}_{++} + [t(1-t)]^{1/2}[e^n_{+-}e^{n+1}_{-+} + e^n_{-+}e^{n+1}_{+-}] \quad (7.10.19)$$

where e^n_{ij} are matrix units in the nth factor of $\bigotimes_{\mathbb{Z}} M_2$, and $t(1-t) = Q^{-1} = \tau$. The identification is as follows. Consider the Potts partition function on the square lattice L of Fig 7.12(a):

$$Z = \sum_\sigma \exp\left(K \sum_{i,j} \delta(\sigma_i, \sigma_j)\right) \quad (7.10.20)$$

where the Hamiltonian $H(\sigma) = -\sum_{i,j} J\delta(\sigma_i, \sigma_j)$ is summed over nearest neighbours in L, and the state σ is a choice of colours in an indexing set I of cardinality Q. We write $\exp K\delta = 1 + \nu\delta$, $\nu = e^K - 1$, and

$$Z = \sum_\sigma \prod_{i,j} (1 + \nu\delta(\sigma_i, \sigma_j)) = \sum_\sigma \sum_G \prod_{(i,j)\in G} \nu\delta(\sigma_i, \sigma_i) \quad (7.10.21)$$

and the final summation is over all subsets G of the set of edges $L^{(1)}$ of L. On each such subgraph G of L, only states which are constant need be considered, so that $Z = \sum_G q^c \nu_1^{\ell_1} \nu_2^{\ell_2}$ where c is the number of components of G, and $\nu_i = e^{K_i} - 1$ are the horizontal and vertical weights for $i = 1, 2$, and there are ℓ_1 and ℓ_2 horizontal and vertical edges in G respectively. This decomposition has an interpretation in terms of the medial lattice L' of Fig. 7.12(b). Shade the regions of the dual lattice L', as in Fig. 7.12(b), in a chessboard manner, so that the outermost infinite region is unshaded. Alternatively, replace each local shading of Fig. 7.13(a) by the over-undercrossing picture of Fig. 7.13(b) to obtain a knot diagram as in Fig. 7.12(c). Then when an edge (i, j) is in a graph G we join up the shaded region, and when it is not we separate them, as in Fig. 7.14(a), so that the medial lattice becomes decomposed into shaded regions, each one corresponding to an original component of the graph G. Putting orientations on the resulting boundary polygons leads to eight possible local configurations as in Fig 7.15 but only six possible local configurations after recomposing the vertex in Fig

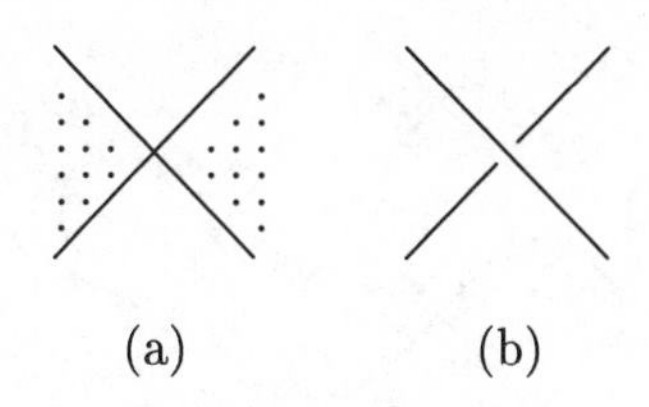

FIG. 7.13. From shading to crossing

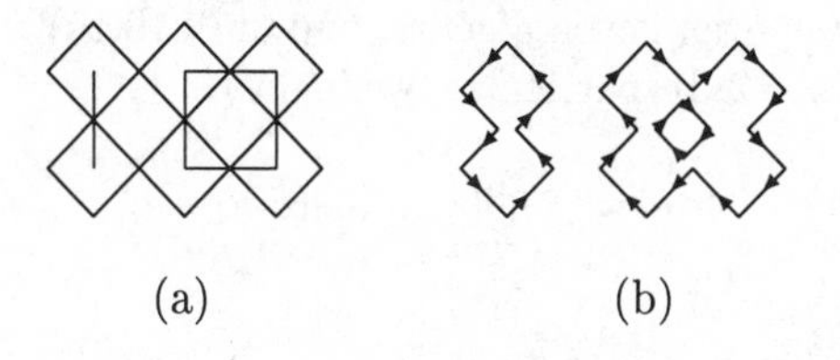

FIG. 7.14. From subgraph G of L to configuration in 6-vertex model on L'

7.16. Thus the Potts partition function on L is equivalent to that of a six-vertex model on L', with states being distributions of arrow orientations, labelled $+, -$ on edges, such that each internal vertex has two incoming and two outgoing arrows with suitable Boltzmann weights. The six-vertex Boltzmann weights, or face operators, are indeed given by the representation (7.10.19).

This representation exists for all $\tau^{-1} = Q > 0$ (not necessary integral Q) but is a $*$-representation only if $Q \geq 4$. It was independently rediscovered by (Pimsner and Popa 1986a) who used it to express the Kramers–Wannier high temperature–low temperature duality $\kappa_Q : e_i \to e_{i+1}$ as a non-commutative Bernoulli shift in F_2 if $Q \geq 4$, i.e. $\left((\bigotimes_{\mathbb{Z}} M_2)''^{\mathrm{T}}, \sigma\right) = \{\{e_n\}'', \kappa\}$, where the von Neumann algebra $(\bigotimes_{\mathbb{Z}} M_2)''$ is generated in the Powers trace with weights $\{t, 1-t\}$, and σ arises from the bilateral shift on $\bigotimes_{\mathbb{Z}} M_2$.

Remark We have already noted in (6.5.1) and (6.5.2) how to identify the even

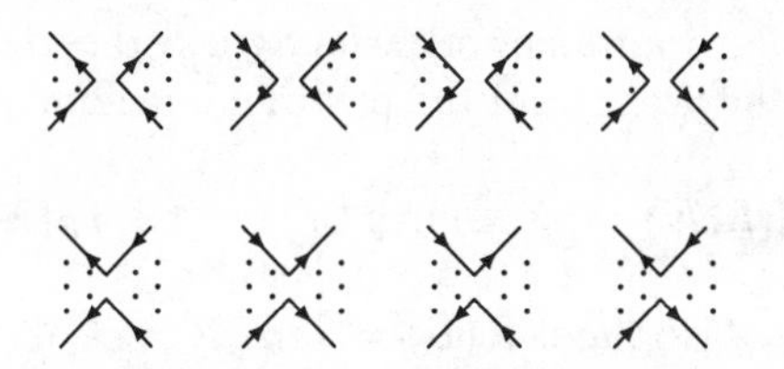

FIG. 7.15. Eight possible local configurations

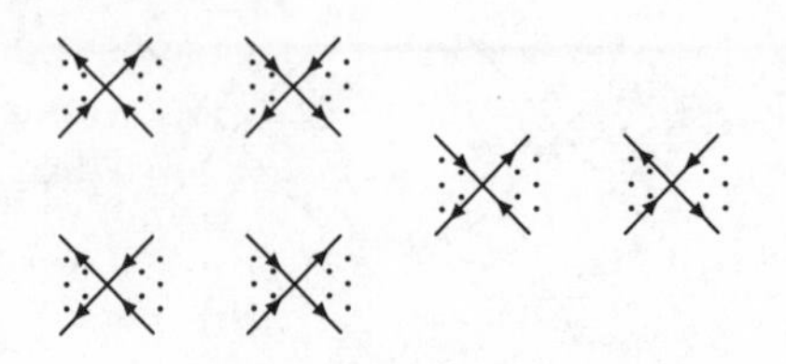

FIG. 7.16. Six-vertex model configurations

Pauli algebra with the even Fermi algebra, and in (7.10.10) with the Temperley–Lieb algebra for $\tau^{-1} = 2$. In particular, we identify

$$e_{2j-1} = c_j^* c_j \qquad e_{2j} = \left[1 + (c_j - c_j^*)(c_{j+1} + c_{j+1}^*)\right]/2 \tag{7.10.22}$$

and

$$c_j^* c_j = e_{2j-1}\,, \qquad c_j^* c_{j+1} = e_{2j-1}(1 - 2e_{2j})e_{2j+1}\,. \tag{7.10.23}$$

We also have already noted in Example 2.23 that the gauge invariant Fermion or observable algebra could be described in terms of generators and relations. This means that we can map the observable algebra into the Temperley–Lieb algebra by sending

$$c_j^* c_j \to e_{2j-1}\,, \quad c_j^* c_{j+1} \to e_{2j-1}(1 - \tau^{-1} e_{2j})e_{2j+1}$$

since if $\{e_i\}$ satisfy (7.10.12) and (7.10.13) then

$$g_j \equiv e_{2j-1}\,, \qquad v_j \equiv e_{2j-1}(1 - \tau^{-1} e_{2j})e_{2j+1} \tag{7.10.24}$$

satisfy the defining relations for the observable algebra.

7.11 High temperature–low temperature duality

We systematically look at high temperature–low temperature duality, first in the classical commutative framework of correlation functions relating $\langle\ \rangle_\beta^\pm$, $\beta > \beta_c$ with $\langle\ \rangle_{\beta^*}^{\text{free}}$ where $\beta^* < \beta_c$ is the dual inverse temperature, and then in the non commutative Pauli algebra framework relating $\varphi_\beta^\pm$, $\beta > \beta_c$ with φ_{β^*}, $\beta^* < \beta_c$.

Consider a finite lattice Λ, and the partition function

$$Z(\Lambda, K) = \textstyle\sum_{X\subset\Lambda} \exp\left(\sum_B \beta K(B)\sigma_B(X)\right)$$

where $H(\sigma) = -\sum_A J_A\sigma_A$, and $K(B) = \beta J_B$.

We let $\mathcal{B}$, the *bonds* of the interaction, denote the support of K, i.e $\mathcal{B} = \{A \subset \Lambda \mid K(A) \neq 0\}$. In the Ising model we will usually just have nearest neighbour bonds, and possibly singleton bonds at the boundary (with a magneti field H, $b^{\pm} = \pm\sum_{x\in\partial\Lambda} H\sigma_x$, and then let $H \to \infty$ to get $\pm$ boundary conditions). For the moment, however, we can work more generally. To obtain a high temperature expansion of a partition function, we introduce the homomorphism $\pi : P(\mathcal{B}) \to \overline{\mathcal{B}} \subset P(\Lambda)$, where $\overline{\mathcal{B}}$ is the *interaction subgroup* generated by the bonds $\mathcal{B}$, and $\pi(B_1, \ldots, B_n) = B_1 \cdots B_n$, where $P(\Lambda)$, $P(\mathcal{B})$ are groups under symmetric difference as in Section 7.2. We let $\mathcal{K} = \operatorname{Ker}\pi$, the *high temperature group* or group of *closed graphs*. Thus a set of bonds $\beta = (B_1, \ldots, B_n)$ lies in $\mathcal{K}$ if and only if, for each $x \in \Lambda$, there is an even number of bonds B_i in β containing x. Then using $\exp(K\sigma) = \cosh K[1 + \sigma \tanh K]$ we find the *high temperature expansion*:

$$Z(\Lambda, K) = 2^{|\Lambda|} \prod_{B\in\mathcal{B}} \cosh K(B) \sum_{\beta\in\mathcal{K}} \prod_{A\in\beta} \tanh K(A) \tag{7.11.1}$$

To obtain the low temperature expansion, we need another group homomorphism, this time $\theta : P(\Lambda) \to P(\mathcal{B})$, defined by $X \to \{B \in \mathcal{B} \mid \sigma_X(B) = -1\}$. The *low temperature group* Γ is the image of θ, or the group of low temperature graphs. The high and low temperature groups are not unrelated. In fact $\langle\beta, \theta(X)\rangle_{\mathcal{B}} = \langle\pi(\beta), X\rangle_{\Lambda}$ where $\langle Y, X\rangle = \sigma_Y(X)$, $\langle\beta, \beta'\rangle = \sigma_\beta(\beta')$ are the identifications $P(\Lambda)^\wedge \cong P(\Lambda)$, $P(\mathcal{B})^\wedge \cong P(\mathcal{B})$ as in Section 7.2. Thus $\Gamma = \mathcal{K}^\perp$ (indeed $\Gamma \cong P(\Lambda)/\operatorname{Ker}\theta \cong \overline{\mathcal{B}} \cong P(\mathcal{B})/\mathcal{K} \cong \mathcal{K}^\perp$). We let $\mathcal{S} = \operatorname{Ker}\theta$, the internal symmetry group $\{S \subset \Lambda \mid \sigma_S(B) = +1, \forall B \in \mathcal{B}\}$. Then the *low temperature expansion* is

$$\begin{aligned} Z(\Lambda, K) &= \prod_{B\in\mathcal{B}} e^{K(B)} \sum_{X\subset\Lambda} \prod_{A\in\mathcal{B}} e^{-K(A)(1-\sigma_A(X))} \\ &= \prod_{B\in\mathcal{B}} e^{K(B)} |\mathcal{S}| \sum_{\beta\in\Gamma} \prod_{A\in\beta} e^{-2K(A)} . \end{aligned} \tag{7.11.2}$$

Correlation functions can be expressed as ratios of partition functions, where we change the interaction K on certain bonds. This can be done in two ways, to produce order or disorder variables σ and μ respectively. We can change an interaction $K \to -K$ or $K \to K + i\pi$. These procedures do not commute (which will result in μ and σ not being local with respect to each other) but are dual. Given a lattice Λ of Fig 7.17(a), we form the dual Λ^*, as in Fig. 7.17(c), essentially by introducing for each edge B between nearest neighbours of Λ a dual edge B^* of Λ^*, orthogonal and intersecting B, and for each point bond in B (we are really only interested in the boundary of Λ), the dual B^* is the edge in Λ^* closest to B). Then define an interaction K^* on the dual bonds $\mathcal{B}^*$ by

$$\sinh 2K^*(B^*) \sinh 2K(B) = 1\,, \text{ or } \tanh K^*(B^*) = e^{-2K(B)} .$$

Then if we reverse the sign of K along the bonds of $\beta \in P(\mathcal{B})$:

$$K_\beta(B) = \begin{cases} K(B) & B \notin \beta \\ -K(B) & B \in \beta \end{cases} \tag{7.11.3}$$

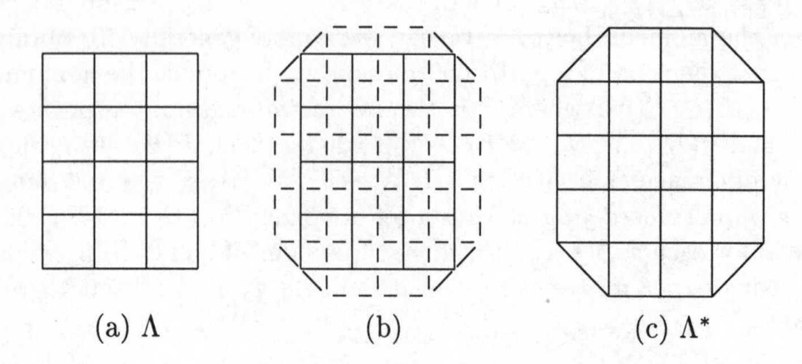

FIG. 7.17. From lattice Λ to dual lattice Λ^*

then in the dual picture, we have made the change $K^* \to K^* + i\pi/2$ on the dual bonds of $\beta^* \in P(\mathcal{B}^*)$:

$$K^*_{\beta^*}(B^*) = \begin{cases} K^*(B^*) & B^* \notin \beta^* \\ K^*(B^*) + i\pi/2 & B^* \in \beta^* \end{cases}. \tag{7.11.4}$$

The action $K^* \to K^* + i\pi/2$ on a set of bonds B means introducing a factor $e^{i\pi/2}\sigma_{B^*} = i\sigma_{B^*}$ into the partition function. Thus if $\beta^* \in P(\mathcal{B}^*)$ is a set of bonds we can write correlation functions as ratios of partition functions:

$$\left\langle \prod_{B^* \in \beta^*} \sigma_{B^*} \right\rangle_{\Lambda^*, K^*} = (-i)^{|\beta|} \frac{Z(\Lambda^*, K^*_{\beta^*})}{Z(\Lambda^*, K^*)}, \quad \beta^* \in P(\mathcal{B}^*) \tag{7.11.5}$$

To take this duality further and relate the high temperature expansion for Λ^* with the low temperature expansion for Λ, note that the duality map $B \to B^*$ on bonds, which extends to a homomorphism of $P(\mathcal{B})$ into $P(\mathcal{B}^*)$, gives a bijection of the low temperature group Γ for $(\Lambda, \mathcal{B})$ onto the high temperature group $\mathcal{K}^*$ for $(\Lambda^*, \mathcal{B}^*)$. Then using $\tanh K^*(B^*) = e^{-2K(B)}$, the high temperature expansion for Λ^* and the low temperature expansion for Λ, we get

$$Z(\Lambda, K) = |\mathcal{S}| 2^{-|\Lambda^*|} \prod_{B \in \mathcal{B}} e^{K(B)} \prod_{B^* \in \mathcal{B}^*} [\cosh K^*(B^*)]^{-1} Z(\Lambda^*, K^*) \tag{7.11.6}$$

Taking the ratio of this formula with the corresponding one when $K \to K_\beta$ for some $\beta \in P(\mathcal{B})$ we get

$$Z(\Lambda, K_\beta)/Z(\Lambda, K) = (-i)^{|\beta|} Z(\Lambda^*, K^*_{\beta^*})/Z(\Lambda^*, K^*). \tag{7.11.7}$$

Comparing this with (7.11.5) and letting $K(B) \to \infty$ on the one point boundary bonds, we see

$$\left\langle \prod_{B^*\in\beta^*} \sigma_{B^*} \right\rangle_{\Lambda^*,K^*}^{\text{free}} = (-1)^{|\beta|} Z^+(\Lambda, K_\beta)/Z^+(\Lambda, K) \tag{7.11.8}$$

where the left hand side is computed using free boundary conditions and the right hand side via + boundary conditions. Now $\langle\ \rangle^{\text{free}}$ is an even state, and we claim that if $A \subset \Lambda^*$ is a set of an even number of sites then we can write $\sigma_A = \prod_{B^*\in\beta^*} \sigma_{B^*}$ for a suitable choice of bonds β^*. We can start with A consisting of two sites a and b and a path $a = \sigma_1, \ldots, \sigma_{n+1} = b$ of nearest neighbour sites from a to b. Then $\sigma_A = \prod_{i=1}^n (\sigma_i \sigma_{i+1})$ so that $\langle \sigma_a \sigma_b \rangle$ can be computed from $Z(\Lambda^*, K_{\beta^*})/Z(\Lambda^*, K)$ by making the changes $K^* \to K^* + i\pi/2$ along the path $\beta^* = \{(\sigma_i, \sigma_{i+1})\}$ from a to b, and is clearly independent of the choice of path. Similarly we can represent the quotient $Z(\Lambda, K_\beta)/Z(\Lambda, K)$ as correlations of a dual variable μ called a disordered variable. (At $K \equiv 0$, infinite temperature, $\langle \mu_r \mu_s \rangle = 1$.) Again take sites r, s of Λ^*, a path $r = \mu_1, \mu_2, \ldots, \mu_{n+1} = s$ from r to s, and then make the change from $K \to -K$ along the bonds of Λ crossed by this path. This gives a correlation function $\langle \mu_r \mu_s \rangle$ which is independent of the path since if we wish to flip a path ⌝↓ to └→ in a small square, then this can be achieved by reversing the sign in the centre site of this square ⊡ (a site of Λ) which leaves the partition function invariant. In this way we can form correlation functions $\langle \mu_A \rangle$, with an interaction K of an arbitrary set of even sites of Λ^*, which by the duality of (7.11.7) is essentially $\langle \sigma_A \rangle$ (at the dual temperature or interaction K^*). Mixed correlation functions of an even number of σs and an even number of μs can be computed by using suitable combinations of the two kinds of interaction changes for bonds say in (Λ, K). Some care is needed, however, due to a sign ambiguity formed when deforming a μ-path through a spin variable σ-site.

In the algebraic picture reversing the signs $K \to -K$ along the vertical bonds in a row amounts to introducing a factor $\sigma_z = \begin{pmatrix} 0 & 1 \\ 1 & 0 \end{pmatrix}$ at each site of the path, (see (7.3.11) and (7.4.9)) so that $\mu_{(j,0)}\mu_{(k,0)}$ is related to $\prod_{i=j}^{k-1} \sigma_z^i$. We examine this duality transformation more closely in the algebraic setting.

In Section 7.7 it was shown that if β, β' are on the same side of the inverse critical temperature (i.e. either $\beta, \beta' > \beta_c$, or $\beta, \beta' < \beta_c$) then there exists an automorphism ν of the Pauli algebra A^P such that $\varphi_{\beta'} = \varphi_\beta \circ \nu$. In fact we have first that $E_\beta = W_\beta^* E_0 W_\beta$, and $\varphi_{\beta'}^F = \varphi_\beta^F \circ \tau(W_\beta^* W_{\beta'})$ on A^F, where φ_β^F is the quasi-free state φ_{E_β}. In Section 7.7 it is shown that if β and β' are on the same side of the critical temperature, then $\tau(W_\beta^* W_{\beta'})|_{A_+}$ extends to a graded automorphism ν of A^P such that $\varphi_{\beta'} = \varphi_\beta \circ \nu$. Now suppose that $\beta < \beta_c, \beta' > \beta_c$. We now claim that if $(\pi_\beta^P, H_\beta^P, \Omega_\beta^P)$ (usually we will drop the subscript β) is the GNS triple for φ_β on A^P, then the isomorphism

$$\pi_\beta^P \circ \tau(W_\beta^* W_{\beta'}) : A_+ \to \pi_\beta^P(A_+) \tag{7.11.9}$$

extends to an endomorphism

$$\nu : A^P \to \pi_\beta^P(A_+)'' \tag{7.11.10}$$

such that

$$\varphi_{\beta'}^+ = \varphi_\beta \circ \nu \tag{7.11.11}$$

where $\varphi_\beta^\pm$ are the pure components of φ_β.

First, recall that the GNS triple (π^P, H^P, Ω^P) of φ_β on A^P can be described (see the Remark on page 275) in terms of the GNS triple (π^F, H^F, Ω^F) of φ_β^F on A^F as follows. Take a representation π of $\hat{A}$ on $H^F \oplus H^F$ by

$$\pi(a) = \pi^F(a) \oplus \pi^F(\theta_-(a)), \quad a \in A^F \tag{7.11.12}$$

$$\pi(T) = \begin{pmatrix} 0 & 1 \\ 1 & 0 \end{pmatrix}. \tag{7.11.13}$$

Then we can identify H^P with $\overline{\pi^F(A_+^F)\Omega^F} \oplus \overline{\pi^F(A_-^F)\Omega^F}$, Ω^P with $\Omega^F \oplus 0$, and π^P with the restriction of π from $\hat{A}$ on $H^P \oplus H^F$ to A^P on H^P.

For brevity, we write ω for $W_\beta^* W_{\beta'}$. Then

$$\theta_- \tau(\omega) \theta_- \tau(\omega)^{-1} = \tau(\theta_- \omega \theta_- \omega^*) = \tau(-X). \tag{7.11.14}$$

If $X = -\theta_- \omega \theta_- \omega^*$, then by the computations of Theorem 7.13, $1 + X$ is trace class, and $\det(-X) = -1$. Hence by Theorem 6.15, $\tau(X)$ is inner and odd, i.e. it can be implemented by a unitary u in A_-^F. But φ_β^F is an even state, and so $\theta = \tau(-1)$ is implemented by a self adjoint unitary r on H^F with $r\Omega^F = \Omega^F$, and $\pi^F\theta(\cdot) = r\pi^F(\cdot)r$. Hence $\tau(\theta_- \omega \theta_- \omega^*)$ is implemented by $r\pi^F(u)$ in φ_β^F. Next note that $[E_\beta, \theta_-]$ is trace class for $\beta \neq \beta_c$ (Lemma 7.12) and so $\varphi_\beta^F \sim \varphi_\beta^F \circ \theta_-$ Theorem 6.15. Hence θ_- is implemented in π^F by a unitary $v = v_\beta$. Since $\theta_-^2 = 1$ and π^F is irreducible we can arrange $v^2 = 1$. We note that v_β is even for $\beta < \beta_c$ Now

$$\begin{aligned} \theta_- E_\beta \theta_- &= \theta_- W_\beta^* E_0 W_\beta \theta_- = \theta_- W_\beta^* \theta_- E_0 \theta_- W_\beta \theta_- \quad \text{as } [E_0, \theta_-] = 0 \\ &= \theta_- W_\beta^* \theta_- W_\beta E_\beta W_\beta^* \theta_- W_\beta \theta_- \end{aligned}$$

or

$$\varphi_\beta^F \circ \theta_- = \varphi_\beta^F \circ \tau(\theta_- W_\beta^* \theta_- W_\beta).$$

Thus θ_- is implemented by the implementer of $\tau(W_\beta^* \theta_- W_\beta \theta_-)$, which is even for $\beta < \beta_c$ [Theorem 7.13]. (We note, in passing, that $\varphi_\beta \sim \varphi_\beta \circ \theta_-$ for $\beta < \beta_c$ for

$$\pi^P(\theta_-(a + Tb)) = \mathrm{Ad}(v \oplus v)\pi^P(a + Tb), \quad a \in A_+^F, b \in A_-^F.)$$

So for $\beta < \beta_c$, $\beta' > \beta_c$ and $\tau = \tau(W_\beta^* W_{\beta'})$, we have in the representation π^F of φ_β^F on A^F:

$$\pi^F \theta \theta_- \tau \theta_- \tau^{-1}(\cdot) = \mathrm{Ad}(r\pi^F(u))\pi^F(\cdot) \tag{7.11.15}$$

$$\pi^F\theta_-(\cdot) = \mathrm{Ad}(v)\pi^F(\cdot) \tag{7.11.16}$$

where u is odd (in A_-^F), and v is a self adjoint even unitary on H^F. If $\gamma = \theta_-\tau\theta_-\tau^{-1}$, then $\gamma\theta_-\gamma\theta_- = 1$, and so identifying u with $\pi^F(u)$, we have

$$ruvruv\pi^F(x)vu^*rvu^*r = \pi^F(x)\,, \quad x \in A^F\,. \tag{7.11.17}$$

By irreducibility of π^F, we have that $ruvruv$ is scalar. By rotating u we can arrange that $ruvruv = 1$, so that ruv and vru are both self adjoint unitaries. Then $X = vru \oplus ruv$ is a self adjoint unitary which satisfies

$$\pi(\tau\theta_-(x)) = X\pi(\tau(x))X\,, \quad x \in A^F\,, \tag{7.11.18}$$

as

$$\pi^F(\tau\theta_-(x) = v\pi^F(\theta_-\tau\theta_-\tau^{-1}(\tau(x))v) = vru\pi^F(\tau(x))u^*rv$$

and

$$\begin{aligned}\pi^F(\theta_-\tau\theta_-(x)) &= \pi^F(\theta_-\tau\theta_-\tau^{-1}(\tau(x))) \\ &= ru\pi^F(\tau(x))u^*r = ruv\pi^F(\theta_-\tau(x))vu^*r \quad x \in A^F\,.\end{aligned}$$

Define $\nu : A \to B(H^F \oplus H^F)$ by

$$\nu(a + Tb) = \pi(\tau(a)) + X\pi(\tau(\beta)) \quad a, b \in A^F\,. \tag{7.11.19}$$

Then ν is a $*$-homomorphism because

$$\begin{aligned}[\nu(Tb)]^* = [X\pi(b)]^* &= \pi(\tau(b^*))X = X\left(\pi\tau\theta_-(b^*)\right) \\ &= \nu(T\theta_-(b^*)) = \nu(b^*T) = [\nu(Tb)^*]\,,\end{aligned}$$

and

$$\begin{aligned}&\nu\left[(a_1 + Tb_1)(a_2 + Tb_2)\right] = \nu(a_1a_2 + \theta_-(b_1)b_2 + Tb_1a_2 + T\theta_-(a_1)b_2) \\ &= \pi\tau(a_1a_2) + \pi\tau(\theta_-(b_1)b_2) + X\pi\tau(b_1a_2) + X\pi(\tau(\theta_-(a_1)b_2)) \\ &= \pi\tau(a_1)\pi\tau(a_2) + X\pi\tau(b_1)X\pi\tau(b_2) + X\pi\tau(b_1)X\pi\tau(a_2) + \pi\tau(a_1)X\pi\tau(b_2) \\ &= \left[\pi\tau(a_1) + X\pi\tau(b_1)\right]\left[\pi\tau(a_2) + X\pi\tau(b_2)\right] = \nu(a_1 + Tb_1)\nu(a_2 + Tb_2)\end{aligned}$$

for $a_1, a_2, b_1, b_2 \in A^F$. Since A^P leave H^P invariant, we have by restriction a representation v of A^P on H^P, by *even* operators. We finally claim that $\varphi_{\beta'}^+ = \varphi_\beta \circ \nu$, but first recall that $\varphi_{\beta'}^+$ can be constructed as follows (Remark on page 275). If $\beta' > \beta_c$ then $\pi_\beta^F\theta_- = v_{\beta'}\pi_{\beta'}^F(\cdot)v_{\beta'}$ where $v_{\beta'}$ is a self adjoint odd unitary (cf. $v_\beta, \beta < \beta_c$). Then

$$\varphi_{\beta'}^+(a + Tb) = \varphi_{\beta'}(a) + \langle v_{\beta'}\pi_{\beta'}^F(b)\Omega_{\beta'}^F, \Omega_{\beta'}^F\rangle \tag{7.11.20}$$

for $a \in A_+^F$, $b \in A_-^F$.

We have

$$\pi^F_\beta \tau\theta_-(x) = v_\beta r u \pi^F \tau(x) u^* r v^*_\beta\,, \quad x \in A^F$$

and so identifying $\pi^F_\beta \circ \tau$ with $\pi^F_{\beta'}$ we can identify $v_{\beta'}$ with $\pm v_\beta r u$, since both are self adjoint unitaries and π^F is irreducible. By appropriate labelling, we can take $v_{\beta'} = v_\beta r u$.

Then for $x = a + Tb$, $a \in A^F_+$, $b \in A^F_-$, we have

$$\begin{aligned}\varphi_\beta(\nu(x)) &= \langle \nu(x)\Omega^P_\beta, \Omega^P_\beta\rangle = \langle [\pi\tau(a) + X\pi(\tau(b))]\Omega^P_\beta, \Omega^P_\beta\rangle \\ &= \langle [\pi_\beta\tau(a) + v_\beta r u \pi_\beta(\tau(b))]\Omega^F_\beta, \Omega^F_\beta\rangle \\ &= \varphi_{\beta'}(a) + \langle v_{\beta'}\pi^F_{\beta'}(b)\Omega^F_{\beta'}, \Omega^F_{\beta'}\rangle = \varphi^+_{\beta'}(x)\,.\end{aligned}$$

We have thus shown:

Theorem 7.20 *For each $\beta < \beta_c$, $\beta' > \beta_c$ the isomorphism $\pi^P_\beta \tau(W^*_\beta W_{\beta'})$: $A_+ \to \pi^P_\beta(A_+)$ extends to an endomorphism $\nu : A^P \to \pi^P_\beta(A_+)''$ such that $\varphi^+_{\beta'} = \varphi_\beta \circ \nu$.*

7.12 Analyticity of correlation functions

Here we apply the automorphism method described in Section 7.7 to proving analyticity of correlation functions.

Theorem 7.21 *If F is a local observable, then $\langle F\rangle^\pm_\beta$ are real analytic in β, J_1, J_2 for $\beta \neq \beta_c$.*

Here $\langle\cdot\rangle^\pm_\beta$ are the equilibrium states on the classical Ising model computed using $\pm$ boundary conditions. Thus $\langle\cdot\rangle^\pm_\beta = \langle\cdot\rangle_\beta$ for any boundary condition for $\beta < \beta_c$, and is the unique equilibrium state, whilst for $\beta > \beta_c$, $\langle\cdot\rangle^\pm_\beta$ are distinct extremal equilibrium states, and any equilibrium state is a convex combination of $\langle\cdot\rangle^\pm_\beta$. This follows from (Messager and Miracle-Sole 1975) who analysed the structure of translation invariant equilibrium states, and (Aizenmann 1980), (Higuchi 1979) who showed that any equilibrium state on the two dimensional Ising model is automatically translation invariant. We do not need to use this deep result, although note that it can be proven directly (Gallavotti *et al.* 1973) that

$$\langle\cdot\rangle_\beta = \frac{1}{2}\left(\langle\cdot\rangle^+_\beta + \langle\cdot\rangle^-_\beta\right) \tag{7.12.1}$$

for free or periodic boundary conditions.

We also have from Theorem 7.6 a decomposition

$$\varphi_\beta = \frac{1}{2}\left(\varphi^+_\beta + \varphi^-_\beta\right) \tag{7.12.2}$$

of the state φ_β on A^P for $\beta > \beta_c$ into inequivalent pure states $\varphi^\pm_\beta$, which arises from considering the thermodynamic limit of Gibbs states, computed using periodic or free boundary conditions. Moreover

$$\mu(F) = \varphi_\beta(F_\beta)$$

where μ is the state on $C(P)$ obtained from periodic or free boundary conditions. If we put $\mu_\pm(F) = \varphi^\pm_\beta(F_\beta)$, it is not clear yet that $\mu_\pm$ are equilibrium states on $C(P)$, indeed extremal ones corresponding to $\pm$ boundary conditions, or that they are indeed states, i.e. $\mu_\pm(F^*F) \geq 0$ as $F \to F_\beta$ is not positive. We first show that $\mu_\pm$ are symmetric i.e. $\mu_\pm(F^*) = \overline{\mu_\pm(F)}$, $F \in C(P)$. For this it is convenient to introduce a new symmetry j on A^P obtained by the infinite tensor product of complex conjugation on M_2, i.e.

$$j\left(\sigma^k_\alpha\right) = \sigma^k_\alpha\,, \quad \alpha = x, z\,, \quad j\left(\sigma^k_y\right) = -\sigma^k_y\,.$$

Then j is a complex conjugate linear automorphism of A^P fixing H_1 and H_2 so that $j(F_\beta) = (F^*)_\beta$, where $F^*(\xi) = \overline{F(\xi)}$ as usual.

At this point it is convenient to observe the following which is a consequence of the fact to be proven in the next section (see the Remark on page 363) that $\{F_\beta \mid F \text{ local}\}$ is dense in A^P. If for a given state φ of A^P, we define $\nu(F) = \varphi(F_\beta)$ for F local in $C(P)$, then ν is symmetric (i.e. $\overline{\nu(F)} = \nu(F^*)$ for all F local) if and only if φ is *j-symmetric* (i.e. $\overline{\varphi(a)} = \varphi \circ j(a)$ for all $a \in A^P$). Since μ is indeed a state on $C(P)$, then φ_β is j-symmetric and so we have

$$\varphi_\beta(a) = \varphi_\beta \circ j(a^*) = \left(\varphi^+_\beta\left(j(a^*)\right) + \varphi^-_\beta\left(j(a^*)\right)\right)/2\,, \quad a \in A^P\,.$$

Then $a \to \varphi^\pm_\beta\left(j(a^*)\right)$ are mutually disjoint states which must coincide with the pair $\varphi^\pm_\beta$ – but we must decide whether $\varphi^\pm_\beta$ are j-symmetric or whether j interchanges them. For $\beta = \infty$, we see that the former case holds by inspection on the product states $\varphi^\pm_\infty$, and then use the automorphism method of Section 7.7 observing that $j\nu_\beta = \nu_\beta j$. Note that as $j\theta_- = \theta_- j$ we can extend j to $\hat{A}$ by $j(T) = T$. Indeed then $j|_{A^F}$ is the Bogoliubov conjugate linear automorphism $B(h) \to B(Ch)$, where C is the conjugation $f \oplus g \to f^* \oplus g^*$ on $\ell^2 \oplus \ell^2$. Noting that C commutes with θ_- and W_β in (7.7.21) we see that $j(v_\beta) = cv_\beta$ for some $c \in \mathbb{T}$. But from $j^2 = 1$ we must have $c^2 = 1$, and from $c = j(v_\beta)v^*_\beta$ and analyticity we must have $c \equiv 1$. Hence j commutes with ν_β as defined in (7.7.27), and so $\varphi^\pm_\beta = \varphi^\pm_\infty \circ \nu_\beta$ are j-symmetric. Consequently, $F \to \mu_\pm(F) \equiv \varphi^\pm_\beta(F_\beta)$ are symmetric. We wish to identify $\mu_\pm$ with the classical $\pm$ equilibrium states on $C(P)$.

We first show that the decompositions of (7.12.1) and (7.12.2) into pure phases coincide.

Lemma 7.22 *With a suitable choice of labelling in* (7.12.1), *we have*

$$\langle F\rangle^\pm_\beta = \varphi^\pm_\beta(F_\beta)\,. \tag{7.12.3}$$

for all local observables F, and all $\beta > \beta_c$.

Proof We have $\varphi_\beta = \varphi_\beta^\pm$ on A_+, and $\langle F\rangle_\beta = \langle F\rangle_\beta^\pm$ for all even F. Hence if F is a local even classical observable, then

$$\langle F\rangle_\beta^\pm = \varphi_\beta^\pm(F) \tag{7.12.4}$$

Let τ denote the action of $\mathbb{Z}^\nu$ on $\mathbb{Z}^\nu$ by translation, $\nu = 1, 2$, and also denote the induced action of $\mathbb{Z}$ on $\bigotimes_{\mathbb{Z}} M_2 = A^P$ by τ and that of $\mathbb{Z}^2$ on $\bigotimes_{\mathbb{Z}^2} \mathbb{C}^2 = C(P)$ also by τ. Then it is easy to see that if G and H are local classical observables, then

$$\left[G\tau_{(0,n)}(H)_\beta\right] - G_\beta\tau_n(H_\beta) \to 0 \tag{7.12.5}$$

in norm as $n \to \infty$. Now if G, H are odd, we have by clustering of the pure states $\varphi_\beta^\pm$ that

$$\varphi_\beta^+(G_\beta\tau_n(H_\beta)) \to \varphi_\beta^+(G_\beta)\varphi_\beta^+(H_\beta) \tag{7.12.6}$$

as $n \to \infty$. Moreover, by clustering of $\langle\cdot\rangle_\beta^\pm$, (which is a consequence of GKS inequalities, see Section 7.2) we see

$$\langle G\tau_{(0,n)}(H)\rangle_\beta^+ \to \langle G\rangle_\beta^+\langle H\rangle_\beta^+\,, \tag{7.12.7}$$

as $n \to \infty$. But $G_{\tau(0,n)}(H)$ is even, if G, H are odd, and so

$$\langle G\tau_{(0,n)}(H)\rangle_\beta^+ = \varphi_\beta^+\left(\left[G\tau_{(0,n)}H\right]_\beta\right)\,, \tag{7.12.8}$$

by (7.12.4). Hence by (7.12.5)–(7.12.8) we have

$$\langle G\rangle_\beta^+\langle H\rangle_\beta^+ = \varphi_\beta^+(G_\beta)\varphi_\beta^+(H_\beta) = \langle G\rangle_\beta^-\langle H\rangle_\beta^-\,, \tag{7.12.9}$$

for all G, H odd. Now $\langle\cdot\rangle_\beta^+$ is not even, and so there exists real H odd such that $\langle H\rangle_\beta^+ \neq 0$. Hence from (7.12.9) with $G = H$, we have $\varphi_\beta^+(H_\beta) = \langle H\rangle_\beta^+$ or $\langle H\rangle_\beta^-$ as by the discussion before this lemma, $\mu_\pm$ are symmetric, i.e. $\varphi_\beta^\pm(H_\beta)$ are real, if H is real. Suppose

$$\varphi_\beta^+(H_\beta) = \langle H\rangle_\beta^+ \neq 0\,, \tag{7.12.10}$$

then dividing (7.12.9) by (7.12.10) we see that

$$\langle G\rangle_\beta^+ = \varphi^+(G_\beta)\,, \tag{7.12.11}$$

for all local odd G. The lemma follows by combining this with (7.12.4). □

Lemma 7.23 *For any local observable F, F_β is real analytic in β, J_1, J_2.*

Proof In the notation of Section 7.3, $F_\beta = \lim_{m\to\infty} F_{\beta M}$. It is enough to show that $F_{\beta M}$ is constant in M for large M, which is clear. □

Lemma 7.24 *The automorphisms $\{\nu_\beta \mid \beta \neq \beta_c\}$ of Theorem 7.6 on A^P can be chosen so that ν_β is real analytic in β, J_1, J_2 (if $\beta \neq \beta_c$).*

Proof If u is a unitary on $l^2 \oplus l^2$ such that $\Gamma u = u\Gamma$, and $\theta_- u\theta_- u^* - 1$ is small in trace class norm, $\det(\theta_- u\theta_- u^*) = 1$, by continuity of the determinant, for $\theta_- u\theta_- u^* - 1$ sufficiently small. Hence taking logarithms we can write $\theta_- u\theta_- u^* = e^{iL}$, where $\|L\| < \pi$, and L is self adjoint, trace class and commutes with Γ. Hence $Q(u) = e^{ib(L)/2}$ satisfies

$$\tau(\theta_- u\theta_- u^*) = \mathrm{Ad}(Q(u)), \tag{7.12.12}$$

on A^F. It is enough to consider $\beta > \beta_c$, and let $V_\beta = W^* W_\beta$. We claim that for each $\beta > \beta_c$, there exists v_β in A^+ such that

$$\tau(\theta_- V_\beta \theta_- V_\beta^*) = \mathrm{Ad}(v_\beta), \tag{7.12.13}$$

$$v_\beta \theta_-(v_\beta) = 1, \tag{7.12.14}$$

and v_β is analytic in β, J_1, J_2 for $\beta > \beta_c$. First, fix $\beta_0 > \beta_c$, and choose any v_{β_0} in A_+^F such that

$$\tau\left(\theta_- V_{\beta_0} \theta_- V_{\beta_0}^*\right) = \mathrm{Ad}\left(v_{\beta_0}\right) \tag{7.12.15}$$

$$v_{\beta_0} \theta_- \left(v_{\beta_0}\right) = 1. \tag{7.12.16}$$

Now,

$$\theta_- V_\beta \theta_- V_\beta^* = \left(\theta_- V_{\beta_0} \theta_- V_{\beta_0}^*\right) \left[V_{\beta_0} \left(\theta_- V_{\beta_0}^* V_\beta \theta_- V_\beta^* V_{\beta_0}\right) V_{\beta_0}^*\right]. \tag{7.12.17}$$

Hence we could define in a neighbourhood of β_0,

$$v_\beta = v_{\beta_0} \tau\left(V_{\beta_0}\right) \left[Q\left(V_{\beta_0}^* V_\beta\right)\right], \tag{7.12.18}$$

which satisfies (7.12.13) and (7.12.14), and is real analytic in β, J_1, J_2 using (7.7.36), (7.7.42) and (7.7.43). Now if v_β is a solution of (7.12.13) and (7.12.14), the only other solution is $-v_\beta$. Hence by connectedness and analyticity, we can thus define $v_\beta \in A_+^F$ such that (7.12.3) and (7.12.4) hold for $\beta > \beta_c$, and v is real analytic. Hence

$$\nu_\beta(a + Tb) = \tau(V_\beta)(a) + T v_\beta \tau(V_\beta)(b), \tag{7.12.19}$$

$a \in A_+^F$, $b \in A_-^F$ defines an analytic family $\{\nu_\beta \mid \beta \neq \beta_c\}$ of automorphisms on A^P. □

Proof of Theorem 7.21 By Lemmas 7.23, 7.24 we see that

$$\varphi_\beta^+(F_\beta) = \varphi_\infty^+ \nu_\beta(F_\beta) \tag{7.12.20}$$

is real analytic for $\beta > \beta_c$, for any local classical observable F. But if $F = \sigma^{00}$, the classical observable of evaluating a configuration at the lattice site $(0,0)$, then $\langle\sigma^{00}\rangle^+_\beta > 0 > \langle\sigma^{00}\rangle^-_\beta$, for $\beta > \beta_c$. We see from Lemma 7.22, that for each β, either $\langle F\rangle^+_\beta = \varphi^+_\beta(F_\beta)$, for all local F, or $\langle F\rangle^+_\beta = \varphi^-_\beta(F_\beta)$, if we define $\varphi^\pm$ as in (7.12.18). By analyticity of (7.12.18), and evaluating at $\beta = \infty$, we see that defining $\varphi^\pm_\beta = \varphi^\pm_\infty \circ \nu_\beta$, is consistent with $\langle F\rangle^\pm = \varphi^\pm_\beta(F_\beta)$, if $\varphi^+ = \bigotimes_{-\infty}^{\infty} \omega_{\left(\begin{smallmatrix}1\\0\end{smallmatrix}\right)}$. ◻

7.13 Magnetization

We show how to combine the computation of the subspace $E_\beta \wedge (1 - \theta_- E_\beta \theta_-)$, $\beta > \beta_c$ in Proposition 7.18, and the identification of the phases in the classical and quantum models of Lemma 7.22 to derive the spontaneous magnetization.

Let $\beta > \beta_c$, and let (π^P, H^P, Ω^P) be the GNS triple for φ_β on A^P described in terms of the GNS triple (π^F, H^F, Ω^F) of ω_β on A^F as in Section 6.8. If v_β is a self adjoint odd unitary on H^F implementing θ_- in π^F, then as in the Remark on page 275 we have

$$\varphi^+_\beta(Tb) = \langle v_\beta \pi^F(b)\Omega^F, \Omega^F\rangle\,, \quad b \in A^F_- . \tag{7.13.1}$$

In particular, taking the classical variable σ^{00} so that $(\sigma^{00})_\beta = \sigma^0_x = T(c_0 + c_0^*)$, we have

$$\langle\sigma^{00}\rangle^+ = \langle v_\beta(c_0 + c_0^*)\Omega, \Omega\rangle \tag{7.13.2}$$

where for convenience we have dropped the superscript F, and representation π^F. To compute the right hand side of (7.13.2) we need the following lemma:

Lemma 7.25

$$\langle v_\beta B(h)\Omega, \Omega\rangle = \langle v_\beta B(E_\beta \wedge (1 - \theta_- E_\beta \theta_-)h)\Omega, \Omega\rangle \quad \text{for } h \in l^2 \oplus l^2 .$$

Proof Since φ^F_β is a Fock state, $B((1-E)f)\Omega = 0$, for all $f \in l^2 \oplus l^2$. Hence

$$\begin{aligned}
&\langle v_\beta B(h)\Omega, \Omega\rangle = \langle v_\beta B(Eh)\Omega, \Omega\rangle = \langle B(\theta_- Eh)v_\beta\Omega, \Omega\rangle \\
&= \langle v_\beta\Omega, B(\Gamma\theta_- Eh)\Omega\rangle = \langle v_\beta\Omega, B(E\Gamma\theta_- Eh)\Omega\rangle = \langle B(\Gamma E\Gamma\theta_- Eh)v_\beta\Omega, \Omega\rangle \\
&= \langle v_\beta B(\theta_-\Gamma E\Gamma\theta_- Eh)\Omega, \Omega\rangle = \langle v_\beta B((1 - \theta_- E\theta_-))Eh\Omega, \Omega\rangle .
\end{aligned}$$

Thus by iteration

$$\begin{aligned}
\langle v_\beta B(h)\Omega, \Omega\rangle &= \langle v_\beta B(((1 - \theta_- E\theta_-)E)^k h)\Omega, \Omega\rangle \\
&= \langle v_\beta B(E \wedge (1 - \theta_- E\theta_-)h)\Omega, \Omega\rangle
\end{aligned}$$

as $\lim_{k\to\infty}[(1 - \theta_- E\theta_-)E]^k = (1 - \theta_- E\theta_-) \wedge E$ by Exercise 1.31. But from Proposition 7.18, we know that the subspace $E_\beta \wedge (1 - \theta_- E_\beta\theta_-)$ is one dimensional and spanned by a unit vector h_β, such that $\langle h_\beta, \Gamma h_\beta\rangle = 0$. Hence

$$\langle v_\beta B(h)\Omega, \Omega\rangle = \langle h, h_\beta\rangle\langle v_\beta B(h_\beta)\Omega, \Omega\rangle \tag{7.13.3}$$

for all $h \in l^2 \oplus l^2$. ◻

Let $E = E_\beta$, $E' = \theta_- E_\beta \theta_-$, $E_{\pi/2} = E \wedge (1 - E') + (1 - E) \wedge E'$ which is of rank 2, spanned by h_β and Γh_β. Let U be the unitary on $l^2 \oplus l^2$ which is the identity on $E_{\pi/2}^\perp$, and interchanges h_β and Γh_β. Then it commutes with Γ, and $E \wedge (1 - UE'U) = 0$. We are thus in the situation of Section 6.9, where in the notation of that section, $P = E$, $P' = UE'U$, $n = 0$, $\hat{R}(P', P) = \exp iH(P'/P)$ moves P to P':

$$\hat{R}(P', P)PR'(P'/P)^* = P' \tag{7.13.4}$$

$$QB(h)Q^* = B(\hat{R}(P'/P)h) \tag{7.13.5}$$

$$\varphi_P(Q) = (\det \cos\alpha)^{\frac{1}{4}} \tag{7.13.6}$$

where $Q = \exp ib(H(P'/P))/2$, $\sin\alpha = |P - P'|$. Now $i(B(h_\beta) - B(\Gamma h_\beta))$ is a self adjoint unitary implementing $\tau(-U)$, and so $v_\beta i(B(h_\beta) - B(\Gamma h_\beta))$ implements $\tau(-\theta_- U)$. Now

$$\theta_- U\hat{R}(P'/P)E\hat{R}(P'/P)^* U\theta_- = E \tag{7.13.7}$$

and so $\tau(-\theta_- U\hat{R}(P'/P))$ leaves E invariant and so is unitarily implementable by Q_0 in φ_E leaving the vacuum invariant. Thus for some $\gamma \in \mathbb{T}$

$$v_\beta(B(h_\beta) - B(\Gamma h_\beta))Q = \gamma Q_0$$

we have

$$\begin{aligned}\gamma\varphi_E(Q^*) &= \varphi_E\left[v_\beta(B(H_\beta) - B(\Gamma h_\beta))\right] \\ &= \varphi_E(v_\beta B(h_\beta)), \quad \text{as } E\Gamma h_\beta = 0\,.\end{aligned} \tag{7.13.8}$$

Then by (7.13.2), (7.13.3), (7.13.6), (7.13.8):

$$\langle\sigma^{00}\rangle^+ = \langle h, h_\beta\rangle\langle v_\beta B(h_\beta)\Omega, \Omega\rangle = \langle h, h_\beta\rangle\gamma\varphi_E(Q^*) = \langle h, h_\beta\rangle\gamma(\det\cos\alpha)^{\frac{1}{4}}$$

where $h = (e_0, e_0)$, $e_0 = (\delta_{i0})_{i=-\infty}^{\infty}$.

We compute this explicitly. If we rotate with $2^{-\frac{1}{2}}\begin{pmatrix}1 & 1\\ 1 & -1\end{pmatrix}$ as in the proof of Proposition 7.18, h_β is the vector of (7.8.8), and $h = (\sqrt{2}e_0, 0)$. We will verify here that (7.8.8) is a unit vector. Now by (7.7.34) and (7.7.35), we can write $e^{i\delta} = f_+ f_-$ where

$$f_- = \left[(1 - z_2 z_1^* e^{i\omega})/(1 - z_1^* e^{i\omega}/z_2)\right]^{\frac{1}{2}}$$

$$f_+ = \left[(1 - z_1^* e^{-i\omega}/z_2)/(1 - z_2 z_1^* e^{-i\omega})\right]^{\frac{1}{2}}$$

have analytic continuations to the interior, exterior of the unit disc respectively. Thus $\left[q_- e^{-i\delta} q_-\right]^{-1} = f_- q_- f_+$ by the theory of Wiener–Hopf or Toeplitz operators. But $q_- f_+ e_0 = e_0$ (so that (7.8.8) is a unit vector) and $\langle f_- e_0, e_0\rangle = 1$. Thus $\langle h, h_\beta\rangle = 1$.

Next

$$\cos^2\alpha = 1 - \sin^2\alpha = (1-(P-P')^2)(1-E_{\pi/2})$$
$$= \left[1-(E-\theta_- E\theta_-)^2\right](1-E_{\pi/2}).$$

Taking $v = 2^{-\frac{1}{2}}\begin{pmatrix}1 & 1\\ 1 & -1\end{pmatrix}$ as usual, we have

$$1 - v(E-\theta_- E\theta_-)^2 v = 1 - \frac{1}{4}\begin{pmatrix}0 & U_\beta^* - \theta_- U_\beta^* \theta_- \\ U_\beta - \theta_- U_\beta \theta_- & 0\end{pmatrix}^2$$

$$= \begin{pmatrix} q_- U_\beta^* q_- U_\beta q_- + (1-q_-)U_\beta^*(1-q_-)U_\beta(1-q_-) & 0 \\ 0 & q_- U_\beta q_- U_\beta^* q_- + (1-q_-)U_\beta(1-q_-)U_\beta^*(1-q_-) \end{pmatrix}.$$

Thus, restricting to $1-E_{\pi/2}$, the orthogonal complement of h_β and Γh_β, and writing $M = \{(q_- e^{-i\delta} q_-)^{-1} e_0\}^\perp = \left[\mathrm{Ker}(q_- U_\beta q_-)^{-1}\right]^\perp$, we have

$$\det\cos^2\alpha = \left[\det\left[q_- U_\beta^* q_- U_\beta q_-\right]\big|_M\right]^4 = \left|\det(q_- e^{i\delta} q_-)\right|^4.$$

Thus the magnetization is

$$\langle\sigma^{00}\rangle^+ = \left|\det(q_- e^{i\delta} q_-)\right|^{\frac{1}{2}}. \tag{7.13.9}$$

where we have ignored the phase γ, since the magnetization is positive. The magnetization can be computed via Szego's theorem (Widom 1976) as

$$\exp\left(\textstyle\sum_{m=0}^{\infty} m k_m k_{-m}\right) \tag{7.13.10}$$

where k_m are the Fourier coefficients of δ, as in (7.7.35), $k_0 = 0$. Substituting, we get

$$\begin{aligned}\textstyle\sum_{m=0}^{\infty} m k_m k_{-m} &= -\textstyle\sum_{m=1}^{\infty} m(1/2m)^2\left[(z_1^*/z_2)^n - (z_2 z_1^*)^n\right]^2\\ &= -1/4\textstyle\sum_{m=1}^{\infty}(1/m)\left((z_1^*/z_2)^{2n} - 2(z_1^*)^{2n} + (z_2 z_1^*)^{2n}\right)\\ &= \frac{1}{4}\log\left\{\frac{(1-(z_1^*/z_2)^2)(1-(z_2 z_1^*)^2)}{(1-(z_1^*)^2)}\right\}\\ &= (1/4)\log\left\{1-(\sinh 2K_1 \sinh 2K_2)^{-2}\right\}.\end{aligned}$$

Thus

$$m^* = \langle\sigma^{00}\rangle^+ = \left\{1-(\sinh 2K_1 \sinh 2K_2)^{-2}\right\}^{\frac{1}{8}} \tag{7.13.11}$$

7.14 Ground states of the one dimensional quantum system arising from the Ising dynamics

If T_M is the transfer matrix at inverse temperature β, we now consider the dynamics on the Pauli algebra A^P obtained from $\lim_{M\to\infty} T_M^{-it}(\cdot)T_M^{it}$, and study the ground states as β varies. The limit will clearly exist on the even algebra A_+ (or even A^F) by quasi-free considerations. The existence of the limit is not clear at all on the Pauli algebra, so we use the automorphism method developed in Section 7.7 (Lemma 7.12) to extend the dynamics from A_+ to A^P. Define $\alpha_z^M \in \mathrm{Aut}(\hat{A})$ by

$$\alpha_z^M = T_M^{-iz}(\cdot)T_M^{iz} \tag{7.14.1}$$

which leaves A invariant. Indeed $\alpha_z|_{A^F}$ is the Bogoliubov automorphism $\tau\left(e^{-2izH_M}\right)$, so that at least on the Fermion algebra

$$\alpha_t^F = \tau\left(e^{-2itH}\right) = \lim_{M\to\infty} \alpha_t^M \tag{7.14.2}$$

exists for real t and defines a dynamical system $\left(A, \alpha_t^F\right)$.

The $*$-algebra generated by $\{B(h)\}$ is the local algebra $A_{(0)}^F$. We call the $*$-algebra generated by $B(h)$, with h of finite support, the *strictly local* elements $A_{(00)}^F$ of A^F. The strictly local elements of $\hat{A}$ are defined to be $A_{(00)}^F + A_{(00)}^F T = A_{(0)}^P + A_{(0)}^P T$, where $A_{(0)}^P = \bigcup A_n^P$.

Lemma 7.26 *(1) The action $t \in \mathbb{R} \to \alpha_t^F \in \mathrm{Aut}(A^F)$ extends uniquely to a continuous action $\hat{\alpha}$ on $\hat{A}$.*

(2) For any strictly local a in $\hat{A}$, $\alpha_t(a)$ is entire analytic in t.

Proof By the automorphism method of Section 7.7, we need to consider the implementability of the quasi-free automorphism $\alpha_t^F\theta_-\alpha_{-t}^F\theta_- = \tau(e^{-2itH}\theta_- e^{2itH}\theta_-)$ on A^F. The estimates of Lemma 7.14 show that $e^{-2itH}\theta_- e^{2itH}\theta_- - 1$ is trace class, and by continuity of the determinant, $\det\left(e^{-2itH}\theta_- e^{2itH}\theta_-\right) \equiv 1$ since it is so at $t = 0$. Thus $\alpha_t^F\theta_-\alpha_{-t}^F\theta_-$ is implementable by an even v_t, which can be chosen so that $v_t(\theta_-(v_t)) = 1$, and

$$\hat{\alpha}_t(a + Tb) = \alpha_t^F(a) + v_t T\alpha_t^F(b)\,, \quad a, b \in A^F\,, \tag{7.14.3}$$

defines an automorphism $\hat{\alpha}_t$ of $\hat{A}$ extending α_t^F on A^F. The problem is to show that v_t can be chosen continuously so that $\hat{\alpha}_t$ is an action. Uniqueness will follow from continuity.

If $\hat{\alpha}_t'$ is another extension of α_t then $v_t' \equiv \hat{\alpha}_t'(T)T$ also implements $\alpha_t^F\theta_-\alpha_{-t}^F\theta_-$ so that $c_t \equiv v_t' v_t^* \in \hat{A} \cap A' = \mathbb{C}$. Then as $\theta = \lim_{m\to\infty} \mathrm{Ad}\prod_{-m}^{m} \sigma_z^j$, and σ_z^j are even, c_t is even. Let $\hat{\theta}_-$ be the dual automorphism of $\hat{A} = \hat{\theta}_-(a_1 + a_2 T) = a_1 - a_2 T$, $a_i \in A$. Then as $\hat{\theta} = \lim \mathrm{Ad}\,\sigma_x^m \sigma_x^{-m}$, and since $\sigma_x^m\sigma_x^{-m}$ is even, $\hat{\theta}_-(c_t) = c_t$, i.e. $c_t \in A$. Hence $c_t \in A_+ \cap A_+' = \mathbb{C}$. Thus $\hat{\alpha}_t'(T) = c_t\hat{\alpha}_t(T)$ so that $c_t^2 = 1$ as T is self adjoint. By continuity $c_t = 1$.

By the estimates of Lemma 7.14, $\delta = \theta_- H\theta_- - H$ is trace class, and $\Delta = b(\delta)/2 \in A^F_+$. Then

$$\mathrm{Ad}(v_t^*)\alpha_t = \theta_-\alpha_t\theta_- \tag{7.14.4}$$

where

$$v_z = \textstyle\sum_{n=0}^{\infty}(iz)^n \displaystyle\int_0^1 \cdots \int_0^{t_{n-1}} dt_1 \cdots dt_n\, \Delta_{t_1 z}\cdots\Delta_{t_n z} \tag{7.14.5}$$

and $\Delta_z = \alpha_z(\Delta) = b(e^{-zH}\delta e^{zH})/2$. Then

$$v_t^* = \textstyle\sum_{n=0}^{\infty}(-it)^n \displaystyle\int_0^1 \cdots \int_0^{t_{n-1}} dt_1 \cdots dt_n \Delta_{t_n t}\cdots\Delta_{t_1 t}\,.$$

Now $d/dt\, v_t = v_t\Delta_t$, $d/dt\, v_t^* = -\Delta_t v_t^*$ so that $v_t v_t^* = 1$ by differentiation, and $v_t^* v_t = 1$ by straightforward expansion and examination of the coefficient of t^n. Similar considerations using $\theta_-\Delta_t = \theta_-\alpha_t\theta_-\Delta$ show that $v_t^* = \theta_-(v_t)$. It then follows from the uniqueness that the continuous family we constructed $\hat{\alpha}_t$ is indeed a one-parameter group as $\hat{\alpha}_{t+s}\hat{\alpha}_{-s}$ is also a continuous extension of $\hat{\alpha}_t$. □

If we write

$$T_z^{1M} = T_z^1 = \exp\left(-izK_1^* \textstyle\sum_{j=-M}^{M} \sigma_z^j\right) \tag{7.14.6}$$

$$T_z^{2M} = T_z^2 = \exp\left(-izK_2 \textstyle\sum_{j=-M}^{M-1} \sigma_x^j \sigma_x^{j+1}/2\right) \tag{7.14.7}$$

then the original transfer matrix is $T_M = T_i^2 T_i^1 T_i^2$. We can define $\alpha_z^{\beta jM} = \mathrm{Ad}(T_z^j)$ $j = 1, 2$, and $\alpha_z^{\beta M} = \alpha_z^{\beta 2M}\alpha_z^{\beta 1M}\alpha_z^{\beta 2M}$ on $\hat{A}$. Then $\alpha_z^{\beta 1} = \lim_{M\to\infty}\alpha_z^{\beta jM}$ exist and define automorphisms of the local algebra $A^P_{(0)} = \cup A^P_n$ where $A^P_n = C^*(\sigma^j_\alpha \mid |j| \le n)$ and on the local algebra $A^F_{(0)}$ generated by $\{B(h)\}$. Indeed for $a \in A^P_n$:

$$\alpha_z^{\beta 1}(a) = \alpha_z^{\beta 1n}(a) \in A^P_n\,, \quad \alpha_z^{\beta 2}(a) = \alpha_z^{\beta 2(n+1)}(a) \in A^P_{n+1} \tag{7.14.8}$$

and

$$\alpha_z^{\beta 1}(B(h)) = \lim\nolimits_{M\to\infty} B(e^{-2izK_1^* H_1^M} h) = B(e^{-2izK_1^* H_1} h) \tag{7.14.9}$$

$$\alpha_z^{\beta 2}(B(h)) = \lim\nolimits_{M\to\infty} B(e^{-izK_2 H_2^M} h) = B(e^{-izK_2 H_2} h) \tag{7.14.10}$$

i.e. $\alpha_z^{\beta 1} = \tau(e^{-2izK_1^* H_1})$, $\alpha_z^{\beta 2} = \tau(e^{-izK_2 H_2})$ on $A^F_{(0)}$. In this notation if

$$F = \textstyle\prod_{j=1}^{n}\sigma(k_j, I_j) \quad k_1 < k_2 < \cdots < k_n\,, \quad \sigma(k, I) = \prod_{i\in I}\sigma_{ki}\,,$$

then

$$F_{\beta M} = T_M^{k_1}\sigma_x(I_1)T_M^{k_2-k_1}\sigma_x(I_2)T_M^{k_3-k_2}\cdots\sigma_x(I_n)T_M^{-k_n}$$

$$= \left\{ \left(\alpha_i^{\beta M}\right)^{k_1} \sigma_x(I_1) \right\} \left\{ \left(\alpha_i^{\beta M}\right)^{k_2} \sigma_x(I_2) \right\} \cdots \left\{ \left(\alpha_i^{\beta M}\right)^{k_n} \sigma_x(I_n) \right\}$$

s local by (7.14.8), where $\sigma_x(I) = \prod_{j \in I} \sigma_x^j$. Moreover, $F_\beta = \lim_{M \to \infty} F_{\beta M}$ where

$$F_\beta = \left\{ \left(\alpha^\beta\right)^{k_1} \sigma_x(I_1) \right\} \left\{ \left(\alpha^\beta\right)^{k_2} \sigma_x(I_2) \right\} \cdots \left\{ \left(\alpha^\beta\right)^{k_n} \sigma_x(I_n) \right\} .$$

Then F_β is linear in F, $1_\beta = 1$ is covariant with respect to translation $\tau_{(0,n)}F)_\beta = \tau_n(F_\beta)$, and satisfies the clustering property

$$\lim_{n \to \infty} \left\| (F_1 \tau_{0,n)}(F_2))_\beta - (F_1)_\beta \tau_n((F_2)_\beta) \right\| = 0 , \tag{7.14.11}$$

where $\tau_{(0,n)}$ is the lattice translation automorphism $\sigma_{ij} \to \sigma_{i,j+n}$ of the C^*-lgebra A generated by σ_{ij} and τ_n is the lattice transformation automorphism f A^P. For the grading θ of A^P and the grading θ of the abelian C^*-algebra A enerated by σ_{ij} determined by $\theta(\sigma_{ij}) = -\sigma_{ij}$, we have $(\theta(F))_\beta = \theta(F_\beta)$.

The states $\varphi_\beta^\pm$ of A^P are lattice translation invariant and pure by Section .7. Hence they have the clustering property

$$\lim_{n \to \infty} \varphi_\beta^\pm (a \tau_n(b)) = \varphi_\beta^\pm(a) \varphi_\beta^\pm(b) . \tag{7.14.12}$$

Proposition 7.27 *For local $a \in A^P$, $\alpha_i^\beta(a) = \alpha^\beta(a)$.*

Proof As in (7.14.8) and (7.14.10) $\lim_{M \to \infty} \alpha_z^{\beta j M}$ exist on $A_{(0)}^F$ and $A_{(0)}^P$, for $= 1, 2$. Taking the analytic continuation of the quasi-free dynamics α_t^F given y (7.14.2) and using (7.14.10) we have

$$\alpha_i^\beta(B(h)) \equiv \lim_{M \to \infty} \alpha_i^{\beta M}(B(h)) = \alpha_i^{\beta 2} \alpha_i^{\beta 1} \alpha_i^{\beta 2}(B(h)) .$$

But $T = (c_1 + c_1^*)\sigma_x^1 = B(e_1 \oplus e_1)\sigma_x^1$ so that $\alpha_i^{\beta 2} \alpha_i^{\beta 1} \alpha_i^{\beta 2}(T)$ is well defined. Let

$$w \equiv \alpha_i^{\beta 2} \left(\alpha_i^{\beta 1} \left(\alpha_i^{\beta 2}(T) T \right) \alpha_i^{\beta 1}(T) T \right) \alpha_i^{\beta 2}(T) T \in A_+ . \tag{7.14.13}$$

s $\alpha_i^{\beta j}(T)T \in A_+$, and $\alpha_i^{\beta j}$ (where defined) leaves A_+ invariant. Since $T^2 = 1$, $^{-1} = T \alpha_i^{\beta 2} \alpha_i^{\beta 1} \alpha_i^{\beta 2}(T)$. Then

$$\begin{aligned} w B(h) w^{-1} &= \alpha_i^{\beta 2} \alpha_i^{\beta 1} \alpha_i^{\beta 2} \left(T \left\{ \alpha_{-i}^{\beta 2} \alpha_{-i}^{\beta 1} \alpha_{-i}^{\beta 2} (T B(h) T) \right\} T \right) \\ &= B \left(e^{K_2 H_2} e^{2 K_1^* H_1} e^{K_2 H_2} \theta_- e^{-K_2 H_2} e^{-2 K_1^* H_1} e^{-K_2 H_2} \theta_- h \right) \\ &= B \left(e^{2H} \theta_- e^{-2H} \theta_- h \right) = v_i B(h) v_i^{-1} . \end{aligned} \tag{7.14.14}$$

The last equality follows from the analytic continuation of

$$B(e^{-2izH}\theta_- e^{2izH}\theta_- h)v_z = v_z B(h)$$

from $z = t$ real to $z = i$. Therefore, $w^{-1}v_i$ is the centre of A_+^F, and so $w =$ cv_i, for some scalar c. We have $(v_z T)^2 = v_z T v_z T = v_z \theta_-(v_z) = 1$, and $(wT)^2 =$ $\alpha_i^{\beta 2}\alpha_i^{\beta 1}\alpha_i^{\beta 2}(T)^2 = 1$ due to $T^2 = 1$. Thus $c^2 = 1$ and we obtain $w = \pm v_i$. We the note that v_i given by (7.14.5) and w given by (7.14.13) are both analytic in K and K_2, $K_2 > K_1^*$. In the first case this is clear from the expression for v_i and th uniform continuous dependence of the Hamiltonian H on (K_1^*, K_2) as in the proc of Theorem 7.6 (see the comment after (7.7.14)). In the second case, we conside expression (7.14.11) for w and again write $T = B(h)\sigma_x^1$. Then (7.14.10) sho that $\alpha_i^{\beta 2}\alpha_i^{\beta 1}\alpha_i^{\beta 2}(B(h))$ is entire analytic in (K_1^*, K_2) and similarly (7.14.8) sho that $\alpha_i^{\beta 2}\alpha_i^{\beta 1}\alpha_i^{\beta 2}(\sigma_x^1)$ is entire analytic in (K_1^*, K_2). To deduce that the + sig in $w = \pm v_i$ persists (away from criticality) it only remains to explicitly comput as usual for $K_1^* = 0$ (and $K_2 = 0$). If $K_1^* = 0$ then $H = K_2 H_2$, and $\alpha_t^\beta = \alpha_2^\beta$ by (7.14.1), (as $M \to \infty$) and (7.14.10). Moreover, $\lim_{\to}$ Ad $(\exp itK_2H_2)$ exist on A_n^P and $B(h)$ by (7.14.8) and (7.14.10) respectively and so defines a strongl continuous one-parameter group of $*$-automorphisms $\alpha_t^{\beta 2}$ of $\hat{A}$. By the uniquenes part of Lemma 7.26 (1), we have $\hat{\alpha}_t^\beta = \alpha_{2t}^{\beta 2}$. But $v_i = \hat{\alpha}_i^\beta(T)T$ from (7.14.3) an $w = \alpha_{2i}^\beta(T)T$ from (7.14.13) so that $v_i = w$. If $K_2 = 0$, $\hat{\alpha}_t^\beta = \alpha_t^{\beta 1}$ fixes T an $v_i = w = 1$. [

Let (H, π, Ω) be the GNS triple of a pure ground state φ of (A, α). Let U_φ b the one-parameter group of unitaries implementing α in π and fixing the vacuun with infinitesimal generator L. Consider the following properties:

(P) positivity $L \geq 0$;

(ND) non-degeneracy dim $E^L(0) = 1$;

(G) gap property $\sigma(L) \cap (0, m) = \varnothing$ for some $m > 0$ and m is not in the poin spectrum of L.

Property (P) holds for all ground states, whilst (ND) will hold for $\varphi_\beta^\pm$ fc $\beta > \beta_c$, and φ_β for $\beta \leq \beta_c$. Suppose φ is a pure ground state disjoint from an other pure ground state. Then (ND) clearly holds, as any $\xi \in E^L(0)H$ gives pure ground state $\psi = \langle \pi(\cdot)\xi, \xi \rangle$ for (A, α), and $\psi \to \mathbb{C}\xi$ determines the ra uniquely when π is irreducible.

Proposition 7.28 *(a) The states φ_β and $\varphi_\beta^\pm$ are ground states of (A^P, α_t^β). $\beta < \beta_c$, φ_β is the unique ground state. If $\beta > \beta_c$, $\varphi_\beta^\pm$ are the only pure groun states and are mutually disjoint.*

(b) For $\beta > \beta_c$, the pure ground states $\varphi_\beta^\pm$ satisfy the gap property (G) *wit $m = 4(K_2 - K_1^*)$.*

(c) For $\beta < \beta_c$ the pure ground state φ_β satisfies the gap property (G) *wit $m = 2(K_1^* - K_2)$.*

Proof By Section 7.7, H does not have an eigenvalue 0 if $K_1^*, K_2 \neq 0$ an indeed has continuous spectrum

$$\sigma(H) = [-K_2 - K_1^* - |K_2 - K_1^*|] \cup [|K_2 - K_1^*|, K_2 + K_1^*] \,.$$

Then $L^F, L_+^F = L^F|_{\text{even}}$ the generators of the second quantization of e^{itH} on $F(EH)$, and its even restriction satisfy (P), (ND), (G) with $m = n\,|K_2 - K_1^*|$, where $n = 2, 4$ respectively. Thus by Proposition 6.37, α_t^F has an unique ground state φ_E on A^F, given by the spectral projection $E = E^H\{-\infty, 0\}$ for H, and moreover that its restriction to A_+ is the unique ground state of $(A_+, \alpha_t|_{A_+})$. If a state φ is invariant under a dynamics α_t, let $L(\varphi, \alpha_t)$ denote the generator of the corresponding unitary group, which implements the dynamics in the GNS triple. Then with $\alpha_t = \tau(e^{iHt})$ we identify

$$L^F = L(\varphi_E, \alpha_t)\,, \quad L_+^F = L(\varphi_E|_{A_+}, \alpha_t|_{A_+})\,.$$

In the range $\beta > \beta_c$, by the Remark on page 275 and Lemma 6.36, we identify the GNS Hilbert space and cyclic vectors of $(A^P, \alpha_t^\beta, \varphi_\beta^\pm)$ with those of $(A_+, \alpha_t^\beta, \varphi_\beta) = (A_+^F, \alpha_t^F, \varphi_E)$ and identifying the representations of $A_+^F \subset A^P$. Thus we identify $L(\varphi_\beta^\pm, \alpha_t)$ with L_+^F, the restriction of $L(\varphi_E, \alpha_t)$ to the even subspace. In particular $\varphi_\beta^\pm$ are ground states of (A^P, α_t^P).

In the range $\beta < \beta_c$ again using the Remark on page 275, $\varphi_{E_\pm}$ are α_t invariant and in $H_\pm$, we identify $L(\varphi_E, \alpha_t)$ with L^F for the Fock state φ_E of A^F. Under this identification the unitary group on $\hat{H}_1$ generated by α_t on A^P is the unitary group for $(\hat{A}, \hat{\alpha}_t, \hat{\varphi}_{E_\pm})$. Thus the infinitesimal generator satisfies (P), (ND), (G) with $m = 2(K_1^* - K_2)$. Then φ_β is a ground state and is the unique even ground state of (A^P, α_t^P) as $(A_+, \alpha_t|_{A_+})$ has a unique ground state, and the restriction of any ground state of A^P to A_+ is clearly a ground state.

Suppose φ is an extremal ground state of (A^P, α_t), $\beta \neq \beta_c$. Then as α_t is even, $(\varphi + \varphi \circ \theta)/2$ is an even ground state of (A^P, α). Thus certainly in the case when $\beta < \beta_c$, when there is an unique even ground state, we have $(\varphi + \varphi \circ \theta)/2 = \varphi_\beta^P$ and since φ_β^P is pure, we have $\varphi = \varphi_\beta^P$. If $\beta > \beta_c$, then by Theorem 6.35(4), we have $\varphi \not\sim \varphi \cdot \theta$ as $\varphi|_{A_+}$ is the unique ground state of A_+ hence pure. Thus from the two decompositions into disjoint pure states: $(\varphi + \varphi \cdot \theta)/2 = (\varphi_\beta^+ + \varphi_\beta^-)/2$ we must conclude that $\varphi = \varphi_\beta^+$ or φ_β^-. (Note that π_φ must be unitarily equivalent to π_β^+ or π_β^- by disjointness, and there is an unique ground state φ_β^+ (respectively φ_β^-) whose GNS representation is π_β^+ (respectively π_β^-) by (ND)). We then conclude that φ_β is the unique ground state of (A, α_t^P) if $\beta < \beta_c$, and that $\varphi_\beta = (\varphi_\beta^+ + \varphi_\beta^-)/2$, if $\beta > \beta_c$ where $\varphi_\beta^\pm$ are the only pure ground states of (A^P, α_t^P) and are disjoint. □

Theorem 7.29 *Let*

$$\psi = \begin{cases} \langle\,\rangle\,, & m = 2\,(K_1^* - K_2) \text{ for } \beta < \beta_c\,, \\ \langle\,\rangle_\pm\,, & m = 4\,(K_2 - K_1^*) \text{ for } \beta > \beta_c\,. \end{cases}$$

Then

$$\lim_{l \to \infty} e^{|l|m} \left[\psi(F_1 \tau_{(l,0)}(F)) - \psi(F_1)\psi(F_2)\right] = 0 \qquad (7.14.15)$$

for any local F_1 and F_2, and for any $\varepsilon > 0$ there exists $F_{\varepsilon 1}^{\pm}$ and $F_{\varepsilon 2}^{\pm}$ such that

$$\lim_{l \to \pm\infty} e^{|l|(m+\varepsilon)} \left|\psi(F_{\varepsilon 1}^{\pm}\tau_{(l,0)}(F_{\varepsilon 2}^{\pm})) - \psi(F_{\varepsilon 1}^{\pm})\psi(F_{\varepsilon 2}^{\pm})\right| = \infty .$$

Proof Consider the linear map Φ from observables to states taking

$$F = \prod_{i=1}^{n} \sigma(k_i, I_i), \quad k_1 < k_2 < \cdots < k_n \tag{7.14.16}$$

to

$$\Phi(F) = e^{-k_1 L}\pi(\sigma_x(I_1))e^{-(k_2-k_1)L} \cdots \pi(\sigma_x(I_n))\Omega .$$

Let $F \to F^+$ be the conjugate linear automorphism of $C(P(\mathbb{Z}^2))$ given by $\sigma(k, I) \to \sigma(-k, I)$. If F_1 is a linear combination of the form (7.14.16) with $N_1 \geq k_i$, and F_2 another linear combination with $N_2 \leq k_i$ where N_1, N_2 are fixed, put $N = N_1 - N_2$ and $\Phi_1 = \Phi(F_1^+)$, $\Phi_2 = \Phi(F_2)$. Then

$$\begin{aligned} &e^{lm}\left(\psi(F_1\tau_{(l,0)}F_2) - \psi(F_1)\psi(F_2)\right) \\ &= e^{Nm}\left\langle (1 - E_L(0))e^{-(l-N)(L-m)}\Phi_2, \Phi_1 \right\rangle . \end{aligned}$$

Since $L - m$ is positive on $(1 - E_L(0))$ without point spectrum at 0 we have $\lim_{l \to +\infty}(1 - E_L(0))e^{-(l-N)(L-m)} = 0$ and (7.14.15) follows.

We will show below, that the set of states $\Phi(F)$ as F varies over the observables in (7.14.16) with $0 \leq k_1 < \cdots < k_n$ is total. So take some such F so that $E_L(m, m + \varepsilon/2)\Phi(F) \neq 0$. Then for $l \geq 0$

$$\begin{aligned} \psi\left(F^+\tau_{(l,0)}F\right) &= \langle e^{-lL}\Phi(F), \Phi(F)\rangle \geq \langle e^{-lL}E_L(m, m+\varepsilon/2)\Phi(F), \Phi(F)\rangle \\ &\geq e^{-l(m+\varepsilon/2)} \left\|E_L(m, m+\varepsilon/2)\Phi(F)\right\|^2 . \end{aligned}$$

Hence

$$\lim_{l \to \infty} e^{l(m+\varepsilon)}\psi\left(F^+\tau_{(l,0)}F\right) = +\infty .$$

We next claim that A_0 the C^*-algebra generated by $\{\alpha_{ik}(\sigma_x^j) \mid k, j \in \mathbb{Z}\}$ is A^P. Now $\sigma_x^j \in A_0$, so it is enough to show $\sigma_y^j \in A_0$. Now $\alpha_z^{\beta M2} = \mathrm{Ad}(T_z^{2M})$ leaves A_0 invariant. Then $\alpha_i(\sigma_x^j) = \alpha_i^{\beta 2}\alpha_i^{\beta 1}\alpha_i^{\beta 2}(\sigma_x^j)$ by the consistency of Proposition 7.27, and is local by (7.14.8). Then $\alpha_i^{\beta 1}\alpha_i^{\beta 2}(\sigma_x^j) = \alpha_{-i}^{\beta 2}\alpha_i(\sigma_x^j) \in A_0$ as A_0 is invariant under $\alpha_{-i}^{\beta 2}$. But $\alpha_2^{\beta 2}(\sigma_x^j) = \sigma_x^j$, we get $\alpha_i^{\beta 1}(\sigma_x^j) \in A_0$. Thus $\alpha_i^{\beta 1}(\sigma_x^j) = \mathrm{Ad}(e^{K_1^*\sigma_z^j})\sigma_x^j = e^{2K_1^*\sigma_z^j}\sigma_x^j$. Hence $e^{2K_1^*\sigma_z^j} = \alpha_i^{\beta 1}(\sigma_x^j)\sigma_x^j \in A_0$. Since $K_1^* \neq 0$, we find $\sigma_z^j \in A_0$.

It remains to show that the vectors $\Phi(F)$ as F varies over the observables in (7.14.16) with $0 \leq k_1 < \cdots < k_n$ is total. So suppose $\langle \Phi(F), \xi\rangle = 0$ for all such F and some ξ. Then

$$G(z_1, \ldots, z_n) = \langle e^{iz_1 L}\pi(\sigma_x(I_1))e^{iz_2 L} \cdots \pi(\sigma_x(I_n)\Omega, \xi\rangle \tag{7.14.17}$$

is a bounded continuous function for $\mathrm{Im}\, z_j \geq 0$, and holomorphic in any subset of variables $\mathrm{Im}\, z_j > 0$, since φ is a ground state. But $G = 0$ for $z_j = ik_j$ where

$k_1, k_2 - k_1, \ldots, k_n - k_{n-1}) \in \mathbb{N}^n$. Hence by an application of Carleson's theorem (Boas 1954), $G = 0$. Then by the density of $C^*(\alpha_{ik}(\sigma_x^j))$ we see $\xi = 0$, i.e. the vectors $\Phi(F)$ where F is as in (7.14.16) with $0 \leq k_1 < \cdots < k_n$ are total. □

Remark Note that $\mathbb{Z}^2$ clustering, namely

$$\lim_{(l,n)\to\infty} \psi(F_1 \tau_{(l,n)}(F_2)) = \psi(F_1)\psi(F_2) \tag{7.14.18}$$

for any continuous functors F_1 and F_2 on configuration space $\{\pm 1\}^{\mathbb{Z}^2}$ was shown in Section 7.2 from the GKS inequalities ($\psi = \langle\,\rangle_\pm$ for $\beta < \beta_c$). In the C^*-setting it follows from the above Proposition for the $l \to \infty$ and from (7.14.15) for $n \to \infty$. These methods can be combined to yield (7.14.18).

Also note that our (conjugate linear) automorphism $F \to F^+$ on classical observables is reflection in the y-axis of the plane $\mathbb{Z}^2$. Then for a function F of the right half plane $(F^+F)_\beta = (F^+)_\beta F_\beta$ and $(F^+)_\beta = F_\beta^*$, and we see (in the set up of Section 7.12) that $\varphi_\mu(F_\beta^* F_\beta) = \mu(F^+F)$, so that positivity of φ_μ is intimately related to *reflection positivity* of μ, namely $\mu(F^+F) \geq 0$, for functions F of the right half plane.

We finally note that the density of $\{F_\beta \mid F \text{ local}\}$ in A^P is a consequence of the density of $C^*(\alpha_{ik}(\sigma_x^j))$ in A^P as in the proof of Theorem 7.29.

Exercise 7.1 Consider the XY-Hamiltonian $H = H(a,b)$:

$$-J\left\{\sum_{j=a}^{b-1}\left[(1+\gamma)\sigma_x^j\sigma_x^{j+1} + (1-\gamma)\sigma_y^j\sigma_y^{j+1}\right] + 2\lambda\sum_{j=a}^{b}\sigma_z^j\right\} \tag{7.14.19}$$

where $J > 0$, and γ, λ are real.

(a) Show that under the Jordan–Wigner transformation H becomes

$$2J\left\{\sum_{j=a}^{b-1}\left[(c_j^*c_{j+1} + c_{j+1}^*c_j) + \gamma(c_j^*c_{j+1}^* + c_{j+1}c_j)\right] - \lambda\sum_{j=a}^{b}(2c_j^*c_j - 1)\right\}. \tag{7.14.20}$$

(b) Show that $\alpha_t = \lim_{N\to\infty} \operatorname{Ad} e^{iH(-N,N)t}$ exists on $\hat{A}$ such that $\alpha_t|_{A^F} = (e^{itK})$ where $K = 2J\begin{pmatrix} U + U^* - 2\lambda & \gamma(U - U^*) \\ -\gamma(U - U^*) & -(U + U^* - 2\lambda)\end{pmatrix}$, U is the bilateral shift on ℓ^2, and $\alpha_t(T) = v_t T$, where

$$\begin{aligned} v_t &= \sum_{n=0}^{\infty} i^n \int_0^t \cdots \int_0^{t_{n-1}} dt_1 \cdots dt_n \alpha_{t_1}(\Delta)\cdots\alpha_{t_n}(\Delta) \\ \Delta &= -4J\left\{(1+\gamma)\sigma_x^0\sigma_x^1 + (1-\gamma)\sigma_y^0\sigma_y^1\right\} = H - \theta_- H \theta_- . \end{aligned}$$

Let E_+, E_-, E_0 be the spectral projectors of K for $(0,\infty)$, $(-\infty, 0)$, $\{0\}$ respectively.

(c) Show that if $(\lambda,\gamma) \neq (0,\pm 1)$ that K has no point spectrum, and if $(\lambda,\gamma) = (0,1)$ that $\sigma(K) = \{\pm 4J\}$.

(d) Show that if either $|\lambda| = 1$, $\gamma \neq 0$ or $|\lambda| < 1$, $\gamma = 0$ then $E_+ - \theta_- E_+ \theta_-$ is not Hilbert–Schmidt.

(e) Show that if either $|\lambda| > 1$ or $(\lambda, \gamma) = (0, \pm 1)$ then $E_+ - \theta_- E_+ \theta_-$ is trac class and $\sigma(E_+, \theta_- E_+ \theta_-) = 1$.

(f) Show that if $|\lambda| < 1$, $\gamma \neq 0$ then $E_+ - \theta_- E_+ \theta_-$ is trace class an $\sigma(E_+, \theta_- E_+ \theta_-) = -1$.

(g) If $(\lambda, \gamma) \neq (0, \pm 1)$, show that $\varphi^P_{E_+}$ is the unique even ground state fc (A^P, α_t).

(h) If $(\lambda, \gamma) = (0, \pm 1)$ then all even ground states of (A^P, α_t) are given by conve combinations $\alpha_0 \varphi^P_{E_+} + \sum \alpha_j \varphi^P_{h_j}$ where $\alpha_j \geq 0$, $\sum \alpha_j = 1$, and $h_j = E^{4J} h_j$ ar mutually orthogonal, where E^{4J} is the eigenprojection of K for eigenvalue 4J and φ_h is the state of A^F in the notation of Theorem 6.38.

(i) If $|\lambda| \geq 1$ or if $\gamma = 0$, $|\lambda| < 1$ show that $\varphi^P_{E_+}$ is the unique ground state fc (A^P, α_t).

(j) If $|\lambda| < 1$, $\gamma \neq 0$ and $(\lambda, \gamma) \neq (0, \pm 1)$, show that φ^P_E is a mixture of two pur states $\varphi^{P\pm}_{E_+}$ which are the unique extremal ground states for (A^P, α_t).

(k) If $(\lambda, \gamma) = (0, \pm 1)$ show that φ^P_E (respectively φ_h) is a mixture of two pu states $\varphi^{P\pm}_E$ (respectively $\varphi^\pm_h$) such that $\{\varphi^{P\pm}_{E_+}, \varphi^\pm_h \mid h\}$ are the unique extrem ground states for (A^P, α_t).

(l) Show that if a is a local observable in A^P and $\varphi_{\lambda,\gamma} = \varphi^P_{E_+}$, that $\varphi_{\lambda,\gamma}(a)$ real analytic in (λ, γ) for $|\lambda| > 1$, and $\varphi^\pm_{\lambda,\gamma}(a)$ in (λ, γ) for $|\lambda| < 1$, $\gamma \neq 0$.

(m) Compute the one-point functions:

$$\varphi^\pm(\sigma^j_x) = \begin{cases} \pm(1-\lambda^2)^{\frac{1}{8}} \left[4\gamma/(1+\gamma^2)\right]^{\frac{1}{4}} & \gamma > 0 \\ 0 & \gamma < 0 \end{cases}$$

$$\varphi^\pm(\sigma^j_y) = \begin{cases} 0 & \gamma > 0 \\ \pm(1-\lambda^2)^{\frac{1}{8}} \left[4\gamma/(1+\gamma^2)\right]^{\frac{1}{4}} & \gamma < 0 \end{cases} .$$

7.15 Spontaneously broken symmetry and the distribution of prim numbers

The phase transition in the two dimensional Ising model is an example of spo taneous symmetry breaking. The $\mathbb{Z}/2$ action on spins, interchanging + and - leaves invariant the unique equilibrium state above the critical temperature b interchanges the two extremal phases $\langle\ \rangle_+$ and $\langle\ \rangle_-$ (or the ground states φ and φ_- on the Pauli algebra) below the critical temperature. In particular th extremal states are not invariant under the $\mathbb{Z}/2$ symmetry group although th dynamics on the Pauli algebra is $\mathbb{Z}/2$ invariant and we speak of a spontaneous broken symmetry.

Another example of spontaneously broken symmetry involving KMS stat rather than ground states appears in a C^*-dynamical system related to the di tribution of prime numbers, which we now describe. This section is not se contained.

Consider the group

$$\Gamma = P_{\mathbb{Q}}^{+} = \left\{ \begin{pmatrix} 1 & b \\ 0 & a \end{pmatrix} \mid b \in \mathbb{Q}, a \in \mathbb{Q}_{+}^{\times} \right\}$$

ith subgroup

$$\Gamma_0 = P_{\mathbb{Z}}^{+} = \left\{ \begin{pmatrix} 1 & m \\ 0 & 1 \end{pmatrix} \mid m \in \mathbb{Z} \right\}.$$

hen Γ_0 is almost normal in Γ in the sense that the orbits of Γ_0 acting on he left on Γ/Γ_0 are finite ; i.e. the cardinality $L(\gamma)$ of the image of $\Gamma_0\gamma\Gamma_0$ in $/\Gamma_0$ is finite for each $\gamma \in \Gamma$. The C^*-algebra $A = C_{\mathbb{Q}}$ is the completion in the egular representation of the Hecke algebra $H(\Gamma, \Gamma_0)$, the convolution algebra f complex valued functions on $\Gamma_0\backslash\Gamma/\Gamma_0$ with finite support. Then if χ_X is the lement corresponding to a coset $X \in \Gamma_0\backslash\Gamma/\Gamma_0$, we have presentation of A in erms of the generators μ_n, $n \in \mathbb{N}^{\times}$, $e(\gamma)$, $\gamma \in \mathbb{Q}/\mathbb{Z}$ (identified with $\chi_{X_n}, \chi_{X_\gamma}$ for he classes of

$$X_n = \begin{pmatrix} 1 & 0 \\ 0 & n \end{pmatrix} \quad \text{and} \quad X_\gamma = \begin{pmatrix} 1 & \gamma \\ 0 & 1 \end{pmatrix}$$

espectively) and relations:

a) $\mu_n^*\mu_n = 1$, for all n

b) $\mu_{nm} = \mu_n\mu_m$, for all n, m

c) $e(0) = 1$, $e(\gamma)^* = e(-\gamma)$, $e(\gamma_1 + \gamma_2) = e(\gamma_1)e(\gamma_2)$, for all $\gamma, \gamma_1, \gamma_2$

d) $\mu_n e(\gamma)\mu_n^* = n^{-1}\sum_{n\delta=\gamma} e(\delta)$, for all n, γ.

he vector state φ on A'' given by the vector of the coset Γ_0 in the regular epresentation has a modular automorphism group σ_t^{φ} so that $\sigma_t = \sigma_{-t}^{\varphi}$ restricts o a strongly continuous one-parameter group of automorphisms of A satisfying

$$\sigma_t(f)(\gamma) = \left(L(\gamma)/L(\gamma^{-1})\right)^{-it} f(\gamma)$$

or $f \in H(\Gamma, \Gamma_0)$, $\gamma \in \Gamma_0\backslash\Gamma/\Gamma_0$, or

$$\sigma_t(\mu_n) = n^{it}\mu_n\,, \quad \sigma_t(e(\gamma)) = e(\gamma)\,, \quad n \in \mathbb{N}^{\times}, \gamma \in \mathbb{Q}/\mathbb{Z}\,.$$

n fact $C^*(\mathbb{Q}/\mathbb{Z})$ generated by the image of e is the fixed point algebra A^σ, under he dynamics σ, the centralizer of the state φ.

Relation (d) says that the semigroup $\mathbb{N}^{\times}$ acts on $C^*(\mathbb{Q}/\mathbb{Z}) = \overline{\text{n}}\,\{e(\gamma) \mid \gamma \in \mathbb{Q}/\mathbb{Z}\}$ by endomorphisms α_n:

$$\alpha_n(e(\gamma)) = n^{-1} \sum_{n\delta=\gamma} e(\delta)$$

o that $A = \overline{\text{lin}}\,\{\mu_m e(\gamma)\mu_n^* \mid \gamma \in \mathbb{Q}/\mathbb{Z}, (m,n) = 1\}$ is identified with the crossed roduct $C^*(\mathbb{Q}/\mathbb{Z}) \rtimes_\alpha \mathbb{N}^{\times}$. The dual action of the dual of the Grothendieck group $\mathbb{Q}_+^{\times} = \mathbb{N}^{\times}(\mathbb{N}^{\times})^{-1}$ on A is

$$\hat{\alpha}_\psi\left(\mu_m e(\gamma)\mu_n^*\right) = \psi(m/n)\mu_m e(\gamma)\mu_n^*\,,$$

$\psi \in \hat{\mathbb{Q}}_+^\times$, $m,n \in \mathbb{N}^\times$, $\gamma \in \mathbb{Q}/\mathbb{Z}$, so that σ_t is a dense one-parameter subgrou of the dual action. Averaging over the compact dual action gives a conditiona expectation $E : A \to C^*(\mathbb{Q}/\mathbb{Z})$,

$$E\left(\mu_m e(\gamma)\mu_n^*\right) = \delta_{mn}\mu_m e(\gamma)\mu_n^*\,.$$

The finite *adele ring* $\mathcal{A}$ of $\mathbb{Q}$ is the restricted product of $(\mathbb{Q}_p)_p$ with respec of $(\mathbb{Z}_p)_p$. (If $p \in \mathcal{P}$, the primes, $\mathbb{Q}_p$ is the completion of $\mathbb{Q}$ with respect to the p adic metric, containing the p-adic integers $\mathbb{Z}_p$ as an open and compact subring. Then $\mathcal{R} = \prod_p \mathbb{Z}_p$ is open and compact in $\mathcal{A}$, and $\mathcal{A}/\mathcal{R} \cong \mathbb{Q}/\mathbb{Z}$. Moreove $\mathbb{Z}_p^\times$, the multiplicative group of $\mathbb{Z}_p$, is an open compact subgroup of $\mathbb{Q}_p^\times$ th multiplicative group of $\mathbb{Q}_p$. Then $\mathcal{A}^\times$, the restricted product of $\left(\mathbb{Q}_p^\times\right)_p$ wit respect to $\left(\mathbb{Z}_p^\times\right)_p$ is the *idele group* of $\mathbb{Q}$, $\mathbb{Q}_+^\times$ is discrete in $\mathcal{A}^\times$ and $W = \mathcal{A}^\times/\mathbb{Q}_+^\times \cong$ $\mathcal{R}^\times$ the *idele class group* is compact and can be identified with $\mathrm{Aut}(\mathbb{Q}/\mathbb{Z}) =$ $\mathrm{proj\,lim}_m \mathrm{Aut}\left(m^{-1}\mathbb{Z}/\mathbb{Z}\right)$.

Recall that $C_\mathbb{Q}$ was obtained in the regular representation from the commu tant of $P_\mathbb{Q}^+$ in $\ell^2(\Delta)$ where $\Delta = P_\mathbb{Q}^+/P_\mathbb{Z}^+$. But $\Delta = P_\mathbb{Q}^+/P_\mathbb{Z}^+$ and $\Delta_1 = P_\mathcal{A}/P_\mathcal{R}$ ar $\Gamma = P_\mathbb{Q}^+$ equivariant, so that the commutants $\Gamma'_\Delta = \Gamma'_{\Delta_1}$ are identified, indee with the commutant of $\overline{\Gamma} = \mathcal{A} \rtimes \mathbb{Q}_+^\times$ acting on Δ_1. Since $\overline{\Gamma}$ is normal in $P_\mathcal{A}$ $\mathrm{Ad}(u)$, $u \in P_\mathcal{A}$ leaves $\overline{\Gamma}'$ invariant with $\mathrm{Ad}(u) = 1$ on Γ'_{Δ_1} for $u \in \overline{P_\mathbb{Q}^+}$. Thus th quotient $W = \mathcal{A}^\times/\mathbb{Q}_+^\times \cong P_\mathcal{A}/\overline{P_\mathbb{Q}^+}$ acts on $P_\mathbb{Q}^{+\prime}$, and leaves $C_\mathbb{Q}$ globally invarian defining an action θ of W on $C_\mathbb{Q}$. Indeed

$$\theta_u(e(\gamma)) = e(u^{-1}\cdot\gamma)\,, \qquad \theta_u(\mu_n) = \mu_n\,. \qquad (7.15.1$$

The action θ of W on $C_\mathbb{Q}$ preserves the state φ and commutes with the dy namics σ. Indeed this is the natural action of $W = \mathrm{Aut}(\mathbb{Q}/\mathbb{Z})$ on $C^*(\mathbb{Q}/\mathbb{Z}) \rtimes_\alpha \mathbb{N}$ since W on $\mathbb{Q}/\mathbb{Z}$ commutes with the action of α. The fixed point algebr $C_\mathbb{Q}^W = C^*(\mu_n \mid n \in \mathbb{N}^\times) = C^*(\mathbb{N}^\times)$. The conditional expectation E_W $C^*(\mathbb{Q}/\mathbb{Z}) \to C^*(\mathbb{Q}/\mathbb{Z})^W \equiv B_{N^\times} = \overline{\mathrm{span}}\,\{\alpha_n(1) = \mu_n\mu_n^* \mid n \in \mathbb{N}^\times\}$ is

$$E_W(e(\gamma)) = \prod_{p|b}\left(\frac{p}{p-1}\alpha_{p^{k_p}}(1) - \frac{1}{p-1}\alpha_{p^{k_p-1}}(1)\right)$$

if $\gamma = a/b$, with $(a,b) = 1$, and $b = \prod p^{k_p}$ as a product of powers of primes.

The actions of θ of W and $\hat{\alpha}$ of $\hat{\mathbb{Q}}_+^\times$ commute and we have a commutin diagram

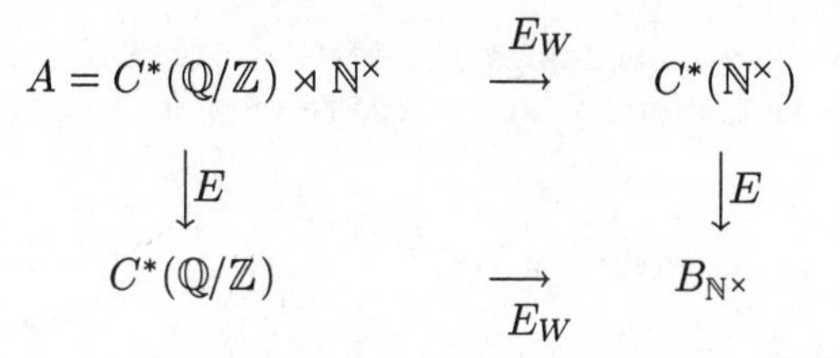

Since the W action on A commutes with the dynamics, there is an induced dynamical system on the W-fixed point algebra A^W whose analysis is immediate and hence determines the structure of the W-invariant KMS$_\beta$ states of (A, σ). In particular there is only one such W-invariant KMS$_\beta$ state of (A, σ) for each β.

Here $B_{\mathbb{N}^\times}$ is the C^*-subalgebra of $\ell^\infty(\mathbb{N}^\times)$ generated by the characteristic functions of $n\mathbb{N}^\times$ identified with $\mu_n\mu_n^*$, and the crossed product $B_{\mathbb{N}^\times} \rtimes \mathbb{N}^\times$ ($\subset C^*(\mathbb{Q}/\mathbb{Z}) \rtimes \mathbb{N}^\times$) is identified with the Toeplitz C^*-algebra $C^*(\mathbb{N}^\times) = C^*(\mu_n \mid n) = \overline{\text{lin}}\,\{\mu_m\mu_n^* \mid m, n \in \mathbb{N}^\times\} = \mathcal{T}(\mathbb{N}^\times) = \bigotimes_{p\in\mathcal{P}} \mathcal{T}_p$, where each $\mathcal{T}_p$ is a standard Toeplitz algebra. The restriction of σ to $\bigotimes_{p\in\mathcal{P}} \mathcal{T}_p$ is a product action $\otimes_p \sigma_t^p$, where σ_t^p is a gauge action $\sigma_t^p(\mu) = p^{it}\mu$. For each $0 < \beta < \infty$, there is a unique KMS state at inverse temperature β, φ_β. Here $\varphi_\beta = \otimes\varphi_\beta^p$, where φ_β^p is the state on the standard Toeplitz algebra $\mathcal{T}$ so that

$$\varphi_\beta^p(\mu^m\mu^{*n}) = (1 - p^{-\beta})p^{-\beta m}\delta_{mn}\,. \tag{7.15.2}$$

(If ψ is a KMS state on $\mathcal{T} = C^*(\mu)$, at inverse temperature β for the dynamics $\sigma_t(\mu) = p^{it}\mu$, implemented by $F(p^{it}) = \oplus_{m=0}^{\infty} \otimes^m p^{it}$ in Fock space, then

$$\psi(\mu^m\mu^{*n}) = \psi\left(\mu^{*n}\sigma_{-i\beta}(\mu^m)\right) = p^{-\beta m}\psi(\mu^{*n}\mu^m)\,.$$

But $\psi(\mu^m) = \psi(\sigma_t(\mu^m)) = p^{imt}\psi(\mu^m)$ so that $\psi(\mu^m) = 0$ if $m \neq 0$. Hence ψ is unique and is φ_β^p as in (7.15.2).) For $\beta > 1$, φ_β is type I_∞, given by normalizing $\text{Trace}(\cdot\, e^{-\beta H})$, and is factorial of type III_1 for $0 < \beta \le 1$, given by the Dixmier trace $\text{Tr}_\omega(\cdot\, e^{-H})$ for $\beta = 1$. The type of the state follows from the theory of ITPFI states (Araki and Woods 1968) – see also (Boca and Zaharescu, preprint 1996) for an analysis of product states over a subset of the primes which for example yields some III_0 factors.

Then the KMS structure of $(C_{\mathbb{Q}}, \sigma)$ is as follows:

(a) For $0 < \beta \le 1$, there exists an unique KMS state φ_β at inverse temperature β. The state φ_β generates the hyperfinite factor of type III_1, and its restriction to $C^*(\mathbb{Q}/\mathbb{Z})$ is given by the positive definite function on $\mathbb{Q}/\mathbb{Z}$:

$$n \to \prod_{p\in\mathcal{P},\, p|b} b^{-\beta}(1 - p^{\beta-1})(1 - p^{-1})^{-1} \tag{7.15.3}$$

where $n = a/b$, $a, b \in \mathbb{Z}$, $(a, b) = 1$, $b > 0$.

(b) For $\beta > 1$, KMS states at inverse temperature β form a simplex whose extreme points $\varphi_{\beta,\chi}$ are type I_∞ factor states and are identified with the compact group W, or complex embeddings $\chi : \mathbb{Q}^{\text{cycl}} \to \mathbb{C}$ of the subfield $\mathbb{Q}^{\text{cycl}}$ of $\mathbb{C}$ generated by the roots of unity, and whose restrictions to $C^*(\mathbb{Q}/\mathbb{Z})$ are given by

$$\varphi_{\beta,\chi}(e(\gamma)) = \zeta(\beta)^{-1} \sum_{n=1}^{\infty} n^{-\beta} \chi(\gamma)^n$$

where the normalization ζ is the Riemann zeta function. The unique W-invariant KMS state $\int_W \varphi_{\beta,\chi}\, d\chi$ for $\beta > 1$ is again given by (7.15.3).

In the low temperature region, $\beta > 1$

$$\nu_\beta = \zeta(\beta)^{-1} \textstyle\sum_n n^{-\beta} \mu_n^*(\cdot)\mu_n = \displaystyle\prod_{p \in \mathcal{P}} \left(\frac{1 - p^{-\beta}}{1 - p^{-\beta} \mu_p^*(\cdot)\mu_p} \right)$$

defines a completely positive unital map on A taking the ground states of (A, σ) to the KMS$_\beta$ states so that

$$\varphi_{\beta,\chi} = \varphi_{\infty,\chi} \circ \nu_\beta$$

where $\varphi_{\infty,\chi}(e(\gamma)) = \chi(\gamma)$, cf. Theorem 7.6 in the Ising case.

7.16 Notes

The presentation here of classical statistical mechanics in Section 7.1 and in 7.2 follows that of lecture courses given by J.T. Lewis at Dublin since 1972. Standard texts on statistical mechanics are (Griffiths 1972), (Ruelle 1972) (Thompson 1979) (Baxter 1982), (Simon 1994), (Bratteli and Robinson 1987) and (Bratteli and Robinson 1981) for quantum statistical mechanics. The presentation of the GKS inequalities Theorem 7.1 is taken from (Griffiths 1972) (following simplifications of (Ginibre 1969)) which has the details of the claim in Section 7.2 that the free energy $\lim_{\to} f(\Lambda)$ exists through arbitrary shapes Λ, not necessarily of size 2^n.

The two-dimensional Ising model with nearest neighbour interactions was shown in (Onsager 1941) to have a phase transition. There is a critical (inverse) temperature β_c such that the spontaneous magnetization $m^* = 0$ for $\beta < \beta_c$, and $m^* \neq 0$ for $\beta > \beta_c$. (Peierls 1936) showed the existence of a spontaneous magnetization, for low enough temperatures. (Kramers and Wannier 1941) showed by the high temperature–low temperature duality that if there exists an unique critical temperature T_c then it satisfies $\sinh 2\beta_c T = 1$, where $\beta_c = 1/kT_c$. (Onsager 1949) computed the spontaneous magnetization as

$$m^* = \begin{cases} [1 - (\sinh 2\beta J)^{-4}]^{\frac{1}{8}} & T < T_c \\ 0 & T > T_c \end{cases} .$$

(Yang 1952) verified this formula. Moreover for $\beta < \beta_c$, a unique equilibrium state exists, whilst for $\beta > \beta_c$, there are exactly two distinct extremal equilibrium states (Aizenmann 1980), (Higuchi 1979). (Schultz *et al.* 1964) reformulated the Onsager–Kaufman transfer matrix treatment, (Kaufman 1949), (Kaufman and Onsager 1949), using a Fermion algebra A^F.

The even subalgebras of A^P and A^F are canonically isomorphic, even though the algebras A^P and A^F themselves are not (for the system infinitely extended to all directions), and so one has a correspondence between even states on the Pauli spin algebra A^P and even states on the Fermion algebra A^F. For a two dimensional lattice, infinitely extended in all directions, the Gibbs state in the thermodynamic limit induces a pure (hence primary) Fock state on the Fermion algebra A^F (Pirogov 1972), (Lewis and Sisson 1975) , (Sisson 1975). (Lewis and Sisson 1975) discussed how the phase transition manifests itself by a jump in the index of a Fredholm operator associated with the Fock state. Subsequently, (Lewis and Winnink 1979) showed that the phase transition also reveals itself in the restricted state on the Fermion algebra $A^F([1,\infty))$ of a half-line (regarded as a subalgebra of A^F). The restricted state is a non-Fock quasi-free state. It is primary for $\beta > \beta_c$, and non-primary for $\beta < \beta_c$. Again this involves the computation of an index (mod 2) of a Fredholm operator. For a half lattice, the Pauli spin algebra $A^P([1,\infty))$ is canonically isomorphic to $A^F([1,\infty))$, and (Kuik 1986) showed that (for periodic boundary conditions) the thermodynamic limit of the Gibbs state induces precisely the above restricted state on $A^P([1,\infty))$. Hence for a half lattice, the manifestation of the phase transition is apparent. In the case of a two dimensional lattice, infinitely extended to all directions (in contrast to a half lattice), the state for the Fermion algebra is a pure (Fock state) for all β. (Araki and Evans 1983) showed that the phase transition manifests itself in the state for the Pauli spin algebra, which is primary for $0 \leq \beta \leq \beta_c$ and non-primary for $\beta > \beta_c$ by Theorem 7.6 (b). Subsequently, (Evans and Lewis 1986) constructed the automorphisms $\{\nu_\beta \mid \beta \neq \beta_c\}$ of Theorem 7.6 (a). Sections 7.8, 7.9, 7.11 and 7.12 are mainly taken from (Carey and Evans 1988) whilst the results of Section 7.14 are from (Araki 1986). Analyticity of correlation functions, Theorem 7.21, is due to (Araki 1986); the proof given here is taken from (Carey and Evans 1988). We refer to (Bratteli and Robinson 1981) for the details of the proofs of the properties of the co-cycle v_t which we outlined in the proof of Lemma 7.26.

(McKean 1964) showed that the high temperature–low temperature duality of Kramers and Wannier could be interpreted as a Poisson summation formula for certain finite groups. The treatment of high temperature–low temperature duality in the classical setting in Section 7.11 is based on (Kadanov and Ceva 1971), (Gruber *et al.* 1977) and (Palmer and Tracy 1981). The results of Exercise 7.1 are from (Araki 1984), (Araki and Matsui 1985), (Araki and Matsui 1986). The treatment of the identification in Section 7.10 of (Temperley and Lieb 1971) of the Potts model and six-vertex model is from (Baxter 1982). The embedding of the observable algebra, the gauge invariant Fermion algebra in a Temperley–Lieb algebra is from (Connes and Evans 1989).

The Q-state Potts model appears to exhibit different behaviour, as to whether $Q \leq 4$ or $Q > 4$. For large Q the Q-different translation invariant equilibrium states which exist by a Peierls argument at low temperatures, persists up to a critical temperature T_c, where there is at least one other state, of complex chaos, so that there are at least $Q+1$ different states at T_c. Above T_c it would appear that only the chaotic state survives. That this description exhausts all translation invariant equilibrium states has been shown in (Kotecky and Shlosman 1985), (Martroisan 1986). It would appear that the Q-state Potts model has a first-order phase transition for $Q > 4$, in that the internal energy U (the first derivative of the free energy) is continuous (Baxter *et al.* 1978), cf. the case $Q = 2$, the Ising model which has a second order phase transition, Section 7.5.

The algebra $\{W_i\}$ of (7.10.5) features in the chiral Potts model, (Baxter *et al* 1988), (Baxter 1988), (Baxter 1989a), (Baxter 1989b), (Capel and Perk 1977a), (Capel and Perk 1977b), (Capel and Perk 1978), (Capel and Perk 1980) including the non-hermitian operator

$$H = -\textstyle\sum_{i=1}^{r-1}\alpha_j W_{2j} - \sum_{i=1}^{r}\gamma_j W_{2j-1}\,. \tag{7.16.1}$$

We emphasized the computation of thermodynamic quantities in the Ising model (with small reference to the hard square and ABF models). We can refer to (Baxter 1982), (Au-Yang and Perk 1989) for reviews and references on the calculations in other integrable models. The Bethe ansatz methods, which we have not touched upon at all, are a vital part in the theory.

The introductions to (Perk 1979),(Perk 1981),(Au-Yang and Perk 1985), (Au-Yang and Perk 1987), (Au-Yang and Perk 1995) survey the literature on Ising and XY-correlation functions, and the difference and differential equations they give rise to. As is hinted at in (7.12.9) and section 7.12, two-point Ising correlation functions can be computed via Toeplitz determinants whose size depends on the distance between sites. The analysis of the asymptotic behaviour for large separation of spins is most intricate at criticality, see e.g. (Au-Yang and Perk 1987). (Away from criticality, the rapid exponential decay of correlations facilitates computation.)

The example in Section 7.15 of spontaneously broken symmetry related to the distribution of prime numbers is due to (Bost and Connes 1995), (see also (Laca) and (Laca and Raeburn)).

The Hecke algebras which are deformations of the group algebra of the symmetric group S_n (cf. Section 11.6 and (11.7.32)–(11.7.34)) arise as follows. Suppose q is the power of a prime, and $\mathbb{F}_q$ the finite field with q elements. Let $\Gamma = SL_n(\mathbb{F}_q)$ and Γ_0 the subgroup of upper triangular matrices. Writing $\Gamma = \Gamma_0 S_n \Gamma_0$, we have a basis of $H(\Gamma,\Gamma_0)$ given by the characteristic functions $T_\omega \equiv \chi_{\Gamma_0\omega\Gamma_0}$, $\omega \in S_n$. Then $H(\Gamma,\Gamma_0)$ is generated by $T_i \equiv T_{(i,i+1)}$, corresponding to the transpositions $(i, i+1)$ with relations

$$\begin{gathered}(T_i+1)(T_i-q)=0\\ T_iT_j = T_jT_i\,, \quad |i-j|>1\,,\\ T_iT_{i+1}T_i = T_{i+1}T_iT_{i+1}\,.\end{gathered}$$

8

CONFORMAL FIELD THEORY

8.1 Introduction

From a lattice model one may obtain a field theory by taking a continuum or scaling limit; letting the lattice spacing $\varepsilon \to 0$, whilst simultaneously approaching the critical temperature. As the scale or correlation length becomes infinite one obtains a scale invariant or conformal invariant theory. (Belavin *et al.* 1980) suggested that the scale invariance at a critical point is enhanced to conformal invariance or using conformal invariance to understand scale invariance of special critical points. Conformal invariance is understood via representations of the Virasoro algebra, and a knowledge of the representations that appear tells us something about the statistical mechanical model, and the nature of the critical point. In this chapter we look at some of the basic principles of the theory, particularly with respect to those aspects of most interest to the operator algebraic approach to statistical mechanics and quantum field theory. We look in detail at the Ising model where many, but of course not all, the features are present and can be analysed explicitly. This means that we also emphasize the viewpoint of conformal field theory from critical statistical mechanical models, although there are other important aspects to the theory, particularly because conformal field theories describe the ground states of string theory.

In Section 8.2 we give the necessary background from two dimensional conformal field theory, in particular the formalism of the operator product expansion, which details how primary fields (those quantum fields which transform well under conformal transformations) can be combined and decomposed into combinations of other primary fields. This is analogous to Clebsch–Gordan rules for products of group representations, indeed the operator product coefficients are like the connection information of Section 8.2 with embeddings of AF algebras, and the parallels with subfactor theory will be particularly brought out in Chapter 9.

In Section 8.3 we begin to analyse the Ising model in some detail, particularly through the corner transfer matrix method of Baxter, and begin to see which representations of the Virasoro algebra appear at criticality. Characters of chiral algebras appear naturally, and these characters are analysed in more detail in Section 8.8 where we consider the modular invariance of partition functions of conformal field theories on the torus. The continuum limit of the Ising model is handled in Section 8.4, where the primary fields and critical exponents are discussed. If we ignore much of the information in the operator product expansion and only consider the multiplicity, much as one does with group representations when one concentrates on the Littlewood–Richardson coefficients instead of the Clebsch–Gordan coefficients, then the multiplicities define a fusion rule algebra FRA. This is analysed in Sections 8.5 and 8.6, for the Ising model, and again in Section 8.7 more generally where we analyse the Moore–Seiberg axioms for a rational conformal field theory with a finite number of primary fields. In Section 8.8 we describe the role of affine Lie algebras, current algebras or Kač–Moody algebras in conformal field theory, the Sugawara construction and how it can be used via the coset construction to yield the interesting representations of the Virasoro algebra, and connections with operator algebra theory and statistical mechanics, in particular how branching coefficients are related to the computations of one point functions in the corner transfer matrix method. Finally in Section 8.9 (see also the notes to this chapter) we consider the question of describing modular invariant partition functions.

In comparison with the rest of this book we are cavalier in this chapter with aspects of rigour, especially in terms of domains and convergence, and manipulation of operator valued distributions, (there are some exceptions — e.g. the Ising analysis of Sections 8.3 and 8.6). The object here is however to provide a link between the operator algebra treatments of statistical mechanics in Chapter 7 and of subfactor theory and TQFT (topological quantum field theory) of Chapters 9 onwards, and conformal field theory. We do not attempt to be exhaustive nor provide any treatment of quantum groups, and only a superficial discussion on link invariants can be found in this book.

8.2 Primary fields, operator product expansion, and correlation functions

We have a vector space $\mathcal{F}$ of local quantum fields $\phi(z, \overline{z})$ with space–time dependence $z = x + it$, which act on a Hilbert space H of states. Fields (operators) and states (vectors) are often interchanged. Indeed, there is an isomorphism from $\mathcal{F}$ into (a dense subspace of) H. This is the Reeh–Schlieder theorem. Given a field ϕ in $\mathcal{F}$ we form a vector $v(\phi)$ in H by

$$v(\phi) = \lim{}_{z,\overline{z}\to 0}\phi(z, \overline{z})v(0) \tag{8.2.1}$$

where $v(0)$, the vacuum, is separating and cyclic for the algebra generated by $\mathcal{F}$. We will begin our discussion of the background to CFT with the field picture.

The fields ϕ are defined on a d-dimensional space–time M^d with metric g i.e. at each point x of M^d, we have an inner product on $\mathbb{R}^d$ (the tangent space at x) given by $\langle v, w\rangle = \sum g_{\mu\nu}v^\mu w^\nu \equiv g_{\mu\nu}v^\mu w^\nu$, e.g. Euclidean space $\mathbb{R}^d$, $g = 1_d$ Minkowski space $\mathbb{R}^{p+q}$, $g = \eta = 1_p \oplus (-1_q)$. We use the metric g to lower

	infinitesimal	exponentiated	Ω
translation	$\varepsilon^\mu = a^\mu$ $a \in \mathbb{R}^d$	$x'^\mu = x^\mu + a^\mu$	1
Lorentz transformations	$\varepsilon^\mu = \omega^\mu{}_\nu x^\mu$ ω anti-symmetric	$x'^\mu = \Lambda^\mu{}_\nu x^\nu$ $\Lambda \in SO(p,q)$	1
scale transformation	$\varepsilon^\mu = \lambda x^\mu$ $\lambda \in \mathbb{R}$	$x'^\mu = \lambda x^\mu$	λ^{-2}
special conformal	$\varepsilon^\mu = b^\mu x^2 - 2x^\mu b \cdot x$	$x'^\mu = \frac{x^\mu + b^\mu x^2}{1+2b\cdot x + b^2 x^2}$	$(1 + 2b \cdot x$ $+ b^2 x^2)^2$

Table 8.1

indices on vectors v^μ to form co-vectors $v_\mu = g_{\mu\nu} v^\mu$, so that $\langle v, w\rangle = v_\mu w^\mu$, identifies the dual of the space of vectors with co-vectors. We raise indices from co-vectors to vectors with $g^{\mu\nu} \equiv \left(g^{-1}\right)_{\mu\nu}$. For each coordinate transformation (diffeomorphism) $f : M^d \to M^d$ we can transform the metric $g \to f_*(g) \equiv g \circ f^{-1} \equiv g'$ so that $f = (M^d, g) \to (M^d, g')$ is isometric, but the identity map $(M^d, g) \to (M^d, g')$ is not necessarily so. If the identity map $(M^d, g) \to (M^d, g')$ induces, for each $x \in M^d$, simply a scalar change $\Omega(x) \neq 0$ on the metric, then f is said to be a *conformal transformation.* Equivalently in the Euclidean case, the map $f = (M^d, g) \to (M^d, g)$ preserves angles (where the angle between two vectors v and w in $\mathbb{R}^d$ is defined by $\langle v, w\rangle/(v^2 w^2)^{1/2}$). The set of conformal transformations is the *conformal group*, a subgroup of the co-ordinate transformations.

We first consider infinitesimal coordinate transformations $x'^\mu = x^\mu + \varepsilon^\mu$ which scale the metric $g = \eta$ of signature p, q. These are the infinitesimal generators of the conformal group. Then with $\partial_v = \partial/\partial x^\nu$, $d\varepsilon_\mu = \partial_\nu \varepsilon_\mu \partial x^\nu$,

$$dx'^2 = dx^2 + (\partial_\nu \varepsilon_\mu + \partial_\mu \varepsilon_\nu)\, dx^\mu\, dx^\nu\,.$$

For this to be a multiple $\Omega(x)$ of dx^2 we need $\partial_\nu \varepsilon_\mu + \partial_\mu \varepsilon_\nu$ to be multiple of $\eta_{\mu\nu}$

$$\partial_\nu \varepsilon_\mu + \partial_\mu \varepsilon_\nu = (2/d)(\partial \cdot \varepsilon)\eta_{\mu\nu} \tag{8.2.2}$$

where the scalar is obtained by considering traces (multiplying by $\eta^{\mu\nu}$ and summing). From (8.2.2), we derive

$$\left(\eta_{\mu\nu}\partial_\theta \partial^\theta + (d-2)\partial_\mu \partial_\nu\right)(\partial \cdot \varepsilon) = 0\,. \tag{8.2.3}$$

For $d > 2$, equations (8.2.2) and (8.2.3) ensure that all third derivatives $\partial_\alpha \partial_\beta \partial_\gamma \varepsilon^\mu$ vanish and so ε^μ is quadratic in x^λ. Such quadratic infinitesimal transformations can be listed as in Table 8.1 (even for $d = 2$). The first two classes of transformations, translations and orthogonal transformations generate the *Poincaré group* $SO(p,q) \times \mathbb{R}^d$ of dimension $d(d-1)/2 + d$. Together with the dilations of dimension 1 they generate the *Weyl group.* The special conformal transforma-

	one-parameter subgroup	X	ℓ
translation	$\begin{pmatrix} 1 & \tau \\ 0 & 1 \end{pmatrix}$	$\begin{pmatrix} 0 & -1 \\ 0 & 0 \end{pmatrix}$	$\frac{d}{du} = (1-\cos\theta)\frac{d}{d\theta}$
rotations	$\begin{pmatrix} \cos\frac{\tau}{2} & \sin\frac{\tau}{2} \\ -\sin\frac{\tau}{2} & \cos\frac{\tau}{2} \end{pmatrix}$	$\frac{1}{2}\begin{pmatrix} 0 & -1 \\ 1 & 0 \end{pmatrix}$	$\frac{(u^2+1)}{2}\frac{d}{du} = \frac{d}{d\theta}$
dilatations	$\begin{pmatrix} e^{\frac{\tau}{2}} & 0 \\ 0 & e^{-\frac{\tau}{2}} \end{pmatrix}$	$\frac{1}{2}\begin{pmatrix} -1 & 0 \\ 0 & 1 \end{pmatrix}$	$u\frac{d}{du} = -\sin\theta\frac{d}{d\theta}$
special conformal	$\begin{pmatrix} 1 & 0 \\ \tau & 0 \end{pmatrix}$	$\begin{pmatrix} 0 & 0 \\ -1 & 0 \end{pmatrix}$	$-u^2\frac{d}{du} = -(1+\cos\theta)\frac{d}{d\theta}$

Table 8.2

tions can be expressed (also infinitesimally) as $x'/x'^2 = x/x^2 + b$, composed of translations and inversions (the *inversion* is $x \to x/x^2$), of dimension d. Adding this gives a total of $(d+1)(d+2)/2$ dimensions, and indeed they generate a group isomorphic to $SO(p+1, q+1)$. In three dimensions or more, the conformal group is finite dimensional. However, in two dimensions, the conformal group is infinite dimensional. In the Euclidean case $g = 1$, (8.2.3) reduces to the Cauchy–Riemann equations $\partial_1\varepsilon_1 = \partial_2\varepsilon_2$, $\partial_1\varepsilon_2 = -\partial_2\varepsilon_1$. Switching from $\mathbb{R}^2$ to the complex plane $z, \overline{z} = x^1 \pm ix^2$, $\varepsilon, \overline{\varepsilon} = \varepsilon^1 \pm i\varepsilon^2$ the conformal transformations consists of maps $z \to f(z)$, $\overline{z} \to \overline{f}(\overline{z})$, where f is holomorphic, and $\overline{f}$ is antiholomorphic, with $\Omega = |\partial f/\partial z|^2$. A basis for infinitesimal transformations would be $z \to z - z^{n+1}$, $\overline{z} \to \overline{z} - \overline{z}^{n+1}$. On functions, these give the generators of the vector fields $\ell_n = -z^{n+1}\partial_z$, $\overline{\ell}_n = -\overline{z}^{n+1}\partial_{\overline{z}}$, with commutation relations

$$[\ell_m, \ell_n] = (m-n)\ell_{m+n}\,, \quad [\overline{\ell}_m, \overline{\ell}_n] = (m-n)\overline{\ell}_{m+n}\,.$$

Only $\ell_0 = -z\partial_z = w\partial_w$, $\ell_1 = -z^2\partial_z = -\partial_w$, $\ell_{-1} = -\partial_z = -w^2\partial_w$ are globally defined on the Riemann sphere $S^2 = \mathbb{C} \cup \{\infty\}$ (under the transformation $z = -1/w$), and they generate the global conformal group or restricted conformal transformations. The other vector fields $\{\ell_n : n \neq 0, \pm 1\}$ do not generate invertible holomorphic transformations $z \to f(z)$ of $\mathbb{C} \cup \{\infty\}$. Corresponding to Table 8.1 in two dimensional Euclidean space we have Table 8.2. A one-parameter group on points with infinitesimal generator X yields a differential operator ℓ on functions: $e^{t\ell} f(w) = f(e^{-tX}w)$ with $\ell f(w) = (d/dt)|_{t=0}(e^{-tX}w)(df/dw)$. In this way we identify with the translation, rotation, dilation and special conformal one-parameter groups, the vector fields as in the fourth column. We have used a variable u which is the Cayley transform of z. Under the Cayley transform $u = e^{i\theta} = (z-i)(z+i)^{-1}$ which maps $\mathbb{R}, i, -i$, to $\mathbb{T}\backslash\{1\}, 0, \infty$, we identify $[(u^2+1)/2]d/du = d/d\theta$, rotation through θ (with $0, \infty$ the fixed points corresponding to the zeros $i, -i$ of w^2+1). Then $\ell_0 = -zd/dz$, $\ell_1 = -z^2d/dz$, $\ell_{-1} = -d/dz$ are related to the corresponding operators in the u variable via:

$$\begin{aligned} d/du &= (1-\cos\theta)\, d/d\theta = i\ell_0 - i(\ell_1 + \ell_{-1}/2 \\ ud/du &= -\sin\theta\, d/d\theta = (\ell_1 - \ell_{-1})/2 \\ u^2 d/du &= (1+\cos\theta)\, d/d\theta = i\ell_0 - i(\ell_1 + \ell_{-1})/2 \end{aligned}$$

In a compactified Minkowski space $M = S^1 \times S^1$ the conformal groups becomes $G = \operatorname{Diff} S^1 \times \operatorname{Diff} S^1$ or two commuting copies of Diff S^1, left and right. Let us look separately at each copy for the time being. We thus have the *Witt algebra*, the Lie algebra of vector fields on S^1 (or $\mathbb{C}$) generated by $\ell_m = -z^{m+1}d/dz$. However it is not representations of Diff S^1 but projective ones which interest us, if in quantum theory it is vector rays which are used, not the vectors themselves. Thus, it is not the Witt algebra which we require but an extension, a larger algebra, called the *Virasoro algebra*, Vir, generated by L_m together with a central element c, satisfying

$$[L_m, L_n] = (m-n)L_{m+n} + \delta_{m,-n}(m^3 - m)c/12\,. \tag{8.2.4}$$

The Virasoro algebra is the unique non-trivial central extension of the Witt algebra, in the following sense. If $[\,,]_f$ is a bracket on Witt $\oplus$ $\mathbb{C}c$, where c is a central element $[L_m, c]_f = 0$ and

$$[L_m, L_n]_f = (m-n)L_{m+n} + f(L_m, L_n)\, c$$

for a bilinear map f, then there is a linear map a on the Witt algebra, $\lambda \in \mathbb{C}$ such that $f(L_m, L_n) = \lambda b(L_m, L_n) + a([L_m, L_n])$ where $b(\ell_m, \ell_n) = \delta_{m,-n}(m^3 - m)/12$. Then $L_m \to L_m + a(L_m)\lambda c$ will identify Witt$\oplus\mathbb{C}c$ with Vir. See Exercise 8.1.

We expect the state space H which carries representations of Vir and $\overline{\text{Vir}}$ to split as

$$H = \bigoplus\nolimits_{i,j} H_i \otimes H_{\bar{\jmath}} \tag{8.2.5}$$

where H_i (respectively $H_{\bar{\jmath}}$) carries an irreducible representation of Vir (respectively $\overline{\text{Vir}}$). In the Minkowski setting, $L_0 + \overline{L}_0$ is the energy operator, which from physical considerations is expected to be bounded below, indeed L_0 and $\overline{L}_0$ are separately bounded below, being independent. Thus putting $H = H_i$, for fixed i, we are led to a highest weight representation, where there is a vector $v_h \in H$, and a scalar $h \in \mathbb{R}$ representing the lowest energy, such that

$$L_0 v_h = h v_h\,,$$

and where the central element c is scalar, from irreducibility. The vacuum vector v_0 is assumed to be regular at $z \to 0$ when acted upon by $T(z) = \sum_{-\infty}^{\infty} z^{-n-2} L_n$, equivalently $L_n v_0 = 0$ for $n \geq -1$. In particular, as $L_{0,\pm 1} v_0 = 0$, the vacuum is $SL(2,\mathbb{R})$ invariant. Since $L_0 L_n v_h = (h-n) L_n v_h$ it follows from (8.2.4), that if the energy is bounded below by h, then:

$$L_n v_h = 0 \quad \text{for } n > 0\,.$$

Moreover H is spanned by

$$L_{-n_1} \cdots L_{-n_m} v_h : n_1 \geq \cdots \geq n_m > 0 \tag{8.2.6}$$

which may not be linearly independent, and is called a *Verma module* $V(h,c)$. The linear span of vectors in (8.2.6) for a fixed m, called the *level*, forms a subspace H^m, where L_0 has eigenvalue $h + |I|$, if $|I| = n_1 + \cdots + n_m$, on (8.2.6). Indeed from the minimal energy property, we see that L_0 has no other eigenvalues. Thus H is graded as $H = \bigoplus_{m=0}^{\infty} H^m$. States of the form $L_{-n_1} \cdots L_{-n_m} v_h$ with $n_1 \geq \cdots \geq n_m > 0$ are called *descendent states.* In the (generic) case where descendent states are linearly independent, the dimension of H^m is $p(m)$ the partitions of m. It is convenient to write $L_I = L_{-n_1} \cdots L_{-n_m}$, if $I = \{n_1, \ldots, n_m\}$ with $n_1 \geq \cdots \geq n_m > 0$.

When switching from the vector picture to fields, the highest weight vectors give a distinguished class of fields, the *primary fields* which transform as

$$\phi(z,\overline{z}) = (f'(z))^h \left(\overline{f}'(z)\right)^{\overline{h}} \phi\left(f(z), \overline{f}(\overline{z})\right) \tag{8.2.7}$$

under the conformal transformation $z \to f(z)$. In particular the identity operator 1 is a primary field, corresponding to the vacuum. Equivalently, considering the infinitesimal version of (8.2.7) and writing $T(z) = \sum_{-\infty}^{\infty} z^{-n-2} L_n$, primary fields are those that transform as

$$T(z)\phi(w,\overline{w}) = \frac{h}{(z-w)^2}\phi(w,\overline{w}) + \frac{1}{(z-w)}\partial_w \phi(w,\overline{w}) + \text{term regular in } z-w \tag{8.2.8}$$

i.e. a singularity is at worst $(z-w)^{-2}$, and analogously for $\overline{T}(\overline{z})$. However, T itself is not a primary field as

$$T(z) = f'(z)^2 T(f(z)) + \frac{c}{12}\{f(z), z\}\,, \tag{8.2.9}$$

where $\{f(z), z\}$ is the *Schwarzian derivative* $f'''/f' - (3/2)(f''/f')^2$, but T is still *quasi-primary* in that the transformation rule (8.2.7) holds for the Möbius group transformations, as the Schwarzian vanishes for $SL(2,\mathbb{R})$. Thus $T(w)$, as a field, has a worse singularity than $(z-w)^{-2}$ when acted on by $T(z)$:

$$T(z)T(w) = \frac{c/2}{(z-w)^4} + \frac{2T(w)}{(z-w)^2} + \frac{\partial T(w)}{(z-w)}\,. \tag{8.2.10}$$

From (8.2.8) we have

$$[L_n, \phi(w)] = \int \frac{dz}{2\pi i} T(z)\phi(w) = h(n+1)w^n\phi(w) + w^{n+1}\partial\phi(w)\,. \qquad (8.2.11)$$

In particular $[L_n, \phi(0)] = 0$, $n > 0$ and so (see (8.2.1)) the state $v_h = \phi(0)v_0$ is indeed a highest weight vector:

$$L_0 v_h = h v_h\,, \quad L_n v_h = 0 \quad n > 0\,.$$

This provides the map from primary fields to the the highest weight vectors.

We now start to discuss the unitary theories, where there is a positive definite inner product. This is natural if we wish to relate to our operator algebra upbringing.

In a unitary theory $L_n^* = L_{-n}$. This is the case if we wish to identify $T^*(z)$ with $T(1/\overline{z})1/\overline{z}^4$. The adjoint of an Euclidean field $A(z,\overline{z})$ is $A(1/\overline{z}, 1/z)\overline{z}^{-2h}z^{-2\overline{h}}$, where the factors $\overline{z}^{-2h}z^{-2\overline{h}}$ are required by conformal considerations ((8.2.7) for $z \to 1/z$) and $z \to 1/z^*$ reverses time $\tau \to -\tau$ in the Minkowski space cylinder picture. A closed string sweeps out as time changes a surface homeomorphic to a cylinder, described say by a compact space coordinate σ and time τ. Mapping to the plane by $z \to e^{\tau + i\sigma}$, takes time development in the cylinder to the radial direction in the plane. Unitarity of the field theory $L_n^* = L_{-n}$ corresponds to reflection positivity of the original statistical mechanical model (cf. reflection positivity corresponds to positivity of the quantum state in Section 7.14).

Unitarity imposes severe restrictions on the possible values of c and h and linear dependence of descendent states, especially for $c < 1$. From $c/2 = \langle [L_2, L_{-2}]\, v_0, v_0\rangle = \langle L_2 L_2^* v_0, v_0\rangle \geq 0$, we immediately have $c \geq 0$. However, $c = 0$ even forces $L_n \equiv 0$, for *all* n. Indeed, from

$$\langle L_{-n}^* L_{-n} v_0, v_0\rangle = \langle [L_n, L_{-n}]\, v_0, v_0\rangle = (2nh + c(n^3 - n)/12)\langle v_0, v_0\rangle \qquad (8.2.12)$$

for $n > 0$, we would see $L_{-n}v_0 = 0$ for all n at least if $h = 0$. However, if $h \neq 0$, the determinant of the inner products of the two vectors $L_{-2n}v_h$, $L_{-n}^2 v_h$ is $4n^3h^2(4h - 5n)$, which is negative for $n >> 0$. Note that (8.2.12) shows that $h \geq 0$ by considering $n = 1$.

Let us consider what happens if the descendent states are not linearly independent so that $\dim H^m < p(m)$, for some $m = 1, 2, \ldots$. In the case of level $m = 1$, $L_{-1}v = 0$ so that $2L_0 v = [L_1, L_{-1}]\, v = 0$ and so $h = 0$, and v is the unique vacuum vector. In the case of level $m = 2$, then $L_{-2}v$ and $L_{-1}^2 v$ are linearly dependent:

$$aL_{-2}v + bL_{-1}^2 v = 0\,. \qquad (8.2.13)$$

Assuming $L_{-1}v \neq 0$, then by applying L_1 and L_2 in turn and using the Virasoro relations, we find $a \neq 0$, say $a = 1$, then $b = -3/2(2h+1)$, and $c = 2h(5 -$

	$\frac{1}{2}$	0	$\frac{3}{2}$	$\frac{7}{16}$	0	3	$\frac{7}{5}$	$\frac{2}{5}$	0
	$\frac{1}{16}$	$\frac{1}{16}$	$\frac{3}{5}$	$\frac{3}{80}$	$\frac{1}{10}$	$\frac{13}{8}$	$\frac{21}{40}$	$\frac{1}{40}$	$\frac{1}{8}$
	0	$\frac{1}{2}$	$\frac{1}{10}$	$\frac{3}{80}$	$\frac{3}{5}$	$\frac{2}{3}$	$\frac{1}{15}$	$\frac{1}{15}$	$\frac{2}{3}$
$q\uparrow$			0	$\frac{7}{16}$	$\frac{3}{2}$	$\frac{1}{8}$	$\frac{1}{40}$	$\frac{21}{40}$	$\frac{13}{8}$
$\vec{p}$						0	$\frac{2}{5}$	$\frac{7}{5}$	3
	$m=3, c=\frac{1}{2}$		$m=4, c=\frac{7}{10}$			$m=5, c=\frac{4}{5}$			
	Ising model		Hard square						

Table 8.3 *Values of h_{pq} for $m = 3, 4, 5$*

$8h)/(2h+1)$. In general, we have to compute the determinants of inner products between descendant states at a fixed level to understand linear dependence. At level 2, we have

$$\begin{pmatrix} \langle L_{-2}v, L_{-2}v\rangle & \langle L_{-2}v, L_{-1}^2 v\rangle \\ \langle L_{-1}^2 v, L_{-2}v\rangle & \langle L_{-1}^2 v, L_{-1}^2 v\rangle \end{pmatrix} = \begin{pmatrix} 4h + c/2 & 6h \\ 6h & 4h(1+2h) \end{pmatrix}$$

which has determinant

$$2(16h^3 - 10h^2 + 2h^2c + hc) = 32(h - h_{1,1}(c))(h - h_{1,2}(c))(h - h_{2,1}(c))$$

where $h_{1,1}(c) = 0$ and $h_{1,2}, h_{2,1} = \left[(5-c) \pm \sqrt{(1-c)(25-c)}\right]/16$ respectively. At level N, the Kač value (Kač 1990) of the determinant of inner product of the $P(N)$ vectors is

$$\alpha_N \prod_{pq\leq N} (h - h_{p,q}(c))^{P(N-pq)}\,. \tag{8.2.14}$$

Here α_N is a constant independent of c and h and if $m = \left(-1 \pm \sqrt{(25-c)/(1-c)}\right)/2$, i.e. $c = 1 - 6/m(m+1)$, then

$$h_{pq} = \left[((m+1)p - mq)^2 - 1\right]/4m(m+1)\,. \tag{8.2.15}$$

Note that h is symmetric under the interchange, $p \to m - p$, $q \to m + 1 - p$, the flip in each variable. Thus there are $m(m-1)/2$ values of $h_{p,q}$ if q is integral in the range $1 \leq q \leq m$. For example see Table 8.3. To avoid double counting, take either those h-values on or below the $q = p$ diagonal, or those where $p+q$ is even. An admissible value (8.2.15) yields a primary state $v(h_{p,q})$, with L_0 eigenvalue $h_{p,q}$ which has a null descendent (a singular vector of zero length or orthogonal to every vector – hence zero for a positive definite inner product) at level pq; e.g.

when $N = 2$, the factor $h - h_{1,1} = h - 0$ in the Kač determinant arises because $L_{-1}v$ itself vanishes. In general the exponent of the factor $(h-h_{p,q})^{p(N-pq)}$ arises in (8.2.14) because if a null state $v(h+n)$ occurs at level n $(= pq)$, then $L_I v(h+n)$ yields $p(N-n)$ null descendents at level N if $|I| = N - n$.

By an analysis of the zeros of the Kač determinant, (Friedan *et al.* 1984) have shown that for $0 < c < 1$, unitary highest weight representations of the Virasoro algebra can at most occur for $c = 1-6/m(m+1)$ for $m = 3, 4\ldots$, with $m(m-1)/2$ possible values of h given by (8.2.15). The analysis of the Kač determinant does not eliminate anything in the range $(c \geq 1,\ h \geq 0)$. The curves $h = h_{p,q}(c)$ are drawn in the (c,h) plane for $pq \leq N$. Across the curves, a single eigenvalue changes sign. Increasing N, and introducing more curves allows more and more (c,h) values to be discarded until we reach the discrete set of (c,h) values for $0 < c < 1$. We shall discuss existence of unitary representations especially for $c < 1$, and the coset space construction in Section 8.8, following (Goddard *et al.* 1986) showing that representations occur at precisely these discrete values.

We have discussed the discrete series $c < 1$ in terms of singular or null vectors. In the field picture there are corresponding singular or null fields. Indeed taking vacuum expectation values $\langle\ \rangle$ of fields, and using the commutation relation (8.2.11), we obtain the relation

$$\begin{aligned}\langle L_{-n}\phi_1(z_1,\overline{z}_1)\phi_2(z_2,\overline{z}_2)\cdots\rangle &= \mathcal{L}_{-n}\langle\phi_1(z_1,\overline{z}_1)\phi_2(z_2,\overline{z}_2)\cdots\rangle \\ &\quad + \langle\phi_1(z_1,\overline{z}_1)\phi_2(z_2,\overline{z}_2)\cdots L_{-n}\rangle \qquad (8.2.16)\end{aligned}$$

where

$$\mathcal{L}_{-n} = \sum_{j\neq 1}\left[\frac{(1-n)h_j}{(z_1-z_j)^n} - \frac{1}{(z_1-z_j)^{n-1}}\frac{\partial}{\partial z_j}\right] \qquad (8.2.17)$$

which shows that $\phi_{p,q}$ is annihilated by differential operator of order pq. In some cases such differential equations can be solved to compute correlation functions, as we shall explicitly see in Section 8.5 for the Ising model.

By looking in more detail at the structure of null or singular vectors and their levels, their explicit expressions as descendent states and their embedding relations (in terms of the associated modules) one can derive expansions for the characters $\chi_h = \mathrm{tr}\, q^{(L_0-c/24)}$ of the Verma module $V(c,h)$ (Rocha-Caridi 1984), (Kent 1991). The insertion of the $-c/24$ correction facilitates modular transformation properties and arises from conformal transformation of the plane to a cylinder (8.10.1). If η is the Dedekind function, the character of the Verma module when all $p(n)$ descendant states at level n are linearly independent is given by

$$q^{-c/24+h}\sum_{n=0}^{\infty}p(n)q^n = q^{-c/24+h}\prod_{n=1}^{\infty}(1-q^n)^{-1} = q^{-(c-1)/24}\eta(q)^{-1}q^h \qquad (8.2.18)$$

A null vector ξ at level pq means we have over-counted in (8.2.18), so we must modify (omitting the overall factor $q^{-(c-1)/24}\eta(q)^{-1}$) $q^{h_{p,q}}$ to $q^{h_{p,q}} - q^{h_{p,q}+pq}$. Since $h_{p,q} + pq = h_{m+p,m+1-q} = h_{p,-q}$ the module given by ξ itself has a null vector at level $(m+p)(m+1-q)$, and we must compensate again by replacing $q^{h_{p,q}} - q^{h_{p,-q}}$ by $q^{h_{p,q}} - q^{h_{p,-q}}(1-q^{(m+p)(m+1-q)})$. Repeating this process *ad infinitum*, and by a careful analysis of the null vectors, and how the various modules embed, we arrive at the character formula:

$$\begin{aligned}\chi_{h_{p,q}} &= \chi_{p,q}(q) = \operatorname{tr} q^{(L_0 - c/24)} \\ &= q^{-(c-1)/24}\eta(q)^{-1}\sum_{k\in\mathbb{Z}}\left(q^{h_{2mk+p,q}} - q^{h_{2mk+p,-q}}\right). \end{aligned} \tag{8.2.19}$$

We will look at some examples. In the case $m=2$, $c=0$, $h=0$ with a one dimensional Verma module this is precisely the Jacobi triple product identity:

$$\chi_{1,1}(q) = \left(\prod_{i=1}^{\infty}(1-q^i)\right)^{-1}\sum_{k\in\mathbb{Z}}(-1)^k q^{(3k^2+k)/2} = 1. \tag{8.2.20}$$

The first two interesting cases $m=3$ and $m=4$, which will correspond to the Ising and hard square models respectively. For brevity, we will omit the factor $q^{-c/24}$.

$$m = 3, c = 1/2$$

$$\chi_0 = \chi_{1,1}(z) = q^{1/24}\eta(q)^{-1} \sum_{\substack{n\in\mathbb{Z}\\ n=0,1 \bmod 4}} (-1)^n q^{\frac{3n^2+n}{4}},$$

$$\chi_{1/2} = \chi_{2,1}(z) = q^{1/24}\eta(q)^{-1} \sum_{\substack{n\in\mathbb{Z}\\ n=2,3 \bmod 4}} (-1)^{n+1} q^{\frac{3n^2+n}{4}},$$

$$\chi_{1/16} = \chi_{2,2}(z) = q^{1/24}\eta(q)^{-1}\sum_{k\in\mathbb{Z}}(-1)^k q^{3k^2+k}.$$

Jacobi's triple product identity (8.2.20) yields

$$\begin{aligned}\chi_{1,1} \pm \chi_{2,1} &= \prod_{n\in\mathbb{N}-\frac{1}{2}}(1\pm q^n), \chi_{2,2} = q^{\frac{1}{16}}\prod_{n\in\mathbb{N}}(1+q^n), \\ \chi_{1,1} = \chi_0 &= \frac{1}{2}\left\{\prod_{n=1}^{\infty}\left(1+q^{n-\frac{1}{2}}\right) + \prod_{n=1}^{\infty}\left(1-q^{n-\frac{1}{2}}\right)\right\} \\ &= 1 + q^2 + q^3 + 2q^4 + \cdots, \\ \chi_{2,1} = \chi_{1/2} &= \frac{1}{2}\left\{\prod_{n=1}^{\infty}\left(1+q^{n-\frac{1}{2}}\right) - \prod_{n=1}^{\infty}\left(1-q^{n-\frac{1}{2}}\right)\right\} \\ &= q^{\frac{1}{2}}\left(1+q+q^2+q^3+\cdots\right), \\ \chi_{2,2} = \chi_{1/16} &= q^{\frac{1}{16}}\prod_{n=1}^{\infty}(1+q^n) = q^{\frac{1}{16}}(1+q+q^2+2q^3+\cdots). \end{aligned}$$

In the next section, we will derive these characters for $c=1/2$ from the corner transfer matrices of the Ising model.

$$m = 4, c = 7/10$$

$$\chi_0 = \chi_{1,1}(z) = q^{\frac{1}{24}}\eta(q)^{-1} \sum_{\substack{n\in\mathbb{Z}\\ n=0,3 \bmod 4}} (-1)^n q^{\frac{5n^2+n}{4}},$$

$$\chi_{3/2} = \chi_{3,1}(z) = q^{\frac{1}{24}}\eta(q)^{-1} \sum_{\substack{n\in\mathbb{Z} \\ n=1,2 \bmod 4}} (-1)^{n+1} q^{\frac{5n^2+n}{4}},$$

$$\chi_{1/10} = \chi_{3,3}(z) = q^{\frac{1}{10}+\frac{1}{24}}\eta(q)^{-1} \sum_{\substack{n\in\mathbb{Z} \\ n=0,1 \bmod 4}} (-1)^{n} q^{\frac{5n^2+3n}{4}},$$

$$\chi_{3/5} = \chi_{3,2}(z) = q^{\frac{1}{10}+\frac{1}{24}}\eta(q)^{-1} \sum_{\substack{n\in\mathbb{Z} \\ n=2,3 \bmod 4}} (-1)^{n+1} q^{\frac{5n^2+3n}{4}},$$

$$\chi_{7/16} = \chi_{2,1}(z) = q^{\frac{7}{16}+\frac{1}{24}}\eta(q)^{-1}\textstyle\sum_{k\in\mathbb{Z}}(-1)^k q^{-(5k^2+3k)},$$

$$\chi_{3/80} = \chi_{2,2}(z) = q^{\frac{3}{80}+\frac{1}{24}}\eta(q)^{-1}\textstyle\sum_{k\in\mathbb{Z}}(-1)^k q^{5k^2+3k}.$$

Using Jacobi's identity (8.2.20), we obtain

$$\chi_{1,1} \pm \chi_{3,1} = \prod_{n=0}^{\infty} \frac{\left(1 \pm q^{5n+\frac{5}{2}}\right)\left(1 \pm q^{5n+\frac{3}{2}}\right)\left(1 \pm q^{5n+\frac{7}{2}}\right)}{(1-q^{5n+2})(1-q^{5n+3})}$$

$$\chi_{3,3} \pm \chi_{3,2} = q^{\frac{1}{10}} \prod_{n=0}^{\infty} \frac{\left(1 \pm q^{5n+\frac{5}{2}}\right)\left(1 \pm q^{5n+\frac{1}{2}}\right)\left(1 \pm q^{5n+\frac{9}{2}}\right)}{(1-q^{5n+1})(1-q^{5n+4})}$$

$$\chi_{2,1} = q^{\frac{7}{16}} \prod_{n=0}^{\infty} \frac{(1+q^{5n+1})(1+q^{5n+4})(1+q^{5n+5})}{(1-q^{5n+2})(1-q^{5n+3})} \qquad (8.2.21)$$

$$\chi_{2,2} = q^{\frac{3}{80}} \prod_{n=0}^{\infty} \frac{(1+q^{5n+2})(1+q^{5n+3})(1+q^{5n+5})}{(1-q^{5n+1})(1-q^{5n+4})}. \qquad (8.2.22)$$

In the next section we will also look at the transfer matrix method for the Ising and hard square models and pick up such Virasoro characters. However, the best way to understand one-point functions in such A_n models of (Andrews *et al.* 1984), and generalizations to higher rank is through characters of Kač–Moody or affine algebras, i.e. via a larger chiral algebra as we shall see in Section 8.8. It may be that in some situations one is interested in a larger algebra $\mathcal{W} \times \overline{\mathcal{W}}$ for which one has to find new generators and relations for a chiral algebra $\mathcal{W}$ containing Vir, or that the left and right chiral algebras $\mathcal{W}_L \times \overline{\mathcal{W}}_R$ do not match although both contain Vir (cf. heterotic theories (Gannon and Ho-Kim 1994)).

In terms of our discussion of (quasi-)primary fields the decomposition (8.2.5) now becomes

$$\mathcal{F} = \textstyle\bigoplus_{i,\bar{i}} [\phi_i] \otimes [\phi_{\bar{i}}] \qquad (8.2.23)$$

where ϕ_i is a primary field and $[\phi_i]$, the space of descendents of ϕ_i, has a basis of quasi-primary fields. We expect that the primary fields form a closed algebra in the sense that the product of primary fields ϕ_i and ϕ_j can be expressed in terms of primary fields:

$$\phi_i(z)\phi_j(w) = \textstyle\sum_k c_{ij}{}^k(z,w)\phi_k(w)\,. \tag{8.2.24}$$

This equality is meant in the sense that one has equality of correlation functions, and for the radially ordered product $|z| > |w|$. (That is the radially ordered product is $\phi_i(z)\phi_j(w)$ for $|z| > |w|$, or $\phi_j(w)\phi_i(z)$ for $|w| > |z|$, which corresponds to time ordering in the Minkowski cylinder $\tau + i\sigma$ or string setting.) Conformal invariance (the fields are primary) yield constraints on the form that the coefficients can take. Indeed

$$c_{ij}{}^k(z,w) = C_{ij}{}^k(z-w)^{h_k-h_i-h_j}\,. \tag{8.2.25}$$

Although a product of fields such as (8.2.24) is usually understood asymptotically as $z \to w$, (8.2.24) and (8.2.25) are valid more generally under the presence of conformal invariance. The operator product expansion $\phi_i(z)\phi_j(w)$, $|z| > |w|$ is expanded in powers of w/z, and in a conformal field theory, after expansion and summation one obtains an analytic function of (z,w). Then the dependence $|z| > |w|$ can be ignored and the result is independent of the ordering $\phi_i(z)\phi_j(w)$ or $\phi_j(w)\phi_i(z)$.

The coefficients $c_{ij}{}^k$ or $C_{ij}{}^k$ are analogous to Clebsch–Gordan coefficients of (3.5.5). There for unitary representations of groups we wrote $\pi_i \otimes \pi_j = u^*(\oplus_k \pi_k)u$, for a unitary intertwiner. The summation is with some multiplicity $N_{ij}{}^k$, the Littlewood–Richardson coefficient, of the representation π_k in the tensor product $\pi_i \otimes \pi_j$. As for group representations in (3.5.4), we can also in (8.2.24) ignore most of the information in $\{C_{ij}{}^k\}$ and concentrate on the multiplicity $N_{ij}{}^k$ of the primary field ϕ_k (or its descendents) in the operator product $\phi_i\phi_j$, written

$$\phi_i * \phi_j = \textstyle\sum_k N_{ij}{}^k \phi_k\,. \tag{8.2.26}$$

We will return to this fusion rule algebra (FRA) shortly in Section 8.7. The content of the operator product expansion with the identity field $\phi_0 = 1$, is trivial:

$$C_{i0}{}^k = N_{i0}{}^k = \delta_i^k\,, \quad C_{0j}{}^k = N_{0j}{}^k = \delta_j^k\,.$$

It is clear moreover that $C_{ij}{}^k = 0$ implies $N_{ij}{}^k = 0$, but the converse is also true. It is also hypothesised that to each field ϕ_i there is a unique *conjugate* field ϕ_j such that the OPE of ϕ_i and ϕ_j contains the identity field. We write $j = i^\vee$, so that $i \to i^\vee$ is an involutive bijection on the primary fields, with $0^\vee = 0$. Note that the operator product of primary fields cannot (unlike the group case) be interpreted using tensor products, because the central charge is additive on tensor products (see also (8.10.35)).

Some information about correlation functions can simply be deduced from conformal invariance by moving some, in fact at most three points, to some

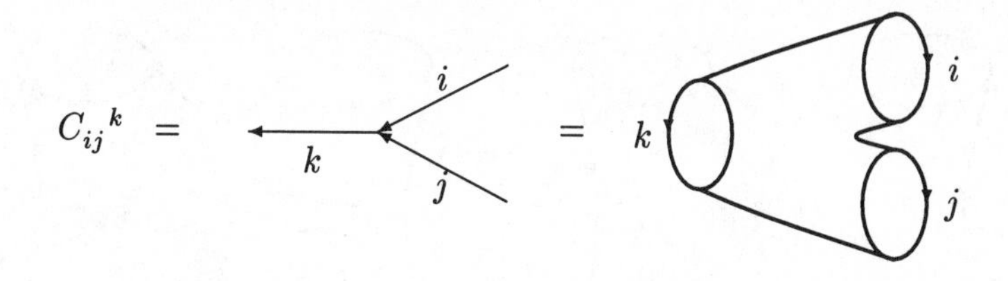

FIG. 8.1. Operator product expansion coefficient

prescribed positions, say 0,1,∞, and then eliminating three variables and working with cross ratios $z_{1234} = z_{12}z_{34}/z_{23}z_{41}$, where $z_{ij} = z_i - z_j$:

$$\langle \phi_1(z_1, \overline{z}_1) \cdots \phi_n(z_n, \overline{z}_n) \rangle = \prod_{1 \leq i < j \leq n} (z_i - z_j)^{h_{ij}} (\overline{z}_i - \overline{z}_j)^{\overline{h}_{ij}} F(z_{ijkl}, \overline{z}_{ijkl}) .$$

Here $h_{ij} = h_{ji}$, and $\sum_{i \neq j} h_{ij} = 2h_j$ and similarly for $\overline{h}_{ij}$. So for n point functions where $n \leq 4$,we have particularly simple forms, which will often be the starting points for more detailed computations of correlation functions (e.g in Section 8.4 using null differential equations):

$$\begin{aligned} &\langle \phi_i(z, \overline{z}) \phi_j(w, \overline{w}) \rangle \\ &= \textstyle\sum_k C_{ij}{}^{k} \overline{C}_{ij}{}^{k} (z - w)^{h_k - h_i - h_j} (\overline{z} - \overline{w})^{\overline{h}_k - \overline{h}_i - \overline{h}_j} \langle \phi_k(w, \overline{w}) \rangle \\ &= (z - w)^{-2h_i} (\overline{z} - \overline{w})^{-2\overline{h}_i} \delta_{ij^\vee} \end{aligned}$$

as $\langle \phi_k \rangle = \delta_{k0}$. Often we suppress the $\overline{z}, \overline{w}$ variables, as in

$$\begin{aligned} \langle \textstyle\prod_{i=1}^{3} \phi_{n_i}(z_i) \rangle &= C z_{12}^{h_3 - h_1 - h_2} z_{23}^{h_1 - h_2 - h_3} z_{13}^{h_2 - h_1 - h_3} \\ \langle \textstyle\prod_{i=1}^{4} \phi_{n_i}(z_i) \rangle &= F(z_{1234}) \textstyle\prod_{i<j} z_{ij}^{-h_i - h_j + h} , \quad h = \textstyle\sum_{i=1}^{4} h_i / 3 . \end{aligned}$$

With our usual matrix convention (Section 2.6) we write $C_{ij}{}^{k}$ as a φ^3 *graph* as in Fig. 8.1. Due to the symmetry

$$C_{ij}{}^{k} = C_{ji}{}^{k}$$

which is implicit in the definition of the radially ordered product, we can also not only rotate the graph in the plane, but invert it, i.e. regard it as a figure in three dimensional space – a three-holed sphere with labellings i, j, k of the boundary given by a right handed corkscrew rule (Fig. 8.1). Matrix multiplication for trivalent graphs is as usual, with summation over internal edges. In the trinion picture, we form punctured Riemann surfaces. Matrix multiplication corresponds to sewing circles on the boundary of neighbouring trinions together – and summing – as we see in the example of Fig. 8.2.

Associativity of the operator product expansion means that when we compute the four-point function $\langle \phi_n(z_1, \overline{z}_1) \phi_m(z_2, \overline{z}_2) \phi_l(z_3, \overline{z}_3) \phi_k(z_4, \overline{z}_4) \rangle$ in different

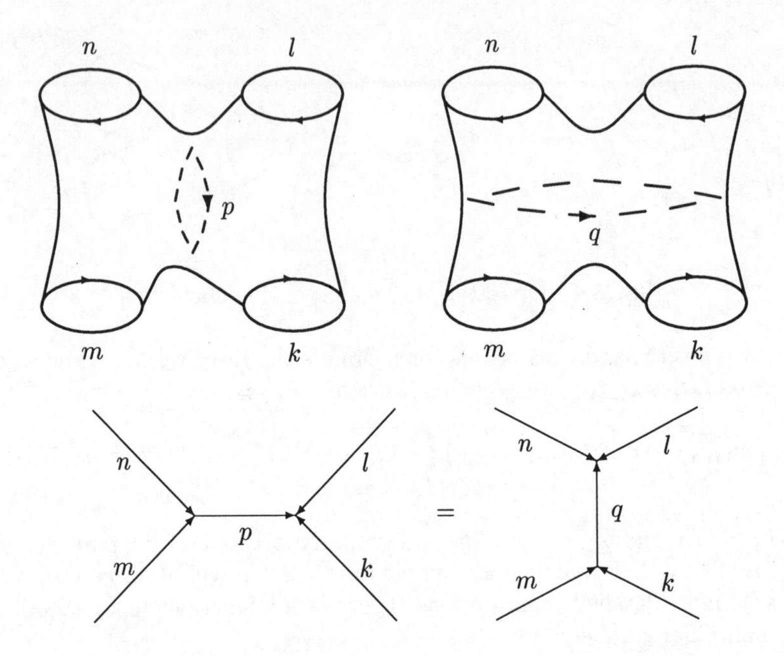

FIG. 8.2. Associativity of operator product expansion

ways, by coupling terms in different orders, then we always get the same result. This yields constraints such as

$$\begin{aligned}&\textstyle\sum_p C_{nm}{}^p C_{lkp} \mathcal{F}^{lk}_{nm}(p|x) \overline{\mathcal{F}}^{lk}_{nm}(p|\overline{x})\\ &= \textstyle\sum_q C_{nlq} C_{mk}{}^q \mathcal{F}^{mk}_{nl}(q|1-x) \overline{\mathcal{F}}^{mk}_{nl}(q|1-\overline{x})\end{aligned} \tag{8.2.27}$$

expressed graphically as in Fig. 8.2. On the left hand side we have coupled first n, m and l, k according to the operator product expansion (8.2.24) taking $z_1 \to z_2$, $z_3 \to z_4$ in terms of p, summed over primary and secondary fields. The terms $\mathcal{F}^{lk}_{nm}(p|x)$, $\overline{\mathcal{F}}^{lk}_{nm}(p|\overline{x})$, are the *conformal blocks* – the contribution from the conformal family $[\phi_p]$ where for simplicity we have taken $(z_1, z_2, z_3, z_4) = (0, x, 1, \infty)$. The conformal blocks satisfy certain differential equations arising from null states cf. (8.2.17). If these can be solved (see Section 8.5 for the Ising case) then they can be fed back into (8.2.27) to obtain constraints on the $C_{ij}{}^k$. Such bootstrap methods can be used to compute $C_{ij}{}^k$ in some circumstances from the consistency equations (8.2.27), cf. computation of connections in Section 11.5. Solving the differential equations satisfied by the four-point correlation functions rather than the conformal blocks can also lead to solutions of the operator product coefficients $C_{ij}{}^k$, as we shall see in Section 8.5 for the Ising model. We will in Section 8.9 introduce other quantum fields – cur-

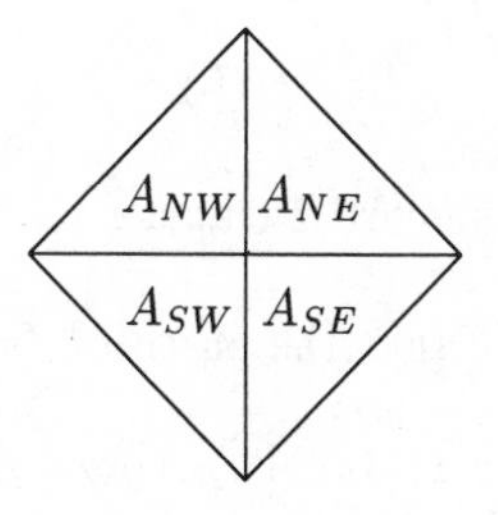

FIG. 8.3. The four corner transfer matrices

rent algebras – and then write the Virasoro generators as quadratic expressions in these fields. Null expressions there will also give rise to differential equations – the Knizhnik–Zamolodchikov equation – which can also be sometimes explicitly solved to obtain information about correlation functions.

Exercise 8.1 If A is a Lie algebra, let $Z^2(A)$, the 2-*cocycles* of A, be the bilinear maps f on A satisfying (a) $f(a_1, a_2) = -f(a_2, a_1)$, (b) $f(a_1, [a_2, a_3]) + f(a_2, [a_3, a_1]) + f(a_3, [a_1, a_2]) = 0$, for $a_i \in A$. The 2-*coboundaries* $B^2(A)$ are $f \in A^*$, $f(a_1, a_2) = f[a_1, a_2]$, and the second cohomology group of A is $H^2(A) = Z^2(A)/B^2(A)$. Show that if W is the Witt algebra, then $H^2(W) = \mathbb{C}b$, where $b(\ell_m, \ell_n) = \delta_{m,-n}(m^3 - m)/12$.

8.3 Corner transfer matrices and Virasoro characters in the Ising model

In Chapter 7, we made extensive use of the transfer matrix between strips to analyse the Ising model and its equilibrium states and magnetisation. Here we use the transfer matrix formalism, on a triangle shaped region, rather than a rectangle. These are the *corner transfer matrices.* We find that the $c = 1/2$ characters of the Virasoro algebra appear from the spectra of corner transfer matrices at criticality.

Take a finite square lattice subdivided as in Fig. 8.3, and let A_{SE}, A_{NE}, A_{NW}, A_{SW}, the corner transfer matrices, be the partition function of each corner. More precisely, we are orienting the lattice $\mathbb{Z}^2$ as in Fig. 7.3, and the boundary values of the finite region are specified, say, either free or fixed (+ or $-$). (If we think of the Ising model as an IRF model, interaction round a face model, with diagonal interactions as in Fig. 7.9 on lattice oriented as in Fig. 8.4(a), then the system decouples into two systems, each on a lattice oriented as in Fig. 7.3 or Fig. 8.4(b)). Then the corner transfer matrix A_{SE} is defined to be the partition function of the SE corner, with boundary values on the cut edges as in Fig. 8.4(c). Similarly

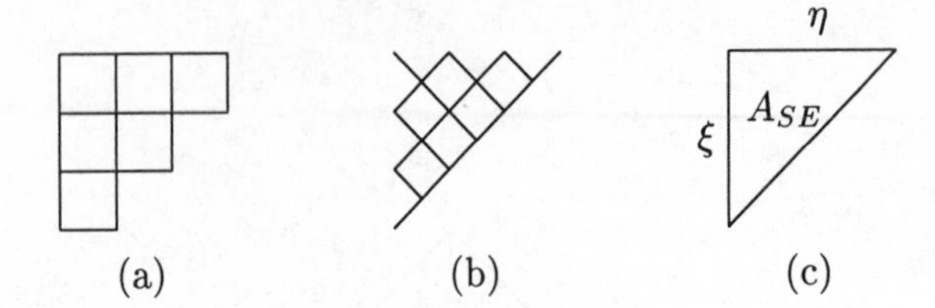

FIG. 8.4. The corner transfer matrix: $(A_{SE})_{\xi\eta}$

for the other three corners, so that the partition function of the square is

$$Z = \operatorname{tr} A_{SE} A_{NE} A_{NW} A_{SW} \,. \tag{8.3.1}$$

As in Section 7.4, we write

$$(A_{SE})_{\xi\eta} = V_N W_N V_{N-1} \cdots W_2 V_1 W_2 \cdots W_{N-1} V_{N-1} W_N V_N \,.$$

Here, up to scalars which we ignore for the time being,

$$V_k = \exp K_2 \left(\textstyle\sum_{j=k}^{N-1} \sigma_x^j \sigma_x^{j+1} + \varepsilon \sigma_x^N \right) \quad W_k = \exp K_1^* \left(\textstyle\sum_{j=k}^{N} \sigma_z^j \right) , \tag{8.3.2}$$

come from interactions along $NE - SW$ and $NW - SE$ diagonals respectively where $\varepsilon = 0$ for free boundary conditions and $\varepsilon = +$ or $-$ for fixed $+$ and fixed $-$ boundary values respectively. If we write $\sinh 2K_1 = \tan 2u$, then

$$\exp(2K_1) = (1 + \sin 2u)/\cos 2u \,, \quad \exp(2K_1) - 1 = \sqrt{2} \sin(\lambda - u)/\sin u \,.$$

Here $\lambda = \pi/4$. At criticality $K_1^* = K_2$, $\sinh 2K_1 \sinh 2K_2 = 1$, so that $\sinh 2K_2 = \cot 2u$, and interchanging K_1 and K_2 is effected by replacing u with $\lambda - u$. Then $A_{NW}(u) = A_{SE}(u)$, $A_{NE}(u) = A_{SW}(u) = A_{SE}(\pi/4 - u)$, and we write $A = A_{SE}$. If $\sigma \equiv 2e - 1$ is a self adjoint unitary

$$\exp K_2 \sigma = s \left(1 + (\exp(2K_2^*) - 1)^{-1} \, 2e \right)$$

(cf. (7.3.11)) where s is a scalar. Thus, again ignoring such scalars for the time being, we write

$$V_k = \prod_{j \geq k} \left(1 + \frac{\sin u}{\sin(\lambda - u)} \sqrt{2} e_{2j} \right) , \quad W_k = \prod_{j=k}^{N} \left(1 + \frac{\sin u}{\sin(\lambda - u)} \sqrt{2} e_{2j-1} \right)$$

with $e_{2j-1} = (1 + \sigma_z^j)/2$, $e_{2j} = (1 + \sigma_x^j \sigma_x^{j+1})/2$, $j < N$ and $e_{2N} = (1 \pm \sigma_x^N)/2$ for $\pm$ boundary conditions; and for free boundary conditions the product in V_k is for $j < N$. Taking linear approximations for u small,

$$A_{SE}(u) \approx \prod (1 + u e_i) \approx 1 - u \mathcal{H} \tag{8.3.3}$$

where

$$\mathcal{H} = -\sum_j 2(j-1)e_j$$

and the summation is over $1 \leq j \leq 2N-1$ free, $1 \leq j \leq 2N$ fixed. We denote these Hamiltonians by $\mathcal{H}_0$, $\mathcal{H}_\pm$ respectively and study their spectra.

The e_i satisfy the Temperley–Lieb relations, and as in Sections 6.4 and 7.5 we can make a Jordan–Wigner transformation to fermion operators. In the case of free boundary conditions, the Hamiltonian $\mathcal{H}_0$ is even with respect to the usual parity operator $\bigotimes_j \sigma_z^j$, and we can identify the Temperley–Lieb operators $\{e_1, e_2, \ldots, e_{2N-1}\}$ with $e_{2n} = c_n^* c_n$, $e_{2n+1} = [1 + (c_n^* - c_n)(c_{n+1} + c_{n+1}^*)]/2$ to get

$$\mathcal{H}_0 = \sum_{j=1}^N 4(j-1)\left(c_j^* c_j - 1/2\right) - \sum_{j=1}^{N-1}(2j-1)(c_j - c_j^*)(c_{j+1} + c_{j+1}^*)\,.$$

The $\pm$ boundary Hamiltonians $\mathcal{H}_\pm$ are not even with respect to the usual parity operator $\bigotimes_j \sigma_z^j$, due to the term σ_x^N. However if we use the Kramers–Wannier duality where we transform $\{e_2, e_3, \ldots, e_{2N}\}$ into Fermi operators, then $\mathcal{H}_\pm$ become even (indeed the parity operator is now σ_z^1, cf. (7.9.15)):

$$\mathcal{H}_\pm = \sum_{j=1}^N 2(2j-1)\left(c_j^* c_j - 1/2\right) - \sum_{j=1}^{N-1} 2j(c_j - c_j^*)(c_{j+1} + c_{j+1}^*)\,.$$

We can write (cf. (7.4.19)–(7.4.21)) $\mathcal{H} = b(H)$, where

$$H = \frac{1}{2}\begin{pmatrix} -W^* - W & W^* - W \\ W - W^* & W^* + W \end{pmatrix} + \begin{pmatrix} L & 0 \\ 0 & -L \end{pmatrix}$$

in $A(\ell^2[1,N] \oplus \ell^2[1,N])$, and

$$(Wf)_j = \begin{cases} 2j f_{j+1} & \text{free} \\ (2j-1) f_{j+1} & \text{fixed} \end{cases} \qquad (Lf)_j = \begin{cases} 2(j-1) f_j & \text{free} \\ (2j-1) f_j & \text{fixed} \end{cases}\,.$$

Conjugating by $2^{-1/2}\begin{pmatrix} 1 & 1 \\ 1 & -1 \end{pmatrix}$ (cf. proof of 7.18) we have

$$H = \begin{pmatrix} L & -W^* \\ -W & L \end{pmatrix}\,. \tag{8.3.4}$$

If we look for a diagonal form with new fermion operators, g_j:

$$\mathcal{H} = \sum_{j=1}^N 4\varepsilon_j g_j^* g_j$$

then the partition function of the complete square lattice is, from (8.3.3) $A(u)A(\pi/4-u)A(u)A(\pi/4-u) \approx (1-\pi H/2)$, with eigenvalues $\lambda_j = \exp(-2\pi\varepsilon_j)$. Reinserting the scalar factors which we had abandoned so far, what we wish to find is a diagonal form of the corner transfer matrix as

$$A = \Lambda_{\max} \exp\left(-\sum_{j=1}^{N} 2\pi\varepsilon_j g_j^* g_j\right) \tag{8.3.5}$$

with $\Lambda_{\max}$ denoting the largest eigenvalue.

Replacing u by $-u$, replaces K_i with $-K_i$. Thus up to a scalar, $A(u)$ and $A(-u)$ are inverses of each other by (8.3.2):

$$A(u)A(-u) = (2\cot 2u)^{N_1} \tag{8.3.6}$$

where N_1 is the number of interactions of type K_1. We can also effect the change $K_i \to -K_i$ in V and W (for free boundary conditions) by replacing $\sigma_z^j \to -\sigma_z^j$, $\sigma_x^j\sigma_x^{j+1} \to -\sigma_x^j\sigma_x^{j+1}$. This can be achieved by conjugating with the spin reversal operator

$$R = \sigma_x^1 \sigma_y^2 \sigma_x^3 \sigma_y^4 \cdots .$$

Thus

$$RA(u)R^* = A(-u) . \tag{8.3.7}$$

In terms of fermion operators, R is up to a scalar $\prod(c_j^* - c_j)$ so that $Rc_j^*R^* = c_j$ up to a phase. Since H is real symmetric, the diagonalizing Bogoliubov transformation is real orthogonal and so $Rg_j^*R^* = g_j$ up to a phase. Hence $Rg_j^*g_jR^* = 1 - g_j^*g_j$, and from (8.3.6) and (8.3.7)

$$(2\cot 2u)^{N_1} = \Lambda^2 \prod_{j=1}^{N} \lambda_j .$$

We first compute λ_j, and then use the last relation to obtain Λ. Let $\mathcal{M} = \mathcal{M}_n \in M_n(\mathbb{C})$ be $\mathcal{M}_{ij} = i\delta_{i,j+1} + j\delta_{i+1,j}$. Then

$$\mathcal{H}_0 = \frac{1}{2}\left[0 \oplus \mathcal{M}_{2N-1}\right] , \quad \mathcal{H}_\pm = \frac{1}{2}\mathcal{M}_{2N} \tag{8.3.8}$$

e.g. for $N = 2$:

$$\mathcal{H}_0 = \frac{1}{2}\begin{bmatrix} 0&0&0&0 \\ 0&0&1&0 \\ 0&1&0&2 \\ 0&0&2&0 \end{bmatrix} \qquad \mathcal{H}_\pm = \frac{1}{2}\begin{bmatrix} 0&1&0&0 \\ 1&0&2&0 \\ 0&2&0&3 \\ 0&0&3&0 \end{bmatrix} .$$

The characteristic polynomial $\mathcal{P}_n$ for $\mathcal{M}_n$ satisfies

$$\mathcal{P}_0(\lambda) = 1 , \quad \mathcal{P}_1(\lambda) = \lambda , \quad \mathcal{P}_{n+1}(\lambda) = \lambda\mathcal{P}_n(\lambda) - n^2\mathcal{P}_{n-1}(\lambda)$$

which are the recursion relations for the *Meixner polynomials* of the second kind (Chihara 1978). Meixner polynomials have the integral representation (Pesche and Truong 1987)

$$\mathcal{P}_n(\lambda) = \frac{n! i^{-n} \cosh(\pi\lambda/2)}{\pi} \int_{-\infty}^{\infty} \frac{\tanh^n x \exp(i\lambda x)\, dx}{\cosh x} . \tag{8.3.9}$$

The lowest-order steepest descent approximation is

$$\mathcal{P}_n(\lambda) = \frac{n! i^{-n} \cosh(\pi\lambda/2)}{\pi} [\exp \psi(x_-, \lambda) + \exp \psi(x_-, \lambda)] \tag{8.3.10}$$

where $\psi(x, \lambda) = \varphi(x, \lambda) - (1/2)\log[-2\varphi''(x, \lambda)]$, φ is the phase function from the integral representation (8.3.9):

$$\varphi(x, \lambda) = i\lambda x + n \log \sinh x - (n+1) \log \cosh x$$

with $x_\pm$ the two stationary points $\varphi'(x_\pm, \lambda) = 0$, namely

$$\tanh x_\pm = \frac{i\lambda \pm [4n(n+1) - \lambda^2]^{\frac{1}{2}}}{2(n+1)}. \tag{8.3.11}$$

The positive zeros in the approximation (8.3.10) are in the range $0 < \alpha < 1$, if $\alpha = \lambda[4n(n+1)]^{-1/2}$, and are given by

$$f(\alpha_j) = \begin{cases} (2j-1)/n & j = 1, \ldots, \quad n/2, \quad n \text{ even} \\ 2j/n & j = 1, \ldots, \quad [n/2], \quad n \text{ odd} \end{cases}$$

if $f(\alpha) = 2\mathrm{Im}\, \psi(x_+, \lambda)/\pi n$. For small α, $f(\alpha) \approx 2\alpha/\pi \log(\alpha/2)$, so that for fixed λ, large n:

$$\mathcal{P}_n(\lambda) \approx \begin{cases} \sin[(\lambda/2)\log n] & n \text{ even} \\ \cos[(\lambda/2)\log n] & n \text{ odd} \end{cases}.$$

Re-inserting the factor 1/2 in the Hamiltonians $\mathcal{H}$ of (8.3.8), the roots of $\mathcal{P}_n$ now yield the low lying energies asymptotically as

$$\varepsilon_j \approx \begin{cases} \pi j/\log N & \text{for } \mathcal{H}_0 \\ \pi(j - \frac{1}{2})/\log N & \text{for } \mathcal{H}_\pm \end{cases}. \tag{8.3.12}$$

We rewrite our inversion information as

$$\log \Lambda_{\max} = \frac{1}{2} N_1 \log 2 \cot 2u + 2u \textstyle\sum_{j=1}^{N-1} \varepsilon_j .$$

Taking the $N-1$ solutions ε_j to (8.3.12) we need to estimate

$$\textstyle\sum_{j=1}^{N-1} \varepsilon_j = [4N(N+1)]^{\frac{1}{2}} \sum_{j=1}^{N-1} f^{-1}(j/N) = N[4N(N+1)]^{\frac{1}{2}} T\left[f^{-1}(x)\right] \tag{8.3.13}$$

by an endpoint trapezoidal quadrature rule for the integral of f^{-1} over the interval [0,1]. The work of (Lyness and Ninham 1967) gives asymptotic expansions for the errors for integration with singularities of the form $xh(x)/\log x$ where h is analytic as

$$T\left[xh(x)/\log x\right] - \int_0^1 xh(x)/\log x \, dx = \sum_{k \geq 0} \frac{a_k}{N^{2+k} \log N} + \frac{b_k}{N^{k+1}}$$

with a leading correction $a_0/N^2 \log N$ where $a_0 = \zeta(-1)h(0) = -h(0)/12$. Reinserting into (8.3.13), which has a N^2 factor, gives us the asymptotic form for free boundary conditions as

$$\log \Lambda_{\max} \approx A_0 N^2 + B_0 N + C_0 + \pi/24 \log N\,.$$

The procedure is similar for fixed boundary conditions except that with the corresponding solutions to (8.3.12), one needs a midpoint rather than an endpoint rule for the errors. In this case $a_0 = (2^{-1}-1)\zeta(-1)h(0) = -h(0)/24$ which gives the asymptotic form for free boundary conditions as

$$\log \Lambda_{\max} \approx A_\pm N^2 + B_\pm N + C_\pm - \pi/48 \log N\,.$$

From the representation

$$A(u) = \Lambda_{\max} \exp \textstyle\sum_j \log \lambda_j g_j^* g_j \tag{8.3.14}$$

we get

$$\operatorname{trace} A(u) = \Lambda_{\max} \textstyle\prod_j (1+\lambda_j)\,. \tag{8.3.15}$$

Ignoring the non-universal part of the asymptotics of $\Lambda_{\max}$, and keeping only $\Lambda_0 \approx q^{1/24}$, $\Lambda_\pm \approx q^{-1/48}$, in the free, fixed cases respectively, where $q = \exp(-1/2 \log N)$, and $\lambda_j = \exp(-2\pi\varepsilon_j) \approx q^j$, $q^{j-1/2}$ respectively, we obtain from (8.3.15) in the free case the Virasoro character $\chi_{1/16}$:

$$\chi_{1/16} = q^{\frac{1}{24}} \textstyle\prod_{n=1}^\infty (1+q^n) = q^{-\frac{1}{48}+\frac{1}{16}}(1+q+q^2+2q^3+\cdots)\,.$$

In the fixed $\pm$ boundary case the other two characters for $c = 1/2$, appear, namely χ_0 and $\chi_{1/2}$. In the case of $\pm$ boundary conditions, the centre spin σ_1^z is the parity operator so that the CTM decomposes into an even and odd part according as to whether the central spin at the origin is aligned or non-aligned with the boundary spins. Decomposing $\prod(1+\lambda_j)$, where j describes parity, we get the generating functions $[\prod(1+\lambda_j) \pm \prod(1-\lambda_j)]/2$, for the even and odd eigenspaces of the CTM. Inserting $\Lambda_{\max} = \Lambda_\pm \approx q^{-1/48}$, $\lambda_j \approx q^{j-1/2}$ these are just the characters χ_0 and $\chi_{1/2}$ of the $h = 0, 1/2$ representations of the Virasoro algebra:

$$\begin{aligned}
\chi_0 &= \frac{1}{2} q^{-\frac{1}{48}} \left\{ \textstyle\prod_{n=1}^\infty (1+q^{n-\frac{1}{2}}) + \prod_{n=1}^\infty (1-q^{n-\frac{1}{2}}) \right\} \\
&= q^{-\frac{1}{48}}(1+q^2+q^3+2q^4+\cdots)
\end{aligned} \tag{8.3.16}$$

$$\begin{aligned}
\chi_{1/2} &= \frac{1}{2} q^{-\frac{1}{48}} \left\{ \textstyle\prod_{n=1}^\infty (1+q^{n-\frac{1}{2}}) - \prod_{n=1}^\infty (1-q^{n-\frac{1}{2}}) \right\} \\
&= q^{-\frac{1}{48}+\frac{1}{2}}(1+q+q^2+q^3+\cdots)\,.
\end{aligned} \tag{8.3.17}$$

We have found the three $c = 1/2$ characters of the Virasoro algebra for $h = 0, 1/2, 1/16$, but can we write down explicit representations of the Virasoro algebra associated with these characters? Let us first consider the cases of χ_0 and $\chi_{1/2}$, where from (8.3.14) and (8.3.15)

$$\chi_0 + \chi_{1/2} = q^{-1/48}\prod_{j=1}^{\infty}(1 + q^{j-1/2}) = q^{-1/48}\operatorname{tr} q^{L_0} \tag{8.3.18}$$

if we put

$$L_0 = \sum_{j=1}^{\infty}(j - 1/2)g_j^* g_j = \sum_{r\in\mathbb{N}-1/2} r b_r^* b_r \tag{8.3.19}$$

and $b_r = g_{r+1/2}$, for $r \in \mathbb{N} - 1/2$. In order to get a self dual representation of the CAR, we put

$$b_{-r} = b_r^*\,, \quad r \in \mathbb{N} - 1/2$$

so that we have a family $\{b_r : r \in \mathbb{Z} + 1/2\}$ of operators satisfying the self dual CAR

$$\{b_r, b_s\} = \delta_{r,-s}\,, \quad b_r^* = b_{-r}\,, \quad r, s \in \mathbb{Z} + 1/2\,. \tag{8.3.20}$$

Then we can write

$$L_0 = \frac{1}{2}\sum_{r\in\mathbb{Z}+1/2} r \,:\, b_r^* b_r \,: \tag{8.3.21}$$

if we put $: b_r^* b_r \,:= b_r^* b_r$ if $r > 0$, and $-b_r b_r^*$ if $r < 0$. If we let $K_{NS} = \ell^2(\mathbb{Z} + 1/2)$ with distinguished basis $\{e_r \mid r \in \mathbb{Z} + 1/2\}$, and conjugation $\Gamma e_r = e_{-r}$, then we can identify b_r with $B(e_r)$ in the self dual CAR algebra over (K_{NS}, Γ). The subscript NS comes from the Fermion fields of (Neveu and Schwarz 1971) in string theory. Then $S_{NS} = \sum_{r\in\mathbb{N}-1/2} e_{-r,-r}$ $(e_{r,s} = e_r \otimes \overline{e_s})$, is a basis projection, and : : above is just the Wick ordering of Section 6.6:

$$: B(f)B(g) \,:= B(f)B(g) - \varphi_{S_{NS}}(B(f)B(g))$$

and

$$L_0 = dQ_{S_{NS}}(\mu_0) \tag{8.3.22}$$

if μ_0 is the diagonal matrix $-\sum_{r\in\mathbb{Z}+1/2} r e_{r,r}$. Comparing with the Witt algebra $\ell_m = -z^m(z\,d/dz + m/2)$, we put $\mu_m = -\sum_{r\in\mathbb{Z}+1/2}(r + m/2)e_{r+m,r}$, and

$$L_m = dQ_{S_{NS}}(\mu_m) = \frac{1}{2}\sum_{r\in\mathbb{Z}+1/2}(r - m/2) \,:\, b_{r-m}^* b_r \,:\,. \tag{8.3.23}$$

The Virasoro relation

$$[L_m, L_n] = (m - n)L_{m+n} + \frac{1}{24}\delta_{m,-n}(m^3 - m) \tag{8.3.24}$$

with $c = 1/2$, is then a consequence of the cocycle relation for dQ and the Schwinger term $S_{S_{NS}}(\mu_m, \mu_n)$. We should identify, according to (8.3.16) and (8.3.17), the $h = 0$ and $h = 1/2$ state spaces with the even and odd tensors in the Fock space of S_{NS}. Indeed we can take $v(0) = \Omega$, $v(1/2) = b_{1/2}^*\Omega$ as $L_0\Omega = 0$, $L_0 b_{1/2}^*\Omega = (1/2)b_{1/2}^*\Omega$.

Similarly, to describe the $h = 1/16$ representation, we can write

$$\chi_{1/16} = q^{1/24}\prod_{n=1}^{\infty}(1 + q^n) = q^{1/48}\operatorname{tr} q^{L_0}$$

where

$$L_0 = \sum_{j=1}^{\infty} j g_j^* g_j + 1/16 \,. \tag{8.3.25}$$

We extend to a self dual family by putting $b_n = g_n$, for $n \in \mathbb{N}$,

$$b_{-n}^* = b_n \,, \qquad n \in \mathbb{N} \tag{8.3.26}$$

and adjoin b_0 so that we have a family $\{b_n \mid n \in \mathbb{Z}\}$ of operators satisfying the self dual CAR

$$\{b_m, b_n\} = \delta_{m,-n} \,, \quad b_m^* = b_{-m} \,, \quad m \in \mathbb{Z} \,.$$

To do this let $K_R = \ell^2(\mathbb{Z})$, with distinguished basis $\{e_n \mid n \in \mathbb{Z}\}$ and conjugation $\Gamma e_n = e_{-n}$, and identify b_n with $B(e_n)$. The subscript R comes from the Fermion fields of (Ramond 1971) in string theory. Then $S_R = e_{0,0}/2 + \sum_{n\in\mathbb{N}} e_{-n,-n} \in Q(K_R, \Gamma)$, where $F_R = S_R - e_{0,0}/2$ is a partial basis projection, and we can write

$$L_0 = \frac{1}{2}\sum_{n\in\mathbb{Z}} n : b_n^* b_n : +1/16 = dQ_{S_R}(\mu_0) + 1/16 \tag{8.3.27}$$

with normal ordering with respect to S_R, and μ_0 is $\sum_{n\in\mathbb{Z}} n e_{n,n}$. We should then put $\mu_m = \sum_{n\in\mathbb{Z}}(n - m/2)e_{n-m,n}$ and

$$L_m = dQ_{S_R}(\mu_m) + \delta_{m,0}/16 = \frac{1}{2}\sum_{n\in\mathbb{Z}}(n - m/2) : b_{n-m}^* b_n : +\delta_{m,0}/16 \,. \tag{8.3.28}$$

Again the Virasoro relation (8.3.24) with $c = 1/2$ is a consequence of the cocycle relation for dQ and the Schwinger term $S_{S_R}(\mu_m, \mu_n)$. This time the Hilbert space of the state φ_{S_R} decomposes to give two copies of the $h = 1/16$ representation of the Virasoro algebra with cyclic vectors $v(1/16) = \Omega_R$, $v'(1/16) = b_0\Omega_R$. (Unlike the Fock state $\varphi_{S_{NS}}$, the Ramond state φ_{S_R} is not pure (see the proof of Theorem 6.40), and decomposes to give two irreducible representations of the Fermi algebra with cyclic vectors $2^{-1/2}(1 \pm b_0)\Omega_R$.)

Thus at criticality, the Virasoro characters naturally appear in the corner transfer matrices with the modular parameter $q = \exp(-1/2 \log N)$ being related to the size N of the finite geometry. Virasoro characters also appear off criticality in the corner transfer matrices with the modular parameter q being related to the perturbation from the critical temperature. We will look at this in some detail in the hard square model to compute one-point functions. However, we note that in the Ising model, these methods enable us to compute the magnetisation of σ_z^{11} at the origin which for $\pm$ boundary conditions, the low temperature region, becomes the parity operator (cf. (7.9.17))

$$\sigma_1^z = \begin{bmatrix} 1 & 0 \\ 0 & -1 \end{bmatrix} \otimes \begin{bmatrix} 1 & 0 \\ 0 & -1 \end{bmatrix} \otimes \cdots \otimes \begin{bmatrix} 1 & 0 \\ 0 & -1 \end{bmatrix} \otimes \cdots \,. \tag{8.3.29}$$

The Hamiltonian is now

$$\mathcal{H} = \sum_{j=1}^{N} 2(2j-1)\left(c_j^* c_j - 1/2\right) - k\sum_{j=1}^{N-1} 2j(c_j - c_j^*)(c_{j+1} + c_{j+1}^*),$$

where $k^{-1} = \sinh 2K_1 \sinh 2K_2$. The product of the corner transfer matrix becomes $A_{SE}A_{NE}A_{NW}A_{SW} = \exp(-2\lambda\mathcal{H})$, where

$$\exp[-2\lambda\mathcal{H}] = \begin{bmatrix} 1 & 0 \\ 0 & q^{1/2} \end{bmatrix} \otimes \begin{bmatrix} 1 & 0 \\ 0 & q^{1+1/2} \end{bmatrix} \otimes \cdots \otimes \begin{bmatrix} 1 & 0 \\ 0 & q^{n-1/2} \end{bmatrix} \otimes \cdots$$

so that the magnetisation is

$$m^* = \mathrm{tr}(\sigma_z^1 \exp[-2\lambda\mathcal{H}])/\mathrm{tr}(\exp[-2\lambda\mathcal{H}]) = \prod_{n=1}^{\infty} \frac{(1-q^{n-1/2})}{(1+q^{n-1/2})}. \tag{8.3.30}$$

Here q has to be derived from k, but elliptic function identities will relate the expression in (8.3.30) back with $(1-k^2)^{1/8}$ as in (7.13.11). Under the transformation $q^{1/2} = e^{\pi i \tau} \to p = e^{-\pi i/\tau}$ (i.e. $\tau \to -1/\tau$) then

$$m^* = \sqrt{2}p^{1/8} \prod_{n=1}^{\infty} \frac{(1+p^{2n})}{(1-p^{2n-1})}$$

we see the behaviour at criticality ($q = 1, p = 0$) and find a critical exponent of $\beta = 1/8$, $m^* \sim |T - T_c|^{1/8}$, cf. Section 7.13.

Virasoro characters for central charge $c = 7/10$ appear in the corner transfer matrix formalism for the hard square model. This is the next case after the Ising model $m = 3$, namely $m = 4$ amongst the ABF models which are based on the Dynkin diagrams A_m. We have already noted in Section 7.1 that the constraints $\sigma_k = 0, 1$, $\sigma_k\sigma_{k+1} = 0$, can be represented by the graph of Fig. 7.6. The only transition not allowed is 1 to 1. We follow the corner transfer method outlined above for the Ising model but being even more cavalier about rigour and away from criticality. To keep the discussion simple, we will first perform with corner transfer matrices, for the Ising model in a framework which has no specific reference to fermions, but which will generalize to other models, including the ABF models in one direction and the eight vertex model in another.

We begin by looking at the integrability condition of commuting transfer matrices. Recall that the row to row transfer matrix T is the partition function of the strip of Fig. 8.5. For two such matrices, with horizontal periodic conditions but with different parameters, to commute it is clearly sufficient (by a graphical argument) that there exists a third family of invertible Boltzmann weights such that we have the local condition of Fig. 8.6

The lattice $\mathbb{Z}^2$ is oriented as in Fig. 8.4(a), and we take the Ising Boltzmann weights as in Fig. 8.7, so that the diagonal interaction parameters are L_1 and K_1, and the vertices $a, b, c, d \in \{\pm 1\}$. Suppose we are given two of the weights

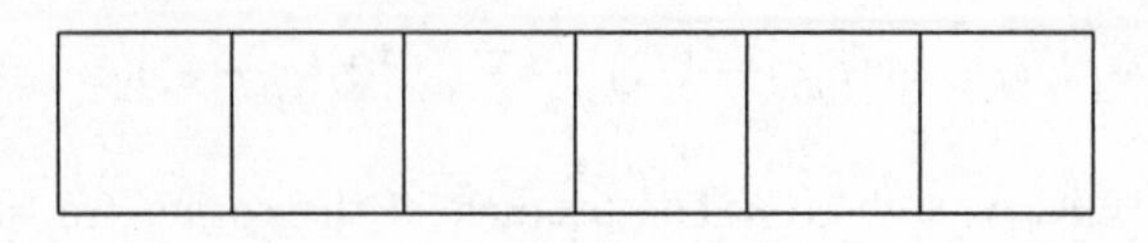

FIG. 8.5. Row to row transfer matrix T

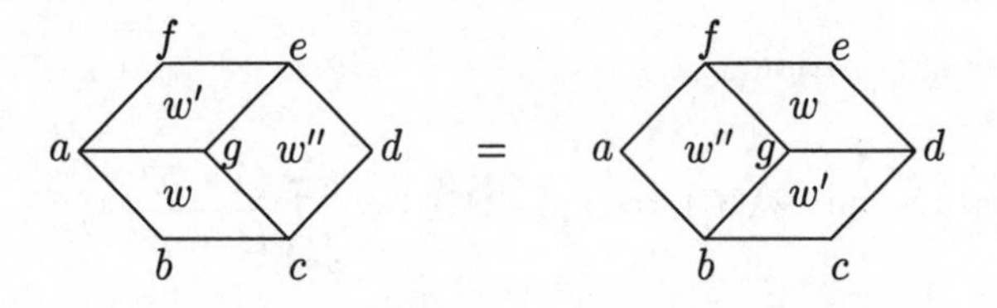

FIG. 8.6. Yang–Baxter equation

w and w' in Fig. 8.6, and we try to find a third w'' such that the equality of partition functions in Fig. 8.6 holds. This integrability condition is the Yang–Baxter equation. Suppose that w' and w'' have diagonal interaction parameters (L_2, K_2) and (L_3, K_3) respectively. In our Ising situation the hexagon to hexagon relation of Fig. 8.6 splits into two equations of the form of Fig. 8.8, a *star–triangle relation* , where α is a constant. Writing

$$\begin{aligned} c &= \cosh(L_1 + L_2 + L_3) \\ c_i &= \cosh(-L_i + L_j + L_k)\,, \quad j, k \neq i \end{aligned}$$

there are only four equations now:

$$\begin{aligned} 2c &= \alpha \exp(K_1 + K_2 + K_3) \\ 2c_1 &= \alpha \exp(K_1 - K_2 - K_3) \\ 2c_2 &= \alpha \exp(-K_1 + K_2 - K_3) \\ 2c_3 &= \alpha \exp(-K_1 - K_2 + K_3) \end{aligned}$$

which yield

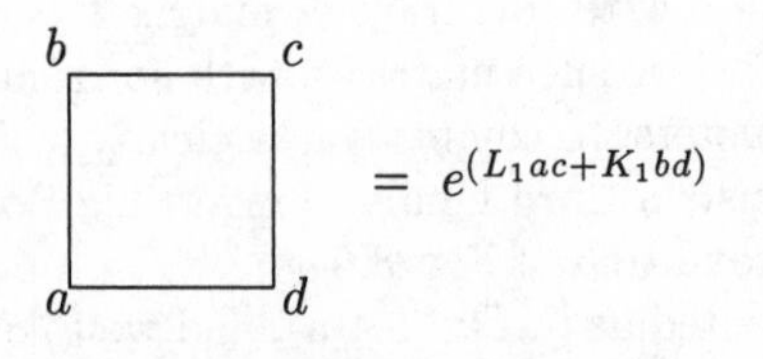

FIG. 8.7. Ising Boltzman weights

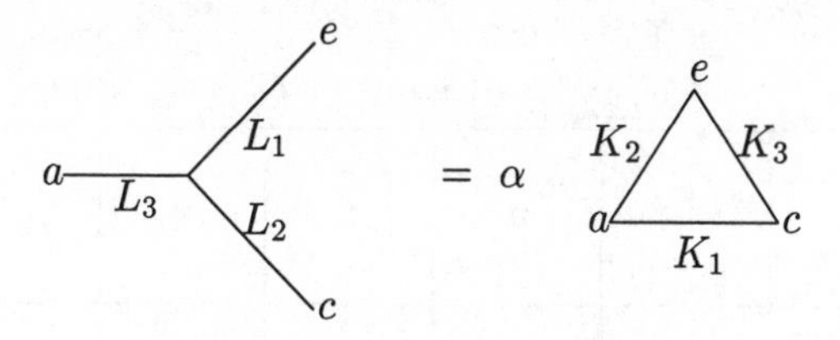

FIG. 8.8. Star–triangle relation

$$\frac{cc_1}{c_2c_3} = e^{4K_1}$$

and so

$$e^{4K_1} - 1 = (\sinh 2L_2 \sinh 2L_3)/c_2c_3 .$$

Then

$$\begin{aligned}
\frac{\sinh 2L_2 \sinh 2L_3}{2(cc_1c_2c_3)^{1/2}} &= \frac{(e^{4K_1}-1)c_2c_3}{2(cc_1c_2c_3)^{1/2}} \\
&= \frac{e^{4K_1}-1}{2}\left(\frac{c_2c_3}{cc_1}\right)^{1/2} \\
&= (e^{4K_1}-1)/2e^{2K_1} = \sinh 2K_1 .
\end{aligned}$$

Hence $\sinh 2K_1 \sinh 2L_1 = (\sinh 2L_1 \sinh 2L_2 \sinh 2L_3)/2(cc_1c_2c_3)^{1/2} \equiv k^{-1}$ is symmetric in L_1, L_2 and L_3. Hence the star–triangle relation implies that $\sinh 2K_1 \sinh 2L_1 = k^{-1}$ is a constant.

A natural way to parameterize K, L for fixed k is through elliptic functions (or the uniformising substitution of (Onsager 1941)). Consider the three (modified) elliptic functions, the theta functions:

$$H(u) = 2q^{1/4} \sinh u \prod_{n=1}^{\infty} (1 - q^{2n}e^{2u})(1 - q^{2n}e^{-2u})$$

$$\Theta(u) = \prod_{n=1}^{\infty} (1 - q^{2n-1}e^{2u})(1 - q^{2n-1}e^{-2u})$$

and the Jacobian elliptic function:

$$\mathrm{snh}(u) = H(u)/\Theta(u) .$$

The entire functions H and Θ are doubly periodic, when $q = e^{-2\tau}$:

$$\begin{aligned}
H(u) &= -H(u + i\pi) = -qe^{-2u}H(u + 2\tau) \\
\Theta(u) &= \Theta(u + i\pi) = -qe^{-2u}\Theta(u + 2\tau) .
\end{aligned}$$

Then we set

σ_3' σ_2' $\sigma_1' = \sigma_1$ σ_2 σ_3

u	u	v	v
u	u	v	v

FIG. 8.9. Partition function $\psi_{\sigma\sigma'}$

$$e^{-2K} = \text{snh}(u) \qquad e^{-2L} = \text{snh}(\lambda - u) \tag{8.3.31}$$

Then for fixed q, $k^{-1} = \sinh 2K \sinh 2L$ is constant as u varies, (with $0 < u < \tau$ to ensure positive Boltzmann weights) provided the crossing parameter λ satisfies $k\text{snh}^2\lambda = 1$.

Then a solution to the Yang–Baxter equation of Fig. 8.6 is $w = w[u]$, $w' = w'[u+u'']$, $w'' = w''[u'']$, i.e. using the same q, but taking u, $u+u''$, and u'' in turn for the spectral parameter u of the above weights. More generally, two transfer matrices T and T' of the row in Fig. 8.5, with horizontal periodic boundary conditions but where the spectral parameter u_j can vary along the row, will commute as long as the differences of the corresponding $u_j - u_j'$ is constant independent of j. The row to row transfer matrices T, where u_i varies, but the differences $u_i - u_j$ is constant, will form a commuting family. The significance of this for the corner transfer matrix philosophy is that in particular, the normalized eigenvectors of T itself will only depend on differences $u_i - u_j$.

We now return to our corner matrix formalism. Again let A_{SE}, A_{NE}, A_{NW}, A_{SW} denote the corner transfer matrices for the triangular regions of Fig. 8. and Fig. 8.4, with some specified boundary conditions. Let us first concentrate on a half plane as in Fig. 8.9 and use the spectral parameter u and v in the SW and SE corners respectively. We first use A_{SW} and A_{SE} to compute its partition function ψ in Fig. 8.10. Here σ and σ' denote the boundary conditions of the top row, and we have some specific ground state boundary conditions in mind along the diagonal boundaries. Now in the infinite volume limit, we can also think of the partition function as being computed through composing strips, in particular through the row to row transfer matrix T as in Fig. 8.11. Here we have taken horizontal cylindrical boundary conditions, and summed over the vertical column boundary values α, again σ and σ' denote the boundary conditions of the top row, and some specified ground state values s and s' along the bottom row. We start with the second interpretation. By the observation regarding eigenvectors of transfer matrices as only depending on differences, the maximal eigenvector X of T depends only on the difference $u - v$, and so

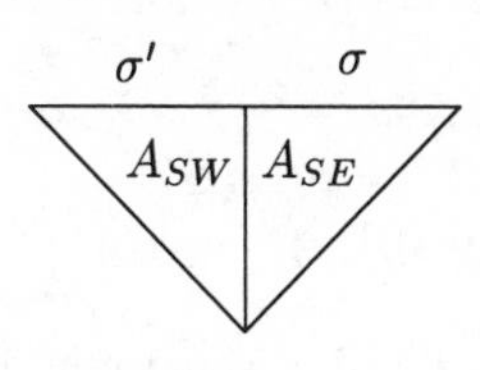

FIG. 8.10. Partition function $\psi_{\sigma\sigma'} = [A_{SW}(u)A_{SE}(v)]$

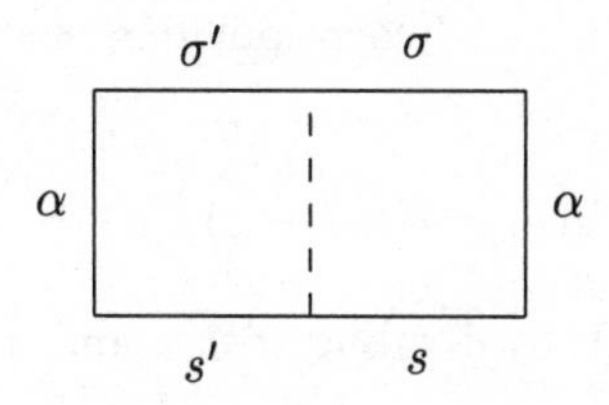

FIG. 8.11. Partition function $\psi_{\sigma\sigma'} = (T^r)_{\sigma\sigma',ss'}$

$$\psi_{\sigma\sigma} = \tau(u,v)\,[X(u-v)]_{\sigma\sigma'}$$

where $\tau(u,v)$ is some scalar normalization. We then go back to the first formulation involving corner transfer matrices to get:

$$A_{SW}(u)A_{SE}(v) = \tau(u,v)X(u-v)\,. \tag{8.3.32}$$

We began with some finite region. To take the infinite limit, having fixed some boundary conditions along the boundary of the plane $\sigma = s$, $\sigma' = s'$, we should renormalize our $A_{SW}(u)$ and $A_{SE}(v)$ by dividing by these ground state values as we take the infinite lattice limit. Suppose now that we specify our ground state boundary values to be the vacuum 0 configuration. Then since not only the Boltzmann weights, but the ground state is invariant under reflection through $NW - SE$ and $NE - SW$ diagonals then the operators A_{SE}, A_{NE}, A_{NW}, A_{SW} are all symmetric and $A_{SE} = A_{NW}$, $A_{NE} = A_{SW}$. Moreover we see from our Boltzmann weights that rotating by $\pi/2$ from A_{SE} to A_{NE} is effected by replacing a spectral parameter v by $\lambda - v$, i.e. there is simple relation between A_{SE} and A_{NE} (and hence with A_{NW} and A_{SW}):

$$A_{NE}(v) = A_{SE}(\lambda - v) = A_{SW}(v) = A_{NW}(\lambda - v)\,.$$

This is referred to as *crossing symmetry.* Then writing the previous relation (8.3.32) in terms of a single matrix or operator, say A_{SE}, we have

$$A(u)A(v) = \tau(u, \lambda - v)X(u + v - \lambda)\,.$$

Then X can be eliminated to give

$$\tau(v, \lambda - u)A(u)A(v) = \tau(u, \lambda - v)A(v)A(u)\,. \qquad (8.3.33$$

Diagonalizing $A(u)$ say and assuming some non-degeneracy of the eigenvalue we are led to the consistency relations

$$\tau(v, \lambda - u) = \tau(u, \lambda - v)\,, \qquad A(u)A(v) = A(v)A(u)\,.$$

This means that the family $A(u), X(v)$ is simultaneously diagonable with larges eigenvalue $a(u)$ say for $A(u)$, and corresponding eigenvalue $x(u)$ of $X(u)$. Th relation (8.3.33) now yields

$$a(u)a(v) = \tau(u, \lambda - v)x(u + v - \lambda)$$

and renormalizing A and X by dividing with a and x respectively we get

$$A(u)A(v) = X(u + v - \lambda)\,.$$

With some smoothness we would then be led to exponential behaviour of th eigenvalues of $A(u)$:

$$A(u) = [m_j \exp(-\alpha_j u)\, \delta_{ij}]_{i,j}\,.$$

Much of the discussion so far is very general. To get the exponents α_j and henc compute some of the thermodynamic functions, we will have to use the specia form of our Boltzmann weights, or at least in some limiting situation – expectin the coefficients α_j to be essentially integral. In fact the complex periodicity c the elliptic functions means that $\exp(-\alpha_j u)$ has a complex period, so that α must be essentially integral. In the limit $L_1 \to \infty$ with K_1 fixed (i.e. $k \to 0$, wit u fixed) the Boltzmann weight becomes diagonal:

$$\begin{matrix} b & \square & c \\ a & & d \end{matrix} = e^{K_1 bd}\delta_{ac} \qquad (8.3.34$$

so that the corresponding diagonal A is immediate from Fig 8.12:

$$A_{\sigma,\sigma'} = \exp\{-u(\sigma_1\sigma_3 + 2\sigma_2\sigma_4 + \cdots + m\sigma_m\sigma_{m+2} + \cdots)\}\,\delta_{\sigma,\sigma'}\,. \qquad (8.3.35$$

Arguing with the rigidity of the α_j (j is now identified with σ), this gives the di agonal form of $A(u)$, even when $k \neq 0$, and so we can compute the magnetizatio as

$$\langle\sigma_1\rangle = \mathrm{tr}\,\sigma_1 A(u)A(\lambda - u)A(u)A(\lambda - u)/\mathrm{tr}\,A(u)A(\lambda - u)A(u)A(\lambda - u)$$

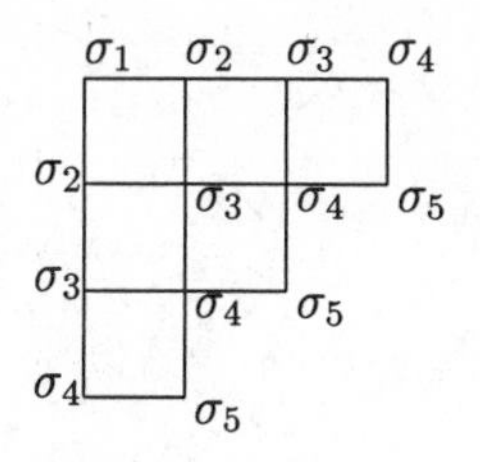

FIG. 8.12. Diagonal corner transfer matrix

$$= \operatorname{tr} \sigma_1 A(2\lambda)/\operatorname{tr} A(2\lambda) . \tag{8.3.36}$$

Ve could normalize A so that the largest eigenvalue is 1 by replacing each $_m\sigma_{m+2}$ by $(\sigma_m\sigma_{m+2} - 1)$. If we write $\mu_j = \sigma_j\sigma_{j+2}$, and identify with a Pauli ıatrix, we are back in the situation of (8.3.30) with $\sigma_1 = \mu_1\mu_3\mu_5 \cdots$ the parity perator.

However, this method generalizes to the eight vertex and ABF models. The ight-vertex model has Boltzmann weights

$$\begin{array}{cc} b & c \\ a & d \end{array} \; = e^{(Lac+Kbd+Mabcd)} \tag{8.3.37}$$

ith the same configuration space as the Ising model, $a, b, c, d \in \{\pm 1\}$, where $I = 0$. In this case, the corresponding Yang–Baxter equation implies that

$$\Delta = -\sinh 2K \sinh 2L - \tanh 2M \cosh 2K \cosh 2L$$

constant, and a suitable parameterization of weights is again via (8.3.31) to- ether with $e^{2M} = \operatorname{snh}(\lambda)$. Again in the limiting case, the diagonal form is nmediate and in fact of the same form, so that the magnetisation can again be xpressed as (8.3.30) where now $q = e^{-\lambda}$.

In the hard square case (and indeed the ABF models on A_n, or even more enerally on $\mathcal{A}^{(n)}$), one can again parameterize solutions of the Yang–Baxter quation by elliptic functions. There are some complications, however, due to enormalizations in the crossing symmetry, and ground states may not necessarily e invariant under rotation through $\pi/2$. In one physical region, region III, where he ground state boundary condition is $\sigma_j \equiv 0$, the limiting case of the Boltzmann reight is the diagonal:

$$\begin{array}{cc} b & c \\ a & d \end{array} \; = r^{b+d-2a} e^{-3u(a-bd)} \delta_{ac} \tag{8.3.38}$$

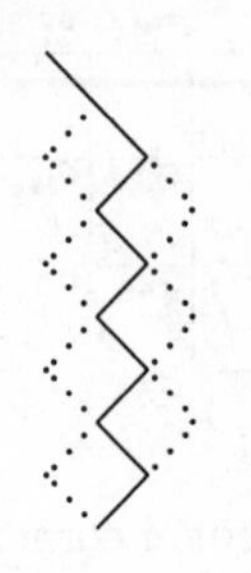

FIG. 8.13. Path space for one-point functions in hard square model

and so the corresponding diagonal A is immediate:

$$\begin{aligned} A_{\sigma\sigma'} &= r^{\sigma_1} e^{-3u(\sigma_2+2\sigma_3+3\sigma_4+\ldots+m\sigma_{m-1})} e^{3u(\sigma_1\sigma_3+2\sigma_2\sigma_4+\ldots+m\sigma_m\sigma_{m+2})} \delta_{\sigma,\sigma'} \\ &\equiv r^{s(\sigma)} e^{-3un(\sigma)} \delta_{\sigma,\sigma'} \,. \end{aligned} \tag{8.3.39}$$

Here $s(\sigma) = \sigma_1$, and $n(\sigma) = \sum_{k\geq 1} k(\sigma_k - \sigma_k\sigma_{k+2})$, and the indices i, j need t be identified with σ, σ'. The magnetisation (cf. (8.3.34)) becomes

$$\begin{aligned} \langle \sigma_1 \rangle &= \sum_{\sigma} \sigma_1 r^{2\sigma_1} q^{n(\sigma)} \Big/ \sum r^{2\sigma_1} q^{n(\sigma)} \\ &= r^2 F(1)/[F(0) + r^2 F(1)] \end{aligned} \tag{8.3.40}$$

where for $j = 0, 1$, $F(j) = \sum_{\sigma} q^{n_j(\sigma)}$ and from (8.3.39) we have put

$$n_j(\sigma) = \textstyle\sum_{k\geq 1}(k + j - 1)(\sigma_{k+1} - \sigma_k\sigma_{k+2}) = \sum_{k\geq 1}(k + j - 1)|\sigma_k - \sigma_{k+2}|/2$$

and $q = e^{-6u}$. The summation is over all sequences $\mathcal{P} = \{(\sigma_k \mid k = 1, 2, \ldots)$ $\sigma_1 = 1, \sigma_k = 0, 1, \sigma_k\sigma_{k+1} = 0$, and $\sigma_m = 0$ for all large $m\}$, all paths on th Bratteli diagram $\hat{A}_4$ in Fig. 8.13 which are eventually on the central line, startin at $*$. By solving relevant recursion relations:

$$\begin{aligned} F(0) &= \sum_{k\geq 0} \frac{q^{k(3k-1)/2}}{(1-q)(1-q^2)\cdots(1-q^k)(1-q)(1-q^3)\cdots(1-q^{2k-1})}, \\ F(1) &= \sum_{k\geq 0} \frac{q^{3k(k+1)/2}}{(1-q)(1-q^2)\cdots(1-q^k)(1-q)(1-q^3)\cdots(1-q^{2k+1})} \,. \end{aligned} \tag{8.3.41}$$

Moreover by classical identities of Rogers (Slater 1951) it is known that

$$\sum_{k\geq 0} q^{k(3k-1)/2}/(1-q)(1-q^2)\cdots(1-q^k)(1-q)(1-q^3)\cdots(1-q^{2k-1})$$

$$= \prod_{n\geq 0, n\neq 0, \pm 4 \bmod n} (1-q^n)^{-1} \tag{8.3.42}$$

$$\sum_{k\geq 0} q^{3k(k+1)/2}/(1-q)(1-q^2)\cdots(1-q^k)(1-q)(1-q^3)\cdots(1-q^{2k+1})$$

$$= \prod_{n\geq 0, n\neq 0, \pm 2 \bmod n} (1-q^n)^{-1}\,.$$

By (8.3.41) and (8.3.42) the characters χ^4_{22} and χ^4_{21} can be expressed as

$$\chi^4_{22} = \prod_{n\geq 0}(1+q^{5n+2})(1+q^{5n+3})(1+q^{5n+5})/(1-q^{5n+1})(1-q^{5n+4})\,,$$

$$\chi^4_{21} = \prod_{n\geq 0}(1+q^{5n+1})(1+q^{5n+4})(1+q^{5n+5})/(1-q^{5n+1})(1-q^{5n+4})\,.$$

Writing $(1+z) = (1-z^2)/(1-z)$, we see that

$$\chi^4_{22} = \prod_{n\geq 0}(1-q^{10n+4})(1-q^{10n+6})(1-q^{10(n+1)})/\prod_{n\geq 1}(1-q^n)$$

$$= \prod_{n\geq 0, n\neq 0, \pm 4 \bmod 10} (1-q^n)^{-1}\,, \tag{8.3.43}$$

$$\chi^4_{21} = \prod_{n\geq 0}(1-q^{10n+2})(1-q^{10n+8})(1-q^{10(n+1)})/\prod_{n\geq 1}(1-q^n)$$

$$= \prod_{n\geq 0, n\neq 0, \pm 2 \bmod 10} (1-q^n)^{-1}\,.$$

Hence combining (8.3.41), (8.3.42), (8.3.43) we have

$$F(0) = \prod_{n\geq 0, n\neq 0, \pm 4 \bmod 10} (1-q^n)^{-1}$$

$$= \sum_{k\geq 0} \frac{q^{k(3k-1)/2}}{(1-q)(1-q^2)\cdots(1-q^k)(1-q)(1-q^3)\cdots(1-q^{2k-1})} = \chi^4_{22}$$

and similarly

$$F(1) = \prod_{n\geq 0, n\neq 0, \pm 2 \bmod 10} (1-q^n)^{-1}$$

$$= \sum_{k\geq 0} \frac{q^{3k(k+1)/2}}{(1-q)(1-q^2)\cdots(1-q^k)(1-q)(1-q^3)\cdots(1-q^{2k+1})} = \chi^4_{21}\,.$$

We concentrate on

$$\sum_\sigma q^{n_j(\sigma)} = \chi_{2,2-j} \quad j = 0, 1\,.$$

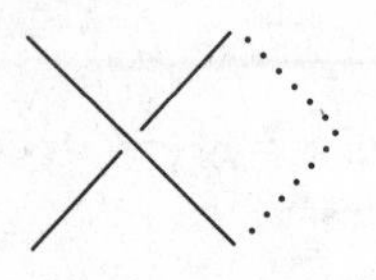

FIG. 8.14. E_j, j odd; for $c = 7/10$

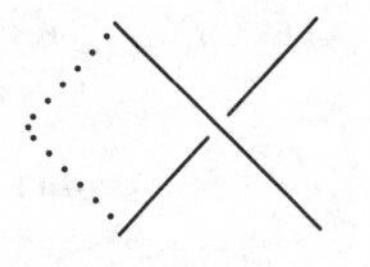

FIG. 8.15. E_j, j even; for $c = 7/10$

We are thus guided to define projections E_j living in $A[j-1, j+1]$, by projections on the two paths of Fig. 8.14 if j is odd, or projections on the two paths of Fig. 8.15 if j is even, and if $\mathrm{H}_{\mathcal{P}}$ denotes the Hilbert space $\ell^2(\mathcal{P})$, to make the following identifications:

$$L_0 = \sum_{j \geq 0} j E_j/2 + 7/16 \qquad (r,s) = (2,1)\,, \tag{8.3.44}$$

$$L_0 = \sum_{j \geq 0} (j-1) E_j/2 + 3/80 \quad (r,s) = (2,2)\,. \tag{8.3.45}$$

Although it is now clear, from dimensional considerations, that the Virasoro algebra acts on $\mathrm{H}_{\mathcal{P}}$ with central charge $c = 7/10$, for $(r,s) = (2,2)$ and $(2,1)$, it still remains to properly understand how $L_n, n \neq 0$ operate. The representation $(2,2)$ of the Virasoro algebra with central charge $c = 7/10$ generate the five other representations with the same central charge, through taking operator product expansions.

This should be compared with (8.3.25), for $c = 1/2$, $h = 1/16$:

$$L_0 = \textstyle\sum_{j=1}^{\infty} j g_j^* g_j + 1/16\,.$$

We write $g_j^* g_j = 1 - e_{2j-1} = E_j$, where e_n are the Temperley–Lieb operators as in Section 7.9, and E_j is the projection on the two paths of Fig. 8.16 in $A[j-1, j+1]$, so that

$$L_0 = \textstyle\sum_{j=1}^{\infty} j E_j + 1/16 \tag{8.3.46}$$

acts on $\ell^2(P)$ where P is now the space of paths which are essentially on the bold path of Fig. 8.17. Similarly for $h = 0$.

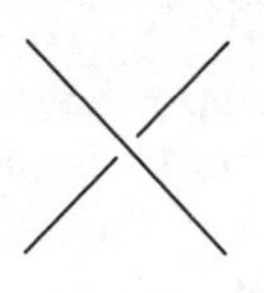

FIG. 8.16. E_j; for $c = 1/2$

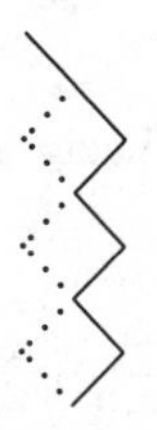

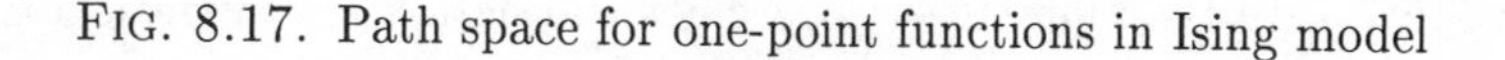

FIG. 8.17. Path space for one-point functions in Ising model

In Section 8.8, we will see that local state probabilities can be described by Virasoro characters in a coset setting of Kač–Moody algebras.

8.4 Ising model continuum limit

In the last section, we obtained the $c = 1/2$ Virasoro characters χ_0, $\chi_{1/2}$, $\chi_{1/16}$ from the critical Ising model. We now want to try and identify the primary fields in a more concrete way, and to get more specific forms of the Virasoro representations in terms of the fermion picture of the Ising model already introduced in Chapter 7.

In the Ising model one can reformulate the Pauli picture $\bigotimes_{\mathbb{Z}} M_2$ with fermions using the Jordan–Wigner transformation:

$$p_j = \textstyle\prod_{r<j} \sigma_z^r \sigma_x^j, \quad q_j = \prod_{r<j} \sigma_z^r \sigma_y^j . \tag{8.4.1}$$

Then

$$\sigma_x^j p_i = \begin{cases} p_i \sigma_x^j & i \le j \\ -p_i \sigma_x^j & i > j \end{cases}$$

and

$$\sigma_x^j q_i = \begin{cases} q_i \sigma_x^j & i < j \\ -q_i \sigma_x^j & i \ge j \end{cases} . \tag{8.4.2}$$

(To avoid confusion, we now write T for the temperature and V for the transfer matrix.) Then in the scaling or continuum limit one obtains the disorder variable μ as $T \downarrow T_c$:

$$V^n \sigma_x^m V^{-n} \to \sigma(a) \tag{8.4.3}$$

as $(m,n) \sim (a^0, a^1) = a$ in the sense that $\varepsilon = |T - T_c| \to 0$, $\varepsilon m = a^0$, $\varepsilon n = \sqrt{-1}a^1$. Similarly one obtains the order variable μ as $T \uparrow T_c$. Moreover the free fermions give rise to a two component Majorana fermion field satisfying the Dirac equation:

$$\left.\begin{matrix} V^n p_m V^{-n} \\ V^n q_m V^{-n} \end{matrix}\right\} \to \psi(x) = \begin{pmatrix} \psi_+(x) \pm \psi_-(x) \\ -i\psi_+(x) \pm i\psi_-(x) \end{pmatrix} \tag{8.4.4}$$

with

$$\psi_\pm(x)\sigma(a) = \begin{cases} \sigma(a)\psi_\pm(x) & x^1 < a^1 \\ -\sigma(a)\psi_\pm(x) & x^1 > a^1 \end{cases} \tag{8.4.5}$$

and the signs reversed for μ.

To be more specific about this we begin with the transfer matrix diagonalized as

$$(2\sinh 2K_1)^{-\frac{2M+1}{2}} V = \left\{ \exp - \frac{1}{2M+1} \left(\sum_\mu \gamma(\theta_\mu)\psi^*(\theta_\mu)\psi(\theta_\mu) - \frac{2M+1}{2} \right) \right\}$$

in Section 7.4. In the limit $M \to \infty$, we get translation in the vertical transfer direction

$$P^0 = \int_{-\pi}^{\pi} \frac{d\theta}{2\pi} \gamma(\theta)\psi^*(\theta)\psi(\theta) \tag{8.4.6}$$

and translation in the horizontal spatial direction

$$P^1 = \int_{-\pi}^{\pi} \frac{d\theta}{2\pi} \theta\psi^*(\theta)\psi(\theta)\,. \tag{8.4.7}$$

In the range $T > T_c$, we have already considered in Section 7.7 (especially (7.7.34

$$e^{2i\vartheta(w)} = \frac{(1 - z_2 z_1^* e^{-iw})(1 - z_2 e^{iw}/z_1^*)}{(1 - z_2 z_1^* e^{iw})(1 - z_2 e^{-iw}/z_1^*)} \tag{8.4.8}$$

where $z_i = \tanh K_i = e^{2K_i^*}$, $z_i^* = \tanh K_i^* = e^{-2K_i}$. Letting $\alpha_1 = z_2 z_1^*$, $\alpha_2 = z_2/z_1^*$ and expanding about the critical temperature, write

$$\alpha_1 = \alpha + \varepsilon m'\,, \quad \alpha_2 = 1 + \varepsilon m \tag{8.4.9}$$

with m, m', α constants and $\varepsilon = 0$ signifying the critical temperature. We write $C_i = \cosh 2K_i$, $C_i^* = \cosh 2K_i^*$, $S_i = \sinh 2K_i$, $S_i^* = \sinh 2K_i^*$, and again expand around criticality:

$$S_2^*, S_1 = \frac{2\sqrt{\alpha}}{1-\alpha} + \varepsilon \frac{(1+\alpha)}{\sqrt{\alpha}(1-\alpha)^2}(m' \pm \alpha m) + O(\varepsilon^2)$$

$$C_2^*, C_1 = \frac{1+\alpha}{1-\alpha} + \varepsilon \frac{2}{(1-\alpha)^2}(m' \pm \alpha m) + O(\varepsilon^2)$$

and insert in (7.7.5)–(7.7.7) to get

$$e^{\pm i\vartheta(\theta)} \sinh \gamma(\theta) = \varepsilon \frac{2\sqrt{\alpha}}{1-\alpha}(\mu \pm \theta/\varepsilon) + O(\varepsilon^2)$$
$$\sinh \gamma(\theta) = \varepsilon \frac{2\sqrt{\alpha}}{1-\alpha}\sqrt{\mu^2 + (\theta/\varepsilon)^2} + O(\varepsilon^2) .$$

It is convenient to set $\kappa = 2\sqrt{\alpha}/(1-\alpha)$, $p^1 = \theta/\varepsilon$, and

$$u^{\pm 1} = \frac{\sqrt{\mu^2 + (p^1)^2} \pm p^1}{\mu}, \quad \psi(u) = \sqrt{\frac{\sinh \gamma(\theta)}{\kappa}}\psi(\theta), \quad \text{for } u > 0$$
$$u^{\pm 1} = \frac{-\sqrt{\mu^2 + (p^1)^2} \pm p^1}{\mu}, \quad \psi(u) = \sqrt{\frac{\sinh \gamma(\theta)}{\kappa}}\psi^*(-\theta), \quad \text{for } u < 0 .$$

Then considering $\varepsilon \to 0$ in (8.4.6) and (8.4.7) we have

$$\langle \psi(u), \psi(u) \rangle = 2\pi |u| \delta(u+u)$$
$$e^{-nP^0} = \exp\left[ix^0 \int_0^\infty \frac{du}{2\pi|u|} p^0 \psi(-u)\psi(u)\right]$$
$$e^{-imP^1} = \exp\left[-ix^1 \int_0^\infty \frac{du}{2\pi|u|} p^1 \psi(-u)\psi(u)\right]$$

where $p^0 = \pm\sqrt{\mu^2 + (p^1)^2}$ if $u \gtrless 0$.

The spin operators $s_{mn} = V^n \sigma_x^m V^{-n}$ are implementers of Bogoliubov automorphisms (see (8.4.2) and Sections 6.7 and 6.9). As such, they can (cf. Section 6.7) be put in a canonical normal ordered form. These forms will then be amenable to taking the limit $M \to \infty$. In this way (Sato *et al.* 1977–1980), we can write

$$s_{mn} = (1 - S_1^2 S_2^2)^{\frac{1}{8}} : \psi_{0,mn} e^{\rho_{mn}/2} :$$

where

$$\psi_{0,mn} = S_2^{-\frac{1}{2}} \int_{-\pi}^{\pi} \underline{d\theta} \left(e^{-im\theta - n\gamma(\theta)}\psi^*(\theta) + e^{im\theta + n\gamma(\theta)}\psi(\theta)\right)$$
$$\rho_{mn} = \int \int_{-\pi}^{\pi} \underline{d\theta}\, \underline{d\theta'}\, (\psi^*(\theta)\psi(\theta)) \begin{pmatrix} R_{mn}^{--}(\theta,\theta') & R_{mn}^{-+}(\theta,\theta') \\ R_{mn}^{+-}(\theta,\theta') & R_{mn}^{++}(\theta,\theta') \end{pmatrix} \begin{pmatrix} \psi^*(\theta') \\ \psi(\theta') \end{pmatrix}$$

$$R_{mn}^{\sigma\sigma'}(\theta,\theta') = \frac{2\sinh\left((\sigma\gamma(\theta) - \sigma'\gamma(\theta'))/2\right)}{1 - e^{-i(\sigma\theta + \sigma'\theta' - i0)}} \times e^{im(\sigma\theta + \sigma'\theta') + (n+\frac{1}{2})(\sigma\gamma(\theta) + \sigma'\gamma(\theta'))} \quad (\sigma, \sigma' = \pm)$$

$$\underline{d\theta} = d\theta/2\pi \sinh \gamma(\theta).$$

The factors $(1 - S_1^2 S_2^2)^{1/8}$ arise from expectation values of the tail $\prod_{r=-\infty}^{0} \sigma_z^r$, cf. Section 7.13, and the correction scalar factors in Chapter 6, e.g. (6.7.21). In the continuum limit, $T \downarrow T_c$, the spin operators s_{mn} become

$$\sigma(a) =: \psi_0(a) e^{\rho_F(a)/2} : \tag{8.4.10}$$

where

$$\rho_F(a) = \int_{-\infty}^{\infty} \int_{-\infty}^{\infty} \frac{\psi(u)}{2\pi|u|} \frac{\psi(u')}{2\pi|u'|} \frac{-i(u-u')}{u+u'-i0} \times e^{-im(a^-(u+u')+a^+(u^{-1}+u'^{-1}))} du\, du'$$

$$\psi_0(a) = \int_{-\infty}^{\infty} \frac{du}{2\pi|u|} e^{-im(a^- u + a^+ u^{-1})} \psi(u), \quad a^{\pm} = (a^0 \pm a^1)/2.$$

Moreover the fermion variables

$$\begin{aligned} p_{mn} &= V^n p_m V^{-n} = e^{-nP^0 - imP^1} p_{00} e^{nP^0 + imP^1} \\ &= \int_{-\pi}^{\pi} \frac{d\theta}{2\pi} \left[e^{-n\gamma(\theta) - im\theta + i\vartheta(\theta)} \psi^*(\theta) + e^{n\gamma(\theta) + im\theta - i\vartheta(\theta)} \psi(\theta) \right], \\ q_{mn} &= V^n q_m V^{-n} = e^{-nP^0 - imP^1} q_{00} e^{nP^0 + imP^1} \\ &= \int_{-\pi}^{\pi} \frac{d\theta}{2\pi} \left[-e^{-n\gamma(\theta) - im\theta - i\vartheta(\theta)} \psi^*(\theta) - e^{n\gamma(\theta) + im\theta + i\vartheta(\theta)} \psi(\theta) \right] \end{aligned}$$

scale as $\varepsilon \to 0$:

$$\sqrt{\frac{2}{m\varepsilon}} \begin{pmatrix} p_{mn} \\ q_{mn} \end{pmatrix} \to \begin{pmatrix} \psi_+(x) + \psi_-(x) \\ -i\psi_+(x) + i\psi_-(x) \end{pmatrix}$$

where $\psi_{\pm}$ are the Majorana fields:

$$\psi_{\pm}(x) = \sqrt{\frac{m}{2}} \int_{-\infty}^{\infty} \frac{du}{2\pi|u|} \sqrt{0+iu}^{\pm 1} e^{im(x^- u + x^+ u^{-1})} \psi(u). \tag{8.4.11}$$

To take the continuum limit from above the critical temperature we need to use Kramers–Wannier high temperature–low temperature duality of Section 7.11. Putting $\bar{s}_m = s_{-M}^{-1} s_m = \sigma_x^{-M} \sigma_x^m$, $\bar{s}_{mn} = V^n \bar{s}_m V^{-n}$ amounts to setting plus boundary conditions $s_{-M} = 1$, and we obtain the even order operator μ rather than the odd σ:

$$T \uparrow T_c : \qquad \mu(a) =: e^{\rho_F(a)/2} : \tag{8.4.12}$$

This time the fermion variables scale as

$$\sqrt{\frac{2}{m\varepsilon}}\begin{pmatrix} p_{mn} \\ q_{mn} \end{pmatrix} \to \begin{pmatrix} \psi_+(x) - \psi_-(x) \\ -i\psi_+(x) - i\psi_-(x) \end{pmatrix}$$

where $\psi_\pm$ are as before in (8.4.11).

The question we are asking is what conformal field theory arises in the Ising model at criticality? What is the central charge, and what are the primary fields? We have seen in Section 8.3 by considering the corner transfer matrices, a central charge $c = 1/2$. According to the abstract theory of Section 8.2, there are three potential primary fields with $h = 0, 1/2, 1/16$. The first is of course the conformal weight of the identity field 1. In general, when taking variables $\phi(x)$ defined on the lattice, scaling to a field $\phi_i = \phi(r)$ in the continuum limit we expect

$$\langle\phi(0)\phi(r)\rangle \sim r^{-2\Delta_i} \tag{8.4.13}$$

with $\Delta_i = h_i + \overline{h}_i$, and spin $s_i = h_i - \overline{h}_i$. As we have already noted in Sections 7.3 and 7.5 typically, for large x one has in a lattice model the behaviour

$$\langle\sigma_0\sigma_x\rangle - \langle\sigma_0\rangle^2 \sim |x|^{-\eta'} \exp -|x|/\xi\,,$$

where the scale factor ξ is the *correlation length* ξ. The correlation length ξ diverges at criticality, where (assuming $\langle\sigma_0\rangle = 0$ at T_c) there is a power law

$$\langle\sigma_0\sigma_x\rangle \sim |x|^{-\eta}$$

and η is a *critical exponent*, computed to be 1/4 in the Ising model. Thus in the continuum limit, the spin σ_{mn} scales to a field $\sigma(a)$, with conformal weights $(h, \overline{h}) = (1/16, 1/16)$. The asymptotics of the energy $\varepsilon_x = \sigma_x\sigma_{x'}$ (where x' is a nearest neighbour, or better as an average over all nearest neighbours) on the lattice is computed from four-point functions for σ:

$$\langle\varepsilon_0\varepsilon_x\rangle = \langle\sigma_0\sigma_{0'}\sigma_x\sigma_{x'}\rangle \sim |x|^{-2}$$

which in the continuum limit is then identified with the $(1/2, 1/2)$ field.

8.5 Ising model fusion rule algebra

In this section we consider the fusion rules satisfied by the primary fields of the critical Ising model. We will do this in the setting of conformal field theory, and then in the next section in our C^*-algebraic framework of endomorphisms on the fermion algebra.

We first compute operator product coefficients by solving the differential equations satisfied by correlation functions

$$G^{(2M,2N)} = \langle\sigma(z_1, \overline{z}_1)\cdots\sigma(z_{2M}, \overline{z}_{2M})$$

$$\times\mu(z_{2M+1},\overline{z}_{2M+1})\cdots\mu(z_{2M+2N},\overline{z}_{2M+2N})\rangle\,.$$

The energy ε and spin (order/disorder) σ,μ have null descendents at level 2, (8.2.13), and so as we have seen in (8.2.16), G satisfies the differential equations:

$$\left\{\frac{4}{3}\frac{\partial^2}{\partial z_i^2}+\sum_{j\neq i}\left[\frac{1/16}{(z_i-z_j)^2}+\frac{1}{(z_i-z_j)}\frac{\partial}{\partial z_j}\right]\right\}G^{2M,2N}=0$$

$(i=1,\ldots,2M+2N)$, and similarly replacing z_i by $\overline{z}_i$. By a Möbius transformation, we can transform G, for $M=2$, $N=0$ as

$$G=(z_{14}z_{23})^{-\frac{1}{8}}(\overline{z}_{13}\overline{z}_{23})^{-\frac{1}{8}}H(x,\overline{x})$$

where x is the cross-ratio $z_{12}z_{34}/z_{23}z_{41}$, $z_{ij}=z_i-z_j$. Writing $H=[x(1-x)\overline{x}(1-\overline{x})]^{-1/8}f$, we get

$$\left[x(1-x)\frac{\partial^2}{\partial x^2}+\left(\frac{1}{2}-x\right)\frac{\partial}{\partial x}+\frac{1}{16}\right]f=0$$

or

$$\left(\frac{\partial^2}{\partial\theta^2}+\frac{1}{4}\right)f=0$$

after the change of variable $x=\sin^2\theta$, $\overline{x}=\sin^2\overline{\theta}$. Two solutions are obtained:

$$f_{1,2}(x)=\left(1\pm\sqrt{1-x}\right)^{\frac{1}{2}}\,. \tag{8.5.1}$$

Thus the single-valued correlation G has a decomposition in terms of multi-valued f_1,f_2:

$$G=\left|\frac{z_{23}z_{41}}{z_{12}z_{34}z_{13}z_{42}}\right|^{\frac{1}{4}}\sum_{i,j=1}^{2}a_{ij}f_i(x)f_j(\overline{x})\,. \tag{8.5.2}$$

Having taken $\overline{x}=x^*$, we need $a_{ij}=a\delta_{ij}$ for some scalar a in order that G be single valued. Such four-point functions can be used to compute fusion coefficients C in the operator product expansion of two fields. We have for example from (8.2.25):

$$\sigma(z_1,\overline{z}_1)\sigma(z_2,\overline{z}_2)=\frac{1}{z_{12}^{\frac{1}{8}}\overline{z}_{12}^{\frac{1}{8}}}+C_{\sigma\sigma\varepsilon}z_{12}^{\frac{3}{8}}\overline{z}_{12}^{\frac{3}{8}}\varepsilon(z_2,\overline{z}_2)\,.$$

Using also a similar expression for $\varepsilon(z_2,\overline{z}_2)\varepsilon(z_4,\overline{z}_4)$, this yields

$$G\sim\frac{1}{|z_{12}|^{\frac{1}{4}}|z_{34}|^{\frac{1}{4}}}+C^2_{\sigma\sigma\varepsilon}\frac{|z_{12}|^{\frac{3}{4}}|z_{34}|^{\frac{3}{4}}}{|z_{24}|^2}+\cdots$$

which needs to be compared with (8.5.2) when $a_{ij} = a\delta_{ij}$, $x \to 0$. The first and second terms yield $a = 1/2$, $C_{\sigma\sigma\varepsilon} = 1/2$ respectively (replacing ε by $-\varepsilon$ if necessary), so that

$$G = \frac{1}{2}\left|\frac{z_{23}z_{41}}{z_{12}z_{34}z_{13}z_{42}}\right|^{\frac{1}{4}} \left(|1+\sqrt{1-x}| + |1-\sqrt{1-x}|\right) . \tag{8.5.3}$$

If we write $\mathcal{F}(h_i, h_j, x) = \mathcal{F}^{ii}_{ii}(j|x)$ (where the latter notation for conformal blocks was used in (8.2.27)) then the two solutions f_1, f_2 to (8.5.1) are the conformal blocks $\mathcal{F}(1/16, 0, x)$, $\mathcal{F}(1/16, 1/2, x)$ respectively and the decomposition (8.5.3) of the four-point function is

$$G = \mathcal{F}(1/16, 0, x)\overline{\mathcal{F}}(1/16, 0, x) + \mathcal{F}(1/16, 1/2, x)\overline{\mathcal{F}}(1/16, 1/2, x) .$$

Thus two conformal families, with dimensions 0 and 1/2 contribute to the operator product expansion $\sigma\sigma$, the identity 1 and the energy density field ε respectively. These operator product coefficients are consistent with the fusion rules:

$$[\sigma][\sigma] = 1 + [\varepsilon] , \quad [\sigma][\varepsilon] = [\varepsilon][\sigma] = [\sigma] , \quad [\varepsilon][\varepsilon] = 1 . \tag{8.5.4}$$

The above differential equation methods for three-point functions does not determine the OPE coefficients C_{ijk} in the more general setting of (8.2.15), but can be used to obtain information on the fusion coefficients N_{ijk}, to obtain the fusion rules:

$$[\phi_{p_1,q_1}]\,[\phi_{p_2,q_2}] = \sum\,[\phi_{p_3,q_3}] \tag{8.5.5}$$

when the summation is over $|p_1 - p_2| + 1 \leq p_3 \leq \min(p_1 + p_2 - 1, 2m - 1 - p_1 - p_2)$, $|q_1 - q_2| + 1 \leq q_3 \leq \min(q_1 + q_2 - 1, 2m + 1 - q_1 - q_2)$. For example

$$(\mathcal{L}_{-2} - 3/(2h_{2,1}+1)\mathcal{L}^2_{-1})\,\langle \phi_{2,1}(z_1)\phi_{p,q}(z_2)\phi_{p',q'}(z_3)\rangle = 0$$

yields $q' = q$, $p' = p \pm 1$ by considering the most singular term in the OPE of $\phi_{2,1}(z_1)\phi_{p,q}(z_2)$ which yields constraints on which dimensions can appear there. Thus in the Ising case, $h_{1,1} = h_{2,3} = 0$, $h_{2,1} = h_{1,3} = 1/2$, $h_{1,2} = h_{2,2} = 1/16$ with primary fields $1 = \phi_{1,1} = \phi_{2,3}$, $\varepsilon = \phi_{2,1} = \phi_{1,3}$, $\sigma = \phi_{1,2} = \phi_{2,2}$. The null differential equation for $\sigma = \phi_{1,2}$ leads to a shift operator; $\phi_{12}\phi_{21}$ is a combination of $\phi_{2,0}$ and $\phi_{2,2}$ with dimensions 63/48 and 1/16 respectively and similarly for $\sigma = \phi_{2,1}$ yields that $\phi_{12}\phi_{21}$ is a combination of $\phi_{0,2}$ and $\phi_{2,2}$ with dimensions 35/48 and 1/16 respectively. Comparing these expansions leads us to $[\sigma][\varepsilon] = [\sigma]$. Similarly expanding $\phi_{13}\phi_{13} = \phi_{21}\phi_{21}$ leads to a truncation from above: $[\varepsilon][\varepsilon] = [1]$. Starting with the fields $\phi_{p,q}$, $0 < p < m+1$, $0 < q < m$ the degenerate fields with $p \geq m+1$ or $q \geq m$ do not appear in the fusion rules.

We end our heuristic discussion of Ising fusion rules, and set this in a rigorous C^*-algebraic formulation in the next section.

8.6 C^*-algebraic approach to the Ising fusion rules

The Ising fusion rules can be concretely constructed in the fermion formalism using Bogoliubov endomorphisms. In the algebraic approach, where for each *sector* we have a representation $\pi_i : A \to B(H)$ of the C^*-algebra A and a ground representation π_0 and endomorphisms ρ_i of A such that $\pi_i = \pi \circ \rho_i$ then the multiplication becomes

$$[\pi_i] \times [\pi_j] = [\pi \circ \rho_i \circ \rho_j] \tag{8.6.1}$$

which then has to be decomposed as a sum of equivalence classes $[\pi_j]$. The considerations will be facilitated through having a notion of localized endomorphisms of von Neumann algebras, to obtain the Ising fusion rules:

$$[\rho_{1/2}]^2 = [\rho_0] + [\rho_1]\,, \quad [\rho_{1/2}][\rho_1] = [\rho_1][\rho_{1/2}] = [\rho_{1/2}]\,, \quad [\rho_1]^2 = [\rho_0] \tag{8.6.2}$$

where we have written $[\rho_i] = [\pi_0 \circ \rho_i]$, $i = 0, 1/2, 1$, cf. (8.5.4).

In the critical Ising model we obtained Majorana fermions, operators satisfying the canonical anti-commutation relations: $\{b_a, b_c\} = \delta_{a,-c}$, and the real Majorana condition: $b_a^* = b_{-a}$, $a, c \in \mathbb{Z} + 1/2$ (NS, Neveu–Schwarz), or $a, c \in \mathbb{Z}$ (R, Ramond). We work with the self dual CAR formalism introduced in Chapter 6 and already heavily utilized in Chapter 7 for the Ising model. More precisely, let $\mathrm{Maj}_{\mathrm{NS}}$, $\mathrm{Maj}_{\mathrm{R}}$ be the self dual CAR algebras over $L^2(S^1, \Gamma)$, and conjugation $\Gamma f = \overline{f}$, with distinguished bases $\{e_r \mid r \in \mathbb{Z} + 1/2\}$ NS; $\{e_n \mid n \in \mathbb{Z}\}$ R where $e_a(z) = z^a$, $a \in (1/2)\mathbb{Z}$ and we will write e_{ab} for $e_a \otimes \overline{e}_b$. Then $b_a = B(e_a)$ satisfy the required relations. To describe these as fields on the circle, we write as in the algebraic picture of Section 8.4, with x real, $z = e^{2\pi i x}$ in S^1:

$$b(x) = \textstyle\sum_s e^{-2\pi i s x} b_s = \sqrt{z} B(z)\,,$$

where $s \in \mathbb{Z} + 1/2$ (NS), $\mathbb{Z}$ (R). Then the canonical anti-commutation relations $\{b_a, b_c\} = \delta_{a,-c}$ translate as

$$[B(f)^*, B(g)] = \langle g, f\rangle = \int_{S^1} \frac{dz}{2\pi i z} g(z)\overline{f(z)}\,.$$

where $B(f) = \int \left(2\pi i z^{1/2}\right)^{-1} f(z) B(z) dz$. The symmetric or Majorana condition $b_a^* = b_{-a}$ becomes

$$B(z)^* = zB(z)\,, \quad B(f)^* = B(\Gamma f)$$

where $\Gamma f(z) = \overline{f(z)}$ for $f \in L^2(S^1)$ and we also have the periodicity

$$b(x+1) = \begin{cases} -b(x) & \mathrm{NS} \\ +b(x) & \mathrm{R} \end{cases}, \quad B(e^{2\pi i} z) = \begin{cases} +B(z) & \mathrm{NS} \\ -B(z) & \mathrm{R} \end{cases}.$$

It is more convenient for us to handle both situations simultaneously in the direct sum of two copies of the self dual CAR algebra over $L^2(S^1)$. We will work with the even part A of the Majorana algebra $\mathrm{Maj} \equiv \mathrm{Maj}_{\mathrm{NS}} \oplus \mathrm{Maj}_{\mathrm{R}}$ (with centre generated by $Z = (-1) \oplus 1$), which by Exercise 8.3 can be identified with the universal algebra generated by local algebras $A(I) \cong A(L^2(I), \Gamma)^+$, for open subintervals I of S^1. (Since the collection of intervals is not directed, the universal algebra is not the even part of $A(L^2(S^1)))$. A has a two-dimensional centre generated by a self adjoint unitary Z, and A is generated by Z and $A(I_\zeta)$, the C^*-algebra of a punctured circle $I_\zeta = S^1 \backslash \{\zeta\}$ for any $\zeta \in S^1$.

When these Ramond and Neveu–Schwarz fermions appeared in Section 8.3 from the corner transfer matrix method, they were in specific ground state representations satisfying

$$\pi_{\mathrm{NS}}(b_r)\Omega_{\mathrm{NS}} = 0\,, \quad r > 0\,, \quad r \in \mathbb{Z} + \tfrac{1}{2}$$
$$\pi_{\mathrm{R}}(b_n)\Omega_R = 0\,, \quad n > 0\,, \quad n \in \mathbb{Z}$$

and the orthogonality relation $\langle \pi_R(b_0)\Omega_{\mathrm{R}}, \Omega_{\mathrm{R}}\rangle = 0$. The NS-representation is irreducible, the R-representation decomposes into two irreducible representations $(H_{\mathrm{R}}^{\pm}, \pi_{\mathrm{R}}^{\pm})$ with cyclic vectors $\Omega_{\mathrm{R}}^{\pm} = 2^{-1/2}(\Omega_{\mathrm{R}} \pm \pi_{\mathrm{R}}(b_0)\Omega_{\mathrm{R}})$ respectively, so that $\pi_{\mathrm{R}}(b_0)\Omega_{\mathrm{R}}^{\pm} = \pm 2^{-1/2}\Omega_{\mathrm{R}}^{\pm}$. The triples $(H_{\mathrm{NS}}, \pi_{\mathrm{NS}}, \Omega_{\mathrm{NS}})$, $(H_{\mathrm{R}}, \pi_{\mathrm{R}}, \Omega_{\mathrm{R}})$ are the GNS triples of the quasi-free states $\varphi_{S_{\mathrm{NS}}}, \varphi_{S_{\mathrm{R}}}$, where

$$S_{\mathrm{NS}} = \textstyle\sum_{r \in \mathrm{N}-1/2} e_{-r,-r}\,, \qquad S_{\mathrm{R}} = e_{0,0}/2 + \sum_{n \in \mathrm{N}} e_{-n,-n}$$
$$\omega_{\mathrm{X}} = \langle \pi_{\mathrm{X}}(\cdot)\Omega_{\mathrm{X}}, \Omega_{\mathrm{X}}\rangle = \varphi_{S_{\mathrm{X}}}\,, \quad \mathrm{X} = \mathrm{R} \text{ or NS}\,.$$

Here S_{NS} is a basis projection but S_{R} is not. So $\pi_{\mathrm{NS}} = \pi_{S_{\mathrm{NS}}}$ is an irreducible Fock representation, and $\pi_{\mathrm{R}} = \pi_{S_{\mathrm{R}}}$ is reducible on $H_{\mathrm{R}}^{+} \oplus H_{\mathrm{R}}^{-}$, $H_{\mathrm{R}}^{\pm}$ has cyclic vectors $\Omega_{\mathrm{R}}^{\pm} = (1 \pm b_0)\Omega_{\mathrm{R}}/\sqrt{2}$ (see the proof of Theorem 6.40).

On the even observable algebra $A = \mathrm{Maj}^+$, the NS-representation decomposes into two inequivalent irreducible representations

$$\pi_{\mathrm{NS}}|_A = \pi_0 \oplus \pi_1\,, \quad H_{\mathrm{NS}} = H_0 \oplus H_1 \tag{8.6.3}$$

on even and odd tensors respectively, whilst the R-representation splits into a sum of two equivalent irreducibles:

$$\pi_{\mathrm{R}}|_A = \pi_{1/2} \oplus \pi'_{1/2}\,, \quad H_{\mathrm{R}} = H_{1/2} \oplus H'_{1/2} \tag{8.6.4}$$

again on even and odd spaces respectively. Eventually we will extend π_J, $J = 0, 1/2, 1$ to $R(I) = \pi_0(A(I))''$, for $I \in \mathcal{I}_\zeta$, the set of open intervals such that ζ does not belong to their closures.

Define the isometries W_J, $J = 0, 1/2, 1$ on $K = L^2(S^1) \oplus L^2(S^1)$ by

$$W_0 = \begin{pmatrix} 1 & 0 \\ 0 & 1 \end{pmatrix}, \quad W_{1/2} = \begin{pmatrix} 0 & V_{1/2} \\ V'_{1/2} & 0 \end{pmatrix}, \quad W_1 = \begin{pmatrix} V_1 & 0 \\ 0 & V'_1 \end{pmatrix}$$
$$V_{1/2} = i(e_{1/2,0} - e_{-1/2,0})/\sqrt{2} + i\textstyle\sum_{n=1}^{\infty}(e_{n+1/2,n} - e_{-n-1/2,-n})$$

$$V'_{1/2} = i\sum_{n=1}^{\infty}(e_{n,n-1/2} - e_{-n,-n+1/2}), \quad V'_1 = e_{0,0} - \sum_{n=1}^{\infty}(e_{n,n} + e_{-n,-n})$$
$$V_1 = e_{1/2,-1/2} + e_{-1/2,1/2} - \sum_{n=1}^{\infty}(e_{n+1/2,n+1/2} - e_{-n-1/2,-n-1/2}).$$

These induce endomorphisms ρ_J on Maj by $\rho_J\left(\hat{B}(f \oplus g)\right) = \hat{B}(W_J(f \oplus g))$, if $\hat{B}(f \oplus g) = B(f) \oplus B(g)$, so that $\rho_0 = \mathrm{id}$,

$$\rho_{1/2}(b_a) = \begin{cases} ib_{a+1/2} & a \geq 1/2 \\ i(b_{1/2} - b_{-1/2})/\sqrt{2} & a = 0 \\ -ib_{a-1/2} & a \leq -1/2 \end{cases} \qquad \rho_{1/2}(Z) = -Z$$

$$\rho_1(b_a) = \begin{cases} -b_a & a \neq 0, \pm 1/2 \\ b_{-a} & a = 0, \pm 1/2 \end{cases} \qquad \rho_1(Z) = Z.$$

Putting $\mathrm{R} = \sqrt{2}b_0 + b_{1/2} + b_{-1/2}$, a self adjoint odd unitary in Maj, then $\rho_1 = \mathrm{Ad}(\mathrm{R})$ is inner on Maj, so that

$$\pi_{\mathrm{X}} \circ \rho_1 \cong \pi_{\mathrm{X}}, \quad \mathrm{X} = \mathrm{NS} \text{ or } \mathrm{R},$$

and, restricting to the even observable algebra,

$$\pi_0 \circ \rho_1 \cong \pi_1, \quad \pi_1 \circ \rho_1 \cong \pi_0, \quad \pi_{1/2} \circ \rho_1 \simeq \pi_{1/2}.$$

We verify from the definitions that

$$V^*_{1/2} S_{\mathrm{NS}} V_{1/2} = S_{\mathrm{R}}, \quad V'^*_{1/2} S_{\mathrm{R}} V'_{1/2} = S_{\mathrm{NS}}.$$

Now Ker $V^*_{1/2}$ is spanned by $e_{1/2} + e_{-1/2}$, and so $\mathrm{Ker}\, V^*_{1/2} \cap \mathrm{Ran}\, S_{\mathrm{NS}} = 0$, i.e. $m_{V_{1/2}} = 1$, $n_{V_{1/2}} = 0$. Hence by Section 6.9, $\pi_{S_{\mathrm{NS}}} \circ \tau(V_{1/2}) \cong \pi_{S_{\mathrm{R}}}$. Indeed as $n_{V_{1/2}} = 0$, $\pi_{S_{\mathrm{NS}}} \circ \tau(V_{1/2}) \cong \pi_{V^*_{1/2} S_{\mathrm{NS}} V_{1/2}} = \pi_{S_{\mathrm{R}}}$ by Proposition 6.39, i.e.

$$\pi_{\mathrm{NS}} \circ \rho_{1/2} \cong \pi_{\mathrm{R}}.$$

In the notation of the proof of Theorem 6.40, $S_{\mathrm{R}} = F_{\mathrm{R}} + E_{\mathrm{R}}/2 = (1 + F_{\mathrm{R}} - \Gamma F_{\mathrm{R}} \Gamma)/2$, where $F_{\mathrm{R}} = \sum_{n \in \mathbb{N}} e_{-n,-n}$, $E_{\mathrm{R}} = e_{0,0}$. Thus $\pi^{\pm}_{S_{\mathrm{R}}} = \pi_{(F_{\mathrm{R}}, \pm e_0)}$ are the irreducible components of $\pi_{S_{\mathrm{R}}}$ given by (6.11.3). Restricting to the even algebra as in Corollary 6.41, both the representations $\pi^{\pm}_{S_{\mathrm{R}}} = \pi_{(F_{\mathrm{R}}, \pm e_0)}$ become equivalent to $\pi_{1/2}$, we have

$$\pi_0 \circ \rho_{1/2} \cong \pi_1 \circ \rho_{1/2} \cong \pi_{1/2}.$$

Now $V'^*_{1/2} V^*_{1/2} S_{\mathrm{NS}} V_{1/2} V'_{1/2} = S_{\mathrm{NS}}$, and $\mathrm{Ker}\, V'^*_{1/2} V^*_{1/2}$ is spanned by $e_+ \equiv e_{1/2}$, and $e_- \equiv \Gamma e_+ = e_{-1/2}$, of which e_- lies in the range of S_{NS}, $\mathrm{Ker}\, V'^*_{1/2} V^*_{1/2} \cap \mathrm{Ran}\, S_{\mathrm{NS}} = \mathbb{C}e_-$, i.e. $m_{V'_{1/2} V_{1/2}} = 2$, $n_{V'_{1/2} V_{1/2}} = 1$. Hence by Proposition 6.39, $\pi_{S_{\mathrm{NS}}} \circ \tau(V_{1/2} V'_{1/2})$ is a sum of two copies of $\pi_{V'^* V S_{\mathrm{NS}} V V'} = \pi_{S_{\mathrm{NS}}}$, i.e.

$$\pi_{\mathrm{NS}} \circ \left(\rho_{1/2}\right)^2 \cong \pi_{\mathrm{NS}} \oplus \pi_{\mathrm{NS}}.$$

Restricting to the even algebra as in Corollary 6.41, we deduce

$$\pi_0 \circ \left(\rho_{1/2}\right)^2 \cong \pi_1 \circ \left(\rho_{1/2}\right)^2 \cong \pi_0 \oplus \pi_1$$

The endomorphisms $\rho_{1/2}, \rho_1$ are not localized, i.e. there is no non-empty interval I' such that they are trivial on $A(I')$. However, some care is needed with the concept of locality as in the present set-up, the vacuum representation π_0 is not faithful since we do not have Haag duality in the sense that $A(I')' \neq A(I)''$ if I' is the open complement of any (open) interval I in the circle, even though there is equality in π_0. So an endomorphism on ρ of $\mathcal{A}$ is *localized* in an interval I, $(I' \neq \varnothing)$, if

$$(\mathrm{a})\, \rho|_{A(I')} = \mathrm{id}\,, \quad (\mathrm{b})\, \rho(A(I_0)) \subset A(I_0)\,, \quad I \subset I_0\,, \quad (\mathrm{c})\, \rho(Z) = \pm Z\,.$$

If $A(I')' = A(I)''$, then (a) would imply (b). In terms of the punctured algebra, we say an endomorphism ρ of $A(I_\zeta)$ is *localized* in an interval I, $(I' = \varnothing)$ if

$$\rho|_{A(I_1)} = \mathrm{id}, I_1 \in \mathcal{I}_\zeta, I_1 \cap I = \varnothing.$$

We can modify the construction to get local endomorphisms satisfying the same fusion relations. By conformal transformations, it is enough to consider $I = \left\{z = e^{i\phi} \mid -\pi/2 < \phi < \pi/2\right\}$. Take $I_\pm = \left\{z \in e^{i\phi} \mid \pi/2 < \phi < \pi\right\}$ so that $L^2(S^1) = L^2(I_-) \oplus L^2(I) \oplus L^2(I_+)$ with orthogonal projections p_-, p, p_+ respectively. Define for $a \in (1/2)\mathbb{Z}$: $f_a(z) = \sqrt{2} z^{2a} \chi_I$, so that $\{f_r \mid r \in \mathbb{Z} + 1/2\}$, $\{f_n \mid n \in \mathbb{Z}\}$ are two orthonormal bases for $L^2(I)$, and write $f_{a,b} = f_a \otimes \overline{f}_b$.
Then define isometries

$$V = p_- - p_+ + i(f_{1/2,0} - f_{-1/2,0})/\sqrt{2} + i\textstyle\sum_{n=1}^\infty (f_{n+1/2,n} - f_{-n-1/2,-n})$$
$$V' = p_- - p_+ + i\textstyle\sum_{n=1}^\infty (f_{n,n-1/2} - f_{-n,-n+1/2})\,.$$

Fix the endomorphisms $\rho_{1/2}^{\mathrm{loc}} = \tau(V)$ or $\tau(V')$, $\rho_{1/2}^{\mathrm{loc}}(Z) = -Z$, ρ_1^{loc} is the automorphism $\rho_1^{\mathrm{loc}} = \tau(h \otimes \overline{h} - 1) = \mathrm{Ad}\, B(h)$ for real (i.e. Γ-invariant) $h \in L^2(I)$, $\|h\| = \sqrt{2}$, $\rho_1^{\mathrm{loc}}(Z) = Z$. Then we claim the endomorphisms ρ_0^{loc}, $\rho_{1/2}^{\mathrm{loc}}$, ρ_1^{loc} obey the same Ising fusion rules and are localized in I. In this case, by a conformal transformation, we can arrange for the localization regions of ρ_1^{loc} and $\rho_{1/2}^{\mathrm{loc}}$ to be disjoint so that the representatives commute.

Now $1 - VV^*$ is the rank one projection on $f_{1/2} + f_{-1/2}$, so that $V^* S_{\mathrm{NS}} V - (V^* S_{\mathrm{NS}} V)^2$ is of rank one. Hence the Powers–Størmer inequality (Exercise 4.10) for positive operators A, B: $\|A - B\|_2^2 \leq \|A^2 - B^2\|_1$ shows that $(V^* S_{\mathrm{NS}} V)^{1/2} - V^* S_{\mathrm{NS}} V$ is Hilbert–Schmidt. But $V^* S_{\mathrm{NS}} V - S_{\mathrm{R}}$ is Hilbert–Schmidt (Exercise 8.3 (i)) and $S_{\mathrm{R}}^{1/2} - S_{\mathrm{R}} = (1/\sqrt{2} - 1/2) e_{0,0}$ is of rank one. Hence $(V^* S_{\mathrm{NS}} V)^{1/2} - S_{\mathrm{R}}^{1/2}$ is Hilbert–Schmidt, and so the quasi-free states $\varphi_{S_{\mathrm{NS}}} \circ \tau(V) = \varphi_{V^* S_{\mathrm{NS}} V}$ and $\varphi_{S_{\mathrm{R}}}$ are quasi-equivalent (Exercise 6.11). Now $m_V = 1$, $n_V = 0$ as $\mathrm{Ker}\; V^* = \mathbb{C}(f_{1/2} + f_{-1/2})$, (and $m_{V'} = 1$, $n_{V'} = 0$ as $\mathrm{Ker}\; V'^* = \mathbb{C} f_0$) so that

$$\pi_{S_{\mathrm{NS}}} \circ \tau(V) \cong \pi^+ \oplus \pi^- \cong \pi_{S_{\mathrm{NS}}} \circ \tau(V')$$

where π^+, π^- are the disjoint non-Fock representations of Theorem 6.40. On the even algebra, the representations $\pi_\pm = \pi_{(F_{\mathrm{R}},\pm e_0)}$ become equivalent and as $\pi_{S_{\mathrm{R}}}$ restricted to the even algebra becomes $\pi_{1/2} \oplus \pi'_{1/2}$, two copies of $\pi_{1/2} \cong \pi'_{1/2}$, then restrictions of $\pi_\pm$ are $\pi_{1/2}$, and $\pi^+_{S_{\mathrm{NS}}} \circ \tau(U) \cong \pi^-_{S_{\mathrm{NS}}} \circ \tau(U)$, for $U = V$ or V'. Consequently

$$\pi_0 \circ \rho^{\mathrm{loc}}_{1/2} \cong \pi_1 \circ \rho^{\mathrm{loc}}_{1/2} \cong \pi_{1/2}\,. \tag{8.6.5}$$

Now $1 - U^{2*}U^2$ is of rank two, and so by a similar argument to the above, using the fact that $U^{2*}S_{\mathrm{NS}}U^2 - S_{\mathrm{NS}}$ is Hilbert–Schmidt, we get $(U^{2*}S_{\mathrm{NS}}U^2)^{1/2} - S_{\mathrm{NS}}$ is Hilbert–Schmidt. Thus the quasi-free states $\varphi_{S_{\mathrm{NS}}} \circ \tau(U^2) = \varphi_{U^{2*}S_{\mathrm{NS}}U^2}$ and $\varphi_{S_{\mathrm{NS}}}$ are quasi-equivalent (Exercise 6.11). Moreover $m_U = 2$, (Ker$(V^*)^2$ is spanned by $f_{1/2} + f_{-1/2}$ and $V(f_{1/2} + f_{-1/2})$, Ker$(V')^*$ spanned by f_0 and $V'f_0$) so that again by Theorem 6.40,

$$\pi_{S_{\mathrm{NS}}} \circ \tau(U)^2 \cong \pi_{S_{\mathrm{NS}}} \oplus \pi_{S_{\mathrm{NS}}}$$

and by restriction to the even algebra

$$\pi_0 \circ \rho^{\mathrm{loc}}_{1/2}\rho^{\mathrm{loc}}_{1/2} \oplus \pi_1 \circ \rho^{\mathrm{loc}}_{1/2}\rho^{\mathrm{loc}}_{1/2} \cong \pi_0 \oplus \pi_1 \oplus \pi_0 \oplus \pi_1\,.$$

By (8.6.5) we have $\pi_0 \cong \pi_1 \circ \rho^{\mathrm{loc}}_{1/2}$ and so $\pi_0 \circ (\rho^{\mathrm{loc}}_{1/2})^2 \cong \pi_1 \circ (\rho^{\mathrm{loc}}_{1/2})^2$. This forces

$$\pi_0 \circ \left(\rho^{\mathrm{loc}}_{1/2}\right)^2 \cong \pi_0 \oplus \pi_1\,. \tag{8.6.6}$$

If I is an interval with non-empty open complement, the local von Neumann algebras $R(I) = \pi_0(A(I))''$, will be used to construct quasi-local algebras A_ζ the norm closure of $\bigcup_{I \in \mathcal{I}_\zeta} R(I)$. We put $R_\zeta(I') = \vee\{R(I_0) \mid I_0 \in \mathcal{I}_\zeta,\, I_0 \cap I = \varnothing\} \subset R(I')$. In fact equality here holds, Haag duality on the punctured circle:

$$R(I)' = R(I') = R_\zeta(I')\,.$$

Since we already have $R_\zeta(I') \subset R(I') \subset R(I)'$ it is enough to show $\pi_0(B(f)B(g)) \in R_\zeta(I')$ if f, g are supported on I'. Let $I_\pm$ be the connected components of $I'\backslash\{\zeta\}$, with characteristic functions $\chi_\pm$, and put $f_\pm = \chi_\pm f$, $g = \chi_\pm g$. Then $\pi_0(B(f)B(g)) = \sum_{\sigma,\sigma'} B(f_\sigma)B(g_{\sigma'})$, and clearly $\pi_0(B(f_\sigma)B(f_\sigma)), \pi_0(B(g_\sigma), B(g_\sigma)) \in R_\zeta(I')$. Take functions $h^\pm_n$ of unit length, whose supports are in $I_\pm$ decreasing to ζ and a weak limit point $\lambda = \lim_{n\to\infty} \pi_0(B(h^+_n)B(h^-_n)) \in R(I')$. But $\pi(B(h^+_n)B(h^-_n)) \in R(I_1)'$ for $I_1 \in \mathcal{I}_\zeta$, and so λ is a scalar by irreducibility – which we can arrange to be non-zero. Hence $\pi_0(B(f_+)B(g_-)) = \lambda^{-1}\lim_{n\to\infty}(B(f_+)B(h^n_+)B(h^n_-)B(f_-)) \in R_\zeta(I')$.

The homomorphisms π_J on $A(I)$ extend to $R(I) = \pi_0(\mathcal{A})''$, i.e. $\pi_J|_{A(I)} \approx \pi_0|_{A(I)}$, $J = 0, 1/2, 1$, $I \in \mathcal{I}_\zeta$. When $J = 1$, $\pi^-_{S_{\mathrm{NS}}} \cong \pi^+_{S_{\mathrm{NS}}} \circ \mathrm{Ad}(V)$ on $A(L^2(S'), \Gamma)^+$, where $V = \sqrt{2}\psi(h)$. When we restrict to $A(L^2(I), \Gamma)^+$ we can replace the odd V by the even $\sqrt{2}\psi(h)\sqrt{2}\psi(h')$ where $h' \in L^2(I_0)$ is real, $\|h'\| = 1$,

$I_0 \in \mathcal{I}_\zeta$, $I_0 \cap I = \varnothing$. Then $\pi^-_{S_{\rm NS}}|_{A(L^2(I),\Gamma)^+} \approx \pi^+_{S_{\rm NS}}|_{A(L^2(I),\Gamma)^+}$. For $J = 1/2$, $I \in \mathcal{I}_\zeta$ we have $\pi_{S_{\rm NS}}|_{A(L^2(I),\Gamma)} \approx \pi_{S_{\rm R}}|_{A(L^2(I),\Gamma)}$ because $(P_I S_{\rm NS} P_I)^{1/2} - (P_I S_{\rm R} P_I)^{1/2}$ is Hilbert–Schmidt (since $P_I S_{\rm NS} P_I - P_I S_{\rm R} P_I$ is trace class, Exercise 8.5) and $\Omega_{S_{\rm NS}}$, $\Omega_{S_{\rm R}}$ remain cyclic for $A(L^2(I),\Gamma)$, (Exercise 8.4 and Exercise 6.7(a)). Restricting to the even algebra, we have by (8.6.3) and (8.6.4) that

$$\pi_0 \oplus \pi_1 \approx \pi_{1/2} \oplus \pi_{1/2} \quad \text{on } A(L^2(I),\Gamma)^+ .$$

Thus since we already know $\pi_0 \approx \pi_1$ on $A(L^2(I),\Gamma)^+$ we have

$$\pi_{1/2} \approx \pi_0 \quad \text{on } A(L^2(I),\Gamma)^+ .$$

We can thus extend π_J to $R(I) = \pi_0(A(I))''$, $A(I) = A(L^2(I),\Gamma)^+$, such that $\pi_J \cong \rho_J^{\rm loc}$ on $A(I_\zeta)$. Thus $\rho_J^{\rm loc}$ extend to A_ζ and satisfy

$$\begin{array}{ll} \rho_J^{\rm loc}|_{R(I_1)} = id & I_1 \in \mathcal{I}_\zeta\,,\ I_1 \cap I = \varnothing \\ \rho_J^{\rm loc}(R(I_0)) \subset R(I_0) & I_0 \in \mathcal{I}_\zeta\,,\ I \subset I_0 \end{array}$$

by extension of the corresponding properties on $A(I_0), A(I_1)$.

We can then compute the Fusion rules of $[\rho_J]$, the equivalence classes of localized endomorphisms which are unitarily equivalent to $\rho_J^{\rm loc}$ in the vacuum representation. Such fusion rules are relatively easy to handle, as if ρ_a, ρ_b are endomorphisms of A_ζ, localized in I_a, I_b respectively, then we can choose an interval $I \in \mathcal{I}_\zeta$, $I_a \cup I_b \subset I$. Then ρ_a, ρ_b are localized in I, and if they are unitarily equivalent on A_ζ, then the conjugating unitary lies in $R(I)$ ($\subset A_\zeta$) by Haag duality, $\rho_a = \mathrm{Ad}(U)\rho_b$. If ρ_c is a third localized endomorphism, then $\rho_c\rho_a \cong \rho_c\rho_b$ via $\rho_c(U)$, which is well defined. Thus by (8.6.6) we have

$$\left[\rho_{1/2}^2\right] = [\rho_0] + [\rho_1] \, .$$

From $\pi_1 \circ \rho_{1/2}^{\rm loc} \cong \pi_{1/2}$ and (8.6.5) we have $\pi_0 \circ \rho_1^{\rm loc}\rho_{1/2}^{\rm loc} \cong \pi_1 \circ \rho_{1/2}^{\rm loc} \cong \pi_{1/2}$ and $\rho_1^{\rm loc}\rho_{1/2}^{\rm loc} = \rho_{1/2}^{\rm loc}\rho_1^{\rm loc}$ if the localization regions of $\rho_1^{\rm loc}$ and $\rho_{1/2}^{\rm loc}$ are disjoint. Thus

$$\left[\rho_{1/2}\rho_1\right] = \left[\rho_1\rho_{1/2}\right] = \left[\rho_{1/2}\right] .$$

The final Ising fusion rule is immediate, as $\left(\rho_1^{\rm loc}\right)^2 = id$:

$$\left[\rho_1^2\right] = [\rho_0] \, . \tag{8.6.7}$$

Exercise 8.2 Let A be the self dual CAR algebra $A(L^2(S^1),\Gamma)$ over $L^2(S^1)$ with conjugation $\Gamma(f) = \overline{f}$. On $L^2(S^1) \oplus L^2(S^1)$ define a field $\hat{B}(f_{\rm NS} \oplus f_{\rm R}) = B(f_{\rm NS}) \oplus B(f_{\rm R})$. If I is an open interval of S^1, and $\zeta \in I$, then let I_1, I_2 be disjoint intervals such that $I = I_1 \cup \{\zeta\} \cup I_2$, with characteristic functions χ_j. If $f \in L^2$ let $\hat{f} = f \oplus f$, $f_j = \chi_j f$. Define

$$B_I(f,g) = \hat{B}(\hat{f}_1)\hat{B}(\hat{g}_1) + \hat{B}(\hat{f}_2)\hat{B}(\hat{g}_2) + Z\left[\hat{B}(\hat{f}_1)\hat{B}(\hat{f}_2) + \hat{B}(\hat{f}_2)\hat{B}(\hat{g}_1)\right]$$

for $f, g \in L^2(S^1)$ and $Z = (-1) \oplus 1 \in A \oplus A$. If $\zeta \notin I$, take $I = I_1$, $I_2 = \varnothing$ in the above formula. Show that:

(i) B_I is complex bilinear
(ii) $2B_I(f,f) = \langle f, \Gamma f\rangle I$
(iii) $2B_I(f,g)B_I(g,h) = \langle g, \Gamma g\rangle B_I(f,h)$,
(iv) $B_I(f,g)^* = B_I(\Gamma g, \Gamma f)$
(v) $B_I(f,g) = B_J(f,g)$, $I \subset J$, $f, g \in L^2(I)$.

Conversely, suppose A is the universal C^*-algebra generated by the range of maps B_I on $L^2(I)^2$ satisfies (i)–(v). Let $I_1 \cap I_2 = \varnothing$, $S^1 = J_+ \cup J_-$, where I_1, I_2, J_+, J_- are four intervals, where J_+ and J_- contain both of I_1 and I_2, one from the left side and one from the right side. Let $f_j \in L^2(I_j)$, $\|f_j\|^2 = 2$. Show that $Z = B_{J_+}(f_1, f_2)B_{J_-}(f_2, f_1)$ is independent of the choices, and is a self adjoint unitary generating the centre of A. Show that for every $\zeta \in S^1$, the universal algebra A is generated by $A(\mathcal{I}_\zeta)$ and Z.

Exercise 8.3 The following operators are Hilbert–Schmidt: (i) $V^*S_{\mathrm{NS}}V - S_{\mathrm{R}}$, (ii) $VS_{\mathrm{NS}}V^* - S_{\mathrm{R}}$, (iii) $V'^*S_{\mathrm{NS}}V' - S_{\mathrm{R}}$, (iv) $V'S_{\mathrm{NS}}V'^* - S_{\mathrm{R}}$.

Exercise 8.4 Let P_I denote the projection on $L^2(I) \subset L^2(S^1)$, $F_{\mathrm{R}} = \sum_{n=1}^{\infty} e_{-n,-n}$, $P_I^{(0)} = P_I - \langle e_0, P_I e_0\rangle^{-1} P_I e_{0,0} P_I$ the projection on $L_0^2(S^1) = (F_{\mathrm{R}} + \Gamma F_{\mathrm{R}}\Gamma)L^2(S^1)$. Show that

$$(1 - P_I)L^2(S^1) \cap S_{\mathrm{NS}}L^2(S^1) = \{0\}\,, (1 - P_I^{(0)})L_0^2(S^1) \cap P_{\mathrm{R}}L_0^2(S^1) = \{0\}\,.$$

Exercise 8.5 Show that $P_I S_{\mathrm{NS}} P_I - P_I S_{\mathrm{R}} P_I$ is trace class, for $I \in \mathcal{I}_\zeta$, and the cut of the half-odd integer basis functions $e_r(z) = z^r$, $r \in \mathbb{Z} + 1/2$ is at $z = \zeta$.

8.7 Moore–Seiberg axioms and fusion rule algebras

We have seen in Section 8.5 that n-point correlation functions can be expressed as finite sums of products of holomorphic and anti-holomorphic functions called *chiral blocks* (or *conformal blocks* when the chiral algebra is the Virasoro algebra). The chiral blocks are multi-valued functions, but the correlation functions are genuine single valued functions. The chiral blocks are holomorphic sections of a vector bundle over a Riemann surface M_g, of genus g. From conformal invariance (null vectors) the chiral blocks are solutions of differential equations (Ward identities, e.g. Knizhnik–Zamolodchikov equation) which restrict the dimension of the vector bundle.

We draw a n-point chiral block $\mathcal{F}_{\phi_{i_1},\dots,\phi_{i_n}} = \mathcal{F}_{i_1,\dots i_n}$, a vector, on a Riemann surface with n punctures, the latter labelled by primary fields or representations of the chiral algebra (e.g. Virasoro or Kač–Moody).

Sewing together two Riemann surfaces along a common portion of their boundaries leads to a map by summation over the labellings of the sewn holes as in Fig. 8.2.

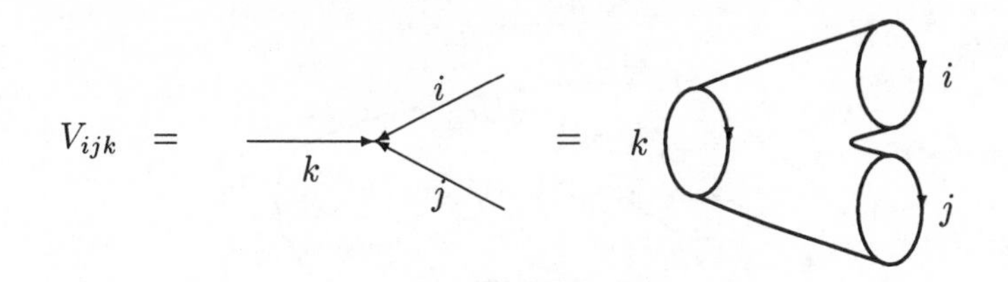

FIG. 8.18. Chiral block V_{ijk}

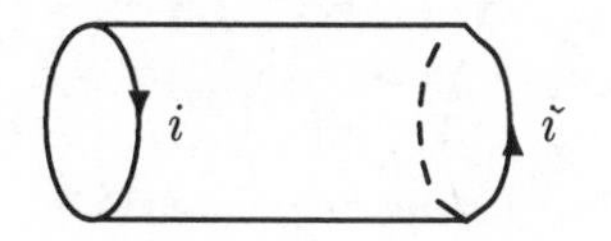

FIG. 8.19. Charge conjugation $C_{ij} = \delta_i^{j^\vee}$

Different sewing sequences will lead to different bases of chiral blocks for a given surface, but they yield the same correlation functions due to associativity of the operator product. This means (using also Exercise 8.6) that there are linear transformations (duality matrices) between bases of chiral blocks on a given surface obtained by different sewing. This is the basic axiom of duality. The Moore–Seiberg axioms give a finite number of generators and relations (called *polynomial equations*) for these duality matrices (cf. Reidemeister moves for links, Section 12.1, Markov moves for braids, Section 12.1, Alexander moves for manifolds Section 12.2 as simplified sets of generators and relations).

First the Riemann surfaces themselves are built up of trinions. Using the conjugation $i \to i^\vee$ we can take the three labellings oriented the same, (or inward) according to the graphical picture of Fig. 8.18. The dimension of the associated space being $N_{ijk} \equiv \dim V_{i,j,k}$. Composing with Fig. 8.19 the charge conjugation $C_{ij} = \delta_i^{j^\vee}$ allows us to reverse an orientation, and lower indices to obtain Fig. 8.1.

We could use, as in Section 8.2, φ^3 graphs instead of Riemann surfaces as in the examples of Fig. 8.20. Each trinion or sphere with three punctures corresponding to an operator product expansion, and a choice of couplings in an operator product expansion is a sewing sequence. If these are the generators, then we need to know the relations.

The two sewing procedures (Fig. 8.2) on the four-punctured sphere yield a *fusing matrix* F to relate the corresponding vectors as in Fig. 8.21. Similarly, two adjacent arms of Fig. 8.22 can be twisted, with a choice of orientation, to yield the *braiding matrix* $B(+)$ and similarly $B(-)$ for the opposite braiding.

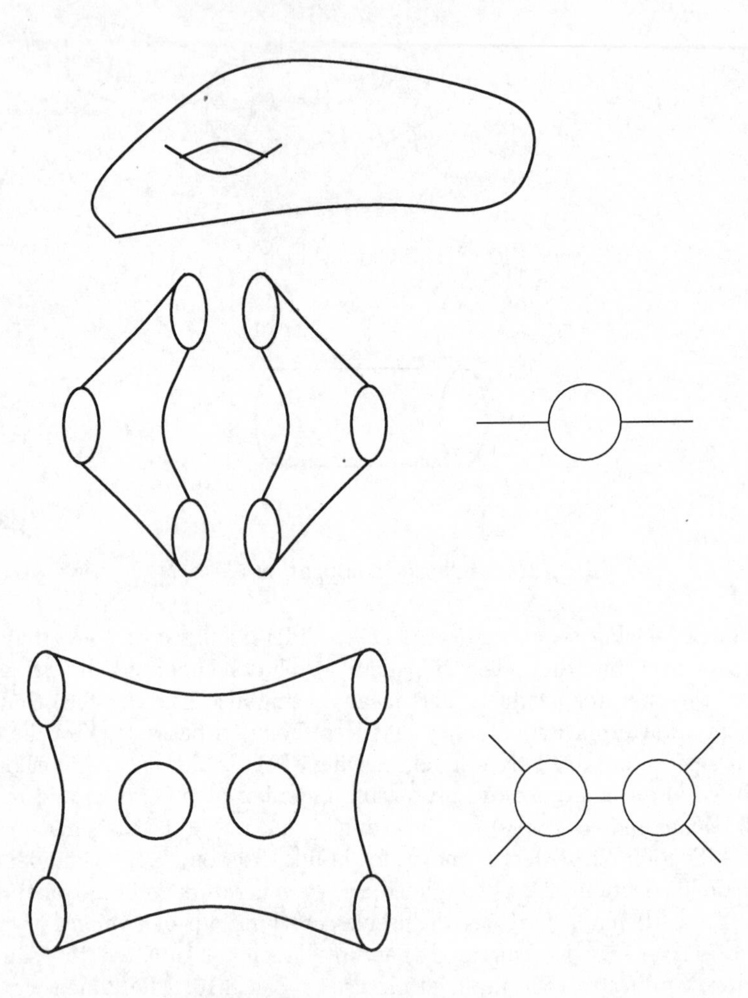

FIG. 8.20. Riemann surfaces and φ^3 graphs

The set of braiding operations $B(\pm)$ contain the more specialized ones, $\Omega(\pm)$ as in Fig. 8.23. whilst B can be recovered from F and Ω as in Fig. 8.24. Duality gives the following hexagon and pentagon identities of Fig. 8.25 and Fig. 8.26:

$$\Omega_{lk}^{m}(\varepsilon)F_{mn}\begin{bmatrix} j & k \\ i & l \end{bmatrix}\Omega_{jk}^{i}(\varepsilon)=\sum_{r}F_{mr}\begin{bmatrix} j & l \\ i & k \end{bmatrix}\Omega_{kr}^{i}(\varepsilon)F_{rn}\begin{bmatrix} k & j \\ i & l \end{bmatrix} \tag{8.7.1}$$

$$F_{sp}\begin{bmatrix} i & j \\ m & n \end{bmatrix}F_{nq}\begin{bmatrix} p & k \\ m & l \end{bmatrix}=\sum_{r}F_{nr}\begin{bmatrix} j & k \\ s & l \end{bmatrix}F_{sq}\begin{bmatrix} i & r \\ m & l \end{bmatrix}F_{rp}\begin{bmatrix} i & j \\ q & k \end{bmatrix}. \tag{8.7.2}$$

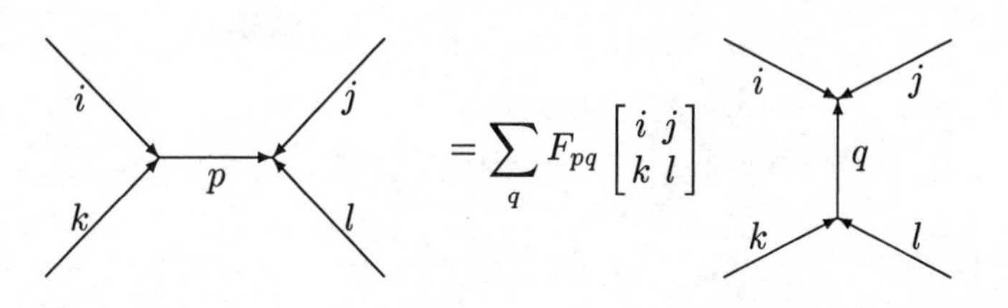

FIG. 8.21. Fusing matrix F

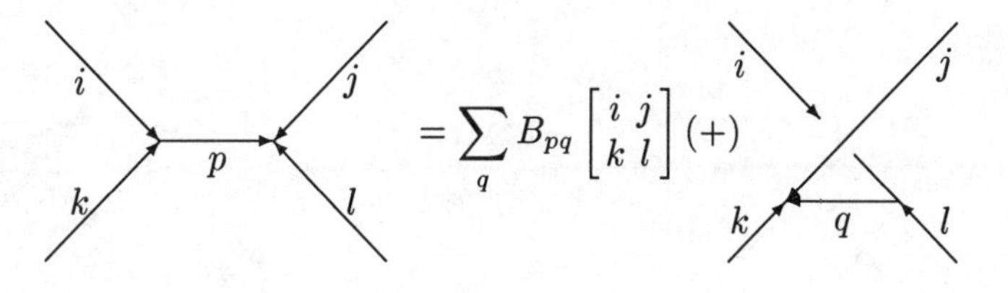

FIG. 8.22. Braiding matrix $B(+)$

This completely describes the picture for genus zero generators and relations. To handle higher genus, it is remarkable that only some generators and relations for genus one are sufficient. For this, we need the representations of the modular group through S and T matrices. These will act on one-point genus zero functions i.e. $\bigoplus_i V_{ji}{}^{i}$, and satisfy the modular relations

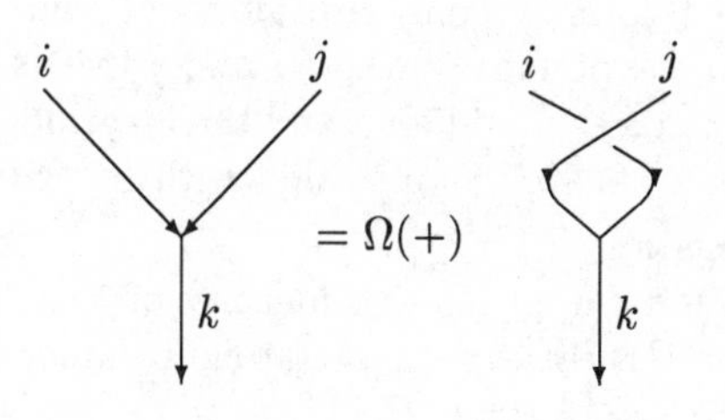

FIG. 8.23. Specialized braiding matrix $\Omega(+)$

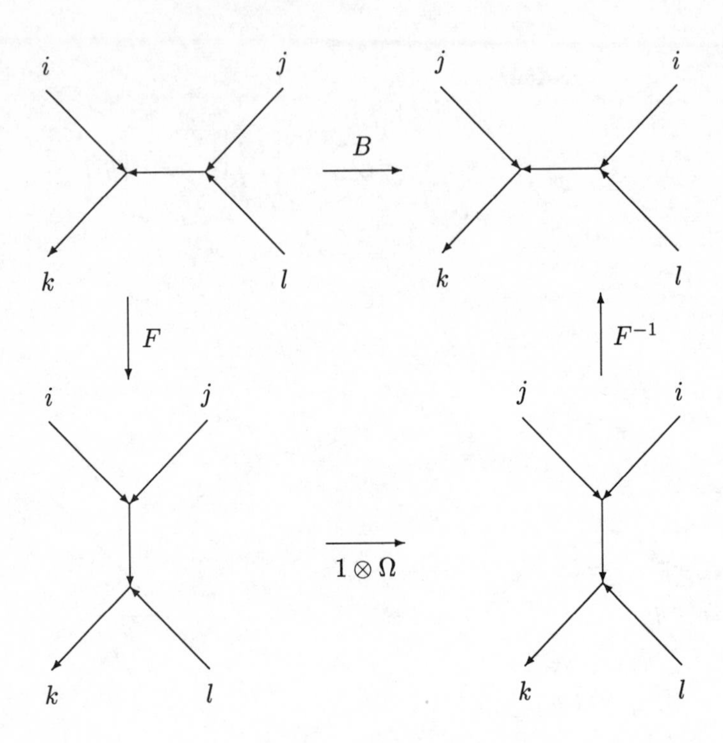

FIG. 8.24. $B(\varepsilon) = F^{-1}(1 \otimes \Omega(\varepsilon))F$

$$STS = T^{-1}ST^{-1} . \tag{8.7.3}$$

We can relate $S(j)$ of Fig. 8.27 to a braiding matrix, Θ by

$$S^2(j) = \bigoplus_i \Theta^i_{ji}(-) \tag{8.7.4}$$

where $\Theta^i_{jk}(\pm) : V^i_{jk} \to V^{k^\vee}_{ji^\vee}$ is not only a braiding of arms of i and k, but also involves a half twist on the boundary where i and j live as in Fig. 8.28.

The S matrix interchanges meridian a and the longitude b, but one has to be careful with orientation $a \to -b$, $b \to a$ so that with $S = S(0)$, $S^2 : a \to -a, b \to -b$ is the charge conjugation $C_{ij} = \delta^{j^\vee}_i$.

The basic idea behind the genus one identity of Fig. 8.29 is that we twist the arms j_1 and j_2 (via Ω), then transport around the a cycle, the meridian or alternatively in the $b = SaS^{-1}$ cycle, the longitude.

The way we have presented the Moore–Seiberg axioms of RCFT is convenient for our subfactor setting of Chapter 13. In the genuine RCFT setting, we can say a bit more. When we look at the occurrence of ϕ_i in the OPE $\phi_j\phi_k$ and in $\phi_k\phi_j$, we see that terms

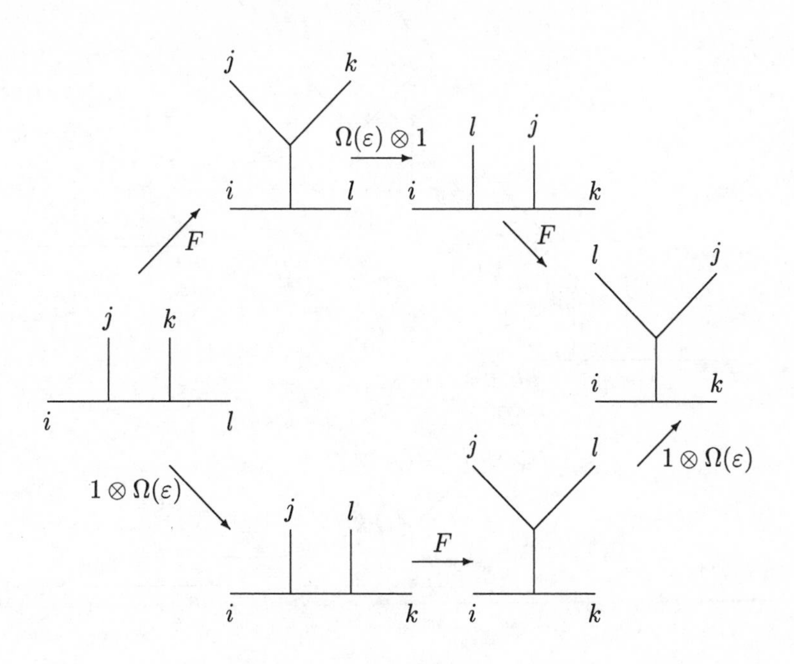

FIG. 8.25. Hexagon relation $F(\Omega(\varepsilon)\otimes 1)F = (1\otimes\Omega(\varepsilon))F(1\otimes\Omega(\varepsilon))$

$$(z-w)^{h_i-h_j-h_k}, \quad (w-z)^{h_i-h_j-h_k}$$

need to be compared. This leads to a choice of expression $(-1)^{h_i-h_j-h_k}$, and with choice of basis for V^i_{jk}, V^i_{kj} which are identified, we find

$$\Omega^i_{jk}(\pm) = \varepsilon^i_{jk} e^{\pm i\pi(h_i-h_j-h_k)}$$

where $\varepsilon = \pm 1$. Note that by topological invariance $\Omega^2 = \Omega^i_{jk}(+)\Omega^i_{kj}(+)$ gives ull kinks or type I twists on the i, j, k arms as in Fig. 8.30 which is given by a phase $e^{2\pi i(h_i-h_j-h_k)}$. Thus taking $j=k=0$, we have Fig. 8.31.

We can now summarize the Moore–Seiberg axioms of RCFT, with the following data, the duality matrices and a complete set of relations for all surfaces, expressed by some of genus zero and one alone.

Data:

1. A finite index set I and a bijection $i \to i^\vee$.
2. Finite dimensional vector spaces V^i_{jk}.
3. Braiding matrices $\Theta^i_{jk}(\pm) : V^i_{jk} \cong V^{k^\vee}_{ji^\vee}$, $\Omega^i_{jk}(\pm) : V^i_{jk} \cong V^i_{kj}$.

   Fusing matrices $F\begin{bmatrix} i & j \\ k & l \end{bmatrix} : \bigoplus_r V^k_{ir}\otimes V^r_{jl} \to \bigoplus_s V^k_{sl}\otimes V^s_{ij}$.

   Modular automorphisms $T, S(j) : \bigoplus_i V^i_{ij} \cong \bigoplus_i V_{ji}{}^i$.

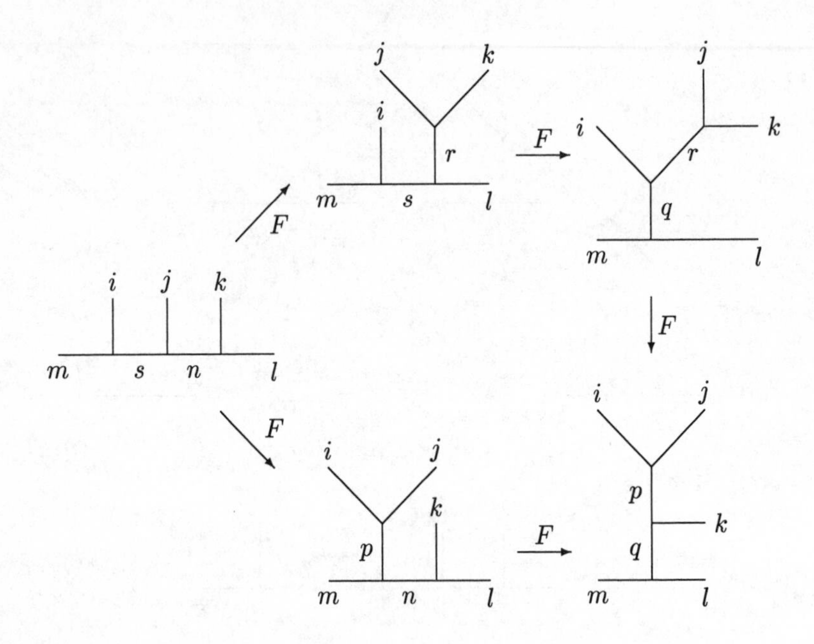

FIG. 8.26. Pentagon identity $P_{23}F_{13}F_{12} = F_{23}F_{12}F_{23}$

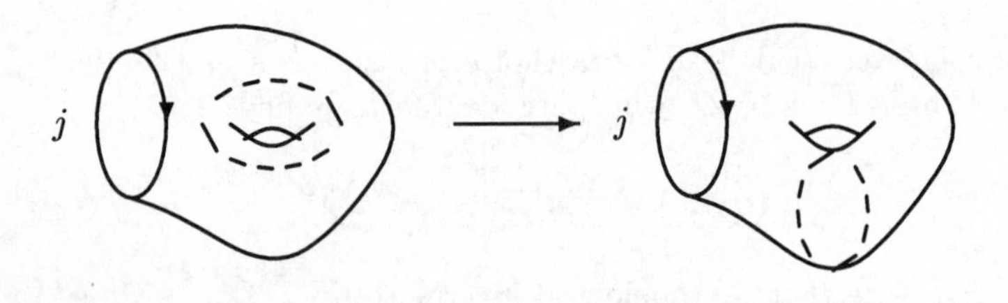

FIG. 8.27. $S(j) : \bigoplus_i V^i_{ji} \to \bigoplus_i V^i_{ji}$

Duality data – polynomial equations:

1. $i^{\vee\vee} = i$.
2. $V^i_{0j} \cong \delta_{ij}\mathbb{C}$, $V^0_{ij} \cong \delta_{ij^\vee}\mathbb{C}$, $V^i_{jk} \cong V^{k^\vee}_{ji^\vee}$, $(V^i_{jk})^\vee \cong V^{i^\vee}_{j^\vee k^\vee}$.
3. $\Omega^2(+) = \Omega^i_{jk}(+)\Omega^i_{kj}(+)$ is scalar
   T on V^i_{ji} is a scalar independent of i.
4. genus zero relations:
   hexagon relations: $F(\Omega(\varepsilon) \otimes 1)F = (1 \otimes \Omega(\varepsilon))F(1 \otimes \Omega(\varepsilon))$.

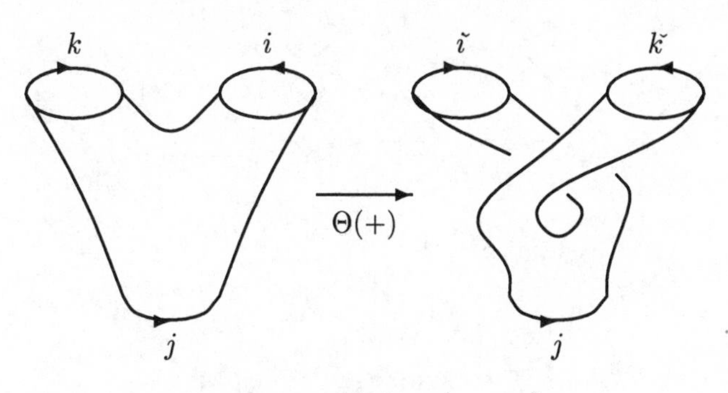

FIG. 8.28. $\Theta(+) : V^{i}_{jk} \to V^{\check{i}}_{j\check{k}}$

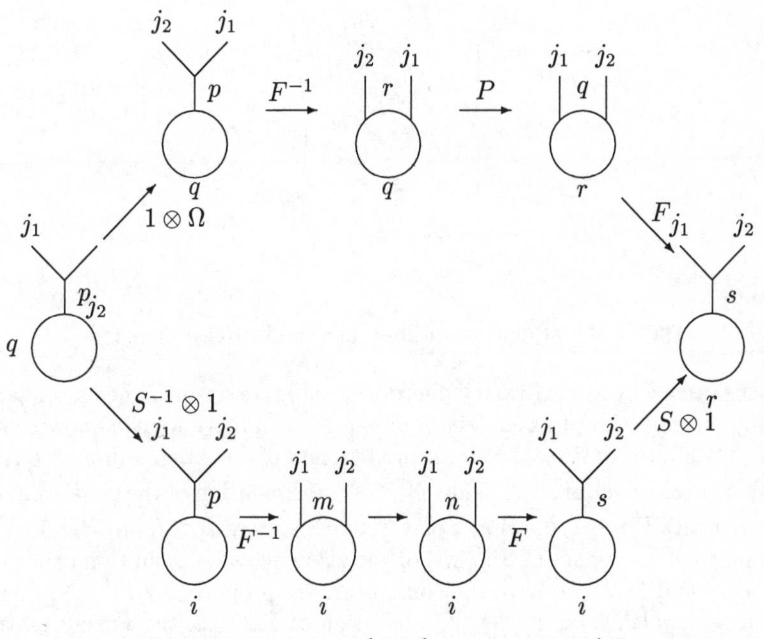

$$(S \otimes 1)F(1 \otimes \Theta(-)\Theta(+))F^{-1}(S^{-1} \otimes 1) = FPF^{-1}(1 \otimes \Omega(-))$$

FIG. 8.29. Genus one identity

pentagon identity: $F_{23}F_{12}F_{23} = P_{23}F_{13}F_{12}$.

5. genus one relations:
   $S^2(j) = \bigoplus_i \Theta^i_{ji}(-)$
   $S(j)T(j)S(j) = T(j)^{-1}S(j)T(j)^{-1}$
   $(S \otimes 1)F(1 \otimes \Theta(-1)\Theta(+))F^{-1}(S^{-1} \otimes 1) = PFF^{-1}(1 \otimes \Omega(-))$

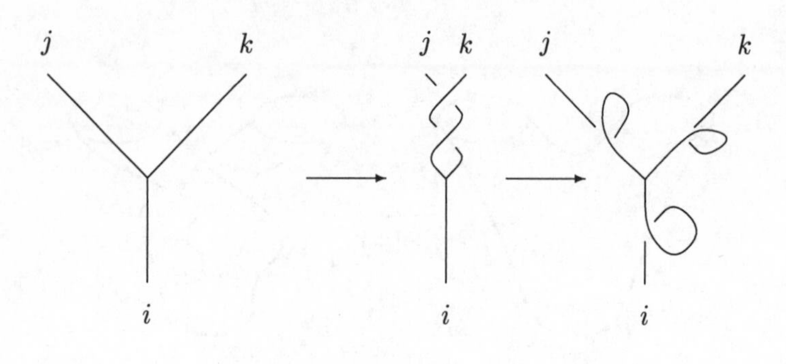

FIG. 8.30. $\Omega^i_{jk}(+)\Omega^i_{kj}(+)$

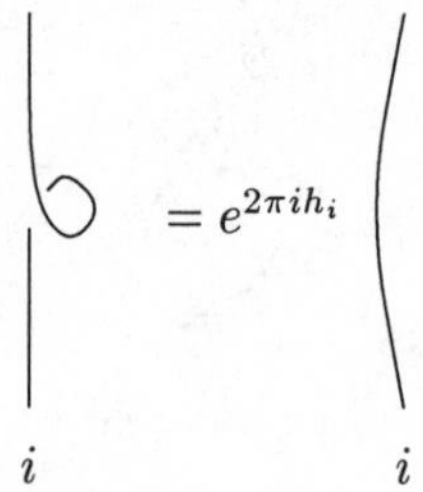

$= e^{2\pi i h_i}$

i i

FIG. 8.31. Removal of kink and conformal dimension

In the same way as for Bratteli diagrams, where we simplify the data of an embedding, a connection, to a matrix of integral multiplicity we can, as we had begun in (8.2.26), to consider the algebraic structure of the integral dimensions ${N_{ij}}^k$ – the fusion rule algebra FRA. Thus ${N_{ij}}^k$ is the dimension of the conformal block associated with Fig. 8.1, or a space of intertwiners, from $H_i \otimes H_j$ to H_k: ${N_{ij}}^k \in \mathbb{N}$. By topological invariance of Fig. 8.1, or the existence of Ω as an isomorphism between ${V_{ij}}^k$ and ${V_{ji}}^k$, we have the commutativity property: ${N_{ij}}^k = {N_{ji}}^k$. Again from the equivalence of the sewing operation of Fig. 8.2, the Fusing matrix F or Fig. 8.21, we have associativity: $\sum {N_{ij}}^m {N_{km}}^l = \sum {N_{ik}}^m {N_{jm}}^l$. The primary field 1 gives an identity for the algebra, corresponding to an index 0: ϕ_0 is an identity for $\mathcal{F}$, so that ${N_{0i}}^j = \delta^j_i$. Conjugation of primary fields, yields an automorphism of the fusion rules, i.e. an involutive bijection $i \to i^\vee$ of the basis preserving the fusion rules ${N_{ij}}^k = {N_{i^\vee j^\vee}}^{k^\vee}$, and $C_{ij} = {N_{ij}}^0$ (see Fig. 8.18) the *charge conjugation* which can be used to raise and lower indices, so that N_{ijk} of Fig. 8.18 is totally symmetric in all three indices. Thus a *fusion ring* or *fusion rule algebra FRA* is a ring over $\mathbb{Z}$ with distinguished basis $x_0, x_1, x_2, \ldots, x_n$ such that

$$x_i x_j = \textstyle\sum_k {N_{ij}}^k x_k$$

satisfy

(a) integrality: ${N_{ij}}^k \in \mathbb{N}$;
(b) commutativity: ${N_{ij}}^k = {N_{ji}}^k$;
(c) associativity: ${N_{ij}}^m {N_{mn}}^l = {N_{ik}}^m {N_{mj}}^l$;
(d) unital: ${N_{0j}}^k = \delta_{jk}$;
(e) conjugation: bijective involution $i \to i^\vee$ such that ${N_{ij}}^0 = \delta_{ij^\vee}$;
(f) ${N_{i^\vee j^\vee}}^{k^\vee} = {N_{ij}}^k$.

Then $C_{ij} = {N_{ij}}^0 = \delta_{ji^\vee}$ is the charge conjugation and satisfies $C^2 = 1$.

A particular representation of a FRA is through the fusion matrices $(N_i)_j^k = {N_{ij}}^k$, which by associativity and commutativity of the original fusion ring, are themselves commutative:

$$N_i N_j = N_j N_i \tag{8.7.5}$$

and satisfy the same fusion rules:

$$N_i N_j = \textstyle\sum_k {N_{ij}}^k N_k \tag{8.7.6}$$

From total symmetry of $N_{ijk} = {N_{ij}}^{k^\vee}$, and (d) we see that $N_i^t = N_{i^\vee}$. Hence the normal matrices N_j are simultaneously diagonable by a matrix S, with eigenvalues algebraic integers $\{\lambda_i^j \mid i\}$ respectively. Then $x_i \to \lambda_l^i$ are one dimensional representations of the fusion ring

$$\lambda_l^i \lambda_l^j = \textstyle\sum_k {N_{ij}}^k \lambda_l^k \,. \tag{8.7.7}$$

The N_i are linearly independent, as if $\sum_i a_i N_i = 0$, then $\sum_i a_i {N_{ij}}^k = 0$ for all j, k in particular by (d), $a_i \equiv 0$. Writing $N_i = \sum \lambda_k^i e_k$ where $e_k = e_{kk}$ is a diagonal matrix unit, we have $e_k = \sum \nu_r^k N_r$, where $M = [\lambda_j^i]$ and $N = [\nu_j^i]$ are mutually inverse matrices. Hence

$$N_i N_{j^\vee} = \textstyle\sum_k \lambda_k^i \lambda_k^{j^\vee} e_k = \sum_{k,r} \lambda_k^i \overline{\lambda_k^j} \nu_r^k N_r = \sum_r {N_{ij^\vee}}^r N_r \,,$$

where we have used $\lambda_k^{j^\vee} = \overline{\lambda_k^j}$ by $N_j^* = N_{j^\vee}$. Hence by linear independence of N_r we have

$$\textstyle\sum_k \lambda_k^i \overline{\lambda_k^j} \nu_k^0 = {N_{ij^\vee}}^0 = \delta_{ij} \,. \tag{8.7.8}$$

From this and $NM = 1$, it follows that

$$\nu_i^j = \nu_0^i \overline{\lambda_j^i} \,,$$

and hence $\nu^i = \nu_0^i \neq 0$. Then from (8.7.8) we deduce

$$\textstyle\sum_k \lambda_i^k \overline{\lambda_j^k} = (\nu^i)^{-1} \delta_{ij} \,.$$

Consequently $(\nu^i)^{-1} = \sum_k |\lambda_i^k|^2 > 0$, and so $\left[\sqrt{\nu^j}\lambda_j^i \mid i,j\right] = S$ is unitary and simultaneously diagonalizes N_i.

Inverting the matrix λ_l^k on the right hand side of (8.7.7) we get the *Verlinde formula*

$$N_{ij}{}^k = \textstyle\sum_l \nu^l \lambda_l^i \lambda_l^j \overline{\lambda_l^k} = \sum_\ell S_{il} S_{jl} \overline{S}_{kl} / S_{0l} \tag{8.7.9}$$

with $S_{ij} = \sqrt{\nu^j}\lambda_j^i$, $S_{0l} = \sqrt{\nu^l}$, $\lambda_l^i = S_{il}/S_{0l}$. In the case when our FRA comes from a CFT, the matrix S is indeed the same as the modular transformation $S = S(0) : \tau \to -1/\tau$ or rotation through $\pi/2$ on the two torus. Then we would have $S^2 = C$ which is equivalent to S being symmetric, since S is unitary. Note that

$$S^* = CS = SC$$

follows from our previous considerations as

$$S_{ij}^* = \overline{S_{ji}} = \sqrt{\nu^i}\,\overline{\lambda_i^j} = \sqrt{\nu^i}\lambda_i^{j^\vee} = S_{j^\vee i}\,.$$

Exercise 8.6 If $\sum_{i=1}^N f_i g_i^* = \sum_{i=1}^M h_i k_i^*$ in a $*$-algebra, where each set $\{f_i\}$ $\{g_i\}$, $\{h_i\}$, $\{k_i\}$ is separately linearly independent, then $N = M$ and $f_i = \sum_j A_{ij} h_i$, $g_i = \sum_j A_{ij}^{-1*} k_j$ for an invertible matrix A.

Exercise 8.7 Show that in the setting of the Ising model fusion rules with $\varepsilon, \sigma, 1$ we can find a gauge choice such that the polynomial equations can be solved to yield

$$F\begin{pmatrix}\sigma & \varepsilon\\ \sigma & \varepsilon\end{pmatrix} = F\begin{pmatrix}\varepsilon & \sigma\\ \varepsilon & \sigma\end{pmatrix} = F\begin{pmatrix}\varepsilon & \varepsilon\\ \sigma & \sigma\end{pmatrix} = F\begin{pmatrix}\sigma & \sigma\\ \varepsilon & \varepsilon\end{pmatrix} = 1$$

$$F\begin{pmatrix}\varepsilon & \varepsilon\\ \varepsilon & \varepsilon\end{pmatrix} = 1\,, \quad F\begin{pmatrix}\sigma & \sigma\\ \sigma & \sigma\end{pmatrix} = 2^{-1/2}\begin{pmatrix}1 & 1\\ 1 & -1\end{pmatrix}$$

$$F\begin{pmatrix}\sigma & \varepsilon\\ \varepsilon & \sigma\end{pmatrix} = -1\,, \qquad B\begin{pmatrix}\sigma & \sigma\\ \sigma & \sigma\end{pmatrix} = e^{-i\pi/8}2^{-1/2}\begin{pmatrix}1 & i\\ i & 1\end{pmatrix}$$

8.8 Current algebras and the Sugawara construction

In the critical Ising model we obtained Majorana fermions, operators satisfying the canonical anti-commutation relations and the real Majorana condition:

$$\{b_a, b_c\} = \delta_{a,-c}\,, \quad b_a^* = b_{-a}$$

$a, c \in \mathbb{Z} + 1/2$ (NS, Neveu–Schwarz), or $a, c \in \mathbb{Z}$ (R, Ramond). To describe these as fields on the circle, we write as in the algebraic picture of Section 8.4, with real, $z = e^{2\pi i x}$ in S^1:

$$B(x) = \textstyle\sum_s e^{-2\pi i s x} b_s = \sqrt{z}\, b(z)\,,$$

where $s \in \mathbb{Z} + 1/2$ NS, or $s \in \mathbb{Z}$ R.

We then wrote the Virasoro generators, for central charge $c = 1/2$, as quadratic expressions

$$L_m = dQ_{S_X}(\ell_m) + \varepsilon_X \delta_{m,0}/16 \tag{8.8.1}$$

where X is NS or R, $\varepsilon_{NS} = 0$, $\varepsilon_R = 1$. We can also produce, by this second quantization procedure, currents $J(f)$ which are also quadratic in the fields $B(f)$. These currents themselves can be used to represent Virasoro generators, quadratic now in the currents, and can be regarded as giving a chiral algebra, the current or Kač–Moody algebra, larger than the Virasoro chiral algebra. To begin this path, let us start to consider the operator product expansion of the fermion fields. We can make the following computation keeping only the singular contribution as $z \to w$:

$$B(z)B(w) = \begin{cases} \dfrac{-1}{z-w} & \text{NS} \\ \\ \dfrac{-1}{z-w}\dfrac{1}{2}\left(\sqrt{\dfrac{z}{w}} + \sqrt{\dfrac{w}{z}}\right) & \text{R} \end{cases} \quad . \tag{8.8.2}$$

Here we need to compute some infinite sums according as

$$\frac{1}{w}\sum_{s>0}\left(\frac{z}{w}\right)^{s-1/2} = \begin{cases} \dfrac{1}{z-w} & \text{NS} \\ \\ \dfrac{-1}{z-w}\sqrt{\dfrac{z}{w}} & \text{R} \end{cases} \quad . \tag{8.8.3}$$

The current J, quadratic in the fields B, is defined to be the operator product expansion of $B(z)B(z)$ with the singularity removed. This is the procedure of normal ordering, which we have encountered in Chapter 6, especially in the context of Section 6.9 in implementing unbounded quasi-free derivations by unbounded operators. Thus

$$J(z) = \frac{1}{2} : B(z)B(z) : . \tag{8.8.4}$$

If we take Fourier components of $J(z)$ and write

$$j(x) = zJ(z) = \textstyle\sum_{n\in\mathbb{Z}} J_n z^{-n} \tag{8.8.5}$$

where $z = e^{2\pi i x}$, then (8.8.4) is equivalent to

$$J_m = \textstyle\sum_n : b^*_{m+n} b_n : = dQ(t_m) \tag{8.8.6}$$

where

$$t_m = \sum_n e_{m+n,n} \,. \tag{8.8.7}$$

Then because of the Schwinger term (Sections 6.7 and 6.9) we have a representation of the Heisenberg algebra:

$$[J_m, J_n] = n\delta_{m,-n}\,, \quad J_m^* = J_{-m}\,. \tag{8.8.8}$$

To understand the operator product expansion JJ, we first need that of JB:

$$J(z)B(w) = (z-w)^{-1}B(w)\,. \tag{8.8.9}$$

It then follows that

$$J(z)J(w) = -(z-w)^{-1}J(w)\,. \tag{8.8.10}$$

The Virasoro algebra can now be recovered from the currents through the quadratic expressions:

$$T_{\text{Sug}}(z) = \,:J(z)J(z):\, \tag{8.8.11}$$

To distinguish this from the Virasoro expressions derived as quadratic expressions in fields $B(w)$, we introduce the Sugawara subscript. We shall see that expressions such as (8.8.11), which are quartic in the fields, can be identical with expressions like that of (8.8.1) which are quadratic in the fields. In Fourier modes $T_{\text{sug}}(z) = \sum_n L_n^{\text{sug}} z^{-n-2}$ this is

$$L_n^{\text{sug}} = \sum_n :J_{n+m}J_{-n}:$$

where $:J_mJ_n: = J_mJ_n$, $m<0$, or J_nJ_m, if $m \geq 0$.

These ideas can be generalized if we start with d free Majorana fermions B^i, $i = 1, \ldots, d$:

$$B^i(z) = \sum z^{-s-1/2} b_s^i\,, \quad \left\{b_s^i, b_t^j\right\} = \delta^{ij}\delta_{s,-t} \tag{8.8.12}$$

(and $s,t \in \mathbb{Z}+1/2$, NS, or $\mathbb{Z}$, R) to construct current algebra relations (8.8.23) generalizing (8.8.8) and representations of the Virasoro algebra (8.8.13), and (8.8.17) generalizing (8.8.1). Moreover, starting from the more general current algebra relations (8.8.23), we can generalize the Sugawara construction (from (8.8.13) and (8.8.17) to (8.8.24) and (8.8.30)). The *energy–momentum tensor*, cf. (8.8.1), is

$$T(z) = \frac{1}{2}\sum_i :B^i(z)\partial B^i(z): + \frac{\varepsilon}{16}dz^{-2} \tag{8.8.13}$$

where $\varepsilon = 0, 1$ NS, R respectively, with B^i a primary field of conformal dimension $h(B) = 1/2$. Indeed we first compute the operator product expansion, cf. (8.2.8):

$$T(z)B^j(w) = \frac{1}{2}(z-w)^{-2}B^j(w) + (z-w)^{-1}\partial B^j(w)\,. \tag{8.8.14}$$

Moreover

$$T(z)T(w) = (d/4)(z-w)^{-4} + 2(z-w)^{-2}T(w) + (z-w)^{-1}\partial T(w) \qquad (8.8.15)$$

so that $T(z) \equiv \sum z^{-n-2}L_n$ indeed gives a representation of the Virasoro algebra with central charge cf. (8.2.10):

$$c = \frac{d}{2} = \frac{1}{2}(\text{number of fermions})\,. \qquad (8.8.16)$$

In particular

$$L_0 = \textstyle\sum_{i=1}^{d}\sum_{n>0} n b^i_{-n} b^i_n + \varepsilon_X d/16 \qquad (8.8.17)$$

where $\varepsilon_X = 0, 1$ NS, R respectively.

The current j in (8.8.5) and (8.8.8) generates a Heisenberg Lie algebra. Recall that an algebra is a vector space with a bilinear operation. If we drop the associativity axiom, a *Lie algebra* is an algebra g, i.e. a vector space with a bilinear operation $[\,,\,] : g \times g \to g$, which satisfies the following properties:

(a) *anti-symmetry* $[x, x] = 0$, $x \in g$;
(b) *Jacobi identity* $[x,[y,z]] + [y,[z,x]] + [z,[x,y]] = 0$, $x, y, z \in g$.

An associative algebra A is a Lie algebra for commutators or the Lie bracket $[a, b] \equiv ab - ba$. Taking a basis $\{T^a \mid a = 1, \ldots, \dim g\}$ for g, writing $[T^a, T^b] = f^{ab}{}_c T^c$, then (a), (b) are equivalent (at least if the underlying field is $\mathbb{R}$ or $\mathbb{C}$ – of characteristic different from 2 to

(a′) $f^{ab}{}_c = -f^{ba}{}_c$;
(b′) $f^{ab}{}_c f^{cd}{}_e + f^{da}{}_c f^{cb}{}_e + f^{bd}{}_c f^{ca}{}_e = 0$.

The *loop algebra* $g_{\text{loop}} = g \otimes \mathbb{C}[z, z^{-1}]$ has a basis $t^a_n \equiv T^a \otimes z^n$, with bracket

$$[t^a_m, t^b_n] = f^{ab}{}_c t^c_{m+n}\,. \qquad (8.8.18)$$

The loop algebra contains g as a subalgebra $T^a \to T^a_0$ whilst z^n will be identified with the shift of (8.8.6). Then g_{loop} may be thought of as the Lie algebra of the loop group LG of smooth maps $S^1 \to G$ (under pointwise multiplication) if g is the Lie algebra of a compact connected Lie group G, with generators $\{T^a\}$. The group of smooth maps $S^1 \to S^1$ (under composition) acts on the maps from S^1 to G and so the Witt algebra acts on the loop algebra through:

$$[\ell_m, t^a_n] = -n t^a_{m+n}\,. \qquad (8.8.19)$$

Although a simple complex Lie algebra g cannot have non-trivial central extensions, g_{loop} has an unique non-trivial central extension $\hat{g}$ satisfying

$$[T^a_m, T^a_n] = f^{ab}{}_c T^c_{m+n} + m\delta_{m+n,0}\kappa^{ab} k\,. \qquad (8.8.20)$$

Here κ is the *Killing form*

$$\kappa^{ab} = \frac{1}{I_{\rm ad}} {\rm tr\ ad}\ T^a {\rm ad}\ T^b = \frac{1}{I_{\rm ad}} f^{ae}{}_c f^{bc}{}_e \tag{8.8.21}$$

$\mathrm{ad}(x) = [x, \cdot] : g \to g$ is the *adjoint representation*, and $I_{\rm ad}$ is a normalizing constant. The Killing form raises or lowers indices, $f^{abc} = f^{ab}{}_d \kappa^{dc}$, and a basis of g may be chosen so that f^{abc} is anti-symmetric. The infinite dimensional Lie algebra $h = \hat{g}$ is called an *affine Kač–Moody algebra.* It is also standard notation to write $\overline{h}$ for the underlying finite dimensional Lie algebra g contained in h, so that $\hat{\overline{h}} = h$.

Unitary representations of an affine algebra $\hat{g}$ are those for which $T^{a^*}_n = T^a_{-n}$ (so that in particular, with $n = 0$, it means that we have chosen self adjoint generators $(T^a)^* = T^a$ for g itself). In the current algebra formulation:

$$j^a(x) = zJ^a(z) = \textstyle\sum_{n\in\mathbb{Z}} T^a_n z^{-n} \tag{8.8.22}$$

where x is real and $z = e^{2\pi i x}$ on the circle as usual, the commutation relations (8.8.20) correspond to the current algebra or canonical commutation relations:

$$[j^a(x), j^b(y)] = \frac{1}{2\pi i} \kappa^{ab} K \delta'(x-y) + f^{ab}{}_c j^c(x) \delta(x-y) . \tag{8.8.23}$$

Just as for a $U(1)$-current algebra, we can form an energy–momentum tensor quadratic in the currents in the *Sugawara construction* (an analogue of a quadratic Casimir operator (8.8.27) which we will discuss in a moment):

$$T(z) = \xi \kappa_{ab} : J^a(z) J^b(z) : \tag{8.8.24}$$

where ξ is some suitable scaling constant to be determined. In fact we have

$$: J^a(z) J^b(z) := \frac{1}{2} f^{ab}{}_c J^c(z) + \frac{\kappa^{ab}}{\xi d} T(z) , \tag{8.8.25}$$

a decomposition into an anti-symmetric part proportional to the structure constants of the finite dimensional Lie algebra g (underlying the affine Lie algebra $\hat{g}$), and a symmetric part proportional to the Killing form of the same algebra g. This form of the operator product of the currents in terms of a Lie algebra has been argued on general grounds by (Zamolodchikov 1985) in a unitary conformal field theory. In particular anti-symmetry follows from the fermion relations and the Jacobi identity from associativity of the operator product. Contracting with the Killing form picks up the symmetric part as in (8.8.24). We have put the proportionality constant as $1/\xi d$ where d is the dimension of g. We compute

$$J^a(z) T(w) = (z-w)^{-2} J^a(w) , \tag{8.8.26}$$

which leads us to take $\xi = 1/(2k + Q)$. Here

$$Q \delta^{cd} = f^{abc} f^{abd} \tag{8.8.27}$$

s the *quadratic Casimir* (cf. (8.8.24)) of the adjoint representation. (The quadratic Casimir of a representation π is $C_\pi = \sum \kappa_{ab}\pi(T^a)\pi(T^b)$). Then we ompute further:

$$T(z)J^a(w) = (z-w)^{-2}J^a(w) + (z-w)^{-1}\partial J^a(w) \tag{8.8.28}$$

$$T(z)T(w) = \frac{kd}{(2k+Q)}(z-w)^{-4} + 2(z-w)^{-2}T(w) + (z-w)^{-1}\partial T(w)\,.$$

o that the central charge is forced to be (cf. (8.2.10)):

$$c = \frac{2kd}{2k+Q} = \frac{k^\vee d}{k^\vee + g^\vee}\,. \tag{8.8.29}$$

The quadratic Casimir of the adjoint representation depends on a normalization or f^{abc}, but, if ψ is the largest root then $g^\vee = Q/\psi^2$ is an invariant, *the dual Coxeter number*. Moreover $k^\vee = 2k/\psi^2$ is the *level*, and d is the dimension of g.

We explain briefly some of the notions from Lie algebra theory used here. A basis for (a simple finite dimensional Lie algebra) g will consist of a maximal commuting set of Hermitian generators H^i, $i = 1, 2, \ldots, r$, where r is the *rank* of g, so that $[H^i, H^j] = 0$, together with step operators E^α such that $[H^i, E^\alpha] = \alpha^i E^\alpha$. The vectors $\alpha = (\alpha^i)$ are called *roots*, and a basis α_j, $j = 1, \ldots, r$ of *simple roots* may be chosen so that every root is either a positive or negative linear combination of simple roots. (The simple *coroots* are $2\alpha_i/\alpha_i^2$, with dual basis the *fundamental weights* Λ_i). From the *Cartan matrix* $K_{ij} \equiv 2\alpha_i \cdot \alpha_j/\alpha_j^2$ form a *Dynkin diagram*, the graph with vertices the simple roots, and $K_{ij}K_{ji}$ edges from i to j if $\alpha_i^2 > \alpha_j^2$. Finite dimensional simple Lie algebras are classified by their Dynkin diagrams. These together with their dual Coxeter numbers are

$$(A_r, sl(r+1), r+1)\,,\quad (B_r, so(2r+1), 2r-1)\,,\quad (C_r, sp(r), r+1)\,,$$
$$(D_r, so(2r), 2r-2)\,,\ (E_6, 12)\,,\ (E_7, 18)\,,\ (E_8, 30)\,,\ (F_4, 9)\,,\ (G_2, 4)\,,$$

see Figs 10.8, 10.9, 10.10, 10.11, and 10.12). We denote the Lie algebra (e.g. $su(2)$) of the Lie group ($SU(2)$) in lower case. In general there are at most two root lengths, long and short. The *simply-laced* cases are those having only one length, the *A-D-E* series. Here $r \le c \le d$, where r is the rank of g, and $c = r$ if and only if the level $k^\vee = 1$, and g is simply-laced, i.e. $g = su(r+1)$, $so(2r)$, E_6, E_7, or E_8 the *A-D-E* series. In particular we are restricted to $c \ge 1$ for the time being with the Sugawara construction.

Taking Laurent components $T(z) = \sum_{n\in\mathbb{Z}} z^{-n-2}L_n$, the components L_n in the Sugawara construction (8.8.24) are

$$L_n = \frac{1}{\psi^2(k^\vee + g^\vee)}\sum \kappa_{ab} : T^a_n T^b_{m-n} : \tag{8.8.30}$$

$$: T^a_n T^b_m := \begin{cases} T^a_n T^b_m & \text{for} \quad n < 0 \\ T^b_m T^a_n & \text{for} \quad n \ge 0 \end{cases}\,. \tag{8.8.31}$$

As for the Virasoro algebra, we will be interested in representations of $\hat{g}\oplus\mathbb{C}L_0$ (cf projective representations of $L(G)\rtimes\mathrm{Rot}(S^1)$, where $\mathrm{Rot}(S^1)$ ($\subset \mathrm{Diff}^+(S^1)$) is th rotation group) which are *highest weight representations* in that L_0 is bounde below so that by the commutation relation $[L_0, T_n^a] = -nT_n^a$ (cf. (8.8.19)), th generator T_n^a, $n > 0$ annihilates any eigenvector of the lowest eigenvalue – th space of vacuum vectors. The vacuum vectors will be invariant under g ($\subset \hat{g}$ and form an irreducible representation. The highest weight representations o the Kač–Moody algebra are classified by irreducible representations of g and a level $k^\vee$. For unitary irreducible highest weight representations the possibilitie are severely restricted (just as the c-values were severely restricted for unitar highest weight irreducible representations). Indeed $k^\vee$ must be integral, and fo a given value of the level, there are only a finite number of admissible (vacuun vector) irreducible representations of g. For example, in the case of $g = su(N)$, al irreducible representations are parameterized by partitions or sequences $\{\lambda_1 \geq \lambda_2 \geq \cdots \geq \lambda_N = 0\}$ of integers (so for $su(2)$ one has for each integer $j \geq 0$, a representation of dimension $j+1$). Then the admissible ones at level $k^\vee = l$ ar those partitions where $\lambda_1 \leq l$.

For example, for $su(2)$, there are $l+1$ admissible representations labelled b $\{0, 1, 2, \ldots, l\}$ at level l, or in terms of half integers $\{0, 1/2, 1, \ldots, l/2\}$. If $I^\pm$, I are the normalized generators of $su(2)$,

$$[I^+, I^-] = 2I^3\,, \qquad [I^3, I^\pm] = \pm I^\pm\,,$$

then in a finite dimensional representation of $su(2)$, the spectrum of $2I^3$ is in tegral. In the normalization where $\psi^2 = 2$, so that $k^\vee = k$ is integral, w need to take generators J^1, J^2, J^3, with structure constants $f^{ijk} = \sqrt{2}\varepsilon^{ijk}$ where $J^1 = \sqrt{2}(I^+ + I^-)$, $J^2 = \sqrt{2}i(I^- - I^+)$, $J^3 = \sqrt{2}I^3$. But the $\hat{I}^\pm = (J^1_{\mp 1} \pm J^2_{\mp 1})/\sqrt{2}$, $\hat{I}^3 = J^3_0/\sqrt{2} - k/2$ form an $su(2)$ triplet, so that k i necessarily integral. In an irreducible representation of $su(2)^\wedge$, the vacuum vec tors form an irreducible representation of $su(2)$, labelled by a half integer j, th spin-j representation, where I^3 has spectrum

$$\{-j, -j+1, \ldots, j\}\,.$$

Taking the unit vector v for the eigenvalue j and using unitarity:

$$0 \leq \left\langle \hat{I}^- I^+ v, v \right\rangle = \left\langle \left[\hat{I}^-, I^+\right] v, v \right\rangle = \left\langle \left(-I^3 + k\right) v, v \right\rangle = -2j + k$$

we see that $j \leq k/2$.

When $g = g_1 \oplus g_2 \oplus \cdots$ is a sum of simple Lie algebras, we take $L^g =$ $L^{g_1} + L^{g_2} + \cdots$, so that the relevant central charge is $c_g = c_{g_1} + c_{g_2} + \cdots$. I particular for abelian g, $f^{ab}{}_i = 0$ so that $Q = 0$ and $c = d = r$.

The Sugawara construction allows us to form an energy–momentum tenso for an affine subalgebra h of g, T_h and then of the coset $T_{g/h} \equiv T_g - T_h$ wit central charge $c_{g/h} = c_g - c_h$ which allowed (Goddard *et al.* 1986) to realize th

(c, h) range with $c < 1$. More precisely, if we have a basis T^a, $a = 1, \ldots, \dim g$ for g with the first $\dim h$ elements a basis for h, then

$$[L^g_m, T^a_n] = -nT^a_{m+n}\,, \quad a = 1, \ldots, \dim g\,, \tag{8.8.32}$$
$$[L^h_m, T^b_n] = -nT^b_{m+n}\,, \quad b = 1, \ldots, \dim h\,, \tag{8.8.33}$$

(cf. (8.8.19)). Hence by subtraction:

$$\left[L^{g/h}_m, T^b_n\right] = 0\,, \quad b = 1, \ldots, \dim h \tag{8.8.34}$$

if $L^{g/h}_m \equiv L^g_m - L^h_m$ (alternatively subtract (8.8.28) for g, h) and so

$$\left[L^{g/h}_n, L^h_n\right] = 0\,, \text{ and } [L^g_m, L^g_n] = [L^h_m, L^h_n] + \left[L^{g/h}_m, L^{g/h}_n\right]\,. \tag{8.8.35}$$

Consequently $L_m \to L^{g/h}_m$ gives a unitary representation of the Virasoro algebra with

$$c_{g/h} = c_g - c_h = \frac{2k \dim g}{2k + Q_g} - \frac{2k \dim h}{2k + Q_h}\,, \tag{8.8.36}$$

where Q_g, Q_h denote quadratic Casimirs of the adjoint-representations of g and h respectively. The $c < 1$ range, namely $1 - 6/(l+2)(l+3)$, $l = 0, 1, 2, \ldots$, and corresponding h values (8.2.15) of Section 8.2, can be realized from the diagonal embedding

$$su(2)_{l+1} \subset su(2)_l \oplus su(2)_1 \tag{8.8.37}$$

where subscripts denote the corresponding levels used in the Sugawara construction as $c_{su(2)} = 3l/(l+2)$ for $su(2)_l$.

In (8.8.4) and (8.8.6) we obtained $U(1)$-current algebra relations, or a representation of $u(1)^\wedge$ via Majorana fermions $B(z)$, as quadratic expressions in the fermions. We can generalize this to an affine algebra $\hat{g}$. Beginning with a family of such fields acted on by an orthogonal representation π of a simple Lie algebra g, currents, quadratic in the fields, will yield a representation of $\hat{g}$. More precisely if we have fermions B^i where i belongs to the indexing set of the representation π, then

$$J^a(z) = \tfrac{1}{2}\textstyle\sum_{i,j} \pi(T^a)_{ij} \; :B^i(z)B^j(z): \tag{8.8.38}$$
$$\text{i.e. } J^a_m = dQ(\pi(T^a) \otimes t_m)$$

satisfy the commutation or current algebra relations:

$$[J^a_m, J^a_n] = f^{ab}{}_c J^c_{m+n} + m\delta_{m+n,0}\kappa^{ab}k^\vee\,. \tag{8.8.39}$$

Here the level is given by $k^\vee = I_\pi/\psi^2$ where I_π is the *Dynkin index* relating the Killing forms, $\operatorname{tr}\pi(T^a)\pi(T^b) = I_\pi \kappa^{ab}$, i.e. $I_\pi = C_\pi d_\pi / d$. The operator product expansion JJ is computed as in the single index case via:

$$J^a(z)B^i(w) = (z-w)^{-1}\sum_j \left[\pi(T^a)_{ij}\right] B^j(w)\,. \qquad (8.8.40)$$

to yield

$$J^a(z)J^b(w) = (z-w)^{-2}\kappa^{ab}k - (z-w)^{-1}f^{ab}{}_c J^c(w)\,. \qquad (8.8.41)$$

For non-orthogonal representations we use Dirac fermions and the dF quantization of Chapter 6 rather than Majorana ones and dQ. That is we have a family of fields ψ_i satisfying the CAR (6.4.1) and (6.4.2), with $(\psi_i)^* = \psi^{*i}$. Define the quadratic current fields:

$$J^a(z) = \frac{1}{2}\sum_{i,j}\left(:\psi^{*i}(z)\pi(T^a)_i{}^j\psi_j(z) + \psi_i(z)\pi^+(T^a)^i{}_j\psi^{*j}(z):\right)$$

which satisfy the current algebra relations at level $k^\vee = (I_\pi + I_{\pi^+})/\psi^2 = 2I_\pi/\psi^2$. We however now have the extra field in the Dirac case:

$$J^0(z) = \sum_j \; :\psi_j(z)\psi^{*j}(z): \qquad (8.8.42)$$

with operator products

$$J^0(z)J^0(w) = d_\pi(z-w)^{-2}\,, \quad J^0(z)J^a(w) = 0\,, \qquad (8.8.43)$$

giving a representation of $\hat{g} \oplus u(1)$.

Remark In Section 8.1 we discussed characters $\operatorname{Tr} q^{L_0}$ of the Virasoro algebra and their computation for the discrete series. In the corresponding characters of unitary irreducible highest weight representations of affine Lie algebras $\hat{g}$, we also take account of the generators of the Cartan subalgebra of g. In the case of $su(2)^\wedge$ at level l, this amounts to considering

$$\chi^l_j(\theta,\tau) = q^{-c/24}\operatorname{Tr}\left(q^{L_0}e^{i\theta J^3_0}\right) \qquad (8.8.44)$$

where $j \in \{0, 1/2, \ldots, l/2\}$, $q = e^{2\pi i\tau}$, $z = e^{2\pi i\theta}$. These characters can be computed in terms of theta functions:

$$\Theta_{k,m}(z,q) = \sum_{\gamma\in\mathbb{Z}+k/2m} q^{m\gamma^2}\left(z^{-m\gamma} - z^{m\gamma}\right). \qquad (8.8.45)$$

It is possible (Kač and Petersen 1984) to show that

$$\chi^l_j(z,q) = \Theta_{2j+1,l+2}(z,q)/\Theta_{1,2}(z,q)\,.$$

The modular group leaves the linear span of characters invariant. Indeed, using the modular properties of the theta functions, we can explicitly compute the action of T and S as

$$\tau \to -1/\tau: \; S^{(l)}_{ij} = \left(\frac{2}{l+2}\right)^{1/2} \sin\frac{\pi(2i+1)(2j+1)}{l+2} \qquad (8.8.46)$$

$$\tau \to -1/\tau : \; T_{ij}^{(l)} = \exp\left\{2\pi\sqrt{-1}\left[\frac{(2j+1)^2}{4(l+2)} - \frac{1}{8}\right]\right\}\delta_{i,j} . \tag{8.8.47}$$

We start with the GKO coset construction for the diagonal embedding g containing h: $su(2)_l \oplus su(2)_1 \supset su(2)_{l+1}$. The tensor product of highest weight modules V for $su(2)$ gives a representation for g which decomposes into irreducible modules for h as

$$V_{j_1,l} \otimes V_{j_2,1} = \sum_{j_3} \Omega_{j_1 j_2 j_3} \otimes V_{j_3,l+1} . \tag{8.8.48}$$

Here Ω is the space of highest weight vectors of spin j_3 for the diagonal and is the irreducible Virasoro module so that in terms of characters

$$\chi_{j_1}^{l}(z,q)\chi_{j_2}^{1}(z,q) = \sum_{j_3} b_{j_1 j_2 j_3}(q)\chi_{j_3}^{l+1}(z,q) , \tag{8.8.49}$$

where b is the Virasoro character (independent of z):

$$b_{j_1 j_2 j_3}(q) = \mathrm{Trace}_{\Omega_{j_1 j_2 j_3}} q^{L_0 - c/24} , \tag{8.8.50}$$

which is non-zero only for $j_1 + j_2 + j_3$ integral.

It is the expressions

$$P(j_3) = b_{j_1 j_2 j_3}(q)\chi_{j_3}^{l+1}/\chi_{j_1}^{l}\chi_{j_2}^{1}$$

which are normalized by (8.8.49) so that $\sum_{j_3} P(j_3) = 1$, which give the local state probability $P(\sigma_{00} = j_3)$ in the ABF A_{l-1} model with respect to a ground state given by (j_1, j_2). See Section 8.3 for the A_4 hard square case (Andrews *et al.* 1984), (Date *et al.* 1987).

From (8.8.49) we can see, using (8.8.46) for level l and $l+1$, that the Virasoro characters for $c = 1 - 6/(l+2)(l+3)$ ($m = l+2$) transform under $S : \tau \to -1/\tau$ as

$$S_{rs}^{pq} = \left[\frac{8}{(l+2)(l+3)}\right]^{1/2} (-1)^{2(r+s)(p+q)} \sin\left(\frac{\pi(2r+1)p}{l+2}\right) \sin\left(\frac{\pi(2s+1)q}{l+3}\right) .$$

The $c < 1$ series, and corresponding Virasoro characters associated with the coset theory $su(2)_l \oplus su(2)_1 \supset su(2)_{l+1}$ clearly generalizes to $su(N)_l \oplus su(N)_1 \supset su(N)_{l+1}$. The modular invariance properties of these families of characters in the Kač–Moody and coset theories will be discussed in the next section. The corresponding coset characters (as in (8.8.50)) arise in the computations of the one-point function of restricted SOS models associated to $su(N)_k$ just as we saw in Section 8.3 for the Ising and hard square models.

We have now described various constructions of a symmetry algebra, and now have to ask what is the relation between them. In some situations, two constructions of the symmetry algebra with different physical input can give rise to the same modules. We have constructions using free fermions. Moreover in the

WZW context, where there is one affine algebra $\hat{g}$ and a representation of the Virasoro algebra, the symmetry algebra is their semi-direct product. To yield the same modules, the level of the affine algebra module π and the central charge should coincide, i.e. we need a WZW theory at level $k^\vee = I_\pi/\psi^2$, and central charge $k^\vee d/(k^\vee + g^\vee)$ should be half the number of fermions $d_\pi/2$.

To provide the same Virasoro module, the central charges should coincide – but by consideration of the difference with zero central charge this should be a sufficient condition for the two constructions of the energy–momentum tensor (one quadratic in fermions as in (8.8.13) or (8.8.17), and the Sugawara expression (8.8.24) or (8.8.30) quadratic in currents, hence quartic in fermions) should coincide. We have already seen from Section 8.2 that unitary modules with zero central charge are trivial.

If we take the Sugawara construction (8.8.24), quadratic in currents

$$T_{\text{sug}} = \sum_{a,b} \frac{1}{\psi^2(k^\vee + g^\vee)} \kappa_{ab} : J^a(z) J^b(z) : \tag{8.8.51}$$

and substitute the currents quadratic in free-fermions:

$$J^a(z) = \frac{1}{2} \sum_{i,j} \pi(T^a)_{ij} : B^i(z) B^j(z) : \tag{8.8.52}$$

we get

$$T_{\text{sug}} = \eta T + T_q \tag{8.8.53}$$

where

$$\eta = \frac{2C_\pi}{I_\pi + C_\theta} = \frac{2d}{d_\pi(1 + I_\theta/I_\pi)} \tag{8.8.54}$$

θ is the adjoint representation, T is the free fermion energy–momentum tensor (8.8.13) arising from the normal ordering in (8.8.51), and T_q is a quartic remainder

$$T_q = \frac{\kappa_{ab}}{4(I_\pi + C_\theta)} \sum_{i,j,k,l} \pi(T^a)_{ij} \pi(T^b)_{kl} : B^i(z) B^j(z) B^k(z) B^l(z) : .$$

Now $\eta = 1$ is equivalent to the central charges of T_{sug} and T coinciding:

$$\frac{k^\vee d}{k^\vee + g^\vee} = \frac{1}{2} d_\pi \tag{8.8.55}$$

and as we have already remarked, this implies $T_q = 0$. From the anti-symmetry of $: B^i B^j B^k B^l :$, this is equivalent to

$$\kappa_{ab} \left\{ \pi(T^a)_{ij} \pi(T^b)_{kl} + \pi(T^a)_{jk} \pi(T^b)_{il} + \pi(T^a)_{ik} \pi(T^b)_{lj} \right\} = 0 \,. \tag{8.8.56}$$

This has an algebraic interpretation, which is why we persisted in writing the expansion (8.8.53). If π is the adjoint module, θ, then the relation (8.8.56) is the Jacobi identity for g, when $\theta(T^a)_{ij} = f^{ai}{}_j$, the structure constants. In general, (8.8.56) has an interpretation as a Jacobi identity. Indeed the equivalent conditions (8.8.55) and (8.8.56) are also equivalent to the cyclic identity for the Riemann tensor of a symmetric space. (Goddard *et al.* 1985) showed that they are equivalent to the existence of a group $H \supset G$, such that H/G is a symmetric space whose tangent space generators transform under G in the same way as the fermions. If we write h as the Lie algebra of the compact Lie group H, with $h = h_+ \oplus h_-$ graded so that $[h_\sigma, h_{\sigma'}] \subset h_{\sigma\sigma'}$ and $h_+ = g$. Letting t^i be generators of h_-, $[T^a, T^b] = f^{ab}{}_c T^c$, $[T^a, t^i] = -x^{ai}{}_j t^j$, $[t^i, t^j] = -y^{ij}{}_a T^a$. From $\mathrm{tr}\ \mathrm{ad}\, t^j(t^i)T^a + \mathrm{tr}\ \mathrm{ad}(t^j)(T^a) = 0$, we have $y^{ija} = x^{aij}$. For convenience, choose a basis so that $\kappa(h) = \mathrm{id}$. We want to put $x^a{}_{ij} = \pi(T^a)_{ij}$. Then the Jacobi identity for t^i, t^j, t^k reduces to (8.8.56) as $[[t^i, t^j], t^k] = x^a{}_{ij} x^a{}_{kl} t^l$, the cyclic identity for the (Riemannian) tensor $x_{ij}{}^a x^a{}_{kl}$.

As an example consider the non-self adjoint vector representation of $su(n)$, and take the corresponding Sugawara construction, so that $c = n-1$, as $d = n^2 - 1$, $g^\vee = 1$, $k^\vee = 1$. However, if we throw in the $u(1)$ current J^0 of (8.8.42) to $su(n) \oplus u(1) = u(n)$, we get a total central charge $(n-1) + 1 = n$. Thus by (8.8.55) the Sugawara construction for $u(n)$ coincides with the free-fermion construction (with n complex fermions).

Another manifestation of quantum equivalence, the coincidence of different constructions of modules, is the possibility that the Sugawara constructions coincide when we have an embedding $h \subset g$ of affine Lie algebras. To have the same central charge in (8.8.29), we need

$$\frac{k^\vee(g)d(g)}{k^\vee(g) + g^\vee(g)} = \frac{k^\vee(h)d(h)}{k^\vee(h) + g^\vee(h)} \tag{8.8.57}$$

or equivalently $L_{g/h} = L_g - L_h = 0$. Identifying $k^\vee(g)/k^\vee(h)$ as the *Dynkin index* $I = I_{h\subset g}$ (= the ratio $I_{\pi \circ i}/I_\pi$ for any representation π of g and the inclusion $i : h \subset g$), this equality of central charges is equivalent to the equality

$$k^\vee(f) = \frac{1}{d(h) - d(g)} \left[I^{-1} d(g) g^\vee(h) - d(h) g^\vee(g) \right] . \tag{8.8.58}$$

In particular what is required is that the right hand side is integral. Such an embedding $h \subset g$ is called a *conformal embedding.* This is equivalent to the situation when irreducible highest weight $g^\wedge$-modules decompose into a *finite* number of irreducible highest weight $h^\wedge$-modules.

As an example, when h is a maximal abelian Cartan subalgebra of a simple g, we have already noted (cf. (8.8.29)) that $c_g = c_h$ or equivalently $L^g = L^h$ if and only if g is simply laced (i.e. *A*-*D*-*E*) and we are using a level one representation of $g^\wedge$. Hence the level one irreducible highest weight $g^\wedge$-module decomposes into a single level one irreducible highest weight $h^\wedge$-module. In this case, the vertex operator construction recovers the Kač–Moody generators of $\hat{g}$ from those of $\hat{h}$.

8.9 Modular invariance

It is argued on physical grounds that the partition function $Z(\tau)$ in a conformal field theory should be invariant under re-parameterization of the torus by $SL(2,\mathbb{Z})$. In the string theory formulation, modular invariance is essentially built into the definition of the partition function $\int D\gamma\, D\phi\, \exp(-S(\phi))$. Here S is the Euclidean action on the fields ϕ, and $D\gamma$ is the path integral over all two-dimensional metrics. In Chapter 7, we wrote the partition function as an average over $e^{-\beta H}$, where H is the Hamiltonian, now $L_0+\overline{L}_0-c/12$ (cf. the shift by $c/24$ arising from mapping the Virasoro algebra on the plane to a cylinder (8.10.1)). We now also have a momentum P $(= L_0 - \overline{L}_0)$ describing evolution along the closed string, so taking both evolutions into account, we first compute

$$Z(\tau) = \operatorname{Tr} e^{-\beta H} e^{i\eta P} = \operatorname{Tr}\left(e^{i2\pi\tau(L_0-c/24)}e^{-i2\pi\overline{\tau}(\overline{L}_0-c/24)}\right). \tag{8.9.1}$$

Here $2\pi i\tau = -\beta + i\eta$ parameterizes the metric of the torus, and we then have to average over τ. If we choose one τ from each orbit under the action of $PSL(2,\mathbb{Z})$, and integrate, we have implicitly assumed $Z(\tau)$ is modular invariant

$$Z(\tau) = Z\left((a\tau+b)/(c\tau+d)\right)\,, \quad a,b,c,d\in\mathbb{Z}\,, \quad ad-bc=1\,.$$

(Nahm 1991) has argued the case for modular invariance in terms of the chiral algebra and its representations, rather than a functional integral setting.

Although we restrict attention to partition functions (i.e. correlation functions with no field inserted) on a torus only, such modular invariants may not be relevant to string theory if they could not be consistently formulated on arbitrary Riemann surfaces – the world sheet swept out by a closed string. One would look for consistency or factorization in deforming a Riemann surface to those of lower genus, and look for sufficient conditions to ensure modular invariance of all correlation functions through a few elementary considerations at low genus. Nahm, in his considerations, uses the behaviour of families of correlation functions to deduce modular invariance of the partition functions themselves.

From a Hilbert space decomposition according to the Virasoro chiral algebra representation (8.2.5) the partition function (8.9.1) decomposes as

$$Z = \textstyle\sum_{i,j} Z_{ij}\chi_i\overline{\chi}_j \tag{8.9.2}$$

where χ_j is the character $\operatorname{Tr} q^{(L_0-c/24)}$, $q = e^{2\pi i\tau}$. The problem we wish to address here, is the classification of expressions (8.9.2) with positive integral coefficients, with the normalization $Z_{00} = 1$, which are modular invariant. The modular group has two generators T and S. The action $T : \tau \to \tau+1$ on characters is simple:

$$\chi_s(\tau+1) = e^{2\pi i(h_s-c/24)}\chi_s(\tau)$$

represented by a diagonal matrix

$$T_{rs} = \exp 2\pi i(h_s - c/24)\delta_{rs}\,.$$

The action of S is $\tau \to -1/\tau$, and as usual $S^2 = C = [\delta_{r,s^\vee}]$ representing time inversion with $C^2 = 1$. Even though iterating $\tau \to -1/\tau$ gives the identity, we have reversed the orientation.

We will discuss modular invariant partition functions $\sum_{i,j} Z_{ij}\chi_i\overline{\chi}_j$ and chiral algebra characters χ_i, for both the Virasoro algebra and Kač–Moody algebras. Modular invariants may appear in various guises as with diagonal invariants, charge conjugation, outer automorphisms (e.g through simple currents), conformal embeddings and more exceptional ones with *ad hoc* constructions (e.g. automorphisms of fusion rules of an extended algebra of Kač–Moody algebra unrelated to simple currents). The classification of all affine $SU(2)$ and $SU(3)$ invariants is complete as we shall describe here. It is an intriguing open problem to classify rational conformal field theories with $c \geq 1$, rational with respect to an extended algebra. Staying with a smaller given chiral algebra such as the Virasoro or Kač–Moody algebra forces us to have an infinite number of primary fields.

It is clear that the diagonal $\sum |\chi_i|^2$ is modular invariant when the left and right chiral algebras coincide. If however $\mathcal{W}_L \neq \mathcal{W}_R$, then even the existence of a single invariant may not be clear (Gannon and Ho-Kim 1994). We will restrict attention to the case when $\mathcal{W}_L = \mathcal{W}_R$. Indeed with respect to a correct, possibly larger, chiral algebra $\mathcal{W}$, all modular invariant partition functions possess this interpretation as a diagonal invariant (possibly twisted with an automorphism as in (8.9.5) below). Note that from (8.7.7) for $l = 0$, and $\lambda^i_j = S_{ij}/S_{0j}$, we have $\lambda^i_0 \geq 1$ (with equality if and only if i is a *simple current*, in the sense that it has an unique fusion rule with other primary fields). For a maximal chiral algebra, uniqueness of the purely anti-holomorphic primary fields says that $Z_{i0} = Z_{0i} = \delta_{i0}$ for a modular invariant Z. With a unitary S-matrix, invariance $SZ = ZS$ says $\lambda^i_0 = \sum_j Z_{ij}\lambda^j_0$, so that as $\lambda^i_0 \geq 1$, Z must be a permutation matrix. The characters χ_i of the larger symmetry algebra may be expressed in terms of those ψ_j of the given algebra, $\chi_i = \sum_j \psi_j$, so that

$$Z = \sum_i \left|\sum_j \psi_j\right|^2. \tag{8.9.3}$$

These are called *integer spin invariants.*

Before we go any further we should consider some examples. Consider the $su(2)$ Kač–Moody modular invariants. At level $k = 16$ we have three modular invariants from Table 8.4 as in Table 8.5. Expanding the second expression of Table 8.4 (writing $Z = \sum_i |\chi_i|^2 + \text{remainder}$), the diagonal terms $\sum_i |\chi_i|^2$ only appear for $i = 1, 3, 5, 7, 9, 11, 13, 15, 17$, which are the Coxeter exponents of D_{10}. Similarly expanding the third expression, only the Coxeter exponents of E_7 appear in the diagonal terms, whilst the Coxeter exponents of A_{17} of course appear in the first invariant, the complete diagonal invariant. (The Coxeter exponents of a Dynkin diagram are the integers $\{m_j\}$ such that $\{2\cos(\pi m_j/h)\}$ are the eigen-

$k \geq 1$	$\sum_{\lambda=1}^{k+1} \lvert\chi_\lambda\rvert^2$	A_{k+1}
$k = 4\rho$, $\rho \geq 1$	$\sum_{\lambda\,\mathrm{odd}=1,\lambda\neq 2\rho+1}^{4\rho+1} \lvert\chi_\lambda\rvert^2 + 2\lvert\chi_{2\rho+1}\rvert^2$ $+\sum_{\lambda\,\mathrm{odd}=1}^{2\rho-1}\left(\chi_\lambda\chi^*_{4\rho+2-\lambda} + c.c.\right)$ $= \sum_{\lambda\,\mathrm{odd}=1}^{2\rho-1} \lvert\chi_\lambda + \chi_{4\rho+2-\lambda}\rvert^2 + 2\lvert\chi_{2\rho+1}\rvert^2$	$D_{2\rho+2}$
$k = 4\rho - 2$, $\rho \geq 2$	$\sum_{\lambda\,\mathrm{odd}=1}^{4\rho-1} \lvert\chi_\lambda\rvert^2 + \lvert\chi_{2\rho}\rvert^2$ $+\sum_{\lambda\,\mathrm{even}=2}^{2\rho-2}\left(\chi_\lambda\chi^*_{4\rho-\lambda} + c.c.\right)$	$D_{2\rho+1}$
$k+2=12$	$\lvert\chi_1+\chi_7\rvert^2 + \lvert\chi_4+\chi_8\rvert^2 + \lvert\chi_5+\chi_{11}\rvert^2$	E_6
$k+2=18$	$\lvert\chi_1+\chi_{17}\rvert^2 + \lvert\chi_5+\chi_{13}\rvert^2 + \lvert\chi_7+\chi_{11}\rvert^2$ $+\lvert\chi_9\rvert^2 + [(\chi_3+\chi_{15})\chi_9^* + c.c.]$	E_7
$k+2=30$	$\lvert\chi_1+\chi_{11}+\chi_{19}+\chi_{29}\rvert^2$ $+\lvert\chi_7+\chi_{13}+\chi_{17}+\chi_{23}\rvert^2$	E_8

Table 8.4 *$su(2)^\wedge$ modular invariants*

Modular invariant	Exponents	Graph
$\lvert\chi_1\rvert^2 + \lvert\chi_2\rvert^2 + \cdots + \lvert\chi_{17}\rvert^2$	$1, 2, 3, \ldots, 17$	A_{17}
$\lvert\chi_1+\chi_{17}\rvert^2 + \lvert\chi_3+\chi_{15}\rvert^2 + \lvert\chi_5+\chi_{13}\rvert^2$ $+\lvert\chi_7+\chi_{11}\rvert^2 + 2\lvert\chi_9\rvert^2$	$1, 3, 5, 7, 9, 11, 13, 15, 17$	D_{10}
$\lvert\chi_1+\chi_{17}\rvert^2 + \lvert\chi_5+\chi_{13}\rvert^2 + \lvert\chi_7+\chi_{11}\rvert^2$ $+\lvert\chi_9\rvert^2 + (\chi_3+\chi_{15})\chi_9^* + \chi_9(\chi_3+\chi_{15})^*$	$1, 5, 7, 9, 11, 13, 17$	E_7

Table 8.5 *$su(2)^\wedge_{k=16}$ modular invariants*

values for the incidence matrix where h is the Coxeter number of the graph). In this way, all affine $su(2)$ modular invariants are described by A-D-E graphs. The correspondence between a graph Γ and a modular invariant is that the Coxeter exponents of the graph Γ appear in the labelling of the diagonal part of the corresponding modular invariant.

In the subfactor theory only A-D-E Dynkin diagrams with D_{even} and E_{even} appear as the principal graphs of subfactors with index less than four. In the rational conformal field theory of $su(2)$ Kač–Moody WZW models described by A-D-E Dynkin diagrams, there is a degeneracy, so that only D_{even} and E_{even}

need be counted. For example in the case of $k = 16$, the modular invariant for E_7 reduces to that of D_{10} under the simple interchange of blocks χ_9 and $\chi_3 + \chi_{15}$. Note that there are two kinds of modular invariants here:

$$\sum |\chi_i|^2 \quad \text{type I} \tag{8.9.4}$$

$$\sum \chi_i \overline{\chi_{\sigma(i)}} \quad \text{type II} \tag{8.9.5}$$

where σ is a permutation of the extended fusion rules. The exceptional invariant E_6 is obtained from the conformal embedding $SU(2) \subset SO(5)$:

$$\left(A_1^{(1)}\right)_{k^\vee=10} \subset \left(B_2^{(1)}\right)_{k^\vee=1} .$$

In general a conformal embedding $g \subset h$ will rewrite an invariant of h as an invariant in terms of g (cf. (8.9.3)), which will be an integer spin invariant if one starts with the diagonal invariant of h. Note that with a conformal embedding, the irreducible h-modules decompose into a finite number of irreducible g-modules. Invariants can also be obtained from conformal embeddings $g_1 \oplus g_2 \subset h$ by contracting an invariant for h with one for g_2 to obtain one for g_1.

The E_8-modular invariant is similarly described via the conformal embedding $SU(2) \subset G_2$:

$$\left(A_1^{(1)}\right)_{k^\vee=28} \subset \left(G_2^{(1)}\right)_{k^\vee=1} .$$

We will see that the D modular invariants, and the D subfactors are $\mathbb{Z}/2$ orbifolds of the A modular invariants, and A subfactors respectively.

The generalization to $SU(N)$ is as follows. The admissible representation of the Kač–Moody algebra $su(N)^\wedge$ at level k, with fundamental weights Λ_i, are the vertices of

$$P_{++}^{(n)} = \{\lambda = \textstyle\sum_{i=1}^{N-1} \lambda_i \Lambda_i \mid \lambda_i \geq 1, \sum_{i=1}^{N-1} \lambda_i \leq n-1\}$$

where $n = k + N$ is the *altitude*. A complete list of all $su(3)^\wedge$ modular invariants is described in Table 8.6 (Gannon 1994) with associated graphs of Fig. 8.18, Fig. 8.32, Fig. 8.33, Fig. 8.34, Fig. 8.35, Fig. 8.36, Fig. 8.37 (Di Francesco and Zuber 1990b). (In Table 8.6, $P_{++}^n(0)$ refers to the vertices of $\mathcal{A}^{(n)}$ of colour 0, cf. Section 13.3). We should also throw in their conjugations Z^c, $(Z^c)_{\lambda\mu} = Z_{C\lambda,\mu}$ – although $\mathcal{D}^{(6)} = \mathcal{D}^{(6)c}$, $\mathcal{D}^{(9)} = \mathcal{D}^{(9)c}$, $\mathcal{E}^{(12)} = \mathcal{E}^{(12)c}$, $\mathcal{E}^{(24)} = \mathcal{E}^{(24)c}$. In analogy with the A-D-E classification, we label these $\mathcal{A}$ (the diagonal invariants), $\mathcal{D}$ (the orbifold invariants) and the exceptional $\mathcal{E}$ invariants. We will discuss orbifold modular invariants in some detail below and return to them again in Chapter 13 in the subfactor setting. These graphs listed are three-colourable and have the same exponents as the Kač–Moody modular invariant. Again the exponents of the graph are its eigenvalues, see (11.7.39) for a description of the $\mathcal{A}^{(n)}$ eigenvalues in terms of the vertices of $\mathcal{A}^{(n)}$. The exponents of the modular invariant are

Graphs	Altitude	Exponents	Type	Modular invariant
$\mathcal{A}^{(n)}$	n	$P_{++}^{(n)}$	I	$A^{(n)}$
$\mathcal{D}^{(n)}$	n $(= 0 \bmod 3)$	$P_{++}^{(n)}(0)$, $(n/3, n/3)$ triple	I	$D^{(n)}$
$\mathcal{D}^{(n)*}$	n $(= 0 \bmod 3)$	$(j,j), (n-2j,j), (j,n-2j)$ $1 \leq j \leq [(n-1)/2]$		$D^{(n)c}$
	n $(\neq 0 \bmod 3)$	$P_{++}^{(n)}(0)$		$D^{(n)}$
	n $(\neq 0 \bmod 3)$	$(j,j); 1 \leq j \leq [(n-1)/2]$		$D^{(n)c}$
$\mathcal{E}^{(8)}$	8	(1,1),(6,1),(1,6),(3,3),(3,2),(2,3) (4,1),(3,4),(1,3),(1,4),(4,3),(3,1)	I	$E^{(8)}$
$\mathcal{E}_1^{(12)}$	12	(1,1),(10,1),(1,10), (5,5),(5,2),(2,5) and twice (3,3),(3,6),(6,3)	I	$E^{(12)}$
$\mathcal{E}_2^{(12)}$				
$\mathcal{E}_3^{(12)}$				
$\mathcal{E}_4^{(12)}$	12	(1,1),(10,1),(1,10),(5,5),(5,2), (2,5),(3,3),(3,6),(6,3) and twice (4,4)	II	$E_{MS}^{(12)c}$
$\mathcal{E}_5^{(12)}$	12	(1,1),(10,1),(1,10),(5,5),(5,2), (2,5),(4,1),(7,4),(1,7),(1,4), (7,1),(4,7),(3,3),(3,6),(6,3) and twice (4,4)	II	$E_{MS}^{(12)}$
$\mathcal{E}^{(24)}$	24	(1,1),(22,1),(1,22),(5,5),(14,5), (5,14),(7,7),(10,7),(7,10), (11,11),(11,2),(2,11),(7,1), (16,7),(1,16),(1,7),(16,1), (7,16),(5,8),(11,5),(8,11), (8,5),(11,8),(5,11)	I	$E^{(24)}$

Table 8.6 *$su(3)^\wedge$ modular invariants*

the labels (from the vertices of $\mathcal{A}^{(n)}$) of the diagonal terms. In some cases more than one graph may initially be associated with the same modular invariant and there may be difficulty in associating a graph with some invariants – unlike the $su(2)$ case. In Section 11.7 we will discuss the question of assigning IRF models to these height models (i.e. Boltzmann weights) at least at criticality and associated subfactors.

The charge conjugation C is an example of an automorphism invariant which of course is only non-diagonal exactly for non-self-conjugate theories. Charge conjugation gives a new non-diagonal invariant only for E_6, D_{odd} and $SU(N)$, $N > 2$.

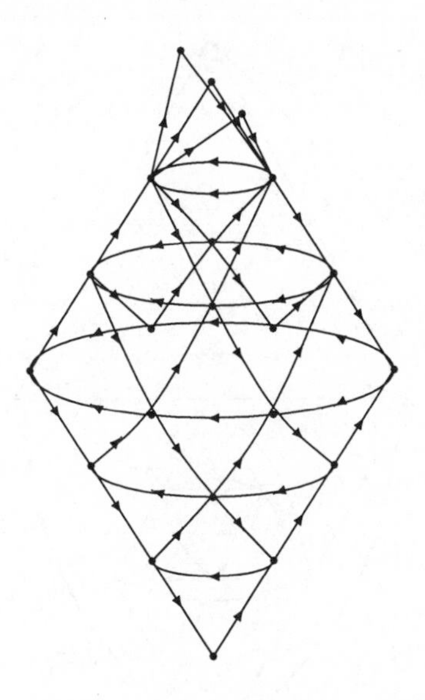

FIG. 8.32. Orbifold graph $\mathcal{D}^{(12)}$

All level 1 $\hat{g}$ modular invariants are known (Degiovanni 1990), (Gannon 1993), as are all level 2 and level 3 $su(N)^\wedge$ modular invariants (Gannon 1997). The level 2 (respectively level 3) exceptional invariants exist at $N = 10, 16, 28$, (respectively $N = 5, 9, 21$). The $\hat{g}$ modular invariants which reflect the symmetries of the Dynkin diagram of g have been classified in (Gannon, preprint 1995). They are very rare, and include only E_7 of the three exceptional invariants in the $su(2)$ case.

We can also consider new modular invariants built up from given ones with the aid of some additional elementary properties of the monodromy of some distinguished primary fields. Most notably one uses the *simple currents* which have unique fusion rules with other primary fields and give automorphisms of the fusion rules (Schellekens and Yankielowicz 1989). A primary field J is *simple* if given any primary field ϕ_i, there is a unique primary field ϕ_j such that the operator product expansion $J\phi_i$ contains ϕ_j alone. In particular the conjugate field $J^\vee$ acts as an inverse $JJ^\vee = J^\vee J = 1$, and with a rational theory the infinite sequence $J, J^2, J^3, \ldots$ is actually a finite set, and so using the inverse $J^\vee$ there must exist an N, and hence a smallest N, such that $J^N = 1$. It is convenient to write $J_i = J^i$, $i = 0, 1, 2, \ldots, N-1$ which are distinct simple currents, and then $J_i^\vee = J^{N-i}$. The complete set of simple currents will form a group $(\mathbb{Z}/N_1) \times (\mathbb{Z}/N_2) \times (\mathbb{Z}/N_3) \times \cdots \times (\mathbb{Z}/N_k)$ called the *centre* of the CFT, which for an $SU(N)$ Kač–Moody theory will indeed be identified with the centre of the $SU(N)$ Lie group. We will describe how to get new modular invariants from simple currents. In the case of $su(2)$ and $su(3)$ we get the modular invariants

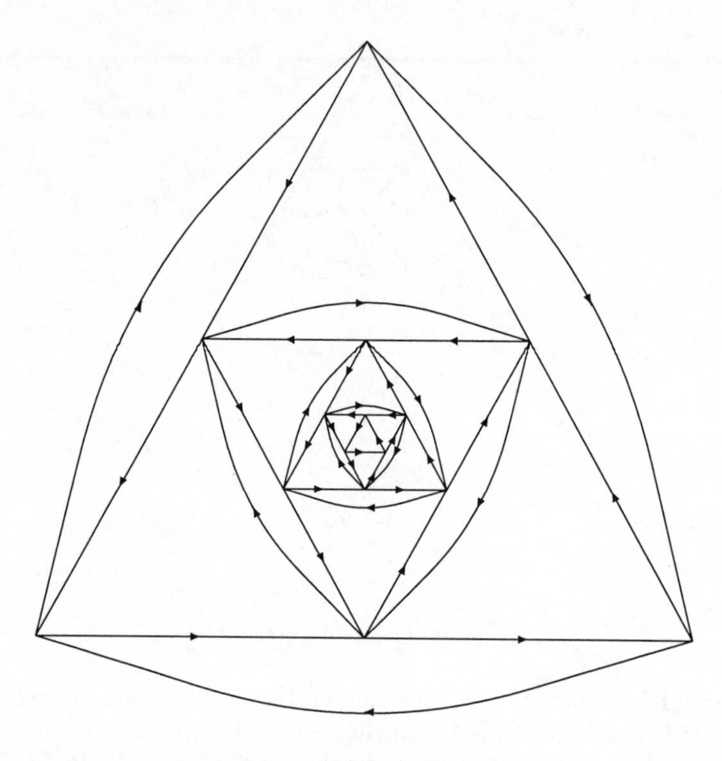

FIG. 8.33. Conjugate orbifold invariant graph $\mathcal{D}^{(12)*}$

for the D and $\mathcal{D}$ graphs as $\mathbb{Z}/2$ and $\mathbb{Z}/3$ orbifolds of the regular graphs.

This is part of the orbifold theory in the modular invariant setting where we divide out by a symmetry group acting on the underlying structures with fixed points which lead to new structures (cf. the order–disorder high temperature – low temperature duality in the Ising model, in Chapter 7, the symmetries on the Pauli algebra which lead to non-AF fixed point algebras, the non-commutative toroidal orbifolds, and the Cuntz algebra orbifolds mentioned in the notes for Chapter 3). The new modular invariants constructed in this way would correspond to WZW models on the orbifold formed by dividing the group manifold by the centre. We will return to orbifolds in the subfactor setting in Chapter 13.

If j represents a simple current J $(= \phi_j)$, then $N_{jlk} = \delta_{k^\vee,jl}$. Hence the Verlinde formula (8.7.9) implies

$$\textstyle\sum_b (S_{jb}S_{lb}/S_{0b})S^*_{bk} = \delta_{k,jl} \tag{8.9.6}$$

and so by unitarity of the S matrix: $S_{b,jl} = (S_{jb}/S_{0b})S_{lb}$. Thus recursively and using symmetry of the S matrix we have $S_{j^\alpha a,b} = (S_{jb}/S_{0b})^\alpha\, S_{ab}$ and

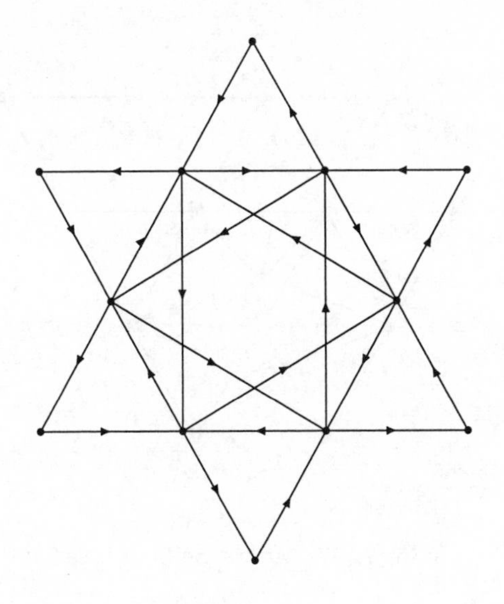

FIG. 8.34. $\mathcal{E}^{(8)}$; $su(3)_5 \subset su(6)_1$

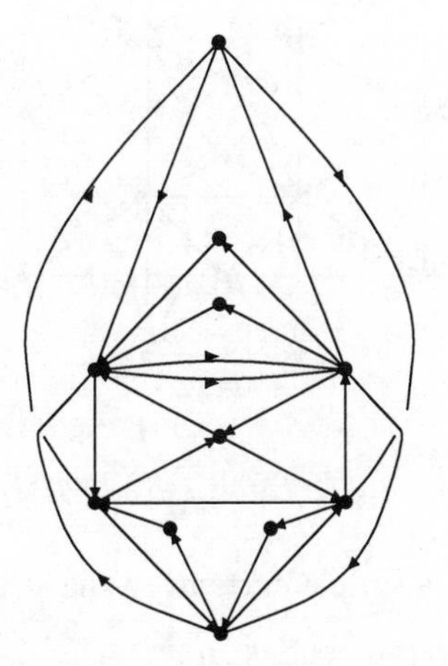

FIG. 8.35. $\mathcal{E}_4^{(12)}$; Conjugate Moore–Seiberg invariant

$$S_{j^\alpha a, j^\beta b} = (S_{jj}/S_{0j})^{\alpha\beta} (S_{00}/S_{0j})^{\alpha\beta} (S_{bj}/S_{b0})^{\alpha} (S_{aj}/S_{a0})^{\beta} S_{ab} \,. \qquad (8.9.7)$$

From the modular group relation (8.7.3) and the Verlinde formula (8.7.9) we have

$$S_{ab}/S_{00} = \textstyle\sum_m (S_{0m}/S_{00}) \, N_{ab}{}^m e^{2\pi i (h_a + h_b - h_m)} \,.$$

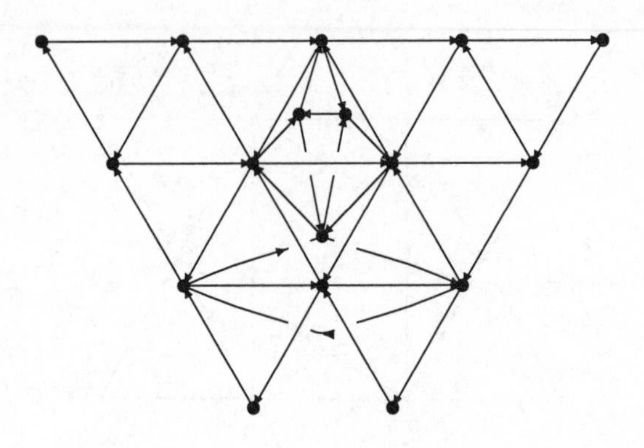

FIG. 8.36. $\mathcal{E}_5^{(12)}$; Moore–Seiberg invariant

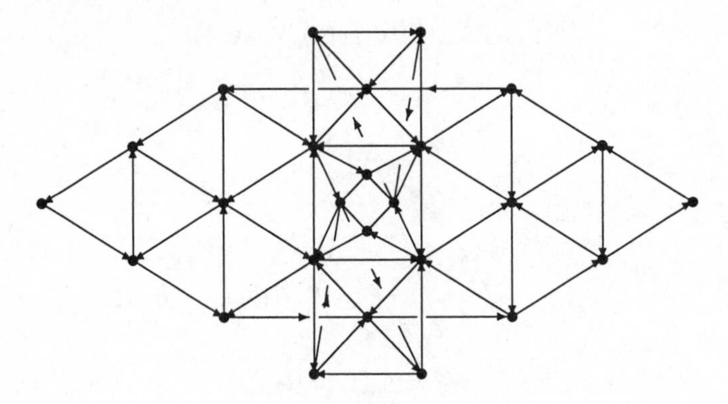

FIG. 8.37. $\mathcal{E}^{(24)}$; $su(3)_{21} \subset E_{7,1}$

In particular, when $b = j$, a simple current, we have

$$S_{aj}/S_{0j} = \exp 2\pi i(h_a + h_j - h_{ja}) \,. \tag{8.9.8}$$

Substituting (8.9.8) into (8.9.7) and with $Q(a)$ the charge $(h_a + h_j - h_{ja}) \bmod \mathbb{Z}$ of the field ϕ_a with respect to the simple current J $(= \phi_j)$, and $h_{j^n} = rn(N-n)/2N \bmod \mathbb{Z}$, where r an integer is the mondromy parameter, we get

$$S_{j^\alpha a, j^\beta b} = \exp\left[2\pi i \beta Q(a)\right] \exp\left[2\pi i \alpha Q(b)\right] \exp(2\pi i \alpha\beta r/N) S_{ab} \,.$$

Orthogonality $S_{j^\alpha a, j^\beta b} S^*_{j^\gamma c, j^\beta b} = \delta_{ac}\delta_{\alpha\gamma}$ when a, c are orbit representatives, and the sum b over orbit representatives, and β over the suitable orbit which is a divisor of N yields

$$\sum_{(\beta,b)} \exp\left[2\pi i\beta Q(a)\right] \exp\left[2\pi i\alpha Q(b)\right] \exp(2\pi i\alpha\beta r/N) \\ \times \exp\left[-2\pi i\gamma Q(b)\right] \exp\left[-2\pi i\beta Q(c)\right] \exp(-2\pi i\beta\gamma r/N) S_{ab} S^*_{bc} = \delta_{ac}\delta_{\alpha\gamma} .$$

This is enough to get our new non-diagonal modular invariants. We write

$$Z_{j^\alpha a, j^\beta b} = \sum_{l=1}^N \delta_{ab}\delta_{\beta,\alpha+l} \frac{1}{N} \sum_{p=1}^N \delta_{ab} \exp\left[2\pi i p\left(Q(a) + \frac{2\alpha+l}{2N} r\right)\right]$$

and compute how it transforms under S:

$$S_{AB} Z_{BC} S^*_{CD} = \sum_{\beta,b} \exp[2\pi i\beta Q(a)] \exp[2\pi i\alpha Q(b)] \exp[2\pi i\alpha\beta r/N] S_{ab} S^*_{bd} \\ \times \sum_{l=1}^N \sum_{p=1}^N \exp[-2\pi i(\beta+l)Q(d)] \exp[-2\pi i(\delta-p)Q(b)] \\ \exp(-2\pi i(\delta-p)(\beta+l)r/N) \exp(-2\pi i prl/2N) .$$

We can now use the orthogonality $S_{ab}S^*_{bd} = \delta_{ad}$ to deduce

$$S_{AB} Z_{BC} S^*_{CD} = \sum_{l=1}^N \sum_{p=1}^N \delta_{ad}\delta^N_{\alpha,\delta-p} \exp\left[-2\pi i l\left(Q(d) + \frac{(2\delta-p)r}{2N}\right)\right] = Z_{AD} .$$

where $\delta^N_{\alpha,\delta-p} = 1$ if $\alpha = \delta - p \bmod N$ and 0 otherwise.

In order to be able to interchange the roles of l and p as we have done, we have to sum over $1 \le l \le N$ even though the actual orbits may be less than N, i.e. we have to overcount with certain multiplicities. This formulation presented so far works for N either even or odd. When N is odd, there are possible reformulations. If N is odd, since r is defined mod N, we can take r to be even. In this case,

$$\delta^1\left(Q(a) + \frac{2h+l}{2N} r\right) = \frac{1}{N} \sum_{p=1}^N \delta_{ab} \exp\left[2\pi i p\left(Q(a) + \frac{2\alpha+l}{2N} r\right)\right] .$$

where $\delta^1(x) = 1$ if $x \in \mathbb{Z}$ and 0 otherwise. Using $Q(j^\alpha a) = Q(a) + r\alpha/N$, this gives us

$$Z_{AB} = \sum_{l=1}^N \delta(B, j^l A)\delta^1(Q(A) + Q(B)) .$$

An exotic example is provided by $(G_2)_q$. The graphs $\mathcal{A}^{(n)}$ in this case are obtained from the integrable representations of the Kač–Moody algebra $G^\wedge_{2,k}$ at level $k = n - 4$ (where 4 is the dual Coxeter number of G_2), with vertices

$$P^{(n)}_{++} = \{\lambda = \lambda_1\Lambda_1 + \lambda_2\Lambda_2 \mid \lambda_i \ge 1, \lambda_1 + 2\lambda_2 \le n-1\}.$$

The fundamental representation $f = (2,1)$ has the property that the fusion diagram at level k is a truncation of the fundamental fusion diagram for G_2

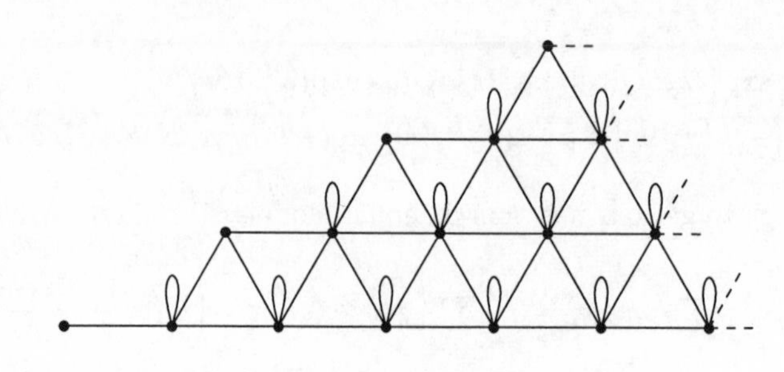

FIG. 8.38. $\mathcal{A}^{(n)}$ for G_2

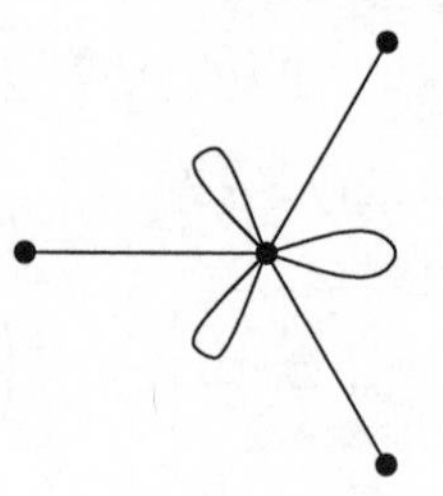

FIG. 8.39. $\mathcal{E}^{(7)}$, exceptional G_2 invariant

representations, (whilst the other $(1,2)$ does not). The matrix $A_{\lambda,\mu} = N^{\mu}_{f,\lambda}$ has eigenvalues

$$\beta^\lambda = \frac{S_{f\lambda}}{S_{*\lambda}} = 1 + 2\cos\left(\frac{2\pi}{3n}\lambda_1\right) + 2\cos\frac{2\pi}{3n}(\lambda_1 + 3\lambda_2) + 2\cos\frac{2\pi}{3n}(2\lambda_1 + 3\lambda_2)$$

for $\lambda \in P^{(n)}_{++}$. Apart from the regular representation of $G_{2,k}$ fusion rules, there are three exceptional ones with exponents:

$$\begin{array}{ll}
\mathrm{Exp}(E^{(7)}) = \{(1,1),(2,2),(3,1)_1,(3,1)_2\}, & \text{level } 3,\\
\mathrm{Exp}(E^{(8)}) = \{(1,1),(4,1),(1,2),(5,1),(2,2)_1,(2,2)_2\}, & \text{level } 4,\\
\mathrm{Exp}(E^{(8)c}) = \{(1,1),(3,1),(2,2),(4,1),(3,2)\}, & \text{level } 4,
\end{array}$$

(Figures 8.38, 8.39, 8.40 and 8.41), which yield the known $G_2{}^\wedge$ modular invariants:

$$Z^{\mathcal{A}(n)} = \sum_{\lambda \in P^{(n)}_{++}} |\chi_\lambda|^2,$$

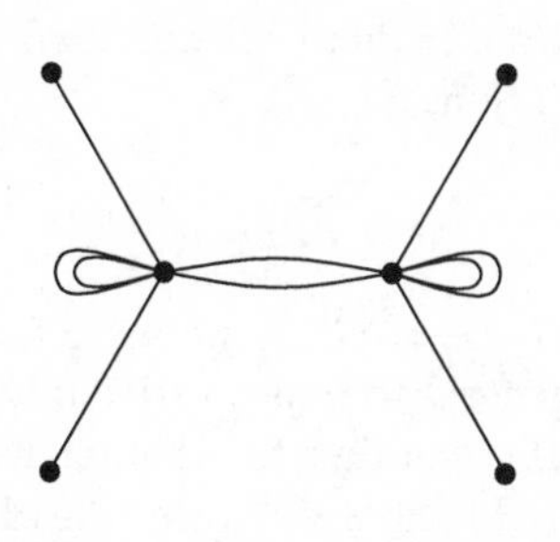

FIG. 8.40. $\mathcal{E}^{(8)}$, exceptional G_2 invariant

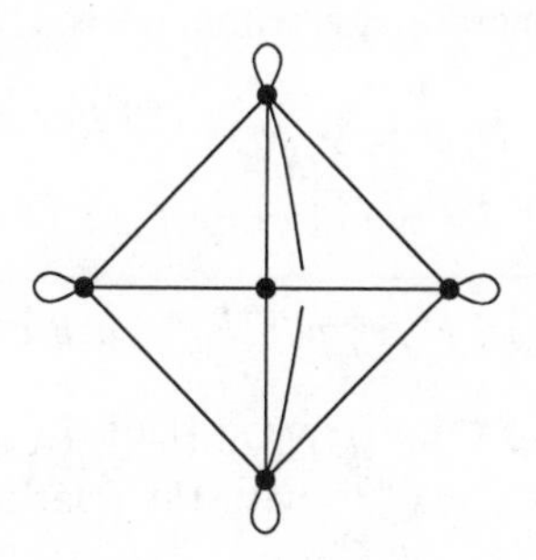

FIG. 8.41. $\mathcal{E}^{(8)*}$, exceptional G_2 invariant

$$
\begin{aligned}
Z^{E(7)} &= |\chi_{(1,1)} + \chi_{(2,2)}|^2 + 2|\chi_{(3,1)}|^2, \\
Z^{E(8)} &= |\chi_{(1,1)} + \chi_{(4,1)}|^2 + |\chi_{(1,2)} + \chi_{(5,1)}|^2 + 2|\chi_{(2,2)}|^2, \\
Z^{E(8)c} &= |\chi_{(1,1)}|^2 + |\chi_{(4,1)}|^2 + |\chi_{(2,2)}|^2 + |\chi_{(3,1)}|^2 + |\chi_{(3,2)}|^2 + |\chi_{(3,2)}|^2 + \\
&\quad \chi_{(1,2)}\chi_{(5,1)} + \text{c.c.}.
\end{aligned}
$$

We discuss further the connections between graphs, modular invariants and their conformal blocks. A modular invariant is labelled by exponents, a subset of the exponents of the diagonal invariant. We first seek a graph, whose exponents or eigenvalues match those of a given modular invariant. In the case of $su(3)^\wedge$ invariants, such graphs, are found (Di Francesco and Zuber 1990a) in Figs 8.32–8.37 associated with the modular invariants of Table 8.6. There may as in the case of the $\mathcal{E}$ invariant be more than one candidate for a given modular invariant. Unlike the $su(2)$ case, we will still have to find Boltzmann weights associated with these candidate target graphs, to turn them into statistical mechanical models, but we postpone that question. However, suppose our graph G has a distinguished vertex $*$, (which we could select from the vertices which minimize the Perron–Frobenius weight). If there are only two vertices f and $\overline{f}$ connected to $*$ we try to solve the relations

$$GN_a = \sum_{b \in G^{(0)}} G_{ab} N_b$$

subject to $N_* = 1$, $N_f = G$, $N_{\overline{f}} = G^t$. If we solve for the fusion matrices ${N_{ab}}^c$, it may not be possible in general to find positive integral values. We distinguish two cases type I or type II according to whether it is possible to get positive integral fusion rules or not. In the $su(2)$ case, the A_n, D_{even}, E_{even} are type I, the D_{odd}, E_7 are type II. In the $su(3)$ case we have labelled in Table 8.6, these graphs which have such extremal vertices $f, \overline{f}$, and hence are of a definite type. It would appear empirically, that type I Kač–Moody modular invariants are related to type I graphs.

When G has non-degenerate spectrum, we can without ambiguity use the Verlinde formula

$${N_{ab}}^c = \sum_\lambda \psi_a^\lambda \psi_b^\lambda \overline{\psi_c^\lambda} / \psi_*^\lambda \,. \tag{8.9.9}$$

Associated with the FRA (of a type I graph) is a *C-algebra.* We put

$${p_{ab}}^c = (\psi_a^* \psi_b^* / \psi_c^* \psi_*^*) \, {N_{ab}}^c \quad a, b, c \in G^{(0)} \,.$$

The representation $N_a N_b = \sum_c {N_{ab}}^c N_c$ of the FRA gives a representation $x_a = (\psi_a^* / \psi_*^*) \, N_a$ of a C-algebra $\mathcal{G} = G^{(0)}$, with the relations

(i) $x_a x_b = \sum {p_{ab}}^c x_c$, ${p_{ab}}^c = {p_{ba}}^c$;
(ii) x_* is a unit, i.e. ${p_{*b}}^c = \delta_{bc}$;
(iii) ${p_{ab}}^c \in \mathbb{R}$;
(iv) ${p_{ab}}^* = \delta_{ab} k_a$, for $k_a > 0$;
(v) $x_a \to k_a$ gives a representation of the algebra.

If we simultaneously diagonalize the commuting matrices, $p_a = [(\psi_a^* / \psi_*^*) \, {p_{ab}}^c]_{b,c}$, with eigenvectors $[E_\lambda \mid \lambda \in \mathcal{G}^*]$,

$$p_a = \theta^{-1} \sum_{\lambda \in \mathcal{G}^*} \pi_{a\lambda} E_\lambda \tag{8.9.10}$$

where $\theta = \sum_a (\psi_a^* / \psi_*^*)^2$, then $\{\theta E_\lambda \mid \lambda \in \mathcal{G}^*\}$ form another C-algebra $\mathcal{G}^*$, the *dual C-algebra*

$$\theta E_\lambda \circ \theta E_\mu = \sum {q_{\lambda\mu}}^\nu \theta E_\nu \tag{8.9.11}$$

for the Hadamard product of matrices $[A \circ B]_{i,j} = A_{ij} B_{ij}$, with dual coefficients

$${q_{\lambda\mu}}^\nu = (\psi_*^\lambda \psi_*^\mu / \psi_*^\nu \psi_*^*) \, {M_{\lambda\mu}}^\nu \tag{8.9.12}$$

in terms of the dual Verlinde formula:

$${M_{\lambda\mu}}^\nu = \sum_{a \in \mathcal{G}} \psi_a^\lambda \psi_a^\mu \overline{\psi_a^\nu} / \psi_a^* \,. \tag{8.9.13}$$

In all known type I cases, the M values are all non-negative real numbers. It has also been empirically observed that the conformal block structure is obtained from a subset T of $\mathcal{G}$, which contains $*$, invariant under σ, and defines a C-subalgebra in that if $i,j \in T$, $c \in \mathcal{G}$, $N_{ij}{}^c \neq 0$ then $c \in T$. If the structure constants p, and Krein parameters q are positive on $\mathcal{G}$, $\mathcal{G}^*$ respectively we can form a quotient $\mathcal{G}/T$ with generators $x_{T_\alpha} = \omega^{-1}\sum_{a\in T_\alpha} x_a$, where $T_* = T$, $T_\alpha, \alpha \in T^* \subset \mathcal{G}^*$ are the equivalence classes for $a \sim c$, $a, c \in \mathcal{G}$, if and only if there is $i \in T$ such that $p_{ai}{}^c \neq 0$, and $\omega = \sum_{i\in T}(S_i^*/S_*^*)^2$. The dual T^* of the C-algebra T is a subalgebra of $\mathcal{G}^*$ with generators $\{E_\alpha \mid \alpha \in T^*\}$. The quotient C-algebra $(\mathcal{G}/T)^* = \mathcal{G}^*/T^*$ is generated by $\theta E_{T_i^*} = k^{-1}\sum_{\lambda\in T_i^*} E_\lambda$, where $\{T_i^* \mid i \in T\}$ is a partition of $\mathcal{G}^*$, arising from the equivalence relation $\lambda \approx \nu$, $\lambda, \nu \in \mathcal{G}^*$, if and only if $q_{\lambda\alpha}{}^\nu \neq 0$ for some α in T^*, and $k = \sum_{\lambda\in T^*}(\psi_1^\lambda/\psi_*^*)^2$. A suitable choice of T which has the correct fusion rules of the Kač–Moody theory in each of the examples (Table 8.4 for $su(2)$, Table 8.6 for $su(3)$, and Figs 8.38, 8.39, 8.40, 8.41 for G_2) yields not only the Kač–Moody modular invariants (8.9.14) but the coset ones (8.9.15) as well:

$$Z^{KM(G)}_{\lambda\overline{\lambda}} = \sum_{i\in T}\delta_{\lambda\in T_i^*}\delta_{\overline{\lambda}\in T_i^*} \tag{8.9.14}$$

$$Z^{\mathrm{coset}\,(A,A,G)}_{(r,s),(\overline{r},\overline{s})} = \frac{1}{N}\delta_{r,\overline{r}}\sum_{i\in T}\delta_{s\in T_i^*}\delta_{\overline{s}\in T_i^*}\,. \tag{8.9.15}$$

The partitions $\{T_i^* \mid i \in T\}$ give the extended blocks, and can be computed from

$$\delta_{\lambda\approx\mu} = \sum_{i\in T}\sigma_i^\lambda\overline{\sigma_i^\mu}$$

where $\sigma_i^\lambda = \psi_i^\lambda/\left(\sum_{k\in T}|\psi_k^\lambda|^2\right)^{1/2}$. For example for the D_{10} invariant, $T = \{1,3,5,7,9,\overline{9}\}$ provides the partition $T_1^* = \{1,17\}$, $T_3^* = \{3,15\}$, $T_5^* = \{5,13\}$, $T_7^* = \{7,11\}$, $T_9^* = \{9\}$, $T_{\overline{9}}^* = \{9\}$ of $\mathrm{Exp}(D_{10})$, the conformal blocks of D_{10}.

8.10 Notes

We have followed the expositions of (Cardy 1990), (Fuchs 1992), (Ginsparg 1990), (Goddard and Olive 1988), (Ruijsenaars 1987), (Schellekens and Yankielowicz 1989) in our presentation.

The form of the singular Virasoro vectors, which we mentioned in Section 8.2, was given in general by (Kent 1991), who also showed how to derive the embeddings of Verma modules, which then immediately gives the characters of the unitary highest weight irreducible representations of the Virasoro algebra. (Rocha-Caridi 1984) had obtained this form for the characters after the derivation of the embeddings in a less concrete fashion.

The corner transfer matrix method of Section 8.3 was initiated by Baxter — see (Baxter 1982). The Ising model analysis of Section 8.3, and the derivation of the $c = 1/2$ Virasoro characters are from (Davies and Pearce 1990), (Davies 1988). The CTM analysis of the hard square model is from (Baxter 1981), (Baxter 1982), with the comments on the path algebra expressions (8.3.44), (8.3.45) from (Connes and Evans 1989).

The Yang–Baxter equation appears in three guises, namely the spin (Fig. 8.8), vertex or S-matrix (Fig. 13.6) and IRF (Fig. 11.76) formulations, which are essentially equivalent as we have hinted at in this book – see (Au-Yang and Perk 1989) for a detailed discussion on the equivalences between such formulations. In the spin language of statistical mechanics the star–triangle relation (Fig. 8.8) has its origins in the work of (Onsager 1941) in the integrability of the Ising model through the commutativity of the transfer matrices and through relating the Ising models on hexagonal honeycomb lattices and triangular lattices, although it already appears in the theory of electrical networks (Kennelly 1899). (McGuire 1964) used the vertex formulation of Fig. 13.6 to describe consistency of factoring an n-body S-matrix into two-body contributions, where the lines correspond to world lines of particles, which (Yang 1967), (Yang 1968) used in nested Bethe ansatz calculations. Baxter also found the vertex formulation as a consistency condition for constructing partition functions (cf. Section 13.2) invariant under type III moves. The references to (Au-Yang and Perk 1989) contain an indication of the diversity of solutions and use put to the Yang–Baxter equation in various physical contexts. For our purposes it is enough to draw attention to the generalizations of (Andrews *et al.* 1984), (Date *et al.* 1987) and the chiral Potts model of the Ising model. The relevance of the former to $SU(N)_k$ subfactors is discussed in Chapter 13. The chiral N-state Potts model has spins in $\mathbb{Z}/N$ at each site of (oriented) $\mathbb{Z}^2$, but the Boltzman weights W corresponding to an edge a, b depends only on $a-b$, but $W(a-b) \neq W(b-a)$ (if $N \neq 2$). A solution to the Yang–Baxter equation was found (Baxter *et al.* 1988),(Au-Yang and Perk 1989) parameterized by curves of genus $N^2(N-2)$ or $(N-1)(N-2)/2$ at criticality cf. 1, 0 respectively in the Ising model. (Au-Yang and Perk 1989) and (Au-Yang and Perk 1995) discuss the historical origin of the equation through regarding the Yang–Baxter equation as a quantum version of the Lax pair integrability condition – although it can now be understood in a quantum group context (Bazhanov and Stroganov 1990).

Section 8.4 is from (Sato *et al.* 1977–1980). (Barouch *et al.* 1973), (Wu *et al.* 1976) found that the scaling limit of the two-point correlation functions at criticality could be expressed in terms of a classical function, the Painlevé transcendent of the third kind. These functions arose classically from monodromy-preserving solutions of second order linear differential equations. (Sato *et al* 1977–1980) pursued this, computing the scaled n-point functions in the Ising model, and relating a wide range of quantum field theory problems and solutions to monodromy-preserving deformation theory for linear (ordinary and partial) differential equations. At criticality the Ising model gives rise to a two component Majorana fermion field ψ on two dimensional space–time satisfying the Dirac equation. The two-dimensional Dirac equation (with m as in (8.4.9))

$$i\frac{\partial v^+}{\partial x^0} = \frac{\partial v^-}{\partial x^1} + mv^- \, ; \quad i\frac{\partial v^-}{\partial x^0} = -\frac{\partial v^+}{\partial x^1} + mv^+$$

is obtained in the scaling limit from a two dimensional difference equation –

whilst the fields $: e^{\rho_F(a)} :$ can be obtained starting from the Dirac equation. The Euclidean continuation of the correlation functions

$$\langle\psi(x)\mu(a_1)\mu(a_2)\cdots\mu(a_\nu)\cdots\mu(a_n)\rangle/\langle\mu(a_1)\cdots\mu(a_n)\rangle$$

give rise to a basis of a n-dimensional space of multi-valued solutions to the Euclidean–Dirac equation, with branch points at $a_1,\ldots,a_n \in \mathbb{R}^2$, whose monodromy is determined by the commutation relation (8.4.5). (McCoy and Perk 1987) found that a Painlevé V formulation is more appropriate for taking the conformal limit. More precisely, the scaled two-point correlation function is the tau function of a Painlevé V equation above and below the critical temperature. (In the Painlevé III formulation, some extra factors have to be incorporated into the correlation functions to obtain tau functions). Then at criticality, this nonlinear differential equation reduces to the second order linear differential equation of conformal field theory for level two null fields (Section 8.5).

(Cardy 1986) argued from conformal invariance for a critical statistical system, that the central charge c may be computed from the asymptotics of the partition function Z on a periodic $N\times M$ square lattice $1 << N << M$. Such general arguments allows the central charge and indeed conformal dimensions to be determined in the Q-state Potts models ($Q\le 4$), ABF models and some six-vertex models (de Vega and Karowski 1987), (Karowski 1988). Indeed, if T and $\hat{T}$ are the $M\times 1$ and $1\times N$ transfer matrices in orthogonal directions with eigenvalues λ_i and $\hat{\lambda}_i$ respectively and maximal eigenvalues $\lambda_{\max}$ and $\hat{\lambda}_{\max}$ respectively, then (Cardy 1986) argues from conformal invariance that

$$\lambda_{\max}\approx\exp\left(-Mf+\frac{\pi}{6}c\right)$$

and hence for $1 << N << M$

$$Z\approx\exp(-NMf)\textstyle\sum_i\left(\hat{\lambda}_i/\hat{\lambda}_{\max}\right)^N$$

$$Z\approx\exp\left(-NMf+\frac{M}{N}\frac{\pi}{6}c\right).$$

The six-vertex model has bond configurations indicated by an arrow on each edge of a square lattice, with the constraint that at each vertex there are precisely two incoming and two outgoing arrows with vertex weights as given in Fig. 8.42. (This is a limiting case of the eight-vertex model of Section 8.3, when K, L, M become infinite with finite differences.) The partition function for the Q-state Potts model on a cylinder can be written in terms of that of a six-vertex model on a cylinder with a seam where there are extra Boltzmann weights, for $Q = 4\cos^2\gamma$, γ is as in the six-vertex weights above, and $\gamma=\pi/(m+1)$, $m = 3,5,\infty$ will correspond to $Q = 2,3,4$. To have isotropic weights we need $(m+1)\theta=\pi/2$. Using Bethe ansatz methods and Cardy's formulae, (de Vega and Karowski 1987), (Karowski 1988) compute the central charge in the six-vertex

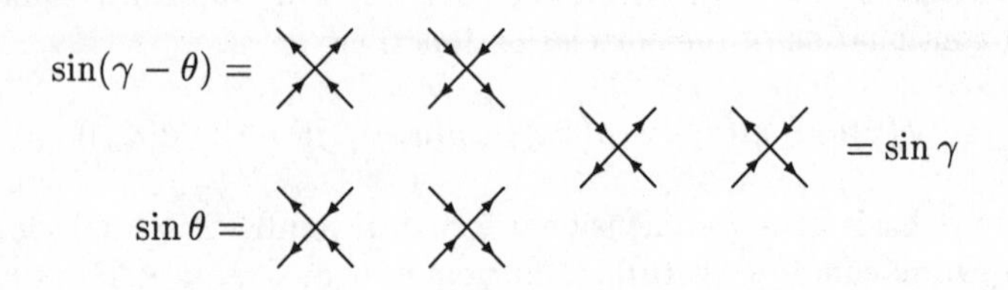

FIG. 8.42. Six-vertex model Boltzman weights

($\gamma = \pi/(m+1)$, $m = 1, 2, 3, \ldots$) and Q-state Potts models, $Q = 2, 3, 4$, to be $c = 1$ and $c = 1 - 6/m(m+1)$, $m = 3, 5, \infty$, and that conformal dimensions can be identified with $h_{p,q}(m)$, $m = 3, 5$.

We can get a better understanding (Cardy 1987) of the correlation length by taking a conformal transformation of the plane onto a cylinder $\mathbb{R} \times [0, L]$:

$$w \equiv u + iv = (L/2\pi) \log z \,.$$

Then according to the transformation law (8.2.9) and computing the relevant Schwarzian we have

$$T(w)_{\text{cylinder}} = (2\pi/L)^2 \left(z^2 T(z)_{\text{plane}} - \frac{c}{24} \right) \tag{8.10.1}$$

and applying this same conformal transformation to two-point correlation functions:

$$\begin{aligned} &\langle \phi(u_1, u_1) \phi(u_2, u_2) \rangle \\ &= (2\pi/L)^{2x} \left[2 \cosh \left(2\pi(u_1 - u_2)/L \right) - 2 \cos \left[2\pi(v_1 - v_2)/L \right] \right]^{-x} \end{aligned}$$

in agreement with (8.4.13) on the plane for $|\omega_1 - \omega_2| << L$, whereas for $u_1 - u_2 >> L$ we have exponential decay:

$$\langle \phi(u_1, u_2) \phi(u_2, u_2) \rangle \sim (2\pi/L)^{2x} \, e^{-(2\pi x/L)(u_1 - u_2)} \,.$$

Thus we can identify $2\pi L/x$ with the correlation length ξ. The shift $-c/24$ in (8.10.1) plays a role when we consider partition functions on a torus in Section 8.9 and in particular how they behave under modular transformations.

The differential equation approach of Section 8.5 to the Ising fusion rules is from (Belavin *et al.* 1980) whilst the C^*-setting of Section 8.6 is from (Mack and Schomerus 1990a) with localized endomorphisms from (Böckenhauer 1996a).

We have identified the critical Ising model with $c = 1/2$ conformal field theory, and $1 = \phi_{1,1}$, $\varepsilon = \phi_{2,1}$, $\sigma = \phi_{2,2}$ as identity, energy and magnetization primary fields respectively. Other critical statistical mechanical models have been identified with the FQS discrete series and some of the primary fields identified. Physically, the possible values of h determine the scaling dimension, and hence the critical exponents. Thus comparing with known exponents, one can match Table 8.3 with the Ising model ($m = 3$), tricritical Ising model ($m = 4$), three-state Potts model and $m \geq 5$ with the ABF models (Friedan *et al.* 1984), (Huse 1984). In the ABF A_n series, $\phi_{1,3}$ corresponds to an energy like variable, and $\phi_{r,r}$ are order variables. The scaling limit of the Q state Potts model is only conformal invariant for $Q \leq 4$. Different critical models appear to give rise to the same central charge and scaling dimensions (cf. when subfactors give rise to the same principal graph), but possibly different operator product expansions (cf. different connections or paragroup structures in the subfactor setting).

We have matched Table 8.3 with the Potts and ABF models. On the other hand we have already noted how the Potts model and ABF models are related to the Jones classification, via the transfer matrix method. It is therefore natural to ask for a direct link between the discrete sequence of index of subfactors and the discrete sequence of unitary irreducible highest weight representations of the Virasoro algebra. In 1983, the connection was noticed, using this transfer matrix formalism, between subfactors on the one hand and statistical mechanics on the other. Namely, that the same Temperley–Lieb algebra that appeared in the Potts model was now appearing in Jones' work on subfactors of the hyperfinite II_1-factor in the theory of von Neumann algebras (Jones 1983) and that the representation of the Temperley–Lieb algebra in the Pauli algebra found by (Pimsner and Popa 1986a) was the same as that obtained by (Temperley and Lieb 1971) when they showed that the Potts model at criticality was equivalent to an ice-type or six-vertex model. Using their representation, Pimsner and Popa could identify the shift $\kappa : e_n \to e_{n+1}$ as a non-commutative Bernoulli shift. In the statistical mechanical picture the shift κ is Kramers–Wannier high temperature–low temperature duality – see the remarks of Onsager on page 123 of (Onsager 1941). These connections between statistical mechanics and von Neumann algebras were made more intriguing by the discovery in 1984 by Jones (Jones 1985) of the one-variable knot polynomial, using the Markov trace on $C^*(e_n)$ and the representation of the braid group. Thus one immediately had the connection between statistical mechanics and links, in particular partition functions of statistical mechanical models and invariants of links. Later the puzzle arose to understand the connection between the discrete sequence of representations of the Virasoro algebra with central charge c less than one, and the discrete sequence of values of the index of subfactors less than four.

The situation has been clarified by (Wassermann 1995), (Wassermann (in preparation a)) – see (Jones 1995). As we have seen in Section 8.2, the positive energy irreducible representations π of a loop group LG are given by a level l and certain allowable irreducible representations of G. Restricting to loops $L_I G$ concentrated on an interval $I \subset S^1$ $\{f \in LG \mid f(x) = 1, x \notin I\}$, we get a subfactor $\pi(L_I G)'' \subset \pi(L_{I^c} G)'$, if I^c is the complementary interval, of type III_1 and of finite index – e.g. of index $4\cos^2 \pi/(l+2)$ in the case of the fundamental representation of $SU(2)$ and level l. The use of positive energy representations, e.g. those of $U(1)$-current algebra (Buchholz *et al.* 1988), and $SU(N)$-current algebras (Wassermann (in preparation a)) provide examples of the algebraic quantum field theory framework of DHR (Doplicher *et al.* 1971,1974).

The $U(1)$-current algebra is described by operators J_n, $n \in \mathbb{Z}$ satisfying the commutation relations

$$[J_m, J_n] = m\delta_{n+m}\,, \qquad J_n^* = J_{-n}$$

and

$$[J_m, H] = mJ_m$$

for the Sugawara Hamiltonian

$$H = \frac{1}{2}J_0^2 + \textstyle\sum_{n=1}^{\infty} J_{-n}J_n \tag{8.10.2}$$

which generates time development $\alpha_t = \mathrm{Ad}(e^{iHt})$. The Bose-field operators

$$J(f) = \textstyle\sum_{n\in\mathbb{Z}} f_{-n}J_n \tag{8.10.3}$$

for $f(z) = \sum f_n z^{-n}$ satisfy the canonical commutation relations

$$[J(f), J(g)] = \int \frac{dz}{2\pi i} f'(z)g(z) \equiv a(f,g)\,, \tag{8.10.4}$$

or in exponential form:

$$W(f) \equiv \exp iJ(f) \tag{8.10.5}$$

$$W(f)W(g) = \exp\left(-\tfrac{1}{2}a(f,g)\right) W(f+g)\,. \tag{8.10.6}$$

The generating functionals (cf. Exercises in Section 4.4),

$$\exp\left(igf_0 - \frac{1}{2}\sum_{n=1}^{\infty} n|f_n|^2\right) \tag{8.10.7}$$

for $g \in \mathbb{R}$ generate the positive energy representations (π_g, H_g, Ω_g) of the CCR algebra $\mathcal{A} = C^*(W(f) \mid f)$ where $\pi_g(J_n)\Omega_g = 0$ for $n > 0$, and $\pi_g(J_0)\Omega_g = g\Omega_g$. (i.e. lowest weight representations).

The representations π_g can be obtained from the vacuum representation π_0 via automorphism γ_g so that $\pi_g = \pi_0 \circ \gamma_g$, where $\gamma_g(W(f)) = e^{igf_0}W(f)$, $\gamma_g(J_n) = J_n + \delta_{n0}g$. More generally, if q is a charge distribution function on the circle S^1 with $zq(z)$ real then

$$\gamma(J(z)) = J(z) + q(z) \tag{8.10.8}$$

preserves the commutation relations (8.10.4), if

$$J(z) = \textstyle\sum_{n\in\mathbb{Z}} J_n z^{-n-1} \tag{8.10.9}$$

and on Weyl operators acts as

$$\gamma(W(f)) = e^{iq(f)}W(f) \tag{8.10.10}$$

where $q(f) = \int q(z)f(z)dz/2\pi i$, and on the Sugawara Hamiltonian (8.10.2) as

$$\gamma(H) = H + \sum J_{-n}q_n + \sum q_{-n}q_n/2 \tag{8.10.11}$$

if $q(z) = \sum q_n z^{-n-1}$. In particular, if $q(z) = gz^{-1}$, then we recover γ_g as before.

The equivalence class $[\pi_\gamma]$ of the morphism $\pi_\gamma = \pi_0 \circ \gamma$, the *superselection sector*, depends only on the charge the equivalence class $[\gamma]$ of the endomorphism γ up to inner unitary equivalence. This unitary equivalence class of $[\gamma]$ is determined by the value $\int q(z)dz/d\pi i = q_0$, see (8.10.12). In the general DHR framework, the local observables generate the observable algebras A, the Hilbert space of states decomposes as $H = \bigoplus H_g$, where each H_g is A-invariant, but with possible multiplicity and then the superselection sector is the subrepresentation on H_g. The problem is to construct additional field operators, relatively local with respect to observables, making transitions between sectors such that H is generated by the vacuum vector Ω in the vacuum sector H_0 (cf. the vertex operator construction). In the philosophy of DHR, the fields are to be constructed from local morphisms, γ, so that π_γ on H_γ is unitarily equivalent to $\pi_0 \circ \gamma$ on the vacuum sector H_0.

Returning to the $U(1)$ case, let us work with the universal state space $H = \bigoplus_{g\in\mathbb{R}} H_g$ where we identify $H_g = H_0$, $\pi = \oplus_g \pi_0 \circ \gamma_g$, Γ_g is the charge shift on H namely $(\Gamma_g\varphi)_h = \varphi_{h-g}$, $\varphi \in \bigoplus_g H_g$ and $U(t) = \oplus_g U_0(t)$ if $U_0(t)$ is the time translation on H_0, coming from the Sugawara Hamiltonian or rotation on the circle, $\alpha_t(W(f)) = W(f_t)$, if $f_t(z) = f(e^{-it}z)$.

If we had unitary field operators ψ_ρ for any charge distribution ρ satisfying (8.10.16), then the cocycle $z(\rho_1, \rho_2) = \psi_{\rho_1}\psi_{\rho_2}\psi^*_{\rho_1+\rho_2}$ lies in $\mathcal{Z}(A'') \subset B(H)$. It can be made scalar valued for a suitable choice of fields ψ_ρ. If ρ_1, ρ_2 have the same charge density, say g, then $\rho_2 - \rho_1 = iv'$ for some v. Then

$$Ad(\psi_{\rho_1}\psi^*_{\rho_2}) = \gamma_{\rho_1-\rho_2} = Ad(W(v)), \tag{8.10.12}$$

so that $W(v)^*\psi_{\rho_1}\psi^*_{\rho_2} \in \mathcal{Z}(A'')$. In particular, for ρ $(= \rho_2)$ given with charge g, and setting $\rho_1(z) = gz^{-1}$ we are led to try

$$\psi_\rho = W(\tilde{\rho})\Gamma_g \tag{8.10.13}$$

since Γ_g $(\equiv \psi_{\rho_1})$ implements $\gamma_{\rho_1} = \gamma_{\rho_g}$, with $\tilde{\rho}$ a solution to

$$i(\rho(z) - g/z) = \tilde{\rho}' \tag{8.10.14}$$

i.e.

$$\tilde{\rho} = i\sum_{n\neq 0} \rho_{-n}z^n/n + l(\rho) \tag{8.10.15}$$

where $l(\rho)$ is a scalar, and $\rho = \sum_n \rho_{-n} z^{n-1}$, $\rho_0 = g$. The choice of ψ given in (8.10.13) will satisfy

$$\psi_\rho W = \gamma_\rho(W)\psi_\rho, \quad W \in A \tag{8.10.16}$$

but the cocycle $(\rho_1, \rho_2) \longrightarrow \psi_{\rho_1}\psi_{\rho_2}\psi^*_{\rho_1+\rho_2}$ will be scalar valued if l is linear. Any other choice of fields is of the form $\eta(\rho)e^{i\lambda(\rho)Q}y(g)\psi_\rho$ where η is an arbitrary phase factor, λ is a real valued continuous linear functional and y, satisfying the cocycle relation $y(g_1)\gamma_{-g_1}(g_2) = y(g_1 + g_2)$ in the unitary group of $\mathcal{Z}(A'')$ depending only on the charge, is a *Klein transformation.*

We will pin ρ down by considering whether the vacuum can be separating for (local) field algebras, and to simplify commutation relations between ψ_{ρ_1} and ψ_{ρ_2} when ρ_1, ρ_2 are fields localized in disjoint intervals I_1 and I_2. First, consider, for such ρ_1, ρ_2 the complex valued commutator

$$\psi_{\rho_1}\psi_{\rho_2}\psi^*_{\rho_1}\psi^*_{\rho_2} = z(\rho_1,\rho_2)z(\rho_2,\rho_1)^* = \exp(-a(\tilde{\rho_1},\tilde{\rho_2}) - il(g_1\rho_2 - g_2\rho_1))$$

where a is as in (8.10.4). If $\zeta \in S^1$, let $\log_\zeta z$ be the branch of the logarithm function with cut along $\mathbb{R}_+\zeta$. Then we will take, for our functional,

$$l_\zeta(\rho) = \int \frac{dz}{2\pi}\rho(z)\log_\zeta(z)\,.$$

Then if $\zeta \in S^1\backslash(I_1 \cup I_2)$, where I_1 and I_2 are disjoint intervals, the functional

$$\mathcal{T} = -a(\tilde{\rho_1},\tilde{\rho_2}) - l_\zeta(g_1\rho_2 - g_2\rho_1) \tag{8.10.17}$$

is seen to be a topological invariant, i.e. depends only on the relative position of I_1, I_2 and ζ and the charges g_1, g_2 of ρ_1 and ρ_2. In fact, $\mathcal{T} = \pm i\pi g_1 g_2$ with the $+$ sign if I_1, I_2 are positively oriented. So if we put

$$\psi^\zeta_\rho = \eta_\zeta(\rho)W(\tilde{\rho_\zeta})\Gamma_g \tag{8.10.18}$$

where $\eta_\zeta(\rho) = \exp(ig\int dz\rho(z)\log_\zeta(z)/4\pi)$, $\tilde{\rho_\zeta}$ is as in (8.10.15) with l_ζ chosen for l, then the field operators satisfy the abelian local braiding relations:

$$\psi^\zeta_1\psi^\zeta_2 = e^{\pm i\pi g_1 g_2}\psi^\zeta_2\psi^\zeta_1 \tag{8.10.19}$$

if the localization regions I_1 and I_2 are disjoint and the $+$ sign if I_1, ζ, I_2 are in clockwise order. Moreover we have the fusion rule:

$$\psi^\zeta_{\rho_1}\psi^\zeta_{\rho_2} = e^{\mathcal{T}/2}\psi^\zeta_{\rho_1+\rho_2} \tag{8.10.20}$$

where $\mathcal{T}$ is imaginary (cf. (8.10.17)) and $\lambda \longrightarrow \psi^\zeta_{\lambda\rho}$ is additive on $\mathbb{R}$, so that $(\psi^\zeta_\rho)^* = \psi^\zeta_{-\rho}$.

Some care is needed in trying to piece together the field algebras $\mathcal{F}_\zeta(I)$ $I \subset S^1\backslash\{\zeta\}$ generated by ψ^ζ_ρ for ρ supported in I , as ζ varies. If ρ is supported on an interval $I \subset S^1\backslash\{\zeta_1,\zeta_2\}$, then

$$\psi_\rho^{\zeta_2}(\psi_\rho^{\zeta_1})^* = e^{-\sigma i\pi g^2} e^{\sigma 2\pi i g Q} \tag{8.10.21}$$

where $\sigma \in \{0,1\}$ depending on the relative position of I, ζ_1 and ζ_2. Unless we restrict values of the charge Q then in general the transition functions 8.10.21 between fields $\psi_\rho^{\zeta_1}$ and $\psi_\rho^{\zeta_2}$ are a non-trivial way of gluing together field algebras $\mathcal{F}_\zeta(I)$, $I \subset S^1\backslash\{\zeta\}$, which are local Minkowski space field algebras (where ζ is identified with the projection of a point at space-like infinity of the space–time manifold $\mathbb{R}\times S^1$), $\mathcal{F}(I) = \cup_\zeta \mathcal{F}_\zeta(I)$. Non trivial elements of the centre of $\mathcal{F}(I)$ (cf. Section 8.5) are an obstruction to the Reeh–Schlieder property of the vacuum being separating for the field algebra $\mathcal{F}(I)$, but this is not so for the algebras $\mathcal{F}_\zeta(I)$ on the punctured circle.

The fields transform under time as

$$U(t)\psi_\rho^\zeta U(t)^* = \left(\eta_\zeta(\rho)/\eta_\zeta(\rho^t)\right)\exp(-iQ\mathcal{J})\psi_{\rho^t}^\zeta\,,$$

where $\rho^t(z) = e^{-it}\rho(e^{-it}z)$, $\eta_\zeta(\rho)/\eta_\zeta(\rho^t) = \exp(ig\mathcal{J})$ and $\mathcal{J} \equiv \int dz(\rho(z) - \rho^t(z))\log_\zeta(z)/2\pi = gt$ for t small enough to ensure that ρ^t is supported in $S^1\backslash\{\zeta\}$. Then $U(t)\psi_t^\zeta U(t)^* = e^{ig^2t/2}e^{-igtQ}\psi_{\rho^t}^\zeta$. Renormalizing, i.e. replacing $U(t)e^{Q^2t/2}$ by $U(t)$, we have $U(t)\psi_t^\zeta U(t)^* = \psi_{\rho^t}^\zeta$ for small t, and the spectrum of the generator of time evolution is $g^2/2 + \mathbb{Z}_+$ on H_g.

To link up with the discussion of Section 8.2, we take $\psi_\rho^{\zeta=-1}$ as the support of ρ shrinks to a point. If we renormalize and set

$$\phi_{g,\lambda}(t) = \exp\{-g^2(\sum|(\tilde{\delta_\lambda})_m|^2/m - \int dz\delta_\lambda(z)\log z)/2\}\alpha_t\psi_{g,\delta_\lambda}^{\zeta=-1}$$

so that $\langle\phi_{g,\lambda}(0)\Omega, \Gamma_g\Omega\rangle = 1$, where $\delta_\lambda(z)$ goes to the Dirac measure $\delta(z-1)$, as $\lambda \searrow 0$. Then with $\psi_g(z) \equiv z^{-g^2}\phi_g(z)$, where $\phi_g = \lim_{\lambda\searrow 0}\phi_{g,\lambda}$ as a distribution, then n-point functions of the primary fields ψ_g satisfy

$$\langle\psi_{g_1}(z_1)\cdots\psi_{g_n}(z_n)\rangle = \prod_{j<k}(z_j - z_k)^{g_j g_k}\delta_{\Sigma g_j} \tag{8.10.22}$$

for $|z_1| > |z_2| > \cdots > |z_n|$.

In a general field theory locality means $\mathcal{R}(I_1) \subset \mathcal{R}(I_2)'$ if $I_1 \cap I_2 = \varnothing$ (disjointness of intervals in the compactified picture corresponds to space–time separation in Minkowski space). However, usually $\mathcal{R}(I) \subset \mathcal{R}(I')'$ will be a strict inclusion, if I' is the (space like) complement. When equality holds, and this is usually in the vacuum representation, an observable which commutes with observables measurable or localized in I' is an observable localized in I — this is *Haag duality*. (In its weakest form duality means $\pi(\mathcal{R}(I))'' = \pi(\mathcal{R}(I'))' \cap \pi(\mathcal{R})''$ in the C^*-situation.) In the case of the observable algebras $\mathcal{A}(I)$ in the vacuum representation, Haag duality is closely related to the geometric realization of the Tomita modular operators. The existence of continuous implementability of conformal transformations $SU(1,1)$ forces $\mathcal{A}(I) = \mathcal{A}(\overline{I})$ for $\overline{I}$ the closure of an interval I. Indeed if γ

in $SU(1,1)$ is such that $\gamma I \subset \overline{I}$, then $U(\gamma)\mathcal{A}(\overline{I})U(\gamma)^{-1} \subset \mathcal{A}(I)$, so that by taking such a sequence γ converging to 1, we get $\mathcal{A}(\overline{I}) = \mathcal{A}(I)$ as the local algebras are already weakly closed. In particular to show Haag duality $\mathcal{A}(I_+) = \mathcal{A}(I'_+)'$ for the open upper semi-circle I_+ with I_- the interior of $I'_+ = S^1\backslash I_+$, i.e. the image of I_+ under the flip j, we need $\mathcal{A}(I_+) = \mathcal{A}(I_-)'$. In fact the Tomita theory operators $S = J\Delta^{1/2}$ have geometrical interpretations: J is the second quantization of the flip or conjugation j on the circle, $\Delta^{1/2}$ is that of $i \in SU(2)$, an analytic continuation of $SU(1,1)$, with $i^2 = -1$. Thus Haag duality is a consequence of Tomita theory $\mathcal{A}(I_-) = J\mathcal{A}(I_-)J = \mathcal{A}(jI_-) = \mathcal{A}(I_+)$.

We turn to the question of extending the observable algebras generated by fields with certain local commutation relations. The relation (8.10.21) was an obstruction to regarding the extension as living on the circle rather than the punctured circle $S^1\backslash\{\zeta\}$. However, if $g^2 = 2N$, $N \in \mathbb{N}$, then the first phase factor in (8.10.21) disappears and we can eliminate the second by taking a quotient forcing $e^{2\pi i g Q} = 1$, or restricting to the subspace of the universal state space H where $e^{2\pi i g Q} = 1$. Then form the observable $\mathcal{A}_N(I)$ as the $*$-algebra generated by ψ_ρ the class of ψ_ρ^ζ for $\zeta \notin I$, where ρ is supported in I with charge $(2N)^{1/2}$. The extension is maximal if and only if N is not divisible by the square of $M \in \mathbb{N}$, $M \neq 1$. If $M^2|N$, then the fields $\psi^\zeta_{\rho/M}$ are local and the fusion rule $(\psi^\zeta_{\rho/M})^M = \psi^\zeta_\rho$ holds.

To describe the superselection sectors of the $\mathcal{A}_N$ theory, we take $V = \exp(2\pi i Q/(2N)^{1/2})$ which generates the centre of $\mathcal{A}_N$. Here $V^{2N} = 1$, so that V has $2N$ eigenvalues $e^{\pi i\nu/N}$, $\nu = 0, \ldots, 2N-1$ which will label the superselection sectors. The fields ψ^ζ_g with charge $g = (2N)^{1/2}$ are not relatively local with respect to observables in $\mathcal{A}_N$, but making a Klein transformation we obtain fields $\phi^\zeta_g = e^{2\pi i Q(2N)^{1/2}}\psi^\zeta_g$ which have this property and generate a field algebra $\mathcal{F}_N$. Since $V\phi^\zeta_g = e^{i\pi/N}\phi^\zeta_g V$, V will take a sector H_ν to $H_{\nu+1}$, induce an action of $\mathbb{Z}_{2N}$ on $\mathcal{F}_N$ such that $\mathcal{F}_N^{\mathbb{Z}_{2N}} = \mathcal{A}_N$. This is an example where all the endomorphisms γ describing sectors are automorphisms, which occurs precisely when there is an abelian gauge group G (here $G = \mathbb{Z}_{2N}$, or in the case of $\mathcal{A}$ itself $G = \mathbb{R}$, $(\gamma = \gamma_g, g \in \mathbb{R})$), so that $\mathcal{A} = \mathcal{F}^G$. In general when the situation is described by endomorphisms which are not all automorphisms, we are in the realm of the dual space of compact groups (Doplicher and Roberts 1989) or paragroups (cf. Ising model example of Section 8.5, or the $SU(N)_k$ setting).

For example, $\mathcal{A}_1$ is the $su(2)^\wedge$ Kač-Moody algebra at level 1, (identify $J(z) = \sqrt{2}J_3(z)$ in the notation of Section 8.7), $\mathcal{A}_2$ is generated by the even polynomials of a free complex Fermion field, $\mathcal{A}_3$ appears in the $\mathbb{Z}/4$-parafermion current algebra of (Zamolodchikov and Fateev 1985), $\mathcal{A}_4$ is the even subalgebra of $\mathcal{A}_1$ (where J_1, J_2 are odd, J_3 even), and not maximal. The extensions $\mathcal{A}_N$ also describe the coset models $so(4N)_1/so(2N)_2$ (Goddard *et al.* 1986).

The irreducible representations of $\mathcal{A}_N$ are labelled by $\nu = 0, \ldots, 2N-1$ where the corresponding partition function $Z_\nu(\beta,\mu) = \mathrm{tr}\ e^{-\beta(H+\mu Q)}$. Now $-\partial/\partial\beta \log Z = \omega_{\beta,\mu}(H+\mu Q)$, where the regular extremal KMS states on (A, α_t) at inverse temperature β are given by

$$\omega_{\beta,\mu}(W(u)) = \exp(i\mu u_0 - 1/2\sum_{n=1}^{\infty} n\coth(\frac{n\beta}{2}|\dot{u}_n|^2)) \tag{8.10.23}$$

o that as $\beta \to \infty$ we recover the pure ground states of (8.10.7). Then using 8.10.23) it follows that

$$Z_n(\beta,\mu) = e^{2\pi i\tau/24}\frac{1}{\eta(t)}\Theta_{2n,2N}(\tau,\zeta,0) \equiv K_n(\tau,\zeta,N) \tag{8.10.24}$$

ith Dedekind η function of (8.2.18), and classical theta function Θ (cf. (8.7.45)), $^{2\pi i\tau} = e^{-\beta}$, $e^{2\pi i\zeta} = e^{-\beta\mu}$. The modular group $SL(2,\mathbb{Z})$ acts on the $2N$ dimen-ional space spanned by $e^{2\pi i u}K_n(\tau,\rho)$ where T $((\tau,\zeta,u) \longrightarrow (\tau+1,\zeta,u))$ acts iagonally by $\exp i\pi(n^2/2N - 1/12)$ and S $((\tau,\zeta,u) \longrightarrow (-1/\tau,-\zeta/\tau,u-\zeta^2/\tau))$ as matrix $[e^{-inm}/(2N)^{1/2}]$. In the maximal case, where N is a product of istinct primes, there are precisely two modular invariant partition functions $\sum Z_{mn}K_m\bar{K}_n$ for each partition of N into coprimes and no other.

Moving from $U(1)$ to $SU(N)$, we take the corresponding positive energy epresentations at a fixed level, say l, of $su(N)^\wedge$, see the comment on page 432 btained by decomposing the level l representation π of $LSU(N)$ in the Hardy pace fermion Fock state of $L^2(S^1,\mathbb{C}^N\otimes\mathbb{C}^l)$. If I is an interval in S^1, let $M(I)$ e the von Neumann algebra generated by the fermion operators on $L^2(I) =$ $^2(I;\ \mathbb{C}^N\otimes\mathbb{C}^l)$ acting on the Hardy Fock state on $L^2(S^1)$. Then if I is a proper on-empty interval in S^1, we can, by a conformal transformation in $SU(1,1)$, ssume it is the upper semicircle I_+. Then (cf. Exercise 6.7 and 8.4), the fermion ock vacuum vector Ω is cyclic and separating for $M(I_+)$, and as in the boson tuation already discussed, the modular conjugation J and modular operator for $M(I_+)$ with respect to Ω have geometric interpretations in terms of PCT onjugation and an analytic continuation of $SU(1,1)$. In particular $M(I_-)$ is the graded) commutant of $M(I_+)$ so that the even part of $M(I_-)$ is the commutant f the even part of $M(I_+)$.

The modular unitary group Δ^{it} is the second quantization of a geometrically escribed unitary group. By the spectral theory of the latter, Δ^{it} is ergodic in xing only $\mathbb{C}\Omega$, and since Ω is separating for $M(I_+)$, the modular automorphism $_t = \Delta^{it}(\cdot)\Delta^{-it}$ is ergodic, fixing only $\mathbb{C}$. Consequently $M(I_+)$ must be a type I factor. (In the semi-finite case the modular group is inner.)

Now again let L_IG denote the subgroup of LG of loops supported in I. Since e positive energy representation π of $LSU(N)$ in the Hardy Fock space of $^2(S^1,\mathbb{C}^N\otimes\mathbb{C}^l)$ is by even operators (cf. the quantization dQ is in terms of en quadratic expressions), we deduce that $\pi(L_{I_+}SU(N))''(\subset M(I_+)_+)$ and $(L_{I_-}SU(N))''(\subset M(I_-)_+)$ commute, i.e. $\pi(L_{I_+}SU(N))'' \subset \pi(L_{I_-}SU(N))'$. oreover since the modular operators of $M(I_\pm)$ are geometric they leave $(L_{I_\pm}SU(N))''$ globally invariant for which by the Reeh–Schlieder theorem, the acuum vector is separating and cyclic. Then by (Takesaki 1972) the modu-r operators for $M(I_+) \supset \pi(L_{I_+}SU(N))''$ can be identified. In particular we gain have an ergodic action of the modular group on the subalgebra so that

$\pi(L_{I_\pm}SU(N))''$ are type III factors and hence so are $\pi_i(L_{I_\pm}SU(N))''$, where π_i are the level k irreducible subrepresentations of π, with $\pi_i(L_{I_+}SU(N))'' \subset \pi_i(L_{I_-}SU(N))'$. Since π is a type III factor representation, all the subrepresentations π_i of $L_ISU(N)$ are equivalent. Since J is essentially the flip j on the circle, and identifying via (Takesaki 1972), we again deduce Haag duality $\pi_0(L_{I_+}SU(N))'' = \pi_0(L_{I_-}SU(N))'$ in the vacuum sector.

If X is an A–B bimodule, where A, B are von Neumann algebras, we let

$$\mathcal{X} = \mathrm{Hom}_{B^{\mathrm{opp}}}(H_0, X)$$

where (H_0, Ω) is a vacuum representation of B, i.e. B acts on H_0, where Ω is cyclic and separating. Then $x \longrightarrow x\Omega$, $\mathcal{X} \longrightarrow X$ allows us to switch from an operator or a field to a state or a vector. In this setting the relative tensor product (Definition 9.61) of bimodules can be formulated using the Reeh–Schlieder identification between states and operators. Then if Y is a B–C bimodule, where C is a third von Neumann algebra, and

$$\mathcal{Y} = \mathrm{Hom}_B(H_0, Y)$$

then the four-point function

$$\langle x_1 \otimes y_1, x_2 \otimes y_2 \rangle = \langle x_2^* x_1 y_2^* y_1 \Omega, \Omega \rangle \qquad (8.10.25)$$

gives an identification of the completion of $\mathrm{Hom}_{B^{\mathrm{opp}}}(H_0, X) \otimes \mathrm{Hom}_B(H_0, Y)$ (modulo null vectors, or vectors of zero length) with the bimodule $X \otimes_B Y$.

In particular, suppose X and Y are two positive energy representations of $SU(N)$ at level l, so that each has an $L_{I_+}SU(N) \times L_{I_+}SU(N)$ action or is an M–M bimodule if $M = \pi_0(L_{I_-}SU(N))''$, where π_0 is the vacuum representation. Let

$$\mathcal{X} = \mathrm{Hom}_{L_{I_-}SU(N)}(H_0, X)$$
$$\mathcal{Y} = \mathrm{Hom}_{L_{I_+}SU(N)}(H_0, Y).$$

Then $\mathcal{X} \otimes \mathcal{Y}$ defines via (8.10.25), $X \otimes_N Y$ which has a natural M–M bimodule structure or an $L_{I_+}SU(N) \times L_{I_+}SU(N)$ action.

Now let p_i and p_j be projections on irreducible summands H_i and H_j of the level l representation $\pi_P^{\otimes l}$, when π_P is the level 1 representation of $LSU(N)$ in the Hardy Fock space of $L^2(S^1, \mathbb{C}^N)$ as above. (All irreducible positive energy representations of level l appear in this way, cf. all irreducible representations of $SU(N)$ appear in the tensor powers of the vector representation.) Fix an equivariant embedding of $\mathbb{C}^N$ in $\mathbb{C}^N \otimes \mathbb{C}^l$, so that if $f \in L^2(S^1, \mathbb{C}^N) \subset L^2(S^1, \mathbb{C}^N \otimes \mathbb{C}^l)$ we can compress to a field operator $\phi(f) = \phi_{ij}(f) \equiv p_i a(f) p_j \in \mathrm{Hom}(H_i, H_j)$. Then $\|\phi(f)\| \leq \|f\|_2$ and satisfies the covariance condition

$$\pi_i(g)\phi(f)\pi_j(g)^* = \phi(g \cdot f), \quad g \in L(G) \rtimes \mathrm{Rot}(S^1)$$

This gives a primary field, an intertwiner between $\mathcal{V} \otimes H_i$ and H_j, where $\mathcal{V} = C^\infty(S^1, V)$ is an ordinary representation of $LSU(N)$ and V the fundamental representation of $SU(N)$. More generally, a primary field can be defined as intertwiner, not necessarily bounded, between $\mathcal{V}^0 \otimes H_i^0$ and H_j^0, where $\mathcal{V}$ is an ordinary representation and the superscript 0 refers to finite energy vectors. Then $\mathcal{V}^0(n) = z^{-n} \otimes V$, V is the *charge* and, at the infinitesimal level, the intertwining property of the primary field $\phi(v,n) = \phi(z^n v)$ is

$$[T_n, \phi(v,m)] = \phi(T \cdot v, m+n), \; [D, \phi(v,m)] = -m\phi(v,m). \qquad (8.10.26)$$

Let $\phi(a,n) : H_j^0 \to H_k^0$ and $\phi(b,m) : H_k^0 \to H_i^0$ be primary fields of charges V_2 and V_3 respectively. Set $V_4 = H_j(0)$, $V_1 = H_i(0)$ the lowest energy spaces, and consider the four-point function putting three of the coordinates at $(0,1,\infty)$ as in page 384, Section 8.2:

$$F(z) = \langle f(z) v_2 \otimes v_3 \otimes v_4, v_1 \rangle = \sum_{n \geq 0} \langle \phi(v_2, n) \phi(v_3, -n) v_4, v_1 \rangle \, z^n \qquad (8.10.27)$$

where $v_j \in V_j$.

The generator D of the rotation group is determined by the Sugawara formula (8.7.30). So inserting $D = L_0 - h$ in the four-point function (8.10.27) (cf. (8.2.16)) and using the commutation relations we obtain the Knizhnik–Zamolodchikov equation:

$$(N+l) df/dz = \left(\frac{\Omega_{34} - (\Delta_k - \Delta_3 - \Delta_4)/2}{z} + \frac{\Omega_{23}}{z-1} \right) f(z);$$

cf. (Knizhnik and Zamolodchikov 1984), (Tsuchiya and Kanie 1988). Here $\Omega_{ij} = -\sum_a \pi_i(T^a)\pi_j(T^a)$, and π_i, $i = 1,2,3,4$ are the natural commuting actions of $su(N)$ on $\mathrm{Hom}_{SU(N)}(V_2 \otimes V_3 \otimes V_4, V_1)$, and T^a is an orthonormal basis for $su(N)$.

So if U, V are either $\mathbb{C}^N$ or its dual, and

$$F_k(z) = F_k^{U,V}(z) = \sum_{n \geq 0} \langle \phi_{ik}^U(u,n) \phi_{kj}^V(v,-n) \xi, \eta \rangle \qquad (8.10.28)$$

then the renormalized $f_k(z) = z^{\lambda_k} F_k(z)$ satisfies the KZ equation

$$(N+l) \frac{df}{dz} = \frac{\Omega_{vj}}{z} f(z) + \frac{\Omega_{uv}}{z-1} f(z) \qquad (8.10.29)$$

where $\lambda_k = (\Delta_k - \Delta_v - \Delta_j)/2(N+l)$ is the value of $\Omega_{vj}/(N+l)$ on $V_k \subset V \otimes V_j$. Then if $\nu_h = (\Delta_n - \Delta_u - \Delta_j)/2(N+l)$ is the value of $\Omega_{uj}/(N+l)$ on $V_h \subset U \otimes V_j$, $G_h = G_h^{V,U} = F_h^{U,V}$, the renormalized $g_h(z) = z^{\nu_h} G_h(z)$ satisfy the same KZ

$$(N+l) \frac{dg}{dz} = \frac{\Omega_{uj}}{z} g(z) + \frac{\Omega_{uv}}{z-1} g(z). \qquad (8.10.30)$$

Then $h(z) = g(z^{-1})$ satisfies

$$(N+l)\frac{dh}{dz} = \frac{(-\Omega_{uj} - \Omega_{uv})}{z}h(z) + \frac{\Omega_{uv}}{z-1}h(z) \tag{8.10.31}$$

and $\Omega_{uv} + \Omega_{vj} + \Omega_{ju} = (\Delta_i - \Delta_u - \Delta_v - \Delta_j)/2 \equiv \mu$ on $\mathrm{Hom}_{SU(N)}(U \otimes V \otimes V_j, V_i)$. Thus $\mu_h \equiv \mu - \nu_h = (\Delta_i - \Delta_v - \Delta_h)/2(N+l)$ is the value of $(N+l)^{-1}(\Omega_{vj} + \Omega_{uv})$ on $V_h \subset U \otimes V_j$ and $g_h(z) = z^{\mu_h} G_h(z^{-1})$ satisfies the KZ equation (8.10.29) expanded around ∞.

In the two cases (either U and V the same or conjugate), the KZ equation has the form

$$\frac{df}{dz} = \frac{A}{z}f + \frac{B}{1-z}f$$

where $B = -\alpha + \beta Q$, $\beta \neq 0$ and Q a rank one idempotent and A, $B - A$ are in generic position with respect to Q which allows a simultaneous rational canonical form and generalized hypergeometric equations.

Then the transport coefficients c_{kh} relating solutions $f_k(z)$ at 0 to $g_h(z^{-1})$ at ∞, $f_k(z) = \sum c_{kh} g_h(z^{-1})$ will provide braiding relations:

$b_{gf} a^*_{g_1 f} = \sum \nu_h a^*_{hg} b_{hg_1}$, with $\nu_h \neq 0$ if $h > g, g_1$ is permissible

$b_{gf} a_{fh} = \sum \mu_{f_1} a_{gf_1} b_{f_1 h}$ with $\mu_{f_1} \neq 0$ if $h < f_1 < g$

if we set $a_{gf} = a^{\square}_{gf} = \phi^{\square}_{gf}(e_{-\alpha}F)$, $\alpha = (\Delta_g - \Delta_f - \Delta_{\square})/2(N+l)$, $(a^*_{gf} = a^{\overline{\square}}_{fg} = \phi^{\overline{\square}}_{fg}(e_\alpha F^*))$, $F \in L^2(I)$ and b in terms of J in $S^1 \backslash \{1\}$ anticlockwise after I. Here $e_\mu(e^{i\theta}) = e^{i\theta\mu}$, taking the cut in the circle at $z = 1$ and $h > g$ means h is obtained from g by adding one box.

The operator $a^*_{\square 0} a_{\square 0} \in B(H_0) \cap (L_{I_-})' = \pi_0(L_{I_+})''$ by Haag duality, so by local equivalence can be transported by $\pi_f : \pi_0 \left(L_{I_+}\right)'' \to \pi_f \left(L_{I_+}\right)''$. Indeed using the above braiding relations:

$$\pi_f\left(a^*_{\square 0} a_{\square 0}\right) = \sum \lambda_g a^*_{gf} a_{gf}, \text{ with } \lambda_g > 0\,. \tag{8.10.32}$$

This latter expression means that if H_f is any irreducible positive energy representation at level l then

$$H_{\square} \otimes_N H_f \cong \oplus_{g>f} H_g \tag{8.10.33}$$

as positive energy bimodules where g runs over all permissible labels descending from f through adding one box.

In a similar way we can at least deduce the partial fusion rules

$$H_{[k]} \otimes H_f \cong \oplus_g H_g \tag{8.10.34}$$

where $[k]$ is the Young tableaux of the anti-symmetric representation with one column of k-boxes, and the summation is over some set of allowable Young tableaux which describe level k-representations of positive energy representations of $LSU(N)$, consistent with classical fusion rules of adding k boxes to f for the representation theory of $SU(N)$. Indeed, it is enough to work with elements $a_P = \phi_{f_k f_{k-1}}(a_k) \cdots \phi_{f_1 f_0}(a_1)$, $a_i \in L^2(I, V)$, for paths P of permissible elements $f = f_0 < f_1 < \cdots < f_k = g$, where $\phi_{gf}(a)$ is the vector vertex operator $\phi^{\square}_{gf}(e_{-\alpha_{gf}} a)$ where the phase factor has been incorporated as before, since the set of vectors $a_P H_f$ (as a_i varies) is dense in $\oplus_{f_k > \cdots > f_1 > f} H_{f_k}$, i.e. we do not need more general primary field operators which may be unbounded. Then if P_0 is the standard path $0 < \square = [1] < \cdots < [k]$, then (8.10.32) generalizes to

$$\pi_f(b^*_{P_0} b_{P_0}) = \sum_{g = f_k > \cdots f_0 = f} (\sum_P \lambda_P(g) b^*_P)(\sum_P \lambda_P(g) a_P)$$

from which (8.10.34) follows.

This is enough to determine the fusion rules completely (cf. p 618, Section 11.7) and identify them with the Verlinde formula (Goodman and Wenzl 1990):

$$H_f \otimes_N H_g = \oplus N^h_{fg} \det(\sigma_h) H_{h'},$$

where if N^h_{fg} are the classical fusion rules for $SU(N)$, h ranges over those permissible elements such that there exists $\sigma_h \in \Lambda_0 \rtimes S_N$ ($\Lambda_0 = \{(N+l)(m_i) \mid \sum m_i = 0\} \subset Z^N$), such that $h' = \sigma_h(h+\delta) - \delta$ is permissible if $\delta = (N-1, N-2, \ldots, 0)$. Note that the fusion rules are abelian because rotation through π on S^1 interchanges I_+ with I_- and identifies $Y \otimes_N X$ with $X \otimes_N Y$ in an $LSU(N)$ covariant way.

For expositions on algebraic quantum field theory and interaction with conformal field theory, subfactor theory etc, see (Kastler 1990). It is argued quite generally in the algebraic setting of Section 8.5, that if ρ is an endomorphism localized in $I \in \mathcal{I}_\zeta$, there is a unitary $\varepsilon_\rho \in R(I) \cap \rho^2(A_\zeta)'$ such that

$$\varepsilon_\rho \rho(\varepsilon_\rho) \varepsilon_\rho = \rho(\varepsilon_\rho) \varepsilon_\rho \rho(\varepsilon_\rho)\,.$$

In fact $\varepsilon_\rho = U^{-1}\rho(U)$ where U is unitary such that $\mathrm{Ad}\, U \circ \rho$ on A_ζ is localized in $I_0 \in \mathcal{I}_\zeta$, $I_0 \subset I'$. Then $\sigma_i \to \rho^{i-1}(\varepsilon_\rho)$ is a representation of the braid group (see Section 12.1). Moreover there should be a left inverse for ρ, a completely positive unital map Φ_ρ such that $\Phi_\rho \circ \rho = id$, and

$$\Phi_\rho(\varepsilon_\rho) = \omega_\rho / d_\rho\,,$$

where d_ρ is the statistical dimension (the square root of a Jones index), and a phase $\omega_\rho = e^{2\pi i h}$ where h is the lowest energy for L_0 in a representation of the Virasoro algebra. In the case of $\rho = \rho_{1/2}$, we find

$$\varepsilon = \frac{\omega_\rho}{\sqrt{2}}\,((1 \pm i) \mp 2ip)$$

where p is the projection generating the commutant $\rho^2(A_\xi)'$, $p = B(e_+)B(e_-)$, $e_\pm$ span the kernel of $(V^*)^2$, $\Gamma e_+ = e_-$. Using Φ_ρ, the completely positive quasi-

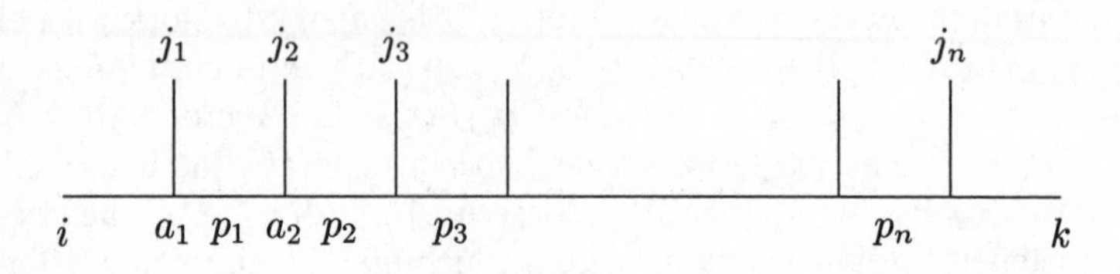

FIG. 8.43. Conformal block

free expectation $A(V^*)$, we obtain the statistical dimension $d_\rho = \sqrt{2}$. Here with $\rho_{1/2}$ being the Ramond sector, $h = 1/16$.

Section 8.7 is from (Moore and Seiberg 1989a), with the treatment on abstract fusion rules from (Caselle *et al.* 1990).

The trinions, and sewing can be described in a picture of intertwiners, chiral vertex operators from $H_i \otimes H_j$ to H_k, and their composition. As we have already said the tensor product of representations of a chiral algebra $\mathcal{A}$ needs some care as we wish to stay with the same central charge. Define $\Delta_{z,0}$ from $\mathcal{A}$ into $\mathcal{A} \otimes \mathcal{A}$ for the Virasoro algebra by

$$\Delta_{z,0}(L_n) = \begin{cases} 1 \otimes L_n + (z^{n-1} L_{-1} + \cdots + L_n) \otimes 1 & n \geq 1 \\ 1 \otimes L_n + (z^{n+1} L_{-1} + z^n (n+1) L_0 + \cdots) \otimes 1 & n < -1 \end{cases} \tag{8.10.35}$$

which can be used to define a representation of $\mathcal{A}$ on $H_j \otimes H_k$. The *chiral vertex operators* of type $\begin{pmatrix} i \\ j\ k \end{pmatrix}$ are then the linear maps $\Phi = \Phi_z : H_j \otimes H_k$ into H_i which intertwine the actions of $\mathcal{A}$, and satisfy $(d/dz)\Phi = \Phi(L_{-1} \otimes 1)$ (cf. (8.10.26)). Writing $\Phi = \begin{pmatrix} i \\ j\ k \end{pmatrix}$, for convenience, a conformal block can be written as

$$\begin{aligned} F(z_1, \ldots z_n) = \Big\langle i\alpha \Big| \begin{pmatrix} i \\ j_1 p_1 \end{pmatrix}_{z_1, a_1} (\beta_1 \otimes \circ) \begin{pmatrix} p_1 \\ j_2 p_2 \end{pmatrix}_{z_2, a_2} \\ \cdots \begin{pmatrix} p_n \\ j_n k \end{pmatrix}_{z_n, a_n} (\beta_n \otimes \circ) \Big| k, \gamma \Big\rangle \end{aligned}$$

or diagrammatically as in Fig. 8.43.

Symmetric spaces, which we mentioned in Section 8.8, have been classified in the mathematical literature (Helgason 1978). The interpretation of the quantum equivalence between Sugawara and free fermion constructions in terms of symmetric spaces of (Goddard *et al.* 1985) can be used to classify conformal embeddings (Arcuri *et al.* 1987).

We refer to (Fuchs 1992), (Gannon, preprint 1995) for reviews on the classification of modular invariants, which we started discussing in Section 8.9. The discussions relating modular invariants, graphs and C-algebras is based on (D

Francesco and Zuber 1990a), (Di Francesco and Zuber 1990b) and (Zuber 1990).

For each fixed level, there are only finitely many modular invariants (Gannon 1993). This follows from the constraint $S_{0\lambda} \geq S_{00} > 0$, which forces

$$1 = Z_{00} = \sum_{\lambda\mu} S_{0\lambda} Z_{\lambda\mu} S_{\mu 0} \geq S_{00}^2 \sum Z_{\lambda\mu} \, .$$

Thus $\sum_{\lambda\mu} Z_{\lambda\mu} \leq 1/S_{00}^2$, and since each $Z_{\lambda\mu}$ is positive and integral, there are only finitely many possibilities for each such matrix entry.

The *A-D-E* classification of $su(2)$ modular invariants is due to (Cappelli *et al.* 1987b). The identification of the E_6 and E_8 $su(2)^\wedge$ modular invariants in terms of conformal embeddings is due to (Bouwknegt and Nahm 1987). (Gannon 1994) completed the $su(3)$ classification by showing that Table 8.6, together with the conjugates, is complete. The $su(3)$ modular invariants were found as follows. The $D^{(n)}$ series for $n = 0$ mod 3 is in (Bernard 1987); the $D^{(n)}$ series for $n \neq 0$ mod 3 was found in (Altschuler *et al.* 1988); the exceptionals $E^{(8)}$, $E^{(24)}$, and the single invariant $E^{(12)}$ corresponding to the graphs $E_1^{(12)}$, $E_2^{(12)}$, $E_3^{(12)}$ were found by (Christe and Ravanini 1989); the invariant $E_{MS}^{(12)}$ corresponding to $E_5^{(12)}$ was found by (Moore and Seiberg 1989b) (the second Moore–Seiberg exceptional $E_{MS}^{(12)c}$, is the conjugate of the authentic Moore–Seiberg invariant $E_{MS}^{(12)}$).

Gannon has a programme to classify modular invariants. The first step in the classification programme is to identify all permutation or automorphism invariants. If we impose in addition to modular invariance, the requirement that

$$Z_{0\lambda} = \delta_{0\lambda}$$

then such an invariant is basically an automorphism of the fusion ring. There is a permutation σ of $\mathcal{A}^{(n)}$ such that $Z_{\lambda\mu} = \delta_{\lambda,\sigma\mu}$ (cf. the statement of Section 8.9 that in a maximal chiral extension, all modular invariants are such automorphism invariants). This classification has been accomplished in the simple case as well as for $su(n_1)\oplus\cdots\oplus su(n_s)$, with no exceptional invariants in the latter situation.

The next step in the programme is to weaken the previous constraint to

$$Z_{0\lambda} \neq 0 \quad \text{implies } \lambda = J0 \text{ for some simple current } J \tag{8.10.36}$$

or equivalently

$$Z_{0\lambda} \neq 0 \quad \text{implies } S_{\lambda,0} = S_{0,0} \, . \tag{8.10.37}$$

These are the $\mathcal{ADE}_7$ invariants and include the simple current invariants, where

$$Z_{\mu\lambda} \neq 0 \quad \text{implies} \quad \lambda = J\mu \tag{8.10.38}$$

for some simple current J.

The final step in the programme is essentially to find the final constraints on states λ coupled to 0, for which $Z_{0\lambda} \neq 0$. It would appear that, with few exceptions, (8.10.36) is almost always satisfied.

The existence of conformal embeddings means that $su(n)_{n-2}$, $su(n)_n$ $su(n)_{n+2}$ will all have invariants violating (8.10.36), and so will not be com pletely covered by the $\mathcal{ADE}_7$ classification. Of course there will be similar in variants for other algebras and levels. So the preliminary task in the third par of the programme is to handle these sorts of generic non-$\mathcal{ADE}_7$ invariants. Thi is realistic because conformal embeddings (and their rank-level dual partners are extremely well-behaved. All of the invariants coming directly from conforma embeddings are explicitly known, since they correspond to level 1 invariants o larger algebras. The classification of conformal embeddings was done simultane ously by (Bais and Bouwknegt 1987) and (Schellekens and Warner 1986). Th $su(n)_k$ conformal embeddings are

$$su(n)_{n-2} \subset su(n(n-1)/2)_1 \text{ (only gives exceptionals for } n \geq 5);$$
$$su(n)_n \subset so(n^2-1)_1 \text{ (only gives exceptionals for } n \geq 4);$$
$$su(n)_{n+2} \subset su(n(n+1)/2)_1 \text{ (only gives exceptionals for } n \geq 3);$$
$$su(2)_{10} \subset so(5)_1\,; \quad su(2)_{28} \subset G_{2,1}\,;$$
$$su(3)_9 \subset E_{6,1}\,; \quad su(3)_{21} \subset E_{7,1}\,;$$
$$su(4)_8 \subset so(20)_1\,; \quad su(6)_6 \subset sp(20)_1\,; \quad su(8)_{10} \subset so(70)_1\,.$$

For $su(2)$, the automorphism invariants are the A-series for all k, and the D series for $k = 2$ mod 4. For $su(3)$, they are $A^{(n)}$ and $A^{(n)c}$ for all n, and $D^{(n)}$ an $D^{(n)c}$ for $n \neq 0$ mod 3. The automorphism invariants for $su(n_1) \oplus \cdots \oplus su(n_r$ were classified in (Gannon 1995). There are no exceptional invariants as w have already mentioned. The automorphism invariants for B_r, C_r, D_r, and th exceptional algebras were classified in (Gannon *et al.* 1996). There are man exceptional invariants: B_r and D_r at $k = 2$, for most r; E_8 at $k = 4$; F_4 at $k = 3$ and G_2 at $k = 4$.

The $\mathcal{ADE}_7$ invariants, i.e. the ones satisfying (8.10.36) will of course includ all the automorphism invariants. For $su(2)$, they are the A-series, the D-series and the E_7 exceptional (hence the name $\mathcal{ADE}_7$). In the $su(3)$ case, the $\mathcal{A}$ invari ant is of course $A^{(n)}$, $A^{(n)c}$, the $\mathcal{D}$ invariants are the $D^{(n)}$, and $D^{(n)c}$ and th $\mathcal{E}_7$ invariants are the two invariants of Moore and Seiberg $E^{(12)}_{MS}$ and $E^{(12)c}_{MS}$ The other invariants $E^{(8)}$, $E^{(12)}$, $E^{(24)}$ correspond to conformal embedding $su(3)_5 \subset su(6)_1$, $su(3)_9 \subset E_{6,1}$, $su(3)_{21} \subset E_{7,1}$ respectively (cf. E_6 and E for $su(2)$).

The $\mathcal{ADE}_7$ invariants have been classified for all $su(n)$ in (Gannon, preprin 1995). There are exceptionals at: $A_{1,16}$ (the E_7 exceptional of Cappelli et al) $A_{2,9}$ (the Moore–Seiberg ones); $A_{3,8}$; $A_{4,5}$; $A_{7,4}$; $A_{8,3}$; $A_{15,2}$. The $\mathcal{ADE}_7$ invari ants for the remaining simple algebras is still under investigation with only D outstanding. Ignoring the automorphism invariants, the exceptionals occur a $C_{r,k}$ when $rk = 16$; $B_{r,2}$, $D_{r,2}$, $B_{r,8}$, $D_{r,8}$ for most r; and $D_{r,k}$ for $rk = 32$, $k >$ (although there may exist other exceptionals for D_r).

The (Bernard 1987) paper anticipated the simple current invariants Schellekens and collaborators realized the generality of the construction. Th simple current invariants for $su(2)$ – i.e. the ones which obey condition (8.10.38) – are the A- and D-series. For su(3), the simple current invariants are again the A- and D-series (but not their conjugations). Simple currents have a meaning for any RCFT. Schellekens and collaborators classified all such invariants, up to a mild condition on the S-matrix: the last paper in the series was (Kreuzer and Schellekens 1994). (For WZW theories their S-matrix condition will almost always be satisfied – an exception is $su(2)$ at level 2.) Schellekens et al. found that there are no exceptionals obeying (8.10.36) (assuming their S-matrix condition). This is a major result, but (8.10.36) is a strong constraint – i.e. many invariants will turn out to obey it but to verify that they will is very difficult. In practice the Gannon programme does not reduce a problem to the Schellekens classification without first completing the entire classification.

The mechanisms for generating non-exceptional invariants are simple currents; conjugation; and (for semi-simple algebras) permutations of identical summand algebras at identical levels (all three of these appear in (Gannon 1995)). The standard mechanisms for generating exceptionals are conformal embeddings; the Galois method (this method is first defined in (Fuchs *et al.* 1994), but new exceptionals were first found by this method in (Fuchs *et al.* 1995)) and rank-level duality (first discussed in (Walton 1989) but exploited more systematically in (Verstegen 1990) and (Verstegen 1991). The conformal embedding exceptionals include the E_6, E_8 ones for $su(2)$. Examples of the Galois ones are $B_{r,2}$ and $D_{r,2}$. Examples of rank-level duality ones are most of the level 2 and 3 exceptionals for $su(n)$. Not all exceptionals are constructible from these methods: e.g. the E_7 exceptional of $su(2)$ requires more convoluted methods, and as far as it seems known the $E_{8,4}$ exceptional automorphism invariants can only be found by direct construction – i.e. it is extremely exceptional. The $\mathcal{ADE}_7$ exceptionals seem to be the most exceptional, in this sense.

The work of Gannon on the classification problem uses none of the machinery developed by the proof of (Cappelli *et al.* 1987b). Their idea was to solve the modular invariance condition explicitly, and then impose the remaining conditions. In general this is not realistic, even for $su(3)$, – as became apparent with (Bauer and Itzykson 1990). The main contribution of Gannon to the problem was to come up with the three-step approach (i.e. first, do the automorphism invariants; then do the remaining $\mathcal{ADE}_7$ invariants; and then finally solve the constraints for the possible λ with $Z_{\lambda,0} \neq 0$, and handle these exceptional levels and algebras more explicitly). The idea is that almost every algebra and level will necessarily satisfy (8.10.36) and hence be an $\mathcal{ADE}_7$ invariant, and that the $\mathcal{ADE}_7$ invariants would be too much to handle without the stepping stone of the more tractable automorphism invariants. Computer checks indicate that the following three constraints reduce almost every choice of algebra and level to (8.10.36).

First, there is the Galois condition arising from a Galois symmetry of the S-matrices. The matrix entries of S lie in a cyclotomic extension of $\mathbb{Q}$. For any element σ of the corresponding Galois group there is a permutation of the weights of $\mathcal{A}^{(n)}$ and a sign map $\varepsilon : \mathcal{A}^{(n)} \to \{\pm 1\}$ such that $\sigma(S_{\lambda\mu}) = \varepsilon_\sigma(\lambda) S_{\sigma\lambda,\mu}$, for all

$\lambda, \mu \in \mathcal{A}^{(n)}$. In particular this forces $Z_{\lambda\mu} = \varepsilon_\sigma(\lambda)\varepsilon_\sigma(\mu)Z_{\sigma\lambda,\sigma\mu}$ so that $Z_{\lambda\mu} \neq 0$ implies that

$$\varepsilon_\sigma(\lambda) = \varepsilon_\sigma(\mu) \tag{8.10.39}$$

for all such σ. Such constraints can be quite severe, e.g. for $su(2)$: when k is odd $Z_{ab} \neq 0$ forces $b = a$ or $b = Ja$ (which equals $k - a$). For even k, there are some extra possible pairs. Similarly for $su(3)$, if $Z_{\lambda\mu} \neq 0$ then $\lambda = C^\alpha J^\beta \mu$, where C is conjugation and J is a simple current when $k + 3$ is coprime to 6. For other k there are a few other possible pairings (λ, μ) such that (8.10.39) holds.

Next, there is T-invariance:

$$(\lambda + \rho)^2 \equiv \rho^2 \mod 2n, \tag{8.10.40}$$

where ρ, the Weyl vector, is half the sum of the positive roots or the sum of the fundamental weights Λ_i. Finally,

$$\sum_\lambda Z_{0,\lambda} S_{\lambda,\mu} \geq 0 \text{ for all } \mu, \tag{8.10.41}$$

from the identity $ZS = SZ$ evaluated at $(0, \mu)$, using positivity of $S_{0,\lambda}$ and $Z_{\lambda,\mu}$. Condition (8.10.41) is quite strong: e.g. the left-hand side will not even be real usually, but is not as easy to use as the Galois condition. There are other constraints, but these are the main ones. The weight 0 is usually much better behaved than the others and (8.10.41) only holds for it. The Galois condition was first noticed independently in (Gannon 1993) and (Ruelle *et al.* 1993). The positivity condition (8.10.41) was first used in (Gannon 1996). The reference for the Galois symmetry is (Coste and Gannon 1994). For the special case of affine algebras, the Galois symmetry can be directly seen by looking at the Kač-Peterson formula for the S-matrix, and there is a geometric interpretation for the permutation $\sigma(\lambda)$ and the parity $\varepsilon_\sigma(\lambda)$.

Note that a complete classification of $su(n_1)_k \oplus su(n_1)_l \oplus su(n_1)_{k+l}$ could imply by (Gannon and Walton 1995) a complete classification of the coset theories $su(n_1)_k \oplus su(n_1)_l / su(n_1)_{k+l}$. The $su(2)$ coset with $l = 1$, and k arbitrary was done by (Cappelli *et al.* 1987b). The $su(2)$ coset for any k, l not both even with $\gcd(k+2, l+2)$, $\gcd(k+2, l)$ and $\gcd(k, l+2)$ all ≤ 3, and the $su(3)$ coset for $l = 1$, and k arbitrary, was treated in (Gannon and Walton 1995).

Given a modular invariant partition function, there is then the problem of associating a conformal field theory or a critical statistical mechanical model within the same universality class, i.e. at least with the correct critical exponents. We will discuss this briefly in Section 11.7, when we seek to assign Boltzmann weights (at least at criticality, with no spectral parameter) for lattice models built up as in Section 7.1 from the graphs associated to modular invariants. This also brings us to the question of associating subfactors with modular invariants and vice versa. Diagonal invariants correspond to Wenzl or WZW subfactors or loop group subfactors $\pi\left(L_{I_+}(G)\right)'' \subset \pi\left(L_{I_-}(G)\right)'$, and similarly for their orbifolds. Conformal embedding $g \subset h$ invariants correspond to inclusions $\pi\left(L_I(G)\right)'' \subset$

$\pi(L_I(H))''$ (in the vacuum representation), see (Xu, in press a) for the E_6, E_8 $su(2)$ modular invariants and $E^{(8)}$, $E^{(12)}$, $E^{(24)}$ $su(3)$ invariants for example with principal graphs, E_6, E_8, $\mathcal{E}^{(6)}$, $\mathcal{E}_1^{(12)}$, $\mathcal{E}^{(24)}$ respectively.

The quantum group $su(N)_q$ and its commutant in the tensor product of the fundamental representation in $\otimes M_N$, the representations of the Hecke algebra where the anti-symmetrizer vanishes, has analogues for $Sp(2N)_q$ with the Birman–Murakami–Wenzl algebra (Birman and Wenzl 1989), (Murakami 1987), replacing the Hecke algebra, and an additional relation replacing the vanishing of the anti-symmetrizer.

9

SUBFACTORS AND BIMODULES

9.1 Introduction

V. F. R. Jones initiated subfactor theory in (Jones 1983) and it has since revolutionized the theory of von Neumann algebras. Furthermore, the discovery of the Jones polynomial, a link invariant, in (Jones 1985) has revealed a truly surprising relation between the operator algebra theory and low-dimensional topology and strengthened links with mathematical physics. We will deal with some of the progress which has been made in this theory in the last decade. In this chapter, we start with the subfactor theory from today's viewpoint based on bimodules. A bimodule in our setting is a Hilbert space with left and right actions of two von Neumann algebras. The importance of bimodule theory in von Neumann algebras was first emphasized by A. Connes in his unpublished manuscript on correspondences, and a detailed study of the bimodule theory was given by S. Popa in his unpublished manuscript 'Correspondences'. It was A. Ocneanu who first realized the importance of bimodules in the Jones theory of subfactors in (Ocneanu 1988), and he obtained several fundamental theorems in the theory, but unfortunately his results have been scattered among his several informal notes and unpublished manuscripts. We will present the basic theory in this chapter as a preliminary to the remainder of this book.

A basic observation of Connes was that a bimodule over a von Neumann algebra gives a correct analogue of a unitary representation of a compact group. In the subfactor theory, the analogy of Table 9.1 plays an important role. We will explain this analogy in detail in this chapter. In Section 9.2, we cover the basic theory of modules over II_1 factors. The two basic notions are dimension (the coupling constant in Section 5.5) and basis. We explain the basic part of the Jones theory of subfactors in Section 9.3, where the fundamental definitions are introduced. In Section 9.4, we study basic constructions in finite dimensional cases and in Section 9.5, we study the Temperley–Lieb algebras. In Section 9.6, the fundamental tools to study subfactors are introduced. In Section 9.7, we define the relative tensor product of bimodules. This is one of the most fundamental operations in the entire theory – analogous to the operator product of

Table 9.1 *Analogy between representations and bimodules*

representation	bimodule
direct sum	direct sum
tensor product	relative tensor product
dimension	(Jones index)$^{1/2}$
Frobenius reciprocity	Frobenius reciprocity
fundamental representation	${}_M M_N$

conformal field theory, which we discussed in Chapter 8. Section 9.8 deals with Frobenius reciprocity for bimodules. This is an analogue of Frobenius reciprocity in the representation theory and it plays a very important role in the subfactor theory.

In this chapter and later on, elements in von Neumann algebras are denoted by lower case roman letters such as $a, b, x, y, \ldots$, elements in Hilbert spaces are denoted by lower case Greek letters such as $\xi, \eta, \zeta, \ldots$, bimodules over von Neumann algebras are denoted by capital roman letters such as $A, B, C, D, X, Y, \ldots$, and intertwiners between bimodules are denoted by small Greek letters such as $\sigma, \rho, \tau, \ldots$.

9.2 Dimension of modules and basis

Before we work on bimodules over von Neumann algebras, we need to work on modules over von Neumann algebras. For simplicity, our von Neumann algebras in this chapter (with the exception of 9.4) will always be II_1 factors. First we introduce the notion of a module over a II_1 factor and its dimension. We will then introduce the notion of a basis of a module over a II_1 factor and prove an existence theorem for bases.

We start with some definitions.

Definition 9.1 *Let M be a II_1 factor. The* opposite algebra *M^{opp} of M is defined as follows. The algebra M^{opp} is the same as M as a Banach space and has the same $*$-operation. The product for $x, y \in M^{\mathrm{opp}}$ is defined by yx which means the product in M. That is, we change the order of multiplication. Of course, $(M^{\mathrm{opp}})^{\mathrm{opp}} = M$.*

Definition 9.2 *Let M be a II_1 factor and H a Hilbert space. We say H is a left [resp. right]* M-module *if we have a representation π of M [resp. M^{opp}] on H. (That is, π is a σ-weakly continuous unital $*$-homomorphism of M [resp. M^{opp}] into $B(H)$.) We simply write $x\xi$ [resp. ξx] if H is a left [resp. right] M-module, $x \in M$, and $\xi \in H$.*

Note that for $x, y \in$M, we have $(xy)\xi = x(y\xi)$ if ξ is in a left M-module and $\xi(xy) = (\xi x)y$ if ξ is in a right M-module. Because M is a factor, Ker π is 0 as a σ-weakly closed proper two-sided ideal, hence the action of M [resp. M^{opp}] is always faithful. (See the remark after Definition 5.15. Because we did not prove this simplicity of II_1 factors, we may impose this faithfulness as a hypothesis — but it is in fact automatic.) We slightly extend the definition in Section 5.5 next.

Definition 9.3 *Let M be a II_1 factor and H a left [resp. right] M-module. The dimension $\dim_M H$ [resp. $\dim H_M$] is defined to be the coupling constant (in Section 5.5) of M [resp. M^{opp}] in H.*

Recall that the dimension is a real number in $[0, \infty]$. We are interested in positive finite dimensions.

For the discussion on basic properties of modules, we deal only with left modules for simplicity, but right modules can be handled in the same way. Let M be a II_1 factor with trace tr and H be a left M-module in the following.

Definition 9.4 *We say that a vector $\xi \in H$ is M-*bounded *if there exists a positive constant C_ξ such that*

$$\|x\xi\| \leq C_\xi \|x\|_2, \quad \text{for all } x \in M,$$

where $\|x\|_2 = \mathrm{tr}(x^*x)^{1/2}$. We write H^{bdd} for the set of all the M-bounded vectors in H.

Lemma 9.5 *The space H^{bdd} is a dense linear subspace of H and invariant under the actions of M and M'.*

Proof It is obvious that H^{bdd} is a linear subspace and invariance under M' is easy. Because tr is a normal state on M, we have a family $\{\xi_j\}_{j\in J}$ in H such that $\mathrm{tr}(x) = \sum_j \langle x\xi_j, \xi_j\rangle$. (See the comment after Theorem 5.9.) Each ξ_j is in H^{bdd} because $\langle x\xi_j, x\xi_j\rangle \leq \mathrm{tr}(x^*x)$. Let e be the projection onto the closure of H^{bdd}. Because $M'H^{\mathrm{bdd}} \subset H^{\mathrm{bdd}}$, we get $e \in M'' = M$. The identity $(1-e)\xi_j = 0$ for all ξ_j implies that $\mathrm{tr}(1-e) = 0$ and that $e = 1$.

For $\xi \in H^{\mathrm{bdd}}$, $x, y \in$M, we have

$$\begin{aligned} 0 \leq \|yx\xi\|^2 &\leq C_\xi^2 \mathrm{tr}(x^*y^*yx) = C_\xi^2 \mathrm{tr}(yxx^*y^*) \\ &\leq C_\xi^2 \|x\|^2 \mathrm{tr}(yy^*) \leq C_\xi^2 \|x\|^2 \|y\|_2^2. \end{aligned}$$

Thus we have $MH^{\mathrm{bdd}} \subset H^{\mathrm{bdd}}$. □

Example 9.6 Let $H = L^2(M)$ with the left multiplication by M. Of course, $\dim_M H = 1$. We get $H^{\mathrm{bdd}} = M$, where M is regarded as a subspace of $L^2(M)$, as follows. It is clear that $M \subset H^{\mathrm{bdd}}$. Conversely, suppose we have $\xi \in H^{\mathrm{bdd}}$. Then the map $x \in M \mapsto x\xi \in L^2(M)$ is extended to a bounded linear map T on $L^2(M)$ by the M-boundedness of ξ. Because T is in $M' = JMJ$, we have an operator $y \in M$ such that T is the right multiplication by y by Proposition 5.30. Then it is clear that the vector ξ is equal to $\hat{y}$, which is $y \in M$ regarded as an element of $L^2(M)$.

Next we introduce an M-valued inner product on H^{bdd} as follows.

Definition 9.7 *For $\xi, \eta \in H^{\mathrm{bdd}}$, the element $\langle \xi, \eta \rangle_M \in M$ is defined by the formula* $\mathrm{tr}(x\langle \xi, \eta \rangle_M) = \langle x\xi, \eta \rangle$ *for all $x \in M$.*

Note that existence of $\langle \xi, \xi \rangle_M$ follows from Theorem 5.20 because

$$|\langle x^*x\xi, \xi \rangle| \le \|x\xi\| \cdot \|x\xi\| \le C_\xi^2 \|x\|_2^2$$

holds for all $x \in M$. The existence of $\langle \xi, \eta \rangle_M$ follows by polarization. The uniqueness of $\langle \xi, \eta \rangle_M$ is clear.

If H is a *right* M-module, we similarly define an inner product $\langle \xi, \eta \rangle_M^\circ$ by $\mathrm{tr}(\langle \xi, \eta \rangle_M^\circ a) = \langle \eta a, \xi \rangle$ for $a \in M$. It is also easy to verify the following basic properties.

Proposition 9.8 *We have the following identities for $x \in M$, $\xi, \xi', \eta \in H^{\mathrm{bdd}}$, $\lambda, \mu \in \mathbb{C}$.*

1. $\langle \lambda\xi + \mu\xi', \eta \rangle_M = \lambda\langle \xi, \eta \rangle_M + \mu\langle \xi', \eta \rangle_M$.
2. $\langle \xi, \eta \rangle_M = \langle \eta, \xi \rangle_M^*$.
3. $\langle \xi, \xi \rangle_M \ge 0$.
4. $\langle x\xi, \eta \rangle_M = x\langle \xi, \eta \rangle_M$.
5. $\langle \xi, x\eta \rangle_M = \langle \xi, \eta \rangle_M x^*$.

Example 9.9 Let p be a non-zero projection in M. Then $H = L^2(M)p$ is a left M-module. Then as in Example 9.6, we get $H^{\mathrm{bdd}} = Mp$. For $xp, yp \in Mp$, we easily get $\langle \hat{x}p, \hat{y}p \rangle_M = xpy^*$.

Example 9.10 Let N be a II_1 factor contained in M. Then $L^2(M)$ becomes a left N-module by the ordinary left multiplication of N. If $x, y \in M$, then $\hat{x}, \hat{y} \in {}_N L^2(M)^{\mathrm{bdd}}$ and we get $\langle \hat{x}, \hat{y} \rangle_N = E_N(xy^*)$, where E_N is the conditional expectation of M onto N with respect to the trace. If we regard $L^2(M)$ as a right N-module, then the inner product is given by $\langle \hat{x}, \hat{y} \rangle_N^\circ = E_N(x^*y)$. Note that $\langle \cdot\ ,\ \cdot \rangle_N^\circ$ is conjugate linear in the first variable and linear in the second variable. (It will follow from Theorem 9.48 that we have the equality ${}_N L^2(M)^{\mathrm{bdd}} = {}_M L^2(M)^{\mathrm{bdd}} = M$, if $\dim_N L^2(M) < \infty$. This is not true if $\dim_N L^2(M) = \infty$.)

Definition 9.11 *A finite set $\{\lambda_j\}_{j=1,2,\ldots,n}$ in H^{bdd} is called a* basis *of H if it satisfies $\xi = \sum_{j=1}^n \langle \xi, \lambda_j \rangle_M \lambda_j$ for all $\xi \in H^{\mathrm{bdd}}$. A basis for a right M-module is also defined similarly.*

Theorem 9.12 *Suppose that* $\dim_M H < \infty$. *Then H has a basis.*

Proof By Definition 5.31, we know that H is isomorphic to $L^2(M) \oplus \cdots \oplus L^2(M) \oplus L^2(M)p$ as a left M-module, where p is a non-zero projection of M, $\dim_M H = n + \mathrm{tr}(p)$, and we have n copies of $L^2(M)$. So we identify H with $L^2(M) \oplus \cdots \oplus L^2(M) \oplus L^2(M)p$.

To get a basis $\{\lambda_1, \ldots, \lambda_n\}$, we set $\lambda_1 = (\hat{1}, 0, \ldots, 0, 0)$, $\ldots$, $\lambda_n = (0, 0, \ldots, \hat{1}, 0)$, $\lambda_{n+1} = (0, 0, \ldots, 0, \hat{p})$. □

We then have the following corollary.

0 1 2 3 4 5 6 7

FIG. 9.1. Possible index values

Corollary 9.13 *For a basis* $\{\lambda_j\}_j$ *for* ${}_M H$, *we have* $\dim {}_M H = \sum_j \langle \lambda_j, \lambda_j \rangle$.

Example 9.14 For a left M-module $L^2(M)$, the singleton $\{\hat{1}\}$ is clearly a basis, but so is the set $\{\hat{p}, \widehat{1-p}\}$ for a projection $p \in M$ with $p \neq 0, 1$. Thus the number of the elements of a basis is not unique.

9.3 Jones index, basic construction and Pimsner–Popa basis

We introduce basic notions and properties of the Jones theory of subfactors in this section.

In the rest of this book, M will be a factor and N a von Neumann subalgebra of M which is also a factor. Such an N is called a *subfactor*. The factors M and N will be of type II_1 in the rest of the book unless explicitly stated otherwise. We define the Jones index of a subfactor as follows.

Definition 9.15 *The* Jones index *of N in M, denoted by $[M : N]$, is defined to be* $\dim {}_N L^2(M)$, *where N acts on $L^2(M)$ by the left multiplication.*

The name "index" and its notation come from an analogy with the index of a subgroup in a group as illustrated in Example 9.20 below. In general we have $[M : N] \in [1, \infty]$, but we are interested in the finite index case, so we assume $[M : N] < \infty$ in the rest of this book unless explicitly stated otherwise.

From a viewpoint of K-theory (in Chapter 3), the condition $[M : N] < \infty$ means that M is a finitely generated projective module over N. Then the inclusion $N \subset M$ gives an element of $K_0(N) = \{r \in \mathbb{R} \mid r \geq 0\}$, which is the Jones index. See Exercise 3.9 and Definition 5.31.

At first sight, it is not clear at all which values are realized by an index of a subfactor. The following surprising result is due to Jones and was the starting point of the entire theory. It is surprising because the index values are restricted while $\dim(e)^{-1}$ takes continuous values.

Theorem 9.16 *For a subfactor $N \subset M$, the index value $[M : N]$ belongs to the set*

$$\{4\cos^2 \frac{\pi}{n} \mid n = 3, 4, 5, \ldots\} \cup [4, \infty]$$

and all the values in this set are realized.

Note that the above set looks like Fig. 9.1. A proof of this theorem will be given in three parts. Here we give a proof of a realization of the values bigger than or equal to four. The restriction on the values below four will be proved in Section 10.3 (Corollary 10.7) (analogous to the restriction of the values of the central charge $c < 1$ (Friedan *et al.* 1984)), and the realization of the values below four will be proved in Section 11.5 (Corollary 11.23) (analogous to the realization of discrete values of the central charge $c < 1$ (Goddard *et al.* 1986)).

Proof [Realizations of the values above four]. Let R be the hyperfinite II_1 factor and $\lambda \in [4, \infty)$. We choose a number $t \in (0,1)$ and a projection p in R so that $1/t + 1/(1-t) = \lambda$ and $\mathrm{tr}(p) = t$. Both pRp and $(1-p)R(1-p)$ are hyperfinite II_1 factors, so by the uniqueness of the hyperfinite II_1 factor (Theorem 5.18), we have an isomorphism θ from pRp to $(1-p)R(1-p)$. (See Exercise 9.1.) Let $M = R$ and $N = \{x + \theta(x) \mid x \in pRp\}$. This N is clearly a subfactor of M, and we compute the index. Because $p \in N' \cap M$, we get

$$\begin{aligned}\dim{}_N L^2(M) &= \dim{}_N pL^2(M) + \dim{}_N(1-p)L^2(M)\\ &= \dim{}_{pNp} pL^2(M) + \dim{}_{(1-p)N(1-p)}(1-p)L^2(M)\\ &= \dim{}_{pRp} pL^2(R) + \dim{}_{(1-p)R(1-p)}(1-p)L^2(R)\\ &= t^{-1} + (1-t)^{-1} = \lambda.\end{aligned}$$

□

We make another definition and give some related basic properties.

Definition 9.17 *For a subfactor N of M, we define the* relative commutant *of N in M to be $N' \cap M = \{x \in M \mid yx = xy$, for all $y \in N\}$.*

Proposition 9.18 1. *For any left M-module H with $\dim{}_M H < \infty$, we get $[M : N] = \dim{}_N H / \dim{}_M H$.*

2. *For factors $N \subset P \subset M$ with $[M : N] < \infty$, we get*

$$[M : N] = [M : P][P : N].$$

3. *The relative commutant $N' \cap M$ is finite dimensional.*

Proof 1. We may assume that H is of the form $L^2(M) \oplus \cdots \oplus L^2(M) \oplus L^2(M)p$, where we have n copies of $L^2(M)$. We have $\dim{}_N L^2(M)p = \mathrm{tr}_{N'}(p) \dim{}_N L^2(M) = \mathrm{tr}(p) \dim{}_N L^2(M)$ by Theorem 5.33 (3). Thus

$$\begin{aligned}\dim{}_N H &= \dim{}_N L^2(M) + \cdots + \dim{}_N L^2(M) + \dim{}_N L^2(M)p\\ &= n[M : N] + \mathrm{tr}(p)[M : N] = [M : N] \dim{}_M H.\end{aligned}$$

2. By 1, we get

$$\begin{aligned}\dim{}_P L^2(M) \dim{}_N L^2(P) &= \dim{}_P L^2(M) \dim{}_N L^2(M) / \dim{}_P L^2(M)\\ &= \dim{}_N L^2(M).\end{aligned}$$

3. Let p_j, $j = 1, \ldots, k$, be mutually orthogonal non-zero projections in $N' \cap M$ with $\sum_{j=1}^k p_j = 1$. Then we have

$$[M : N] = \dim{}_N L^2(M) = \sum_{j=1}^k \dim{}_N p_j L^2(M) = \sum_{j=1}^k \dim{}_{p_j N p_j} p_j L^2(M)$$

$$\geq \sum_{j=1}^{k} \dim_{p_j M p_j} p_j L^2(M) = \sum_{j=1}^{k} 1/\mathrm{tr}_M(p_j),$$

where we used Theorem 5.33 (4). Because $\sum_{j=1}^{k} \mathrm{tr}(p_j) = 1$, we get the inequality $[M : N] \geq k^2$, which shows that $N' \cap M$ is finite dimensional. (See Exercise 9.2.) □

Note that the above proof of (3) also shows the following corollary.

Corollary 9.19 *If a subfactor N satisfies $[M : N] < 4$, then it has a trivial relative commutant, $N' \cap M = \mathbb{C}$.*

Here we present the most fundamental example.

Example 9.20 Let G be a finite group of order n and H a subgroup of order m. Suppose G acts on a II_1 factor R outerly. (If R is the hyperfinite II_1 factor, then an outer action of any finite group on R exists. See Section 6.7.) Set $N = R$ and $M = R \rtimes G$. By Exercise 5.2, M is a II_1 factor. Then as a left N-module, $L^2(M) = L^2(N)u_{g_1} \oplus L^2(N)u_{g_2} \oplus \cdots \oplus L^2(N)u_{g_n}$, where $G = \{g_1, g_2, \dots, g_n\}$ and u_g denotes the implementing unitary for $g \in G$. Thus we get $[M : N] = n$.

Furthermore, set $P = R \rtimes H$. By $[M : N] = n$ and $[P : N] = m$ and Proposition 9.18 (2), we get $[M : P] = n/m$, that is, $[R \rtimes G : R \rtimes H] = [G : H]$. This shows the name index is justified in analogy with the index of a subgroup. Our basic idea is that a general subfactor arises from a certain generalization of a group – a *paragroup*.

We next introduce the basic construction.

Definition 9.21 *Let M, N act on $L^2(M)$ by the left multiplication. The conditional expectation $E_N : M \to N$ naturally extends to a projection of $L^2(M)$ onto $L^2(N)$. We denote this projection by e_N and call it the* Jones projection. *We define M_1 to be the von Neumann algebra generated by M and e_N on $L^2(M)$. This is also written as $\langle M, e_N \rangle$. (The symbol $\langle \cdot \rangle$ means the von Neumann algebras generated by the operators in the bracket.) We call this construction of M_1 the* basic construction. *The sequence of three factors $N \subset M \subset M_1$ is called* standard. *If a sequence $N \subset M \subset P$ has an isomorphism Φ from P to M_1 which is identity on M, then this is also called standard.*

We list the basic properties of M_1. Recall that the index $[M : N]$ has been assumed to be finite.

Proposition 9.22 1. *For $x \in M$, we have $e_N x e_N = E_N(x) e_N$.*

2. *$M \cap \{e_N\}' = N$.*
3. *$J e_N = e_N J$.*
4. *$M_1 = J N' J$, and hence M_1 is also a II_1 factor.*
5. *$[M_1 : M] = [M : N]$.*

6. *The $*$-algebra*

$$\{a + \sum_{j=1}^{n} a_j e_N b_j \mid a, a_j, b_j \in M, n = 1, 2, \ldots\}$$

is strongly dense in M_1.

Proof 1. Let both sides act on a vector $\hat{y}$ for $y \in M$, to obtain $E_N(x)\widehat{E_N(y)}$.

2. For $x \in N$, $y \in M$, we have $e_N x \hat{y} = \widehat{E_N(xy)} = x\widehat{E_N(y)} = x e_N \hat{y}$, thus $N \subset \{e_N\}'$.

Suppose $x \in M \cap \{e_N\}'$. Then applying $x e_N = e_N x$ to the vector $\hat{1}$, we get $\hat{x} = \widehat{E_N(x)}$, which means $x = E_N(x) \in N$.

3. For $x \in M$, we get

$$J e_N \hat{x} = J\widehat{E_N(x)} = \widehat{E_N(x)^*} = \widehat{E_N(x^*)} = e_N \widehat{x^*} = e_N J \hat{x}.$$

4. Take the commutant of the identity in 2 to get $\langle M', e_N \rangle = N'$. By applying $J \cdot J$, we get $\langle M, e_N \rangle = J N' J$.

5. We have the following:

$$[M_1 : M] = [J M_1 J : J M J] = [N' : M'] = \frac{\dim_{M'} L^2(M)}{\dim_{N'} L^2(M)} = [M : N]$$

by (4) and Theorem 5.33 (4).

6. The set is really a $*$-subalgebra of M_1 by 1. Then the density follows from the definition of M_1. □

By (4), we know that M_1 has the unique trace extending tr on M. This extension is also denoted by tr. A trace satisfying the relation in the following Proposition is called a *Markov* trace. The name Markov originally came from Markov's theorem on knots and braids (see Section 12.1), but it can be also regarded as describing a (non-commutative) Markov chain.

Proposition 9.23 *For $x \in M$, we get $\mathrm{tr}(x e_N) = [M : N]^{-1}\mathrm{tr}(x)$. Thus $E_M(e_N) = [M : N]^{-1}$.*

Proof First, note that we have

$$1 = \dim_N e_N(L^2(M)) = \mathrm{tr}_{N'}(e_N)[M : N]$$

by Theorem 5.33 (3). Thus we get $\mathrm{tr}(e_N) = \mathrm{tr}_{JN'J}(e_N) = [M : N]^{-1}$.

For $x \in N$, the map $x \mapsto \mathrm{tr}(x e_N)$ is a non-normalized trace. By the uniqueness of the trace on N, we get $\mathrm{tr}(x e_N) = [M : N]^{-1}\mathrm{tr}(x)$ for $x \in N$.

For $x \in M$, we have

$$\mathrm{tr}(x e_N) = \mathrm{tr}(e_N x e_N) = \mathrm{tr}(E_N(x) e_N)$$

$$= [M : N]^{-1}\mathrm{tr}(E_N(x)) = [M : N]^{-1}\mathrm{tr}(x),$$

where we used Proposition 9.22 (1). The identity $E_M(e_N) = [M : N]^{-1}$ then follows from the definition of the conditional expectation. □

Definition 9.24 *We set $e_1 = e_N$. After making a basic construction to get $M_1 = \langle M, e_1\rangle$, we make another basic construction for $M \subset M_1$ to get $M_2 = \langle M_1, e_2\rangle$, and continue to get $M_n = \langle M_{n-1}, e_n\rangle$. The sequence*

$$N \subset M \subset M_1 \subset M_2 \subset M_3 \subset \cdots$$

is called the Jones tower.

We have the following theorem due to Jones. Relations (1) and (2) are called the *Jones relations* or the *Temperley–Lieb–Jones relations*, and the trace in Relation (3) produced the Jones polynomial, a polynomial invariant of links. (See Section 12.1.)

Theorem 9.25 *For the Jones projections $e_1, e_2, \ldots$, we have the following:*

1. $e_j e_k = e_k e_j$, *if* $|j - k| \geq 2$.
2. $e_j e_{j\pm 1} e_j = [M : N]^{-1} e_j$.
3. *For $x \in M_j$, we get $\mathrm{tr}(x e_{j+1}) = [M : N]^{-1}\mathrm{tr}(x)$, where* tr *is the unique extension of the original trace to M_{j+1}.*

Proof 1. This is trivial by Proposition 9.22 (2).

2. We have $E_M(e_N) = [M : N]^{-1}$ by Proposition 9.23. Thus

$$e_j e_{j-1} e_j = E_{M_{j-2}}(e_{j-1}) e_j = [M : N]^{-1} e_j$$

by Proposition 9.22 (1). This implies that

$$([M : N]^{1/2} e_j e_{j-1})([M : N]^{1/2} e_j e_{j-1})^* = e_j,$$

thus

$$([M : N]^{1/2} e_j e_{j-1})^*([M : N]^{1/2} e_j e_{j-1}) = [M : N] e_{j-1} e_j e_{j-1}$$

is also a projection. Applying the trace to the inequality

$$[M : N] e_{j-1} e_j e_{j-1} \leq e_{j-1},$$

we get $\mathrm{tr}(e_j) \leq \mathrm{tr}(e_{j-1})$ because $\mathrm{tr}(e_{j-1} e_j e_{j-1}) = \mathrm{tr}(e_j e_{j-1} e_j)$, but here we have an equality. By the faithfulness of the trace, we get $[M : N] e_{j-1} e_j e_{j-1} = e_{j-1}$.

3. By Proposition 9.22 (5), we get $[M_j : M_{j-1}] = [M : N]$. Then the identity is clear by Proposition 9.23. □

The right N-module $L^2(M)$ has a basis, but now we would like to choose a basis within M, not in its completion. Such a basis will be called a Pimsner–Popa basis. We need some preliminaries.

Lemma 9.26 *For any $x \in M_1$, there exists $y \in M$ satisfying $xe_N = ye_N$.*

Proof If such a y exists, then $E_M(xe_N) = y[M : N]^{-1}$ implies that y has to be equal to $[M : N]E_M(xe_N)$. So we prove that $xe_N = [M : N]E_M(xe_N)e_N$ for all $x \in M_1$. Direct computation shows that this is true for x of the form $a + \sum_{j=1}^{n} a_j e_N b_j$, $a, a_j, b_j \in M$. Then density in Proposition 9.22 (6) and strong continuity of E_M proves the desired identity. □

Theorem 9.27 *Let $[M : N] = n + \alpha$ with n an integer and $\alpha \in [0, 1)$. There exist $m_1, m_2, \ldots, m_{n+1} \in M$ with the following properties:*

1. $E_N(m_j^* m_k) = 0$ *if* $j \neq k$.
2. $E_N(m_j^* m_j) = 1$ *if* $j \leq n$ *and* $E_N(m_{n+1}^* m_{n+1}) = p$, *where p is a projection in N with* $\mathrm{tr}(p) = \alpha$.
3. $\sum_{j=1}^{n+1} m_j e_N m_j^* = 1$.
4. $\sum_{j=1}^{n+1} m_j m_j^* = [M : N]$.
5. *For any $x \in M$, we have* $x = \sum_{j=1}^{n+1} m_j E_N(m_j^* x)$.

A system $\{m_1, m_2, \ldots, m_{n+1}\}$ given by this theorem is called a *Pimsner–Popa basis*.

Proof Choose mutually orthogonal projections $p_1, p_2, \ldots, p_{n+1} \in M_1$ so that $\mathrm{tr}(p_j) = 1/[M : N]$ for $j = 1, 2, \ldots, n$ and $\mathrm{tr}(p_{n+1}) = 1 - n/[M : N]$. We have partial isometries $v_j \in M_1$ so that $v_j v_j^* = p_j$, $v_j^* v_j = e_N$ $(j = 1, 2, \ldots, n)$, and $v_{n+1}^* v_{n+1} \leq e_N$. By Lemma 9.26, there exist $m_j \in M$, $j = 1, 2, \ldots, n+1$, satisfying $v_j = v_j e_N = m_j e_N$. Then $v_j^* v_k = 0$ for $j \neq k$ implies $0 = e_N m_j^* m_k e_N = E_N(m_j^* m_k) e_N$, hence $E_N(m_j^* m_k) = 0$, which is the identity in (1). For $j = 1, 2, \ldots, n$, we get $E_N(m_j^* m_j) e_N = e_N m_j^* m_j e_N = v_j^* v_j = e_N$. This and a similar formula for $j = n + 1$ give the identities in (2).

For (3), we have

$$\sum_{j=1}^{n+1} m_j e_N m_j^* = \sum_{j=1}^{n+1} v_j^* v_j = \sum_{j=1}^{n+1} p_j = 1.$$

Applying E_M to this formula, we get (4).

For $x \in M$, we have

$$xe_N = 1 \cdot xe_N = \sum_{j=1}^{n+1} m_j e_N m_j^* x e_N = \sum_{j=1}^{n+1} m_j E_N(m_j^* x) e_N.$$

Applying E_M to this formula, we get $x = \sum_{j=1}^{n+1} m_j E_N(m_j^* x)$, as desired for (5). □

The following is a stronger form of Proposition 9.22 (6).

Proposition 9.28 *The $*$-algebra*

$$\{\sum_{j=1}^{n} a_j e_N b_j \mid a_j, b_j \in M, n = 1, 2, \ldots\}$$

is equal to M_1.

Proof Choose a Pimsner–Popa basis $\{m_1, m_2, \ldots, m_n, m_{n+1}\}$. Let x be an element in M_1. By Theorem 9.27 (3), we get $x = x1 = \sum_{j=1}^{n+1} x m_j e_N m_j^*$. By Lemma 9.26, there exists $y_j \in M$ satisfying $y_j e_N = x m_j e_N$. Then we get $x = \sum_{j=1}^{n+1} y_j e_N m_j^*$. □

Exercise 9.1 Let R be a hyperfinite II_1 factor and p a non-zero projection in R. Prove that pRp is also hyperfinite.

Exercise 9.2 Let R be a von Neumann algebra and suppose that the maximum number of mutually orthogonal non-zero projections in R is k. Prove that R is finite dimensional.

9.4 Finite dimensional cases

In this section, we study the basic extension of finite dimensional operator algebras to illustrate the ideas we have so far developed for II_1 factors.

A trace on a finite dimensional matrix algebra $A = \mathrm{End}(\ell^2(X))$ is uniquely determined up to a scalar multiple of the canonical un-normalized trace:

$$(x_{\mu\nu}) = x \to \sum_{\mu} x_{\mu\mu} = \sum_{\mu\nu} f_{\mu\nu} x f_{\nu\mu}. \tag{9.4.1}$$

Thus a trace on a multi-matrix algebra $A = \bigoplus_i \mathrm{End}(\ell^2(X_i))$ is determined by a sequence (s_i), indexed by the minimal central projections of A, called a trace vector:

$$\mathrm{tr}(\textstyle\bigoplus_i x_i) = \sum_i s_i \mathrm{tr}_i(x_i) \tag{9.4.2}$$

if tr_i is the canonical trace (9.4.1) on $A_i = \mathrm{End}(\ell^2(X_i))$. Indeed $s_i = \mathrm{tr}\, f_i$, if f_i is a minimal projection in A_i. The trace is positive if $s_i \geq 0$ for all i and faithful if $s_i > 0$ for all i.

Consider an embedding of a finite dimensional C^*-algebra A in another B with embedding matrix $\Lambda = (\lambda_{ji})$ where the rows j (respectively columns i) are indexed by the minimal central projections of B (respectively A). Suppose we have traces tr_1 and tr_2 on A, B respectively represented by trace vectors s, t respectively. The traces are compatible under the embedding described by Λ if and only if

$$t\Lambda = s \tag{9.4.3}$$

Suppose that the traces tr_1 and tr_2 are both positive and faithful on A and B respectively. In particular the embedding of A in B is injective, and we denote

FIG. 9.2. A Bratteli diagram

either trace by tr for convenience. Regarding A as a subspace of B for the inner product defined by the trace ($\langle x, y\rangle = \text{tr } y^*x, \quad x, y \in B$) let $E_A = E$ denote the orthogonal projection of B onto A, the *conditional expectation* of A onto B relative to the trace, i.e. for $b \in B$, $E_A(b)$ is the unique element $b' \in A$ such that tr $b'a =$ tr ba for all $a \in A$.

We can regard E as an operator e in End(B), regarding B as a Hilbert space. Then $C = \langle B, e\rangle$ denotes the C^*-subalgebra of End(B) generated by B (acting by left multiplication) and e, and is finite dimensional, being a subalgebra of the finite dimensional matrix algebra End(B). This construction is the Jones basic extension of B by A:

$$A \subset B \subset \langle B, e\rangle \,. \tag{9.4.4}$$

We will determine the structure of the inclusion $B \subset \langle B, e\rangle$. Before we do that it is illuminating to look at an example. Consider the inclusion $\mathbb{C} \to M_n$ represented by the Bratteli diagram as in Fig. 9.2.

We can identify End(M_n) as $M_n \otimes M_n^{\text{opp}}$, where M_n^{opp} is the opposite algebra by

$$x \otimes y(z) = xzy, \quad x, y, z \in M_n \tag{9.4.5}$$

and we can identify M_n^{opp} with M_n via the transpose map. Thus End(M_n) is identified with $M_n \otimes M_n$. The projection $E_{\mathbb{C}}$ of M_n onto $\mathbb{C}$ is the normalized trace itself:

$$E_{\mathbb{C}}(x) = \frac{1}{n} \sum f_{\mu\nu} x f_{\nu\mu} \tag{9.4.6}$$

which under the above identification End(M_n) $\cong M_n \otimes M_n$ yields

$$e = \frac{1}{n} \sum f_{\mu\nu} \otimes f_{\mu\nu} = \frac{1}{n} \sum f_{\mu\mu,\nu\nu} \tag{9.4.7}$$

in the path algebra of $\mathbb{C} \to M_n \to M_n \otimes M_n$ as in Fig. 9.5, where we have identified by reflection the unit of $\mathbb{C}$ with that of $M_n \otimes M_n$, and the edges of Path(p, q) with Path(q, p). In this example, we claim $\langle M_n, e\rangle =$ End(M_n) $\cong M_n \otimes M_n$. To see this first compute $f_{\bar{\alpha},\alpha} e$ as in Fig. 9.4. Then

$$f_{\bar{\alpha},\alpha} e f_{\bar{\beta},\beta} = n f_{\bar{\alpha}\alpha,\bar{\beta}\beta} \tag{9.4.8}$$

and in particular $\langle M_n, e\rangle \cong M_n \otimes M_n$ and in the path picture of Fig. 9.5, we get the identity of Fig. 9.3.

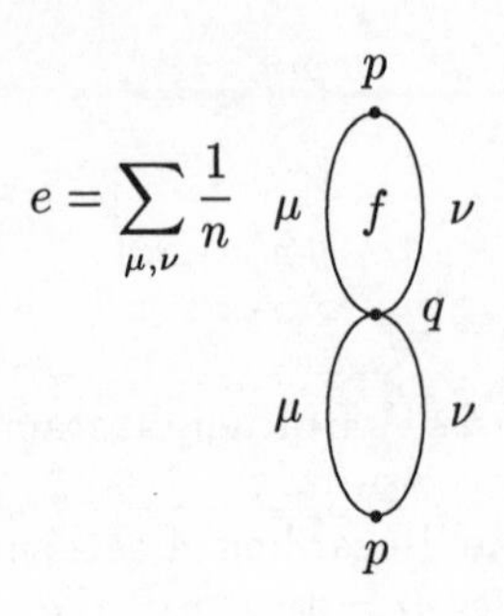

FIG. 9.3. Jones projection

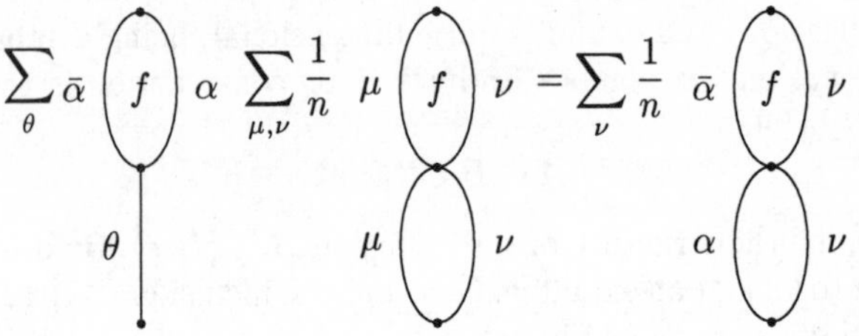

FIG. 9.4. $f_{\bar{\alpha},\alpha} e$

Next consider the embedding $A = M_k$ in $B = M_n \otimes M_k$ obtained by tensoring the previous embedding $\mathbb{C} \to M_n$ with M_k. Then $E_A : M_n \otimes M_k \to 1 \otimes M_k \cong M_k$ can be identified with $E_{\mathbb{C}} \otimes 1_k$. Moreover as before we identify $\mathrm{End}(M_n \otimes M_k)$ with $M_n \otimes M_k \otimes M_n \otimes M_k$ which we can identify with $M_n \otimes M_n \otimes M_k \otimes M_k$. In this way $e_A \in \mathrm{End}(M_n \otimes M_k)$ is identified with $e_{\mathbb{C}} \otimes 1_k \otimes 1_k$, and so $\langle M_n \otimes M_k, e_{M_k} \rangle$ is identified with $\langle M_n \otimes 1 \otimes M_k \otimes 1, e_{\mathbb{C}} \otimes 1 \otimes 1 \otimes 1) \rangle$ which is $M_n \otimes M_n \otimes M_k \otimes 1$. Thus $C = \langle B, e \rangle$ can be identified with $M_n \otimes M_n \otimes M_k$, with $e = e_{\mathbb{C}} \otimes 1$ given

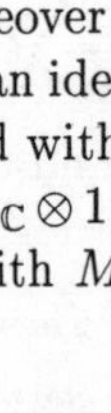

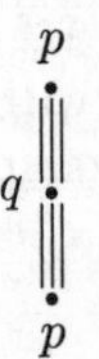

FIG. 9.5. A Bratteli diagram for a basic extension

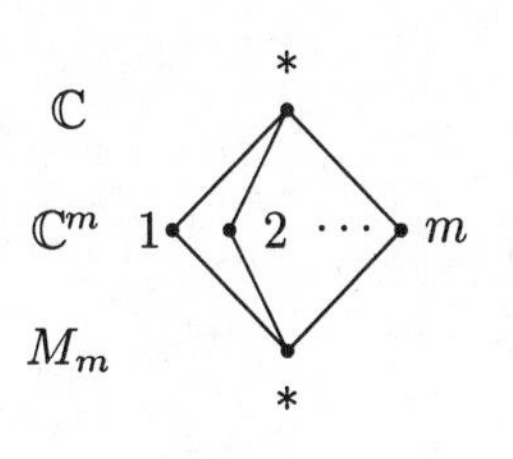

FIG. 9.6. Basic extension $\mathbb{C} \subset \mathbb{C}^m \subset M_m$

as in Fig. 9.3 where p, q, r are now the units of M_k, $M_n \otimes M_k$, $M_n \otimes M_n \otimes M_k$ respectively, with the Bratteli diagram of their embeddings again described by Fig. 9.5.

Note that in $\mathrm{End}(M_n \otimes M_k)$, we have identified the commutant of the right action of $A = M_k$ with C, and the multiplicity of the embedding of A in B is the same as that of B in C. This is a general feature. If $A \subset B \subset \langle B, e\rangle$ is an extension of arbitrary finite dimensional C^*-algebras then we can identify C with $\mathrm{End}_A(B)$, the endomorphisms of B which commute with the right action of A, and the graph of the multiplicity of the embedding of B in C is obtained by reflecting that A in B.

To switch between left and right actions it is convenient to use the involution J in the trace of B, namely the conjugate linear unitary on B, given by the involution $Jb = b^*$, $b \in B$. This defines a unitary as $\mathrm{tr}\ bb^* = \mathrm{tr}\ b^*b$. If π, ρ are the left and right actions of B on itself given by $\pi(b)a = ba$, $\rho(b)a = ab$, $a, b \in B$, then $J\pi(a)J = \rho(a^*)$, $\pi(B)' = \rho(B)$, $\rho(B)' = \pi(B)$ where the commutants are computed in $\mathrm{End}(B)$. Then $ebe = E(b)e$, $b \in B$, $\{e\}' \cap B = A$, $\langle B', e\rangle = A'$, $JeJ = e$. Consequently $J\pi(A)'J = \langle \pi(B), e\rangle$, i.e. $\pi(A)' = J\langle \pi(B), e\rangle J = \langle\pi(B)', JeJ\rangle$. Hence $J\pi(A)'J = (J\pi(A)J)' = \rho(A)'$ and $\mathrm{End}_{\rho(A)}(B) = JA'J = \langle B, e\rangle$.

In particular we see that the centre of $C = \langle B, e\rangle$ is isomorphic to the centre of A', which is identical with the centre of A. Thus we identify the minimal central projections of $C = \langle B, e\rangle$ with those of A, under the identification $p_i \leftrightarrow Jp_iJ$. To complete the picture of the Bratteli diagram of the embedding B in $\langle B, e\rangle$ we must compute the matrix of multiplicities of the embedding. The embedding matrix of B in $\langle B, e\rangle$ is in fact the same as that of A in B under the identification $p \leftrightarrow JpJ$ of the minimal projections in A with those of $\langle B, e\rangle$. We will prove this by using e and path matrix units for B to build a path model for $\langle B, e\rangle$, whilst simultaneously identifying e in the path model for $\langle B, e\rangle$. Before we do that it is illuminating to look at a further example, the embedding $\mathbb{C} \to \mathbb{C}^m$ with trace vectors s, (t_i) respectively, with $s = t_1 + t_2 + \cdots + t_m$ as in (9.4.3). Then the path algebra for $\mathbb{C} \to \mathbb{C}^m \to \mathrm{End}(\mathbb{C}^m)$ is given by reflection as in Fig. 9.6.

The expectation $E : \mathbb{C}^m \to \mathbb{C}$ is the re-normalized trace

$$E(x_1, x_2, \ldots, x_m) = \left(\textstyle\sum_i t_i x_i\right)/s = \left(\textstyle\sum_i t_i x_i/s\right)(1, 1, \ldots, 1) \qquad (9.4.9)$$

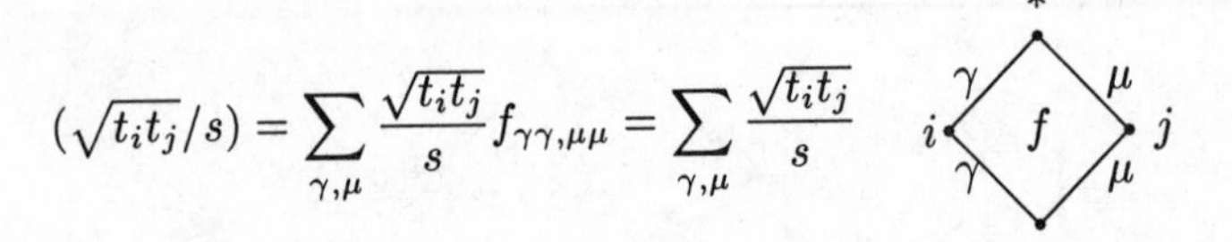

FIG. 9.7. A Jones projection e

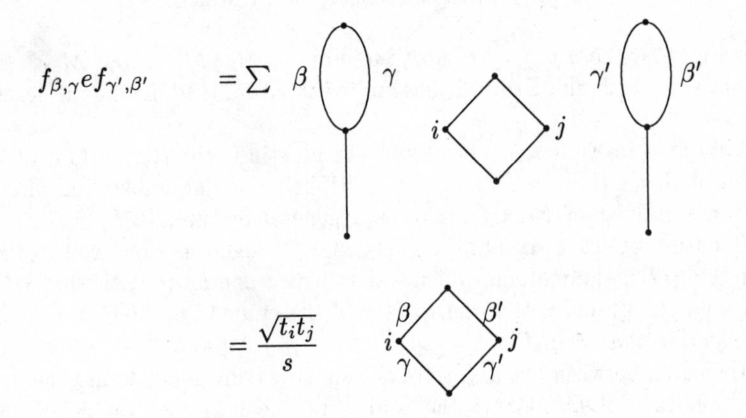

FIG. 9.8. $f_{\beta,\gamma} e f_{\gamma',\beta'} = ((t_i t_j)^{1/2}/s) f_{\beta\gamma,\beta'\gamma'}$

We then need to compute the matrix of E with respect to an orthonormal basis for $\mathbb{C}^m$. Since $(1,0,\ldots,0)$, $(0,1,0,\ldots,0)$, $\ldots$ have norms $\sqrt{t_1}, \sqrt{t_2}, \ldots, \sqrt{t_m}$ respectively, the matrix of E with respect to the unit vectors $(1/\sqrt{t_1}, 0, \ldots, 0)$, $(0, 1/\sqrt{t_2}, \ldots, 0)$, $\ldots$ is (reflecting edges) as in Fig. 9.7 with the summation over $\gamma \in \text{Path}(*, i)$, $\mu \in \text{Path}(*, j)$. Moreover for $\beta, \gamma \in \text{Path}(*, i)$, $\gamma', \beta' \in \text{Path}(*, j)$, we have the identity of Fig. 9.8 and so

$$f_{\beta\gamma,\beta'\gamma'} = \frac{s}{\sqrt{t_i t_j}} f_{\beta,\gamma} e f_{\gamma',\beta'} \tag{9.4.10}$$

Thus not only do we see from (9.4.10) (and similarly (9.4.8)) that $BeB = \langle B, e\rangle = \text{End}(B)$, but we can use this formula to define matrix units for $\langle B, e\rangle$ in terms of e and those of B. Moreover one can invert (9.8) to get a path representation (9.7) for e. This carries over to the general situation of the basic extension of finite dimensional C^*-algebras, but to get the complete picture, we should first tensor Fig. 9.6 with M_k to obtain Fig. 9.9.

Identifying $e = e_{M_k}$ with $e_{\mathbb{C}} \otimes 1_{M_k}$, we see that

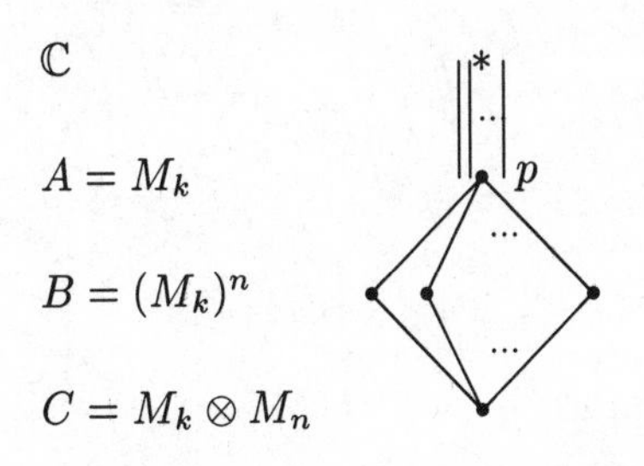

FIG. 9.9. Basic extension $\mathbb{C} \subset M_k \subset (M_k)^n \subset M_k \otimes M_n$

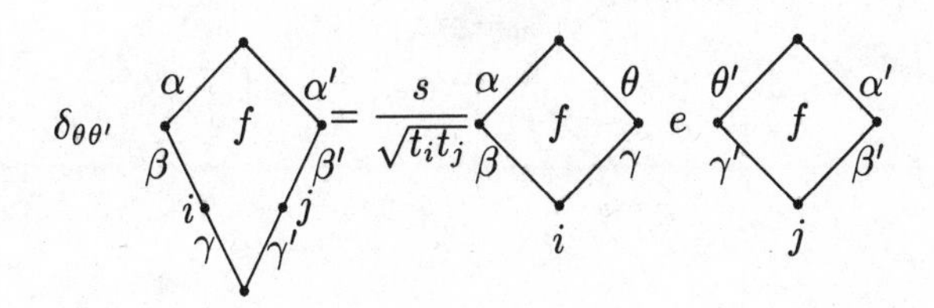

FIG. 9.10. $\delta_{\theta,\theta'} f_{\alpha\beta\gamma,\alpha'\beta'\gamma'} = (s/(t_i t_j)^{1/2}) f_{\alpha\beta,\theta\gamma} e f_{\theta'\gamma',\alpha'\beta'}$

$$\frac{s}{\sqrt{t_i t_j}} f_{\alpha\beta,\theta\gamma} e f_{\theta'\gamma',\alpha'\beta'} = \delta_{\theta,\theta'} f_{\alpha\beta\gamma,\alpha'\beta'\gamma'} \tag{9.4.11}$$

or

$$\begin{aligned} &\frac{s}{\sqrt{t_i t_j}} f_{\alpha\theta} \otimes f_{\beta\gamma}(1 \otimes e) f_{\theta'\alpha'} \otimes f_{\gamma'\beta'} \\ &= \frac{s}{\sqrt{t_i t_j}} \delta_{\theta\theta'} f_{\alpha\alpha'} \otimes (f_{\beta\gamma} e f_{\gamma'\beta'}) \\ &= \delta_{\theta\theta'} f_{\alpha,\alpha'} \otimes f_{\beta\gamma,\beta'\gamma'} = \delta_{\theta\theta'} f_{\alpha\beta\gamma,\alpha'\beta'\gamma'} \end{aligned}$$

for $\theta, \theta', \alpha, \alpha' \in \text{Path}(*, p)$, $\beta, \gamma \in \text{Path}(p, i)$, $\beta', \gamma' \in \text{Path}(p, j)$, i, j are minimal projections in B. Equation (9.4.11) or Fig. 9.10 are what we can use to directly define matrix units in $\langle B, e\rangle$.

Consider thus the case of a unital embedding of a multi-matrix algebra A in B, with multiplicity graph Λ and compatible faithful trace vectors s, t respectively. Graphically, the embedding $i : A \to B$ is as in Fig. 9.11.

The adjoint E is thus as in Fig. 9.12 since $f_{\zeta,\zeta'}$, $f_{\zeta\eta,\zeta'\eta'}$ (matrix units in A and B respectively) have norms s_k , t_i respectively if k, i are the terminal vertices of $\zeta, \zeta\eta$ respectively. We claim that the product in Fig. 9.13 vanishes unless $\theta = \theta'$

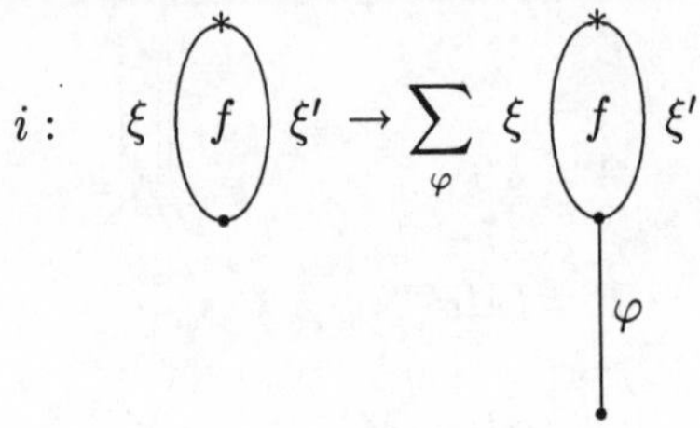

FIG. 9.11. Embedding $i : A \to B$

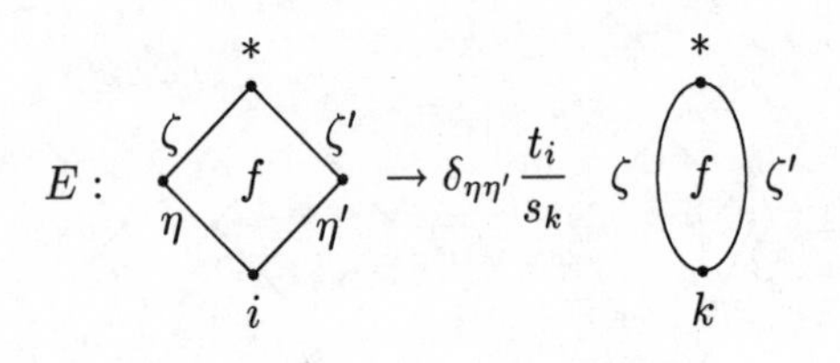

FIG. 9.12. Conditional expectation $E : B \to A$

and is then independent of θ. Here $\alpha\beta, \theta\gamma \in \text{Path}(*, i)$, $\theta'\gamma', \alpha'\beta' \in \text{Path}(*, j)$, $\theta \in \text{Path}(*, k)$ and i, j are minimal projections in B, and k a minimal projection in A. This claim is a consequence of:

Lemma 9.29 *In the above situation:*

$$(f_{\alpha\beta,\theta\gamma} e f_{\theta'\gamma',\alpha'\beta'}) f_{\zeta\eta,\zeta'\eta'} = \delta_{\theta\theta'} \delta_{\gamma'\eta'} \delta_{\alpha'\beta',\zeta\eta} \frac{t_i}{s_k} f_{\alpha\beta,\zeta'\gamma}$$

if $\theta' \in \text{Path}(*, i), \theta'\gamma' \in \text{Path}(*, k)$.

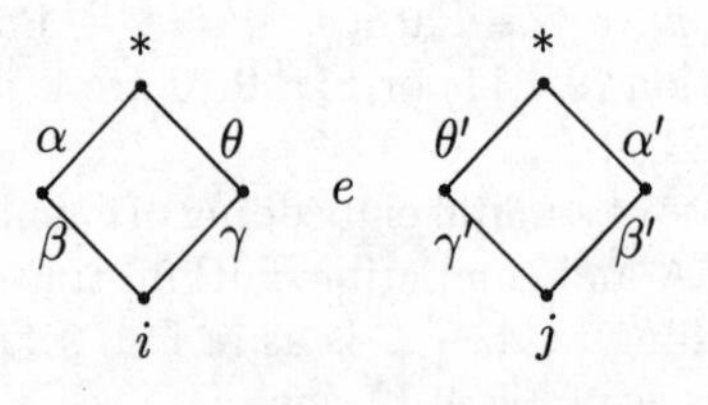

FIG. 9.13. $f_{\alpha\beta,\theta\gamma} e f_{\theta'\gamma',\alpha'\beta'}$

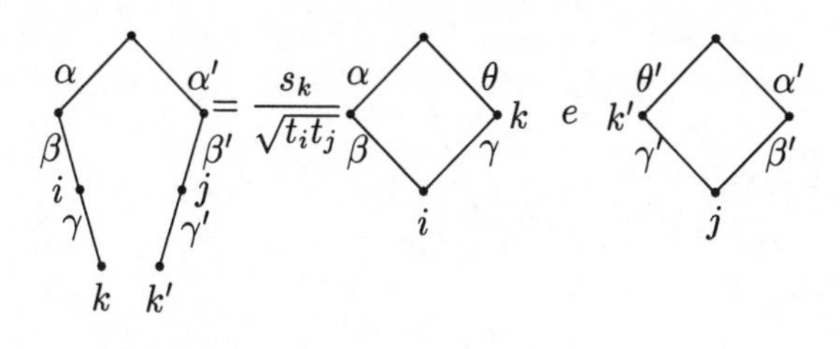

FIG. 9.14. $h_{\alpha\beta\gamma,\alpha'\beta'\gamma'} = (s_k/(t_it_j)^{1/2}) f_{\alpha\beta,\theta\gamma} e f_{\theta\gamma',\alpha'\beta'}$

Proof Use $ef_{\theta'\gamma',\alpha'\beta'} f_{\zeta\eta,\zeta'\eta'} = \delta_{\alpha'\beta',\zeta\eta} E(f_{\theta'\gamma',\zeta'\eta'})$ and the formula in Fig. 9.12 for E. □

Now consider the path algebra $\mathbb{C} \to A \to B \to C$ where we take C to be the algebra obtained by reflecting the graph of A in B. For $\alpha\beta\gamma \in \mathrm{Path}(*,k)$, $\alpha'\beta'\gamma' \in \mathrm{Path}(*,k')$ where k,k' are minimal projections in C (identified with one in A by reflection), $\alpha\beta \in \mathrm{Path}(*,i)$, $\alpha'\beta' \in \mathrm{Path}(*,j)$, i,j minimal projections in B, take any pair of edges θ,θ' in $\mathrm{Path}(*,k), \mathrm{Path}(*,k')$ respectively (where now k,k' are identified with minimal projections in B). We define $h_{\alpha\beta\gamma,\alpha'\beta'\gamma'}$ in $\langle B,e\rangle$ by

$$h_{\alpha\beta\gamma,\alpha'\beta'\gamma'} = \frac{s_k}{\sqrt{t_it_j}} f_{\alpha\beta,\theta\gamma} e f_{\theta\gamma',\alpha'\beta'} \tag{9.4.12}$$

which is graphically represented as in Fig. 9.14.

Then Lemma 9.29 shows that $h_{\alpha\beta\gamma,\alpha'\beta'\gamma'}$ vanishes unless $\theta = \theta'$ (in particular, unless $k = k'$) and is then independent of the choice of θ. We claim that

$$f_{\alpha\beta\gamma,\alpha'\beta'\gamma'} \to h_{\alpha\beta\gamma,\alpha'\beta'\gamma'} \tag{9.4.13}$$

is a representation from the path model for the algebra C into $\langle B,e\rangle$ and is in fact an isomorphism.

Proposition 9.30 *There is a $*$-isomorphism (9.4.13) from the path model for $\mathbb{C} \to A \to B \to C$ where the multiplicity graph of B in C is the reflection of that of A in B into the extended algebra $\langle B,e\rangle$ such that the diagram in Fig. 9.15 commutes under the natural inclusions of B in C and $\langle B,e\rangle$. Moreover we have the representation of the Jones projection e of Fig. 9.16 in the path model for $A \to B \to C$.*

Proof We have graphical identities as in Fig. 9.17. Hence

$$h_{\alpha\beta\gamma,\alpha'\beta'\gamma'} h_{\zeta\eta\psi,\zeta'\eta'\psi'} = \delta_{\alpha'\beta'\gamma',\zeta\eta\psi} h_{\alpha\beta\gamma,\zeta'\eta'\psi'} \tag{9.4.14}$$

This ensures that $f \to h$ is a representation of the path model for the algebra C, which is faithful because of the identity of Fig. 9.18. To show that this ho-

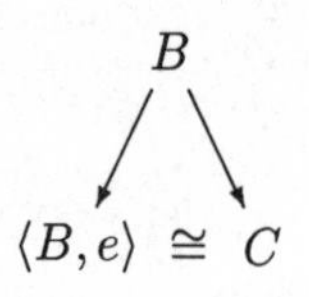

FIG. 9.15. Basic extension identification

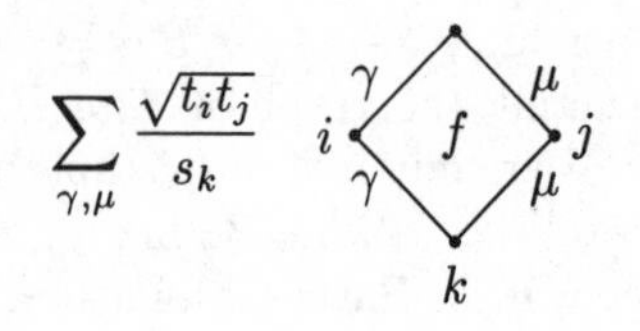

FIG. 9.16. Jones projection e

momorphism is onto $\langle B, e\rangle$ we need to express e and matrix units in B in terms of hs. We have

$$
\begin{aligned}
e &= \sum_{\alpha,\beta\alpha',\beta'} f_{\alpha\beta,\alpha\beta} e f_{\alpha'\beta',\alpha'\beta'} \\
&= \sum_{\alpha,\beta,\beta'} f_{\alpha\beta,\alpha\beta} e f_{\alpha\beta',\alpha\beta'}, \quad \text{by Lemma 9.32} \qquad (9.4.15) \\
&= \sum_{\alpha,\beta,\beta'} \frac{\sqrt{t_i t_j}}{s_k} h_{\alpha\beta\beta,\alpha\beta'\beta'} \qquad (9.4.16)
\end{aligned}
$$

Moreover we have the identity as in Fig. 9.19 using Lemma 9.29. Thus (9.15) holds, and $BeB = \langle B, e\rangle$ can be identified with C. □

Remark We have derived, using the matrix unit picture, a representation for the extended algebra $\langle B, e\rangle$ and e as in Fig. 9.16. It is instructive to put this in the face operator picture. The projection e is represented in Fig. 9.16 as a sum of rank one projections, for which we already have the graphical picture in Fig 2.9. Define vectors graphically as in Fig. 9.20 where $j = s(\mu)$, $i = r(\mu) = s(\nu)$, $k = r(\nu)$, so that we get Fig. 9.21. (In Section 11.7 we will renormalize e to an operator U, where $U^2 = \beta U$). Then (9.4.12) has the graphical explanation as in Fig. 9.22.

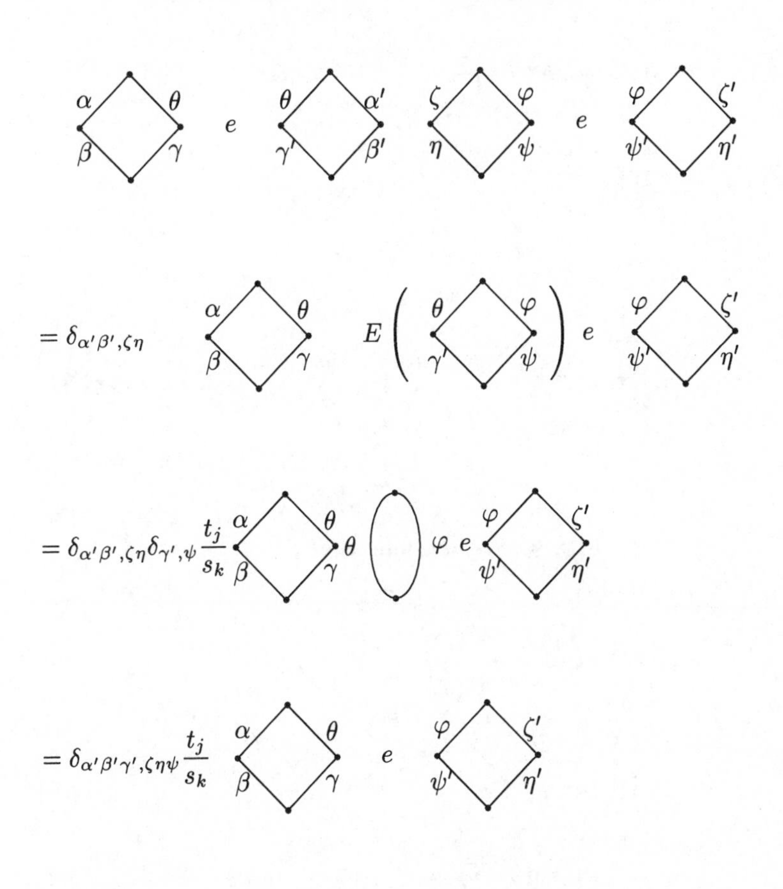

FIG. 9.17. Derivation of $h_{\alpha\beta\gamma,\alpha'\beta'\gamma'}h_{\zeta\eta\psi,\zeta'\eta'\psi'} = \delta_{\alpha'\beta'\gamma',\zeta\eta\psi}h_{\alpha\beta\gamma,\zeta'\eta'\psi'}$

Remark We have indeed shown that the multiplicity graph of the embedding of B' in A' (which is the same as that of $JB'J = B$ in $JA'J$) is obtained by reflection of that of A in B. This is indeed true in general for an arbitrary unital representation of B on a finite dimensional Hilbert space V. If $B \subset \mathrm{End}(V)$ is a unital embedding, and B is a factor or simple, then by Proposition 9.30,

$$(V, B, B') \cong (V_1 \otimes V_2, \mathrm{End}(V_1) \otimes 1_{V_2}, 1_{V_2} \otimes \mathrm{End}(V_2))$$

and $\mathrm{End}(V) \cong B \otimes B'$. This can all be represented by a *Murray diagram* in Fig. 9.23. The tensor product $V_1 \otimes V_2 = \oplus_{(i,j)\in\Lambda}\mathbb{C}$ using a rectangular lattice Λ of height $\dim V_1$ and width $\dim V_2$, so that B acts on columns only and B' on rows only. A projection $p \in B$ is represented (at least in a suitable basis) by chopping down a column as in Fig. 9.24 into the enclosed rectangle which represents $pV_1 \otimes V_2$. Here we have identified $p \in \mathrm{End}(V_1)$ with $p \otimes 1_{V_2}$. We write $B_p = pBp$, and identify $B'_p = pB'p$ with $(B_p)'$ and B'. Similarly, if $p \in B, q \in B'$

$$\|h_{\alpha\beta\gamma,\alpha'\beta'\gamma'}\|^2 = \mathrm{tr}\; h_{\alpha\beta\gamma,\alpha'\beta'\gamma'} h_{\alpha'\beta'\gamma',\alpha\beta\gamma} = \mathrm{tr}\; h_{\alpha\beta\gamma,\alpha\beta\gamma}$$

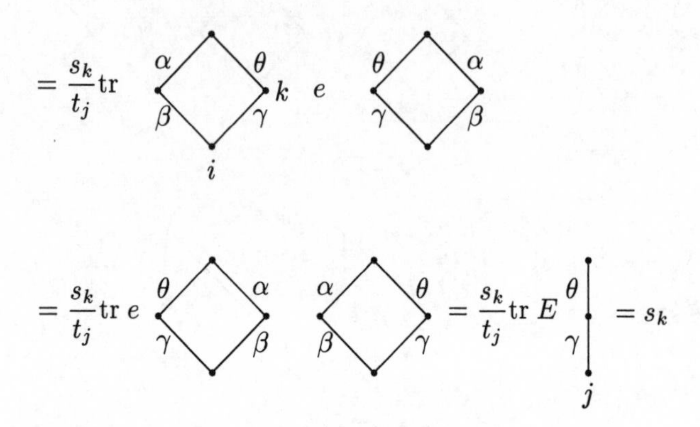

FIG. 9.18. Faithfulness of $f \mapsto h$

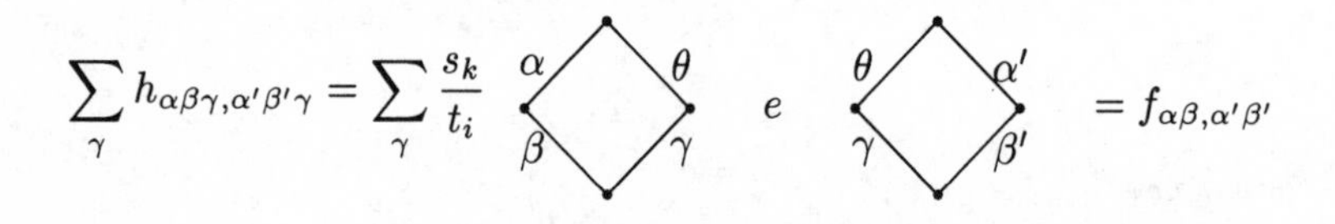

FIG. 9.19. $\sum_\gamma h_{\alpha\beta\gamma,\alpha'\beta'\gamma} = f_{\alpha\beta,\alpha'\beta'}$

are projections, then the enclosed rectangle in Fig. 9.25 represents

$$(pqV, B_{pq}, (B')_{pq}) = (pV_1 \otimes qV_2, \mathrm{End}(pV_1) \otimes 1_{qV_2}, 1_{pV_2} \otimes \mathrm{End}(qV_2)).$$

More generally, let B be a finite dimensional C^*-algebra with minimal central projections $\{q_i\}$. If $B \subset \mathrm{End}(V)$ is a unital embedding, we can represent

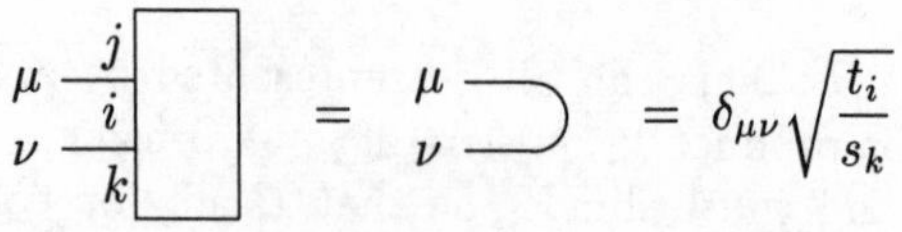

$$= \delta_{\mu\nu}\sqrt{\frac{t_i}{s_k}}$$

FIG. 9.20. Face vector

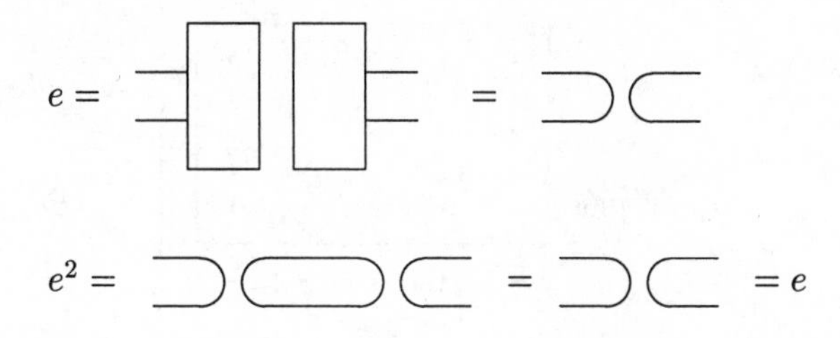

FIG. 9.21. Face operator $e = e^2$

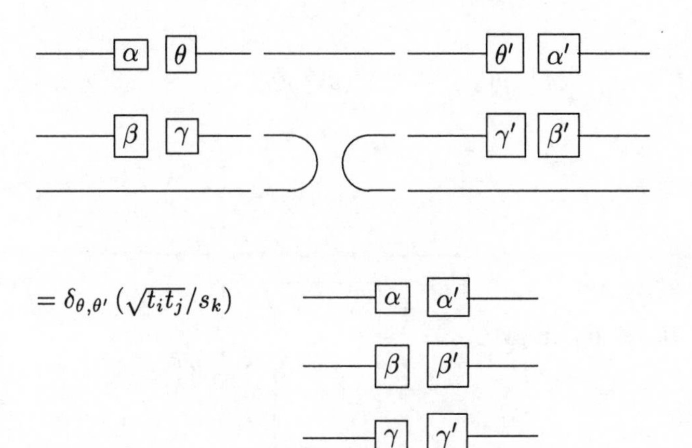

FIG. 9.22. $h_{\alpha\beta\gamma,\alpha'\beta'\gamma'} = (s_k/(t_i t_j)^{1/2}) f_{\alpha\beta,\theta\gamma} e f_{\theta\gamma',\alpha'\beta'}$

$$(V, B, B') = \bigoplus_j (q_j V, Bq_j, (Bq_j)' \cap \mathrm{End}(q_j V))$$

by a sequence of Murray diagrams $(B')_{q_j} \cong B'_{q_j} \cap \mathrm{End}(q_j V)$ placed as in the

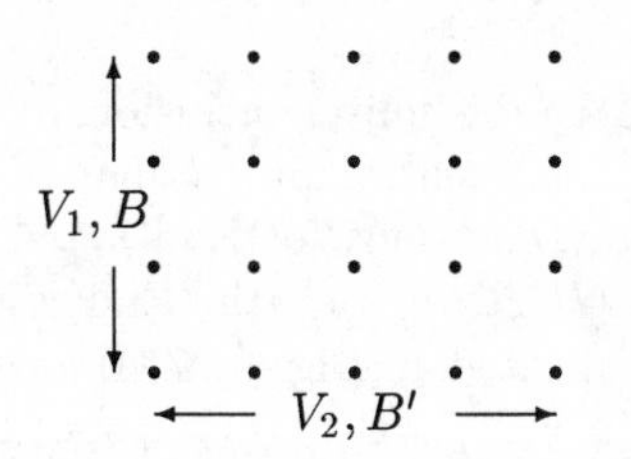

FIG. 9.23. $(V_1 \otimes V_2, \mathrm{End}(V_1) \otimes 1_{V_2}, 1_{V_2} \otimes \mathrm{End}(V_2))$

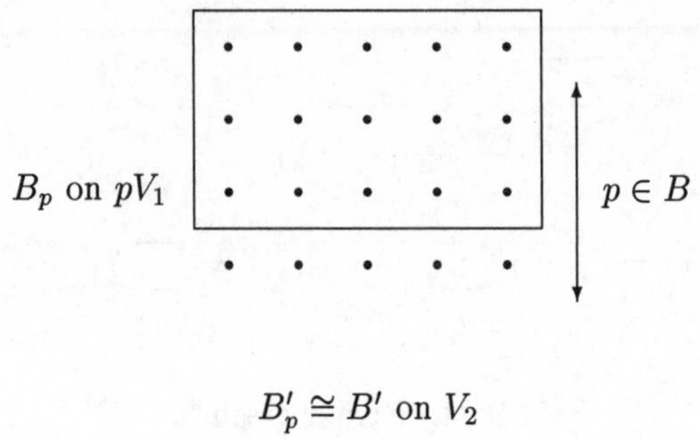

FIG. 9.24. $(pV_1 \otimes V_2, \mathrm{End}(pV_1) \otimes 1, 1_{pV_1} \otimes \mathrm{End}(V_2))$

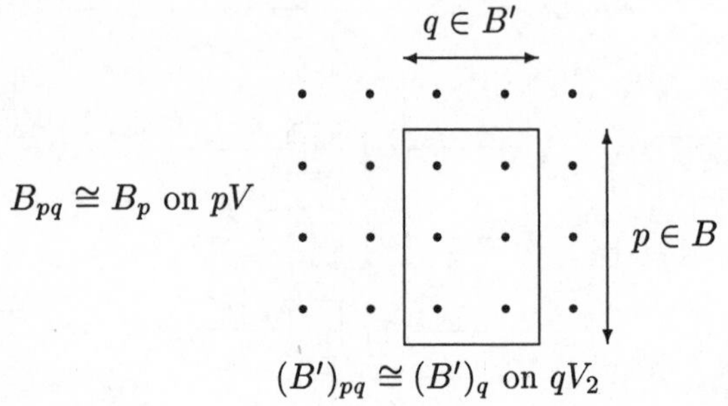

FIG. 9.25. Murray diagram for $(pqV, B_{pq}, (B')_{pq})$

above example. B consists of column transformations, and B' those on rows. The centre $B \cap B'$ consists of transformations constant on each rectangle. If p, q are projections in B, B' respectively, then $(pqV, B_{pq}, (B')_{pq} = (B_{pq})' \cap \mathrm{End}(pqV))$ is represented by chopping down each rectangle representing (B, B') in Fig. 9.23 as in Fig.9.25.

Now let $A \subset B \subset \mathrm{End}(V)$ be unital inclusions of finite dimensional C^*-algebras, with $\{p_i\}$ the minimal central projections of A so that $p_i q_j \in A' \cap B$. Then we have two Murray diagrams for $\mathrm{End}(p_i q_j V)$ arising from chopping down those for (A, A') and (B, B') in the ith $(\mathrm{End}(p_i V))$ and jth $(\mathrm{End}(q_j V))$ rectangles respectively as indicated in Fig. 9.27. Here the extended rectangles represent $(p_i V, A_{p_i}, (A')_{p_i}), (q_j V, B_{q_j}, (B')_{q_j}(q_j))$ respectively. Then if λ_{ji} denotes the multiplicity of A_{p_i} in B_{q_j}, μ_{ij} the multiplicity of B'_{q_j} in A'_{p_i}, and $m_i^2 = \dim(A_{p_i})$, $n_j^2 = \dim(B_{q_j})$ we see since the areas of the enclosed squares are both $\dim(p_i q_j V)$ that

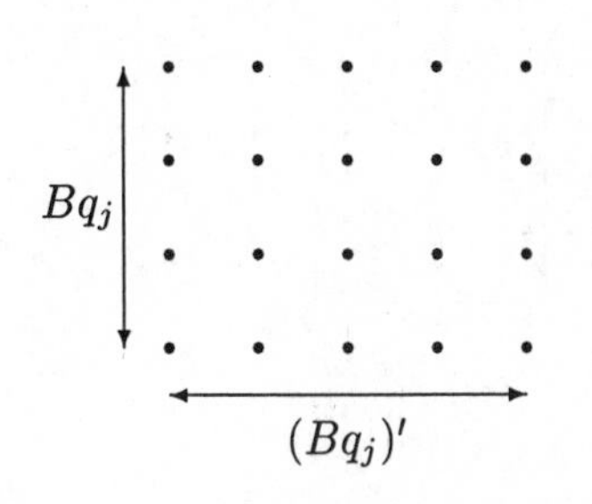

FIG. 9.26. Murray diagram for (V, B, B')

$$m_i(\mu_{ij}n_j) = (\lambda_{ji}m_i)n_j\,.$$

Hence $\mu_{ij} = \lambda_{ji}$, i.e. the multiplicity graph of A in B is the transpose of that of B' in A'.

We next study Pimsner–Popa basis in finite dimensional cases. Let $A \subset B$ be a unital inclusion of finite dimensional C^* -algebras with multiplicity graph Λ and compatible faithful trace vectors s, t respectively. The derivation of the Bratteli diagram of the inclusion $B \subset \langle B, e\rangle$ in the above not only enables us to determine the path representation for e and show $\langle B, e\rangle = BeB$ but also gives us a basis – a Pimsner–Popa basis – for B as an A-module.

In the notation as above, we have the identity of Fig. 9.28 if we set λ_μ as in Fig. 9.29. Here $\mu = \alpha\beta\gamma \in \mathrm{Path}(*, k)$ is an arbitrary path in the Bratteli diagram of $\mathbb{C} \to A \to B \to C$, and as usual θ is a random edge such that $\alpha\beta, \theta\gamma \in \mathrm{Path}(*, i)$ in the path space for $\mathbb{C} \to A \to B$. If we have λ_μ, λ_ν as in Fig. 9.29, then we can compute $\lambda_\mu^*\lambda_\nu$ and $E(\lambda_\mu^*\lambda_\nu)$ as in Figures 9.30 and 9.31 respectively. Here f_μ is a projection in A, and $\lambda_\mu f_\mu = \lambda_\mu$

Let $A \subset B$ be a unital embedding of C^* -algebras and $E : B \to A$ a conditional expectation – a faithful completely positive unital projection. Let $\mathcal{E} = {}_BB$ as a left B-module with inner product

$$\langle b, b'\rangle = E(b'^* b). \tag{9.4.17}$$

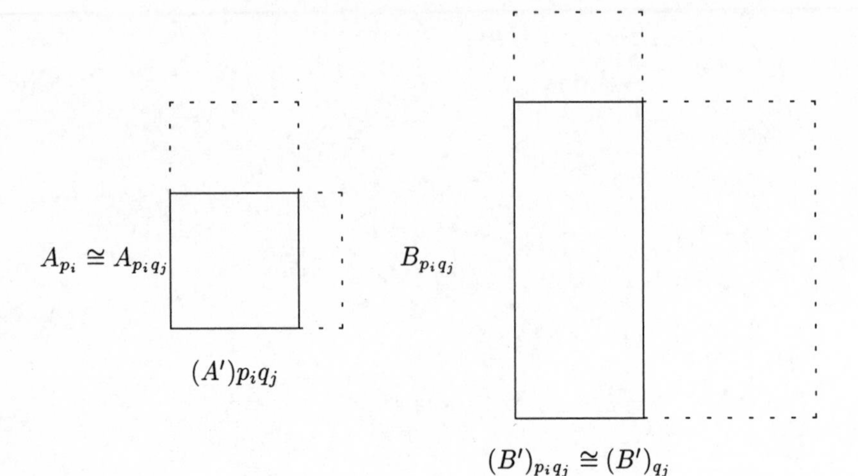

FIG. 9.27. Murray diagrams for $\mathrm{End}(p_i q_j V)$; $p_i \in A \cap A'$, $q_j \in B \cap B'$

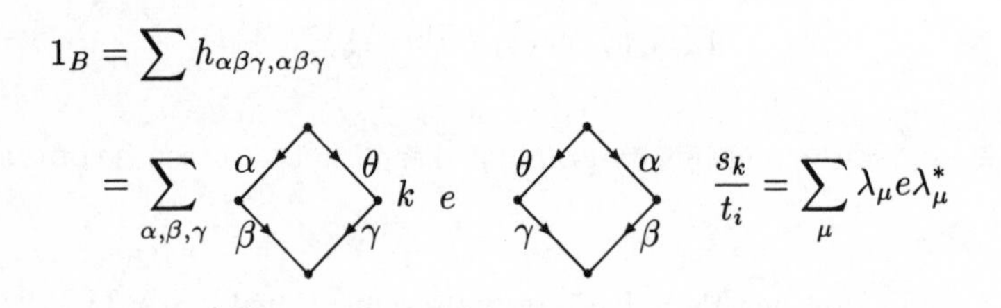

FIG. 9.28. Pimsner–Popa basis relation $1 = \sum \lambda_\mu e \lambda_\mu^*$

Then B acts on $\mathcal{E}$ by left multiplication and E defines a self-dual map – a projection – e in $\mathcal{L}(\mathcal{E})$, such that

(9.4.18) $E(b)e = ebe$

(9.4.19) $a \to ae$ gives an isomorphism (which we will abbreviate as $A \cong Ae$) of C^*-algebras.

Moreover, as above, we have $A = B \cap \{e\}'$. More precisely let π denote the action of B on $\mathcal{E}$ by $\pi(a)b = \alpha\beta$, and $\Omega = 1$ the unit vector in $\mathcal{E}$, then $eb\Omega = E(b)\Omega$, $\pi E(b)e = e\pi(b)e$, $b \in B$. Then π defines a $*$-representation of B on $\mathcal{E}$ and e a projection $e = e^2 = e^*$ in $\mathcal{L}(\mathcal{E})$ because

$$\langle \pi(b)a, \pi(b)a' \rangle = E(a^*b^*ba) \leq \|b\|^2 f(a^*a) = \|b\|^2 \langle a, a \rangle \qquad (9.4.20)$$

$$\langle \pi(b)a, a' \rangle = E(a'^*ba) = E((ba')^*a) = \langle a', \pi(b)a \rangle \qquad (9.4.21)$$

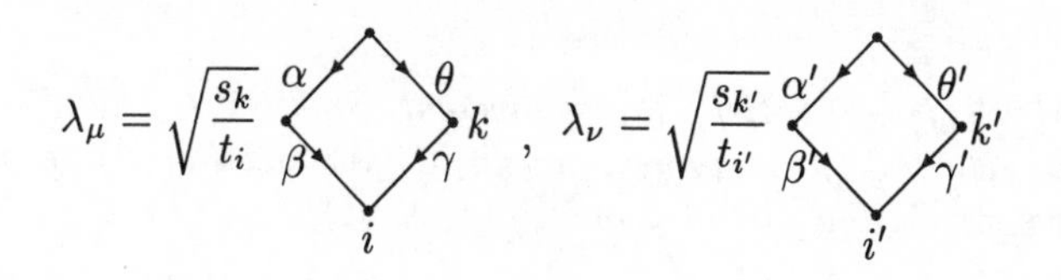

FIG. 9.29. Pimsner–Popa basis; λ_μ, λ_ν

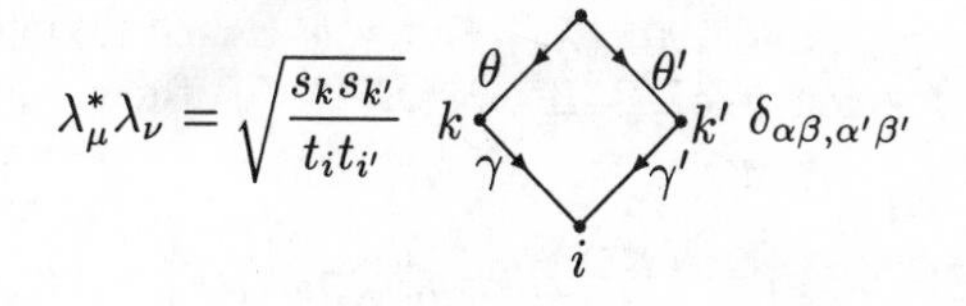

FIG. 9.30. $\lambda_\mu^* \lambda_\nu$

$$\langle ea, eb \rangle = E(b^*)E(b) \leq E(b^*b) = \langle b, b \rangle \text{ for } b, a \in B \tag{9.4.22}$$

where we have used the Schwarz inequality for completely positive maps, and $e^2 = e$ as $E^2 = E$ and $e^* = e$ as E is symmetric, $E(b^*) = E(b)^*$.

Lemma 9.31 *Let $A \subset B$ be an inclusion of unital C^*-algebras, $E : B \to A$ a conditional expectation, and $\{\lambda_\mu\}$ a finite set of elements of B.*

The following conditions are equivalent:

1. $b = \sum_\mu \lambda_\mu E(\lambda_\mu^* b)$, *all* $b \in B$
2. $b = \sum_\mu E(b\lambda_\mu)\lambda_\mu^*$, *all* $b \in B$.

In which case $\sum_\mu \lambda_\mu \lambda_\mu^*$ lies in the centre of B and is independent of the choice of $\{\lambda_\mu\}$ satisfying the above conditions, and we denote this by Index E. If there exists a projection e satisfying (9.4.18) and (9.4.19) then the above conditions are implied by the following:

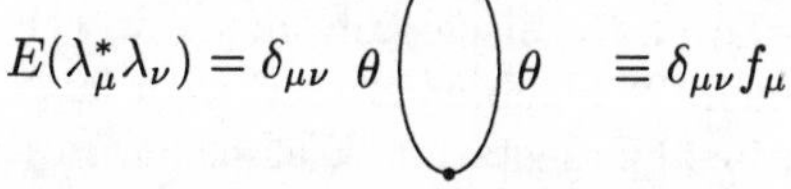

$$E(\lambda_\mu^* \lambda_\nu) = \delta_{\mu\nu} \;\; \theta \bigcirc \theta \;\; \equiv \delta_{\mu\nu} f_\mu$$

FIG. 9.31. $E(\lambda_\mu^* \lambda_\nu) = \delta_{\mu\nu} f_\mu$

$$1 = \sum_\mu \lambda_\mu e \lambda_\mu^*. \tag{9.4.23}$$

Let $\{\lambda_\mu\}_{\mu \in I}$, $\{\gamma_\nu\}_{\nu \in J}$ satisfy the equivalent conditions 1 and 2. Then $(E(\lambda_\mu \lambda_{\mu'}^*))$ and $(E(\gamma_\nu \gamma_{\nu'}^*))$ are projections in $M_I(A)$ and $M_J(A)$ respectively, which are equivalent via the partial isometry $(E(\lambda_\mu \gamma_\nu^*))$.

Proof $1 \Leftrightarrow 2$ by taking adjoints. Suppose $\{\lambda_\mu\}$, $\{\gamma_\nu\}$ satisfy 1 and 2. Then for $b \in B$

$$\sum_\mu b\lambda_\mu \lambda_\mu^* = \sum_\mu \sum_\nu \gamma_\nu E(\gamma_\nu^* b \lambda_\mu^*)\lambda_\mu = \sum_\nu \sum_\mu \gamma_\nu E(\gamma_\nu^* b \lambda_\mu^*)\lambda_\mu = \sum \gamma_\nu \gamma_\nu^* b$$

by applying 1 to $\{\gamma_\nu\}$ and $b\lambda_\mu$, and 2 to $\{\lambda_\mu\}$ and $\gamma_\nu^* b$. Hence taking $b = 1$, we see that $\sum \lambda_\mu \lambda_\mu^* = \sum \gamma_\nu \gamma_\nu^* \in$ centre of B. If $v = (E(\lambda_\mu \gamma_\nu^*))$ then

$$vv^* = \left(\sum_\nu E(\lambda_\mu \gamma_\nu^*) E(\gamma_\nu \lambda_{\bar\mu}^*)\right) = \left(\sum_\nu E(\lambda_\mu \gamma_\nu^* E(\gamma_\nu \lambda_{\bar\mu}^*))\right) = (E(\lambda_\mu \lambda_{\bar\mu}^*))$$

from which (b) follows. To complete (a), let e be a projection as stated. Then if (9.4.23) holds, and $b \in B$:

$$be = 1 \cdot be = \sum \lambda_\mu e \lambda_\mu^* be = \sum \lambda_\mu E(\lambda_\mu^* b)e.$$

Hence $b = \sum \lambda_\mu E(\lambda_\mu^* b)$. □

Thus from a basis $\{\lambda_\mu\}$ we can construct a projection, and hence an element of $D_\infty(A)$ and $K_0(A)$ which is independent of the choice of basis. It is convenient to introduce matrix algebra $M_\alpha(A)$, where $a \in D_\infty(A)$. For this let B be any unital C^*-algebra, and p a projection in a matrix algebra over B, say $p \in M_r(A)$. Define $M_p(A) = M_r(A)_p$. If moreover $q \in M_s(A)$ and p and q are Murray–von Neumann equivalent via a partial isometry v, $p = v^*v$, $q = vv^*$ (v is an $s \times r$ matrix over A) then $x \to vxv^*$ gives an isomorphism between $M_p(A)$ and $M_q(A)$. Hence $M_p(A)$ only depends on the Murray–von Neumann equivalence class of p, which we denote by $M_{[p]}(A)$. If we have cancellation of projections in $K_0(A)$ in the sense of Proposition 3.4(c), we can thus define $M_\alpha(A)$ for $\alpha \in K_0(A)_+$. In particular, if $M(\mathbf{n}) = M(n_1) \oplus \cdots \oplus M(n_p)$ is a multi-matrix algebra, $\mathbf{m} = (m_1, \ldots, m_p) \in \mathbb{N} = K_0(M(n))_+$, then $M_{\mathbf{m}}(\mathbf{n}) \cong M(\mathbf{mn})$, where $\mathbf{mn} = (m_1 n_1, \ldots, m_p n_p)$. If N is a II_1 von Neumann algebra with finite dimensional centre, we can define $M_\tau(N)$ for $\tau \in \mathbb{R}^n = K_0(N)$, if $n =$ dimension of the centre of N.

Let us inspect what the above discussion means in the case of a unital inclusion of finite dimensional C^*-algebras $A \subset B$ with faithful trace vectors s, t and corresponding conditional expectation E as usual. In this case we have the identity of Fig. 9.32, where q_i are the minimal central projections of B, or we can write Index $E = \Lambda s/t$. The matrix $\left(E(\lambda_\mu^* \lambda_\nu)\right) = (\delta_{\mu\nu} f_\mu) \equiv f$ and $[f] = \left[\sum_j \lambda_{ji}\right]$

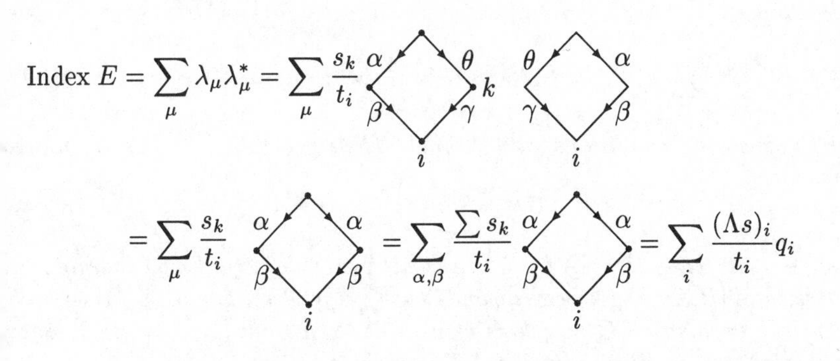

FIG. 9.32. Index $E = \Lambda s/t$

n $K_0(A)_+ = \bigoplus_i \mathbb{N}$, where $\Lambda = (\lambda_{ji})$ is the multiplicity matrix of the embedding of A in B. Thus Index(E) is a scalar β if and only if $\Lambda s = \beta t$, i.e. as $s = \Lambda^t t$, if and only if $\Lambda\Lambda^t t = \beta t$ for some $\beta > 0$. We claim that this is the case if and only if there exists a Markov τ-trace on $\langle B, e\rangle$. We will prove the existence of a Markov trace if Index E is scalar – which is all we will use. However, it is interesting to discuss all this in the more general context of C^*-index theory.

Let B be a unital C^*-algebra, and $\langle B, e\rangle$ the C^*-algebra generated by B and a single projection e. A *Markov τ-trace* or *Markov trace* on $\langle B, e\rangle$ is a faithful normalized trace tr such that

$$\text{tr } be = \tau \text{tr } b, \quad b \in B. \tag{9.4.24}$$

If such a trace exists on B, then it is uniquely determined, at least on BeB, by its restriction to B and $\tau = \text{tr } e \in (0, 1]$. If $\{\lambda_\mu\}$ is a Pimsner–Popa basis as in(9.4.23), and tr is a τ-Markov trace on $\langle B, e\rangle$, then

$$\begin{aligned} 1 = \text{tr} \sum \lambda_\mu e \lambda_\mu^* &= \text{tr } e \sum \lambda_\mu^* \lambda_\mu = \tau \text{tr} \sum \lambda_\mu^* \lambda_\mu \\ &= \tau \sum \text{tr } \lambda_\mu \lambda_\mu^* = \tau \text{tr Index } E \end{aligned} \tag{9.4.25}$$

and so

$$\tau^{-1} = \text{tr (Index } E). \tag{9.4.26}$$

Lemma 9.32 *Let $A \subset B$ be an unital inclusion of C^*-algebras, and $E : A \to B$ a faithful conditional expectation, with e a projection in a C^*-algebra containing B such that $E(b)e = ebe$, $b \in B$, and $A \cong Ae$. Let $\{\lambda_\mu\}$ be a finite set of elements of B.*

(a) The following conditions are equivalent:

1. *$E(\lambda_\mu \lambda_\nu^*) = \delta_{\mu\nu} f_\mu$ for some family of projections f_μ in A.*
2. *$\lambda_\mu e$ are partial isometries with orthogonal ranges.*

If 1 and 2 hold, and if there exists a Markov τ-trace on $\langle B, e\rangle$, then we hav the following:

$$\sum \lambda_\mu e \lambda_\nu^* = 1 \text{ if and only if } \sum \operatorname{tr} f_\mu = \tau^{-1}. \quad (9.4.27$$

(b) Suppose the conditions in 1–2 hold. Then every $b \in B$ has an uniqu decomposition

$$b = \sum \lambda_\mu b_\mu \quad (9.4.28$$

where $b_\mu \in f_\mu A$. Thus $B \cong \bigoplus f_\mu A$, as a right A-module, is finitely generated an projective, and $[B]$ in $K_0(A)$ corresponds to $\sum_\mu [f_\mu]$. The map $b_1 \otimes_A b_2 \mapsto b_1 e b$ gives an isomorphism of C^*-algebras between $B \otimes_A B$ and $\langle B, e\rangle = BeB$, wher multiplication and involution in $B \otimes_A B$ are given by

$$(b_1 \otimes_A b_2)^* = b_2^* \otimes_A b_1^*, \quad (b_1 \otimes_A b_2)(b_1' \otimes_A b_2') = b_1 E(b_2 b_1') \otimes_A b_2'. \quad (9.4.29$$

In particular $1 \in B$ is identified with $\sum_\mu \lambda_\mu \otimes_A \lambda_\mu^*$.

Proof (a) If (1) holds then $e\lambda_\mu^* \lambda_\nu e = \delta_{\mu\nu} f_\mu e$, so that $\lambda_\nu e$ is a family of partia isometries with orthogonal ranges. Conversely, if (2) holds, then

$$E(\lambda_\mu^* \lambda_\mu)^2 e = e\lambda_\mu^* \lambda_\mu e \lambda_\mu^* \lambda_\mu e = e\lambda_\mu^* \lambda_\mu e = E(\lambda_\mu^* \lambda_\mu) e.$$

Hence by $A \cong Ae$, we see that $E(\lambda_\mu^* \lambda_\mu)$ is a projection f_μ. Similarly $E(\lambda_\mu^* \lambda_\nu) =$ $\delta_{\mu\nu} f_\mu$. Next (9.4.27) follows from

$$\begin{aligned} \operatorname{tr}(1 - \sum \lambda_\mu e \lambda_\mu^*) &= 1 - \tau \sum \operatorname{tr} \lambda_\mu^* \lambda_\mu \\ &= 1 - \tau \sum \operatorname{tr} E(\lambda_\mu^* \lambda_\mu) = 1 - \tau \sum \operatorname{tr} f_\mu. \end{aligned}$$

(b) If (1) and (2) hold then by Lemma 9.31 we have $b = \sum \lambda_\mu E(\lambda_\mu^* b) =$ $\sum \lambda_\mu b_\mu$ where $b_\mu \in f_\mu A$. If $b = \sum \lambda_\nu b_\nu$ where $b_\mu \in f_\mu A$, then $E(\lambda_\mu^* b) =$ $\sum_\nu E(\lambda_\mu^* \lambda_\nu) b_\nu = f_\nu b_\nu = b_\nu$ and so uniqueness follows. Any elemen of $B \otimes_A B$ has a decomposition $\sum \lambda_\mu z_{\mu\nu} \otimes \lambda_\nu^*$ where $z_{\mu\nu} \in f_\mu A f_\nu$. I $\sum \lambda_\mu z_{\mu\nu} e \lambda_\nu^* = 0$ then premultiplying by $e\lambda_{\bar\mu}^*$ and postmultiplying by $\lambda_{\bar\nu} e$ w get $\sum_{\mu,\nu} E(\lambda_{\bar\mu}^* \lambda_\mu) e z_{\mu\nu} E(\lambda_\nu^* \lambda_{\bar\nu}) = 0$. Hence $z_{\bar\mu\bar\nu} e = f_{\bar\mu} z_{\bar\mu\bar\nu} f_{\bar\nu} e = 0$ and so $z_{\bar\mu\bar\nu} = 0$ This establishes the isomorphism between $B \otimes_A B$ and BeB, the $*$-algebr generated by e and B. We need to show that the latter is complete. To se this let $\bar e$ be the projection in $\mathcal{L}(\mathcal{E})$ defined as in (9.4.18)–(9.4.22). Then w have an isomorphism between $B \otimes_A B$ and $B\bar e B \subset \mathcal{L}(\mathcal{E})$. But $B\bar e B = \mathcal{F}(\mathcal{E})$ the ideal of finite rank operators on $\mathcal{E}$, which contains $1_\mathcal{E} = \sum \lambda_\mu \bar e \lambda_\mu^*$. Henc $B\bar e B = \mathcal{F}(\mathcal{E}) = \mathcal{K}(\mathcal{E}) = \mathcal{L}(E)$, which is complete. Hence in general we se from $BeB \cong B \otimes_A B \cong B\bar e B$ that BeB is complete, and is indeed $\langle B, e\rangle$, th C^*-algebra generated by B and e. □

Lemma 9.33 *Let $A \subset B$ be a unital inclusion of C^*-algebras, $E : B \to A$ faithful conditional expectation, and $\{\lambda_\mu\}$ a Pimsner–Popa basis for E. The*

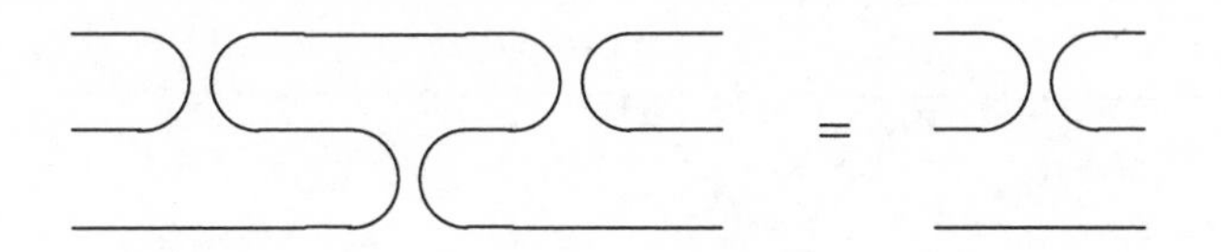

FIG. 9.33. $U_1U_2U_1 = U_1$

$\hat{E} : B \otimes_A B \to B$ defined by $\hat{E}(a \otimes b) = (\text{Index}\, E)\, ab$ is a faithful conditional expectation. If e is a projection in a C^*-algebra containing B, such that $E(b)e = ebe$, $b \in B$ and $A \cong Ae$, such that (1) and (2) hold in Lemma 9.32, then $\{\lambda_\mu e \text{Index}\,(E)^{1/2}\} \subset BeB \cong B \otimes_A B$ is a Pimsner–Popa basis for $\hat{E}$, with Index $\hat{E} = \sum \lambda_\mu e \text{Index}\,(E) e \lambda_\mu^*$. Thus Index $\hat{E} =$ Index E if Index E is in A.

Suppose in the context of Lemma 9.33 (e.g. back in the finite dimensional context) that ω is trace on B such that $\omega = \omega \circ E$. Then $\bar{\omega}(a \otimes b) = \bar{\omega}(aey) = \omega(ab)$ is a trace on $B \otimes_A B \cong BeB$. If Index(E) is a scalar, then $(\text{Index(E)}^{-1})\bar{\omega}$ is a Markov trace on $\langle B, e \rangle$, as $\bar{\omega}(b) = \sum \bar{\omega}(b\lambda_\mu e \lambda_\mu^*) = \sum \omega(b\lambda_\mu\lambda_\mu^*) = (\text{Index}\,\text{E})\omega(\text{b})$, so that $\bar{\omega}(ae) = \bar{\omega}(ae1) = \omega(a)$.

Remark Now summarizing the finite dimensional situation $A \subset B$, where $s = \Lambda^t t$ and $\Lambda s = \beta t$, i.e. we have a Markov trace, we can iterate the basic construction of $B \subset C = \langle B, e \rangle$, to get a sequence of projections $e = e_1, e_2, \ldots$ such that $e_2e_1e_2 = \tau e_1$, $e_1e_2e_1 = \tau e_1$ etc. (cf. proof of Proposition 9.23). The Bratteli diagrams of the inclusions are obtained by reflecting that of the previous stage, beginning with that of $A \subset B$, and the projections e_i have path algebra representations as in Fig. 9.16. Using $U_i = e_i/\sqrt{\tau}$, and Fig. 9.21, the relation $U_1U_2U_1 = U_1$ has the graphical representation of Fig. 9.33.

9.5 Dimension groups of Temperley–Lieb algebras

Consider a locally finite graph Γ with distinguished vertex $*$ and incidence matrix Δ. Markov traces on the AF algebra $A(\Gamma, *)$ can be used to compute its dimension group $K_0(A(\Gamma, *))$, at least for certain classes of graphs. It is then possible to discuss embeddings $A(\Gamma_1, *_1)$ in $A(\Gamma_2, *_2)$ of AF algebras associated with graphs Γ_1 and Γ_2 from a dimension group point of view.

From the remark above we have the following. A Markov trace Tr on $A(\Gamma, *)$ is given from a solution $\{\phi_v > 0 \mid v \in \Gamma^{(0)}\}$ to

$$x\phi_v = \textstyle\sum_w \Delta(v, w)\phi_w, \quad \phi_* = 1. \tag{9.5.1}$$

For any such family $\{\phi_v\}$ there is a representation of the Temperley–Lieb algebra, which is generated by the projections satisfying the relations in Theorem 9.25, in $A(\Gamma, *)$ by projections $\{e_i \mid i \in \mathbb{N}\}$ which satisfy

$$e_n e_{n\pm1} e_n = \tau e_n; \quad e_n e_m = e_m e_n, \quad |m - n| > 1\,, \tag{9.5.2}$$

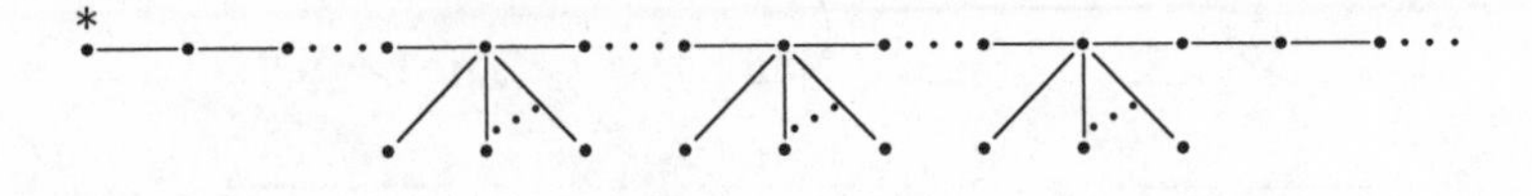

FIG. 9.34. Graph Γ

$p-1 \quad p \quad p+1 \quad p+r-2$

$0 \quad 1 \quad 2 \quad p-2$

$\bar{p}$

$\overline{p+q-3}$

$\overline{p+q-2}$

FIG. 9.35. $T_{p,q,r}, p, q < \infty, r \leq \infty$

$$\mathrm{Tr}(ye_m) = \tau \mathrm{Tr} y, \quad y \in A(\Gamma, *)_n \tag{9.5.3}$$

where $\tau = x^{-2}$.

For certain graphs Γ, we can find rational functions $\{\phi_v \mid v \in \Gamma^{(0)}\}$ in ar indeterminate x such that (9.5.1) holds. After a change of variable, $t = x^{-2}$ these functions can be used to generate or describe $K_0(A(\Gamma, *))$. The graph we consider are those as in Fig. 9.34, consisting of one (possibly infinite) linea branch A_m, $m \leq \infty$, or $A_{\infty,\infty}$, together with a finite number of branches o length one.

These graphs are not entirely pathological as they conveniently describe th limit points of the set of norms of graphs (Shearer 1989). For more complicate graphs, coming from $SU(N)$ counterparts with $N > 2$, as in Section 13.3, w would need functions ϕ_v of more than one indeterminate. We will be most par ticularly concerned with the T-shaped graph in Fig. 9.35.

Then for $q = 1, 2$ and $\alpha = (v, m)$ for the vertex $v \in T^{(0)}$ at level m in th Bratteli diagram $\hat{T}^{(0)}$ we will assign using ϕ_v, a polynomial Q_α, which will giv a map $\alpha \to Q_\alpha$ from projections (at level m of the Bratteli diagram) in $A(\Gamma, *$ into $\mathbb{Z}[t]$. This will identify $K_0(A(\Gamma, *))$ with $\mathbb{Z}[t]/\langle Q_r \rangle$, where $Q_\infty = 0$, and Q_r i essentially the characteristic polynomial of the graph $T_{p,q,r}$ and $\langle Q_r \rangle$ is the idea generated by Q_r. Thus for $r = \infty$, $K_0(A(T_{p,q,r}))$ as a group, does not depend o p or on the graph. This dependence does, however, appear when we look at th ordering in K_0. Then

$$K_0(A(T_{p,q,r},*)) = \begin{cases} \{0\} \cup \{P \mid P \in \mathbb{Z}[t], P(t) > 0, t \in [0,\gamma)\} & r = \infty \\ \{0\} \cup \{P + \langle Q \rangle \mid P \in \mathbb{Z}[t], P(\gamma) > 0\} & r < \infty \end{cases}$$

where $\gamma = \gamma_{p,q,r} = \|T_{p,q,r}\|^{-2}$, $\gamma_{p,q} = \gamma_{p,q,\infty}$ and $\|\Gamma\| = \|\Delta\|$.

For the graph A_∞ the solution $\{\phi_v \mid v \in A_\infty^{(0)}\}$ to equation (9.5.1) is given by the *Chebyshev polynomials of the second kind* $S_n \in \mathbb{Z}[x]$ satisfying

$$S_r = xS_{r-1} - S_{r-2}, \quad S_0(x) = 1, \quad S_1(x) = x \tag{9.5.4}$$

$$S_n(x) = \sum_{k, 0 \le 2k \le n} (-1)^k \; {}^{(n-k)}C_k \; x^{n-2k} \tag{9.5.5}$$

$$S_n(x) = \prod_{k=1}^{n} (x - 2\cos(k\pi/(n+1))) \tag{9.5.6}$$

$$S_n(t + t^{-1}) = (t^{n+1} - t^{-(n+1)})/(t - t^{-1}) \tag{9.5.7}$$

$$S_n(2\cos\theta) = \sin(n+1)\theta / \sin\theta. \tag{9.5.8}$$

The first few are

$$s_1 = x\,, \quad s_2 = x^2 - 1\,, \quad s_3 = x^3 - 2x\,, \quad s_4 = x^4 - 3x^2 + 1\,.$$

Moreover if we extrapolate (9.5.4) to be defined for negative r so that $S_{-n}(x) = -S_{n-2}(x)$, then for $n \ge 0$:

$$x^{n+1} = \sum_{r=0}^{n} {}^nC_r \; S_{n-2r+1}(x). \tag{9.5.9}$$

Let $\{\phi_v\}$ be a family of rational functions associated to a graph Γ, satisfying (9.5.1). Then we define, for $v \in \Gamma^{(0)}$:

$$Q_v(t) = x^{-d(v)}\phi_v(x) \tag{9.5.10}$$

where $t = x^{-2}$. For $\Gamma = A_\infty$, $Q_r = P_r$, where $P_r \in \mathbb{Z}[t]$, $r = 0, 1, 2, \ldots$, are defined by

$$P_r(t) = x^{-r}S_r(x) \tag{9.5.11}$$

$t = x^{-2}$, and are the polynomials:

$$P_r = P_{r-1} - tP_{r-2}, \quad P_0(t) = 1, \quad P_{-1}(t) = 0\,. \tag{9.5.12}$$

The first few are

$$P_1 = 1\,, \quad P_2 = 1 - t\,, \quad P_3 = 1 - 2t\,, \quad P_4 = 1 - 3t + t^2\,.$$

Lemma 9.34 *Let $\{Q_v\}$ be the rational functions associated with the graph $T_{p,q,\infty}$, $p \ge q \ge 1$. Then*

$$\{t \ge 0 \mid Q_v(t) \ge 0, \text{ for all } v \in T^{(0)}_{p,q,\infty}\} = [0, \gamma_{p,q}], \tag{9.5.13}$$

Proof For $q = 1$, $T_{p,1,\infty} = A_\infty$, and so we see from (9.5.6) and (9.5.11) that

$$\{t \geq 0 \mid P_n(t) > 0, n = 0, 1, 2, \ldots\} = [0, 1/4] \tag{9.5.14}$$

We refer to (Evans and Gould 1994a) for the general case. □

We associate to each vertex (v, n) of $\hat{T}_{p,q,\infty}$, the polynomial

$$Q_{(v,n)}(t) = t^{(n-d(v))/2} Q_v(t) \tag{9.5.15}$$

where d is the distance function on $\hat{T}_{p,q,\infty}$. Thus our notation is consistent with the embedding of $T_{p,q,\infty}$ in $\hat{T}_{p,q,\infty}$. Now let $\langle Q_{p+r-1}\rangle$ be the ideal in $\mathbb{Z}[t]$ generated by $Q_{(p+r-1,0)} = Q_{p+r-1}$ where $Q_\infty = 0$, and denote by $\overline{\mathbb{Z}[t]}$ the quotient ring $\mathbb{Z}[t]/\langle Q_{p+r-1}\rangle$. For $f(t) \in \mathbb{Z}[t]$, we write $\bar{f}$ for its image $f + I$ in $\mathbb{Z}[t]/\langle Q_{p+r-1}\rangle$. The set $\{1, \bar{t}, \ldots, \bar{t}^{d-1}\}$ forms a basis of $\overline{\mathbb{Z}[t]} = \mathbb{Z}[\bar{t}]$, considered as a $\mathbb{Z}$-module where $d = \deg Q_{p+r-1}$ if $r < \infty$, and $d = \infty$ if $r = \infty$.

Also for a connected graph Γ, and $\eta \in \Gamma^{(0)}$, we let $P^k(\eta)$ denote the set of paths of length k from η. For $i \in P^k(\eta)$, let $r(i)$ denote the endpoint. Note that since Γ is connected there is at least one $i \in P^2(\eta)$ such that $r(i) = \eta$.

Lemma 9.35 *Consider the graph $T_{p,2,r}$, for $p \geq 2$, with associated polynomials $\{\bar{Q}_{(v,n)} \mid (v,n) \in \hat{T}^{(0)}_{(p,2,r)}\}$.*

(a) The polynomials $\{\bar{Q}_{(v,n)}\}$ satisfy the following splitting rules:

$$\bar{Q}_{(v,n)} = \textstyle\sum_w \Delta(v,w) \bar{Q}_{(w,n+1)}.$$

(b) The linear span over $\mathbb{Z}$ of the polynomials $\{\bar{Q}_{(v,n)} \mid (v,n) \in \hat{T}^{(0)}_{p,2,r}\}$ is $\overline{\mathbb{Z}[t]}$.

(c) If p is even, then for each fixed k, the set of polynomials associated with level $2k+1$ of $\hat{T}_{p,2,r}$ are linearly independent over $\mathbb{Z}$. If p is odd, those associated with level $2k$ are linearly independent over $\mathbb{Z}$.

(d) We have

$$(1-t)\bar{Q}_{(v,n)} = \sum_{i \in P^2(v)}{}' \bar{Q}_{(r(i),n+2)}$$

where $\sum'$ means that we omit the vertex v exactly once from the summation.

Proof (a) is clear.

(b) and (c): Suppose p is even, then there are $k+1$ vertices on level $2k+1$ of $\hat{A}_\infty$. We prove that if the monomials $\{t^j\}_{j=0}^k$ are in the linear span over $\mathbb{Z}$ of the polynomials associated with level $2k+1$ of $\hat{T}_{p,2,\infty}$ (for level 1 this is clear), then $\{t_j\}_{j=0}^{k+1}$ are in the corresponding linear span for level $2k+3$ of $\hat{A}_\infty$. This is enough to establish (b) and (c) by induction. Suppose that for level $2k+1$ we have

$$t^j = \textstyle\sum_l a_{jl} Q_{(2l+1,2k+1)}, \quad j = 0, 1, 2, \ldots, k \tag{9.5.16}$$

where $a_{jl} \in \mathbb{Z}$. Then using the splitting rules, we express each polynomial $Q_{(2l+1,2k+1)}$ in (9.5.16) as a sum of the polynomials associated with level $2k+2$,

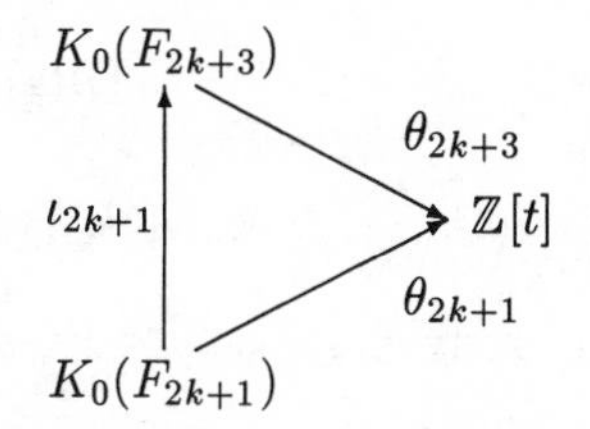

FIG. 9.36. Composition of $K_0(A(T_{p,2,\infty}))$

then we repeat to get each of those on level $2k+2$, as a sum of those on level $2k+3$. We also have $t^{k+1} = Q_{(1,2k+3)}$ and so the set $\{t_j\}_{j=0}^{k+1}$ is in the desired linear span.

(d) If $(u,v) \in T^{(0)}_{p,2,r}$, we write $u \sim v$ if $(u,v) \in T^{(1)}_{p,2,r}$. Then from the definitions of the polynomials ϕ_v we have

$$\phi_2\phi_v = (x^2-1)\phi_v = x(\sum_{u\sim v}\phi_u) - \phi_v = \sum_{u\sim v}\sum_{w\sim u}\phi_w - \phi_v = \sum_{w\in P^2(v)}{}'\phi_w.$$

Now $Q_v(t) = x^{-d(v)}\phi_v(x)$, where $t = x^{-2}$, and so

$$\begin{aligned} Q_2Q_{(v,n)} &= t^{(n-d(v))/2}Q_2Q_v = x^{-n-2}\phi_2\phi_v = \sum_{w\in P^2(v)}{}'\phi_w \\ &= x^{-n-2}\sum_w{}' x^{d(w)}Q_w(t) = \sum_w{}' Q_{(w,n+2)}(t). \end{aligned}$$

□

Let $F_n = A(T_{p,2,\infty})_n$ denote the finite dimensional algebra corresponding to level n of the Bratteli diagram of $A(T_{p,2,\infty})$ and $j_n : F_n \to F_{n+1}$ the corresponding inclusion. Thus if p is even $K_0(F_{2k+1}) \cong \mathbb{Z}^{k+1}$, and we can define maps $\theta_{2k+1} : K_0(F_{2k+1}) \to \mathbb{Z}[\bar{t}]$ by

$$\theta_{2k+1}[(a_i)_{i=0}^k] = \textstyle\sum_{j=0}^k a_j\bar{Q}_{(2j+1,2k+1)} \tag{9.5.17}$$

which are injective by Lemma 9.35 (c). It follows from Lemma 9.35 (a) that the diagram in Fig. 9.36 is commutative, where $\iota_{2k+1} = K_0(j_{2k+2}\circ j_{2k+1})$. Hence using Lemma 9.35 (b) we have an isomorphism

$$\theta_\infty : \varinjlim K_0(F_{2k+1}) \cong K_0(A(T_{p,2,\infty})) \to \mathbb{Z}[\bar{t}] \tag{9.5.18}$$

The case p odd is similar – we have an isomorphism:

$$\theta_\infty : \varinjlim K_0(F_{2k}) \cong K_0(A(T_{p,2,\infty})) \to \mathbb{Z}[\bar{t}]. \tag{9.5.19}$$

We now characterize the elements in $\mathbb{Z}[\bar{t}]$ corresponding to the positive cone of $K_0(A(T_{p,2,r}))$. We let

$$R = \mathrm{lin}_{\mathbb{Z}}\{\bar{Q}_{(v,n)}\} = \mathbb{Z}[\bar{t}] \cong K_0(A(T_{p,2,r})) \tag{9.5.20}$$

$$R_+ = \mathrm{lin}_{\mathbb{N}}\{\bar{Q}_{(v,n)}\} \cong K_0(A(T_{p,2,r}))_+. \tag{9.5.21}$$

Then R is a ring with an additively closed subset R_+ satisfying

$$R_+ - R_+ = R, \quad R_+ \cap (-R_+) = \{0\}, \tag{9.5.22}$$

so that considered as a group, (R, R_+) is partially ordered. Since (R, R_+) is a dimension group it is unperforated, and $1 = \bar{1} = \bar{Q}_{(0,0)}$ is an order unit in (R, R_+), i.e. for all $P \in R_+$, there exists $N \in \mathbb{N}$ such that $N \cdot 1 - P \in R_+$. Later we will show that $R_+ R_+ \subset R_+$, so that (R, R_+) is in fact a partially ordered commutative ring. Let $\bar{Q} \in R_+$, then $\bar{Q} = \sum a_{(r,n)} \bar{Q}_{(r,n)}$, with $a_{(r,n)} \in \mathbb{N}$. Thus either $Q = 0$ or

$$Q(t) > 0 \quad \text{for } t \in (0, \gamma_{p,2}].$$

We will show that this property characterizes R_+, and consequently that $K_0(A(T_{p,2,r}))$ is an ordered ring.

Lemma 9.36 *Let C be a subset of $\mathbb{Z}[\bar{t}]$, with $\bar{1} \in C$ such that $(\mathbb{Z}[\bar{t}], C)$ is an ordered group and $\bar{t}C \subset C$, $(\bar{1} - \bar{t})C \subset C$. Then every normalized extremal state on $(\mathbb{Z}[\bar{t}], C)$ is a ring homomorphism.*

Proof Let ϕ be a normalized (i.e $\phi(\bar{1}) = 1$) extremal state. Put $\varphi(\bar{P}) = \phi(\bar{t}\bar{P})$, for $\bar{P} \in \mathbb{Z}[\bar{t}]$. Then since $\bar{t}C \subset C$, φ is a state on $(\mathbb{Z}[\bar{t}], C)$. Suppose $\bar{P} \in C$, then $(\bar{1} - \bar{t})\bar{P} \geq 0$, i.e. $\bar{P} \geq \bar{t}\bar{P}$ and so $\phi(\bar{P}) \geq \phi(\bar{t}\bar{P}) = \varphi(\bar{P})$. Hence φ is a state dominated by ϕ, but ϕ is extremal and so there exists a real non-negative number λ such that $\lambda\phi = \varphi$. Thus since ϕ is normalized we have

$$\lambda = \lambda\phi(\bar{1}) = \varphi(\bar{1}) = \phi(\bar{t} \cdot \bar{1}) = \phi(\bar{t})\,. \tag{9.5.23}$$

It follows inductively from (9.5.23) that $\phi(\bar{t}^n) = \lambda^n$ for all $n \geq 0$, hence $\phi(\bar{P}) = P(\lambda)$ for all $\bar{P} \in \mathbb{Z}[\bar{t}]$. □

Corollary 9.37 *Let $p \geq q$, $q = 1, 2$. If ϕ is a normalized extremal state on $K_0(A(T_{p,q,r}))$, then $\phi(P) = P(\lambda)$ for some λ and all $P \in K_0(A(T_{p,2,r})) \cong \mathbb{Z}[\bar{t}]$. If $r = \infty$, $\lambda \in [0, \gamma_{p,q}]$, whilst $\gamma = \gamma_{p,q,r}$ if $r < \infty$.*

Proof It is clear from the definition of R_+ that $tR_+ \subset R_+$; moreover $(1 - \bar{t})R_+ \subset R_+$ by Lemma 9.35 (d). Since $\bar{t} \in R_+$, we have $\phi(\bar{t}) = \lambda \geq 0$. Moreover $\bar{Q}_v \in R_+$ for all $v \in T^{(0)}_{p,q,r}$, and so $\phi(Q_v) = Q_v(\lambda) \geq 0$. Hence $\lambda \in [0, \gamma_{p,q}]$ by Lemma 9.34. □

Theorem 9.38 *For $p \geq q$, $q = 1, 2$ we have $K_0(A(T_{p,q,r})) \cong \mathbb{Z}[t]/\langle Q_{p+r-1}\rangle$ and $K_0(A(T_{p,q,r}))_+$ corresponds under this isomorphism to the set*

$$\{0\} \cup \{Q \in \mathbb{Z}[t] \mid Q(\lambda) > 0 \text{ for } \lambda \in (0, \gamma_{p,q}]\} \qquad r = \infty\,,$$

$$\{0\} \cup \{Q \in \mathbb{Z}[\bar{t}] \mid Q(\gamma_{p,q,r}) > 0\} \qquad r < \infty\,.$$

Proof $r = \infty$. It remains to show that if $Q \in \mathbb{Z}[t]$ and $Q(\lambda) > 0$ for $\lambda \in (0, \gamma_{p,q}]$, then Q is a linear combination of the polynomials $\{Q_{(v,n)} \mid (v,n) \in \hat{T}^{(0)}_{p,2,\infty}\}$ with non-negative integer coefficients. We can write $Q(t) = t^k P(t)$, for some non-negative integer k, where $P(\lambda) > 0$ for all $\lambda \in [0, \gamma_{p,q}]$. Let ϕ be any normalized extremal state on $K_0(A(T_{p,q,\infty}))$. Then by Corollary 9.37, $\phi(P) = P(\lambda)$ for some $\lambda \in [0, \gamma_{p,q}]$, and so $\phi(P) > 0$. It follows from (Goodearl and Handelman 1976) that P is in $K_0(A(T_{p,q,\infty}))_+$ and hence so is $Q = t^k P$.

Suppose now $r < \infty$. Now $(\phi_v(\beta_{p,q,r}))_{v \in T^{(0)}_{p,q,r}}$ is a Perron–Frobenius eigenvector for the incidence matrix Δ of $T_{p,q,r}$, where $\beta_{p,q,r} = \|T_{p,q,r}\|$ is the Perron–Frobenius eigenvalue, and so $\phi_v(\beta_{p,q,r}) > 0$ for each $v \in T^{(0)}_{p,q,r}$. Then it follows from the definition of $Q_v(t)$, that $Q_v(\gamma_{p,q,r}) > 0$ for each $v \in T^{(0)}_{p,q,r}$. Hence the positive cone $\overline{R}_+$ is contained in the set $\{\overline{Q}; Q \in \mathbb{Z}[t], Q(\gamma_{p,q,r}) > 0\} \cup \{0\}$. Now suppose that $\overline{Q} \neq 0$ is an element of this set, then if ϕ is the extremal state on $(\overline{R}, \overline{R}_+)$ we have $\phi(\overline{Q}) = Q(\gamma_{p,q,r}) > 0$, and so by (Goodearl and Handelman 1976) it follows that $\overline{Q} \in \overline{R}_+$. □

In particular we have $K_0(A(A_n)) \cong \mathbb{Z}[t]/\langle P_n\rangle$, where $P_\infty = 0$. The positive cone may be identified with the set

$$\{\bar{Q} = Q + \langle P_n\rangle \mid Q(\gamma) > 0\} \cup \{0\}$$

where $\gamma = \|A_n\|^{-2} = (2\cos(\pi/(n+1)))^{-2}$ if $n < \infty$ or $\{Q \in \mathbb{Z}[t] \mid Q(\lambda) > 0, \lambda \in (0, 1/4]\} \cup \{0\}$ if $n = \infty$.

Corollary 9.39 *$K_0(A(T_{p,q,r}))$ is a partially ordered ring, for $p \geq 1$.*

In particular, Corollary 9.39 says that there exist non-negative integers $a^{\gamma}_{\alpha\beta}$ for $\alpha, \beta, \gamma \in \hat{T}^{(0)}_{p,2,r}$ such that

$$\bar{Q}_\alpha \bar{Q}_\beta = \textstyle\sum_\gamma a^{\gamma}_{\alpha\beta} \bar{Q}_\gamma\,. \tag{9.5.24}$$

A special case of this has already been verified in Lemma 9.35(d).

Suppose that $\Gamma = T_{p,q,r}$ ($q = 1, 2$), and $\|\Gamma\| \geq 2$. Then since $K_0(A(\Gamma)) \cong \mathbb{Z}[t]/\langle Q_{p+r-1}\rangle$ the quotient map of $\mathbb{Z}[t]$ onto $\mathbb{Z}[t]/\langle Q_{p+r-1}\rangle$ gives a surjective ring homomorphism $\pi : K_0(A(A_\infty)) \to K_0(A(\Gamma))$. Since the positive cone of $K_0(A(\Gamma))$ can be identified with the set $\{\bar{\bar{Q}} = Q + \langle Q_{p+r-1}\rangle \mid Q(\gamma) > 0\} \cup \{0\}$ where $\gamma = \|\Gamma\|^{-2} \in (0, 1/4]$, the map π is positive, and since $f \in \operatorname{Ker} \pi$ implies

$f(\gamma)=0$, $K_0(A(A_\infty))_+ \cap \operatorname{Ker}\pi = \{0\}$, and $\pi(K_0(A(A_\infty))_+) \subset K_0(A(\Gamma))_+$. Since $K_0(A(\Gamma))_+$ is generated by the polynomials

$$\{\bar{\bar{Q}}_{(v,m)} = Q_{(v,m)} + \langle Q_{p+r-1}\rangle \mid (v,m) \in \hat{T}^{(0)}_{p,q,r}\}$$

there are non-negative integers $a_{m\alpha}$, $\alpha \in \hat{T}^{(0)}_{p,q,r}$ such that

$$\bar{\bar{P}}_m = \textstyle\sum_\alpha a_{m\alpha}\bar{\bar{Q}}_\alpha, \tag{9.5.25}$$

for $m = 0, 1, 2, \ldots$.

Now suppose that $\|\Gamma\| < 2$, so $\Gamma = A_k, D_k, E_6, E_7, E_8$ as in Figs 10.8–10.12, and $\|\Gamma\| = 2\cos(\pi/(n+1))$ for some $n \geq 2$. Now $K_0(A(\Gamma)) \cong \mathbb{Z}[t]/\langle Q_{p+r-1}\rangle$. One can check, by comparing Coxeter exponents of the graphs A_n, and Γ, that $Q_{p+r-1} \mid P_n$, and hence $\langle Q_{p+r-1}\rangle \supset \langle P_n\rangle$. Thus the map $\pi : \mathbb{Z}[t]/\langle P_n\rangle \to \mathbb{Z}[t]/\langle Q_{p+r-1}\rangle$ given by $\pi(\bar{f}) = \bar{\bar{f}}$ is a surjective ring homomorphism, and its kernel is the ideal in $\mathbb{Z}[t]/\langle P_n\rangle$ generated by Q_{p+r-1}. And so from the identification of the positive cones, we have a positive map $\pi : K_0(A(A_n)) \to K_0(A(\Gamma))$ such that $K_0(A(A_n))_+ \cap \operatorname{Ker}\pi = \{0\}$, and $\pi(K_0(A(A_n))_+) = K_0(A(\Gamma))_+$. Hence there are non-negative integers $a_{m\alpha}$, $\alpha \in \hat{T}^{(0)}_{p,q,r}$, such that (9.5.25) holds for $m = 0, 1, \ldots, n-1$. The graphs E_6, E_7, E_8 have norm $2\cos\pi/m$, with $m = 12, 18, 30$ respectively. Thus we have embeddings of $A(A_{11})$, $A(A_{17})$, $A(A_{29})$ in $A(E_6)$, $A(E_7)$, and $A(E_8)$ respectively. The graph E_9 has norm 2, and so we have an embedding of $A(A_\infty)$ in $A(E_9)$. We will discuss liftings of these maps between dimension groups and intertwining graphs in more detail in Section 11.7.

9.6 Higher relative commutants and commuting squares

We are now back in the set up of II_1 factors $N \subset M$. Proposition 9.18 (2) and Proposition 9.22 (5) imply $[M_k : N] = [M : N]^{k+1} < \infty$. Thus for all k, we know that $N' \cap M_k$ is finite dimensional by Proposition 9.18 (3). We make the following definitions.

Definition 9.40 *The increasing sequence of the finite dimensional algebras $\{N' \cap M_k\}_k$ is called the tower of* higher relative commutants.

Definition 9.41 *Draw the Bratteli diagram of the tower of the higher relative commutants $\{N' \cap M_k\}_k$. We will later prove in Section 10.2 that the Bratteli diagram at each step consists of the reflection of the previous step and a (possibly empty) new graph. (See Fig. 9.37 for example.) The "new graph" put together is called the* principal graph *of the subfactor $N \subset M$. (The thick part is the principal graph in Fig. 9.37.) We also get a graph for the tower $\{M' \cap M_k\}_k$. This second "new" graph is called a* dual principal graph. *The (dual) principal graph can be infinite in general, but if they are finite, we say that the subfactor $N \subset M$ has* finite depth. *(We will prove in the Remark in page 529 that if the principal graph or the dual principal graph is finite, then the other is automatically finite.)*

Finite depth will be the subfactor analogue of rationality in conformal field theory. (See Chapter 8.)

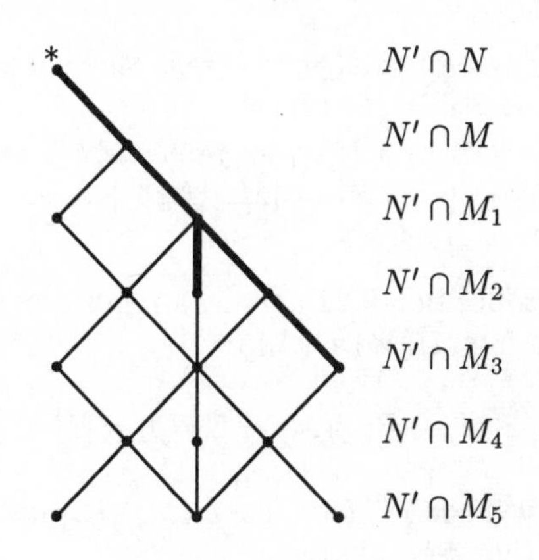

FIG. 9.37. Example of a principal graph

For a subfactor $N \subset M$, the basic construction gives a bigger algebra. We now want to make a construction giving a smaller algebra in the converse direction.

Proposition 9.42 *For $N \subset M$, there exists a subfactor P of N such that $P \subset N \subset M$ is standard.*

Proof Choose a projection $p \in$M with $\operatorname{tr}(p) = [M : N]^{-1}$. Then both M and N act on $L^2(M)p$ by the left multiplication. Since we have $\dim {}_N L^2(M)p = \dim {}_N L^2(M)\operatorname{tr}(p) = 1$ by Theorem 5.33 (3), we can identify $L^2(M)p$ with $L^2(N)$ as left N-modules. With this identification, we may think that M acts on $L^2(N)$. On $L^2(N)$, we have a natural conjugation J, and we set $P = JM'J$, where M' is the commutant of M in $B(L^2(N))$. By Proposition 9.22 (4), this gives the desired P. □

Definition 9.43 *We call the construction of P in Proposition 9.42 a* downward basic construction. *By repeating this procedure, we can construct a decreasing sequence of II_1 factors as follows.*

$$\cdots \subset N_3 \subset N_2 \subset N_1 \subset N \subset M.$$

We call such a sequence a tunnel.

An important point is that the downward basic construction depends on a projection p and is not canonical and hence a tunnel is not unique, but we have uniqueness of a tunnel in a weaker sense. We give a precise meaning of this in Proposition 9.45.

First we need a preliminary.

Proposition 9.44 *Let ϕ be a map from the unitary group $\mathcal{U}(M)$ of M to the set of projections* $\mathrm{Proj}(M_1)$ *in M_1 defined by $\phi(u) = ue_N u^*$ for $u \in \mathcal{U}(M)$. Then the following hold.*

1. *The map ϕ induces a bijection from the quotient $\mathcal{U}(M)/\mathcal{U}(N)$ onto $\{p \in \mathrm{Proj}(M_1) \mid E_M(p) = [M : N]^{-1}\}$.*
2. *The map ϕ induces a bijection from the quotient $\mathcal{N}(N)/\mathcal{U}(N)$ onto $\{p \in \mathrm{Proj}(N' \cap M_1) \mid E_M(p) = [M : N]^{-1}\}$, where $\mathcal{N}(N)$ denotes the set of normalizers of N in M.*

Proof 1. By Proposition 9.22 (2) and Proposition 9.23, the map ϕ induces a well-defined map from $\mathcal{U}(M)/\mathcal{U}(N)$ to

$$\{p \in \mathrm{Proj}(M_1) \mid E_M(p) = [M : N]^{-1}\}.$$

Suppose $ue_N u^* = ve_N v^*$ for $u, v \in \mathcal{U}(M)$. Then $v^*u \in N$ by Proposition 9.22 (2). This means that ϕ is injective.

Choose a projection $p \in \mathrm{Proj}(M_1)$ with $E_M(p) = [M : N]^{-1}$. Because $\mathrm{tr}(p) = [M : N]^{-1} = \mathrm{tr}(e_N)$, we get a unitary $v \in M_1$ with $ve_N v^* = p$ in a II_1 factor M_1. (Note that we have $p \sim e_N$ and $1 - p \sim 1 - e_N$.) By Lemma 9.26, we have an element $u \in M$ with $ue_N = ve_N$, which implies $ue_N u^* = p$. By applying E_M to this formula, we get $uu^* = 1$ by Proposition 9.23. This means u in a II_1 factor M is unitary, hence ϕ is surjective.

2. First note that if $u \in \mathcal{N}(N)$, then $ue_N u^*$ is indeed in $N' \cap M_1$ by Proposition 9.22 (2).

The injectivity of ϕ has already been proved in (1). For the surjectivity, choose a projection $p \in \mathrm{Proj}(N' \cap M_1)$ with $E_M(p) = [M : N]^{-1}$. We have $u \in \mathcal{U}(M)$ with $\phi(u) = p$ by 1. Commutativity of $ue_N u^*$ and N implies commutativity of e_N and u^*Nu. Proposition 9.22 (2) then implies $u^*Nu \subset N$. Because $[M : u^*Nu] = [M : N]$, we get equality here. □

We now have uniqueness of the downward basic construction in the following weaker sense.

Proposition 9.45 *Suppose that for a subfactor $N \subset M$, we have two downward basic constructions $P \subset N \subset M$ and $Q \subset N \subset M$. Then there exists a unitary u in N satisfying $uPu^* = Q$.*

Proof We have $M = \langle N, e_P \rangle = \langle N, e_Q \rangle$. Because $E_N(e_P) = E_N(e_Q) = [M : N]^{-1}$ by Proposition 9.23, we get a unitary $u \in N$ satisfying $ue_P u^* = e_Q$. By Proposition 9.22 (2), we get

$$Q = N \cap \{e_Q\}' = N \cap \{ue_P u^*\}' = u(N \cap \{e_P\}')u^* = uPu^*.$$

□

By Proposition 9.45 and induction, we immediately get the following.

Corollary 9.46 *Let*

$$N_k \subset \cdots \subset N_3 \subset N_2 \subset N_1 \subset N \subset M$$

and

$$P_k \subset \cdots \subset P_3 \subset P_2 \subset P_1 \subset N \subset M$$

be two choices of tunnel. Then there exists a unitary $u \in N$ *such that* $P_j = uN_ju^*$ *for* $j = 1, 2, \ldots, k$.

We give an abstract characterization of the basic construction.

Proposition 9.47 *Let* P *be a* II_1 *factor containing* M *and a projection* $f \in N' \cap P$ *with* $E_M(f) = [M : N]^{-1} = [P : M]^{-1}$. *Then there is an isomorphism* $\Phi : P \to M_1$ *satisfying* $\Phi(x) = x$ *for* $x \in M$ *and* $\Phi(f) = e_N$.

Proof Apply the downward basic construction to $M \subset P$ to get a subfactor Q of M so that $P = \langle M, e_Q \rangle$. Because $E_M(e_Q) = [P : M]^{-1} = E_M(f)$, we get a unitary $v \in M$ satisfying $ve_Qv^* = f$ by Proposition 9.44 (1). Apply $\mathrm{Ad}(v)$ to $Q \subset M \subset P = \langle M, e_Q \rangle$, then we get $vQv^* \subset M \subset \langle M, f \rangle = P$. We have $N \subset M \cap \{f\}' = vQv^*$ and $[M : N] = [M : vQv^*]$, thus $N = vQv^*$. So we are done. □

Theorem 9.48 *For* $x \in M_+$, *we have* $E_N(x) \geq (1/[M : N])x$. *For a subfactor* N *of a* II_1 *factor* M *(possibly with infinite index), the constant* $1/[M : N]$ *is the biggest possible constant in this inequality, where we use the convention* $1/\infty = 0$.

Proof For simplicity, we assume that $[M : N] < \infty$ in the following proof. For a proof in the general case, see (Pimsner and Popa 1986a).

We perform a downward basic construction to get $P \subset N \subset M$ standard. Choose an $x \in M_+$ and write x as y^*y with $y = \sum_{j=1}^{n} a_j e_P b_j$ $(a_j, b_j \in N)$, by Proposition 9.28. We have

$$x = \sum_{j,k=1}^{n} b_j^* e_P a_j^* a_k e_P b_k = \sum_{j,k=1}^{n} b_j^* E_P(a_j^* a_k) e_P b_k,$$

by Proposition 9.22 (1). Define an element A by

$$A = \left(E_P(a_j^* a_k)\right)_{jk} \in M_n(\mathbb{C}) \otimes P.$$

Set

$$B = \begin{pmatrix} b_1 \\ b_2 \\ \vdots \\ b_n \end{pmatrix}, \quad E = \begin{pmatrix} e_P & 0 & \cdots & 0 \\ 0 & e_P & \cdots & 0 \\ \vdots & \vdots & \ddots & \vdots \\ 0 & 0 & \cdots & e_P \end{pmatrix} \in M_n(\mathbb{C}) \otimes M.$$

By complete positivity of E_P, we know that A is positive and hence there exists $C \in M_n(\mathbb{C}) \otimes P$ with $C^*C = A$. (The complete positivity follows from the positivity of the conditional expectation onto $M_n(\mathbb{C}) \otimes P$.) Then we get

$$x = B^*C^*CEB = B^*C^*ECB \le B^*C^*CB = [M : N]E_N(x),$$

which is the desired inequality.

Conversely, suppose that there exists $\lambda \ge 0$ such that $E_N(x) \ge \lambda x$ for all $x \in M_+$. Setting $x = e_P \in M_+$ in the inequality, we get $[M : N]^{-1} \ge \lambda$. □

The inequality in Theorem 9.48 is called the *Pimsner–Popa inequality.* We next study the iteration of the basic construction.

Theorem 9.49 *Let $N \subset M \subset M_1 \subset M_2 \subset \cdots$ be a Jones tower. Then $N \subset M_{k-1} \subset M_{2k-1}$ is standard with the Jones projection equal to*

$$[M : N]^{k(k-1)/2}(e_k e_{k-1} \cdots e_1)(e_{k+1}e_k \cdots e_2) \cdots (e_{2k-1}e_{2k-2} \cdots e_k).$$

Proof We give a proof for the case $k = 2$ for simplicity. The other cases can be proved similarly by induction.

Set $f = [M : N]e_2e_1e_3e_2$. By Theorem 9.25, we get $f = f^* = f^2$. By Proposition 9.22 (2), we know that $f \in N' \cap M_3$. By Proposition 9.47, it is enough to prove $E_{M_1}(f) = [M_3 : M_1]^{-1} = [M : N]^{-2}$ and hence it is enough to have $\mathrm{tr}(fx) = [M : N]^{-2}\mathrm{tr}(x)$ for $x \in M_1$. This follows from the following computation using Proposition 9.22 (1) and Proposition 9.23.

$$\begin{aligned} \mathrm{tr}(fx) &= [M : N]\mathrm{tr}(e_2e_1e_3e_2x) \\ &= [M : N]\mathrm{tr}(e_2xe_2e_1e_3) = \mathrm{tr}(e_2xe_2e_1) \\ &= \mathrm{tr}(E_M(x)e_2e_1) = [M : N]^{-1}\mathrm{tr}(e_1E_M(x)) \\ &= [M : N]^{-2}\mathrm{tr}(E_M(x)) = [M : N]^{-2}\mathrm{tr}(x). \end{aligned}$$

□

Proposition 9.50 *Let*

$$\cdots N_3 \subset N_2 \subset N_1 \subset N \subset M \subset M_1 \subset M_2 \subset \cdots$$

be the Jones tower and a choice of tunnel. Then we have a family of anti-isomorphism $\Phi_k : N_k' \cap M \to M' \cap M_{k+1}$, $k = 0, 1, \ldots$, such that Φ_{k+1} is an extension of Φ_k. (Here we put $N_0 = N$.)

Proof We choose a tunnel and regard N_k as subalgebras of $B(L^2(M))$. Let J be the canonical conjugation on $L^2(M)$. We set $Q_k = JN_{k-1}'J$ for $k = 1, 2, \ldots$. Then we can naturally identify M_k with Q_k, $k = 1, 2, \ldots$, so that we realize the Jones tower on $L^2(M)$. Then the map Φ_k is defined by $\Phi_k(x) = Jx^*J$. □

We now start a new important topic concerning commuting squares of inclusions of von Neumann algebras. Commuting squares will be used to approximate inclusions of infinite dimensional operator algebras with those of finite dimensional algebras.

Proposition 9.51 *Let*

$$\begin{matrix} K_0 & \subset & K_1 \\ \cap & & \cap \\ K_2 & \subset & K_3 \end{matrix}$$

be four von Neumann algebras with a faithful normal finite trace tr *on* K_3. *Then the following conditions are equivalent:*

1. $E_{K_1}(K_2) \subset K_0$.
2. $E_{K_2}(K_1) \subset K_0$.
3. $E_{K_1}E_{K_2} = E_{K_0}$.
4. $E_{K_2}E_{K_1} = E_{K_0}$.
5. $E_{K_2}E_{K_1} = E_{K_1}E_{K_2}$ *and* $K_0 = K_1 \cap K_2$.
6. *For* $x \in K_2$, *we have* $E_{K_0}(x) = E_{K_1}(x)$.
7. *For* $x \in K_1$, *we have* $E_{K_0}(x) = E_{K_2}(x)$.

Proof It is clear that condition (3) implies condition (6), and condition (6) implies condition (1).

Suppose we have condition (1). For $x \in K_3$, we get $E_{K_1}(E_{K_2}(x)) \in K_0$, thus for $a \in K_0$ we get

$$\mathrm{tr}(E_{K_1}(E_{K_2}(x))a) = \mathrm{tr}(E_{K_2}(x)a) = \mathrm{tr}(xa) = \mathrm{tr}(E_{K_0}(x)a).$$

This implies condition (3). Thus conditions (1), (3), (6) are equivalent.

Suppose we have condition (3). Consider the identity $E_{K_1}E_{K_2} = E_{K_0}$ on $L^2(K_3)$ and take the adjoints of the both sides. Then we get condition (5).

Assume condition (5) next. We know that

$$E_{K_1}(E_{K_2}(x)) = E_{K_2}(E_{K_1}(x)) \in K_1 \cap K_2 = K_0$$

for any $x \in K_3$. Then

$$\mathrm{tr}(E_{K_1}(E_{K_2}(x))a) = \mathrm{tr}(E_{K_2}(x)a) = \mathrm{tr}(xa) = \mathrm{tr}(E_{K_0}(x)a)$$

implies condition (3). Thus conditions (1), (3), (5), (6) are equivalent.

Similarly, we can prove that conditions (2), (4), (5), (7) are equivalent. □

Definition 9.52 *If four algebras*

$$\begin{matrix} K_0 & \subset & K_1 \\ \cap & & \cap \\ K_2 & \subset & K_3 \end{matrix}$$

satisfy one of the equivalent conditions in Proposition 9.51, we say that they form a commuting square.

For our purpose, the most important commuting squares are given by the following.

Proposition 9.53 *For a subfactor $N \subset M$, choose a subalgebra S of N. Then*

$$\begin{array}{ccc} S' \cap N & \subset & S' \cap M \\ \cap & & \cap \\ N & \subset & M \end{array}$$

is a commuting square.

Proof For $x \in S' \cap M$, it is clear that we have $E_N(x) \in S' \cap N$, because $S \subset N$. Thus we get condition (2) of Proposition 9.51. □

Corollary 9.54 *The following is a sequence of commuting squares.*

$$\begin{array}{ccccccc} M' \cap M & \subset & M' \cap M_1 & \subset & M' \cap M_2 & \subset & \cdots \\ \cap & & \cap & & \cap & & \\ N' \cap M & \subset & N' \cap M_1 & \subset & N' \cap M_2 & \subset & \cdots \end{array}$$

Proof By the Proposition 9.53, the conditional expectations from $M' \cap M_1$ onto $M \cap M$ and from $N' \cap M_1$ onto $N' \cap M$ are both restrictions of that from M_1 onto M. □

9.7 Bimodules and relative tensor products

In this section, we study the basic properties of bimodules as tools to study subfactors.

Definition 9.55 *Let P, Q be II_1 factors. A Hilbert space X is called a P–Q* bimodule *if it is a left P-module and a right Q-module and the left P-action and the right Q-action commute. We write ${}_PX_Q$ for this bimodule in order to express the II_1 factors explicitly.*

Basic examples of bimodules in the subfactor theory are $L^2(M_k)$ regarded as an M-M bimodule. For simplicity, we write ${}_MM_M$, ${}_NM_M$, and so on for ${}_ML^2(M)_M$, ${}_NL^2(M)_M$, and so on respectively. That is, we drop the symbol $L^2(\cdot)$.

Definition 9.56 *A bimodule ${}_PX_Q$ is said to be of* finite type *if* $\dim {}_PX < \infty$ *and* $\dim X_Q < \infty$. *We also set* $[X] = [{}_PX_Q] = \dim {}_PX \dim X_Q$ *and call this the* Jones index *of the bimodule.*

Note that the commutativity of the left and right actions implies that $P \subset Q'$ is a subfactor of type II_1, where Q' is the commutant of the right Q-action in $B(X)$. (Thus a bimodule of finite type naturally gives a subfactor of finite index.) Then by Theorem 5.33 (2) and Proposition 9.18 (1), we get $[{}_PX_Q] = [Q' : P]$, thus the name Jones index is justified.

Proposition 9.57 *If ${}_PX_Q$ is of finite type, then a vector in X is P-bounded if and only if it is Q-bounded.*

Proof We choose a basis of X as a left P-module as in the proof of Theorem 9.12. Then it is easy to see that the vectors in the basis are P'-bounded, hence Q-bounded. If $\xi \in X$ is P-bounded, it is a P-linear combination of this basis, so ξ is also Q-bounded. The converse direction holds by symmetry. □

So we just write X^{bdd} for P-bounded (or Q-bounded) vectors in ${}_P X_Q$ of finite type.

Definition 9.58 *Let X be a P-Q bimodule. We define a* conjugate *bimodule $\bar{X} = {}_Q\bar{X}_P$, which is the conjugate Hilbert space $\bar{X}$ with the actions $b \cdot \bar{\xi} \cdot a = \overline{a^* \cdot \xi \cdot b^*}$, where $a \in P, b \in Q, \xi \in X$.*

Example 9.59 We determine the conjugate bimodule of ${}_N M_M$. We can define a map $\Phi : {}_M\bar{M}_N \to {}_M M_N$ by $\Phi(\bar{x}) = x^*$. (Strictly speaking, we have to take the completion, so we use $\hat{x}$ instead of x and extend Φ to the entire space.) This Φ gives an isomorphism of bimodules, so we can identify the conjugate of ${}_N M_M$ with ${}_M M_N$.

We can verify the following identity by a direct computation.

Proposition 9.60 *Let X be a P-Q bimodule. For ${}_Q\bar{X}_P$, we have*

$$\langle \bar{\xi}, \bar{\eta} \rangle_Q = \langle \xi, \eta \rangle_Q^{\circ},$$

where $\xi, \eta \in X^{\mathrm{bdd}}$.

Our next aim is to introduce a correct notion of a 'product' for bimodules. We will make a relative tensor product of a P-Q bimodule X and a Q-R bimodule Y over a II_1 factor Q.

First we make an algebraic tensor product $X^{\mathrm{bdd}} \otimes Y^{\mathrm{bdd}}$ and put an inner product there by $\langle \xi \otimes \eta, \xi' \otimes \eta' \rangle = \langle \xi \langle \eta, \eta' \rangle_Q, \xi' \rangle$. This is a (possibly degenerate) inner product on the space, so we form a Hilbert space by completion after taking a quotient by the kernel. This becomes naturally a P–R bimodule. (This Hilbert space is non-zero. See Proposition 9.63.)

Definition 9.61 *The above bimodule is called a* relative tensor product *of ${}_P X_Q$ and ${}_Q Y_R$ over Q and denoted by ${}_P X \otimes_Q Y_R$. The image of an algebraic tensor product $\xi \otimes \eta$ in ${}_P X \otimes_Q Y_R$ is denoted by $\xi \otimes_Q \eta$.*

It is easy to see $\xi a \otimes_Q \eta = \xi \otimes_Q a\eta$ for $a \in Q$, $\xi \in X$, $\eta \in Y$. It is also easy to see that the relative tensor product operation is associative.

By direct computations, we can prove the following proposition easily, because $X^{\mathrm{bdd}} \otimes_Q Y^{\mathrm{bdd}}$ is dense in $(X \otimes_Q Y)^{\mathrm{bdd}}$.

Proposition 9.62 *Suppose we have bimodules ${}_P X_Q$, ${}_Q Y_R$ with* $\dim {}_P X < \infty$, $\dim {}_Q Y < \infty$. *If $\{\lambda_j\}_j$ is a basis of ${}_P X$ and $\{\mu_k\}_k$ is a basis of ${}_Q Y$, then $\{\lambda_j \otimes_Q \mu_k\}_{j,k}$ is a basis of ${}_P(X \otimes_Q Y)$.*

Proposition 9.63 *Suppose we have bimodules ${}_P X_Q$, ${}_Q Y_R$ with* $\dim {}_P X < \infty$, $\dim {}_Q Y < \infty$. *Then we have* $\dim {}_P(X \otimes_Q Y) = \dim {}_P X \dim {}_Q Y$.

Proof Set $\dim_Q Y = n + \alpha$, with $n \in \mathbb{N}$ and $0 \le \alpha < 1$. We can identify ${}_QY$ with $L^2(Q) \oplus \cdots \oplus L^2(Q) \oplus L^2(Q)p$ where we have n copies of $L^2(Q)$ and p is a projection in Q with $\mathrm{tr}(p) = \alpha$. Then $X \otimes_Q Y$ is naturally identified with $X \oplus \cdots \oplus X \oplus Xp$. We then get

$$\begin{aligned}\dim_P(X \otimes_Q Y) &= \dim_P X + \cdots + \dim_P X + \dim_P(Xp)\\ &= n \dim_P X + \mathrm{tr}_{P'}(p) \dim_P X = (n+\alpha)\dim_P X\\ &= \dim_P X \dim_Q Y.\end{aligned}$$

□

As a corollary, we have the following.

Corollary 9.64 *Suppose we have bimodules ${}_PX_Q$, ${}_QY_R$ of finite type. Then* $[{}_PX_Q][{}_QY_R] = [{}_PX \otimes_Q Y_R]$.

The following proposition is the key to the bimodule technique in subfactor theory (cf. Lemma 9.32).

Proposition 9.65 *There are isomorphisms:*

$${}_M(M_k)_M \cong {}_M(M \otimes_N M \otimes_N \cdots \otimes_N M)_M,$$

where we have $k+1$ copies of M.

Proof We define a map $\Phi : M \otimes_N M \to M_1$ by

$$\Phi(x \otimes_N y) = [M : N]^{1/2} x e_N y.$$

(Note that $\Phi(1 \otimes_N 1)$ is not equal to $1 \in M_1$.)

We have

$$\begin{aligned}&\langle x_1 \otimes_N y_1, x_2 \otimes_N y_2\rangle = \langle x_1 \langle y_1, y_2\rangle_N, x_2\rangle\\ &= \langle x_1 E_N(y_1 y_2^*), x_2\rangle = \mathrm{tr}(x_2^* x_1 E_N(y_1 y_2^*))\end{aligned}$$

and

$$\begin{aligned}&[M:N]\langle x_1 e_N y_1, x_2 e_N y_2\rangle = [M:N]\mathrm{tr}(y_2^* e_N x_2^* x_1 e_N y_1)\\ &= [M:N]\mathrm{tr}(x_1 e_N y_1 y_2^* e_n x_2^*) = [M:N]\mathrm{tr}(x_2^* x_1 E_N(y_1 y_2^*) e_N)\\ &= \mathrm{tr}(x_2^* x_1 E_N(y_1 y_2^*))\end{aligned}$$

by Proposition 9.22 (1) and Proposition 9.23. This means Φ is an isometry.

By Proposition 9.28, Φ is also surjective. It is easy to see that Φ induces a bimodule isomorphism ${}_M(M_1)_M \cong {}_M(M \otimes_N M)_M$.

For M_2, we have

$$M_2 \cong M_1 \otimes_M M_1 \cong (M \otimes_N M) \otimes_M (M \otimes_N M) \cong M \otimes_N M \otimes_N M.$$

The general case follows by induction. □

9.8 Frobenius reciprocity for bimodules

We will now introduce Frobenius reciprocity for bimodules as an analogue to Frobenius reciprocity for group representations. This Frobenius reciprocity plays the key role in the analogy between bimodules and representations of compact groups in Table 9.1. This is necessary for understanding the higher relative commutants of subfactors in terms of bimodules and intertwiners. See the Remark after Corollary 9.72 for its relation to classical Frobenius reciprocity.

Definition 9.66 *Let ${}_PX_Q$, ${}_PY_Q$ be bimodules with P, Q II_1 factors. We denote by* $\mathrm{Hom}({}_PX_Q, {}_PY_Q)$ *the set of all bounded linear operators from X to Y commuting with the left action of P and the right action Q. If $Y = X$, we write* $\mathrm{End}({}_PX_Q)$ *for* $\mathrm{Hom}({}_PX_Q, {}_PX_Q)$. *If* $\mathrm{End}({}_PX_Q) = \mathbb{C}$, *we say that ${}_PX_Q$ is* irreducible.

Example 9.67 Consider the example of ${}_NM_M$ arising from a subfactor $N \subset M$. Then $\mathrm{End}({}_NM_M)$ is the space of the left multiplications by the elements in $N' \cap M$. Hence this bimodule is irreducible if and only if the subfactor has trivial relative commutant.

We now list the basic properties of bimodules. In the first place:

Proposition 9.68 *If the bimodules ${}_PX_Q$ and ${}_PY_Q$ are of finite type, then* $\mathrm{Hom}({}_PX_Q, {}_PY_Q)$ *is finite dimensional.*

Proof It is enough to prove that $\mathrm{End}({}_PX_Q \oplus {}_PY_Q)$ is finite dimensional, because $\mathrm{Hom}({}_PX_Q, {}_PY_Q)$ is a subspace. Because ${}_PX_Q \oplus {}_PY_Q$ is clearly of finite type, the space $\mathrm{End}({}_PX_Q \oplus {}_PY_Q)$, which is the intersection of the commutants of the right and left actions, is finite dimensional by Proposition 9.18 (3). □

Definition 9.69 *Let ${}_PX_Q$, ${}_QY_R$, ${}_PZ_R$ be three irreducible bimodules of finite type. Define the intertwiner space as* $\mathcal{H}^Z_{X,Y} = \mathrm{Hom}({}_PX \otimes_Q Y_R, {}_PZ_R)$ *and so on. For $\sigma, \rho \in \mathcal{H}^Z_{X,Y}$, we have a natural inner product $\langle \sigma, \rho \rangle = \sigma\rho^* \in$* $\mathrm{End}({}_PZ_R) = \mathbb{C}$. *For an intertwiner $\sigma \in \mathcal{H}^Z_{X,Y}$, we define the right hand side* Frobenius dual *$\sigma^Y \in \mathcal{H}^X_{Z,\bar{Y}}$ by*

$$\sigma^Y(\zeta \otimes_R \bar{\eta}) = (\dim X_Q)^{1/2}(\dim Z_R)^{-1/2}\pi_r(\eta)^*(\sigma^*(\zeta)), \quad \zeta \in Z, \eta \in Y,$$

where π_r denote the right tensor multiplication $\pi_r(\eta) : \xi \in X \mapsto \xi \otimes \eta \in X \otimes_Q Y$. Similarly, we define the left hand side Frobenius dual ${}^X\sigma \in \mathcal{H}^Y_{\bar{X},Z}$ for $\sigma \in \mathcal{H}^Z_{X,Y}$ by

$${}^X\sigma(\bar{\xi} \otimes_P \zeta) = (\dim {}_QY)^{1/2}(\dim {}_PZ)^{-1/2}\pi_l(\xi)^*(\sigma^*(\zeta)), \quad \xi \in X, \zeta \in Z,$$

where π_l denote the left tensor multiplication.

Then we have the following.

Proposition 9.70 *For $\sigma \in \mathcal{H}^{Z}_{X,Y}$, define $\hat{\sigma} : {}_P X_Q \to {}_P Z \otimes_R \bar{Y}_Q$ by*

$$\hat{\sigma}(\xi) = \sum_j \sigma(\xi \otimes_Q \eta_j) \otimes_R \bar{\eta}_j,$$

where $\{\eta_j\}_j$ is a basis of Y_R. Then we have

$$\sigma^Y = (\dim X_Q)^{1/2} (\dim Z_R)^{-1/2} \hat{\sigma}^*.$$

Proof For $\xi \in X^{\mathrm{bdd}}, \eta \in Y^{\mathrm{bdd}}, \zeta \in Z^{\mathrm{bdd}}$, we get

$$\begin{aligned}
\langle \hat{\sigma}^*(\zeta \otimes_R \bar{\eta}), \xi \rangle &= \sum_j \langle \zeta \otimes_R \bar{\eta}, \sigma(\xi \otimes_Q \eta_j) \otimes_R \bar{\eta}_j \rangle \\
&= \sum_j \langle \zeta \langle \bar{\eta}, \bar{\eta}_j \rangle_R, \sigma(\xi \otimes_Q \eta_j) \rangle \\
&= \sum_j \langle \zeta \langle \eta, \eta_j \rangle^{\circ}_R, \sigma(\xi \otimes_Q \eta_j) \rangle \\
&= \sum_j \langle \zeta, \sigma(\xi \otimes_Q \eta_j) \langle \eta_j, \eta \rangle^{\circ}_R \rangle \\
&= \langle \zeta, \sigma(\xi \otimes_Q \eta) \rangle \\
&= \langle \sigma^*(\zeta), \pi_r(\eta)(\xi) \rangle \\
&= \langle \pi_r(\eta)^* \sigma^*(\zeta), \xi \rangle,
\end{aligned}$$

by Proposition 9.60, and thus we are done. □

Next we show that Frobenius reciprocity holds.

Theorem 9.71 *The Frobenius duality map $\cdot^Y : \mathcal{H}^{Z}_{X,Y} \to \mathcal{H}^{X}_{Z,\bar{Y}}$ is a conjugate linear isomorphism preserving norms.*

Proof First we prove $(\sigma^Y)^{\bar{Y}} = \sigma$. By Proposition 9.70, we have

$$\begin{aligned}
(\sigma^Y)^{\bar{Y}}(\xi \otimes_Q \bar{\bar{\eta}}) &= (\dim Z_R)^{1/2} (\dim X_Q)^{-1/2} \pi_r(\bar{\eta})^* (\sigma^Y)^*(\xi) \\
&= \pi_r(\bar{\eta})^* \hat{\sigma}(\xi).
\end{aligned}$$

Then for any $\zeta \in Z^{\mathrm{bdd}}$, we have

$$\begin{aligned}
\langle \pi_r(\bar{\eta})^* \hat{\sigma}(\xi), \zeta \rangle &= \sum_j \langle \sigma(\xi \otimes_Q \eta_j) \otimes_R \bar{\eta}_j, \zeta \otimes_R \bar{\eta} \rangle \\
&= \sum_j \langle \sigma(\xi \otimes_Q \eta_j) \langle \eta_j, \eta \rangle^{\circ}_R, \zeta \rangle \\
&= \langle \sigma(\xi \otimes_Q \eta), \zeta \rangle.
\end{aligned}$$

Thus we have $(\sigma^Y)^{\bar{Y}} = \sigma$. Next we show $\|\sigma\|^2 = \|\sigma^Y\|^2$, where the norm come from the inner product as in Definition 9.69. We choose bases $\{\xi_j\}_j$ for X_Q $\{\eta_k\}_k$ for Y_R, and $\{\zeta_m\}_m$ for Z_R. Then

$$\begin{aligned}
\dim X_Q \|\sigma^Y\|^2 &= \sum_j \langle \sigma^Y (\sigma^Y)^* \xi_j, \xi_j \rangle \\
&= \frac{\dim X_Q}{\dim Z_R} \sum_{j,k,l} \langle \sigma(\xi_j \otimes_Q \eta_k) \otimes_R \bar{\eta}_k, \sigma(\xi_j \otimes_Q \eta_l) \otimes_R \bar{\eta}_l \rangle \\
&= \frac{\dim X_Q}{\dim Z_R} \sum_{j,k,l} \langle \sigma(\xi_j \otimes_Q \eta_k) \langle \eta_k, \eta_l \rangle_R^\circ, \sigma(\xi_j \otimes_Q \eta_k) \rangle \\
&= \frac{\dim X_Q}{\dim Z_R} \sum_{j,l} \langle \sigma(\xi_j \otimes_Q \eta_l), \sigma(\xi_j \otimes_Q \eta_l) \rangle \\
&= \frac{\dim X_Q}{\dim Z_R} \sum_{j,l,m} \langle \sigma(\xi_j \otimes_Q \eta_l), \zeta_m \langle \zeta_m, \sigma(\xi_j \otimes_Q \eta_l) \rangle_R^\circ \rangle \\
&= \frac{\dim X_Q}{\dim Z_R} \sum_{j,l,m} \langle \sigma(\xi_j \otimes_Q \eta_l) \langle \sigma(\xi_j \otimes_Q \eta_l), \zeta_m \rangle_R^\circ, \zeta_m \rangle \\
&= \frac{\dim X_Q}{\dim Z_R} \sum_{j,l,m} \langle \sigma(\xi_j \otimes_Q \eta_l) \langle \xi_j \otimes_Q \eta_l, \sigma^*(\zeta_m) \rangle_R^\circ, \zeta_m \rangle \\
&= \frac{\dim X_Q}{\dim Z_R} \sum_m \langle \sigma\sigma^*(\zeta_m), \zeta_m \rangle \\
&= \|\sigma\|^2 \dim X_Q,
\end{aligned}$$

where we used Corollary 9.13 and Proposition 9.62. Thus we are done. □

Corollary 9.72 *Let ${}_P X_Q$ and ${}_Q Y_P$ be irreducible bimodules. The number of copies of ${}_P P_P$ in ${}_P X \otimes_Q Y_P$ as a direct summand is 0 or one. This number is one if and only if ${}_P X_Q$ is isomorphic to ${}_P \bar{Y}_Q$.*

Proof The number of copies is given by the dimension of $\mathcal{H}^P_{X,Y}$. By Frobenius reciprocity, Theorem 9.71, this is isomorphic to

$$\mathcal{H}^X_{P,\bar{Y}} = \mathrm{Hom}({}_P \bar{Y}_Q, {}_P X_Q).$$

Because X, Y are irreducible, the dimension of this intertwiner space is 0 or one, and it is one if and only if ${}_P X_Q$ is isomorphic to ${}_P \bar{Y}_Q$. □

Remark We explain the term Frobenius reciprocity. In the subfactor situation for $N \subset M$, we typically have $P = N$, $Q = N$, $R = M$, and ${}_Q Y_R = {}_N M_M$. Suppose we have ${}_N X_N$. Then for the tensor product ${}_N X \otimes_N M_M$, we have a right action of M, which is bigger than N. That is, we extend the algebra acting on the right. This is regarded as an analogue of the induced representation. The dimension of $\mathcal{H}^Z_{X,M}$ counts the multiplicity of Z in the analogue of the induced representation $X \otimes_N M$. If we start with ${}_N Z_M$, the tensor product ${}_N Z \otimes_M M_N$ is just an ${}_N Z_N$, which means the bimodule given by restricting the right action to N. This is regarded as an analogue of the restriction of a representation, and the dimension of $\mathcal{H}^X_{Z,\bar{M}}$ counts the multiplicity of X in the analogue of the restricted representation. (Note that the conjugate bimodule of ${}_N M_M$ is ${}_M M_N$, as in Example 9.59.) Frobenius reciprocity here, as in the case of group representations, gives that these two dimensions are the same.

For $\sigma \in \mathcal{H}^Z_{X,Y}$, we next define the conjugate intertwiner $\bar{\sigma} \in \mathcal{H}^{\bar{Z}}_{\bar{Y},\bar{X}}$ by $\bar{\sigma}(\bar{\eta} \otimes_Q \bar{\xi}) = \overline{\sigma(\xi \otimes_Q \eta)}$. With this definition and a similar computation to the above proof of Theorem 9.71, we get a more general Frobenius reciprocity as follows.

Theorem 9.73 *For the symmetric group on the set of three von Neumann algebras $\{P, Q, R\}$, Frobenius duality gives an action of S_3 by isomorphisms between the following spaces:*

$$\mathcal{H}^Z_{X,Y}, \mathcal{H}^{\bar{Y}}_{\bar{Z},X}, \mathcal{H}^{\bar{X}}_{Y,\bar{Z}}, \mathcal{H}^{\bar{Z}}_{\bar{Y},\bar{X}}, \mathcal{H}^{Y}_{\bar{X},Z}, \mathcal{H}^{X}_{Z,\bar{Y}}.$$

Here the even permutations act as unitaries and the odd permutations as bijective conjugate linear isometries.

Proof We have to show that the relations in S_3 are preserved.

For $\sigma \in \mathcal{H}^Z_{X,Y}$, we have two ways to get an intertwiner in $\mathcal{H}^{\bar{X}}_{Y,\bar{Z}}$. One is $(\sigma^Y)^-$ and the other is ${}^Y\bar{\sigma}$. We will show that these two intertwiners are equal.

For $\bar{\xi} \in \bar{X}$ and a basis $\{\eta_j\}_j$ of Y_R, we have

$$\begin{aligned}
\overline{(\sigma^Y)^*}(\bar{\xi}) &= \overline{(\sigma^Y)^*(\xi)} \\
&= \sqrt{\frac{\dim X_Q}{\dim Z_R}}\,\overline{\hat{\sigma}(\xi)} \\
&= \sqrt{\frac{\dim X_Q}{\dim Z_R}}\,\overline{\sum_j \sigma(\xi \otimes_Q \eta_j) \otimes_R \bar{\eta}_j} \\
&= \sqrt{\frac{\dim X_Q}{\dim Z_R}} \sum_j \eta_j \otimes_R \overline{\sigma(\xi \otimes_Q \eta_j)},
\end{aligned}$$

where we used Proposition 9.70. Similarly, we also get

$$({}^{\bar{Y}}\bar{\sigma})^*(\bar{\xi}) = \sqrt{\frac{\dim X_Q}{\dim Z_R}} \sum_j \eta_j \otimes_R \overline{\sigma(\xi \otimes_Q \eta_j)},$$

thus we have $\overline{(\sigma^Y)^*}(\bar{\xi}) = ({}^{\bar{Y}}\bar{\sigma})^*(\bar{\xi})$.

Similarly, for $\sigma \in \mathcal{H}^Z_{X,Y}$, we have two ways to get an intertwiner in $\mathcal{H}^{\bar{Y}}_{\bar{Z},X}$. One is ${}^Z(\sigma^Y)$ and the other is $({}^X\sigma)^-$. We show that these two intertwiners are the same.

Note that

$$\pi_l(\zeta)^*(\zeta' \otimes_R \bar{\eta}) = \langle \zeta, \zeta' \rangle^\circ_R \bar{\eta}$$

for $\zeta, \zeta' \in Z^{\text{bdd}}$, $\bar{\eta} \in \bar{Y}$ by the definition of the inner product in the relative tensor product. Then for a basis $\{\eta_j\}_j$ of Y_R, we get

$$
\begin{aligned}
&{}^{Z}(\sigma^{Y})(\bar{\zeta} \otimes_P \xi) \\
&= \sqrt{\frac{\dim {}_R\bar{Y}}{\dim {}_P X}} \pi_l(\zeta)^*((\sigma^Y)^*(\xi)) \\
&= \sqrt{\frac{\dim {}_R\bar{Y}}{\dim {}_P X}} \pi_l(\zeta)^* \sqrt{\frac{\dim X_Q}{\dim Z_R}} \sum_j \sigma(\xi \otimes_Q \eta_j) \otimes_R \bar{\eta}_j \\
&= \sqrt{\frac{\dim {}_R\bar{Y}}{\dim {}_P X}} \sqrt{\frac{\dim X_Q}{\dim Z_R}} \sum_j \langle \zeta, \sigma(\xi \otimes_Q \eta_j) \rangle_R^\circ \bar{\eta}_j \\
&= \sqrt{\frac{\dim {}_R\bar{Y}}{\dim {}_P X}} \sqrt{\frac{\dim X_Q}{\dim Z_R}} \sum_j \overline{\eta_j \langle \eta_j, \pi_l(\xi)^*(\sigma^*(\zeta)) \rangle_R^\circ} \\
&= \sqrt{\frac{\dim {}_R\bar{Y}}{\dim {}_P X}} \sqrt{\frac{\dim X_Q}{\dim Z_R}} \overline{\pi_l(\xi)^*(\sigma^* \zeta)} \\
&= \sqrt{\frac{\dim {}_R\bar{Y}}{\dim {}_P X}} \sqrt{\frac{\dim X_Q}{\dim Z_R}} \sqrt{\frac{\dim {}_P Z}{\dim {}_Q Y}} \overline{{}^{X}\sigma(\bar{\xi} \otimes_P \zeta)} \\
&= \sqrt{\frac{\dim {}_R\bar{Y}}{\dim {}_P X}} \sqrt{\frac{\dim X_Q}{\dim Z_R}} \sqrt{\frac{\dim {}_P Z}{\dim {}_Q Y}} ({}^{X}\sigma)^{-}(\bar{\zeta} \otimes_P \xi),
\end{aligned}
$$

where we used Definition 9.69 and Proposition 9.70. (We have also used the fact $\langle \zeta, \sigma(\xi \otimes_Q \eta_j) \rangle_R^\circ = \langle \sigma^*(\zeta), \xi \otimes_Q \eta_j \rangle_R^\circ$, which is easy to show.) In the last expression, the coefficient

$$
\sqrt{\frac{\dim {}_R\bar{Y}}{\dim {}_P X}} \sqrt{\frac{\dim X_Q}{\dim Z_R}} \sqrt{\frac{\dim {}_P Z}{\dim {}_Q Y}}
$$

is 1 if there exists a non-zero $\sigma \in \mathcal{H}_{X,Y}^{Z}$ because Frobenius reciprocity gives an isometric isomorphism by Theorem 9.71. □

9.9 Notes

In Section 9.2, we followed the treatment of an unpublished manuscript 'An invariant coupling between 3-manifolds and subfactors, with connections to topological and conformal quantum field theory' of A. Ocneanu (1991), and also used (Yamagami 1993a).

Sections 9.3 and 9.6 are mainly taken from the fundamental paper (Jones 1983). (We are also influenced by the presentation in the book (Goodman *et al.* 1989).) The basic construction was first introduced by (Skau 1977) for a different purpose and has been also used in (Christensen 1979). Theorem 9.16 and Theorem 9.25 were the major results in (Jones 1983). The Jones relations in Theorem 9.25 had been found in an entirely different context by (Temperley and Lieb 1971) – see Chapter 7, so the relations also have the name Temperley–Lieb–Jones. Lemma 9.26, Theorem 9.27, and Proposition 9.28 first appeared in (Pimsner and Popa 1986a).

The exposition of Section 9.4 is taken from unpublished lectures of Evans at RIMS, Kyoto University on AF algebras of Statistical Mechanics in 1990–91. The notion of an index in the C^*-context, and Lemmas 9.32–9.33 are from (Watatani 1990). The idea of the Murray diagram is from (Murray 1990). Section 9.5 is from (Evans and Gould 1994a). The notion of the principal graph was introduced by V. F. R. Jones.

The downward basic construction was first introduced by (Jones 1983), and its properties, Propositions 9.44, Proposition 9.45 and Corollary 9.46 was proved in (Pimsner and Popa 1986a). The Pimsner–Popa inequality, Theorem 9.48, was also proved by (Pimsner and Popa 1986a), and Proposition 9.47, Theorem 9.49, and Proposition 9.50 were proved by (Pimsner and Popa 1988). Their direct aim in (Pimsner and Popa 1986a) was to clarify a relation between the Jones index and the entropy introduced in (Connes and Størmer 1975). This work on entropy has been very useful in a recent series of work on the analytic classification of subfactors by Popa. The notion of commuting square was introduced by (Popa 1983b) in a different context. Here we followed the treatment of Chapter 4 of (Goodman *et al.* 1989).

The relative tensor product in Section 9.7 was first introduced by A. Connes in his unpublished manuscript and was published in (Sauvageot 1983). A detailed account of the bimodule theory has been given in the unpublished manuscript 'Correspondences' (1986) of S. Popa. The main source of Sections 9.7 and 9.8 is again Ocneanu's unpublished manuscript "An invariant coupling between 3-manifolds and subfactors, with connections to topological and conformal quantum field theory". We also used (Yamagami 1993a) and the unpublished notes of Yamagami 'A report on Ocneanu's lectures', which was a preliminary version of (Yamagami 1993a).

10

AXIOMATIZATION OF PARAGROUPS

10.1 Introduction

In this chapter, we study Ocneanu's paragroups. A *paragroup* is a certain quantization of an ordinary (finite) group which Ocneanu introduced in order to characterize commuting squares arising as higher relative commutants of subfactors as in Corollary 9.54. (A paragroup is a different kind of quantization from a quantum group, though it has a close relation to a quantum group.) We list combinatorial axioms for paragroups in a slightly different formulation from the one in (Ocneanu 1988) and show that we get data satisfying these axioms from subfactors. Our object is a triple $(\mathcal{G}, \mu, W)$ consisting of a connected unoriented finite graph $\mathcal{G}$, a positive real-valued function μ on the set of vertices of the graph $\mathcal{G}$, and a connection W on the graph $\mathcal{G}$. Our formal setting is very close to that in the theory of exactly solvable lattice models of Chapters 7 and 8, as explained in Section 11.9.

In Section 10.2, we express the higher relative commutants of subfactors in terms of bimodules with the techniques of Chapter 9. Section 10.3 deals with the rather easy axiom of unitarity of connections. We will study the renormalization axiom in Section 10.4. This expresses a certain type of symmetry of the system, and will play an important role in the following chapters. Section 10.5 studies the flatness axiom. In actual computations and applications, this is the most important axiom. The abstract definition of a paragroup is completed in this section. The most fundamental examples of paragroups are presented in Section 10.6. These really are finite groups, and we will show how to regard them as paragroups.

Section 10.7 explains Longo's sector technique for subfactors of type III. This section is not self-contained, and is mainly for operator algebraists. In many aspects, Ocneanu's bimodule approach and Longo's sector approach are parallel, but Longo's method does have some advantage, as we will see in Section 12.9.

10.2 Higher relative commutants and bimodules

Our first aim is to represent the higher relative commutants in terms of bimodules and intertwiners. In this section we assume that the subfactor $N \subset M$ has finite index and finite depth.

Lemma 10.1 *We have isomorphisms* $\mathrm{End}({}_N(M_k)_N) \cong N' \cap M_{2k+1}$ *and* $\mathrm{End}({}_N(M_k)_M) \cong N' \cap M_{2k}$ *respectively. These spaces are finite dimensional.*

Proof We can naturally identify $\mathrm{End}({}_N(M_k)_N)$ with $N' \cap M_{2k+1}$ by Theorem 9.49 and Proposition 9.22 (4). It is finite dimensional by Proposition 9.18 (3). The other case is similarly proved. □

By Proposition 9.65, we know that

$$ {}_M(M_k)_M \cong {}_M(M \otimes_N M \otimes_N \cdots \otimes_N M)_M, $$

where we have $k+1$ copies of M. Thus we can identify the tower of the higher relative commutants

$$ N' \cap N = \mathbb{C} \subset N' \cap M \subset N' \cap M_1 \subset N' \cap M_2 \subset N' \cap M_3 \subset \cdots $$

with

$$ \begin{aligned} &\mathrm{End}({}_N N_N) \subset \mathrm{End}({}_N M_M) \subset \mathrm{End}({}_N M_N) \\ &\subset \mathrm{End}({}_N M \otimes_N M_M) \subset \mathrm{End}({}_N M \otimes_N M_N) \subset \cdots, \end{aligned} $$

by Lemma 10.1. Here it is easy to see that the embeddings in the latter increasing sequence, which are compatible with the former increasing sequence, are given by the two kinds of maps. One is by the trivial inclusion $\mathrm{End}({}_N {M_k}_M) \subset \mathrm{End}({}_N {M_k}_N)$ and the other $\mathrm{End}({}_N {M_k}_N) \subset \mathrm{End}({}_N M_k \otimes_N M_M)$ is $x \mapsto x \otimes_N \mathrm{id}_M$.

With these observations, the Bratteli diagram of the higher relative commutants and the principal graph are expressed inductively for $n = 0, 1, 2, \ldots$ in terms of bimodules as follows.

First we have a single vertex representing a bimodule ${}_N N_N$ for the case $n = 0$. We mark this bimodule with the symbol $*$. Then repeat the following two procedures for $n = 0, 1, 2, \ldots$ alternately.

If n is odd, then we set $k = (n-1)/2$ and take an irreducible decomposition of ${}_N(M_k)_M$ as a bimodule, and assign a vertex to each irreducible component. (We have only finitely many components – otherwise, we would have an infinite dimensional $\mathrm{End}_N(M_k)_M$, which would contradict Lemma 10.1.) We draw edges from each ${}_N X_N$ in the previous step $n-1$ to ${}_N Y_M$ in the current step with the number of the edges equal to the multiplicity of ${}_N Y_M$ in ${}_N X \otimes_N M_M$.

If n is even, then we set $k = (n-2)/2$ and take an irreducible decomposition of ${}_N(M_k)_N$ as a bimodule at each step, and assign a vertex to each irreducible component. We draw edges from ${}_N Y_M$ in the previous step $n-1$ to ${}_N X_N$ in the current step with the number of edges equal to the multiplicity of ${}_N X_N$ in ${}_N Y \otimes_M M_N = {}_N Y_N$.

These procedures yield the desired Bratteli diagram. The fact mentioned in Definition 9.41 that each step of the Bratteli diagram consists of the reflection of the previous step together with a new part means that the multiplicity of ${}_NY_M$ in ${}_NX \otimes_N M_M$ is equal to the multiplicity of ${}_NX_N$ in ${}_NY \otimes_M M_N = {}_NY_N$. This is a consequence of Frobenius reciprocity, Theorem 9.71. We can thus interpret the principal graph introduced in Definition 9.41 as the graph describing the branching rule of the relative tensor products of ${}_NN_N$ with ${}_NM_M$, ${}_MM_N, \ldots$, from the right. In this interpretation, the name *induction–restriction graph* is sometimes used for the principal graph.

We have a similar description of the dual principal graph by looking at the sequence

$$ {}_MM_M, {}_MM_N, {}_MM \otimes_N M_M, {}_MM \otimes_N M_N, \ldots $$

of bimodules. This way of interpreting the (dual) principal graph, however, does not give a strong enough invariant.

As a next step, we take intertwiners corresponding to downward paths on the Bratteli diagram with finite length starting from $*$. Let ${}_NX_N$, ${}_NY_M$ be irreducible bimodules. If the multiplicity of ${}_NY_M$ in ${}_NX \otimes_N M_M$ is m, then the dimension of $\mathrm{Hom}({}_NX \otimes_N M_M, {}_NY_M)$ is also m. Also note that for elements σ, ρ in this Hom space, we have an inner product $\langle \sigma, \rho \rangle = \sigma\rho^* \in \mathrm{End}({}_NY_M) = \mathbb{C}$ by irreducibility of Y. We take an orthonormal basis $\{\sigma_1, \sigma_2, \ldots, \sigma_m\}$ of $\mathrm{Hom}({}_NX \otimes_N M_M, {}_NY_M)$ with respect to this inner product. Note that each σ_j is a co-isometry by the definition of the inner product. By Frobenius reciprocity, Theorem 9.71, the dimension of $\mathrm{Hom}({}_NY \otimes_M M_N, {}_NX_N)$ is also m and we have an orthonormal basis $\{\sigma'_1, \sigma'_2, \ldots, \sigma'_m\}$, where $\sigma'_j \in \mathrm{Hom}({}_NY \otimes_M M_N, {}_NX_N)$ is defined by $\sigma'_j = \sigma_j^M$ as a Frobenius dual as in Definition 9.69. (Note that the orthogonality also follows from Theorem 9.71.) Similarly, an orthonormal basis $\{\rho_j\}$ of $\mathrm{Hom}({}_NY \otimes_M M_N, {}_NX_N)$ produces an orthonormal basis $\{\rho'_j\}$ of $\mathrm{Hom}({}_NX \otimes_N M_M, {}_NY_M)$, and we can show that $\sigma''_j = \sigma_j$. In the following procedure, we assign an intertwiner to each edge in the Bratteli diagram.

1. Let $k = 0$, and start with a bimodule ${}_NN_N$ at level 0.
2. Depending on the parity of k, we form the tensor product of the bimodules at level k with ${}_NM_M$ or ${}_MM_N$ from the right, and take the bimodules at level $k+1$ by irreducible decompositions.
3. To an edge appearing in the Bratteli diagram from the level k to $k+1$ which is a reflection of an edge in the previous level, we assign σ', where σ is the intertwiner assigned to the edge in the previous step. To the other edges, we assign new intertwiners so that they form an orthonormal basis.
4. Increase k by 1 and go to Step 2.

Note that because we assume finite depth, the principal graph is finite, which means that we will have no new bimodules any more at some level in Step 2. Next we assign intertwiners to downward paths of finite length from $*$ on the Bratteli diagram. For example, take a path of length 3, and suppose that the three edges are labelled by the intertwiners $\sigma_1 : {}_NN \otimes_N M_M \to {}_NX_{1M}$,

$\sigma_2 : {}_N X_1 \otimes_M M_N \to {}_N X_{2N}$, $\sigma_3 : {}_N X_2 \otimes_N M_M \to {}_N X_{3M}$ from the top to the bottom. We assign to this path the composition

$$\sigma_3(\sigma_2 \otimes \mathrm{id}_{{}_N M_M})(\sigma_1 \otimes \mathrm{id}_{{}_M M_N} \otimes \mathrm{id}_{{}_N M_M}) : \\ {}_N N \otimes_N M \otimes_M M \otimes_N M_M \to {}_N X_{3M}$$

of these three intertwiners. A general path is dealt with similarly.

We then take a pair of downward paths σ_+, σ_- of the same length on the Bratteli diagram starting from $*$ with same endpoint. Identifying the paths with the compositions of intertwiners, we assign the composition $\sigma_+^* \sigma_-$ of the intertwiners to the pair (σ_+, σ_-). If σ_+, σ_- have length 3 as in the above example, this composition is in

$$\mathrm{End}({}_N N \otimes_N M \otimes_M M \otimes_N M_M) = \mathrm{End}({}_N M_{1M}) = N' \cap M_2.$$

In general, if a pair (σ_+, σ_-) has length k, the composition $\sigma_+^* \sigma_-$ of the intertwiners gives a partial isometry in a system of matrix units of $N' \cap M_{k-1}$, and we know that $N' \cap M_{k-1}$ is spanned by such pairs (σ_+, σ_-) by counting the dimension. We next look at the embedding $N' \cap M_{k-1} \subset N' \cap M_k$ in this setting, and see that a pair (σ_+, σ_-) of length k is embedded from $N' \cap M_{k-1}$ into $N' \cap M_k$ as $\sum_\rho (\sigma_+ \cdot \rho, \sigma_- \cdot \rho)$ of length $k+1$. Here ρ means a path of length 1 from the endpoint of $\sigma_\pm$, and $\sigma_\pm \cdot \rho$ is a concatenation of paths. (Note that in the concatenation of paths, we first draw the path $\sigma_\pm$ and then underneath draw the path ρ, which gives the reversed order of the compositions of the corresponding intertwiners.) In this way, we get a series of specific bases for the increasing sequence of the finite dimensional algebras $\{N' \cap M_k\}_k$. Such a pair (σ_+, σ_-) and such a sequence of finite dimensional algebras are called *string* and *string algebras* respectively.

We naturally have a similar description of $\{M' \cap M_k\}_k$ as string algebras.

10.3 Connections and unitarity

In this section, we start to axiomatize paragroups and show how to get data satisfying these axioms from a subfactor $N \subset M$. Recall that the subfactor $N \subset M$ has finite index and finite depth. In the rest of this chapter, we also require the trivial relative commutant condition $N' \cap M = \mathbb{C}$. This is not really necessary, but this condition is desirable for technical simplicity.

The increasing sequence $\{N' \cap M_k\}_k$ considered in Section 10.2 is not enough to recover the original subfactor $N \subset M$. (For example, two subfactors $N \subset N \rtimes (\mathbb{Z}/2 \oplus \mathbb{Z}/2)$ and $N \subset N \rtimes \mathbb{Z}/4$, where the two groups act as outer actions, are not isomorphic although they have the same towers of higher relative commutants. See Section 10.6.) We also take the following increasing sequences:

$$\begin{array}{ccccccc} \mathrm{End}({}_N N_N) & \subset & \mathrm{End}({}_N M_M) & \subset & \mathrm{End}({}_N M_N) & \subset & \cdots \\ \cap & & \cap & & \cap & & \\ \mathrm{End}({}_M M_N) & \subset & \mathrm{End}({}_M M \otimes_N M_M) & \subset & \mathrm{End}({}_M M \otimes_N M_N) & \subset & \cdots . \end{array}$$

Iere the horizontal embeddings are as in Section 10.2, and the vertical em-
eddings are given by $\mathrm{End}({}_N M_N) \ni \sigma \mapsto \mathrm{id}_{{}_M M_N} \otimes \sigma \in \mathrm{End}({}_M M \otimes_N M_N)$,
or example. By an argument similar to the one in Section 10.2, we can
dentify the second line in the above sequence with the increasing sequence
$M' \cap M_1 \subset M' \cap M_2 \subset M' \cap M_3 \subset \cdots$. The first line was identified with
$N' \cap N \subset N' \cap M \subset N' \cap M_1 \subset \cdots$, and the vertical embeddings show that
he image of the algebra $N' \cap M_k$ by the embedding into $M' \cap M_{k+2}$ in the
econd line is $M_1' \cap M_{k+2}$. (In order to verify this, it is enough to show that
$\mathrm{d}_{{}_M M_N} \otimes \sigma$ commutes with the left multiplication of e_1 on ${}_M M \otimes_N M_N$, for
xample. This commutativity follows from Proposition 9.65.) Thus the above
equence is identified with the following:

$$
\begin{array}{ccccccc}
M_1' \cap M_1 & \subset & M_1' \cap M_2 & \subset & M_1' \cap M_3 & \subset & \cdots \\
\cap & & \cap & & \cap & & \\
M' \cap M_1 & \subset & M' \cap M_2 & \subset & M' \cap M_3 & \subset & \cdots .
\end{array}
$$

The string algebra method in Section 10.2 produces a specific basis for each
inite dimensional algebra here. (That is, in the Bratteli diagram of the above
equence, a downward or right-bound path with finite length from $*$ corresponds
o a composition of intertwiners. Note that the Bratteli diagram of the increasing
equence $\{N' \cap M_k\}_k$ is now written horizontally, though it was written verti-
ally in Section 10.2.) Now we have several natural bases for the same algebra.
'or example, $\mathrm{End}({}_M M \otimes_N M_N)$ in the above sequence has three paths from
$\mathrm{End}({}_N N_N) = \mathbb{C}$;

$$
\begin{aligned}
&\mathrm{End}({}_N N_N) \subset \mathrm{End}({}_N M_M) \subset \mathrm{End}({}_N M_N) \subset \mathrm{End}({}_M M \otimes_N M_N),\\
&\mathrm{End}({}_N N_N) \subset \mathrm{End}({}_N M_M) \subset \mathrm{End}({}_M M \otimes_N M_M) \subset \mathrm{End}({}_M M \otimes_N M_N),\\
&\mathrm{End}({}_N N_N) \subset \mathrm{End}({}_M M_N) \subset \mathrm{End}({}_M M \otimes_N M_M) \subset \mathrm{End}({}_M M \otimes_N M_N).
\end{aligned}
$$

'hese three sequences give three different bases of $\mathrm{End}({}_M M \otimes_N M_N)$. The pro-
edure of changing these bases is determined by a *connection*, which will com-
lete the data for the paragroup, see also Section 3.5. Take the first two of the
bove three sequences of inclusions. For paths of length three, the first parts
$\mathrm{End}({}_N N_N) \subset \mathrm{End}({}_N M_M)$ are the same, so the difference arises from whether
he path passes through the upper-right corner or the lower-left corner in the
ext block. Take a path σ_1 from an upper-left vertex (N-M bimodule) A to an
pper-right vertex (N-N bimodule) B, a path σ_2 from B to an lower-right ver-
ex (M-N bimodule) D, a path σ_3 from an upper-left vertex (N-M bimodule)
A to a lower-left vertex (M-M bimodule) C, and a path σ_4 from C to a lower-
ight vertex (M-N bimodule) D. Then the compositions $\sigma_2(\mathrm{id}_{{}_M M_N} \otimes \sigma_1)$ and
$\sigma_4(\sigma_3 \otimes \mathrm{id}_{{}_M M_N})$ are both in $\mathrm{Hom}({}_M M \otimes_N A \otimes_M M_N, D)$, and if we let each σ_j
ary, the two ways of composing give two orthonormal bases of this Hom space.
ssign a complex number

$$\sigma_4(\sigma_3 \otimes \mathrm{id}_{{}_M M_N})(\mathrm{id}_{{}_M M_N} \otimes \sigma_1)^* \sigma_2^* \in \mathbb{C}$$

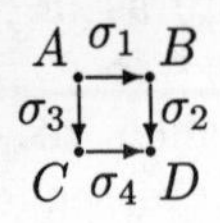

FIG. 10.1. The connection $\sigma_4(\sigma_3 \otimes \mathrm{id}_{{}_M M_N})(\mathrm{id}_{{}_M M_N} \otimes \sigma_1)^* \sigma_2^* \in \mathbb{C}$

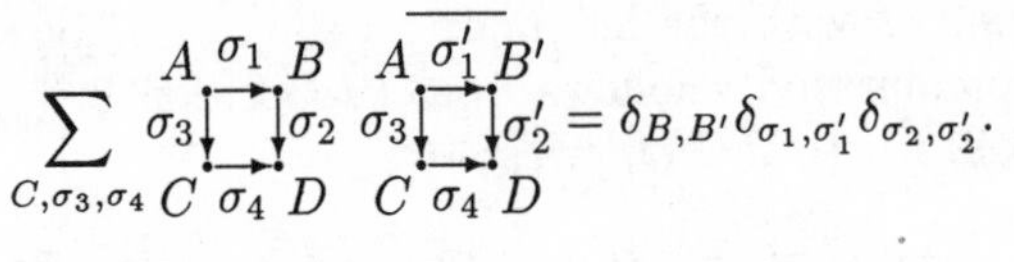

$$\sum_{C,\sigma_3,\sigma_4} \begin{array}{c} A \;\sigma_1\; B \\ \sigma_3 \square \sigma_2 \\ C \;\sigma_4\; D \end{array} \; \overline{\begin{array}{c} A \;\sigma_1'\; B' \\ \sigma_3 \square \sigma_2' \\ C \;\sigma_4\; D \end{array}} = \delta_{B,B'} \delta_{\sigma_1,\sigma_1'} \delta_{\sigma_2,\sigma_2'}.$$

FIG. 10.2. Unitarity of the connection

to the square determined by A, B, C, D and $\sigma_1, \sigma_2, \sigma_3, \sigma_4$, and denote this number by the diagram in Fig. 10.1.

We have two kinds of diagrams because the upper-left vertex is an N-N bimodule or an N-M bimodule. In both cases, we get the identity in Fig. 10.2, which is called the *unitarity* property.

(The number of paths from A to D of length 2 is the same for paths through the upper-right corner and those through the lower-left corner.) Looking at the Bratteli diagram of the above sequence from the left to the right, we notice that the graphs at each step consist of the reflection (with respect to a vertical axis) of the previous step and a new part. The finite depth assumption means that no new parts appear after some step, so finally we have a combination of four kinds of graphs as in Fig. 10.3. Here we have four connected graphs $\mathcal{G}_0, \mathcal{G}_1, \mathcal{G}_2, \mathcal{G}_3$ where $\mathcal{G}_j$ and $\mathcal{G}_{j+1}$ have common vertices V_{j+1} for $j \in \mathbb{Z}/4$. We also denote the union of the four graphs by $\mathcal{G}$. An example showing how the vertices of $\mathcal{G}$ are connected is in Fig. 10.4 and Fig. 10.5. Figure 10.4 is a two-dimensional picture of the graph cut at vertices V_0 and Fig. 10.5 is a three-dimensional picture. Note that in this section and the following, pictures like Fig. 10.3 really mean pictures as in Figs 10.4 and 10.5.

We say that V_0 is the set of *even* vertices of $\mathcal{G}_0$ and $\mathcal{G}_3$, V_1 the set of *odd* vertices of $\mathcal{G}_0$ and $\mathcal{G}_1$, V_2 the set of *even* vertices of $\mathcal{G}_1$ and $\mathcal{G}_2$, and V_3 the set of *odd* vertices of $\mathcal{G}_2$ and $\mathcal{G}_3$. (That is, the graphs $\mathcal{G}_j$ are *bipartite* in the sense that the vertices are divided into the even and odd sets and only even vertices and

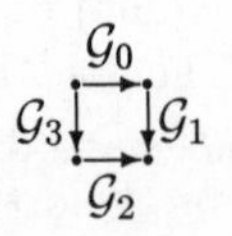

FIG. 10.3. Four graphs for a paragroup

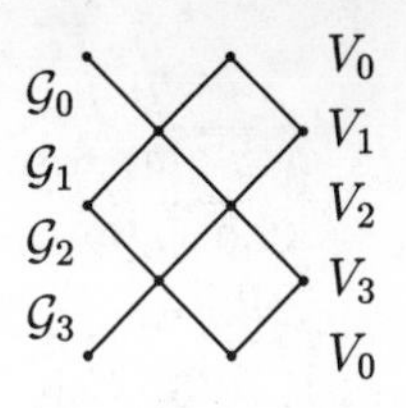

FIG. 10.4. Example of four graphs for a paragroup

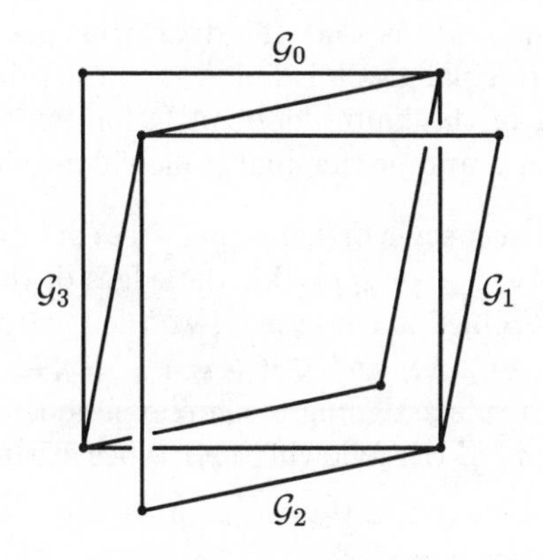

FIG. 10.5. Example of four graphs for a paragroup

odd vertices are connected.)

In Fig. 10.3, the vertices V_0 correspond to the N-N bimodules appearing in the irreducible decompositions of $M \otimes_N \cdots \otimes_N M$, V_1 to the N-M bimodules, V_2 to the M-M bimodules, and V_3 to the M-N bimodules. That is, the graph $\mathcal{G}_0$ is determined by how an N-N bimodule X is decomposed when ${}_N M_M$ is tensored from the right as in ${}_N X \otimes_N M_M = \bigoplus_j n_j {}_N(Y_j)_M$, but taking the conjugate bimodules of both sides, we get ${}_M M \otimes_N \bar{X}_N = \bigoplus_j n_j {}_M(\bar{Y}_j)_N$, which gives the decomposition rule for left tensor multiplication by ${}_M M_N$. This shows that the graphs $\mathcal{G}_0$ and $\mathcal{G}_3$ are both isomorphic to the principal graph of $N \subset M$. Similarly, both $\mathcal{G}_1$ and $\mathcal{G}_2$ are isomorphic to the dual principal graph. (But we do not require these isomorphisms as a part of the axioms. Instead, these isomorphisms are consequences of the axioms.)

Remark Choose an N-N bimodule X corresponding to a vertex in V_0 of the graph and consider how many edges it has adjacent to vertices in V_1. The number of these edges is equal to the number of irreducible N-M bimodules (with multiplicity counted) in ${}_N X \otimes_N M_M$. Because the Jones index value of each bimodule is bigger than or equal to 1, the number of edges is bounded by $[{}_N X \otimes_N M_M] = [M : N][{}_N X_N]$. This is true even without the finite depth

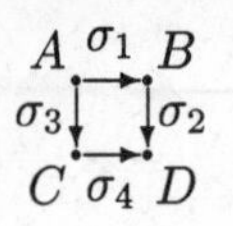

FIG. 10.6. Cell

assumption.

Suppose that the principal graph of $N \subset M$ is finite. Then we have only finitely many vertices in V_1. With the above remark, we then have only finitely many vertices in V_2, which means that the dual principal graph is also finite. Conversely, if the dual principal graph is finite, then the principal graph is finite too. So in the definition of the finite depth in Definition 9.41, it is enough to assume that the principal graph *or* the dual principal graph is finite.

We have already chosen specific orthonormal bases of intertwiners of ${}_M M \otimes_N X_N \to {}_M Y_N$ and ${}_M M \otimes_N X_M \to {}_M Y_M$ for the edges of the vertical graphs $\mathcal{G}_3$ and $\mathcal{G}_1$ respectively. By Frobenius reciprocity, we then get orthonormal bases of intertwiners ${}_N M \otimes_M Y_N \to {}_N X_N$ and ${}_N M \otimes_M Y_M \to {}_N X_M$ respectively.

In this way, we have a rule assigning a complex number to a square made of the four edges $\sigma_1, \sigma_2, \sigma_3, \sigma_4$ of the following four types arising from the graph $\mathcal{G}$.

1. $\sigma_1 \in \mathcal{G}_0, \sigma_2 \in \mathcal{G}_1, \sigma_3 \in \mathcal{G}_3, \sigma_4 \in \mathcal{G}_2$.
2. $\sigma_1 \in \mathcal{G}_0, \sigma_2 \in \mathcal{G}_3, \sigma_3 \in \mathcal{G}_1, \sigma_4 \in \mathcal{G}_2$.
3. $\sigma_1 \in \mathcal{G}_2, \sigma_2 \in \mathcal{G}_1, \sigma_3 \in \mathcal{G}_3, \sigma_4 \in \mathcal{G}_0$.
4. $\sigma_1 \in \mathcal{G}_2, \sigma_2 \in \mathcal{G}_3, \sigma_3 \in \mathcal{G}_1, \sigma_4 \in \mathcal{G}_0$.

Such a square is called a *cell.* (See Fig. 10.6.)

The assignment of complex numbers to cells is called a *connection* and denoted by W. We will also use the diagram in Fig. 10.6 to denote this number. We will list axioms of the graph $\mathcal{G}$ and a connection W. We sometimes drop the labels of edges or vertices if no confusion arises. We also use the convention that if $(\sigma_1, \sigma_2, \sigma_3, \sigma_4)$ does not make a square, the symbol in Fig. 10.6 denotes the number 0.

Axiom 1. (Unitarity) *Choose four edges $\sigma_1, \sigma_2, \sigma_1', \sigma_2'$ of the graph $\mathcal{G}$ with one of the following properties.*

1. $\sigma_1, \sigma_1' \in \mathcal{G}_0, \sigma_2, \sigma_2' \in \mathcal{G}_1$.
2. $\sigma_1, \sigma_1' \in \mathcal{G}_0, \sigma_2, \sigma_2' \in \mathcal{G}_3$.
3. $\sigma_1, \sigma_1' \in \mathcal{G}_2, \sigma_2, \sigma_2' \in \mathcal{G}_1$.
4. $\sigma_1, \sigma_1' \in \mathcal{G}_2, \sigma_2, \sigma_2' \in \mathcal{G}_3$.

Suppose that there exist edges σ_3, σ_4 of $\mathcal{G}$ such that $\sigma_1, \sigma_2, \sigma_3, \sigma_4$ makes a cell. Then we have the identity as in Fig. 10.7.

We have already seen that this axiom is satisfied by our connection arising from a subfactor.

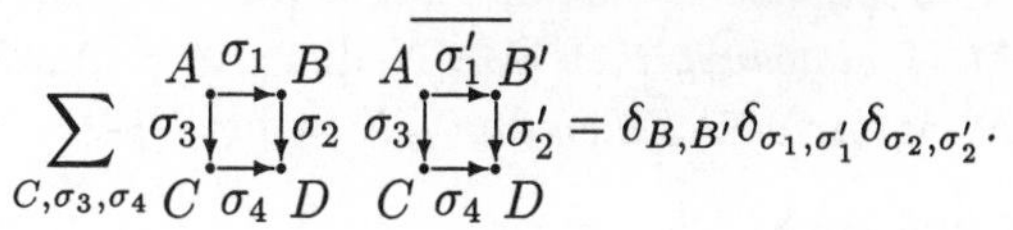

$$\sum_{C,\sigma_3,\sigma_4} \begin{array}{ccc} A & \xrightarrow{\sigma_1} & B \\ \sigma_3\downarrow & & \downarrow\sigma_2 \\ C & \xrightarrow[\sigma_4]{} & D \end{array} \quad \overline{\begin{array}{ccc} A & \xrightarrow{\sigma_1'} & B' \\ \sigma_3\downarrow & & \downarrow\sigma_2' \\ C & \xrightarrow[\sigma_4]{} & D \end{array}} = \delta_{B,B'}\delta_{\sigma_1,\sigma_1'}\delta_{\sigma_2,\sigma_2'}.$$

FIG. 10.7. Unitarity of a connection

Definition 10.2 *Each vertex of the graph $\mathcal{G}$ represents a bimodule. We assign the positive number $\mu(X) = [X]^{1/2}$ to each vertex representing a bimodule X. (Here the symbol $[X]$ denotes the Jones index of a bimodule, as in Definition 9.56.)*

There are two distinguished vertices $*_N \in V_0$ and $*_M \in V_2$ corresponding to ${}_N N_N$ and ${}_M M_M$ respectively. (They correspond to $*$ of the principal graph and $*$ of the dual principal graph.) The function μ assigns a positive number greater than or equal to 1 to each vertex of $\mathcal{G}$ and satisfies $\mu(*_N) = \mu(*_M) = 1$.

We will soon need a classical theorem of Perron–Frobenius, which we state in terms of graphs, since it is the form we need in this book. (This form is actually weaker than the ordinary form of the theorem, but it is sufficient for our purposes. See (Gantmacher 1960) for a proof.)

Theorem 10.3 *Let Γ be a finite connected bipartite graph. We denote the set of even [resp. odd] vertices of Γ by I [resp. J]. Let $\Delta = (\Delta_{ij})_{i\in I, j\in J}$ be the incidence matrix of Γ defined by*

$$\Delta_{ij} = \text{the number of edges connecting } i \text{ and } j.$$

Then we have the following.

1. *There exist a map $\mu : I \cup J \to \mathbb{R}_+$ and $\beta > 0$ such that $\Delta^t(\mu|_I) = \beta\mu|_J$ and $\Delta(\mu|_J) = \beta\mu|_I$.*
2. *The number β in 1 is unique and the vector μ in 1 is unique up to a positive scalar.*
3. $\bigcap_{k=1}^{\infty} (\Delta\Delta^t)^k(\mathbb{R}_+^I) = \mathbb{R}_+ \mu|_I$.
4. *For any vector $v \in \mathbb{R}_+^I$,*

$$\lim_{k\to\infty} \frac{(\Delta\Delta^t)^k v}{\|(\Delta\Delta^t)^k v\|} = \frac{\mu|_I}{\|\mu|_I\|}.$$

We call β the *Perron–Frobenius eigenvalue* of Γ and μ the *Perron–Frobenius eigenvector* of Γ.

We need the following proposition.

Proposition 10.4 *For each bimodule arising from a subfactor $N \subset M$ with finite index and finite depth as above, we have the following.*

1. *If X is an N-N bimodule, then* $\dim {}_N X = \dim X_N$.

2. *If X is an N-M bimodule, then* $\dim {}_N X = [M:N] \dim X_M$.
3. *If X is an M-N bimodule, then* $[M:N] \dim {}_M X = \dim X_N$.
4. *If X is M-M bimodule, then* $\dim {}_M X = \dim X_M$.

Proof For ${}_N X_N$ and ${}_M Y_N$, set

$$n_{X,Y} = \dim \mathrm{Hom}({}_M Y_N, {}_M M \otimes_N X_N) = \dim \mathrm{Hom}({}_N X_N, {}_N M \otimes_M Y_N),$$

where we used Frobenius reciprocity, Theorem 9.71. Then we have the following four identities by Proposition 9.63.

$$\begin{aligned}[M:N]^{1/2} \dim {}_N X &= [M:N]^{1/2} \dim {}_M M \otimes_N X \\ &= \sum_Y n_{X,Y} ([M:N]^{1/2} \dim {}_M Y),\end{aligned}$$

$$\begin{aligned}[M:N]^{1/2} ([M:N]^{1/2} \dim {}_M Y) &= \dim {}_N M \otimes_M Y \\ &= \sum_X n_{X,Y} \dim {}_N X,\end{aligned}$$

$$\begin{aligned}[M:N]^{1/2} \dim X_N &= [M:N]^{-1/2} \dim M \otimes_N X_N \\ &= \sum_Y n_{X,Y} ([M:N]^{-1/2} \dim Y_N),\end{aligned}$$

$$\begin{aligned}[M:N]^{1/2} ([M:N]^{-1/2} \dim Y_N) &= \dim M \otimes_M Y_N \\ &= \sum_X n_{X,Y} \dim X_N.\end{aligned}$$

These mean that the assignment of $\dim_N X$ to X and $[M:N]^{1/2} \dim_M Y$ to Y and the assignment of $\dim X_N$ to X and $[M:N]^{-1/2} \dim Y_N$ to Y are both the Perron–Frobenius eigenvectors for the matrix $[n_{X,Y}]_{X,Y}$. Because $\dim_N N = \dim N_N = 1$, we get $\dim {}_N X = \dim X_N$ and

$$[M:N] \dim {}_M Y = \dim Y_N,$$

by Theorem 10.3 (2).

The other two cases are proved similarly. □

Corollary 10.5 *Suppose that $N \subset M$ is a subfactor of finite index and finite depth. Let X, Y be bimodules of the same kind arising from $N \subset M$ as above. Then we have $[X+Y]^{1/2} = [X]^{1/2} + [Y]^{1/2}$.*

Proof Suppose X and Y are M-N bimodules. We have

$$[X+Y] = \dim {}_M X \dim X_N + \dim {}_M Y \dim X_N$$

$$+\dim {}_M X \dim Y_N + \dim {}_M Y \dim Y_N,$$

and

$$([X]^{1/2} + [Y]^{1/2})^2 = \dim {}_M X \dim X_N + \dim {}_M Y \dim Y_N \\ +2\sqrt{\dim {}_M X \dim X_N \dim {}_M Y \dim Y_N}.$$

Thus the desired equality is equivalent to

$$(\sqrt{\dim {}_M Y \dim X_N} - \sqrt{\dim {}_M X \dim Y_N})^2 = 0.$$

This is satisfied because Proposition 10.4 implies

$$\frac{\dim {}_M X}{\dim X_N} = \frac{\dim {}_M Y}{\dim Y_N}.$$

The other three cases are proved in the same way. □

Then we can prove that μ gives the Perron–Frobenius eigenvectors of the incidence matrices of these four graphs with the eigenvalue $[M:N]^{1/2}$, as follows. For example, if X is an N-N bimodule X, then we get $[M:N]^{1/2}\mu(X) = \sum_{{}_N Y_M} \mu(Y)$. Here the summation $\sum_{{}_N Y_M}$ is over Y which are connected to X with the principal graph and $\mu(Y)$ is with multiplicity n if the number of edges connecting X and Y is n. We can prove this identity by applying $[\cdot]^{1/2}$ to both sides of ${}_N X \otimes_N M_M = \sum {}_N Y_M$ using Corollary 10.5. Thus the following axiom is satisfied for our graph $\mathcal{G}$ with $\beta = [M:N]^{1/2}$. (Recall that our graph $\mathcal{G}$ is finite.)

Axiom 2. (Harmonicity) *The map μ satisfies $\mu(*_N) = \mu(*_M) = 1$ for the two distinguished vertices $*_N \in V_0$ and $*_M \in V_2$ in the graph $\mathcal{G}$. We have the following identities for some positive number β:*

$$\beta\mu(X) = \sum_{Y \in V_k} n_{X,Y}\mu(Y),$$

where X is in V_j, $k = j \pm 1 \in \mathbb{Z}/4$, and $n_{X,Y}$ is the number of edges connecting X and Y.

It is possible that we have a function μ on an infinite graph and it determines an eigenvector with positive entries for the incidence matrix of the graph with a positive eigenvalue. We call such an eigenvector and an eigenvalue a *Perron–Frobenius eigenvector* and a *Perron–Frobenius eigenvalue.* Note that if the connected graph $\mathcal{G}$ is finite, the eigenvalue β and the function μ giving the eigenvector are determined uniquely by Theorem 10.3. For this reason, this μ is often called the *Perron–Frobenius weight.*

With this harmonic axiom, we can prove the restriction on possible index values in Theorem 9.16 as follows. First, we cite the following theorem of Kronecker. (See Chapter 1 of (Goodman *et al.* 1989) for a proof, for example.)

FIG. 10.8. Dynkin diagram A_n, Coxeter number $n+1$

FIG. 10.9. Dynkin diagram D_n, Coxeter number $2n-2$

Theorem 10.6 *If a Perron–Frobenius eigenvalue is less than 2, then the graph must be one of Dynkin diagrams A_n, D_n, E_6, E_7, E_8 in Figs 10.8–10.12, and the eigenvalue is of the form $2\cos\pi/k$ with $k = 3, 4, \ldots$. This integer k is determined by the graph, and called the* Coxeter number *– see Section 8.7.*

Corollary 10.7 *If the index value $[M : N]$ of a subfactor is below four, it must be of the form $4\cos^2\pi/n$ with $n = 3, 4, \ldots$.*

This should be compared with the discrete series $c = 1 - 6/m(m+1)$ of highest weight irreducible representations of the Virasoro algebra for central charge $c < 1$. See Section 8.2.

We also have the following consequence of the harmonic axiom and restriction on the index values below 4, because $\mu(*) = 1$ and $\mu(X) = [X]^{1/2} \geq 1$ for a general X. This is often called the *$2\cos\pi/n$-rule.*

Corollary 10.8 *If a finite graph G is a principal graph of a subfactor, then the vertex $*$ is one of the vertices with the lowest Perron–Frobenius eigenvector entries. Furthermore, if we normalize the lowest Perron–Frobenius eigenvector entries to be 1, then all the other entries must be in the set $\{2\cos\pi/n \mid n = 3, 4, \ldots\} \cup [2, \infty)$.*

For example, we can conclude using Corollary 10.8 that the Dynkin diagram D_5 is not the principal graph of any subfactor as follows.

Write the Perron–Frobenius eigenvector entries as in Fig. 10.13. The number $(1 + \sqrt{2}/2)\sqrt{2 - \sqrt{2}}$ is approximately equal to 1.30656 and is in the interval

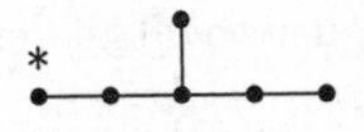

FIG. 10.10. Dynkin diagram E_6, Coxeter number 12

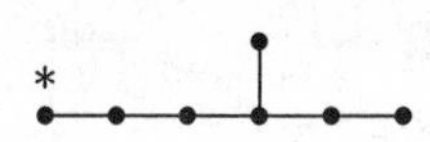

FIG. 10.11. Dynkin diagram E_7, Coxeter number 18

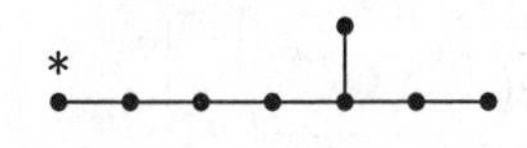

FIG. 10.12. Dynkin diagram E_8, Coxeter number 30

$(1, \sqrt{2})$. It means that the leftmost vertex is $*$, if D_5 is the principal graph of a subfactor, but then the two rightmost vertices have entries which do not fall in the set of allowed index values. This means that D_5 cannot be the principal graph of any subfactor. This $2\cos\pi/n$-rule is often useful to reject a graph as a principal graph. However, we note that a graph which passes this test is not necessarily the principal graph of some subfactor. The Dynkin diagrams D_{2n+1}, $n > 2$, are such examples. We will discuss this in more detail in Sections 11.5, 12.3, and 13.4.

We next look at the four bimodules $*_N = {}_N N_N \in V_0$, ${}_N M_M \in V_1$, $*_M = {}_M M_M \in V_2$, ${}_M M_N \in V_3$. The trivial relative commutant condition $N' \cap M = \mathbb{C}$ implies the irreducibility of the bimodules ${}_N M_M$ and ${}_M M_N$ as in Example 9.67. Thus the number of edges in $\mathcal{G}_0$ connected to $*_N$ is one, and the edge has ${}_N M_M$ as the other endpoint which is also the endpoint of the only edge in $\mathcal{G}_1$ connected to $*_M$. By axiomatizing this situation, we have the following axiom.

Axiom 3. (Initialization) *We have vertices ${}_N{**}_M \in V_1$ and ${}_M{**}_N \in V_3$ satisfying the following properties.*

1. *The graph $\mathcal{G}_0$ has only one edge connected to the vertex $*_N \in V_0$ and the other endpoint of the edge is ${}_N{**}_M$.*

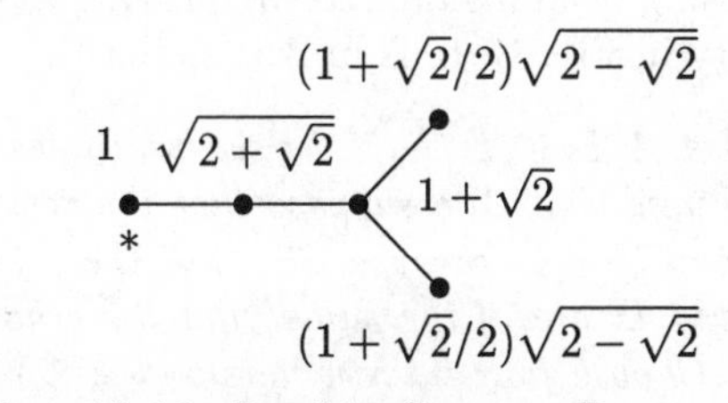

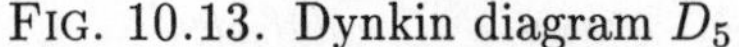
FIG. 10.13. Dynkin diagram D_5

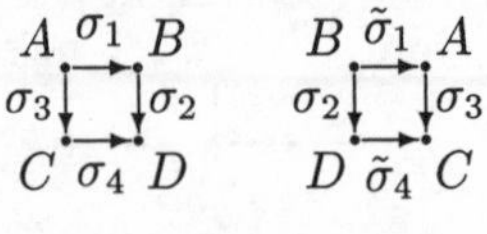

$$\begin{array}{ccc} A & \xrightarrow{\sigma_1} & B \\ {\scriptstyle\sigma_3}\downarrow & & \downarrow{\scriptstyle\sigma_2} \\ C & \xrightarrow[\sigma_4]{} & D \end{array} \qquad \begin{array}{ccc} B & \xrightarrow{\tilde{\sigma}_1} & A \\ {\scriptstyle\sigma_2}\downarrow & & \downarrow{\scriptstyle\sigma_3} \\ D & \xrightarrow[\tilde{\sigma}_4]{} & C \end{array}$$

FIG. 10.14. Related connections

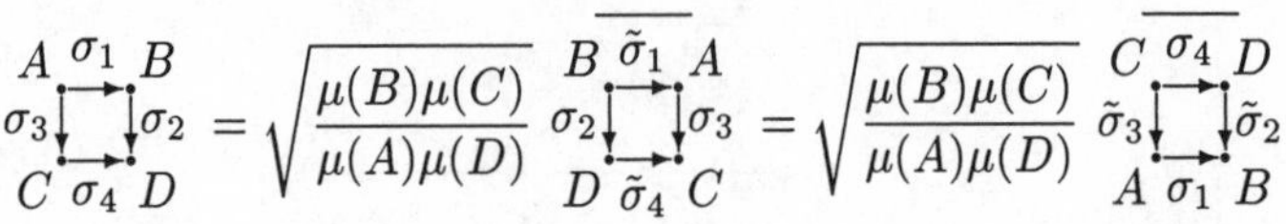

$$\begin{array}{ccc} A & \xrightarrow{\sigma_1} & B \\ {\scriptstyle\sigma_3}\downarrow & & \downarrow{\scriptstyle\sigma_2} \\ C & \xrightarrow[\sigma_4]{} & D \end{array} = \sqrt{\frac{\mu(B)\mu(C)}{\mu(A)\mu(D)}}\ \overline{\begin{array}{ccc} B & \xrightarrow{\tilde{\sigma}_1} & A \\ {\scriptstyle\sigma_2}\downarrow & & \downarrow{\scriptstyle\sigma_3} \\ D & \xrightarrow[\tilde{\sigma}_4]{} & C \end{array}} = \sqrt{\frac{\mu(B)\mu(C)}{\mu(A)\mu(D)}}\ \overline{\begin{array}{ccc} C & \xrightarrow{\sigma_4} & D \\ {\scriptstyle\tilde{\sigma}_3}\downarrow & & \downarrow{\scriptstyle\tilde{\sigma}_2} \\ A & \xrightarrow[\sigma_1]{} & B \end{array}}$$

FIG. 10.15. Renormalization of a connection

2. *The graph $\mathcal{G}_3$ has only one edge connected to the vertex $*_N \in V_0$ and the other endpoint of the edge is ${}_M{*}{*}_N$.*
3. *The graph $\mathcal{G}_1$ has only one edge connected to the vertex $*_M \in V_2$ and the other endpoint of the edge is ${}_N{*}{*}_M$.*
4. *The graph $\mathcal{G}_2$ has only one edge connected to the vertex $*_M \in V_2$ and the other endpoint of the edge is ${}_M{*}{*}_N$.*

10.4 Renormalization of connections

In this section, we study the renormalization axiom. Consider the two connection values in Fig. 10.14. Here, the two edges $\tilde{\sigma}_1$ and $\tilde{\sigma}_4$ denote the reversed edges given by Frobenius reciprocity from σ_1 and σ_4 respectively. Because Frobenius reciprocity was given by an explicit formula, it is natural to expect that the two connection values in Fig. 10.14 have a simple relation. The following axiom, the renormalization axiom, gives this relation. The two axioms unitarity and renormalization together are often called *bi-unitarity.* The renormalization axiom is sometimes called the crossing symmetry, whose name comes from the theory of solvable lattice models. (See Chapters 7 and 8.)

Axiom 4. (Renormalization) *We have the identities as in Fig. 10.15, for a cell given by $(\sigma_1, \sigma_2, \sigma_3, \sigma_4)$.*

We prove that the first identity of this renormalization axiom is satisfied by our connection arising from a subfactor by proving the following stronger form. The second identity in Fig. 10.15 is proved similarly.

Theorem 10.9 *Let A, B, C, D, X, Y be six bimodules arising from a subfactor $N \subset M$ as in Section 10.2. We suppose that the relative tensor products $X \otimes A \otimes Y$, $X \otimes B$, and $C \otimes Y$ are possible, and the four bimodules $X \otimes A \otimes Y$, $X \otimes B$, $C \otimes Y$, and D are of the same kind. (We have dropped subscripts for $\otimes$ for simplicity.) Choose four intertwiners $\sigma_1 : A \otimes Y \to B$, $\sigma_2 : X \otimes B \to D$, $\sigma_3 : X \otimes A \to C$, and $\sigma_4 : C \otimes Y \to D$ and set*

$$W = \sigma_4(\sigma_3 \otimes \mathrm{id}_Y)(\mathrm{id}_X \otimes \sigma_1)^* \sigma_2^* \in \mathrm{End}(D) = \mathbb{C},$$

and

$$W' = \sigma_4^Y(\sigma_2 \otimes \mathrm{id}_{\bar{Y}})(\mathrm{id}_X \otimes \sigma_1^Y)^* \sigma_3^* \in \mathrm{End}(C) = \mathbb{C}.$$

Then we have

$$W' = \sqrt{\frac{\mu(A)\mu(D)}{\mu(B)\mu(C)}}\bar{W}.$$

Proof For simplicity, we assume that all the bimodules are N-N bimodules. (The other cases are proved in the same way.) We choose a basis $\{\gamma_j\}_j$ for C_N, a basis $\{\eta_k\}_k$ for Y_N, and a basis $\{\delta_l\}_l$ for D_N. Then we have

$$\begin{aligned}
&(\dim C_N)W' \\
&= \sum_j \langle \sigma_4^Y(\sigma_2 \otimes \mathrm{id}_{\bar{Y}})(\mathrm{id}_X \otimes \sigma_1^Y)^* \sigma_3^* \gamma_j, \gamma_j \rangle \\
&= \sqrt{\frac{\mu(A)\mu(C)}{\mu(B)\mu(D)}} \sum_{j,k,k'} \langle ((\mathrm{id}_X \otimes \sigma_1)(\sigma_3^*\gamma_j \otimes \eta_k) \otimes \bar{\eta}_k, \sigma_2^*\sigma_4(\gamma_j \otimes \eta_{k'}) \otimes \bar{\eta}_{k'} \rangle \\
&= \sqrt{\frac{\mu(A)\mu(C)}{\mu(B)\mu(D)}} \sum_{j,k} \langle ((\mathrm{id}_X \otimes \sigma_1)(\sigma_3^*\gamma_j \otimes \eta_k), \sigma_2^*\sigma_4(\gamma_j \otimes \eta_k) \rangle \\
&= \sqrt{\frac{\mu(A)\mu(C)}{\mu(B)\mu(D)}} \sum_{j,k,l} \langle \sigma_2(\mathrm{id}_X \otimes \sigma_1)(\sigma_3^*\gamma_j \otimes \eta_k), \delta_l \langle \delta_l, \sigma_4(\gamma_j \otimes \eta_k)\rangle_N^\circ \rangle \\
&= \sqrt{\frac{\mu(A)\mu(C)}{\mu(B)\mu(D)}} \sum_{l} \langle \sigma_2(\mathrm{id}_X \otimes \sigma_1)(\sigma_3^* \otimes \mathrm{id}_Y)\sigma_4^*\delta_l, \delta_l \rangle \\
&= \sqrt{\frac{\mu(A)\mu(C)}{\mu(B)\mu(D)}} (\dim D_N)\bar{W},
\end{aligned}$$

which implies the conclusion. Here we have used

$$\sqrt{\frac{\dim A_N \dim D_N}{\dim B_N \dim C_N}} = \sqrt{\frac{\mu(A)\mu(D)}{\mu(B)\mu(C)}},$$

which follows from Proposition 10.4 and the definition of μ in Definition 10.2. □

We will also use a standard convention as in Fig. 10.16 for the symbols denoting connection values. Note that for simplicity we have dropped the labels of edges.

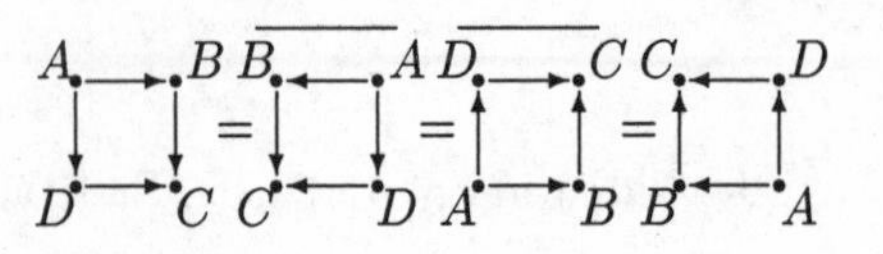

FIG. 10.16. Conventions for connections

10.5 Flatness of connections

In this section, we study the final axiom for a paragroup. In order to explain this axiom, we need to introduce an analogue of the partition function in statistical mechanics. (See Chapters 7 and 8.)

We again study the problem of the bases changes of the same End space. Take a partial isometry

$$(\sigma_{1,+} \cdot \sigma_{2,+} \cdot \sigma_{3,+}, \sigma_{1,-} \cdot \sigma_{2,-} \cdot \sigma_{3,-})$$

in a matrix unit of $\mathrm{End}({}_M M \otimes_N M_N)$ corresponding to the embeddings

$$\mathrm{End}({}_N N_N) \subset \mathrm{End}({}_N M_M) \subset \mathrm{End}({}_N M_N) \subset \mathrm{End}({}_M M \otimes_N M_N).$$

We express this operator in terms of a basis corresponding to the embeddings

$$\mathrm{End}({}_N N_N) \subset \mathrm{End}({}_M M_N) \subset \mathrm{End}({}_M M \otimes_N M_M) \subset \mathrm{End}({}_M M \otimes_N M_N).$$

We need a coefficient in the expansion

$$\sum c_{\sigma_{1,+}\cdot\sigma_{2,+}\cdot\sigma_{3,+},\rho_{1,+}\cdot\rho_{2,+}\cdot\rho_{3,+}} \bar{c}_{\sigma_{1,-}\cdot\sigma_{2,-}\cdot\sigma_{3,-},\rho_{1,-}\cdot\rho_{2,-}\cdot\rho_{3,-}} \times (\rho_{1,+} \cdot \rho_{2,+} \cdot \rho_{3,+}, \rho_{1,-} \cdot \rho_{2,-} \cdot \rho_{3,-})$$

with respect to the latter basis

$$\{(\rho_{1,+} \cdot \rho_{2,+} \cdot \rho_{3,+}, \rho_{1,-} \cdot \rho_{2,-} \cdot \rho_{3,-})\},$$

and this coefficient

$$c_{\sigma_{1,+}\cdot\sigma_{2,+}\cdot\sigma_{3,+},\rho_{1,+}\cdot\rho_{2,+}\cdot\rho_{3,+}}$$

is given as in Fig. 10.17.

In Fig. 10.17, the summation in the left hand side is over all τ connecting the endpoint of σ_1 and the endpoint of ρ_2, and the right hand side is defined by the left hand side. (That is, we make a product of the connections for each configuration of the inside of the square and a summation of the products over all the configurations. See Fig. 10.18 below for a more general setting.) We have a similar formula for

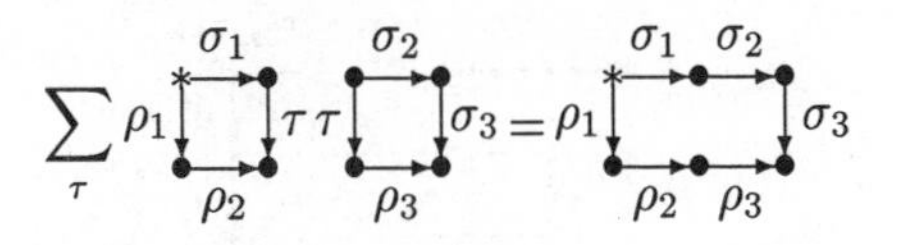

FIG. 10.17. Partition functions of connections

$$c_{\sigma_{1,-}\cdot\sigma_{2,-}\cdot\sigma_{3,-},\rho_{1,-}\cdot\rho_{2,-}\cdot\rho_{3,-}}$$

and longer paths are dealt with similarly. Note that these coefficients are similar to the partition functions in statistical mechanics as in Chapters 7 and 8 — except that we do not expect positivity here.

Two bimodules $*_N = {}_N N_N$, $*_M = {}_M M_M$ play the role of an identity in the tensor product operations. We write $*$ for one of the two possibilities. It corresponds to the initial vertex $*$ of the (dual) principal graph. In the above, we had a partition function for rectangles with size $1 \times n$. Here we have more general rectangles. For this purpose, we consider the following double sequence of the End spaces.

$$\begin{array}{ccccccc}
\mathrm{End}({}_N N_N) & \subset & \mathrm{End}({}_N M_M) & \subset & \mathrm{End}({}_N M_N) & \subset \cdots \\
\cap & & \cap & & \cap & \\
\mathrm{End}({}_M M_N) & \subset & \mathrm{End}({}_M M \otimes_N M_M) & \subset & \mathrm{End}({}_M M \otimes_N M_N) & \subset \cdots \\
\cap & & \cap & & \cap & \\
\mathrm{End}({}_N M_N) & \subset & \mathrm{End}({}_N M \otimes_N M_M) & \subset & \mathrm{End}({}_N M \otimes_N M_N) & \subset \cdots \\
\cap & & \cap & & \cap & \\
\vdots & & \vdots & & \vdots &
\end{array}$$

Here the vertical embedding is given by composition of the identity intertwiner corresponding to tensoring ${}_M M_N$ or ${}_N M_M$ from the left. For example, the first entry $\mathrm{End}({}_N M_N)$ in the third line and the third entry $\mathrm{End}({}_N M_N)$ in the first line are of course isomorphic, but they are not identified, because they are embedded into the third entry $\mathrm{End}({}_N M \otimes_N M_N)$ of the third line as different subalgebras. By generalizing the argument identifying the End spaces with the higher relative commutants in the first two lines, we can identify this double sequence with $\{N_k' \cap M_{l-1}\}$ for a tunnel $\{N_k\}$. (For the first two lines, we had $M_1' \cap M_k$ and $M' \cap M_k$ in Section 10.2. Applying the map $x \mapsto J_N J_M x J_M J_N$, we get $N' \cap M_{k-2}$ and $N_1' \cap M_{k-2}$.)

For each algebra in the double sequence, we have several matrix unit systems. In order to move between different systems of bases, we apply the same method as above. Here we need the vertical Frobenius reciprocity which changes the orientations of the arrows $A \to C$, $B \to D$ in Fig. 10.6. The partition function for a general rectangle with size $m \times n$ gives the coefficient in the change of bases.

Based on this observation, we introduce a partition function with fixed boundary conditions in the paragroup theory as follows. We choose one of $*_N$ and $*_M$ and write $*$ for it. We choose four sequences of edges $(\sigma_1, \sigma_2, \ldots, \sigma_{2n})$,

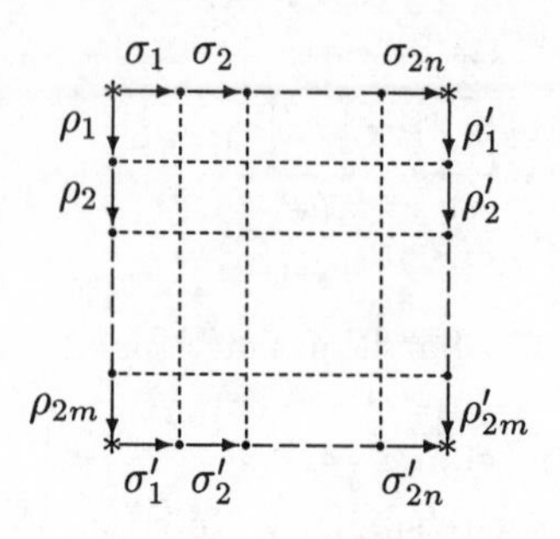

FIG. 10.18. Configuration

$(\sigma'_1, \sigma'_2, \ldots, \sigma'_{2n})$, $(\rho_1, \rho_2, \ldots, \rho_{2m})$, and $(\rho'_1, \rho'_2, \ldots, \rho'_{2m})$ with the following properties.

1. The edges $\sigma_1, \sigma_{2n}, \sigma'_1, \sigma'_{2n}, \rho_1, \rho_{2m}, \rho'_1, \rho'_{2m}$ have $*$ as one of the endpoints.
2. The edges σ_j and σ_{j+1} have a common endpoint for $j = 1, \ldots, 2n-1$.
3. The edges σ'_j and σ'_{j+1} have a common endpoint for $j = 1, \ldots, 2n-1$.
4. The edges ρ_k and ρ_{k+1} have a common endpoint for $k = 1, \ldots, 2m-1$.
5. The edges ρ'_k and ρ'_{k+1} have a common endpoint for $k = 1, \ldots, 2m-1$.
6. If $* = *_N$, then the edges σ_j and σ'_j are in $\mathcal{G}_0$ for $j = 1, \ldots, 2n$.
7. If $* = *_N$, then the edges ρ_k and ρ'_k are in $\mathcal{G}_3$ for $k = 1, \ldots, 2n$.
8. If $* = *_M$, then the edges σ_j and σ'_j are in $\mathcal{G}_2$ for $j = 1, \ldots, 2n$.
9. If $* = *_M$, then the edges ρ_k and ρ'_k are in $\mathcal{G}_1$ for $k = 1, \ldots, 2n$.

Then we have the following definition of the *partition function* with fixed boundary condition. First, the dotted lines inside the diagram in Fig. 10.18 means we complete the interior of the large rectangle with edges from the graph $\mathcal{G}$ and each choice of internal edges is called a *configuration.* Figure 10.18 is the product of the $4nm$ connection values of a chosen configuration. Then Fig. 10.19 is the sum of the values in Fig. 10.18 over all the possible configurations, which is a direct analogue of a partition function with fixed boundary values $\sigma, \sigma', \rho, \rho'$ in solvable lattice model theory as in Chapters 7 and 8. Obviously, we can define a partition function in the case in which all four corner vertices are not necessarily $*$.

With this definition, we can write the following flatness axiom.

Axiom 5. (Flatness) *We choose $*_N$ or $*_M$ and denote the vertex by $*$. Then we have the identity as in Fig. 10.20 for all edges $\sigma_j, \sigma'_j, \rho_k, \rho'_k$ as above.*

We prove that our connection W arising from the subfactor $N \subset M$ satisfies this axiom.

Theorem 10.10 *The connection W arising from a subfactor $N \subset M$ satisfies the flatness axiom.*

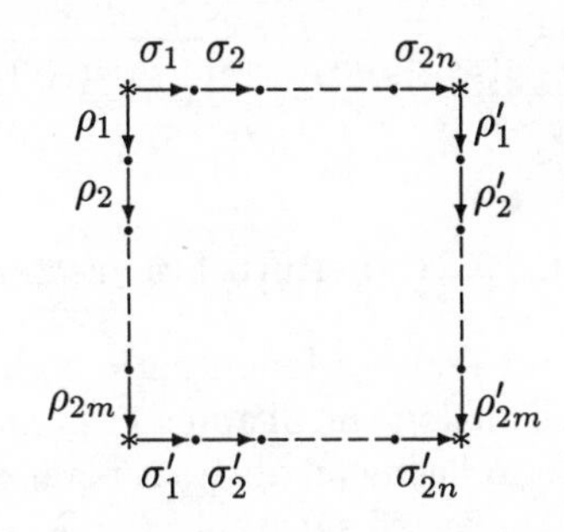

FIG. 10.19. Partition function of connections

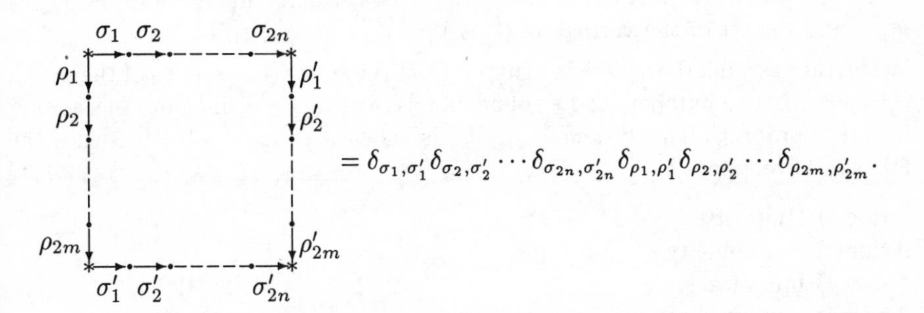

$$= \delta_{\sigma_1,\sigma'_1}\delta_{\sigma_2,\sigma'_2}\cdots\delta_{\sigma_{2n},\sigma'_{2n}}\delta_{\rho_1,\rho'_1}\delta_{\rho_2,\rho'_2}\cdots\delta_{\rho_{2m},\rho'_{2m}}.$$

FIG. 10.20. Flatness property

Proof First suppose that $* = *_N$, $\sigma_j = \sigma'_j$ for $j = 1, \ldots, 2n$ and $\rho_k = \rho'_k$ for $k = 1, \ldots, 2m$. Recall that a concatenation of edges represents a composition of intertwiners as in Section 10.2.

Because all four corners are $*$, the definition of composition of intertwiners implies that the intertwiner through the upper-right corner and that through the lower-left corner both give

$$\begin{aligned}&(\rho_{2m}(\mathrm{id}_M \otimes \rho_{2m-1})\cdots(\mathrm{id}_M \otimes \cdots \otimes \mathrm{id}_M) \otimes \rho_1) \otimes \mathrm{id}_N \\ \otimes\ &(\sigma_{2n}(\sigma_{2n-1} \otimes \mathrm{id}_M)\cdots\sigma_1 \otimes (\mathrm{id}_M \otimes \cdots \otimes \mathrm{id}_M)).\end{aligned}$$

That is, the rectangle diagram has value 1 as the inner product of an intertwiner from

$${}_N M \otimes_M M \otimes_N \cdots \otimes_M M_N$$

to ${}_N N_N$ with itself. If we change part of the left vertical path or a bottom right-bound path, then the unitarity implies that the new diagram has value 0.

The case $* = *_M$ is handled in the same way. □

The term *flat connection* comes from an interpretation of this condition as a discrete analogue of the flatness of a connection in differential geometry in the

$$W^{\#}\begin{pmatrix} A \overset{\rho_1}{\longrightarrow} B \\ \rho_3 \downarrow \quad \downarrow \rho_2 \\ C \underset{\rho_4}{\longrightarrow} D \end{pmatrix} = \sum_{\sigma_1,\sigma_2,\sigma_3,\sigma_4} u(\rho_3,\sigma_3)u(\rho_4,\sigma_4)\overline{u(\rho_1,\sigma_1)u(\rho_2,\sigma_2)}\, W \begin{pmatrix} A \overset{\sigma_1}{\longrightarrow} B \\ \sigma_3 \downarrow \quad \downarrow \sigma_2 \\ C \underset{\sigma_4}{\longrightarrow} D \end{pmatrix}$$

FIG. 10.21. Perturbed connection

sense that 'parallel transport' along a loop from $*$ does not change the form of a loop. We will discuss this analogy in Chapter 11 in more detail.

We are now finally able to define an abstract paragroup. Suppose we have a graph $\mathcal{G}$ consisting of four connected subgraphs $\mathcal{G}_0$, $\mathcal{G}_1$, $\mathcal{G}_2$, $\mathcal{G}_3$ with the following properties.

1. The sets of the edges of $\mathcal{G}_j$, $j = 0, 1, 2, 3, 4$, are disjoint.
2. The set of the vertices of the graph $\mathcal{G}$ is a disjoint union of V_0, V_1, V_2, V_3 and the set of the vertices of $\mathcal{G}_j$ is $V_j \cup V_{j+1}$ for $j \in \mathbb{Z}/4$.

We further assume that there is a map μ from the set of the vertices of the graph $\mathcal{G}$ to positive real numbers and a connection W abstractly defined on cells arising from the graph $\mathcal{G}$. The system $(\mathcal{G}, \mu, W)$ is called a *paragroup* if it satisfies the following five axioms:

Axiom 1 Unitarity.
Axiom 2 Harmonicity.
Axiom 3 Initialization.
Axiom 4 Renormalization.
Axiom 5 Flatness.

The definition of the connection W involved choices of orthonormal basis of intertwiners, so the paragroup given by a subfactor $N \subset M$ is not unique. In order to eliminate this inconvenience, we now define an equivalence relation between two paragroups satisfying the above five axioms.

Definition 10.11 *Let $(\mathcal{G}, \mu, W)$ be a system satisfying the above five axioms.* A perturbation u *of the connection W is a set of unitary matrices $(u(\sigma,\rho))_{\sigma,\rho}$ associated with each pair of adjacent vertices X, Y of $\mathcal{G}$, where σ, ρ are edges of $\mathcal{G}$ connecting X and Y. We require that $u(\tilde{\sigma}, \tilde{\rho}) = \overline{u(\sigma,\rho)}$. The perturbed connection $W^{\#}$ is defined by the formula in Fig. 10.21. A choice of such a unitary u is called a* gauge choice.

The paragroups $(\mathcal{G}, \mu, W)$ and $(\mathcal{G}, \mu, W^{\#})$ are called equivalent. We simply write $(\mathcal{G}, \mu, W)$ for the equivalence class. Two paragroups $(\mathcal{G}, \mu, W)$ and $(\mathcal{G}', \mu', W')$ are called isomorphic if there is a perturbation $W^{\#}$ of W and a graph isomorphism $\theta : \mathcal{G} \to \mathcal{G}'$ such that θ maps the vertices $*_N$, ${}_N{*}{*}_M$, ${}_M{*}{*}_N$, $*_M$ of $\mathcal{G}$ to the corresponding vertices of $\mathcal{G}'$ and we have $\mu' \cdot \theta = \mu$ and $W' \cdot \theta = W^{\#}$.

At the end of this section, we show a related property.

Proposition 10.12 *Suppose that the partition function of size $2m \times 2n$ in Fig. 10.22 has a non-zero value. Then we have $X = \bar{Y}$.*

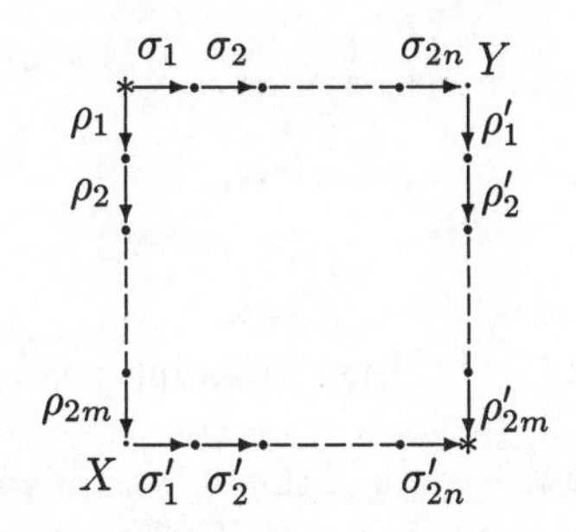

FIG. 10.22. Partition function having $*$ at two corners

Proof For this partition function, we compare two ways of decomposition of

$$_N M \otimes_M M \otimes \cdots \otimes_M M_N.$$

Then Frobenius reciprocity, Theorem 9.71, gives the conclusion easily. □

10.6 Paragroups arising from finite groups

The name 'paragroup' comes from the fact that it is regarded as a certain quantization of a Galois group. An analogue of the classical Galois theory has been studied in the theory of operator algebras for a long time. In its simplest form, it recovers a finite group G from an inclusion $R \subset R \rtimes G$ or $R^G \subset R$, where the group G acts on a II_1 factor R as an outer action. It was Ocneanu's basic idea that once we translate this method into the bimodule language, we can use the same bimodule method with general subfactors and obtain some structure analogous to groups. This was the origin of the paragroup theory. So we should be able to recover the group structure as a paragroup by looking at subfactors $R \subset R \rtimes G$ or $R^G \subset R$. We show how to do this in this section. There are several other methods to get paragroups from group actions on factors as in (Bisch and Haagerup 1996), (Kosaki and Yamagami 1992), (Wassermann 1988), but the method here is the most fundamental among those.

Suppose we have an outer action α of a finite group G on a II_1 factor R. (Recall that any finite group G has an outer action on the hyperfinite II_1 factor by results in Section 6.7.) Then the crossed product construction gives a larger von Neumann algebra $R \rtimes G$. This $R \rtimes G$ is also a II_1 factor, and we get $[R \rtimes G : R] = |G| < \infty$ as in Example 9.20. By constructing a paragroup with $N = R$, $M = R \rtimes G$, we get N-N bimodules parametrized by the elements of G, M-M bimodules parametrized by the elements of $\hat{G}$, a single M-N bimodule, and a single N-M bimodule as follows.

Let $G = \{g_1, g_2, \ldots, g_n\}$. As in Example 9.20, we get $M = \bigoplus_{j=1}^n N u_{g_j}$, which is the irreducible decomposition as N-N bimodules. (That is, each $N u_{g_j}$ is an irreducible N-N bimodule and different g_j give non-isomorphic bimodules. The right action on $N u_{g_j}$ is given by $x u_{g_j} \cdot y = x \alpha_{g_j}(y) u_{g_j}$. See Exercise 10.1.)

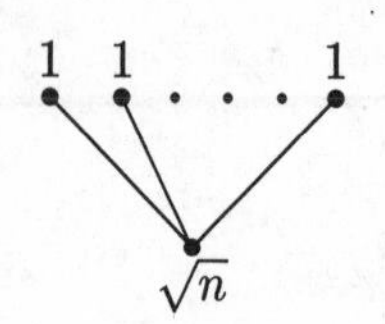

FIG. 10.23. The principal graph for $R \subset R \times G$

This means that the principal graph of this subfactor with the Perron–Frobenius eigenvector μ contains the graph in Fig. 10.23. This graph shows that our μ gives the correct eigenvector with correct eigenvalue, so we conclude that this is really the principal graph. That is, the principal graph has one edge from the single N-M bimodule to each bimodule labelled by each element of G.

We next study M-M bimodules by looking at the irreducible decomposition of $M \otimes_N M$. For each irreducible unitary representation σ of G, we set

$$\xi^\sigma_{ij} = \sum_{g \in G} \overline{\sigma_{ij}(g^{-1})} u_{g^{-1}} \otimes_N u_g \in M \otimes_N M, \quad i, j = 1, \ldots, \dim \sigma.$$

For $a, b \in N$, we get

$$\begin{aligned} \langle a u_h \xi^\sigma_{ij}, b u_k \xi^\rho_{lm} \rangle &= \delta_{hk} \operatorname{tr}(ab^*) \sum_{g \in G} \sigma_{ij}(g^{-1}) \overline{\rho_{lm}(g^{-1})} \\ &= n \operatorname{tr}(ab^*) \delta_{hk} \delta_{\sigma,\rho} \delta_{il} \delta_{jm}. \end{aligned}$$

Thus $\bigoplus_{\sigma,i,j} M \xi^\sigma_{i,j}$ is an orthogonal decomposition of $M \otimes_N M$ by the Peter–Weyl theorem. (The Peter–Weyl theorem is used to show that this sum really spans $M \otimes_N M$.) Because we have

$$\begin{aligned} \xi^\sigma_{ij} x &= x \xi^\sigma_{ij}, \quad x \in N, \\ \xi^\sigma_{ij} u_h &= u_h \sum_l \overline{\sigma_{il}(h)} \xi^\sigma_{lj}, \quad h \in G, \end{aligned}$$

we know that $X^\sigma_j = \sum_i M \xi^\sigma_{ij}$ is an M-M bimodule. We have $M \otimes_N M = \bigoplus_{\sigma,j} X^\sigma_j$ as a decomposition of M-M bimodules. By computations based on Schur's lemma, we can deduce that the bimodules X^σ_j's are irreducible and that X^σ_j and X^ρ_l are isomorphic if and only if $\rho = \sigma$. In this way, we can conclude that the dual principal graph of $N \subset M$ with the Perron–Frobenius eigenvector contains the graph in Fig. 10.24, where the number of edges between X^σ_j and ${}_M M_N$ is $\dim \sigma$.

Again we conclude that the graph in Fig. 10.24 is really the dual principal graph by checking the Perron–Frobenius eigenvalue. That is, the dual principal graph has edges from the unique M-N bimodule to each isomorphism class of bimodules labelled by an element in $\hat{G}$ with multiplicity equal to the dimension of the representation.

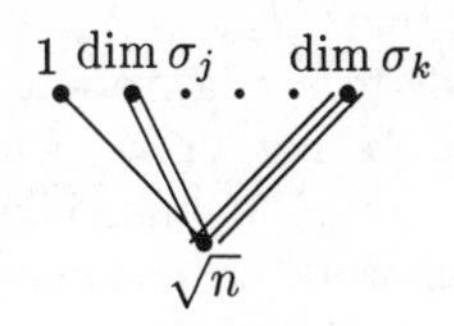

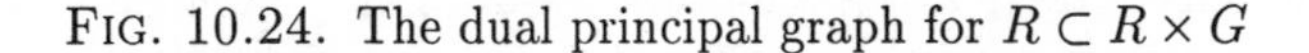

FIG. 10.24. The dual principal graph for $R \subset R \times G$

$$= \sigma_{ij}(g), \quad g \in G, \sigma \in \hat{G}.$$

FIG. 10.25. Connection for $R \subset R \times G$

By choosing an appropriate intertwiner to each edge, we conclude that the flat connection of this paragroup is given by Fig. 10.25. (Choose $\xi^{\sigma}_{li} \in X^{\sigma}_i$ and trace the images of this element in the bimodules appearing in Fig. 10.25.)

In short, any finite group can appear as a 'quantized Galois group' over the hyperfinite II_1 factor.

Exercise 10.1 Let α be an automorphism of a II_1 factor N. Define a bimodule ${}_N(N_\alpha)_N$ by $x \cdot \xi \cdot y = x\xi\alpha(y)$. Prove that this bimodule is isomorphic to ${}_N N_N$ if and only if α is inner.

Prove that $\overline{{}_N(N_\alpha)_N}$ is isomorphic to ${}_N(N_{\alpha^{-1}})_N$.

Prove that ${}_N(N_\alpha \otimes_N N_\beta)_N$ is isomorphic to ${}_N(N_{\alpha\beta})_N$.

Exercise 10.2 Let G, H be two finite groups. Construct the paragroups corresponding to these groups as above. Prove that if the two paragroups are isomorphic, then G, H are also isomorphic as groups.

10.7 Paragroups based on sector approach for properly infinite subfactors

Based on ideas in quantum field theory, (Longo 1989) and (Longo 1990) established a theory of sectors. We will explain that Longo's sector theory essentially produces the same objects as paragroups by replacing bimodules by sectors of type III factors. This method requires more preliminaries on operator algebraic side, but once the basic theory has been established, working with sectors is often easier than with bimodules. Furthermore, the sector method has a real advantage over the bimodule method, as we will see in Section 12.9. This section is *not* logically self-contained. We will present the method based on work of M. Izumi, H. Kosaki and R. Longo without proofs.

First we study a case of a II_1 factor M. Suppose we have an endomorphism ρ of M. Then we can define an M-M bimodule M_ρ by

$$x \cdot \xi \cdot y = x\xi\rho(y), \quad x, y \in M, \xi \in L^2(M).$$

An easy computation shows that $M_\rho \otimes_M M_\sigma \cong M_{\rho\sigma}$, that is, a composition of endomorphisms corresponds to a relative tensor product of bimodules. (See Exercise 10.1.) For a II_1 factor M, not all M-M bimodules arise in this way, but for a type III factor M, then any M-M bimodule arises from an endomorphism of M in this way. So we can work on endomorphisms instead of bimodules. A natural equivalence relation '$\sim$' is defined for endomorphisms of M so that $\rho \sim \rho'$ if and only if there exists a unitary $u \in M$ satisfying $u\rho(x)u^* = \rho'(x)$ for all $x \in M$. We call an equivalence class of endomorphisms a *sector*, and denote the set of sectors by $\mathrm{Sect}(M)$.

We also have a natural notion of a direct sum of sectors and use the symbol $+$ for a direct sum. For $\rho \in \mathrm{Sect}(M)$, we can define $\bar{\rho}$ by $\rho\bar{\rho} = \gamma$, where γ is Longo's canonical endomorphism as in (Longo 1987). (The canonical endomorphism γ of M is defined by $\gamma(x) = J_N J_M x J_M J_N$, where J_M and J_N are the modular conjugation of M and N respectively realized on the *same* Hilbert space using a cyclic separating vector for both M and N. See Section 5.6 for basic notions in the Tomita–Takesaki theory. The fact that there exists a simultaneous cyclic and separating vector for both M and N is not trivial at all. Such a vector is constructed in a rather transcendental way.) This $\bar{\rho}$ plays the role of a conjugate bimodule in the sector theory. We have $\overline{\rho_1 + \rho_2} = \bar{\rho}_1 + \bar{\rho}_2$ and $\overline{\rho_1\rho_2} = \bar{\rho}_2\bar{\rho}_1$.

Each sector ρ has a statistical dimension $d(\rho) \in [1, \infty]$, which is $[M : \rho(M)]_0$, where the symbol $[\cdot : \cdot]_0$ means the minimal index in the sense of (Hiai 1988). (The definition of the Jones index was extended to arbitrary subfactors, not necessarily of type II_1, by (Kosaki 1986). Later (Hiai 1988) introduced the notion of a minimal index.) The set $\mathrm{Sect}(M)$ has an additive and multiplicative structure given by direct sum and composition. The map $d : \mathrm{Sect}(M) \to [1, \infty]$ preserves these two operations. (The multiplicativity of the statistical dimension d was proved by (Kosaki and Longo 1992), (Longo 1992).)

We have a notion of intertwiner as follows. For two sectors ρ, ρ', we say that $x \in M$ is an intertwiner between ρ and ρ' if $\rho(y)x = x\rho'(y)$ for all $y \in M$. We say that a sector is *irreducible* if the scalars are the only intertwiners between ρ and itself. This condition is equivalent to the subfactor $\rho(M) \subset M$ having trivial relative commutant. We have an analogue of Frobenius reciprocity, Theorem 9.73, for sectors. (See (Longo 1990) and related papers.)

In Section 10.2, we worked on the higher relative commutants for the Jones tower, but here it is more natural to work with a tunnel. Let $N \subset M$ be a subfactor of type III with finite index and finite depth. (Type III subfactors also have a basic construction, so we can define the finite depth condition just as in the type II_1 case.) We also assume that $N \cong M$. (This is not a strong restriction. See (Longo 1992), for example.) We choose an arbitrary isomorphism ρ from M onto N, then it is an endomorphism of M so gives a sector. We use the same letter ρ for this sector.

We start with the identity endomorphism $1 = \mathrm{id}_M$, and multiply ρ and $\bar{\rho}$ alternatively on the right to get a series of sectors,

$$1, \rho, \rho\bar{\rho}, \rho\bar{\rho}\rho, \rho\bar{\rho}\rho\bar{\rho}, \cdots.$$

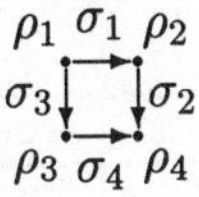

$$\begin{array}{ccc} \rho_1 & \sigma_1 & \rho_2 \\ \sigma_3 \downarrow & \square & \downarrow \sigma_2 \\ \rho_3 & \sigma_4 & \rho_4 \end{array}$$

FIG. 10.26. Cell of sectors

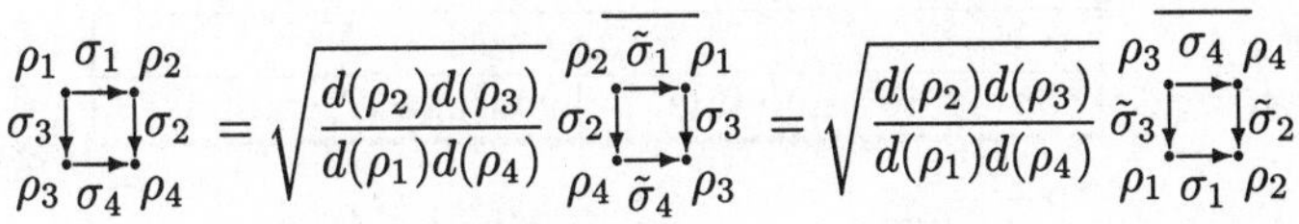

$$\begin{array}{ccc} \rho_1 & \sigma_1 & \rho_2 \\ \sigma_3 \downarrow & \square & \downarrow \sigma_2 \\ \rho_3 & \sigma_4 & \rho_4 \end{array} = \sqrt{\frac{d(\rho_2)d(\rho_3)}{d(\rho_1)d(\rho_4)}}\ \overline{\begin{array}{ccc} \rho_2 & \tilde{\sigma}_1 & \rho_1 \\ \sigma_2 \downarrow & \square & \downarrow \sigma_3 \\ \rho_4 & \tilde{\sigma}_4 & \rho_3 \end{array}} = \sqrt{\frac{d(\rho_2)d(\rho_3)}{d(\rho_1)d(\rho_4)}}\ \overline{\begin{array}{ccc} \rho_3 & \sigma_4 & \rho_4 \\ \tilde{\sigma}_3 \downarrow & \square & \downarrow \tilde{\sigma}_2 \\ \rho_1 & \sigma_1 & \rho_2 \end{array}}$$

FIG. 10.27. Renormalization for sectors

Then we decompose these sectors into irreducibles, and assign a vertex to each irreducible sector, and draw edges between them in a similar way to that in Section 10.2. Then we get the dual principal graph. (Because we work with endomorphisms, we have a tunnel instead of a tower. So we have the dual principal graph instead of the principal graph here.) The vertices of the graph are labelled by irreducible sectors and they have a positive number determined by the statistical dimension d. This gives the Perron–Frobenius eigenvector with the eigenvalue $[M : N]_0^{1/2}$. For the principal graph, we use

$$1, \bar{\rho}, \bar{\rho}\rho, \bar{\rho}\rho\bar{\rho}, \bar{\rho}\rho\bar{\rho}\rho, \cdots.$$

To get a cell, we multiply on the left by $\bar{\rho}$ and ρ alternately. Again from a composition of four intertwiners, we get a connection for each cell as in Fig. 10.26, where ρ_j are irreducible sectors and σ_j are intertwiners.

The unitarity axiom holds just as in the bimodule case. As seen above, in the bimodule language, harmonicity follows from the additivity and the multiplicativity of the statistical dimension. The initialization axiom is satisfied as in the bimodule case. For renormalization, we have the identity of Fig. 10.27, which is due to (Izumi 1993b). Here the symbol $\tilde{}$ denotes the Frobenius dual for intertwiners in the sector setting. The flatness axiom holds as in the bimodule case. In this way, we get an abstract paragroup also in the sector approach.

Finally, we have Table 10.1 for a correspondence between several concepts in bimodule theory and sector theory.

10.8 Notes

The reasoning in Section 10.2 is a detailed version of the 'encoding argument' in (Ocneanu 1988). The presentation here is based on (Kawahigashi 1996).

The axioms for paragroups were introduced in (Ocneanu 1988). Here we have slightly changed the axioms so that we can use a similar set of axioms in more general situations in Chapter 11. For example, we do not have the *contragredient map* of (Ocneanu 1988) here. Our convention of normalizing constants of

Table 10.1 *Correspondence Table between bimodules and sectors*

bimodule	sector
direct sum	direct sum
relative tensor product	composition
(Jones index)$^{1/2}$	statistical dimension
Frobenius reciprocity	Frobenius reciprocity
${}_M M_N$	$\rho : M \to N$
conjugate bimodule	conjugate sector

connections is slightly different from the original formulation in (Ocneanu 1988), but compatible with more recent papers on paragroups.

The $2\cos\pi/n$-rule first implicitly appeared in (Popa 1990a), and first explicitly appeared and was used as a rule to reject some graphs in (Izumi 1991). The proof of Theorem 10.9 and the proof of flatness are taken from A. Ocneanu's unpublished manuscript 'An invariant coupling between 3-manifolds and subfactors, with connections to topological and conformal quantum field theory'. (See also (Yamagami 1993a).) The definitions of equivalence and isomorphism of connections are taken from (Ocneanu 1988). (A similar equivalence had been considered in (Andrews *et al.* 1984) in the context of solvable lattice models.) The example in Section 10.6 is given in (Ocneanu 1991). (Also see (Yamagami 1993a).) Section 10.7 is based on (Izumi 1991), (Kosaki and Longo 1992), (Longo 1989), (Longo 1990), and (Longo 1992). The renormalization axiom for sectors has appeared in section 2 of (Izumi 1993b).

11

STRING ALGEBRAS AND FLAT CONNECTIONS

11.1 Introduction

We introduced a notion of commuting squares in Section 9.6, and will now characterize finite dimensional commuting squares in Section 11.2.

In Section 11.3, we give a construction of subfactors from certain commuting squares. This is in the converse direction of the construction of a paragroup from a subfactor in Chapter 10, but the construction of a subfactor in this section starts with properties weaker than the full set of axioms of paragroups. (The paragroup of the subfactor may not then be the data that we started with.)

In Section 11.4, we give one method to compute the principal graph of a subfactor constructed in Section 11.3, known as Ocneanu's compactness argument. In general, it is hard to carry out this method in concrete examples (i.e. solve the relevant equations), but this is still the only general method and is useful to eliminate many possibilities. Together with Popa's deep results on generating properties for subfactors, the result in this section gives a bijective correspondence between a paragroup and a subfactor of the hyperfinite II_1 factor with trivial relative commutant, finite index and finite depth.

We will determine all paragroups for subfactors with index less than 4 in Section 11.5, which should be compared with the classification in conformal field theories with central charge $c < 1$ – as in the classification of affine $SU(2)$ modular invariant partition functions on the torus in Section 8.8. First, we determine the bi-unitary connections on the Dynkin diagrams, and will give a general method, called the *triple point obstruction*, due to Ocneanu. We follow Izumi's method to check flatness of the connections.

We will give another method to construct subfactors with finite index and finite depth from the Dynkin diagrams. This was introduced in the text (Goodman *et al.* 1989). It depends on having a pair of Dynkin diagrams with the same Coxeter number – if the diagrams are identical, we are back in the situation of the construction of subfactors with index less than four. We will compute the principal graphs of these subfactors with a simplified version of the method of (Okamoto 1991). In Section 11.7, we study the GHJ construction of subfactors

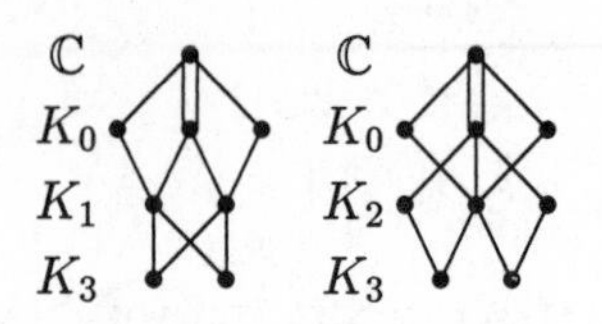

FIG. 11.1. Bratteli diagrams

in a more general setting (corresponding to the higher rank case of looking at $\hat{g}$ modular invariants, particularly for $g = su(3)$.)

So far, we have worked only on subfactors with finite depth. In Section 11.8, we will discuss subfactors with infinite depth. S. Popa introduced a notion of 'strong amenability' to characterize a good class of subfactors with infinite depth. We will discuss how to get a 'paragroup with strong amenability'. This section is not self-contained. (Strictly speaking, from this generalized viewpoint, paragroups we have considered so far should be called *finite paragroups.*)

Finally in Section 11.9, we discuss parallels between paragroup theory and the theory of exactly solvable lattice models, and the role of the Yang–Baxter equation.

11.2 Characterization of finite dimensional commuting squares

In Corollary 9.54, we have shown that higher relative commutants give rise to commuting squares. Because the connection introduced in Section 10.3 contains complete information on the higher relative commutants, the commuting square condition should be expressed in terms of connections. More generally, we will characterize general finite dimensional commuting squares in terms of generalized *connections.*

Let K_0, K_1, K_2, K_3 be a quadruple of finite dimensional C^*-algebras and suppose that we have a trace tr on K_3.

$$\begin{array}{ccc} K_0 & \subset & K_1 \\ \cap & & \cap \\ K_2 & \subset & K_3 \end{array} \tag{11.2.1}$$

We study when this square is a commuting square. First we draw two Bratteli diagrams for two inclusions $\mathbb{C} \subset K_0 \subset K_1 \subset K_3$ and $\mathbb{C} \subset K_0 \subset K_2 \subset K_3$. See Fig. 11.1 for example. We recall from Chapters 2 and 3 (especially Sections 2.6, 2.7, 2.9, 3.4, and 3.5) that for these two diagrams, we have *string algebra* expressions. That is, for each diagram, we have the following procedure.

Definition 11.1 *We put the label $*$ to the vertex at the top level corresponding to $\mathbb{C}$. Take a pair $\sigma = (\sigma_+, \sigma_-)$ of downward paths on the Bratteli diagram with the same length and the same endpoints. (Such a pair is called a* string*. We denote the length of σ, the endpoint of σ, and the starting point of σ by $|\sigma|$, $r(\sigma)$,*

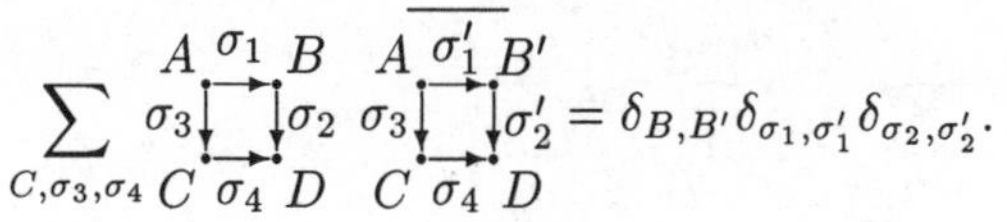

$$\sum_{C,\sigma_3,\sigma_4} \overline{\begin{array}{ccc} A & \sigma_1 & B \\ \sigma_3 & \square & \sigma_2 \\ C & \sigma_4 & D \end{array}} \begin{array}{ccc} A & \sigma_1' & B' \\ \sigma_3 & \square & \sigma_2' \\ C & \sigma_4 & D \end{array} = \delta_{B,B'}\delta_{\sigma_1,\sigma_1'}\delta_{\sigma_2,\sigma_2'}.$$

FIG. 11.2. Unitarity (1)

$s(\sigma)$ respectively.) We define the multiplication and the $*$-structure for strings as follows:

$$(\sigma_+, \sigma_-) \cdot (\rho_+, \rho_-) = \delta_{\sigma_-,\rho_+}(\sigma_+, \rho_-),$$
$$(\sigma_+, \sigma_-)^* = (\sigma_-, \sigma_+).$$

Clearly strings of length 1, 2, and 3 give systems of matrix units for K_0, K_1, K_3 and K_0, K_2, K_3.

A better way of looking at this is as follows. We assign a Hilbert space to each vertex of the Bratteli diagram so that the dimension of the Hilbert space is given by the number at the vertex in the Bratteli diagram. We identify each downward path from $*$ on the Bratteli diagram with an embedding of $\mathbb{C}$ corresponding to $*$ into the Hilbert space corresponding to the endpoint of the path so that the ranges of different paths are orthogonal. Then the string $\sigma = (\sigma_+, \sigma_-)$ is identified with a partial isometry $\sigma_+\sigma_-^*$ on the Hilbert space corresponding to the endpoint of the paths $\sigma_\pm$. The above multiplication rule and the $*$-operation are compatible with this identification. Furthermore, the embedding is given as

$$(\sigma_+, \sigma_-) \mapsto \sum_{|\rho|=1} (\sigma_+ \cdot \rho, \sigma_- \cdot \rho) \tag{11.2.2}$$

in this identification (cf. (2.9.11)).

For K_3, we have two systems of matrix units for the same algebra. We study the relation between the two (cf. Section 3.5).

Let $\mathcal{G}_0, \mathcal{G}_1, \mathcal{G}_2, \mathcal{G}_3$ be the Bratteli diagrams for the inclusions $K_0 \subset K_1$, $K_1 \subset K_3$, $K_2 \subset K_3$, and $K_0 \subset K_2$ respectively. Then these four graphs combine to form a single graph $\mathcal{G}$ as in Fig. 10.3. Each string is a rank one partial isometry as above, so we can conclude that the change of basis between the two systems of matrix units is given by a unitary matrix, each entry corresponding to a cell as in Fig. 10.6, where $\sigma_1 \in \mathcal{G}_0, \sigma_2 \in \mathcal{G}_1, \sigma_3 \in \mathcal{G}_3, \sigma_4 \in \mathcal{G}_2$. (In the above identification of paths and embeddings, Fig. 10.6 denotes a complex number with composition of four operators.) Unitarity of the matrix is expressed graphically as follows.

Choose four edges σ_1, σ_2, σ_1', σ_2' of the graph $\mathcal{G}$ so that $\sigma_1, \sigma_1' \in \mathcal{G}_0$, $\sigma_2, \sigma_2' \in \mathcal{G}_1$. Suppose moreover that there exist edges σ_3, σ_4 of $\mathcal{G}$ such that σ_1, σ_2, σ_3, σ_4 makes a cell. Then we have the identity of Fig. 11.2.

Similarly, choose four edges $\sigma_3, \sigma_4, \sigma_3', \sigma_4'$ of the graph $\mathcal{G}$ so that $\sigma_3, \sigma_3' \in \mathcal{G}_3, \sigma_4, \sigma_4' \in \mathcal{G}_2$. Suppose again that there exist edges σ_1, σ_2 of $\mathcal{G}$ such that

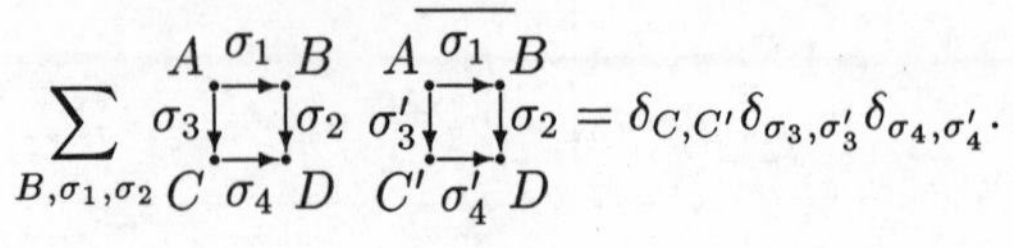

$$\sum_{B,\sigma_1,\sigma_2} \sigma_3 \overset{A \;\sigma_1\; B}{\underset{C\;\sigma_4\;D}{\square}} \sigma_2 \; \overline{\sigma_3' \overset{A\;\sigma_1\;B}{\underset{C'\;\sigma_4'\;D}{\square}} \sigma_2} = \delta_{C,C'}\delta_{\sigma_3,\sigma_3'}\delta_{\sigma_4,\sigma_4'}.$$

FIG. 11.3. Unitarity (2)

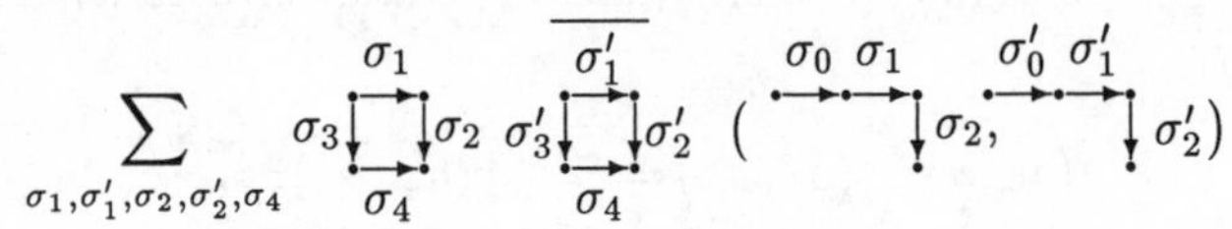

$$\sum_{\sigma_1,\sigma_1',\sigma_2,\sigma_2',\sigma_4} \sigma_3 \overset{\sigma_1}{\underset{\sigma_4}{\square}} \sigma_2 \; \overline{\sigma_3' \overset{\sigma_1'}{\underset{\sigma_4}{\square}} \sigma_2'} \; \left(\overset{\sigma_0\;\sigma_1}{\longrightarrow} \downarrow \sigma_2, \overset{\sigma_0'\;\sigma_1'}{\longrightarrow} \downarrow \sigma_2' \right)$$

FIG. 11.4. Change of bases

$\sigma_1, \sigma_2, \sigma_3, \sigma_4$ makes a cell. Then we have the identity of Fig. 11.3.

The choice of this unitary matrix is not unique, but it is possible to prove that the freedom, a *gauge choice*, is unique up to isomorphism as in Definition 10.11. (We do not need this fact in the rest of this book.)

Take a string $x = (\sigma_0 \cdot \sigma_3, \sigma_0' \cdot \sigma_3') \in K_2$, where σ_3, σ_3' are on $\mathcal{G}_3$. We check Condition 6 in Proposition 9.51, where $\mu(A)$ denotes the trace value of a minimal projection in the direct summand corresponding to the vertex A. We have the expression

$$\sum_{\sigma_1} \delta_{\sigma_3,\sigma_3'} \frac{\mu(r(\sigma_3))}{\mu(s(\sigma_3))} (\sigma_0 \cdot \sigma_1, \sigma_0' \cdot \sigma_1)$$

for $E_{K_0}(x)$ embedded into K_1 by Lemma 9.29. Next we embed x into K_3 and then change the basis to get the formula in Fig. 11.4. Applying E_{K_1} to the expression in Fig. 11.4, we get the formula in Fig. 11.5. Comparing the two expressions, we conclude that the commuting square condition is equivalent to the condition that for all choices of $\sigma_1, \sigma_1', \sigma_3, \sigma_3'$, we have the identity of Fig. 11.6.

Theorem 11.2 *The square (11.2.1) is a commuting square if and only if the identity in Fig. 11.6 holds for the corresponding connection.*

The identity in Fig. 11.6 means that a certain matrix given by normalized entries of the original connection is an isometry. In many interesting cases, this

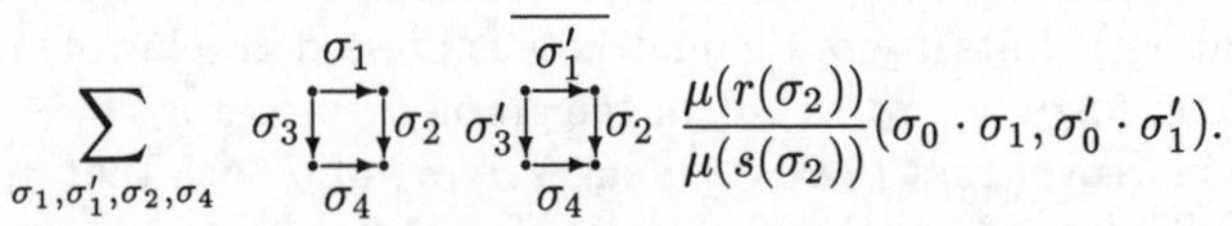

$$\sum_{\sigma_1,\sigma_1',\sigma_2,\sigma_4} \sigma_3 \overset{\sigma_1}{\underset{\sigma_4}{\square}} \sigma_2 \; \overline{\sigma_3' \overset{\sigma_1'}{\underset{\sigma_4}{\square}} \sigma_2} \; \frac{\mu(r(\sigma_2))}{\mu(s(\sigma_2))} (\sigma_0 \cdot \sigma_1, \sigma_0' \cdot \sigma_1').$$

FIG. 11.5. The image of the conditional expectation

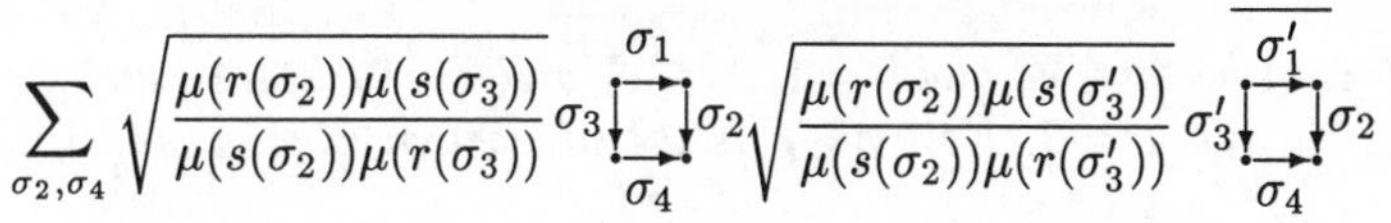

$$\sum_{\sigma_2,\sigma_4} \sqrt{\frac{\mu(r(\sigma_2))\mu(s(\sigma_3))}{\mu(s(\sigma_2))\mu(r(\sigma_3))}} \; \sigma_3 \underset{\sigma_4}{\overset{\sigma_1}{\square}} \sigma_2 \sqrt{\frac{\mu(r(\sigma_2))\mu(s(\sigma_3'))}{\mu(s(\sigma_2))\mu(r(\sigma_3'))}} \; \overline{\sigma_3' \underset{\sigma_4}{\overset{\sigma_1'}{\square}} \sigma_2}$$

$$= \delta_{\sigma_3,\sigma_3'}\delta_{\sigma_1,\sigma_1'}.$$

FIG. 11.6. The commuting square condition

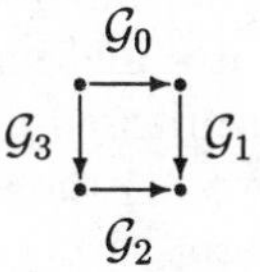

$$\mathcal{G}_3 \underset{\mathcal{G}_2}{\overset{\mathcal{G}_0}{\square}} \mathcal{G}_1$$

FIG. 11.7. Four graphs

isometric matrix is a square matrix and hence is unitary. In such a case, we say that the original commuting square is a *symmetric commuting square.*

11.3 String algebra construction of subfactors

We will construct a subfactor from an abstract bi-unitary connection in this section. Our initial data is as follows.

First, we have a graph $\mathcal{G}$ consisting of four finite subgraphs $\mathcal{G}_0, \mathcal{G}_1, \mathcal{G}_2, \mathcal{G}_3$ as in Fig. 11.7. Again $\mathcal{G}_j$ and $\mathcal{G}_{j+1}$ have common vertices V_{j+1} for $j \in \mathbb{Z}/4$. We assume that the graphs $\mathcal{G}_0$ and $\mathcal{G}_2$ are connected, but we *do not assume* that $\mathcal{G}_1$ and $\mathcal{G}_3$ are connected. We suppose that $\mathcal{G}_0$ has more than one edge.

Definition 11.3 *We can define a cell for this graph $\mathcal{G}$ as in Section 10.3, and we call an assignment of a complex number to each cell a* connection. *We assume that each vertex A has assigned to it a strictly positive number $\mu(A)$. Our assumptions on the connection and μ are as follows.*

1. *Unitarity, as in Axiom 1.*
2. *Harmonicity in the following sense. We have two positive numbers β, β' with the following property. We have*

$$\beta\mu(X) = \sum_{Y \in V_k} n_{X,Y}\mu(Y),$$

for the four cases $X \in V_0$ and $k = 3$, $X \in V_1$ and $k = 2$, $X \in V_2$ and $k = 1$, and $X \in V_3$ and $k = 0$. We also have

$$\beta' \mu(X) = \sum_{Y \in V_k} n_{X,Y} \mu(Y),$$

for the four cases $X \in V_0$ and $k = 1$, $X \in V_1$ and $k = 0$, $X \in V_2$ and $k = 3$, and $X \in V_3$ and $k = 2$. Here $n_{X,Y}$ is the number of edges connecting X and Y.

3. *Renormalization, as in Axiom 4.*

If our connection satisfies the above three conditions, we call it a bi-unitary connection.

Note that here the terminologies 'connection' and 'bi-unitary connection' are used in a more general setting than in Chapter 10. This abstract bi-unitary connection characterizes a symmetric commuting square as in Section 11.2.

We fix a vertex in V_0 and call it $*$ and normalize μ so that $\mu(*) = 1$. First, we use only the graph $\mathcal{G}_0$. For an oriented edge σ in the graph, we denote by $s(\sigma)$ the starting point of σ and by $r(\sigma)$ the endpoint of σ. (The letters s and r denote the source and the range respectively as in the groupoid language of the notes to Chapter 2.) We now define the string algebra formally as in Chapter 2.

Definition 11.4 *An oriented* path *on $\mathcal{G}_0$ is a succession of edges: $\sigma = (\sigma_1, \sigma_2, \ldots, \sigma_n)$, where $r(\sigma_j) = s(\sigma_{j+1})$ for $j = 1, 2, \ldots, n-1$. We write $\tilde{\sigma}$ for the edge σ with the reversed orientation. A* string *on $\mathcal{G}$ is a pair of paths, $\rho = (\rho_+, \rho_-)$, with $s(\rho_+) = s(\rho_-) = *$, $r(\rho_+) = r(\rho_-)$, and $|\rho_+| = |\rho_-|$, where $|\cdot|$ denote the length of a path. We also use the symbols $r(\cdot)$, $|\cdot|$ for strings.*

Define an algebra $\mathrm{String}^{(n)}\mathcal{G}_0$ with the linear basis of the n-strings, that is, the strings with length n. We define a multiplication and a $*$-operation by

$$(\rho_+, \rho_-) \cdot (\sigma_+, \sigma_-) = \delta_{\rho_-, \sigma_+} (\rho_+, \sigma_-),$$
$$(\rho_+, \rho_-)^* = (\rho_-, \rho_+),$$

which makes $\mathrm{String}^{(n)}\mathcal{G}_0$ into a finite dimensional C^*-algebra.

We next embed $\mathrm{String}^{(n)}\mathcal{G}_0$ into $\mathrm{String}^{(n+k)}\mathcal{G}_0$ for $k > 0$. The embedding $i_n^{n+k} : \mathrm{String}^{(n)}\mathcal{G}_0 \to \mathrm{String}^{(n+k)}\mathcal{G}_0$ by

$$i_n^{n+k}(\rho_+, \rho_-) = \sum_{|\sigma|=k} (\rho_+ \cdot \sigma, \rho_- \cdot \sigma),$$

where $\rho_\pm \cdot \sigma$ denotes the concatenation of the paths and the summation is over all σ for which such a concatenation is possible as in (2.9.11).

We introduce a trace tr on $\bigcup_{n=0}^\infty \mathrm{String}^{(n)}\mathcal{G}_0$. We define $\mathrm{tr} : \mathrm{String}^{(n)}\mathcal{G}_0 \to \mathbb{C}$ by $\mathrm{tr}((\rho_+, \rho_-)) = \delta_{\rho_+, \rho_-} \beta'^{-|\rho|} \mu(r(\rho))$. This trace tr is compatible with the embedding i_n^{n+k}, that is, we have $\mathrm{tr}(i_n^{n+k}(\rho)) = \mathrm{tr}(\rho)$ by the harmonic condition, Definition 11.3 (2). We know that this tr is the only trace on $\bigcup_{n=0}^\infty \mathrm{String}^{(n)}\mathcal{G}_0$, because any trace has to come from a Perron–Frobenius eigenvector on the graph by Theorem 10.3 (3).

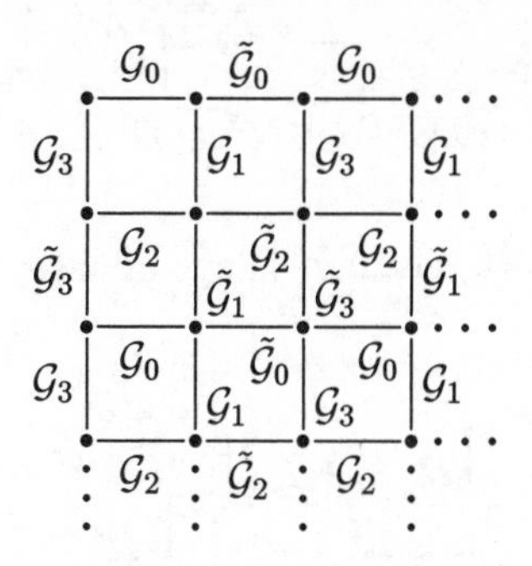

FIG. 11.8. Nested graphs

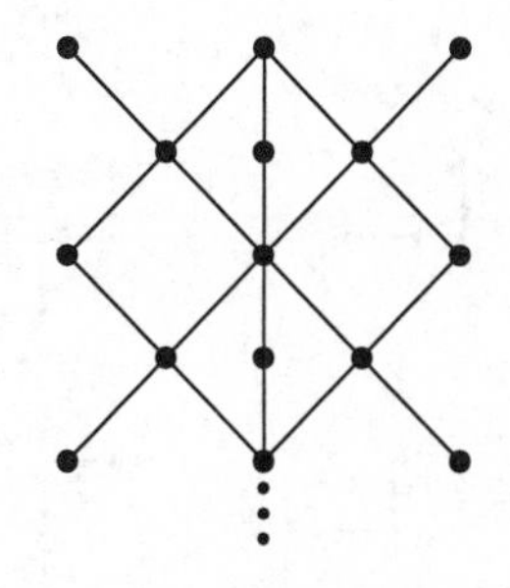

FIG. 11.9. Path space from $\mathcal{G}_0$

By completing the infinite dimensional algebra $\bigcup_{n=0}^{\infty} \mathrm{String}^{(n)}\mathcal{G}_0$ in the GNS-representation with respect to the trace, we get a hyperfinite II_1 factor because of the uniqueness of the trace. That is, we get the same von Neumann algebras from any graph by Theorem 5.18. (Note that at finite levels, the algebra $\mathrm{String}^{(n)}\mathcal{G}_0$ is not simple unless the graph $\mathcal{G}_0$ is of very special form, but we get a simple algebra after passing to the 'limit' as $n \to \infty$.)

We now use all four graphs. We make the nest of graphs as in Fig. 11.8 by reflecting graphs. (The symbols $\tilde{\mathcal{G}}_j$ mean the reflected graphs.)

For example, if the graph $\mathcal{G}_0$ is the Dynkin diagram E_6, Fig. 10.10, the top part $\mathcal{G}_0, \tilde{\mathcal{G}}_0, \mathcal{G}_0, \tilde{\mathcal{G}}_0, \ldots$ looks like Fig. 11.9.

The string algebra $\mathrm{String}^{(n)}\mathcal{G}_0$ is defined with paths on this graph in Fig. 11.9. We now imitate the construction on the string algebra on this nest of four graphs. That is, we start at the upper left corner of the graphs in Fig. 11.8, and go to the right or downward direction step by step. In this way, we can define a path on the four graphs, and form string algebras. (Note that our $*$ is in V_0.) We write $A_{k,l}$ for the string algebra given by going to the right for l steps and to the downward direction for k steps. However, we have several ways

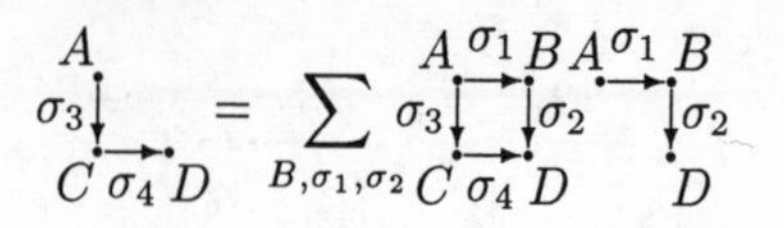

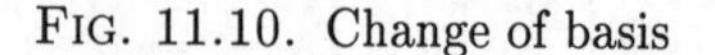

FIG. 11.10. Change of basis

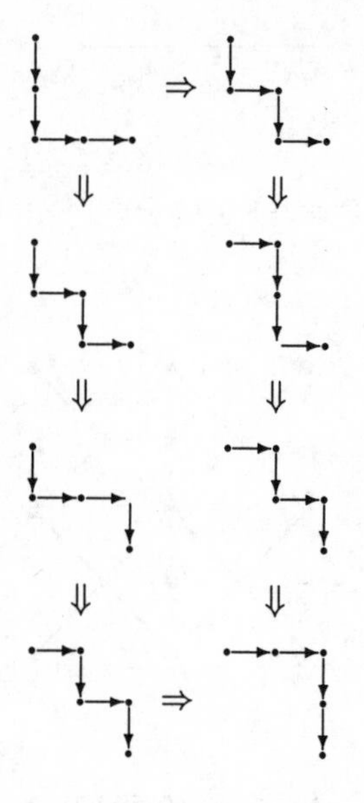

FIG. 11.11. Compatibility of changes of bases

to go to the right for l steps and downward for k steps. For example, $A_{1,1}$ has two representations; in one, we go to the right first and then downward, and in the other we go downward first and to the right next. In general, we have ${}_{k+l}C_k$ representations for $A_{k,l}$. In order for our definitions to be consistent, we need to give identifications between these algebras. Recall that the multiplication in the string algebra was given through rank-one partial isometries, so it is enough for us to give unitary identification among paths in order to give identifications of the finite dimensional C^*-algebras. This is given by the bi-unitary connection. (cf. (3.5.9)) That is, we use the rule of local changes of bases as in Fig. 11.10. (This makes sense because of the unitarity condition, Definition 11.3 (1).)

In general, we repeat this local change to make the identification. Note that as in Fig. 11.11, repetition of these local changes is compatible, that is, the result does not depend on the order of local changes.

In this way, we have a well-defined double sequence of finite dimensional C^*-algebras $\{A_{k,l}\}_{k,l=0,1,2,\ldots}$. Note that we have embeddings $A_{k,l} \subset A_{k+1,l}$ and

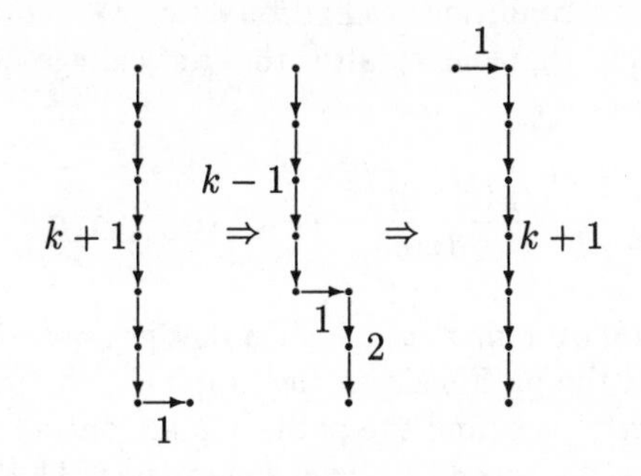

FIG. 11.12. Basis change for the Jones projection

$A_{k,l} \subset A_{k,l+1}$ as in (11.2.2). We also note that we have a compatible trace tr on this double sequence. That is, we set

$$\mathrm{tr}((\rho_+, \rho_-)) = \delta_{\rho_+,\rho_-} \beta^{-k} \beta'^{-l} \mu(r(\rho)),$$

for $(\rho_+, \rho_-) \in A_{k,l}$. As before, this is a well-defined trace on $\bigcup_{k,l} A_{k,l}$. We define $A_{k,\infty}$ to be the weak closure of $\bigcup_{l=1}^{\infty} A_{k,l}$ in the GNS-representation with respect to the trace tr. As before, each $A_{k,\infty}$ is a hyperfinite II_1 factor. Then following Fig. 9.16, we make the definition.

Definition 11.5 *We define the (vertical)* Jones projection $e_k \in A_{k+1,0}$ *as follows:*

$$e_k = \sum_{\sigma,\gamma,\delta} \frac{1}{\beta} \frac{\sqrt{\mu(r(\gamma))\mu(r(\delta))}}{\mu(r(\sigma))} (\sigma \cdot \gamma \cdot \tilde{\gamma}, \sigma \cdot \delta \cdot \tilde{\delta}),$$

where the summation is over all σ, γ, δ *such that* $|\sigma| = k - 1$, $s(\sigma) = *$, *and* $|\gamma| = |\delta| = 1$, *and all the paths* σ, γ, δ *are on the graph* $\mathcal{G}_3$.

We have the adjective 'vertical' because the formula is given as a vertical string. We can similarly define the horizontal Jones projections in $A_{0,k+1}$.

We have the following proposition.

Proposition 11.6 *The string* e_k *in Definition 11.5 satisfies* $e_k = e_k^* = e_k^2$.

Proof It is trivial that $e_k = e_k^*$. The identity $e_k = e_k^2$ also follows by a direct computation (cf. Section 9.4, e.g. the Remark on page 501) with the harmonic condition, Definition 11.3 (2). □

In Definition 11.5, the Jones projection here has no relation *a priori* to the Jones projection in Section 9.6, but later we will identify these two kinds of the Jones projections in Theorem 11.9. (See also Section 9.4.)

Now we embed e_k into $A_{k+1,1}$. First our e_k is represented as a string going downward $k + 1$ steps first and then going to the right for one step. Then we change the basis as in Fig. 11.12, where the numbers denote the length of paths.

The renormalization condition, Definition 11.3 (3), and the coefficient of e_k in Definition 11.5 imply that our e_k after the first change of basis in Fig. 11.1 is expressed as follows:

$$e_k = \sum_{\sigma',\sigma,\gamma,\delta} \frac{1}{\beta} \frac{\sqrt{\mu(r(\gamma))\mu(r(\delta))}}{\mu(r(\sigma))} (\sigma' \cdot \sigma \cdot \gamma \cdot \tilde{\gamma}, \sigma' \cdot \sigma \cdot \delta \cdot \tilde{\delta}),$$

where the summation is over all $\sigma', \sigma, \gamma, \delta$ such that $|\sigma'| = k-1, |\sigma| = 1, s(\sigma') = *$ and $|\gamma| = |\delta| = 1$, and the path σ' is on the graph $\mathcal{G}_3$, the path σ is on $\mathcal{G}_2$ or $\mathcal{G}$ depending on the parity of k, and the paths γ, δ are on the graph $\mathcal{G}_1$.

Furthermore, after the second basis change in Fig. 11.12 e_k is expressed as follows:

$$e_k = \sum_{\sigma',\sigma,\gamma,\delta} \frac{1}{\beta} \frac{\sqrt{\mu(r(\gamma))\mu(r(\delta))}}{\mu(r(\sigma))} (\sigma' \cdot \sigma \cdot \gamma \cdot \tilde{\gamma}, \sigma' \cdot \sigma \cdot \delta \cdot \tilde{\delta}),$$

where the summation is over all $\sigma', \sigma, \gamma, \delta$ such that $|\sigma'| = 1, |\sigma| = k-1, s(\sigma') = *$ and $|\gamma| = |\delta| = 1$, and the path σ' is on the graph $\mathcal{G}_0$ and all the paths σ, γ, δ are on the graph $\mathcal{G}_1$.

By repeating this procedure, if we embed e_k into $A_{k+1,l}$, we get the following expression for e_k:

$$e_k = \sum_{\sigma',\sigma,\gamma,\delta} \frac{1}{\beta} \frac{\sqrt{\mu(r(\gamma))\mu(r(\delta))}}{\mu(r(\sigma))} (\sigma' \cdot \sigma \cdot \gamma \cdot \tilde{\gamma}, \sigma' \cdot \sigma \cdot \delta \cdot \tilde{\delta}),$$

where the summation is over all $\sigma', \sigma, \gamma, \delta$ such that $|\sigma'| = l, |\sigma| = k-1, s(\sigma) = *$ and $|\gamma| = |\delta| = 1$, and the path σ' is on the graph $\mathcal{G}_0$ and all the paths $\sigma, \gamma,$ are on the graph $\mathcal{G}_3$ or $\mathcal{G}_1$ depending on the parity of l.

All the above expressions, the formulae for e_k have formally the same expression. We represent this fact by saying that *the Jones projections do not change their form by changes of basis.*

Because we have a trace, we also have conditional expectations. An easy computation shows the following lemma – the first part of which is Lemma 9.29

Lemma 11.7 1. *The conditional expectation $E_k : A_{k,0} \to A_{k-1,0}$ is given by the following:*

$$E_k((\sigma_+ \cdot \rho_+, \sigma_- \cdot \rho_-)) = \delta_{\rho_+,\rho_-} \frac{\mu(r(\rho_+))}{\beta\mu(r(\sigma_+))} (\sigma_+, \sigma_-),$$

where paths $\sigma_\pm$, $\rho_\pm$ are on the graph $\mathcal{G}_3$ and satisfy $|\sigma_\pm| = k-1$, $|\rho_\pm| = 1$.

2. *For $x \in A_{k,0}$, we have $e_k x e_k = E_k(x) e_k$.*

3. *For $x \in A_{k,0}$, we have $\mathrm{tr}(x e_k) = \mathrm{tr}(x)/\beta^2$. Hence $E_{A_{k,0}}(e_k) = 1/\beta^2$.*

By the renormalization condition (Definition 11.3 (3)) and Theorem 11.2, we know that

$$\begin{array}{ccc} A_{k,l} & \subset & A_{k,l+1} \\ \cap & & \cap \\ A_{k+1,l} & \subset & A_{k+1,l+1} \end{array}$$

is a commuting square for any $k, l \geq 0$.

We would like to compute the index $[A_{k,\infty} : A_{k-1,\infty}]$ and describe the basic construction of the subfactor $A_{k-1,\infty} \subset A_{k,\infty}$. We first need the following lemma.

Lemma 11.8 *For sufficiently large l, we have $A_{k+1,l} = \langle A_{k,l}, e_k \rangle$.*

Proof For sufficiently large l, the Bratteli diagram for the inclusion $A_{k,l} \subset A_{k+1,l}$ is a reflection of that for $A_{k-1,l} \subset A_{k,l}$. In this case, we prove $A_{k+1,l} = A_{k,l}e_kA_{k,l}$. Note that it is clear that $A_{k+1,l} \supset A_{k,l}e_kA_{k,l}$. Take any string $z = (\sigma_+ \cdot \rho_+, \sigma_- \cdot \rho_-) \in A_{k+1,l}$ with $|\sigma_\pm| = k$, $|\rho_\pm| = 1$. We have some path τ with $|\tau| = k-1$ and $r(\tau) = r(\rho_\pm)$, because of the assumption on the Bratteli diagram. Then we know that the product

$$(\sigma_+ \cdot \rho_+, \tau \cdot \tilde{\rho}_+ \cdot \rho_+)e_k(\tau \cdot \tilde{\rho}_- \cdot \rho_-, \sigma_- \cdot \rho_-)$$

is equal to z up to a positive scalar, thus we are done. $\square$

Choose an arbitrary $y \in A_{k,l}$ and express it as $y = \sum_{j=1}^n a_j e_{k-1} b_j$, $(a_j, b_j \in A_{k-1,l})$ by Lemma 11.8. Setting $x = y^*y$, we apply an argument in the proof of Theorem 9.48. We have

$$\begin{aligned} x &= \sum_{j,m=1}^n b_j^* e_{k-1} a_j^* a_m e_{k-1} b_m \\ &= \sum_{j,m=1}^n b_j^* E_{A_{k-2,l}}(a_j^* a_m) e_{k-1} b_m, \end{aligned}$$

by Lemma 11.7 (2). Define an element A by

$$A = \left(E_{A_{k-2,l}}(a_j^* a_m)\right)_{jm} \in M_n(\mathbb{C}) \otimes A_{k-2,l}.$$

Set

$$B = \begin{pmatrix} b_1 \\ b_2 \\ \vdots \\ b_n \end{pmatrix}, \quad E = \begin{pmatrix} e_{k-1} & 0 & \cdots & 0 \\ 0 & e_{k-1} & \cdots & 0 \\ \vdots & \vdots & \ddots & \vdots \\ 0 & 0 & \cdots & e_{k-1} \end{pmatrix} \in M_n(\mathbb{C}) \otimes A_{k,l}.$$

We know that A is positive and hence there exists $C \in M_n(\mathbb{C}) \otimes A_{k-2,l}$ with $C^*C = A$. Then we get

$$x = B^*C^*CEB = B^*C^*ECB \leq B^*C^*CB = \beta^2 E_{A_{k-1,l}}(x),$$

by Lemma 11.7 (3).

This shows that we have $E_{A_{k-1,\infty}}(x) \geq x/\beta^2$ for any positive $x \in A_{k,\infty}$. Since Lemma 11.7 (3) gives $E_{A_{k-1,\infty}}(e_{k-1}) = 1/\beta^2$, we get $[A_{k,\infty} : A_{k-1,\infty}] = \beta^2$ by Theorem 9.48. Then Proposition 9.47 shows that $A_{k-1,\infty} \subset A_{k,\infty} \subset A_{k+1,\infty}$ is identified with the basic construction of $A_{k-1,\infty} \subset A_{k,\infty}$ for $k \geq 1$. Thus we have proved the following theorem.

Theorem 11.9 *The string algebra construction from a bi-unitary connection as in Definition 11.3 gives a subfactor $A_{0,\infty} \subset A_{1,\infty}$ with the Jones index β^2. Its Jones tower is given by*

$$A_{0,\infty} \subset A_{1,\infty} \subset A_{2,\infty} \subset A_{3,\infty} \subset A_{4,\infty} \subset \cdots.$$

If the graphs $\mathcal{G}_1$ and $\mathcal{G}_3$ are also connected, we can construct a subfactor $A_{\infty,0} \subset A_{\infty,1}$ in a symmetric way. It is not clear at all what kind of relations we have between the two subfactors $A_{0,\infty} \subset A_{1,\infty}$ and $A_{\infty,0} \subset A_{\infty,1}$ in general. It was shown in (Sato, in press a) that if one of the two has a finite depth, so does the other. A characterization of two subfactors arising in this way was given in (Sato 1997b) and (Sato, in press b) in terms of systems of bimodules.

Here we also derive a useful formula for the Jones projections on the string algebras. Fix a (possibly extended) Dynkin diagram A_n $(1 \leq n \leq \infty)$. Let β be the Perron–Frobenius eigenvalue of the graph if $n < \infty$, and β be any real number with $\beta \geq 2$ if $n = \infty$. Let μ_k be the entry of the eigenvector of the graph of the kth vertex corresponding to the eigenvalue β. We normalize μ_k so that $\mu_1 = 1$. Then we have the Jones projections $\{e_k\}_k$ defined as above in the string algebra of A_n starting from the first vertex $*$. (Here we need only a single sequence of string algebras, not a double sequence.) Let f_k be the largest projection orthogonal to $e_1, e_2, \ldots, e_{k-1}$. We then have a projection p with $f_k = f_{k+1} + p$, as in Fig. 11.13. Because f_{k+1} is orthogonal to e_k, we have $pe_kp = f_ke_kf_k$. From the definition of the Jones projection in Definition 11.5, we get

$$pe_kp = \frac{\mu_{k+1}}{\beta\mu_k}p.$$

Thus we get the following theorem.

Theorem 11.10 *In the above setting, we have*

$$f_{k+1} = f_k - \frac{\beta\mu_{k-1}}{\mu_k} f_k e_k f_k.$$

We complete this section by discussing the *generating property* of S. Popa. Suppose we start with a subfactor $N \subset M$ of the hyperfinite II$_1$ factor with finite index and finite depth. We apply the construction of a paragroup as in Chapter 10 and then apply the string algebra construction in this chapter to the paragroup. Then what we have is a subfactor

$$\overline{\bigcup_{n=0}^{\infty} N' \cap M_n} \subset \overline{\bigcup_{n=0}^{\infty} N_1' \cap M_n},$$

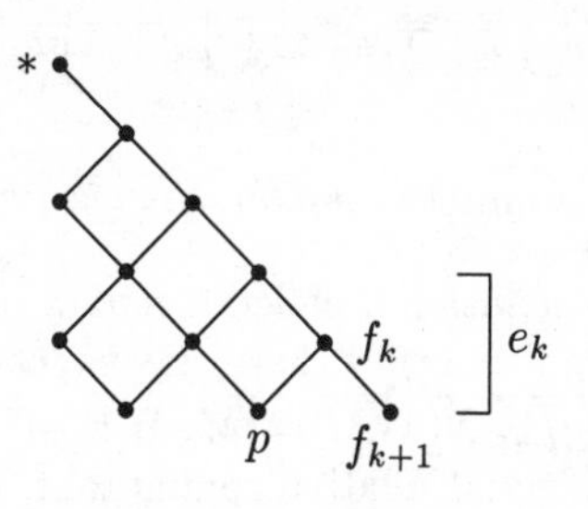

FIG. 11.13. The projection f_k

where $\overline{}$ denotes the weak closure in the GNS-representation with respect to the trace. We would like to know the relation between this subfactor and the original subfactor. A series of deep theorems in (Popa 1990a), (Popa 1994a), (Popa 1994b) answers this natural and difficult question in more general situations; the two subfactors are essentially the same – strictly speaking, these two subfactors are anti-isomorphic. (An anti-isomorphism is the same as an isomorphism except that it reverses the order of the multiplication.)

Popa has the following three definitions.

Definition 11.11 *We say that a subfactor $N \subset M$ with finite index is* extremal *if $E_{N'\cap M}(e) \in \mathbb{C}$ for a Jones projection $e \in M$.*

It is clear that the trivial relative commutant condition implies extremality. It has been shown by (Pimsner and Popa 1988) that extremality is equivalent to the condition that the family of maps $\{\Phi_k\}_k$ in Proposition 9.50 is trace-preserving.

Definition 11.12 *Let Γ be the incidence matrix of the principal graph of a subfactor $N \subset M$ with finite index. We write Γ^t for its transpose matrix. We say that the subfactor $N \subset M$ is* amenable *if $[M : N] = \|\Gamma^t\Gamma\|$, where the right hand side means the operator norm of the matrix on ℓ^2*

Actually, this is one of the several equivalent definitions of amenability. See (Popa 1994a) for the other equivalent conditions. (The name amenability comes from an analogy with the amenability of groups in (Greenleaf 1969). One of Popa's conditions, Theorem 4.1.1. (iv) in (Popa 1994a), looks like a quantized version of the group amenability.)

Definition 11.13 *An amenable subfactor $N \subset M$ is called* strongly amenable *if $\bigvee_{n=1}^{\infty} M' \cap M_n$ is a factor. (This $\bigvee$ notation means the weak closure of $\bigcup_{n=1}^{\infty} M' \cap M_n$ in the GNS-representation with respect to the trace.)*

Then the main theorem in (Popa 1994a) is as follows.

Theorem 11.14 *Let $N \subset M$ be an extremal subfactor of the hyperfinite II_1 factor. It is strongly amenable if and only if the inclusion*

$$\overline{\bigcup_{n=0}^{\infty} N' \cap M_n} \subset \overline{\bigcup_{n=0}^{\infty} N_1' \cap M_n}$$

is anti-isomorphic to the original subfactor $N \subset M$.

The condition in the conclusion is called a *generating property*. This is known to be equivalent to the condition that there exists a choice of tunnel $\{N_k\}_k$ with $\overline{\bigcup_{k=1}^{\infty} N_k' \cap N} = N$ and $\overline{\bigcup_{k=1}^{\infty} N_k' \cap M} = M$. If a subfactor $N \subset M$ has finite index, finite depth and a trivial relative commutant, then the harmonic axiom and the factoriality of the string algebra completion imply strong amenability. Thus Theorem 11.14 implies that we have the generating property in this case.

This theorem of Popa is one of the deepest in the theory of von Neumann algebras, and is beyond the scope of this book, where the emphasis is on the combinatorial rather than analytic aspects of subfactor theory. So we do not try to give an explanation here, but we will give a brief discussion of the meaning of this theorem in the case of finite depth. (See the above-mentioned papers or the monograph (Popa 1995a) for more details.)

The above theorem of Popa says that the paragroup is a complete invariant for irreducible subfactors of the hyperfinite II_1 factor with finite index and finite depth. A basic idea of the 'quantized Galois theory' is to regard a subfactor $N \subset M$ as an inclusion of N into a 'crossed product of N with an action of a paragroup', as explained in Section 10.6. Of course, this does not make sense literally, because we do not have a definition of an 'action of a paragroup', but still this viewpoint is conceptually useful. From this viewpoint, the construction of a paragroup from a subfactor is regarded as a procedure to find an algebraic object acting on N, and this procedure does not extract information about the 'action' itself, except for its 'outerness'. For example, if a subfactor $N \subset M$ is indeed of a form $N \subset N \rtimes G$, then this procedure finds what the group G is, and the only information we get about the action is that it is outer. That is, any detailed information about the action is lost in this procedure. It thus seems rather strange that we can recover the subfactor only from this information about the group without knowing about the action, but it is known by (Jones 1980b) that an outer action of a finite group G on the hyperfinite II_1 factor is unique up to conjugacy, so it is indeed possible to recover the crossed product subfactor only from the information on the group. In this sense, the above theorem of Popa is regarded as a certain uniqueness theorem for an 'outer action' of a paragroup on the hyperfinite II_1 factor, which generalizes the theorem of Jones.

11.4 Ocneanu's compactness argument and flatness of connections

Consider a subfactor $A_{0,\infty} \subset A_{1,\infty}$ constructed with a bi-unitary connection on four graphs as in Section 11.3. This is the most fundamental way to construct a subfactor, so we would like to compute the paragroup arising from this subfactor. That is, we would like to compute the higher relative commutants $A_{0,\infty}' \cap A_{k,\infty}$. This is one of the most natural and important problems in subfactor theory. At

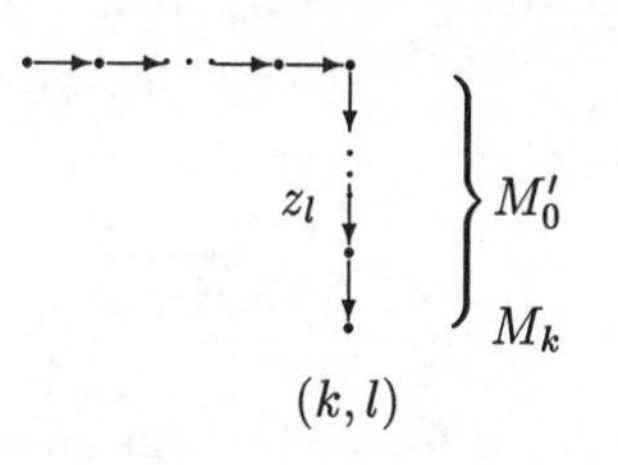

FIG. 11.14. A string expression for z_l

least at an abstract level, we can give a complete answer to this problem. This is due to Ocneanu, and we explain his method in this section.

We start with an arbitrary element from $A'_{0,\infty} \cap A_{k,\infty}$. Since this relative commutant is contained in a von Neumann algebra arising as a weak closure of an infinite dimensional algebra, it seems hard to describe this object in terms of strings. Our aim in this section is to show that this element in the relative commutant indeed comes from a string in a finite dimensional algebra and has a very natural expression as a *flat field of strings.*

Let $z^0 \in A'_{0,\infty} \cap A_{k,\infty}$ and set $z_l = E_{A_{k,l}}(z^0)$. Then $z_l \in A'_{0,l} \cap A_{k,l}$ as in Fig. 11.14. Let d be the smallest number such that the Bratteli diagram for $A_{0,d} \subset A_{0,d+1}$ is a reflection of that for the previous step $A_{0,d-1} \subset A_{0,d}$. Note that if $2l \geq d$, then all the algebras $A'_{0,2l} \cap A_{k,2l}$ are canonically isomorphic, because all of them have bases expressed as strings on the graph $\mathcal{G}_3$. (This is the key point of the entire argument.) Let A be a (finite dimensional) model C^*-algebra isomorphic to these algebras, and we fix a natural isomorphism ϕ_{2l} from $A'_{0,2l} \cap A_{k,2l}$ to A. That is, A is realized as a string algebra on the graph $\mathcal{G}_3$ with length k and with the initial vertices in V_0. (We allow any vertex in V_0 as the initial vertex of a string. So this algebra A is a direct sum of string algebras beginning at one of the vertices in V_0.) The operator norm $\|\phi_{2l}(z_{2l})\|$ is equal to $\|z_{2l}\|$, which is bounded by $\|z^0\|$ because of a property of the conditional expectation. So the set $\{\phi_{2l}(z_{2l})\}_l$ is bounded in the finite dimensional algebra A. By compactness, we have a subsequence $\{l_j\}_j$ such that $\phi_{2l_j}(z_{2l_j}) \to z$, $\phi_{2l_j+2}(z_{2l_j+2}) \to z'$ for some $z, z' \in A$ as $j \to \infty$. (Here we have used the compactness in an essential way. This is the reason this proof is called the compactness argument.)

We also note that if $2l \geq d$, then all the algebras $A'_{0,2l} \cap A_{k,2l+2}$ are canonically isomorphic. We denote the finite dimensional model C^*-algebra isomorphic to these algebras by $\tilde{A}$, and we fix a natural isomorphism $\tilde{\phi}_{2l}$ from $A'_{0,2l} \cap A_{k,2l+2}$ to $\tilde{A}$.

Consider the diagram in Fig. 11.15 representing the algebra $\tilde{A}$ as a direct sum of string algebras beginning at one of the vertices in V_0, where different string expressions are identified using the connection. That is, we have two As vertically corresponding to two embeddings of A into $\tilde{A}$, and two 2-string algebras horizontally. We regard this as a model of the embedding $A'_{0,2l} \cap A_{k,2l} \subset A'_{0,2l} \cap A_{k,2l+2}$. Because $\|z_{2l} - z_{2l+2}\|_2 \to 0$ as $l \to \infty$, we know that the two strings

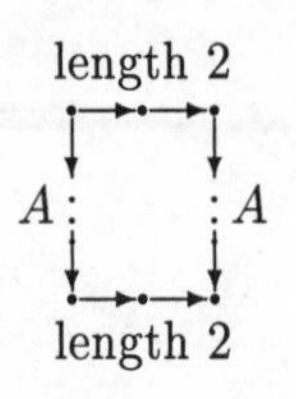

FIG. 11.15. The algebra $\tilde{A}$

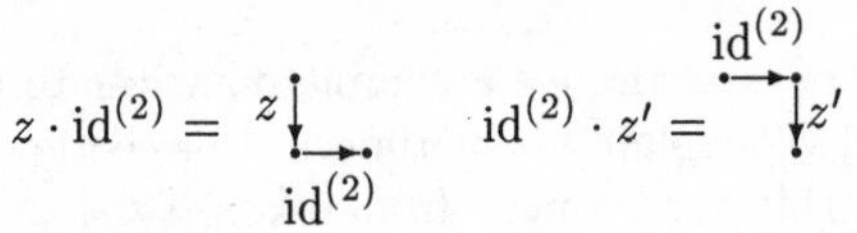

$z \cdot \mathrm{id}^{(2)} =$ z $\mathrm{id}^{(2)}$ $\qquad \mathrm{id}^{(2)} \cdot z' =$ $\mathrm{id}^{(2)}$ z'

FIG. 11.16. $z \cdot \mathrm{id}^{(2)}$ and $\mathrm{id}^{(2)} \cdot z'$

in Fig. 11.16 are identified using the connection as limits of the subsequences. (Note that different ls give different L^2-norms on $\tilde{A}$ via $\tilde{\phi}_{2l}$, but that does not matter in the limit because the l^2-norms converge as $l \to \infty$ by Theorem 10.3 (4), which implies that the limit of the trace value of a minimal projection from the vertex X to the vertex Y in $\tilde{A}$ is a scalar multiple of $\mu(X)\mu(Y)$.)

We now have $z \cdot \mathrm{id}^{(2)} = \mathrm{id}^{(2)} \cdot z'$, where $\mathrm{id}^{(2)}$ denotes the string $\sum_{|\sigma|=2}(\sigma, \sigma)$ on the graph $\mathcal{G}_0$ and $\cdot$ means a concatenation of strings. (Note that this equality means that they are identified using the connection.) We want to show that z and z' are the same as strings in A. (That is, they are identified without using the connection.) Let e be the Jones projection in the upper horizontal string algebra in the above diagram, which is defined as in Definition 11.5 for horizontal strings. (We have several starting vertices in V_0 now, but it does not matter here. We use the same formula for several initial vertices.) Because the Jones projection does not change its form when identified using the connection, this e has the same form as the original e in the lower horizontal string algebras after the identification with W in Fig. 11.16, that is, $e \cdot \mathrm{id}^{(k)} = \mathrm{id}^{(k)} \cdot e$, where $\mathrm{id}^{(k)}$ denotes the string $\sum_{|\sigma|=k}(\sigma, \sigma)$ on the graph $\mathcal{G}_3$ and $\cdot$ means a concatenation of strings again. (This equality again means that they are identified using the connection.)

On the other hand, we can easily show the identification $z' \cdot e = e \cdot z'$ using the connection by induction on k. This proof is the same as in the proof that the Jones projections do not change their form in the identifications using the connection. This implies

$$
\begin{aligned}
&(z \cdot \mathrm{id}^{(2)}) \times (\mathrm{id}^{(k)} \cdot e) = (\mathrm{id}^{(2)} \cdot z') \times (e \cdot \mathrm{id}^{(k)}) \\
&= e \cdot z' = z' \cdot e = (z' \cdot \mathrm{id}^{(2)}) \times (\mathrm{id}^{(k)} \cdot e),
\end{aligned}
$$

where '$\times$' means multiplication in the string algebra and '$\cdot$' means concatenation of strings. (Note that the product $\times$ generalizes an ordinary matrix multiplica-

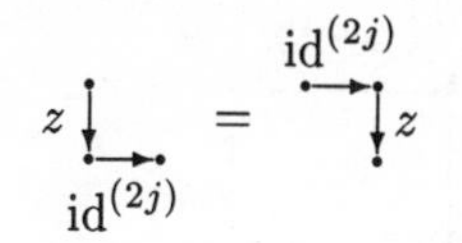

FIG. 11.17. $z \cdot \mathrm{id}^{(2j)} = \mathrm{id}^{(2j)} \cdot z$

tion and the concatenation $\cdot$ generalizes a tensor product of matrices.) Taking a conditional expectation to the left vertical algebra A in Fig. 11.15 with respect to the limit trace on $\tilde{A}$, we get $z = z'$ in A. This means that for any j we get the identity as in Fig. 11.17, where we identify two strings using the connection.

Let $z(X)$ be a component of z with initial vertex $X \in V_0$ and regard $z(*)$ as an element in $A_{k,0}$. (Note that we have fixed $*$.) We say that $\{z(X)\}_X$ is *flat* when the identity in Fig. 11.17 holds. (As we have seen, it is enough to have the identity for $j = 1$.) Then the identity in Fig. 11.17 shows that $\lim_{j\to\infty} \|z_{2l_j} - z(*)\|_2 = 0$, because of the definition of z and the fact that the trace on $\tilde{A}$ given with $\tilde{\phi}_{2l}$ from the trace on $A_{k,\infty}$ converges to the limit trace on $\tilde{A}$ described as above as l goes to ∞. This then means that $z^0 = z(*) \in A_{k,0}$.

Conversely, suppose we have a field of strings $\{z(X)\}_X$ which is flat in the sense that it satisfies the identity in Fig. 11.17. Then by the same argument as above, we conclude that $z(*) \in A_{k,0}$ is in $A'_{0,\infty} \cap A_{k,\infty}$.

Thus we have proved the following theorem.

Theorem 11.15 *In the string algebra construction from a bi-unitary connection as in Definition 11.3, the higher relative commutants $A'_{0,\infty} \cap A_{k,\infty}$ are given by flat fields of k-strings on $\mathcal{G}_3$ in the above sense. In particular, $A'_{0,\infty} \cap A_{k,\infty} \subset A_{k,0}$.*

The argument used in the above proof is called Ocneanu's *compactness argument.* This argument itself is very important and powerful as well as the Theorem. For example, see (Sato, in press a), (Sato 1997b) for applications of the argument and the notion of a flat field.

To use this abstract criterion to compute higher relative commutants can be highly non-trivial even for small graphs as we shall see in Section 11.5 for Dynkin diagrams.

The fact that the Jones projections do not change their forms by basis changes means that the Jones projections give a flat families of strings. We say the *Jones projections are flat.* This means that $e_1, e_2, \ldots, e_k \in A'_{0,\infty} \cap A_{k+1}$.

Definition 11.16 *We say that the bi-unitary connection as in Definition 11.3 is* flat with respect to $* \in V_0$ *if for any element $x \in A_{k,0}$, there exists a flat field $\{z(X)\}_X$ of k-strings on $\mathcal{G}_3$ such that $z(*) = x$. By the above argument, this condition is equivalent to the condition that any element $x \in A_{k,0}$ commutes with $A_{0,\infty}$.*

We will study a relation between flatness in Definition 11.16 and flatness in Axiom 5.

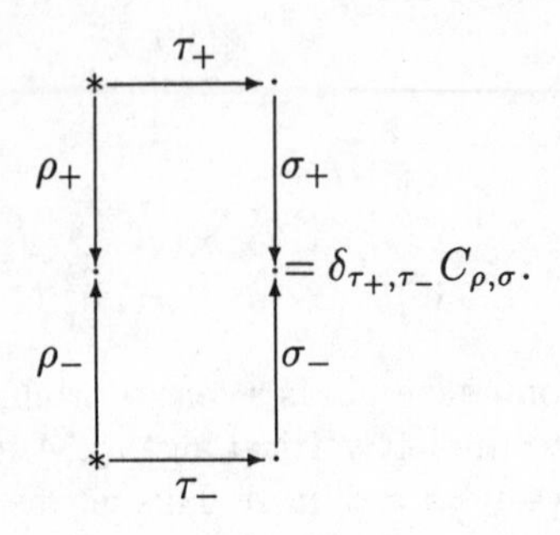

FIG. 11.18. Flatness condition (2)

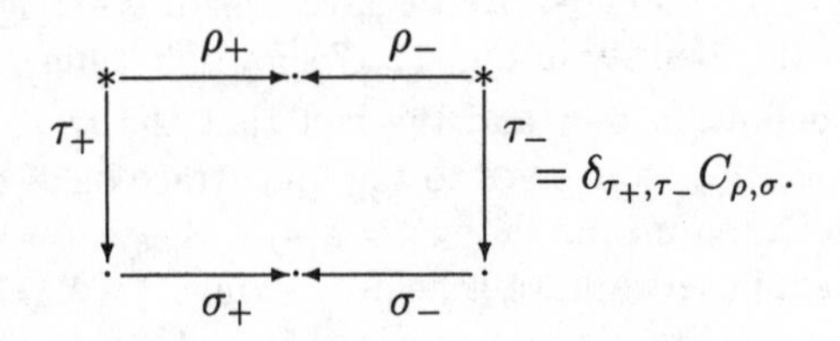

FIG. 11.19. Flatness condition (3)

Theorem 11.17 *In the string algebra construction from a bi-unitary connection as in Definition 11.3, the following conditions are equivalent:*

1. *Any two elements $x \in A_{k,0}$ (in the vertical string algebra) and $y \in A_{0,l}$ (in the horizontal string algebra) commute.*
2. *For each vertical string $\rho = (\rho_+, \rho_-) \in A_{k,0}$, we have the identity of Fig 11.18, where $C_{\rho,\sigma} \in \mathbb{C}$ depends only on ρ and $\sigma = (\sigma_+, \sigma_-)$.*
3. *For each horizontal string $\rho = (\rho_+, \rho_-) \in A_{0,k}$, we have the identity of Fig 11.19, where $C_{\rho,\sigma} \in \mathbb{C}$ depends only on ρ and $\sigma = (\sigma_+, \sigma_-)$.*
4. *For any horizontal paths σ_+, σ_- and vertical paths ρ_+, ρ_- with all the sources and ranges equal to $*$, we have the identity of Fig. 11.20.*

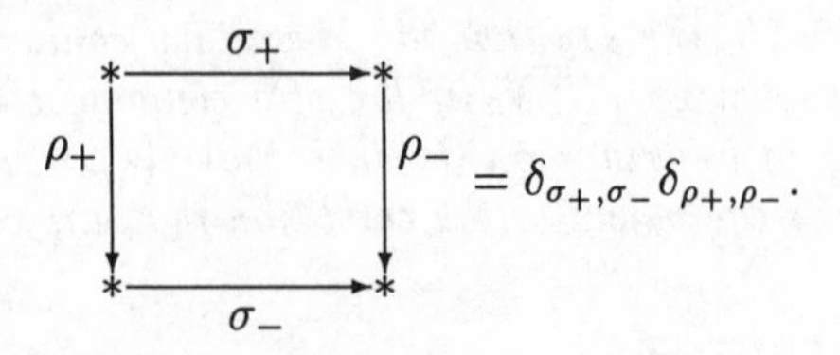

FIG. 11.20. Flatness condition (4)

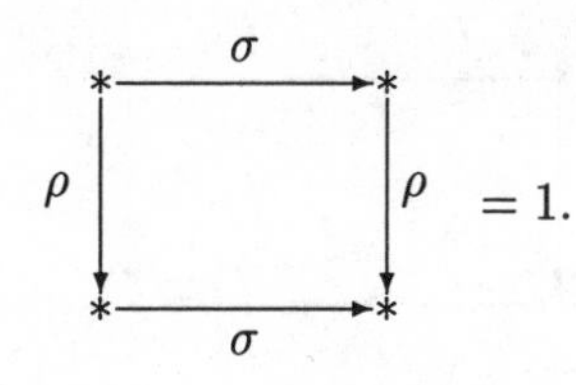

FIG. 11.21. Flatness condition (5)

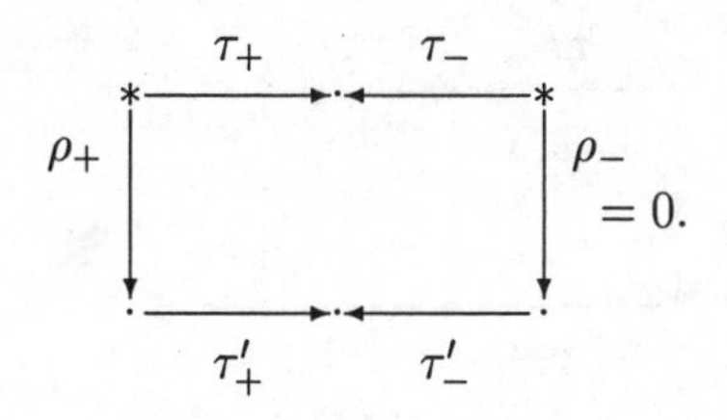

FIG. 11.22. A flatness identity

5. *For any horizontal path σ and vertical path ρ with $s(\sigma) = r(\sigma) = s(\rho) = r(\rho) = *$, we have the identity of Fig. 11.21.*

Proof (1)⇔(2). Let $m = |\sigma| = |\rho|$ in (2). Condition (2) means $\rho \in A_{k,m} \cap A'_{0,m}$ after the identification using the connection. Because k, m are arbitrary, we get the equivalence of (1) and (2).

(1)⇔(3). Same as above.

(2),(3)⇒(4). Suppose $\rho_+ \neq \rho_-$. Because $s(\sigma_+) = r(\sigma_+) = s(\sigma_-) = r(\sigma_-) = *$, we can write $\sigma_+ = \tau_+ \cdot \tilde{\tau}_-$ and $\sigma_- = \tau'_+ \cdot \tilde{\tau}'_-$ with $s(\tau_+) = s(\tau_-) = s(\tau'_+) = s(\tau'_-) = *$. Then by (3), we get the identity in Fig. 11.22. This implies the left hand side of the identity in (4) is 0. Similarly, if $\sigma_+ \neq \sigma_-$, then the formula is 0.

Suppose $\sigma_+ = \sigma_-$ and $\rho_+ = \rho_-$. Write $\sigma_+ = \sigma_- = \tau_+ \cdot \tilde{\tau}_-$ with $s(\tau_+) = s(\tau_-) = *$ as above and set $\tau = (\tau_+, \tau_-)$. Define C as in Fig. 11.23. By (3), this C does not depend on $\rho_+ (= \rho_-)$. Let $p = \sum_{\pi_1,\pi_2} (\pi_1 \cdot \pi_2, \pi_1 \cdot \pi_2)$, where the summation is over all the vertical paths π_1 and horizontal paths π_2 with $s(\pi_1) = r(\pi_1) = s(\pi_2) = *$, $r(\pi_2) = r(\tau_+)$, $|\pi_1| = |\rho_+|$ and $|\pi_2| = |\tau_+|$. This p

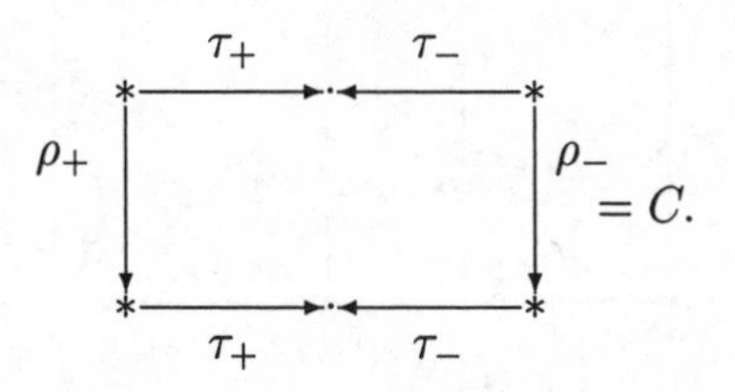

FIG. 11.23. A flatness identity

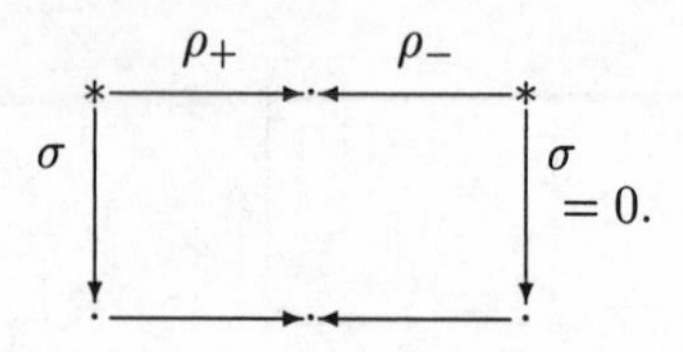

FIG. 11.24. A flatness identity

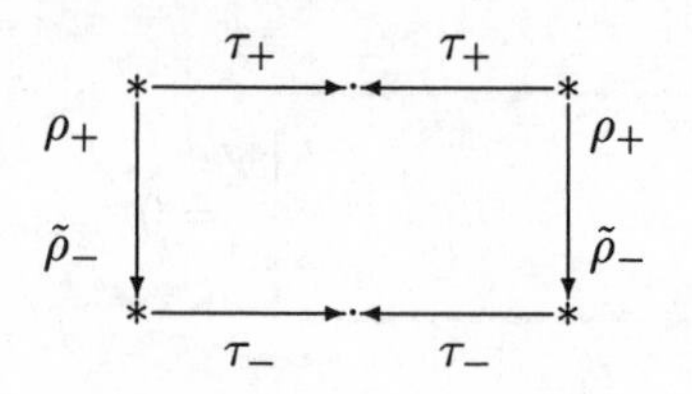

FIG. 11.25. A partition function

is a projection commuting with ρ and we know that $p\tau = C\sum_\pi(\pi,\pi)\cdot\tau$, where the summation is over all the vertical paths π with $s(\pi) = r(\pi) = *$, $|\pi| = |\rho_+|$. Because τ is a partial isometry, we get $|C| = 1, 0$. Because C does not depend on the choice of $\rho_+ = \rho_-$, we set $\rho_+ = \rho_- = \sigma\cdot\tilde{\sigma}$, where σ is an arbitrary vertical path with $2|\sigma| = |\rho_+|$. Then we get $C \geq 0$, hence $C = 1, 0$. If $C = 0$, we get the identity in Fig. 11.24 for all σ with $|\sigma| = |\rho_+|/2$ and $s(\sigma) = *$. This means that a non-zero element τ is identified with 0 using the connection, which is a contradiction. Thus we get $C = 1$, which implies Condition (4).

(4)$\Rightarrow$(2). Suppose $\tau_+ \neq \tau_-$ in Condition (2). Then applying Condition (4) to the diagram in Fig. 11.25, we get the left hand side of the formula in (2) is 0 for all σ_+, σ_-. Now we fix $\rho_\pm$, τ, τ' with $s(\rho_\pm) = s(\tau) = s(\tau') = *$, $r(\tau) = r(\tau')$. Then it is enough to show the identity in Fig. 11.26. Expanding the left hand

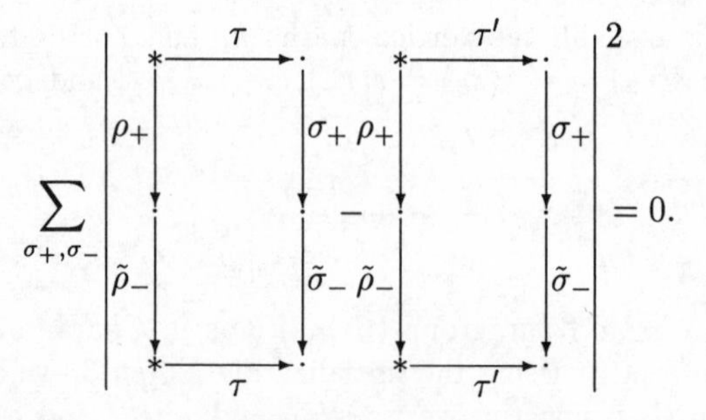

FIG. 11.26. A flatness identity

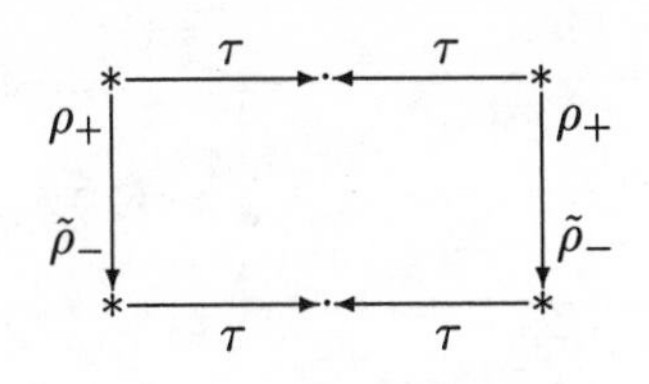

FIG. 11.27. A partition function equal to 1

side, we get $1+1-1-1=0$. Indeed, for example, the first 1 is obtained as in Fig. 11.27.

(4)⇒(5). Trivial.

(5)⇒(4). Fix ρ_+, σ_-. By unitarity, the sum of the squares of the absolute values of the left hand side of the formula in Condition (4) for all σ_+, ρ_- is 1. By Condition (5), all the terms except for $\sigma_+ = \sigma_-$ and $\rho_- = \rho_+$ are zero. □

Note that we have assumed only $\mathcal{G}_0$ and $\mathcal{G}_2$ are connected, so the roles of vertical and horizontal strings are not symmetric, strictly speaking. But in many cases, all the four graphs are connected, and then of course the roles of vertical and horizontal strings are symmetric.

Definition 11.18 *Suppose we have a bi-unitary connection which is flat with respect to $*$. (Thus we have all five conditions in Theorem 11.17.) In Condition 2 of Theorem 11.17, we fix an identical pair of paths $\tau_+ = \tau_-$. Then we have a map $\rho \mapsto \sum_\sigma C_{\rho,\sigma}\sigma$. (The image is in the string algebra with the initial vertex $r(\tau_\pm)$.) This is a $*$-isomorphism from $A_{k,0}$ into the k-string algebra with the initial vertex $r(\tau_\pm)$. We call this map the* horizontal parallel transport *from $*$ to $r(\tau_\pm)$. We similarly define* vertical parallel transport. *These two together are called* parallel transports.

Remark In Axiom 5, we have flatness with respect to both $*_N$ and $*_M$. Note that in the above Definition, we use the name flat connection for ones satisfying Condition 4 of Theorem 11.17 only for one $*$, while we need the flatness for both $*_N$ and $*_M$ to get a paragroup.

Remark Condition (4) of Theorem 11.17 can be written as in Fig. 11.28, where $\rho = (\rho_+, \rho_-)$ and $\sigma = (\sigma_+, \sigma_-)$. This means the string ρ does not change its form in the parallel transport from $*$ to $*$. This is the reason Ocneanu calls this condition flatness in analogy to the flatness in differential geometry. (That is, a loop does not change its form in a parallel transport.) Note that flatness depends on the choice of $*$.

Suppose we have a paragroup $(\mathcal{G}, \mu, W)$ as in Chapter 10 and construct a string algebra from it. Using the initialization axiom 3, we define σ_0 to be the unique edge in $\mathcal{G}_0$ connecting vertices $*_N$ and ${}_N{*}{*}_M$ and σ_1 to be the unique edge in $\mathcal{G}_1$ connecting vertices ${}_N{*}{*}_M$ and $*_M$. We take the sequence of string algebras on the graph $\mathcal{G}_1$ starting with $*_M$ and identify them with

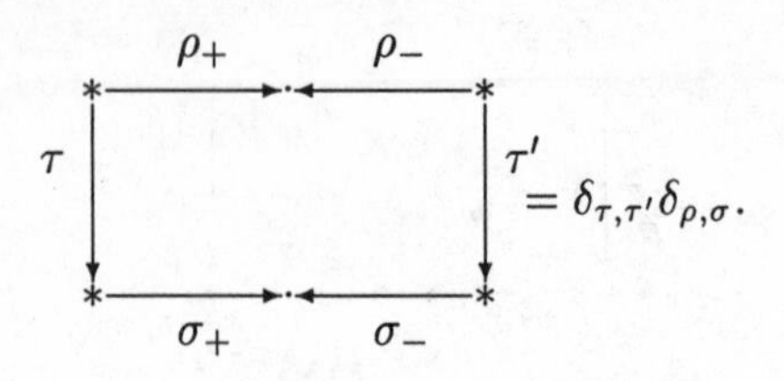

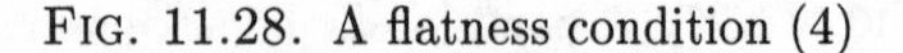

FIG. 11.28. A flatness condition (4)

$$\begin{array}{ccc} A & \sigma_1 & B \\ \sigma_3 & \square & \sigma_2 \\ C & \sigma_4 & D \end{array} = \overline{\begin{array}{ccc} A & \sigma_3 & C \\ \sigma_1 & \square & \sigma_4 \\ B & \sigma_2 & D \end{array}}$$

FIG. 11.29. Opposite connection

$$A_{0,0} \subset A_{0,1} \subset A_{1,1} \subset A_{2,1} \subset A_{3,1} \subset \cdots$$

by identifying σ_0 and σ_1. (This is possible by the initialization axiom.) We now label a sequence of string algebras on the graph $\mathcal{G}_1$ starting with $*_M$ as

$$A_{-1,1} \subset A_{0,1} \subset A_{1,1} \subset A_{2,1} \subset A_{3,1} \subset \cdots.$$

Then we can also define a sequence

$$A_{-1,1} \subset A_{-1,2} \subset A_{-1,3} \subset A_{-1,4} \subset A_{-1,5} \subset \cdots$$

as the string algebras on the graph $\mathcal{G}_2$ starting with $*_M$, and as before, we have a well-defined double sequence $\{A_{kl}\}_{kl}$ for $k \geq -1$, $l \geq 0$, $(k,l) \neq (-1,0)$ using the connection. By repeating similar arguments with the initialization axiom, we can construct a well-defined double sequence $\{A_{kl}\}_{kl}$ of string algebras for $k+l \geq 0$. We also extend the definition of $A_{k,\infty}$ to the negative k. Using the flatness axiom and Theorem 11.17 for both $*_N$ and $*_M$, we know that for $k \geq -l \geq -m$, $x \in A_{k,l}$ and $y \in A_{-l,m}$, we have $xy = yx$. In particular, we have $A_{k,0} = A'_{0,\infty} \cap A_{k,\infty}$ and $A_{k,1} = A'_{-1,\infty} \cap A_{k,\infty}$ for $k \geq 0$. Thus the paragroup for $A_{0,\infty} \subset A_{1,\infty}$ is given as follows.

We first switch $\mathcal{G}_0$ with $\mathcal{G}_3$, and $\mathcal{G}_1$ with $\mathcal{G}_2$. The map μ is retained, and the new connection is defined as in Fig. 11.29, where the left hand side of the identity denotes the value of the new connection and the right hand side is for the original value.

We call this new paragroup the *opposite* paragroup of the original. Clearly, the opposite of the opposite paragroup is the original paragroup. By Proposition 9.50, mutually opposite paragroups give mutually anti-isomorphic subfactors.

We can also define the *dual* paragroup which is given by replacing $\mathcal{G}_0$, $\mathcal{G}_1$, $\mathcal{G}_2$, $\mathcal{G}_3$ by $\mathcal{G}_2$, $\mathcal{G}_3$, $\mathcal{G}_0$, $\mathcal{G}_1$ respectively and using the same connection. It is clear that the dual paragroup corresponds to $M \subset M_1$ while the original paragroup corresponds to $N \subset M$.

So far, we have proved the following theorem.

Theorem 11.19 *If we construct a subfactor $N \subset M$ from a paragroup with the string algebra method, the paragroup arising from this subfactor is the opposite paragroup of the original one.*

We give the following application.

Theorem 11.20 *Let $N \subset M$ be a subfactor arising from a paragroup as in Section 11.3. Let M_∞ be the GNS-completion of $\bigcup_{n=1}^\infty M_n$ with respect to the trace. Then we have $(N' \cap M_\infty)' \cap M_\infty = N$.*

Proof Choose an x in $N' \cap M_\infty$ and set $x_k = E_{M_k}(x)$. Then we have $x_k \in N' \cap M_k = A_{k+1,0}$ by Theorem 11.19. Because the limit of $\{x_k\}_k$ in the weak operator topology is x, we know that $N' \cap M_\infty$ is the weak closure of $\bigcup_{k=1}^\infty A_{k+1,0}$ in M_∞. Choose y in $(N' \cap M_\infty)' \cap M_\infty$ and set $y_k = E_{A_{\infty,k}}(y)$, where $A_{\infty,k}$ is the weak closure of $\bigcup_{j=1}^\infty A_{jk}$ in M_∞. It is clear that the limit of $\{y_k\}_k$ in the weak operator topology is y. By a similar argument to the above, we get $y_k \in A'_{\infty,0} \cap A_{\infty,k} = A_{0,k}$ by changing the role of the horizontal and vertical directions. We then get $y \in N$.

The other inclusion is trivial. □

We now discuss the conditions in Theorem 11.17. Condition (1) of the Theorem involves operators in infinite dimensional algebras, and Conditions (2)–(5) involve infinitely many diagrams of partition functions. With this observation, it seems that we need to verify infinitely many identities to check flatness, but this turns out to be not the case.

We define the *depth* of a graph with $*$ to be the largest distance from $*$ to a vertex on the graph. By the proof of Lemma 11.8, we know that if $k \geq d_{\mathcal{G}_3}$, then $A_{k+1,0}$ is generated by $A_{k,0}$ and the kth vertical Jones projection, where $d_{\mathcal{G}_j}$ is the depth of the graph $\mathcal{G}_j$. (A similar statement holds for $A_{0,k}$.) Because the vertical Jones projections always commute with horizontal strings, in order to verify flatness, it is enough to check the identities in (2) for the diagrams of size less than $2d_{\mathcal{G}_3} \times d_{\mathcal{G}_0}$, and similarly it is enough to check the identities in (4) for the diagrams of size less than $2d_{\mathcal{G}_3} \times 2d_{\mathcal{G}_0}$. This means that whether a given connection on four finite graphs is flat or not can be determined by finitely many computations. However, note that for a Dynkin diagram such as E_6, E_7, or E_8, these computations can be formidable.

We discuss an example of a flat connection corresponding to finite group case of Section 10.6.

Example 11.21 Consider the connection corresponding to a finite group G defined in Fig. 10.25. Because it gives a paragroup, we already know that this connection satisfies the flatness axiom. Here we try to verify flatness with respect to $* = 1 \in G$ by a direct computation. Take the 2×2-diagram as in Fig. 11.30.

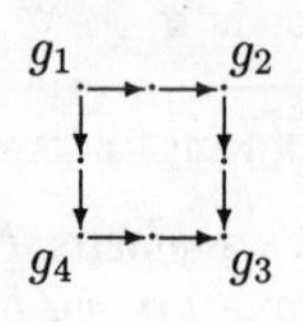

FIG. 11.30. A 2×2-diagram

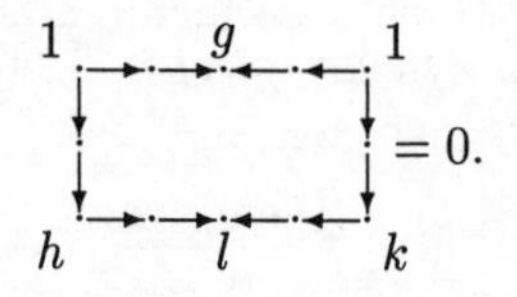

FIG. 11.31. Flatness in the group case

If we fix a centre vertex σ for this diagram, the sum of the products of four connection values for all the configurations is given by

$$\begin{aligned}
&\sum_{i,j,k,l} \sigma_{ij}(g_1)\overline{\sigma_{kj}(g_2)}\sigma_{kl}(g_3)\overline{\sigma_{il}(g_4)} \cdot \frac{|\sigma|}{n} \\
&= \sum_{i,j,k,l} \sigma_{ij}(g_1)\sigma_{jk}(g_2^{-1})\sigma_{kl}(g_3)\sigma_{li}(g_4^{-1}) \cdot \frac{|\sigma|}{n} \\
&= \frac{|\sigma|}{n}\mathrm{Tr}(\sigma(g_1 g_2^{-1} g_3 g_4^{-1})).
\end{aligned}$$

Now we let σ vary, then we get the value

$$\sum_{\sigma} \frac{|\sigma|}{n}\mathrm{Tr}(\sigma(g_1 g_2^{-1} g_3 g_4^{-1})) = \delta_{g_1 g_2^{-1} g_3 g_4^{-1},1}$$

for the above diagram by the orthogonality relations in the representation theory of finite groups. Because our graph has depth 2, we only need to check the identities in Fig. 11.31 for all $g, h, k, l \in G$ with $h \neq k$ in order to verify flatness. (This corresponds to Condition (3) in Theorem 11.17.)

This easily follows from the above computation for 2×2-diagrams. A similar computation also works for the $*$ of the dual principal graph. Thus we get flatness by direct computations.

11.5 Flat connections on the Dynkin diagrams

In this section, we will determine all the paragroups corresponding to subfactors with index less than four. First note that the Perron–Frobenius eigenvalue of the incidence matrix of the graph must be less than 2, hence each graph is one of

Table 11.1 *Dynkin diagrams with eigenvalues*

The Perron–Frobenius eigenvalue	Dynkin diagrams
$2\cos\pi/(2n+1)$	A_{2n}
$2\cos\pi/(2n-2)$, $n \neq 7, 10, 16$	A_{2n-3}, D_n
$2\cos\pi/12$	A_{11}, D_7, E_6
$2\cos\pi/18$	A_{17}, D_{10}, E_7
$2\cos\pi/30$	A_{29}, D_{16}, E_8

the Dynkin diagrams of type A, D, E by Theorem 10.6. By Corollary 10.8, the vertex $*$ of each of the Dynkin diagrams must be as in Figs 10.8–10.12.

Let $\mathcal{G}_0$ be one of the A-D-E Dynkin diagrams. All the other graphs $\mathcal{G}_1, \mathcal{G}_2, \mathcal{G}_3$ must have the same Perron–Frobenius eigenvalues. (See Table 11.1.) First we look at the graph $\mathcal{G}_3$. The number of even vertices of $\mathcal{G}_3$ must be the same as that of $\mathcal{G}_0$. Furthermore, the entries of the Perron–Frobenius eigenvector of $\mathcal{G}_0$ must match those of $\mathcal{G}_3$. These conditions, given an inspection of A-D-E diagrams as the only possibilities, forces $\mathcal{G}_3$ to be the same graph as $\mathcal{G}_0$.

The $2k+1$-string algebras on the graph $\mathcal{G}_0$ starting with $*_N \in V_0$ must be isomorphic to the $2k+1$-string algebras on the graph $\mathcal{G}_1$ starting with $*_M \in V_2$, because both are isomorphic to $N' \cap M_{2k}$. This forces $\mathcal{G}_1$ to be a copy of $\mathcal{G}_0$. This also forces the graphs $\mathcal{G}_0$ and $\mathcal{G}_1$ to be connected trivially at V_1. The same argument works for $\mathcal{G}_2$ and $\mathcal{G}_3$.

Similarly, we can represent $N' \cap M_{2k-1}$ both as the $2k$-string algebra on the graph $\mathcal{G}_0$ starting with $*_N \in V_0$ and that on $\mathcal{G}_3$ starting with $*_N$. This forces that the graphs $\mathcal{G}_0$ and $\mathcal{G}_3$ to be copies of each other and connected trivially at V_0. The same argument works for $\mathcal{G}_1$ and $\mathcal{G}_2$. Thus we know that all the four graphs are the same, and the vertices are matched in a trivial way.

We next determine all bi-unitary connections on this graph and will later determine which of them are flat.

We fix one of the A-D-E Dynkin diagrams and identify all four graphs $\mathcal{G}_0, \mathcal{G}_1, \mathcal{G}_2, \mathcal{G}_3$. We write down an explicit formula for connections as follows. Recall that μ is given as the Perron–Frobenius eigenvector of the incidence matrix of the graph. We then have the harmonicity axiom because we identify the four graphs. We set $\varepsilon = \sqrt{-1}\exp\left(\pi\sqrt{-1}/2N\right)$, where N is the Coxeter number of the Dynkin diagram. We define a connection on the diagram as in Fig. 11.32. (This is natural from the braid viewpoint. See Section 11.9.) Because we now identify the four graphs, the formula in Fig. 11.32 assigns a complex number to each cell. Note that we do not have to specify edges because the graph has no multiple edges.

We first see that this indeed gives a bi-unitary connection. We check the unitarity axiom as in Fig. 11.33.

Suppose $A \neq D$ first. Then the left hand side of the identity in Fig. 11.33 is equal to $\sum_B \delta_{BC}\varepsilon\delta_{BC'}\bar{\varepsilon} = \delta_{CC'}$.

Suppose $A = D$ next. Then the left hand side of the identity in Fig. 11.33 is equal to the following:

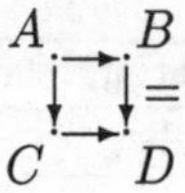

$$\begin{array}{c} A \quad B \\ \boxed{} \\ C \quad D \end{array} = \delta_{BC}\varepsilon + \sqrt{\frac{\mu(B)\mu(C)}{\mu(A)\mu(D)}}\delta_{AD}\bar{\varepsilon}.$$

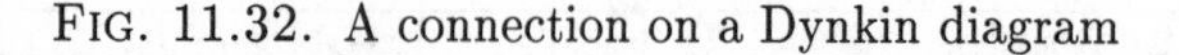

FIG. 11.32. A connection on a Dynkin diagram

$$\sum_B \begin{array}{c} A \quad B \\ \boxed{} \\ C \quad D \end{array} \overline{\begin{array}{c} A \quad B \\ \boxed{} \\ C' \quad D \end{array}} = \delta_{C,C'}.$$

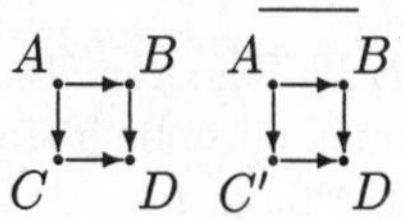

FIG. 11.33. Unitarity condition

$$\begin{aligned} &\sum_B \delta_{BC}\varepsilon\delta_{BC'}\bar{\varepsilon} + \sum_B \delta_{BC}\varepsilon^2 \frac{\mu(B)^{1/2}\mu(C')^{1/2}}{\mu(A)} \\ &\qquad + \sum_B \frac{\mu(B)^{1/2}\mu(C)^{1/2}}{\mu(A)}\delta_{BC'}\bar{\varepsilon}^2 + \sum_B \frac{\mu(C)^{1/2}\mu(C')^{1/2}\mu(B)}{\mu(A)^2} \\ &= \delta_{CC'} + \frac{\mu(C)^{1/2}\mu(C')^{1/2}}{\mu(A)}(\varepsilon^2 + \bar{\varepsilon}^2 + \frac{\sum_B \mu(B)}{\mu(A)}) \\ &= \delta_{CC'} + \frac{\mu(C)^{1/2}\mu(C')^{1/2}}{\mu(A)}(\varepsilon^2 + \bar{\varepsilon}^2 + \beta) \\ &= \delta_{CC'}. \end{aligned}$$

Here we used $\varepsilon^2 + \bar{\varepsilon}^2 + \beta = 0$, which follows from the definition of ε. The other identity of the unitarity axiom holds similarly.

It is trivial from Fig. 11.32 that the renormalization axiom holds. So Fig. 11.32 indeed defines a bi-unitary connection. It is clear that the same computation as above works after we replace ε by $\bar{\varepsilon}$. Thus we have two bi-unitary connections on each of the Dynkin diagrams. We next determine all bi-unitary connections up to isomorphism.

First, we consider the A-diagrams. We work on the following example of A_5. We now distinguish the four graphs $\mathcal{G}_0, \mathcal{G}_1, \mathcal{G}_2, \mathcal{G}_3$ and label the vertices as in Fig. 11.34. Note that $\mu(X_1) = \mu(X_3) = \mu(X_6) = \mu(X_8) = 1$, $\mu(X_2) = \mu(X_7) = 2$, $\mu(X_4) = \mu(X_5) = \mu(X_9) = \mu(X_{10}) = \sqrt{3}$.

We first look at the connection value in Fig. 11.35. By the unitarity axiom, we know that the absolute value of this connection value is 1. By the freedom of the gauge choices, we can and do set this value to be 1. Note that gauges for the four edges appearing in Fig. 11.35 have been fixed and we cannot change them later. Then the renormalization axiom implies the identity in Fig. 11.36.

Similarly, we can set the connection value as in the first identity in Fig. 11.37. Note that gauge choices have been made for the edge connecting X_1 and X_4 and

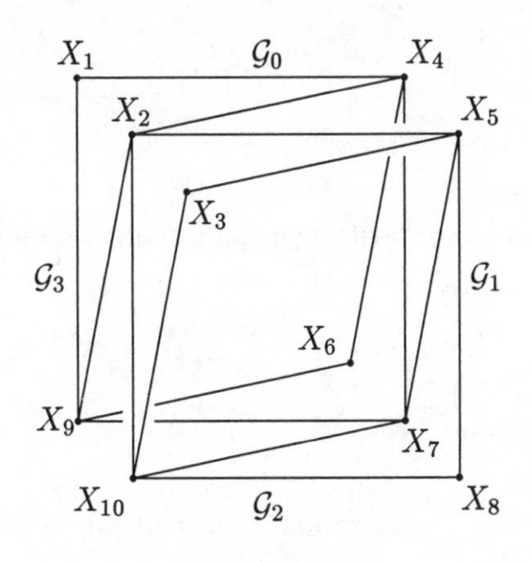

FIG. 11.34. The Dynkin diagram A_5

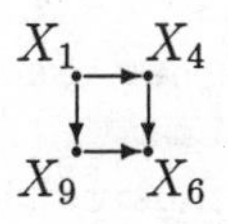

FIG. 11.35. Connection value

the edge connecting X_1 and X_9 in Fig. 11.37, but we still have a freedom of gauge choice for the other two edges. We have used this freedom of the two choices. Using the renormalization axiom again, we get the second identity in Fig. 11.37. We similarly have identities in Fig. 11.38 after similar gauge choices. Figs 11.36–11.38 and the unitarity axiom applied to a 2×2-matrix imply the first identity in Fig. 11.39 and then the renormalization axiom implies the second identity in Fig. 11.39. Similarly, by gauge choices, the unitarity axiom and the renormalization axiom, we get identities in Figs 11.40, 11.41. Again by the unitarity axiom applied to a 2×2-matrix, we get the identity in Fig. 11.42.

It is clear that we can repeat this procedure of gauge choices, the renormalization axiom, and the unitarity axiom applied to a 2×2-matrix. Thus on A_n, we have at most one isomorphism class of bi-unitary connections. We have already seen that a bi-unitary connection does exist on the A_n graphs, so we conclude that on each of the graphs A_n, there is one and only one bi-unitary connection

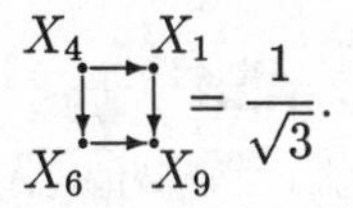

FIG. 11.36. A renormalization identity

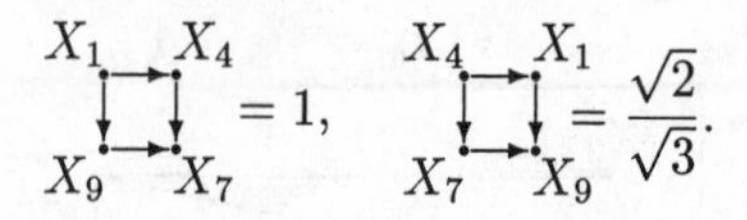

FIG. 11.37. Unitarity and renormalization identities

X_2 X_4 / X_9 X_6 $= 1,\quad$ X_4 X_2 / X_6 X_9 $= \dfrac{\sqrt{2}}{\sqrt{3}}.$

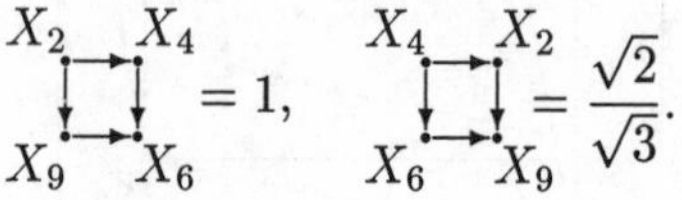

FIG. 11.38. Unitarity and renormalization identities

up to isomorphism. (It means in particular that a change of ε to $\bar{\varepsilon}$ as above gives a connection isomorphic to the original connection.)

Next we study the case of D_n, E_6, E_7, E_8. We apply the same procedure as above by starting the vertex $*$. Then we can continue the same process until we meet the triple point of the graph.

We label vertices around the triple point as in Fig. 11.43. By appropriate choices of gauges, we have a 3×3-matrix as follows, which must be unitary by the unitarity axiom.

$$\begin{pmatrix} \dfrac{\mu(B_1)}{\mu(A)} & \dfrac{\sqrt{\mu(B_1)\mu(B_2)}}{\mu(A)} & \dfrac{\sqrt{\mu(B_1)\mu(B_3)}}{\mu(A)} \\ \dfrac{\sqrt{\mu(B_1)\mu(B_2)}}{\mu(A)} & x & y \\ \dfrac{\sqrt{\mu(B_1)\mu(B_3)}}{\mu(A)} & y & z \end{pmatrix},$$

where $|y| = \sqrt{\mu(B_2)\mu(B_3)}/\mu(A)$, $x, y, z \in \mathbb{C}$. We want to determine $x, y, z \in \mathbb{C}$ so that the above matrix is a unitary. Set $x_j = \mu(B_j)/\mu(A)$ for $j = 1, 2, 3$. Then we get $|x| = \sqrt{1 - x_2(x_1 + x_3)}$, $|y| = \sqrt{x_2 x_3}$ by unitarity. Because the other condition for unitarity is

$$x_1\sqrt{x_2} + \sqrt{x_2}x + \sqrt{x_3}y = 0,$$

the number of equivalence classes of possible unitary matrices is two if and only if we have

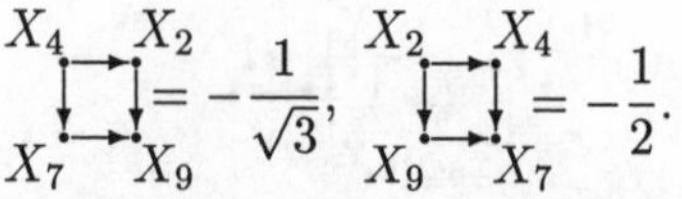

X_4 X_2 / X_7 X_9 $= -\dfrac{1}{\sqrt{3}},\quad$ X_2 X_4 / X_9 X_7 $= -\dfrac{1}{2}.$

FIG. 11.39. Unitarity and renormalization identities

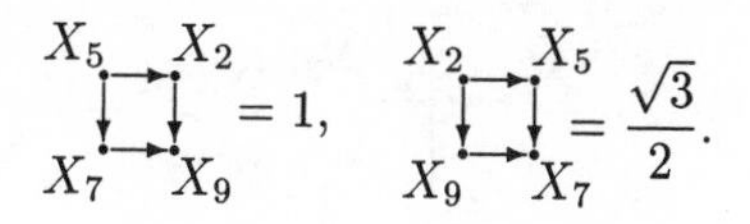

FIG. 11.40. Unitarity and renormalization identities

X_4 X_2 X_7 X_{10} $= 1,$ X_2 X_4 X_{10} X_7 $= \frac{\sqrt{3}}{2}.$

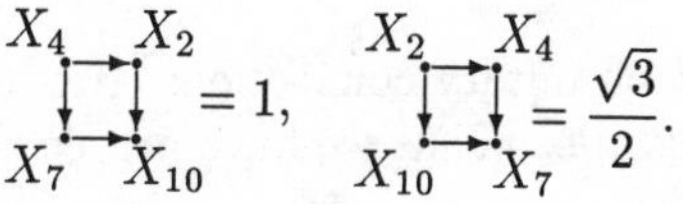

FIG. 11.41. Unitarity and renormalization identities

$$\sqrt{1 - x_2(x_1 + x_3)} + x_3 > x_1 > |x_3 - \sqrt{1 - x_2(x_1 + x_3)}|.$$

For E_6, E_7, E_8, we can check this inequality by direct computation. For D_n, we can set $\mu(B_1) = \mu(B_2) = 1/2$, $\mu(A) = \beta/2$, $\mu(B_3) = \beta^2/2 - 1$, where β is the Perron–Frobenius eigenvalue of the D_n, that is, $\beta = 2\cos(\pi/(2n-2))$. The above inequality in this case is

$$\sqrt{1 - \frac{1}{\beta}\left(\frac{1}{\beta} + \frac{\beta^2 - 2}{\beta}\right)} + \frac{\beta^2 - 2}{\beta} > \frac{1}{\beta} > \frac{\beta^2 - 3}{\beta},$$

and this is valid because $2 < \beta^2 < 4$.

These show that for each of the graphs D_n, E_6, E_7, E_8 we have either two equivalence classes of bi-unitary connections or no bi-unitary connection at all. Because we already know the existence of one bi-unitary connection, we conclude that each of the above graphs indeed has two equivalence classes of bi-unitary connections.

Moreover, in the D_n case, we have the following procedure. We first start with $*$ and apply the above method of determining a connection as in the A_n diagrams until we meet the triple point. We can choose a real connection up to this point. Then at the triple point, we have a 3×3-unitary matrix as above, but now the $(1,1)$-entry is different from the above. We get $x = \bar{y} = z$ in the above 3×3-matrix. Thus two connections are isomorphic up to the graph automorphism of D_n switching the two endpoints. Thus each of the graph D_n has one and only

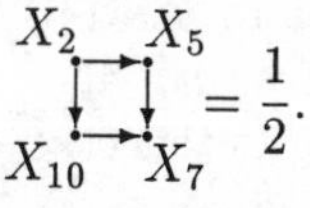

X_2 X_5 X_{10} X_7 $= \frac{1}{2}.$

FIG. 11.42. A unitarity identity

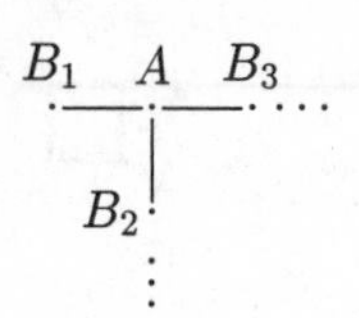

FIG. 11.43. The triple point of the graphs D, E

one isomorphism class of bi-unitary connections. It is clear that this argument does not work for the E graphs. Thus we have proved the following theorem.

Theorem 11.22 *Each graph A_n has precisely one bi-unitary connection up to isomorphism. Each graph D_n has precisely one connection up to isomorphism. Each of the graphs E_6, E_7, E_8 has precisely two connections up to isomorphism.*

As soon as we have a bi-unitary connection, we can apply the string algebra construction to get a subfactor. Thus we get the following, which gives a proof of the realization part of Theorem 9.16. We list it as a corollary here.

Corollary 11.23 *Each of the values $4\cos^2 \pi/n$, with $n = 3, 4, \ldots$, is realized as an index value $[M : N]$ of a subfactor N of a hyperfinite II_1 factor M.*

Remark If we apply the above argument of the 3×3-unitary matrix to a graph with Perron–Frobenius eigenvalue bigger than 2 and without a cycle of length 4, it is easy to see that we have no solutions of x, y, z. It means that we have no bi-unitary connection on a graph with Perron–Frobenius eigenvalue bigger than 2 and without a cycle of length 4 if we put the same four graphs together in a trivial way. This fact is often called Ocneanu's *triple point obstruction.*

Remark If the graph is one of the extended Dynkin diagrams $E_6^{(1)}$, $E_7^{(1)}$, $E_8^{(1)}$ as in Figs 11.55–11.57, the above method still applies. In these cases, we have only one bi-unitary connection on each of the graphs up to isomorphism. This method, however, does not apply to the $D_n^{(1)}$ diagrams. See (Izumi and Kawahigashi 1993).

The initialization axiom holds trivially for these bi-unitary connections. Now we want to determine whether the bi-unitary connections we have found for the Dynkin diagrams are flat or not. (Note that the vertex $*$ is already determined by Corollary 10.8.) Our aim is to prove the following theorem.

Theorem 11.24 *The bi-unitary connections on A_n, D_{2n}, E_6, E_8 are flat. The bi-unitary connections on D_{2n+1}, E_7 are not flat.*

This gives a complete solution to the problem which of principal graphs appear for subfactors with index less than 4.

First we consider the case of A_n. In this case, we know that the string algebra $A_{k,0}$ is generated by the Jones projections $e_1, e_2, \ldots, e_{k-1}$ by the proof of Lemma 11.8. Because the Jones projections always commute with $A_{0,l}$, we get flatness by Theorem 11.17. (In general, we need to verify flatness for both $*_N$ and $*_M$,

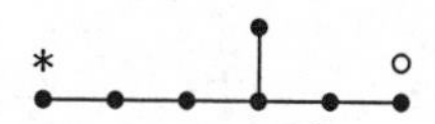

FIG. 11.44. Dynkin diagram E_7

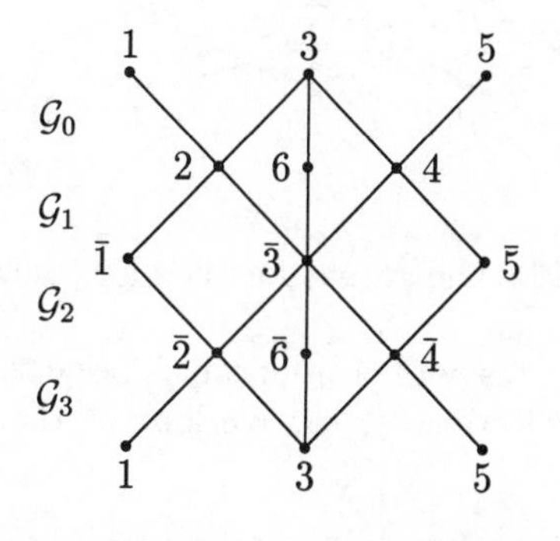

FIG. 11.45. The Dynkin diagram E_6

but now the two cases are the same.) We have thus proved Theorem 11.24 in the case of A_n. Thus we have a series of subfactors of the hyperfinite II_1 factor with principal graph A_n. These are often called the *Jones subfactors* because they were first constructed in (Jones 1983).

Next we consider the case of E_7. In this case, the vertex marked with $\circ$ in Fig. 11.44 has the Perron–Frobenius eigenvector violating the $2\cos\pi/n$-rule (Corollary 10.8). So the graph E_7 cannot have a flat connection as in the case of D_5 we discussed.

Next we give a proof of flatness of the bi-unitary connections on E_6. This is the method of (Izumi 1994a). Similar, but more complicated, computations give a proof of flatness of the E_8 case. We omit the proof for E_8 and refer readers to (Izumi 1994a). It is possible to work on the D_n cases with this method (with induction on n) as pointed out in (Izumi 1994a), but we will give a different proof of (non-)flatness for the D_n cases in Section 13.4, so we also omit the proof for the D_n cases here.

For the case of E_6, we label vertices as in Fig. 11.45.

We define the two strings $p, q \in A_{3,0}$ by

$$p = (1 \to \bar{2} \to 3 \to \bar{4}, 1 \to \bar{2} \to 3 \to \bar{4}),$$
$$q = (1 \to \bar{2} \to 3 \to \bar{6}, 1 \to \bar{2} \to 3 \to \bar{6}).$$

By the flatness of the vertical Jones projections and the proof of Lemma 11.8,

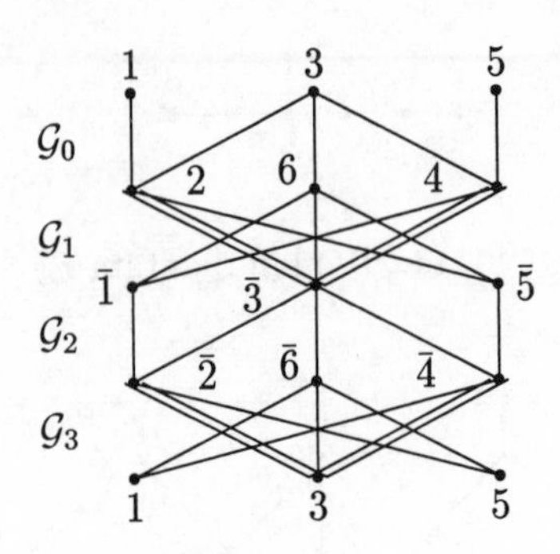

FIG. 11.46. The graphs for the new connection

we know that $p+q$ commutes with $A_{0,l}$ for any l, because $1-p-q$ is expressed in terms of the Jones projections e_1, e_2. We look at the following sequence of commuting squares:

$$\begin{array}{ccccccc} (p+q)A_{0,0} & \subset & (p+q)A_{0,1} & \subset & (p+q)A_{0,2} & \subset \cdots \\ \cap & & \cap & & \cap & \\ (p+q)A_{3,0}(p+q) & \subset & (p+q)A_{3,1}(p+q) & \subset & (p+q)A_{3,2}(p+q) & \subset \cdots \end{array} \tag{11.5.1}$$

Note that it is easy to see that they indeed form commuting squares by using Condition (7) of Proposition 9.51. Let f_k be the kth horizontal Jones projection in $(p+q)A_{0,k+1}$. Then for $x \in (p+q)A_{0,k}$, we have

$$f_k x f_k = E_{(p+q)A_{0,k-1}}(x) f_k = E_{(p+q)A_{3,k-1}(p+q)}(x) f_k.$$

In the series (11.5.1) of commuting squares, the horizontal Jones projections on the top line are embedded as the horizontal Jones projections on the second line. Then the characterization of commuting squares in Section 11.2, the expression of the horizontal Jones projections, and the proof of Lemma 11.8 imply that the sequence (11.5.1) can be regarded as the top two lines of the double sequence of string algebras constructed with a bi-unitary connection on the four graphs. It is easy to see that the four graphs are connected as in Fig. 11.46, where the vertices are labelled by numbers $1, 2, 3, 4, 5, 6, \bar{1}, \bar{2}, \bar{3}, \bar{4}, \bar{5}, \bar{6}$. (See Example 11.25 for this type of computation.)

We know that we have a bi-unitary connection on Fig. 11.46 and want to compute it explicitly. We will make the computation easier by appropriate gauge choices. Our claim is that we get the following sixteen formulae after appropriate gauge choices. Some unimportant terms are left blank and entries marked with $*$ denote non-zero entries.

$$ {}^{1}\square_{\bar{3}} = \begin{array}{c} \\ \bar{6} \\ \bar{4} \end{array} \begin{array}{c} \begin{array}{cc} 2_1 & 2_2 \end{array} \\ \begin{pmatrix} 1 & 0 \\ 0 & 1 \end{pmatrix} \end{array}, \tag{11.5.2} $$

$$ \begin{array}{ccc} 1 & & 2 \\ & \square & \\ \bar{4} & & \bar{5} \end{array} = 1, \tag{11.5.3} $$

$$ {}^{3}\square_{\bar{1}} = \begin{array}{c} \\ \overline{2_1} \\ \overline{2_2} \end{array} \begin{array}{c} \begin{array}{cc} 4 & 6 \end{array} \\ \begin{pmatrix} 1 & 0 \\ 0 & 1 \end{pmatrix} \end{array}, \tag{11.5.4} $$

$$ {}^{3}\square_{\bar{3}} = \begin{array}{c|ccccc} & 2_1 & 2_2 & 4_1 & 4_2 & 6 \\ \hline \overline{2_1} & 0 & * & & 0 & * \\ \overline{2_2} & & 0 & 0 & & 0 \\ \overline{4_1} & 0 & & & 0 & \\ \overline{4_2} & & 0 & 0 & * & 0 \\ \bar{6} & 0 & & * & 0 & \end{array}, \tag{11.5.5} $$

$$ {}^{3}\square_{\bar{5}} = \begin{array}{c} \\ \overline{4_1} \\ \overline{4_2} \end{array} \begin{array}{c} \begin{array}{cc} 2 & 6 \end{array} \\ \begin{pmatrix} 1 & 0 \\ 0 & 1 \end{pmatrix} \end{array}, \tag{11.5.6} $$

$$ {}^{5}\square_{\bar{3}} = \begin{array}{c} \\ \bar{2} \\ \bar{6} \end{array} \begin{array}{c} \begin{array}{cc} 4_1 & 4_2 \end{array} \\ \begin{pmatrix} 1 & 0 \\ 0 & 1 \end{pmatrix} \end{array}, \tag{11.5.7} $$

$$ \begin{array}{ccc} 5 & & 4 \\ & \square & \\ \bar{2} & & \bar{1} \end{array} = 1, \tag{11.5.8} $$

$$ {}^{2}\square_{\bar{2}} = \begin{array}{c} \\ \overline{3_1} \\ \overline{3_2} \end{array} \begin{array}{c} \begin{array}{cc} 3_1 & 3_2 \end{array} \\ \begin{pmatrix} 0 & \\ & 0 \end{pmatrix} \end{array}, \tag{11.5.9} $$

$$ {}^{2}\square_{\bar{6}} = \begin{array}{c|cc} & 1 & 3 \\ \hline \bar{3}_1 & 1 & 0 \\ \bar{3}_2 & 0 & \end{array}, \tag{11.5.10} $$

$$ {}^{2}\square_{\bar{4}} = \begin{array}{c|ccc} & 1 & 3_1 & 3_2 \\ \hline \bar{3}_1 & 0 & 0 & \\ \bar{3}_2 & * & & 0 \\ \bar{5} & & * & 0 \end{array}, \tag{11.5.11} $$

$$ {}^{6}\square_{\bar{2}} = \begin{array}{c|cc} & 3_1 & 3_2 \\ \hline \bar{1} & 0 & 1 \\ \bar{3} & * & 0 \end{array}, \tag{11.5.12} $$

$$ \begin{array}{ccc} 6 & & 3 \\ & \square & \\ \bar{3} & & \bar{6} \end{array} = 1, \tag{11.5.13} $$

$$ {}^{6}\square_{\bar{4}} = \begin{array}{c|cc} & 3_1 & 3_2 \\ \hline \bar{3} & & 0 \\ \bar{5} & 0 & 1 \end{array}, \tag{11.5.14} $$

$$ {}^{4}\square_{\bar{2}} = \begin{array}{c|ccc} & 3_1 & 3_2 & 5 \\ \hline \bar{1} & * & 0 & \\ \bar{3}_1 & & 0 & * \\ \bar{3}_2 & 0 & & 0 \end{array}, \tag{11.5.15} $$

$$ {}^{4}\square_{\bar{6}} = \begin{array}{c|cc} & 3 & 5 \\ \hline \bar{3}_1 & * & 0 \\ \bar{3}_2 & 0 & 1 \end{array}, \tag{11.5.16} $$

$$ {}^{4}\square_{\bar{4}} = \begin{array}{c|cc} & 3_1 & 3_2 \\ \hline \bar{3}_1 & & 0 \\ \bar{3}_2 & 0 & \end{array}. \tag{11.5.17} $$

The meaning of the above formulae is as follows. We fix the upper left corne and the lower right corner of each cell. We let the upper right corner and th lower left corner change and then get a matrix. On the right hand sides of th

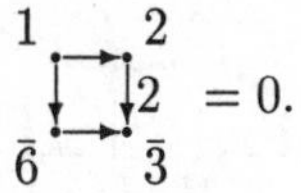

$$\begin{array}{ccc} 1 & & 2 \\ & \square & 2 \\ \bar{6} & & \bar{3} \end{array} = 0.$$

FIG. 11.47. An example of a connection value

equations above, we have these matrices. We label multiple edges by 1,2. For example, the $(1,2)$-entry of the matrix in Equation (11.5.2) means that we have the identity in Fig. 11.47. We put the labelling of the multiple edges as a subscript of the label of the vertices, to keep the notation as simple as possible.

A proof of the above claim is as follows.

By appropriate gauge choices, we obtain identities (11.5.3), (11.5.8), (11.5.13), (11.5.2), (11.5.4), (11.5.6), (11.5.7) in this order. We obtain identity (11.5.10) from identity (11.5.2) using the renormalization axiom and the unitarity axiom. In the same way, we obtain identities (11.5.12), (11.5.14), (11.5.16) from identities (11.5.4), (11.5.6), (11.5.7) respectively. In this procedure, we obtained new 0-entries at four places. With the renormalization axiom, we get four more 0-entries in identity (11.5.5).

With the renormalization axiom applied to identities (11.5.2) and (11.5.6), we get identity (11.5.11) in the following form first, where the entries $*$ denote some non-zero values as usual:

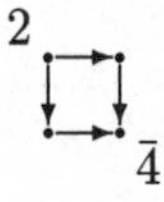

$$\begin{array}{cc} 2 & \\ & \square \\ & \bar{4} \end{array} = \begin{array}{c} \\ \overline{3_1} \\ \overline{3_2} \\ \bar{5} \end{array} \begin{array}{c} \begin{array}{ccc} 1 & 3_1 & 3_2 \end{array} \\ \begin{pmatrix} 0 & & \\ * & & \\ & * & 0 \end{pmatrix} \end{array}.$$

Together with the unitarity axiom, this implies identity (11.5.11). Similarly, we get identity (11.5.15) from identities (11.5.4) and (11.5.7). In these steps, we obtained four new 0-entries, and the renormalization axiom gives four 0-entries in identity (11.5.5). For identity (11.5.5), we now have the following form:

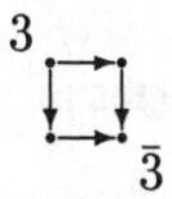

$$\begin{array}{cc} 3 & \\ & \square \\ & \bar{3} \end{array} = \begin{array}{c} \\ \overline{2_1} \\ \overline{2_2} \\ \overline{4_1} \\ \overline{4_2} \\ \bar{6} \end{array} \begin{array}{c} \begin{array}{ccccc} 2_1 & 2_2 & 4_1 & 4_2 & 6 \end{array} \\ \begin{pmatrix} x_1 & * & & 0 & * \\ & y_1 & 0 & & 0 \\ 0 & & & y_2 & \\ & 0 & x_2 & * & 0 \\ 0 & & * & 0 & \end{pmatrix} \end{array}. \tag{11.5.18}$$

The unitarity axiom implies $x_1 = x_2 = 0$. These imply identities (11.5.9) and (11.5.17) with the renormalization axiom and the unitarity. Then we get $y_1 = y_2 = 0$ by looking at identities (11.5.9) and (11.5.17) together with the renormalization axiom. Thus we have proved all 16 identities.

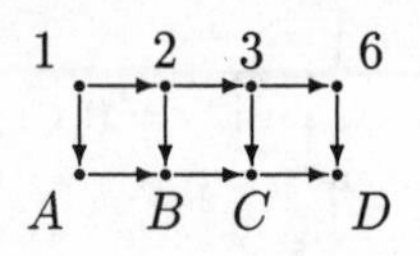

FIG. 11.48. An E_6-partition function

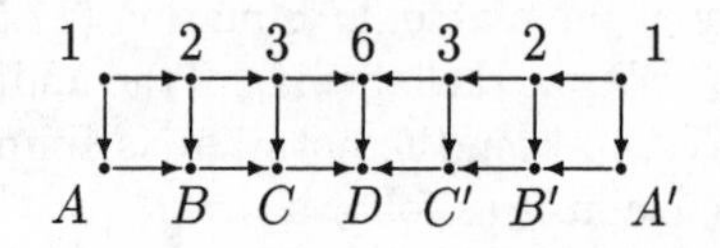

FIG. 11.49. Another E_6-partition function

We now study the double sequence of string algebras arising from this connection. We set

$$x = (1 \to 2 \to 3 \to 6, 1 \to 2 \to 3 \to 6) \in A_{0,3}\,.$$

Our aim is to prove that this string x commutes with $(p+q)A_{3,0}(p+q) \cong \mathbb{C}\oplus\mathbb{C}$ For this purpose, we look at the partition function in Fig. 11.48.

We investigate when the partition function in Fig. 11.48 is nonzero. We classify cases according to the choice of the corner vertex D.

1. $D = \bar{1}$. In this case, in order to get a non-zero value, we must have $C = \overline{2_2}$ Then we must have $B = \overline{3_1}$ and hence $A = \bar{6}$.

2. $D = \bar{3}$. We have three possibilities $\overline{2_1}$, $\overline{4_1}$, $\bar{6}$ for C. In order to get a non-zero value, we must have $B = \overline{3_2}, \bar{5}$ in each of the three cases. Then $A = \bar{4}$.

3. $D = \bar{5}$. In this case, in order to get a non-zero value, we must have $C = \overline{4_2}$ Then we must have $B = \overline{3_1}$ and hence $A = \bar{6}$.

With this observation, we next look at the partition function in Fig. 11.49 We can conclude that this partition function has value 0 if A and A' are different Then the equivalence between Conditions 1 and 3 in Theorem 11.17 implies tha projections $p, q \in (p+q)A_{3,0}(p+q)$ commute with any $(p+q)A_{0,l}$, because th algebras $(p+q)A_{0,l}$ are generated by the horizontal Jones projections and a singl element x.

This means that the projections p, q in (11.5.1) commute with any $(p+q)A_{0,}$ which is isomorphic to $A_{0,l}$ and hence that p, q are in the higher relative commutants of the subfactor given by the bi-unitary connection on the Dynkin diagram E_6. This finally means that the bi-unitary connection on E_6 is flat because Condition (1) in Theorem 11.17 holds. Thus we have proved Theorem 11.24 in the case of E_6.

Remark Suppose that the connections on E_7 were flat. Then for any projection $p \in A_{k,0}$ of the double sequence of string algebras, we would find a bi-unitary connection on the four graphs by looking at the algebras $pA_{k,l}p$ as above, but it turns out that we cannot find the four graphs consistently in the above way. This gives another proof of non-flatness of the connections on E_7. The same method applies to the connections on D_{2n+1}.

Exercise 11.1 Prove the flatness of the connections on D_{2n} with the above method.

In the case of E_7, we already know that the two connections are not flat, but they do produce subfactors with index $4\cos^2 \pi/18$ as in Section 11.3. The principal graphs of the subfactors must be either A_{17} or D_{10}, but it is not clear which the right ones are. The above method (with more computations) shows that D_{10} is the correct principal graph for both connections. See (Evans and Kawahigashi 1994b) for the computations.

Remark In the cases of E_6 and E_8, we have two bi-unitary connections, and they are the complex conjugates of each other. (Note that if we have a paragroup, then taking the complex conjugate of the connection values gives another paragroup. This new paragroup may or may not be isomorphic to the original paragroup.)

We study what relation there is between two subfactors arising from mutually complex conjugate connections. In the string algebra construction of a subfactor, we have the transpose operation of (finite dimensional) matrices on each A_{kl}. We regard this operation as a map from one system of the double sequence of string algebras with a connection to another system with the complex conjugate connection. Then it is easy to see that this map is compatible with the identifications under the connections. This means that the two subfactors are mutually anti-isomorphic, because the transpose operation reverses the order of the multiplication. Thus the two subfactors with the principal graph E_6 (or E_8) arising from the two conjugate connections are not isomorphic, but mutually anti-isomorphic.

We next explain the situation at and beyond the index value 4. The cases of index 4 are still similar to the above except for that we now have infinite graphs. As an analogue of Theorem 10.6, if the Perron–Frobenius eigenvector is equal to 2, we have the following graphs (*extended Dynkin diagrams*); three infinite graphs $A_{\infty,\infty}$ (Fig. 11.50), A_∞ (Fig. 11.51), D_∞ (Fig. 11.52), two infinite series $A_{2n-1}^{(1)}$ (Fig. 11.53), $D_n^{(1)}$ (Fig. 11.54), and three exceptional graphs $E_6^{(1)}$ (Fig. 11.55), $E_7^{(1)}$ (Fig. 11.56), $E_8^{(1)}$ (Fig. 11.57).

When the principal graph is one of the three infinite graphs, it is known that we have only one isomorphism class for the double sequence of higher relative commutants. (See (Popa 1990c), (Popa 1994a), and an argument in Section 11.8.) Then the main theorem in (Popa 1994a) implies that for each of the three infinite graphs, we have only one isomorphism class of subfactors of the hyperfinite II_1 factor.

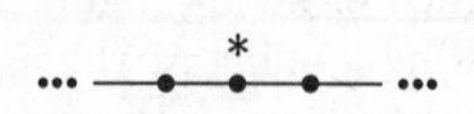

FIG. 11.50. Extended Dynkin diagram $A_{\infty,\infty}$

FIG. 11.51. Extended Dynkin diagram A_{∞}

Next consider the $A_{2n-1}^{(1)}$ and $D_n^{(1)}$ cases. If the principal graph is one of these, we can show that the dual principal graph is the same and the four graphs are connected trivially as above. We then get a one-parameter family of bi-unitary connections in each case. We can prove that only n connections [resp. $n-2$ connections] satisfy the flatness axiom for $A_{2n-1}^{(1)}$ [resp. $D_n^{(1)}$]. Popa's generating theorem again implies that the number of isomorphism classes of subfactors of the hyperfinite II_1 factor with principal graph $A_{2n-1}^{(1)}$ [resp. $D_n^{(1)}$] is n [resp. $n-2$]. It is also possible to reduce this problem to a classical problem of classifying dihedral group actions on the hyperfinite II_1 factor which has been solved by (Jones 1980b). See (Popa 1990c) and (Izumi and Kawahigashi 1993) for more details on $A_{2n-1}^{(1)}$ and $D_n^{(1)}$.

In the case of $E_6^{(1)}$, $E_7^{(1)}$, $E_8^{(1)}$, we again conclude that all four graphs are the same and connected trivially. In each case, we have only one bi-unitary connection. It can be shown that these are flat. Existence of subfactors with these principal graphs, hence existence of paragroup structures on these graphs, was first proved in 4.7.d of (Goodman *et al.* 1989). It is again possible to reduce this case to a problem of classifying actions of finite groups such as the tetrahedron group, the cube group, and the icosohedron group. This is due to M. Izumi.

At the end of this section, we discuss what is known for the case index bigger than 4. Recently (Haagerup 1994) has shown a list of candidates of principal graphs for the index range $(4, 3+\sqrt{3}]$ by a subtle combinatorial argument. In

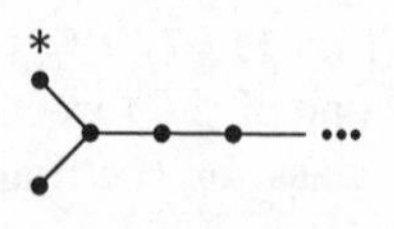

FIG. 11.52. Extended Dynkin diagram D_{∞}

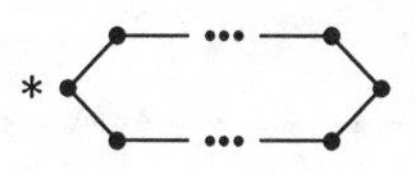

FIG. 11.53. Extended Dynkin diagram $A_{2n-1}^{(1)}$ with $2n$ vertices

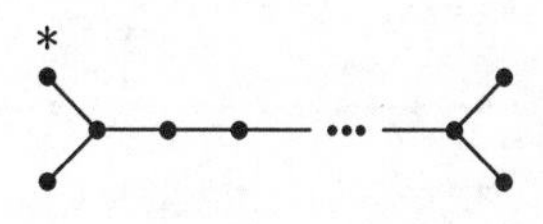

FIG. 11.54. Extended Dynkin diagram $D_n^{(1)}$ with $n+1$ vertices

particular, he has shown that if the index value is between 4 and $(5+\sqrt{13})/2$, then the principal graph must be A_∞. Haagerup–Schou and Ocneanu constructed many subfactors of the hyperfinite II_1 factors with trivial relative commutants with indices in this range with the method described in Section 11.3. (See (Schou 1990).) The result of (Cvetković *et al.* 1982) implies that the smallest index value above 4 arising from the method in Section 11.3 is $4.026\cdots$ and this value is the square of the Perron–Frobenius eigenvalue of the graph in Fig. 11.58. (See also Appendix I of (Goodman *et al.* 1989).) Ocneanu's construction of a subfactor with trivial relative commutant and this index value based on the graph in Fig. 11.58 is described in (Schou 1990). That is, Ocneanu constructed a bi-unitary connection on the four graphs in the setting of Section 11.3, where the two graphs $\mathcal{G}_1$ and $\mathcal{G}_3$ are taken to be the graph in Fig. 11.58, and the other graphs $\mathcal{G}_0$ and $\mathcal{G}_2$ are different. Because we know that the principal graph is A_∞, his bi-unitary connection does not satisfy the flatness property for any choice of $*$. According to a recent announcement in (Popa 1991), this is the smallest possible value of the Jones index of a subfactor of the hyperfinite II_1 factor with trivial relative commutant.

S. Popa also proved that any value bigger than 4 can be realized as a Jones index of some subfactor with trivial relative commutant of a II_1 factor which is not hyperfinite in (Popa 1993). Thus, if we drop the hyperfiniteness condition, we

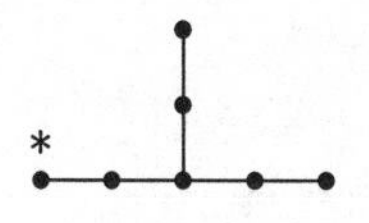

FIG. 11.55. Extended Dynkin diagram $E_6^{(1)}$

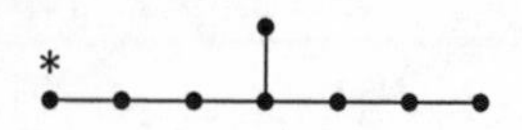

FIG. 11.56. Extended Dynkin diagram $E_7^{(1)}$

FIG. 11.57. Extended Dynkin diagram $E_8^{(1)}$

have a complete answer of possible values of the Jones index of general subfactors with trivial relative commutant.

It has been shown in (Haagerup 1994) that the index value $(5+\sqrt{13})/2$ is realized for a subfactor of a hyperfinite II_1 factor with trivial relative commutant and finite depth. (Also see (Asaeda and Haagerup, preprint 1998).) This is called the *Haagerup subfactor.* He has also proved that at this index value, we have only two, mutually dual, (finite) paragroups. The pair of the principal and dual principal graphs in this case is given as in Figs 11.59 and 11.60. The four graphs are now connected as in Fig. 11.61, where the top four vertices are naturally identified with the bottom four vertices. It is a good exercise to verify that we have only one bi-unitary connection up to isomorphism on the system of these four graphs. Haagerup's result means that this connection satisfies the flatness axiom.

Furthermore, a recent work of (Asaeda and Haagerup, preprint 1998) has proved that we have a subfactor of finite depth with index $(5+\sqrt{17})/2$ corresponding to the graphs (3) in page 36 of (Haagerup 1994). This is called the *Asaeda–Haagerup subfactor.* This paragroup and its dual are the only ones with this index value. (The paper (Asaeda and Haagerup, preprint 1998) has a general theory useful for this kind of existence problem, but it is too technical to be presented here.)

A recent numerical computation in (Ikeda, preprint 1997) further suggests that the connection for graphs (2) with $n=7$ in page 36 of (Haagerup 1994) is also flat. It is fairly easy to write down the connection for this graph and the

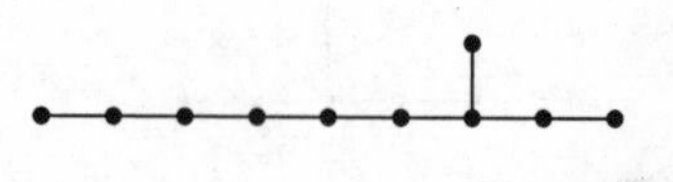

FIG. 11.58. Ocneanu's graph

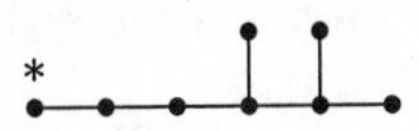

FIG. 11.59. The principal graph of Haagerup

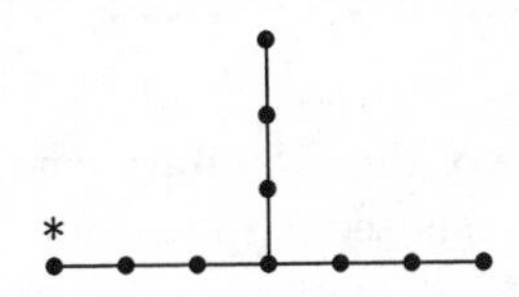

FIG. 11.60. The dual principal graph of Haagerup

index value is is $4.377\cdots$. This, together with the above examples of Asaeda and Haagerup, suggests that we have many 'exotic' paragroups which do not come from ordinary quantum groups.

Recent work of (Bisch and Haagerup 1996) shows that such a listing of all the possible graphs is unlikely to be possible for index values slightly larger, such as around 6, because combinatorial complexity increases exponentially.

Exercise 11.2 Suppose we have a sequence of projections $\{e_j\}_{j\geq 0}$ satisfying the Temperley–Lieb–Jones relations in the following sense, where $\beta = 2\cos(\pi/(n+1))$ with $n = 2, 3, \ldots$.

1. $e_j e_k = e_k e_j$, if $|j-k| \geq 2$.
2. $e_j e_{j\pm 1} e_j = \beta^{-2} e_j$.
3. We have a trace on the algebras generated by e_js and for $x \in \langle e_0, e_1, \ldots, e_k\rangle$ we have $\mathrm{tr}(x e_{k+1}) = \beta^{-2}\mathrm{tr}(x)$.

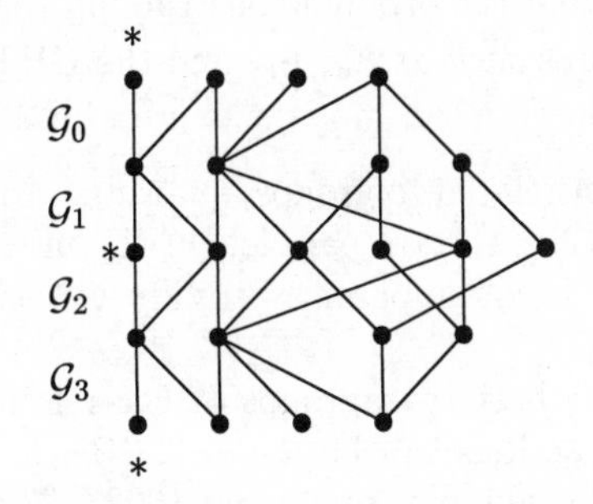

FIG. 11.61. The way the four graphs are connected

Let M be the weak closure of the algebra generated by $\{e_j\}_{j\geq 0}$ in the GNS-representation with respect to the trace and N be its von Neumann subalgebra generated by $\{e_j\}_{j\geq 1}$.

Prove that M, N are both hyperfinite II_1 factors and identify this subfactor $N \subset M$ with the Jones subfactors with principal graph A_n.

Exercise 11.3 Verify that the vertex marked with $\circ$ in Fig. 11.44 violates the $2\cos\pi/n$-rule (Corollary 10.8).

Exercise 11.4 Find all the bi-unitary connections on the extended Dynkin diagrams $D_n^{(1)}$.

Exercise 11.5 Prove flatness of the bi-unitary connections on the Dynkin diagram E_8 with the method of this section for computing the principal graph of the E_6 connections.

Exercise 11.6 Prove that the subfactors arising from the bi-unitary connections on the Dynkin diagram E_7 have principal graph D_{10} with the above method.

Exercise 11.7 Construct a bi-unitary connection on the graphs $\mathcal{G}_j$ where the two graphs $\mathcal{G}_1$ and $\mathcal{G}_3$ are taken to be the graph in Fig. 11.58, and $\mathcal{G}_0$ and $\mathcal{G}_2$ are some different graphs.

Exercise 11.8 Compute the bi-unitary connection on the graphs in Fig. 11.61.

11.6 The Goodman–de la Harpe–Jones subfactors

The point in the discussion at the end of the previous section is that the unitarity and the renormalization axioms are rather easy to satisfy while the flatness axiom is much harder to satisfy. In other words, it is much easier to construct a subfactor (of the hyperfinite II_1 factor with trivial relative commutant) for a given index value than to construct or determine the paragroup for that index value. Indeed, the known paragroups fall in the following five categories:

1. classical cases arising from group/Hopf algebra actions;
2. 'diagonal' paragroups arising from Wess–Zumino–Witten models such as $SU(n)_k$ (Section 13.3);
3. paragroups arising from the orbifold construction (Section 13.4);
4. exceptional paragroups such as E_6, E_8, and the GHJ construction;
5. genuine exceptions.

Categories (1)–(4) seem related to groups, quantum groups, Hopf algebras in some way and categories (2)–(4) have parallels in the modular invariant theory of Section 8.8, whilst (5) seems apparently totally unrelated to the theory of modular invariants.

We have seen the most basic paragroups of Case 1 in Section 10.6. Other constructions from group actions can be found in (Bisch and Haagerup 1996), (Goodman *et al.* 1989), (Kosaki and Yamagami 1992), (Wassermann 1988). See (Longo 1994a), (Szymanski 1994) for Hopf algebra cases. We will discuss Cases 2 and 3 in Chapter 13.

The Haagerup example mentioned at the end of the previous section is one example in Case 5. Here we discuss a series of examples for Case 4 which has appeared in (Goodman *et al.* 1989), and consider further examples in the next section.

Let G be one of the A-D-E Dynkin diagrams. Then choose an arbitrary vertex and mark it with $*_G$. (This is not necessarily $*$ in the principal graph.) Label the l-string algebra on the graph G with the initial vertex $*_G$ as

$$A_{1,0} \subset A_{1,1} \subset A_{1,2} \subset A_{1,3} \subset \cdots.$$

Note that we have a unique trace on $\bigcup_l A_{1,l}$ as in Section 11.3. Let G' be the Dynkin diagram of type A with the same Coxeter number as G. Mark one of its two endpoints with $*_{G'}$. We label the l-string algebra on the graph G' with the initial vertex $*_{G'}$ as

$$A_{0,0} \subset A_{0,1} \subset A_{0,2} \subset A_{0,3} \subset \cdots.$$

Note again that we have a unique trace on $\bigcup_l A_{0,l}$. In both series of string algebras, we can define (horizontal) Jones projections. Because $A_{0,l}$ is generated by the first $l-1$ Jones projections as in the proof of Lemma 11.8, we get an embedding of $A_{0,l}$ into $A_{1,l}$ by identifying the Jones projections on G' and G. In this way, we get the following sequence of embeddings and it is easy to see that they make a series of commuting squares. (Use Condition (7) of Proposition 9.51.)

$$\begin{array}{ccccccccc} A_{0,0} & \subset & A_{0,1} & \subset & A_{0,2} & \subset & A_{0,3} & \subset & \cdots \\ \cap & & \cap & & \cap & & \cap & & \\ A_{1,0} & \subset & A_{1,1} & \subset & A_{1,2} & \subset & A_{1,3} & \subset & \cdots \end{array} \qquad (11.6.1)$$

Then it is easy to see that this series gives the top two lines of the double sequence of string algebras arising from a bi-unitary connection as in Definition 11.3 as in the case of the sequence (11.5.1). Thus we get a subfactor $A_{0,\infty} \subset A_{1,\infty}$ as in Section 11.3. We call these subfactors the *Goodman–de la Harpe–Jones (GHJ) subfactors* because this construction first appeared in (Goodman *et al.* 1989).

Example 11.25 It is illuminating to work on a concrete example, so let G be E_6 with the choice of $*_G = C_1$ and then we have $G' = A_{11}$ with $*_{G'} = B_1$ as in Fig. 11.62. (It turns out that this is the 'most interesting' choice of a pair of A-D-E graphs with marked vertices.)

Then in the notation of Definition 11.3, we have $\mathcal{G}_0 = A_{11}$ and $\mathcal{G}_2 = E_6$ with $V_0 = \{B_1, B_3, B_5, B_7, B_9, B_{11}\}$, $V_1 = \{B_2, B_4, B_6, B_6, B_{10}\}$, $V_2 = \{C_2, C_4, C_6\}$, $V_3 = \{C_1, C_3, C_5\}$. In order to determine the other two graphs $\mathcal{G}_1$ and $\mathcal{G}_3$, we write the two Bratteli diagrams of the two series of the string algebras as in Fig. 11.63.

From Fig. 11.63, we can compute the graphs $\mathcal{G}_1$, $\mathcal{G}_3$ as follows. By looking at level 0 (the top level), we know that the vertex B_1 is connected to C_1 with a single edge on $\mathcal{G}_3$. By looking at level 2 next, we also know that the vertex

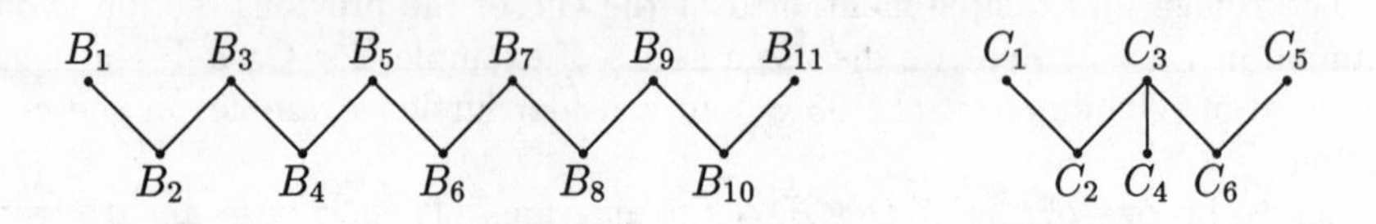

FIG. 11.62. The vertices of A_{11} and E_6

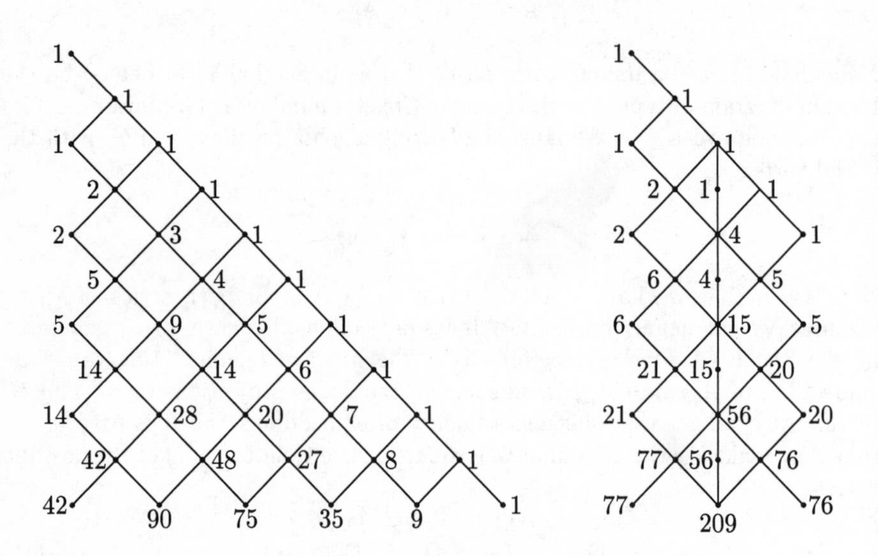

FIG. 11.63. The Bratteli diagrams for A_{11} and E_6

B_3 is connected to C_3 with a single edge on $\mathcal{G}_3$. By looking at level 3 next, we also know that the vertex B_5 is connected to C_3 with a single edge on $\mathcal{G}_3$. By repeating this procedure, we get the graph $\mathcal{G}_3$ as in Fig. 11.64. (It is not clear *a priori* that we get a finite graph in this procedure, and if we get an infinite graph here, our method in Sections 11.3 and 11.4 does not work, but it turns out that we always have a finite graph in this procedure. This is a special property of the *A*-*D*-*E* Dynkin diagrams. See the next section for more discussion on this point.)

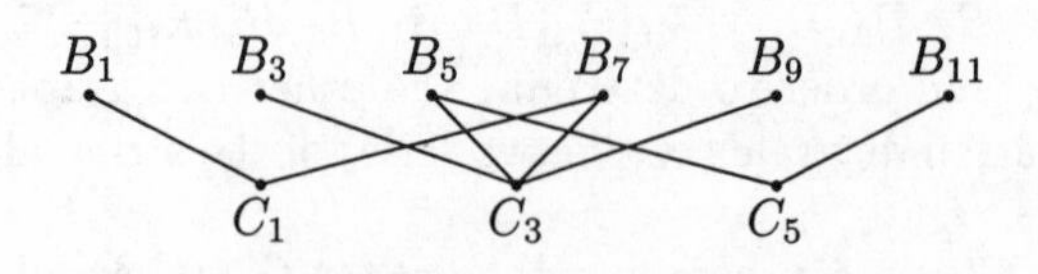

FIG. 11.64. The graph $\mathcal{G}_3$

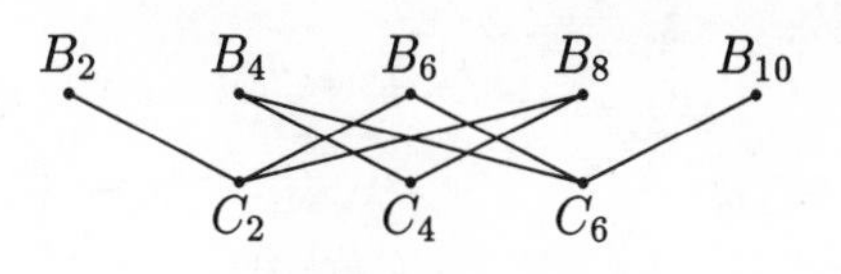

FIG. 11.65. The graph $\mathcal{G}_1$

In a similar process, we also get the graph $\mathcal{G}_1$ as in Fig. 11.65. We now know that we have a bi-unitary connection on the system of these four graphs. (The Perron–Frobenius eigenvector of $\mathcal{G}_0 = G'$ is normalized so that we have $\mu(*_{G'}) = 1$, and that of $\mathcal{G}_2 = G$ is normalized so that the Perron–Frobenius eigenvalue of $\mathcal{G}_3$ is equal to $\mu(*_G)$.) We do not have to write down the values of the connections explicitly, though it is possible by solving the bi-unitarity relations.

We apply the construction in Section 11.3 to get a subfactor. The index of this subfactor is equal to the Perron–Frobenius eigenvalue of the graphs $\mathcal{G}_3$, $\mathcal{G}_1$, and it is $3+\sqrt{3}$ in the above example of E_6. This value is the smallest above 4 among the values given in this construction. This is why we said that the choice of E_6 and $*$ as above is the most interesting. For the other values, see (Goodman *et al.* 1989).

We want to compute the principal graphs of these GHJ subfactors. Because the algebras $A_{0,l}$ are generated by the (horizontal) Jones projections, any element in $A_{0,l}$ commutes with $A_{k,0}$ for any k, which means that our bi-unitary connection is flat with respect to $*_{G'}$. Thus Theorem 11.15 implies that the higher relative commutants of $A_{0,\infty} \subset A_{1,\infty}$ are given by

$$A_{0,0} \subset A_{1,0} \subset A_{2,0} \subset A_{3,0} \subset \cdots.$$

That is, the principal graph of the subfactor is the graph $\mathcal{G}_3$ and thus this subfactor has finite depth, whenever we start with one of the *A*-*D*-*E* Dynkin diagrams and a choice of $*$. In particular, the principal graph of the subfactor in Example 11.25 arising from E_6 is the graph in Fig. 11.64. (This index value $3+\sqrt{3}$ is at one endpoint of the interval of Haagerup's consideration mentioned at the end of the previous section.)

Remark If we choose G to be D_n and $*$ to be its vertex with the lowest Perron–Frobenius eigenvector entry, then the graph G' is A_{2n-3}. In this case, the graph $\mathcal{G}_1$ and $\mathcal{G}_3$ are both disjoint unions of copies of the Dynkin diagram A_3, so this gives an example of the bi-unitary connection as in Definition 11.3 where we have disconnected graphs. These subfactors have index 2.

It is important to note that the dual principal graph of the subfactor in Example 11.25 arising from E_6 is *not* the graph in Fig. 11.65. Indeed, the graph in Fig. 11.65 violates the $2\cos\pi/n$-rule (Corollary 10.8), so it cannot be a (dual) principal graph of any subfactor. In general, it is a hard problem to determine

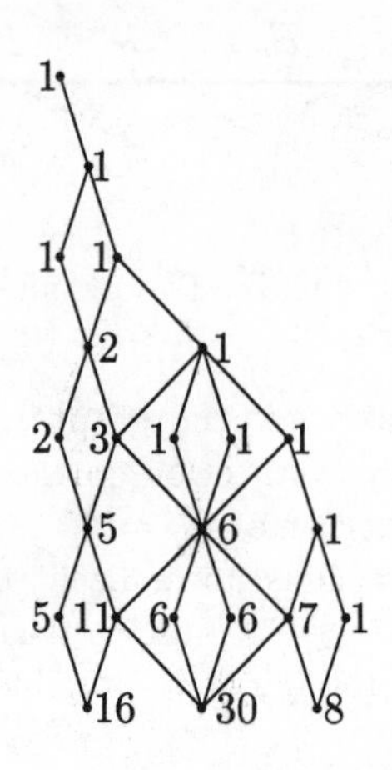

FIG. 11.66. The Bratteli diagram for $N' \cap M_k$

the dual principal graph of a subfactor constructed as above, but we can solve it for the subfactor $N \subset M$ in Example 11.25 as follows.

We first draw the Bratteli diagram for the higher relative commutants $N' \cap M_k$ as in Figure 11.66.

Then the odd levels of this tower must coincide with the odd levels of the other higher relative commutants $M' \cap M_k$ as in the argument for the cases with index less than 4. (The top level is counted as level 0 here.) This fact determines the dual principal graph step by step, and we conclude that the dual principal graph is the same as principal graph in Fig. 11.64. (In general, this method does not determine the dual principal graph from the principal graph uniquely. We happen to be fortunate here.)

At the end of this section, we discuss the case $G = A_n$ in the above construction of the GHJ subfactors. We choose $*_G$ to be the mth vertex from the end. Then it turns out that the index value of the resulting subfactor is $\left(\sin^2 m\pi/(n+1)\right) / \left(\sin^2 \pi/(n+1)\right)$. This subfactor is described naturally as follows.

We start with the double sequence $\{A_{kl}\}_{kl}$ of string algebras arising from the flat connection on the Dynkin diagram A_n in Section 11.5. Then we have a minimal projection p corresponding to the mth vertex in $A_{m-1,0} = A'_{0,\infty} \cap A_{m-1,\infty}$. Then the sequence

$$\begin{array}{ccccccc} pA_{0,0} & \subset & pA_{0,1} & \subset & pA_{0,2} & \subset & \cdots \\ \cap & & \cap & & \cap & & \\ pA_{m-1,0}p & \subset & pA_{m-1,1}p & \subset & pA_{m-1,2}p & \subset & \cdots \end{array}$$

gives the same sequence as in (11.6.1) for $G = A_n$ and $*_G$ to be the mth vertex of A_n. Thus the resulting subfactor is identified with $pN \subset pM_{m-2}p$, where

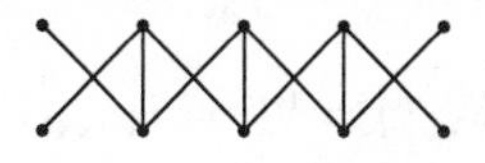

FIG. 11.67. The principal graph for the case of A_9, $m = 3$

the subfactor $N \subset M$ has the principal graph A_n and the minimal projection $p \in N' \cap M_{m-2}$ corresponds to the mth vertex of A_n as above. That is, the GHJ subfactors for A_n graphs are given as cut-down subfactors of the basic construction of subfactors with principal graph A_n.

We show as an example the principal graph for the case $n = 9$, $m = 3$ in Fig. 11.67. The index of this subfactor is $(\sin^2 3\pi/10)/(\sin^2 \pi/10)$. This is also the subfactor corresponding to the N-N bimodule represented by the third vertex of A_9.

Exercise 11.9 Compute the principal graphs of the other GHJ subfactors constructed from Dynkin diagrams.

11.7 Intertwining Verlinde formulae

We rephrase the GHJ models in a setting which allows us to understand higher rank possibilities ($SU(N) : N > 2$), beginning with a systematic way of understanding the intertwining or embedding matrices between a Dynkin diagram and A_n diagrams in terms of intertwining Verlinde formulae.

The Ising model based on the graph A_3 can be generalized to models based on the $\mathcal{A}^{(n)}$ graphs with vertices $P_{++}^{(n)}$,, the set of level k $(= N - n)$ of the Kač–Moody algebra $A_{N-1}^{(1)} = su(N)^\wedge$. The associated connection is a degenerate case of the Boltzmann weights of an integrable model. We regard the GHJ subfactors as embeddings of the path algebras of the face operators (in this case the Temperley–Lieb algebra) in other path algebras of bipartite graphs. The generalization will be to embed $SU(N)_q$ face algebras into path algebras, such as those on N-colourable graphs, but before doing so let us look again at the $N = 2$ case.

We have a Dynkin diagram G, embedded in the $A_{n-1} = \mathcal{A}^{(n)}$ graph abbreviated as A, via an intertwining matrix V:

$$AV = VG. \tag{11.7.1}$$

The labels of the orthogonal eigenvectors $\{\psi^\rho\}$ of G are called the exponents of G, and form a subset of the exponents of A (with multiplicity), with eigenvectors $\phi^\rho = (S_{\rho i})$. Then any linear combination of $\phi^\rho \otimes \overline{\psi^\rho}$ will intertwine A and G. We consider the *intertwining Verlinde formula*

$$V_{\lambda a}{}^{b} = \textstyle\sum_\rho \psi_a^\rho \bar{\psi}_b^\rho \phi_\lambda^\rho / \phi_*^\rho = \sum_\rho \psi_a^\rho \bar{\psi}_b^\rho S_{\lambda\rho}/S_{*\rho}, \tag{11.7.2}$$

where ${V_{*a}}^b = \delta_{ab}$, ${V_{1a}}^b = G_{ab}$. However, with this approach, and again in the higher rank case, it is not clear that the ${V_{\lambda a}}^b$ are positive and integral. Integrality will follow from

$$[{V_{\lambda a}}^b] = V_\lambda = S_{\lambda-1}(G), \tag{11.7.3}$$

which is a consequence of the intertwining relation, and where S_n are the Chebyshev polynomials of the second kind.

Let Γ be a locally finite bi-partite graph with incidence matrix Δ. In Section 11.6 we showed simply by considering dimension groups that there are embeddings of the Temperley–Lieb algebra $A(A_n)$ in $A(\Gamma)$, where n is finite if $\|\Gamma\| < 2$, and $n = \infty$ otherwise. Also from Section 11.6, we see that at the dimension group level this is related to the question of expressing the generators of $K_0(A(A_\infty))$ essentially the Chebyshev polynomials of the second kind S_n (associated with the graph A_∞) as linear combinations of the rational functions $\{\phi_v \mid v \in \Gamma^{(0)}\}$ (associated with the graph Γ), with non-negative integer coefficients.

Let $V(\Gamma)$ denote the free module over $\mathbb{Z}$, generated by the vertices of Γ identifying an element $a \in V(G)$ as $a = (a_v)$, $a_v \in \mathbb{Z}$, $v \in \Gamma^{(0)}$. We let $V(\Gamma)_+$ be the positive cone generated by the vertices. The incidence matrix Δ acts on $V(\Gamma)$. We seek a sequence $a_n = (a_{nv})$ in $V(\Gamma)_+$ with

$$S_n = \textstyle\sum_{v\in\Gamma^{(0)}} a_{nv}\phi_v, \quad n = 0, 1, 2, \dots . \tag{11.7.4}$$

We claim that the a_n can be found from the recurrence relations for (S_n) and (ϕ_v) as follows. In the first place $S_0 = \phi_* = 1$, and so we can take $a_{0v} = \delta_{v*}$. Next we use the recurrence relations

$$xS_n = \Delta_{A_\infty} S_n, \quad x\phi_v = \Delta_\Gamma \phi_v \tag{11.7.5}$$

to see that if $a_0, \dots, a_n$ are given, then

$$xS_n = \textstyle\sum_v a_{nv} x\phi_v = \sum_v a_{nv}(\sum_w \Delta_{vw}\phi_w) = \sum_w(\sum_v \Delta_{vw} a_{nv})\phi_w \tag{11.7.6}$$

and so

$$S_{n+1} = xS_n - S_{n-1} = \textstyle\sum_w(\sum_v \Delta_{vw} a_{nv} - a_{n-1w})\phi_w \, . \tag{11.7.7}$$

Thus we can take

$$a_{n+1} = \Delta a_n - a_{n-1}, \quad \text{for } n \geq 0 \tag{11.7.8}$$

with $a_{-1} = 0$.

This calculation (11.7.8) is illustrated in Fig. 11.68 for $\Gamma = E_\infty$. The numbers appearing in level n of $\hat{E}_\infty$ give the coefficients a_{nv} of the polynomial ϕ_v (corresponding to vertex v in $E_\infty^{(0)}$) in the expansion of the Chebyshev polynomial S_n. One obtains the coefficients at level $n+1$ by pushing forward on $\hat{E}_\infty$ those on level n to level $n+1$, then adding the values obtained at each vertex of level $n+1$, and then subtracting the number from the same vertex on level $n-1$. The question

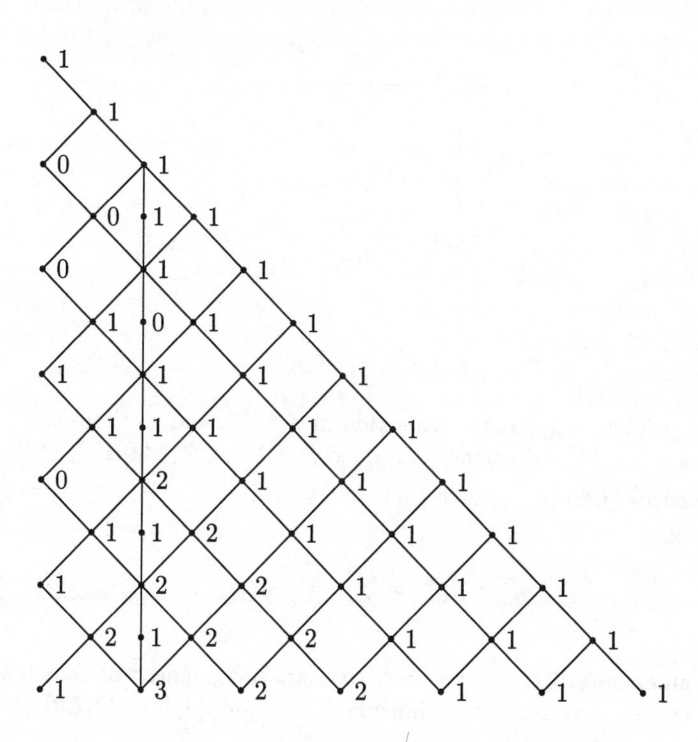

FIG. 11.68. The Bratteli diagram for $A(A_\infty) \subset A(E_\infty)$

then arises of whether a_n, as defined above, lie in $V(\Gamma)_+$ (i.e. whether $a_{nv} \geq 0$ for all n, v). If a_n in $V(\Gamma)$ satisfy $a_{-1} = 0$, $a_1 = 1$, and $a_{n+1} = \Delta a_n - a_{n-1}$, then a_n satisfy the same recurrence relation as the Chebyshev polynomials S_n and so $a_n = S_n(\Delta)a_0$ for $n \geq 0$. It is thus important to know when $S_n(\Delta) \geq 0$. Here a linear map, $T = [T_{vw}]$, between $V(\Gamma_1)$ and $V(\Gamma_2)$ is said to be positive, written $T \geq 0$ if T maps $V(\Gamma_1)_+$ into $V(\Gamma_2)_+$, or equivalently $T_{vw} \geq 0$, for all v, w.

Lemma 11.26 *Suppose that Γ_1 and Γ_2 are locally finite connected graphs with distinguished vertices $*_1$ and $*_2$ respectively. Let $(e_n)_{n\in\mathbb{N}}$, $(f_n)_{n\in\mathbb{N}}$ denote canonical families of Jones projections with common parameter β^{-2} in $A(\Gamma_1)$, and $A(\Gamma_2)$ respectively, and $\pi : A(\Gamma_1) \to A(\Gamma_2)$ a unital embedding such that:*

(a) The diagram in Fig. 11.69 commutes for all n, where $\pi_n = \pi|_{A(\Gamma_1)_n}$, and h_n, k_n are standard inclusions.

(b) $\mathrm{tr}_2 \cdot \pi_n = \mathrm{tr}_1$, where tr_i is a Markov trace on $A(\Gamma_i)$, $i = 1, 2$.

(c) $\pi(e_n) = f_n$ for all $n \geq 1$, (so $\pi_{n+1}(e_n) = f_n$).

Then there exists a positive linear map $A : V(\Gamma_1) \to V(\Gamma_2)$ such that: (1) $A\Delta_1 = \Delta_2 A$, (2) A has no rows, or columns zero, (3) $A*_1 = *_2$.

Proof Let p_i^n denote a minimal projection in $A(\Gamma_1)_n$ corresponding to vertex

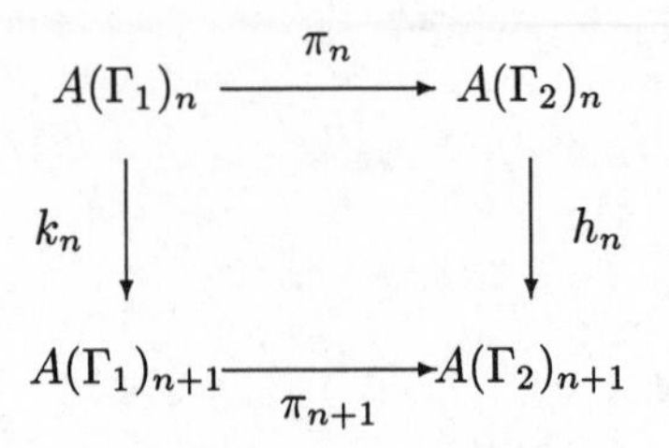

FIG. 11.69.

(i, n) of $\hat{\Gamma}_1$. Then $\pi_n(p_i^n)$ is a projection in $A(\Gamma_2)_n$, and so there are families of equivalent minimal projections $\{q_{j,k(j)}^n \mid k(j) = 1, \ldots, b_{ji}^n\}$ in $A(\Gamma_2)_n$ corresponding to vertices (j, n) in $\hat{\Gamma}_2$, such that

$$\pi_n(p_i^n) = \sum_j \sum_{k(j)=1}^{b_{ji}^n} q_{j,k(j)}^n. \tag{11.7.9}$$

The numbers $\{\beta_{ji}^n\}_j$ are non-negative, are independent of the choice of p_i^n, and are not all zero, since π_n is injective. Now multiplying (11.7.9) by f_{n+1}, we have

$$f_{n+1}\pi_n(p_i^n) = \sum_j \sum_{k(j)=1}^{b_{ji}^n} f_{n+1}q_{j,k(j)}^n,$$

but by (a), and (c)

$$f_{n+1}\pi_n(p_i^n) = \pi_{n+2}(e_{n+1})\pi_n(p_i^n) = \pi_{n+2}(e_{n+1}p_i^n),$$

and so we have

$$\pi_{n+2}(e_{n+1}p_i^n) = \sum_j \sum_{k(j)=1}^{b_{ji}^n} f_{n+1}q_{j,k(j)}^n. \tag{11.7.10}$$

Then by the Markov property of tr_1 and tr_2, we have $\mathrm{tr}_1(e_{n+1}p_i^n) = \beta^{-2}\mathrm{tr}_1(p_i^n)$, and $\mathrm{tr}_2(f_{n+1}q_{j,k(j)}^n) = \beta^{-2}\mathrm{tr}_2(q_{j,k(j)}^n)$. Hence $e_{n+1}p_i^n$ is a minimal projection in $A(\Gamma_1)_{n+2}$ corresponding to vertex $(i, n+2)$ of $\hat{\Gamma}_1$, and $f_{n+1}q_{j,k(j)}^n$ is a minimal projection in $A(\Gamma_2)_{n+2}$ corresponding to vertex $(j, n+2)$ of $\hat{\Gamma}_2$. It follows from (11.7.9) and (11.7.10) that the coefficients occurring in the decomposition of a minimal projection as in (11.7.9) corresponding to vertex (i, n) of $\hat{\Gamma}_1$, $n \geq 1$, is independent of the level n.

Now put $A = (b_{j,i})_{i\in\Gamma_1^{(0)}, j\in\Gamma_2^{(0)}}$, then since $A(\Gamma_1)_0 \cong \mathbb{C} \cong A(\Gamma_2)_0$, and $\pi_0 : A(\Gamma_1)_0 \to A(\Gamma_2)_0$ we see that $A*_1 = *_2$. Note that since π is unital, the rows of A are non-zero. We now show that $A\Delta_1 = \Delta_2 A$. Let $\Delta_k(n)$, $k = 1, 2$, be the finite submatrix of Δ_k, whose rows and columns are labelled by vertices $v \in \Gamma_k^{(0)}$, $k = 1, 2$, with $d(v) \leq n + 1$. Similarly, let $A(n)$ denote the finite submatrix of A whose rows are labelled by $j \in \Gamma_2^{(0)}$ with $d(j) \leq n + 1$, and whose columns are labelled by $i \in \Gamma_1^{(0)}$ with $d(i) \leq n + 1$.

It follows from (a) that for each n we have

$$K_0(h_n)K_0(\pi_n) = K_0(\pi_{n+1})K_0(k_n) . \tag{11.7.11}$$

But it is immediate from the definition of $A(\Gamma_1)$ and $A(\Gamma_2)$ that for n even, $K_0(k_n)$ is the submatrix of $\Delta_1(n)$ mapping even vertices to odd ones. Similar remarks apply to $K_0(k_n)$, $K_0(\pi_n)$, and $K_0(\pi_{n+1})$. Thus (11.7.11) implies that $\Delta_2(n)A(n-1) = A(n)\Delta_1(n)$ holds for every n. Hence $\Delta_2 A = A\Delta_1$. □

Remark When Γ is finite one can always define canonical families of Jones projections in $A(\Gamma)$ as in Definition 11.5. Note that condition (c) in Lemma 11.26 will imply that $\|\Gamma_1\| = \|\Gamma_1\| = \beta$, when Γ_1 and Γ_2 are finite. For general Γ, if there is a positive solution to equation (9.5.1) for some x, then one can define a positive Markov trace and a canonical family of Jones projections in $A(\Gamma)$ with parameter x^{-2} – see the Remark on page 490.

Proposition 11.27 *Suppose that Γ is a locally finite connected graph, with distinguished vertex $*$. Let $\{f_n\}_{n\in\mathbb{N}}$ be the canonical family of Jones projections in $A(\Gamma)$. Define m as follows. For $\|\Gamma\| < 2$, m is given by $\|\Gamma\| = 2\cos(\pi/(m+1))$. For $\|\Gamma\| \geq 2$, take $m = \infty$. Then we can identify $A(A_m)$ with the algebra generated by $\{1, f_1, f_2, \ldots\}$. If we define $\pi : A(A_m) \to A(\Gamma)$ by $\pi(1) = 1$, $\pi(f_n) = f_n$, then π is a unital embedding, and there exists a positive linear map $A : V(A_m) \to V(\Gamma)$ such that: (a) $A\Delta(A_m) = \Delta(\Gamma)A$. (b) A has no rows, or columns zero. (c) $A*_1 = *$, where $*_1 = 0$ is the distinguished vertex of A_m. Let $a_0 = *$, and define $a_k \in V(\Gamma)$ by $a_k = S_k(\Delta(\Gamma))a_0$, $k = 1, 2, 3, \ldots$. Then for $\|\Gamma\| < 2$ we have $A = (a_0, a_1, \ldots, a_{m-1})$, and for $\|\Gamma\| \geq 2$ we have $A = (a_0, a_1, a_2, \ldots)$.*

Proof Now $\pi : A(A_m) \to A(\Gamma)$ defined by $\pi(1) = 1$, $\pi(f_n) = f_n$ is a unital embedding which clearly satisfies the condition of Lemma 11.26 with $*_1 = 0$, and $*_2 = *$ varying over Γ. Hence when m is finite there exists $A = (a_0, a_1, \ldots, a_{m-1})$ with the required properties. Now $A\Delta(A_m) = (a_1, a_0 + a_2, \ldots, a_{m-2})$. Thus $\Delta(\Gamma)A = A\Delta(A_m)$ implies that $a_1 = \Delta(\Gamma)a_0$, $a_{m-2} = \Delta(\Gamma)a_{m-1}$, and $a_{k-1} + a_{k+1} = \Delta(\Gamma)a_k$ for $k = 1, \ldots, m-2$, i.e. $a_k = S_k(\Delta(\Gamma))a_0$, where $a_0 = *$, $k = 1, \ldots, m-1$. The $m = \infty$ case is similar. □

Corollary 11.28 *Let Γ be a connected, locally finite graph with incidence matrix Δ. Suppose that there is a positive solution to equation (9.5.1) for some x. Then if $\|\Gamma\| = 2\cos(\pi/(m+1))$, we have $S_k(\Delta) \gneq 0$, for $k = 0, 1, \ldots, m-1$, and $S_m(\Delta) = 0$. If $\|\Gamma\| \geq 2$, then $S_k(\Delta) \gneq 0$ for $k = 0, 1, 2, \ldots$.*

Proof This follows immediately from Proposition 11.27. □

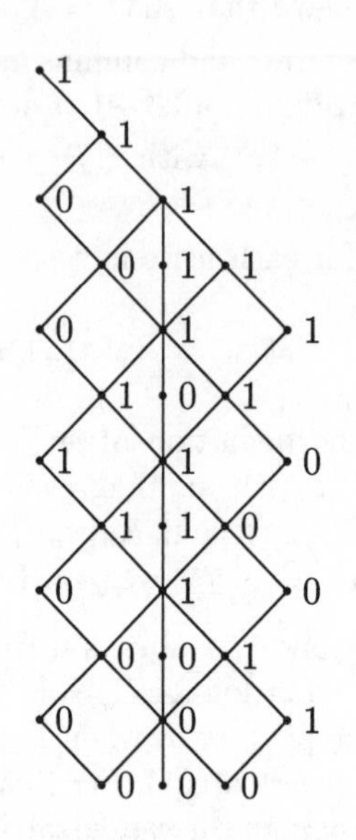

FIG. 11.70. $A(A_{11}) \subset A(E_6)$

In view of the remarks at the beginning of this section, (9.5.10), and (9.5.15), it is clear that the dimension groups maps of the discussion at the end of Section 9.5 lift to the maps at the algebra level given by Proposition 11.27. In fact one can obtain expansions (9.5.25) of $\overline{\overline{P_m}}$ by taking the vectors $a_m = (a_{m\alpha})_\alpha \in V(\Gamma)$ of Proposition 11.27 as coefficients. Note also that the $\overline{\overline{Q_\alpha}}$ that occur in this expansion with non-zero coefficients are associated with level m of $\hat{T}^p_{q,r}$. We illustrate this in Figs 11.70,11.71,11.81 for the cases $\Gamma = E_6, E_7, E_8$. The graphs E_6, E_7, E_8 have norm $2\cos(\pi/m)$, with $m = 12, 18, 30$ respectively. Thus we have embeddings of $A(A_{11})$, $A(A_{17})$, $A(A_{29})$ in $A(E_6)$, $A(E_7)$, and $A(E_8)$ respectively. The graph E_9 has norm 2, and so we have an embedding of $A(A_\infty)$ in $A(E_9)$.

In a rational conformal field theory with primary fields $\{\phi_i\}$ we have the operator product expansion

$$\phi_j * \phi_k = \textstyle\sum_i N_{jk}{}^i \phi_i$$

where the non-negative integers $N_{jk}{}^i$ are the fusion rules. Since this product is associative, the algebra generated by the $\{\phi_i\}$ with this product will have a representation as linear operators. In fact the matrices $N_j = (N_{jk}{}^i)_{k,i}$ will afford such a representation, i.e. $N_j N_k = \sum N_{jk}{}^i N_i$.

The matrices N_j are symmetric, mutually commuting, non-negative, and integral. Thus they will generate a commutative ring of symmetric matrices over $\mathbb{Z}$, which when equipped with the positive cone $\mathrm{lin}_{\mathbb{N}}\{N_i\}$ will be a partially ordered

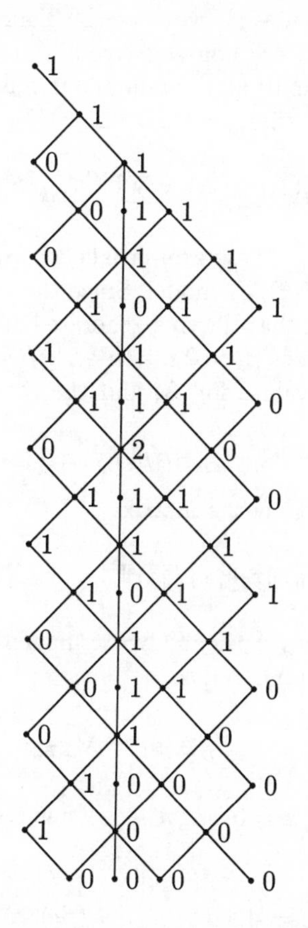

FIG. 11.71. $A(A_{17}) \subset A(E_7)$

subring of $M_n(\mathbb{Z})$ where n is the number of primary fields.

We will now show that for certain graphs Γ, one can use the rational functions $\{\phi_v \mid v \in \Gamma^{(0)}\}$ defined in Section 9.5 to construct such families of matrices, and that the ring they generate is $\mathbb{Z}[\Delta]$, where Δ is the incidence matrix of Γ. It is then an easy matter to find a unitary matrix S that diagonalizes the fusion rules.

Example 11.29 Let $\Gamma = A_n$, then the incidence matrix is given by

$$\begin{pmatrix} 0\,1 & & & \\ 1\,0\,1 & & & \\ \;\;1\,0 & & & \\ & & \ddots & 1 \\ & & 1 & 0 \end{pmatrix}. \tag{11.7.12}$$

Now since $\|\Gamma\| = 2\cos(\pi/n+1)$, we know by Lemma 11.26 that the matrices $S_j(\Delta)$, $j = 0, 1, \ldots, n-1$, are non-negative and non-zero. It is clear also that they are symmetric and mutually commute. It follows immediately that they satisfy the formal product rule:

$$S_j(\Delta)S_k(\Delta) = \textstyle\sum_{i=0}^{n-1} N_{jk}{}^{i} S_i(\Delta) \tag{11.7.13}$$

where $S_j(\Delta) = (N^i_{jk})^{n-1}_{k,i=0}$. These are, in fact, the same rules that occur in the Wess–Zumino–Witten $SU(2)_{n-1}$ model. Since the characteristic polynomial for Δ is $S_n(x)$, it is easy to see that the ring generated over $\mathbb{Z}$ by the $\{S_j(\Delta)\}$ is $\mathbb{Z}[\Delta]$. Now $S_n(x)$ has distinct zeros $\beta_k = 2\cos((k+1)\pi/(n+1))$, $k = 0, 1, 2, \ldots, n-1$. Thus these are the eigenvalues for Δ, and the corresponding eigenvectors are given by

$$x_k = (S_0(\beta_k), S_1(\beta_k), \ldots, S_{n-1}(\beta_k))^t$$

for $k = 0, 1, 2, \ldots, n-1$. Thus the matrix

$$S = (x_0/\|x_0\|, x_1/\|x_1\|, \ldots, x_{n-1}/\|x_{n-1}\|)$$

is unitary, and diagonalizes Δ, i.e. $S^*\Delta S = \mathrm{diag}(\beta_0, \beta_1, \ldots, \beta_{n-1})$. The x_k will also be eigenvectors of $S_j(\Delta)$, and so

$$S_j(\Delta)x_k = \lambda^j_k x_k$$

where $\lambda^j_k = S_j(\beta_k)$, for $j, k = 0, \ldots, n-1$. Thus the same matrix S will diagonalize each $S_j(\Delta)$, and

$$S^* S_j(\Delta) S = \mathrm{diag}(S_j(\beta_0), S_j(\beta_1), \ldots, S_j(\beta_{n-1})).$$

Note also that by (11.7.13) the eigenvalues λ^j_k satisfy

$$\lambda^i_k \lambda^j_k = \textstyle\sum_l N_{ij}{}^l \lambda^l_k$$

for $k = 0, 1, \ldots, n-1$, and thus determine the n one-dimensional representations of the fusion rule algebra. Now using (9.5.8) and that $\beta_k = 2\cos((k+1)\pi/(n+1))$ it follows that

$$\lambda^j_k = S_j(\beta_k) = \sin((j+1)(k+1)\pi/(n+1))/\sin((k+1)\pi/(n+1))$$

and

$$S_{jk} = (S_j(\beta_k)/\|x_k\|) = \sqrt{(2/(n+1))}\sin((j+1)(k+1)\pi/(n+1))$$

for $j, k = 0, 1, \ldots, n-1$. Thus the matrix $S = [S_{jk}]$ is symmetric, and since it is also unitary, it satisfies the relation $S^2 = I$.

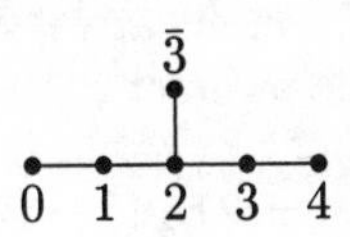

FIG. 11.72. Dynkin diagram E_6

Example 11.30 Let $\Gamma = E_6$, then the incidence matrix Δ is

$$\begin{pmatrix} 0&1&0&0&0&0\\ 1&0&1&0&0&0\\ 0&1&0&1&0&1\\ 0&0&1&0&1&0\\ 0&0&0&1&0&0\\ 0&0&1&0&0&0 \end{pmatrix}$$

where we have ordered the vertices as $0, 1, 2, 3, 4, \bar{3}$ (see Fig. 11.72). The characteristic polynomial for Δ is $x^6 - 5x^4 + 5x^2 - 1$, hence Δ is invertible. Recall from Section 9.5 the rational functions associated with the graph E_∞. For the subgraph E_6 they are

$$\phi_0 = 1, \qquad \phi_1 = x, \qquad \phi_2 = x^2 - 1,$$

$$\phi_3 = x^3 - 3x + x^{-1}, \quad \phi_4 = x^4 - 4x^2 + 2, \quad \phi_{\bar{3}} = x - x^{-1}.$$

Now since Δ satisfies its own characteristic polynomial we have $\Delta^{-1} = \Delta^5 - 5\Delta^3 + 5\Delta$. Thus the matrices $\phi_v(\Delta)$, $v \in E_6^{(0)}$ are contained in, and actually generate $\mathbb{Z}[\Delta]$. They are given below:

$$\phi_0(\Delta) = I, \qquad \phi_1(\Delta) = \Delta,$$

$$\phi_2(\Delta) = \begin{pmatrix} 0&0&1&0&0&0\\ 0&1&0&1&0&1\\ 1&0&2&0&1&0\\ 0&1&0&1&0&1\\ 0&0&1&0&0&0\\ 0&1&0&1&0&0 \end{pmatrix}, \quad \phi_3(\Delta) = \begin{pmatrix} 0&0&0&1&0&0\\ 0&0&1&0&1&0\\ 0&1&0&1&0&1\\ 1&0&1&0&0&0\\ 0&1&0&0&0&0\\ 0&0&1&0&0&0 \end{pmatrix},$$

$$\phi_4(\Delta) = \begin{pmatrix} 0&0&0&0&1&0\\ 0&0&0&1&0&0\\ 0&0&1&0&0&0\\ 0&1&0&0&0&0\\ 1&0&0&0&0&0\\ 0&0&0&0&0&1 \end{pmatrix}, \quad \phi_{\bar{3}}(\Delta) = \begin{pmatrix} 0&0&0&0&0&1\\ 0&0&1&0&0&0\\ 0&1&0&1&0&0\\ 0&0&1&0&0&0\\ 0&0&0&0&0&1\\ 1&0&0&0&1&0 \end{pmatrix}.$$

Thus they constitute a family of non-negative, symmetric, mutually commuting integral matrices. They satisfy the product rule $\phi_v(\Delta)\phi_w(\Delta) = \sum_u N_{vw}{}^u \phi_u(\Delta)$, where $\phi_v(\Delta) = (N_{vw}{}^u)$.

The eigenvalues of Δ are $\beta_k = 2\cos(k\pi/12)$, $k = 1,4,5,7,8,11$, so $\beta_1 = -\beta_{11} = \sqrt{2+\sqrt{3}}$, $\beta_4 = -\beta_8 = 1$, and $\beta_5 = -\beta_7 = \sqrt{2-\sqrt{3}}$. The corresponding eigenvectors are $x_k = (\phi_0(\beta_k), \phi_1(\beta_k), \phi_2(\beta_k), \ldots, \phi_{\bar{3}}(\beta_k))^t$.

The eigenvalues are distinct, so the matrix

$$S = (x_1/\|x_1\|, x_4/\|x_4\|, x_5/\|x_5\|, x_7/\|x_7\|, x_8/\|x_8\|, x_{11}/\|x_{11}\|)$$

is unitary and

$$S^*\Delta S = \mathrm{diag}(\beta_1, \beta_4, \beta_5, \beta_7, \beta_8, \beta_{11}).$$

One also has $\phi_v(\Delta)x_k = \lambda_v^k x_k$ where $\lambda_v^k = \phi_v(\beta_k)$, $v \in E_6^{(0)}$, $k = 1,4,5,7,8,11$, and

$$S^*\phi_v(\Delta)S = \mathrm{diag}(\phi_v(\beta_1), \ldots, \phi_v(\beta_{11})).$$

Example 11.31 Let $G = SU(2)$, $\pi : G \to M_2$ the standard representation $H = \mathbb{T}$, so that $(\pi|\mathbb{T})(z) = \begin{pmatrix} z & 0 \\ 0 & \bar{z} \end{pmatrix}$, $z \in \mathbb{T}$, and $\alpha_g = \bigotimes \mathrm{Ad}\,\pi(g)$, the action of $SU(2)$ on $\bigotimes_{\mathbb{N}} M_2$, so that $(\bigotimes_{\mathbb{N}} M_2)^{\mathbb{T}} \supset (\bigotimes_{\mathbb{N}} M_2)^{SU(2)}$ as in Section 3.5.

The embedding $A_m^{SU(2)} \to A_m^{\mathbb{T}}$ is determined by (3.5.4). We have $\chi_0|\mathbb{T} = \sigma_0$, $\chi_1|\mathbb{T} = \sigma_1 + \sigma_{-1}$, and $\chi_2|\mathbb{T} = [\chi_1^2 - \chi_0]|\mathbb{T} = (\sigma_1 + \sigma_{-1})^2 - \sigma_0 = \sigma_2 + \sigma_0 + \sigma_{-2}$. Inductively from

$$\chi_{i+1} = \chi_1\chi_i - \chi_{i-1} \tag{11.7.14}$$

we see

$$\chi_m|\mathbb{T} = \sigma_m + \sigma_{m-2} + \cdots + \sigma_{-m+2} + \sigma_{-m}\,. \tag{11.7.15}$$

From (11.7.14) we see that $\chi_i = S_i(x) \in \mathbb{Z}[x]$, if $x = \chi_1$, where $\{S_i\}$ are the Chebyshev polynomials of the second kind, the polynomials associated with the graph A_∞. Similarly the polynomials $\{\phi_i\}$ associated with the graph $A_{\infty,\infty}$ may be obtained from the irreducible characters of $\mathbb{T}$ by making the substitution $u = \sigma_1$, $v = \sigma_{-1}$, so that

$$\phi = \begin{cases} u^i & i \geq 0 \\ v^i & i < 0 \end{cases} \tag{11.7.16}$$

where $u + v = \chi_1$ and $uv = 1$. Then (11.7.15) becomes

$$S_m = \phi_m + \phi_{m-2} + \cdots + \phi_{-m}\,. \tag{11.7.17}$$

The coefficients of $\{\phi_i\}$ appearing in the decomposition of $\{S_m\}$ are non-negative, since they arise from the decomposition of group characters. Alternatively, as $\|A_{\infty,\infty}\| = 2$, we know from Corollary 11.28 that $S_m(\Delta) \geq 0$ for all m, where $\Delta = \Delta(A_{\infty,\infty})$, which gives independent confirmation of this fact. Also by Proposition

11.27, the matrix $R = [R_{ij}]$ satisfying $R\Delta_{\Gamma_\chi} = \Delta_{\Gamma_\sigma} R$ with $R_{ij} \geq 0$, and $R* = *$ is

$$R = [a_0, a_1, \ldots] = \begin{pmatrix} \cdot\ \cdot\ 0 \cdot \cdot \\ \cdot\ \cdot\ 0\ 1 \cdot \cdot \\ \cdot\ 0\ 1\ 0 \cdot \cdot \\ 0\ 1\ 0\ 1 \cdot \cdot \\ 1\ 0\ 1\ 0 \cdot \cdot \\ 0\ 1\ 0\ 1 \cdot \cdot \\ \cdot\ 0\ 1\ 0 \cdot \cdot \\ \cdot\ \cdot\ 0\ 1 \cdot \cdot \\ \cdot\ \cdot\ 0 \cdot \cdot \end{pmatrix} \tag{11.7.18}$$

where $a_0^T = (0, \ldots 0, 1, 0 \ldots) = (\delta_{i0})$, $a_m = S_m(\Delta(A_{\infty,\infty}))a_0$.

We now define $P_i \in \mathbb{Z}[\tau]$ by

$$P_i(\tau) = x^{-i} S_i(x) \tag{11.7.19}$$

if $\tau = x^{-2}$, $i = 0, 1, 2, \ldots$ so that P_i are the Jones polynomials (9.5.12). Polynomials $P_{(i,n)}$ associated with the vertices of $\hat{A}_\infty$ (see Fig. 2.24), the Bratteli diagram for $A(A_\infty) \cong (\otimes M_2)^{SU(2)}$, are then defined by

$$P_{(i,n)}(\tau) = \tau^{(n-i)/2} P_i(\tau) = x^{-n} S_i(x)\,. \tag{11.7.20}$$

Similarly define $Q_{(i,n)}$ for $(i, n) \in \hat{A}^{(0)}_{\infty,\infty}$ (see Fig. 2.40) by

$$Q_{(i,n)}(t) = x^{-n} \phi_i\,. \tag{11.7.21}$$

If $u/x = t$, $v/x = 1 - t$, and $i > 0$ then

$$Q_{(i,n)}(t) = u^i/x^n = (u^i/x^i)(1/x^{n-i}) = t^i \tau^{(n-i)/2}\,.$$

But $\tau = x^{-2} = (u/x)(v/x) = t(1-t)$, and so

$$Q_{(i,n)}(t) = t^{(n+i)/2}(1-t)^{(n-i)/2} \in \mathbb{Z}[t]\,, \tag{11.7.22}$$

which are the polynomials that appear in Theorem 3.12. The methods of Section 9.5 show that

$$K_0\left((\otimes M_2)^{SU(2)}\right) \cong \mathbb{Z}[\tau] = \mathrm{lin}_{\mathbb{Z}} \left\{P_{(i,n)}(\tau)\right\} \tag{11.7.23}$$

with positive cone $K_0\left((\otimes M_2)^{SU(2)}\right)_+$ identified with

$$\begin{gathered} \{0\} \cup \{f \in \mathbb{Z}[\tau] \mid f(\lambda) > 0\,, \lambda \in (0, \tfrac{1}{4}]\} = \mathrm{lin}_{\mathbb{N}} \left\{P_{(i,n)}(\tau)\right\}\,, \\ K_0\left((\otimes M_2)^{\mathbb{T}}\right) \cong \mathbb{Z}[t] \cong \mathrm{lin}_{\mathbb{Z}} \left\{Q_{(i,n)}(t)\right\} \end{gathered} \tag{11.7.24}$$

with positive cone $K_0\left((\otimes M_2)^{\mathbb{T}}\right)_+$ identified with

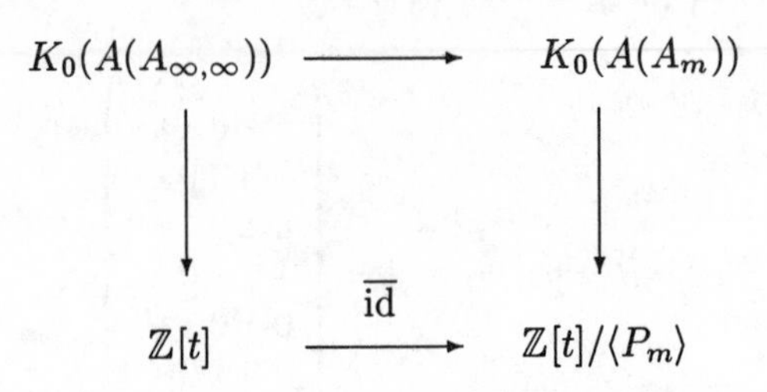

FIG. 11.73. Embedding the GICAR algebra in a Temperley–Lieb algebra

$$\{0\} \cup \{f \in \mathbb{Z}[t] \mid f(t) > 0\,, t \in (0,1)\} = \mathrm{lin}_{\mathbb{N}}\left\{Q_{(i,n)}(t)\right\}.$$

Now since $\tau = t(1-t)$, there is an inclusion

$$K_0((\otimes M_2)^{SU(2)}) \to K_0\left((\otimes M_2)^{\mathbb{T}}\right)$$

given by $f(\tau) \to g(t)$, where $g(t) = f(t(1-t))$. This map is clearly positive, i.e. if $f \in \mathbb{Z}[\tau]$, $f(\lambda) > 0$ for $\lambda \in (0,1/4]$, then $g(\eta) = f(\eta(1-\eta)) > 0$ for $\eta \in (0,1)$. This implies that we can express the polynomials $P_{(i,n)}$ as non-negative integer linear combinations of the polynomials $Q_{(j,m)}$. In fact using (11.7.17) we have

$$P_n(\tau) = \textstyle\sum_{k=0}^{n} t^k (1-t)^{n-k} \tag{11.7.25}$$

so that

$$\begin{aligned} P_{(i,n)}(\tau) = \tau^{(n-i)/2} P_i(\tau) &= \textstyle\sum_{k=0}^{n} t^{(n+2k-i)/2}(1-t)^{(n-(2k-i))/2} \\ &= \textstyle\sum_{k=0}^{i} Q_{(2k-i,n)}(t)\,. \end{aligned} \tag{11.7.26}$$

The inclusion $(\bigotimes_{\mathbb{N}} M_2)^{SU(2)} \subset (\bigotimes_{\mathbb{N}} M_2)^{\mathbb{T}}$ is immediate. However, we saw in Section 7.9 that we could embed the GICAR $(\bigotimes_{\mathbb{N}} M_2)^{\mathbb{T}}$ in a Temperley–Lieb algebra (e.g. $(\bigotimes_{\mathbb{N}} M_2)^{SU(2)}$). In fact the gauge invariant observable algebra $(\otimes_{\mathbb{Z}} M_2)^{\mathbb{T}} \cong A(A_{\infty,\infty})$ was embedded in $A(A_m)$ for $3 \le m \le \infty$. Now $K_0(A(A_{\infty,\infty})) \cong \mathbb{Z}[t]$, with the positive cone identified with

$$\{0\} \cup \{P \in \mathbb{Z}[t] \mid P(t) > 0\,, t \in (0,1)\}\,.$$

We have shown in Theorem 9.38 that $K_0(A(A_m)) \simeq \mathbb{Z}[t]/\langle P_m\rangle$, where $P_\infty = 0$, and positive cone ordered by $4\cos^2 \pi/(m+1)$ when $m < \infty$, and $(0,1/4]$ when $m = \infty$. We now assert that we have a commutative diagram as given by Fig. 11.73 for $3 \le m \le \infty$, where the vertical maps are isomorphisms and $\overline{\mathrm{id}}$ is the quotient map induced by the identity map on $\mathbb{Z}[t]$. This means that the Powers trace on $(\bigotimes_{\mathbb{N}} M_2)^{\mathbb{T}}$, the restriction of $\bigotimes_{\mathbb{N}} \mathrm{tr}\left(\left[\begin{pmatrix} t & 0 \\ 0 & 1-t \end{pmatrix} \cdot\right]\right)$, where $t(1-t) = \tau$, is taken to the Markov trace on the Temperley–Lieb algebra.

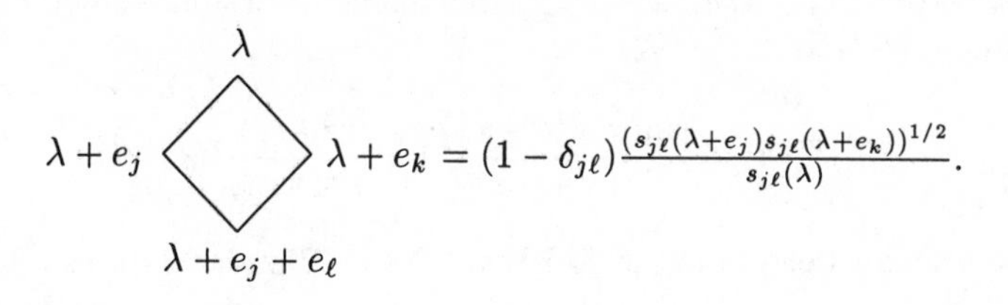

FIG. 11.74. $SU(N)_l$ face operator U_i

To check this assertion, it is convenient to label paths in $\hat{A}_{\infty,\infty}$ by sequences Ω in $\{-1,+1\}$, so that $/, \backslash$ correspond to $-1,+1$ respectively. In the notation of Section 2.9 let g_i be the projection in $C(\Omega) \subset A(A_{\infty,\infty})$ given by the characteristic function of $\{x \in \Omega : x(i) = 1\}$, and v_i the partial isometry in $A(A_{\infty,\infty})$ with initial support $g_{i+1}(1-g_i)$, and final support $g_i(1-g_{i+1})$. Thus v_i is the flip given by Fig. 2.42. The embedding of $A(A_{\infty,\infty})$ in $A(A_m)$, $1 \le m \le \infty$, is given by

$$\begin{aligned} g_i &\to e_{2i-1} \\ v_i &\to e_{2i-1}(1-\tau^{-1}e_{2i})e_{2i+1} \end{aligned} \tag{11.7.27}$$

(see the Remark on page 343). The class of $g_1 \cdots g_r(1-g_{r+1})\cdots(1-g_{r+s})$ in $K_0(A(A_{\infty,\infty})) \cong \mathbb{Z}[t]$ is $t^r(1-t)^s$ by the identification of Theorem 3.12 and (11.7.22). We need to compute the class $R_{r,s}$ of

$$e_1e_3\cdots e_{2r-1}(1-e_{2r+1})(1-e_{2r+3})\cdots(1-e_{2(r+s)+1})$$

in $K_0(A(A_m)) \cong \mathbb{Z}[t]/\langle P_m\rangle$. We need to show that

$$R_{r,s} = t^r(1-t)^s\,. \tag{11.7.28}$$

To see this, we first note $R_{r,0} = t^r$ as $e_1e_3\cdots e_{2r-1}$ corresponds to the extreme left hand path in $\hat{A}_m$ (Fig. 2.43). Then (11.7.28) follows inductively on s, using Lemma 9.35(d).

In the bipartite or $SU(2)$ case, it is relatively easy to put the Temperley–Lieb algebra into a path algebra of a such a graph. In the higher rank case, it is by no means obvious at all, whether one can embed an $SU(N)_q$ face algebra, $N > 2$, into a path algebra of a given graph Γ and whether the obvious candidate for an intertwiner V, is the intertwining Verlinde formula, or the analogue of the Chebyshev polynomials applied to $V(\cdot)$, is a matrix with positive integral entries. One does not have a complete picture even when $N = 3$, or a list of all possible 3-colourable graphs which permit embeddings.

The critical Ising weights on A_3 generalize to $\mathcal{A}^{(n)}$ as in Fig. 11.74. Here

$s_{ij}(\lambda) = \sin[\pi(e_i - e_j)\cdot\lambda/n]$, $\lambda \in P_{++}^{(n)}$, and the inner product is determined by $e_j \cdot e_k = \delta_{jk} - 1/N$. Then

$$X_i(u) = \frac{\sin(\pi(1/n - u))}{\sin(\pi/n)} 1 + \frac{\sin(\pi u)}{\sin(\pi/n)} U_i$$

are the Boltzmann weights which satisfy the Yang–Baxter relation:

$$X_i(v)X_{i+1}(v+u)X_i(u) = X_{i+1}(u)X_i(v+u)X_{i+1}(v),$$

due to the following relations between U_i (cf. Section 7.9) where $\beta = 2\cos(\pi/n)$:

$$U_i^2 = \beta U_i\,, \tag{11.7.29}$$

$$[U_i, U_j] = 0\,, \quad |i-j| > 1 \tag{11.7.30}$$

$$U_iU_{i+1}U_i - U_i = U_{i+1}U_iU_{i+1} - U_{i+1}\,. \tag{11.7.31}$$

Note that the above formula for U_i reduces to Fig. 11.32 when $N \to 2$. These relations define a Hecke algebra. If we put $\beta = 2\cos(\pi/n) = q + q^{-1}$, $q = e^{i\pi/n}$,

$$g_j = q^{-1} - U_j = \lim_{u\to -i\infty} 2i\sin(\pi/n)e^{-i\pi u}X_j(u) \tag{11.7.32}$$

then (11.7.29)–(11.7.31) are equivalent to

$$(q^{-1} - g_j)(q + g_j) = 0 \tag{11.7.33}$$

$$[g_i, g_j] = 0\,, \quad |i-j| > 1 \tag{11.7.34}$$

$$g_ig_{i+1}g_i = g_{i+1}g_ig_{i+1}\,. \tag{11.7.35}$$

When $q = 1$, this reduces to the (group algebra of) the permutation group as (11.7.33) degenerates to $g_j^2 = 1$ (so that g_j represents a transition $(j, j+1)$). By Weyl, the representation of the permutation group in $\bigotimes_{\mathrm{N}} M_N$, is the fixed point algebra of the product action of $SU(N)$. Deforming this, there is a representation of the Hecke algebra in $\bigotimes_{\mathrm{N}} M_N$, whose commutant is a representation of a deformation of $SU(N)$, the quantum group $SU(N)_q$. Here the labels of irreducible representations are given by $\mathcal{A}$, i.e. Young tableaux of at most $N-1$ rows, but when q is a root of unity $e^{2\pi i/n}$ we have the further constraint of $\mathcal{A}^{(n)}$, i.e. at most $l = n - N$ columns, where l is the level. When $N = 2$, we have already come across this representation of the Hecke algebra in $\bigotimes_{\mathrm{N}} M_2$, via the Pimsner–Popa Temperley–Lieb representation of the operators e_i of Section 7.8, now written βU_i, and so in this representation there are further relations, indeed not only does (11.7.29) hold but $U_iU_{i+1}U_i = U_i$ (the Temperley–Lieb relation). More generally, when embedding in $\bigotimes_{\mathrm{N}} M_N$, with the commutant of $SU(N)_q$, there are further relations as follows. If $\rho = \prod_{j\in I_\rho} \tau_j$ is a minimal representation of a permutation as a product of transpositions, the operator $g_\rho = \prod_{j\in I_\rho} g_j$ is well defined as a consequence of the braid relation (11.7.29). The extra relation is then the vanishing of the q-antisymmetrizer

$$\sum_{\rho\in S^*_{N+1}}(-q)^{|I_\rho|}g_\rho = 0\,, \tag{11.7.36}$$

e.g. for $N = 3$ we need

$$(U_i + U_{i+1} - U_{i+2}U_{i+1}U_i)\,(U_{i+1}U_{i+2}U_{i+1} - U_{i+1}) = 0\,. \tag{11.7.37}$$

What we require are embeddings of the algebra generated by $U_1, U_2, \ldots$ where $\beta = q + q^{-1}$ subject to the constraint (11.7.36) depending on N.

The embedding or intertwining matrix V that we are looking for, will (as in (11.7.1) for $N = 2$) intertwine the $\mathcal{A}^{(n)}$ graph, again abbreviated to A, with the secondary G matrix, embedding the A path algebra in the one for G. Again V will be given by an intertwining Verlinde formula.

First however let us concentrate on the A matrix alone. Here we have the Verlinde formula

$$N_{\nu\lambda}{}^{\mu} = \sum_\rho S_{\nu\rho}S_{\lambda\rho}\overline{S_{\mu\rho}}/S_{*\rho} \tag{11.7.38}$$

where $*$ is the apex $\sum_i \Lambda_i$ of the Weyl alcove $\mathcal{A}^{(n)}$ (see Section 8.7). If f is the fundamental representation (i.e. $f = \Lambda_1$) then the matrices N^μ describe the fusion rules of the level $k = n - N$ representations of $su(N)^\wedge$ (which we know from Section 8.9 to be labelled by the vertices of $\mathcal{A}^{(n)}$),

$$\nu * \lambda = \sum_\mu N_{\nu\lambda}{}^{\mu}\mu$$

with $N_{f\lambda}{}^{\mu} = A_{\lambda\mu}$. The S matrix diagonalizes the fusion rules, and

$$\gamma^\rho = S_{f\rho}/S_{*\rho} = \sum_{j=1}^{N}\exp(2\pi i(e_j\cdot\rho)/n) \tag{11.7.39}$$

are the eigenvalues of the A-matrix, with eigenvectors $\{S_{\mu\rho} \mid \mu\} = \phi^\rho$, with largest eigenvalue $\gamma = \gamma^* = S_{N-1}(\beta)$ when $\rho = *$ and where S_N is the Chebyshev polynomial of the second kind of degree N (9.5.4), and Perron–Frobenius eigenvector $\phi = \phi^*$ with components

$$\phi_\mu = \prod_{\alpha>0}\sin((\pi/n)\alpha\cdot\mu)/\sin((\pi/n)\alpha\cdot *)$$

where the product runs over the positive roots of $SU(N)$.

Next we consider these concepts in the context of a secondary graph G. Taking a cue from the graphs which appear when looking at $su(3)^\wedge$ modular invariants, we have a tri-colourable graph whose eigenvalues are a subset of those of the regular $\mathcal{A}^{(n)}$ graph, say ϕ^ρ_f/ϕ^ρ_1 where ρ runs over a subset Exp(G) of the vertices of $\mathcal{A}^{(n)}$, and let $\psi^\rho = \{\psi^\rho_a\}$ be corresponding eigenvectors (with possible degeneracies). Here we are implicitly taking $G = \sum(\phi^\rho_f/\phi^\rho_*)\psi^\rho\otimes\overline{\psi^\rho}$ to be normal. We can again form the intertwining Verlinde formula:

$$V_\nu = \sum_\rho\,(\phi^\rho_\nu/\phi^\rho_*)\,\psi^\rho\otimes\overline{\psi^\rho} \tag{11.7.40}$$

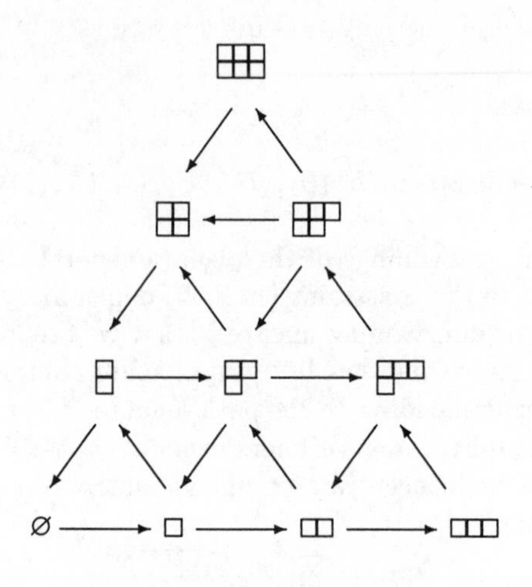

FIG. 11.75. $SU(3)_3 = \mathcal{A}^{(6)}$

so that $V_f = G$, and start to worry about the positivity and integrality of its matrix entries, intertwining property. To address integrality first, we seek polynomials to replace the Chebyshev polynomials in (11.7.3), namely a polynomial ϕ_ν indexed by the vertices of $\mathcal{A}^{(\infty)}$ such that

$$x\phi_\nu = \sum_\mu A(\nu,\mu)\phi_\mu\,, \qquad y\phi_\nu = \sum_\mu A^t(\nu,\mu)\phi_\mu\,.$$

Thus ϕ_ν is a polynomial in two variables because now A is not symmetric unlike the $SU(2)$ case. Indexing with Young tableaux as in Fig. 11.75 on page 610, the first few are

$$\begin{array}{llll} \varnothing = 1 & (1,1) = y & (2,2) = y^2 - x & (3,3) = y^3 - 2xy + 1 \\ (1) = x & (2,1) = xy - 1 & (3,2) = xy^2 - xy - y & \\ (2) = x^2 - y & (3,1) = x^2y - y^2 - x & & \\ (3) = x^3 - 2xy + 1 & & & \end{array} \tag{11.7.41}$$

We want to identify V_ν with $S_\nu(G, G^t)$ (which requires G normal surely) which could at least guarantee integrality of $V_{\nu b}{}^a$ and a lack of dependence on a choice of eigenvectors if there is degeneracy. First it is clear, since N_ν and S_ν are built out of the same fusion rules that $N_\nu = S_\nu(N_f, N_{\overline{f}})$ (which is the equality $V_\nu = S_\nu(G, G^t)$ in the case $A = G$). So if γ_ρ, $\rho \in \mathrm{Exp}(A)$ are the eigenvalues of A, then $S_\nu(\gamma_\rho, \overline{\gamma_\rho})$ are the eigenvalues of N_ν, and comparing with (11.7.38), has to be identified with $\phi_\nu^\rho/\phi_*^\rho$. Thus

$$V_{\nu a}{}^b = \sum_\rho \psi_a^\rho \overline{\psi_b^\rho} S_\nu(\gamma_\rho, \overline{\gamma_\rho}) = \left[S_\nu(G, G^t)\right]_{a,b}\,. \tag{11.7.42}$$

Now that we have the embedding graph V we can try to solve the intertwining Yang–Baxter equation, i.e. find a face operator on the G graph, and an unitary G–A intertwiner satisfying the relation of Fig. 11.75. This can be done immediately, without any further knowledge or hypothesized relations on the unknown G–face operator in the situation of the $\mathcal{D}^{(6)*}$ and $\mathcal{E}^{(8)}$ graphs of Table 8.6 (Di Francesco and Zuber 1990a). However, to proceed further and find solutions as the graphs become more complex, new ideas are needed. This is provided by a consideration of traces and partition functions on the G and A models and how they should be related via the intertwining V_ν matrices. This provides further relations between the unknown face operator and connections which can be solved to provide the required G–A embeddings in a few more examples.

If a is the initial vertex at level 1 and b at level L in the Bratteli diagram of the path algebra $A(G)$ we have the unnormalized trace

$$Z^G_{ab}(x) = \sum x_{\xi,\xi}\,, \quad \text{for } x \in A[G][1,L]\,,$$

and a summation over all paths $\xi = (a_1 = a, a_2, \ldots, a_{L-1}, a_L = b)$ from a to b. Define

$$Z^\mu(x) = \textstyle\sum_b Z^G_{ab}(x)\psi^\mu_b/\psi^\mu_a\,.$$

Since

$$\sum_\lambda \;\gamma \overset{\mu}{\underset{\lambda}{\Diamond}} \gamma \;\; \phi^*_\lambda \;=\; S_{N-2}(\beta)\phi^*_\mu \tag{11.7.43}$$

and

$$\sum_\lambda A_{\mu\lambda}\phi^*_\lambda = S_{N-1}(\beta)\phi^*_\mu \tag{11.7.44}$$

where S is a Chebyshev polynomial, we have that

$$Z^*(xU_{n+1}) = S_{N-2}(\beta)/S_{N-1}(\beta)Z^*(x) \tag{11.7.45}$$

for x any word in $1, U_1, \ldots, U_n$ so that Z^* is the Markov trace (cf. Section 11.3). Thus for the G-graph we would hypothesize that the largest eigenvalue of G coincides with that of A:

$$\textstyle\sum_c G_{bc}\psi^*_c = S_{N-1}(\beta)\psi^*_b$$

which actually follows from (11.7.1) and the uniqueness of the Perron–Frobenius eigenvalue, as long as G is irreducible. Corresponding to (11.7.45), we would want an analogous Markov property in the G-picture:

$$\sum_c \; b \overset{a}{\underset{c}{\Diamond}} b \;\; \psi^*_c \;=\; S_{N-2}(\beta)\psi^*_b. \tag{11.7.46}$$

With these extra equations at our disposal, the face operators for $\mathcal{E}^{(8)}$ can be found (Sochen 1991).

Now inductively on l, if $x = U_{i_1} \cdots U_{i_l}$, $0 \leq i_1 < \cdots < i_l$, it follows from the Markov property, and the vanishing of the anti-symmetrizer that the following identity holds:

$$Z^G_{ab}(x)\psi^\mu_b/\psi^\mu_a = \textstyle\sum_\lambda Z^A_{*\lambda}(x)\phi^\mu_\lambda/\phi^\mu_* .$$

The significance of this formula is that, from the orthogonality $\sum_\mu \psi^\mu_b \overline{\psi^\mu_{b'}} = \delta_{bb'}$, we deduce

$$Z^G_{ab}(x) = \textstyle\sum_\lambda \left(\sum_\mu \left(\overline{\psi^\mu_b}\psi^\mu_a \phi^\mu_\lambda/\phi^\mu_*\right)\right) Z^A_{*\lambda} = \sum_\lambda V_{\lambda a}{}^{b} Z^A_{*\lambda}(x)$$

with $V_{\lambda a}{}^{b}$ given by the intertwining Verlinde formula, and more importantly we have new constraints on the face operators relating Z^G_{ab} with $Z^A_{*\lambda}$. The simplest of these are:

$$\sum_b \; a \underset{b}{\overset{b}{\Diamond}} c \;=\; \sum V_{\lambda a}{}^{c} \; * \underset{\mu}{\overset{\mu}{\Diamond}} \lambda \;=\; G_{ca}\beta \tag{11.7.47}$$

$$\begin{aligned} \sum_{b,c} \; a \underset{b}{\overset{c}{\Diamond\!\!\Diamond}} d \;\; (\text{inner labels } b, c) &= \beta^2\delta_{ad} + V_{(*+\Lambda_1+\Lambda_2),a}{}^{d} \\ &= (\beta^2 - 1)\delta_{ad} + (G^t G)_{ad} \end{aligned} \tag{11.7.48}$$

With the aid of (11.7.47) and (11.7.48) (Sochen 1991) could find the face operators, which satisfied the Hecke algebra relations and the vanishing of (11.7.37) for $\mathcal{E}_1^{(12)}$ and $\mathcal{E}^{(24)}$.

We can then form the GHJ subfactor the inclusion $\mathcal{E} \subset \mathcal{A}$ for these exceptional graphs. Flatness is a consequence of the IYBE as in Section 11.9. Thus as in the A-D-E situation, the principal graph is given by the intertwining matrix – which can be written down via (11.7.42) and (11.7.41).

11.8 Paragroups for strongly amenable subfactors

So far, we have assumed the finite depth condition. In this section, we describe what happens if the subfactor $N \subset M$ has infinite depth (but has finite index and trivial relative commutant). This section is *not* logically self-contained.

We suppose that a subfactor $N \subset M$ has finite index and infinite depth in this section. For technical simplicity, we again assume the trivial relative commutant condition $N' \cap M = \mathbb{C}$. (Or we can assume extremality as in Definition 11.11 more generally.) We still have higher relative commutants and four kinds of bimodules, and get infinite graphs $\mathcal{G}_1, \mathcal{G}_2, \mathcal{G}_3, \mathcal{G}_4$ and a connection W satisfying all five axioms. (The proof of Proposition 10.4 does not work any more because it used the Perron–Frobenius theorem. But we can still deduce the harmonic axiom from the extremality – we do not discuss it here.) We have another condition, *strong amenability* as in Definition 11.13. This property characterizes a good subclass among the subfactors of the hyperfinite II_1 factor with infinite depth in the sense that the inclusion $\overline{\bigcup_{n=0}^{\infty} N' \cap M_n} \subset \overline{\bigcup_{n=0}^{\infty} N_1' \cap M_n}$ is anti-isomorphic to the original subfactor $N \subset M$, as in Theorem 11.14. We can express strong amenability in terms of two extra axioms of paragroups as follows.

In the first place, the algebras $\bigcup_{n=0}^{\infty} N' \cap M_n$ and $\bigcup_{n=0}^{\infty} N_1' \cap M_n$ may not be factors any more, and if they are not factors, we certainly cannot have the above anti-isomorphism. So we have to require factoriality as an axiom.

Axiom 6. (Ergodicity) *The GNS-completions of two algebras $\bigcup_{l=0}^{\infty} A_{0,l}$ and $\bigcup_{l=0}^{\infty} A_{1,l}$ with respect to the trace induced by μ are both factors.*

The other axiom expresses Popa's key notion 'amenability'. Let Γ, Γ' be the incidence matrices of the graphs $\mathcal{G}_0$ and $\mathcal{G}_1$ respectively. (Note that they are infinite graphs.) We also write Γ^t, Γ'^t for their transpose matrices. We mean by $\|\Gamma^t\Gamma\|$, $\|\Gamma'^t\Gamma'\|$ the operator norm of these matrices respectively on ℓ^2.

Because now the graphs are infinite, the eigenvalue β in the harmonic axiom is not determined by the graphs, so we have to regard β as additional data to the graphs. (For example, if the graph Γ is A_∞, then β can be any number greater than or equal to 2.) Then we have the following axiom.

Axiom 7. (Amenability) *In the above setting, we have $\beta^2 = \|\Gamma^t\Gamma\| = \|\Gamma'^t\Gamma'\|$.*

With the method in (Popa 1994a), (Popa 1995b), one can prove that the higher relative commutants of strongly amenable subfactors of the hyperfinite II_1 factor are characterized by Axioms 1–7.

Recently, (Popa 1995b) gave a complete characterization of higher relative commutants of arbitrary subfactors of type II_1 with finite index. (That is, we no longer assume hyperfiniteness.) His axioms are similar to flat connections on infinite graphs, but he directly works on commuting squares instead of connections. See (Popa 1995b) for more details.

11.9 IRF models and the Yang–Baxter equation

In this section, we discuss an analogy between paragroups and exactly solvable lattice models. We will return to this point in the notes to Chapter 13, after we have looked at orbifold subfactors.

In the theory of exactly solvable lattice models (see Chapter 7), there is also a graph and an assignment of a positive number to each cell arising from the graph. (This type of lattice model is called an IRF – Interaction-Round-Faces – model.) We have an analogy between paragroups and IRF models and we list the correspondence table as in Table 11.2. This analogy is especially clear if we have the same four graphs connected trivially for paragroups as in the Dynkin diagram cases, because we can think that the connection is defined on a single graph in such a case.

One important point in the analogy table is that we have no counterpart of the spectral parameter for paragroups. That is, in the theory of IRF models, their

Table 11.2 *Analogy between paragroups and IRF models*

connection	Boltzmann weight
gauge choice	gauge choice
partition function	partition function
unitarity	unitarity
renormalization	crossing symmetry
—	spectral parameter

Boltzmann weight, the counterpart of a connection, depends on the parameter called a spectral parameter (the interaction constants K_1, K_2 in the Ising model, or better still, think of the substitution $k^{-1} = \sinh 2K_1 \sinh K_2$ parametrizing solutions of the Yang–Baxter equation), but we have no such parameter in the paragroup theory. In a sense, the paragroup theory corresponds to the IRF model theory at criticality, and it is closer to rational conformal field theory in this sense. (This will be discussed in detail in Chapter 13.)

In Table 11.2, one notices that important axioms are missing both in the paragroup theory and the IRF model theory. They are the flatness axiom in the paragroup theory and the Yang–Baxter equation in the IRF models. These two axioms are not directly related in general, but they do correspond to each other in certain special cases. We study the case of the Dynkin diagrams A_n in detail here.

We now have a single graph A_n and a connection defined as in Fig. 11.32 for each cell arising from the graph A_n. Then the corresponding IRF models have been found by (Andrews *et al.* 1984), see Chapter 8. The relation between the Boltzmann weight in (Andrews *et al.* 1984) and the flat connection as in Fig. 11.32 is as follows. We get the identity in Fig. 11.32 after letting the spectral parameter in (Andrews *et al.* 1984) go to a certain limit, as in formula (11.7.32). Then we know that the Yang–Baxter equation still holds for the connection on A_n in the following sense.

We have the identity (11.7.35) as in Fig. 11.76, where both sides mean the sum of the products of three connection values over all the possible choices of X_7 for any fixed choice of the vertices X_1, X_2, X_3, X_4, X_5, X_6 on the graph. In the above identity, we used the graphical convention as in Fig. 11.77. We study what this equation means in the paragroup/string algebra theory. We define the *face operator* $F_k \in A_{0,k+1}$ as in Fig. 11.78. (See also (11.7.32) and (7.1.12).) Note that here the coefficient of a string is given by the connection. Comparing the definition of the connection in Fig. 11.32 and Definition 11.5 of the Jones projections, we get the identity $F_k = \varepsilon + \beta\bar{\varepsilon} f_k$, where f_k is the kth horizontal Jones projection in $A_{0,k+1}$. We embed F_k into $A_{1,k+2}$ and make a basis change as before. Then the identities in Fig. 11.79 show that $F_k \in A_{1,k+2}$ is expressed with the same formula as in Fig. 11.78 where the path σ first goes down for one step and then goes to the right direction for $k-1$ steps, and paths $\gamma_\pm$, $\delta_\pm$ are horizontal. (In the identities in Fig. 11.79, we have dropped all labelling for

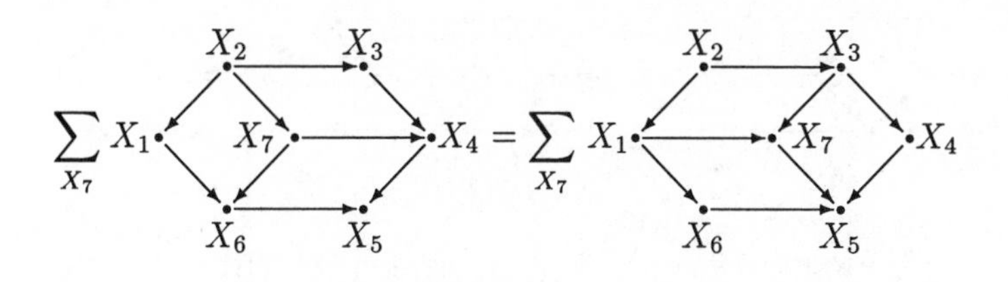

FIG. 11.76. The Yang–Baxter equation

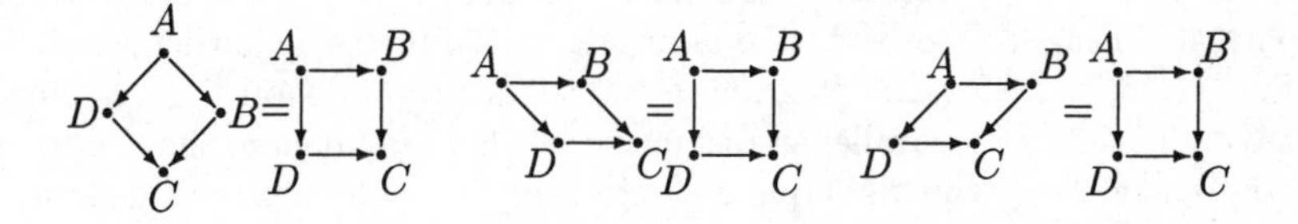

FIG. 11.77. Graphical conventions

simplicity. We use the Yang–Baxter equation in the first identity and unitarity in the second identity.)

What we have so far seen is that we get the flatness of the (horizontal) Jones projections from the Yang–Baxter equation. We can reverse this argument, so we know that *the Yang–Baxter equation in (Andrews et al. 1984) at criticality is equivalent to the flatness of the Jones projections on the Dynkin diagram A_n*. A similar relation holds between the Yang–Baxter equation in (Jimbo *et al.* 1987), (Jimbo *et al.* 1988) and flatness of the face operators in Wenzl's construction of subfactors arising from Hecke algebras of type A_n (Wenzl 1988a). See (Evans and Kawahigashi 1994a) for more details. In this book, these subfactors of Wenzl will be treated as quantum $SU(n)_k$ subfactors in 13.3 in connection with rational conformal field theory.

Note that in the case of the Dynkin diagrams of type D and E, we still have flatness of the Jones projections, and they imply the Yang–Baxter equation of the connections as in Fig. 11.76. Also note that we have the Yang–Baxter equations even for D_{2n+1} and E_7 for which the flatness axiom fails.

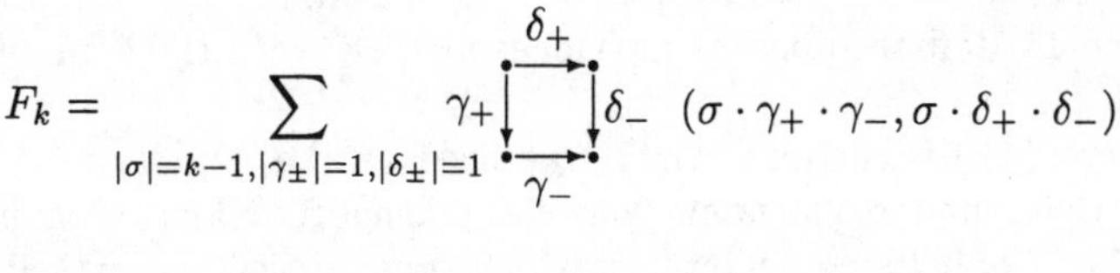

$$F_k = \sum_{|\sigma|=k-1,|\gamma_\pm|=1,|\delta_\pm|=1} \gamma_+ \overset{\delta_+}{\underset{\gamma_-}{\square}} \delta_- \quad (\sigma\cdot\gamma_+\cdot\gamma_-, \sigma\cdot\delta_+\cdot\delta_-)$$

FIG. 11.78. A face operator

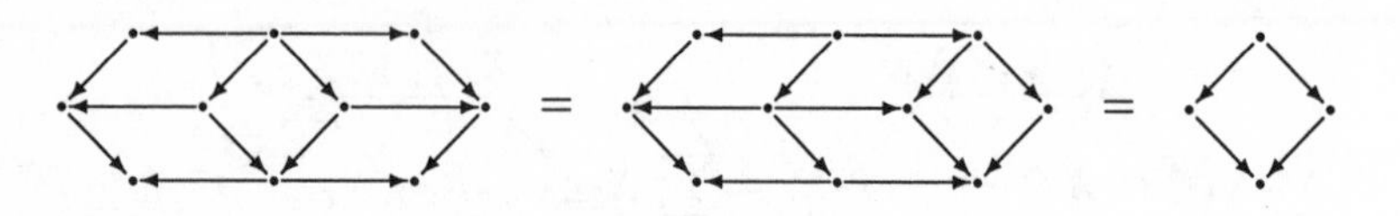

FIG. 11.79. $F_k \in A_{0,k+1} \mapsto A_{1,k+2}$ via YBE

Finally, we discuss the Yang–Baxter equation for the subfactor with index $3+\sqrt{3}$ in Example 11.25. In the embedding of the string algebras on A_{11} to those on E_6, we have flatness of the horizontal Jones projections. As above, this implies the Yang–Baxter equation as in Fig. 11.76 in the following sense. Now in Fig. 11.76, the left parallelogram in the left hexagon denotes the connection for A_{11}, the right two parallelograms in the left hexagon denote the connections for the embedding, the right parallelogram in the right hexagon denotes the connection for E_6, and the left two parallelograms in the right hexagon denote the connections for the embedding. That is, this equation involves three different kind of connections. This is the *intertwining Yang–Baxter equation*, IYBE.

11.10 Notes

The results in Section 11.2 are due to unpublished work of U. Haagerup and J. Schou (1988) and the dissertation of (Schou 1990). Here our treatment is closer to the unpublished manuscript 'Graph geometry, quantized groups and nonamenable subfactors' (Lake Tahoe Lectures, 1989) of A. Ocneanu. The string algebras were first defined in (Evans 1985a) (with the name path algebras) and were later rediscovered by (Ocneanu 1988), (Sunder 1987). Theorem 11.10 is due to (Wenzl 1987).

Results in Sections 11.3 and 11.4 are mainly taken from (Ocneanu 1991). Ocneanu's compactness argument first appeared in (Ocneanu 1991).

The deep theorem of S. Popa on the generating property mentioned at the end of Section 11.3 was first proved in the finite depth case in (Popa 1990a). He later proved the general theorem in (Popa 1994a) and gave a different proof in (Popa 1994b) including certain non-hyperfinite cases. In (Popa 1995a), he generalized the method to the type III subfactors.

The formula for $e_1 \vee \cdots \vee e_{n-1}$ in Theorem 11.10 is due to (Wenzl 1987); the derivation here from the path algebra setting is from (Evans and Gould 1994b).

Theorem 11.19 first appeared in (Ocneanu 1988) with the name 'Range Theorem'.

Condition (4) of Theorem 11.17 was used as the definition of flatness by (Ocneanu 1988) under the name 'parallel transport axiom'. Condition (2) of Theorem 11.17 and its equivalence to Condition (4) was mentioned there, too. Condition (1) of Theorem 11.17 is explicitly made first in (Kawahigashi 1995a) and was inserted in the lecture notes (Ocneanu 1991) by the recorder, the second

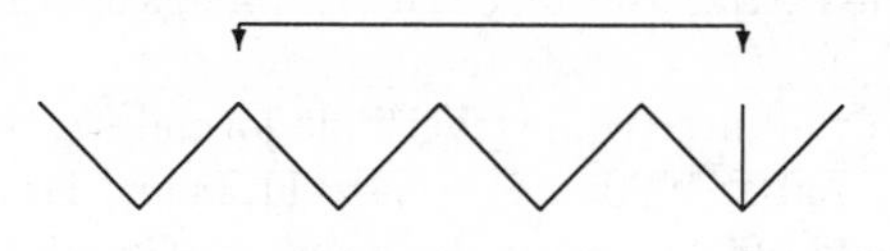

FIG. 11.80. Flip $\chi_9 \leftrightarrow \chi_3 + \chi_{15}$ on D_{10}

author of this book. The proof of Theorem 11.17 here is taken from (Kawahigashi 1995a).

Theorems 11.22 and 11.24 first appeared in (Ocneanu 1988). An outline of his proof of Theorem 11.22 was given in (Ocneanu 1988), (Ocneanu 1991). Here we followed (Ocneanu 1991) and (Kawahigashi 1995a). Our proof of Theorem 11.24 here is based on the method of (Izumi 1994a). (A subfactor with principal graph E_6 was first constructed by (Bion-Nadal 1992).) Another method based on explicit computations of partition functions was later found by (Hayashi 1993). The construction in Exercise 11.2 is the original one in (Jones 1983). Exercise 11.3 first appeared in (Izumi 1991). Exercises 11.4–11.6 are taken from (Izumi and Kawahigashi 1993), (Izumi 1994a), (Evans and Kawahigashi 1994b) respectively.

In Section 8.9, we noted that the flip $\chi_9 \leftrightarrow \chi_3 + \chi_{15}$ on the extended blocks identifies the two modular invariants for D_{10} and E_7. This led Zuber to conjecture that the subfactor built from the E_7 connection

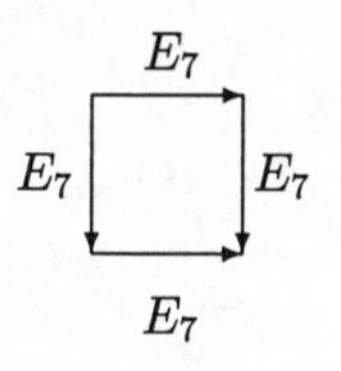

is the D_{10} subfactor (i.e. the flat part of E_7 is D_{10}). This conjecture was verified by (Evans and Kawahigashi 1994b); indeed using a commuting square or connection

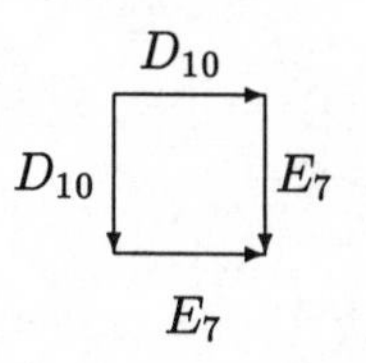

where the two D_{10}s are glued together by a non-trivial identification corresponding exactly to the interchange $\chi_9 \leftrightarrow \chi_3 + \chi_{15}$ of blocks as in Fig. 11.80.

The construction of subfactors in Section 11.6 first appeared in (Goodman *et al.* 1989), and the principal graphs were computed first in (Okamoto 1991). Here we followed the method of (Kawahigashi 1992b). The dual principal graph of the subfactor with index $3 + \sqrt{3}$ in Example 11.25 was first computed by U.

Haagerup (unpublished). Here we followed (Kawahigashi 1992b), (Kawahigashi 1995c).

The discussion on the Intertwining Verlinde formula in Section 11.7 is based on (Di Francesco and Zuber 1990a). Corollary 11.28 and Examples 11.29–31 are from (Evans and Gould 1994a). A similar result to Corollary 11.28 is obtained in (de la Harpe and Wenzl 1987) in the case when Γ is finite, and $\|\Gamma\| \geq 2$. Here they show that $S_k(\Delta)$ occurs as an inclusion matrix for a different pair of algebras, and make essential use of the basic construction.

The dimension groups of the AF algebras $(\otimes M_2)^{\mathbb{T}}$ and $(\otimes M_2)^{SU(2)}$ were first characterized by (Renault 1980) and (Wassermann 1981) respectively.

The Goodman–de la Harpe–Jones construction has been recently generalized in (Xu 1997). Xu's construction gives other subfactors in Case 4 of the list in page 590. Xu discusses a relation of his construction to conformal embedding of $SU(3)_k$ with $k = 5, 9, 21$, while the original Goodman–de la Harpe–Jones construction is related to conformal embeddings of $SU(2)_k$. (See Section 8.8 for conformal embeddings.)

The discussion in Section 11.9 is taken from (Evans and Kawahigashi 1994a), see also (Roche 1990). The validity of the intertwining Yang–Baxter equation at the end of Section 11.9 was first proved in (Roche 1990) by direct computations. The more general proof here is due to V. F. R. Jones.

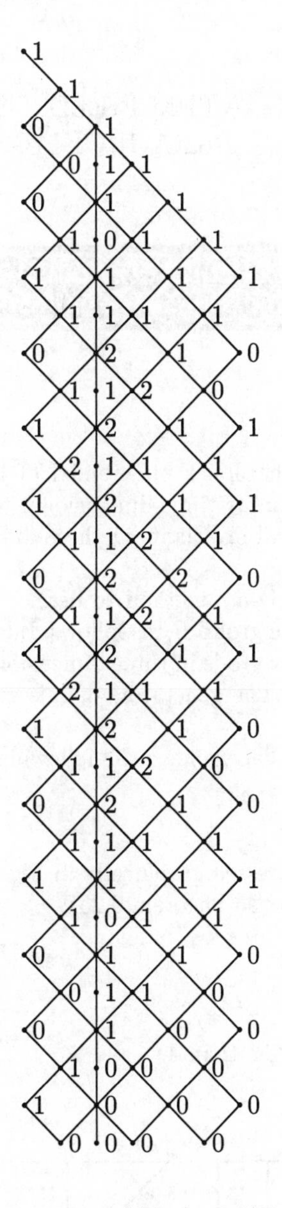

FIG. 11.81. $A(A_{29}) \subset A(E_8)$

12

TOPOLOGICAL QUANTUM FIELD THEORY (TQFT) AND PARAGROUPS

12.1 Introduction

It was the discovery of the Jones polynomial of links in (Jones 1985) which opened an entirely new era for three-dimensional topology. In this chapter, we study 'quantum' topological invariants in three-dimensional topology using the subfactor theory.

We first explain the original method to define the Jones polynomial in this section. We define the braid group B_n as follows. Intuitively, this is defined as the set of n-strands between two rods in three dimensional space with multiplication defined by concatenation. The generators of this group are $\sigma_1, \sigma_2, \ldots, \sigma_{n-1}$, as defined in Fig. 12.1.

Then it is easy to see that we have the following relations:

1. $\sigma_j \sigma_k = \sigma_k \sigma_j$, if $|j-k| \geq 2$;
2. $\sigma_j \sigma_{j+1} \sigma_j = \sigma_{j+1} \sigma_j \sigma_{j+1}$.

The second identity is expressed graphically in Fig. 12.2.

Recall the Jones relations (Theorem 9.25):

1. $e_j = e_j^* = e_j^2$;
2. $e_j e_k = e_k e_j$, if $|j-k| \geq 2$;
3. $e_j e_{j\pm 1} e_j = (4\cos^2 \pi/m)^{-1} e_j$,

for subfactors with index less than 4.

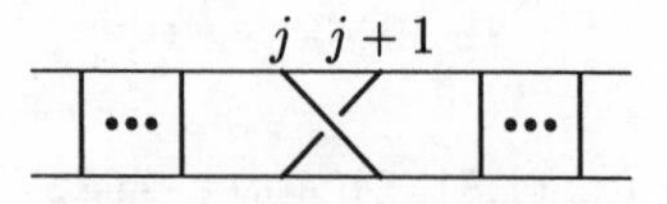

FIG. 12.1. The generator σ_j

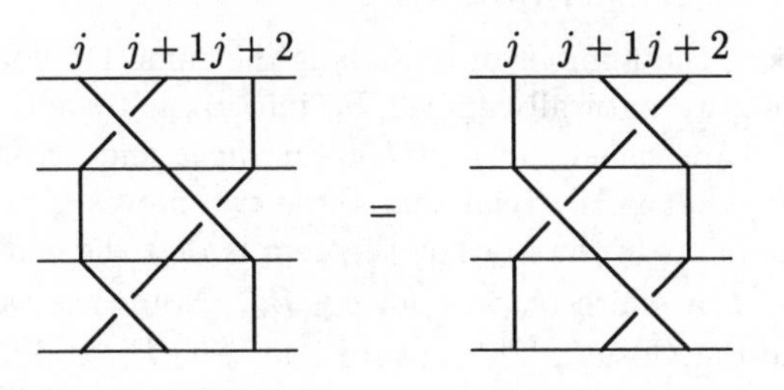

FIG. 12.2. $\sigma_j\sigma_{j+1}\sigma_j = \sigma_{j+1}\sigma_j\sigma_{j+1}$

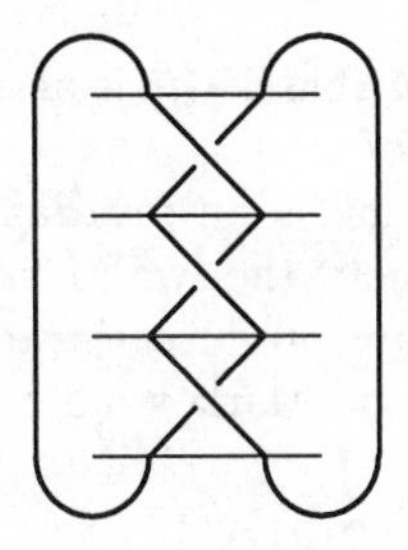

FIG. 12.3. The closure of σ_1^3 is a trefoil

Clearly, these Jones relations are similar to the above braid relations. (This similarity was first pointed out to V. F. R. Jones by D. Hatt and P. de la Harpe.) We set $\sqrt{t} = \exp(\pi i/m)$ so that we have $(4\cos^2 \pi/m)^{-1} = t(1+t)^{-2}$. We also put $g_j = \sqrt{t}(te_j - (1-e_j))$. Then we have the following relations:

1. $g_j g_k = g_k g_j$, if $|j-k| \geq 2$;
2. $g_j g_{j+1} g_j = g_{j+1} g_j g_{j+1}$.

Thus we have a representation $r_m : \sigma_j \mapsto g_j$ of the braid group B_n into the von Neumann algebra generated by the Jones projections $\{e_j\}_j$.

If we have a braid $b \in B_n$, we can connect the top and the bottom of the kth strand for each k trivially to get a link. We call this link the *closure* of the braid b and denote it by $\hat{b}$. (The example of Fig. 12.3 yields a trefoil knot.) A classical theorem of (Alexander 1930) says that any link can be obtained as a closure of some braid. It is possible that different braids give the same closure, but it is known by a classical theorem of Markov when two braids give the same closure. That is, different braids give the same closure if and only if they are equivalent in the equivalence relation generated by the following two types of moves:

1. for $b, c \in B_n$, we have $b \sim cbc^{-1}$;

2. for $b \in B_n \subset B_{n+1}$, we have $b \sim b\sigma_n^{\pm 1}$.

In the first relation, the number of strands is the same for both braids, but in the second relation, we naturally embed B_n into B_{n+1} by adding a new strand to the right. They are called the *first Markov move* and *second Markov move* respectively. (Note that it is trivial that these two moves give braids with the same closure. The point of the Markov theorem is that the converse holds.)

Suppose a link L is expressed as $\hat{b}$ for $b \in B_n$. (Note that we have to specify B_n when we take the closure. The closure $\hat{b}$ for $b \in B_n$ is different from $\hat{b}$ for the same b in B_{n+1} – the latter has one extra component.) Then we define a (Laurent) polynomial $V_L(t)$ of $\sqrt{t}$ by

$$V_L(t) = \left(-\frac{1+t}{\sqrt{t}}\right)^{n-1} \operatorname{tr}(r_m(b)).$$

(Here $\sqrt{t}$ is a number $\exp \pi i/m$, thus $V_L(t)$ is also a number, but we can regard $V_L(t)$ as a Laurent polynomial of $\sqrt{t}$.)

We show that this polynomial is an invariant of the link. It is enough to prove that $V_L(t)$ is invariant under the two Markov moves. The trace property $\operatorname{tr}(r_m(b)) = \operatorname{tr}(r_m(cbc^{-1}))$ easily produces invariance under the first Markov move. Invariance under the second Markov move follows from the Markov trace property. For example, we have

$$\begin{aligned}
&\operatorname{tr}(r_m(b)g_n) \\
&= \operatorname{tr}(r_m(b)\sqrt{t}(te_n - (1-e_n))) \\
&= \sqrt{t}\operatorname{tr}(r_m(b)(-1+(1+t)e_n)) \\
&= \sqrt{t}(-\operatorname{tr}(r_m(b)) + (1+t)\operatorname{tr}(r_m(b)e_n)) \\
&= \sqrt{t}\operatorname{tr}(r_m(b))\left(-1+\frac{t}{t+1}\right) \\
&= \frac{-\sqrt{t}}{t+1}\operatorname{tr}(r_m(b)).
\end{aligned}$$

This $V_L(t)$ is called the *Jones polynomial* of the link L. For example, a trefoil is expressed as $\sigma_1^{3\wedge}$. Thus its Jones polynomial is equal to

$$\begin{aligned}
&-\frac{1+t}{\sqrt{t}}\operatorname{tr}(g_1^3) \\
&= -\frac{1+t}{\sqrt{t}}\operatorname{tr}(\sqrt{t}^3(te_1 - (1-e_1))^3) \\
&= -t(1+t)\operatorname{tr}((-1+(1+t)e_1)^3) \\
&= -t(1+t)\operatorname{tr}(t^3 e_1 + e_1 - 1) \\
&= -t(1+t)\left((t^3+1)\frac{t}{(t+1)^2} - 1\right)
\end{aligned}$$

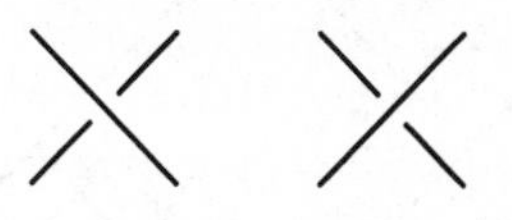

FIG. 12.4. Crossings

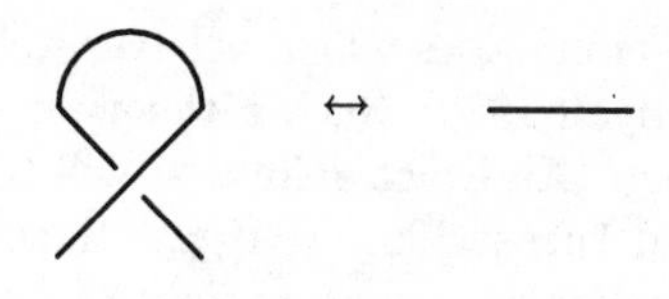

FIG. 12.5. The Reidemeister move of type I

$$= -t^2(t^2 - t + 1) + t + t^2$$
$$= -t^4 + t^3 + t.$$

In the above definition of the Jones polynomial, we first represented a link in a specific way as the closure of a braid. Then we defined a polynomial based on this representation, and later proved that the polynomial does not depend on specific representations, and hence defined an invariant of the link. This is a general procedure in combinatorial constructions in the recent theory of quantum invariants of invariants of links and manifolds.

E. Witten predicted, based on physical arguments, topological invariants of three dimensional manifolds related to the Jones polynomial in (Witten 1989a). In this theory, we have an assignment of Hilbert spaces to compact three di-

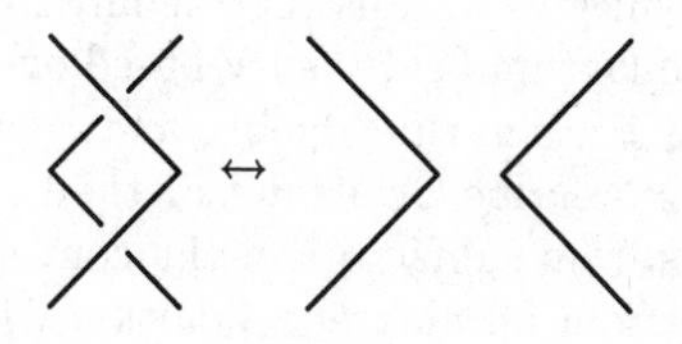

FIG. 12.6. The Reidemeister move of type II

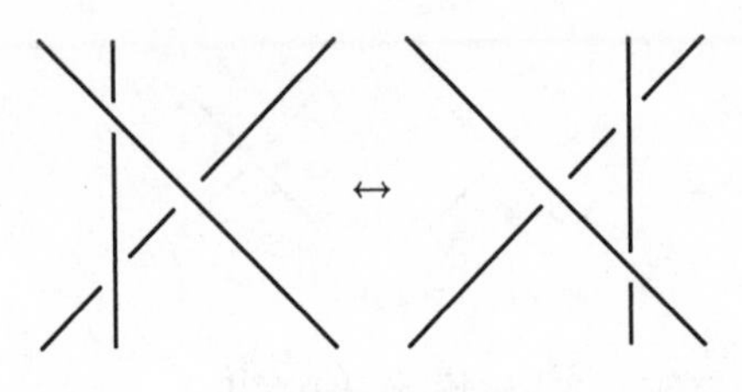

FIG. 12.7. The Reidemeister move of type III

mensional manifolds with boundaries which behaves well with respect to certain axioms formulated by (Atiyah 1989). Such a theory in general is called a *topological quantum field theory.* (An exact definition will be given in Section 12.2.)

It was (Reshetikhin and Turaev 1991) who gave the first rigorous construction of a three-dimensional topological quantum field theory. In their construction, they realize a three-manifold with a surgery along a link in S^3, and use a theorem of (Kirby 1978) to show topological invariance. (The theorem of Kirby plays the role of the Markov theorem in telling us which links give rise to the same manifold via surgery.) This method has been generalized by several authors. There is another construction of three dimensional topological quantum field theory based on triangulation due to (Turaev and Viro 1992). It is this method that we describe in this chapter in connection with subfactor theory.

Both methods of Reshetikhin–Turaev and Turaev–Viro are quite general and depend on initial data. An example of such initial data is given by the quantum group $U_q(sl_2(\mathbb{C}))$ at certain roots of unity. In this case, the Reshetikhin–Turaev invariant and the Turaev–Viro invariant are both complex number valued invariants of closed three-manifolds. It has been shown by (Turaev, preprint 1991) that the latter is the square of the absolute value of the former for $q = \exp(\pi i/r)$. With this theorem, the Reshetikhin–Turaev invariant looks more fundamental than the Turaev–Viro invariant, but the Turaev–Viro method does have an advantage over the Reshetikhin–Turaev method, because it requires only a 'lower symmetry' in the definition. It is this point that plays a key role in connection to the paragroup theory.

In Section 12.2, we will give an abstract definition of a Turaev–Viro three dimensional topological quantum field theory based on triangulation of oriented manifolds with an abstract fusion rule algebra and quantum $6j$-symbols.

Sections 12.3 and 12.4 describe Ocneanu's method to get fusion rule algebras and quantum $6j$-symbols from subfactors in the converse direction. We explain how to get a paragroup from an abstract quantum $6j$-symbol in Section 12.5. The results in Sections 12.3–12.5 mean that the notion of a paragroup can be also formulated in terms of a (graded) fusion algebra with quantum $6j$-symbols. This formulation is conceptually clearer and matches several constructions in this and the next chapters well. In Section 12.6, we construct Ocneanu's tube algebra, finite dimensional C^*-algebras, from a paragroup. We discuss its topological and operator algebraic meaning. For this purpose, we also describe Ocneanu's subfactor construction of an asymptotic inclusion. Analogues of the S and T-matrices and the Verlinde identity in the subfactor theory are presented in Section 12.7. It is explained in Section 12.8 that the results in Sections 12.6 and 12.7 are naturally interpreted in the framework of the 'quantum double' construction for paragroups. We show in Section 12.9 that Longo's sector theory in Section 10.7 gives an alternative method to the bimodule theory and that it does have some advantage over the bimodule theory.

12.2 3-dimensional TQFT of Turaev–Viro type based on triangulation

We construct in this section a three dimensional topological quantum field theory, abbreviated as TQFT$_3$, based on the Turaev–Viro triangulation method.

As initial data of this general construction, we need an abstract *fusion rule algebra* and *quantum 6j-symbols*.

Definition 12.1 *Suppose we have a finite set $\{X_i\}_{i\in I}$. If the free $\mathbb{C}$-module $\bigoplus_{i\in I}\mathbb{C}X_i$ has a C^*-algebra structure with the following properties, we call this algebra a* fusion rule algebra.

1. *We have a set of non-negative integers N_{XY}^Z for $X, Y, Z \in \{X_i\}_{i\in I}$ so that $XY = \sum_Z N_{XY}^Z Z$.*
2. *We have a bijective map $^-$ on $\{X_i\}_{i\in I}$ so that the $*$-operation is given by $X_i \mapsto \bar{X}_i$.*
3. *We have $X_0 \in \{X_i\}_{i\in I}$ which gives the identity of the C^*-algebra.*
4. *For $X, Y, Z \in \{X_i\}_{i\in I}$, we have*

$$N_{XY}^Z = N_{\bar{Z}X}^{\bar{Y}} = N_{Y\bar{Z}}^{\bar{X}} = N_{\bar{Y}\bar{X}}^{\bar{Z}} = N_{\bar{X}Z}^{Y} = N_{Z\bar{Y}}^{X}.$$

Note that we do *not* assume that fusion rule algebras are commutative as in Chapter 8. In the following, we also need the following definition of the abstract intertwiner space.

Definition 12.2 *Suppose we have a fusion rule algebra as in Definition 12.1. We say that the fusion rule algebra has* intertwiner spaces *if the following hold:*

1. *For each $X, Y, Z \in \{X_i\}_{i\in I}$, we have a Hilbert space $\mathcal{H}_{XY}^Z$ with dimension N_{XY}^Z. This Hilbert space is called an intertwiner space.*
2. *For any $X, Y, Z \in \{X_i\}_{i\in I}$, the symmetric group on the set $\{X, Y, Z\}$ gives an action of S_3 by isomorphisms between the following spaces:*

$$\mathcal{H}_{XY}^Z, \mathcal{H}_{\bar{Z}X}^{\bar{Y}}, \mathcal{H}_{Y\bar{Z}}^{\bar{X}}, \mathcal{H}_{\bar{Y}\bar{X}}^{\bar{Z}}, \mathcal{H}_{\bar{X}Z}^{Y}, \mathcal{H}_{Z\bar{Y}}^{X}.$$

Here the even permutations act as unitaries and the odd permutations act as bijective conjugate linear isometries. (Note that these spaces have the same dimension by Definition 12.1 (4).)

A classical example of a fusion rule algebra with intertwiner spaces is given by a representation ring of a finite group. From our viewpoint, the following is a fundamental example containing this classical case.

Example 12.3 Let $N \subset M$ be a subfactor with finite depth. Let I be the set of the irreducible N-N bimodules arising from the subfactor as in the definition of the paragroup in Section 10.2. With the relative tensor product and the conjugate operation, they make a fusion rule algebra. The intertwiner spaces are given as $\mathcal{H}_{X,Y}^{Z}$ as in Theorem 9.73.

With Example 12.3 in mind, we use the same notation as in Theorem 9.73 for isomorphisms in Definition 12.2. For example, the image of $\sigma \in \mathcal{H}_{X,Y}^{Z}$ in $\mathcal{H}_{Z,\bar{Y}}^{X}$ under the isomorphism is denoted by σ^{Y} and the image of $\sigma \in \mathcal{H}_{X,Y}^{Z}$ in $\mathcal{H}_{\bar{Y},\bar{X}}^{\bar{Z}}$ under the isomorphism is denoted by $\bar{\sigma}$.

Definition 12.4 *Let $A, B, C, D, X, Y \in \{X_i\}_{i\in I}$. Suppose that we have a sesquilinear form on $\mathcal{H}_{XA}^{B} \otimes \mathcal{H}_{BY}^{D}$ and $\mathcal{H}_{AY}^{C} \otimes \mathcal{H}_{XC}^{D}$ denoted by*

$$Z(A, B, C, D, X, Y \mid \sigma_1, \sigma_2, \sigma_3, \sigma_4)$$

for $\sigma_1 \in \mathcal{H}_{XA}^{B}$, $\sigma_2 \in \mathcal{H}_{BY}^{D}$, $\sigma_3 \in \mathcal{H}_{AY}^{C}$, $\sigma_4 \in \mathcal{H}_{XC}^{D}$ and that we have a map $[\,\cdot\,]$ from $\{X_i\}_{i\in I}$ to the set of positive real numbers. We assume that the map $[\,\cdot\,]$ satisfies the following:

1. $[X] = [\bar{X}]$ *for any* $X \in \{X_i\}_{i\in I}$;
2. $[X]^{1/2}[Y]^{1/2} = \sum_{Z\in\{X_i\}} N_{XY}^{Z}[Z]^{1/2}$ *for any* $X, Y, Z \in \{X_i\}_{i\in I}$.

This $Z(A, B, C, D, X, Y \mid \sigma_1, \sigma_2, \sigma_3, \sigma_4)$ is called a quantum $6j$-symbol *if it satisfies the three properties; tetrahedral symmetry (Definition 12.5), unitarity (Definition 12.6), and the pentagon identity (Definition 12.7).*

Note that when we use the word 'quantum' here, we have an axiomatic combinatorial approach and do not have a deformation parameter q as in the quantum group theory. In a purely combinatorial approach to the quantum $6j$-symbols, the values $[X]$ do not have to be positive in general, but we require this positivity here because this condition is natural from the viewpoint of the subfactor theory, and positive inner product spaces.

Definition 12.5 *We first draw a labelled tetrahedron as in Fig. 12.8 for*

$$Z(A, B, C, D, X, Y \mid \sigma_1, \sigma_2, \sigma_3, \sigma_4).$$

(That is, six oriented edges are labelled with A, B, C, D, X, Y and four faces are labelled with $\sigma_1, \sigma_2, \sigma_3, \sigma_4$.) We rotate this tetrahedron and/or take its mirror image, and get a new labelled tetrahedron. (One example is given in Fig. 12.9. There are 24 such labelled tetrahedra where 12 of them are mirror images of the original tetrahedron.) Then we reverse orientations of some edges so that the orientations match those in the original Fig. 12.8. When we reverse an orientation of an edge, we put a bar on its label, and relabel faces according to the

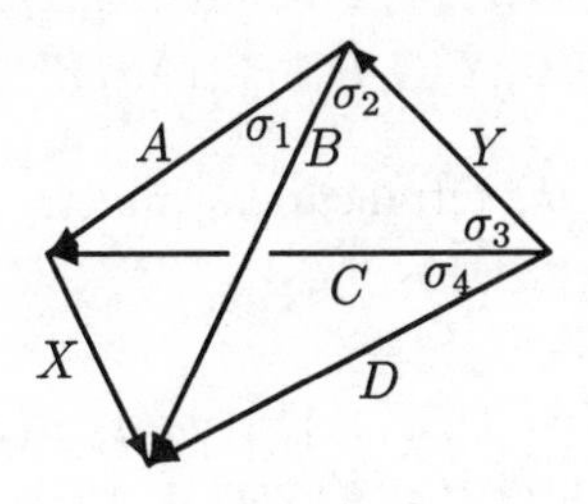

FIG. 12.8. A labelled tetrahedron

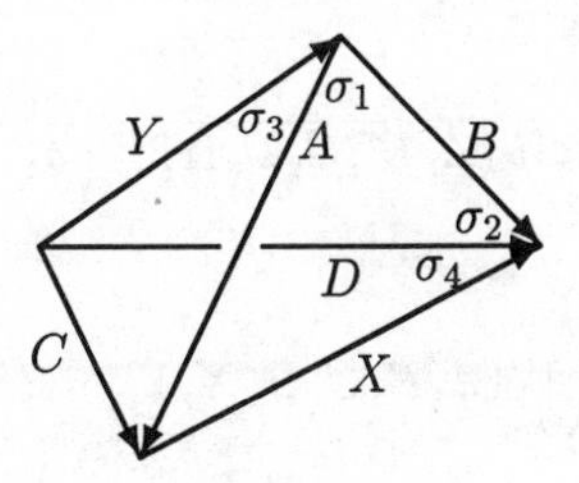

FIG. 12.9. A rotated labelled tetrahedron

isomorphisms in Definition 12.2. (In the case of Fig. 12.9, the resulting relabelling is given in Fig. 12.10.)

Then we require that the value of Z for the relabelled figure is the same as [resp. complex conjugate of] the value of the original Z when we change the original figure without [resp. by] taking a mirror image.

In the case of the above example, this means the identity

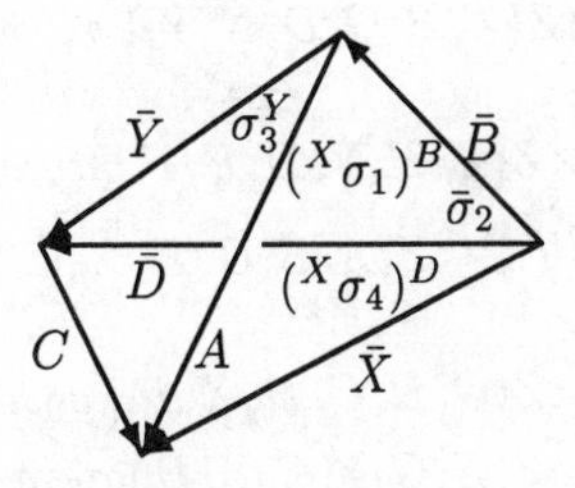

FIG. 12.10. A labelled tetrahedron rotated and relabelled

$$\begin{aligned}&Z(A,B,C,D,X,Y \mid \sigma_1,\sigma_2,\sigma_3,\sigma_4)\\&= Z(\bar{Y},A,D,\bar{X},C,\bar{B} \mid \sigma_3^Y,({}^X\sigma_1)^B,\bar{\sigma}_2,({}^X\sigma_4)^D).\end{aligned}$$

This property is called the tetrahedral symmetry *of quantum 6j-symbols.*

Definition 12.6 *Setting*

$$\begin{aligned}&W(A,B,C,D,X,Y \mid \sigma_1,\sigma_2,\sigma_3,\sigma_4)\\&= [B]^{1/4}[C]^{1/4}Z(A,B,C,D,X,Y \mid \sigma_1,\sigma_2,\sigma_3,\sigma_4),\end{aligned}$$

we require the following two identities for $A,B,C,D,X,Y \in \{X_i\}_{i\in I}$.

$$\begin{aligned}&\sum_{C,\sigma_3,\sigma_4}\Big(W(A,B,C,D,X,Y \mid \sigma_1,\sigma_2,\sigma_3,\sigma_4)\\&\qquad\times\overline{W(A,B',C,D,X,Y \mid \sigma_1',\sigma_2',\sigma_3,\sigma_4)}\Big)\\&= \delta_{(B,\sigma_1,\sigma_2),(B',\sigma_1',\sigma_2')},\\&\sum_{B,\sigma_1,\sigma_2}\Big(W(A,B,C,D,X,Y \mid \sigma_1,\sigma_2,\sigma_3,\sigma_4)\\&\qquad\times\overline{W(A,B,C',D,X,Y \mid \sigma_1,\sigma_2,\sigma_3',\sigma_4')}\Big)\\&= \delta_{(C,\sigma_3,\sigma_4),(C',\sigma_3',\sigma_4')}.\end{aligned}$$

(In the first identity, we take the summation so that σ_3 and σ_4 form orthonormal bases of $\mathcal{H}_{AY}^C$ and $\mathcal{H}_{XC}^D$ respectively. In the second identity, we take the summation so that σ_1 and σ_2 form orthonormal bases of $\mathcal{H}_{XA}^B$ and $\mathcal{H}_{BY}^D$ respectively.) This property is called unitarity *of the quantum 6j-symbols.*

Definition 12.7 *For $A,B,C,D,E,F,G,H,X,Y \in \{X_i\}_{i\in I}$, we require the following identity.*

$$\begin{aligned}&\sum_{\sigma_2}\big(Z(A,B,C,D,X,Y \mid \sigma_1,\sigma_2,\sigma_3,\sigma_4)\\&\qquad\times Z(E,H,Y,D,B,F \mid \sigma_5,\sigma_6,\sigma_7,\sigma_2)\big)\\&= \sum_{G,\sigma_8,\sigma_9,\sigma_{10}}\big([G]^{1/2}Z(A,B,G,H,X,E \mid \sigma_1,\sigma_5,\sigma_8,\sigma_9)\\&\qquad\times Z(G,H,C,D,X,F \mid \sigma_9,\sigma_6,\sigma_{10},\sigma_4)\\&\qquad\times Z(E,G,Y,C,A,F \mid \sigma_8,\sigma_{10},\sigma_7,\sigma_3)\big),\end{aligned}$$

where we have $\sigma_1 \in \mathcal{H}_{XA}^B$, $\sigma_2 \in \mathcal{H}_{BY}^D$, $\sigma_3 \in \mathcal{H}_{AY}^C$, $\sigma_4 \in \mathcal{H}_{XC}^D$, $\sigma_5 \in \mathcal{H}_{BE}^H$, $\sigma_6 \in \mathcal{H}_{HF}^D$, $\sigma_7 \in \mathcal{H}_{EF}^Y$, $\sigma_8 \in \mathcal{H}_{AE}^G$, $\sigma_9 \in \mathcal{H}_{XG}^H$, $\sigma_{10} \in \mathcal{H}_{GF}^C$, and the summations are taken so that σ_2, σ_8, σ_9 and σ_{10} form orthonormal bases of $\mathcal{H}_{BY}^D$, $\mathcal{H}_{AE}^G$,

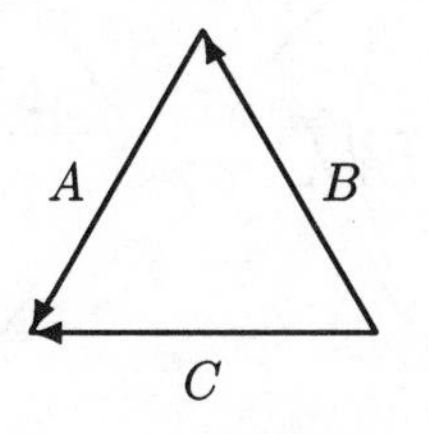

FIG. 12.11. A labelled triangle

$\mathcal{H}_{XG}^{H}$ and $\mathcal{H}_{GF}^{C}$ respectively. This property is called the pentagon identity *of the quantum $6j$-symbols.*

We now fix a fusion rule algebra with intertwiner spaces and quantum $6j$-symbols. We will construct a three dimensional topological quantum field theory (TQFT$_3$) from this data as follows.

First we assume that we have a compact oriented triangulated three-manifold P without boundary. We call this triangulation T. That is, we think of P as being assembled from finitely many tetrahedra. Let C_e be the set of all the possible assignments of $X \in \{X_i\}_{i\in I}$ to each edge appearing in the triangulation T. Fix an element c in C_e. Then each triangle in the triangulation is labelled as in Fig. 12.11. Then we assign σ_j to this labelled triangle so that the σ_j make an orthonormal basis of $\mathcal{H}_{AB}^{C}$. Let $C_f(c)$ be the set of all the possible assignments of σ_j to each triangle appearing in the triangulation T. We define $w = \sum_{i\in I}[X_i]$. For $c \in C_e$, we set $[c]$ to be the product of all $[X_j]$ appearing in c (with multiplicity counted). We set a to be the number of vertices appearing in the triangulation T. Fix an element $d \in C_f(c)$. We have labelled tetrahedra, so each tetrahedron is assigned a number given by the quantum $6j$-symbols. This number is well defined because of the tetrahedral symmetry. We define $Z(d)$ to be the product of these values over all the tetrahedra.

We then make the following definition.

Definition 12.8 *In the above setting, we set*

$$Z(P,T) = w^{-a} \sum_{c\in C_e} \sum_{d\in C_f(c)} [c]^{1/2} Z(d). \tag{12.2.1}$$

This is called a simplicial state sum.

We will often simply call a *state sum* instead of a simplicial state sum. The above definition of the simplicial state sum uses a specific triangulation T. Our aim is to prove that it does not depend on T and hence defines a complex valued topological invariant of a manifold P. (Conceptually, this construction is similar to that of the Jones polynomial in Section 12.1 – we first give a definition

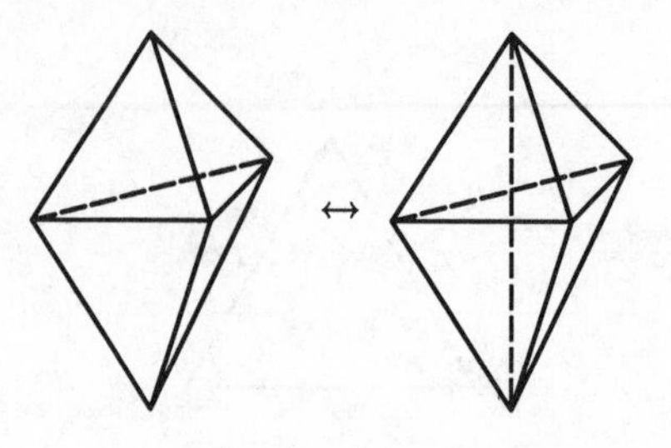

FIG. 12.12. Move I

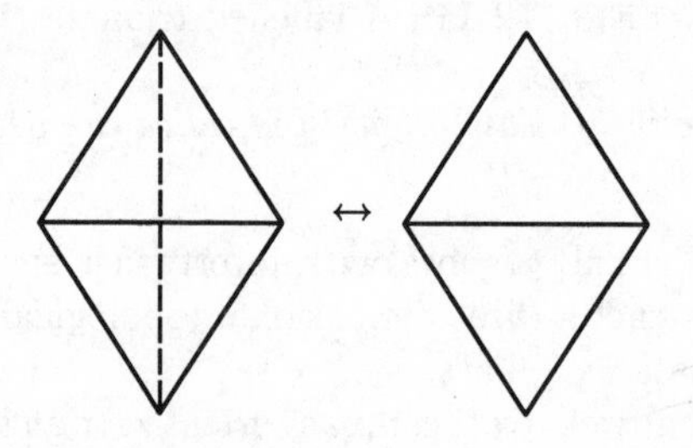

FIG. 12.13. Move II

depending on a specific representation of a topological object and later prove that it does not depend on representations.)

For this purpose, we need to know how different triangulations of the same manifold are related. There is such a characterization due to (Pachner 1978), which plays the same role the Markov theorem did in the Jones polynomial. The Pachner theorem is based on classical theorems of Alexander and Newman and holds in any dimension. We will use it in the following form in three dimensions.

Theorem 12.9 *Two triangulations of the same closed three-manifold can be transformed from one to the other by a finite sequence of moves from the following list of transformations and their inverses.*

1. *(Move I) We have two tetrahedra sharing one common triangle. Then we split these into three tetrahedra sharing one common triangle pairwise as in Fig. 12.12.*
2. *(Move II) We have two tetrahedra sharing two common triangles. Then these collapse into two triangles as in Fig. 12.13.*
3. *(Move III) We have two tetrahedra sharing three common triangles. Then these collapse into one triangle as in Fig. 12.14.*

Thus we only need to prove that the definition in (12.2.1) is invariant under these three local moves.

The invariance under Move I is exactly the pentagon identity. (Note that the term $[c]^{1/2}$ in 12.2.1 is inserted so that the factor $[G]^{1/2}$ on the right hand side of the pentagon identity cancels out in Move I.) In the case of Move II, what we

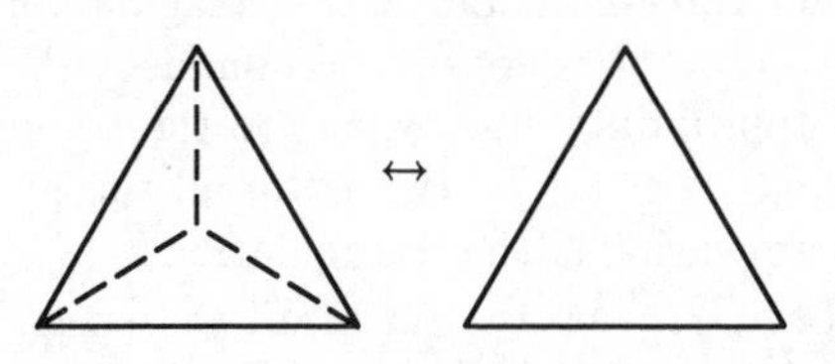

FIG. 12.14. Move III

have to prove is that the value for the tetrahedra is equal to the Kronecker δ of two pairs of labelled triangles. This is exactly unitarity. We only have to check the coefficients $[B]^{1/4}[C]^{1/4}$ in the definition of W and the coefficients $[X]^{1/2}$ in that of $[c]^{1/2}$ cancel out.

To study Move III, we need a lemma.

Lemma 12.10 *For any $Z \in \{X_i\}_{i\in I}$, we have*

$$w = [Z]^{-1/2} \sum_{X,Y\in\{X_i\}} N_{X,Y}^{Z}[X]^{1/2}[Y]^{1/2}.$$

Proof By $N_{XY}^{Z} = N_{Z\bar{Y}}^{X}$, we get

$$\begin{aligned}
&[Z]^{-1/2} \sum_{X,Y\in\{X_i\}} N_{X,Y}^{Z}[X]^{1/2}[Y]^{1/2} \\
&= [Z]^{-1/2} \sum_{Y\in\{X_i\}} [Y]^{1/2} \sum_{X\in\{X_i\}} N_{Z\bar{Y}}^{X}[X]^{1/2} \\
&= [Z]^{-1/2} \sum_{Y\in\{X_i\}} [Y]^{1/2}[Z]^{1/2}[Y]^{1/2} \\
&= \sum_{Y\in\{X_i\}} [Y] = w.
\end{aligned}$$

□

Then invariance under Move III is again by unitarity.

Thus we have proved that the formula (12.2.1) defines a topological invariant of an oriented closed three-manifold P. We write $Z(P)$ for this value. Furthermore, the tetrahedral symmetry, Definition 12.5, implies that if we take a mirror image of P, all the values for the tetrahedra become complex conjugate. So we have proved the following theorem.

Theorem 12.11 *The formula 12.2.1 defines a complex-valued topological invariant for oriented closed three-manifolds from quantum 6j-symbols. If we take a mirror image of the manifold, then the resulting value of the invariant is the complex conjugate.*

We next study the case of a manifold with boundary.

Let P be a compact three-manifold with triangulation T. Let ϕ be an assignment of $X \in \{X_i\}_{i \in I}$ to the edges on the boundary ∂P in the triangulation and an assignment of appropriate intertwiners to the triangles on the boundary ∂P in the triangulation T. Let C_e^ϕ be the set of all the possible assignments of $X \in \{X_i\}_{i \in I}$ to edges appearing in the triangulation T of P which extend ϕ. Fix an element c in C_e^ϕ. Let $C_f^\phi(c)$ be the set of all the possible assignments of σ_j to each triangle appearing in the triangulation T which extend ϕ. (For triangles which are not on the boundary, we again assign σ_j to each labelled triangle so that they form an orthonormal basis.) For $c \in C_e^\phi$, we set $[c]$ to be the product of all $[X_j]$ appearing in the edges which are not on the boundary in c and all the $[X_j]^{1/2}$ appearing in the edges which are on the boundary in c. Let a' be the total number of vertices on the boundary ∂P in the triangulation T and $a + a'$ be the number of vertices in the triangulation T of P. Fix an element $d \in C_f^\phi(c)$. We have labelled the tetrahedra, so each tetrahedron has a value given by the quantum $6j$-symbols. This number is well defined because of the tetrahedral symmetry. We define $Z(d)$ to be the product of these values over all tetrahedra.

We then make the following definition.

Definition 12.12 *In the above setting, we put*

$$Z_\phi(P, T) = w^{-a-a'/2} \sum_{c \in C_e^\phi} \sum_{d \in C_f^\phi(c)} [c]^{1/2} Z(d). \tag{12.2.2}$$

Note that the weights for the vertices and edges on the boundary are given so that their weights are square roots of those of the vertices and edges inside the manifold. As before, we can prove that this $Z_\phi(P, T)$ does not depend on the triangulation T as long as we fix the triangulation on the boundary ∂P and the assignment ϕ on the boundary ∂P. We write $Z_\phi(P)$ for this value. (Instead of the classical theorem of Alexander mentioned above, we need to appeal to a new relative version with boundary, due to (Turaev and Viro 1992).)

We next construct a topological quantum field theory from the above data. For a triangulated surface S, we define a Hilbert space $C(S)$ as follows. We consider all assignments of $\{X_i\}_{i \in I}$ to the edges in the triangulation and all assignments of σ_j to labelled triangles in the triangulation. After fixing an assignment of $\{X_i\}_{i \in I}$ to the edges, the set of all assignments of σ_j to all the labelled triangles has a natural Hilbert space structure arising from that in Definition 12.2. (This is the tensor product of all the Hilbert spaces associated with labelled triangles.) Then we define $C(S)$ to be the direct sum of these Hilbert spaces over all possible assignments of $\{X_i\}_{i \in I}$ to the edges in the triangulation.

Choose a cobordism $W = (P, i_+, i_-)$ between triangulated surfaces S_+, S_-. That is, P is a compact three-manifold and $i_\pm : S_\pm \to \partial P$ are embeddings with $\partial P = i_+(S_+) \cup i_-(S_-)$ and $i_+(S_+) \cap i_-(S_-) = \varnothing$. We can define a linear map $\Phi_W : C(S_+) \to C(S_-)$ so that we have $(\Phi_W(\alpha), \beta) = Z_{\alpha \cup \beta}(P)$ for $\alpha \in C(S_+)$, $\beta \in C(S_-)$. Cobordisms can be considered as morphisms of a category whose objects are surfaces. The composition of two cobordisms $W_1 = (P_1, i_1 : S_1 \to \partial P_1, i_2 : S_2 \to \partial P_1)$ and $W_2 = (P_2, i_2 : S_2 \to \partial P_2, i_3 : S_3 \to \partial P_2)$ is given as $W_2 \cdot W_1 = (P_1 \cup P_2, i_1, i_3)$, given by gluing P_1 and P_2 along S_2. Then Definition 12.12 implies $\Phi_{W_2 \cdot W_1} = \Phi_{W_2} \cdot \Phi_{W_1}$.

In this definition, the map Φ for a trivial cobordism $\mathrm{id}_S = (S \times [0,1], S \times \{0\}, S \times \{1\})$ may not be an identity. To eliminate this difficulty, we introduce a new Hilbert space defined by $Q(S) = C(S)/\mathrm{Ker}(\Phi_{\mathrm{id}_S})$. Any cobordism $W = (P, i_+ : S_+ \to \partial P, i_- : S_- \to \partial P)$ is homeomorphic to $W \cdot \mathrm{id}_{S_+}$ and this implies $\mathrm{Ker}(\Phi_{\mathrm{id}_{S_+}}) \subset \mathrm{Ker}(\Phi_W)$. Thus the map $\Phi_W : C(S_+) \to C(S_-)$ induces a map from $Q(S_+)$ to $Q(S_-)$. We denote this map by Ψ_W. By the definition, Ψ_{id_S} is injective, and then the identity $\Psi_{\mathrm{id}_S} \cdot \Psi_{\mathrm{id}_S} = \Psi_{\mathrm{id}_S}$ implies $\Psi_{\mathrm{id}_S} = \mathrm{id}$ on $Q(S)$. Thus we have obtained the following two conditions:

1. $\Psi_{\mathrm{id}_S} = \mathrm{id}$ on $Q(S)$;
2. $\Psi_{W_2 \cdot W_1} = \Psi_{W_2} \cdot \Psi_{W_1}$.

Furthermore, it is easy to see that the definition of $Q(S)$ does not depend on the triangulation of S. Thus the maps $S \mapsto Q(S)$ and $W \mapsto \Psi_W$ give a functor from the category of cobordism of surfaces to the category of finite dimensional Hilbert spaces.

Definition 12.13 *A functor as above is called a* three dimensional topological quantum field theory. *It will be abbreviated as* $TQFT_3$.

There are more general definitions of topological quantum field theories. (See (Atiyah 1989) for example.) Here we required that the spaces $Q(S)$ have Hilbert space structures and are finite dimensional, and we constructed the functor based on triangulation. When we want to specify these conditions explicitly, we say that our topological quantum field theory is a three dimensional unitary rational simplicial topological quantum field theory. The adjectives *unitary*, *rational*, and *simplicial* refer to the Hilbert space structure, finite dimensionality, and the method based on triangulation, respectively. See (Turaev 1994) for more details.

Remark It is easy to define an equivalence relation of quantum $6j$-symbols based on gauge choice as in Definition 10.11 for connections. It is clear that equivalent quantum $6j$-symbols give the same TQFT3.

A. Ocneanu has recently proved that for a given fusion rule algebra, we have only finitely many equivalence classes of quantum $6j$-symbols. His main argument shows that a 'small perturbation' of quantum $6j$-symbols does not change the equivalence class of the quantum $6j$-symbols. As a corollary, this implies that we have only finitely many equivalence classes of flat connections on a given finite graph, using the results in Sections 12.4 and 12.5.

Exercise 12.1 Prove that $Z(S^3) = 1/w$ for the three dimensional sphere S^3.

Exercise 12.2 Prove that $Z(S^2 \times S^1) = 1$ for the product of the two and one dimensional spheres.

12.3 Fusion rule algebras for subfactors

We introduce a fusion rule algebra from a subfactor which has finite index and finite depth.

Let $N \subset M$ be a subfactor with finite index and finite depth. We have a finite system of four kinds of irreducible bimodules arising from ${}_N M_M$ as in Section 10.2. When we take two bimodules from the system, we can take a relative tensor product as in Section 9.7 as long as the algebra acting from the right on the first bimodule coincides with the algebra acting from the left on the second bimodule. The resulting bimodule decomposes into a finite direct sum of bimodules in the system by the finite depth condition. We call this multiplication rule the *fusion rule* of the subfactor.

If we restrict our attention to the set of N-N bimodules in the system, the free $\mathbb{C}$-module over the system naturally gives a fusion rule algebra in the sense of Definition 12.1, where the product is given by the relative tensor product and the involution $*$ is given by the conjugate operation on bimodules. If we consider all four kinds of bimodules, our algebra has a restricted multiplication. We will study this kind of *graded fusion rule algebra* in more detail in Section 12.5.

Example 12.14 Let us consider the example of a paragroup arising from a finite group as in Section 10.6. In this example, the N-N bimodules are labelled by elements of the finite group G and their multiplication comes from the multiplication in the group. So in this case, the fusion rule algebra is nothing but the group algebra $\mathbb{C}[G]$. Note that this is a non-commutative algebra if the group G is non-commutative.

The principal graph of a subfactor gives partial information of the fusion rule. That is, the graph gives us the multiplication rule by ${}_N M_M$. In general, this information does not determine the entire fusion rule – two groups with the same order give the same principal graph. In some cases, the principal graph gives complete information of the fusion rule, and in other cases, we can tell that a certain graph cannot appear as a principal graph of any subfactor just by looking at the partial information of the fusion rule.

For example, suppose that the D_5 Dynkin diagram were the principal graph of some subfactor $N \subset M$. (We have already seen that this is impossible after Corollary 10.8.) We label bimodules as in Fig. 12.15. At the beginning of Section 11.5, we saw that the dual principal graph must also be D_5, and then the labelling is as in Fig. 12.16. From the labelling in Fig. 12.15, we get the following fusion rules:

$$
\begin{aligned}
{}_N N \otimes_N M_M &= {}_N M_M; \\
{}_N X \otimes_N M_M &= {}_N M_M \oplus {}_N Y_M \oplus {}_N Z_M; \\
{}_N M \otimes_M M_N &= {}_N N_N \oplus {}_N X_N; \\
{}_N Y \otimes_M M_N &= {}_N X_N; \\
{}_N Z \otimes_M M_N &= {}_N X_N.
\end{aligned}
$$

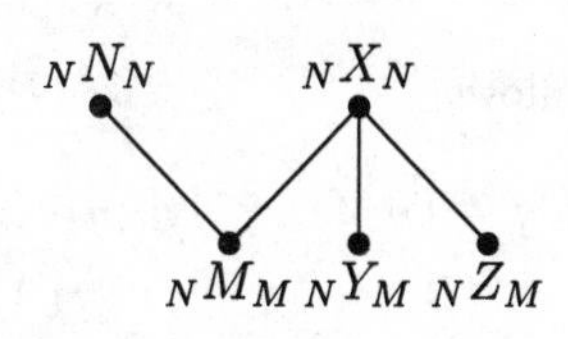

FIG. 12.15. Dynkin diagram D_5

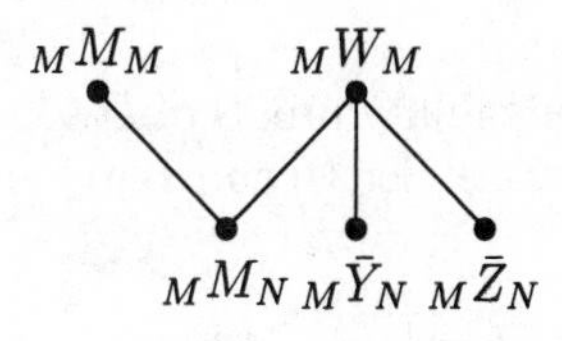

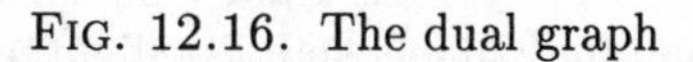

FIG. 12.16. The dual graph

Similarly, from Fig. 12.16, we get the following fusion rules:

$$\begin{aligned}
{}_MM \otimes_M M_N &= {}_MM_N;\\
{}_MW \otimes_M M_N &= {}_MM_N \oplus {}_M\bar{Y}_N \oplus {}_M\bar{Z}_N;\\
{}_MM \otimes_N M_M &= {}_MM_M \oplus {}_MW_M;\\
{}_M\bar{Y} \otimes_N M_M &= {}_MW_M;\\
{}_M\bar{Z} \otimes_N M_M &= {}_MW_M.
\end{aligned}$$

We then have

$$\begin{aligned}
{}_NM_M \oplus {}_NY_M \oplus {}_NZ_M &= {}_NM \otimes_M \bar{W}_M\\
&= {}_NM \otimes_M (M \otimes_N Y)_M\\
&= {}_N(M \otimes_M M)_N \otimes_N Y_M\\
&= ({}_NN_N \oplus {}_NX_N) \otimes_N Y_M\\
&= {}_NY_M \oplus {}_NX \otimes_N Y_M.
\end{aligned}$$

This implies

$$ {}_NX \otimes_N Y_M = {}_NM_M \oplus {}_NZ_M. $$

This, together with ${}_NN \otimes_N Y_M = {}_NY_M$ and Frobenius reciprocity (Theorem 9.71), implies

$$\begin{aligned}
{}_NM \otimes_M \bar{Y}_N &= {}_NX_N,\\
{}_NY \otimes_M \bar{Y}_N &= {}_NN_N,
\end{aligned}$$

$$_N Z \otimes_M \bar{Y}_N = {_N X_N}.$$

Then we can compute as follows:

$$\begin{aligned} 2_N X_N &= ({_N M_M} \oplus {_N Z_M}) \otimes_M \bar{Y}_N \\ &= ({_N X} \otimes_N Y_M) \otimes_M \bar{Y}_N \\ &= {_N X} \otimes_N ({_N Y} \otimes_M \bar{Y}_N) \\ &= {_N X} \otimes_N N_N \\ &= {_N X_N}, \end{aligned}$$

which is a contradiction. So we have obtained another proof of the impossibility of D_5 as a principal graph.

It is possible to give impossibility proofs of D_{2n+1} and E_7 along this line with more complicated computations. See (Izumi 1991) and (Sunder and Vijayarajan 1993) for more details.

12.4 Quantum $6j$-symbols from subfactors

We next produce quantum $6j$-symbols from a subfactor with finite index and finite depth. Our quantum $6j$-symbols here are slightly more general than those in Section 12.2.

Let $N \subset M$ be a subfactor with finite index and finite depth. We then have a finite system of four kinds of bimodules, which form a fusion rule algebra as in Section 12.3. (We use the convention that when we form a product, we always consider only pairs of bimodules for which the relative tensor product is possible.) As in Definition 9.69, we have intertwiner spaces associated with the fusion rule algebra.

Choose four algebras $P, Q, R, S \in \{M, N\}$, six bimodules ${_Q A_R}$, ${_P B_R}$, ${_Q C_S}$, ${_P D_S}$, ${_P X_Q}$, ${_R Y_S}$ from the finite system of bimodules, and four intertwiners $\sigma_1 \in \mathcal{H}^B_{X,A}$, $\sigma_2 \in \mathcal{H}^D_{B,Y}$, $\sigma_3 \in \mathcal{H}^C_{A,Y}$, $\sigma_4 \in \mathcal{H}^D_{X,C}$. Then the composition $\sigma_2(\sigma_1 \otimes 1_Y)(1_X \otimes \sigma_3)^* \sigma_4^*$ gives an endomorphism of ${_P D_S}$, which is a scalar by irreducibility.

Definition 12.15 *We set*

$$W(A, B, C, D, X, Y \mid \sigma_1, \sigma_2, \sigma_3, \sigma_4) = \sigma_2(\sigma_1 \otimes 1_Y)(1_X \otimes \sigma_3)^* \sigma_4^*,$$

and define the quantum $6j$-symbol Z by

$$\begin{aligned} &Z(A, B, C, D, X, Y \mid \sigma_1, \sigma_2, \sigma_3, \sigma_4) \\ &= [B]^{-1/4}[C]^{-1/4} W(A, B, C, D, X, Y \mid \sigma_1, \sigma_2, \sigma_3, \sigma_4). \end{aligned}$$

By direct computation based on Frobenius reciprocity, we get the following identities:

$$Z(A, B, C, D, X, Y \mid \sigma_1, \sigma_2, \sigma_3, \sigma_4)$$

$$= Z(\bar{A},\bar{C},\bar{B},\bar{D},\bar{Y},\bar{X},|\,\bar{\sigma}_3,\bar{\sigma}_4,\bar{\sigma}_1,\bar{\sigma}_2);$$
$$= \frac{Z(\bar{A},X,Y,D,B,C\mid \sigma_1^A,\sigma_4,{}^A\sigma_3,\sigma_2)}{Z(A,B,C,D,X,Y\mid \sigma_1,\sigma_2,\sigma_3,\sigma_4)};$$
$$= \frac{Z(B,A,D,C,\bar{X},Y\mid {}^X\sigma_1,\sigma_3,\sigma_2,{}^X\sigma_4)}{Z(A,B,C,D,X,Y\mid \sigma_1,\sigma_2,\sigma_3,\sigma_4)}.$$

Here the first and the third identities are easy. For the second identity, we need to prove

$$(\sigma_1^A\otimes 1_C)(1_B\otimes {}^A\sigma_3)^* = \frac{[X]^{1/4}[Y]^{1/4}}{[B]^{1/4}[C]^{1/4}}(1_X\otimes\sigma_3)(\sigma_1^*\otimes 1_Y).$$

This is proved as follows for bounded vectors $\beta\in B$, $\eta\in Y$, with Proposition 9.70 and Definition 9.69:

$$\begin{aligned}
&(\sigma_1^A\otimes 1_C)(1_B\otimes {}^A\sigma_3)^*(\beta\otimes\eta)\\
&= \frac{(\dim_R Y)^{1/4}}{(\dim_Q C)^{1/4}}\sum_i(\sigma_1^A\otimes 1_C)(\beta\otimes\bar{\alpha}_i\otimes\sigma_3(\alpha_i\otimes\eta))\\
&= \frac{(\dim X_Q)^{1/4}(\dim_R Y)^{1/4}}{(\dim B_R)^{1/4}(\dim_Q C)^{1/4}}\sum_i \pi_r(\alpha_i)^*\sigma_1^*(\beta)\otimes\sigma_3(\alpha_i\otimes\eta)\\
&= \frac{[X]^{1/4}[Y]^{1/4}}{[B]^{1/4}[C]^{1/4}}\sum_i(1_X\otimes\sigma_3)(\pi_r(\alpha_i)\pi_r(\alpha_i)^*\sigma_1^*(\beta)\otimes\eta)\\
&= \frac{[X]^{1/4}[Y]^{1/4}}{[B]^{1/4}[C]^{1/4}}(1_X\otimes\sigma_3)(\sigma_1^*\otimes 1_Y)(\beta\otimes\eta),
\end{aligned}$$

where $\{\alpha_i\}_i$ is a basis of A as a left module. These three formulae correspond exactly to the *tetrahedral symmetry* in Definition 12.5.

Because the quantum $6j$-symbols here are defined in terms of intertwiners, unitarity in Definition 12.6 also holds trivially. The other property we have to verify is the pentagon identity in Definition 12.7, but this requires some argument. So we first introduce another important property, *flatness*, which is a slight generalization of the flatness in Theorem 10.10.

Suppose we have a large diagram as in Fig. 12.17 with fixed boundaries. That is, each edge on the boundary has an assignment of intertwiners, each vertex on the boundary has an assignment of bimodules, and each row and column of the diagram has an assignment of bimodule tensor products from the left and the right respectively. Then the value assigned to this diagram is defined as the sum over all configurations of the products of all values of W (as in Definition 12.6) in each configuration. This is a direct generalization of partition functions for ordinary connections in Section 10.5. We can also regard this value as an inner product of two intertwiners; one is given by the composition of intertwiners at the bottom boundary and those at the left boundary, and the other is given by

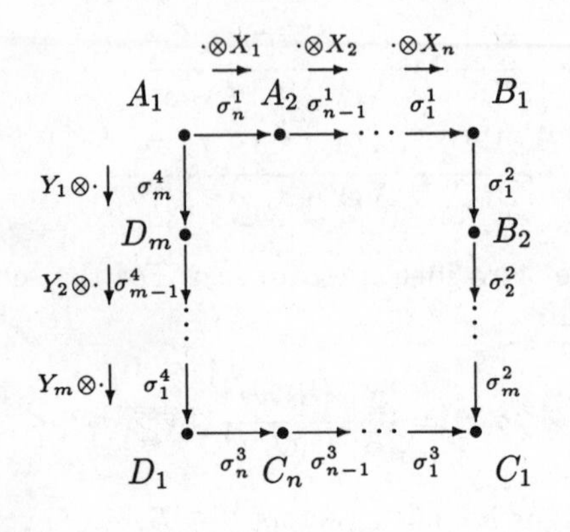

FIG. 12.17. Partition function

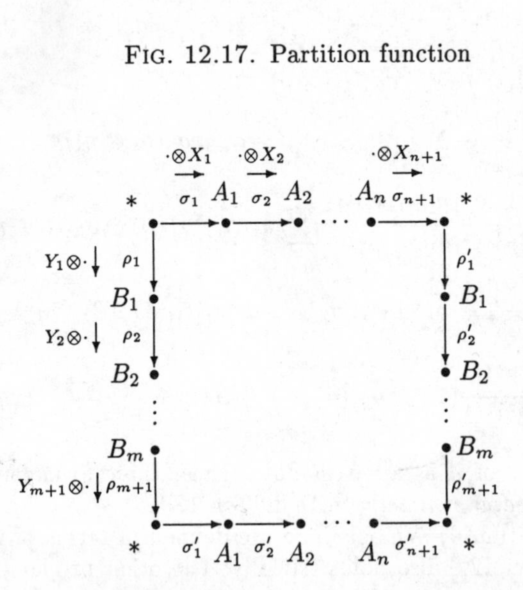

FIG. 12.18. Flatness

the composition of those at the right boundary and the top boundary.

Another important property of quantum $6j$-symbols is as follows.

Theorem 12.16. (Flatness) *For the diagram as in Fig. 12.18 with* ${}_N{*}_N$ *or* ${}_M{*}_M$ *at all the four corners, we get the value*

$$\delta_{\sigma_1,\sigma_1'}\delta_{\sigma_2,\sigma_2'}\cdots\delta_{\sigma_{n+1},\sigma'_{n+1}}\delta_{\rho_1,\rho_1'}\delta_{\rho_2,\rho_2'}\cdots\delta_{\rho_{m+1},\rho'_{m+1}}.$$

For the proof it is enough, because of unitarity, to prove that the value is 1 if $\sigma_j = \sigma_j'$ for $j = 1, 2, \ldots, n+1$ and $\rho_k = \rho_k'$ for $k = 1, 2, \ldots, m+1$. This is proved

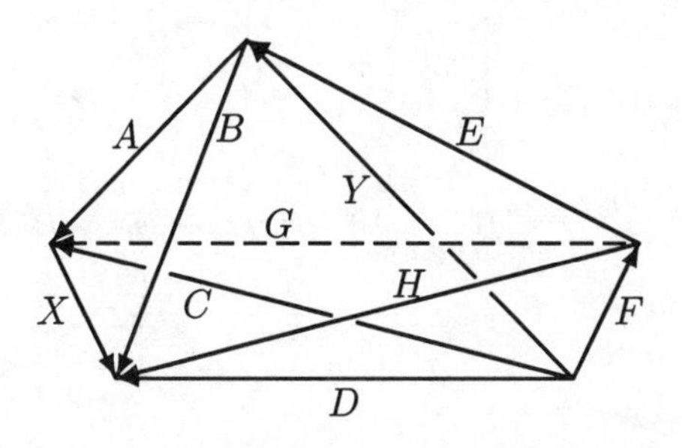

FIG. 12.19. Two divisions into tetrahedra

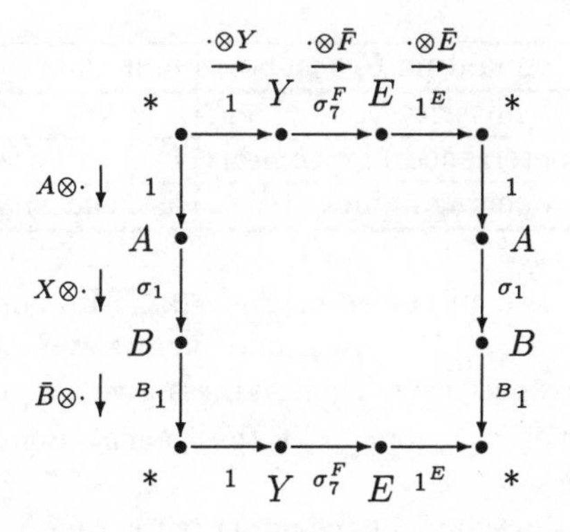

FIG. 12.20. A 3×3-partition function

exactly as in the proof of Theorem 10.10.

We next prove the pentagon identities (Definition 12.7) from Theorem 12.16. For simplicity, we concentrate only on the N-N bimodules in the system and ignore the other three kinds of bimodules for a while.

We label edges and triangles as in Fig. 12.19. On the one hand, this picture consists of two tetrahedra; one has edges labelled by A, B, C, D, X, Y and the second by E, H, Y, D, B, F. On the other hand, this picture consists of three tetrahedra; one has edges labelled by A, B, G, H, X, E, another by G, H, C, D, X, F, and the third by E, G, Y, C, A, F.

We also label 10 triangles by intertwiners $\sigma_1 \in \mathcal{H}_{XA}^B$, $\sigma_2 \in \mathcal{H}_{BY}^D$, $\sigma_3 \in \mathcal{H}_{AY}^C$, $\sigma_4 \in \mathcal{H}_{XC}^D$, $\sigma_5 \in \mathcal{H}_{BE}^H$, $\sigma_6 \in \mathcal{H}_{HF}^D$, $\sigma_7 \in \mathcal{H}_{EF}^Y$, $\sigma_8 \in \mathcal{H}_{AE}^G$, $\sigma_9 \in \mathcal{H}_{XG}^H$, $\sigma_{10} \in \mathcal{H}_{GF}^C$.

Then we apply flatness to the 3×3-diagram in Fig. 12.20 to get the value 1. Here the symbol 1 means a trivial intertwiner such as

$$n \otimes_N \eta \in {}_N N \otimes_N Y_N \mapsto n\eta \in {}_N Y_N.$$

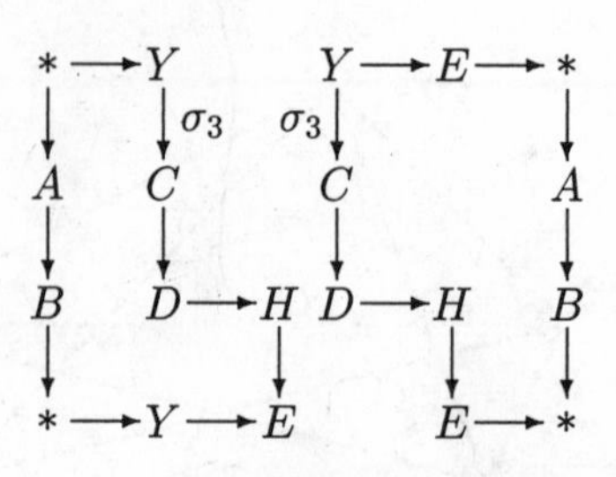

FIG. 12.21. Pentagon identity

Table 12.1 *Relation between connections, quantum 6j-symbols, and bimodules*

connection	quantum $6j$-symbol	bimodule property
unitarity	unitarity	—
renormalization	tetrahedral symmetry	Frobenius reciprocity
flatness	pentagon identity	associativity

Now cut Fig. 12.20 into two pieces as in Fig. 12.21. In this cutting, we label the path from $*$ to $Y, C, D, H, E, *$ by ρ and denote the values of the above two diagrams by a_ρ, $\bar{b}_\rho$ respectively. Then flatness gives $\sum_\rho a_\rho \bar{b}_\rho = 1$, unitarity implies $\sum_\rho |a_\rho|^2 = \sum_\rho |b_\rho|^2 = 1$, and so by the Cauchy–Schwarz inequality, we get $a_\rho = b_\rho$ for all ρ.

In Fig. 12.21, look at the square at the upper left corner first. The intertwiner from A to C must be σ_3 to get a non-zero value, and in this case, we get the value 1. The square at the lower left corner has a similar property, and in this case, we get the value $[D]^{1/4}[B]^{-1/4}[Y]^{-1/4}$ because of the tetrahedral symmetry (Definition 12.5). We can compute the values for the upper right and lower right squares for the right diagram in Fig. 12.21 similarly. We still have four squares left in Fig. 12.21. Writing down the identity $a_\rho = b_\rho$ explicitly, we get the pentagon identity as in Definition 12.7. That is, the left hand side of the pentagon identity corresponds to the left piece in Fig. 12.21, and the right hand side to the right piece.

We thus have the correspondences of Table 12.1 between bi-unitary connections and quantum $6j$-symbols. We will study the entries in the last column in more detail in the next section.

A subfactor arising from a finite group G, as in Section 10.6, gives the following example.

Example 12.17 For a subfactor arising from a finite group G as in Section 10.6, we have the fusion rule algebra as in Example 12.14. In this example, the quantum $6j$-symbols for N-N bimodules are easily written.

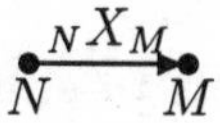

$$\underset{N}{\bullet}\xrightarrow{{}_N X_M}\underset{M}{\bullet}$$

FIG. 12.22. Compatible assignment of a bimodule to an edge

Note that each N-N bimodule is labelled by a group element $g \in G$. So we will identify bimodules and group elements. Then the dimension of the intertwiner space $\mathcal{H}^k_{gh}$ is Kronecker delta, $\delta_{gh,k}$. For all $g \in G$, we have $[g] = 1$, and the only non-trivial quantum $6j$-symbols are of the following form:

$$Z(h, gh, hk, ghk, g, k) = 1, \quad g, h, k \in G.$$

Here we have dropped the intertwiners σ_j because their spaces are all one-dimensional.

We now recall that we have four kinds of bimodules. We will show that we can define the TQFT_3 using all four kinds of bimodules. For simplicity, we suppose that our compact three-manifold P with triangulation T has no boundary.

We assign a label N or M arbitrarily to each vertex of the triangulation T, and denote this labelling by L. When we label an edge by a bimodule, the bimodule must be compatible with L. See Fig. 12.22 for an example. We can prove topological invariance of the state sum in 12.2.1 exactly as before, but we now also want to prove that this state sum does not depend on L either. The key move for this argument is Move III in Theorem 12.9. In Move III, one vertex disappears, and the labelling of this vertex can be either N or M. It is enough to prove that one can change the labelling of one vertex without changing the state sum. First we take a finer triangulation if necessary so that we may regard a neighbourhood of the vertex as a Euclidean ball. Then by repeated use of Move I, we can decrease the number of edges connected to the vertex down to 3. Then finally by Move III, we can eliminate the vertex. So we can conclude that the labelling of this vertex does not affect the value $Z(P, T, L)$.

In particular, we can use the labelling assigning N to all the vertices. In this case, we use only N-N bimodules. Similarly, we can use only M-M bimodules. These two methods produce the same value for any closed three-manifold.

Exercise 12.3 Fix a finite group G and consider the fusion rule algebra and quantum $6j$-symbols arising from G as in Examples 12.14 and 12.17. For the topological invariant $Z(P)$ defined with these as in Definition 12.8, show that

$$Z(P) = |\mathrm{Hom}(\pi_1(P), G)|/|G|,$$

where $\pi_1(P)$ is the fundamental group of the manifold P and $\mathrm{Hom}(\pi_1(P), G)$ is the set of homomorphisms from $\pi_1(P)$ to G.

12.5 From quantum $6j$-symbols to subfactors

In this section, we construct (a family of) subfactors with finite index and finite depth from a system of quantum $6j$-symbols, as a converse operation to the construction in Section 12.4.

In order to recover a paragroup from a fusion rule algebra with a system of quantum $6j$-symbols, our fusion rule algebra has to be graded. So our data is as follows.

Definition 12.18 *We have a* graded fusion rule algebra $\mathcal{A}$ *as follows. The associative $\mathbb{C}$-algebra $\mathcal{A}$ is spanned by finitely many X_i as a $\mathbb{C}$-vector space. Each X_i has a left attribution and a right attribution, and there are two possible attributions, denoted by A and B. By the notation ${}_AX_{iB}$ we mean that X_i has the left attribution A and the right attribution B. Each X_j has its conjugate $\bar{X}_j$ among the X_i, and the conjugate operation interchanges right and left attributions. We require $\bar{\bar{X}}_j = X_j$. For X and Y among the X_i, the product is given by $X \cdot Y = \sum_Z N_{X,Y}^Z Z$, where $N_{X,Y}^Z$ is a non-negative integer and Z is among the X_i. If the right attribution of X and the left attribution of Y are different, the product is 0. If Z has non-zero $N_{X,Y}^Z$, its left attribution is the same as that of X and its right attribution is the same as that of Y. We also require*

$$N_{X,Y}^Z = N_{\bar{Z},X}^{\bar{Y}} = N_{Y,\bar{Z}}^{\bar{X}} = N_{\bar{Y},\bar{X}}^{\bar{Z}} = N_{\bar{X},Z}^{Y} = N_{Z,\bar{Y}}^{X}.$$

We further require that we have two identities ${}_A1_A$ and ${}_B1_B$ with ${}_A\bar{1}_A = {}_A1_A$ and ${}_B\bar{1}_B = {}_B1_B$. The identities satisfy ${}_A1_A \cdot X = X$ if the left attribution of X is A. Similarly, we have $X \cdot {}_A1_A = X$, ${}_B1_B \cdot X = X$, and $X \cdot {}_B1_B = X$, if X has an appropriate attribution in each formula. Each X_i has an assignment $[X_i]$ of a positive value. The vector $\sum_{{}_AX_{iB}} [X_i]^{1/2} X_i$ is a simultaneous eigenvector for left multiplications by ${}_AX_{j}{}_A$. Similar statements hold for all types of attributions.

We then assume that this graded fusion rule algebra has a system of quantum $6j$-symbols as in Definition 12.4. (When we form intertwiner spaces, we consider only pairs for which the multiplication is possible with the correct matching of the attributes.)

In many natural examples of quantum $6j$-symbols arising from a quantum group or a rational conformal field theory, we do not have two-sided labelling of attribution as above. In such a case, we can use a trivial modification of the following construction, as we will see later.

We first choose an element Y from the fusion rule algebra. Suppose first that Y has the left attribution A and the right attribution B. Define the bipartite graphs $\mathcal{G}_0^0$, $\mathcal{G}_1^0$, $\mathcal{G}_2^0$, $\mathcal{G}_3^0$ as follows. Even vertices of $\mathcal{G}_0^0$ and $\mathcal{G}_3^0$ are given by ${}_AX_A$ in the fusion rule algebra, odd vertices of $\mathcal{G}_0^0$ and $\mathcal{G}_1^0$ by ${}_AX_B$, odd vertices of $\mathcal{G}_3^0$ and $\mathcal{G}_2^0$ by ${}_BX_A$, and even vertices of $\mathcal{G}_1^0$ and $\mathcal{G}_2^0$ by ${}_BX_B$. The number of edges for $\mathcal{G}_0^0$ between ${}_AX_A$ and ${}_AZ_B$ is given by $N_{X,Y}^Z$. Similarly, we use the numbers

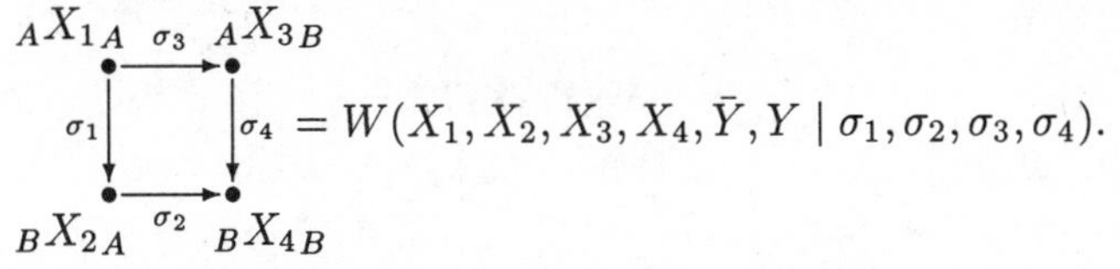

$$\begin{array}{ccc} {}_A X_{1A} & \xrightarrow{\sigma_3} & {}_A X_{3B} \\ \sigma_1 \downarrow & & \downarrow \sigma_4 \\ {}_B X_{2A} & \xrightarrow{\sigma_2} & {}_B X_{4B} \end{array} = W(X_1, X_2, X_3, X_4, \bar{Y}, Y \mid \sigma_1, \sigma_2, \sigma_3, \sigma_4).$$

FIG. 12.23. A connection from a $6j$-symbol

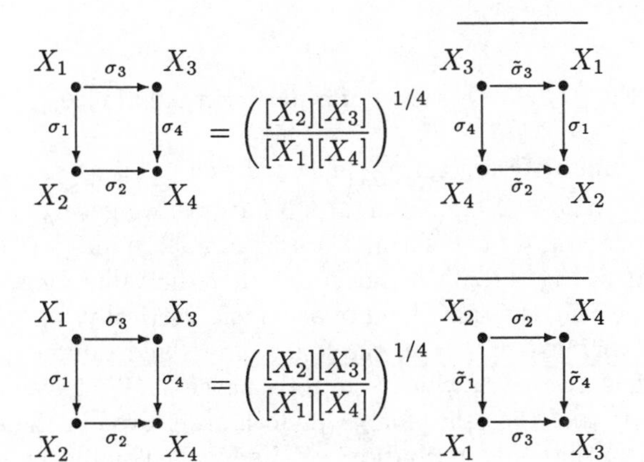

FIG. 12.24. Renormalization

$N^Z_{\bar{Y},X}$, $N^Z_{\bar{Y},X}$, $N^Z_{X,Y}$ for $\mathcal{G}^0_3$, $\mathcal{G}^0_1$, $\mathcal{G}^0_2$ respectively. We then look at the connected component containing ${}_A*_A$, and label the resulting four bipartite graphs as $\mathcal{G}_0$, $\mathcal{G}_1$, $\mathcal{G}_2$, $\mathcal{G}_3$. Note that $\mathcal{G}_1$ and $\mathcal{G}_2$ always contain ${}_B*_B$. For each set of edges with the same endpoints, we label them by an orthonormal basis of the corresponding intertwiner space. We then define a connection as in Fig. 12.23. We also define other types of connections using the renormalization axiom as in Fig. 12.24.

In order to prove that this data gives a paragroup, we have to show unitarity and flatness, because we have already defined the connection so that the renormalization axiom holds. Unitarity for the connection is an immediate consequence of the tetrahedral symmetry and unitarity of the quantum $6j$-symbols. So our aim is to prove flatness. Note that the proof of the pentagon identity from the special type of flatness for 3×3-diagrams in Section 12.4 also works in the converse direction. So we now have 3×3-flatness from the pentagon identity. We want to deduce ordinary flatness as in the flatness axiom.

We use a graphical method as follows. (In the following diagrams, we omit labelling of edges and vertices for simplicity. Our convention is that parallel edges from $*$ or to $*$ denote the same choices of intertwiners.) First we split the 3×3-diagram in Section 12.4 into two pieces as in Fig. 12.25. As in Section 12.4, we label the cutting path by ρ and denote the values of the two diagrams by a_ρ and

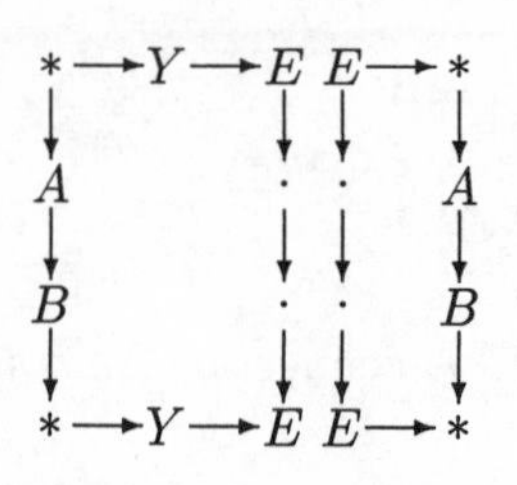

FIG. 12.25. Division of a partition function

$\bar{b}_\rho$ respectively. By unitarity, we get $\sum_\rho |a_\rho|^2 \le 1$ and $\sum_\rho |b_\rho|^2 \le 1$. With these and the formula $\sum_\rho a_\rho \bar{b}_\rho = 1$ expressing 3×3-flatness, we get $a_\rho = b_\rho$ for all ρ again. With this relation, we can shrink the size of a large diagram one step at a time. That is, as in Fig. 12.26, we can prove that the value for a diagram of $3 \times n$-size is 1. Repeating the same kind of argument vertically, we can conclude that a diagram of any size with $*$ at the four corners and same configurations on parallel edges has a value 1, which is exactly flatness. (We have two kinds of flatness, one for ${}_A*_A$ and the other for ${}_B*_B$. Both are proved in the same way.)

The case where Y has the attribution ${}_BY_A$ is handled similarly. If Y has an attribution ${}_AY_A$, then we modify the above construction as follows. For the four corners of the graphs $\mathcal{G}_0, \mathcal{G}_1, \mathcal{G}_2, \mathcal{G}_3$, we now use only ${}_AX_A$. Then all the rest is the same as before. The case where Y has the attribution ${}_BY_B$ is handled similarly.

Furthermore, if the graphs $\mathcal{G}_0^0$, $\mathcal{G}_1^0$, $\mathcal{G}_2^0$, $\mathcal{G}_3^0$ are connected for a certain choice of ${}_AY_B$ or ${}_BY_A$, then all the elements of the fusion rule algebra are identified with bimodules arising from a subfactor. Thus we have proved the following.

Theorem 12.19 *A system of quantum $6j$-symbols on a graded fusion rule algebra produces a family of subfactors. If we get connected graphs in the above construction for some choice in the fusion rule algebra, the quantum $6j$-symbols arise from a subfactor.*

Remark Suppose we have a subfactor $N \subset M$ with finite depth. We then have a fusion rule algebra and quantum $6j$-symbols arising from this subfactor. Then by choosing any bimodule from this system and applying the above method, we obtain some paragroup.

Suppose that we choose a bimodule ${}_NY_M$ appearing in the irreducible decomposition of ${}_NM_{kM}$, for example. Then we have a projection p from $L^2(M_k)$ onto Y and this gives a minimal projection in $N' \cap M_{2k}$. Let R be the von Neumann algebra given by the right action of M on Y. Since the left action of N on Y is identified with Np on $L^2(M_k)$, we have inclusions of II_1 factors, $Np \subset p(M_{2k})p \subset R$. By comparing the Jones indices, we conclude that $R = p(M_{2k})p$. That is, the subfactor $pN \subset p(M_k)p$ is the one associated with the bimodule ${}_NY_M$.

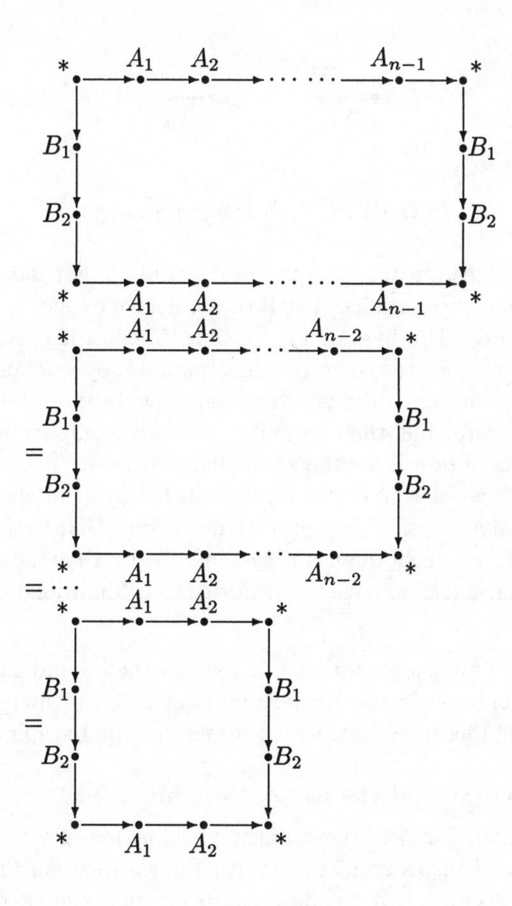

FIG. 12.26. Shrinking partition functions

Using an argument in Section 5 of (Evans and Kawahigashi 1995b), we can show that the paragroup arising from ${}_N Y_M$ as above corresponds to this subfactor $pN \subset p(M_k)p$.

Remark Suppose we have no attribution in the fusion rule algebra. Then we choose a Y from the fusion rule algebra, and proceed as in the above case where Y has the same right and left attributions. Then the same argument works.

The results in the previous section and this section mean that the the axioms for paragroup are essentially equivalent to those for quantum $6j$-symbols on four kinds of bimodules. (Strictly speaking, we have to choose a generator Y to get a paragroup from quantum $6j$-symbols. The paragroup does depend on the choice of Y. In this sense, different paragroups correspond to the same quantum $6j$-symbols.) From a conceptual viewpoint, quantum $6j$-symbols on bimodules are

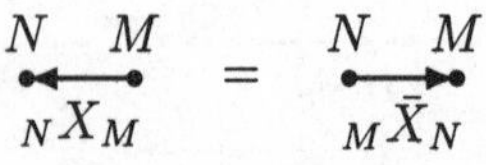

$$\underset{{}_N X_M}{\overset{N \quad M}{\bullet\!\!\longleftarrow\!\!\bullet}} = \underset{{}_M \bar{X}_N}{\overset{N \quad M}{\bullet\!\!\longrightarrow\!\!\bullet}}$$

FIG. 12.27. Labelling of an edge

better to handle as seen in this and the next chapters, but flat connections are better for concrete computations. Recall that we have explicitly determined flat connections on the Dynkin diagrams in Section 11.5, but it is very hard to write down explicit formulae for the corresponding quantum $6j$-symbols. The quantum $6j$-symbols corresponding to the A_n-diagrams have been obtained by (Kirillov and Reshetikhin 1988), but their formulae are very complicated, and no such formulae have been obtained for the D, E-diagrams.

It is easy to get a flat connection from quantum $6j$-symbols because we just use a part of the given data. The opposite direction is harder because we want to recover the full system of quantum $6j$-symbols from partial data given by a connection. The flatness axiom can be regarded as a condition ensuring that this recovery is possible.

Exercise 12.4 Fix any subfactor with index less than 4 and choose any vertex of its principal graph. As in the Remark on page 644, we then get a paragroup from the corresponding bimodule. Compute its principal graph.

12.6 Tube algebras and the asymptotic inclusion

We start with a subfactor N of the hyperfinite II_1 factor M which is constructed from a paragroup as in Section 11.3. (With Popa's theorem (Theorem 11.14), this requirement is equivalent to the assumption that the subfactor has finite index, finite depth, and trivial relative commutant. Actually, we can drop the trivial relative commutant condition for this chapter without any change of the arguments.) In this case, we can naturally identify $N \subset M$ with $\overline{\bigcup_{n=0}^{\infty} N_n' \cap N} \subset \overline{\bigcup_{n=0}^{\infty} N_n' \cap M}$, where $\overline{}$ denotes the weak closure.

We denote by $\mathcal{M}$ the finite system of four kinds of bimodules arising from the subfactor as in Section 11.2. We also have intertwiner spaces and a system of quantum $6j$-symbols as in Section 12.4.

We label graphical objects such as vertices, edges, and triangles with algebras (N or M), bimodules in $\mathcal{M}$, and intertwiners respectively. Consider an oriented triangulated surface (possibly with boundaries). We put a label of an algebra N or M to each vertex. We next assign a label of a bimodule to an oriented edge so that it matches the labelling of the vertices as in Fig. 12.22. A labelled edge is identified with the edge with reversed orientation and labelling by the conjugate bimodule, as in Fig. 12.27.

Next we assign an intertwiner to a triangle with labelled edges so that the intertwiner maps from a tensor product of two of the three labelled bimodules to

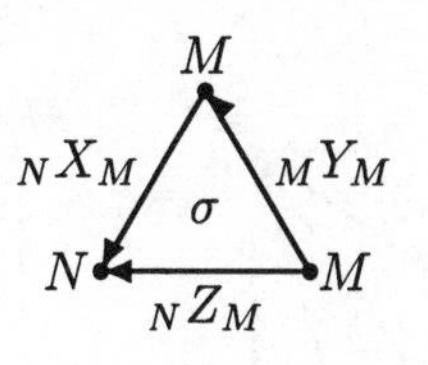

FIG. 12.28. Labelling of a triangle

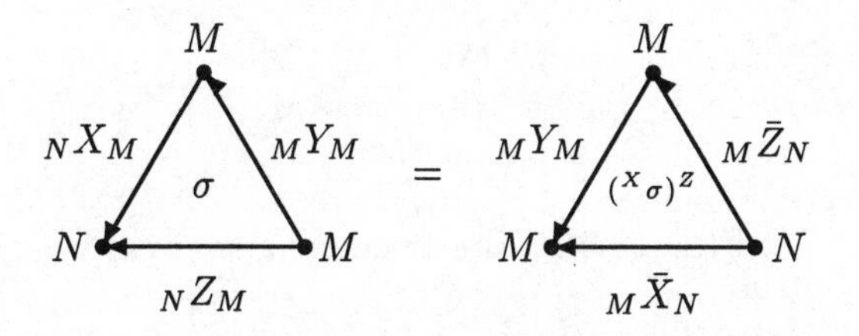

FIG. 12.29. Identification of labelled triangles

the other as in Fig. 12.28, where $\sigma \in \mathrm{Hom}({}_N X \otimes_M Y_M, {}_N Z_M)$. After rotating Fig. 12.28 by $2\pi/3$ in the counterclockwise orientation, we get the right diagram in Fig. 12.29. We identify the labelled diagram in Fig. 12.28 with the right labelled diagram in Fig. 12.29 using Frobenius reciprocity (Theorem 9.73). We can also take the mirror image of Fig. 12.28, and then make the identification as in Fig. 12.30, again using Frobenius reciprocity (Theorem 9.73).

Next we assign a finite dimensional Hilbert space to each oriented triangulated surface S with labelled boundary ∂S. This is an extension of the definition of $Q(S)$ in Section 12.2. The only difference now is that S has a boundary. Let $C(S)$ be the finite dimensional Hilbert space of all possible labellings of the triangulated S that are compatible with the labels on ∂S. (It has a Hilbert space structure as in the definition of $C(S)$ in Section 12.2.) Let IS be the quotient space of $I \times S$ with respect to the relation $(t, x) \sim (s, x)$ for any $t, s \in I$, $x \in \partial S$, where I is the unit interval $[0, 1]$. This is a three-manifold with boundary. Using

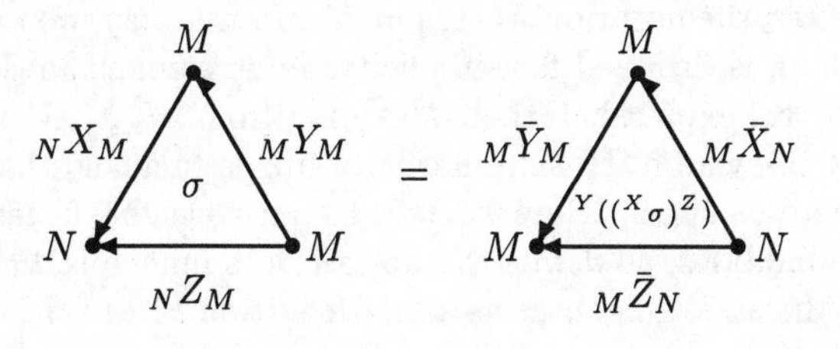

FIG. 12.30. Identification of mirror images

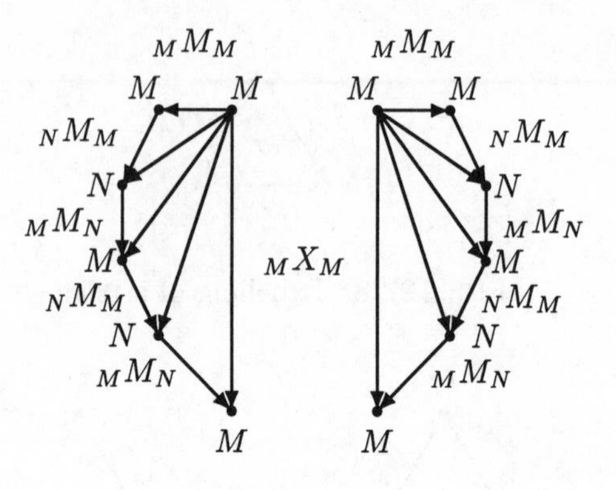

FIG. 12.31. Intertwiners on triangles

this, we can define the Hilbert space $Q(S)$ associated with S as in Section 12.2. We will write H_S for $Q(S)$ in the rest of this chapter.

Next we modify the string algebra construction in Section 11.3 so that the algebra is expressed in terms of two dimensional pictures instead of one dimensional pictures. First note that M^{opp} is naturally identified with $M' \cap M_\infty = \overline{\bigcup_k \text{End}({}_M M_{k\,M})}$.

Consider Fig. 12.31. Putting labels of intertwiners on Fig. 12.31, the left hand side of the picture gives an intertwiner in

$$\text{Hom}({}_M X_M, {}_M M \otimes_N M \otimes_M M \otimes_N M \otimes_M M_M)$$

and the right hand side of the picture one in

$$\text{Hom}({}_M M \otimes_N M \otimes_M M \otimes_N M \otimes_M M_M, {}_M X_M).$$

Thus by composing these two intertwiners, we get an element in

$$\text{End}({}_M M \otimes_N M \otimes_M M \otimes_N M \otimes_M M_M) = M'_{-4} \cap M,$$

where we have used the notation $M_{-k-1} = N_k$. In this way, we regard Fig. 12.32, where the labelling is dropped for simplicity, as an element in $M'_{-4} \cap M$.

Note that in this expression, the embedding from $M'_{-4} \cap M$ into $M'_{-5} \cap M$ is given by Fig. 12.33, where the summation over σ is taken so that the σ form an orthonormal basis, as usual. (Here we have to put a weight, in terms of the Jones indices of the bimodules, to define the appropriate inner product as in the case of the trace for the string algebras as after Definition 11.4. With this weight, our embedding here becomes compatible with the inner product.) Note that with the 'limit' of this embedding, we get M as noted at the beginning of this section. In this graphical expression, we have a natural multiplication structure, which is just a rewriting of the multiplication structure of the string algebra in Definition 11.1.

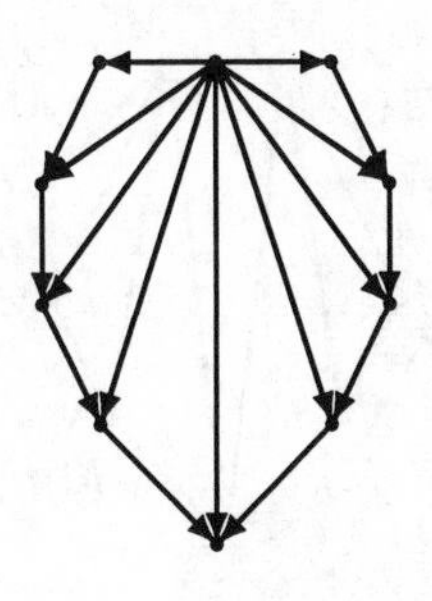

FIG. 12.32. An element in $M'_{-4} \cap M$

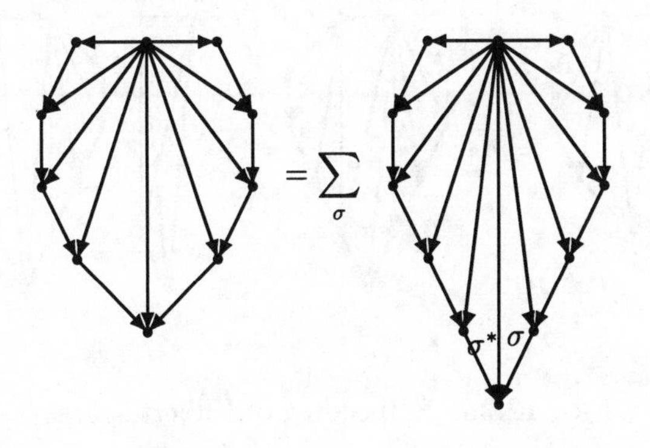

FIG. 12.33. Embedding of $M'_{-4} \cap M$ into $M'_{-5} \cap M$

We next introduce a similar graphical expression of M-M bimodules in the system $\mathcal{M}$. The bimodule is given by an increasing sequence of finite dimensional Hilbert spaces. An element of the finite dimensional Hilbert space is given by Fig. 12.34 with labels of intertwiners. We regard this element as a map from ${}_M M \otimes_N M \otimes_M M \otimes_N M_M$ to ${}_M M \otimes_N M \otimes_M M \otimes_N M \otimes_M X_M$. Note that we can also regard this element as a map from ${}_M M \otimes_N M \otimes_M M \otimes_N M \otimes_M \bar{X}_M$ to ${}_M M \otimes_N M \otimes_M M \otimes_N M_M$ by Frobenius reciprocity, Theorem 9.73. (This way of regarding the picture as a map is more appropriate for computations of the M-valued inner product on this bimodule.) The embedding is given by Fig. 12.35, where the labelling of vertices and edges is dropped again for simplicity. Here the coefficient in the right hand side is the W which appears in the Definition 12.15 of the quantum $6j$-symbol. On each finite dimensional Hilbert space, we

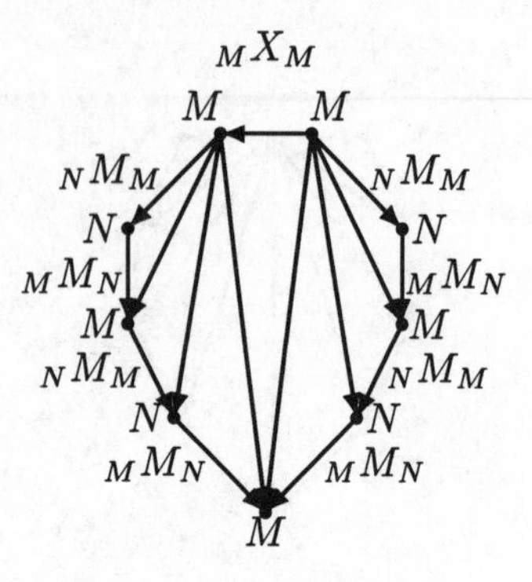

FIG. 12.34. A Hilbert space making a bimodule

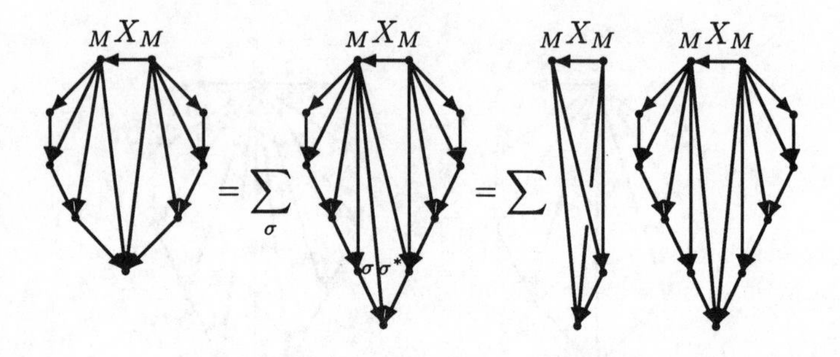

FIG. 12.35. Embedding of Hilbert spaces

have natural left and right actions of $M'_{-k} \cap M$, so by taking the completion of the union with respect to a natural inner product, we get an M-M bimodule. We call it a *surface bimodule* and denote this by A_X. (See (Asaeda and Haagerup, preprint 1998) for a more general construction of open string bimodules.)

Recall that we have the change of basis rule (Fig. 12.36) for paths in terms of the connections in the string algebra construction. Our new change of basis rule

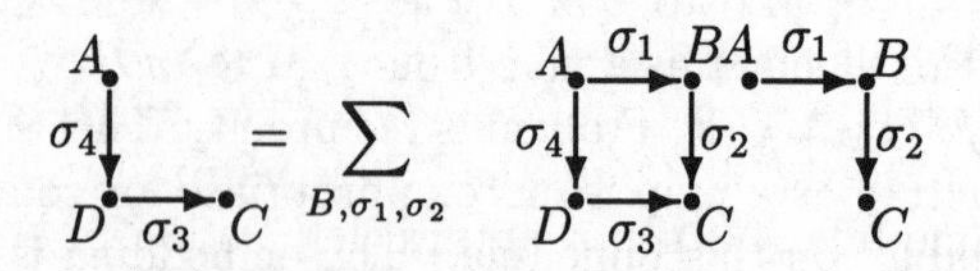

FIG. 12.36. Basis change in a string algebra

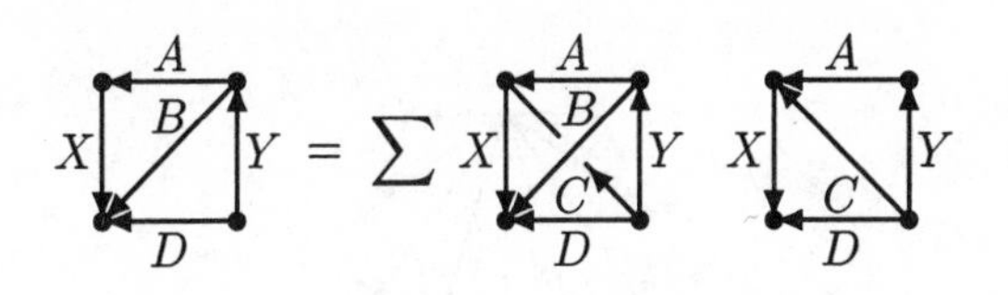

FIG. 12.37. Basis change in a surface bimodule

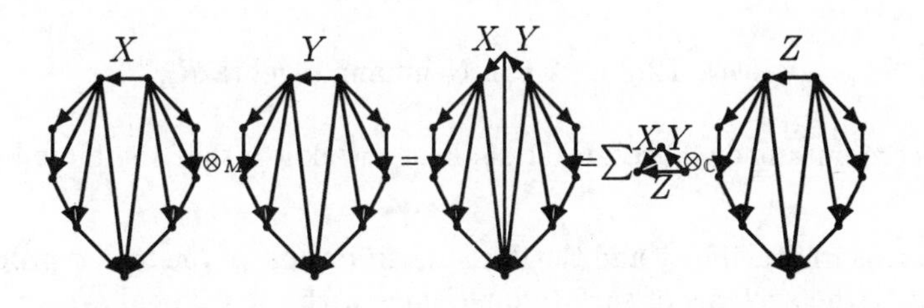

FIG. 12.38. A decomposition of a bimodule

(Fig. 12.37) for surfaces in terms of W is essentially a rewriting of Fig. 12.36.
We claim that these surface bimodules are irreducible.

Theorem 12.20 *The bimodule A_X is irreducible.*

Proof First, note that we have the graphical decomposition rule (Fig. 12.38) of tensor products of the surface bimodules as appropriate bimodules, where the symbol $\otimes_{\mathbb{C}}$ in the last part means an ordinary tensor product of Hilbert spaces. Here the first equality actually defines an isomorphism from $A_X \otimes_M A_Y$. (We can show that this indeed defines an isomorphism by computing the M-valued inner product.) The triangle on the right hand sides denotes a Hilbert space with dimension N_{XY}^Z. That is, $A_X \otimes_M A_Y = \sum_Z N_{X,Y}^Z A_Z$. (At this point, it is not known that this is the *irreducible* decomposition.)

Let μ_X be the square root of the Jones index $[A_X]$ corresponding to the bimodule A_X. Note that A_* for $X = * = {}_M M_M$ is clearly isomorphic to ${}_M M_M$ and hence $\mu_* = 1$. Because of $\mu_X \mu_Y = \sum_Z N_{X,Y}^Z \mu_Z$, $\mu_* = 1$, and uniqueness of the Perron–Frobenius eigenvector (Theorem 10.3 (2)), we get $\mu_X = [X]^{1/2}$ for each M-M bimodule X in the system $\mathcal{M}$. Let R_X be the von Neumann algebra corresponding to Fig. 12.39. (Note that R_* is naturally identified with M.)

Then the bimodule A_X has a natural left action of R_X in addition to the left and right actions of M. We have a natural embedding of M into R_X with index $[X]$, and the left action of R_X and the right action of M on A_X commute. By computing the index, we know that the commutant of the right M action on A_X is the left R_X action. The natural embedding of M into R_X has a trivial

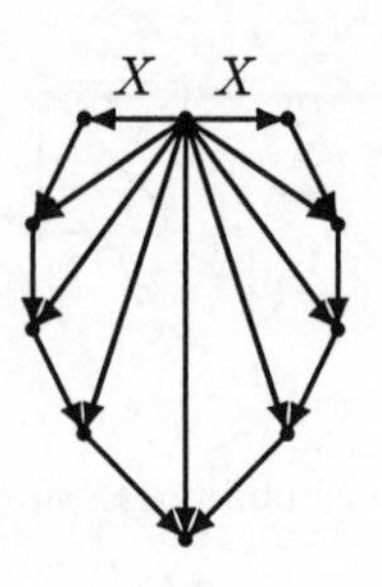

FIG. 12.39. A von Neumann algebra R_X

relative commutant by Theorem 11.15, hence we know that the bimodule A_X is irreducible. □

With this irreducibility and the decomposition rule of the tensor products, we can identify the systems of surface bimodules with the original system of M-M bimodules in $\mathcal{M}$.

In the study of the Hilbert spaces associated with surfaces with labelled boundaries as above, we encounter the problem of labelling circles which are parts of boundaries of surfaces in a more natural way. For this purpose, we will introduce a new algebra, the *tube algebra* of Ocneanu, and show that its centre gives a natural labelling of circles in the TQFT_3.

Definition 12.21 *Take M-M bimodules X, X', Y, Y', A, A', A'' in the system $\mathcal{M}$. We define a finite dimensional C^*-algebra* Tube $\mathcal{M}$

$$\text{Tube } \mathcal{M} = \bigoplus_{X,Y,A} \text{Hom}(X \otimes A, A \otimes Y),$$

with the product defined graphically as in Fig. 12.40. In the expression of σ in Fig. 12.40, the top and the bottom edges are labelled by the same bimodule A, and the square is regarded as a tube under this identification. The meaning of the bracket in the second line of Fig. 12.40 is as follows. Each element in the bracket is regarded as a tube labelled by an intertwiner. We identify the boundaries of the two tubes by gluing the two circles labelled with X and the two circles labelled with Y' respectively. In this way, we obtain a two-dimensional torus labelled by an intertwiner. We then regard this as a solid torus with boundary labelled by an intertwiner. By Definition 12.12, this three-dimensional manifold with labelled boundary gives rise to a complex number which is the value of the bracket. The name tube algebra comes from this graphical expression. (See Fig. 12.41.)

In the Turaev–Viro type TQFT_3 arising from the system $\mathcal{M}$ as in Sections 12.2 and 12.4, we study the Hilbert space $H_{S^1\times S^1}$ associated with the two dimensional torus $S^1 \times S^1$, which was denoted by $Q(S^1 \times S^1)$ in Section 12.2. Note

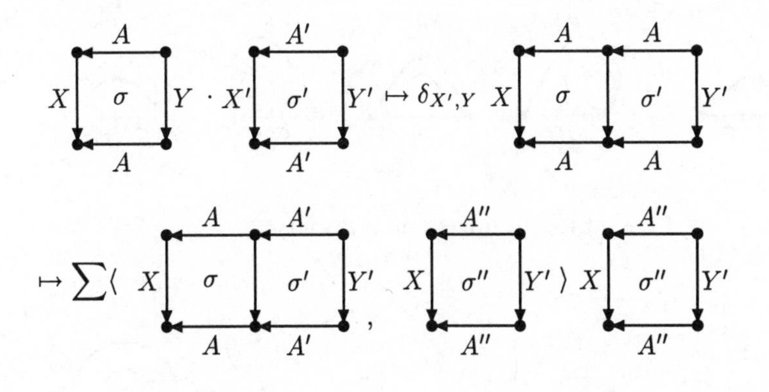

FIG. 12.40. Product in the tube algebra

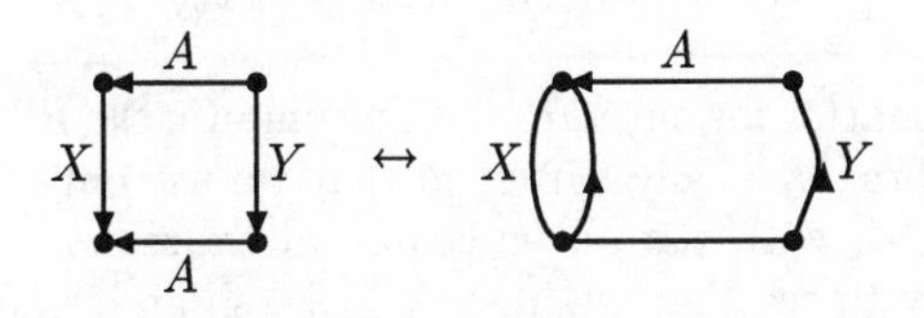

FIG. 12.41. An element of the tube algebra

that by definition, the topological invariant of the three dimensional torus in this TQFT_3 gives the dimension of this Hilbert space. For the study of this space, we can fix any triangulation of the two dimensional torus, so we choose the one as in Fig. 12.42. In Fig. 12.42, the top and the bottom edges are identified, and the left and right edges are identified to give a torus. This is not a triangulation in the standard sense in combinatorial topology, because it has only one vertex,

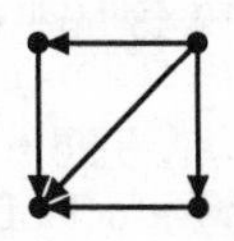

FIG. 12.42. Triangulation of $S^1 \times S^1$

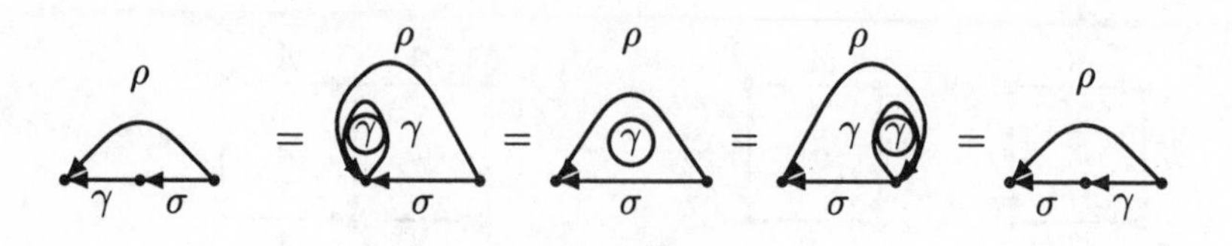

FIG. 12.43. Commutativity

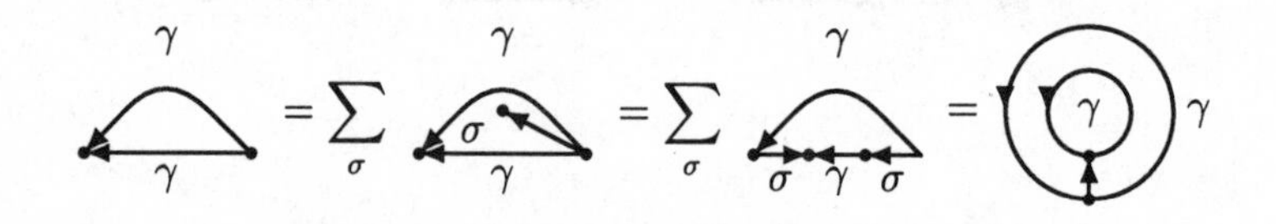

FIG. 12.44. An element in $H_{S^1 \times S^1}$

but that does not matter for our purpose. To compute the Hilbert space $H_{S^1 \times S^1}$, we assign an algebra M to any vertex so that we use only M-M bimodules in $\mathcal{M}$. If $\sigma \in \text{Tube}\,\mathcal{M}$, with the above graphical expression, is in the centre of Tube $\mathcal{M}$, it is easy to see that we have $X = Y$, so an element in the centre of Tube M naturally gives an element in $H_{S^1 \times S^1}$ as the same intertwiner on the same picture in $H_{S^1 \times S^1}$ modulo the degenerate vectors. The next theorem says that this gives an isomorphism.

Theorem 12.22 *The centre of the tube algebra* Tube $\mathcal{M}$ *is identified with the Hilbert space* $H_{S^1 \times S^1}$ *associated with the Turaev–Viro type* $TQFT_3$ *arising from* $\mathcal{M}$.

Proof This is proved graphically as follows. Because it is easier to draw two dimensional pictures than three dimensional pictures, we have drawn Figs 12.43 and 12.44 in one dimension less. (Each figure represents a number.) To get the correct pictures in three-dimensions, it is enough to multiply Figs 12.43 and 12.44 with S^1. The interiors of the small inner discs labelled with γ in the second, third, and fourth pictures of Fig. 12.43 and in the fourth picture of Fig. 12.44 are all empty.

Let γ be a unit vector in $H_{S^1 \times S^1}$. The meaning of Fig. 12.43 is as follows. The left picture denotes the inner product of ρ and $\gamma\sigma$. This picture represents a solid torus, a three dimensional manifold, whose boundary is labelled with ρ, γ, σ, and thus gives a complex number as a topological invariant of a three dimensional manifold with a labelled boundary. The next picture represents a product of two topological invariants of three dimensional manifolds with labelled boundaries. One of the two values is 1, because it is given as the inner product

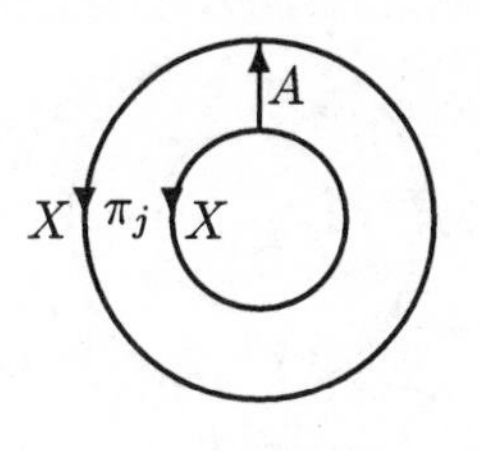

FIG. 12.45. π_j, a minimal central projection of Tube $\mathcal{M}$

of γ and itself, and the other is the original value as in the left picture. We now regard this value as a summation of products of two topological invariants of three dimensional manifolds with labelled boundaries over unit vectors γ_0 in an orthonormal basis of $H_{S^1\times S^1}$ containing γ. This is possible because one of the two invariants is δ_{γ,γ_0}, so the summation can be reduced to a single term as given above. We now drop one torus, which is a common boundary of two three dimensional manifolds above, because it is labelled with γ_0 as above and we have a summation over γ_0, and thus get the next picture. A similar argument gives the rest of the transformation in Fig. 12.43. Then the entire figure shows that the unit vector γ in $H_{S^1\times S^1}$ commutes with any σ in the tube algebra, because $\gamma\sigma$ and $\sigma\gamma$ have the same inner products with any $\rho \in H_{S^1\times S^1}$.

Conversely, let γ be a central element in the tube algebra. The left picture in Fig. 12.44 gives a self-inner product of γ on the tube, which is again given as the topological invariant of a three dimensional manifold with boundary labelled with γ and γ. The next picture represents a summation of topological invariants of a three dimensional manifold with boundary labelled with $\gamma, \gamma, \sigma, \sigma$, with the summation over unit vectors σ in an orthonormal basis of the Hilbert space for the tube. In the next picture, we cut the tube labelled with σ and use the commutativity $\sigma\gamma = \gamma\sigma$. Finally, by a similar argument, we get the last picture which gives a self-inner product of γ in $H_{S^1\times S^1}$. The identity of the two inner product values means that γ is in $H_{S^1\times S^1}$. (Note that γ is a linear combination of intertwiners from $X\otimes A$ to $A\otimes Y$ for various X, A, B in the system $\mathcal{M}$.)

□

We denote the set of the minimal central projections of Tube $\mathcal{M}$ by $\{\pi_1, \ldots, \pi_n\}$. We label circles, which are parts of boundaries of surfaces, with π_j as follows. First we rewrite the picture of π_j as in Fig. 12.45. (Actually, the annulus is labelled with intertwiners and a summation over A, X, which we have dropped for simplicity.) By labelling a circle, which is a part of the boundary of a surface S, with π_j, we mean the following. We thicken the circle so that we have an annulus which has the circle as a common boundary with S. Then we have a new surface, part of which is labelled with intertwiners. We can still assign a Hilbert space to this partially labelled surface just as above. Then the product rule $\pi_j\pi_k = \delta_{jk}\pi_j$ and the minimality of π_j imply that the dimension of

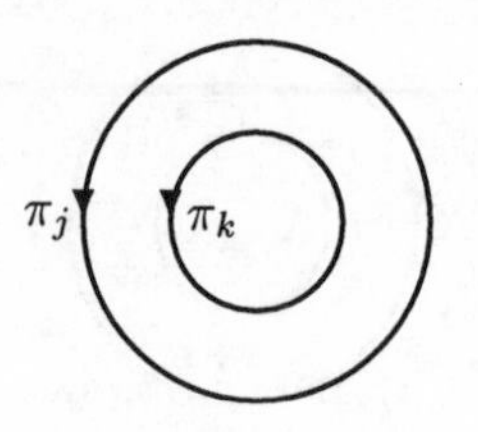

FIG. 12.46. An annulus with boundary labelled with π_j, π_k

the Hilbert space for the annulus in Fig. 12.46 is δ_{jk}.

We will give a more operator algebraic meaning to the above definitions. First, we need to introduce Ocneanu's *asymptotic inclusion.*

Definition 12.23 *Let $N \subset M$ be a subfactor with finite index. Let M_∞ be the closure of $\bigcup_{n=0}^{\infty} M_n$ in the GNS-representation with respect to the trace. We call the inclusion $M \vee (M' \cap M_\infty) \subset M_\infty$ the* asymptotic inclusion *constructed from $N \subset M$. (Here $M \vee (M' \cap M_\infty)$ means the von Neumann algebra generated by M and $M' \cap M_\infty$.)*

This definition is quite general, but we are interested in the case where $N \subset M$ is a subfactor of the hyperfinite II_1 factor with finite index, finite depth, and trivial relative commutant. In such a case, we can prove that the asymptotic inclusion is again a subfactor of the hyperfinite II_1 factor with finite index, finite depth, and a trivial relative commutant as follows.

We may assume that the subfactor $N \subset M$ is constructed from a paragroup and the corresponding double sequence $\{A_{kl}\}_{kl}$ of the string algebras as in Section 11.3. By replacing the paragroup with its dual, we may assume that M and $M' \cap M_\infty$ are given by $\bigvee_{n=0}^{\infty} A_{0,n}$ and $\bigvee_{n=0}^{\infty} A_{n,0}$. Then M_∞ is given by $\bigvee_{n=0}^{\infty} A_{n,n}$. Then it is trivial that $M \vee (M' \cap M_\infty)$ and M_∞ are hyperfinite II_1 factors. The relative commutant is

$$(M \vee (M' \cap M_\infty))' \cap M_\infty = (M' \cap M_\infty)' \cap (M' \cap M_\infty) = \mathbb{C}.$$

It is also easy to see that the commuting squares

$$\begin{array}{ccc} A_{n,0} \vee A_{0,n} & \subset & A_{n+1,0} \vee A_{0,n+1} \\ \cap & & \cap \\ A_{n,n} & \subset & A_{n+1,n+1} \end{array}$$

approximates the asymptotic inclusion. (Use the inclusion $A_{n,0} \vee A_{0,n+1} \subset A_{n,n+1}$ to see the commuting square condition.) This is again of the form arising from a bi-unitary connection as in Section 11.3. That is, we have four graphs as in Fig. 12.47, where the sets V_i of vertices are as in Section 10.3.

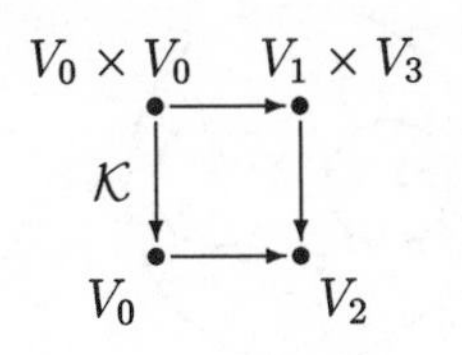

FIG. 12.47. The four graphs for the asymptotic inclusion

The number of edges from (X_1, X_2) to X on the graph $\mathcal{K}$ is given by the multiplicity of the bimodule X in $X_1 \otimes_M X_2$. (We can verify this fact by identifying $A_{n,0} \vee A_{0,n}$ and $A_{n,n}$ with appropriate End spaces.) It is easy to see that the Perron–Frobenius eigenvector μ' of the graph $\mathcal{K}$ in Fig. 12.47 is given by $\mu'((X_1, X_2)) = \mu(X_1)\mu(X_2)$ for $(X_1, X_2) \in V_0 \times V_0$ and $\mu'(X) = \sqrt{\sum_{Y \in V_0} \mu(Y)^2}\mu(X)$ for $X \in V_0$, and the Perron–Frobenius eigenvalue is the square root of $\sum_{Y \in V_0} \mu(Y)^2$. (Use Lemma 12.10.) Thus the index of the asymptotic inclusion is equal to $\sum_{Y \in V_0} \mu(Y)^2$, called the *global index* of $N \subset M$. Note that this is equal to w defined before Definition 12.8 for the system of N-N (or M-M) bimodules arising from the subfactor $N \subset M$. (Several recent results suggest that the global index is the correct way to measure the size of a paragroup. Note that if a paragroup arises from an outer action of a finite group, then the global index of the corresponding subfactor is just the order of the finite group.) So far we have proved the following. We will later prove that this asymptotic inclusion has finite depth in Theorem 12.29.

Theorem 12.24 *If $N \subset M$ is a subfactor of the hyperfinite II_1 factor with finite index, finite depth, and a trivial relative commutant, then the asymptotic inclusion $M \vee (M' \cap M_\infty) \subset M_\infty$ is again a subfactor of the hyperfinite II_1 factor with finite index and a trivial relative commutant. Its index is given by the global index of the original subfactor.*

We also note the following easy observation.

Proposition 12.25 *We have the following identity:*

$$\sum_{X \in V_0} \mu(X)^2 = \sum_{X \in V_1} \mu(X)^2 = \sum_{X \in V_3} \mu(X)^2.$$

Proof We denote the vectors $\{\mu(X)\}_{X \in V_0}$ and $\{\mu(X)\}_{X \in V_1}$ by a and b respectively. Then by the harmonic axiom, we have

$$\langle \beta a, a \rangle = \langle Gb, a \rangle = \langle b, G^t a \rangle = \langle b, \beta b \rangle,$$

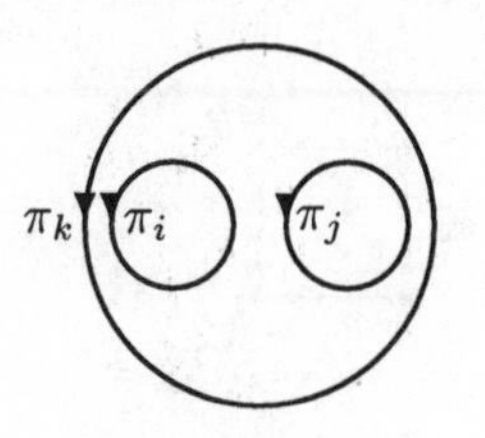

FIG. 12.48. A partially labelled surface

where $G = [G_{jk}]_{j\in V_0,k\in V_1}$ is the incidence matrix for the graph $\mathcal{G}_0$ and G^t is the transpose matrix of G. This shows the first identity and we have the other identity similarly. □

Our next aim is to introduce the 'convolution product' in the centre of the tube algebra and identify it with the fusion rule of M_∞-M_∞ bimodules of the asymptotic inclusion. In particular, we will know then that the tensor product operation of the M_∞-M_∞ bimodules arising from the asymptotic inclusion is commutative. This is not trivial at all from the definition of the asymptotic inclusion – even in the group case.

We first introduce a new commutative multiplication $*$, a *convolution*, on $H_{S^1\times S^1}$ as follows. Let N_{ij}^k be the dimension of the Hilbert space associated with the partially labelled surface in Fig. 12.48. (Here this picture is a disc with two discs removed from the interior and has three boundary components.) We define $\pi_i * \pi_j = \sum_k N_{ij}^k \pi_k$ and extend this linearly to $H_{S^1\times S^1}$. It is easy to see that this multiplication is associative. Commutativity follows because switching π_i and π_j in Fig. 12.48 gives a homeomorphic picture. We call the algebra $H_{S^1\times S^1}$ with this convolution product the *asymptotic fusion rule algebra* of the original system $\mathcal{M}$ of bimodules.

In order to deal with bimodules arising from the asymptotic inclusion $M \vee (M' \cap M_\infty) \subset M_\infty$, we first express the algebras $M \vee (M' \cap M_\infty)$ and M_∞ graphically. Because $M' \cap M_\infty$ is naturally identified with M^{opp}, we can identify the algebra $M \vee (M' \cap M_\infty)$ with Fig. 12.49. Here the bottom copy for M is the same as above and the top copy is for M^{opp}. We have the same embedding rule as before. The algebra M_∞ is expressed as $\bigvee_{k,l} A_{kl}$, so we have Fig. 12.50 for M_∞.

The upward and downward embeddings are expressed as in Fig. 12.51, where the arrow in the middle step indicates a change of bases. Next the algebra M_∞ as an $M \vee (M' \cap M_\infty)$-$M_\infty$ bimodule is identified with Fig. 12.52. Again this has two directions of embedding as in Fig. 12.51.

With the above preliminaries, we now deal with bimodules arising from the asymptotic inclusion $M \vee (M' \cap M_\infty) \subset M_\infty$. First, we handle $M \vee (M' \cap M_\infty)$-$M \vee (M' \cap M_\infty)$ bimodules. They are expressed as in Fig. 12.53, where X and Y are any pair of M-M bimodules of the original system $\mathcal{M}$

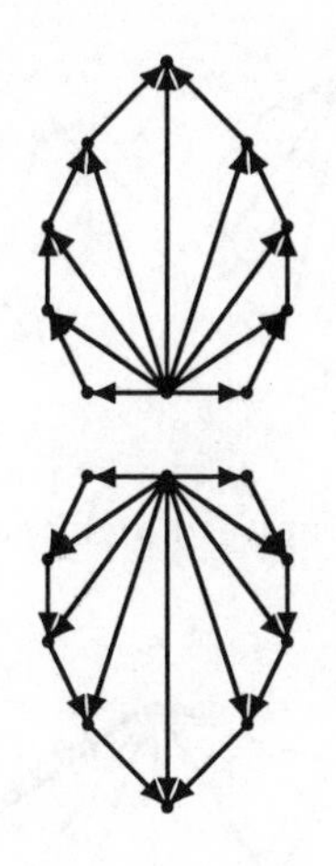

FIG. 12.49. The algebra $M \vee (M' \cap M_\infty)$

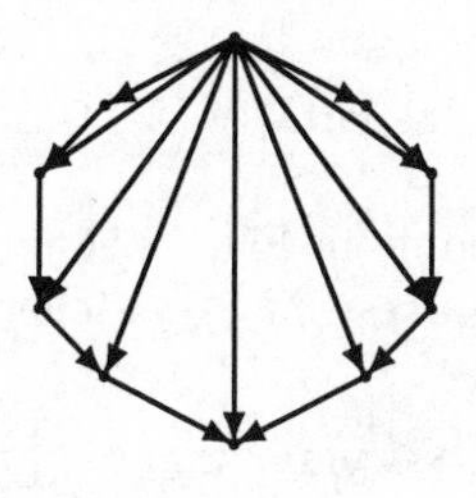

FIG. 12.50. The algebra M_∞

and this picture again has two directions of embeddings. We denote this bimodule by $B_{X,Y}$. This is essentially just a pair of M-M bimodules, and the tensor product structure is easy to see. In particular, this bimodule is irreducible as an $M \vee (M' \cap M_\infty)$-$M \vee (M' \cap M_\infty)$ bimodule.

Next we multiply the $M \vee (M' \cap M_\infty)$-$M \vee (M' \cap M_\infty)$ bimodule in Fig. 12.53 by ${}_{M\vee(M'\cap M_\infty)}M_{\infty M_\infty}$ from the right. Then we have the $M \vee (M' \cap M_\infty)$-$M_\infty$ bimodule as in Fig. 12.54, which is decomposed as in Fig. 12.55. We claim that this is really an irreducible decomposition. Note that the dimension of the Hilbert space associated to the left triangle in Fig. 12.55 is $N^Z_{X,\bar{Y}}$.

Theorem 12.26 *An $M \vee (M' \cap M_\infty)$-$M_\infty$ bimodule given by Fig. 12.56 is irreducible.*

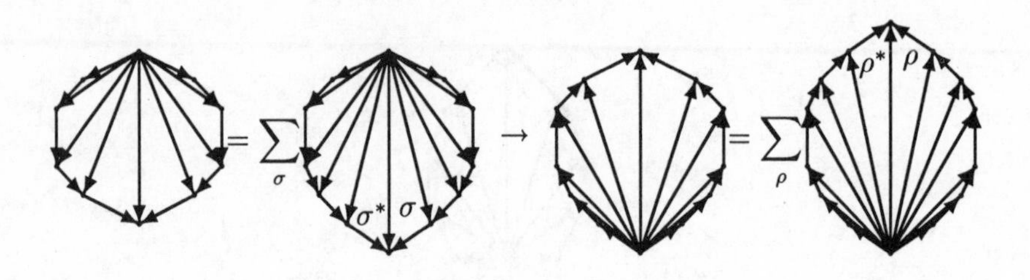

FIG. 12.51. The upward and downward embedding

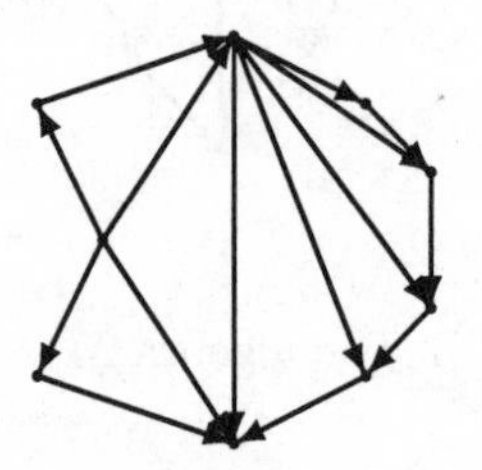

FIG. 12.52. M_∞ as an $M \vee (M' \cap M_\infty)$-$M_\infty$ bimodule

Proof We denote the bimodule in Fig. 12.56 by B_Z for an M-M bimodule Z in $\mathcal{M}$. Note that $B_* = {}_{M\vee(M'\cap M_\infty)}M_{\infty M_\infty}$. Then a graphical inspection easily produces

$$B_Z \otimes_{M_\infty} M_{\infty M\vee(M'\cap M_\infty)} = \bigoplus_{X,Y} N_{X,\bar{Y}}^Z B_{X,Y},$$

which is the irreducible decomposition.

Let the B_j be the irreducible $M \vee (M' \cap M_\infty)$-$M_\infty$ bimodules in the system, and suppose we have $B_Z = \oplus_j n_j^Z B_j$. Suppose we also have

$$B_{X,Y} \otimes_{M\vee(M'\cap M_\infty)} M_{\infty M_\infty} = \sum_j n_j^{X,Y} B_j$$

as the irreducible decomposition. Then the above irreducible decomposition of $B_Z \otimes_{M_\infty} M_{\infty M\vee(M'\cap M_\infty)}$ and the formula

$$B_{X,Y} \otimes_{M\vee(M'\cap M_\infty)} M_{\infty M_\infty} = \sum_j n_j^{X,Y} B_j$$

with Frobenius reciprocity, Theorem 9.73, give $N_{X,\bar{Y}}^Z = \sum_j n_j^Z n_j^{X,Y}$. We set $\mu_j = [B_j]^{1/2}$ and $\mu_Z = [B_Z]^{1/2}$. Then Fig. 12.55 shows that

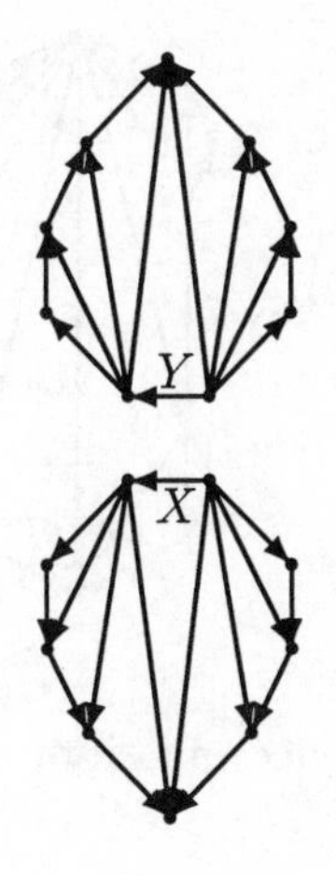

FIG. 12.53. An $M \vee (M' \cap M_\infty)$-$M \vee (M' \cap M_\infty)$ bimodule

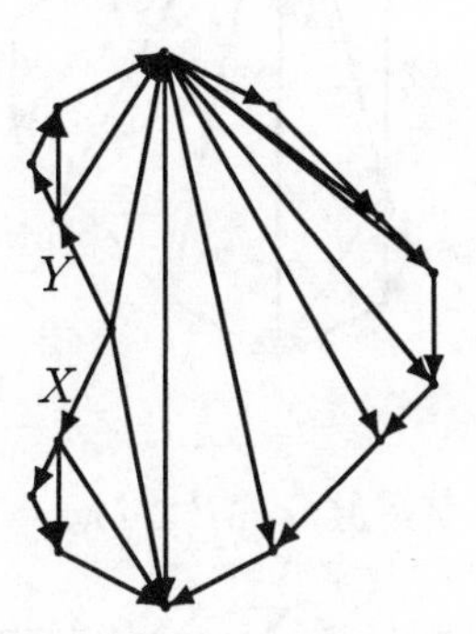

FIG. 12.54. The resulting bimodule of the product

$$\sum_j n_j^{X,Y} \mu_j = [B_{X,Y}]^{1/2} [{}_{M\vee(M'\cap M_\infty)} M_\infty {}_{M_\infty}]^{1/2} = \sum_Z N_{X,\bar{Y}}^Z \mu_Z .$$

We also have

$$\begin{aligned}\sum_Z N_{X,\bar{Y}}^Z \mu_Z &= \sum_{Z,j,k} (n_k^Z n_k^{X,Y})(n_j^Z \mu_j) \\ &= \sum_j (\sum_{Z,k} n_k^Z n_k^{X,Y} n_j^Z) \mu_j\end{aligned}$$

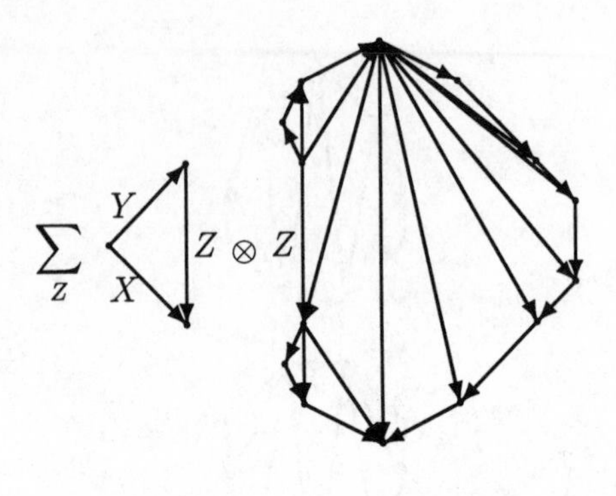

FIG. 12.55. A decomposition of a bimodule

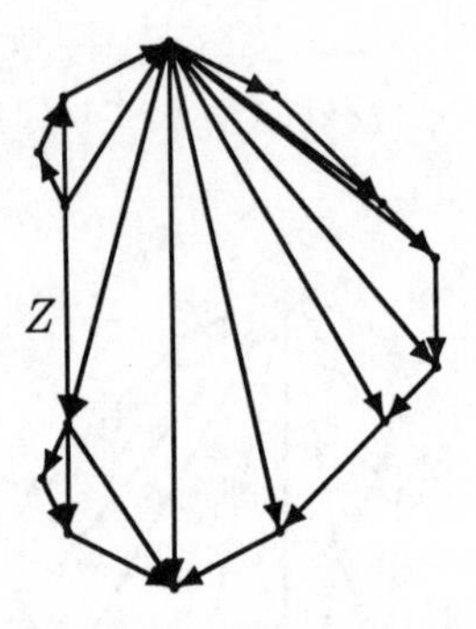

FIG. 12.56. An $M \vee (M' \cap M_\infty)$-$M_\infty$ bimodule

$$\geq \sum_j \Big(\sum_Z (n_j^Z)^2 n_j^{X,Y}\Big)\mu_j.$$

We fix X, Y. Note that for any j, we have Z with $n_j^Z \geq 1$. We thus get

$$\sum_Z (n_j^Z)^2 n_j^{X,Y} = n_j^{X,Y}$$

for all j.

We next fix j and choose X, Y so that $n_j^{X,Y} \geq 1$. Then we have $\sum_Z (n_j^Z)^2 = 1$, that is, for any j we have only one Z with $n_j^Z \geq 1$ and we have $n_j^Z = 1$ for this Z.

We next fix Z and choose X, Y with $N_{X,\bar{Y}}^Z \geq 1$. Since we have $N_{X,\bar{Y}}^Z = \sum_j n_j^Z n_j^{X,Y}$, we choose j so that we have $n_j^Z \geq 1$. These imply $n_j^{X,Y} \geq 1$. Then we have

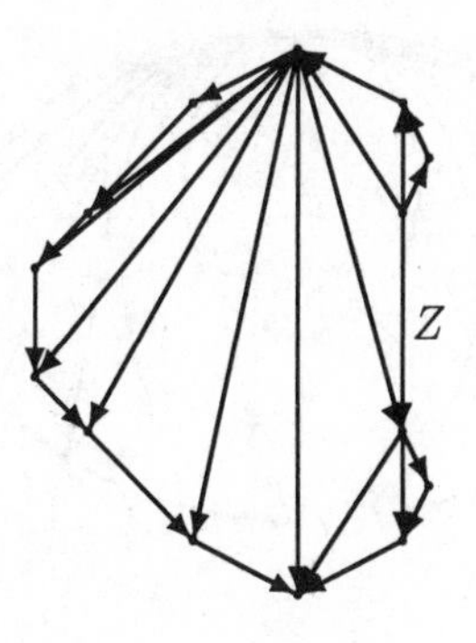

FIG. 12.57. An M_∞-$M \vee (M' \cap M_\infty)$ bimodule

$$n_j^{X,Y} = \sum_{k,Z'} n_k^{Z'} n_k^{X,Y} n_j^{Z'} = (\sum_k n_k^Z n_k^{X,Y}) n_j^Z,$$

where we first have $\leq$ for the first equality, but we can conclude that the equality holds from the above equality for $\sum_j n_j^{X,Y} \mu_j$. This implies $n_k^Z n_k^{X,Y} = 0$ for all $k \neq j$. Since we have $n_k^Z \leq n_k^{X,Y}$ by $N_{X,\bar{Y}}^Z \geq 1$, this implies the irreducible decomposition of B_Z has just one summand B_j. This means each B_Z is irreducible. □

This allows us to describe the principal graph of the asymptotic inclusion as follows. We have a system of $M \vee (M' \cap M_\infty)$-$M \vee (M' \cap M_\infty)$ bimodules $B_{X,Y}$ labelled by X, Y in $\mathcal{M}$ and a system of $M \vee (M' \cap M_\infty)$-$M_\infty$ bimodules B_Z labelled by Z in $\mathcal{M}$. We have the *fusion graph* whose vertices are bimodules $B_{X,Y}$ and B_Z and which has $N_{X,\bar{Y}}^Z$ edges between $B_{X,Y}$ and B_Z. The connected component of the fusion graph containing the vertex $B_{*,*}$ is the principal graph of the subfactor $M \vee (M' \cap M_\infty) \subset M_\infty$, where $*$ denotes the identity bimodule ${}_M M_M$. This proves that the asymptotic inclusion has finite depth, as mentioned above.

Next we have to deal with M_∞-M_∞ bimodules. First we have a graphical description for an M_∞-$M \vee (M' \cap M_\infty)$ bimodule $\bar{B}_Z$ as in Fig. 12.57.

We form a tensor product $\bar{B}_Z \otimes_{M \vee (M' \cap M_\infty)} B_{Z'}$. First we have a graphical description as in Fig. 12.58. Here the interior of the inner disk is empty. We need an irreducible decomposition of this M_∞-M_∞ bimodule. Denote by $N_{Z,Z'}^{\pi_j}$ the dimension of the Hilbert space associated with the annulus in Fig. 12.59. Here the boundary of this annulus consists of two circles. The inner circle has two components and they are labelled by Z and Z' respectively. The outer circle is labelled by π_j. First we put an element $1 = \sum_j \pi_j$ into the empty disk in Fig. 12.58, which gives an isomorphism from the bimodule in Fig. 12.58 to a direct sum of bimodules labelled by π_j. Then we can interchange the intertwiner on the inner annulus in Fig. 12.58 and the intertwiner π_j, since π_j is in the centre

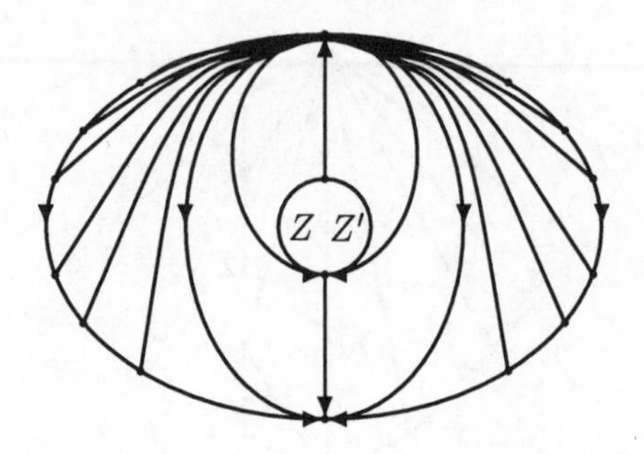

FIG. 12.58. The M_∞-$M \vee (M' \cap M_\infty)$ bimodule $\bar{B}_Z \otimes_{M\vee(M'\cap M_\infty)} B_{Z'}$

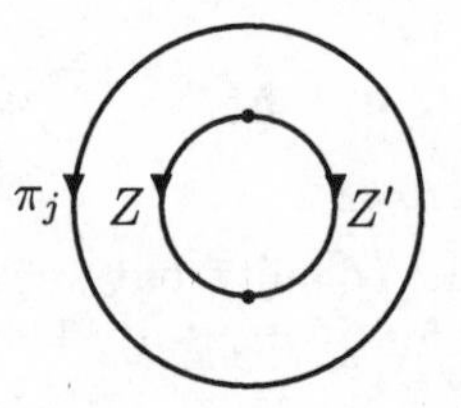

FIG. 12.59. A Hilbert space associated with an annulus

of the tube algebra and we get the decomposition of the M_∞-M_∞ bimodule as in Fig. 12.60, where the label π_j is attached to the inner annulus of the right hand side diagram.

We denote this M_∞-M_∞ bimodule on the right hand side of Fig. 12.60 by B_{π_j} and claim that the decomposition in Fig. 12.60 is the irreducible decomposition.

Theorem 12.27 *Each B_{π_j} is an irreducible M_∞-M_∞ bimodule.*

Proof A graphical expression yields the identity

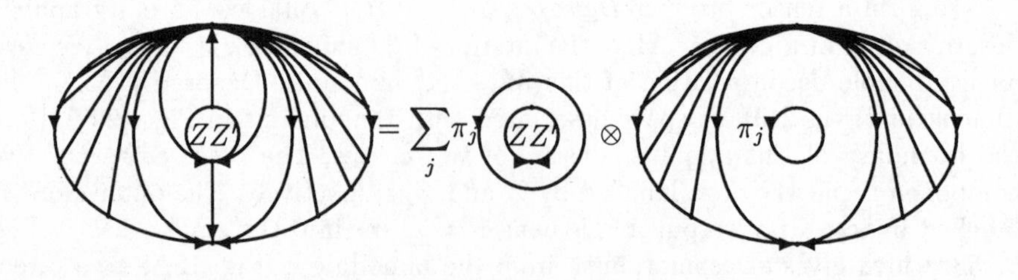

FIG. 12.60. A decomposition of an M_∞-M_∞ bimodule

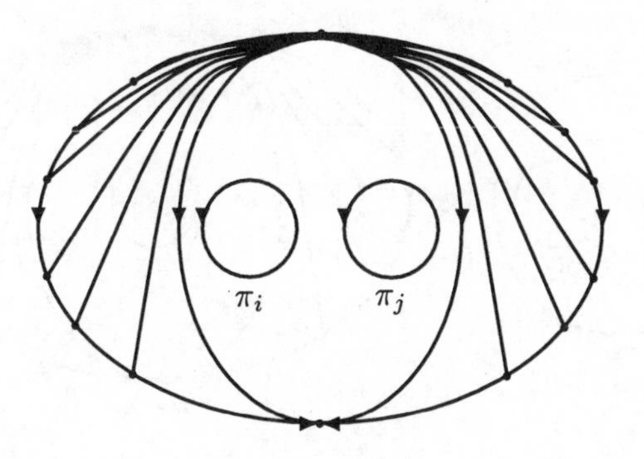

FIG. 12.61. The M_∞-M_∞ bimodule $B_{\pi_i} \otimes_{M_\infty} B_{\pi_j}$

$$B_{\pi_j} \otimes_{M_\infty} M_{\infty M \vee (M' \cap M_\infty)} = \bigoplus_Z N_{Z,*}^{\pi_j} \bar{B}_Z,$$

which is the irreducible decomposition as an M_∞-$M \vee (M' \cap M_\infty)$ bimodule. Then the same argument as in the proof of Theorem 12.26 works. □

It is easy to see that the system of bimodules B_{π_j} is closed under the tensor product of M_∞-M_∞ bimodules. We call this system the *tube system* and denote it by $\mathcal{M}_\infty$. The tensor product in the tube system is given by the following theorem.

Theorem 12.28 *In $\mathcal{M}_\infty$, we have*

$$B_{\pi_i} \otimes_{M_\infty} B_{\pi_j} = \bigoplus_k N_{ij}^k B_{\pi_k}.$$

Proof The bimodule $B_{\pi_i} \otimes_{M_\infty} B_{\pi_j}$ is expressed as in Fig. 12.61. Then this decomposes as in Fig. 12.62. □

In this way, we can identify the fusion rule algebra of the tube system with the asymptotic fusion rule algebra. We now have a graph $\mathcal{G}$ with vertices labelled by B_Z and B_{π_j} with $N_{Z,*}^{\pi_j}$ edges between vertices represented by B_Z and B_{π_j}. The above theorem shows that the connected component of the graph $\mathcal{G}$ containing B_* is the dual principal graph of the asymptotic inclusion $M \vee (M' \cap M_\infty) \subset M_\infty$. If the fusion graph is connected, then the graph $\mathcal{G}$ is also connected. In particular, the above theorem shows that the tensor product operation of the M_∞-M_∞ bimodules arising from the asymptotic inclusion is commutative, as mentioned above. Thus we have proved the following theorem.

Theorem 12.29 *The tube system $\mathcal{M}_\infty$ of M_∞-M_∞ bimodules with the relative tensor product is isomorphic to the asymptotic fusion rule algebra of $H_{S^1 \times S^1}$ with the convolution product $*$. The tube system contains the M_∞-M_∞ bimodules*

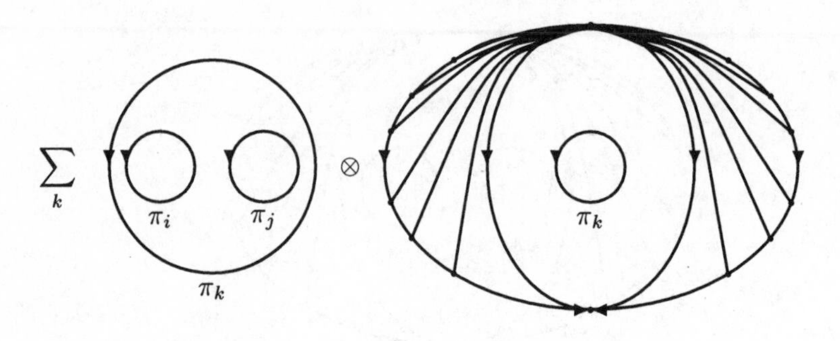

FIG. 12.62. Decomposition of a bimodule

arising from the asymptotic inclusion $M \vee (M' \cap M_\infty) \subset M_\infty$, and if the fusion graph of the original subfactor $N \subset M$ is connected, all the M_∞-M_∞ bimodules in the tube system arise in this way.

Thus if the fusion graph is connected, we have an identification between the natural basis of the Hilbert space $H_{S^1 \times S^1}$ of the TQFT_3 arising from the original subfactor $N \subset M$, the minimal central projections in the tube algebra Tube $\mathcal{M}$ of the original subfactor $N \subset M$, and the irreducible M_∞-M_∞ bimodules arising from the asymptotic inclusion $M \vee (M' \cap M_\infty) \subset M_\infty$ by Theorems 12.22 and 12.29. If the fusion graph is not connected, some bimodules in the tube system are not given by the basic constructions from the asymptotic inclusion.

Exercise 12.5 Suppose we have a two-sided sequence of projections $\{e_j\}_{j\in\mathbb{Z}}$ satisfying the Temperley–Lieb–Jones relations in the following sense, where $\beta = 2\cos(\pi/(n+1))$ with $n = 2, 3, \ldots$.

1. $e_j e_k = e_k e_j$, if $|j-k| \geq 2$.
2. $e_j e_{j\pm1} e_j = \beta^{-2} e_j$.
3. We have a trace on the algebras generated by the e_j and for $x \in \langle e_j, e_{j+1}, \ldots, e_k\rangle$, we have $\text{tr}(xe_{k+1}) = \beta^{-2}\text{tr}(x)$.

Let M be the weak closure of the algebra generated by $\{e_j\}_{j\in\mathbb{Z}}$ in the GNS-representation with respect to the trace and N be its von Neumann subalgebra generated by $\{e_j\}_{j\neq0}$. Prove that M, N are both hyperfinite II_1 factors and identify this subfactor $N \subset M$ with the asymptotic inclusion of the subfactor of the hyperfinite II_1 subfactor with principal graph A_n. Also compute the index and the principal graph.

Exercise 12.6 Consider the fusion rule algebra and the quantum $6j$-symbols arising from a finite group G as in Examples 12.14 and 12.17. Regard these as a system of M-M bimodules and quantum $6j$-symbols arising from M-M bimodules. Prove that the minimal central projections in the corresponding tube algebra are labelled by pairs (A, σ), where A is a conjugacy class in G and σ is an irreducible representation of the stabilizer of a representative element in A.

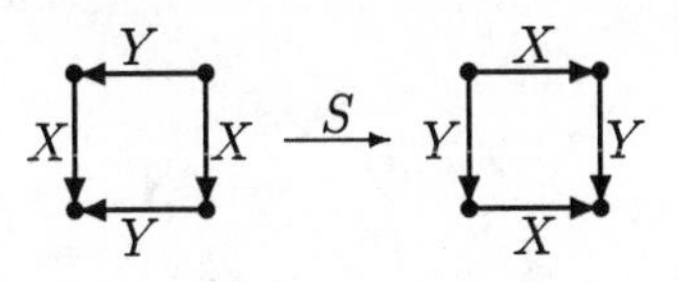

FIG. 12.63. The S-matrix

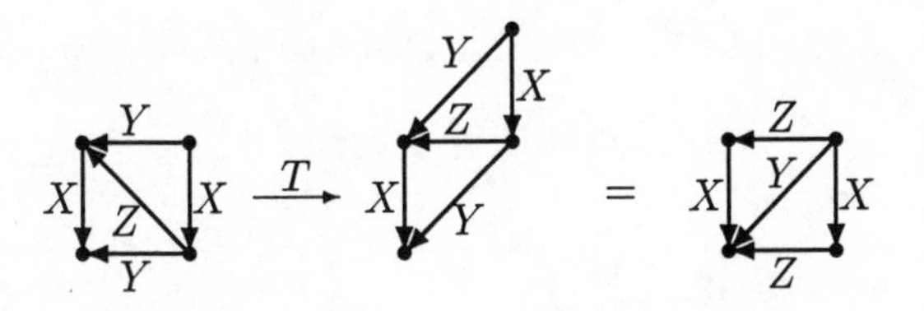

FIG. 12.64. The T-matrix

12.7 S-matrix, T-matrix, and the Verlinde identity

We introduce the S and T-matrices for our tube system $\mathcal{M}_\infty$, and prove several analogues of the theorems in RCFT due to (Vafa 1988), (Verlinde 1988a). That is, we define the S-matrix and the T-matrix so that the S-matrix diagonalizes the fusion rule algebra of the M_∞-M_∞ bimodules of the asymptotic inclusion, the T-matrix is diagonal, and its diagonal entries are roots of unity.

We define the operations S, T on $H_{S^1\times S^1}$ graphically as in Fig. 12.63 and Fig. 12.64.

It is easy to see that these two operators are unitary on $H_{S^1\times S^1}$. Note that the operation T is a 'twist' of a tube. These give a unitary representation of the group $SL(2,\mathbb{Z})$ generated by the following two matrices, denoted by S and T again:

$$S = \begin{pmatrix} 0 & -1 \\ 1 & 0 \end{pmatrix}, \quad T = \begin{pmatrix} 1 & 1 \\ 0 & 1 \end{pmatrix}.$$

Theorem 12.30 *For $\sigma, \rho \in H_{S^1\times S^1}$, we have*

$$S(0)\cdot S(\sigma * \rho) = S(\sigma)\cdot S(\rho),$$

where the product $\cdot$ denotes the multiplication in the tube algebra and 0 denotes the identity with respect to the convolution $*$. That is, the S-matrix diagonalizes the asymptotic fusion rule algebra.

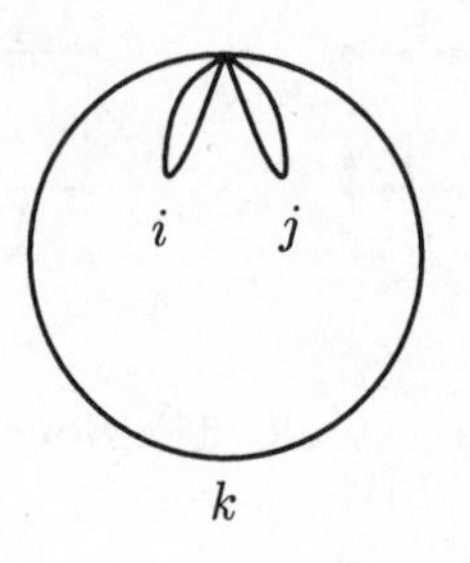

FIG. 12.65. The product in the tube algebra

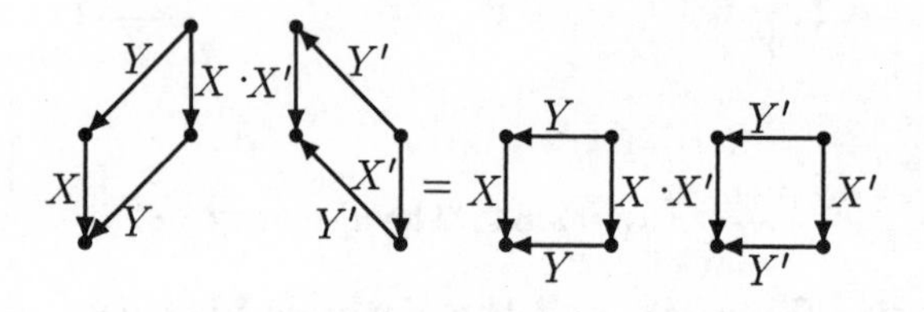

FIG. 12.66. An action of the T-matrix

Proof By linearity, we may assume ρ, σ to be i, j, minimal central projections in the tube algebra. Then for another minimal central projection k in the tube algebra, the fusion rule coefficient N_{ij}^k is the dimension of the Hilbert space in Fig. 12.48. (We have dropped the letter π here.) Then this dimension is given by the coefficient of Sk in the tube algebra product $(Si)\cdot(Sj)$, up to a normalization constant, because we can can compute the dimension as the topological invariant of the three-manifold with boundary given as the product of S^1 and Fig. 12.65.

We can determine the normalization constant by setting $i = 0$, and get the formula as desired. □

This is an analogue of the celebrated Verlinde formula, (8.7.9).

Again from Fig. 12.66, we have the following identity.

$$(T\sigma \cdot T^*\rho, \tau) = (\sigma \cdot \rho, \tau), \quad \text{for all } \tau.$$

Thus we have

$$T\sigma \cdot T^*\rho = \sigma \cdot \rho,$$

which implies the matrix T is diagonal to the natural basis $\{\pi_j\}$ of the Hilbert space $H_{S^1 \times S^1}$. That is, we have a scalar $t_j \in \mathbb{C}$ with $T\pi_j = t_j \pi_j$. It is possible

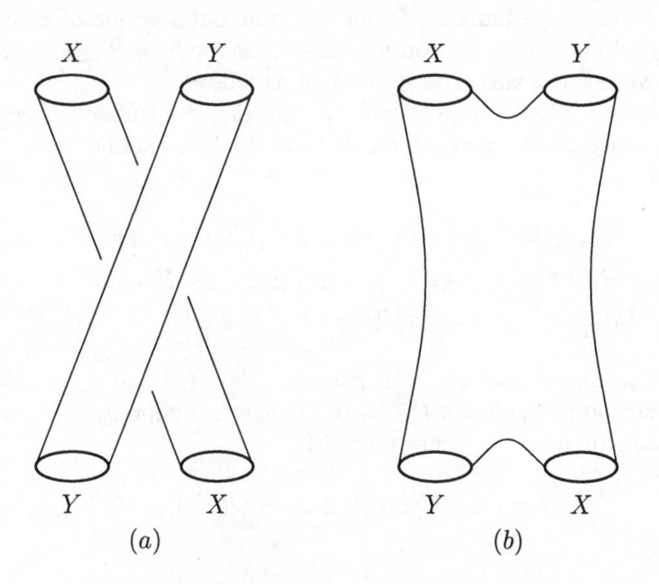

FIG. 12.67. Braiding on the tube system

to prove that the diagonal entries of the T-matrix are roots of unity as in (Vafa 1988).

At the end of this section, we see that we have a natural braiding structure on the tube system as described in Fig. 12.67. The meaning of the Figure is as follows. We choose X, Y from the tube system. What we need as a candidate for braiding is an element in $\mathrm{Hom}(X \otimes Y, Y \otimes X)$. We can put intertwiners corresponding to the labels X, Y on the two tubes on the left hand picture of Fig. 12.67. This gives an element in the Hilbert space corresponding to the left hand picture of Fig. 12.67. There is a natural sesquilinear form on the two Hilbert spaces on the two pictures in Fig. 12.67, which is obtained by putting picture (a) inside picture (b). (In this way, we have a three-dimensional manifold whose boundary is a union of the two pictures in Fig. 12.67.) This sesquilinear form gives an element in the Hilbert space corresponding to picture (b) of Fig. 12.67 from the element of the other Hilbert space. This is the element we need for braiding. It is easy to see that this construction satisfies all the necessary properties for braiding. A more detailed discussion on braiding can be found in (Ocneanu 1994).

12.8 Group case and the quantum double construction

The construction of the asymptotic inclusion is a way to get a new paragroup from a given paragroup. We explain why this passage can be regarded as a paragroup analogue of the quantum double construction of (Drinfeld 1987) by looking at the group case in detail, so start with a finite group G. This section is *not* self-contained.

Let R be the hyperfinite II_1 factor and α an outer action of a finite group G on R. As in Section 10.6, we consider the subfactor $N = R \subset M = R \rtimes_\alpha G$. We explicitly write down this crossed product as follows.

Suppose R acts on a Hilbert space H. On the new Hilbert space $\ell^2(G, H)$ of all the H-valued ℓ^2 functions on G, we have the following actions:

$$\begin{aligned}(u_g\xi)(h) &= \xi(g^{-1}h), \quad g, h \in G, \xi \in \ell^2(G, H)\\ (v_g\xi)(h) &= \xi(hg), \quad g, h \in G, \xi \in \ell^2(G, H)\\ (\pi(x)\xi)(h) &= \alpha_{h^{-1}}(x)\xi(h), \quad x \in R, h \in G, \xi \in \ell^2(G, H).\end{aligned}$$

Then the crossed product M is generated by $\pi(x)$ and u_g, and is identified with the fixed point algebra of $R \otimes M_G(\mathbb{C})$ under the action $\beta_g = (\alpha_g \otimes \mathrm{id}) \cdot \mathrm{Ad}(v_g)$. Then we can compute the Jones tower of

$$N = (R \otimes \ell^\infty(G))^\beta \subset M = (R \otimes M_G(\mathbb{C}))^\beta$$

as

$$\begin{aligned}&(R \otimes \ell^\infty(G))^\beta \subset (R \otimes M_G(\mathbb{C}))^\beta\\ &\subset (R \otimes M_G(\mathbb{C}) \otimes \ell^\infty(G))^{\beta \otimes \mathrm{Ad}v} \subset (R \otimes M_G(\mathbb{C}) \otimes M_G(\mathbb{C}))^{\beta \otimes \mathrm{Ad}v}\\ &\subset (R \otimes M_G(\mathbb{C}) \otimes M_G(\mathbb{C}) \otimes \ell^\infty(G))^{\beta \otimes \mathrm{Ad}v \otimes \mathrm{Ad}v} \subset \cdots.\end{aligned}$$

We set

$$P = M_G(\mathbb{C}) \otimes M_G(\mathbb{C}) \otimes M_G(\mathbb{C}) \otimes \cdots,$$

which is the closure in the GNS-representation with respect to the trace, and hence a hyperfinite II_1 factor. We also set an outer action γ of G by

$$\gamma_g = \mathrm{Ad}v_g \otimes \mathrm{Ad}v_g \otimes \mathrm{Ad}v_g \otimes \cdots.$$

Then we have

$$\begin{aligned}M_\infty &= \bigvee_n (R \otimes M_G(\mathbb{C}) \otimes \cdots \otimes M_G(\mathbb{C}))^{\beta \otimes \mathrm{Ad}v \otimes \cdots \otimes \mathrm{Ad}v}\\ &= (R \otimes M_G(\mathbb{C}) \otimes P)^{\beta \otimes \gamma},\end{aligned}$$

where we can change the order of taking $\bigvee$ and the fixed point algebra in the second equality because the group G is finite. We know that

$$M_G(\mathbb{C}) \otimes \cdots \otimes M_G(\mathbb{C})^{\mathrm{Ad}v \otimes \cdots \otimes \mathrm{Ad}v}$$

is contained in the higher relative commutants $M' \cap M_{2k}$. Because we already know the principal graph, and hence the dimension of the higher relative commutants as in Section 10.6, we can conclude that the above algebra is indeed the higher relative commutant. We thus know that $M' \cap M_\infty = P^\gamma$. We set $Q = R \otimes P$ and define an action of $G \times G$ by $\beta \otimes \gamma$. Then we conclude that the asymptotic inclusion $M \vee (M' \cap M_\infty) \subset M_\infty$ is given by $Q^{G \times G} \subset Q^G$, where G is embedded into $G \times G$ by $g \mapsto (g, g)$.

We want to know fusion rule algebras of the M_∞-M_∞ bimodules arising from the asymptotic inclusion. This corresponds to the TQFT$_3$ arising from the quantum $6j$-symbols in 12.17. This is a special case of TQFT$_3$ considered in (Dijkgraaf and Witten 1990), and the fusion rule algebra there was described by (Dijkgraaf *et al.* 1989). It was later noticed by (Dijkgraaf *et al.* 1990) that this fusion rule algebra is given by the duals of the quantum double of the function algebra over G, in the sense of (Drinfeld 1987). Thus Theorem 12.29 implies that the fusion rule for the subfactor $Q^{G \times G} \subset Q^G$ should be also described by the quantum double, if the fusion graph of $\hat{G}$ is connected. This has been directly verified in the Example of Section 4 of (Kosaki *et al.* 1997). Because a general paragroup is a 'quantization' of a finite group, we can say that the asymptotic inclusion gives the *quantum double of a paragroup.*

A general paragroup is more like a general Hopf algebra than a quantum group, so that in general the fusion rule algebra is not commutative and we have no analogue of the R-matrix. We can eliminate these inconveniences by passing to the 'quantum double'. The construction in Section 13.5 will further clarify this idea.

Conceptually, the quantum 'double' construction pairs the original object with something 'dual' to the original object, and then appeals to some machinery to produce a higher symmetry. In the ordinary quantum double construction, the 'dual' object is the dual Hopf algebra and the 'machinery' changes the product structure as in (Drinfeld 1987). In the asymptotic inclusion, the 'dual' object is an opposite algebra $M^{\mathrm{opp}} \cong M' \cap M_\infty$, and the 'machinery' allows us to pass from the system of the $M \vee (M' \cap M_\infty)$-$M \vee (M' \cap M_\infty)$ bimodules to that of the M_∞-M_∞ bimodules. (Recall that the system of the $M \vee (M' \cap M_\infty)$-$M \vee (M' \cap M_\infty)$ bimodules consist of a pair, one from the original system and one from its 'opposite'.)

We can also regard the Turaev–Viro type TQFT$_3$ based on triangulation as a kind of the quantum double construction. If the original data, the quantum $6j$-symbols, have a 'sufficiently high symmetry', then the resulting TQFT$_3$ splits as a tensor product of the Reshetikhin–Turaev type TQFT$_3$ and its complex conjugate as shown by (Turaev 1994). This is natural because it is expected that if the original object has enough symmetry from the beginning, then the 'quantum double construction' should just give a pair consisting of the original object and a certain kind of 'dual'.

Ocneanu also has this type of splitting result in (Ocneanu 1994), in the presence of a non-degenerate braided system of bimodules. This non-degenerate bimodule assumption makes mathematically precise the notion of a 'high enough symmetry'.

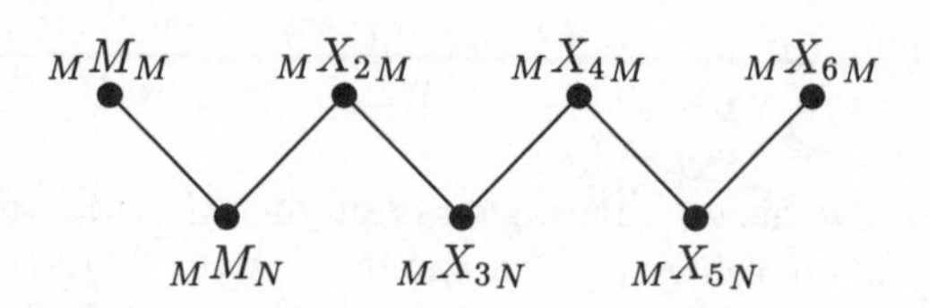

FIG. 12.68. Dynkin diagram A_7 labelled with bimodules

12.9 Quantum $6j$-symbols and self-conjugate sectors

We study a TQFT$_3$ based on Longo's sector approach in Section 10.7. This section is *not* self-contained.

In (Turaev and Viro 1992), there are explicit computations of the topological invariants for the quantum $6j$-symbols of (Kirillov and Reshetikhin 1988) arising from the quantum group $U_q(sl_2(\mathbb{C}))$. Conceptually, this should correspond to the subfactor with principal graph A_n when $q = \exp(\pi i/(n+1))$, so we might want to conclude that the TQFT$_3$ arising from subfactors with principal graph A_n is the same as the one in (Turaev and Viro 1992), but this is not the case. This is because our fusion rule algebra is *graded* as in Section 12.3 while there is no such grading in (Kirillov and Reshetikhin 1988). Let us take a more concrete example. Suppose we have the dual principal graph A_7 and the vertices are labelled with bimodules as in Fig. 12.68.

Then we cannot multiply X_{3N} and $_M X_5$ because the algebras N and M do not match, while the corresponding multiplication is possible in the quantum group setting. Thus our quantum $6j$-symbols have less data than in the quantum $6j$-symbols in the quantum group theory. This was noted by T. Kohno.

We want to find a method to eliminate this difficulty. First, we see that a naive approach based on an endomorphism picture does not work. That is, in the case of subfactors with principal graph A_n, we have a natural endomorphism σ sending $e_j \mapsto e_{j+1}$ from $M = \langle e_1, e_2, e_3, \ldots \rangle$ onto $N = \langle e_2, e_3, e_4, \ldots \rangle$. So we can transform an M-N bimodule into an M-M bimodule by changing the right action by this endomorphism. Suppose that this gave a desired fusion rule algebra. The bimodule $_M M_N$ is now changed into an M-M bimodule M_σ in the notation in Section 10.7. This should be self-conjugate, so the product with itself, M_{σ^2} should give $_M M_M \oplus {_M X_{2M}}$. But the irreducible decomposition of M_{σ^2} as an M-M bimodule is given by $(Me_1)_{\sigma^2} \oplus M(1-e_1)_{\sigma^2}$ and they do not have the correct dimensions as left M-modules, and we reach a contradiction. This remark is due to M. Izumi.

What we should instead do here is to use endomorphisms in a more sophisticated way, that is, in a type III setting.

We start with a paragroup which is isomorphic to its dual. Then a recent method in (Izumi 1993b) based on Cuntz algebra endomorphisms gives a subfactor $N \subset M$ of type III. We can then choose an endomorphism ρ of M onto N

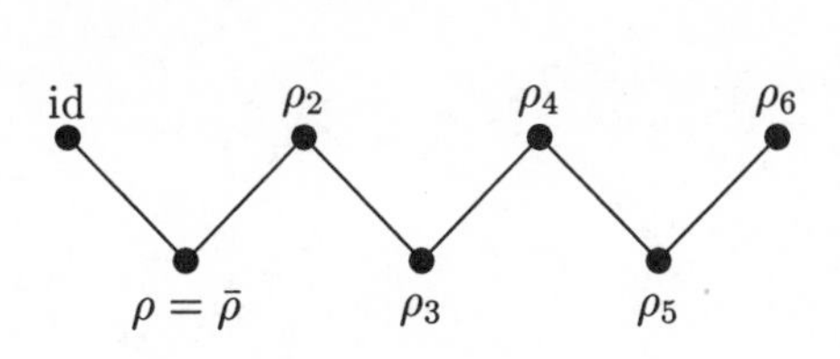

FIG. 12.69. Dynkin diagram A_7 labelled with sectors

so that we have $\rho = \bar{\rho}$ as sectors. Then the condition $\rho = \bar{\rho}$ insures that we have a finite system of sectors which is closed under conjugation and composition. For example, in the case of the Dynkin diagram A_7, the dual principal graph is labelled as in Fig. 12.69. We can now multiply ρ_3 and ρ_5 and the result is $\rho_2 + \rho_4$. Note that we can multiply ρ_3 and ρ_5 as compositions of endomorphisms without the condition $\rho = \bar{\rho}$, but then we could have no reason why the result would be expressed in terms of ρ_j in the system.

12.10 Notes

The definition of the Jones polynomial in Section 12.1 follows the original definition in (Jones 1985).

A link diagram is a projection of a link onto a two dimensional plane, composed of arcs with possible crossings containing two strands only, one over and one under as in Fig. 12.3. The Jordan curve theorem ensures that a diagram in the plane with no crossings is equivalent to the unknot. In general it is difficult to decide whether two diagrams represent knots which can be continuously deformed from one to the other. This is the case, if and only if one can go from one diagram to another via a sequence of local moves, the *Reidemeister moves* of types I, II, III. To actually find a sequence of moves in a particular example may not be easy. Reidemeister's theorem does not help us, at this stage, to decide whether the trefoil is knotted; or whether its left and right handed versions are equivalent (Fig. 12.3).

In some sense, the type I move is more subtle than the other two although apparently simpler at first sight. If we thicken the strings to turn them into ribbons, then two link diagrams on the plane represent the same ribbon links in three dimensional Euclidean space if and only if one diagram can be transformed to another via at most type II or III moves, in which case we say the diagrams are *regularly isotopic*. It is convenient, when introducing invariants for links (cf. also RCFT in Chapter 13) to first understand invariants of regular isotopy and then keep track of them under type I moves.

Many link invariants can be obtained through the notion of a *state* on a diagram. By a state we mean here a choice or distribution of symbols on the diagram, on the components of the string's arcs, vertices or regions of the plane (cf. state in statistical mechanical models, e.g. in the Ising model – a distribution of + and − over vertices). Placing symbols on arcs or string components is

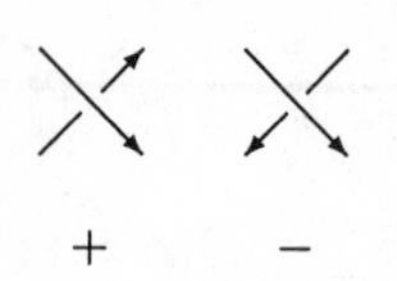

FIG. 12.70. Positive and negative crossings

akin to colouring. With two colours on strings, this is equivalent to choosing an orientation on each component. For an oriented diagram, one can distinguish between two kinds of crossings; a crossing is positive if as one approaches the crossing along the underpass, then the overcrossing is from left to right, negative if from right to left as in Fig. 12.70. Then it can be readily verified that the *writhe* $w(K)$ of an oriented link diagram K:

$$w(K) = \#(\text{positive crossings}) - \#(\text{negative crossings}) \tag{12.10.1}$$

is an invariant of regular isotopy. It is also useful to discuss regular isotopy in terms of *framed* links. A *framing* f of an (unoriented) link L is an assignment of an integer to each component. To incorporate the framing into a link diagram D for L, choose any orientation of the link. If D_S is the diagram of a component S of the link diagram, then we say that the diagram D represents (L, f) if $f(S) = w(D_S)$ for each component S. We can easily find a diagram to represent a framed link by introducing extra kinks if necessary. Then if D, D' are link diagrams representing the framed link (L, f) and (L', f'), D, D' are regularly isotopic in $\mathbb{R}^2 \cup \{\infty\}$ if and only if (L, f) and (L', f') are *ambient* isotopic (i.e. there is a frame-preserving ambient isotopy between L and L'). Any closed oriented three-manifold can be obtained by surgery on a *framed* (unoriented) link (L, f).

Colouring arcs is another possibility of assigning states to a link diagram. The three arcs of a trefoil in Fig. 12.3 can be coloured with three distinct colours. At any crossing, all three colours appear, but under type I moves crossings where only one colour is used will appear. A link diagram is said to be 3-*colourable* if the arcs can be coloured with three distinct colours such that at any crossing either three colours or one colour is used. The property of being 3-colourable is invariant under all three Reidemeister moves for knot diagrams (but not for links). Hence a trefoil is knotted as the unknot is not 3-colourable. This idea is not in itself sufficient to distinguish the trefoil from its mirror image, but a different and more sophisticated notion of state is required, which gives rise to the Jones polynomial (Jones 1985) via the bracket polynomial of (Kauffman 1986). A state will now be a choice of removing a crossing by cutting and re-tieing the strings as in Fig. 12.71 (cf. Fig. 7.15).

A state then gives a collection of δ non-intersecting circles in the plane, where a certain number α of a regions have been joined (as in the second option of Fig. 12.71), and a certain number β of b regions have been joined (as in the first

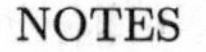

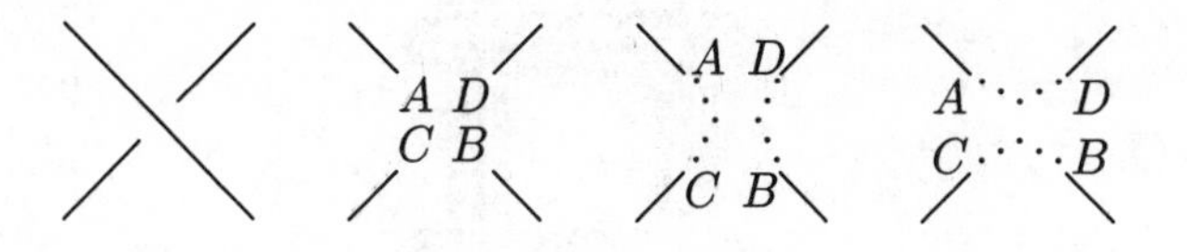

FIG. 12.71. State choices

option of Fig. 12.71). (In Fig. 7.13 (a), the a [resp. b] region corresponds to the shaded [resp. unshaded] region.) Multiplying the weights a and b together according to their multiplicity and weighting each component by d, we define the *bracket polynomial*

$$\langle K \rangle = \textstyle\sum_{\sigma} a^{\alpha} b^{\beta} d^{\delta} / d \tag{12.10.2}$$

with the summation over all states (Kauffman 1986). The bracket polynomial satisfies the recursion relation under local changes:

$$\langle \times \rangle \;=\; a\langle \asymp \rangle \;+\; b\langle \supset\subset \rangle \tag{12.10.3}$$

and is normalized to be 1 on the unknot.

Braid diagrams are a decorative feature in Celtic art. Around the sixth to seventh Century AD, plaitwork, the regular weaving pattern as in Fig. 7.12(c) developed in Europe incorporating more intricate braided patterns. In (Allen 1899), we find a description of the *breaks* used to decorate a plait (Fig. 12.71 appears explicitly in (Allen 1899)) through removing crossings:

If we now wish to make a break in the plait, any two of the cords are at the point where they cross each other, leaving four loose ends, $ABCD$. To make a break,the loose ends are joined together in pairs. This can be done in two ways only: (1) A can be joined to C, and D to B, forming a vertical break: or (2) A can be joined to D, and C to B, forming a horizontal break.

Allen decomposed the resulting braids into tangles (which he called 'knots') and identified eight fundamental tangles as generators of nearly all the braid diagrams which appeared in Celtic art.

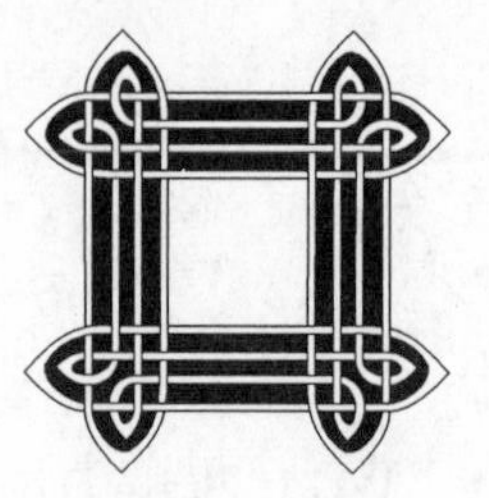

The bracket polynomial is invariant under type II moves if $ab = 1$, $a^2+b^2+d = 0$, and then automatically invariant under type III moves. In this the type I moves involve factors of $-a^{\pm 3}$. The writhe is exactly what one needs to keep track of type I moves, so that the product

$$V(L) = (-a)^{-3w(D)}\langle D\rangle/(-a^2 - a^{-2})$$

is an invariant of ambient isotopy, i.e. a link invariant. With the change of variable $t = a^{-4}$, this becomes the Jones polynomial $V_K(t)$ (a Laurent polynomial in $\sqrt{t}$) which is uniquely determined by the normalization $V(\bigcirc) = 1$ and the recursion relation:

$$t^{-1}V(\times) - tV(\times) = (t^{\frac{1}{2}} - t^{-\frac{1}{2}})\, V(\asymp)\,. \tag{12.10.4}$$

The *skein relation* (12.10.4) can be obtained from the bracket polynomial (12.10.2) and its variant:

$$\langle \times \rangle = a\langle \asymp \rangle + a^{-1}\langle \supset\subset \rangle \tag{12.10.5}$$

$$\langle \times \rangle = a^{-1}\langle \asymp \rangle + a\langle \supset\subset \rangle \tag{12.10.6}$$

and so eliminating $\langle \supset\subset \rangle$ gives

$$a\,\langle \times \rangle - a^{-1}\langle \times \rangle = (a^2 - a^{-2})\langle \asymp \rangle$$

which, when decorated with orientation and writhes, yields (12.10.4) when $t = a^{-4}$.

The definition of the TQFT_3 in Section 12.2 is taken from (Turaev and Viro 1992). The most detailed reference on various constructions and properties of topological quantum field theories is (Turaev 1994). Our definitions of abstract fusion rule algebra and quantum $6j$-symbols follow the unpublished manuscript 'An invariant coupling between three-manifolds and subfactors, with connections to topological and conformal quantum field theory' (1991) of A. Ocneanu, and are more general than those in (Turaev and Viro 1992) in some points. Exercises 12.1 and 12.2 are from (Turaev and Viro 1992).

Most of the material in Sections 12.3 and 12.4 appeared first in Ocneanu's unpublished manuscript mentioned above. It was (Pasquier 1988) who first noticed that the graphs D_{2n+1}, E_7 do not have consistent fusion rules, but we need to argue further to prove that they do not arise as principal graphs of subfactors because we have a $\mathbb{Z}/2$-grading of vertices in the paragroup theory unlike the setting of Pasquier. In the subfactor theory, the impossibility of D_{2n+1}, E_7 as principal graphs in this method were first published by (Izumi 1991) and (Sunder and Vijayarajan 1993) independently.

The result in Section 12.5 was claimed by Ocneanu in several talks. The proof here is taken from (Evans and Kawahigashi 1995b). The computations in Exercise 12.4 were first explicitly done by (Izumi 1991).

The asymptotic inclusion was first introduced in (Ocneanu 1988) with further results in (Ocneanu 1991). Most of the material in Section 12.6 were given by Ocneanu in his seminar talks at the University of California, Berkeley in July, 1993. We became aware of them through handwritten notes of the talks by one of the participants. The presentation here follows (Evans and Kawahigashi 1995a). A similar construction to the asymptotic inclusion for sectors has been introduced by (Longo and Rehren 1995). See (Masuda 1997) for the relation between the two constructions. The figures in the proof of Theorem 12.22 were taken from Ocneanu's unpublished manuscript 'Low dimensional operator cohomology – preliminary course notes –' (1993). See (Ocneanu 1994) for more related results. Theorem 5.1 in (Evans and Kawahigashi 1995a), corresponding to Theorem 12.30 here, was incorrect because the normalizing coefficient was missing there. The definition of a surface bimodule has already appeared in (Ocneanu 1988) under the name 'open string bimodule' in a slightly different form. The index value in Exercise 12.5 was first computed in (Choda 1989).

The results in Section 12.7 are due to Ocneanu and were shown in his lectures at Collège de France in 1991. We learnt them through handwritten notes of the lectures by a participant, which were distributed by Ocneanu. The presentation here follows (Evans and Kawahigashi 1995a). The construction of the braiding at the end of Section 12.7 is due to (Ocneanu 1994).

Ocneanu, in talks around the period 1991–95, presented the basic idea that we can regard the asymptotic inclusion as a subfactor analogue of the quantum double construction of (Drinfeld 1987). The method to compute the asymptotic inclusion of a finite group crossed product in Section 12.8 is due to M. Izumi.

The result in Section 12.9 is new and relies on (Izumi 1993b).

13

RATIONAL CONFORMAL FIELD THEORY (RCFT) AND PARAGROUPS

13.1 Introduction

Conformal quantum field theory has experienced much interaction with various fields of mathematics in the last ten years. We will explain here relations between the combinatorial aspects of (rational) conformal field theory (RCFT) and the combinatorial aspects of paragroup theory.

In Section 13.2, we present the most fundamental examples of RCFT, the Wess–Zumino–Witten models. We will construct a paragroup from an RCFT in Section 13.3. If we start with the Wess–Zumino–Witten models $SU(n)_k$, we get the series of Jones–Wenzl subfactors. In Section 13.4, we introduce the orbifold construction for paragroups. In particular, we will prove (non-)flatness of the connections on the Dynkin diagrams D_n, which was left open in Section 11.5. Section 13.5 deals with a converse construction of an RCFT from a paragroup.

13.2 Wess–Zumino–Witten models and graphical representation

We discuss examples satisfying the Moore–Seiberg axioms arising from a connected, simply connected, compact simple Lie group G and an integer k called a *level.* Then a set of combinatorial data satisfying the Moore–Seiberg axioms can be constructed for a pair of (G, k) as in (Witten 1989b). This is often called the WZW model G_k. Here we introduce some graphical representation of the data.

We have a *commutative* fusion rule algebra $\bigoplus_{i\in I} \mathbb{C}i$. We put a label with an element in I to each edge and to each region of an oriented knotted planar graph with at most trivalent vertices and crossings. We put a labelling as in Fig. 13.1 only when $N_{ij}^k > 0$.

For the braiding and fusing matrix B and F, we have graphical representations as in Figs 13.2, 13.3. Special forms of the fusing matrices are also written as in Fig. 13.4.

Then unitarity, the Yang–Baxter equation, and the braiding–fusion relation are represented graphically as in Figs 13.5, 13.6, and 13.7 respectively, where we omitted labelling and orientation of edges for simplicity. Note that Figs 13.5,

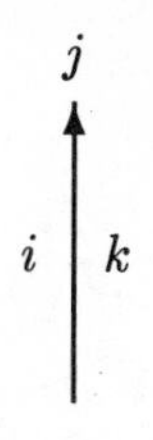

FIG. 13.1. Labelling of an edge and regions

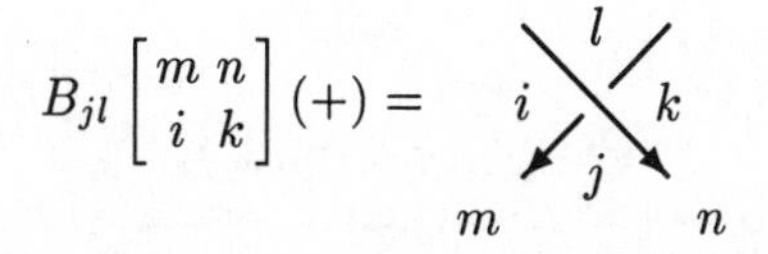

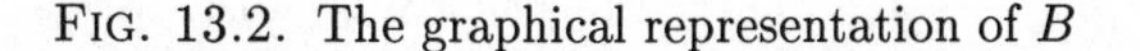
FIG. 13.2. The graphical representation of B

13.6 represent Reidemeister moves of of type II, III as in Figs 12.6, 12.7. (As usual, we can reverse the orientation of an edge by changing its label from i to $\bar{\imath}$ or $i^\vee$.)

These three graphical identities mean that we can associate a numerical value with a diagram up to *regular isotopy.* (See Notes to Chapter 12 for a discussion on regular isotopy.) In order to get topological invariance, we need another relation in addition to the regular isotopy. That relation is described in Fig. 13.8, but our graphical representation is not invariant this time, and we get an additional coefficient here.

Using the S-matrix, we can define $F_j = S_{0j}/S_{00} > 0$ for $j \in I$, where 0

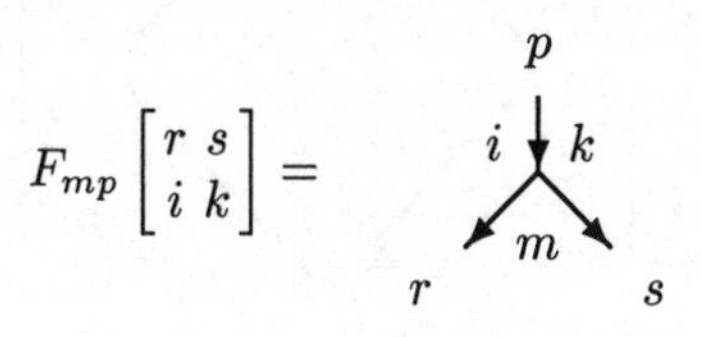

FIG. 13.3. The graphical representation of F

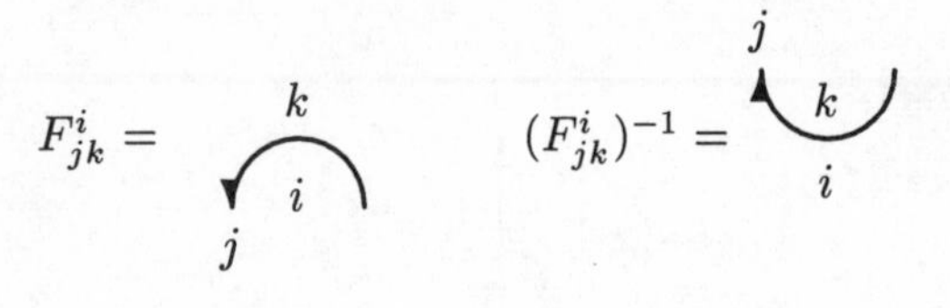

FIG. 13.4. Special cases of F

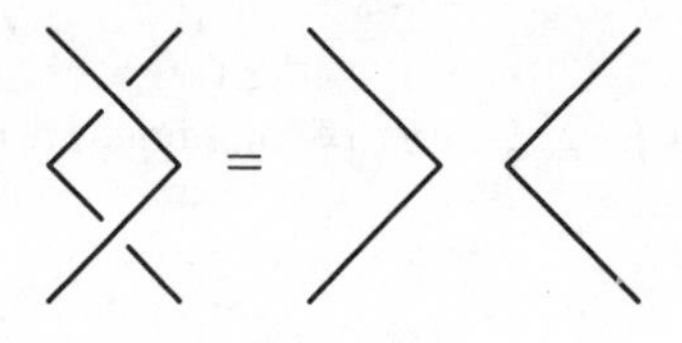

FIG. 13.5. Unitarity

denotes the identity of the fusion rule algebra. Then it has been shown in (Degiovanni 1992), (Moore and Seiberg 1989a), (Turaev and Wenzl 1993), (Witten 1989b) that we may assume the relations as in Figs 13.1, 13.9, 13.10, and 13.11 with various orientations. (Note that the diagram in Fig. 13.11 is a tetrahedron, but this diagram is 'dual' to those in Chapter 12 in the sense that the roles of vertices and faces are now interchanged.)

Take any labelled finite planar graph. We assign a number to such a graph as follows. We fix a time direction of on the plane and slice the graph into finitely many parts or time periods so that each time slice only contains a graph of the type of Figs 13.2, 13.3, or 13.9. (This is always possible by a small change of the time direction if necessary.) In the example in Fig. 13.12, the trefoil is split according to seven time periods.

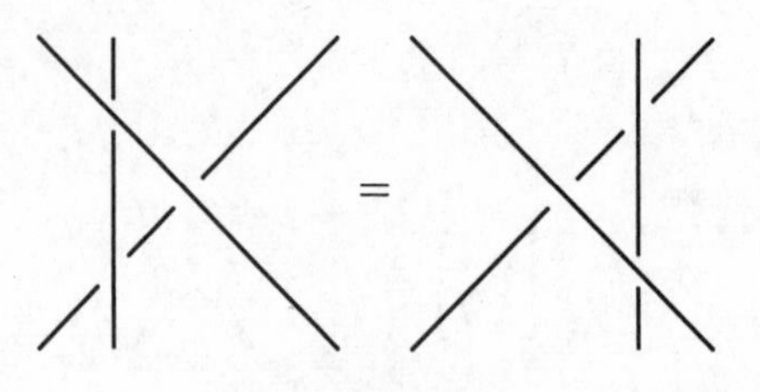

FIG. 13.6. The Yang–Baxter equation

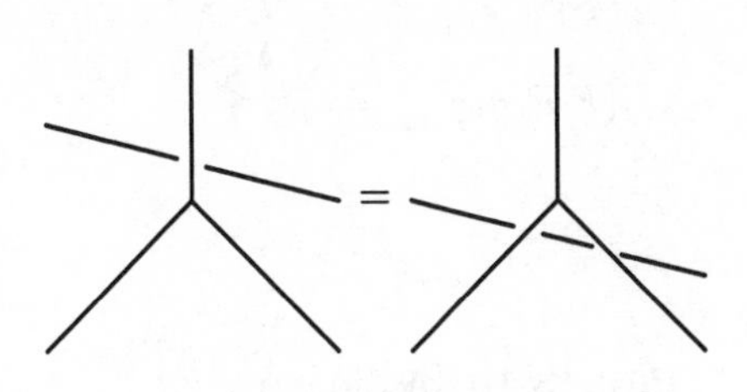

FIG. 13.7. The braiding-fusion relation

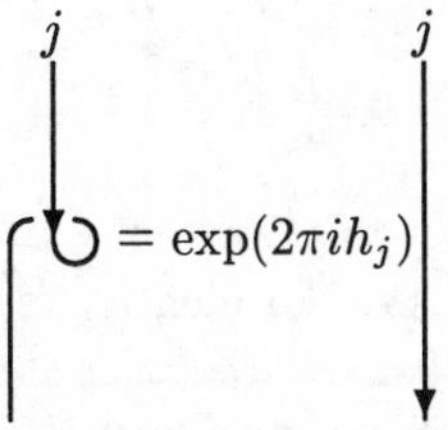

$j \qquad j$

$= \exp(2\pi i h_j)$

FIG. 13.8. Twist and a conformal dimension

Then each time period gives a number (depending on parameters) as in Figs 13.2, 13.3, 13.4 above, and the total graph gives a numerical value as their product over all time periods. If we have a partially labelled graph, we consider all possible labellings for unlabelled parts, and then take a sum of all the values of the labelled graphs. In this procedure, we always assign a label 0 to the unbounded region outside the graph.

Strictly speaking, we also have to specify intertwiners as in the case of the quantum $6j$-symbols in Section 12.2 because N_{ij}^k is in general bigger than 1, but we omit them in our graphical representation for simplicity, as in many treatments of this topic.

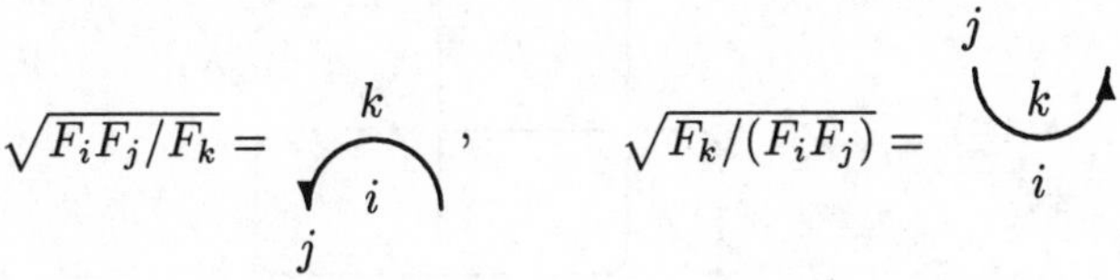

$\sqrt{F_i F_j / F_k} =$ k, i, j , $\qquad \sqrt{F_k / (F_i F_j)} =$ j, k, i

FIG. 13.9. Identities for F_j

$$(F_iF_jF_k)^{1/4} = \quad \begin{matrix} i \quad j \quad k \\ \bar{i} \; \bar{k} \\ 0 \end{matrix}$$

FIG. 13.10. An identity for F_j

$$B_{jl}\begin{bmatrix} m & n \\ i & k \end{bmatrix}(+) = \frac{1}{(F_iF_kF_mF_n)^{1/4}} \quad \begin{matrix} 0 \\ l \\ i \quad k \\ j \\ m \qquad n \end{matrix}$$

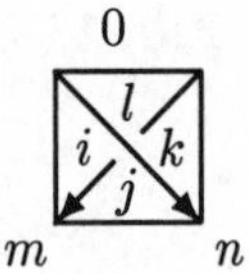

FIG. 13.11. An identity for B_{jl}, F_j

13.3 Construction of paragroups from RCFT

Based on a graphical representation in Section 13.2, we construct a paragroup from a WZW model G_k. Our construction here is similar to that in Section 12.5.

Fix an element x in I. We will then define a paragroup, which will be isomorphic to its dual paragroup. First we define the graphs $\mathcal{G}_j$, $j = 0, 1, 2, 3$, which are connected as in Fig. 10.3.

Definition 13.1 *Let $V_0^0 = V_1^0 = V_2^0 = V_3^0 = I$. We define four graphs $\mathcal{G}_0^0, \mathcal{G}_1^0, \mathcal{G}_2^0, \mathcal{G}_3^0$ as follows, where $\mathcal{G}_j^0$ has $V_j^0 \cup V_{j+1}^0$ as the set of its vertices for $j \in \mathbb{Z}/4$.*

1. *For $j \in V_0^0$, $k \in V_1^0$, we draw edges of $\mathcal{G}_0^0$ connecting j, k with multiplicity N_{jx}^k.*

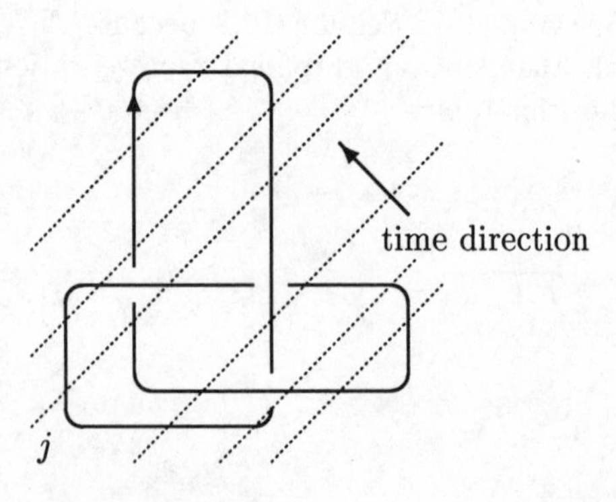

FIG. 13.12. Time direction and slices

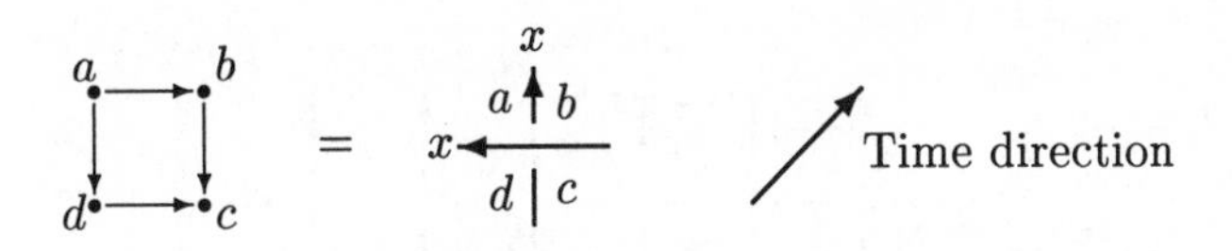

FIG. 13.13. Definition of a connection

2. *For $j \in V_1^0$, $k \in V_2^0$, we draw edges of $\mathcal{G}_1^0$ connecting j, k with multiplicity $N_{\bar{x}j}^k$.*
3. *For $j \in V_3^0$, $k \in V_2^0$, we draw edges of $\mathcal{G}_2^0$ connecting j, k with multiplicity N_{jx}^k.*
4. *For $j \in V_0^0$, $k \in V_3^0$, we draw edges of $\mathcal{G}_3^0$ connecting j, k with multiplicity $N_{\bar{x}j}^k$.*

We define $\mathcal{G}_j$ to be a connected component of $\mathcal{G}_j^0$ so that the $\mathcal{G}_j$ together make the connected component of the $\mathcal{G}_j^0$ containing $0 \in V_0^0$. We define V_j as the intersection of V_j^0 and the set of vertices of $\mathcal{G}_j$.

Then we define a connection on these graphs as in Fig. 13.13. If $N_{ax}^b > 0$, for example, we actually have labels for the upward arrow in the right hand side of the formula in Fig. 13.13. In such a case, we use the same labels for the arrow from a to b on the left hand side of the formula in Fig. 13.13. However, we drop such labellings in our diagrams for simplicity.

Setting $\mu(a) = F_a$, the harmonic axiom holds. The unitarity axiom and the renormalization axiom are trivial consequences of unitarity and tetrahedral symmetry of the braiding matrix. The initialization axiom is trivial. Thus the only non-trivial axiom we have to verify is flatness. We have to compute the partition function with $*$ at all the four corners. In the flatness axiom, we may assume $\sigma_j = \sigma'_j$ and $\rho_k = \rho'_k$, and then we have to prove that the value of the partition function is 1. By the definition of the connection, this problem is the same as showing that the value assigned to the labelled diagram in Fig. 13.14 is 1, where some labellings are dropped for simplicity. (In the example in Fig. 13.14, this diagram has 24 connection values and corresponds to a 4×6-partition function.)

In order to compute this value, we modify the diagram to the one in Fig. 13.15. This means that we have multiplied the partition function value with a positive scalar. Now we can modify Fig. 13.15 to the one in Fig. 13.16 without changing its numerical value. The diagram in Fig. 13.16 no longer has any crossings, so this value is a positive scalar, and we can easily check that it is equal to the the inverse of the positive scalar we multiplied in the passage from Fig. 13.14 to Fig. 13.15. This means that the original numerical value in Fig. 13.14 is 1, which exactly proves the flatness axiom.

In this way, we can construct a large family of paragroups. Their (dual) principal graphs are given as the graphs of fusion rules of irreducible representations of G with truncation at level k, i.e. the graphs $\mathcal{A}^{()}$ of the statistical mechanical

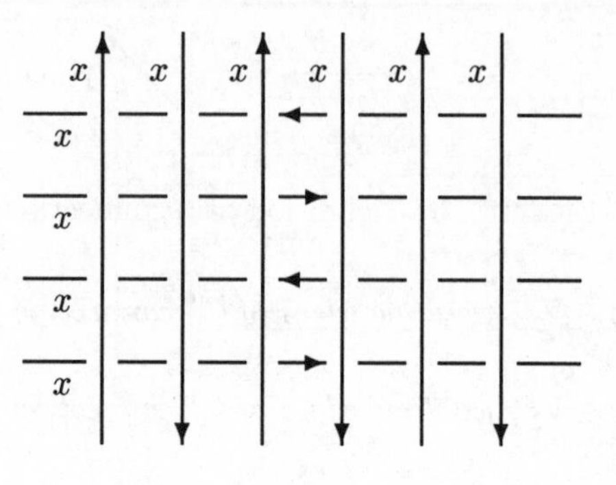

FIG. 13.14. A partition function for the flatness axiom

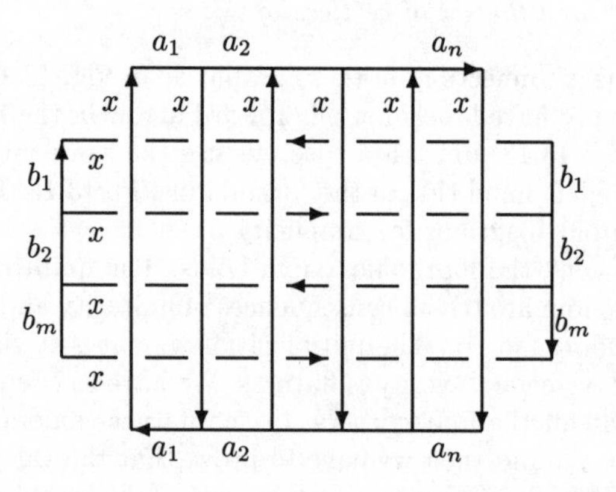

FIG. 13.15. Scalar multiple of the partition function

models of Chapter 7 (see also Section 11.7), or the positive energy, irreducible highest-weight representations of the loop group LG at level k of Chapter 8.

If we start with $SU(2)_k$ and the canonical choice of the generator of the fusion rule algebra, we get the principal graph A_{k+1}. (Note that we have the graph A_∞ for the irreducible representations of $SU(2)$. After the truncation, we get A_{k+1}.) So it coincides with the paragroup considered in Section 11.5. If we choose another x in the fusion rule algebra, we get a paragroup given by the basic constructions of the A_{k+1} subfactor.

We list some examples of the principal graphs for $SU(3)_k$ with the choice of the generator of the fusion rule algebra in Fig. 13.17. (Note that the dual principal graphs are the same, because the paragroup is isomorphic to its dual.) The

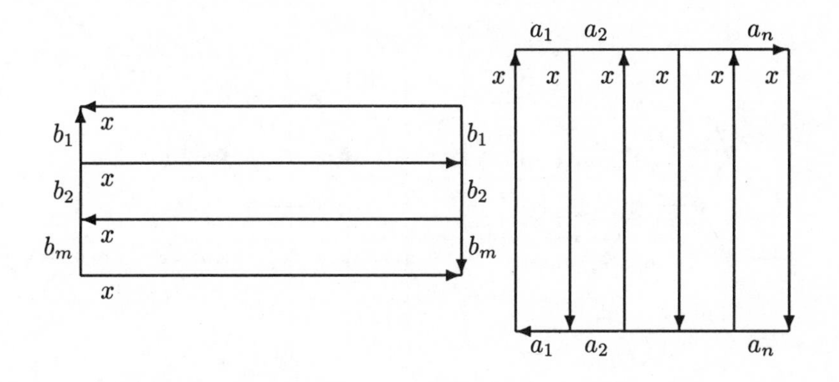

FIG. 13.16. Use of regular isotopy invariance

subfactors corresponding to these paragroups were first constructed by (Wenzl 1988a) with a more direct method based on Hecke algebras of type A. So these series of subfactors are often called the *Jones–Wenzl subfactors.* Here we assumed the framework of the WZW-model, so we constructed paragroups more easily.

Here we give an explanation on how to get these graphs. Take $SU(3)_6$ as an example. First we look at the irreducible representations of $SU(3)$ which make up the top graph in Fig. 13.18. Here the vertex marked with $*$ is the trivial representation and the one with $\circ$ is the generator of the system. The arrows give fusion rules given by multiplication by the generator. This is an infinite graph and we cut this graph at level 6 to get the graph at lower left in Fig. 13.18. The $\mathbb{Z}/3$-grading on the vertices are written on the boundary of the large triangle. (An arrow goes from a grading j to $j+1$ mod 3.) Finally, we leave only the edges connecting vertices with grading 0 and 1 and forget the orientations of these edges to get the graph at lower right in Fig. 13.18, which is the principal graph of the paragroup corresponding to $SU(3)_6$. The index value we have for $SU(n)_k$ is $(\sin^2 n\pi/(n+k))/(\sin^2 \pi/(n+k))$.

Remark These subfactors corresponding to $SU(n)_k$ have the same index values as those considered at the end of Section 11.6 for the A_n diagrams. These subfactors are, however, mutually non-isomorphic, because the principal graphs are different. Compare Figs 11.67 and 13.17.

Exercise 13.1 Find principal graphs for paragroups corresponding to $SU(4)_k$ for $k = 4, 5, 6, 7, 8$.

Exercise 13.2 For $SU(n)_k$, we can get a finite oriented graph whose vertices are given by I and whose edges are given by multiplication by the generator, as in Fig. 13.18. We can build a string algebra from $*$ on this oriented graph as in Section 11.2. Let g_j be the jth face operator on this string algebra arising from the braiding matrix as in Section 11.9. First prove the following three relations:

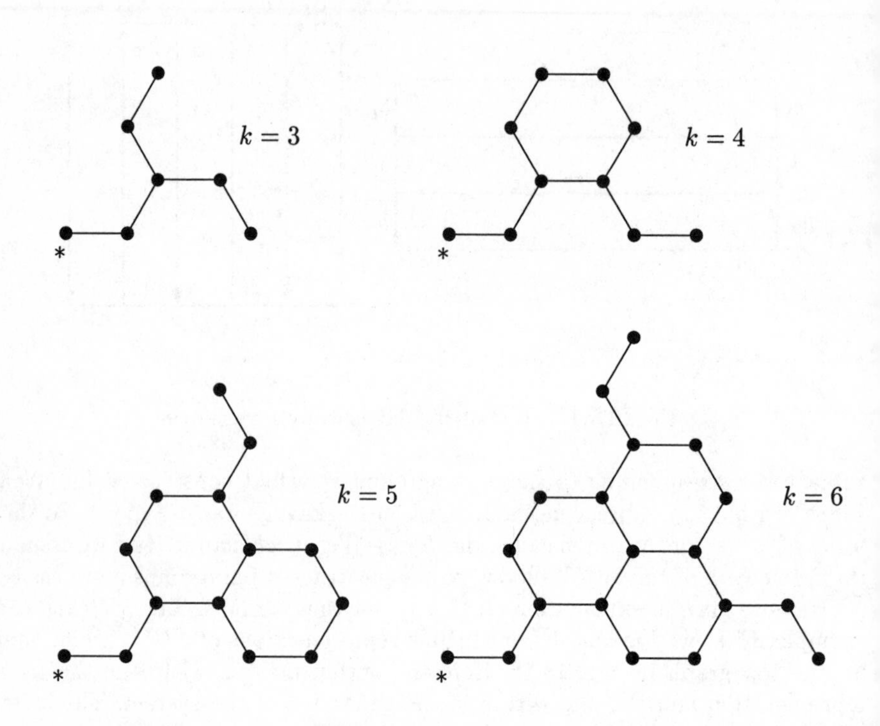

FIG. 13.17. Examples of principal graphs for $SU(3)_k$

1. $g_j g_{j+1} g_j = g_{j+1} g_j g_{j+1}$;
2. $g_j g_k = g_k g_j$, if $|j-k| \geq 2$;
3. $g_j^2 = (q-1)g_j + q$, for some root q of 1.

Then prove that the string algebra is generated by $\{g_j\}_{j\geq 1}$.

Let M be the weak closure of the algebra generated by $\{g_j\}_{j\geq 1}$ in the GNS-representation with respect to the canonical trace on the string algebra and N be its von Neumann subalgebra generated by $\{g_j\}_{j\geq 2}$.

Prove that M, N are both hyperfinite II_1 factors and identify this subfactor $N \subset M$ with the subfactor corresponding to the $SU(n)_k$ paragroup described above.

Exercise 13.3 Using Exercise 13.2 and a double sequence of commuting squares, construct a series $\{g_j\}_{j\in\mathbb{Z}}$ of face operators satisfying the same three relations as above.

Let M be the weak closure of the algebra generated by $\{g_j\}_{j\in\mathbb{Z}}$ in the GNS-representation with respect to the trace and N be its von Neumann subalgebra generated by $\{g_j\}_{j\neq 1}$.

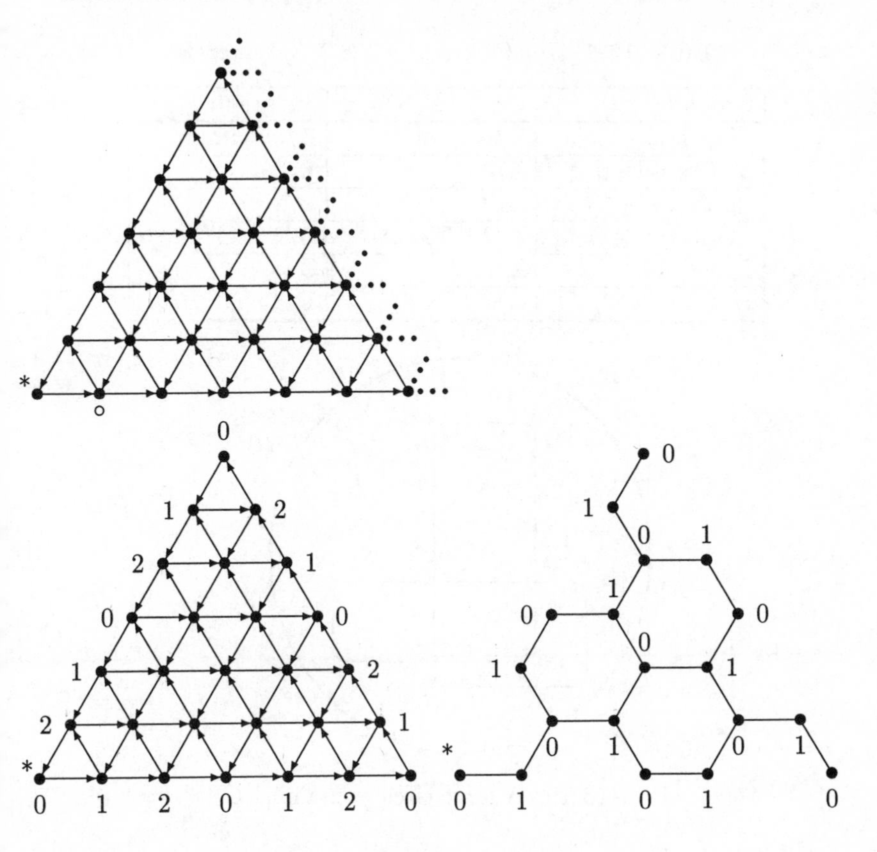

FIG. 13.18. Principal graph for $SU(3)_6$

Prove that M, N are both hyperfinite II_1 factors and identify this subfactor $N \subset M$ with the asymptotic inclusion of the subfactor of the hyperfinite II_1 ubfactor corresponding to the $SU(n)_k$ paragroup. (Use the method in Section 2.6.) Also compute the index and the principal graph.

3.4 Orbifold construction and conformal dimension

Ve next introduce the orbifold construction of subfactors in the setting in Section 3.3. The basic idea is to take a 'quotient' of a paragroup with a certain symmetry n a non-trivial way. As in Section 13.2, we fix a connected, simply connected, ompact and simple Lie group G and a level k. We assume that the group G has non-trivial centre $Z(G)$ and choose a non-trivial cyclic subgroup Z of $Z(G)$. This means that we have one of the entries in Table 13.1.

Then an element $z \in Z$ acts on the index set I of the fusion rule algebra and ve have the following as in Chapter 8.

$$N_{ij}^{k} = N_{iz(j)}^{z(k)}, \quad N_{ij}^{z(k)} = N_{z(i)j}^{k}, \quad S_{0j} = S_{0z(j)}\,.$$

Table 13.1 *Group G, center $Z(G)$, subgroup Z*

Group G	Centre $Z(G)$	Subgroup Z
$SU(n)$, $n \geq 2$	$\mathbb{Z}/n$	$\mathbb{Z}/m$ with $m \mid n$
$SO(2n+1)$, $n \geq 2$	$\mathbb{Z}/2$	$\mathbb{Z}/2$
$Sp(2n)$, $n \geq 2$	$\mathbb{Z}/2$	$\mathbb{Z}/2$
$SO(2n)$, $n \geq 2$	$\mathbb{Z}/2 \times \mathbb{Z}/2$ or $\mathbb{Z}/4$	$\mathbb{Z}/2$ or $\mathbb{Z}/4$
E_6	$\mathbb{Z}/3$	$\mathbb{Z}/3$
E_7	$\mathbb{Z}/2$	$\mathbb{Z}/2$

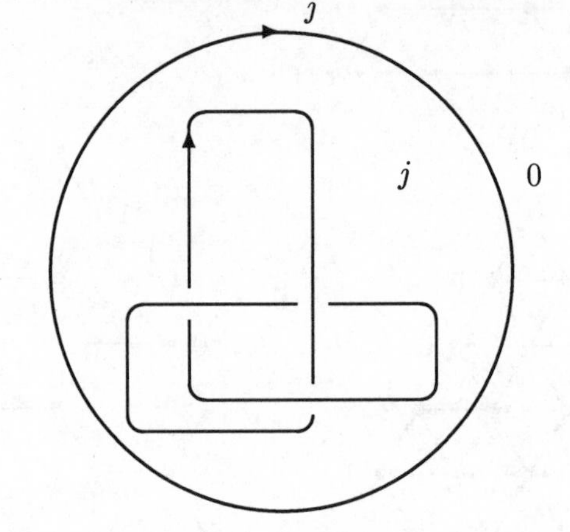

FIG. 13.19. A large circle containing G

These in particular imply $z(0)j = z(j)$.

Definition 13.2 *We say that Z* preserves the grading *if $z(0) \in V_0$ in the construction in Definition 13.1 for all $z \in Z$.*

We assume that the condition in Definition 13.2 holds. Then we need the following lemma.

Lemma 13.3 *Let H be a finite graph labelled as in Section 13.2 with the unbounded outside region labelled by 0. Let H^j be the same graph with the unbounded outside region labelled by j. Then the two labelled graphs are assigned the same numerical value.*

Proof We put a large circle labelled with j on a diagram of H so that it contains H as in Fig. 13.19.

On one hand, the value for this new graph is the product of F_j and the value for the graph H^j, and on the other hand, we can split this new graph into a disjoint union of H and the circle, and then the resulting value is the product of F_j and the value for the graph H. Since F_j is not zero, we get the desired conclusion. □

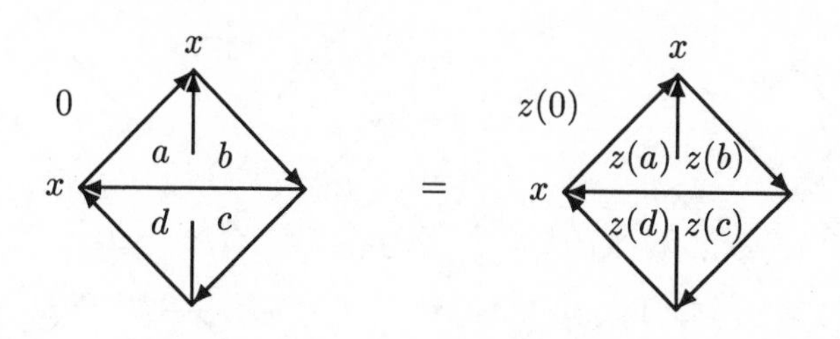

FIG. 13.20. Invariance of the connection

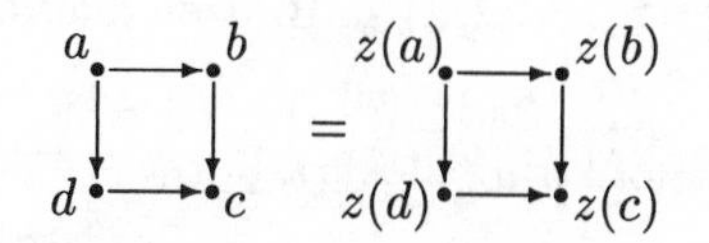

FIG. 13.21. Invariance of the connection

Applying Lemma 13.3 to the left hand diagram of Fig. 13.20, we get the identity in Fig. 13.20. By dividing both sides by the same scalar, we get the identity in Fig. 13.21, where we have dropped edge labellings again for simplicity.

We then set $V_0^0 = \{z(0) \mid z \in Z\}$. We modify the construction of the double sequence of string algebras in Section 11.3 as follows. Instead of using just $*$ as the starting point, we use all the vertices in V_0^0 as possible starting points as in Fig. 13.22 (see Example 2.22). Then we use the graph $\mathcal{G}_j$ as in Section 11.3 to get the double sequence $\{A_{kl}\}_{kl}$. So A_{00} is a direct sum of copies of $\mathbb{C}$, the number of copies being the order of V_0^0. In other words, in Definition 11.1 of a string (σ_+, σ_-) with $s(\sigma_+) = s(\sigma_-) = *$, $r(\sigma_+) = r(\sigma_-)$, $|\sigma_+| = |\sigma_-|$, we now drop the requirement $s(\sigma_+) = s(\sigma_-) = *$ and instead impose $s(\sigma_+), s(\sigma_-) \in V_0^0$. The multiplication and the $*$-operation are defined by the same formula as in Definition 11.1.

To make the construction clearer, consider the following example. Let G be $SU(2)$ and take the level k to be 6. Then as noted at the end of Section 13.3, the graph we get is A_7. The ordinary Bratteli diagrams for

$$A_{00} \subset A_{01} \subset A_{02} \subset \cdots$$

for the construction in Section 11.3 would be the one in the left half of of Fig. 13.22, but now our new construction gives the Bratteli diagram in the right half of Fig. 13.22.

Furthermore, just as in the construction preceding Theorem 11.19 in Section 11.4, we can extend the definition of the string algebras so that we have the

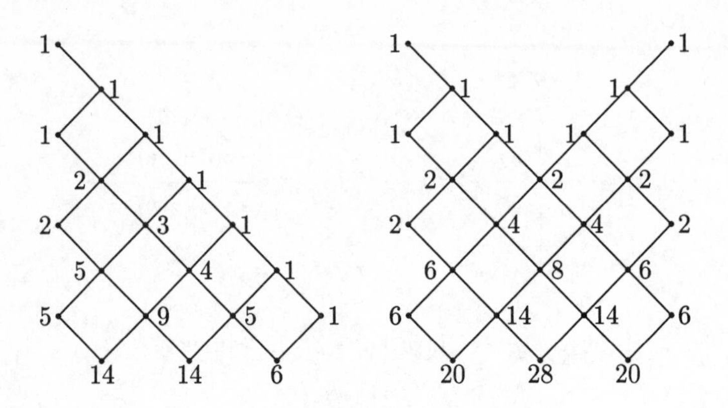

FIG. 13.22. The Bratteli diagrams

double sequence $\{A_{kl}\}_{kl}$ of string algebras for $k+l \geq 0$. We have the following theorem.

Theorem 13.4 *The subfactor $A_{0,\infty} \subset A_{1,\infty}$ constructed above with multi-starting points has the same higher relative commutants as the subfactor constructed from the paragroup in an ordinary way as in Section 11.3. Hence they are isomorphic by Theorem 11.14.*

Proof We can apply Theorem 11.15 to the subfactor $A_{0,\infty} \subset A_{1,\infty}$. Then both $A'_{0,\infty} \cap A_{k,\infty}$ and $A'_{1,\infty} \cap A_{k,\infty}$ are the same as the higher relative commutants of the subfactor constructed from the paragroup in the standard way. □

The above theorem means that our new construction here realizes the same subfactor as before in a 'non-canonical' way, but our new construction does have an advantage over the standard realization as follows.

Because the subgroup Z of the centre of G acts on vertices and the edges of the graph $\mathcal{G}_0$, it acts on each A_{kl} as algebra automorphisms. By the identity in Fig. 13.21, this action is compatible with the identification with the connection, hence we have a well-defined automorphic action of Z on each A_{kl}. This naturally extends to factors $A_{k,\infty}$. We denote this action by α. Note that the action α_z for $z \in Z$ is expressed as follows, where the labelling of edges is dropped for simplicity again:

$$\begin{aligned}&\alpha_z(a_0 \longrightarrow a_1 \longrightarrow \cdots \longrightarrow a_n, b_0 \longrightarrow b_1 \longrightarrow \cdots \longrightarrow b_n)\\ &= (z(a_0) \longrightarrow z(a_1) \longrightarrow \cdots \longrightarrow z(a_n), z(b_0) \longrightarrow z(b_1) \longrightarrow \cdots \longrightarrow z(b_n)).\end{aligned}$$

We now want to take fixed point algebras

$$A^{\alpha}_{k,\infty} = \{x \in A_{k,\infty} \mid \alpha_z(x) = x, \quad \text{for all } z \in Z\}.$$

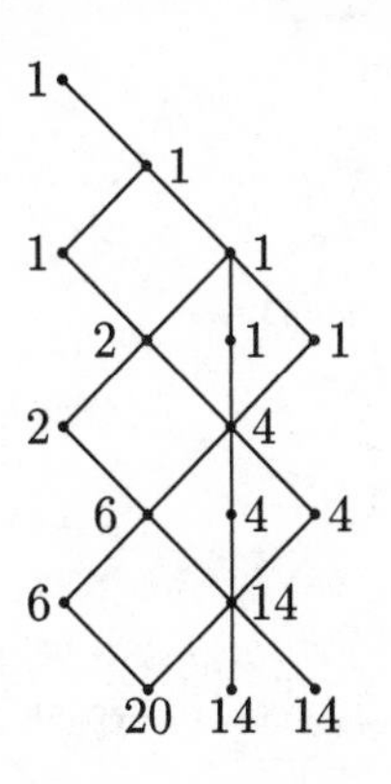

FIG. 13.23. The Bratteli diagrams for B_{kl}; D_5 from A_7

Definition 13.5 *The subfactor* $A^{\alpha}_{0,\infty} \subset A^{\alpha}_{1,\infty}$ *is called an* orbifold subfactor. *This construction is called the* orbifold construction.

Our aim is to compute paragroups of orbifold subfactors. First we set $B_{kl} = A^{\alpha}_{kl}$ for $k + l \geq 0$. Then it is easy to see that the following:

$$\begin{array}{ccc} B_{kl} & \subset & B_{k,l+1} \\ \cap & & \cap \\ A_{kl} & \subset & A_{k,l+1} \end{array}$$

is a commuting square, by comparing the two conditional expectations in the vertical direction. By the symmetry of Definition 11.5, we also know that the horizontal and vertical Jones projections are in B_{kl}. This shows that the double sequence $\{B_{kl}\}_{kl}$ also arises from a bi-unitary connection as in Definition 11.3. (Because α has finite order, we know that the graphs involved for $\{B_{kl}\}_{kl}$ must be finite, by looking at the above commuting square. See our examples which will come later. Also note that $B_{00} = \mathbb{C}$.)

To get a better idea of this procedure, take the example in Fig. 13.22. In this case, the group Z is $\mathbb{Z}/2$ and it acts on the Bratteli diagram in the right half of Fig. 13.22. Then the Bratteli diagram for the sequence of the resulting fixed point algebras

$$B_{00} \subset B_{01} \subset B_{02} \subset \cdots$$

is given as in Fig. 13.23, which is the Bratteli diagram of the string algebras on D_5. If we start with $SU(2)_{2n-4}$ with $Z = \mathbb{Z}/2$, we first get the principal graph A_{2n-3}, and then get the graph D_n after taking the fixed point algebras. Thus we have a bi-unitary connection on D_n in this way, but we have already seen in Section 11.5 that we have only one isomorphism class of bi-unitary connections on D_n. Thus the problem of determining whether the bi-unitary connections on the double sequence $\{B_{kl}\}_{kl}$ are flat contains the problem of deciding flatness of

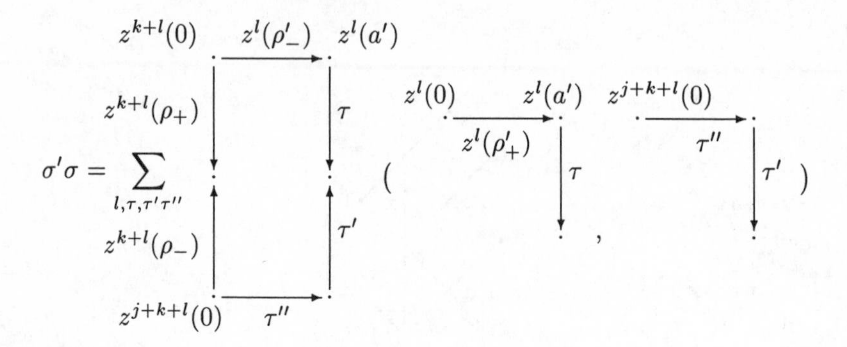

FIG. 13.24. An expression for $\sigma'\sigma$

the bi-unitary connections on the Dynkin diagrams D_n, which was left open in Section 11.5.

We now want to check when any string in $B_{k,0}$ commutes with everything in $B_{0,l}$ for all l. (By self-duality of the original connection, this problem is the same as checking when any string in $B_{k,-1}$ commutes with everything in $B_{1,l}$ for all l.)

Let n be the order of the cyclic group Z and z a generator of Z. We simply write α for α_z. Choose paths $\rho_\pm$ with the same length m on the original graph $\mathcal{G}_3$ with $s(\rho_+) = 0$, $s(\rho_-) = z^j(0)$, and $r(\rho_+) = r(\rho_-) = a$. We set $\sigma = \sum_{l=0}^n \alpha^l((\rho_+, \rho_-)) \in B_{m,0}$. It is easy to see that the algebra $B_{m,0}$ is linearly spanned by the elements of this form. Similarly, choose paths $\rho'_\pm$ with the same length m' on the original graph $\mathcal{G}_0$ with $s(\rho'_+) = 0$, $s(\rho'_-) = z^k(0)$, and $r(\rho'_+) = r(\rho'_-) = a'$ and set $\sigma' = \sum_{l=0}^n \alpha^l((\rho'_+, \rho'_-)) \in B_{0,m'}$. Again, the algebra $B_{0,m'}$ is linearly spanned by the elements of this form. We want to express the identity $\sigma\sigma' = \sigma'\sigma$ in terms of partition functions. First we express $\sigma'\sigma$ as in Fig. 13.24.

We first show that most partition functions in Fig. 13.24 are 0.

Lemma 13.6 *The partition function in Fig. 13.24 is 0 unless $\tau'' = z^{j+l}(\rho'_-)$.*

Proof Suppose $\tau'' \neq z^{j+l}(\rho'_-)$. Then it is enough to prove that the partition function in Fig. 13.25 has value 0, where we reversed the orientations of two paths. (Recall the conventions in Fig. 11.77.) We take a summation of the absolute value squared of the values of the partition functions in Fig. 13.25 over all possible τ, τ'. Then as in the proof of Theorem 11.17, we get the larger diagram as in Fig. 13.26. By invariance of the connection under Z, it is enough to prove the diagram in Fig. 13.27 has value 0, as in the case of Fig. 13.15. (We have dropped some labels again for simplicity. In particular, recall that we have dropped labels for intertwiners.) Then by regular isotopy invariance, we can modify the diagram in Fig. 13.27 to that in Fig. 13.28. Then the assumption $\tau'' \neq z^{j+l}(\rho'_-)$ implies that the value for the diagram in Fig. 13.28 is 0. □

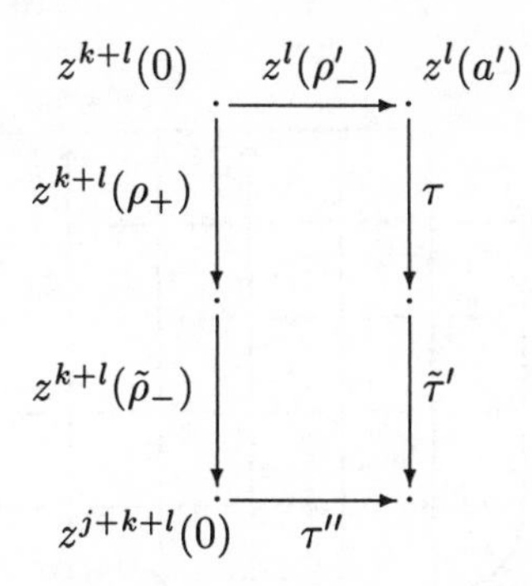

FIG. 13.25. A partition function

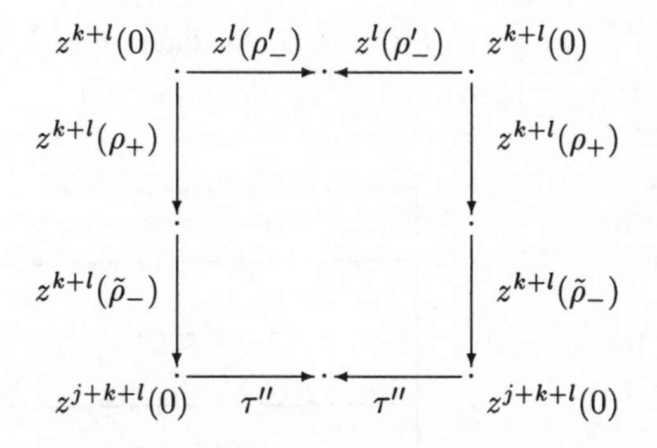

FIG. 13.26. The larger partition function

Next we compute $\sigma\sigma'$ in a similar way, and get the identity in Fig. 13.29. Then as in Lemma 13.6, we can prove that the partition function in Fig. 13.29 is 0 unless we have $\tau'' = z^l(\rho'_+)$. Comparing the identities in Figs 13.24 and 13.29, we conclude that the identity in Fig. 13.30 is necessary and sufficient for the identity $\sigma\sigma' = \sigma'\sigma$.

We take the absolute value squared of the difference of the two partition functions in Fig. 13.30, and then sum over all possible τ, τ'. Then we get four large partition functions as in the proof of Theorem 11.17. The argument in the proof of Lemma 13.6 shows that two of them have value 1. Finally we need to consider a partition function as in Fig. 13.31, and then $\sigma\sigma' = \sigma'\sigma$ for all σ, σ' if and only if the partition function in Fig. 13.31 always has value 1. Here we dropped z^l because of the invariance of the connection under Z. (At first, our condition is that the real part of the value of the partition function in Fig. 13.31 is 1, but unitarity implies that this is indeed equivalent to the condition that the value is 1.)

As in the proof of Theorem 11.17, where we now use the argument in the proof of Lemma 13.6, we can easily show that the value of the partition function

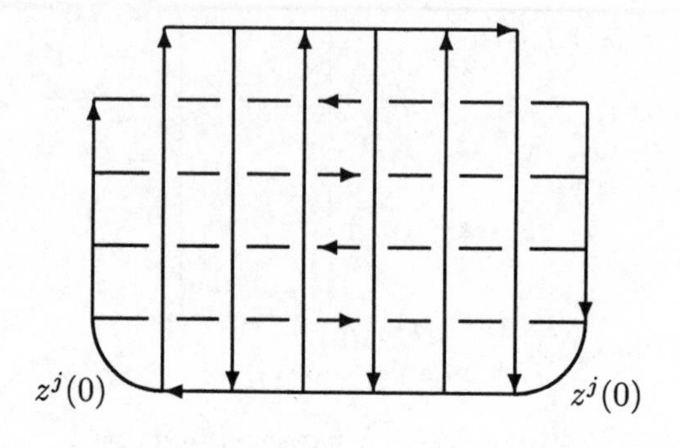

FIG. 13.27. A diagram with value 0

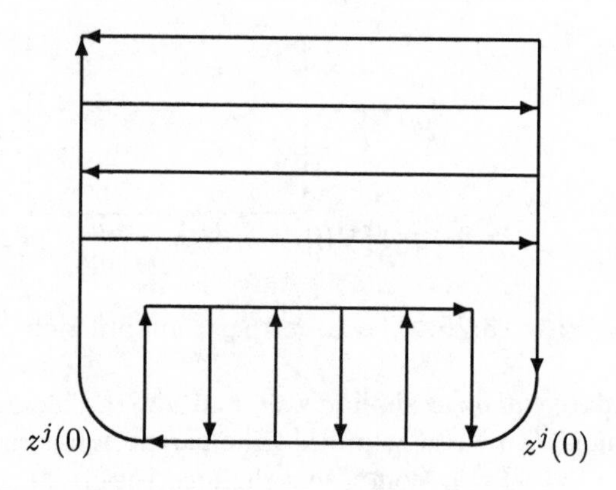

FIG. 13.28. After a regular isotopy move

in Fig. 13.31 does not depend on $\rho_\pm$, $\rho'_\pm$ as long as we fix j, k. (Here the lengths of $\rho_\pm$, $\rho'_\pm$ can vary.) Denote this value by $\lambda(j,k)$. Then again a similar argument, with the invariance of the connection under Z, shows that $\lambda(j,k) = \lambda(1,-1)^{-jk}$. Thus we have reached the conclusion that the bi-unitary connection for the double sequence $\{B_{kl}\}_{kl}$ is flat if and only if the partition function in Fig. 13.31 has value 1 for $j = 1$, $k = -1$. We can reduce the computation of the partition function again to the computation of the value of the labelled graph as in Fig. 13.32.

We again want to simplify the picture, but this value is invariant only under regular isotopy, and this time we have a writhe (see Notes to Chapter 12) for a curve at the lower left corner of Fig. 13.33 arising from the original graph in Fig.

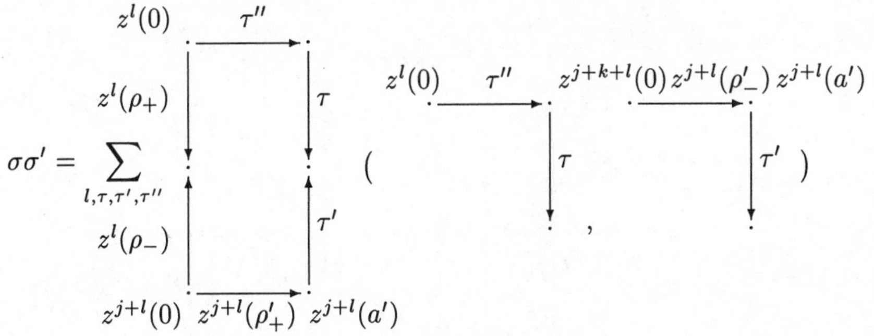

$$\sigma\sigma' = \sum_{l,\tau,\tau',\tau''} \begin{array}{c} z^l(0) \xrightarrow{\tau''} \\ z^l(\rho_+) \downarrow \tau \\ z^l(\rho_-) \uparrow \tau' \\ z^{j+l}(0)\ z^{j+l}(\rho'_+)\ z^{j+l}(a') \end{array} \left(\begin{array}{c} z^l(0) \xrightarrow{\tau''} \\ \downarrow \tau \end{array} , \begin{array}{c} z^{j+k+l}(0)\, z^{j+l}(\rho'_-)\, z^{j+l}(a') \\ \downarrow \tau' \end{array} \right)$$

FIG. 13.29. The expression for $\sigma\sigma'$

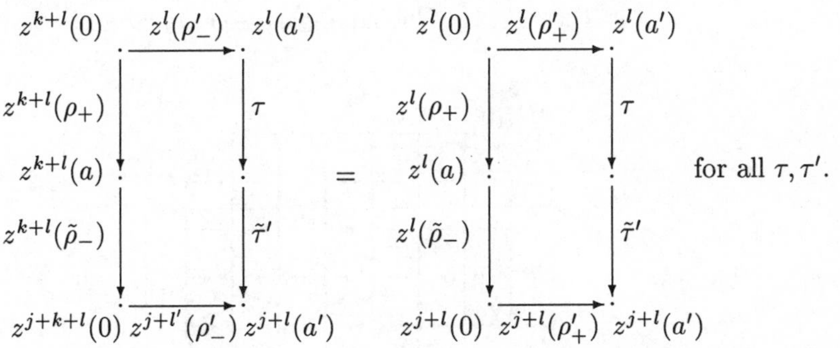

$$\begin{array}{c} z^{k+l}(0) \quad z^l(\rho'_-) \quad z^l(a') \\ z^{k+l}(\rho_+) \downarrow \tau \\ z^{k+l}(a) \\ z^{k+l}(\tilde\rho_-) \downarrow \tilde\tau' \\ z^{j+k+l}(0)\, z^{j+l'}(\rho'_-)\, z^{j+l}(a') \end{array} = \begin{array}{c} z^l(0) \quad z^l(\rho'_+) \quad z^l(a') \\ z^l(\rho_+) \downarrow \tau \\ z^l(a) \\ z^l(\tilde\rho_-) \downarrow \tilde\tau' \\ z^{j+l}(0)\ z^{j+l}(\rho'_+)\ z^{j+l}(a') \end{array} \quad \text{for all } \tau, \tau'.$$

FIG. 13.30. The necessary and sufficient condition for $\sigma'\sigma = \sigma\sigma'$

13.32. If we remove this writhe as in Fig. 13.8, we have a coefficient $\exp(2\pi i h)$, where h is the conformal dimension of $z^{-1}(0)$. Thus we have proved the following theorem.

Theorem 13.7 *The bi-unitary connection for the double sequence $\{B_{kl}\}_{kl}$ in the orbifold construction is flat if an only if the conformal dimensions of $z(0)$, $z \in Z$, are all integers.*

For $SU(n)_k$ with $Z = \mathbb{Z}/n$, it is easy to see that the action of Z preserves the grading in the sense of Definition 13.2 if and only if n divides the level k. We now have to determine when the conformal dimensions of $z(0)$, $z \in Z$, are all integers. In the first row of Table (5.1.30) of (Fuchs 1992), we set $r = n-1$, $i = 1$ and multiply the entry for Δ by k for $SU(n)_k$. Then we know that the conformal dimensions are all integers if and only if $n \mid k$ when n is odd, and $2n \mid k$ when n is even. In particular, the flatness condition for $SU(2)_k$ is $4|k$ and thus the bi-unitary connection on the Dynkin diagram D_n is flat if and only if n is even. This was the case left open in the proof of Theorem 11.24. See (Dijkgraaf and Witten 1990) for discussion of this condition in a physical context.

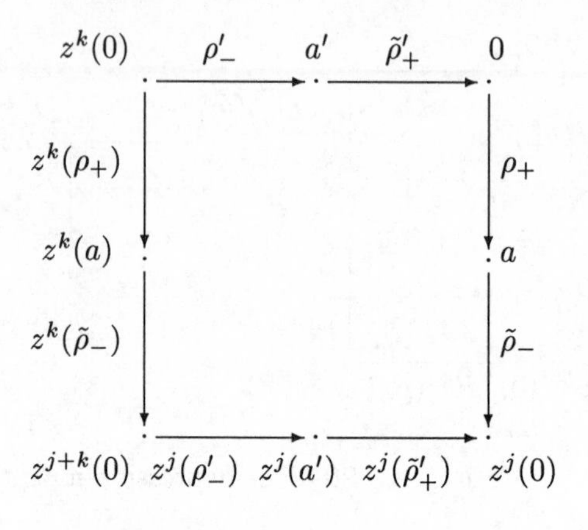

FIG. 13.31. The large partition function

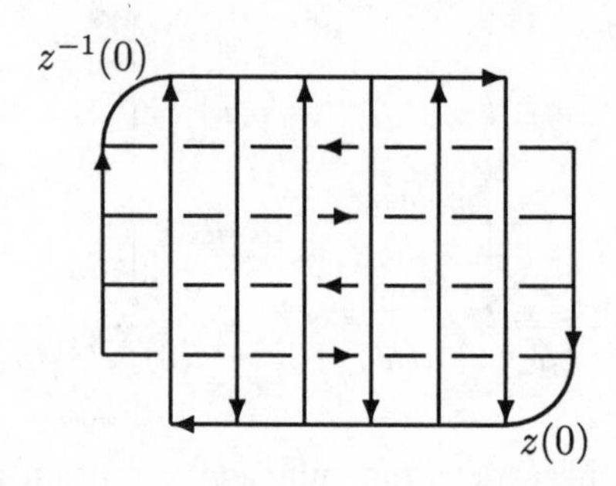

FIG. 13.32. Before a regular isotopy move

Finally, we list some examples of the principal graphs of the orbifold subfactors for $SU(3)_k$ in Fig. 13.34. Note that in this case we have $Z = \mathbb{Z}/3$, and the grading is kept and we have flatness as long as k is a multiple of 3.

13.5 From the asymptotic inclusion to RCFT

It is natural to try to get a system satisfying the Moore–Seiberg axioms of RCFT from a paragroup, as a converse to the construction in Section 13.3. It was A. Ocneanu who realized that we can get such data of RCFT *after* passing to the asymptotic inclusion in Section 12.6 from a subfactor with finite index and finite depth. We will describe the construction in detail. Most of the necessary definitions are already made in Sections 12.6 and 12.7, so we only need to verify the Moore–Seiberg axioms in Section 8.4 based on geometric operations.

Let $N \subset M$ be a subfactor of the hyperfinite II_1 factor with finite index and finite depth. We construct the tube algebra Tube $\mathcal{M}$ as in Section 12.6, and we denote the set of the minimal central projections of Tube $\mathcal{M}$ by $\{\pi_1, \ldots, \pi_n\}$. By

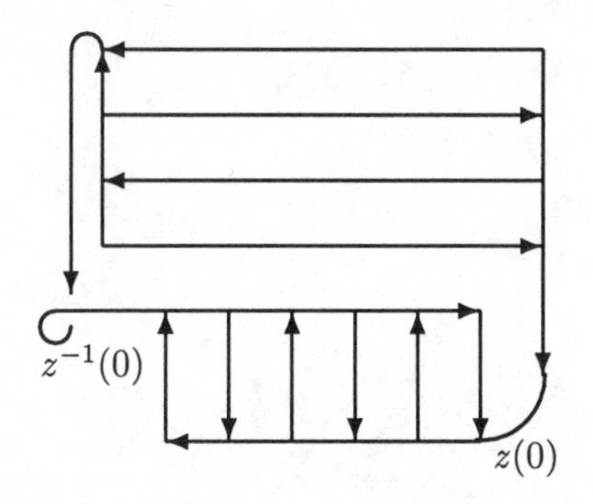

FIG. 13.33. After a regular isotopy move

Theorem 12.22 and Theorem 12.29, this system of minimal central projections is regarded as a system of M_∞-M_∞ bimodules. We set $I = \{\pi_1, \dots, \pi_n\}$, and then this gives a fusion rule algebra in the sense of Definition 12.1. Instead of the notation X_i in Definition 12.1, we will simply write i so that our notation will be compatible with that in RCFT. We denote by 0 the identity element in the fusion rule algebra. It is clear that the finite set I gives a fusion rule algebra in the sense in RCFT. (The involution in I corresponds to the conjugate operation of bimodules, of course.)

We now define data V^i_{jk}, $\Theta^i_{jk}(\pm)$, $\Omega^i_{jk}(\pm)$, $F\begin{bmatrix} j_1 & j_2 \\ i_1 & k_2 \end{bmatrix}$, $S(j)$, and T for the Moore–Seiberg set of axioms for RCFT.

For $i, j, k \in I$, we define the finite dimensional Hilbert space V^i_{jk} as the Hilbert space H_S defined in Section 12.6 for the labelled surface S in Fig. 13.35. We write N^i_{jk} for the dimension of V^i_{jk}.

Next, the operation $\Theta^i_{jk}(+)$ [resp. $\Theta^i_{jk}(-)$] is defined as in Fig. 13.36. An element in V^i_{jk} is represented by a surface in the upper left corner in Fig. 13.36 labelled with intertwiners as in Section 12.6. We first make 180° twists of the two circles labelled with i, k of the labelled surface for V^i_{jk} in the directions of the arrows on the circles. This is the first step in Fig. 13.36. Next we twist the upper right half of the surface in the positive [resp. negative] direction along the axis going through the circle labelled by j while fixing the left bottom circle labelled with j as in the second step in Fig. 13.36. In the third step, we just change orientations of two circles to get labels $\check{k}$ and $\check{\imath}$. The operations $\Theta^i_{jk}(\pm)$ are compositions of these three steps, and they are isomorphisms from V^i_{jk} to $V^{\check{k}}_{j\check{\imath}}$.

We define the operation $\Omega^i_{jk}(+)$ [resp. $\Omega^i_{jk}(-)$] as in Fig. 13.37. We twist the lower half of the surface by 180° in the positive [resp. negative] direction while fixing the top circle labelled with i. They are isomorphisms from V^i_{jk} to V^i_{kj}.

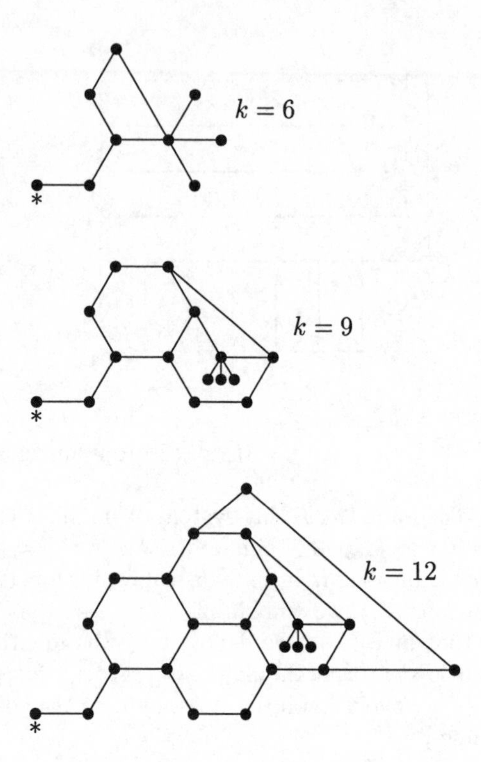

FIG. 13.34. Examples of principal graphs of orbifold subfactors from $SU(3)_k$

The operation $F\begin{bmatrix} j_1 & j_2 \\ i_1 & k_2 \end{bmatrix}$ is an isomorphism from $\bigoplus_r V^{i_1}_{j_1,r} \otimes V^r_{j_2,k_2}$ to $\bigoplus_s V^{i_1}_{s,k_2} \otimes V^s_{j_1,j_2}$ defined as in Fig. 13.38, where we have summations over r and s on the left and right hand sides respectively.

The next operation $S(j)$ on $\bigoplus_i V^i_{ji}$ is defined as a $90°$ rotation as in Fig. 13.39, where two edges with double arrows are identified in both pictures. (Note that the Hilbert space $\bigoplus_i V^i_{ji}$ is graphically represented with a torus with a hole whose boundary circle is labelled with j, as in the left hand side of Fig. 13.39. Also compare this figure with Fig. 12.63.)

Similarly, the operation T on $\bigoplus_i V^i_{ji}$ is defined as in Fig. 13.40, where two edges with the same labels are identified in each diagram. It is easy to see that this operation is independent of j. (Again compare this figure with Fig. 12.64.)

The above gives all the required data. The axioms we need to verify are as follows:

1. $(\check{\imath})\check{} = i$.
2. $V^i_{0j} \cong \delta_{ij}\mathbb{C}$, $V^0_{ij} \cong \delta_{i\check{j}}\mathbb{C}$, $V^i_{jk} \cong V^{\check{k}}_{j\check{\imath}}$, $(V^i_{jk})^* \cong V^{\check{\imath}}_{\check{j}\check{k}}$.
3. $\Omega^i_{jk}(+)\Omega^i_{kj}(+)$ is a multiplication by a phase.

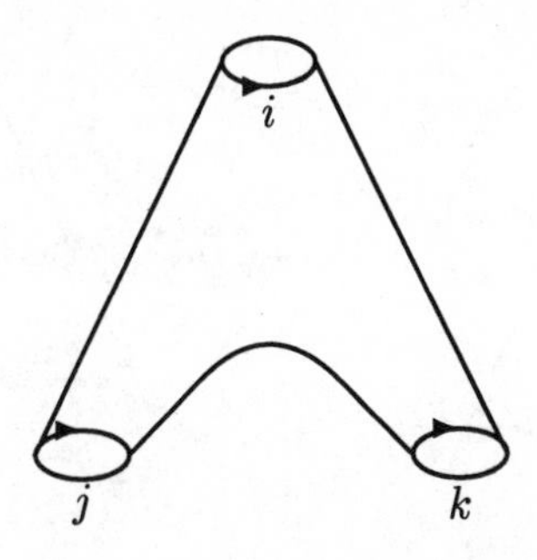

FIG. 13.35. The Hilbert space V^i_{jk}

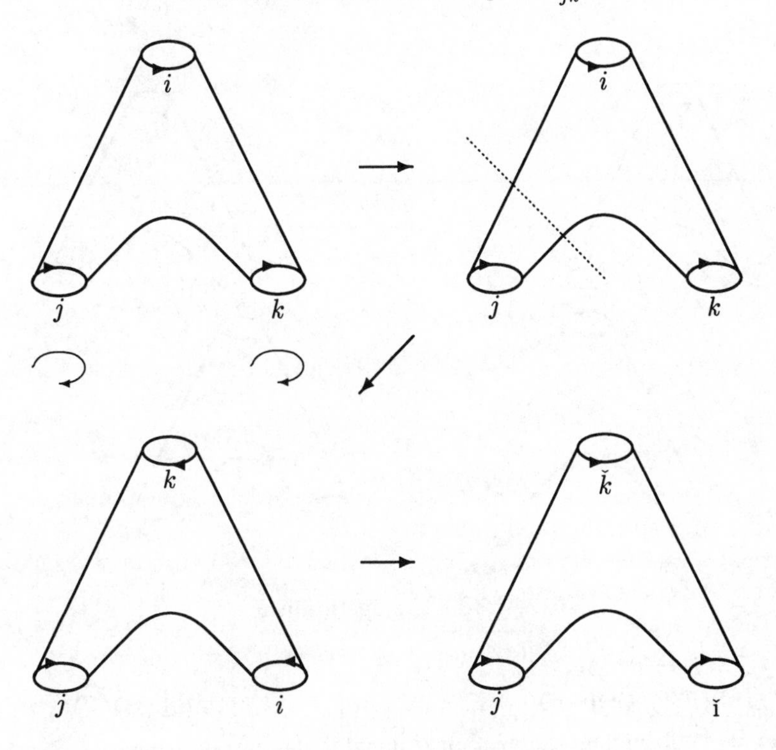

FIG. 13.36. The operations $\Theta^i_{jk}(\pm)$

Similarly, the action of T on V^i_{ji} is a diagonal matrix of phases independent of j.

4. $F(\Omega(\varepsilon) \otimes \mathrm{id})F = (\mathrm{id} \otimes \Omega(\varepsilon))F(\mathrm{id} \otimes \Omega(\varepsilon))$, where $\varepsilon = \pm$.
5. $F_{23}F_{12}F_{23} = P_{23}F_{13}F_{12}$.
6. $S^2(j) = \bigoplus_i \Theta^i_{ji}(-)$.

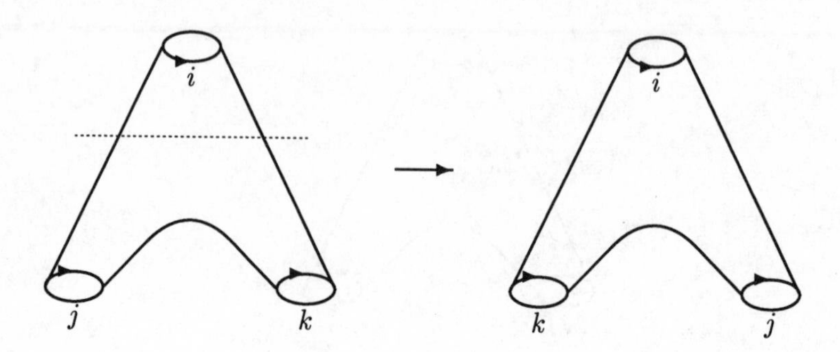

FIG. 13.37. The operations $\Omega^i_{jk}(\pm)$

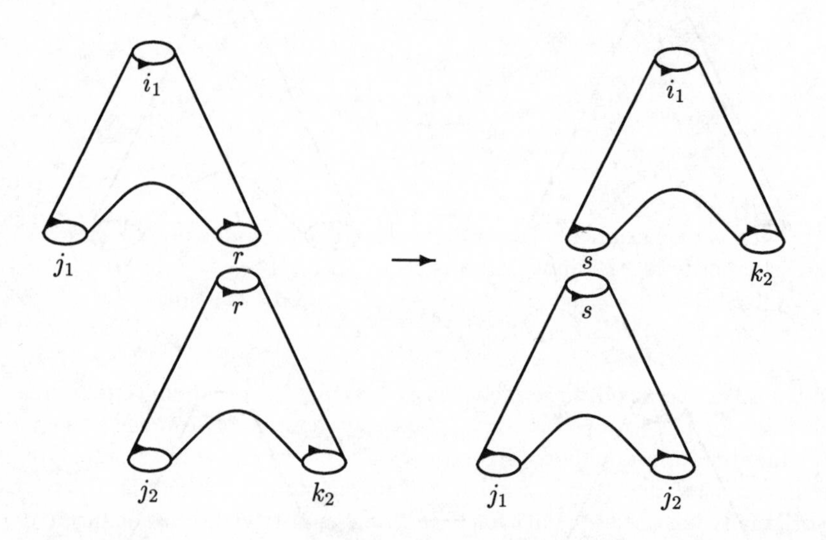

FIG. 13.38. The operation F

7. $S(j)TS(j) = T^{-1}S(j)T^{-1}$.
8. $(S \otimes \mathrm{id})(F(\mathrm{id} \otimes \Theta(-)\Theta(+))F^{-1})(S^{-1} \otimes \mathrm{id}) = FPF^{-1}(\mathrm{id} \otimes \Omega(-))$.

In the above formulae, the symbol P denotes the permutation;

$$P(x \otimes y) = (y \otimes x).$$

We also define P_{23}, F_{12}, F_{23}, F_{13} by

$$P_{23}(x \otimes y \otimes z) = x \otimes z \otimes y, \quad F_{12} = F \otimes \mathrm{id}, \quad F_{23} = \mathrm{id} \otimes F, \quad F_{13} = P_{23}F_{12}P_{23},$$

respectively.

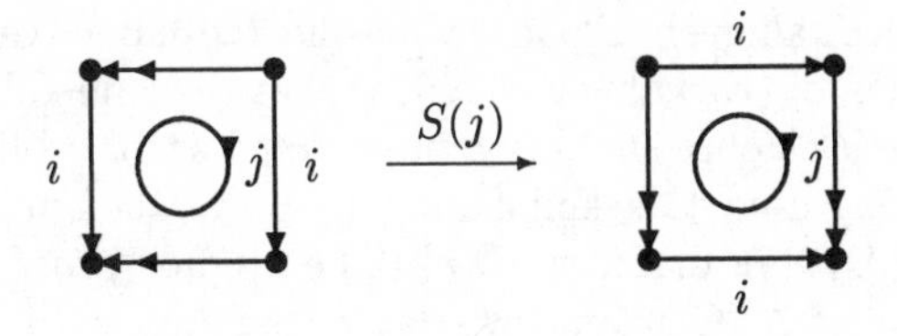

FIG. 13.39. The operation $S(j)$

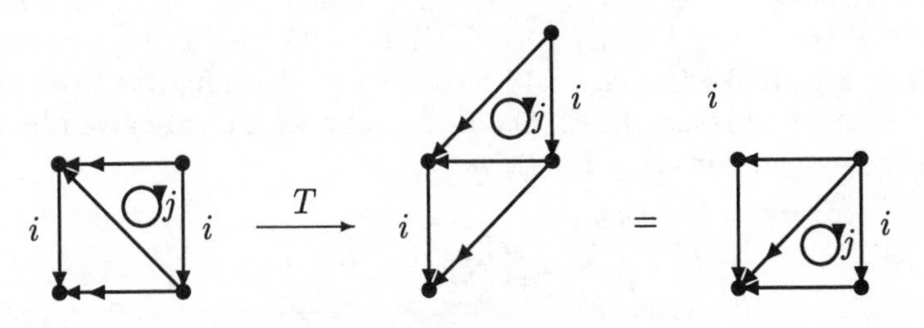

FIG. 13.40. The operation T

By graphical arguments based on the above definitions and Section 12.7, we can easily verify all the above conditions.

We note that the Ω matrix recovers conformal dimensions and the T matrix recovers the central charge c. It is easy to see that our central charge c in this construction is 0. An interpretation of this fact is as follows. As explained in Section 12.8, the asymptotic inclusion is regarded as the 'quantum double construction' for paragroups. So our data in this section corresponds to something *after* the quantum double construction and then we can think that the original 'hidden' central charge is 'cancelled out' in this double procedure. (Note that if we originally have a set of data for RCFT with non-trivial c, the quantum double construction gives a new set of data with $c = 0$.)

13.6 Notes

Our presentation in Section 13.2 followed (Witten 1989b).

The construction of a paragroup from an RCFT in Section 13.3 is due to (de Boer and Goeree 1991). Our notation is slightly different from theirs, and the same as in (Xu 1994). Originally, the paragroups of the subfactors arising from Hecke algebras of type A in (Wenzl 1988a) were found in (Evans and Kawahigashi 1994a) using the solutions of (Jimbo *et al.* 1987), (Jimbo *et al.* 1988) to the Yang–Baxter equation. The original construction in (Wenzl 1988a) is close to the one in Exercise 13.2. The index value in Exercise 13.3 was first computed in (Erlijman 1995). See (Goto, preprint 1996), (Erlijman, preprint 1996) for the identification with the asymptotic inclusion.

The orbifold construction was first realized as a tool for constructing subfactors in (Evans and Kawahigashi 1994a), when the first named author commented on a primitive version of (Kawahigashi 1995a) that the embedding of the A path algebra in the D path algebra should be interpreted as an orbifold construction. [(Kawahigashi 1995a) used this embedding to prove the (non-)flatness of the connections on the Dynkin diagrams D_n.] (The original proof of flatness of the connections on D_{2n} of A. Ocneanu was later communicated to us. It is quite different from the proof here and can be found in the Appendix of (Kawahigashi 1992b).) The real role of the conformal dimension was unnoticed until (Xu 1994). The presentation in Section 13.4 follows (Xu 1994).

Orbifold models have been considered in statistical mechanical models (Fendley 1989), (Fendley and Ginsparg 1989), for $\mathbb{Z}/2$ and in (Di Francesco and Zuber 1990a), (Di Francesco and Zuber 1990b), (Zuber 1990) for $\mathbb{Z}/3$ symmetries of the graphs as in Fig. 10.8 and Fig. 13.18 respectively, obtaining the orbifold graphs of Fig. 10.9 and 8.32 respectively. The Boltzmann weights are given by (Fendley 1989), (Fendley and Ginsparg 1989), as

$$W'\begin{pmatrix} & \alpha \quad \alpha' & \\ & \beta \quad \beta' & \end{pmatrix} = \sum \omega \; W\begin{pmatrix} & \overline{\alpha} \quad \overline{\alpha}' & \\ & \overline{\beta} \quad \overline{\beta}' & \end{pmatrix} \tag{13.6.1}$$

where the weights W' of the orbifold model are given as a linear combination of those weights W of the initial model, with the coefficient ω being square roots, and cubic roots, respectively (and vice versa, with the initial weights W being expressed as a similar combination of the orbital weights W'). We can interpret this orbifold construction via path algebras. In particular, this allows us to easily generalize the weights to the $\mathbb{Z}/N$ orbifolds of the $SU(N)$ models, and derive the Yang–Baxter equation without any pain. The Yang–Baxter equation was checked by hand in the $N = 2, 3$ cases in (Fendley 1989) and (Fendley and Ginsparg 1989), whilst (Di Francesco and Zuber 1990a), page 643 expected off-critical orbifolds for $N > 3$ models as a consequences of identities between theta functions.

Suppose that G is a finite group of symmetries of the graph Γ leaving a subset $\Gamma_0^{(0)}$ of the vertices $\Gamma^{(0)}$ globally invariant. Then there is an induced action of G on strings $\alpha \mapsto g^{-1}\alpha$ and hence on the path algebra $A(\Gamma, \Gamma_0^{(0)})$ (in the notation of Example 2.22) by $(\alpha, \beta) \mapsto (g^{-1}\alpha, g^{-1}\beta)$. The fixed point algebra $A(\Gamma, \Gamma_0^{(0)})^G$ can be described as the path algebra built on an orbifold graph as follows.

For the case $\Gamma = \mathcal{A}^{(n)}$, we define an action of the cyclic group $\mathbb{Z}/N$ as follows. We set $A_0 = *$, and label the other end vertices of the graph $\mathcal{A}^{(n)}$ by $A_1 = A_0 + (n - N)e_1$, $A_2 = A_1 + (n - N)e_2$, ..., $A_{N-1} = A_{N-2} + (n - N)e_{N-1}$. Define a rotation symmetry ρ of the graph $\mathcal{A}^{(n)}$ by

$$\rho(A_j + \sum_k c_k e_k) = A_{j+1} + \sum_k c_k e_{k+1} \tag{13.6.2}$$

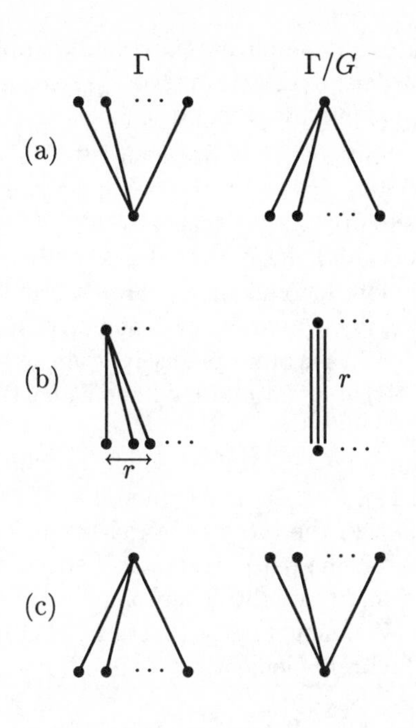

FIG. 13.41. From Γ to the orbifold Γ/G

where the indices are in $\mathbb{Z}/N$ and $c_k \in \mathbb{N}$. Note that $\rho^N = id$. We take $\Gamma_0^{(0)} = \{A_0, \ldots, A_{N-1}\}$, the orbit of $*$ under $\mathbb{Z}/N$.

In this case, the orbit of any vertex is either trivial or can be identified with the group G. Then at any level of the path algebra $A(\Gamma, \Gamma_0^{(0)})$ we have sums of blocks of the following kind. First a singleton B, a full matrix algebra globally invariant under G, and direct sums of matrix algebras $\oplus_{h \in G} A_h$, where A_h is a matrix algebra transformed isomorphically to A_{g+h} under the action of $g \in G$. To determine $[A(\Gamma, \Gamma_0^{(0)})_n]^G$, note that the fixed point algebra of each B is identified with a direct sum of a family $\{B_h \mid h \in G\}$ of isomorphic matrix algebras indexed by G, and the fixed point algebra of $\oplus_{h \in G} A_h$ is identified with A_0. Thus we can construct a graph Γ/G to yield the Bratteli diagram of $A(\Gamma, \Gamma_0^{(0)})$ as follows. We replace each trivial orbit in Γ by a copy of G, and each non-trivial orbit in Γ by a singleton. These are the vertices of the orbifold graph Γ/G. The edges of the orbifold graph are obtained by considering the inclusion of $[A(\Gamma, \Gamma_0^{(0)})_n]^G$ in $[A(\Gamma, \Gamma_0^{(0)})_{n+1}]^G$. There are three cases to consider, as in Fig. 13.41.

In case (a), each vertex of a non-trivial orbit at level n is joined to the trivial orbit by a single edge at level $n+1$ for Γ, which is reversed in the orbifold graph. Case (c) is similar. In case (b), each vertex of a non-trivial orbit at level n is joined to r vertices in a non-trivial orbit at level $n+1$ for Γ. In the orbifold graph, we have r edges joining the trivial orbits.

This gives us the rules for computing the orbifold graph Γ/G, so that we have a filtered embedding $A(\Gamma/G,*) \to A(\Gamma,\Gamma_0^{(0)})^G$ where $* \equiv \Gamma_0^{(0)}/G \in (\Gamma/G)^{(0)}$.

To describe the embedding of the path algebra $A(\Gamma/G)$ in $A(\Gamma)$, through a connection, recall the terminology of Sections 2.9, 3.4 and 3.5. If we have two AF path algebras $A(\lambda)$ and $A(\mu)$ with Bratteli diagrams coming from sequences of graphs $\lambda_1, \lambda_2, \ldots$, and $\mu_1, \mu_2, \ldots$, then any filtered $*$-homomorphism ϕ between path algebras $A(\lambda)$ and $A(\mu)$, filtered in the sense that $A(\lambda)_n$ is taken to $A(\mu)_n$ with this restriction denoted by ϕ_n, are described via connections. Now the $*$-homomorphism ϕ_n is up to an inner automorphism of $A(\mu)_n$ a path endomorphism of the type (2.9.14). More precisely, there is a graph i_n between the nth levels of $A(\lambda)$ and $A(\mu)$, and a unitary intertwiner that implements ϕ_n as in (3.5.9).

If all four graphs $i_n, \lambda_n, \mu_n, j_n$ are the same (as will be the case in the $\mathcal{A}^{(n)}$ models – see Figure 13.18), then an intertwiner in $(\lambda_n j_n, i_n \mu_n)$ is also an element of the path algebra $A(\lambda)$. In the theory of subfactors, the braid element plays a dual role. It may appear in the path algebra as a specialization (see for example Section 11.7) of a Boltzmann weight. There may also be a shift endomorphism κ (related to Kramers–Wannier duality (Section 7.11) of the path algebra, $\kappa : A(\lambda) \to A(\lambda)$ taking the braid element

$$\sigma_i \in A[i-1,i+1] \to \sigma_{i+1} \in A[i,i+2].$$

Indeed, because of the braid relation $\sigma_i\sigma_{i+1}\sigma_i = \sigma_{i+1}\sigma_i\sigma_{i+1}$ we have

$$\kappa(\sigma_i) = \lim_{n\to\infty} \sigma_1\sigma_2\cdots\sigma_n(\sigma_i)\sigma_n^{-1}\cdots\sigma_2^{-1}\sigma_1^{-1} \tag{13.6.3}$$

and in fact we may have

$$\kappa(x) = \lim_{n\to\infty} \sigma_1\sigma_2\cdots\sigma_n(x)\sigma_n^{-1}\cdots\sigma_2^{-1}\sigma_1^{-1} \tag{13.6.4}$$

as in the $\mathcal{A}^{(n)}$ models (see (Evans and Gould 1994a) and (Goodman and Wenzl 1990) for other examples). In this way, the braid elements appear as the intertwiners or connection for the filtered homomorphism κ. Then when we use (3.5.9) to transform in the double complex picture of Section 11.4 an element W such as a Boltzmann weight or a genuine partition function is taken to

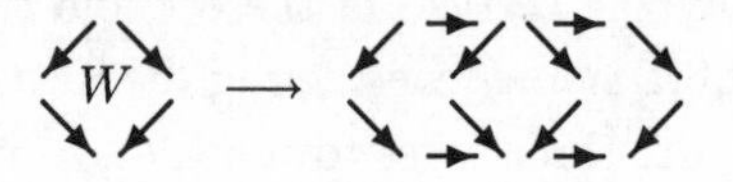

The elements

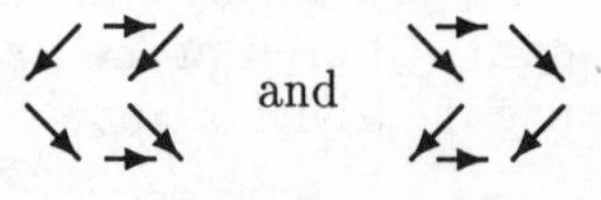

which appear here arise from 'partition functions' of braid elements (with no spectral parameter). In this respect, the expressions

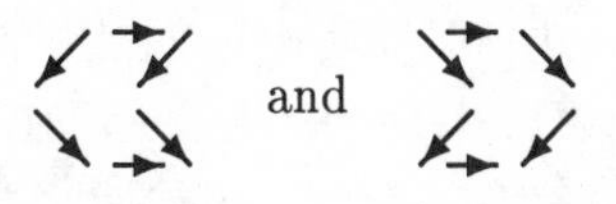

which we compute are not genuine partition functions but intertwiners between the submodel and the model. Note that a unitary matrix cannot in any case have all positive entries except in rather trivial cases. In this regard, the relation between connections and Boltzmann weights needs to be taken with a pinch of salt.

Before we complete the intertwiners between a path algebra and its orbifold model, let us recall the situation for subalgebras of UHF algebras arising as fixed point algebras of limit inner actions. The connections were the Clebsch–Gordan coefficients, see Section 3.5. In a similar way, the connection for orbifold models is obtained as the Clebsch–Gordan coefficients which put the cyclic permutations of Fig. 13.41 in diagonal form. If ω is a primitive Nth root of unity, then

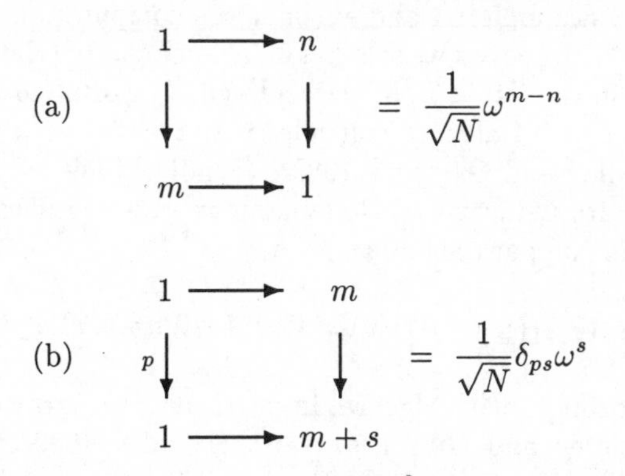

where $m, n \in \mathbb{Z}/N$, $p, s \in \{0, 1, 2, \ldots, r-1\} \subset \mathbb{Z}/N$.

The embedding $A^G \to A$ of limit inner actions of Example 3.10 or graph symmetries of these Notes (e.g. Fig. 13.41) have the property that not only are they filtered but they also satisfy

$$A[\lambda][m,n]^G \to A[\kappa][m,n]\,.$$

To see this, note that the action of G leaves invariant $A[\lambda][m,n]$ (e.g. by (2.9.15)) and

$$A[\lambda][m,n]^G \subset A[\lambda][m,n] \cap A[\kappa][0,n]$$

as $A[\kappa][0,n] = A[\lambda][0,n]^G$. Take $x \in A[\lambda][m,n]^G$, $y \in A[\kappa][0,m] \subset A[\lambda][0,m]$, then $xy = yx$ as $x \in A[\lambda][m,n]$, $y \in A[\lambda][0,m]$. Hence $x \in A[\kappa][0,n] \cap A[\kappa][0,m]' = A[\kappa][m,n]$.

In particular, this means that if $W \in A[\lambda][m-1,m+1]^G$, then the corresponding W' of $A[\lambda][m-1,m+1]$ satisfy

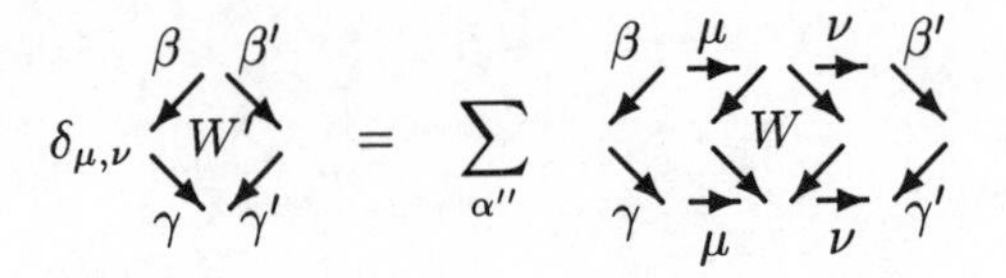

where

and

are the intertwining unitary connections. In the case of Example 3.10, this explains and verifies the vertex-IRF correspondence computations of (Roche 1990) Appendix 4. (Note also that Property T of (Roche 1990) holds since we clearly have commuting squares as the conditional expectation $E_n : A_n \to A_n^G$ is consistently given by $E = \int \alpha_g \, dg$). Also it verifies the corresponding equation in the orbifold model Example 1.1 and explains the computations of (Kawahigashi 1995a), Lemma 5.1. Moreover, we see that $\mathbb{Z}/N$ analogue of (Kawahigashi 1995a), Lemma 5.1 holds in the $(\mathcal{A}^{(n)}, \mathbb{Z}/N)$ case as well. The orbifold weights which we define via the fixed point algebra coincide (via (2.9.15)) with the $\mathbb{Z}/2$ and $\mathbb{Z}/3$ examples of (Fendley and Ginsparg 1989), (Fendley 1989), when $N = 2, 3$, respectively. Also (with G-invariant) Boltzmann weights, it is clear that the Yang–Baxter equation in the path algebra

$$W_i(u)W_{i+1}(u+v)W_i(v) = W_{i+1}(v)W_i(u+v)W_{i+1}(u)$$

also holds in the orbifold path algebra. In particular, we verify directly the computations of (Fendley and Ginsparg 1989), (Fendley 1989) that the orbifold weights defined from (13.6.1) satisfy the Yang–Baxter equation, and immediately generalize from $N = 2, 3$ to $N > 3$ to get the $\mathcal{A}^{(n)}/\mathbb{Z}/N$ models without having to check any identities between theta functions (Di Francesco and Zuber 1990a), page 643.

It was communicated to us by Ocneanu that he has the construction in Section 13.5. (Also see (Ocneanu 1994).) The presentation here is due to us.

14

COMMUTING SQUARES OF II_1 FACTORS

14.1 Introduction

We will study a certain type of commuting squares of II_1 factors in connection to topology, physics, and problems of ordinary subfactors.

In Section 14.2, we introduce a notion of bi-commuting square and study its basic properties. Section 14.3 deals with the intertwining Yang–Baxter equation and the triple sequence of string algebras. We show several examples of commuting squares of II_1 factors arising from RCFT in Section 14.4. We discuss an analogue of the coset construction in RCFT in Section 14.5. As a concrete application of the general machinery, we study the GHJ subfactor introduced in Section 11.6 in the last section of this chapter.

14.2 Bi-commuting squares and the planar basic construction

We first need a definition.

Definition 14.1 *If four II_1 factors $M_{00}, M_{01}, M_{10}, M_{11}$ satisfy the following properties, we say that they make a bi-commuting square of finite index.*

1. *The square*

$$\begin{array}{ccc} M_{00} & \subset & M_{01} \\ \cap & & \cap \\ M_{10} & \subset & M_{11} \end{array}$$

   is a commuting square.

2. $[M_{11} : M_{00}] < \infty$.
3. $[M_{11} : M_{10}] = [M_{01} : M_{00}]$.

Of course, the third condition is equivalent to $[M_{11} : M_{01}] = [M_{10} : M_{00}]$.

Suppose that the square

$$\begin{array}{ccc} M_{00} & \subset & M_{01} \\ \cap & & \cap \\ M_{10} & \subset & M_{11} \end{array}$$

is a commuting square of II_1 factors with $[M_{11} : M_{00}] < \infty$. Let e_{01}, e_{10}, e_{00} be the projections onto $L^2(M_{01}), L^2(M_{10}), L^2(M_{00})$ respectively on $L^2(M_{11})$.

We consider the basic constructions $M_{21} = \langle M_{11}, e_{01}\rangle$, $M_{12} = \langle M_{11}, e_{10}\rangle$, and $M_{22} = \langle M_{11}, e_{00}\rangle$. We can naturally regard M_{12}, M_{21} as subalgebras of M_{22}. We prove several statements under this set of assumptions.

Lemma 14.2 *Under the above assumptions, the following are equivalent.*

1. *The square*
$$\begin{array}{ccc} M_{11} & \subset & M_{12} \\ \cap & & \cap \\ M_{21} & \subset & M_{22} \end{array}$$
is a commuting square.
2. $E_{M_{12}}(e_{01}) = [M_{11} : M_{01}]^{-1}$.
3. $E_{M_{21}}(e_{10}) = [M_{11} : M_{10}]^{-1}$.
4. $E_{M_{12}}(e_{01}) \in \mathbb{C}$.
5. $E_{M_{21}}(e_{10}) \in \mathbb{C}$.

Proof Assume Condition 1 first. Then we get

$$E_{M_{12}}(e_{01}) = E_{M_{11}}(e_{01}) = [M_{11} : M_{01}]^{-1}.$$

We similarly get Condition 3 from Condition 1.

Assume Condition 4. It is enough to prove Condition 1. First note that we get Condition 2 by looking at the trace value. A general element x in M_{21} is of the form $\sum_{j=1}^{n} x_j e_{01} y_j$ with $x_j, y_j \in M_{11}$. Then we clearly have $E_{M_{11}}(x) = E_{M_{12}}(x)$. □

Lemma 14.3 *Under the above assumptions, suppose that we have* $[M_{11} : M_{01}] = [M_{10} : M_{00}]$ *and that* $M_{11} = \overline{M_{10} \cdot M_{01}}^{\sigma\text{-strong}}$, *where*

$$M_{10} \cdot M_{01} = \left\{ \sum_{j=1}^{n} x_j y_j \mid x_j \in M_{10}, y_j \in M_{01}, n \geq 0 \right\}.$$

Then the square

$$\begin{array}{ccc} M_{11} & \subset & M_{12} \\ \cap & & \cap \\ M_{21} & \subset & M_{22} \end{array}$$

is a commuting square.

Proof By Lemma 14.2, it is enough to prove $E_{M_{12}}(e_{01}) = [M_{11} : M_{01}]^{-1}$, that is,

$$\mathrm{tr}(e_{01}(x e_{10} y)) = \frac{1}{[M_{11} : M_{01}]} \mathrm{tr}(x e_{10} y), \quad x, y \in M_{11}.$$

By the second assumption, we may assume that x, y are of the form $x = n_1 m_1$, $y = m_2 n_2$ with $n_1, n_2 \in M_{01}$ and $m_1, m_2 \in M_{10}$.

Then we have

$$\begin{aligned}
\operatorname{tr}(e_{01}(xe_{10}y)) &= \operatorname{tr}(e_{01}n_1m_1e_{10}m_2n_2)\\
&= \operatorname{tr}(n_2n_1e_{01}e_{10}m_1m_2)\\
&= \operatorname{tr}(n_2n_1e_{00}m_1m_2)\\
&= \frac{1}{[M_{11}:M_{00}]}\operatorname{tr}(n_2n_1m_1m_2)\\
&= \frac{1}{[M_{11}:M_{00}]}\operatorname{tr}(xy)\\
&= \frac{1}{[M_{11}:M_{01}][M_{11}:M_{10}]}\operatorname{tr}(xy)\\
&= \frac{1}{[M_{11}:M_{01}]}\operatorname{tr}(xe_{10}y),
\end{aligned}$$

which is the desired conclusion. □

Lemma 14.4 *Let $N \subset M \subset L$ be II_1 factors with $[L : N] < \infty$. Choose a Pimsner–Popa basis $\{u_1, u_2, \ldots, u_n\}$ for $N \subset M$. Then we have*

$$e_M = \sum_{j=1}^{n} u_j e_N u_j^*$$

on $L^2(L)$.

Proof For $x \in L$, we get

$$\begin{aligned}
E_M(x) &= \sum_{j=1}^{n} u_j E_N(u_j^* E_M(x))\\
&= \sum_{j=1}^{n} u_j E_N(E_M(u_j^* x))\\
&= \sum_{j=1}^{n} u_j E_N(u_j^* x).
\end{aligned}$$

Thus we have

$$e_M(\hat{x}) = \widehat{E_M(x)} = \left(\sum_{j=1}^{n} u_j E_N(u_j^* x)\right)\hat{} = \left(\sum_{j=1}^{n} u_j e_N u_j^*\right)\hat{x},$$

which gives the conclusion. □

Lemma 14.5 *Under the above assumptions, we choose Pimsner–Popa bases $\{u_1, u_2, \ldots, u_n\}$ and $\{v_1, v_2, \ldots, v_m\}$ for $M_{00} \subset M_{10}$ and $M_{00} \subset M_{01}$ respectively. We set*

$$a = \sum_{j=1}^{n}\sum_{k=1}^{m} u_j v_k e_{00} v_k^* u_j^* \in M_{22};$$

then a is a projection and

$$\operatorname{tr}(a) = \frac{[M_{10} : M_{00}][M_{01} : M_{00}]}{[M_{11} : M_{00}]}.$$

Proof By Lemma 14.4, we know that $e_{01} = \sum_{k=1}^{m} v_k e_{00} v_k^*$. Thus we have $a = \sum_{j=1}^{n} u_j e_{01} u_j^*$. It is trivial that $a = a^*$ and we have the following identities:

$$\begin{aligned}
a^2 &= \sum_{j=1}^{n} u_j e_{01} u_j^* \sum_{k=1}^{n} u_k e_{01} u_k^* \\
&= \sum_{j=1}^{n}\sum_{k=1}^{n} u_j e_{01} u_j^* u_k e_{01} u_k^* \\
&= \sum_{j=1}^{n}\sum_{k=1}^{n} u_j E_{M_{01}}(u_j^* u_k) e_{01} u_k^* \\
&= \sum_{j=1}^{n}\sum_{k=1}^{n} u_j E_{M_{00}}(u_j^* u_k) e_{01} u_k^* \\
&= \sum_{j=1}^{n} u_j e_{01} u_j^* = a.
\end{aligned}$$

We also have

$$\begin{aligned}
\operatorname{tr}(a) &= \operatorname{tr}\left(E_{M_{11}}\left(\sum_{j=1}^{n}\sum_{k=1}^{m} u_j v_k e_{00} v_k^* u_j^*\right)\right) \\
&= \operatorname{tr}\left(\sum_{j=1}^{n}\sum_{k=1}^{m} u_j v_k E_{M_{11}}(e_{00}) v_k^* u_j^*\right) \\
&= \frac{1}{[M_{11} : M_{00}]}\operatorname{tr}\left(\sum_{j=1}^{n}\sum_{k=1}^{m} u_j v_k v_k^* u_j^*\right) \\
&= \frac{[M_{10} : M_{00}][M_{01} : M_{00}]}{[M_{11} : M_{00}]}.
\end{aligned}$$

□

Lemma 14.6 *Under the above assumptions, we assume that* $M_{11} = \overline{M_{10} \cdot M_{01}}^{\sigma\text{-strong}}$. *Then we have* $[M_{11} : M_{01}] = [M_{10} : M_{00}]$ *and* $[M_{11} : M_{10}] = [M_{01} : M_{00}]$.

Proof We set the operator a as in Lemma 14.5 and will prove that $a = 1$. By the assumption, we get $L^2(M_{11}) = \overline{M_{10} \cdot M_{01}}^{\|\cdot\|_2}$, thus it is enough to show $a\widehat{mn} = \widehat{mn}$ for $m \in M_{10}, n \in M_{01}$. We get

$$\begin{aligned} a\widehat{mn} &= \sum_{j=1}^{n} u_j e_{01} u_j^* \widehat{mn} \\ &= \left(\sum_{j=1}^{n} u_j E_{M_{01}}(u_j^* m) n \right)^{\wedge} \\ &= \left(\sum_{j=1}^{n} u_j E_{M_{00}}(u_j^* m) n \right)^{\wedge} \\ &= m\hat{n} = \widehat{mn}, \end{aligned}$$

which shows $a = 1$.

Lemma 14.5 and the identity $\mathrm{tr}(a) = 1$ implies the conclusion. □

Proposition 14.7 *For the commuting square*

$$\begin{array}{ccc} M_{00} & \subset & M_{01} \\ \cap & & \cap \\ M_{10} & \subset & M_{11} \end{array}$$

of II_1 *factors with* $[M_{11} : M_{00}] < \infty$, *the following are equivalent.*

1. *The following*

$$\begin{array}{ccc} M_{11} & \subset & M_{12} \\ \cap & & \cap \\ M_{21} & \subset & M_{22} \end{array}$$

   is a commuting square.
2. $[M_{11} : M_{01}] = [M_{10} : M_{00}]$.
3. $[M_{11} : M_{10}] = [M_{01} : M_{00}]$.
4. $M_{11} = M_{10} \cdot M_{01} = \{\sum_{j=1}^{n} x_j y_j \mid x_j \in M_{10}, y_j \in M_{01}, n \geq 0\}$.
5. $M_{11} = M_{01} \cdot M_{10}$.
6. $M_{11} = \overline{M_{10} \cdot M_{01}}^{\sigma\text{-strong}}$.
7. $M_{11} = \overline{M_{01} \cdot M_{10}}^{\sigma\text{-strong}}$.

Proof First note that we always have $[M_{11} : M_{01}] \geq [M_{10} : M_{00}]$.

Assume Condition 1. Then we get

$$\begin{aligned} &[M_{11} : M_{01}] = [M_{21} : M_{11}] = [M_{22} : M_{11}]/[M_{22} : M_{21}] \\ &\leq [M_{22} : M_{11}]/[M_{12} : M_{11}] = [M_{11} : M_{00}]/[M_{11} : M_{10}] = [M_{10} : M_{00}], \end{aligned}$$

where we have used Theorem 9.48. Thus Condition 1 implies Condition 2.

Assume Condition 2. By Lemma 14.5, we get $\mathrm{tr}(a) = 1$, which means $a = 1$. Applying $ax = x$ to $\hat{1} \in L^2(M_{11})$, we get

$$\hat{x} = \left(\sum_{j=1}^{n} \sum_{k=1}^{m} u_j v_k e_{00} v_k^* u_j^* x \right)^{\wedge},$$

and this implies

$$x = \sum_{j=1}^{n} \sum_{k=1}^{m} u_j v_k E_{M_{00}}(v_k^* u_j^* x) \in M_{10} \cdot M_{01}.$$

It is trivial that Condition 4 implies Condition 6. Suppose we have Condition 6. By Lemma 14.6, we get Condition 2. Then Lemma 14.3 implies Condition 1.

Thus we have proved equivalence of Conditions 1, 2, 4, 6. The equivalence of Conditions 1, 3, 5, 7 are proved in the same way. □

We will always assume in this chapter that the square

$$\begin{array}{ccc} M_{00} & \subset & M_{01} \\ \cap & & \cap \\ M_{10} & \subset & M_{11} \end{array}$$

satisfies the conditions in Definition 14.1.

For simplicity of notation, we write ${}_{ij}X_{kl}$ for a bimodule ${}_{M_{ij}}X_{M_{kl}}$, where $i, j, k, l = 0, 1$. We also write $\otimes_{ij}$ for the relative tensor $\otimes_{M_{ij}}$ for bimodules and E_{ij} for $E_{M_{ij}}$. We again drop $L^2()$ for simplicity of notation as in Section 9.7. For example, ${}_{00}(M_{00})_{00}$ means the M_{00}-M_{00} bimodule $L^2(M_{00})$.

Lemma 14.8 *Suppose we have a commuting square of II_1 factors as in Definition 14.1. We then have natural isomorphisms $M_{01} \otimes_{00} M_{10} \cong M_{11}$ as M_{01}-M_{10} bimodules, and $M_{10} \otimes_{00} M_{01} \cong M_{11}$ as M_{10}-M_{01} bimodules.*

Proof We define a map $\pi : M_{01} \otimes_{00} M_{10} \to M_{11}$ by $\pi(x \otimes_{00} y) = xy$ for $x \in M_{01}, y \in M_{10}$. For $x_1, x_2 \in M_{01}$ and $y_1, y_2 \in M_{10}$, we get

$$\begin{aligned} \langle x_1 \otimes_{00} y_1, x_2 \otimes_{00} y_2 \rangle &= \langle x_1 E_{00}(y_1 y_2^*), x_2 \rangle \\ &= \mathrm{tr}(x_2^* x_1 E_{00}(y_1 y_2^*)) \\ &= \mathrm{tr}(x_2^* x_1 y_1 y_2^*) \\ &= \langle \pi(x_1 \otimes_{00} y_1), \pi(x_2 \otimes_{00} y_2) \rangle, \end{aligned}$$

where we have used the commuting square condition $E_{00} = E_{10}$ on M_{01}. Thus π extends to an isometry which is surjective by Proposition 14.7. The other isomorphism is proved similarly. □

In many concrete cases, we further assume that $M_{00} \subset M_{11}$ is of finite depth and all the type II_1 factors concerned are hyperfinite. In general, we have the following proposition.

Proposition 14.9 *Suppose $N \subset M$ is a subfactor with finite index and finite depth. Then for any factor P with $N \subset P \subset M$, the subfactors $N \subset P$ and $P \subset M$ have finite depth.*

Proof Trivially, we have ${}_N P \otimes_P M_M \cong {}_N M_M$. By the finite depth condition, we know that we have only finitely many mutually non-isomorphic irreducible N-N bimodules in the decompositions of

$${}_N M \otimes_M M \otimes_N M \otimes_M M \otimes_N \cdots \otimes_M M_N.$$

By replacing ${}_N M_M$ here with ${}_N P \otimes_P M_M$ and using the fact ${}_P M \otimes_M M_P$ contains a copy of the trivial P-P bimodule ${}_P P_P$, we can conclude that

$${}_N P \otimes_P P \otimes_N P \otimes_P P \otimes_N \cdots \otimes_P P_N$$

contains only finitely many irreducible N-N bimodules, which means that $N \subset P$ has finite depth. The other case is proved similarly. □

The converse of the above proposition does not hold. That is, even if both subfactors $N \subset P$ and $P \subset M$ have finite depth, the subfactor $N \subset M$ can be infinite depth, or even non-amenable. See (Bisch and Haagerup 1996) for many concrete examples. In a situation in Definition 14.1, we have the following proposition, which can be regarded as a partial converse to Proposition 14.9.

Proposition 14.10 *Suppose we have one of the following conditions in Definition 14.1. Then the subfactor $M_{00} \subset M_{11}$ has finite depth.*

1. *The subfactors $M_{00} \subset M_{01}$ and $M_{00} \subset M_{10}$ both have finite depth.*
2. *The subfactors $M_{01} \subset M_{11}$ and $M_{10} \subset M_{11}$ both have finite depth.*
3. *The subfactors $M_{00} \subset M_{01}$ and $M_{01} \subset M_{11}$ both have finite depth.*
4. *The subfactors $M_{00} \subset M_{10}$ and $M_{10} \subset M_{11}$ both have finite depth.*

Proof We prove only the first case because the others are similar. By Lemma 14.8, we know that $M_{10} \otimes_{00} M_{01} \cong M_{01} \otimes_{00} M_{10}$ as M_{00}-M_{00} bimodules. Then

$${}_{00}M_{11} \otimes_{00} M_{11} \otimes_{00} \cdots \otimes_{00} (M_{11})_{00}$$

is isomorphic to

$${}_{00}M_{10} \otimes_{00} M_{10} \otimes_{00} \cdots \otimes_{00} M_{10} \otimes_{00} M_{01} \otimes_{00} M_{01} \otimes_{00} \cdots \otimes_{00} (M_{01})_{00}.$$

Then the two finite depth conditions imply that the subfactor $M_{00} \subset M_{11}$ also has finite depth. □

We set $f_1 = e_{10} \in M_{12}$ and $e_1 = e_{01} \in M_{21}$. We define $M_{20} = M_{21} \cap \{f_1\}'$ and $M_{02} = M_{12} \cap \{e_1\}'$. Because

$$\begin{array}{ccc} M_{00} & \subset & M_{01} \\ \cap & & \cap \\ M_{10} & \subset & M_{11} \end{array}$$

is a commuting square, we get $E_{M_{21}}(f_1) = [M_{22} : M_{21}]^{-1}$ and $E_{M_{12}}(e_1) = [M_{22} : M_{12}]^{-1}$. Thus by Proposition 9.44, we know that $M_{20} \subset M_{21} \subset M_{22}$ and $M_{02} \subset M_{12} \subset M_{22}$ are both standard. We now prove that $M_{00} \subset M_{10} \subset M_{20}$ and $M_{00} \subset M_{01} \subset M_{02}$ are both standard. First note that the commuting square condition implies $e_1 f_1 = f_1 e_1$, hence $e_1 \in M_{20}$ and $f_1 \in M_{02}$. By Proposition 9.53, the square

$$\begin{array}{ccc} M_{10} = M_{11} \cap \{f_1\}' & \subset & M_{11} \\ \cap & & \cap \\ M_{20} = M_{21} \cap \{f_1\}' & \subset & M_{21} \end{array}$$

is a commuting square. So we get $E_{M_{10}}(e_1) = E_{M_{11}}(e_1) = [M_{21} : M_{11}]^{-1}$. This implies $[M_{20} : M_{10}] = [M_{21} : M_{11}] = [M_{10} : M_{00}]$ by Theorem 9.48. By Proposition 9.47, we conclude that $M_{00} \subset M_{10} \subset M_{20}$ is standard. Similarly, we can prove that $M_{00} \subset M_{01} \subset M_{02}$ is standard. By repeating this type of argument, we get the following theorem.

Theorem 14.11 *We can construct the double sequence $\{M_{kl}\}_{kl\geq 0}$ and two sequences of projections $\{e_j\}_{j=1,2,\ldots}$, $\{f_j\}_{j=1,2,\ldots}$ so that the following hold.*

1. $M_{kl} \subset M_{k+1,l} \subset M_{k+2,l}$ *is standard.*
2. $M_{kl} \subset M_{k,l+1} \subset M_{k,l+2}$ *is standard.*
3. *The projection e_j is a Jones projection for $M_{j-1,k} \subset M_{j,k}$ for all j, k.*
4. *The projection f_j is a Jones projection for $M_{k,j-1} \subset M_{k,j}$ for all j, k.*

14.3 The intertwining Yang–Baxter equation and triple sequences of string algebras

We would like to construct interesting bi-commuting squares of II_1 factors. As in the construction of subfactors, we use a construction based on finite dimensional objects, and we want to extend the construction with string algebras in Section 11.3 to our new setting.

Our initial data is a system $(\mathcal{G}, \mu, \beta_1, \beta_2, \beta_3, W)$ consisting of a finite connected unoriented graph $\mathcal{G}$, a positive real-valued function μ on the vertices of $\mathcal{G}$, positive numbers β_1, β_2, β_3, and a 'connection' W on the graph $\mathcal{G}$.

We require the following for the graph $\mathcal{G}$. The set of the vertices of the graph $\mathcal{G}$ is a disjoint union of the eight sets ${}_{ij}\mathcal{G}_{kl}$, where

$$\begin{aligned}(i,j,k,l) = {} & (0,0,0,0), (0,0,0,1), (1,1,0,0), (1,1,0,1), \\ & (0,0,1,0), (0,0,1,1), (1,1,1,0), (1,1,1,1).\end{aligned}$$

We write ${}^{i',j'}_{i,j}\mathcal{G}^{k',l'}_{k,l}$ for the graph whose vertices are given as the union ${}_{i,j}\mathcal{G}_{k,l} \cup {}_{i',j'}\mathcal{G}_{k',l'}$ and whose edges are the edges of $\mathcal{G}$ connecting a vertex in ${}_{i,j}\mathcal{G}_{k,l}$ to a vertex in ${}_{i',j'}\mathcal{G}_{k',l'}$, where

$$\begin{aligned}(i',j',k',l',i,j,k,l) = {} & (0,0,0,0,0,0,0,1), (0,0,0,1,1,1,0,1), \\ & (0,0,0,0,1,1,0,0), (1,1,0,0,1,1,0,1),\end{aligned}$$

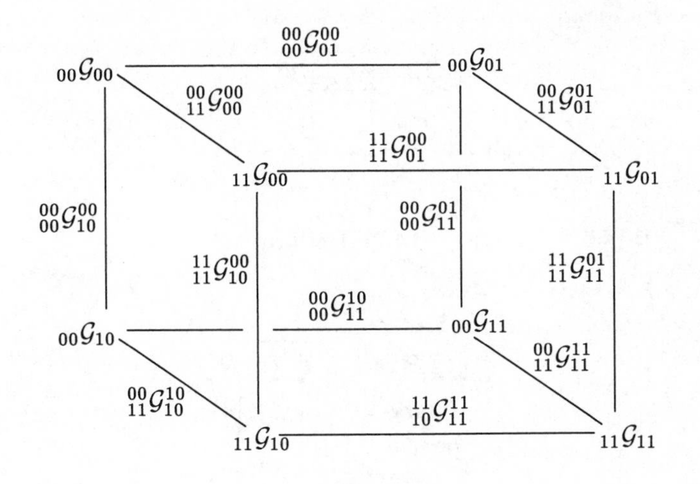

FIG. 14.1. Nested graph $\mathcal{G}$

$$
\begin{gathered}
(0,0,1,0,0,0,1,1),(0,0,1,1,1,1,1,1),\\
(0,0,1,0,1,1,1,0),(1,1,1,0,1,1,1,1),\\
(0,0,0,0,0,0,1,0),(0,0,0,1,0,0,1,1),\\
(1,1,0,0,1,1,1,0),(1,1,0,1,1,1,1,1),\\
(0,0,0,1,0,0,0,0),(1,1,0,1,0,0,0,1),\\
(1,1,0,0,0,0,0,0),(1,1,0,1,1,1,0,0),\\
(0,0,1,1,0,0,1,0),(1,1,1,1,0,0,1,1),\\
(1,1,1,0,0,0,1,0),(1,1,1,1,1,1,1,0),\\
(0,0,1,0,0,0,0,0),(0,0,1,1,0,0,0,1),\\
(1,1,1,0,1,1,0,0),(1,1,1,1,1,1,0,1).
\end{gathered}
$$

(Thus ${}^{i'j'}_{ij}\mathcal{G}^{k'l'}_{kl} = {}^{ij}_{i'j'}\mathcal{G}^{kl}_{k'l'}$ and we have twelve different ${}^{i'j'}_{ij}\mathcal{G}^{k'l'}_{kl}$.) Each graph ${}^{i'j'}_{ij}\mathcal{G}^{k'l'}_{kl}$ may not be connected, but we do *require* that ${}^{00}_{11}\mathcal{G}^{00}_{00}$, ${}^{00}_{11}\mathcal{G}^{01}_{01}$, ${}^{00}_{11}\mathcal{G}^{10}_{10}$, and ${}^{00}_{11}\mathcal{G}^{11}_{11}$ are connected. Note that the graph $\mathcal{G}$ looks like a cube whose eight vertices correspond to the eight sets of vertices and whose twelve edges correspond to the twelve sets of edges.

The first condition corresponds to the harmonic axiom 2. For two vertices $x, y \in \mathcal{G}$, we write $m(x,y)$ for the number of edges of $\mathcal{G}$ connecting x and y.

Axiom 8. (Harmonicity) *We have the following identities.*

$$\beta_1 \mu(x) = \sum_{y \in {}_{ij}\mathcal{G}_{\bar{k}l}} m(x,y)\mu(y), \qquad \textit{for } x \in {}_{ij}\mathcal{G}_{kl},$$

$$\sum_{D,\sigma_3,\sigma_4} \begin{array}{ccc} A & \xrightarrow{\sigma_1} & B \\ {\scriptstyle\sigma_3}\downarrow & & \downarrow{\scriptstyle\sigma_2} \\ D & \xrightarrow[\sigma_4]{} & C \end{array} \;\; \overline{\begin{array}{ccc} A & \xrightarrow{\sigma_1'} & B' \\ {\scriptstyle\sigma_3}\downarrow & & \downarrow{\scriptstyle\sigma_2'} \\ D & \xrightarrow[\sigma_4]{} & C \end{array}} = \delta_{B,B'}\delta_{\sigma_1,\sigma_1'}\delta_{\sigma_2,\sigma_2'}.$$

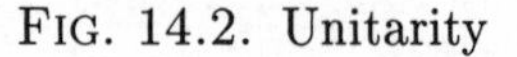

FIG. 14.2. Unitarity

$$\begin{array}{ccc} A & \xrightarrow{\sigma_1} & B \\ {\scriptstyle\sigma_3}\downarrow & & \downarrow{\scriptstyle\sigma_2} \\ D & \xrightarrow[\sigma_4]{} & C \end{array} = \overline{\begin{array}{ccc} A & \xrightarrow{\sigma_3} & D \\ {\scriptstyle\sigma_1}\downarrow & & \downarrow{\scriptstyle\sigma_4} \\ B & \xrightarrow[\sigma_2]{} & C \end{array}}$$

FIG. 14.3. Renormalization (1)

$$\beta_2\mu(x) = \sum_{y \in {}_{ij}\mathcal{G}_{k\bar{l}}} m(x,y)\mu(y), \qquad \textit{for } x \in {}_{ij}\mathcal{G}_{kl},$$

$$\beta_3\mu(x) = \sum_{y \in {}_{\bar{i}\bar{j}}\mathcal{G}_{kl}} m(x,y)\mu(y), \qquad \textit{for } x \in {}_{ij}\mathcal{G}_{kl},$$

where $\bar{i} = 1, 0$, *for* $i = 0, 1$, *respectively.*

We mark a vertex in ${}_{00}\mathcal{G}_{00}$ with $*$ and require $\mu(*) = 1$. A *cell* is a quadruple $(\sigma_1, \sigma_2, \sigma_3, \sigma_4)$ of edges of the graph $\mathcal{G}$ with $s(\sigma_1) = a$, $r(\sigma_1) = b$, $s(\sigma_2) = b$, $r(\sigma_2) = c$, $s(\sigma_3) = a$, $r(\sigma_3) = d$, $s(\sigma_4) = d$, $r(\sigma_4) = c$, where a, b, c, d are four vertices of $\mathcal{G}$ in mutually different ${}_{ij}\mathcal{G}_{kl}$, as in Section 10.3. The complex valued map W is defined on the set of cells and called a *connection.* We also use the graphical notation for $W(\sigma_1, \sigma_2, \sigma_3, \sigma_4)$ as in Section 10.3. We again sometimes drop the labels for edges or vertices if no confusion arises. We also use the convention that if $(\sigma_1, \sigma_2, \sigma_3, \sigma_4)$ is not a cell, the above symbol denotes the number 0. We write $v(\sigma)$ for the set of two vertices of an edge σ of the graph $\mathcal{G}$. The next axiom corresponds to the unitarity axiom 1.

Axiom 9. (Unitarity) *Choose four edges* $\sigma_1, \sigma_2, \sigma_1', \sigma_2'$ *of the graph* $\mathcal{G}$ *with* $v(\sigma_1) = \{a, b\}$, $v(\sigma_2) = \{b, c\}$, $v(\sigma_1') = \{a, b'\}$, $v(\sigma_2') = \{b', c\}$ *so that* b *and* b' *belong to the same* ${}_{ij}\mathcal{G}_{kl}$. *Suppose that there exist edges* σ_3, σ_4 *of* $\mathcal{G}$ *such that* $\sigma_1, \sigma_2, \sigma_3, \sigma_4$ *make a cell. Then we have the identity of Fig. 14.2.*

We also have the following corresponding to the renormalization axiom 4.

Axiom 10. (Renormalization) *For a cell* $(\sigma_1, \sigma_2, \sigma_3, \sigma_4)$, *we have the two identities of Figs 14.3 and 14.4.*

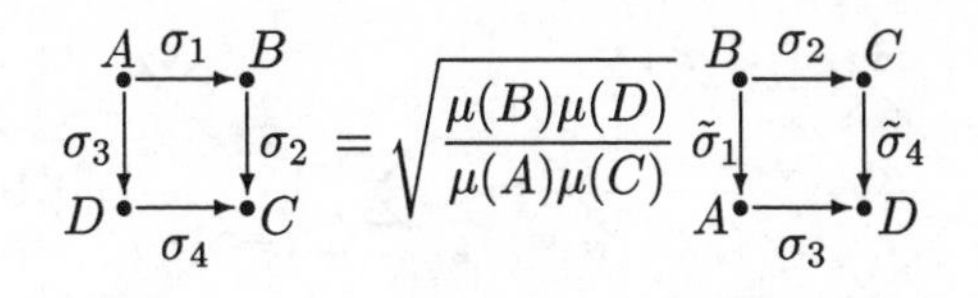

FIG. 14.4. Renormalization (2)

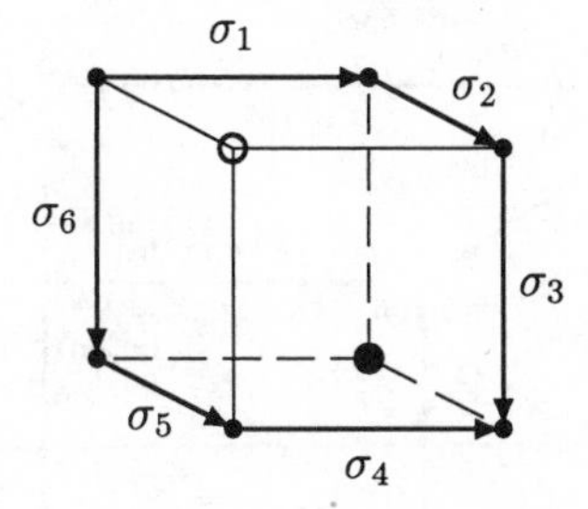

FIG. 14.5. A hexagon in the graph $\mathcal{G}$

We also use the standard convention for the graphic expressions as in Fig. 10.16.

Next we need a new axiom which does not correspond to any axiom in the paragroup case. First we choose six edges $\sigma_1, \sigma_2, \sigma_3, \sigma_4, \sigma_5, \sigma_6$ so that they make a hexagon in the graph $\mathcal{G}$ as in Fig. 14.5. Then we have the following axiom.

Axiom 11. (Intertwining Yang–Baxter Equation) *The identity of Fig. 14.6 holds, where both sides mean the sum of the product of three connection values over all possible internal choices of A_7, σ_7, σ_8, σ_9 for any fixed choice of σ_1, σ_2, σ_3, σ_4, σ_5, σ_6 on the boundary as above. On the left hand side, the vertex A_7 is chosen from the corner marked with $\circ$ in Fig. 14.5, and on the right hand side, it is chosen from the corner $\bullet$.*

In the above identity, we used the same convention as in Fig. 11.77.

We now construct a *triple* sequence of string algebras $\{A_{jkl}\}_{j,k,l\geq 0}$ from the above set of data. Using the starting point $*$, we construct the three sequences of string algebras $\{A_{j00}\}_{j\geq 0}$, $\{A_{0k0}\}_{k\geq 0}$, $\{A_{00l}\}_{l\geq 0}$ starting from $*$ with the graphs ${}^{00}_{11}\mathcal{G}^{00}_{00}$, ${}^{00}_{00}\mathcal{G}^{00}_{10}$, ${}^{00}_{00}\mathcal{G}^{00}_{01}$ respectively. Then use the entire graph $\mathcal{G}$ and the connection W to get the triple sequence as in Section 11.3. Here the intertwining Yang–Baxter equation insures that the identifications with connections are compatible. Note that the first part of the sequence looks as in Figure 14.7. Here the arrows

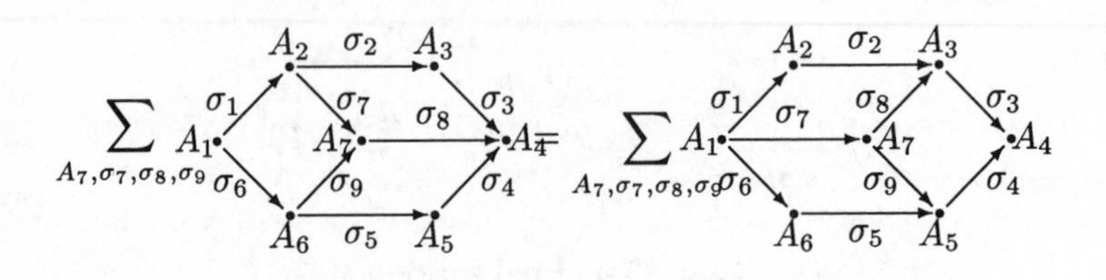

FIG. 14.6. The intertwining Yang–Baxter equation

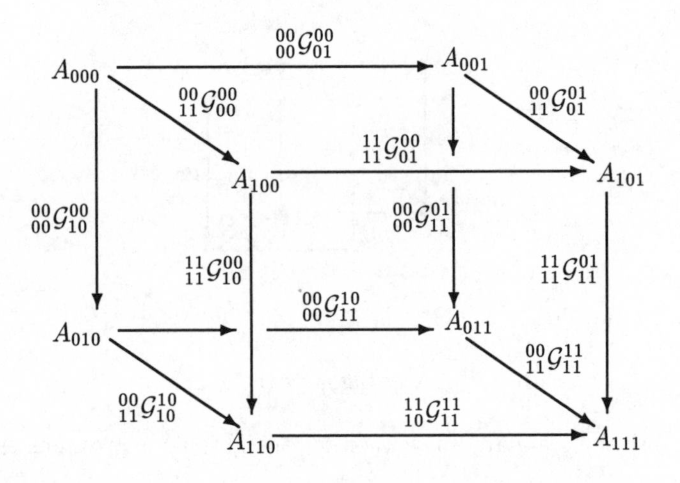

FIG. 14.7. Triple sequences of string algebras

denote embeddings, and we look at the graph from the same direction as in Fig. 14.1.

We define a normalized trace tr on $\cup_{jkl} A_{jkl}$ as follows. For a string $(\sigma_+, \sigma_-) \in A_{jkl}$, we define $\mathrm{tr}((\sigma_+, \sigma_-)) = \beta_1^{-k}\beta_2^{-l}\beta_3^{-j}\mu(r(\sigma_+))$. By the harmonic axiom and the embedding rule of the string algebra, this tr is well defined on $\cup_{jkl} A_{jkl}$ as in Section 11.3. The we define $A_{\infty kl}$ to be the von Neumann algebra obtained by the GNS-completion with respect to tr. We define the vertical Jones projections $e_k \in A_{0,k+1,l}$ and $f_l \in A_{0,k,l+1}$ as in Definition 11.5. We also define the Jones projections p_j in $A_{j+1,0,0}$ similarly. Because the four graphs $^{00}_{11}\mathcal{G}^{00}_{00}$, $^{00}_{11}\mathcal{G}^{11}_{11}$, $^{00}_{11}\mathcal{G}^{01}_{01}$, and $^{00}_{11}\mathcal{G}^{10}_{10}$ are connected, all the von Neumann algebras $A_{\infty kl}$ are hyperfinite II_1 factors.

As in Section 11.3, we can conclude that the square

$$\begin{array}{ccc} M_{00} & \subset & M_{01} \\ \cap & & \cap \\ M_{10} & \subset & M_{11} \end{array}$$

satisfies the conditions in Definition 14.1 and the double sequence M_{kl} of hyperfinite II_1 factors is constructed from the above square as in Theorem 14.11.

In the following sections, we will show how to construct systems of data satisfying the axioms of this section. As in the paragroup case, it is possible to characterize 'canonical' forms of data satisfying the axioms in this section as a paragroup type invariant of bi-commuting squares of II_1 factors of finite index. See (Kawahigashi 1995c) for details.

14.4 RCFT, orbifold construction and commuting squares of II_1 factors

If we want to construct a bi-commuting square of II_1 factors, the easiest way is to use two subfactors $N \subset M$ and $P \subset Q$, and make the following commuting square with the tensor product operation:

$$\begin{array}{ccc} Q \otimes N & \subset & Q \otimes M \\ \cup & & \cup \\ P \otimes N & \subset & P \otimes M \end{array}$$

No surprises, however, occur in this construction and so we need to do something different in order to get more interesting objects. This kind of commuting square should be regarded as trivial. We will show that we can construct a triple sequence of string algebras giving non-trivial commuting squares of II_1 factors from an RCFT with essential use of the Yang–Baxter equation. This is a generalization of the constructions in Sections 13.3 and 13.4. The reason that we can construct such a commuting square from an RCFT is that 'an RCFT has a higher symmetry than an ordinary paragroup'.

We proceed as in Section 13.3. Fix an RCFT. Choose fields x, y, z from the RCFT. The trivial modification of the construction in Section 13.3 gives a graph $\mathcal{G}$, a function μ, numbers β_j, and a connection W. We can show that the harmonic axiom, the unitarity axiom, and the renormalization axiom hold exactly as in Section 13.3. The intertwining Yang–Baxter equation follows from Fig. 14.8.

As a very simple example of this construction, take the Lie group $SU(2)$ and set the level k to be $n-1$. In this case, we have a flat connection on the graphs A_n as in Section 11.5, which satisfies the Yang–Baxter equation. This follows from the flatness of the Jones projections as in Sections 11.3 and 14.3. (Also compare the formula with the one in (Andrews *et al.* 1984).) It is easy to identify this commuting square with the following. Take a subfactor $N \subset M$ of type A_n, and make a basic construction $N \subset M \subset M_1$. Set $\varepsilon = \sqrt{-1}\exp(\pi\sqrt{-1})/2(n+1)$, $\beta = 2\cos(\pi/(n+1))$, and $u = \varepsilon + \bar{\varepsilon}\beta e$, then u is a unitary, and we get a commuting square:

$$\begin{array}{ccc} N & \subset & M \\ \cap & & \cap \\ uMu^* & \subset & M_1 \end{array}$$

Furthermore, it is easy to see that the orbifold construction as in Section 13.4 also works in our setting here. This is because the intertwining Yang–Baxter equation is preserved in the orbifold construction. (Recall that the intertwining Yang–Baxter equation is the compatibility condition for different identifications of strings with the connections.)

We can prove that this data gives non-trivial bi-commuting squares of hyperfinite II_1 factors in the above sense by looking at the Bratteli diagrams for $M'_{00} \cap M_{kl}$. See (Kawahigashi 1995c) for more details.

14.5 Coset construction and subfactors with infinite index

In (de Boer and Goeree 1991), there was a correspondence between constructions in RCFT and subfactor theory as in Table 14.1.

In this table, the tensor product of paragroups should be a construction of $N \otimes P \subset M \otimes Q$ from two subfactors $N \subset M$ and $P \subset Q$. The construction of $N \subset P$ from $N \subset M, M \subset P$ is a 'composition' of subfactors as studied in (Bisch and Haagerup 1996). The orbifold construction has been already studied in Section 13.4. The extended algebra construction is regarded as an 'inverse' of the orbifold construction from our viewpoint.

So in this section, we study the last entry, the coset construction. We regard this problem of finding a subfactor analogue of the coset construction as a purely operator algebraic problem to construct a new subfactor $S' \cap N \subset S' \cap M$ from

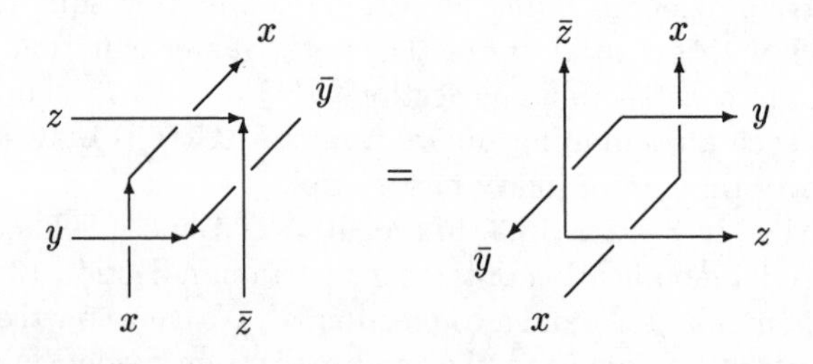

FIG. 14.8. The intertwining Yang–Baxter equation

Table 14.1 *Analogy between RCFT and subfactors*

RCFT	Subfactor
tensor product	$N \subset M, M \subset P \Rightarrow N \subset P$
coset construction	$N \subset M, S \subset N \Rightarrow S' \cap N \subset S' \cap M$
orbifold construction	$N \subset M \Rightarrow N^G \subset M^G$
extended algebras	$N \subset M \Rightarrow N \rtimes G \subset M \rtimes G$

a given subfactor $N \subset M$ for an appropriate choice of $S \subset N$. To get *factors* $S' \cap N$ and $S' \cap M$, the subalgebra S must be a factor. If S is of finite dimension, there is nothing interesting, so we must assume that S is a type II_1 subfactor of N. If $[N : S] < \infty$, $S' \cap N$ is finite dimensional, so we need to assume $[N : S] = \infty$. Thus our problem can be stated as follows. For a given subfactor $N \subset M$, what kind of subfactors $S' \cap N \subset S' \cap M$ arise, when S is a subfactor of N with infinite index such that $S' \cap N$ and $S' \cap M$ are factors? This can be regarded as a classification problem for S (for a fixed $N \subset M$), too, and then the paragroup of $S' \cap N \subset S' \cap M$ is an invariant for S. Note that by the commuting square condition, we get $[S' \cap M : S' \cap N] \leq [M : N]$, and we expect that natural constructions would give an equality here. (The equality does not hold in general. Take a subfactor $P \subset Q$ with $P' \cap Q = \mathbb{C}$, and then set $S = P \otimes \mathbb{C} \subset N = P \otimes P \subset M = Q \otimes P$. Of course, such a construction is not interesting for us.)

First we note the following proposition.

Proposition 14.12 *Let $N \subset M$ be a subfactor with finite index and S be a subfactor of N such that $S' \cap N$ and $S' \cap M$ are type II_1 factors with $[S' \cap M : S' \cap N] = [M : N]$. Then it is possible to choose a tunnel*

$$\cdots \subset N_2 \subset N_1 \subset N \subset M$$

so that $N_j \supset S$ and

$$\cdots \subset S' \cap N_2 \subset S' \cap N_1 \subset S' \cap N \subset S' \cap M$$

is a tunnel for $S' \cap N \subset S' \cap M$. We have $N_j' \cap N \subset (S' \cap N_j)' \cap (S' \cap N)$ and $N_j' \cap M \subset (S' \cap N_j)' \cap (S' \cap M)$.

Proof First choose a projection $e_0 \in S' \cap M$ with $E_{S' \cap N}(e_0) = [S' \cap M : S' \cap N]^{-1} = [M : N]^{-1}$. By the commuting square condition, $E_N(e_0) = [M : N]^{-1}$, so $N_1 = N \cap \{e_0\}'$ gives a downward basic construction $N_1 \subset N \subset M$. Because $e_0 \in S' \cap M$, we get $N_1 \supset S$. Because $S' \cap N_1 = (S' \cap N) \cap \{e_0\}'$, we also know that $S' \cap N_1 \subset S' \cap N \subset S' \cap M$ is a downward basic construction. By repeating this argument, we get the desired tunnel. The last two inclusions are then trivial. □

We call a construction of $S' \cap N \subset S' \cap M$ from $N \subset M$ as above the *coset construction*. The above proposition shows that the coset construction does not decrease the higher relative commutants.

The next lemma gives a basis for the coset construction in our setting.

Lemma 14.13 *Suppose that the following square of hyperfinite II_1 factors satisfies the conditions of Definition 14.1 and that $M_{00} \subset M_{11}$ has finite depth:*

$$\begin{array}{ccc} M_{00} & \subset & M_{01} \\ \cap & & \cap \\ M_{10} & \subset & M_{11} \end{array}$$

We form a double sequence $\{M_{kl}\}_{k,l\geq 0}$ of type II_1 factors from the basic construction and further assume the condition $M'_{00}\cap M_{0l}\subset M'_{10}\cap M_{1l}$ for $l\geq 0$. We define $M_{\infty,l}$ as the GNS-completion of $\bigvee_k M_{kl}$ with respect to the trace. Then

$$M_{\infty,0}\subset M_{\infty,1}\subset M_{\infty,2}\subset M_{\infty,3}\subset\cdots$$

is the Jones tower, and we get two equalities $M'_{\infty,0}\cap M_{\infty,l}=M'_{00}\cap M_{0l}$ and $M'_{\infty,1}\cap M_{\infty,l}=M'_{01}\cap M_{0l}$ for $l\geq 1$, which imply that the subfactors $M_{00}\subset M_{01}$ and $M_{\infty,0}\subset M_{\infty,1}$ are isomorphic by Theorem 11.14.

Proof The Jones projections for the towers

$$M_{k0}\subset M_{k1}\subset M_{k2}\subset M_{k3}\subset\cdots$$

are common for all k. From this fact and the commuting square condition, it is easy to see that

$$M_{\infty,0}\subset M_{\infty,1}\subset M_{\infty,2}\subset M_{\infty,3}\subset\cdots$$

is the Jones tower. Any element in $M'_{00}\cap M_{0l}$ commutes with any element in M_{k0} by $M'_{00}\cap M_{0l}\subset M'_{10}\cap M_{1l}$, so we get the inclusion $M'_{00}\cap M_{0l}\subset M'_{\infty,0}\cap M_{\infty,l}$.

Take any $x\in M'_{\infty,0}\cap M_{\infty,l}$. Set $x_k=E_{M_{kl}}(x)\in M'_{k0}\cap M_{kl}$. The inclusion $M'_{00}\cap M_{0l}\subset M'_{2k,0}\cap M_{2k,l}$ implies $M'_{00}\cap M_{0l}=M'_{2k,0}\cap M_{2k,l}$ because these two higher relative commutants are isomorphic and finite dimensional. So $x_{2k}\in M'_{2k,0}\cap M_{2k,l}=M'_{00}\cap M_{0l}$ implies $x_0=E_{M_{0l}}(x_{2k})=x_{2k}$, and then $x=\lim_k x_{2k}=x_0\in M'_{00}\cap M_{0l}$, which is the converse inclusion.

Next $M'_{01}\cap M_{0l}\subset M'_{00}\cap M_{0l}$ implies $M'_{01}\cap M_{0l}\subset M'_{01}\cap M'_{10}\cap M_{1l}$, which in turn implies $M'_{01}\cap M_{0l}\subset M'_{11}\cap M_{1l}$ by Proposition 14.7. Then the same argument as above shows $M'_{\infty,1}\cap M'_{\infty,l}=M'_{01}\cap M'_{0l}$. □

Lemma 14.14 *Suppose that the following square of hyperfinite II_1 factors satisfies the condition of Definition 14.1 and that $M_{00}\subset M_{11}$ has finite depth.*

$$\begin{array}{ccc} M_{00} & \subset & M_{01}\\ \cap & & \cap\\ M_{10} & \subset & M_{11}\end{array}$$

We form a double sequence $\{M_{kl}\}_{k,l\geq 0}$ of type II_1 factors and further assume the condition $M'_{00}\cap M_{k0}=M'_{01}\cap M_{k1}$ for $k\geq 0$. Set $S=M'_{10}\cap M_{\infty,0}=\bigvee_{k\geq 1}(M'_{10}\cap M_{k0})$. Then $S'\cap M_{\infty,0}=M_{10}$ and $S'\cap M_{\infty,1}=M_{11}$.

Proof Because the subfactor $M_{10}\subset M_{20}$ has finite depth, it is given with the string algebra construction as in Section 11.3 from a paragroup. Then Theorem 11.20 implies $(M'_{10}\cap M_{\infty,0})'\cap M_{\infty,0}=M_{10}$, which means $S'\cap M_{\infty,0}=M_{10}$.

If $x\in M'_{10}\cap M_{k0}$, then $x\in M'_{00}\cap M_{k0}=M'_{01}\cap M_{k1}$, so $x\in M'_{10}\cap M'_{01}\cap M_{k1}=M'_{11}\cap M_{k1}$. If $x\in M'_{11}\cap M_{k1}$, then $x\in M'_{01}\cap M_{k1}=M'_{00}\cap M_{k0}$, so $x\in M'_{10}\cap M_{k0}$. Thus we have proved $M'_{10}\cap M_{k0}=M'_{11}\cap M_{k1}$, and $S=M'_{11}\cap M_{\infty,1}=\bigvee_{k\geq 1}(M'_{11}\cap M_{k1})$. Thus we also get $S'\cap M_{\infty,1}=M_{11}$. □

The above two lemmas immediately show the following theorem which gives a relation between the coset construction and a bi-commuting square of II_1 factors.

Theorem 14.15 *Suppose that the following square satisfies the conditions of Definition 14.1:*

$$\begin{array}{ccc} M_{00} & \subset & M_{01} \\ \cap & & \cap \\ M_{10} & \subset & M_{11} \end{array}$$

We form a double sequence $\{M_{kl}\}_{k,l\geq 0}$ of type II_1 factors as in Theorem 14.11 and further assume the following two conditions:

1. $M'_{00} \cap M_{0l} \subset M'_{10} \cap M_{1l}$ *for* $l \geq 0$;
2. $M'_{00} \cap M_{k0} = M'_{01} \cap M_{k1}$ *for* $k \geq 0$.

Then there exists a subfactor S of M_{00} such that $S' \cap M_{00} \subset S' \cap M_{01}$ is isomorphic to $M_{10} \subset M_{11}$.

We have the following example of Theorem 14.15.

Example 14.16 Form a triple sequence of the string algebras $\{A_{jkl}\}$ with the connection used in the construction of the Goodman–de la Harpe–Jones subfactors as in Section 11.6. (See the last paragraph in Section 11.9.) Because the subfactor $A_{00} \subset A_{01}$ has the Dynkin diagram A_{11} as the principal graph, Condition 1 of Theorem 14.15 is automatically satisfied. We can construct $\{A_{-1,k,l}\}$ for $k \geq 0$, $l \geq 1$ as in Section 11.4, and then the flatness of the Jones projection and Ocneanu's compactness argument imply $M'_{00} \cap M_{k0} = A_{0k0} = A_{-1,k,1} = M'_{01} \cap M_{k1}$, which is Condition 2 of Theorem 14.15. Thus we know that both subfactors of type E_6 can be constructed from the subfactor of type A_{11} with the coset construction.

Similarly, both subfactors of type E_8 can be constructed from the subfactor of type A_{29} with the coset construction. (We can make a similar construction for A_{17}, but then the resulting subfactor is of type D_{10}, not E_7.)

We have another example of the coset construction.

Example 14.17 As in the orbifold construction in Section 13.4, we have an action α of a finite group Z on a subfactor $N \subset M$ obtained from a WZW-model. Form the following commuting square.

$$\begin{array}{ccc} M_{00} = N & \subset & M_{01} = M \\ \cap & & \cap \\ M_{10} = N \rtimes_\alpha Z & \subset & M_{11} = M \rtimes_\alpha Z \end{array}$$

We know that Condition 1 of Theorem 14.15 is satisfied, as we will see in Section 15.6. Condition 2 is also satisfied because both vertical inclusions are crossed products by the same group. Thus the subfactor $N \rtimes_\alpha Z \subset M \rtimes_\alpha Z$, which is conjugate to the orbifold subfactor $N^\alpha \subset M^\alpha$, is obtained from $N \subset M$ by the coset construction.

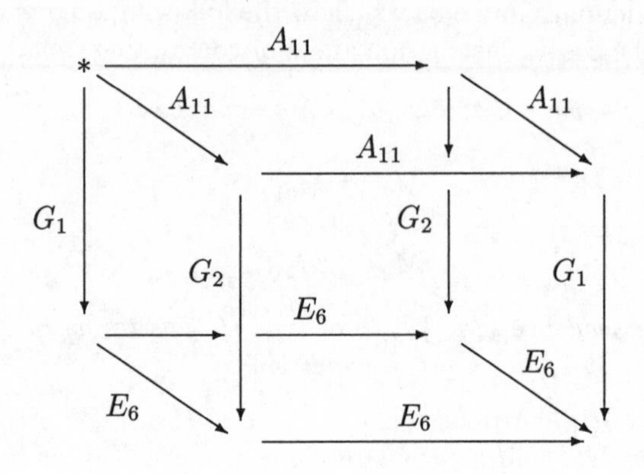

FIG. 14.9. Nested graphs

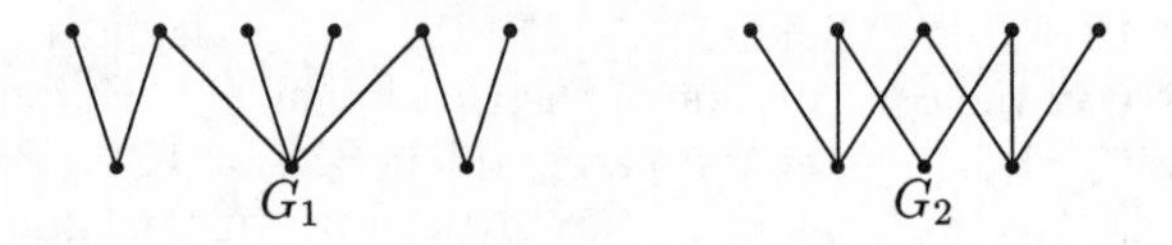

FIG. 14.10. Two graphs intertwining A_{11} and E_6

Thus our orbifold construction in Section 13.4 can also be regarded as a special case of Theorem 14.15.

14.6 The Goodman–de la Harpe–Jones subfactors, fusion algebras, and $\mathbf{TQFT_3}$

In this section, we study Goodman–de la Harpe–Jones subfactors in Section 11.6 based on commuting squares of II_1 factors in Example 14.16. The Goodman–de la Harpe–Jones subfactor with index $3+\sqrt{3}$ has received special attention, because it is one of the few known subfactors with non-integer indices which do not come from Wess–Zumino–Witten models directly and has relatively small index. Again by the last paragraph in Section 11.9, we can make a triple sequence of the string algebras from this system of connections.

In the case of E_6 and A_{11}, our graph, which gives the Bratteli diagrams of the commuting cube, looks like Fig. 14.9. The graphs G_1, G_2 are as in Fig. 14.10. Note that for the top face of the cube in Fig. 14.9, we use the standard connection for A_{11}, for the bottom face of the cube, we use the standard connection for E_6, and for the four side faces, we use the connection appearing in the Goodman–de la Harpe–Jones commuting square in Example 11.25. The explicit formula for the last connection is given in the table on page 418 of (Roche 1990). (Roche verified the intertwining Yang–Baxter equation by direct computation, but it is a consequence of the flatness of the Jones projections as noted above.) From the triple sequence of the string algebras, we get a commuting square

$$\begin{array}{ccc} M_{00} & \subset & M_{01} \\ \cap & & \cap \\ M_{10} & \subset & M_{11} \end{array} \tag{14.6.1}$$

satisfying the conditions in Definition 14.1. Here the subfactor $M_{00} \subset M_{01}$ is of type A_{11}, $M_{10} \subset M_{11}$ is of type E_6, and $M_{00} \subset M_{10}$ and $M_{01} \subset M_{11}$ are the Goodman–de la Harpe–Jones subfactor with index $3 + \sqrt{3}$. (There are two flat connections on E_6, so we have two commuting squares as in (14.6.1) corresponding to the two connections.) The flatness of the Jones projection implies flatness with respect to $*$ in an obvious sense, which is shared by A_{11}, E_6 and G_1, so we conclude that the original connection arising from the Goodman–de la Harpe–Jones commuting square is 'canonical' in the sense it appears as the 'higher relative commutants' $M'_{00} \cap M_{kl}$. The above graph G_2 of Fig. 14.10 cannot be a principal graph of any subfactor because it is rejected by the '$2\cos\pi/n$-rule' (Corollary 10.8), but it does appear as a part of the paragroup type invariant of a commuting square of II$_1$ factors mentioned in the last paragraph of Section 14.3. Similar results hold for the case E_7 and E_8, for which (Di Francesco and Zuber 1990a) make several computations.

We can compute the fusion rule of the N-N bimodules of the Goodman–de la Harpe–Jones subfactor $N \subset M$ with index $3 + \sqrt{3}$ with this observation.

First we establish a general theory. Suppose that we have a bi-commuting square of II$_1$ factors as in Definition 14.1 and that $M_{00} \subset M_{11}$ has finite depth. We construct a triple sequence (X_{jkl}) of bimodules inductively as follows. Let $X_{000} = {}_{00}(M_{00})_{00}$. If k, l are even, then set $X_{j,k,l+1} = X_{jkl} \otimes_{00} (M_{01})_{01}$ and $X_{j,k+1,l} = X_{jkl} \otimes_{00} (M_{10})_{10}$. If k is even and l is odd, then set $X_{j,k,l+1} = X_{jkl} \otimes_{01} (M_{01})_{00}$ and $X_{j,k+1,l} = X_{jkl} \otimes_{01} (M_{11})_{11}$. If k is odd and l is even, then set $X_{j,k,l+1} = X_{jkl} \otimes_{10} (M_{11})_{11}$ and $X_{j,k+1,l} = X_{jkl} \otimes_{10} (M_{10})_{00}$. If k, l are odd, then set $X_{j,k,l+1} = X_{jkl} \otimes_{11} (M_{11})_{10}$ and $X_{j,k+1,l} = X_{jkl} \otimes_{11} (M_{11})_{01}$. If j is even, then set $X_{j+1,k,l} = {}_{11}(M_{11}) \otimes_{00} X_{jkl}$, and if j is odd, then set $X_{j+1,k,l} = {}_{00}(M_{11}) \otimes_{11} X_{jkl}$. With natural isomorphisms

$$\begin{aligned}
{}_{00}(M_{01} \otimes_{01} M_{11})_{11} &\cong {}_{00}(M_{11})_{11} \cong {}_{00}(M_{10} \otimes_{10} M_{11})_{11}, \\
{}_{01}(M_{01} \otimes_{00} M_{10})_{10} &\cong {}_{01}(M_{11})_{10} \cong {}_{01}(M_{11} \otimes_{11} M_{11})_{10}, \\
{}_{10}(M_{10} \otimes_{00} M_{01})_{01} &\cong {}_{10}(M_{11})_{01} \cong {}_{10}(M_{11} \otimes_{11} M_{11})_{01}, \\
{}_{11}(M_{11} \otimes_{10} M_{10})_{00} &\cong {}_{11}(M_{11})_{00} \cong {}_{11}(M_{11} \otimes_{01} M_{01})_{00},
\end{aligned}$$

we know that this construction is compatible. Thus X_{jkl} is an $M_{[j][j]}$-$M_{[k][l]}$ bimodule, where $[j]$ denotes 0 [resp. 1] when j is even [resp. odd].

We set $B_{jkl} = \mathrm{End}(X_{jkl})$, where $\mathrm{End}(X_{jkl})$ means the set of bounded linear maps on the Hilbert space X_{jkl} which commute with the left and right actions of the type II$_1$ factors. Note that this is finite dimensional. If k, l are even, we regard B_{jkl} as a subalgebra of $B_{j,k,l+1}$ with the embedding $\xi \in \mathrm{End}(X_{jkl}) \mapsto \xi \otimes_{00}$

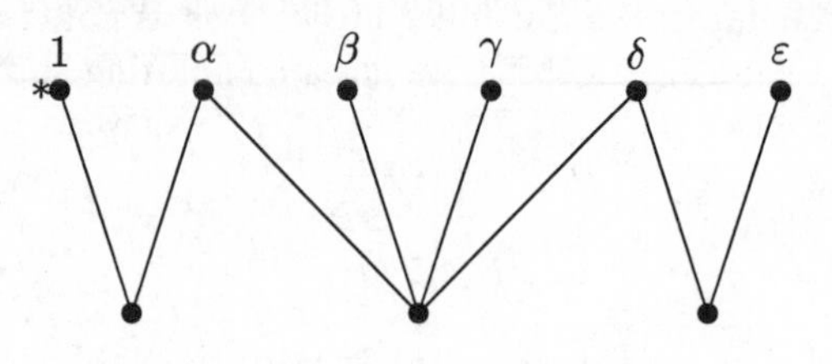

FIG. 14.11. Labelling of vertices

$\mathrm{id}_{00}(M_{01})_{01} \in \mathrm{End} X_{j,k,l+1}$ and define the other embeddings similarly. Because the embeddings are compatible, we get a triple increasing sequence of finite dimensional algebras B_{jkl}. By looking at the isomorphism between each B_{jkl} and each $M'_{-j,-j} \cap M_{kl}$ and the embeddings of the both triple sequences, we can conclude that the sequence (B_{jkl}) and the sequence of the higher relative commutants $(M'_{-j,-j} \cap M_{kl})$ are isomorphic, as in Section 10.2. (Here $M_{-j,-j}$ is a tunnel for $M_{00} \subset M_{11}$.)

We take irreducible decompositions of the bimodules $X_{2j,2k,2l}$ to get a set of (isomorphism classes of) all the M_{00}-M_{00} bimodules arising in this way. Set this to be the vertex set ${}_{00}\mathcal{G}_{00}$. Similarly, we take vertex sets ${}_{00}\mathcal{G}_{01}$, ${}_{00}\mathcal{G}_{10}$, ${}_{00}\mathcal{G}_{11}$, ${}_{11}\mathcal{G}_{00}$, ${}_{11}\mathcal{G}_{01}$, ${}_{11}\mathcal{G}_{10}$, and ${}_{11}\mathcal{G}_{11}$, from the irreducible decompositions of the bimodules $X_{2j,2k,2l+1}$, $X_{2j,2k+1,2l}$, $X_{2j,2k+1,2l+1}$, $X_{2j+1,2k,2l}$, $X_{2j+1,2k,2l+1}$, $X_{2j+1,2k+1,2l}$, and $X_{2j+1,2k+1,2l+1}$ respectively. For an M_{00}-M_{00} bimodule $X \in {}_{00}\mathcal{G}_{00}$ and an M_{11}-M_{00} bimodule $Y \in {}_{11}\mathcal{G}_{00}$, we put edges in ${}^{00}_{11}\mathcal{G}^{00}_{00}$ so that their number is equal to the multiplicity of Y in ${}_{11}(M_{11} \otimes_{00} X)_{00}$. Note that by Frobenius reciprocity, Theorem 9.71, the number of edges is also equal to the multiplicity of X in ${}_{00}(M_{11} \otimes_{11} Y)_{00}$. We similarly construct edges for the other parts of $\mathcal{G}$. Note that ${}^{00}_{11}\mathcal{G}^{00}_{00}$ and ${}^{00}_{11}\mathcal{G}^{11}_{11}$ are connected because these are the principal graph and the dual principal graph of the subfactor $M_{00} \subset M_{11}$. Because the subfactor $M_{00} \subset M_{11}$ is of finite depth, our graph $\mathcal{G}$ is finite. We set $* = {}_{00}({}_{00}M_{00})_{00}$. For a bimodule ${}_{ij}X_{kl}$ in the vertex set of $\mathcal{G}$, we define the function μ by $\mu({}_{ij}X_{kl}) = \sqrt{\dim {}_{ij}X \dim X_{kl}}$, where $i, j, k, l = 0, 1$. Because we have the finite depth condition, we get the harmonic axiom, as in Section 10.3. In this way, we get the Bratteli diagrams for $M'_{00} \cap M_{kl}$ in terms of bimodules.

In the above graph in Fig. 14.9, the even vertices of the A_{11} and those of G_1 are identified, and this identification means that these two systems of bimodules are the same. That is, the fusion rule of the N-N bimodules for the GHJ subfactor is the same as that of the subfactor of type A_{11}. With the labelling of even vertices as in Fig. 14.11, we get the multiplication table as in Table 14.2.

Our commuting cube, however, does not satisfy the counterpart of the initialization axiom, so we cannot compute the dual higher relative commutants $M'_{11} \cap M_{kl}$ directly. We are in the same situation as in Section 11.6, where the general method did not give the dual principal graphs. Our next aim is to compute the Bratteli diagram for this series $M'_{11} \cap M_{kl}$.

Table 14.2 *Multiplication of the N-N bimodules*

$\times$	α	β	γ	δ	ε
α	$1+\alpha+\beta+\gamma+\delta$	$\alpha+\gamma+\delta$	$\alpha+\beta+\delta$	$\alpha+\beta+\gamma+\delta+\varepsilon$	δ
β	$\alpha+\gamma+\delta$	$1+\beta+\delta$	$\alpha+\gamma+\varepsilon$	$\alpha+\beta+\delta$	γ
γ	$\alpha+\beta+\delta$	$\alpha+\gamma+\varepsilon$	$1+\beta+\delta$	$\alpha+\gamma+\delta$	β
δ	$\alpha+\beta+\gamma+\delta+\varepsilon$	$\alpha+\beta+\delta$	$\alpha+\gamma+\delta$	$1+\alpha+\beta+\gamma+\delta$	α
ε	δ	γ	β	α	1

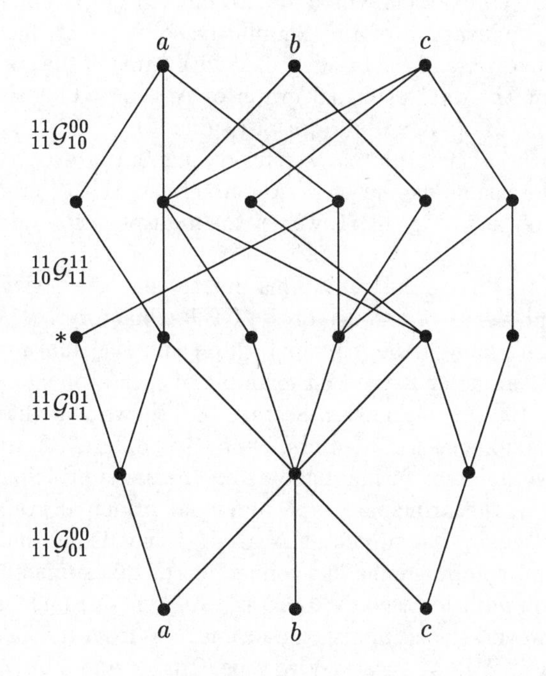

FIG. 14.12. The Bratteli diagrams for the dual principal connection

We look for the graphs $^{11}_{11}\mathcal{G}^{00}_{10}$, $^{11}_{10}\mathcal{G}^{11}_{11}$, $^{11}_{11}\mathcal{G}^{01}_{11}$, and $^{11}_{11}\mathcal{G}^{00}_{01}$. The above observation shows that a connected component of the graph $^{11}_{10}\mathcal{G}^{11}_{11}$ is E_6, and a connected component of the graph $^{11}_{11}\mathcal{G}^{01}_{11}$ is G_1. Because the number of vertices in $_{00}\mathcal{G}_{11}$ is three, the number of vertices in $_{11}\mathcal{G}_{00}$ is also three and two of them have the same weight $\mu(\cdot)$ by the contragredient map. Any connected component of the graph $^{11}_{11}\mathcal{G}^{00}_{01}$ has the Perron–Frobenius eigenvalue $2\cos(\pi/12)$, so each connected component must be one of A_{11}, D_7, and E_6. The above observation on $_{11}\mathcal{G}_{00}$ shows that $^{11}_{11}\mathcal{G}^{00}_{01}$ is connected and equal to E_6, and it implies that $^{11}_{11}\mathcal{G}^{01}_{11}$ is connected and equal to G_1. It then implies that $^{11}_{10}\mathcal{G}^{11}_{11}$ has two connected components and the other is another copy of E_6, and this uniquely determines $^{11}_{11}\mathcal{G}^{00}_{10}$. The graphs obtained are as in Figure 14.12.

Table 14.3 *Multiplication of the M-M bimodules*

$\times$	α	β	γ	δ	ε
α	$1+\alpha+\beta+\gamma+\delta$	$\alpha+\gamma+\delta$	$\alpha+\beta+\delta$	$\alpha+\beta+\gamma+\delta+\varepsilon$	δ
β	$\alpha+\gamma+\delta$	$1+2\beta+\varepsilon$	$\alpha+\delta$	$\alpha+\gamma+\delta$	β
γ	$\alpha+\beta+\delta$	$\alpha+\delta$	$1+2\gamma+\varepsilon$	$\alpha+\beta+\delta$	γ
δ	$\alpha+\beta+\gamma+\delta+\varepsilon$	$\alpha+\gamma+\delta$	$\alpha+\beta+\delta$	$1+\alpha+\beta+\gamma+\delta$	α
ε	δ	β	γ	α	1

This is a non-trivial example where a disconnected graph appears as a part of the paragroup type invariant of a bi-commuting square of II_1 factors. It shows that the three even vertices of E_6 are identified with three of the six even vertices of $\mathcal{G}_1$, which is now the dual principal graph of $N \subset M$. This shows that the fusion algebra of the M-M bimodules for our subfactor $N \subset M$ contains a sub-fusion algebra which is isomorphic to the fusion algebra of even vertices of E_6. This determines the fusion algebra of M-M bimodules. With the same labelling of even vertices of G_1 as in Fig. 14.11, we get the multiplication table as in Table 14.3.

Tables 14.2 and 14.3 are different. This means that the fusion algebras for N-N bimodules and M-M bimodules of the GHJ subfactor $N \subset M$ with index $3+\sqrt{3}$ are different, though the principal graph and the dual principal graph are the same. This subfactor is the first example of such a phenomenon. Recent work of (Haagerup 1994) mentioned in Section 11.5 shows that this example has the smallest index among such subfactors. (Note that different finite groups with the same order give different fusion algebras on the same principal graphs, but these two cannot be the principal graph and dual principal graph of a single subfactor.) In particular, the subfactor $N \subset M$ is not isomorphic to its dual $M \subset M_1$. (M. Izumi computed the flat connection of this subfactor, and it also follows from his computation that $N \subset M$ and $M \subset M_1$ are not isomorphic. It seems difficult, however, to distinguish the fusion rules from the flat connection.)

Furthermore, a TQFT_3 of Turaev–Viro type (Turaev and Viro 1992) based on triangulations arising from subfactors is computed with $6j$-symbols of only N-N bimodules (or M-M bimodules) as noted in Section 12.4. So the TQFT_3 for the GHJ subfactor with index $3+\sqrt{3}$ is the same as that for the subfactor of type A_{11}. Similar results hold for the GHJ subfactors arising from the Dynkin diagrams E_7 and E_8. The TQFT_3s arising from subfactors of type E_6 use the sub-fusion algebras of the fusion algebra of the M-M bimodules of the GHJ subfactor with index $3+\sqrt{3}$ which gives the same TQFT_3 as the N-N bimodules of the same subfactor and as the A_{11} TQFT_3. In this sense, the E_6 TQFT_3s use only partial information of the A_{11} TQFT_3, but it does not necessarily mean that the E_6 TQFT_3s are less interesting. Also note that there are two subfactors of type E_6, and they correspond to two vertices β, γ of the dual principal graph of the GHJ subfactor in Fig. 14.11.

If we make the same construction for the D_n diagrams, we again have a graph D_n as a part of the paragroup type invariant of a bi-commuting square of II_1

factors. The graph D_{2n+1} is impossible as a principal graph, but it does appear as a part of this paragroup type invariant.

14.7 Notes

The notion of bi-commuting squares in Section 14.2 was introduced by (Sano and Watatani 1994) with several equivalent definitions under the name 'commuting and co-commuting square'. A notion of a non-degenerate commuting square in (Popa 1994a) can be regarded as its generalization. Proposition 14.9 was first proved by (Bisch 1994a). Proposition 14.10 was first proved by (Wierzbicki 1994).

The results in Sections 14.3–14.6 are due to (Kawahigashi 1995c). The intertwining Yang–Baxter equation as a compatibility condition of identifications first appeared in (Kawahigashi 1995a). It was shown in (Bisch 1994b) that purely algebraic computations from the principal graph of the GHJ subfactor with index $3 + \sqrt{3}$ are not enough to determine its fusion algebra, since they lead to several distinct fusion algebras, some of which are different from our Tables 14.2 and 14.3. Our method, given in (Kawahigashi 1995c), allows us to determine the right fusion rules of the subfactor. This result has recently been generalized to the other GHJ subfactors by A. Ocneanu.

15

AUTOMORPHISMS OF SUBFACTORS AND CENTRAL SEQUENCES

15.1 Introduction

As seen in Section 13.4, the orbifold construction gives new series of interesting subfactors as simultaneous fixed point algebras of finite group actions on the original subfactors. This construction suggests that automorphisms appearing in the orbifold construction have interesting operator algebraic properties. The aim of this chapter is to show that this is indeed the case. This chapter is mainly for operator algebraists.

After remarkable work by (Connes 1975a), (Connes 1977), there has been extensive work on the automorphisms of operator algebras, and many results on the complete classification of automorphisms and group actions on hyperfinite factors have been obtained as in (Jones 1980b), (Ocneanu 1985), (Jones and Takesaki 1984), (Sutherland and Takesaki 1989), (Kawahigashi *et al.* 1992), (Kawahigashi and Takesaki 1992). In this work, the roles of two classes of automorphisms have been emphasized. One is the class of approximately inner automorphisms and the other is that of the centrally trivial automorphisms. We will study the subfactor versions of these automorphisms.

In Section 15.2, we introduce the Loi invariant for automorphisms of subfactors and explain its important properties, based on work of S. Popa. We introduce strongly outer automorphisms of subfactors in Section 15.3 and give their characterization due to Choda–Kosaki in terms of bimodules. In Section 15.4, we define centrally trivial automorphisms of subfactors and explain their characterization due to S. Popa for strongly amenable subfactors. In this section, we also introduce ultraproduct algebras. The treatment is not logically self-contained. Construction of the central sequence subfactors is introduced in Section 15.5 and we will show that it essentially gives the same paragroup as the asymptotic inclusion introduced in Section 12.6. In Section 15.6, we characterize automorphisms used in the orbifold construction of Section 13.4. We introduce a subfactor analogue of the Connes invariant χ in Section 15.7 and examples related to the orbifold construction. In Section 15.8, we introduce a subfactor analogue of the

Jones invariant κ as an additional invariant to the relative χ of Section 15.7. We compute the invariant in connection to the orbifold construction.

Exercise 15.1 Let M be a II_1 factor and α an automorphism of M. We denote the following subfactor by $M(\alpha)$:

$$\left\{ \begin{pmatrix} x & 0 & 0 \\ 0 & \alpha(x) & 0 \\ 0 & 0 & \alpha(x) \end{pmatrix} \middle| \; x \in M \right\} \subset \left\{ \begin{pmatrix} x_{11} & x_{12} & x_{13} \\ x_{21} & x_{22} & x_{23} \\ x_{31} & x_{32} & x_{33} \end{pmatrix} \middle| \; x_{ij} \in M \right\}.$$

If two subfactors $M(\alpha)$ and $M(\beta)$ are isomorphic, what can we say about α and β?

Exercise 15.2 Let $N \subset M$ be a subfactor with finite index. For $\alpha \in \mathrm{Aut}(M)$ with $\alpha(N) = N$, we denote the following commuting squares of II_1 factors by $C(\alpha)$:

$$\begin{array}{ccc} \left\{ \begin{pmatrix} x & 0 & 0 \\ 0 & \alpha(x) & 0 \\ 0 & 0 & \alpha(x) \end{pmatrix} \middle| \; x \in N \right\} & \subset & \left\{ \begin{pmatrix} x_{11} & x_{12} & x_{13} \\ x_{21} & x_{22} & x_{23} \\ x_{31} & x_{32} & x_{33} \end{pmatrix} \middle| \; x_{ij} \in N \right\}, \\ \cap & & \cap \\ \left\{ \begin{pmatrix} x & 0 & 0 \\ 0 & \alpha(x) & 0 \\ 0 & 0 & \alpha(x) \end{pmatrix} \middle| \; x \in M \right\} & \subset & \left\{ \begin{pmatrix} x_{11} & x_{12} & x_{13} \\ x_{21} & x_{22} & x_{23} \\ x_{31} & x_{32} & x_{33} \end{pmatrix} \middle| \; x_{ij} \in M \right\}. \end{array}$$

If two commuting squares $C(\alpha)$ and $C(\beta)$ are isomorphic, what can we say about α and β?

15.2 The Loi invariant and approximately inner automorphisms of subfactors

We first list some basic definitions and notation.

Definition 15.1 *For a subfactor $N \subset M$, we set*

$$\mathrm{Aut}(M, N) = \{\alpha \in \mathrm{Aut}(M) \mid \alpha(N) = N\},$$
$$\mathrm{Int}(M, N) = \{\mathrm{Ad}(u) \in \mathrm{Aut}(M, N) \mid u \in \mathcal{U}(N)\}.$$

We say that an element in $\mathrm{Aut}(M, N)$ *is an* automorphism of a subfactor $N \subset M$. *We denote the closure of* $\mathrm{Int}(M, N)$ *in* $\mathrm{Aut}(M, N)$ *by* $\overline{\mathrm{Int}}(M, N)$. *An automorphism in* $\overline{\mathrm{Int}}(M, N)$ *is called* approximately inner.

Note that the group $\mathrm{Aut}(M, N)$ has the standard topology where we define $\alpha_i \to \alpha$ if $\|\phi \cdot \alpha_i - \phi \cdot \alpha\| \to 0$ for all $\phi \in M_*$. Our aim in this section is to introduce the Loi invariant for automorphisms of subfactors and characterize its triviality as approximate innerness for strongly amenable subfactors.

Definition 15.2 *Let $N \subset M$ be a subfactor of type II_1 with finite index and $\alpha \in \mathrm{Aut}(M, N)$. Because α automatically commutes with E_N, we can extend the* *automorphism to M_n by setting $\alpha(e_n) = e_n$. We use the same symbol α for this extension. We then restrict this α to the higher relative commutants $\bigcup_n N' \cap M_n$. (Note that $\alpha(N' \cap M_n) = N' \cap M_n$ from the definition.) This restriction is called the* Loi invariant *of the automorphism α. We denote by Φ the map assigning the Loi invariant to an automorphism $\alpha \in$* Aut(M, N).

We write α_k for the restriction of α on $N' \cap M_k$. Then the family $\{\alpha_k\}_k$ satisfies the conditions of the following Definition.

Definition 15.3 *We denote by $\mathcal{G}$ the set of families of automorphisms α_k on $N' \cap M_k$ satisfying the following conditions.*

1. *Each α_k preserves the trace.*
2. *Each α_k preserves the inclusion $M' \cap M_k \subset N' \cap M_k$.*
3. *Each α_{k+1} is an extension of α_k.*
4. *Each α_k satisfies $\alpha_k(e_j) = e_j$ for $j = 1, 2, \ldots, k$.*

It is clear that the set $\mathcal{G}$ has a natural topological group structure. The following proposition is trivial from the definition of Φ.

Proposition 15.4 *The map Φ in Definition 15.2 gives a group homomorphism from* Aut$(M, N)/$Int(M, N) *to $\mathcal{G}$ and it is continuous as a map from* Aut(M, N) *to $\mathcal{G}$.*

We next study the case where a subfactor $N \subset M$ of a hyperfinite II_1 factor is extremal and strongly amenable as in Theorem 11.14. (For example, if $N \subset M$ arises from a paragroup as in Section 11.3, this assumption holds.) In this case, by a comment after Definition 11.11, the family of maps $\{\Phi_k\}_k$ in Proposition 9.50 is trace-preserving. (Note that this property holds trivially if we have the finite depth condition, because we have a unique trace on string algebras arising from a finite graph.) In this case, we represent all the M_k on $L^2(M)$ as in the proof of Proposition 9.50 by setting $M_{k+1} = J N_k' J$ and $e_{2-k} = J e_k J$ for $k = 1, 2, \ldots$, where J means J_M on $L^2(M)$.

Theorem 15.5 *Let $N \subset M$ be an extremal and strongly amenable subfactor of a hyperfinite II_1 factor. Then an automorphism $\alpha \in$* Aut(M, N) *is approximately inner if and only if $\Phi(\alpha)$ is trivial.*

Proof By Proposition 15.4, it is immediate that $\Phi(\alpha)$ is trivial if α is approximately inner. (Here we do not need the strong amenability assumption.)

Now we assume that $\Phi(\alpha)$ is trivial. We may and do assume that the tunnel $\{N_k\}_k$ has a generating property by Theorem 11.14.

Using Proposition 9.44 (1), we get a unitary $u_0 \in N$ with $\alpha(e_0) = u_0 e_0 u_0^*$. Set $\alpha_0 = (\mathrm{Ad} u_0^*)\alpha$. We define a unitary operator U_0 on $L^2(M)$ by $U_0(\hat{x}) = \widehat{\alpha_0(x)}$. It is easy to see that we have $\alpha_0 = \mathrm{Ad}(U_0)$ on M. Recall that the factor M_2 is generated by M and the appropriate Jones projection f corresponding to E_{N_1} by Theorem 9.49. Since α_0 commutes with E_{N_1}, we get $\mathrm{Ad}(U_0)(f) = f$. This implies that the extension of α_0 to M_2 as above also coincides with $\mathrm{Ad}(U_0)$.

The assumption of the triviality of $\Phi(\alpha)$ implies that α_0 is also trivial on $M' \cap M_2$. By $J_M U_0 = U_0 J_M$ and $J_M(N_1' \cap M)J_M = M' \cap M_2$, we also know that α_0 is trivial on $N_1' \cap M$, which implies $\alpha = \mathrm{Ad}(u_0)$ on $N_1' \cap M$.

We next choose a unitary $u_1 \in N_1$ with $\alpha_0(e_{-1}) = u_1 e_{-1} u_1^*$. Then similarly we show that $\alpha = \mathrm{Ad}(u_0 u_1)$ on $N_2' \cap M$. By repeating this procedure, we get $\alpha = \mathrm{Ad}(u_0 u_1 \cdots u_k)$ on $N_{k+1}' \cap M$ for unitaries $u_j \in N_j$. This clearly shows $\alpha \in \overline{\mathrm{Int}}(M, N)$ by the generating property, because we have $\alpha(x) = \lim_{k\to\infty} \mathrm{Ad}(u_0 u_1 \cdots u_k)(x)$ for all x in an L^2-dense subset of M. □

15.3 Strongly outer automorphisms of subfactors

In this section, we will introduce another important class of automorphisms. We make the following definition. (Compare this with Lemma 5.27.)

Definition 15.6 *Let $N \subset M$ be a subfactor with finite index and $\alpha \in \mathrm{Aut}(M, N)$. As in Definition 15.2, we extend α to the Jones tower. We say that α is* strongly outer *if for any n there does not exist a non-zero $a \in M_n$ such that $\alpha(x)a = ax$ for all $x \in M$.*

Our aim is to characterize strong-outerness in terms of bimodules. We first introduce the following notation. Let $X = {}_M X_N$ be an M-N bimodule and α, β be automorphisms of M, N respectively. Then we define M-N bimodules ${}_\alpha X = {}_M({}_\alpha X)_N$ and $X_\beta = {}_M(X_\beta)_N$ by

$$x \cdot \xi \cdot y = \alpha(x)\xi y, \qquad \text{for } \xi \in {}_\alpha X, x \in M, y \in N,$$
$$x \cdot \xi \cdot y = x\xi\beta(y), \qquad \text{for } \xi \in X_\beta, x \in M, y \in N,$$

respectively. We use the same notations for the other three kinds of bimodules in an obvious way. If a bimodule X contains a copy of a bimodule Y as a direct summand, we write $X \succ Y$. The following theorem characterizes strong outerness.

Theorem 15.7 *For $\alpha \in \mathrm{Aut}(M, N)$, the following are equivalent.*

1. *There exists a non-zero element $a \in M_n$ such that $ax = \alpha(x)a$ for all $x \in M$.*
2. ${}_M(M_n)_M \succ {}_M({}_\alpha M)_M$.

Proof First assume Condition 1. If n is even, set $i = n/2$. There exists a non-zero element $a \in M_n$ such that $ax = \alpha(x)a$ for all $x \in M$. We identify a with its left multiplication which is an element of $\mathrm{End}((M_i)_M)$. Since

$$a(x\xi) = \alpha(x)a(\xi), \qquad \text{for all } x \in M, \xi \in L^2(M_i),$$

we can regard this a as an element of $\mathrm{Hom}({}_M(M_i)_M, {}_M({}_\alpha M_i)_M)$. This implies that there exists a non-zero intertwiner

$$\widetilde{a^*} \in \mathrm{Hom}({}_M(M_i) \otimes_M (M_i)_M, {}_M({}_\alpha M)_M) = \mathrm{Hom}({}_M(M_n)_M, {}_M({}_\alpha M)_M),$$

where $\widetilde{a^*}$ means the Frobenius dual of a^*. Hence ${}_M(M_n)_M \succ {}_M({}_\alpha M)_M$ by the irreducibility of the bimodule ${}_M({}_\alpha M)_M$. We can apply a similar argument when n is odd.

Conversely, assume Condition 2 holds. We have ${}_M(M_n)_M \succ {}_M({}_\alpha M)_M$. Then there exists a non-zero intertwiner

$$\sigma \in \mathrm{Hom}({}_M(M_n)_M, {}_M({}_\alpha M)_M) = \mathrm{Hom}({}_M M \otimes_M (M_n)_M, {}_M({}_\alpha M)_M),$$

so we have a non-zero intertwiner $\tilde{\sigma}^* \in \mathrm{Hom}({}_M M_M, {}_M({}_\alpha M_n)_M)$. We define $a = \tilde{\sigma}^*(\hat{1})$. Then we have $a \in {}_M({}_\alpha M_n)_M^{\mathrm{bdd}} = M_n$, and $a \neq 0$, so we have

$$ax = \tilde{\sigma}^*(\hat{1})x = \tilde{\sigma}^*(\hat{1} \cdot x) = \tilde{\sigma}^*(x \cdot \hat{1}) = x \cdot \tilde{\sigma}^*(\hat{1}) = \alpha(x)\tilde{\sigma}^*(\hat{1}) = \alpha(x)a,$$

for all $x \in M$. □

Note that ${}_M({}_\alpha M)_M$ and ${}_M({}_{\alpha'} M)_M$ are isomorphic if and only if there is a unitary $u \in M$ with $\mathrm{Ad}(u) \cdot \alpha = \alpha'$. Thus Condition 2 in the above Theorem determines α only up to inner automorphisms of M. Also note that we can find candidates for α satisfying Condition 2 by looking at the even vertices of the dual principal graph of $N \subset M$ with the Perron–Frobenius weight 1.

It is possible that an automorphism α of M satisfying ${}_M(M_n)_M \succ {}_M({}_\alpha M)_M$ does not preserve N globally, as we will see below. The following theorem gives a characterization of such automorphisms.

Theorem 15.8 *Suppose $\alpha \in \mathrm{Aut}(M)$ satisfies ${}_M(M_n)_M \succ {}_M({}_\alpha M)_M$. Then the following are equivalent.*

1. *There exists a unitary $u \in M$ such that $\mathrm{Ad}u \cdot \alpha \in \mathrm{Aut}(M, N)$.*
2. *There exists $\beta \in \mathrm{Aut}(N)$ such that*

$$_M({}_\alpha M) \otimes_M M_N \cong {}_M M \otimes_N ({}_\beta N)_N.$$

If Condition 2 holds, then ${}_N(M_{n-1})_N \succ {}_N({}_\beta N)_N$ and the unitary $u \in \mathcal{U}(M)$ can be chosen so that $\mathrm{Ad}u \cdot \alpha|_N = \beta$.

Proof First note that

$$_M({}_\alpha M)_N \cong {}_M({}_\alpha M) \otimes_M M_N$$

and

$$_M M \otimes_N ({}_\beta N)_N \cong {}_M(M_{\beta^{-1}})_N.$$

Assume first that Condition 1 holds. We have a unitary $u \in M$ such that $\mathrm{Ad}u \cdot \alpha \in \mathrm{Aut}(M, N)$. Define $\beta \in \mathrm{Aut}(N)$ by $\beta = \mathrm{Ad}u \cdot \alpha|_N$. We extend $(\mathrm{Ad}u \cdot \alpha)^{-1} \in \mathrm{Aut}(M)$ to $L^2(M)$ and denote it by V. Then we have

$$V \in \mathrm{Hom}({}_M({}_\alpha M)_N, {}_M({}_{\mathrm{Ad}(\alpha^{-1}(u)^*)} M_{\beta^{-1}})_N).$$

Indeed,

$$V(x \cdot \xi \cdot y) = \mathrm{Ad}(\alpha^{-1}(u)^*)(x) \cdot V(\xi) \cdot \beta^{-1}(y),$$

for all $x \in M, y \in N, \xi \in {}_M({}_\alpha M)_N$. The map V is a surjective isometry, so we have

$${}_M({}_\alpha M)_N \cong {}_M({}_{\mathrm{Ad}(\alpha^{-1}(u)^*)}M_{\beta^{-1}})_N \cong {}_M(M_{\beta^{-1}})_N.$$

Conversely, assume Condition 2 holds. We have $\beta \in \mathrm{Aut}(N)$ such that ${}_M({}_\alpha M)_N \cong {}_M(M_{\beta^{-1}})_N$. Then there exists a unitary $U \in B(L^2(M))$ such that

$$U(\alpha(x)\hat{a}y) = U(x \cdot a \cdot y) = x \cdot U(\hat{a}) \cdot y = xU(\hat{a})\beta^{-1}(y),$$

for all $x \in M, y \in N, a \in M$. So if we define $\eta = U(\hat{1})$, then we can easily verify $\eta = \hat{v}$ with $v \in \mathcal{U}(M)$ and

$$\begin{aligned}\alpha^{-1}(x)\hat{v} = \alpha^{-1}(x)U(\hat{1}) &= U(x\hat{1}) = U(\hat{x}) \\ &= U(\hat{1}x) = U(\hat{1})\beta^{-1}(x) = \hat{v}\beta^{-1}(x),\end{aligned}$$

for all $x \in N$. So we have

$$\beta^{-1}(x) = v^*\alpha^{-1}(x)v = \mathrm{Ad}(v^*) \cdot \alpha^{-1}(x),$$

for all $x \in N$, and

$$\mathrm{Ad}(\alpha(v)) \cdot \alpha|_N = (\mathrm{Ad}(v^*) \cdot \alpha^{-1})^{-1}|_N = \beta \in \mathrm{Aut}(N).$$

Hence there exists a unitary $u \in \mathcal{U}(M)$ such that $\mathrm{Ad}u \cdot \alpha \in \mathrm{Aut}(M, N)$.

Furthermore, if ${}_M(M_n)_M \succ {}_M({}_\alpha M)_M$, the following holds:

$$\begin{aligned}{}_M(M_n) \otimes_M \overline{{}_M({}_\alpha M)_M} &\cong {}_M(M_{n-1}) \otimes_N M \otimes_M \overline{{}_M({}_\alpha M)_M} \\ &\cong {}_M(M_{n-1}) \otimes_N \overline{{}_M({}_\alpha M) \otimes_M M_N} \\ &\cong {}_M(M_{n-1}) \otimes_N \overline{{}_M M \otimes_N ({}_\beta N)_N} \succ {}_M M_M.\end{aligned}$$

The irreducibility of ${}_M M \otimes_N ({}_\beta N)_N$ implies that

$${}_M(M_{n-1})_N \succ {}_M M \otimes_N ({}_\beta N)_N\,,$$

and similarly we have

$${}_N M \otimes_M (M_{n-1})_N \succ {}_N({}_\beta N)_N\,.$$

Hence we have ${}_N(M_{n-1})_N \succ {}_N({}_\beta N)_N$. □

Remark Consider the Haagerup subfactor with (dual) principal graphs as in Figs 11.59, 11.60. The dual principal graph has three even vertices with the normalized Perron–Frobenius eigenvector entries equal to 1. They give automorphisms of M, but Theorem 15.8 shows that we cannot adjust these automorphisms to be in $\mathrm{Aut}(M,N)$ except for the identity for $*$. This shows that the conditions in Theorem 15.8 do not always hold. This remark was first noted by M. Izumi.

15.4 Centrally trivial automorphisms of subfactors

Definition 15.9 *We say that a norm-bounded sequence* (x_n) *in* N *is* central *with respect to* M *if we have* $\|x_n y - y x_n\|_2 \to 0$ *for all* $y \in M$.

Definition 15.10 *We say that* $\alpha \in \mathrm{Aut}(M,N)$ *is* centrally trivial *if* α *acts trivially on central sequences in* N *with respect to* M. *(That is, the condition means that we have* $\|\alpha(x_n) - x_n\|_2 \to 0$ *for any central sequence* (x_n) *in* N *with respect to* M.*) We denote by* $\mathrm{Ct}(M,N)$ *the subgroup of centrally trivial automorphisms.*

We next introduce ultraproduct von Neumann algebras. (See (McDuff 1970) for more details.) Let ω be a *free ultrafilter* ω over $\mathbb{N}$. That is, ω is a set of subsets of $\mathbb{N}$ with the following properties:

1. $\varnothing \notin \omega$.
2. If $A \in \omega$ and $A \subset B \subset \mathbb{N}$, then $B \in \omega$.
3. If $A, B \in \omega$, then $A \cap B \in \omega$.
4. For any subset A of $\mathbb{N}$, we get $A \in \omega$ or $\mathbb{N} \setminus A \in \omega$.
5. For any finite subset A of $\mathbb{N}$, we get $A \notin \omega$.

A subset ω satisfying Properties 1, 2, 3 is called a filter. Property 4 means maximality of a filter and a maximal filter is called an ultrafilter. Property 5 means that ω is not of the form $\{A \subset \mathbb{N} \mid n \in A\}$ for any n. (Equivalence between Property 5 and this condition follows from the other properties.)

From an operator algebraic viewpoint, a free ultrafilter is understood as follows. The C^*-algebra ℓ^∞ is unital and commutative, so it is of the form $C(X)$ for a compact space X. This space X naturally contains $\mathbb{N}$ and is called the Stone–Čech compactification of $\mathbb{N}$. A free ultrafilter is identified with an elements in $X \setminus \mathbb{N}$. When we have a bounded sequence (λ_n) in $\mathbb{C}$, we regard it as an element in $\ell^\infty = C(X)$ and can then evaluate at ω. We denote this value by $\lim_{n\to\omega} \lambda_n$. Actually, for any sequence (a_n) in a compact space, we have a natural notion of an *ultralimit* $\lim_{n\to\omega} a_n$ in the compact space as follows.

For any sequence (a_n) in a compact space and an element A in a free ultrafilter ω, we set $X_A = \{a_n \mid n \in A\}$. For $A, B \in \omega$, we have $X_A \cap X_B = X_{A\cap B} \neq \varnothing$. This, together with the compactness of the space, implies $X_\omega = \bigcap\{\bar{X}_A \mid A \in \omega\}$ is not empty. Suppose X_ω contains two distinct points a, b. Then we have two open neighbourhoods U, V of a, b respectively with $\bar{U} \cap \bar{V} = \varnothing$. Set $A = \{n \in \mathbb{N} \mid a_n \in U\}$. It is easy to see $A \neq \varnothing$, $A^c \neq \varnothing$. If $A \in \omega$, then we have $X_A \subset U$, hence $\bar{X}_A \subset \bar{U}$, which implies $b \neq \bar{X}_A$ contradicting $b \in X_\omega$. Thus $A^c \in \omega$. Then $X_{A^c} \subset U^c$ and we get $a \notin \bar{X}_{A^c}$, which is again a contradiction. Hence we know that X_ω consists of only one point a. We say that this a is an ultralimit of (a_n)

and write $a = \lim_{n\to\omega} a_n$. Note that for any open neighbourhood U of a, the set $A_U = \{n \in \mathbb{N} \mid a_n \in U\}$ is an element of ω.

Let M be a von Neumann algebra with a faithful trace tr. Let C_ω be an algebra of norm-bounded sequences (x_n) in M with $\lim_{n\to\omega} \|x_n y - y x_n\|_2 = 0$ for all $y \in M$. This is a C^*-subalgebra of $M \otimes \ell^\infty$. For $(x_n) \in M \otimes \ell^\infty$, we set $\mathrm{tr}^\omega(x) = \lim_{n\to\omega} \mathrm{tr}(x)$. This is a state on $M \otimes \ell^\infty$ and C_ω. We denote the image of $M \otimes \ell^\infty$ and C_ω in the GNS-representation with respect to tr^ω by M^ω, M_ω respectively.

Proposition 15.11 *The algebras M^ω and M_ω are von Neumann algebras. The original algebra M is embedded into M^ω as the set of constant sequences, and then we have $M_\omega = M' \cap M^\omega$.*

Proof We first prove that the unit ball of M^ω is complete with respect to the L^2-norm arising from the trace.

Let $\{x_k\}_k$ be a Cauchy sequence in the unit ball of M^ω with respect to the L^2-norm. We may assume that $\|x_{k+1} - x_k\|_2 < 1/2^k$.

We choose a representative of x_k inductively. We represent x_1 by a sequence $(x_1(n))_n$ in the unit ball of M. Suppose we have chosen representatives of $x_1, x_2, \ldots, x_{k-1}$. First we take a sequence $(y(n))_n$ in the unit ball as a representative of x_k. We set

$$A_k = \{n \in \mathbb{N} \mid \|y(n) - x_{k-1}(n)\|_2 < 1/2^k\}.$$

Then we define $x_k(n) = y(n)$ for $n \in A_k$ and $x_k(n) = x_{k-1}(n)$ for $n \notin A_k$. Since A_k is in ω, this $(x_k(n))_n$ is also a representative of x_k in M^ω. For each n, the sequence $\{x_k(n)\}_k$ is a Cauchy sequence in the unit ball of M with respect to the L^2-norm, and thus has a limit $x(n)$. We can easily show that the image x of $(x(n))_n$ in M^ω is the limit of the Cauchy sequence $\{x_k\}_k$.

This shows that M^ω is indeed a von Neumann algebra. The rest is easy. □

Note that if the original algebra M is of finite dimensional, then we have $M = M^\omega$.

Lemma 15.12 *Let M be a II_1 factor, A its von Neumann subalgebra, and x an element in M. Then $E_{A'\cap M}(x)$ is in the L^2-closed convex hull of $\{uxu^* \mid u \in \mathcal{U}(A)\}$.*

Proof Let K be the closed convex hull of $\{uxu^* \mid u \in \mathcal{U}(A)\}$ in $L^2(M)$. The set K has a unique element y with minimum L^2-norm. Since the norm unit ball of M is complete with respect to the L^2-norm, this y is in M, and the uniqueness of y implies that it commutes with any unitary in A. That is, y is in $A' \cap M$. For any element $z \in K$, we see that $E_{A'\cap M}(z) = E_{A'\cap M}(x)$. This shows that $E_{A'\cap M}(x) = y$. □

Corollary 15.13 *Let M be a II_1 factor and x an element in M. If $\mathrm{tr}(x) = 0$ and $\|x\|_2 = \varepsilon > 0$, then we have a unitary $u \in M$ with $\|ux - xu\|_2 \geq \varepsilon/2$.*

Proof Suppose that we have $\|uxu^* - x\|_2 < \varepsilon/2$ for all $u \in \mathcal{U}(M)$ for a contradiction.

Since $\mathrm{tr} = E_{M' \cap M}$, Corollary 15.13 implies that $\|0 - x\|_2 < \varepsilon/2$, which is a contradiction. □

Proposition 15.14 *If M is a II_1 factor, then M^ω is also a II_1 factor.*

Proof Since M^ω has a trace and is infinite dimensional, it is enough to show that the centre is trivial. Take an element $x \in M^\omega \setminus \mathbb{C}$. We will show that we have an element $y \in M^\omega$ with $xy \neq yx$.

We may assume that $x = (x_n)$, $\mathrm{tr}(x_n) = 0$ for all n, and $\lim_{n\to\omega} \|x_n\|_2 = \varepsilon > 0$. Let $A = \{n \in \mathbb{N} \mid \|x_n\|_2 > \varepsilon/2\}$. For each $n \in A$, we use Corollary 15.13 and find a unitary $u_n \in M$ with $\|u_n x_n - x_n u_n\|_2 > \varepsilon/4$. We set $u_n = 1$ for $n \notin A$. Then the element $y = (u_n)_n \in M^\omega$ does not commute with x. □

Proposition 15.15 *If M is a hyperfinite II_1 factor, then M_ω is also a II_1 factor.*

Proof Take an element $x \in M_\omega \setminus \mathbb{C}$. We will show that we have an element $y \in M_\omega$ with $xy \neq yx$. We may assume that $x = (x_n)$, $\mathrm{tr}(x_n) = 0$ for all n, and $\lim_{n\to\omega} \|x_n\|_2 \neq 0$.

We represent M as an infinite tensor product of $M_2(\mathbb{C})$ with respect to the trace. Let R_k be the relative commutant in M of the tensor product of the first k copies of $M_2(\mathbb{C})$.

For each k, we may regard x as an element of R_k^ω. That is, we have a bounded sequence $(x_n^k)_n$ in R_k with $\lim_{n\to\omega} \|x_n^k - x_n\|_2 = 0$. Set $A_0 = \mathbb{N}$ and

$$A_k = \{n \in \mathbb{N} \mid \|x_n^k - x_n\|_2 < 1/2^k\} \cap A_{k-1} \cap [k, \infty)$$

for $k \geq 1$. Then we set $x_n^\infty = x_n^k$ for $n \in A_{k-1} \setminus A_k$ for $k \geq 1$. It is easy to see that $(x_n^\infty)_n$ also represents the element x in M_ω.

For $n \in A_{k-1} \setminus A_k$, we choose a unitary u_n in R_{k-1} with $\|u_n x_n^\infty - x_n^\infty u_n\| \geq \|x_n^\infty\|_2/2$ by Corollary 15.13. It then follows that the sequence $(u_n)_n$ defines an element $y \in M_\omega$ with $xy \neq yx$. □

It is not difficult to see that the above II_1 factors M^ω and M_ω do not act on separable Hilbert spaces. (See (McDuff 1970) for more details on these algebras.)

The idea of the above construction is to form a von Neumann algebra of certain sequences of operators using an ultrafilter. If we have a von Neumann algebra, then we can perform several operations such as a spectral decomposition or a polar decomposition within the algebra, which is a great technical advantage.

This construction is essentially an appeal to the non-standard analysis of (Robinson 1974), though it is a tradition in the theory of operator algebras to use the ultraproduct method without mentioning non-standard analysis. In non-standard analysis, we have an ultrareal number field which contains infinities and infinitesimals. From this viewpoint, we have thrown away operators with infinite norms and ignored infinitesimal differences in $\|\cdot\|_2$ in the above construction. In this way, we get a von Neumann algebra instead of an ultra-von Neumann algebra. When M is infinite dimensional, our new algebra M^ω is much larger than the original M even after throwing away infinities and ignoring infinitesimals. (In non-standard analysis, infinite ultra-integers are called *hyperfinite*. So our *hyperfinite* von Neumann algebra can be characterized by the condition that it is embedded into a hyperfinite dimensional multi-matrix algebra with an infinitesimal error!)

The ultraproduct construction is a kind of 'completion' in the sense that it adds many 'ideal' elements to the original structure. This is given by sequences technically, but we do not get a good conceptual understanding if we always consider an element of the ultraproduct algebra as a sequence. We seldom think of a real number in our mind as a Cauchy sequence of rational numbers!

We give a concrete example on how to think of M_ω as a certain 'completion'. Let M be the hyperfinite II_1 factor and represent it as an infinite tensor product of $M_2(\mathbb{C})$. Let x_n be an operator given by

$$1 \otimes 1 \otimes \cdots \otimes 1 \otimes \begin{pmatrix} 1 & 0 \\ 0 & -1 \end{pmatrix} \otimes 1 \otimes \cdots,$$

where the entry different from 1 is at the nth place. Then the sequence (x_n) gives a non-trivial element in M_ω. We regard the sequence $1, 2, 3, \ldots$ as a positive infinite integer α in the ultrareal number field. (There are many different positive infinite integers. For example, we get a different positive infinite integer $\alpha + 1$ given by the sequence $2, 3, 4, \ldots$.) Then we regard the above element in M_ω as an operator having a non-trivial entry

$$\begin{pmatrix} 1 & 0 \\ 0 & -1 \end{pmatrix}$$

only at the αth place. Roughly speaking, the algebra M_ω is a set of operators having non-trivial entries in the tensor product at the places corresponding to infinite positive integers.

It is possible that $M_\omega = \mathbb{C}$ for certain M. Let F_2 be the free group of two generators a, b. For $g \in F_2$, we have a unitary u_g on $\ell^2(F_2)$ by the left regular representation. The von Neumann algebra M generated by u_gs is a II_1 factor and called the group von Neumann algebra of F_2. A general element in F_2 is expressed as

$$a^{n_1} b^{m_1} a^{n_2} b^{m_2} \cdots a^{n_k} b^{m_k}$$

and a typical element in M is a linear combination of such elements. Passing to the ultraproduct, we have elements such as

$$a^{n_1} b^{m_1} a^{n_2} b^{m_2} \cdots a^{n_\alpha} b^{m_\alpha},$$

where n_j, m_j, α are (possibly infinite) ultra integers. Because there is no relation between a and b, it is not hard to imagine that only linear combinations of such elements commuting with any element in F_2 are scalar multiples of the identity. This suggests that $M_\omega = \mathbb{C}$ and it is indeed possible to give a rigorous proof of this. (See (McDuff 1970) for example.)

We now come back to the subfactor situation. It is easy to see that we have

$$\mathrm{Ct}(M,N) = \{\alpha \in \mathrm{Aut}(M,N) \mid \alpha = \mathrm{id} \text{ on } N^\omega \cap M'\}.$$

(This is an analogue of a corresponding result in the single factor case.) The following characterization was proved by (Popa, preprint 1992). We state this theorem without proof.

Theorem 15.16 *Suppose $N \subset M$ is an extremal and strongly amenable subfactor of the hyperfinite II_1 factor. Then $\alpha \in \mathrm{Aut}(M,N)$ is centrally trivial if and only if α is not strongly outer.*

Remark The above theorem, together with Theorem 15.7, determines a group $\mathrm{Ct}(M,N)/\mathrm{Int}(M,N)$ for an extremal and strongly amenable subfactor of the hyperfinite II_1 factor, but there is one point to be noted. Condition 2 in Theorem 15.7 determines α only up to inner automorphisms of M, but we are often interested in $\mathrm{Ct}(M,N)/\mathrm{Int}(M,N)$, where $\mathrm{Int}(M,N)$ comes from inner automorphisms of N. If M has a non-trivial normalizer u for N, these two classes of inner automorphisms in $\mathrm{Aut}(M,N)$ are different. That is, a unitary $u \in M \setminus N$ with $uNu^* = N$ gives such a difference. Such normalizers can be detected by Proposition 9.44 (2), and many 'interesting' subfactors have no such normalizers. But we have to be careful about this difference if we do have such a normalizer.

Popa also obtained a classification result for group actions. We first need the following definition.

Definition 15.17 *Let G be a discrete group and α an action of G on a von Neumann algebra M. A map u from G to the unitary group of M is called a* unitary cocycle *for α if $u_{gh} = u_g\alpha_g(u_h)$ for $g,h \in G$.*

Two actions α, β of G on M are called conjugate *if there exists an automorphism θ of M such that $\alpha_g = \theta \cdot \beta_g \cdot \theta^{-1}$ for all $g \in G$.*

Two actions α, β of G on M are called cocycle conjugate *if there exists an automorphism θ of M and a unitary cocycle u for α such that $Ad(u_g) \cdot \alpha_g = \theta \cdot \beta_g \cdot \theta^{-1}$ for all $g \in G$.*

For actions on a subfactor $N \subset M$, we have trivial modifications of the above definitions. (A unitary cocycle must have values in the subfactor N.)

S. Popa proved the following theorem in (Popa, preprint 1992) whose proof is beyond the scope of this book.

Theorem 15.18 *Let α, β be actions of a discrete amenable group G on a strongly amenable and extremal subfactor $N \subset M$ of the hyperfinite II_1 factor. If α_g, β_g are strongly outer for all $g \in G \setminus \{1\}$, then the actions α, β are cocycle conjugate if and only if they have the same Loi invariants.*

Compare this with the following theorem in (Ocneanu 1985) in the single factor case, which can now also be proved by Popa's classification theorem on subfactors in (Popa 1994a) if the group G is finitely generated.

Table 15.1 *Analogy between paragroups and flows of weights*

A subfactor $N \subset M$ of type II_1	A type III factor M
paragroup	flow of weights
strong amenability	injectivity
$\overline{\mathrm{Int}}(M,N)$	$\overline{\mathrm{Int}}(M)$
$\mathrm{Ct}(M,N)$	$\mathrm{Ct}(M)$
Loi invariant	Connes–Takesaki module
tower construction M_n	$M \rtimes_\sigma \mathbb{R}$
tunnel construction N_n	centralizer M_ϕ
automorphism in the orbifold construction	modular automorphism σ

Theorem 15.19 *Let α, β be actions of a discrete amenable group G on the hyperfinite II_1 factor. If α_g, β_g are outer for all $g \in G \setminus \{1\}$, then the actions α, β are cocycle conjugate.*

We now discuss the analogy between paragroups and flows of weights of type III factors in the automorphism viewpoint as in Table 15.1.

If we have an automorphism of a type III factor, it naturally induces an action on the flow of weights called the Connes–Takesaki module (Connes and Takesaki 1977). This is similar to the Loi invariant in Definition 15.2 in the following sense. Theorem 15.5 is an analogue of the fact that the kernel of the Connes–Takesaki module is characterized by approximate innerness for hyperfinite type III factors, which was announced in (Connes 1976c) without a proof and proved by (Kawahigashi *et al.* 1992). Also in the course of Loi's proof, we see a similarity between the crossed product by a modular automorphism group and the Jones tower, and between the centralizer and the tunnel construction. That is, the constructions of the crossed product by a modular automorphism group and the Jones tower and are canonical, but the constructions of a centralizer and a tunnel involve non-canonical choices.

Next recall that centrally trivial automorphisms of hyperfinite type III factors are characterized as extended modular automorphisms up to inner perturbation. This was also announced in (Connes 1976c) without a proof and proved in (Kawahigashi *et al.* 1992). If we compare this to Theorem 15.16, we are led to an idea that this subgroup of $\mathrm{Aut}(M,N)/\mathrm{Int}(M,N)$ appearing at the even vertices of the dual principal graph with the normalized Perron–Frobenius eigenvector entries equal to 1 is a discrete analogue of modular automorphism groups of type III factors. From this viewpoint, we can get a natural interpretation of Theorem 15.7 as follows.

Theorem 15.7 means that an automorphism of a subfactor comes from this discrete analogue of the modular automorphism group if and only if it becomes 'almost' inner when extended to the Jones tower. In the case of separable type III factors (Haagerup and Størmer 1990b) characterized pointwise inner automorphisms by proving that an automorphism comes from the extended modular automorphism group if and only if it becomes inner when extended to the crossed product by the modular automorphism group for the dominant weight. So the analogy holds again. Furthermore, in type III subfactor theory, these even vertices correspond to automorphisms appearing in the decomposition of the powers of Longo's canonical endomorphism in (Longo 1987). Longo had already noticed that his canonical endomorphisms were similar to modular automorphism groups. This also supports the above similarity. We believe that further analogies will hold.

Popa's Theorem 15.18 means that centrally free actions of discrete amenable groups on strongly amenable and extremal subfactors of the hyperfinite II_1 factor are classified by the Loi invariant. This is again an analogue of a classification result in hyperfinite type III factors, where centrally free actions of discrete amenable groups on hyperfinite type III factors are classified by the Connes–Takesaki modules as in (Connes 1976c), (Kawahigashi *et al.* 1992), (Ocneanu 1985), (Sutherland and Takesaki 1989).

The basic philosophy in the above type of analogy is that any phenomenon on type III (single) factor theory appears in type II_1 subfactor theory in a discrete form. A mixture of both will appear in type III subfactor theory. The simultaneous generalization of the Connes–Takesaki module and Loi invariant in (Winsløw 1994a) is the first example of such a mixture.

15.5 The central sequence subfactors and asymptotic inclusions

We will next explain how the central sequence structure in a subfactor is related to the asymptotic inclusion introduced in Section 12.6.

In this section, $N \subset M$ is a subfactor of the hyperfinite II_1 factor M constructed from a paragroup as in Section 11.3. Again by Theorem 11.14, this is the same as assuming that $N \subset M$ is a subfactor of the hyperfinite II_1 factor M with finite index, finite depth, and trivial relative commutant. We set $\beta = [M:N]^{1/2}$. Also set γ to be the global index $\sum_{X\in V_0} \mu(X)^2$ as in Theorem 12.24. (Recall Proposition 12.25.) We fix a free ultrafilter ω over $\mathbb{N}$.

By replacing the paragroup with its dual again as in Section 12.6, we may assume that M is given as $\bigvee_{n=0}^{\infty} A_{0,n}$. We study the double sequence $\{A_{k,l}\}_{k+l\geq 0}$ of the string algebras. We also set $M_k = A_{k,\infty} = \bigvee_{l=-k}^{\infty} A_{kl}$, $A_{\infty,l} = \bigvee_{k=-l}^{\infty} A_{k,l} = M'_{-l} \cap M_\infty$, and $A_{\infty,\infty} = M_\infty = \bigvee_{k=0}^{\infty} A_{k,k}$.

Lemma 15.20 *Fix a free ultrafilter ω on $\mathbb{N}$. Set $N_j = M_{-(j+1)}$ for $j \geq -1$ and let $\{e_{-j}\}_{j=0,1,2\ldots}$ be the Jones projections with $e_{-j} \in N'_{j+1} \cap N_{j-1}$. Then*

$$\cdots \subset N_2^\omega \subset N_1^\omega \subset N^\omega \subset M^\omega$$

is a tunnel for the ultraproduct II_1 subfactor $N^\omega \subset M^\omega$ and Jones projections $\{e_{-j}\}_{j=0,1,2\ldots} \subset M^\omega$ satisfy $N_j^\omega \cap \{e_{-j}\}' = N_{j+1}^\omega$.

Proof We prove this by induction on j. It is easy to see that $[N_j^\omega : N_{j+1}^\omega] = [M:N]$ by Theorem 9.48. We get

$$E_{N_{j+1}^\omega}(e_{-j-1}) = [M:N]^{-1} = [N_j^\omega : N_{j+1}^\omega]^{-1}.$$

This identity and $e_{-j} \in (N_{j+1}^{\omega})' \cap N_{j-1}^{\omega}$ imply $N_{j+1}^{\omega} = N_j^{\omega} \cap \{e_{-j}\}'$ as in Proposition 9.45. □

Lemma 15.21 *We have* $M' \cap N^{\omega} = \bigcap_{k=0}^{\infty} M_{-k}^{\omega}$.

Proof By Lemma 15.20, it is enough to prove that $M' \cap N^{\omega} = \{e_0, e_{-1}, e_{-2}, \ldots\}' \cap N^{\omega}$. Because the Jones projections are in M, it is trivial that

$$M' \cap N^{\omega} \subset \{e_0, e_{-1}, e_{-2}, \ldots\}' \cap N^{\omega}.$$

Choose an $x \in \{e_0, e_{-1}, e_{-2}, \ldots\}' \cap N^{\omega}$, and $y \in N_j' \cap M$ for some j. Then $x \in N^{\omega} \cap \{e_0, \ldots, e_{-j+1}\}' = N_j^{\omega}$ by Lemma 15.20. This implies $xy = yx$. By $\bigvee_j (N_j' \cap M) = M$, we get $x \in M'$. □

Lemma 15.22 *There exists a strictly positive constant c such that for any positive element x in $N_j \vee \mathcal{Z}(N_j' \cap M)$, we have $E_{N_j}(x) \geq cx$, where $\mathcal{Z}(N_j' \cap M)$ means the centre of $N_j' \cap M$.*

Proof Let $n^{(j)} = (n_k^{(j)})$ and $p^{(j)} = (p_k^{(j)})$ be the vectors denoting the size and the trace of the minimal projection of each simple summand of $N_j' \cap M$, respectively. There exists j_0 such that $p_k^{(2j+1)} = [M : N]^{j_0 - j} p_k^{(2j_0+1)}$ for $j \geq j_0$ by the finite depth assumption. Because $n^{(j)} \cdot p^{(j)} = 1$, the vector $[M : N]^{j_0 - j} n^{(2j+1)}$ approaches a Perron–Frobenius eigenvector of $\Gamma\Gamma^t$ as $j \to \infty$ by Theorem 10.3 (4), where Γ denotes the inclusion matrix of the principal graph. For $p_k^{(2j)}$, we have a similar result. Then we set $c = \inf_{j,k} n_k^{(j)} p_k^{(j)}$, which is strictly positive. □

Lemma 15.23 *There is a strictly positive constant c such that $E_{N^{\omega} \cap M'}(x) \geq cx$ for any $x \in (M_{\omega})_+$.*

Proof We simply write E for $E_{N^{\omega} \cap M'}$. First note that $E(x) = \lim_{j \to \infty} E_{N_j^{\omega}}(x)$ for $x \in M^{\omega}$ by Lemma 15.21. By Theorem 12.24 and Theorem 9.48, we have a strictly positive constant c' such that $E_{M \vee (M' \cap M_{\infty})}(x) \geq c'x$ for all $x \in (M_{\infty})_+$. Then for the same constant c', we have $E_{M \vee (M' \cap M_{k+1})}(x) \geq c'x$ for all $x \in (M_{k+1})_+$ by the commuting square condition. We then have $E_{(N_j \vee (N_j' \cap M))^{\omega}}(x) \geq c'x$ for all $x \in (M_{\omega})_+ \subset M_+^{\omega}$. For $x \in M_{\omega}$, we get

$$E_{(N_j \vee (N_j' \cap M))^{\omega}}(x) = E_{(N_j \vee \mathcal{Z}(N_j' \cap M))^{\omega}}(x),$$

where we use $(N_j \vee (N_j' \cap M))^{\omega} = N_j^{\omega} \vee (N_j' \cap M)$ which is due to the finite dimensionality of $N_j' \cap M$. Then we get

$$E_{N_j^{\omega}}(x) = E_{N_j^{\omega}}(E_{(N_j \vee \mathcal{Z}(N_j' \cap M))^{\omega}}(x)) \geq c'c''x,$$

where c'' is given by Lemma 15.22. This gives the conclusion. □

Lemma 15.24 *An inclusion $N^{\omega} \cap M' \subset M_{\omega}$ gives a subfactor of type II_1 with finite index.*

Proof By Lemma 15.23 and Proposition 9.48, it is enough to show that $N^\omega \cap M'$ is a factor. We also use Lemma 15.21. First we claim that $x = (x_n)$ is in $\bigcap_k M_{-k}^\omega$ if and only if x is represented with (x_n) such that $F_k = \{n \mid x_n \in M_{-k}\} \in \omega$. It is trivial that such an x is in $\bigcap_k M_{-k}^\omega$. Suppose that $x^0 = (x_n^0) \in M_{-1}^\omega \cap M'$. For all k, there is a sequence $x^k = (x_n^k) \in M_{-k}^\omega$ with $x^k = x^0$ in M_{-1}^ω. Set $F_0 = \mathbb{N}$ and

$$F_k = \{n \mid \|x_n^{k-1} - x_n^k\|_2 < 1/2^k\} \cap F_{k-1} \cap [k, \infty).$$

Then each F_k is in ω. Put $x_n = x_n^k$ for $n \in F_k \setminus F_{k+1}$ and set $x = (x_n) \in N^\omega$. Then on F_k, we get $\|x_n^k - x_n\|_2 \leq 1/2^k$. Thus $\|x^0 - x\| \leq_2 1/2^k$ for all k, which implies $x^0 = x$. Thus we have the claim.

Suppose we have $x = (x_n) \in \mathcal{Z}(N^\omega \cap M')$ with $\mathrm{tr}(x) = 0$. We have to show $x = 0$. We may assume that $\mathrm{tr}(x_n) = 0$ and $\|x_n\|_2 = 1$ for all n and will obtain a contradiction. We choose (x_n) as in the above claim. We have $\bigcap_k F_k = \varnothing$. Then for each $n \in F_k \setminus F_{k+1}$, we choose a unitary $y_n \in M_{-k}$ so that $\|[x_n, y_n]\|_2 \geq 1/2$ by Corollary 15.13. Then the sequence $y = (y_n)$ is in $N^\omega \cap M'$ and $\|[x, y]\|_2 \neq 0$, which is a contradiction. □

The subfactor $N^\omega \cap M' \subset M_\omega$ is called the *central sequence subfactor* of the original subfactor $N \subset M$. Next we have an important technical tool to compute relative commutants of various ultraproduct algebras.

Lemma 15.25. (Central freedom lemma) *Let $L \subset P \subset Q$ be finite von Neumann algebras and L be a hyperfinite factor. Then we get*

$$(L' \cap P^\omega)' \cap Q^\omega = L \vee (P' \cap Q)^\omega.$$

Proof Note that it is trivial that the right hand side is contained in the left hand side.

First we prove the lemma for the case $L = \mathbb{C}$. Take $x = (x_n) \in (P^\omega)' \cap Q^\omega$. We prove that $x = (E_{P' \cap Q}(x_n))$. Suppose not. Then there exists a positive ε and a set $F \in \omega$ such that $\|x_n - E_{P'\cap Q}(x_n)\|_2 \geq \varepsilon$ for all $n \in F$. Moreover, for all $n \in F$, there exists a unitary $y_n \in P$ with $\|y_n\| = 1$ and $\|[x_n, y_n]\|_2 \geq \varepsilon/2$ by Lemma 15.12. Setting $y = (y_n) \in P^\omega$, we get $\|[x, y]\|_2 \geq \varepsilon/2$, which is a contradiction.

Next assume that L is a hyperfinite II_1 factor. Represent $L = \bigotimes_{n\in\mathbb{N}} M_2(\mathbb{C})$ and set $L_m = \bigotimes_{n=1}^m M_2(\mathbb{C})$. We claim that

$$\left(\bigcap_m (L_m' \cap P)^\omega\right)' \cap Q^\omega = \bigvee_m \left(((L_m' \cap P)^\omega)' \cap Q^\omega\right).$$

It is clear that the right hand side is contained in the left hand side. To prove the converse inclusion, suppose that $x = (x_n) \in Q^\omega$ satisfies $x \notin \bigvee_m (((L_m' \cap P)^\omega)' \cap Q^\omega)$. Then there exists a positive ε such that $\|x - E_{((L_m'\cap P)^\omega)' \cap Q^\omega}(x)\|_2 \geq \varepsilon$ for all m. Then we have a unitary $y^m \in (L_m' \cap P)^\omega$ and $\|[x, y^m]\|_2 \geq \varepsilon/2$. Put $F_0 = \mathbb{N}$ and

$$F_m = \{n \mid \|[x_n, y_n^m]\|_2 \geq \varepsilon/2\} \cap F_{m-1} \cap [m, \infty).$$

Each F_m is in ω, and define $y = (y_n)$ with $y_n = y_n^m$ for $n \in F_m \setminus F_{m+1}$. Then $y \in \bigcap_m (L'_m \cap P)^\omega$ and $\|[x, y]\|_2 \geq \varepsilon/2$, which complete the proof of the claim.

Then we have

$$\begin{aligned}
(L' \cap P^\omega)' \cap Q^\omega &= \left(\bigcap_m (L'_m \cap P^\omega)\right)' \cap Q^\omega \\
&= \left(\bigcap_m (L'_m \cap P)^\omega\right)' \cap Q^\omega \\
&= \bigvee_m (((L'_m \cap P)^\omega)' \cap Q^\omega) \\
&= \bigvee_m (\mathbb{C} \otimes (L'_m \cap P)^\omega)' \cap (L_m \otimes (L'_m \cap Q)^\omega) \\
&= \bigvee_m (L_m \vee (P' \cap Q)^\omega) \\
&= L \vee (P' \cap Q)^\omega.
\end{aligned}$$

□

Recall that our double sequence of the higher relative commutants looks like the following.

$$\begin{array}{ccccccc}
 & & & & A_{-n,n} & \subset \cdots \to & A_{-n,\infty} = M_{-n} \\
 & & & & \cap & & \cap \\
 & & & & \vdots & & \vdots \\
 & & & & \cap & & \cap \\
 & & A_{0,0} & \subset \cdots \subset & A_{0,n} & \subset \cdots \to & A_{0,\infty} = M_0 \\
 & & \cap & & \cap & & \cap \\
 & & \vdots & & \vdots & & \vdots \\
 & & \cap & & \cap & & \cap \\
A_{n,-n} & \subset \cdots \subset & A_{n,0} & \subset \cdots \subset & A_{n,n} & \subset \cdots \to & A_{n,\infty} = M_n \\
\cap & & \cap & & \cap & & \cap \\
\vdots & & \vdots & & \vdots & & \vdots \\
\downarrow & & \downarrow & & \downarrow & & \\
A_{\infty,-n} & \subset \cdots \subset & A_{\infty,0} & \subset \cdots \subset & A_{\infty,n} & \subset \cdots &
\end{array}$$

Lemma 15.26 *Let n be a positive even integer. Let $e \in A_{\infty,n}$ be the Jones projection for $A_{\infty,-n} \subset A_{\infty,0}$. Let f be the central support of e in $A'_{\infty,-n} \cap A_{\infty,n} = A_{n,n}$ and $p \in A_{\infty,0} \vee A_{0,n}$ be the Jones projection for $A'_{n,0} \cap A_{\infty,0} \subset A_{\infty,0}$. Then the following hold:*

1. *$p \in A_{n,0} \vee A_{0,n}$;*

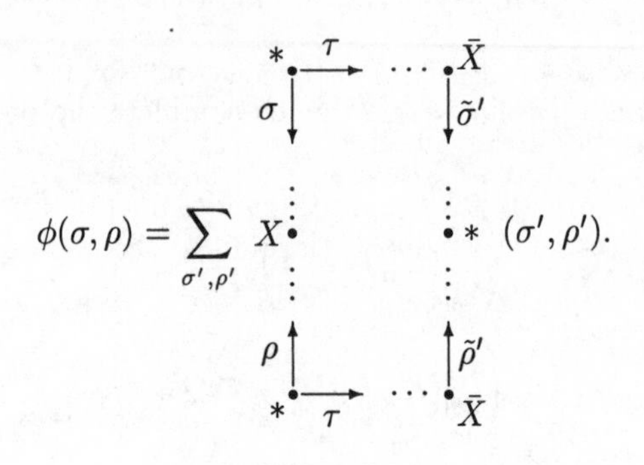

FIG. 15.1. The definition of ϕ_X

2. $ep = e$;
3. $E_{A_{0,n} \vee A_{\infty,0}}(e) = E_{A_{0,n}}(f)p$.

Proof For each $X \in V_0$ and an even integer n, define a map ϕ_X from the vertical string algebra starting from $*$ to X with length n to that starting from $*$ to $\bar{X}$ by Fig. 15.1. By flatness, this is a unital homomorphism from a full matrix algebra to another of the same dimension by Theorem 10.12 and Theorem 11.17, and so is an isomorphism. Using this for all X, we get an automorphism ϕ of the string algebra starting from $*$.

Set

$$p = \sum_{\sigma,\rho} \frac{1}{|\mathrm{Path}_{*,x}(n)|}(\sigma,\rho) \otimes \phi^{-1}(\sigma,\rho) \in A_{n,0} \vee A_{0,n},$$

where both $A_{n,0}$ and $A_{0,n}$ are expressed as vertical string algebras from $*$ with length n after fixing an identification of the graphs $\mathcal{G}_0$ and $\mathcal{G}_3$, and $|\mathrm{Path}_{*,x}(n)|$ denotes the number of paths with length n from $*$ to x. Then a direct computation shows that this p is the correct Jones projection for $A'_{n,0} \cap A_{\infty,0} \subset A_{\infty,0}$. It is also easy to see that $p \in A_{n,0} \vee A_{0,n}$ and we get $ep = e$ because these are both Jones projections and the subalgebra corresponding to e is smaller.

We also have

$$E_{A_{0,n}}(f) = \frac{1}{\beta^n} \sum_{x \in V_0} \frac{1}{\mu(x)} |\mathrm{Path}_{*,x}(n)| \sum_{s(\sigma)=*,r(\sigma)=x} (\sigma,\sigma),$$

where the right hand side is expressed as a horizontal string. By the commuting square condition, in order to prove $E_{A_{0,n} \vee A_{\infty,0}}(e) = E_{A_{0,n}}(f)p$, it is enough to see $\mathrm{tr}(e\tau) = \mathrm{tr}(E_{A_{0,n}}(f)p\tau)$ for $\tau \in A_{n,0} \vee A_{0,n}$ of the form $\tau = (\sigma,\rho) \otimes \phi^{-1}(\sigma',\rho')$, where these are expressed as horizontal strings. For this τ, a direct computation shows that both $\mathrm{tr}(e\tau)$ and $\mathrm{tr}(E_{A_{0,n}}(f)p\tau)$ are equal to $\delta_{\sigma,\sigma'}\delta_{\rho,\rho'}\mu(r(\sigma))\beta^{-3n}$. (For $\mathrm{tr}(e\tau)$, we easily get the Kronecker δs by $ep = e$. Then it is enough to compute $\mathrm{tr}(e(\sigma,\rho))$, which is easy.) □

Remark Note that the algebra $A'_{n,0} \cap A_{\infty,0}$ is not a factor, but this is not significant.

Lemma 15.27 *Let $e \in A_{\infty,n}$ be the Jones projection for $A_{\infty,-n} \subset A_{\infty,0}$. Then this e is also a Jones projection in $A_{n,\infty}$ for $A_{-n,\infty}, \subset A_{0,\infty}$.*

Proof If $n = 1$, this directly follows from the identification of strings explained in Section 11.4. For general n, we use the formula of e in Theorem 9.49 expressed with e_js and reduce the problem to the case $n = 1$. □

Lemma 15.28 *Let n, e, f be as in Lemma 15.26. Let $0 < \varepsilon < 1/3$ and $\bar{e} \in A_{\infty,n}$ be a projection. Suppose we have*

$$\|E_{A_{0,n} \vee A_{\infty,0}}(\bar{e}) - \gamma^{-1}\|_1 < \varepsilon\gamma^{-1},$$
$$\|E_{A_{0,n}}(f) - \gamma^{-1}\| < \varepsilon\gamma^{-1},$$

where $\|x\|_1$ means $\mathrm{tr}(|x|)$. Then for $x \in A_{0,\infty} \cap A'_{0,n}$ with $\|x\| \leq 1$, we get

$$\|\bar{e}x\bar{e} - \bar{e}E_{A_{-n,\infty}}(x)\|_2 < 3\varepsilon^{1/4}.$$

Proof First choose v_is in $A_{\infty,0}$ so that $\bar{e} = \sum_{i=1}^m v_i^* e v_i$. (Write $\bar{e} = \bar{e}1\bar{e}$ and use Theorem 9.27 (3) as in the proof of Proposition 9.28.) By Lemmas 15.26 and 15.27 we have

$$\begin{aligned}
&\|\bar{e}x\bar{e} - \bar{e}E_{A_{-n,\infty}}(x)\|_2^4 \\
&\leq 4\|\bar{e}x\bar{e} - \bar{e}E_{A_{-n,\infty}}(x)\|_1^2 \\
&= 4\left\|\sum_{i=1}^m v_i^* e v_i x\bar{e} - \sum_{i=1}^m v_i^* e v_i E_{A_{-n,\infty}}(x)\right\|_1^2 \\
&= 4\left\|\sum_{i=1}^m v_i^* e x p v_i\bar{e} - \sum_{i=1}^m v_i^* e x e v_i\right\|_1^2 \\
&= 4\left\|\sum_{i=1}^m v_i^* e x(p v_i\bar{e} - e v_i)\right\|_1^2 \\
&\leq 4\mathrm{tr}\left(\sum_{i=1}^m v_i^* e x x^* e v_i\right)\mathrm{tr}\left(\sum_{i=1}^m (\bar{e}v_i^* p - v_i^* e)(p v_i\bar{e} - e v_i)\right) \\
&= 4\mathrm{tr}(\bar{e}E_{A_{-n,\infty}}(xx^*))\mathrm{tr}\left(\sum_{i=1}^m \bar{e}v_i^* p v_i\bar{e} - \bar{e} - \bar{e} + \bar{e}\right) \\
&\leq 4\|\bar{e}\|_1 \mathrm{tr}\left(\bar{e}(\sum_{i=1}^m v_i^* p v_i - 1)\bar{e})\right).
\end{aligned}$$

Next by Lemma 15.26 (3) we have

$$\begin{aligned}
&\gamma^{-1}\left\|\sum_{i=1}^m v_i^* p v_i - 1\right\|_1 \\
&\leq \left\|\gamma^{-1}\sum_{i=1}^m v_i^* p v_i - E_{A_{0,n}}(f)\sum_{i=1}^m v_i^* p v_i\right\|_1 + \left\|E_{A_{0,n}}(f)\sum_{i=1}^m v_i^* p v_i - \gamma^{-1}\right\|_1 \\
&= \left\|\gamma^{-1}\sum_{i=1}^m v_i^* p v_i - E_{A_{0,n}}(f)\sum_{i=1}^m v_i^* p v_i\right\|_1 + \left\|E_{A_{0,n}\vee A_{\infty,0}}(\bar{e}) - \gamma^{-1}\right\|_1 \\
&\leq \varepsilon\gamma^{-1}\left\|\sum_{i=1}^m v_i^* p v_i\right\|_1 + \varepsilon\gamma^{-1}.
\end{aligned}$$

Setting $C = \| \sum_{i=1}^m v_i^* p v_i \|_1$, we obtain

$$C \leq \left\|\sum_{i=1}^m v_i^* p v_i - 1\right\|_1 + 1 \leq \varepsilon C + \varepsilon + 1,$$

which implies $C \leq (1+\varepsilon)/(1-\varepsilon) \leq 1 + 3\varepsilon$. Thus we get

$$\left\|\sum_{i=1}^m v_i^* p v_i - 1\right\|_1 \leq \varepsilon(1+3\varepsilon) + \varepsilon \leq 5\varepsilon\,.$$

With this, we get

$$\|\bar{e}x\bar{e} - \bar{e}E_{A_{-n,\infty}}(x)\|_2^4 \leq 20\varepsilon,$$

which produces the desired estimate. □

Lemma 15.29 *Let $\tilde{M} \subset A_{0,\infty} \vee A_{\infty,0} \subset M_\infty$ be a downward basic construction with the Jones projection $\tilde{e} \in M_\infty$. Then*

$$\begin{array}{ccc}
\tilde{M}^\omega & \subset & (A_{0,\infty} \vee A_{\infty,0})^\omega \\
\cup & & \cup \\
N^\omega \cap M' & \subset & M_\omega
\end{array}$$

is a commuting square.

Proof First note that $N^\omega \cap M' = M^\omega \cap M'_\infty$. Because $\tilde{e} \in M_\infty$, we get $N^\omega \cap M' \subset \tilde{M}^\omega$. Choose a positive $\varepsilon < 1/3$. Choose n_0 so large that if $n > n_0$, we get $\|E_{A_{0,n}\vee A_{\infty,0}}(\tilde{e}) - \gamma^{-1}\|_2 < \varepsilon\gamma^{-1}/2$ and $\|E_{A_{0,n}}(f) - \gamma^{-1}\| < \varepsilon\gamma^{-1}$, and we can find by a standard argument a projection $\bar{e} \in A_{\infty,n}$ with $\|\bar{e} - \tilde{e}\|_2 < \varepsilon\gamma^{-1}/2$, where γ is the global index of $N \subset M$. (We perturb $E_{A_{\infty,n}}(\tilde{e})$ to a projection within $A_{\infty,n}$.) Here f is the central support of the Jones projection for $A_{\infty,-n} \subset A_{\infty,0}$ in $A'_{\infty,-n} \cap A_{\infty,n}$ as in Lemma 15.26. The estimate for f is possible by looking at the explicit expression

$$E_{A_{0,n}}(f) = \frac{1}{\beta^n}\sum_{x\in V_0}\frac{1}{\mu(x)}|\mathrm{Path}_{*,x}(n)| \sum_{s(\sigma)=*,r(\sigma)=x}(\sigma,\sigma)$$

and Theorem 10.3 (4). Choose $x = (x_m) \in M_\omega$ with $\|x\| = 1$. Fix even $n > n_0$ and we may assume that $x_m \in M \cap A'_{0,n}$. Then

$$\begin{aligned}
&\|\tilde{e}E_{\tilde{M}^\omega}(x) - \tilde{e}E_{M^\omega_{-n}}(x)\|_2 \\
&= \|\tilde{e}x\tilde{e} - \bar{e}x\tilde{e} + \bar{e}x\tilde{e} - \bar{e}x\bar{e} + \bar{e}x\bar{e} - \bar{e}E_{M^\omega_{-n}}(x) + \bar{e}E_{M^\omega_{-n}}(x) - \tilde{e}E_{M^\omega_{-n}}(x)\|_2 \\
&\le 3\varepsilon^{1/4} + 3\varepsilon \le 6\varepsilon^{1/4},
\end{aligned}$$

by Lemma 15.28. Because $E_{M^\omega_{-n}}(x)$ converges to $E_{N^\omega \cap M'}(x)$ in L^2-norm as in Lemma 15.21, we get

$$\|\tilde{e}E_{\tilde{M}^\omega}(x) - \tilde{e}E_{N^\omega \cap M'}(x)\|_2 \le 6\varepsilon^{1/4}$$

for all $\varepsilon > 0$, which implies $E_{\tilde{M}^\omega}(x) = E_{N^\omega \cap M'}(x)$. □

Lemma 15.30 *In the above setting,*

$$N^\omega \cap M' \subset M_\omega \subset \langle M_\omega, \tilde{e}\rangle$$

is standard where $\tilde{e}$ is regarded as an element of M^ω_∞. The central sequence subfactor $N^\omega \cap M' \subset M_\omega$ has trivial relative commutant, and the index $[M_\omega : N^\omega \cap M']$ is given by the global index γ.

Proof The only part we have to prove for the basic construction is that the central support q of $\tilde{e}$ in $\langle M_\omega, \tilde{e}\rangle$ is 1, since this implies that the natural map from the basic construction of $N^\omega \cap M' \subset M_\omega$ to $\langle M_\omega, \tilde{e}\rangle$ is a unital isomorphism. By the central freedom lemma (Lemma 15.25), we get $(M_\omega)' \cap M^\omega_\infty = M \vee A^\omega_{\infty,0} \supset M \vee A_{\infty,0}$. Then

$$\gamma^{-1} = E_{(M_\omega)'\cap M^\omega_\infty}(\tilde{e}) = E_{(M_\omega)'\cap M^\omega_\infty}(q\tilde{e}) = qE_{(M_\omega)'\cap M^\omega_\infty}(\tilde{e}) = \gamma^{-1}q.$$

Thus we obtain

$$[M_\omega : N^\omega \cap M'] = E_{M_\omega}(\tilde{e})^{-1} = E_{M_\omega}(E_{M\vee A_{\infty,0}}(\tilde{e}))^{-1} = \gamma.$$

By the central freedom lemma (Lemma 15.25) and its proof, we also get

$$\begin{aligned}
(N^\omega \cap M')' \cap M_\omega &= \left(\bigcap_k M^\omega_{-k}\right)' \cap M^\omega \cap M' \\
&= \bigvee_k (M'_{-k} \cap M) \cap M' = M \cap M' = \mathbb{C}.
\end{aligned}$$

□

Lemma 15.31 *Suppose $P_0 \subset P_1 \subset P_2 \subset Q$ are II_1 factors with $[P_1 : P_0] < \infty$ and $P'_0 \cap P_1 = \mathbb{C}$ and $P_0 \subset P_1 \subset P_2 = \langle P_1, e\rangle$ is standard with the Jones* *projection e in Q. We also assume that $P_1' \cap Q \subset P_0' \cap Q$ are II_1 factors with $[P_0' \cap Q : P_1' \cap Q] = [P_1 : P_0]$. Then*

$$P_2' \cap Q \subset P_1' \cap Q \subset P_0' \cap Q$$

is also standard.

Proof Because

$$\begin{array}{ccc} P_1' \cap Q & \subset & P_0' \cap Q \\ \cup & & \cup \\ P_1' \cap P_2 & \subset & P_0' \cap P_2 \end{array}$$

is a commuting square by Proposition 9.53, we get $E_{P_1' \cap Q}(e) = [P_0' \cap Q : P_1' \cap Q]^{-1}$. This makes

$$P_1' \cap Q \cap \{e\}' \subset P_1' \cap Q \subset P_0' \cap Q$$

standard by Proposition 9.47. □

Lemma 15.32 *We have $M_\infty^\omega \cap (N^\omega \cap M')' = \bigvee_k A_{\infty,k}^\omega$.*

Proof By the central freedom lemma (Lemma 15.25), the right hand side is equal to $\bigvee_k ((M_{-k}^\omega)' \cap M_\infty^\omega)$. It is trivial that this is contained in the left hand side, which is equal to $(\bigcap M_{-k}^\omega)' \cap M_\infty^\omega$. The converse inclusion is proved as in the claim in the last step in the proof of the central freedom lemma (Lemma 15.25). □

Let

$$\begin{aligned} &A_{0,\infty} \vee A_{\infty,0} \subset A_{\infty,\infty} \subset \langle A_{\infty,\infty}, f_1 \rangle \\ &\subset \langle A_{\infty,\infty}, f_1, f_2 \rangle \subset \langle A_{\infty,\infty}, f_1, f_2, f_3 \rangle \subset \cdots \end{aligned}$$

be the Jones tower and $f_1, f_2, \ldots$ be the Jones projections. Set $B_k^l = A_{\infty,k} \vee \{f_1, \ldots, f_l\}$. Then we can show that the following double sequence gives commuting squares.

$$\begin{array}{ccccccccc} \vdots & & \vdots & & \vdots & & \vdots & & \\ A_{0,k} \vee A_{\infty,0} & \subset & A_{\infty,k} & \subset & B_k^1 & \subset & B_k^2 & \subset & \cdots \\ \cap & & \cap & & \cap & & \cap & & \\ A_{0,k+1} \vee A_{\infty,0} & \subset & A_{\infty,k+1} & \subset & B_{k+1}^1 & \subset & B_{k+1}^2 & \subset & \cdots \\ \cap & & \cap & & \cap & & \cap & & \\ \vdots & & \vdots & & \vdots & & \vdots & & \end{array}$$

Furthermore, the vertical limit of the double sequence gives the tower for $A_{0,\infty} \vee A_{\infty,0} \subset A_{\infty,\infty}$. Then we have the following lemma.

Lemma 15.33 1. *We have the following identity:*

$$\bigvee_k (B_k^j)^\omega \cap \left(\bigvee_k (A_{0,k} \vee A_{\infty,0})^\omega \right)' = \langle A_{\infty,\infty}, f_1, \ldots, f_j \rangle \cap (A_{0,\infty} \vee A_{\infty,0})'.$$

2. *The sequence*

$$\bigvee_k (A_{0,k} \vee A_{\infty,0})^\omega \subset \bigvee_k A^\omega_{\infty,k} \subset \bigvee_k (B^1_k)^\omega \subset \bigvee_k (B^2_k)^\omega \subset \cdots$$

gives the Jones tower.

Proof 1. We have

$$\begin{aligned}
&\bigvee_k \left(B^j_k\right)^\omega \cap \left(\bigvee_k (A_{0,k} \vee A_{\infty,0})^\omega\right)' \\
&= \bigvee_k \left(B^j_k\right)^\omega \cap \bigcap_k ((A_{0,k} \vee A_{\infty,0})^\omega)' \\
&= \bigcap_k \left(((A_{0,k} \vee A_{\infty,0})^\omega)' \cap \bigvee_{l\geq k} \left(B^j_l\right)^\omega \right) \\
&= \bigcap_k \bigvee_l \left(((A_{0,k} \vee A_{\infty,0})^\omega)' \cap (B^j_l)^\omega \right) \\
&= \bigcap_k \bigvee_l \left((A_{0,k} \vee A_{\infty,0})' \cap B^j_l \right)^\omega \\
&= \bigcap_k \bigvee_l ((A_{0,k} \vee A_{\infty,0})' \cap B^j_l) \\
&= \left(\bigvee_k B^j_k\right) \cap \left(\bigvee_k (A_{0,k} \vee A_{\infty,0})\right)' \\
&= \langle A_{\infty,\infty}, f_1, \ldots, f_j \rangle \cap (A_{0,\infty} \vee A_{\infty,0})'.
\end{aligned}$$

Here we used the central freedom lemma (Lemma 15.25) and the finite dimensionality of $(A_{0,k} \vee A_{\infty,0})' \cap B^j_l$. This finite dimensionality follows from finite dimensionality of the centre of $A_{0,k} \vee A_{\infty,0}$ and the Pimsner–Popa inequality for the conditional expectation from B^j_l onto $A_{0,k} \vee A_{\infty,0}$.

2. By the same kind of argument as above, we can prove that each $\bigvee_k (B^l_k)^\omega$ is a factor. First we prove that

$$\bigvee_k (A_{0,k} \vee A_{\infty,0})^\omega \subset \bigvee_k A^\omega_{\infty,k} \subset \bigvee_k \left(B^1_k\right)^\omega$$

is standard. (The other parts of the proof are just repetitions of the same arguments.) For the inclusion $\bigvee_k A^\omega_{\infty,k} \subset \bigvee_k (B^1_k)^\omega$, we have the Pimsner–Popa inequality $E(x) \geq \gamma^{-1} x$ because of the commuting square condition. We have $f_1 \in \bigvee_k (B^1_k)^\omega$ with $E_{\bigvee_k A^\omega_{\infty,k}}(f_1) = \gamma^{-1}$. This implies $[\bigvee_k (B^1_k)^\omega : \bigvee_k A^\omega_{\infty,k}] = \gamma$, hence the conclusion by Proposition 9.47. □

Lemma 15.34 *Suppose $P_0 \subset P_1 \subset P_2 \subset Q$ are II_1 factors with $[P_1 : P_0] < \infty$ and $P_0' \cap P_1 = \mathbb{C}$ and $P_0 \subset P_1 \subset P_2$ is standard. We also assume that $P_2' \cap Q \subset P_1' \cap Q \subset P_0' \cap Q$ are II_1 factors with $[P_1' \cap Q : P_2' \cap Q] = [P_1 : P_0]$ and this sequence is standard. Then we can choose a tower*

$$P_0 \subset P_1 \subset P_2 \subset P_3 \subset P_4 \subset \cdots \subset Q$$

so that

$$\cdots P_4' \cap Q \subset P_3' \cap Q \subset P_2' \cap Q \subset P_1' \cap Q \subset P_0' \cap Q$$

is a tunnel.

Proof Set $\lambda = [P_1 : P_0]$. Let $f_1 \in P_2$ be the Jones projection for $P_0 \subset P_1$. Then we claim that f_1 is the Jones projection in $P_0' \cap Q$ for the inclusion $P_2' \cap Q \subset P_1' \cap Q$. Using Proposition 9.53, we can prove that

$$\begin{array}{ccc} P_1' \cap Q & \subset & P_0' \cap Q \\ \cup & & \cup \\ P_1' \cap P_2 & \subset & P_0' \cap P_2 \end{array}$$

is a commuting square. So we get $E_{P_1' \cap Q}(f_1) = \lambda$. Thus f_1 is the Jones projection. Take $f_2 \in P_1' \cap Q$ with $E_{P_2' \cap Q}(f_2) = \lambda$. Then

$$P_2' \cap Q \cap \{f_2\}' \subset P_2' \cap Q \subset P_1' \cap Q$$

is a downward basic construction by Proposition 9.45. We next claim that $P_1 \subset P_2 \subset P_3 = \langle P_2, f_2 \rangle$ is standard. For $x, y \in P_1$, we get

$$f_2(xf_1y)f_2 = xf_2f_1f_2y = \lambda f_2xy = f_2E_{P_1}(xf_1y).$$

Thus it is enough to see that the central support of f_2 in P_3 is 1. (See the proof of Lemma 15.30.) This can be proved as in the proof of Lemma 15.30 with $E_{P_2' \cap Q}(f_2) = \lambda$. It is now trivial that $P_2' \cap Q \cap \{f_2\}' = P_3' \cap Q$. We repeat this procedure to get the conclusion. □

Lemma 15.35 *Set $P_0 = N^\omega \cap M'$ and $P_1 = M_\omega$. We have a Jones tower*

$$P_0 \subset P_1 \subset P_2 \subset P_3 \subset \cdots \subset M_\infty^\omega$$

so that

$$\cdots \subset P_3' \cap M_\infty^\omega \subset P_2' \cap M_\infty^\omega \subset P_1' \cap M_\infty^\omega \subset P_0' \cap M_\infty^\omega$$

is a tunnel.

Proof By Lemma 15.32 and the central freedom lemma (Lemma 15.25), we get that the inclusion $P_1' \cap M_\infty^\omega \subset P_0' \cap M_\infty^\omega$ is given by $\bigvee_k (A_{0,k} \vee A_{\infty,0})^\omega \subset \bigvee_k A_{\infty,k}^\omega$. By Lemma 15.33 (2), this is a subfactor with index γ. We can set $P_2 = \langle P_1, \tilde{e} \rangle$ by Lemmas 15.30 and 15.31. Now we just apply Lemma 15.34 to get the conclusion. □

Lemma 15.36 *The higher relative commutants of M_ω for the subfactor $N^\omega \cap M' \subset M_\omega$ are contained in the higher relative commutants of $A_{0,\infty} \vee A_{\infty,0}$ for the subfactor $A_{0,\infty} \vee A_{\infty,0} \subset A_{\infty,\infty}$ with a trace-preserving injective anti-homomorphism.*

Proof By Lemma 15.35, it is easy to see that the higher relative commutants of M_ω for the subfactor $N^\omega \cap M' \subset M_\omega$ are contained in the higher relative commutants of the tunnel for $\bigvee_k (A_{0,k} \vee A_{\infty,0})^\omega \subset \bigvee_k A^\omega_{\infty,k}$. But this latter subfactor has trivial relative commutant by Lemma 15.33. So the relative commutants for the tunnel are anti-isomorphic to the relative commutants for the tower with a trace preserving anti-isomorphism by Proposition 9.50. We get the conclusion by Lemma 15.33 (1). □

Theorem 15.37 *Let $N \subset M$ be a subfactor of the hyperfinite II_1 factor with finite index, finite depth, and trivial relative commutant. The paragroups of the central sequence subfactor $N^\omega \cap M' \subset M_\omega$ and of the asymptotic inclusion $M \vee (M' \cap M_\infty) \subset M_\infty$ are mutually dual.*

The trivial relative commutant condition here is actually redundant, and can be removed it with a little bit of more work in Lemma 15.26.

It was proved in Lemma 15.36 that the dual canonical commuting square of the central sequence subfactor is contained in the canonical commuting square of the asymptotic inclusion with a trace-preserving injective anti-homomorphism, so we only have to prove the converse inclusion, because the asymptotic inclusion is clearly anti-isomorphic to itself.

We set $P_0 = N^\omega \cap M'$, $P_1 = M_\omega$. We construct the Jones tower $P_0 \subset P_1 \subset P_2 \subset P_3 \subset \cdots$ within M^ω_∞ as in Lemma 15.35. We denote the Jones projection for the subfactor $P_0 \subset P_1$ by $\tilde{e}$. For a general subalgebra R of M^ω_∞, we write R^c for $R' \cap M^\omega_\infty$. By Lemma 15.35, the sequence $\cdots \subset P^c_3 \subset P^c_2 \subset P^c_1 \subset P^c_0$ is a tunnel. We have $P^c_0 = \bigvee_k A^\omega_{\infty,k}$ by Lemma 15.32 and

$$P^c_1 = M \vee (M' \cap M_\infty)^\omega = \bigvee_k (A_{0,k} \vee A_{\infty,0})^\omega$$

by the central freedom lemma (Lemma 15.25). By Theorem 11.20 and the central freedom lemma (Lemma 15.25), we get $P^{cc}_1 = P_1$ and $P^{cc}_0 = \bigcap_{k \geq 0} M^\omega_{-k} = P_0$. Choose a Jones projection $f \in P_1$, i.e. $E_{P_0}(f) = \gamma^{-1}$. We then need the following lemma.

Lemma 15.38 *In the above context, the inclusion $P^c_1 \subset P^c_0 \subset \langle P^c_0, f\rangle$ is standard.*

Proof First note that $E_{P^c_1}(\tilde{e}) = \gamma^{-1}$. Because $P^c_2 \subset P^c_1 \subset P^c_0 = \langle P^c_1, \tilde{e}\rangle$ is standard, a general element of P^c_0 is a linear combination of elements of the form $x\tilde{e}y$ with $x, y \in P^c_1$. For such x, y, we get

$$f(x\tilde{e}y)f = xf\tilde{e}fy = \gamma^{-1}xfy = fE_{P^c_1}(x\tilde{e}y).$$

Thus it is now enough to prove that the central support of f in $\langle P_0^c, f\rangle$ is 1. This is proved as in Lemma 15.30. □

We set $Q_1 = \langle P_0^c, f\rangle$. Then $P_1^c \subset P_0^c \subset Q_1$ and $Q_1^c \subset P_0^{cc} \subset P_1^{cc}$ are both standard. By Lemma 15.34, we can construct a Jones tower

$$P_1^c \subset P_0^c \subset Q_1 \subset Q_2 \subset Q_3 \subset \cdots \subset M_\infty^\omega$$

so that

$$\cdots \subset Q_3^c \subset Q_2^c \subset Q_1^c \subset P_0^{cc} \subset P_1^{cc}$$

is a tunnel. We first have

$$P_0' \cap P_k \subset P_k^{cc} \cap P_0^c = (P_k^c)' \cap P_0^c.$$

With the trace-preserving anti-isomorphism, we can identify $(P_k^c)' \cap P_0^c$ with $P_0^{cc} \cap Q_k$, which is contained in $P_0 \cap Q_k^{cc} = (Q_k^c)' \cap P_0$, which is again contained in $P_0' \cap P_k$ with the trace-preserving anti-isomorphism. This shows that $P_0' \cap P_k = (P_k^c)' \cap P_0^c$. With this, we have proved Theorem 15.37 as desired.

Our next objective is to prove the following theorem for the group case. With the results in Section 12.8 and Theorem 15.37, we know the paragroup of the central sequence subfactor for a subfactor arising from a crossed product by an outer action of a finite group. Here we compute this paragroup directly.

Theorem 15.39 *Let N be a hyperfinite factor of type II_1 and M be the crossed product $N \rtimes G$ of N with an outer action of a finite group G. Then the asymptotic inclusion, $N^\omega \cap M' \subset M_\omega$, is of the form $Q \rtimes G \subset Q \rtimes (G \times G)$, where Q is some factor of type II_1 with an outer action of $G \times G$ and G is embedded into $G \times G$ with a map $g \mapsto (g, g)$.*

We denote the implementing unitaries in M by λ_g, the projections in M_1 corresponding to the group elements by f_g, and the implementing unitaries in M_2 by ρ_g, where g is an element of G. Note that we have relations $\lambda_g f_h \lambda_g^* = f_{gh}$, $\rho_g f_h \rho_g^* = f_{hg^{-1}}$ for $g, h \in G$.

First we claim that

$$N^\omega \cap M' \subset M_\omega \subset \langle M_\omega, f_1\rangle$$

is standard. Note that for $x = (x_n) \in M_\omega$, we get $E_{N^\omega \cap M'}(x) = (E_N(x_n))$ because M is the crossed product by a group action. Thus we have $f_1 x f_1 = E_{N^\omega \cap M'}(x) f_1$ for $x \in M_\omega$. So it is enough (as in Lemma 15.30) to prove that the central support q of f_1 in $\langle M_\omega, f_1\rangle$ is 1. The central freedom lemma (Lemma 15.25) implies $(M_\omega)' \cap M_1^\omega = M \vee (M' \cap M_1)^\omega = M$, so we get $E_{(M_\omega)' \cap M_1}(f_1) = E_m(f_1) = 1/n$, where n is the order of the group G. Then we have $1/n = E_{(M_\omega)' \cap M_1}(f_1) = E_{(M_\omega)' \cap M_1}(qf_1) = q/n$, and the claim is proved.

Next, let $P = N' \cap M_1^\omega$. By the central freedom lemma (Lemma 15.25), we get $P' \cap P = (N \vee (M_1' \cap M_1')^\omega) \cap N' = \mathbb{C}$, so P is a factor. We define an action of $G \times G$ on P by $\mathrm{Ad}(\lambda_g \cdot \rho_h)$ for $(g,h) \in G \times G$ and claim that this action is outer. Suppose that there exists a unitary $U \in N' \cap M_1^\omega$ with $\mathrm{Ad}U = \mathrm{Ad}(\lambda_g \cdot \rho_h)$ for some $(g,h) \in G \times G$. Then $\lambda_g \rho_h U^* \in (N' \cap M_1^\omega)' \cap M_2^\omega = N$, where we used the central freedom lemma (Lemma 15.25) again. This implies that U is in M_2 and hence in $N' \cap M_1^\omega \cap M_2 = N' \cap M_1$. By $\lambda_g \rho_h U^* \in N$ and $U \in M_1$, we get $\rho_h \in M_1$ and hence $h = 1$. By $\lambda_g U^* \in N$, we get $U \in M$ and then $U \in N' \cap M_1$ implies $U \in \mathbb{C}$ and $g = 1$, and thus the outerness claim is proved. We regard G as a subgroup of $G \times G$ with the diagonal embedding as in the theorem.

It is easy to see that M_ω is equal to the fixed point algebra $P^{G \times G}$ with this action. We next claim that $P^G = \langle M_\omega, f_1 \rangle$. The inclusion $\langle M_\omega, f_1 \rangle \subset P^G$ is trivial. Because $[P : P^G] = n$, it is enough to show $[P : \langle M_\omega, f_1 \rangle] \leq n$. By the commuting square condition, we get $[N' \cap M_1^\omega : N' \cap M^\omega] \leq n$, and it is easy to see $[N' \cap M^\omega : M' \cap M^\omega] \leq n$. Thus we have $[N' \cap M_1^\omega : M' \cap M^\omega] \leq n^2$. Because $M' \cap M^\omega \subset \langle M_\omega, f_1 \rangle \subset P$ and $[\langle M_\omega, f_1 \rangle : M_\omega] = n$, we get $[P : \langle M_\omega, f_1 \rangle] \leq n$, as desired.

Consequently, we have proved that the basic construction of the central sequence subfactor is of the form $P^{G \times G} \subset P^G$, which gives Theorem 15.39.

15.6 Automorphisms used in the orbifold construction

Consider the action α of $\mathbb{Z}/2$ on subfactors of type A_{4n-3} used in the orbifold construction in Section 13.4. The Loi invariant is always trivial on subfactors of type A_n, so all automorphisms of these subfactors are approximately inner by Theorem 15.5. Because the simultaneous fixed point algebra gives a different subfactor D_{2n}, this action α is different from the 'standard' action of $\mathbb{Z}/2$, which is of the form $\mathrm{id} \otimes \sigma$ on $N \bar{\otimes} R \subset M \bar{\otimes} R$, where σ is a (unique) outer action of $\mathbb{Z}/2$ on the common hyperfinite II_1 factor R and $N \bar{\otimes} R \subset M \bar{\otimes} R$ is isomorphic to $N \subset M$, because the two subfactors have the same higher relative commutants. We will show that this action α is centrally trivial.

We work in a more general situation. Let α be an automorphism appearing in the orbifold construction for subfactors arising from RCFT as in Section 13.4. Then computations based on topological moves of knotted graphs as in Section 13.4 show the identities in Fig. 15.2 and Fig. 15.3 for any k, where α denotes the induced symmetry of the graph. We can show that the identity in Fig. 15.2 implies triviality of the Loi invariant of α and that the identity in Fig. 15.3 implies central triviality of α as follows.

In Theorem 13.4, we concretely computed the higher relative commutants of the subfactor arising from RCFT with multiple starting vertices. The description there shows that the action α acts trivially on the higher relative commutants. Thus α is approximately inner by Theorem 15.5.

In the context of the double sequence of string algebras with multiple starting vertices in Section 13.4, we can construct a natural tunnel

$$\cdots \supset N_3 \supset N_2 \supset N_1 \supset N \supset M$$

with the same multiple starting vertices as in Section 11.4. (Note that this tunnel

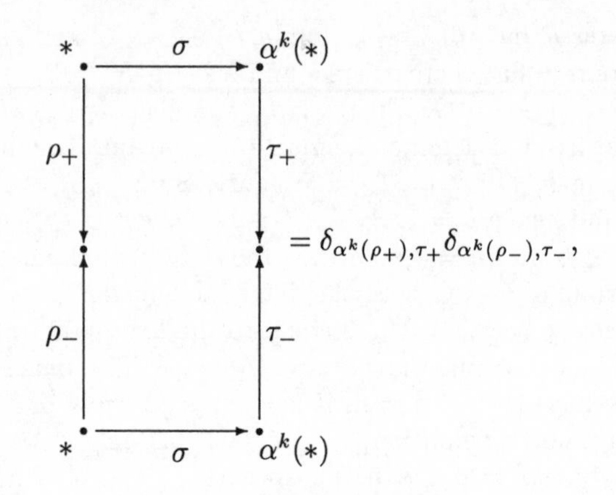

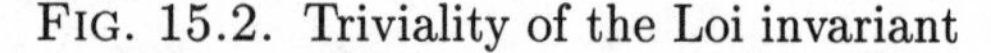

FIG. 15.2. Triviality of the Loi invariant

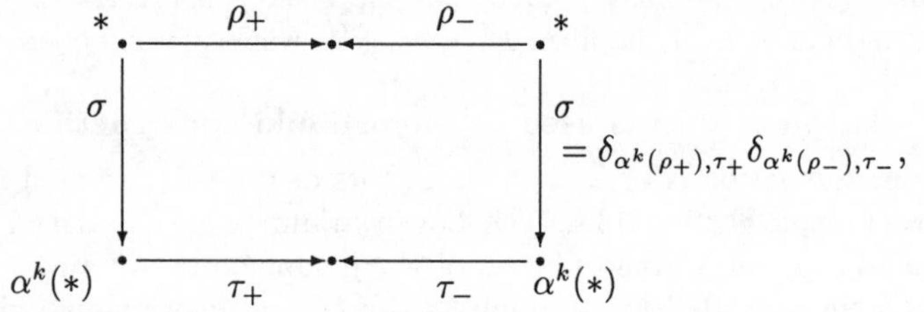

$*$ ρ_+ ρ_- $*$

σ σ $= \delta_{\alpha^k(\rho_+),\tau_+}\delta_{\alpha^k(\rho_-),\tau_-},$

$\alpha^k(*)$ τ_+ τ_- $\alpha^k(*)$

FIG. 15.3. Central triviality

does not have a generating property.) We observe that $N^\omega \cap M' \subset N_n^\omega$. The algebra N_n has minimal projections p_j corresponding to the multiple starting vertices at the first stage. Let $x \in N^\omega \cap M'$. Then we may assume that $x \in \sum_j (p_j N_n p_j)^\omega$ and x is expressed as $\sum_j x_j$ in this decomposition. Now x must commute with N hence with vertical strings with length n, so looking at the vertical parallel transport from the multiple endpoints of the principal graph to the fixed point by the symmetry of the principal graph for a sufficiently large n, we can conclude that $\alpha(x_j) = x_{j+1}$. This means that α acts trivially on $N^\omega \cap M'$.

In this sense, approximate innerness and central triviality appear in a symmetric way, and we get that $\alpha \in \mathrm{Ct}(M, N) \cap \overline{\mathrm{Int}}(M, N)$. Note that we get this regardless of whether the orbifold construction produces a flat connection.

Next we aim to show that the automorphisms appearing in the orbifold construction for subfactors arising from RCFT as in Section 13.4 actually give the entire group of centrally trivial automorphisms of the subfactor up to $\mathrm{Int}(M, N)$ as long as M does not have a non-trivial normalizer of N, which is the case in most examples. For example, in the case of quantum $SU(n)_k$ subfactors arising from the WZW-model $SU(n)_k$ as in Section 13.4 with indices $\sin^2(n\pi/(k+n))/\sin^2(\pi/(k+n))$, we get $\mathrm{Ct}(M, N)/\mathrm{Int}(M, N) = \mathbb{Z}/d$

with $d = (k, n)$ as long as M does not have a non-trivial normalizer of N. This is true because Theorems 15.7 and 15.16 imply that the order of the group $\mathrm{Ct}(M, N)/\mathrm{Int}(M, N)$ is bounded by the number of even vertices of the dual principal graph with normalized Perron–Frobenius weights 1 and this upper bound is now attained. Based on this observation, it is possible to extend the orbifold construction to general subfactors without hyperfiniteness. See (Goto 1994), (Goto 1995).

In the analogy Table 15.1, the orbifold construction in Section 13.4 is regarded as an analogue of the change of the flows of weights in crossed products by the modular automorphism groups as in (Kawahigashi and Takesaki 1992), (Sekine 1990).

We close this section with a remark for operator algebraists. We can verify that the orbifold $\mathbb{Z}/2$ actions on A_{4n-3} and the $\mathbb{Z}/2$ actions flipping the two tails of D_{2n} are in Takesaki duality. This is regarded as an analogue of the duality between the Connes–Takesaki module (Connes and Takesaki 1977) and the Sutherland–Takesaki modular invariant (Sutherland and Takesaki 1989).

15.7 The relative Connes invariant χ

As a relative version of the Connes invariant $\chi(M)$ introduced in (Connes 1975c), we set

$$\chi(M, N) = \frac{\mathrm{Ct}(M, N) \cap \overline{\mathrm{Int}}(M, N)}{\mathrm{Int}(M, N)}.$$

Our first aim is derive an exact sequence for computing $\chi(M \rtimes G, N \rtimes G)$. First we need some lemmas.

Lemma 15.40 *Suppose $N^\omega \cap M'$ is a factor and $\alpha \in \mathrm{Aut}(M, N) \setminus \overline{\mathrm{Int}}(M, N)$. If (x_n) is a bounded sequence in N with $\|x_n \alpha(y) - y x_n\|_2 \to 0$ as $n \to \infty$ for all $y \in M$, then $\|x_n\|_2 \to 0$.*

Proof Put $x = (x_n) \in N^\omega$, then $x\alpha(y) = yx$ for all $y \in M$. Set

$$\sigma(y) = \begin{pmatrix} \alpha(y) & 0 \\ 0 & y \end{pmatrix} \in M^\omega \otimes M_2(\mathbb{C}), \qquad y \in M.$$

Then

$$\begin{pmatrix} 0 & 0 \\ x & 0 \end{pmatrix} \in \sigma(M)' \cap (N^\omega \otimes M_2(\mathbb{C})).$$

We claim that $\sigma(M)' \cap (N^\omega \otimes M_2(\mathbb{C}))$ is a factor. Indeed, suppose

$$\begin{pmatrix} a & b \\ c & d \end{pmatrix} \in \mathcal{Z}(\sigma(M)' \cap (N^\omega \otimes M_2(\mathbb{C}))).$$

Then

$$\begin{pmatrix} 1 & 0 \\ 0 & 0 \end{pmatrix}, \begin{pmatrix} 0 & 0 \\ 0 & 1 \end{pmatrix} \in \sigma(M)' \cap (N^\omega \otimes M_2(\mathbb{C}))$$

implies that $b = c = 0$. Furthermore, $a, d \in \mathcal{Z}(M' \cap N^\omega) = \mathbb{C}$, and we get $a = d$ because

$$\left[\begin{pmatrix} a & 0 \\ 0 & d \end{pmatrix}, \begin{pmatrix} 0 & 0 \\ x & 0 \end{pmatrix}\right] = 0.$$

Thus factoriality is proved and there exists a unitary $u \in N^\omega$ with

$$\begin{pmatrix} 0 & 0 \\ u & 0 \end{pmatrix} \in \sigma(M)' \cap (N^\omega \otimes M_2(\mathbb{C})).$$

Writing $u = (u_n)$, $u_n \in \mathcal{U}(N)$, we get $\lim_{n\to\infty} \mathrm{Ad}(u_n) = \alpha$. □

Lemma 15.41 *Suppose $N^\omega \cap M'$ is a factor and let G be a finite group in $\mathrm{Aut}(M, N)$ with $G \cap \overline{\mathrm{Int}}(M, N) = \{1\}$. If $(x_n) \in (N \rtimes G)^\omega \cap (M \rtimes G)'$, then $(x_n) = (E_{N^G}(x_n))$ in $(N \rtimes G)^\omega$.*

Proof Write $x_n = \sum_{g\in G} x_n^g u_g$, $x_n^g \in N$. For each g, (x_n^g) is a bounded sequence and $\|x_n^g g(a) - a x_n^g\|_2 \to 0$ for all $a \in M$. By Lemma 15.40, we get $\|x_n^g\|_2 \to 0$ for all $g \neq 1$. Since $\|[x_n, u_g]\|_2 \to 0$ and

$$E_{N^G}(x) = \frac{1}{|G|} \sum_{g\in G} g(x), \qquad x \in N$$

we obtain the conclusion. □

Lemma 15.42 *Let $N \subset M$ be a subfactor and $A \subset N$ a subalgebra with the condition $N^\omega \cap M' \subset A^\omega$. Then if $\alpha \in \overline{\mathrm{Int}}(M, N)$, there is a unitary $w \in N$ and a sequence of unitaries z_n of A such that $\alpha = \mathrm{Ad}(w) \lim_{n\to\infty} \mathrm{Ad}(z_n)$.*

Proof For each $\varepsilon > 0$, there is a neighbourhood V_ε of id in $\mathrm{Aut}(M, N)$ such that if $\mathrm{Ad}(u) \in V_\varepsilon$ with $u \in \mathcal{U}(N)$, then $\|u - E_A(u)\|_2 < \varepsilon/6$. (Otherwise, we could get a central sequence contradicting the hypothesis.) By the polar decomposition, we get a unitary $v \in A$ with $\|u - v\|_2 < \varepsilon$.

Given $\alpha \in \overline{\mathrm{Int}}(M, N)$, choose a decreasing sequence U_n of neighbourhoods of α with $U_n^{-1} U_n \subset V_{1/2^n}$ and $\mathrm{Ad}(u_n) \in U_n$ with $\lim_{n\to\infty} \mathrm{Ad}(u_n) = \alpha$. Then $\mathrm{Ad}(u_n^* u_{n+1}) \in V_{1/2^n}$ so there is a unitary $v_n \in A$ with $\|u_n^* u_n + 1 - v_n\|_2 \leq 1/2^n$. This means $\|u_n v_n u_{n+1}^* - 1\|_2 \leq 1/2^n$ so that $\prod_{i=1}^n u_i v_i u_{i+1}^*$ is Cauchy. Let its limit be t. Then

$$\prod_{i=1}^k v_i = u_1^* \left(\prod_{i=1}^k u_i v_i u_{i+1}^* \right) u_{k+1},$$

so that $\lim_{n\to\infty} \mathrm{Ad}(\prod_{i=1}^n v_i)$ exists and is equal to $\mathrm{Ad}(u_1^* t)\alpha$. But each of the v_i is in A so, putting $z_n = \prod_{i=1}^n v_i$ and $w = t^* u_1$, we have the result. □

Now suppose $N^\omega \cap M'$ is a factor and let G be a finite subgroup of $\mathrm{Aut}(M, N)$ with $G \cap \overline{\mathrm{Int}}(M, N) = \{1\}$. We set $K = G \cap \mathrm{Ct}(M, N)$,

$$K^\perp = \{\gamma : G \to \mathbb{T} \mid \gamma \text{ a character vanishing on } K\},$$

and L to be the image of

$$G \cdot \mathrm{Ct}(M,N) \cap \overline{\{\mathrm{Ad}(u) \mid u \in \mathcal{U}(N^G)\}}$$

in $\mathrm{Aut}(M,N)/\mathrm{Int}(M,N)$. Then we have the following.

Theorem 15.43 *In the above context, we have maps $\delta : K^\perp \to \chi(M \rtimes G, N \rtimes G)$ and $\pi : \chi(M \rtimes G, N \rtimes G) \to L$ such that the following sequence is exact:*

$$\{1\} \to K^\perp \to \chi(M \rtimes G, N \rtimes G) \to L \to \{1\}.$$

Proof We define $\delta(\eta) \in \mathrm{Aut}(M,N)$ by $\delta(\eta)(\sum_g a_g u_g) = \sum_g \eta(g) a_g u_g$. By Lemma 15.42, we get $\delta(\eta) \in \mathrm{Ct}(M \rtimes G, N \rtimes G)$.

Next, note that the action of G on $N^\omega \cap M'$ is by outer automorphisms and the identity on K. Hence it gives a stable action of G/K and thus if $\eta \in K^\perp$, $\bar{\eta}$ is a coboundary, where $\bar{\eta}(g) = \overline{\eta(g)}$. (An action is called stable when this kind of 1-cohomology vanishing holds. See Proposition 3.1.2 in (Jones 1980b), for example.) Then we have a unitary $U \in N^\omega \cap M'$ with $g(U) = \bar{\eta}(g)U$. Representing U by a sequence of unitaries, we get unitaries (u_n) in N which are central with respect to M such that $\|g(u_n) - \bar{\eta}(g)u_n\|_2 \to 0$. Then $\mathrm{Ad}(u_n) \to \delta(\eta)$ in $\mathrm{Aut}(M \rtimes G, N \rtimes G)$ by

$$\begin{aligned} \|u_n u_g u_n^* - \eta(g) u_g\|_2 &= \|u_g u_n u_g^* - \bar{\eta}(g) u_n\|_2 \to 0, \\ \|u_n a u_n^* - a\|_2 &\to 0, \qquad a \in M. \end{aligned}$$

It is easy to see that $\delta(\eta) \notin \mathrm{Int}(M,N)$ for $\eta \neq 0$, so this δ is an injective homomorphism from $K^\perp$ to $\chi(M \rtimes G, N \rtimes G)$.

Next we define $\pi : \chi(M \rtimes G, N \rtimes G) \to L$. So let $\alpha \in \mathrm{Ct}(M \rtimes G, N \rtimes G) \cap \overline{\mathrm{Int}}(M \rtimes G, N \rtimes G)$ represent an element of $\chi(M \rtimes G, N \rtimes G)$. By Lemma 15.42, we get $\alpha = \mathrm{Ad}(w) \lim_{n\to\infty} \mathrm{Ad}(z_n)$ with unitaries $w \in N \rtimes G$ and $z_n \in N^G$. We want to define $\pi(\alpha) \in L$ as the image of the automorphism $\theta = \lim \mathrm{Ad}(z_n) \in \mathrm{Aut}(M,N)$.

First let us show that θ is well defined, i.e. if $\alpha = \mathrm{Ad}(y) \lim_{n\to\infty} \mathrm{Ad}(v_n)$ with $y \in N \rtimes G$, $v_n \in N^G$ and $\theta' = \lim_{n\to\infty} \mathrm{Ad}(v_n)$, then $\theta = \theta'$ mod $\mathrm{Int}(M,N)$. But under these circumstances $\theta = \mathrm{Ad}(t)\theta'$ for some unitary $t \in N \rtimes G$. But since θ and θ' leave N globally, so does $\mathrm{Ad}(t)$. Thus t is in the normalizer of N inside $N \rtimes G$. It is not hard to show that all elements of the normalizers are of the form xu_g for some $g \in G$ and $x \in N$. Hence $\theta = \mathrm{Ad}(xu_g)\theta'$ But both θ and θ' are in $\overline{\mathrm{Int}}(M,N)$, so that $\mathrm{Ad}(u_g) \in \overline{\mathrm{Int}}(M,N)$. But by hypothesis $G \cap \overline{\mathrm{Int}}(M,N) = \{1\}$. Hence $g = id$ and $\theta = \theta'$ modulo $\mathrm{Int}(M,N)$.

The next thing to check is that θ is indeed in $G \cdot \mathrm{Ct}(M,N)$. For this, note that, since $\alpha \in \mathrm{Ct}(M \rtimes G, N \rtimes G)$, θ must act trivially on $(N^\omega \cap M')^G$. Since $N^\omega \cap M'$ is a factor and G/K acts on this outerly, it is easy to see that $\theta \in G \cdot \mathrm{Ct}(M,N)$.

At this stage, we have a well-defined map from $\chi(M \rtimes G, N \rtimes G)$ to L. It easily follows that this is a homomorphism. We must now show exactness. We already have $\mathrm{Ker}\, \delta = 0$ so let us show the other three inclusions.

(1) $\mathrm{Im}\,\delta \subset \mathrm{Ker}\,\pi$. We express $\delta(\eta)$ as $\mathrm{Ad}(w)\lim_{n\to\infty}\mathrm{Ad}(z_n)$ with $w \in N \rtimes G$ and $z_n \in N^G$. Because w is a normalizer of N in $N \rtimes G$, we get $w = xu_g$ with $x \in N$. By $G \cap \mathrm{Int}\, b(M,N) = \{1\}$, we get $g = 1$, and then get $\pi(\delta(\eta)) = 0$ by the definition of π.

(2) $\mathrm{Ker}\,\pi \subset \mathrm{Im}\,\delta$. If $\pi([\alpha]) = 0$, then up to an inner automorphism given by a unitary in $N \rtimes G$, α is the identity on M. Then α is a dual automorphism, i.e., $\alpha(\sum_g a_g u_g) = \sum_g \eta(g) a_g u_g$ for some homomorphism $\eta : G \to \mathbb{T}$. But such an η must annihilate $G \cap \mathrm{Ct}(M,N)$ as follows. Since α is given by $\lim_{n\to\infty}\mathrm{Ad}(v_n)$ with $v_n \in N$ central in M and for $g \in \mathrm{Ct}(M,N)$,

$$\|v_n u_g v_n^* - u_g\|_2 = \|u_g v_n^* u_g^* - v_n^*\|_2 \to 0$$

as $n \to \infty$. Hence $\alpha(u_g) = u_g$ and η annihilates $\mathrm{Ct}(M,N) \cap G$.

(3) The last property to show is that π is surjective. We shall construct a section for π as follows. Let $\alpha = \lim_{n\to\infty}\mathrm{Ad}(v_n) \in \mathrm{Aut}(M,N)$ for $v_n \in \mathcal{U}(N^G)$ and $g \in G$ such that $g\alpha \in \mathrm{Ct}(M,N)$. Then since α commutes with G, define $\beta = \mathrm{Aut}(M \rtimes G, N \rtimes G)$ by $\beta(\sum a_g u_g) = \sum_g \alpha(a_g) u_g$. Certainly, if $[\beta] \in \chi(M \rtimes G, N \rtimes G)$, then $\pi([\beta]) = [\alpha]$, because $\beta = \lim_{n\to\infty}\mathrm{Ad}(v_n)$ in $\mathrm{Aut}(M \rtimes G, N \rtimes G)$. It is clear that $\beta \in \overline{\mathrm{Int}}(M \rtimes G, N \rtimes G)$. Moreover, $\mathrm{Ad}(u_g)\beta$ is $g\alpha$ on M, so by Lemma 15.42, $\mathrm{Ad}(u_g)\beta$ acts trivially on all central sequences of $N \rtimes G$ with respect to $M \rtimes G$. Thus $[\beta] \in \chi(M \rtimes G, N \rtimes G)$. □

We have the following concrete example by the arguments of the previous section.

Proposition 15.44 *In the case of hyperfinite subfactors with index less than 4, we have:*

$$\begin{aligned}
\chi(M,N) &= 0, && \text{for } A_{2n}, D_{2n+4}, E_8, \text{ with } n \geq 1\\
\chi(M,N) &= \mathbb{Z}/2, && \text{for } A_{2n+3}, E_6, \text{ with } n \geq 1\\
\chi(M,N) &= \mathbb{Z}/2 \oplus \mathbb{Z}/2, && \text{for } A_3\\
\chi(M,N) &= \mathbb{Z}/3 \oplus \mathbb{Z}/3, && \text{for } D_4.
\end{aligned}$$

The subfactors of type A_3 and D_4 are exceptional because they have non-trivial normalizers. (See the Remark following Theorem 15.16.)

In the case of the quantum $SU(n)_k$ subfactors of Section 13.3 with index $\sin^2(n\pi/(k+n))/\sin^2(\pi/(k+n))$, we considered a $\mathbb{Z}/(k,n)$ action arising from a paragroup symmetry for the orbifold construction in Section 13.4 for the case $k \equiv 0 \pmod n$. These automorphisms are in both $\overline{\mathrm{Int}}(M,N)$ and $\mathrm{Ct}(M,N)$. We also have the following example.

Proposition 15.45 *For the quantum $SU(n)_k$ subfactors with index*

$$\sin^2(n\pi/(k+n))/\sin^2(\pi/(k+n)), \quad N \geq 3$$

in Section 13.3, we get $\chi(M,N) = \mathbb{Z}/d$ with $d = (k,n)$.

Next, we will prove that all the finite abelian groups can be realized as $\chi(M,N)$ for some hyperfinite type II_1 subfactor $N \subset M$ with finite index, finite depth, and trivial relative commutants. We first construct $\mathbb{Z}/d$ with Proposition 15.45 and then get an arbitrary finite abelian group using tensor products. One can show that the tensor product of two centrally trivial automorphisms is centrally trivial and that the tensor product of two approximately inner automorphisms is approximately inner. Thus we have the following.

Theorem 15.46 *Any finite abelian group can be realized as $\chi(M,N)$ for a hyperfinite type II_1 subfactor $N \subset M$ with finite index, finite depth and trivial normalizer.*

It is a very important problem to distinguish non-amenable subfactors with the same indices using $\chi(M,N)$ as an analogue of the Connes construction in (Connes 1975c) for single II_1 factors. Our exact sequence, Theorem 15.43, is expected to be useful for this purpose, but so far, no example has been computed in the non-amenable subfactor case.

Finally, we discuss the problem of finding non-commutative symmetries of paragroups. In the orbifold construction in Section 13.4, the finite group acting on subfactors arising from RCFT was the centre of the Lie group with which we started. In particular, this finite group is always abelian. In general, there seems to be no reason why a symmetry group of a given paragroup is abelian and the problem of finding an interesting non-abelian symmetry has caught the attention of physicists, but our viewpoint of automorphisms suggests that any interesting symmetry should be abelian. That is, $\chi(M,N)$ is always abelian, though $\mathrm{Ct}(M,N)/\mathrm{Int}(M,N)$ is not necessary abelian. (A trivial example is a subfactor $R^G \subset R$ with G finite, non-commutative, and acting on R outerly.) See (Goto 1996) for further discussion on this.

15.8 The relative Jones invariant κ and orbifold subfactors

Our aim in this section is to introduce a relative Jones invariant κ as a finer invariant than the relative Connes invariant χ and identify it with the conformal dimension for the quantum $SU(n)_k$ subfactors in Section 13.3.

Comparing the results in Sections 13.4 and 15.7, we noticed that the relative Connes invariant χ does not see the obstruction for flatness in the orbifold construction. This is rather unsatisfactory because it is a general belief that an algebraic property (such as flatness) should be equivalent to an analytic property of ultraproducts/central sequences as long as we have a certain amenability condition such as strong amenability of subfactors. (We have seen in Sections 15.2 and 15.4 that such a belief can be realized.) Our aim in this section is to clarify the situation by introducing the relative Jones invariant κ as a finer invariant than the relative Connes invariant χ. We give the definition of κ as follows.

Let $N \subset M$ be a subfactor with finite index. Choose $\alpha, \beta \in \mathrm{Ct}(M,N) \cap \overline{\mathrm{Int}}(M,N)$. As $\beta \in \overline{\mathrm{Int}}(M,N)$, we have unitaries $\{u_n\}_n$ in N with $\beta = \lim_{n\to\infty} \mathrm{Ad}(u_n)$. We look at $\alpha \cdot \beta \cdot \alpha^{-1}$. On one hand, this is equal to $\mathrm{Ad}(u) \cdot \beta = \lim_{n\to\infty} \mathrm{Ad}(uu_n)$ for some unitary $u \in N$ because $\chi(M,N)$ is abelian, and on the other hand this is clearly equal to $\lim_{n\to\infty} \mathrm{Ad}(\alpha(u_n))$. These imply that the sequence $\{u_n^* u^* \alpha(u_n)\}_n$ is central in M. We also know that this sequence is a Cauchy sequence by the following estimate:

$$\begin{aligned}\|u_n^* u^* \alpha(u_n) - u_m^* u^* \alpha(u_m)\|_2 &= \|u_m u_n^* u^* \alpha(u_n u_m^*) - u^*\|_2 \\ &\to 0, \quad \text{as } n, m \to \infty,\end{aligned}$$

because of the central triviality of α. Thus there exists a complex number $\kappa(a, \beta)$ with modulus one such that $\lim_{n\to\infty} u_n^* u^* \alpha(u_n) = \kappa(\alpha, \beta)$. A standard argument shows that this number $\kappa(\alpha, \beta)$ does not depend on the choice of u and $\{u_n\}_n$. A direct computation using $\alpha \cdot \beta \cdot \alpha^{-1} = \lim_{n\to\infty} \mathrm{Ad}(uu_n)$ shows that $\kappa(a, \beta) = \kappa(\mathrm{Ad}(v) \cdot \alpha, \mathrm{Ad}(w) \cdot \beta)$ for unitaries $v, w \in N$, thus κ is a well-defined map from $\chi(M, N) \times \chi(M, N)$ to $\mathbb{T}$.

We also define $\kappa(\alpha) = \kappa(\alpha, \alpha)$ for $\alpha \in \chi(M, N)$. The following proposition means that this κ is a quadratic form over $\chi(M, N)$. We omit its proof because we do not use this proposition in this book. See (Chen 1993) for a proof in the single factor case.

Proposition 15.47 1. *For $\alpha \in \chi(M, N)$, we have $\kappa(\alpha) = \kappa(\alpha^{-1})$.*
2. *The form $b_\kappa(\alpha, \beta) = \kappa(\alpha\beta)\overline{\kappa(\alpha)\kappa(\beta)}$ is symmetric and bilinear.*

We can also define κ using ultraproducts as follows. We fix a free ultrafilter ω over $\mathbb{N}$. Let U be a unitary in N^ω implementing an approximately inner automorphism β. Then $a(U) = \kappa(\alpha, \beta) uU$ as above.

Let $SU(n)_k$ be the Wess–Zumino–Witten model for $SU(n)$ with level k. Theorem 15.45 says that $\chi(M, N)$ for the corresponding subfactor is $\mathbb{Z}/d$, where $d = (n, k)$. In these cases, the $\mathbb{Z}/d$ action is realized in a concrete way as in Section 13.4, so we can take $u = 1$ in the above definition. We will compute κ for these subfactors and identify κ with the obstruction for flatness in the orbifold construction. We fix n, k with $d > 1$. We denote the global index of $N \subset M$ by γ as in Theorem 12.24. We denote the $\mathbb{Z}/d$ action on $N \subset M$ by α.

We have the following lemma.

Lemma 15.48 *The global index of the orbifold subfactor $N \rtimes_\alpha \mathbb{Z}/d \subset M \rtimes_\alpha \mathbb{Z}/d$ is given by γ/d if d is odd or $2d$ divides k and by $2\gamma/d$ if d is even and $2d$ does not divide k.*

Proof This follows from Theorem 13.7 and a description of A^α_{kl} in Definition 13.5. □

Our next aim is to compute $(N \rtimes_\alpha \mathbb{Z}/d)^\omega \cap (M \rtimes_\alpha \mathbb{Z}/d)'$ and $(M \rtimes_\alpha \mathbb{Z}/d)^\omega \cap (M \rtimes_\alpha \mathbb{Z}/d)'$ for a fixed free ultrafilter ω over $\mathbb{N}$. We first compute $(N \rtimes_\alpha \mathbb{Z}/d)^\omega \cap M'$ and $(M \rtimes_\alpha \mathbb{Z}/d)^\omega \cap M'$. Choose a sequence of unitaries $\{u_n\}_n$ in N so that $\alpha_1 = \lim_{n\to\infty} \mathrm{Ad}(u_n)$. We denote by U the element in N^ω corresponding to the sequence $\{u_n\}_n$. We denote by u the implementing unitary for α_1 in the crossed product. A general element in $N \rtimes_\alpha \mathbb{Z}/d$ is written as $\sum_{j=0}^{d-1} X_j u^j$ where $X_j \in N^\omega$. Suppose this element commutes with M. Then we have $X_j u^j x = x X_j u^j$ for all

$x \in M$ and $j = 0, 1, \ldots, d-1$. Because $uxu^* = \alpha_1(x) = UxU^*$ for $x \in M$, the condition we have is $X_j U^j \in N^\omega \cap M'$. So a general element in $(N \rtimes_\alpha \mathbb{Z}/d)^\omega \cap M'$ is expressed as $\sum_{j=0}^{d-1} Y_j (U^* u)^j$, where each Y_j is in $N^\omega \cap M'$. Similarly, we know that a general element in $(M \rtimes_\alpha \mathbb{Z}/d)^\omega \cap M'$ is expressed as $\sum_{j=0}^{d-1} Z_j (U^* u)^j$, where each Z_j is in $M^\omega \cap M'$.

Next we compute $(N \rtimes_\alpha \mathbb{Z}/d)^\omega \cap (M \rtimes_\alpha \mathbb{Z}/d)'$. Let $\kappa = \kappa(\alpha_1, \alpha_1)$. Then we have $u(U^* u)u^* = \bar{\kappa}$ and $\kappa^d = 1$. Let l be the minimal positive integer with $\kappa^l = 1$. Note that the quadratic form $\kappa(\cdot, \cdot)$ is trivial if and only if $l = 1$. Then a general element in $(N \rtimes_\alpha \mathbb{Z}/d)^\omega \cap (M \rtimes_\alpha \mathbb{Z}/d)'$ is expressed as $\sum_{j=0}^{d/l-1} Y_j (U^* u)^{jl}$, where each Y_j is in $N^\omega \cap M'$.

We need the following lemma.

Lemma 15.49 *The action* $\mathrm{Ad}(U^* u)$ *gives an automorphism* σ *of* $N^\omega \cap M' \subset M_\omega$ *and* σ^j *is outer for* $j = 1, 2, \ldots, d-1$ *on both factors* $N^\omega \cap M'$ *and* M_ω.

Proof The action $\mathrm{Ad}(u)$ acts trivially on $N^\omega \cap M'$ and freely on M_ω. Because $UNU^* = N$ and $UMU^* = M$, it is clear that σ gives an automorphism of $N^\omega \cap M' \subset M_\omega$.

Suppose that σ^j is inner on $N^\omega \cap M'$ for some j with $0 < j < d$. Then we have a unitary $V \in N^\omega \cap M'$ with $VXV^* = U^{-j} X U^j$ for all $X \in N^\omega \cap M'$. Then we have $U^j V \in (N^\omega \cap M')' \cap N^\omega$. By Lemma 15.21 and the proof of the central freedom lemma (Lemma 15.25), we get $(N^\omega \cap M')' \cap N^\omega$. This means we have a unitary $v \in N$ such that

$$\mathrm{Ad}(v)(x) = \mathrm{Ad}(U^j V)(x) = \mathrm{Ad}(U^j)(x) = \alpha_j(x),$$

for all $x \in M$, which contradicts the freeness of α.

The freeness of σ on M_ω is proved similarly. □

By Lemma 15.49, we get

$$[(N \rtimes_\alpha \mathbb{Z}/d)^\omega \cap (M \rtimes_\alpha \mathbb{Z}/d)' : N^\omega \cap M'] = \frac{d}{l}.$$

It is known that α on M_ω is an outer action, and so $[M_\omega : M_\omega^\alpha] = d$ and

$$[(M \rtimes_\alpha \mathbb{Z}/d)^\omega \cap (M \rtimes_\alpha \mathbb{Z}/d)' : M_\omega^\alpha] = d.$$

(See (Connes 1977).) By the central triviality of α, we get

$$\begin{aligned} N^\omega \cap M' &\subset M_\omega^\alpha \subset (M \rtimes_\alpha \mathbb{Z}/d)^\omega \cap (M \rtimes_\alpha \mathbb{Z}/d)', \\ N^\omega \cap M' &\subset (N \rtimes_\alpha \mathbb{Z}/d)^\omega \cap (M \rtimes_\alpha \mathbb{Z}/d)' \\ &\subset (M \rtimes_\alpha \mathbb{Z}/d)^\omega \cap (M \rtimes_\alpha \mathbb{Z}/d)'. \end{aligned}$$

Because $[M_\omega^\alpha : N^\omega \cap M'] = \gamma/d$, the identity

$$[(M \rtimes_\alpha \mathbb{Z}/d)^\omega \cap (M \rtimes_\alpha \mathbb{Z}/d)' : M_\omega^\alpha]\,[M_\omega^\alpha : N^\omega \cap M']$$

$$= [(M \rtimes_\alpha \mathbb{Z}/d)^\omega \cap (M \rtimes_\alpha \mathbb{Z}/d)' : (N \rtimes_\alpha \mathbb{Z}/d)^\omega \cap (M \rtimes_\alpha \mathbb{Z}/d)']$$
$$\times [(N \rtimes_\alpha \mathbb{Z}/d)^\omega \cap (M \rtimes_\alpha \mathbb{Z}/d)' : N^\omega \cap M'],$$

together with Lemma 15.48 and Theorem 13.7 implies that if the resulting connection in the orbifold construction is flat, then $l = 1$, and that if the resulting connection in the orbifold construction is not flat, then $l = 2$. Thus we have the following theorem.

Theorem 15.50 *For the quantum $SU(n)_k$ subfactor $N \subset M$, which has $\chi(M,N) = \mathbb{Z}/d$ with $d = (k,n)$, the relative Jones invariant κ is trivial if and only if the resulting connection in the $\mathbb{Z}/d$-orbifold construction is flat.*

For other orbifold subfactors arising from a connected, simply connected, compact and simple Lie group G as in Section 13.4, we have a similar result.

15.9 Notes

The basic definitions and results in Section 15.2 are due to (Loi 1996). (We used here towers in Definition 15.2 instead of the tunnels used in (Loi 1996).) Theorems 15.16 and 15.18 were given in (Popa, preprint 1992). Strongly outer automorphisms of subfactors were introduced by (Choda and Kosaki 1994). The same notion was also independently introduced with the name 'properly outer automorphisms' in (Popa 1992). The characterization, Theorem 15.7, is due to (Choda and Kosaki 1994). Theorem 15.8 is due to (Kosaki 1993). Our presentation here follows (Goto 1994). Approximately inner automorphisms and centrally trivial automorphisms of subfactors were introduced by (Loi 1996).

The central sequences have played an important role in classification theory of von Neumann algebras since (Dixmier and Lance 1969). They were used to construct mutually non-isomorphic II_1 factors by (McDuff 1969), (Sakai 1970). The most important technique to use central sequences was established by (Connes 1975a) based on results of (McDuff 1970). The central sequence subfactor was first introduced by (Ocneanu 1988). Theorem 15.37 was given by A. Ocneanu in his Tokyo lectures in 1990 together with sketches of his proofs. The presentation here is based on (Kawahigashi 1993), (Kawahigashi 1995b).

The results in Section 15.6 are due to (Evans and Kawahigashi 1993). Goto's results mentioned here are in (Goto 1994), (Goto 1995). The relative Connes invariant χ in Section 15.7 was introduced in (Kawahigashi 1993). For the treatment of the original Connes invariant in a single factor case, we followed an unpublished manuscript 'Notes on Connes' invariant $\chi(M)$' of V. F. R. Jones and (Chen 1993). An interesting relation between the original χ and the relative χ has been found by (Rădulescu, preprint 1996). The results in Section 15.8 are due to (Kawahigashi 1995b). The invariant κ for a single factor was introduced in (Jones 1980a) and studied in (Chen 1993).

BIBLIOGRAPHY

Aizenmann, M. (1980). Translation invariance and instability of phase co-existence in the two dimensional Ising system. *Communications in Mathematical Physics*, **73**, 83–94.

Akutsu, Y. and Wadati, M. (1987). Knot invariants and critical statistical systems. *Journal of the Physics Society of Japan*, **56**, 839–842.

Albeverio, S. and Høegh-Krohn, R. (1980). Ergodic actions by compact groups on C^*-algebras. *Mathematische Zeitschrift*, **174**, 1–17.

Alexander, J. W. (1930). The combinatorial theory of complexes. *Annals of Mathematics*, **(2) 31**, 294–322.

Allen, G. D., Narcowich, F. J. and Williams, J. P. (1975). An operator version of a theorem of Kolmogorov. *Pacific Journal of Mathematics*, **61**, 305–312.

Allen, J. R. (1899). Celtic crosses of Wales. *Archaeologica Cambrensis.*

Altschuler, D., Lacki, J. and Zaugg, Ph. (1988). The affine Weyl group and modular invariant partition functions. *Physics Letters B*, **205**, 281–284.

Andrews, G. E. (1976). The theory of partitions. In *Encyclopedia of mathematics and its applications* (ed. G.-C. Rota), Addison-Wesley, Reading, Massachusetts.

Andrews, G. E., Baxter, R. J., and Forrester, P. J. (1984). Eight vertex SOS model and generalized Rogers–Ramanujan type identities. *Journal of Statistical Physics*, **35**, 193–266.

Araki, H. (1960). Hamiltonian formalism and the canonical commutation relations in quantum field theory. *Journal of Mathematical Physics*, **1**, 492–504.

Araki, H. (1963). A lattice of von Neumann algebras associated with the quantum theory of a free Bose field. *Journal of Mathematical Physics*, **4**, 1343–1362.

Araki, H. (1968). On the diagonalization of a bilinear Hamiltonian by a Bogoliubov transformation. *Publications of the RIMS, Kyoto University* Ser. A, **4**, 387–412.

Araki, H. (1969). Gibbs states of a one dimensional quantum lattice. *Communications in Mathematical Physics*, **14**, 120–157.

Araki, H. (1970a). On quasi-free states of CAR and Bogoliubov automorphisms. *Publications of the RIMS, Kyoto University*, **6**, 385–442.

Araki, H. (1970b). Factorizable representations of current algebra. *Publications of the RIMS, Kyoto University*, **5**, 361–422.

Araki, H. (1971). On quasi-free states of the canonical commutation relations (II). *Publications of the RIMS, Kyoto University*, **7**, 121–152.

Araki, H. (1972). Normal positive linear mappings of norm 1 from a von Neumann algebra into its commutant and its application. *Publications of the RIMS, Kyoto University*, **8**, 439–469.

Araki, H. (1984). On the XY model on two sided infinite chain. *Publications of the RIMS, Kyoto University*, **20**, 227–296.

Araki, H. (1986). Analyticity of correlation functions for the two dimensional Ising model. *Communications in Mathematical Physics*, **106**, 241–266.

Araki, H. (1987). Canonical anticommutation relations. *Contemporary Mathematics*, **62**, 23–186.

Araki, H. and Barouch, E. (1983). On the dynamics and ergodic properties of the XY-model. *Journal of Statistical Physics*, **31**, 327–345.

Araki, H. and Evans, D. E. (1983). On a C^*-algebra approach to phase transition in the two dimensional Ising model. *Communications in Mathematical Physics*, **91**, 489–503.

Araki, H. and Matsui, T. (1985). Ground states of the XY-model. *Communications in Mathematical Physics*, **101**, 213–245.

Araki, H. and Matsui, T. (1986). Analyticity of ground states of the XY-model. *Letters in Mathematical Physics*, **11**, 87–94.

Araki, H. and Woods, E. J. (1968). A classification of factors. *Publications of the RIMS, Kyoto University*, Ser. A, **4**, 51–130.

Araki, H. and Wyss, W. (1964). Representations of canonical anticommutation relations. *Helvetica Physica Acta*, **37**, 136–159.

Araki, H., Carey, A. L., and Evans, D. E. (1984). On $\mathcal{O}_{n+1}$. *Journal of Operator Theory*, **12**, 247–264.

Arcuri, R. C., Gomes, J. F., and Olive, D. I. (1987). Conformal subalgebras and symmetric spaces. *Nuclear Physics B*, **285**, 327–339.

Arendt, W., Chernoff, P. R., and Kato, T. (1982). A generalisation of dissipativity and positive semigroups. *Journal of Operator Theory*, **8**, 167–180.

Aronszajn, N. (1950). Theory of reproducing kernels. *Transactions of the American Mathematical Society*, **68**, 337–404.

Artin, E. (1947). Theory of braids. *Annals of Mathematics*, **48** 101–126.

Arveson, W. B. (1974). On groups of automorphisms of operator algebras. *Journal of Functional Analysis*, **15**, 217–243.

Arveson, W. B. (1976). An invitation to C^*-algebras. *Graduate Texts in Mathematics*, **39**, Springer, Berlin.

Arveson, W. B. (1977). Notes on extensions of C^*-algebras. *Duke Mathematical Journal*, **44**, 329–355.

Arveson, W. B. (1982). The harmonic analysis of automorphism groups. In *Operator algebras and applications, Part 1, Proceedings of the Symposium on Pure Mathematics*, American Mathematical Society, Providence, **38**, 199–269.

Arveson, W. B. (1989a). Continuous analogues of Fock space. *Memoirs of the American Mathematical Society*, **80**(3).

Arveson, W. B. (1989b). An addition formula for the index of semigroups of endomorphisms of $B(H)$. *Pacific Journal of Mathematics*, **137**, 19–36.

Arveson, W. B. (1989c). Continuous analogues of Fock space III : singular states. *Journal of Operator Theory*, **22**, 165–205.

Arveson, W. B. (1990). Continuous analogues of Fock space II : the spectral C^*-algebra. *Journal of Functional Analysis*, **90**, 138–205.

Arveson, W. B. (1990). Continuous analogues of Fock space IV : essential states. *Acta Mathematica*, **164**, 265–300.

Asaeda, M. and Haagerup, U. (preprint 1998). Exotic subfactors of finite depth with Jones index $(5+\sqrt{13})/2$ and $(5+\sqrt{17})/2$.

Atiyah, M. (1967). *K-theory*. Benjamin, New York.

Atiyah, M. (1970). Global theory of elliptic operators. *Proceedings of an International Conference on Functional Analysis and Related Topics*, University of Tokyo Press, Tokyo.

Atiyah, M. (1989). Topological quantum field theory. *Publications Mathématiques IHES*, **68**, 175–186.

Atiyah, M. and Hirzebruch, F. (1961). Vector bundles and homogeneous spaces. *Proceedings of Symposia on Pure Mathematics*, American Mathematical Society, Providence, **3**, 7–38.

Atiyah, M. and Singer, I. (1963). The index of elliptic operators on compact manifolds. *Bulletin of the American Mathematical Society*, **69**, 422-433.

Atiyah, M. and Singer, I. (1968). The index of elliptic operators I, II, III. *Annals of Mathematics*, **87**, 484–530, 531–545 (with Segal, G.), 546–604.

Atiyah, M. F. and Singer, I. (1969). Index theory for skew-adjoint Fredholm operators. *Publications Mathématiques IHES*, **37**, 305–326.

Aubert, P.-L. (1976). Théorie de Galois pour une W^*-algèbre. *Commentarii Mathematici Helvetici*, **39 (51)**, 411–433.

Au-Yang, H. and Perk, J. H. H. (1985). Ising models and soliton equations. In *Proceeding of the III International Symposium on Selected Topics in Statistical Mechanics, Dubna, August 22–26, 1984 (JINR, Dubna, USSR), Vol. II*, 138–151.

Au-Yang, H. and Perk, J. H. H. (1987). Critical correlations in a Z-invariant inhomogeneous Ising model. *Physica A* **144**, 44–104.

Au-Yang, H. and Perk, J. H. H. (1989). Onsager's star–triangle equation : Master key to integrability. *Proceedings of the Taniguchi Symposium*, Kyoto, October 1988, in *Advanced Studies in Pure Mathematics*, (Kinokuniya-Academic, Tokyo), **19**, 57–94.

Au-Yang, H. and Perk, J. H. H. (1995). The chiral Potts models revisited. *Journal of Statistical Physics*, **78**, 17–78.

Baez, J. C., Segal, I. E. and Zhou, Z. (1992). Introduction to algebraic and constructive quantum field theory, *Princeton University Press*, Cambridge.

Baillet, M., Denizeau, Y., and Havet, J.-F. (1988). Indice d'une esp/'erance conditionnelle. *Composito Mathematica*, **66**, 199–236.

Bais, S. and Bouwknegt, P. (1987). A classification of subgroup truncations of the Bosonic string. *Nuclear Physics B*, **279**, 561–570.

Baker, B. M. (1978, 1979). Free states of the gauge invariant canonical anticommutation relations I, II. *Transactions of the American Mathematical Society*, **237**, 35–61; **254**, 135–155.

Baker, B. M. (1980). A central decomposition. *Journal of Functional Analysis*, **35**, 1–25.

Baker, B. M. and Powers, R. T. (1983a). Product states and C^*-dynamical systems of product type. *Journal of Functional Analysis*, **50**, 229–266.

Baker, B. M. and Powers, R. T. (1983b). Product states of the gauge invariant and rotationally invariant CAR algebras. *Journal of Operator Theory*, **10**, 365–393.

Balslev, E., Manuceau, J. and Verbeure, A. (1968a). Representations of anticommutation relations and Bogoliubov transformations. *Communications in Mathematical Physics*, **8**, 315–326.

Balslev, E., Manuceau, J. and Verbeure, A. (1968b). States on Clifford algebras. *Communications in Mathematical Physics*, **7**, 55–76.

Barouch, E., McCoy, B. M. and Wu, T. T. (1973). Zero field susceptibility of the two-dimensional Ising model near T_c. *Physical Review Letters*, **31**, 1409–1411.

Bashilov, Yu. A. and Pokrovsky, S. V. (1980). Conservation Laws in the quantum version of N-positional Potts model. *Communications in Mathematical Physics*, **76**, 129–141.

Batty, C. J. K. (1978). Dissipative mappings and well-behaved derivations. *Journal of the London Mathematical Society (2)*, **18**, 527–533.

Batty, C. J. K. and Davies, E. B. (1983). Positive semigroups and resolvents. *Journal of Operator Theory*, **10**, 357–363.

Batty, C. J. K. and Robinson, D. W. (1984). Positive one-parameter semigroups on ordered Banach spaces. *Acta Applicandae Mathematicae*, **2**, 221–296.

Batty, C. J. K., Digernes, T. and Robinson, D. W. (1983). Positive semigroups on ordered Banach spaces. *Journal of Operator Theory*, **9**, 371–400.

Bauer, M. and Itzykson, C. (1990). Modular transformations of $SU(N)$ affine characters and their commutant. *Communications in Mathematical Physics*, **127**, 617–636.

Baum, P., Connes, A. and Higson, N. (1994). Classifying space for proper actions and K-theory of group C^*-algebras. *Contemporary Mathematics*, **167**, 241–291.

Baxter, R. J. (1981). Rogers–Ramanujan identities in the Hard Hexagon model. *Journal of Statistical Physics*, **26**, 427–452.

Baxter, R. J. (1982). *Exactly solved models in statistical mechanics.* Academic Press, New York.

Baxter, R. J. (1988). The superintegrable chiral Potts model. *Physics Letters A*, **133**, 185–189.

Baxter, R. J. (1989a). A simple solvable $Z_4(N)$ Hamiltonian. *Physics Letters A*, **140**, 155–157.

Baxter, R. J. (1989b). Superintegrable Chiral Potts model: thermodynamic properties, an 'inverse' model, and a simple associated Hamiltonian. *Journal of Statistical Physics*, **57**, 1–39.

Baxter, R. J., Kelland, S. B. and Wu, F. Y. (1976). Potts model or Whitney Polynomial. *Journal of Physics. A. Mathematical and General*, **9**, 397–406.

Baxter, R. J., Temperley, H. N. V. and Ashley, S. E. (1978). Triangular Potts model and its transition temperature and related models. *Proceedings of the Royal Society of London A*, **358**, 535–559.

Baxter, R. J., Perk, J. H. H. and Au-Yang, H. (1988). New solutions of the star–triangle relations for the chiral Potts model. *Physics Letters A* **128**, 138–142.

Bazhanov, V. V. and Stroganov, Yu. G (1990). Chiral Potts model as a descendent of the six–vertex model. *Journal of Statistical Physics*, **59**, 799–817.

Beggs, E. and Evans, D. E. (1991). The real rank of algebras of matrix valued functions. *International Journal of Mathematics*, **2**, 131–138.

Belavin, A. A., Polyakov, A. M., and Zamolodchikov, A. B. (1980). Infinite conformal symmetry in two-dimensional quantum field theory. *Nuclear Physics B*, **241**, 333–380.

Bellisard, J. and Simon, B. (1982). Cantor spectrum for the almost Mathieu equation. *Journal of Functional Analysis*, **48**, 408–419.

Bendat, J. and Sherman, S. (1955). Monotone and convex operator functions. *Transactions of the American Mathematical Society*, **79**, 58–71.

Berezin, F. A. (1966). A method of second quantization. *Academic Press*, London/New York.

Bernard, D. (1987). String characters from Kač–Moody automorphisms. *Nuclear Physics B*, **288**, 628–648.

Binnenhei, C. (1995). Implementation of endomorphisms of the CAR algebra. *Reviews in Mathematical Physics*, **7**, 833–869.

Binnenhei, C. (1997). On the even CAR algebra. *Letters in Mathematical Physics*, **40**, 91–93.

Bion-Nadal, J. (1992). Subfactor of the hyperfinite II_1 factor with Coxeter graph E_6 as invariant. *Journal of Operator Theory*, **28**, 27–50.

Birman, J. (1974). Braids, links and mapping class groups. *Annals of Mathematical Studies*, **82**.

Birman, J. S. and Wenzl, H. (1989). Braids, link polynomials and a new algebra. *Transactions of the American Mathematical Society*, **313**, 249–273.

Bisch, D. (1990). On the existence of central sequences in subfactors. *Transactions of the American Mathematical Society*, **321**, 117–128.

Bisch, D. (1992). Entropy of groups and subfactors. *Journal of Functional Analysis*, **103**, 190–208.

Bisch, D. (1994a). A note on intermediate subfactors. *Pacific Journal of Mathematics*, **163**, 201–216.

Bisch, D. (1994b). On the structure of finite depth subfactors. In *Algebraic methods in operator theory* (ed. R. Curto and P. E. T. Jørgensen), Birkhäuser, Basel, pp. 175–194.

Bisch, D. (1994c). Central sequences in subfactors II. *Proceedings of the American Mathematical Society*, **121**, 725–731.

Bisch, D. (1994d). An example of an irreducible subfactor of the hyperfinite II_1 factor with rational, non-integer index. *Journal für die Reine und Angewandte Mathematik*, **455**, 21–34.

Bisch, D. (1997). Bimodules, higher relative commutants and the fusion algebra associated to a subfactor. In *Operator algebras and their applications.* Fields Institute Communications, Vol. 13, American Mathematical Society, pp. 13–63.

Bisch, D. (in press). Principal graphs of subfactors with small Jones index. *Mathematische Annalen.*

Bisch, D. and Haagerup, U. (1996). Composition of subfactors: New examples of infinite depth subfactors. *Annales Scientifiques de l'École Normale Supérieur*, **29**, 329–383.

Bisch, D. and Jones, V. F. R. (1997a). Algebras associated to intermediate subfactors. *Inventiones Mathematicae*, **128**, 89–157.

Bisch, D. and Jones, V. F. R. (1997b). A note on free composition of subfactors. In *Geometry and physics, (Aarhus 1995)*, Lecture Notes in Pure and Applied Mathematics, Vol. 184, Marcel Dekker, New York, pp. 339–361.

Bisognano, J. and Wichmann, E. (1975). On the duality condition for a hermitian scalar field. *Journal of Mathematical Physics*, **16**, 985–1007.

Blackadar, B. (1980). A simple C^*-algebra with no non-trivial projections. *Proceedings of the American Mathematical Society*, **78**, 504–508.

Blackadar, B. (1981). A simple unital projectionless C^*-algebra. *Journal of Operator Theory*, **5**, 63–71.

Blackadar, B. (1986). K-theory for operator algebras. *Mathematical Sciences Research Institute Publications*, **5**, Springer, Berlin, Heidelberg.

Blackadar, B. (1990). Symmetries of the CAR algebra. *Annals of Mathematics*, **131**, 589–623.

Blackadar, B. and Cuntz, J. (1982). The structure of stable algebraically simple C^*-algebras. *American Journal of Mathematics*, **104**, 813–822.

Blackadar, B. and Rørdam, M. (1992). Extending states on preordered semigroups and the existence of quasitraces on C^*-algebras. *Journal of Algebra*, **152**, 240–247.

Blackadar, B., Bratteli, O., Elliott, G. A., and Kumjian, A. (1992a). Reduction of real rank in inductive limit C^*-algebras. *Mathematische Annalen*, **292**, 111–126.

Blackadar B., Kumjian A., and Rørdam M.(1992b). Approximately central matrix units and the structure of non-commutative tori. *K-Theory*, **6**, 267–284.

Blattner, R. (1958). Automorphic group representations. *Pacific Journal of Mathematics*, **8**, 665–677.

Boas, R. P. (1954). *Entire functions.* Academic Press, New York.

Boca, F. (preprint 1995). Metric diophantine approximation and non-commutative tori.

Boca, F. P. and Zaharescu, A. (preprint 1996). Factors of type III and the distribution of prime numbers.

Böckenhauer, J. (1996a). Localised endomorphisms of the Chiral Ising model. *Communications in Mathematical Physics*, **177**, 265–304.

Böckenhauer, J. (1996b). An algebraic formulation of level one Wess–Zumino–Witten models. *Reviews in Mathematical Physics*, **8**, 925–947.

Bongaarts, P. J. M. (1970). The electron–positron field, coupled to external electromagnetic potentials as an elementary C^*-algebra theory. *Annals of Physics*,(N.Y.), **56**, 108–138.

Borchers, H. J. (1966). Energy and momentum as observables in quantum field theory. *Communications in Mathematical Physics*, **2**, 49–54.

Borchers, H. J. (1974). Characterisation of inner *-automorphisms of W^*-algebras. *Publications of the RIMS, Kyoto University*, **10**, 11–49.

Bost, J.-B. and Connes A. (1995). Hecke algebras, type III factors and phase transitions with spontaneous symmetry breaking in number theory. *Selecta Math. (New Series)*, **1**, 411–457.

Bott, R. (1959). The stable homotopy of the classical groups. *Annals of Mathematics*, **70**, 313–337.

Bouwknegt, P. and Nahm, W. (1987). Realizations of the exceptional modular invariant $A_1^{(1)}$ partition functions. *Physics Letters B*, **184**, 359–362.

Bozejko, M. and Speicher, R. (1991). An example of generalized Brownian motion. *Communications in Mathematical Physics*, **137**, 519–531.

Bozejko, M. and Speicher, R. (1994). Completely positive maps on Coxeter groups, deformed commutation relations and operator spaces. *Mathematische Annalen*, **300**, 97–120.

Bratteli, O. (1972). Inductive limits of finite dimensional C^*-algebras. *Transactions of the American Mathematical Society*, **171**, 195–234.

Bratteli, O. (1982). A remark on extensions of quasifree derivations on the CAR-algebra. *Letters in Mathematical Physics*, **6**, 499–504.

Bratteli, O. and Kishimoto, A. (1992). Non-commutative spheres, III, Irrational rotations. *Communications in Mathematical Physics*, **147**, 605–624.

Bratteli, O. and Robinson, D. W. (1979). *Operator algebras and quantum statistical mechanics. I.* Springer, Berlin.

Bratteli, O. and Robinson, D. W. (1981). *Operator algebras and quantum statistical mechanics. II.* Springer, Berlin.

Bratteli, O. and Robinson, D. W. (1987). *Operator algebras and quantum statistical mechanics I.* 2nd edition, Springer, Berlin.

Bratteli, O., Elliott, G. A., Herman, R. (1980). On the possible temperatures of a C^*-dynamical system. *Communications in Mathematical Physics*, **74**, 281–295.

Bratteli, O., Elliott, G. A. and Kishimoto, A. (1986a). The temperature state space of a C^*-dynamical system II. *Annals of Mathematics*, **123**, 205–263.

Bratteli, O. and Evans, D. E., Goodman, F. M., and Jørgensen, P. E. T. (1986b). A dichotomy for derivations on $\mathcal{O}_n$. *Publications of the RIMS, Kyoto University*, **22**, 103–117.

Bratteli, O., Elliott, G. A., Evans, D. E., and Kishimoto, A. (1989). Quasi-product actions of a compact abelian group on a C^*-algebra. *Tohoku Journal of Mathematics*, **41**, 133–161.

Bratteli, O., Elliott, G. A., Evans, D. E. and Kishimoto, A. (1991a). On the classification of inductive limits of inner actions of a compact group. In Current topics in operator algebras. (eds. H. Araki *et al.*), World Scientific Press, pp. 13–24.

Bratteli, O., Elliott, G. A., Evans, D. E., and Kishimoto, A. (1991b). Non-commutative spheres, I. *International Journal of Mathematics*, **2**, 139–166.

Bratteli, O., Elliott, G. A., Evans, D. E., and Kishimoto, A. (1992). Non-commutative spheres II: rational rotations. *Journal of Operator Theory*, **27**, 53–85.

Bratteli, O., Kishimoto, A., Rørdam M., and Størmer E. (1993). The crossed product of a UHF algebra by a shift. *Ergodic Theory and Dynamical Systems*, **13**, 615–626.

Bratteli, O., Evans, D. E. and Kishimoto, A. (1993). Almost shift invariant projections in infinite tensor products. In Quantum and non-commutative analysis (ed. H. Araki et al.), Kluwer Academic, pp. 427–434.

Bratteli, O., Evans, D. E. and Kishimoto, A. (1994b). Crossed products of totally disconnected spaces by $\mathbb{Z}_2 \times \mathbb{Z}_2$. *Ergodic Theory and Dynamical Systems*, **13**, 445–484.

Bratteli, O., Elliott, G. A., Evans, D. E., and Kishimoto, A. (1994a). Finite group actions on AF algebras obtained by folding the interval. *K-theory*, **8**, 443–464.

Bratteli, O., Evans, D. E. and Kishimoto, A. (1995). The Rohlin property for quasi-free automorphisms of the Fermion algebra. *Proceedings of the London Mathematical Society*, **71**, 675–694.

Bratteli, O., Elliott, G. A., Evans, D. E., and Kishimoto, A. (to appear a). On the classification of C^*-algebra of real rank zero III: The infinite case. In *Operator algebras and their applications.* Fields Institute Communications, American Mathematical Society.

Bratteli, O., Elliott, G. A., Evans, D. E., and Kishimoto, A. (to appear b). Homotopy of a pair of approximately commuting unitaries in a simple purely infinite unital C^*-algebra. *Journal of Functional Analysis.*

Brauer, R. (1937). On algebras which are connected with the semisimple continuous groups. *Annals of Mathematics*, **38**, 856–872.

Bricmont, J., Kuroda, K., and Lebowitz, J. L. (1985). First order phase transitions in lattice and continuous systems : Extension of Pirogov–Sinai theory. *Communications in Mathematical Physics*, **101**, 501–538.

Brown, L. G. (1977). Stable isomorphism of hereditary subalgebras of C^*-algebras. *Pacific Journal of Mathematics*, **71**, 335–348.

Brown, L. and Pedersen, G. K. (1991). C^*-algebras of real rank zero. *Journal of Functional Analysis*, **99**, 131–149.

Brown, L., Douglas, R. and Fillmore, P. (1973). Unitary equivalence modulo the compact operators and extensions of C^*-algebras. In *Proceedings of a Conference on Operator Theory*, Lecture Notes in Mathematics, **345**, Springer, Berlin, pp. 58–128.

Brown, L., Douglas, R., and Fillmore, P. (1977). Extensions of C^*-algebras and K-homology. *Annals of Mathematics*, **105**, 265–324.

Buchholz, D., Mack, G., and Todorov, I. (1988). The current algebra on the circle as a germ of local field theories. *Nuclear Physics B (Proceedings Supplement)*, **5**, 20–56.

Buchholz, D., Mack, G., and Todorov, I. (1990). Localised automorphisms of the U(1)-current algebra on the circle : an instructive example. In *The algebraic theory of superselection sectors. Introduction and recent results* (ed. D. Kastler), World Scientific, Singapore, pp. 356–378.

Bunce, J. W. and Deddens, J. A. (1975). A family of simple C^*-algebras related to weighted shift operators. *Journal of Functional Analysis*, **19**, 13–24.

Bures, D. H. C. (1963). Certain factors constructed as finite tensor products. *Compositio Mathematica*, **15**, 169–191.

Busby, R. C. (1968). Double centralizers and extensions of C^*-algebras. *Transactions of the American Mathematical Society*, **132**, 79–99.

Capel, H. W. and Perk, J. H. H. (1977a). Autocorrelation function of the x-component of the magnetization in the one-dimensional XY-model. *Physica A*, **87**, 211–242.

Capel, H. W. and Perk, J. H. H. (1977b). Time-dependent xx-correlations in the one-dimensional XY-model. *Physica A*, **89**, 265–303.

Capel, H. W. and Perk, J. H. H. (1978). Transverse correlations in the inhomogeneous XY-model at infinite temperature. *Physica A*, **92**, 163–184.

Capel, H. W. and Perk, J. H. H. (1980). Time- and frequency-dependent correlation functions for the homogeneous and alternating XY-models. *Physica A*, **100**, 1–23.

Cappell, S. E. (1976). On homotopy invariance of higher signatures. *Inventiones Mathematicae*, **33**, 171–179.

Cappelli, A., Itzykson, C. and Zuber, J.-B. (1987a). Modular invariant partition functions in two dimensions. *Nuclear Physics B*, **280**, 445–465.

Cappelli, A., Itzykson, C., and Zuber, J.-B. (1987b). The A-D-E classification of minimal and $A_1^{(1)}$ conformal invariant theories. *Communications in Mathematical Physics*, **113**, 1–26.

Cardy, J. L. (1986). Operator content of two dimensional conformally invariant theories. *Nuclear Physics*, **B270 [FS16]**, 186–204.

Cardy, J. L. (1987). Conformal invariance. In *Phase transitions and critical phenomena* Vol. 11 (eds. C. Domb, J. Lebowitz.) Academic Press, New York, pp. 55–126.

Cardy, J. L. (1990). Conformal invariance and statistical mechanics. In *Champs, cordes et phénomènes critiques (Les Houches, 1988)* (ed. E. Brezin and J. Sinn-Justin), North-Holland, Amsterdam pp. 169–245.

Carey, A. L. (1984). Some infinite dimensional groups and bundles. *Publications of the RIMS, Kyoto University*, **20**, 1103–1117.

Carey, A. L. (1985). Some homogeneous spaces and representations of the Hilbert Lie group $U(H)_2$. *Revue Roumaine de Mathématiques Pures et Appliquées*, **30**, 505–520.

Carey, A. L. and Evans D. E. (1987). On an automorphic action of $U(n, 1)$ on $\mathcal{O}_n$. *Journal of Functional Analysis*, **70**, 90–110.

Carey, A. L. and Evans, D. E. (1988). The operator algebras of the two dimensional Ising model. In *Braids* (ed. J. Birman and A. Libgober), *Contemporary Mathematics*, **78**, 117–165.

Carey, A. L. and Evans D. E. (1990). On algebras almost commuting with Clifford algebras. *Journal of Functional Analysis*, **88**, 279–298.

Carey, A. L. and O'Brien, D. M. (1983). Automorphisms of the infinite dimensional Clifford algebra and the Atiyah Singer mod 2 index. *Topology*, **22**, 437–448.

Carey, A. L. and Ruijsenaars, S. N. M. (1987). On fermion gauge groups, current algebras, and Kač–Moody algebras. *Acta Applicandae Mathematicae*, **10**, 1–86.

Carey, A. L., Hurst, C. A., and O'Brien, D. M. (1972). Automorphisms of the canonical anticommutation relations and index theory. *Journal of Functional Analysis*, **48**, 360–393.

Cartier, P. (1990). Développements récents sur les groupes de tresses. Applications à la topologie et à l'algèbre. *Séminaire Bourbaki, exp. 716, Astérisque*, **189–190**, 17–67.

Caselle, M. and Ponzano, G. (1989). Fusion rule algebras from graph theory. *Physics Letters B*, **224**, 303–308.

Caselle, M., Ponzano, G. and Ravanini, F. (1990). Orthogonal polynomial structures and fusion algebras of rational conformal field theories. *Physics Letters B*, **251**, 260–265.

Ceccherini, T. (1994). Approximately inner and centrally free commuting squares of type II_1 factors and their classification. Ph. D. thesis. University of California, Los Angeles.

Chadam, J. M. (1968). Unitarity of dynamical propagators of perturbed Klein–Gordon equations. *Journal of Mathematical Physics*, **9**, 386–396.

Chen, J. (1993). The Connes invariant $\chi(M)$ and cohomology of groups. Ph. D. thesis. University of California, Berkeley.

Chihara, T. S. (1978). *An introduction to orthogonal polynomials.* Gordon and Breach, New York.

Choda, M. (1989). Index for factors generated by Jones' two sided sequence of projections. *Pacific Journal of Mathematics*, **139**, 1–16.

Choda, M. (1991). Entropy for $*$-endomorphisms and relative entropy for subalgebras. *Journal of Operator Theory*, **25**, 125–140.

Choda, M. (1992). Entropy for canonical shift. *Transactions of the American Mathematical Society*, **334**, 827–849.

Choda, M. (1993). Duality for finite bipartite graphs (with applications to II_1 factors). *Pacific Journal of Mathematics*, **158**, 49–65.

Choda, M. (1994). Square roots of the canonical shifts. *Journal of Operator Theory*, **31**, 145–163.

Choda, M. and Hiai, F. (1991). Entropy for canonical shifts. II. *Publications of the RIMS, Kyoto University*, **27**, 461–489.

Choda, M. and Kosaki, H. (1994). Strongly outer actions for an inclusion of factors. *Journal of Functional Analysis*, **122**, 315–332.

Choi, M.-D. (1972). Positive linear maps on C^*-algebras. *Canadian Journal of Mathematics*, **24**, 520–529.

Choi, M.-D. (1974). A Schwarz inequality for positive linear maps on C^*-algebras. *Illinois Journal of Mathematics*, **18**, 565–574.

Choi, M.-D. (1975). Completely positive linear maps on complex matrices. *Linear Algebra and its Applications*, **10**, 285–290.

Choi, M.-D. (1980). Some assorted inequalities for positive linear maps on C^*-algebras. *Journal of Operator Theory*, **4**, 271–285.

Choi, M.-D. and Effros, E. G. (1976a). The completely positive lifting problem for C^*-algebras. *Annals of Mathematics*, **104**, 585–609.

Choi, M.-D. and Effros, E. G. (1976b). Separable nuclear C^*-algebras and injectivity. *Duke Mathematical Journal*, **43**, 309–322.

Choi, M.-D. and Effros, E. G. (1977). Nuclear C^*-algebras and injectivity, the general case. *Indiana University Journal of Mathematics*, **26**, 443–446.

Choi, M.-D. and Effros, E. G. (1978). Nuclear C^*-algebra and the approximation property. *American Journal of Mathematics*, **100**, 61–79.

Choi, M.-D. and Elliott, G. A. (1990). Density of the self adjoint elements with finite spectrum in an irrational rotation C^*-algebra. *Mathematica Scandinavica*, **67**, 73–86.

Choi, M.-D., Elliott, G. A. and Yui, N. (1990). Gauss polynomials and the rotation algebras. *Inventiones Mathematicae*, **99**, 225–246.

Christe, P. and Ravanini, F. (1989). $G_N \otimes G_L/G_{N+L}$ conformal field theories and their modular invariant partition functions. *International Journal of Modern Physics A*, **4**, 897–920.

Christensen, E. (1978a). Generators of semigroups of completely positive maps. *Communications in Mathematical Physics*, **62**, 167–171.

Christensen, E. (1978b). Extensions of derivations. *Journal of Functional Analysis*, **27**, 234–247.

Christensen, E. (1979). Subalgebras of a finite algebra. *Mathematische Annalen*, **243**, 17–29.

Christensen, E. (1982a). Derivations and their relation to perturbation of operator algebras. *Proceedings of Symposia in Pure Mathematics*, **38**, 261–274.

Christensen, E. (1982b). Extensions of derivations II. *Mathematica Scandinavica*, **50**, 111–122.

Christensen, E. and Evans, D. E. (1979). Cohomology of operator algebras and quantum dynamical semigroups. *Journal of the London Mathematical*

Society, **20**, 358–368.

Coburn, L. A. (1967). The C^*-algebra generated by an isometry I. *Bulletin of the American Mathematical Society*, **73**, 722–726.

Coburn, L. A. (1969). The C^*-algebra generated by an isometry II. *Transactions of the American Mathematical Society*, **137**, 211–217.

Combes, F. (1968). Poids sur une C^*-algèbre. *Journal de Mathématiques Pures et Appliquées*, **47**, 57–100.

Connes, A. (1973). Une classification des facteurs de type III. *Annales Scientifiques de l'École Normale Supérieure*, **6**, 133–252.

Connes, A. (1975a). Outer conjugacy classes of automorphisms of factors. *Annales Scientifiques de l'École Normale Supérieure*, **8**, 383–419.

Connes, A. (1975b). Hyperfinite factors of type III_0 and Krieger's factors. *Journal of Functional Analysis*, **18**, 318–327.

Connes, A. (1975c). Sur la classification des facteurs de type II. *Comptes Rendus de l'Academie des Sciences, Série I, Mathématiques*, **281**, 13–15.

Connes, A. (1975d). A factor not antiisomorphic to itself. *Annals of Mathematics*, **101**, 536–554.

Connes, A. (1976a). Classification of injective factors. *Annals of Mathematics*, **104**, 73–115.

Connes, A. (1976b). Outer conjugacy of automorphisms of factors. *Symposia Mathematica*, **XX**, 149–160.

Connes, A. (1976c). On the classification of von Neumann algebras and their automorphisms. *Symposia Mathematica*, **XX**, 435–478.

Connes, A. (1977). Periodic automorphisms of the hyperfinite factor of type II_1. *Acta Scientiarum Mathematicarum*, **39**, 39–66.

Connes, A. (1978). On the cohomology of operator algebras. *Journal of Functional Analysis*, **28**, 248–253.

Connes, A. (1979). Sur la théorie non commutative de l'integration. *Springer Lecture Notes in Math.*, **725**, 19–143.

Connes, A. (1980a). C^*-algebres et geomètrie différentielle. *Comptes Rendus de l'Academie des Sciences, Série I, Mathématiques*, 559–604.

Connes, A. (1980b). Spatial theory of von Neumann algebras. *Journal of Functional Analysis*, **35** (1980), 153–164.

Connes, A. (1981). An analogue of the Thom isomorphism for crossed products of a C^*-algebra by an action of $\mathbb{R}$. *Advances in Mathematics*, **39**, 311–355.

Connes, A. (1982a). Foliations and Operator Algebras. *Proceedings of Symposia in Pure Mathematics.* (ed. R. V. Kadison), **38**, 521–628.

Connes, A. (1982b). Classification des facteurs. *Proceedings of the Symposia in Pure Mathematics (II)*, **38**, 43–109.

Connes, A. (1985a). Non-commutative differential geometry I-II. *Publications Mathématiques IHES*, **62**, 41–144.

Connes, A. (1985b). Factors of type III_1, property L'_λ and closure of inner automorphisms. *Journal of Operator Theory*, **14**, 189–211.

Connes, A. (1985c). Non commutative differential geometry, Chapter III: De

Rham homology and non commutative algebra. *Publications Mathématiques IHES*, **62**, 257–360.

Connes, A. (1994). *Noncommutative geometry.* Academic Press, New York.

Connes, A. and Evans, D. E. (1989). Embeddings of $U(1)$-current algebras in non-commutative algebras of classical statistical mechanics. *Communications in Mathematical Physics*, **121**, 507–525.

Connes, A. and Higson, N. (1990). Déformations, morphismes asymptotiques et K-théorie bivariante. *Comptes Rendus de l'Académie des Sciences, Série I, Mathématiques*, **311**, 101–106.

Connes, A. and Karoubi, M. (1988). Caractère multiplicatif d'un module de Fredholm. *K-theory*, **2**, 431–463.

Connes, A. and Krieger, W. (1977). Measure space automorphism groups, the normalizer of their full groups, and approximate finiteness. *Journal of Functional Analysis*, **24**, 336–352.

Connes, A. and Rieffel, M. (1985). Yang–Mills for non-commutative tori. *Contemporary Mathematics*, **62**, 237–265.

Connes, A. and Skandalis, G. (1984). The longitudinal index theorem for foliations. *Publications of the RIMS, Kyoto University*, **20**, 1139–1183.

Connes, A. and Størmer, E. (1975). Entropy for automorphisms of II_1 von Neumann algebras. *Acta Mathematica*, **134**, 289–306.

Connes, A. and Takesaki, M. (1977). The flow of weights on factors of type III. *Tohoku Mathematical Journal*, **29**, 73–555.

Cook, J. M. (1951). The mathematics of second quantization. *Proceedings of the National Academy of Sciences of the USA*, **37**, 417–420.

Cooper, J. L. B. (1947). One-parameter semigroups of isometric operators in Hilbert space. *Annals of Mathematics (2)*, **48**, 827–842.

Cooper, J. L. B. (1950a). The characterisation of quantum-mechanical operators. *Proceedings of the Cambridge Philosophical Society*, **46**, 614–619.

Cooper, J. L. B. (1950b). The paradox of separated systems in quantum theory. *Proceedings of the Cambridge Philosophical Society*, **46**, 620–625.

Coste, A. and Gannon, T. (1994). Remarks on Galois symmetry in rational conformal field theories. *Physics Letters B*, **323**, 316–321.

Cuntz, J. (1977). Simple C^*-algebras generated by isometries. *Communications in Mathematical Physics*, **57**, 173–185.

Cuntz, J. (1981a). K-theory for certain C^*-algebras. *Annals of Mathematics*, **113**, 181–197.

Cuntz, J. (1981b). A class of C^*-algebras and topological Markov chains II. Reducible Markov chains and the *Ext* functor for C^*-algebras. *Inventiones Mathematicae*, **63**, 25–40.

Cuntz, J. (1984). K-theory and C^*-algebras. *Lecture Notes in Mathematics*, **1046**, Springer, Berlin, pp. 55–79.

Cuntz, J. and Evans D. E. (1981). Some remarks on the C^*-algebras associated with certain topological Markov chains. *Mathematica Scandinavica*, **48**, 235–240.

Cuntz, J. and Krieger, W. (1980). A class of C^*-algebras and topological Markov chains. *Inventiones Mathematicae*, **56**, 251–268.

Cvetković, D., Doob, M. and Gutman, I. (1982). On graphs whose spectral radius does not exceed $(2+\sqrt{5})^{1/2}$. *Ars Combinatoria*, **14**, 225–239.

Dadarlat, M. (1995). Reduction of dimension three of local spectra of real rank zero C^*-algebras. *Journal für die Reine und Angewandte Mathematik*, **460**, 189–212.

Dadarlat, M. and Gong, G. (in press). A classification result for approximately homogeneous C^*-algebras of real rank zero. *Duke Mathematical Journal.*

Dadarlat, M. and Loring, T. (1996a). Classifying C^*-algebras via ordered mod p K-theory. *Mathematische Annalen*, **305**, 601–616.

Dadarlat, M. and Loring, T. (1996b). A universal multicoefficient theorem for the Kasparov groups. *Duke Mathematical Journal*, **84**, 355–377.

Dadarlat, M. and Nemethi, A. (1990). Shape theory and connective K-theory. *Journal of Operator Theory*, **23**, 207–291.

Dadarlat, M., Nagy, G., Nemethi, A. and Pasnicu, C. (1992). Reduction of topological stable rank in inductive limits of C^*-algebras. *Pacific Journal of Mathematics*, **153**, 267–276.

Date, E., Jimbo, M., Miwa, T., and Okado, M. (1987). Solvable lattice models. *Theta functions – Bowdoin 1987, Part 1*, Proceedings of Symposia in Pure Mathematics Vol. 49, American Mathematical Society, Providence, R.I., pp. 295–332.

Date, E., Jimbo, M., Kuniba, A., Miwa, T., and Okado, M. (1988). Exactly solvable SOS models II: Proof of the star-triangle relation and combinatorial identities. *Advanced Studies in Pure Mathematics*, **16**, 17–122.

David, M. C. (1996). Paragroupe d'Adrian Ocneanu et algèbre de Kač. *Pacific Journal of Mathematics*, **172**, 331–363.

Davidson, K. R. (1996). *C^*-Algebras by example.* Fields Institute monograph.

Davies, B. (1988). Corner transfer matrices for the Ising model. *Physica A*, **154**, 1–20.

Davies, B. and Pearce, P. A. (1990). Conformal invariance and critical spectrum of corner transfer matrices. *Journal of Physics A. Mathematics Gen.*, **23**, 1295–1312.

Davies, E. B. (1972a). Some contraction semigroups in quantum probability. *Z. Wahrscheinlichkeitstheorie und Verw. Gebiete*, **23**, 261–273.

Davies, E. B. (1972b). Diffusion for weakly coupled quantum oscillators. *Communications in Mathematical Physics*, **27**, 309–325.

Davies, E. B. (1974). Markovian master equations. *Communications in Mathematical Physics*, **39**, 91–110.

Davies, E. B. (1976a). Markovian master equations II. *Mathematische Annalen*, **219**, 147–158.

Davies, E. B. (1976b). *Quantum theory of open systems.* Academic Press, New York.

Davies, E. B. (1977). Quantum dynamical semigroups and the neutron diffusion equation. *Reports on Mathematical Physics*, **11**, 169–188.

Davies, E. B. (1978). Irreversible dynamics of infinite Fermion systems. *Communications in Mathematical Physics*, **55**, 231–258.

Davies, E. B. (1980). *One-parameter semigroups*, LMS monographs 15, Academic Press, New York.

de Boer, J. and Goeree, J. (1991). Markov traces and II_1 factors in conformal field theory. *Communications in Mathematical Physics*, **139**, 267–304.

Degiovanni, P. (1990). $\mathbb{Z}/N\mathbb{Z}$ conformal field theories. *Communications in Mathematical Physics*, **127**, 71–99.

Degiovanni, P. (1992). Moore and Seiberg's equations and 3D toplogical field theory. *Communications in Mathematical Physics*, **145**, 459–505.

de la Harpe, P. (1979). Moyennabilité du groupe unitaire et propriété P de Schwartz des algèbres de von Neumann. *Algèbres d'opérateurs (Séminaire, Les Plans-sur-Bex, Suisse 1978)* (ed. P. de la Harpe), *Lecture Notes in Mathematics*, Springer, **725**, 220–227.

de la Harpe, P. and Wenzl, H. (1987). Opérations sur les rayons spectraux de matrices symétriques entières positives. *Comptes Rendus de l'Académie des Sciences, Série I, Mathématiques*, **305**, 733–736.

Dell'Antonio, G. F. (1968). Structure of the algebras of some free systems. *Communications in Mathematical Physics*, **9**, 81–117.

Demoen, B., Vanheuverzwijn, P., and Verbeure, A. (1977). Completely positive maps on the CCR algebra. *Letters in Mathematical Physics*, **2**, 161–166.

Derdinger, R. (1980), (1984). Uber das Spektrum positiver Generatoren. *Mathematische Zeitschrift*, **172**, 281–293; *Phil. Soc.* **96**, 315–322.

Derdinger, R. and Nagel, R. (1979). Der Generator stark stetiger Verbandshalbgruppen aus $c(x)$ unddessen Spektrum. *Mathematische Annalen*, **245**, 159–177.

de Schreye, D. and van Daele, A. (1981). A new technique for dealing with Cuntz algebras. *Mathematische Annalen*, **257**, 397–401.

de Vega, H. J. and Karowski, M. (1987). Conformal invariance and integrable theories. *Nuclear Physics B*, **285**, 619–638.

Di Francesco, P. (1992). Integrable lattice models, graphs, and modular invariant conformal field theories. *International Journal of Modern Physics A*, **7**, 407–500.

Di Francesco, P. and Zuber, J.-B. (1990a). $SU(N)$ lattice integrable models associated with graphs. *Nuclear Physics B*, **338**, 602–646.

Di Francesco, P. and Zuber, J.-B. (1990b). $SU(N)$ lattice integrable models and modular invariance. In *Recent Developments in Conformal Field Theories*, 1989, World Scientific, Trieste, pp. 179–215.

Dijkgraaf, R., and Witten, E. (1990). Topological gauge theories and group cohomology. *Communications in Mathematical Physics*, **129**, 393–429.

Dijkgraaf, R., Vafa, C., Verlinde, E., and Verlinde, H. (1989). The operator algebra of orbifold models. *Communications in Mathematical Physics*, **123**,

485–526.

Dijkgraaf, R., Pasquier, V., and Roche, Ph. (1990). Quasi Hopf algebras, group cohomology and orbifold models. *Nuclear Physics B(Proc. Suppl.)*, **18**, 60–72.

Dijkgraaf, R., Pasquier, V., and Roche, Ph. (1991). Quasi-quantum groups related to orbifold models. *Proceedings of the International Colloquium on Modern Quantum Field Theory*, World Scientific, Singapore, pp. 375–383.

Dixmier, J. (1964). *Les C^*-algebras et leurs représentations*, Gauthier-Villars, Paris.

Dixmier, J. (1967). On some C^*-algebras considered by Glimm. *Journal of Functional Analysis*, **1**, 182–203.

Dixmier, J. (1969). *Les algèbres d'opérateurs dans l'espace Hilbertien. (Algèbres de von Neumann)*, 2nd ed Gauthier Villars, Paris.

Dixmier, J. (1981). *Von Neumann Algebras*, North-Holland, Amsterdam.

Dixmier, J. and Lance, C. (1969). Deux nouveaux facteurs de type II. *Inventiones Mathematicae*, **7**, 226–234.

Dixon, L., Harvey, J. A., Vafa, C., and Witten, E. (1985). Strings on orbifolds. *Nuclear Physics B*, **261**, 678–686.

Dixon, L., Harvey, J. A., Vafa, C., and Witten, E. (1986). Strings on orbifolds, II. *Nuclear Physics B*, **274**, 285–314.

Dobrushin, R. L. (1968). Gibbsian random fields for lattice systems with pairwise interactions. *Functional Analysis and its applications*, **2**, 292–301.

Doplicher, S. and Roberts, J. E. (1989). A new duality theory for compact groups. *Inventiones Mathematicae*, **98**, 157–218.

Doplicher, S., Haag, R. and Roberts, J. (1971,74). Local observables and particle statistics, I, II. *Communications in Mathematical Physics*, **23**, 199–230 and **35**, 49–85.

Douglas, R. G. (1972). *Banach algebra techniques in operator theory*, Academic Press, New York.

Douglas, R. G. (1980). C^*-Algebra Extensions and K-Homology. *Annals of Mathematical Studies*, Princeton University Press, **95**.

Drinfeld, V. G. Quantum groups. In *Proceedings of ICM-86*, Berkeley, 798–820, Springer.

Dubin, D. A. (1974). *Solvable models in algebraic statistical mechanics.* Oxford University Press, Oxford.

Dunford, N. and Schwartz, J. T. (1958). *Linear Operators* Vol. I. Interscience, New York.

Durhuus, B., Jakobsen, H. P., and Nest, R. (1993). Topological quantum field theories from generalized $6j$-symbols. *Reviews in Mathematical Physics*, **5**, 1–67.

Dykema, K. and Nica, A. (1993). On the Fock representation of the q-commutation relations. *Journal für die Reine und Angewandte Mathematik*, **440**, 210–212.

Effros, E. G. (1981). Dimensions and C^*-algebras. *CBMS Lectures*, **46**, American Mathematical Society.

Effros, E. G. (1982). On the structure theory of C^*-algebras: some old and new problems. *Proceedings of Symposia in Pure Mathematics*, **38**, 19–34.

Effros, E. G. and Hahn, F. (1967). Locally compact transformation groups and C^*-algebras. *Memoirs of the American Mathematical Society*, **75**.

Effros, E. G. and Lance, E. C. (1977). Tensor products of operator algebras. *Advances in Mathematics*, **25**, 1–34.

Effros, E. G., Handelman, D. E., and Shen, C. L. (1981). Dimension groups and their affine representations. *American Journal of Mathematics*, **102**, 385–407.

Eilers, S. (1996). A complete invariant for AD algebras with real rank zero and bounded torsion in K_1. *Journal of Functional Analysis*, **139**, 325–348.

Elliott, G. A. (1976). On the classification of inductive limits of sequences of semisimple finite-dimensional algebras. *Journal of Algebra*, **38**, 29–44.

Elliott, G. A. (1979). On totally ordered groups and K_0. In *Ring Theory Conference at Waterloo (1978)*, Lecture Notes in Mathematics, 734, Springer, Berlin, pp. 1–49.

Elliott, G. A. (1984) On the K-theory of the C^*-algebra generated by a projective representation of a torsion free discrete abelian group. In *Operator Algebras and Group Representations 1*, Pitman, London.

Elliott, G. A. (1986). Gaps in the spectrum of an almost periodic Schrödinger operator. In *Geometric methods in operator algebras*, (ed. H. Araki and E. G. Effros), Pitman Research Notes in Mathematics, 123, Longman, London, pp. 181–191.

Elliott, G. A. (1990). Dimension groups with torsion. *International Journal of Mathematics*, **1**, 361–380.

Elliott, G. A. (1993a). On the classification of C^*-algebras of real rank zero. *Journal für die Reine und Angewandte Mathematik*, **443**, 179–219.

Elliott, G. A. (1993b). A classification of certain simple C^*-algebras. In *Quantum and non-commutative analysis* (ed. H. Araki *et al.*), 373–385, Kluwer Academic.

Elliott, G. A. (1993c). Are amenable C^*-algebras classifiable? *Representation Theory of Groups and Algebras. Contemporary Mathematics*, American Mathematical Society, **145**, 423–427.

Elliott, G. A. (1995). The classification problem for amenable C^*-algebras. In *Proceedings of the International Congress of Mathematicians, Zürich 1994*, Birkhaüser, pp. 922–932.

Elliott, G. A. (1997). A classification of certain simple C^*-algebras, II. *Journal of the Ramanujan Mathematical Society*, **12**, 97–134.

Elliott, G. A. and Evans, D. E. (1993). The structure of the irrational rotation C^*-algebra. *Annals of Mathematics*, **138**, 477–501.

Elliott, G. A. and Gong, G. (1996a). On the classification of C^*-algebras of real rank zero, II. *Annals of Mathematics*, **144**, 497–610.

Elliott, G. A. and Gong, G. (1996b). On inductive limits of matrix algebras over the two-torus. *American Journal of Mathematics*, **118**, 263–290.

Elliott, G. A. and Lin, Q. (1996). Cut-down method in the inductive limit decomposition of non-commutative tori. *Journal of the London Mathematical Society*, **54**, 121–134.

Elliott, G. A. and Rørdam, M. (1993). The automorphism group of the irrational rotation C^*-algebra. *Communications in Mathematical Physics*, **155**, 3–26.

Elliott, G. A. and Rørdam, M. (1995). Classification of certain infinite simple C^*-algebras, II. *Commentarii Mathematici Helvetici*, **70**, 615–638.

Elliott, G. A. and Su, H. (to appear). K-theoretic classification for inductive limit $\mathbb{Z}_2$ actions on AF algebras, *Canadian Journal of Mathematics*.

Elliott, G. A. and Thomsen, K. (1994). The state space of the K_0-group of a simple separable C^*-algebra. *Geometric and Functional Analysis*, **4**, 522–538.

Elliott, G. A., Evans, D. E., and Kishimoto, A. (to appear a). Outer conjugacy classes of trace scaling automorphisms of stable UHF algebras. *Mathematica Scandinavica*.

Elliott, G. A., Gong, G., and Li, L. (in preparation). On simple inductive limits of matrix algebras over higher dimensional spaces, II.

Elliott, G. A., Natsume, T. and Nest, R. (1993). The Heisenberg group and K-theory. *K-theory*, **7**, 409–428.

Elliott, G. A., Gong, G., Jiang, X., and Su, H. (to appear b). A classification of simple limits of dimension drop C^*-algebras. In *Operator algebras and their applications* (eds P. A. Fillmore and J. A. Mingo), *Fields Institute Communications*, American Mathematical Society, Providence, Vol. 1, 1996.

Emch, G. G. (1972). *Algebraic methods in statistical mechanics and quantum field theory.* Wiley, New York.

Emch, G. G. (1976). Generalised K-flows. *Communications in Mathematical Physics*, **49**, 191–215.

Emch, G. G., Albeverio, S. and Eckmann, J.–P. (1978). Quasi-free generalized K-flows. *Reports in Mathematical Physics*, **13**, 73–85.

Erlijman, J. (1995). New subfactors from braid group representations. Ph. D. thesis. University of Iowa.

Erlijman, J. (preprint 1996). Two-sided braid subfactors and asymptotic inclusions.

Evans, D. E. (1976a). Positive linear maps on operator algebras. *Communications in Mathematical Physics*, **48**, 15–22.

Evans, D. E. (1976b). On the spectrum of a one-parameter strongly continuous representation. *Mathematica Scandinavica*, **39**, 80–82.

Evans, D. E. (1977). Irreducible quantum dynamical semigroups. *Communications in Mathematical Physics*, **54**, 293–297.

Evans, D. E. (1979a). Completely positive quasi-free maps on the Fermion algebra. In *Proceedings of Conference on Mathematical Problems in Quantum Theory of Irreversible Processes*, Naples. (eds. L. Accardi, V. Gorini, and G.

Parravicini.) pp. 136–162

Evans, D. E. (1979b). Completely positive quasi-free maps on the CAR algebra. *Communications in Mathematical Physics*, **70**, 53–68.

Evans, D. E. (1980a). Dissipators for symmetric quasi free dynamical semigroups on the CAR algebra. *Journal of Functional Analysis*, **37**, 318–330.

Evans, D. E. (1980b). On $\mathcal{O}_n$. *Publications of the RIMS, Kyoto University*, **16**, 915–927.

Evans, D. E. (1980c). A review on semigroups of completely positive maps, in K. Osterwalder ed., Mathematical Problems in Theoretical Physics, SLNP 116, *Springer*, 400–406.

Evans, D. E. (1982a). Gauge actions on $\mathcal{O}_A$. *Journal of Operator Theory*, **7**, 79–100.

Evans, D. E. (1982b). Entropy of automorphisms of AF algebras. *Publications of the RIMS, Kyoto University*, **18**, 1045–1051.

Evans, D. E. (1984a). The C^*-algebras of topological Markov chains. *Tokyo Metropolitan University Lecture Notes.*

Evans, D. E. (1984b) Quantum dynamical semigroups, symmetry groups, and locality. *Acta Applicandae Mathematicae*, **2**, 333–352.

Evans, D. E. (1985a). The C^*-algebras of the two-dimensional Ising model. Springer Lecture Notes in Mathematics, **1136**, 162–176.

Evans, D. E. (1985b). Quasi-product states on C^*-algebras. *Operator algebras and their connections with topology and ergodic theory*, Springer Lecture Notes in Mathematics, **1132**, 129–151.

Evans, D. E. (1989a). Operator algebras and critical phenomenon. In *Proceedings of the Congress of the International Association of Mathematical Physics, Swansea.* (eds. I.M. Davies, A. Truman and B. Simon.) Adam Hilgler, Bristol and New York, pp. 418–429.

Evans, D. E. (1989b). An algebraic approach to critical phenomena. *Leuven Notes in Mathematical and Theoretical Physics, Seies A : Mathematical Physics, Volume 1* (eds. M. Fannes, A. Verbeure) Leuven University Press, 15–30.

Evans, D. E. (1990). C^*-algebraic methods in statistical mechanics and field theory. *International Journal of Modern Physics B*, **4**, 1069–1118.

Evans, D. E. (1992). Some non-commutative orbifolds. In *Ideas and methods in mathematical analysis, stochastics, and applications.* (eds. S. Albeverio, J. E. Fenstad, H. Holden, T. Lindstrom.) Cambridge University Press, pp. 344–364.

Evans, D. E. (1997) The Rohlin property and classification of trace scaling automorphisms of AF algebras. In *Operator Algebras and Quantum Field Theory*, (ed. S. Doplicher et al), 123–135, International Press.

Evans, D. E. and Gould, J. D. (1989). Dimension groups, embeddings and presentations of AF algebras associated to solvable lattice models. *Modern Physics Letters A*, **20**, 1883–1890.

Evans, D. E. and Gould, J. D. (1994a). Dimension groups and embeddings of graph algebras, *International Journal of Mathematics*, **5**, 291–327.

Evans, D. E. and Gould, J. D. (1994b). Presentations of AF algebras associated to T-graphs, *Publications of the RIMS, Kyoto University*, **30**, 767–798.

Evans, D. E. and Hanche-Olsen, H. (1979). The generators of positive semigroups. *Journal of Functional Analysis*, **32**, 207–212.

Evans, D. E. and Høegh-Krohn, R. (1978). Spectral properties of positive maps on C^*-algebras. *Journal of the London Mathematical Society*, **17**, 345–355.

Evans, D. E. and Kawahigashi, Y. (1993). Subfactors and conformal field theory. In *Quantum and non-commutative analysis* (ed. H. Araki *et al.*), Kluwer Academic, pp. 341–369.

Evans, D. E. and Kawahigashi, Y. (1994a). Orbifold subfactors from Hecke algebras. *Communications in Mathematical Physics*, **165**, 445–484.

Evans, D. E. and Kawahigashi, Y. (1994b). The E_7 commuting squares produce D_{10} as principal graph. *Publications of the RIMS, Kyoto University*, **30**, 151–166.

Evans, D. E. and Kawahigashi, Y. (1995a). On Ocneanu's theory of asymptotic inclusions for subfactors, topological quantum field theories and quantum doubles. *International Journal of Mathematics*, **6**, 205–228.

Evans, D. E. and Kawahigashi, Y. (1995b). From subfactors to 3-dimensional topological quantum field theories and back – a detailed account of Ocneanu's theory. *International Journal of Mathematics*, **6**, 537–558.

Evans, D. E. and Kawahigashi, Y. (preprint 1997). Orbifold subfactors from Hecke algebras II – Quantum doubles and braiding.

Evans, D. E. and Kishimoto, A. (1988). Duality for automorphisms on a compact C^*-dynamical system. *Ergodic Theory and Dynamical Systems*, **8**, 173–189.

Evans, D. E. and Kishimoto, A. (1991). Compact group actions on UHF algebras obtained by folding the interval. *Journal of Functional Analysis*, **98**, 346–360.

Evans, D. E. and Kishimoto, A. (1997). Trace scaling automorphisms of certain stable AF algebras. *Hokkaido Mathematical Journal*, **26**, 211–224.

Evans, D. E. and Lewis, J.T. (1976a). Dilations of dynamical semigroups. *Communications in Mathematical Physics*, **50**, 219–227.

Evans, D. E. and Lewis, J.T. (1976b). Completely positive maps on some C^*-algebras, (unpublished).

Evans, D. E. and Lewis, J. T. (1977a). Dilations of irreversible evolutions in algebraic quantum theory. *Communications of the Dublin Institute for Advanced Studies* Ser. A **24**.

Evans, D. E. and Lewis, J. T. (1977b). Some semigroups of completely positive maps on the CCR algebra. *Journal of Functional Analysis*, **26**, 369–377.

Evans, D. E. and Lewis, J. T. (1984). The spectrum of the transfer matrix in the C^*-algebra of the two-dimensional Ising model. *Communications in Mathematical Physics*, **92**, 309–327.

Evans, D. E. and Lewis, J. T. (1986). On a C^*-algebra approach to phase transition in the two-dimensional Ising model II. *Communications in Mathematical Physics*, **102**, 521–535.

Evans, D. E. and Su, H. (in press). K-theoretic classification for certain $\mathbb{Z}_2$-actions on inductive limits of Cuntz algebras. *International Journal of Mathematics.*

Fack, T. and Maréchal, O (1979). Sur la classification des symétries des C^*-algèbres UHF. *Canadian Journal of Mathematics*, **31**, 496–523.

Fack, T. and Marèchal, O. (1981). Sur la classification des automorphisms périodiques des C^*-algèbres UHF. *Journal of Functional Analysis*, **40**, 267–301.

Faddeev, L. (1984). *Integrable models in (1+1)-dimensional quantum field theory* (Lectures in les Houches, 1982), Elsevier Science Publishers, Amsterdam, pp. 563–608.

Fannes, M. and Verbeure, A. (1974). On the time evolution automorphisms of the CCR-algebra for quantum mechanics. *Communications in Mathematical Physics*, **35**, 257–264.

Felder, F., Fröhlich, J., and Keller, G. (1990). On the structure of unitary conformal field theory. II. Representation-theoretic approach. *Communications in Mathematical Physics*, **130**, 1–49.

Feller, W. (1953). On the generation of unbounded semigroups of bounded linear operators. *Annals of Mathematics*, **58**, 166–174.

Fendley, P. (1989). New exactly solvable orbifold models. *Journal of Physics A*, **22**, 4633–4642.

Fendley, P. and Ginsparg, P. (1989). Non-critical orbifolds. *Nuclear Physics B*, **324**, 549–580.

Fenn, R. and Rourke, C. (1979). On Kirby's calculus of links. *Topology*, **18**, 1–15.

Ferdinand, A. and Fisher, M. (1969). Bounded and inhomogeneous Ising models. I. Specific-heat anomaly of a finite lattice. *Physical Review*, **185** (2), 832–846.

Fillmore, P. (1996). *A user's guide to operator algebras.* Canadian Mathematical Society Series of Monographs and Advanced Texts, Wiley, New York.

Ford, G. W., Kač, M. and Mazur, P. (1965). Statistical mechanics of assemblies of coupled oscillators. *Journal of Mathematical Physics*, **6**, 504–515.

Fredenhagen, K. (1977). Implementation of automorphisms and derivations of the CAR-algebra. *Communications in Mathematical Physics*, **52**, 255–266.

Fredenhagen, K. (1994). Superselection sectors with infinite statistical dimension. In *Subfactors – Proceedings of the Taniguchi Symposium, Katata*, (ed. H. Araki, et al.), World Scientific, Singapore, pp. 242–258.

Fredenhagen, K., Rehren, K.-H., and Schroer, B. (1989). Superselection sectors with braid group statistics and exchange algebras. *Communications in Mathematical Physics*, **125**, 201–226.

Fredenhagen, K., Rehren, K.-H., and Schroer, B. (1992). Superselection sectors with braid group statistics and exchange algebras II. Geometric aspects and conformal covariance. *Reviews in Mathematical Physics, Special Issue*, 113–157.

Freyd, P., Yetter, D., Hoste, J., Lickorish, W., Millet, K., and Ocneanu, A. (1985). A new polynomial invariant of knots and links. *Bulletin of the American Mathematical Society*, **12**, 239–246.

Friedan, D., Qui, Z. and Shenker, S. (1984). Conformal invariance, unitarity and critical exponents in two dimensions. *Physical Review Letters*, **52**, 1575–1578.

Friedrichs, K. O. (1953). *Mathematical aspects of the quantum theory of fields.* Wiley-Interscience, New York.

Fröhlich, J. (1988). The statistics of fields, the Yang–Baxter equation, and the theory of knots and links (Cargese Lectures, 1987). In *Non-Pert. Quantum Field Theory*, (ed. G. 't Hooft *et al.*), Plenum, New York.

Fröhlich, J. and Gabbiani, F. (1990). Braid statistics in local quantum theory. *Reviews in Mathematical Physics*, **2**, 251–353.

Fröhlich, J. and Kerler, T. (1993). *Quantum groups, quantum categories and quantum field theory.* Lecture Notes in Mathematics, 1542, Springer, Berlin.

Fröhlich, J. and King, C. (1989a). The Chern–Simons theory and knot polynomials. *Communications in Mathematical Physics*, **126**, 167–199.

Fröhlich, J. and King, C. (1989b). Two-dimensional conformal field theory and three-dimensional topology. *International Journal of Modern Physics A*, **4**, 5321–5399.

Fröhlich, J. and Marchetti, P.-A. (1988). Quantum field theory of anyons. *Letters in Mathematical Physics*, **16**, 347–358.

Fröhlich, J. and Marchetti, P.-A. (1989). Quantum field theories of vortices and anyons. *Communications in Mathematical Physics*, **121**, 177–223.

Fuchs, J. (1992). *Affine Lie algebras and quantum groups*, Cambridge University Press, Cambridge.

Fuchs, J., Gato-Rivera, B., Schellelkens, B., and Schweigert, C. (1994). Modular invariants and fusion rule automorphisms from Galois theory. *Physics Letters B*, **334**, 113–120.

Fuchs, J., Schellelkens, B., and Schweigert, C. (1995). Galois modular invariants of WZW models. *Nuclear Physics B*, **437**, 667–694.

Gabbiani, F. and Fröhlich, J. (1993). Operator algebras and conformal field theory. *Communications in Mathematical Physics*, **155**, 569–640.

Gallavotti, G., Martin-Lof, A. and Miracle-Sole, S (1973). In *Statistical mechanics and mathematical problems* (ed. A. Lenard), Springer, Berlin, Heidelberg, New York.

Gannon, T. (1993). WZW commutants, lattices and level – one partition functions. *Nuclear Physics B*, **396**, 708–736.

Gannon, T. (1994). The classification of affine $SU(3)$ modular invariant partition functions. *Communications in Mathematical Physics*, **161**, 233–264.

Gannon, T. (1995). Symmetries of Kač–Peterson modular matrices of affine algebras. *Inventiones Mathematicae*, **122**, 341–357.

Gannon, T. (preprint 1995). Kač–Peterson, Perron–Frobenius, and classification of conformal field theories.

Gannon, T. (1996). The classification of su(3) modular invariants revisited. *Annales de l'Institut Henri Poincaré*, **65**, 15–55.

Gannon, T. (1997). The level two and three modular invariants of $SU(n)$. *Letters in Mathematical Physics*, **39**, 289–298.

Gannon, T. and Ho-Kim, Q. (1994). The rank – four heterotic modular invariant partition functions. *Nuclear Physics B*, **425**, 319–342.

Gannon, T. and Walton, T. (1995). On the classification of diagonal coset modular invariants. *Communications in Mathematical Physics*, **173**, 175–197.

Gannon, T., Ruelle, Ph., and Walton, M. (1996). Automorphism modular invariants of current algebras. *Communications in Mathematical Physics*, **179**, 121–156.

Gantmacher, F. R. (1960). *The theory of matrices.* Vol. 2. Chelsea, New York.

Gelfand, I. M. and Naimark, M. (1943). On the embedding of normed rings into the ring of operators in Hilbert space. *Rossiiskaya Akademiya Nauk. Matematicheskii Sbornik*, **12**, 197–213.

Gepner, D. and Witten, E. (1986). String theory on group manifolds. *Nuclear Physics B*, **278**, 493–549.

Ginibre, J. (1969). Simple proof and generalisation of Griffiths' second equality. *Physical Review Letters*, **23**, 828–830.

Ginsparg, P. (1990). Applied conformal field theory. In *Champs, cordes et phénomènes critiques (Les Houches, 1988)* (ed. E. Brezin and J. Sinn-Justin), North-Holland, Amsterdam, pp. 1–168.

Giordano, T. (1988). Classification of approximately finite real C^*-algebras. *Journal für die Reine und Angewandte Mathematik*, **385**, 161–194.

Giordano, T., Putnam, I. F. and Skau, C. (1995). Topological orbit equivalence and C^*-crossed products. *Journal für die Reine und Angewandte Mathematik*, **469**, 51–111.

Glimm, J. (1960). On a certain class of operator algebras. *Transactions of the American Mathematical Society*, **95**, 318–340.

Glimm, J. (1961). Type I C^*-algebras. *Annals of Mathematics*, **73**, 572–612.

Glimm, J. and Jaffe, A. (1972). The $(\lambda\phi^4)_2$ Quantum field theory without cutoffs. IV Perturbations of the Hamiltonian. *Journal of Mathematical Physics*, **13**, 1568–1584.

Goddard, P. and Olive, D. (ed.) (1988). Kač–Moody and Virasoro algebras. *Advanced Series in Mathematical Physics*, **3**, World Scientific, Singapore.

Goddard, P., Nahm, W., and Olive, D. (1985). Symmetric spaces, Sugawara's energy momentum tensor in two dimensions and free Fermions. *Physics Letters*, **160B**, 111–116.

Goddard, P., Kent, A., and Olive, D. (1986). Unitary representations of the Virasoro and super-Virasoro algebras. *Communications in Mathematical*

Physics, **103**, 105–119.

Gohberg I. C. and Krein, M. G. (1969). *Introduction to the theory of linear non-self-adjoint operators.* American Mathematical Society. Providence.

Goldman, M. (1960). On subfactors of type II_1. *The Michigan Mathematical Journal*, **7**, 167–172.

Goldschmidt, D. M. and Jones, V. F. R. (1989) Metaplectic link invariants. *Geometriae Dedicata*, **31**, 165–191.

Gong, G. (preprint 1994). On the classification of C^*-algebras of real rank zero and unsuspended E-equivalent types.

Gong, G. (preprint a). On inductive limits of matrix algebras over higher dimensional spaces, Part II.

Gong, G. (preprint b). On simple inductive limits of matrix algebras over higher dimensional spaces.

Gong, G., Jiang, X. and Su, H. Two properties of certain simple C^*-algebras.

Goodearl, K. R. and Handelman, D. E. (1976). Rank functions and K_0 of regular rings. *Journal of Pure and Applied Algebra*, **7**, 195–216.

Goodearl, K. R. and Handelman, D. E. (1987). Classification of ring and C^*-algebra direct limits of finite-dimensional semisimple real algebras. *Memoirs of the American Mathematical Society*, **372**.

Goodman, F. and Nakanishi, T. (1991). Fusion algebras in integrable systems in two dimensions. *Physics Letters B*, **262**, 259–264.

Goodman, F. and Wenzl, H. (1990). Littlewood Richardson coefficients for Hecke algebras at roots of unity. *Advances in Mathematics*, **82**, 244–265.

Goodman, F., de la Harpe, P. and Jones, V. F. R. (1989). *Coxeter graphs and towers of algebras.* MSRI Publications, 14, Springer, Berlin.

Goodman, R. and Wallach, N. (1984). Structure and unitary representations of loop groups and the group of diffeomorphisms of the circle. *Journal der Reine Angewandtte Mathematik*, **347**, 69–133.

Gorini, V., Kossakowski, A. and Sudarshan, E. C. G. (1976). Completely positive dynamical semigroups on N-level systems. *Journal of Mathematical Physics*, **17**, 821–825.

Goto, S. (1994). Orbifold construction for non-AFD subfactors. *International Journal of Mathematics*, **5**, 725–746.

Goto, S. (1995). Symmetric flat connections, triviality of Loi's invariant, and orbifold subfactors. *Publications of the RIMS, Kyoto University*, **31**, 609–624.

Goto, S. (1996). Commutativity of automorphisms of subfactors modulo inner automorphisms. *Proceedings of the American Mathematical Society.* **124**, 3391–3398.

Goto, S. (preprint 1996). Quantum double construction for subfactors arising from periodic commuting squares.

Green, H. S. and Hurst, C. A. (1964). *Order–disorder phenomena.* Wiley Interscience, New York.

Greenleaf, F. P. (1969). *Invariant means on topological groups and their applications*, Van Nostrand, New York.

Greiner, G., Voigt, J. and Wolff, M. (1981). On the spectral bound of the generator of semigroups of positive operators. *Journal of Operator Theory*, **5**, 245–256.

Griffiths, R. B. (1967). Correlations on Ising ferromagnets I, II. *Journal of Mathematical Physics*, **8**, 478–483, 484–489.

Griffiths, R. B. (1972). Rigorous results and theorems. In *Phase transitions and critical phenomena Vol 1*, (eds C. Domb, M. Green. Academic Press, New York), pp. 7–109.

Gruber, C., Hintermann, A. and Merlini, D. (1977). *Group analysis of classical lattice systems*. Lecture Notes in Physics, 60, Springer, Berlin.

Haag, R. (1992). *Local quantum physics*. Springer, Berlin.

Haag, R and Kastler, T. (1964). An algebraic approach to quantum field theory. *Journal of Mathematical Physics*, **5**, 848–861.

Haag, R., Hugenholtz, N. M. and Winnink, M. (1967). On the equilibrium states in quantum mechanics. *Communications in Mathematical Physics*, **5**, 215–236.

Haagerup, U. (1975). Normal weights on W^*-algebras. *Journal of Functional Analysis*, **19**, 302–317.

Haagerup, U. (1983). All nuclear C^*-algebras are amenable. *Inventiones Mathematicae*, **74**, 305–319.

Haagerup, U. (1987). Connes' bicentralizer problem and the uniqueness of the injective factor of type III_1. *Acta Mathematica*, **158**, 95–148.

Haagerup, U. (preprint 1991). Quasi-traces on exact C^*-algebras are traces.

Haagerup, U. (1994). Principal graphs of subfactors in the index range $4 < [M : N] < 3+\sqrt{2}$. In *Subfactors – Proceedings of the Taniguchi Symposium, Katata*, (ed. H. Araki *et al.*), World Scientific, Singapore, pp. 1–38.

Haagerup, U. and Størmer, E. (1990a). Equivalences of normal states on von Neumann algebras and the flow of weights. *Advances in Mathematics* **83**, 180–262.

Haagerup, U. and Størmer, E. (1990b). Pointwise inner automorphisms of von Neumann algebras (with an appendix by Sutherland, C. E.). *Journal of Functional Analysis*, **92**, 177–201.

Haagerup, U. and Winsløw, C. (preprint 1997). The Effros–Maréchal topology in the space of von Neumann algebras.

Hakeda, J. and Tomiyama, J. (1967). On some extension properties of von Neumann algebras. *Tôhoku Mathematical Journal*, **19**, 315–323.

Halmos, P. R. (1967). *A Hilbert space problem book*. Van Nostrand, Princeton.

Hamachi, T. and Kosaki, H. (1988a). Index and flow of weights of factors of type III. *Proceedings of the Japan Academy*, **64A**, 11–13.

Hamachi, T. and Kosaki, H. (1988b). Inclusions of type III factors constructed from ergodic flows. *Proceedings of the Japan Academy*, **64A**, 195–197.

Hamachi, T. and Kosaki, H. (1993). Orbital factor map. *Ergodic Theory and Dynamical Systems*, **13**, 515–532.

Handelman, D. E. (1978). K_0 of von Neumann algebras and AF C^*-algebras. *Quarterly Journal of Mathematics, Oxford*, **(2) 29**, 427–441.

Handelman, D. E. (1985). Positive polynomials and product type actions of compact groups. *Memoirs of the American Mathematical Society*, **320**.

Handelman, D. E. (1987). Extending traces on fixed point C^*-algebras under xerox product type actions of compact Lie groups. *Journal of Functional Analysis*, **72**, 44–57.

Handelman, D. E. and Rossmann, W. (1984). Product type actions of finite and compact groups. *Indiana University of Mathematics*, **33**, 479–509.

Handelman, D. E. and Rossman, W. (1985). Actions of compact groups on C^*-algebras. *Illinois Journal of Mathematics*, **29**, 51–95.

Havet, J.-F. (1976). Espérance conditionnelles permutables à un group d'automorphismes sur une algèbre de von Neumann. *Comptes Rendus de l'Académie des Sciences, Série I, Mathématiques*, **282**, 1095–1098.

Havet, J.-F. (1990). Espérance conditionnelle minimale. *Journal of Operator Theory*, **24**, 33–35.

Hayashi, T. (1993). Quantum group symmetry of partition functions of IRF models and its application to Jones' index theory. *Communications in Mathematical Physics*, **157**, 331–345.

Helgason, S. (1978). *Differential geometry, Lie groups and symmetric spaces.* Academic Press, London.

Herman, R.H. and Jones, V.F.R. (1982). Period two automorphisms of UHF algebras. *Journal of Functional Analysis*, **45**, 169–176.

Herman, R. H. and Jones, V. F. R. (1983). Models of finite group action. *Mathematica Scandinavica*, **52**, 312–320.

Herman, R. H. and Ocneanu, A. (1984). Stability for integer actions on UHF C^*-algebras. *Journal of Functional Analysis*, **59**, 132–144.

Herman, R. H. and Vaserstein, L. N. (1984). The stable rank of C^*-algebras. *Inventiones Mathematicae*, **77**, 553–555.

Herman, R. H., Putnam, I. F. and Skau, C. (1992). Ordered Bratteli diagrams, dimension groups and topological dynamical systems. *International Journal of Mathematics*, **3**, 827–864.

Hiai, F. (1988). Minimizing indices of conditional expectations onto a subfactor. *Publications of the RIMS, Kyoto University*, **24**, 673–678.

Hiai, F. (1990). Minimum index for subfactors and entropy I. *Journal of Operator Theory*, **24**, 301–336.

Hiai, F. (1991). Minimum index for subfactors and entropy, II. *Journal of the Mathematical Society of Japan*, **43**, 347–380.

Hiai, F. (1995). Entropy for canonical shifts and strong amenability. *International Journal of Mathematics*, **6**, 381–396.

Hiai, F. (preprint 1995). Standard invariants for crossed products inclusions of factors.

Hiai, F. and Izumi, M. (preprint 1996). Amenability and strong amenability for fusion algebras with applications to subfactor theory.

Higson, N. (1987). A characterization of KK-theory. *Pacific Journal of Mathematics*, **126**, 253–276.

Higson, N. (1988). Algebraic K-theory of stable C^*-algebras. *Advances in Mathematics*, **67**, 1–140.

Higson, N. (1990). Categories of fractions and excision in K-theory. *Journal of Pure and Applied Algebra*, **65**, 119–138.

Higuchi, Y. (1979). On the absence of non-translationally invariant Gibbs states for the two dimensional Ising model. *Random fields, Colloquia Mathematica Societatis János Bolyai, Eszlergom, Hungary*, **27**.

Hille, E. (1948). *Functional analysis and semigroups.* American Mathematical Society Collected Publications, American Mathematical Society, New York, **31**.

Hille, E. (1952). On the generation of semigroups and the theory of conjugate functions. *Kungl. Fys. Salls. I.Lund Forhand*, **21**, 1–13.

Hille, E. and Philips, R. S. (1957). *Functional analysis and semigroups.* American Mathematical Society Collected Publications, American Mathematical Society, New York, **31**.

Hochstenbach, W. (1975). Feldtheorie mit einem äußeren potential, Thesis, Hamburg.

Høegh-Krohn, R., and Skjelbred, T. (1981). Classification of C^*-algebras admitting ergodic actions of the two-dimensional torus. *Journal der Reine Angewandtte Mathematik*, **328**, 1–8.

Høegh Krohn, R., Landstad, M. and Størmer, E. (1981). Compact ergodic groups of automorphisms. *Annals of Mathematics.* **114**, 75–86.

Hofman, A. J. (1972). On limit points of spectral radii of non-negative symmetric integral matrices. *Lecture Notes in Mathematics*, **303**, 165–172, Springer, Berlin.

Hudson, R. L. and Parthasarathy, K. R. (1986). Unification of Fermion and Boson stochastic calculus. *Communications in Mathematical Physics.* **104**, 457–470.

Hugenholtz, N. M. and Kadison, R. V. (1975). Automorphisms and quasi-free states on the CAR algebra. *Communications in Mathematical Physics.* **43**, 161–177.

Huse, D. (1984). Exact exponents for infinitely many new critical points. *Physics Review B*, **30**, 3908–3915.

Husemoller, D. (1966). *Fibre bundles.* Graduate Texts in Mathematics, 20, Springer, New York.

Ikeda, K. (preprint 1997). Numerical evidence for flatness of Haagerup's connections.

Ingarden, R. S. and Kossakowski, A. (1975). On the connection of non-equilibrium information thermodynamics with non-hamiltonian quantum mechanics of open systems. *Annals of Physics*, **89**, 451–485.

Izumi, M. (1991). Application of fusion rules to classification of subfactors. *Publications of the RIMS, Kyoto University*, **27**, 953–994.

Izumi, M. (1992). Goldman's type theorem for index 3. *Publications of the RIMS, Kyoto University*, **28**, 833–843.

Izumi, M. (1993a). On type III and type II principal graphs for subfactors. *Mathematica Scandinavica*, **73**, 307–319.

Izumi, M. (1993b). Subalgebras of infinite C^*-algebras with finite Watatani indices I. Cuntz algebras. *Communications in Mathematical Physics*, **155**, 157–182.

Izumi, M. (1994a). On flatness of the Coxeter graph E_8. *Pacific Journal of Mathematics*, **166**, 305–327.

Izumi, M. (1994b). Canonical extension of endomorphisms of factors. In *Subfactors – Proceedings of the Taniguchi Symposium, Katata*, (ed. H. Araki, *et al.*), World Scientific, Singapore, pp. 274–293.

Izumi, M. (preprint 1995). Subalgebras of infinite C^*-algebras with finite indices II. Cuntz–Krieger algebras.

Izumi, M. and Kawahigashi, Y. (1993). Classification of subfactors with the principal graph $D_n^{(1)}$. *Journal of Functional Analysis* **112**, 257–286.

Izumi, M. and Kosaki, H. (1996). Finite dimensional Kač algebras arising from certain group actions on a factor. *International Mathematics Research Notices*, **8**, 357–370.

Izumi, M., Longo, R. and Popa, S. (in press). A Galois correspondence for compact groups of automorphisms of von Neumann algebras with a generalization to Kač algebras. *Journal of Functional Analysis.*

Janich, K. (1965). Vectorraumbündel und der Raum der Fredholm Operatoren. *Mathematische Annalen*, **161**, 129–142.

Jiang, X. and Su, H. (preprint 1996). On a simple unital projectionless C^*-algebra.

Jimbo, M. (1986a). A q-difference analogue of $U(g)$ and the Yang–Baxter equation. *Letters in Mathematical Physics*, **102**, 537–567.

Jimbo, M. (1986b). A q-analogue of $U(N+1)$, Hecke algebra and the Yang–Baxter equation. *Letters in Mathematical Physics*, **11**, 247–252.

Jimbo, M., Miwa, T. and Okado, M. (1987). Solvable lattice models whose states are dominant integral weights of $A_{n-1}^{(1)}$. *Letters in Mathematical Physics*, **14**, 123–131.

Jimbo, M., Miwa, T. and Okado, M. (1988). Solvable lattice models related to the vector representation of classical simple Lie algebras. *Communications in Mathematical Physics*, **116**, 507–525.

Johnson, B. E. (1964). An introduction to the theory of centralizers. *Proceedings of the London Mathematical Society (3)*, **14**, 299–320.

Johnson, B. E. (1972). *Cohomology in Banach algebras.* Memoirs of the American Mathematical Society, 127.

Jones, V. F. R. (1980a). A factor anti-isomorphic to itself but without involutory anti-automorphisms. *Mathematica Scandinavica*, **45**, 103–117.

Jones, V. F. R. (1980b). *Actions of finite groups on the hyperfinite type II_1 factor*. Memoirs of the American Mathematical Society, 237.

Jones, V. F. R. (1983). Index for subfactors. *Inventiones Mathematicae*, **72**, 1–15.

Jones, V. F. R. (1985). A polynomial invariant for knots via von Neumann algebras. *Bulletin of the American Mathematical Society*, **12**, 103–112.

Jones, V. F. R. (1986a). A new knot polynomial and von Neumann algebras. *Notices of the American Mathematical Society*, **33**, 219–225.

Jones, V. F. R. (1986b). Braid groups, Hecke algebras and type II_1 factors. (1986). In *Geometric Methods in Operator Algebras* (ed. H. Araki and E. G. Effros), Longman, London, pp. 242–273.

Jones, V. F. R. (1987). Hecke algebra representations of braid groups and link polynomials. *Annals of Mathematics*, **126**, 335–388.

Jones, V. F. R. (1989a). On a certain value of the Kauffman polynomial. *Communications in Mathematical Physics*, **125**, 459–467.

Jones, V. F. R. (1989b). On knot invariants related to some statistical mechanical models. *Pacific Journal of Mathematics*, **137**, 311–334.

Jones, V. F. R. (1990a). Knots, braids and statistical mechanics. In *Advances in differential geometry and topology*, 149–184.

Jones, V. F. R. (1990b). Baxterization. *International Journal of Modern Physics B*, **4**, 701–713.

Jones, V. F. R. (1990c). Notes on subfactors and statistical mechanics. *International Journal of Modern Physics A*, **5**, 441–460.

Jones, V. F. R. (1991a). von Neumann algebras in mathematics and physics. In *Proceedings of ICM-90, Berkeley*, 121–138, Springer, Berlin.

Jones, V. F. R. (1991b). *Subfactors and knots.* CBMS Regional Conference Series in Mathematics, 80.

Jones, V. F. R. (1992). From quantum theory to knot theory and back: a von Neumann algebra excursion. In *American Mathematical Society Centennial Publications*, Vol. II, pp. 321–336.

Jones, V. F. R. (1994). The Potts model and the symmetric group. In *Subfactors – Proceedings of the Taniguchi Symposium, Katata*, (ed. H. Araki *et al.*), World Scientific, New York, pp. 259–267.

Jones, V. F. R. (1995). Fusion en algèbres de von Neumann et groupes de lacets (d'après A. Wassermann). *Seminaire Bourbaki*, **800**, 1–20.

Jones, V. F. R. and Takesaki, M. (1984). Actions of compact abelian groups on semifinite injective factors. *Acta Mathematica*, **153**, 213–258.

Jones, V. and Wassermann, A. (in preparation). Fermions on the circle and representations of loop groups.

Jørgensen, P. E. T. and Werner, R. F. (1994). Coherent states of the q-canonical commutation relations. *Communications in Mathematical Physics*, **164**, 455–471.

Jørgensen, P. E. T., Schmidt, L. M. and Werner, R. F. (1994). q-canonical commutation relation and stability of the Cuntz algebras. *Pacific Journal of Mathematics*, **165**, 131–151.

Kač, V. (1990). *Infinite dimensional Lie algebras*, 3rd edn, Cambridge University Press, Cambridge.

Kač, V. and Petersen, D. H. (1984). Infinite dimensional Lie algebras, theta functions, and modular forms. *Advances in Mathematics*, **53**, 125–264.

Kadanov, L. P. and Ceva, H. (1971). Determination of an operator algebra for the two dimensional Ising model. *Physical Review B*, **3**, 3918–3938.

Kadison, R. V. (1952). A generalized Schwarz inequality and algebraic invariants for operator algebras. *Annals of Mathematics*, **56**, 494–503.

Kadison, R. V. (1965). Transformations of states in operator theory and dynamics. *Topology.* **3** (Supplement 2), 177–198.

Kadison, R. V. (1966). Derivations on operator algebras. *Annals of Mathematics*, **83**, 280–293.

Kadison, R. V. (1975). A note on derivations of operator algebras. *Bulletin of the London Mathematical Society*, **7**, 41–44.

Kadison, R. V. and Ringrose, J. R. (1983). *Fundamentals of the theory of operator algebras*, Vol I. Academic Press, New York.

Kadison, R. V. and Ringrose, J. R. (1986). *Fundamentals of the theory of operator algebras*, Vol II. Academic Press, New York.

Kallman, R. R. (1971). Groups of inner automorphisms of von Neumann algebras. *Journal of Functional Analysis*, **7**, 43–60.

Kaplansky, I. (1951a). A theorem on rings of operators. *Pacific Journal of Mathematics*, **1**, 227–232.

Kaplansky, I. (1951b). The structure of certain operator algebras. *Transactions of the American Mathematical Society*, **70**, 219–255.

Kaplansky, I. (1951c). Group algebras in the large. *Tôhoku Mathematical Journal*, **3**, 249–256.

Kaplansky, I. (1968). *Rings of Operators.* Benjamin.

Kaplansky, I. (1970). *Algebraic and analytic aspects of operator algebras.* CBMS Regional Conference Series in Mathematics, American Mathematical Society, Providence, 1.

Karlin, S. and Studden, W. J. (1966). Tchebycheff systems: with applications in analysis and statistics. *Interscience Publishers, Wiley, New York.*

Karowski, M. (1988). Finite size corrections for integrable systems and conformal properties of six-vertex models. *Nuclear Physics*, **B 300**, 473–499.

Karowski, M. (1989). Conformal quantum field theories and integrable systems. In *Proceedings of the Brasov Summer School, September 1987* (ed. P. Dita *et al.*), Academic Press, New York, pp. 147–161.

Kasparov, G. (1980,1981). The operator K-functor and extensions of C^*-algebras. *Izvestiya Akademii Nauk SSSR, Seriya Matematicheskaya*, **44**, 571–636; *Mathematics USSR - Izvestiya*, **16**, 513–572.

Kasteleyn, P. W. and Fortuin, C. M. (1969). Phase transitions in Lattice systems with random local properties. *Journal of the Physical Society of Japan Supplement*, **26**, 11–14.

Kastler D. (1990). (editor) *The algebraic theory of Superselection sectors. Introduction and recent results.* World Scientific, Singapore.

Kato, A. (1987). Classification of modular invariant partition functions in two dimensions. *Modern Physics Letters A*, **2**, 585–600.

Kato, T. (1966). *Perturbation theory for linear operators.* Springer, Berlin.

Kauffman, L. H. (1986). Chromatic polynomial, Potts model, Jones polynomial. *Lecture notes.*

Kauffman, L. (1987). State models and the Jones polynomial. *Topology*, **26**, 395–407.

Kauffman, L. (1990). An invariant of regular isotopy. *Transactions of the American Mathematical Society*, **318**, 417–471.

Kauffman, L. (1991). *Knots and Physics.* World Scientific, Singapore.

Kauffman, L. and Lins, S. L. (1994). *Temperley–Lieb recoupling theory and invariants of* 3*-manifolds.* Princeton University Press, Princeton.

Kaufman, B. (1949). Crystal Statistics II. *Physical Review*, **76**, 1232–1243.

Kaufman, B. and Onsager, L. (1949). Crystal Statistics III. *Physical Review*, **76**, 1244–1252.

Kawahigashi, Y. (1992a). Automorphisms commuting with a conditional expectation onto a subfactor with finite index. *Journal of Operator Theory*, **28**, 127–145.

Kawahigashi, Y. (1992b). Exactly solvable orbifold models and subfactors. *Functional Analysis and Related Topics*, Lecture Notes in Mathematics, Springer, **1540**, 127–147.

Kawahigashi, Y. (1993). Centrally trivial automorphisms and an analogue of Connes's $\chi(M)$ for subfactors. *Duke Mathematical Journal*, **71**, 93–118.

Kawahigashi, Y. (1994). Paragroups and their actions on subfactors. In *Subfactors – Proceedings of the Taniguchi Symposium, Katata* (ed. H. Araki *et al.*), World Scientific, Singapore, pp. 64–84.

Kawahigashi, Y. (1995a). On flatness of Ocneanu's connections on the Dynkin diagrams and classification of subfactors. *Journal of Functional Analysis*, **127**, 63–107.

Kawahigashi, Y. (1995b). Orbifold subfactors, central sequences and the relative Jones invariant κ. *International Mathematical Research Notices*, pp. 129–140.

Kawahigashi, Y. (1995c). Classification of paragroup actions on subfactors. *Publications of the RIMS, Kyoto University*, **31**, 481–517.

Kawahigashi, Y. (1996) Paragroups as quantized Galois groups of subfactors. *Sugaku Expositions*, **9**, 21–35.

Kawahigashi, Y. (1997a) Classification of approximately inner automorphisms of subfactors. *Mathematische Annalen*, **308**, 425–438.

Kawahigashi, Y. (1997b) Quantum doubles and orbifold subfactors. In *Operator Algebras and Quantum Field Theory*, (ed. S. Doplicher et al), 271–283, International Press.

Kawahigashi, Y. and Takesaki, M. (1992). Compact abelian group actions on injective factors. *Journal of Functional Analysis*, **105**, 112–128.

Kawahigashi, Y., Sutherland, C. E., and Takesaki, M. (1992). The structure of the automorphism group of an injective factor and the cocycle conjugacy of discrete abelian group actions. *Acta Mathematica*, **169**, 105–130.

Kawamuro, K. (preprint 1998). Central sequence subfactors and double commutant properties.

Kazhdan, V. and Lusztig, G. (1994). Tensor structures arising from affine Lie algebras. *IV, Journal of the American Mathematical Society*, **7**, 383–453.

Kelly, D. G. and Sherman, S. (1968). General Griffiths' inequality on correlations in Ising feromagnets. *Journal of Mathematical Physics*, **9**, 466–484.

Kennelly, A. E. (1899). The equivalence of triangles and three pointed stars in conducting networks. *Electrical World and Engineer*, **34**, 413–414.

Kent, A. (1991). Projections of Virasoro singular vectors. *Physics Letters B*, **273**, 56–62.

Kirby, R. (1978). A calculus of farmed links in S^3. *Inventiones Mathematicae*, **45**, 35–56.

Kirby, R. and Melvin, P. (1990). On the 3-manifold invariants of Witten and Reshetikhin–Turaev. *Inventiones Mathematicae*, **105**, 473–545.

Kirchberg, E. (preprint 1994). The classification of purely infinite C^*-algebras using Kasparov's theory. *Fields Institute preprint.*

Kirchberg, E. (1995). Exact C^*-algebras, tensor products and classification of purely infinite algebras. In *Proceedings of ICM-94, Zürich*, 943–954, Birkhäuser.

Kirillov, A. N. and Reshetikhin, N. Yu. (1988). Representations of the algebra $U_q(sl_2)$, q-orthogonal polynomials and invariants for links. *Infinite dimensional Lie algebras and groups* (ed. Kač, V. G.), Advanced Series in Mathematical Physics, vol. 7, pp. 285–339.

Kishimoto, A. (1976). Dissipations and derivations. *Communications in Mathematical Physics*, **47**, 25–32.

Kishimoto, A. (1977). On the fixed point algebra of a UHF algebra under a periodic automorphism of product type. *Publications of the RIMS, Kyoto University*, **13**, 777–791.

Kishimoto, A. (1990). Actions of finite groups on certain inductive limit. *International Journal of Mathematics*, **1**, 267–292.

Kishimoto, A. (1995). The Rohlin property for automorphisms of UHF algebras. *Journal für die Reine und Angewandte Mathematik*, **465**, 183–196.

Kishimoto, A. (1996). The Rohlin property for shifts on UHF algebras and automorphisms of Cuntz algebras. *Journal of Functional Analysis.* **140**, 100–123.

Klaus, M. and Scharf, G. (1977). The regular external field problem in quantum electro-dynamics. *Helvetica Physica Acta*, **50**, 779–802.

Knizhnik, V. and Zamolodchikov, A. (1984). Current algebra and Weiss–Zumino models in two dimensions. *Nuclear Physics B*, **247**, 83–103.

Kohno, T. (1987). Monodromy representations of braid groups and Yang–Baxter equations. *Annales de l'Institut Fourier, Grenoble*, **37**(4), 139–160.

Kohno, T. (1992a). Topological invariants for 3-manifolds using representations of mapping class groups I. *Topology*, **31**, 203–230.

Kohno, T. (1992b). Three-manifold invariants derived from conformal field theory and projective representations of modular groups. *International Journal of Modern Physics*, **6**, 1795–1805.

Kolmogorov, A. N. (1941). Stationary sequences in Hilbert's space. *Byulleten Moskovskogo Gosudarstvennogo Universiteta Matematika No.6*, **2**, 40pp.

Körner, T. W. (1986). *Fourier analysis*. Cambridge University Press, Cambridge.

Kosaki, H. (1986). Extension of Jones' theory on index to arbitrary factors. *Journal of Functional Analysis*, **66**, 123–140.

Kosaki, H. (1989). Characterization of crossed product (properly infinite case). *Pacific Journal of Mathematics*, **137**, 159–167.

Kosaki, H. (1990). Index theory for type III factors. In *Mappings of operator algebras, Proceedings of U.S.-Japan Seminar* (ed. H. Araki and R. V. Kadison), Birkhäuser, pp. 129–139.

Kosaki, H. (1993). Automorphisms in the irreducible decompositions of sectors. In *Quantum and non-commutative analysis* (ed. H. Araki *et al.*), pp. 305–316, Kluwer Academic.

Kosaki, H. (1994a). AFD factor of type III_0 with many isomorphic index 3 subfactors. *Journal of Operator Theory*, **32**, 17–29.

Kosaki, H. (1994b). Some remarks on automorphisms for inclusions of type III factors. in *Subfactors – Proceedings of the Taniguchi Symposium, Katata*, (ed. H. Araki *et al.*), World Scientific, Singapore, pp. 153–171.

Kosaki, H. (1996). Sector theory and automorphisms for factor-subfactor pairs. *Journal of the Mathematical Society of Japan*, **48**, 427–454.

Kosaki, H. and Loi, P. H. (1995). A remark on non-splitting inclusions of type III_1 factors. *International Journal of Mathematics*, **6**, 581–586.

Kosaki, H. and Longo, R. (1992). A remark on the minimal index of subfactors. *Journal of Functional Analysis*, **107**, 458–470.

Kosaki, H. and Yamagami, S. (1992). Irreducible bimodules associated with crossed product algebras. *International Journal of Mathematics*, **3**, 661–676.

Kosaki, H., Munemasa, A., and Yamagami, S. (1997). On fusion algebras associated to finite group actions. *Pacific Journal of Mathematics*, **177**, 269–290.

Kossakowski, A. (1972a). On quantum statistical mechanics of non-hamiltonian systems. *Reports on Mathematical Physics*, **3**, 247–274.

Kossakowski, A. (1972b). On the necessary and sufficient conditions for a generator of a quantum dynamical semigroup. *Bulletin de l'Academie Polonaise des Sciences, ser. Mathématiques, Astronomiques et Physique*, **20**, 1021–1025.

Kossakowski, A. (1973). On the general form of the generator of a dynamical semigroup for the spin 1/2 system. *Bull. Acad. Polon. Sér. Sci. Math.*

Astronom. Phys., **21** 649–653.

Kostov, I. (1988). Free field presentation of the A_n coset models on the torus. *Nuclear Physics B*, **300**, 559–587.

Kotecky, R. and Shlosman, S. B. (1985). First-order phase transitions in large entropy lattice models. *Communications in Mathematical Physics*, **83**, 493–515.

Kramers, H. A. and Wannier, G. H. (1941). Statistics of the two dimensional ferromagnet part 1. *Physical Review*, **60**, 252–262.

Kraus, K. (1971). General state changes in quantum theory. *Annals of Physics*, **64**, 311–335.

Kreuzer, M. and Schellekens, A. N. (1994). Simple currents versus orbifolds with discrete torsion – a complete classification. *Nuclear Physics B*, **411**, 97–121.

Kristensen, P., Mejlbo, L. M., and Thue Poulsen, E. (1967). Tempered distributions in infinitely many dimensions. III. Linear transformations of field operators. *Communications in Mathematical Physics*, **6**, 29–48.

Kuik, R. (1986). On the q-state Potts model by means of non-commutative algebras. Ph. D. Thesis, Groningen.

Kulish, P. and Reshetikhin, N. (1983). Quantum linear problem for the sine-Gordon equation and higher representations. *Journal of Soviet Mathematics*, **23**, 2435–2441.

Kumjian, A. (1988). An involutive automorphism of the Bunce–Deddens algebra. *Comptes Rendus Mathématique Rep. de l'Académie des Sciences, Canada*, **10**, 217–218.

Kumjian, A. (1990). On the K-theory of the symmetrized non-commuting torus. *Comptes Rendus Mathématique Rep. de l'Académie des Sciences, Canada* **12**. 87–89.

Kumjian, A., Pask, D., Raeburn, I. and Renault, J. (preprint 1996). Graphs, groupoids and Cuntz–Krieger algebras.

Kümmerer, B. (1987). Survey on a theory of non-commutative stationary Markov processes. *Quantum Probability and Applications III, Springer Lecture Notes in Mathematics*, **1303**, 154–182.

Kuniba, A., Akutsu, Y. and Wadati, M. (1986). Virasoro algebra, von Neumann algebra and critical eight vertex SOS model. *Journal of Physics Society of Japan*, **55**, 3285–3288.

Kunze, R. A. (1967). Positive-definite operator-valued kernels and unitary representations. *Functional Analysis, Proceedings of Conference*, University of California, Irvine (ed. B. R. Gelbaum), Academic Press, New York.

Labontè , G. (1974). On the nature of 'strong' Bogoliubov transformations for Fermions. *Communications in Mathematical Physics*, **36**, 59–72.

Laca, M. (to appear). Semigroups of *-endomorphisms, Dirichlet series and phase transitions. *Journal of Functional Analysis*.

Laca, M. and Raeburn, I. (to appear). A semigroup crossed product arising in number theory. *Journal of the London Mathematical Society*.

Lance, E. C. (1973). On nuclear C^*-algebras. *Journal of Functional Analysis*, **12**, 157–176.

Lance, E. C. (1982). Tensor products and nuclear C^*-algebras. *Proceedings of Symposia in Pure Mathematics (I)* **38**, 379–400.

Lanford, O. E. and Ruelle, D. (1969). Observables at infinity and states with short range correlations in statistical mechanics. *Communications in Mathematical Physics*, **13**, 194–215.

Lax, P. D. and Phillips, R. S. (1967). *Scattering theory.* Academic Press, New York.

Lewis, J. T. and Sisson, P. N. M. (1975). A C^*-algebra of the 2-dimensional Ising model. *Communications in Mathematical Physics*, **44**, 279–292.

Lewis, J. T. and Thomas, L. C. (1975a). On the existence of a class of stationary quantum stochastic processes. *Annales de l'Institut Henri Poincaré*, Section A, **22**, 241–248.

Lewis, J. T. and Thomas, L. C. (1975b). How to make a heat bath. In *Functional Integration* (ed. A. M. Arthurs), *Proceedings of International Conference*, Cumberland Lodge, London, April 1974, Clarendon Press, Oxford, pp. 97–123.

Lewis, J. T. and Winnink, M. (1979). The Ising model phase transition and the index of states on the Clifford algebra. *Coll. Mathematical Society Janes Bolyai*, **27**, Random Fields, Esztergom, Hungary.

Lickorish, W. (1988). Polynomials for links. *Bulletin of the American Mathematical Society*, **20**, 558–588.

Lieb, E. H. and Ruskai, M. B. (1974). Some operator inequalities of the Schwarz type. *Advances in Mathematics*, **12**, 269–273.

Lin, H. and Phillips, N. C. (1995). Classification of direct limits of even Cuntz-circle algebras. *Memoirs of the American Mathematical Society*, **565**.

Lin, Q. (1996). Cut down method in the inductive limit decomposition of non commutative tori, III : a complete answer in 3-dimensions. *Communications in Mathematical Physics*, **179**, 555–575.

Lindblad, G. (1976a). On the generators of quantum dynamical semigroups. *Communications in Mathematical Physics*, **48**, 119–130.

Lindblad, G. (1976b). Dissipative operators and cohomology of operator algebras. *Letters in Mathematical Physics*, **1**, 219–224.

Lindblad, G. (1976c). Brownian motion of a quantum harmonic oscillator. *Reports on Mathematical Physics*, **10**, 393–406.

Loesch, S. Zhou, Y.-K. and Zuber, J.-B. (preprint 1996). Identities in representations of the Hecke algebra.

Loi, P. H. (1988). On the theory of index and type III factors. Thesis, Pennsylvania State University.

Loi, P. H. (1994a). On the derived tower of certain inclusions of type III_λ factors of index 4. *Pacific Journal of Mathematics*, **165**, 321–345.

Loi, P. H. (1994b). Remarks on automorphisms of subfactors. *Proceedings of the American Mathematical Society*, **121**, 523–531.

Loi, P. H. (preprint 1995). Strongly free actions of finite groups on inclusions of type III_λ factors.

Loi, P. H. (1996). On automorphisms of subfactors. *Journal of Functional Analysis*, **141**, 275–293.

Loi, P. H. (preprint 1996a). A structural result of irreducible inclusions of type III_λ factors.

Loi, P. H. (preprint 1996b). Commuting squares and the classification of finite depth inclusions of AFD type III_λ factors.

Loke, T. (1994). Operator algebras and conformal field theory of the discrete series representations of Diff(S^1). Thesis, University of Cambridge.

Longo, R. (1978) A simple proof of the existence of modular automorphisms in approximately finite dimensional von Neumann algebras. *Pacific Journal of Mathematics*, **75**, 199–205.

Longo, R. (1979). Automatic relative boundedness of derivations in C^*-algebras. *Journal of Functional Analysis*, **34**, 21–28.

Longo, R. (1987). Simple injective subfactors. *Advances in Mathematics*, **63**, 152–171.

Longo, R. (1989). Index of subfactors and statistics of quantum fields, I. *Communications in Mathematical Physics*, **126**, 217–247.

Longo, R. (1990). Index of subfactors and statistics of quantum fields II. *Communications in Mathematical Physics*, **130**, 285–309.

Longo, R. (1992). Minimal index and braided subfactors. *Journal of Functional Analysis*, **109**, 98–112.

Longo, R. (1994a). A duality for Hopf algebras and for subfactors I. *Communications in Mathematical Physics.* **159**, 133–150.

Longo, R. (1994b). Problems on von Neumann algebras suggested by quantum field theory. In *Subfactors – Proceedings of the Taniguchi Symposium, Katata*, (ed. H. Araki *et al.*), World Scientific, Singapore, pp. 233–241.

Longo, R., and Rehren, K.-H. (1995). Nets for subfactors. *Reviews in Mathematical Physics*, **7**, 567–597.

Longo, R., and Roberts, J. E. (1997). A theory of dimension. *K-theory*, **11**, 103–159.

Loring, T. A. (1996). Stable relations II. Corona semi-projectivity and dimension drop C^*-algebras. *Pacific Journal of Mathematics*, **172**, 461–475.

Löwner, K. (1934). Über monotone Matrixfunctionen *Mathematische Zeitschrift*, **38**, 177–216.

Lumer, G. and Philips, R. S. (1961). Dissipative operators in a Banach space. *Pacific Journal of Mathematics*, **11**, 679–698.

Lundberg, L. E. (1976). Quasi-free 'second-quantization'. *Communications in Mathematical Physics*, **50**, 103–112.

Lyness, J. N. and Ninham, B. W. (1967). Numerical quadrature and asymptotic expansions. *Mathematics of Computation*, **21**, 162–178.

McCoy, B. M. and Perk, J. H. H. (1987). Relation of conformal field theory and deformation theory for the Ising model. *Nuclear Physics B*, **285[FS19]**,

279–294.
McCoy, B. and Wu, T. (1972). *The two dimensional Ising model.* Harvard University Press, Cambridge, Massachusetts, 40.
McDuff, D. (1969). Uncountably many II_1 factors. *Annals of Mathematics*, **90**, 372–377.
McDuff, D. (1970). Central sequences and the hyperfinite factor. *Proceedings of the London Mathematical Society*, **21**, 443–461.
McGuire, J. B. (1964). Study of exactly soluble one-dimensional N-body problems. *Journal of Mathematical Physics*, **5**, 622–636.
McKean, H. P. (1964). Kramers–Wannier duality for the two dimensional Ising model as an instance of Poisson's summation formula. *Journal of Mathematical Physics*, **5**, 775–776.
Mack, G. and Schomerus, V. (1990a). Conformal field algebras with quantum symmetry from the theory of superselection sectors. *Communications in Mathematical Physics*, **134**, 139.
Mack, G. and Schomerus, V. (1990b). Endomorphisms and quantum symmetry of the conformal Ising model. In *The algebraic theory of superselection sectors. Introduction and recent results* (ed. D. Kastler), World Scientific, Singapore, pp. 388–427.
Mackey, G. (1957). Borel structures in groups and their duals. *Transactions of the American Mathematical Society*, **85**, 134–165.
Manuceau, J. and Verbeure, A. (1976). Non factor quasi-free states of the CAR algebra. *Communications in Mathematical Physics*, **18**, 319–326.
Manuceau, J., Rocca, F., and Testard, D. (1969). On the product form of quasi-free states. *Communications in Mathematical Physics*, **12**, 43–57.
Manuceau, J., Sirugue, M., Testard, D., and Verbeure, A. (1973). The smallest C^*-algebra for CCR. *Communications in Mathematical Physics*, **32**, 231–243.
Markov, A. (1935). Über die freie Äquivalenz geschlossener Zöpfe. *Rossiiskaya Akademiya Nauk, Matematicheskii Sbornik*, **1**, 73–78.
Martroisan, D. H. (1986). Translation invariant Gibbs states in the q-state Potts model. *Communications in Mathematical Physics*, **105**, 281–290.
Masuda, T. (1997). An analogue of Longo's canonical endomorphism for bimodule theory and its application to asymptotic inclusions. *International Journal of Mathematics*, **8**, 249–265.
Masuda, T. (in press). Classification of actions of discrete amenable groups on strongly amenable subfactors of type III_λ. *Proceedings of the American Mathematical Society.*
Masuda, T. (preprint 1997). Classification of strongly free actions of discrete amenable groups on strongly amenable subfactors of type III_0.
Matsui, T. (1986). Explicit formulas for correlation functions of ground states of the 1 dimensional XY model. *Annales de l'Institut Henri Poincaré*, **45**, 49–59.

Matsui, T. (1987a). On quasi-equivalence of quasi-free states of gauge invariant CAR algebras. *Journal of Operator Theory*, **17**, 281–290.

Matsui, T. (1987b). Factoriality and quasi-equivalence of quasi-free states of $\mathbb{Z}_2$ and $U(1)$ invariant CAR algebras. *Review Roumaine de Mathématiques Pure et Appliqués*, **32**, 693–700

Matsui, T. (1990a). Uniqueness of translationally invariant ground states in quantum spin system. *Communications in Mathematical Physics*, **126**, 453–467.

Matsui, T. (1990b). A link between quantum and classical Potts models. *Journal of Statistical Physics*, **59**, 781–798.

Messager, A. and Miracle-Sole, S. (1975). Equilibrium states of the two-dimensional Ising model in the two-phase region. *Communications in Mathematical Physics*, **40**, 187–196.

Milnor, J. (1963). Morse Theory. *Annals of Mathematical Studies*, **51**, 9.

Miyadera, J. (1952). Generation of a strongly continuous semigroups of operators. *Tôhoku Journal of Mathematics* **4**, 109–114.

Mlak, W. (1966). Unitary dilations in case of ordered groups. *Annales Polonici Mathematici, Polish Academy of Science*, **17**, 321–328.

Montrell, E. and Potts, R. B. and Ward, J. C. (1967). Correlations and spontaneous magnetisation of the two dimensional Ising model. *Journal of Mathematical Physics*, **4**, 308–322.

Moore, C. C. (1967). Invariant measures on product spaces. *Proceedings of the 5th Berkeley Symposium on Mathematical Statistics and Probability*, **II**(II), 447–459.

Moore, G. and Seiberg, N. (1989a). Classical and quantum conformal field theory. *Communications in Mathematical Physics*, **123**, 177–254.

Moore, G. and Seiberg, N. (1989b). Naturality in conformal field theory. *Nuclear Physics B*, **313**, 16–40.

Munemasa, A. and Watatani, Y. (1992). Paires orthogonales de sous-algèbres involutives. *Comptes Rendus de l'Académie des Sciences, Série I, Mathématiques*, **314**, 329–331.

Murakami, H. (1994). Quantum $SU(2)$-invariants dominate Casson's $SU(2)$-invariant. *Mathematical Proceedings of the Cambridge Philosophical Society*, **115**, 253–281.

Murakami, J. (1987). The Kauffman polynomial of links and representation theory. *Osaka Journal of Mathematics*, **24**, 745–758.

Murphy, G. J. (1990). *C^*-algebras and operator theory.* Academic Press, New York.

Murray, F. J. (1990). The rings of operators. *Symposia Mathematica*, **50**, 57–60.

Murray, F. J. and von Neumann, J. (1936). On rings of operators. *Annals of Mathematics*, **37**, 116–229.

Murray, F. J. and von Neumann, J. (1937). On rings of operators II. *Transactions of the American Mathematical Society*, **41**, 208–248.

Murray, F. J. and von Neumann, J. (1943). On rings of operators IV. *Annals of Mathematics*, **44**, 716–808.

Nagel, R. (1981/82). Zur Characterisierung stabiler Operator Halbgruppen. *Semesterbericht Funktionanalysis, Tubingen*, 99–119.

Nahm, W. (1988). Lie group exponents and $SU(2)$ current algebras. *Communications in Mathematical Physics*, **118**, 171–176.

Nahm, W. (1991). A proof of modular invariance. *International Journal of Modern Physics*, **6**, 2837–2845.

Naimark, M. A. (1943a). On a representation of additive operator set function. *C.R. (Doklady) Acad. Sci. URSS(N.S.)*, **41**, 359–361.

Naimark, M. A. (1943b). Positive-definite operator functions on a commutative group. *Bull. Acad. Sci. URSS. Ser. Math. [Izvestia Akad. Nauk SSSR]*, **1**, 237–244.

Naimark, M. A. (1972). *Normed algebras.* Wolters Noordhoff, Groningen.

Nakanishi, T. and Tsuchiya, A. (1992). Level-rank duality of WZW models in conformal field theory. *Communications in Mathematical Physics*, **144**, 351–372.

Natsume, T. (1985). On $K_*(C^*(SL(2,\mathbb{Z})))$. *Journal of Operator Theory*, **13**, 103–118.

Nelson, E. (1960). Analytic vectors. *Annals of Mathematics*, **70**, 572–614.

Nest, R. (1988). Cyclic cohomology of non-commutative tori. *Canadian Journal of Mathematics*, **XL**, 1046–1057.

Nest, R. (preprint 1989). Cohomology of a certain non-commutative orbifold.

Neveu, A. and Schwarz, J. H. (1971). Factorisable dual model of pions. *Nuclear Physics B*, **31**, 86–112.

Nill, F. and Wiesbrock, H.-W. (1995). A comment on Jones inclusions with infinite index. *Reviews in Mathematical Physics*, **7**, 599–630.

Ocneanu, A. (1985). *Actions of discrete amenable groups on factors*, Lecture Notes in Mathematics **1138**, Springer, Berlin.

Ocneanu, A. (1988). Quantized group, string algebras and Galois theory for algebras. In *Operator algebras and applications*, Vol. 2 (Warwick, 1987), (ed. D. E. Evans and M. Takesaki), London Mathematical Society Lecture Note Series Vol. 136, Cambridge University Press, Cambridge, pp. 119–172.

Ocneanu, A. (1991). Quantum symmetry, differential geometry of finite graphs and classification of subfactors. University of Tokyo Seminary Notes 45, (Notes recorded by Kawahigashi, Y.).

Ocneanu, A. (1994). Chirality for operator algebras (recorded by Kawahigashi, Y.). In *Subfactors - Proceedings of the Taniguchi Symposium, Katata*, (ed. H. Araki *et al.*), World Scientific, Singapore, pp. 39–63.

Okamoto, S. (1991). Invariants for subfactors arising from Coxeter graphs. *Current Topics in Operator Algebras*, World Scientific Publishing, 84–103.

Olesen, D. (1974). On norm continuity and compactness of the spectrum. *Mathematica Scandinavica*, **35**, 145–148.

Olesen, D., Pedersen, G. and Takesaki, M. (1980). Ergodic actions of compact abelian groups *Journal of Operator Theory*, **3**, 237–269.

Onsager, L. (1941). Crystal Statistics I. *Physical Review*, **65**, 117–149.

Onsager, L. (1949). Discussion remark (Spontaneous magnetisation of the two-dimensional Ising model, *Nuovo Cimento, Supplement*, **6**, 261–262.

Pachner, U. (1978). Bistellare Aquivalenz kombinatorischer Mannigfaltigkeiten. *Archiv der Mathematik*, **30**, 89–98.

Palmer, J. and Tracy, C. (1981). Two dimensional Ising correlations: convergence of the scaling limit. *Advances in Applied Mathematics*, **2**, 329–388.

Palmer, J. and Tracy, C. (1983). Two dimensional Ising correlations: the SMJ analysis. *Advances in Applied Mathematics*, **4**, 46–102.

Parthasarathy, K. R. (1992). *An introduction to quantum stochastic calculus.* Birkhaüser, Basel.

Parthasarathy, K. R. and Schmidt, K. (1972). Positive-definite kernels, continuous tensor products and central limit theorems of probability theory. *Springer Lecture Notes in Mathematics, Springer, Berlin*, **272**.

Paschke, W. L. (1981). K-theory for actions of the circle group. *Journal of Operator Theory*, **6**, 125–133.

Paschke, W. L. and Salinas, N. (1979). Matrix algebras over $\mathcal{O}_n$. *Michigan Mathematical Journal*, **26**, 3–12.

Pask, D. and Raeburn, I. (1996). On the K-theory of Cuntz–Krieger algebras. *Publications of the RIMS, Kyoto University*, **32**, 415–443.

Pask, D. and Sutherland, C. (1994). Filtered inclusions of path algebras; a combinatorial approach to Doplicher–Roberts duality. *Journal of Operator Theory*, **31**, 99–121.

Pasquier, V. (1987a). Two-dimensional critical systems labelled by Dynkin diagrams. *Nuclear Physics B*, **285**, 162–172.

Pasquier, V. (1987b). Operator content of the ADE lattice models *Journal of Physics A*, **20**, 5707–5717.

Pasquier, V. (1988). Etiology of IRF-models. *Communications in Mathematical Physics*, **118**, 355–364.

Payen, R. (1964). Représentation spectrale des fonctions aléatoires stationaires à valeurs dans un espace de Hilbert. *Comptes Rendus de l'Académie des Sciences, Paris*, **259**, 3929–3932.

Pedersen, G. K. (1979). *C^*-algebras and their automorphism groups.* Academic Press, New York.

Pedersen, G. K. (1989). *Analysis now.* Graduate texts in mathematics. Springer, New York.

Peierls, R. (1936). On Ising's problem of ferromagnetism. *Proceedings of the Cambridge Philosophical Society*, **32**, 477–481.

Perk, J. H. H. (1979). Time-dependent correlations in the one-dimensional XY-model. Ph. D. Thesis. Leiden.

Perk, J. H. H. (1981). Nonlinear partial difference equations for Ising model n-point Green's functions. In *Proceedings of the II International Symposium on Selected Topics in Statistical Mechanics*, Dubna, August 25–29, pp. 165–180.

Peschel, I. and Truong, T. T. (1987). Corner transfer matrices and conformal invariance. *Zeitschrift für Physik B. Condensed Matter*, **69**, 385–391.

Petz, D. (1990). *An invitation to the algebra of canonical commutation relations.* Leuven Notes in Mathematics and Theoretical Physics, 2, Leuven University Press.

Phillips, N. C. (preprint 1994). A classification theorem for nuclear purely infinite simple C^*-algebras.

Phillips, R. S. (1952). On the generation of semigroups of linear operators. *Pacific Journal of Mathematics*, **2**, 343–369.

Pimsner, M. and Popa, S. (1978). The Ext groups of some C^*-algebras considered by J. Cuntz. *Revue Roumaine de Mathématiques Pures et Appliquées*, **23**, 1069–1076.

Pimsner, M. and Popa, S. (1986a). Entropy and index for subfactors. *Annales Scientifiques de l'École Normale Supérieur*, **19**, 57–106.

Pimsner, M. and Popa, S. (1986b). Sur les sous-facteurs d'indice finite d'un facteur de II_1 ayant la propriét T. *Comptes Rendus de l'Académie des Sciences, Série I, Mathématiques*, **303**, 359–361.

Pimsner, M. and Popa, S. (1988). Iterating the basic constructions. *Transactions of the American Mathematical Society*, **310**, 127–134.

Pimsner, M. and Popa, S. (1991). Finite dimensional approximation of pairs of algebras and obstructions for the index. *Journal of Functional Analysis*, **98**, 270–291.

Pimsner, M. and Voiculescu, D. (1980a). Imbedding the irrational rotation C^*-algebra into an AF-algebra. *Journal of Operator Theory*, **4**, 201–210.

Pimsner, M. and Voiculescu, D. (1980b). Exact sequences for K-groups and Ext-groups of certain crossed product C^*-algebras. *Journal of Operator Theory*, **8**, 131–156.

Pirogov, S. (1972). States associated with the two dimensional Ising model. *Theoretical Mathematical Physics*, **11(3)**, 614–617.

Polyakov, A. M. (1980). Critical symmetry of critical fluctuations. *JETP Letters, American Institute of Physics*, **12**, 381–390.

Ponomarenko, O. I. (1956). Correlation kernels of random functions with values in Hilbert space. *Akademiya Nauk Ukraïni. Dopvidi. Matematika.* Sér. A, 35–36.

Popa, S. (1981). On a problem of R.V. Kadison on maximal abelian $*$-subalgebras in finite factors. *Inventiones Mathematicae*, **65**, 269–281.

Popa, S. (1983a). Maximal injective subalgebras in factors associated with free groups. *Advances in Mathematics*, **50**, 27–48.

Popa, S. (1983b). Orthogonal pairs of $*$-subalgebras in finite von Neumann algebras. *Journal of Operator Theory*, **9**, 253–268.

Popa, S. (1986). A short proof of 'injectivity implies hyperfiniteness' for finite von Neumann algebras. *Journal of Operator Theory*, **16**, 261–272.

Popa, S. (1989a). Relative dimension, towers of projections and commuting squares of subfactors. *Pacific Journal of Mathematics* **137**, 181–207.

Popa, S. (1989b). Sousfacteurs, actions des groupes et cohomologie. *Comptes Rendus de l'Académie des Sciences, Série I, Mathématiques*, **309**, 771–776.

Popa, S. (1990a). Classification of subfactors: reduction to commuting squares. *Inventiones Mathematicae*, **101**, 19–43.

Popa, S. (1990b). Some rigidity results in type II_1 factors. *Comptes Rendus de l'Académie des Sciences, Série I, Mathématiques*, **311**, 535–538.

Popa, S. (1990c). Sur la classification des sousfacteurs d'indice fini du facteur hyperfini. *Comptes Rendus de l'Académie des Sciences, Série I, Mathématiques*, **311**, 95–100.

Popa, S. (1991). Subfactors and classification in von Neumann algebras. *Proceedings of the International Congress of Mathematicians, Kyoto 1990*, 987–996.

Popa, S. (1992). On the classification of actions of amenable groups on subfactors. *Comptes Rendus de l'Académie des Sciences, Série I, Mathématiques*, **315**, 295–299.

Popa, S. (preprint 1992). Classification of actions of discrete amenable groups on amenable subfactors of type II.

Popa, S. (1993). Markov traces on universal Jones algebras and subfactors of finite index. *Inventiones Mathematicae*, **111**, 375–405.

Popa, S. (1994a). Classification of amenable subfactors of type II. *Acta Mathematica*, **172**, 352–445.

Popa, S. (1994b). Approximate innerness and central freeness for subfactors : A classification result. In *Subfactors – Proceedings of the Taniguchi Symposium, Katata*, (ed. H. Araki *et al.*), World Scientific, Singapore, pp. 274–293.

Popa, S. (1994c). Classification of subfactors of finite depth of the hyperfinite type III_1 factor. *Comptes Rendus de l'Académie des Sciences, Série I, Mathématiques*, **318**, 1003–1008.

Popa, S. (1994d). Symmetric enveloping algebras, amenability and AFD properties for subfactors. *Mathematical Research Letters*, **1**, 409–425.

Popa, S. (in press). Some ergodic properties for infinite graphs associated to subfactors. *Ergodic Theory and Dynamical Systems.*

Popa, S. (1995a). Classification of subfactors and of their endomorphisms. *CBMS Lecture Notes Series*, **86**.

Popa, S. (1995b). An axiomatization of the lattice of higher relative commutants of a subfactor. *Inventiones Mathematicae*, **120**, 427–446.

Popa, S. (1995c). Free-independent sequence in type II_1 factors and related problems. *Astérisque*, **232**, 187–202

Popa, S. and Wassermann, A. (1992). Actions of compact Lie groups on von Neumann algebras. *Comptes Rendus de l'Académie des Sciences, Série I, Mathématiques*, **315**, 421–426.

Popescu, G. (1989). Isometric dilations for infinite sequences of noncommuting operators. *Transactions of American Mathematical Society*, **316**, 523–536.

Potts, R. (1952). Some generalized order–disorder transformations. *Mathematical Proceedings of the Cambridge Philosophical Society*, **48**, 106–109.

Power, S. C. (1978) Simplicity of C^*-algebras of minimal dynamical systems. *Journal of the London Mathematical Society (2)*, **18**, 534–538.

Powers, R. T. (1967). Representations of uniformly hyperfinite algebras and their associated von Neumann rings. *Annals of Mathematics*, **86**, 138–171.

Powers, R. T. (1987). A non-spatial continuous semigroup of $*$-endomorphisms of $B(H)$. *Publications of the RIMS, Kyoto University*, **23**, 1053–1069.

Powers, R. T. (1988). An index theory for semigroups of endomorphisms of $B(H)$ and type II factors. *Canadian Journal of Mathematics*, **40**, 86–114.

Powers, R. T. (1991). On the structure of continuous spatial semigroups of $*$-endomorphisms of $B(H)$. *International Journal of Mathematics*, **2**, 323–360.

Powers, R. T. (preprint 1994). New examples of continuous spatial semigroups of endomorphisms of $B(H)$.

Powers, R. T. and Sakai, S. (1975). Existence of ground states and KMS states for approximately inner dynamics. *Communications Mathematical Physics*, **39**, 273–288.

Powers, R. T. and Størmer, E. (1970). Free states of the canonical anti-commutation relations. *Communications in Mathematical Physics*, **16**, 1–33.

Pressley, A. and Segal, G. (1986). *Loop groups*. Oxford University Press, Oxford.

Putnam, I. F. (1989). The C^*-algebras associated with minimal homeomorphisms of the Cantor set. *Pacific Journal of Mathematics*, **136**, 329–353.

Putnam, I. F. (1990a). On the topological stable rank of certain transformation group C^*-algebras. *Ergodic Theory and Dynamical Systems*, **10**, 187–207.

Putnam, I. F. (1990b). The invertible elements are dense in the irrational rotation C^*-algebras. *Journal für die Reine und Angewandte Mathematik*, **410**, 160–166.

Rădulescu, F. (1992). Sous-facteurs de $\mathcal{L}(F_\infty)$ d'indice $4\cos^2 \pi/n$, $n \geq 3$. *Comptes Rendus de l'Académie des Sciences, Série I, Mathématiques*, **315**, 37–42.

Rădulescu, F. (1994). Random matrices, amalgamated free products and subfactors of the von Neumann algebra of a free group of noninteger index. *Inventiones Mathematicae*, **115**, 347–389.

Rădulescu, F. (preprint 1996). An invariant for subfactors in the von Neumann algebra of a free group.

Ramond, P. (1971). Dual theory for free fermions. *Physical Review*, **D3**, 2415–2418.

Reed, M. and Simon, B. (1972). *Methods of modern mathematical physics, Vol 1, Functional Analysis.* Academic Press, New York.

Renault, J. (1980). A groupoid approach to C^*-algebras. *Lecture Notes in Mathematics*, **793**.

Reshetikhin, N. Yu. (preprint 1987). Quantized universal enveloping algebras, the Yang–Baxter equation and invariant of links I.

Reshetikhin, N. Yu. and Turaev, V. G. (1991). Invariants of 3-manifolds via link polynomials and quantum groups. *Inventiones Mathematicae*, **103**, 547–598.

Rideau, G. (1968) On some representations of the anticommutation relations. *Communications in Mathematical Physics*, **9**, 229–241.

Riedel, N. (1985). On the topological stable rank of irrational rotation algebras. *Journal of Operator Theory*, **13**, 143–150.

Rieffel, M. A. (1981). C^*-algebras associated with irrational rotations. *Pacific Journal of Mathematics*, **93**, 415–429.

Rieffel, M. A. (1983a). Dimensions and stable rank in the K-theory of C^*-algebras. *Proceedings of the London Mathematical Society*, **46**(3), 301–333.

Rieffel, M. A. (1983b). The cancellation theorem for projective modules over irrational rotation C^*-algebras. *Proceedings of the London Mathematical Society*, **(3) 47**, 285–302.

Riesz, F. and Sz.-Nagy, B. (1955). *Functional Analysis*. New York.

Ringrose, J. R. (1971). *Compact non-self-adjoint operators*. Van Nostrand–Reinhold mathematical studies, 35, Van Nostrand–Reinhold, New York.

Ringrose, J. R. (1972). Cohomology of operator algebras. *Springer Lecture Notes in Mathematics*, **247**, 355–434.

Robinson, A. (1974). *Non-standard Analysis*. 2nd edn., North-Holland, Amsterdam.

Rocca, F., Sirugue, M. and Testard, D. (1969a). Translation invariant quasifree states and Bogoliubov transformations. *Annales de l'Institut Henri Poincaré*, **3**, 247–258.

Rocca, F., Sirugue, M. and Testard, D. (1969b). On a class of equilibrium states under the Kubo–Martin–Schwinger boundary condition. *Communications in Mathematical Physics*, **13**, 317–334.

Rocha-Caridi, A. (1984). Vacuum vector representation of the Virasoro algebra. In *Vertex Operators in Mathematical Physics*. MSRI Publications, 3, Springer, Heidelberg, pp. 451–473.

Roche, Ph. (1990). Ocneanu cell calculus and integrable lattice models. *Communications in Mathematical Physics*, **127**, 395–424.

Rørdam, M. (1993). Classification of inductive limits of Cuntz algebras. *Journal für die Reine und Angewandte Mathematik*, **440**, 175–200.

Rørdam, M. (1995a). Classification of Cuntz–Krieger Algebras. *K-theory*, **9**, 31–58.

Rørdam, M. (1995b). Classification of certain infinite simple C^*-algebras. *Journal of Functional Analysis*, **131**, 415–458.

Rosenberg, J. (1983). C^*-algebras, positive scalar curvature, and the Nokikov conjecture. *Publications Mathématiques IHES*, **58**, 197–212.

Rosenberg, J. (1994). *Algebraic K-theory and its applications*. Springer, Berlin.

Rosenberg, J. and Schochet, C. (1987). The Künneth theorem and the universal coefficient theorem for Kasparov's generalised K-functor. *Duke Mathematical Journal*, **55**, 431–474.

Rössgen, M. and Varnhagen, R. (1995). Steps towards lattice Virasoro algebras : $su(1,1)$. *Physics Letters*, **B 350**, 203–211.

Rosso, M. (1987). Comparison des groupes $SU(2)$ quantique de Drinfeld et de

Woronowicz. *Comptes Rendus de l'Académie des Sciences, Série I, Mathématiques*, **403**, 323–326.

Rosso, M. (1988). Finite dimensional representations of the quantum analog of the enveloping algebra of a complex simple Lie algebra. *Communications in Mathematical Physics*, **117**, 581–593.

Roth, J. P. (1976). Opérateurs dissipatifs et semigroupes dans les espaces de fonctions continues. *Annales de l'Institut Fourier*, **26**, 1–97.

Ruelle, D. (1972). On the use of 'Small External Fields' in the problem of symmetry breakdown in statistical mechanics. *Annals of Physics*, **69**, 364–374.

Ruelle, Ph., Thiran, E. and Weyers, J. (1993). Implications of an arithmetical commutant for symmetry of the modular invariants. *Nuclear Physics B*, **402**, 693–708.

Ruijsenaars, S. N. M. (1977). On Bogoliubov transformations for systems of relativistic charged particles. *Journal of Mathematical Physics*, **18**, 517–526.

Ruijsenaars, S. N. M. (1987). Conformal invariance in 2D and Virasoro algebras.

Sakai, S. (1960). On a conjecture of Kaplansky. *Tôhoku Mathematical Journal*, **12**, 31–33.

Sakai, S. (1966). Derivations of W^*-algebras. *Annals of Mathematics*, **83**, 273–279.

Sakai, S. (1968). Derivations of simple C^*-algebras. *Journal of Functional Analysis*, **2**, 202–206.

Sakai, S. (1970). An uncountable number of II_1 and II_∞ factors. *Journal of Functional Analysis*, **5**, 236–246.

Sakai, S. (1971). *C^*-algebras and W^*-algebras.* Springer, Berlin–Heidelberg–New York.

Sano, T. (1996). Commuting co-commuting squares and finite dimensional Kač algebras. *Pacific Journal of Mathematics*, **172**, 243–253.

Sano, T. and Watatani, Y. (1994). Angles between two subfactors. *Journal of Operator Theory*, **32**, 209–241.

Sato, N. (1997a). Fourier transform for irreducible inclusions of type II_1 factors with finite index and its application to the depth two case. *Publications of the RIMS, Kyoto University*, **33**, 189–222.

Sato, N. (1997b). A relation between two subfactors arising from a non-degenerate commuting square – tensor categories and TQFT's. *International Journal of Mathematics*, **8**, 407–420.

Sato, N. (in press a). A relation between two subfactors arising from a non-degenerate commuting square –An answer to a question raised by V. F. R. Jones. *Pacific Journal of Mathematics.*

Sato, N. (in press b). Constructing a non-degenerate commuting square from equivalent systems of bimodules. *International Mathematics Research Notices.*

Sato, M., Miwa, T. and Jimbo, M. (1977–1980). Holonomic quantum fields. *Publications of the RIMS, Kyoto University*, I (1977), **14**, 223–267; IV (1979),

15, 871–972; V (1980), **16**, 531–584.

Sauvageot, J. L. (1983). Sur le produit tensoriel relatif d'espace de Hilbert. *Journal of Operator Theory*, **9**, 237–252.

Schellekens, A. N. (1990). Conformal field theory for four-dimensional strings. *Nuclear Physics B (Proceedings Supplement)*, **15**, 3–34.

Schellekens, A. N. and Warner, N. P. (1986). Conformal subalgebras of Kač-Moody algebras. *Physical Review D*, **34**, 3092–3096.

Schellekens, A. N. and Yankielowicz, S. (1989). Modular invariants from simple currents. An explicit proof. *Physics Letters B*, **227**, 387–391.

Schor, R. and O'Carroll, M. (1982). The scaling limit and Osterwalder–Schrader axioms for the two dimensional Ising model. *Communications in Mathematical Physics*, **84**, 153–170.

Schou, J. (1990). Commuting squares and index for subfactors. Ph. D. thesis, Odense University.

Schrader, R. and Uhlenbrock, D. A. (1975). Markov structures on Clifford algebras. *Journal of Functional Analysis*, **18**, 369–413.

Schultz, T. D., Mattis, D. C. and Lieb, E. (1964). Two dimensional Ising model as a solvable problem of many Fermions. *Reviews of Modern Physics*, **36**, 856–871.

Schwartz, J. (1967). *W^*-algebras.* Gordon and Breach, New York.

Segal, G. Notes on conformal field theory. Unpublished manuscript.

Segal, I. E. (1947). Irreducible representations of operator algebras. *Bulletin of the American Mathematical Society*, **53**, 73–88.

Segal, I. E. (1961). Foundations of the theory of infinite-dimensional systems, II. *Canadian Journal of Mathematics*, **13**, 1–18.

Segal, I. E. (1962). Mathematical characterization of the physical vacuum for a linear Bose–Einstein field. *Illinois Journal of Mathematics*, **6**, 500–523.

Sekine, Y. (1990). Flows of weights of crossed products of type III factors by discrete groups. *Publications of the RIMS, Kyoto University*, **26**, 655–666.

Shale, D. and Stinespring, W. F. (1964). States on the Clifford algebra. *Annals of Mathematics*, **80**, 365–381.

Shale, D. and Stinespring, W. F. (1965). Spinor representations of infinite orthogonal groups. *Journal of Mathematical Mech.*, **14**, 315–322.

Shearer, J. B. (1989). On the distribution of the maximum eigenvalue of graphs. *Linear Algebra and its Applications*, **114/115**, 17–20.

Simon, B. (1977a). An abstract Kato's inequality for generators of positivity preserving semigroups. *Indiana University Mathematics Journal*, **26**, 1067–1073.

Simon, B. (1977b). Notes on infinite determinants of Hilbert space operators. *Advances in Mathematics*, **24**, 244–273.

Simon, B. (1979). Trace ideals and their applications. *London Mathematical Society Lecture Note Series* **35**, Cambridge University Press, Cambridge.

Simon, B. (1982). Almost periodic Schrödinger operators: a review. *Advances in Applied Mathematics*, **3**, 463–490.

Simon, B. (1994). *Statistical mechanics.* Princeton University Press.

Sirugue, M. and Winnink, M. (1970). Contraints imposed upon a system that satisfies the KMS boundary condition. *Communications in Mathematical Physics*, **19**, 161–168.

Sisson, P. N. M. (1975). A C^*-algebra of the Ising model. Ph. D. thesis. Trinity College, Dublin.

Skau, C. (1977). Finite subalgebras of a von Neumann algebra. *Journal of Functional Analysis*, **25**, 211–235.

Sklyanin, E. (1982). Some algebraic structures connected with the Yang–Baxter equation. *Journal of Soviet Mathematics*, **19**, 1546–1596.

Slater. L. J. (1951). Further identities of the Rogers–Ramanujan type. *Proceedings of the London Mathematical Society*, **54**(2), 147–167.

Slodowy, P. (1980). *Simple singularities and simple algebraic groups*, Lecture Notes in Mathematics, 815, Springer, Berlin.

Slodowy, P. (1983). In *Algebraic Geometry* (ed. J. Dolgachev), Lecture Notes in Mathematics, 1008, Springer, Berlin.

Sochen, N. (1991). Integrable models through representations of the Hecke algebra. *Nuclear Physics B*, **360**, 613–640.

Stinespring, W. F. (1955). Positive functions on C^*-algebras. *Proceedings of the American Mathematical Society*, **6**, 211–216.

Størmer, E. (1963). Positive linear maps on operator algebras. *Acta Mathematica*, **110**, 233–278.

Størmer, E. (1974). Positive linear maps of C^*-algebras. In *Foundations of Quantum Mechanics and Ordered Linear Spaces* (ed. A. Hartkämper and H. Neumann), Advanced Study Institute, Marburg, 1973, Lecture Notes in Physics, 29, Springer, Berlin.

Størmer, E. (1980). Decomposition of positive projections on C^*-algebras. *Mathematische Annalen*, **247**, 21–41.

Strătilă, Ş. (1981). *Modular theory in operator algebras.* Editura Academiei and Abacus Press, Tunbridge Wells.

Strătilă, Ş. and Voiculescu, D. (1978). On a class of KMS states for the unitary group $U(\infty)$. *Mathematische Annalen*, **235**, 87–110.

Strătilă, Ş. and Zsido. L. (1979). *Lectures on von Neumann algebras.* Editura Academiei and Abacus Press, Tunbridge Wells.

Su, H. (1992). Ph. D. thesis. University of Toronto.

Su, H. (1993). K-theoretic classification for certain real rank zero C^*-algebras, In *Quantum and Non-commutative Analysis* (ed. H. Araki), Kluwer Academic Publications.

Su, H. (1995). On the classification of C^*-algebras of real rank zero: inductive limits of matrix algebras over non-Hausdorff graphs. *Memoirs of the American Mathematical Society*, 114 (no. 547).

Su, H. (to appear). K-theoretic classification for certain inductive limit $\mathbb{Z}_2$-actions on real rank zero C^*-algebras. *Transactions of the American Mathematical Society.*

Suciu, I. (1967). Unitary dilation in case of partially ordered groups. *Bull. Acad. Polon. Sér. Sci. Math. Astronom. Phys.*, **15**, 271–275.

Suciu, I. (1973). Dilation and extension theorems for operator-valued mappings (Romanian, English summary). *The theory of operators and operator algebras, Editura Acad. R.S.R., Bucharest*, 93–130.

Sullivan, W. G. (1975). Markov processes for random fields. *Communications of the Dublin Institute for Advanced Studies, Series A*, **23**.

Sunder, V. S. (1987). A model for AF-algebras and a representation of the Jones projections. *Journal of Operator Theory*, **18**, 289–301.

Sunder, V. S. (1991). On commuting squares and subfactors. *Journal of Functional Analysis*, **101**, 286–311.

Sunder, V. S. (1992). II_1 factors, their bimodules and hypergroups. *Transactions of the American Mathematical Society*, **30**, 227–256.

Sunder, V. S. and Vijayarajan, A. K. (1993). On the non-occurrence of the Coxeter graphs β_{2n+1}, E_7, D_{2n+1} as principal graphs of an inclusion of II_1 factors. *Pacific Journal of Mathematics* **161**, 185–200.

Sutherland, C. E. (1994). Finite extensions of flows and subfactors. In *Subfactors – Proceedings of the Taniguchi Symposium, Katata*, (ed. H. Araki *et al.*), World Scientific, Singapore, pp. 172–189.

Sutherland, C. E. and Takesaki, M. (1985). Actions of discrete amenable groups and groupoids on von Neumann algebras. *Publications of the RIMS, Kyoto University*, **21**, 1087–1120.

Sutherland, C. E. and Takesaki, M. (1989). Actions of discrete amenable groups on injective factors of type III_λ, $\lambda \neq 1$. *Pacific Journal of Mathematics*, **137**, 405–444.

Swan, R. G. (1962). Vector bundles and projective modules. *Transactions of the American Mathematical Society*, **105**, 264–277.

Sz.-Nagy, B. (1953). Sur les contractions de l'espace de Hilbert. *Acta Scientiarum Mathematicarum*, **15**, 87–92.

Sz.-Nagy, B. (1955). Prolongements des transformations de l'espace de Hilbert qui sortent de cet espace. *Appendice au livre 'Leçons d'analyse fonctionelle'* par F. Riesz et B. Sz.-Nagy. Akadémai Kiadó, Budapest, 36pp.

Sz.-Nagy, B. and Foias, C. (1970). *Harmonic analysis of operators on Hilbert space.* North–Holland, Amsterdam.

Szymanski, W. (1994). Finite index subfactors and Hopf algebra crossed products. *Proceedings of the American Mathematical Society*, **120**, 519–528.

Takai, H. (1975). On a duality for crossed products of C^*-algebras. *Journal of Functional Analysis*, **19**, 25–39.

Takesaki, M. (1964). On the cross-norm of the direct product of C^*-algebras. *Tôhoku Mathematical Journal*, **16**, 111–122.

Takesaki, M. (1970). *Tomita's theory of modular Hilbert algebras and its applications.* Lecture Notes in Mathematics, 128, Springer, Berlin.

Takesaki, M. (1972). Conditional expectations in von Neumann algebras. *Journal of Functional Analysis*, **9**, 306–321.

Takesaki, M. (1973). Duality for crossed products and the structure of von Neumann algebras of Type III. *Acta Mathematica*, **131**, 249–310.

Takesaki, M. (1979). *Theory of operator algebras I.* Springer, Berlin.

Takesaki, M. (1983). Structure of factors and automorphism groups. *CBMS Regional Conference Series, American Mathematical Society*, **51**.

Taylor, J. L. (1975). Banach Algebras and topology. In *Algebras in analysis* (ed. J.H. Williamson), Academic Press, New York.

Temperley, H. N. V. (1971). Graph-theoretical problems and planar lattices. *Proceedings of the Royal Society, London* **A322**, 251–280.

Temperley, H. N. V. and Lieb. E. H. (1971). Relations between the 'percolation' and 'colouring' problem and other graph-theoretical problems associated with regular planar lattices: some exact results for the 'percolation' problem. *Proceedings of the Royal Society A*, **322**, 251–280.

Thomas, L. C. (1971). Ph. D. Thesis, University of Oxford.

Thompson, C. J. (1979). *Mathematical Statistical Mechanics*, Princeton University Press, Princeton.

Todorov, I. (1986). Infinite dimensional Lie algebras in conformal QFT models. In *Conformal Groups and Related Symmetries* (ed. A. Barat, H. Doebner), Lecture Notes in Physics, 261, Springer, Berlin, pp. 387–443.

Tomiyama, J. (1957). On the projection of norm one in W^*-algebras. *Proceedings of the Japan Academy*, **33**, 608–612.

Tomiyama, J. (1987). *Invitation to C^*-algebras and topological dynamics*, World Scientific, Singapore.

Tsuchiya, A. and Kanie, Y. (1988). Vertex operators on conformal field theory on $\mathbf{P}^1$ and monodromy representations of braid groups. *Advanced Studies in Pure Mathematics*, **16**, 297–372.

Tsuchiya, A., Ueno, K. and Yamada, Y. (1989). Conformal field theory on universal family of stable curves with gauge symmetries. *Advanced Studies in Pure Mathematics*, **19**, 459–566.

Tsui, S. K. (1977). A note on generators of semigroups. *Transactions of the American Mathematical Society*, **66**, 305–310.

Turaev, V. (1988). The Yang–Baxter equation and invariants of links. *Inventiones Mathematicae*, **92**, 527–553.

Turaev, V. G. (preprint 1991). Topology of shadows.

Turaev, V. G. (1992a). Shadow links and face models of statistical mechanics. *Journal of Differential Geometry*, **36**, 35–74.

Turaev, V. G. (1992b). Modular categories and 3-manifold invariants. *International Journal of Modern Physics B*, **6**, 1807–1824.

Turaev, V. G. (1993). Quantum invariants of links and 3-valent graphs in 3-manifolds. *Publications Mathématiques IHES*, **77**, 121–171.

Turaev, V. G. (1994). *Quantum invariants of knots and 3-manifolds.* de Gruyter Studies in Mathematics, Vol. 18.

Turaev, V. G. and Viro, O. Y. (1992). State sum invariants of 3-manifolds and quantum $6j$-symbols. *Topology*, **31**, 865–902.

Turaev, V. G. and Wenzl, H. (1993). Quantum invariants of 3-manifolds associated with classical simple Lie algebras. *International Journal of Mathematics*, **4**, 323–358.

Turaev, V. G. and Wenzl, H. (preprint 1996). Semisimple and modular categories from link invariants.

Vafa, C. (1988). Toward classification of conformal field theories. *Physics Letters B*, **206**, 421–426.

van Daele, A. and Verbeure, A. (1971). Quasi-equivalence of quasi-free states on the Weyl algebra. *Communications in Mathematical Physics*, **21**, 171–191.

Vanheuverzwijn, P. (1977). Generators for quasi-free completely positive semigroups. *Annales de l'Institut Henri Poincaré. Physique Théorique*, **29**, 123–138.

Vaserstein, L. N. (1971). Stable rank of rings and dimensionality of topological spaces. *Functional Analysis and its Applications*, **5**, 102–110.

Verlinde, E. P. (1988a). Fusion rules and modular transformation in $2D$ conformal field theory. *Nuclear Physics B*, **300**, 360–376.

Verlinde, E. P. (1988b). Conformal field theory and its applications to strings. Thesis, University of Utrecht.

Vershik, A. and Kerov, S. V. (1987). Locally semisimple algebras. Combinatorial theory and the K_0-functor. *Journal of Soviet Mathematics*, **38**, 1701–1733.

Verstegen, D. (1990). New exceptional modular invariant partition functions for simple Kač–Moody algebras. *Nuclear Physics B*, **346**, 349–386.

Verstegen, D. (1991). Conformal embeddings, rank-level duality and exceptional modular invariants. *Communications in Mathematical Physics*, **137**, 567–586.

Villadsen, J. (1995). The range of the Elliott invariant. *Journal für die Reine und Angewandte Mathematik*, **462**, 31–55.

Villadsen, J. (in press). Simple C^*-algebras with perforation. *Journal of Functional Analysis.*

Villadsen, J. (preprint 1997). On the stable rank of simple C^*-algebras.

Voiculescu, D. (1985). Symmetries of some reduced free product C^*-algebras. In *Operator algebras and their connections with topology and ergodic theory* (eds. H. Araki, C.C. Moore, S. Stratila, D. Voiculescu) Springer Lecture Notes in Mathematics, **1132**, 556–588.

von Neumann, J. (1935). Charakterisierung des Spektrums eines Integral Operators. *Act. Sci. et Ind.*, **229**.

von Neumann, J. (1940). On rings of operators III. *Annals of Mathematics*, **41**, 94–161.

von Neumann, J. (1961). *Collected works, Vol. III. Rings of operators.* Pergamon Press, Oxford.

Wakui, M. (1992). On Dijkgraaf–Witten invariant for 3-manifolds. *Osaka Journal of Mathematics*, **29**, 675–696.

Walters, S. G. (1995). Inductive limit automorphisms of the irrational rotation C^*-algebra. *Communications in Mathematical Physics*, **171**, 365–381.

Walton, M. (1989). Conformal branching rules and modular invariants. *Nuclear Physics B*, **322**, 775–790.

Walton, M. (1990). Fusion rules of Wess–Zumino–Witten models. *Nuclear Physics B*, **340**, 777–789.

Wassermann, S. (1976). On tensor products of certain group C^*-algebras. *Journal of Functional Analysis*, **23**, 239–254.

Wassermann, A. J. (1981). Automorphic actions of compact groups on operator algebras. Ph. D. thesis. University of Pennsylvania.

Wassermann, A. J. (1988). Coactions and Yang–Baxter equations for ergodic actions and subfactors. In *Operator algebras and applications* (eds. D.E. Evans and M. Takesaki), London Mathematical Society Lecture Notes, **136**, 203–236.

Wassermann, A. J. (1995). Operator algebras and conformal field theory. In *Proceedings of the International Congress of Mathematicians, Zürich 1994*, Birkhaüser, 966–979.

Wassermann, A. J. (in press). Operator algebras and conformal field theory III : Fusion of positive energy representations of $SU(N)$ using bounded operators. *Inventiones Mathematicae.*

Wassermann, A. J. (in preparation a). Operator algebras and conformal field theory II : Fusion for von Neumann algebras and loop groups.

Wassermann, A. J. (in preparation b). Operator algebras and conformal field theory IV : Loop, groups, invariant theory and subfactors.

Wassermann, A. J. (with contributions by V. Jones). (to appear) Lectures on operator algebras and conformal field theory. *Proceedings of Borel Seminar, Bern 1994.*

Watatani, Y. (1981). Toral automorphisms on irrational rotation algebras. *Mathematica Japonica*, **26**, 479–484.

Watatani, Y. (1990). Index for C^*-subalgebras. *Memoirs of the American Mathematical Society*, **424**.

Watatani, Y. (1996). Lattices of intermediate subfactors. *Journal of Functional Analysis*, **140**, 312–334.

Watatani, Y. and Wierzbicki, J. (1995). Commuting squares and relative entropy for two subfactors. *Journal of Functional Analysis*, **133**, 329–341.

Wegge-Olsen, N. E. (1993). *K-Theory and C^*-algebras.* Oxford University Press, Oxford.

Wenzl, H. (1987). On sequences of projections. *Comptes Rendus Mathématiques, La Société Royale du Canada, L'Académie des Sciences*, **9**, 5–9.

Wenzl, H. (1988a). Hecke algebras of type A_n and subfactors. *Inventiones Mathematicae*, **92**, 345–383.

Wenzl, H. (1988b). On the structure of Brauer's centralizer algebras. *Annals of Mathematics*, **128**, 173–193.

Wenzl, H. (1990a). Representations of braid groups and the quantum Yang–Baxter equation. *Pacific Journal of Mathematics*, **145**, 153–180.

Wenzl, H. (1990b). Quantum groups and subfactors of type B, C, and D. *Communications in Mathematical Physics*, **133**, 383–432.

Wenzl, H. (1993). Braids and invariants of 3-manifolds. *Inventiones Mathematicae*, **114**, 235–275.

Wenzl, H. (preprint 1997). C^*-tensor categories from quantum groups.

Widom, H. (1976). Asymptotic behaviour of block Toeplitz matrices and determinants II. *Advances in Mathematics*, **21**, 1–29.

Wierzbicki, J. (1994). An estimate of the depth from an intermediate subfactor. *Publications of the RIMS, Kyoto University*, **30**, 1139–1144.

Wiesbrock, H.-W. (1992). A comment on a recent work of Borchers. *Letters in Mathematical Physics*, **25**, 157–159.

Wiesbrock, H.-W. (1993a). Half-sided modular inclusions of von Neumann algebras. *Communications in Mathematical Physics*, **157**, 83–92.

Wiesbrock, H.-W. (1993b). Symmetries and half-sided modular inclusions of von Neumann algebras. *Letters in Mathematical Physics*, **28**, 107–114.

Wiesbrock, H.-W. (1994). A note on strongly additive conformal field theory and half-sided modular conormal standard inclusions. *Letters in Mathematical Physics*, **31**, 303–307.

Wiesbrock, H.-W. (1995). Superselection structures and localized Connes' cocycles. *Reviews in Mathematical Physics*, **7**, 133–160.

Wiesbrock, H.-W. (preprint 1995). Symmetries and modular intersections of von Neumann algebras.

Winnink, M. (1969). Algebraic aspects of the Kubo–Martin–Schwinger boundary condition. *Cargese Lectures in Physics*, **4**, 235–255, Gordon and Breach, New York.

Winsløw, C. (1993). Strongly free actions on subfactors. *International Journal of Mathematics*, **4**, 675–688.

Winsløw, C. (1994a). Approximately inner automorphisms on inclusions of type III_λ factors. *Pacific Journal of Mathematics*, **166**, 385–400.

Winsløw, C. (1994b). Automorphisms on an inclusion of factors. In *Subfactors – Proceedings of the Taniguchi Symposium, Katata*, (ed. H. Araki *et al.*), World Scientific, Singapore, pp. 139–152.

Winsløw, C. (1995a). Crossed products of II_1-subfactors by strongly outer actions. *Proceedings of the American Mathematical Society*, **347**, 985–991.

Winsløw, C. (1995b). The flow of weights in subfactor theory. *Publications of the RIMS, Kyoto University*, **31**, 519–532.

Winsløw, C. (1997a). Global structure in the semigroup of endomorphisms on a von Neumann algebra. *Journal of Operator Theory*, **37**, 1–16.

Winsløw, C. (1997b). Classification of strongly amenable subfactors of type III_0. *Journal of Functional Analysis*, **148**, 296–313.

Winsløw, C. (in press). A topology on the semigroup of endomorphisms on a von Neumann algebra. *Journal of the Mathematical Society of Japan.*

Winsløw, C. (preprint 1997a). Endomorphisms, automorphisms and subfactors of von Neumann algebras.

Winsløw, C. (preprint 1997b). Convergence and limit range for endomorphisms I.

Witten, E. (1988). Topological quantum field theory. *Communications in Mathematical Physics*, **117**, 353–386.

Witten, E. (1989a). Quantum field theory and Jones polynomial. *Communications in Mathematical Physics*, **121**, 351–399.

Witten, E. (1989b). Gauge theories and integrable lattice models. *Nuclear Physics B*, **322**, 629–697.

Wolfe, J. C. (1975). Free states and automorphisms of the Clifford algebra. *Communications in Mathematical Physics*, **45**, 53–58.

Wu, T. T., McCoy, B. M., Tracy, C. A. and Barouch, E. (1976). Spin–spin correlation functions for the two-dimensional Ising model : Exact theory in the scaling region. *Physical Review B*, **13**, 316–374.

Xu, F. (1994). Orbifold construction in subfactors. *Communications in Mathematical Physics*, **166**, 237–254.

Xu, F. (1996). The flat parts of non-flat orbifolds. *Pacific Journal of Mathematics*, **172**, 299–306.

Xu, F. (1997). Generalized Goodman–Harpe–Jones construction of subfactors I, II. *Communications in Mathematical Physics*, **184**, 475–491, 493–508.

Xu, F. (preprint 1997a). Jones–Wassermann subfactors for disconnected intervals.

Xu, F. (preprint 1997b). A note on quivers with symmetries.

Xu, F. (in press a). New braided endomorphisms from conformal inclusions. *Communications in Mathematical Physics.*

Xu, F. (in press b). Standard λ-lattices from quantum groups. *Inventiones Mathematicae.*

Xu, F. (in press c). Applications of braided endomorphisms from conformal inclusions. *International Mathematics Research Notices.*

Yamagami, S. (1992). Algebraic aspects in modular theory. *Publications of the RIMS, Kyoto University*, **28**, 1075–1106.

Yamagami, S. (1993a). A note on Ocneanu's approach to Jones index theory. *International Journal of Mathematics*, **4** (1993), 859–871.

Yamagami, S. (1993b). Vector bundles and bimodules. In *Quantum and non-commutative analysis*, (ed. H. Araki *et al.*), Kluwer Academic, pp. 321–329.

Yamagami, S. (1994). Modular theory for bimodules. *Journal of Functional Analysis*, **125**, 327–357.

Yamagami, S. (1996a). On unitary representations of compact quantum groups. *Communications in Mathematical Physics*, **167**, 509–529.

Yamagami, S. (1996b). Crossed products in bimodules. *Mathematische Annalen*, **305**, 1–24.

Yang, C. N. (1952). The spontaneous magnetisation of a two–dimensional Ising model. *Physical Review*, **85**, 809–816.

Yang, C. N. (1967). Some exact results for the many-body problem in one dimension with repulsive delta-function interaction. *Physical Review Letters*,

19, 1312–1315.

Yang, C. N. (1968). S matrix for the one-dimensional N-body problem with repulsive or attractive δ-function interaction. *Physical Review*, **168**, 1920–1923.

Yosida, K. (1948). On the differentiability and representation of one-parameter semigroups of linear operators. *Journal of the Mathematical Society of Japan*, **1**, 15–21.

Zabzyk, J. (1975). A note on C_0-semigroups. *Bulletin Acad. Polon. Sci. Ser. Sci. Math. Astr. Phys.*, **23**, 895–898.

Zamolodchikov, A. (1985). Infinite additional symmetries in two-dimensional conformal field theory. *Theoretical Mathematical Physics*, **65**, 1205–1213.

Zamolodchikov, A. B. and Fateev, V. A. (1985) Nonlocal (parafermion) currents in two-dimensional conformal quantum field theory and self-dual critical points in Z_N-symmetric statistical systems. *Soviet Physics JETP*, **62**, 215–225.

Zhang, S. (1990). A property of purely infinite simple C^*-algebras. *Proceedings of the American Mathematical Society*, **109**, 717–720.

Zuber, J. B. (1990). Graphs, algebras, conformal field theories and integrable lattice models. *Nuclear Physics B(Proc. Suppl.)*, **18**, 313–326.

INDEX